Transforming Growth Factor-β in Cancer Therapy, Volume I

CANCER DRUG DISCOVERY AND DEVELOPMENT

Beverly A. Teicher, PhD, *SERIES EDITOR*

Checkpoint Responses in Cancer Therapy, edited by *Wei Dai, 2008*

Cancer Proteomics: *From Bench to Bedside,* edited by *Sayed S. Daoud, 2008*

Transforming Growth Factor-β in Cancer Therapy, *Volume II: Cancer Treatment and Therapy,* edited by *Sonia Jakowlew, 2008*

Transforming Growth Factor-β in Cancer Therapy, *Volume I: Basic and Clinical Biology,* edited by *Sonia Jakowlew, 2008*

Microtubule Targets in Cancer Therapy, edited by *Antonio T. Fojo, 2008*

Antiangiogenic Agents in Cancer Therapy, *Second Edition,* edited by *Beverly A. Teicher and Lee M. Ellis, 2007*

Apoptosis and Senescence in Cancer Chemotherapy and Radiotherapy, *Second Edition,* edited by *David A. Gerwitz, Shawn Edan Holtz, and Steven Grant, 2007*

Molecular Targeting in Oncology, edited by *Howard L. Kaufman, Scott Wadler, and Karen Antman, 2007*

In Vivo Imaging of Cancer Therapy, edited by *Anthony F. Shields and Patricia Price, 2007*

Cytokines in the Genesis and Treatment of Cancer, edited by *Michael A. Caligiuri, Michael T. Lotze, and Frances R. Balkwill, 2007*

Regional Cancer Therapy, edited by *Peter M. Schlag and Ulrike Stein, 2007*

Gene Therapy for Cancer, edited by *Kelly K. Hunt, Stephan A. Vorburger, and Stephen G. Swisher, 2007*

Deoxynucleoside Analogs in Cancer Therapy, edited by *Godefridus J. Peters, 2006*

Cancer Drug Resistance, edited by *Beverly A. Teicher, 2006*

Histone Deacetylases: *Transcriptional Regulation and Other Cellular Functions,* edited by *Eric Verdin, 2006*

Immunotherapy of Cancer, edited by *Mary L. Disis, 2006*

Biomarkers in Breast Cancer: *Molecular Diagnostics for Predicting and Monitoring Therapeutic Effect,* edited by *Giampietro Gasparini and Daniel F. Hayes, 2006*

Protein Tyrosine Kinases: *From Inhibitors to Useful Drugs,* edited by *Doriana Fabbro and Frank McCormick, 2005*

Bone Metastasis: *Experimental and Clinical Therapeutics,* edited by *Gurmit Singh and Shafaat A. Rabbani, 2005*

The Oncogenomics Handbook, edited by *William J. LaRochelle and Richard A. Shimkets, 2005*

Camptothecins in Cancer Therapy, edited by *Thomas G. Burke and Val R. Adams, 2005*

Combination Cancer Therapy: *Modulators and Potentiators,* edited by *Gary K. Schwartz, 2005*

Cancer Chemoprevention, *Volume 2: Strategies for Cancer Chemoprevention,* edited by *Gary J. Kelloff, Ernest T. Hawk, and Caroline C. Sigman, 2005*

Death Receptors in Cancer Therapy, edited by *Wafik S. El-Deiry, 2005*

Cancer Chemoprevention, *Volume 1: Promising Cancer Chemopreventive Agents,* edited by *Gary J. Kelloff, Ernest T. Hawk, and Caroline C. Sigman, 2004*

Proteasome Inhibitors in Cancer Therapy, edited by *Julian Adams, 2004*

Nucleic Acid Therapeutics in Cancer, edited by *Alan M. Gewirtz, 2004*

DNA Repair in Cancer Therapy, edited by *Lawrence C. Panasci and Moulay A. Alaoui-Jamali, 2004*

Hematopoietic Growth Factors in Oncology: *Basic Science and Clinical Therapeutics,* edited by *George Morstyn, MaryAnn Foote, and Graham J. Lieschke, 2004*

Handbook of Anticancer Pharmacokinetics and Pharmacodynamics, edited by *William D. Figg and Howard L. McLeod, 2004*

Anticancer Drug Development Guide: *Preclinical Screening, Clinical Trials, and Approval, Second Edition,* edited by *Beverly A. Teicher and Paul A. Andrews, 2004*

Handbook of Cancer Vaccines, edited by *Michael A. Morse, Timothy M. Clay, and Kim H. Lyerly, 2004*

Drug Delivery Systems in Cancer Therapy, edited by *Dennis M. Brown, 2003*

Oncogene-Directed Therapies, edited by *Janusz Rak, 2003*

Cell Cycle Inhibitors in Cancer Therapy: *Current Strategies,* edited by *Antonio Giordano and Kenneth J. Soprano, 2003*

Chemoradiation in Cancer Therapy, edited by *Hak Choy, 2003*

Fluoropyrimidines in Cancer Therapy, edited by *Youcef M. Rustum, 2003*

Targets for Cancer Chemotherapy: *Transcription Factors and Other Nuclear Proteins,* edited by *Nicholas B. La Thangue and Lan R. Bandara, 2002*

Tumor Targeting in Cancer Therapy, edited by *Michel Pagé, 2002*

Hormone Therapy in Breast and Prostate Cancer, edited by *V. Craig Jordan and Barrington J. A. Furr, 2002*

Tumor Models in Cancer Research, edited by *Beverly A. Teicher, 2002*

Tumor Suppressor Genes in Human Cancer, edited by *David E. Fisher, 2001*

Matrix Metalloproteinase Inhibitors in Cancer Therapy, edited by *Neil J. Clendeninn and Krzysztof Appelt, 2001*

Farnesyltransferase Inhibitors in Cancer, edited by *Saïd M. Sebti and Andrew D. Hamilton, 2001*

Transforming Growth Factor-β in Cancer Therapy, Volume I

Basic and Clinical Biology

Edited by

Sonia B. Jakowlew, PhD

Cancer Training Branch, Office of Centers, Training and Resources, National Cancer Institute, Bethesda, MD

Foreword by

Michael B. Sporn, MD

Department of Pharmacology, Dartmouth Medical School, Hanover, NH

Humana Press

999 Riverview Drive, Suite 208
Totowa, New Jersey 07512

www.humanapress.com

This publication is printed on acid-free paper. ∞

ANSI Z39.48-1984 (American National Standards Institute) Permanence of Paper for Printed Library Materials.

Production Editor: Michele Seugling

Cover design by Nancy Fallatt

For additional copies, pricing for bulk purchases, and/or information about other Humana titles, contact Humana at the above address or at any of the following numbers: Tel: 973-256-1699; Fax: 973-256-8341; or visit our Website: http://humanapress.com

Printed in the United States of America. 10 9 8 7 6 5 4 3 2 1

e-ISBN: 978-1-59745-292-2

Library of Congress Control Number: 2007931768

In Memoriam

All of us in the transforming growth factor-β (TGF-β) research community were deeply saddened to learn of the death of Anita B. Roberts on May 26, 2006. Ironically, Anita died from cancer, a disease that is now closely linked to the peptide growth factor she and her colleagues discovered in the early 1980s and that will forever be associated with her name. As I reflect on her life and her contributions to science, three memories stand out that I believe illustrate the extraordinary person that Anita was.

I first met Anita in Mike Sporn's office at the NCI in the summer of 1984. I was a young scientist working for a biotechnology company in the Bay Area of California. Our research team had discovered two growth factors, designated Chondrogenic Induction Factors A and B (CIF-A and CIF-B), that appeared to be highly related to a growth factor called TGF-β that had been recently purified and characterized by Anita. Several of my colleagues and I met with Mike and Anita to share our data with them and to discuss how we might collaborate to more clearly elucidate the relationships between TGF-β, CIF-A and CIF-B. Eventually we determined that CIF-A was TGF-β1 and CIF-B was TGF-β2. I clearly remember in my first meeting with Anita, her high levels of intensity, inquisitiveness, and energy. I was also struck by her scientific rigor and attention to detail. These were professional characteristics that would become hallmarks of her scientific career during the next 22 years.

A second memory I have of Anita was related to a Wound Healing Society meeting we attended together in the early 1990s in Richmond, Virginia. One evening, several of us ended up in a "drinking establishment" listening to a Blue Grass band. Late in the evening, Anita and I walked back to the hotel where we were staying, when I asked her about her family, since all we ever seemed to talk about was science. Anita told me that she married her husband Bob who she had first met when she was a freshman in high school. It was clear to me that she loved her husband and children very much and was very proud of their accomplishments. It was refreshing for me to see someone who was both successful in the pursuit of their career and, at the same time, so committed to their family life.

One of my final memories of Anita was at the 2005 Keystone Symposium on TGF-β that Anita, Mike Sporn and I had co-organized. Shortly after the three of us had agreed to organize the meeting, Anita was diagnosed with gastric carcinoma. I was stunned by the news and remember lying awake that night trying to imagine how Anita must feel. Given her prognosis, I thought it was unlikely she would survive the next 12 months to even attend the meeting. Much to my amazement, not only did Anita attend, but she sat with me during the entire meeting completely immersed in the science and reminiscing about how the TGF-β field had progressed during the past 25 years. Anita's attendance at that meeting and another TGF-β meeting held in San Diego in February, 2006, symbolized her incredible courage, determination, and perseverance.

Anita was truly a remarkable individual. She exhibited an unusual balance of scientific brilliance and humility. She was a very caring and loving person who changed the lives of many of the people she touched. For those of us working in the field of TGF-β, her death is a terrible loss. On the other hand, Anita would want us to "press on," to explore the unknown and determine whether this growth factor she discovered 25 years ago is a valid therapeutic target for diseases like cancer and pathological fibrosis. This will continue to be a difficult mission, but remembering Anita's personal and professional qualities should fortify our efforts and sustain us when we are discouraged. Although Anita is no longer with us, her indomitable spirit lives on to provide us inspiration and hope.

John M. McPherson, PhD

FOREWORD

Transforming Growth Factor-β in Cancer Therapy, Volume I: Basic and Clinical Biology

The present volume brings together a wealth of information that is fundamental to understanding the role of TGF-β in the pathogenesis, prevention, and treatment of cancer. It is not even 25 years since TGF-β was first isolated and characterized as a dimeric peptide from both human and bovine sources *(1–3)*, but the entire field of TGF-β research has grown and expanded so that it is now a central theme in all of cell biology. There is almost no tissue or organ in the mammalian body in which TGF-β does not play a central role in embryonic differentiation or in adult function, and furthermore, malfunction of the normal physiology of TGF-β can have disastrous consequences in almost all of these sites. Therefore, the present comprehensive review of so many aspects of TGF-β function is a most welcome attempt to bring together a huge body of experimental data that is of the utmost importance in the field of oncology.

There are 45 chapters in this volume, which start with the most fundamental aspects of the molecular biology of TGF-β, with particular emphasis on the critical role of the unique receptors for TGF-β and Smad signaling. Furthermore, a full 13 chapters are devoted to members of the TGF-β superfamily (other than TGF-β itself), including bone morphogenetic proteins, activins, inhibins, and Mullerian inhibiting substances, all of which have multifunctional actions that resemble those of TGF-β and which signal through similar Smad pathways. The concluding 14 chapters deal with specific aspects of TGF-β in inflammation and fibrosis, processes that are intimately involved with the genesis and metastasis of carcinoma.

Thus, the reader of this volume will be treated to an unusual and unique overview of the role of TGF-β in basic and clinical cancer biology. The topic is an extremely complicated one because of the multifunctional nature of TGF-β. As we have noted many times before *(4)*, TGF-β may be considered the prototypical multifunctional signaling molecule. As is the case for all the other peptide growth factors, it is an element of a complex biological signaling language, providing the basis for intercellular (and perhaps even intracellular) communication in higher organisms. Like a symbol or a letter of the alphabet in a language or code, the meaning of the action of TGF-β can only be considered in a cellular context. TGF-β always acts as a member of a set of other signals, and to understand its action, one must always consider the biological context in which it acts. The successful manipulation of TGF-β for control of cancer will depend on understanding this biological context, because TGF-β has unique potential, both to suppress, as well as to enhance, the development and progression of malignancy.

Many problems in TGF-β physiology still remain unanswered *(5)*. One of the most notable problems is the role of the TGF-β that is found intracellularly, most particularly in mitochondria. Mitochondrial TGF-β simply does not fit into the classical receptor/ Smad signaling paradigm, and has unfortunately received relatively little attention.

However, given the emphasis on the role of TGF-β in apoptosis, and the importance of mitochondrial function for control of apoptosis, one may anticipate that this problem, too, will be solved in the not-too-distant future. Yet another intriguing problem for which there is presently almost no hard information is whether TGF-β might be covalently modified, as by alkylation, acylation, phosphorylation, or prenylation, after it is synthesized on the ribosome. Such covalent modification would allow for a great increase in the regulatory activities of TGF-β, as has been shown elegantly for histones, with the development of the "histone code" *(6)*. Thus, the present volume, rich as it is in abundant and important information, may be regarded not only as an excellent summary of the state of the art in TGF-β research, but also a harbinger of significant exciting new developments to come.

Michael B. Sporn, MD
Department of Pharmacology,
Dartmouth Medical School,
Hanover, NH

REFERENCES

1. Frolik CA, Dart LL, Meyers CA, Smith DM, Sporn MB. Purification and initial characterization of a type-beta transforming growth factor from human placenta. Proc Natl Acad Sci USA 1983;80:3676–3680.
2. Assoian RK, Komoriya A, Meyers CA, Sporn MB. Transforming growth factor-beta in human platelets: Identification of a major storage site, purification, and characterization. J Biol Chem 1983;258:7155–7160.
3. Roberts AB, Anzano MA, Meyers CA, Wideman J, Blacher R, Pan Y-C E, Stein S, Lehrman SR, Smith JM, Lamb LC, Sporn MB. Purification and properties of a type beta transforming growth factor from bovine kidney. Biochemistry 1983;22:5692–5698.
4. Sporn MB, Roberts AB. TGF-β: Problems and prospects. Cell Regulation 1990;1:875–882.
5. Sporn MB. The early history of TGF-β, and a brief glimpse of its future. Cytokine Growth Factor Rev 2006;17:3–7.
6. Fischle W, Wang Y, Allis CD. Binary switches and modification cassettes in histone biology and beyond. Nature 2003;425:475–479.

Preface

The transforming growth factor-β (TGF-β) superfamily of secreted signaling proteins shows high conservation among eukaryotes. This superfamily is composed of proteins that regulate cell fate in development and homeostasis including tissue remodeling, histogenesis and maintenance of epithelial homeostasis. TGF-β was originally named for its ability to induce malignant behavior of normal fibroblasts and it was proposed that TGF-β might play a role in uncoupling a cell from normal growth control. Paradoxically, TGF-β exhibits a ubiquitous pattern of expression in normal developing and adult tissues, and unlike most polypeptide growth factors, TGF-β is produced by, and can act on, nearly every cell type. Numerous studies have established that the TGF-β signaling mechanism begins with TGF-β ligand binding to TGF-β receptors that causes receptor serine/threonine kinases to phosphorylate and activate receptor-regulated Smads, and/or initiate non-Smad signaling through activation of mitogen-activated protein kinases, phosphatidylinositol 3-kinase and other mediators. The receptor-regulated Smads heterooligomerize with the common Smad, Smad4, before translocating to the nucleus, where they regulate gene expression. Mutations and epigenetic dysregulation of TGF-β signaling mechanisms occur commonly in major human diseases including cancer, fibrosis, and immune and vascular diseases. The TGF-β signaling system controls a wide range of cellular functions that depend on cell type and physiological or patho-physiological context. In epithelial cells, TGF-β may play several roles including inhibition of cell growth, initiation of apoptosis and induction of epithelial to mesenchymal transition. In contrast, the effects of TGF-β on cellular growth and apoptosis in stromal fibroblasts are minor compared with its potent ability to stimulate cell-matrix adhesion and matrix remodeling and promotion of cell motility. Elucidation of cell type- and context-dependent molecular signaling mechanisms that control the variations in functional specificity of TGF-β signaling is extremely important in understanding key processes that occur in normal development and homeostasis and how these processes change in cancer and disease. With the discovery that TGF-β is a potent growth inhibitor of epithelial cells and the identification of inactivating mutations within the TGF-β signaling pathway in cancer, it has become clear that the TGF-β signaling system is a tumor suppressor pathway in early stages of cancer progression. However, many human cancers show increased expression of TGF-β that is associated with poor patient prognosis and increased frequency of metastasis. The stage-specific duality of multifunctional TGF-β is the emerging paradigm for the role of TGF-β in cancer and disease and the mechanism by which the switch of TGF-β from heroic tumor suppressor to villainous pro-oncogene occurs is the subject of intense investigation. New therapeutic opportunities may emerge from a clearer understanding of the molecular and cellular contexts that permit the tumor suppressor versus oncogenic activities of TGF-β.

It has been nearly 25 years since TGF-β was discovered and several thousand articles have been published about the role of TGF-β in normal and tumor cells. During the same time, there has been a large increase in our understanding of TGF-β in cancer and disease. There has also been a significant change in the general direction of therapeutics' discovery and development. Both protein therapeutics and small molecule therapeutics are now described as being molecularly targeted. It is time to review the field of TGF-β in cancer and cancer therapy in the post-genomic, molecularly targeted era.

As part of the *Cancer Drug Discovery and Development* series, *Transforming Growth Factor-β in Cancer Therapy* attempts to provide an overview of the current status of knowledge about TGF-β in basic and clinical biology and in cancer treatment and therapy. As with other volumes in the series, *Transforming Growth Factor-β in Cancer Therapy* is intended as a resource for both new and experienced investigators who are interested in the fields of polypeptide growth factors, cancer, and human disease, especially since TGF-β regulates and is regulated by a host of growth factors and transcription factors. Persons who are not working directly in the field, but who have a desire to learn about TGF-β and new approaches for using the TGF-β signaling pathway components for treatment and therapy in cancer and disease, will also benefit from this book. Each chapter presents a thorough review of the specific subject matter along with current state-of-the-art information by a leading expert in the field.

Transforming Growth Factor-β in Cancer Therapy is presented in two companion volumes. The aim of Volume I is to provide a compendium of findings about the role of TGF-β in basic and clinical biology. The book examines in detail basic concepts of TGF-β signaling in normal physiology and cancer pathobiology that have been elucidated in the past two and a half decades. Volume I begins with a Foreword authored by the elder statesman, Michael B. Sporn, MD, who along with his colleague, Anita B. Roberts, PhD, originally discovered TGF-β and pioneered studies on its structure, function, expression and signaling. Volume I contains 45 chapters and is divided into three parts. Part I presents basic concepts of TGF-β signaling in normal physiology and cancer pathobiology including the topics TGF-β in homeostasis, latent TGF-β and its activation and availability for interaction with latent TGF-β binding proteins in in vitro and in vivo microenvironments, Smad-dependent and Smad-independent pathways in TGF-β signaling, nucleocytoplasmic shuttling mechanisms of TGF-β, transcriptional regulation of the TGF-β ligand isoforms, the role of the proteasome in controlling TGF-β signaling, crosstalk of TGF-β with other regulatory molecules and signaling pathways, TGF-β signaling in epithelial to mesenchymal transition, mechanisms of TGF-β-induced apoptosis and cell cycle regulation, interaction of TGF-β with matrix metalloproteinases, and interaction of TGF-β with oncogenes like Ras in mouse models of tumorigenesis and their relation to human cancer. In addition to prototypical TGF-β, Part II explores the role of several other members of the TGF-β superfamily in normal and tumor biology including Mullerian inhibiting substances, growth and differentiation factors, bone morphogenetic proteins, activins, inhibins, endoglin and Betaig-h3. A proteomics analysis of TGF-β superfamily members is presented, along with studies of regulation of the TGF-β superfamily by betaglycan and myostatin. Part III examines the importance of TGF-β in inflammation and fibrosis including the roles of triterpenoids, vitamin D, Smads and thrombospondin in TGF-β signaling in the fibrotic response, gene therapy using ultrasound-microbubble-mediated inducible Smad7, negative regulation of signaling by inhibitory Smads, and use of TGF-β peptide inhibitors and TGF-β antisense oligonucleotides for therapy of fibrosis and an overexpressed truncated TGF-β type II receptor that inhibits fibrotic behavior. The aim of companion Volume II of *Transforming Growth Factor-β in Cancer Therapy* is to provide a compendium of findings about the role of TGF-β in cancer treatment and therapy. Volume II begins with a Foreword penned by the elder statesman Carl-Henrik Heldin, PhD, who, along with his pioneering studies on the structure and function of platelet derived growth factor (PDGF) and applications of PDGF to new treatments of cancer, has also conducted seminal studies on mechanisms of TGF-β signaling and its inhibition, and is now applying TGF-β to new regimens for treating cancer. As in Volume I, Volume II of *Transforming Growth Factor-β in Cancer Therapy* is divided into three parts and contains 46 chapters. Part I examines TGF-β in developing and advanced cancers with the role of TGF-β in the development and progression of several different cancer types including cancers of the pancreas, head and neck, cervix, endometrium, lung, bone, breast, esophagus, colon, kidney, brain, prostate and blood. Part II presents an examination of TGF-β in cancer treatment and therapy. Changes in the TGF-β signaling network in human neoplasia are presented, along with the bifunctionality of TGF-β as both a tumor suppressor and a pro-progression factor in metastasis, mutations that have been shown to occur in the TGF-β ligands, TGF-β receptors and Smads in various cancers, and the role of TGF-β in immunity, immune suppression, angiogenesis, hematopoiesis and vascular morphogenesis. In addition, a hierarchial molecular profiling of TGF-β in progressive tumorigenesis and the predictions that follow from such an approach are discussed. The use of adenovirus-mediated gene transfer of TGF-β signal antagonists is discussed, along with the role of TGF-β in DNA damage responses and tumor protection and anti-tumor immunity, manipulation of TGF-β signaling to enhance therapy, activation of multiple protein kinases by TGF-β that may enhance negative growth control, cancer-associated fibroblasts as novel targets in anti-cancer therapy, DNA methylation and histone deacetylation inhibitors as therapeutic agents for reconstitution of TGF-β signaling, and reactive stroma in the evolution of tumors and malignant invasiveness. Part III explores the development of inhibitors of TGF-β signaling for therapy. These inhibitors include activin receptor-like kinase inhibitors as antagonists of TGF-β signaling, soluble TGF-β type II and type III receptors that inhibit tumorigenesis and malignant progression, isoform-specific anti-TGF-β antibodies, small molecule inhibitors of TGF-β type I and type II receptors to reverse epithelial to mesenchymal transition, a TGF-β-related tumor protection strategy to enhance anti-tumor immunity and small-binding peptide aptamers that interfere with TGF-β signaling.

It is not possible to include every important contribution that has been made to our understanding of the role of TGF-β in basic and clinical biology and in cancer treatment and therapy in *Transforming Growth Factor-β in Cancer Therapy* because of space considerations. Each part of Volume I and Volume II of *Transforming Growth Factor-β in Cancer Therapy* could readily be expanded to be a volume in itself. Apologies are extended to those investigators whose important contributions could not be included.

I would like to express my gratitude to all the contributors of *Transforming Growth Factor-β in Cancer Therapy* who responded so promptly to my repeated requests and suggestions during the preparation of their chapters, and for their patience while I collated and edited the volumes. I am very grateful to Beverly A. Teicher, PhD, for her guidance and suggestions as series editor. I am especially grateful to my mentor, colleague and friend, Anita B. Roberts, PhD, the third most cited female researcher in the United States, who despite the illness that, sadly, eventually took her life in May, 2006, gave me many valuable suggestions about the content of *Transforming Growth Factor-β in Cancer Therapy*. John M. McPherson, PhD, has contributed an In Memoriam for Anita B. Roberts that appears at the beginning of both volumes.

This book is dedicated to my husband, Edward W. Maxwell, for his never-ending love, support, encouragement and understanding about the importance and need for this book at this time.

Sonia B. Jakowlew, PhD

Biography

Sonia B. Jakowlew, PhD, is currently a Program Director in the Cancer Training Branch of the National Cancer Institute in Bethesda, Maryland, after being a Principal Investigator in the Cell and Cancer Biology Branch of the National Cancer Institute for over 15 years. Her research focuses on the role of transforming growth factor-β (TGF-β) in normal epithelial homeostasis and carcinogenesis, with emphasis on lung tumorigenesis. Her recent projects have included 1) developing new mouse model systems that are relevant to lung cancer in which the functioning of the TGF-β pathway is experimentally compromised by deletion of TGF-β ligand and downstream components; 2) examining the complex dual role of TGF-β as both a tumor suppressor and oncogene by novel approaches including analysis of chemically-initiated and spontaneous lung tumorigenesis in mutated mice; 3) characterizing molecular mechanisms underlying the changes in cell cycle regulation and growth control in experimental systems in which TGF-β function is compromised; and 4) identifying known and novel genes that are regulated by TGF-β in normal and malignant lung cells and characterizing the differences in their regulation. Dr. Jakowlew conducted research in isolating and characterizing myosin heavy chain mRNA in the embryonic chicken for her dissertation at the Roche Institute of Molecular Biology under the direction of Mohammed A.Q. Siddiqui, PhD, and received her doctorate in biochemistry from Rutgers University in 1981 with highest honors. She completed a post-doctoral research fellowship at the Laboratoire de Genetique Moleculaire des Eucaryotes of the Institute de Chimie Biologique in Strasbourg, France, with Prof. Pierre Chambon where she investigated and identified genes that were differentially responsive to the effects of estradiol hormone in human breast cancer cells. She joined the National Cancer Institute as a Staff Fellow in 1984 in the Laboratory of Chemoprevention headed by Michael B. Sporn, MD, and worked under the direction of Anita B. Roberts, PhD, to identify genes and promoters for multiple TGF-β isoforms, including TGF-β2, TGF-β3 and TGF-β4 in the chicken embryo and to characterize their expression during embryogenesis and development. She has made the cDNAs to these various genes available to investigators in various laboratories throughout the world. Dr. Jakowlew has authored and co-authored over 70 articles in peer-reviewed journals and has over a dozen book chapters to her credit. She has mentored numerous post-doctoral fellows, graduate students, undergraduate students, and high school students, and has received awards for her outstanding mentorship abilities while at the National Cancer Institute.

TABLE OF CONTENTS

In Memoriam .... *v*

Foreword .... *vii*

Preface .... *ix*

Biography .... *xiii*

Contributors .... *xix*

PART I. BASIC CONCEPTS OF TGF-β SIGNALING IN NORMAL PHYSIOLOGY AND CANCER PATHOBIOLOGY

1 Component Hardware for Transforming Growth Factor-β Signal Transduction: *TGF-β Ligands, TGF-β Receptors, and Smads* **3**
Sonia B. Jakowlew

2 TGF-β Signaling in Homeostasis and Cancer **23**
Joan Seoane

3 TGF-β Availability: *Latent TGF-β and Latent TGF-β Binding Proteins* **37**
Vesna Todorovic and Daniel B. Rifkin

4 Biological Functions of Latent TGF-β-Binding Proteins and Activation of TGF-β **57**
Marko Hyytiäinen, Katri Koli, and Jorma Keski-Oja

5 Regulators of Smad2/3 Transcription and Phosphorylation **77**
Azeddine Atfi and Marie-Francoise Bourgeade

6 Interaction of Smad4 and Embryonic Liver Fodrin-β-spectrin in Hyperplasia, Neoplasia, and Tumor Suppression **91**
Sang Soo Kim, Charles Shuman, and Lopa Mishra

7 Regulation of Smad Activity by Phosphorylation **105**
Fang Liu

8 Nuclear Targeting of TGF-β-Activated Smads in Normal and Tumor Biology **125**
Lan Xu

9 Ski, SnoN, and Akt as Negative Regulators of Smad Activity: *Balancing Cell Death and Cell Survival* **139**
Erwan Le Scolan and Kunxin Luo

10 Transforming Growth Factor-β in Cancer Therapy: *Smurfs in TGF-β Signaling and Regulation of Bone Homeostasis and Cancer* **155**
Ying E. Zhang and Motozo Yamashita

11 Involvement of km23 Dynein Light Chains in TGF β Signaling **169**
Qunyan Jin, Guofeng Gao, and Kathleen M. Mulder

12 TGF-β Signaling in Fibroblastic Cells and Oncogenesis **185**
Neil A. Bhowmick

13 Mechanisms of TGF-β-Induced Apoptosis in Cancer Cells **199**
Tae-Aug Kim and Seong-Jin Kim

14 Mechanisms of Cell Cycle Regulation by TGF-β Disabled in Cancer **213**
Leslie I. Gold and Jon Lecanda

15 Transforming Growth Factor-β in Cancer Therapy: *Proteasome in the Control of TGF-β Signaling During Development and Cancer* **243**
Arja Band and Marikki Laiho

16 TGF-β and Smad Signaling in Transcriptome Reprogramming During EMT **259**
Sylvie Thuault, Ulrich Valcourt, Marcin Kowanetz, and Aristidis Moustakas

17 Functional Interactions Between MMPs and TGF-β in Normal and Tumor Tissue **275**
Christina H. Stuelten, Anita B. Roberts, and William G. Stetler-Stevenson

18 Cooperation Between TGF-β and Wnt Pathway Components in Regulating Mesenchymal Stem Cell Function **293**
Irwin M. Liu, Stephen H. Schilling, and Xiao-Fan Wang

19 Interaction of Oncogenic Ras and TGFβ1 Signaling in Cancer: *Lessons From the Multistage Skin Carcinogenesis Model* **305**
Adam Glick, Dong-Soon Bae, Xin Liu, and Kinnimulki Vijayachandra

PART II. TGF-β SUPERFAMILY MEMBERS IN NORMAL AND TUMOR BIOLOGY

20 Proteomics Analysis of TGFβ Superfamily Members **319**
Serhiy Souchelnytskyi

21 Mullerian Inhibiting Substance: *A Potential Nontoxic, Naturally Occurring Therapeutic Agent for Certain Human Cancers* **333**
David T. MacLaughlin and Patricia K. Donahoe

22 Activins, Inhibins, and Bone Morphogenetic Proteins as Modulators and Biomarkers of Prostate Cancer Progression **355**
Tetsu Hayashida, Vandana Gupta, Sam Thiagalingam, and Shyamala Maheswaran

23 Bone Morphogenic Proteins and Angiogenesis in Developing Tumors **373**
George Berberian, Deepak Malhotra, and John E. Langenfeld

24 Histone Methyltransferase and Smad Interactions in BMP Signaling **383**
Jennifer E. Leader, M. Pilar Frontelo, and Robert J. Lechleider

25 Endoglin (CD105): *A Strong Candidate for Immunologic Targeting of Tumor Neovasculature in Human Malignancies* **395**
Ester Fonsatti and Michele Maio

26 Activins and Inhibins in Cancer Progression **411**
Gail P. Risbridger and Christopher Butler

27 Bone Morphogenetic Proteins/Growth Differentiation Factors and Smad Activation in Ovarian Granulosa Cells and Carcinoma **425**
Sabine Mazerbourg

28 Regulation of the Transforming Growth Factor-β Superfamily by Betaglycan **451**
Fernando López-Casillas, M. Magdalena Vilchis-Landeros, José Esparza-López, M. Verónica Ponce-Castañeda, N. Gabriela Velasco-Loyden, Patricia Juárez, Jose Luis Montiel, and Valentín Mendoza

29 Uterine Sensitization Associated Gene-1: *A Bone Morphogenetic Protein Antagonist with a Role in Kidney* **463**
Motoko Yanagita

30 Role of Betaig-h3 Gene in Carcinogenesis **475**
Yongliang Zhao and Tom K. Hei

31 Activins and the Liver **483**
Michael Grusch, Chantal Rodgarkia-Dara, Wilfried Bursch, and Rolf Schulte-Hermann

32 Myostatin Signaling in Normal and Tumor Cells **509**
Xiangyang Zhu, Li-fang Liang, and Ronald L. Stotish

PART III. TGF-β IN INFLAMMATION AND FIBROSIS

33 Triterpenoids, Vitamin D, and TGF-β/Smad Signaling in Inflammation and Cancer **525**
Nanjoo Suh

34 Thrombospondin-Dependent Activation of Latent TGF-β in Fibrosis and Disease **549**
Joanne E. Murphy-Ullrich

35 TGF-β in Pulmonary Fibrosis **569**
Masahito Ebina

36 The Role of TGF-β in Bleomycin Induced Pulmonary Fibrosis **581**
Nasreen Khalil and Robert O'Connor

37 Smads in the Fibrotic Response: *Findings in the Smad3 Knockout Mouse* **595**
Kathleen C. Flanders

38 TGF-β Signaling in Renal Fibrosis and Inflammation: *Gene Therapy Using Ultrasound-Microbubble-Mediated Inducible Smad7* **609**
Hui Y. Lan

39 Transforming Growth Factor-β Peptide Signaling in Pulmonary Development, Bronchopulmonary Dysplasia, Fibrosis, and Emphysema **621**
David Warburton, Wei Shi, Martin Kolb, and Jack Gauldie

40 The Role of TGF-β in Radiation and Chemotherapy Induced Pulmonary Fibrosis: *Inhibition of TGF-β as a Novel Therapeutic Strategy* **629**
Patricia J. Sime, R. Matthew Kottmann, Heather F. Lakatos, and Thomas H. Thatcher

41 Negative Regulation of the TGF-β Family Signal Pathway by Inhibitory Smads and Their Involvement in Cancer and Fibrosis **649**
Susumu Itoh, Hiroyuki Suzuki, Fumiko Itoh, Kohei Miyazono, and Mitsuyasu Kato

42 TGF-β-Related Antifibrotic Strategies in the Eye **663**
M. Francesca Cordeiro, Annelie Maass, Clive Migdal, and Li Guo

43 Transforming Growth Factor-β in Cutaneous Tissue Repair **675**
Jisun Cha and Vincent Falanga

44 Topical Application of TGF-β1 Peptide Inhibitors for the Therapy of Skin Fibrosis **693**
Javier Dotor and José L. Pablos

45 Overexpressed Truncated TGF-β Type II Receptor Inhibits Fibrotic Behavior of Keloid Fibroblasts In Vitro and Experimental Scar Formation In Vivo **703**
Wei Liu, Chekhau Chua, Zhen Gao, Xiaoli Wu, and Yilin Cao

Index **723**

Contributors

Azeddine Atfi, PhD • INSERM U673, Hopital Saint-Antoine, Paris, France

Dong-Soon Bae, PhD • Department of Pharmacology and Physiology, The George Washington University, Washington, DC, USA

Arja Band, PhD • Molecular and Cancer Biology Program, Biomedicum Helsinki, The University of Helsinki, Helsinki, Finland

George Berberian, MD • Department of Surgery, Division of Thoracic Surgery, UMDNJ-Robert Wood Johnson Medical School, New Brunswick, NJ, USA

Neil A. Bhowmick, PhD • Department of Urologic Surgery, Vanderbilt University Medical Center, Nashville, TN, USA

Marie-Francoise Bourgeade, PhD • INSERM U785, Hopital Paul Brousse, Villejuif, France

Wilfried Bursch, PhD • Department of Medicine, Institute of Cancer Research, Unit Toxicology and Prevention, Medical University of Vienna, Vienna, Austria

Christopher Butler, PhD • Center for Urological Research, Monash Institute of Medical Research, Monash Medical Center, Monash University, Melbourne, Clayton, Victoria, Australia

Yilin Cao, MD, PhD • Department of Plastic and Reconstructive Surgery, Shanghai Ninth People's Hospital, Shanghai Jiao Tong University School of Medicine, Shanghai, People's Republic of China

Jisun Cha, MD • Department of Dermatology and Skin Surgery, Roger Williams Medical Center, Providence, RI, USA

Chekhau Chua, MD, PhD • Department of Plastic and Reconstructive Surgery, Shanghai Ninth People's Hospital, Shanghai Jiao Tong University School of Medicine, Shanghai, People's Republic of China

M. Francesca Cordeiro, PhD, MD, MRCP, FRCOphth • Department of Pathology, UCL Institute of Ophthalmology and Western Eye Hospital, London, United Kingdom

Patricia K. Donahoe, MD • Pediatric Surgical Research Laboratory, Massachusetts General Hospital, Boston, MA, USA

Javier Dotor, PhD • CIMA, Universidad de Navarra, Pamplona, Spain

Masahito Ebina, MD, PhD • Department of Respiratory Oncology and Molecular Medicine, Institute of Aging, Development and Cancer, Tohoku University, Sendai, Japan

José Esparza-López, PhD • Departamento de Biología Celular, Instituto de Fisiología Celular, Universidad Nacional Autónoma de México, México City, México

Vincent Falanga, MD, FACP • Department of Dermatology and Skin Surgery, Roger Williams Medical Center, Providence, RI, USA

Kathleen C. Flanders, PhD • Laboratory of Cell Regulation and Carcinogenesis, National Cancer Institute, The National Institutes of Health, Bethesda, MD, USA

Ester Fonsatti, PhD • Division of Medical Oncology and Immunotherapy, Department of Oncology, University Hospital of Siena, Istituto Toscano Tumori, Siena, Italy

M. Pilar Frontelo, PhD • Brookdale Department of Molecular, Cell and Developmental Biology, Mount Sinai School of Medicine, New York, NY, USA

Guofeng Gao, MD • Department of Pharmacology, The Pennsylvania State University College of Medicine, Hershey, PA, USA

Zhen Gao, MD • Department of Plastic and Reconstructive Surgery, Shanghai Ninth People's Hospital, Shanghai Jiao Tong University School of Medicine, Shanghai, People's Republic of China

Jack Gauldie, PhD • Department of Pathology and Medical Medicine, Centre for Gene Therapeutics, McMaster University, Hamilton, Ontario, Canada

Adam Glick, PhD • Center for Molecular Toxicology and Carcinogenesis, The Pennsylvania State University, University Park, PA, USA

Leslie I. Gold, PhD • Departments of Medicine and Pathology, New York University Cancer Institute, New York University School of Medicine, New York, NY, USA

Michael Grusch, PhD • Department of Medicine, Institute of Cancer Research, Medical University of Vienna, Vienna, Austria

Li Guo, MD, PhD • Glaucoma and Retinal Neurodegeneration Research Group, UCL Institute of Opthalmalogy, London, United Kingdom

Vandana Gupta, PhD • Department of Medicine, Genetics Program and Cancer Center, Boston University School of Medicine, Boston, MA, USA

Tetsu Hayashida, MD • Surgical Oncology, Massachusetts General Hospital, Boston, MA, USA

Tom K. Hei, PhD • Center for Radiological Research, Columbia University, New York, NY, USA

Marko Hyytiäinen, PhD • Department of Virology, Biomedicum Helsinki, The University of Helsinki, Helsinki, Finland

Fumiko Itoh, PhD • Department of Experimental Pathology, Graduate School of Comprehensive Human Sciences, University of Tsukuba, Tsukuba, Ibaraki, Japan

Susumu Itoh, PhD • Department of Experimental Pathology, Graduate School of Comprehensive Human Sciences, University of Tsukuba, Tsukuba, Ibaraki, Japan

Sonia B. Jakowlew, PhD • Cancer Training Branch, Office of Centers, Training and Resources, National Cancer Institute, Bethesda, MD, USA

Qunyan Jin, MD • Department of Pharmacology, The Pennsylvania State University College of Medicine, Hershey, PA, USA

Patricia Juárez, MSc • Departamento de Biología Celular, Instituto de Fisiología Celular, Universidad Nacional Autónoma de México, México City, México

Mitsuyasu Kato, MD, PhD • Department of Experimental Pathology, Graduate School of Comprehensive Human Sciences, University of Tsukuba, Tsukuba, Ibaraki, Japan

Jorma Keski-Oja, MD, FRCPC, FCCP • Department of Virology, Biomedicum Helsinki, The University of Helsinki, Helsinki, Finland

Nasreen Khalil, MD, FRCPC, FCCP • Division of Respiratory Medicine, Jack Bell Research Center, The University of British Columbia, Vancouver, British Columbia, Canada

Sang Soo Kim, PhD • Radiation Medicine Branch, National Cancer Center, Goyang, Gyeonggi, Korea

Seong-Jin Kim, PhD • Laboratory of Cell Regulation and Carcinogenesis, National Cancer Institute, The National Institutes of Health, Bethesda, MD, USA

Tae-Aug Kim, PhD • Laboratory of Cell Regulation and Carcinogenesis, National Cancer Institute, The National Institutes of Health, Bethesda, MD, USA

Martin Kolb, MD, PhD • Department of Pathology and Medical Medicine, Centre for Gene Therapeutics, McMaster University, Hamilton, Ontario, Canada

Katri Koli, PhD • Department of Virology, Biomedicum Helsinki, The University of Helsinki, Helsinki, Finland

R. Matthew Kottmann, MD • Division of Pulmonary and Critical Care Medicine, The University of Rochester School of Medicine, Rochester, NY, USA

Marcin Kowanetz, PhD • Ludwig Institute for Cancer Research, Biomedical Center, Uppsala, Sweden, and Genentech, Inc., South San Francisco, CA, USA

Marikki Laiho, PhD • Molecular and Cancer Biology Program, Biomedicum Helsinki, The University of Helsinki, Helsinki, Finland

Heather F. Lakatos, MSc • Division of Pulmonary and Critical Care Medicine, The University of Rochester School of Medicine, Rochester, NY, USA

Hui Y. Lan, MD, PhD • Department of Medicine, The University of Hong Kong, Li Ka Shing Faculty of Medicine, Hong Kong

John E. Langenfeld, MD • Department of Surgery, Division of Thoracic Surgery, UMDNJ-Robert Wood Johnson Medical School, New Brunswick, NJ, USA

Jennifer E. Leader, PhD • Interdisciplinary Program in Tumor Biology, Lombardi Cancer Center, Washington, DC, USA

Jon Lecanda, PhD • Departments of Medicine and Pathology, New York University Cancer Institute, New York University School of Medicine, New York, NY, USA

Robert J. Lechleider, MD • Medical Oncology Research Unit, National Cancer Institute, The National Institutes of Health, Bethesda, MD, USA

Erwan Le Scolan, PhD • Department of Molecular and Cell Biology, University of California, Berkeley, CA, USA

Li-fang Liang, PhD • MetaMorphix, Inc., Beltsville, MD, USA

Fang Liu, PhD • Center for Advanced Biotechnology and Medicine, Susan Lehman Cullman Laboratory for Cancer Research, Department of Chemical Biology, Ernest Mario School of Pharmacy, Rutgers, The State University of New Jersey, Cancer Institute of New Jersey, Piscataway, NJ, USA

Irwin M. Liu, BSc • Department of Pharmacology and Cancer Biology, Duke University Medical Center, Durham, NC, USA

Wei Liu, MD, PhD • Department of Plastic and Reconstructive Surgery, Shanghai Ninth People's Hospital, Shanghai Jiao Tong University School of Medicine, Shanghai, People's Republic of China

Xin Liu, MD, PhD • Penn State Cancer Institute, Hershey, PA, USA

Fernando López-Casillas, MD, PhD • Departamento de Biología Celular, Instituto de Fisiología Celular, Universidad Nacional Autónoma de México, México City, México

Kunxin Luo, PhD • Department of Molecular and Cell Biology, University of California, Berkeley, CA, USA

Annelie Maass, MSc • Glaucoma and Retinal Neurodegeneration Research Group, UCL Institute of Opthalmalogy, London, United Kingdom

David T. MacLaughlin, PhD • Pediatric Surgical Research Laboratory, Massachusetts General Hospital, Boston, MA, USA

Shyamala Maheswaran, PhD • Department of Surgical Oncology, Massachusetts General Hospital, Boston, MA, USA

Michele Maio, MD, PhD • Department of Oncology, Division of Medical Oncology and Immunotherapy, The University Hospital of Siena, Istituto Toscano Tumori, Siena, Italy; Cancer Bioimmunotherapy Unit, Centro di Riferimento Oncologico, Aviano, Italy

Deepak Malhotra, MD • Department of Surgery, Division of Thoracic Surgery, UMDNJ-Robert Wood Johnson Medical School, New Brunswick, NJ, USA

Sabine Mazerbourg, PhD • Université Henri Poincairé-Nancy 1, Faculté des Sciences, Nancy, France

John M. McPherson, PhD • Biological Products R&D, Genzyme Corporation, Framingham, MA, USA

Valentín Mendoza, MD • Departamento de Biología Celular, Instituto de Fisiología Celular, Universidad Nacional Autónoma de México, México City, México

Clive Migdal, MD • Western Eye Hospital, St. Mary's Hospital, London, United Kingdom

Lopa Mishra, MD • Departments of Medicine and Surgery, Lombardi Cancer Center, Georgetown University, Department of Surgery, VA Medical Center, Washington, DC, USA

Kohei Miyazono, MD, PhD • Department of Molecular Pathology, Graduate School of Medicine, University of Tokyo, Tokyo, Japan

Jose Luis Montiel, PhD • Departamento de Biología Celular, Instituto de Fisiología Celular, Universidad Nacional Autónoma de México, México City, México

Aristidis Moustakas, PhD • Ludwig Institute for Cancer Research, Biomedical Center, Uppsala, Sweden

Kathleen M. Mulder, PhD • Department of Pharmacology, The Pennsylvania State University College of Medicine, Hershey, PA, USA

Joanne E. Murphy-Ullrich, PhD • Department of Pathology and the BioMatrix Engineering and Regenerative Medicine Center, The University of Alabama at Birmingham, Birmingham, AL, USA

Robert O'Connor, MD, FRCPC • Department of Pathology, Vancouver General Hospital, Vancouver, British Columbia, Canada

José L. Pablos, MD, PhD • Servicio de Reumatología, Hospital 12 de Octubre, Madrid, Spain

M. Verónica Ponce-Castañeda, MD, PhD • Departamento de Biología Celular, Instituto de Fisiología Celular, Universidad Nacional Autónoma de México, México City, México

Daniel B. Rifkin, PhD • Department of Cell Biology and Medicine, New York University School of Medicine, New York, NY, USA

Gail P. Risbridger, PhD • Center for Urological Research, Monash Institute of Medical Research, Monash Medical Center, Monash University, Melbourne, Clayton, Victoria, Australia

Anita B. Roberts, PhD • Laboratory of Cell Regulation and Carcinogenesis, National Cancer Institute, The National Institutes of Health, Bethesda, MD, USA

Chantal Rodgarkia-Dara, PhD • Department of Medicine, Institute of Cancer Research, Unit Toxicology and Prevention, Medical University of Vienna, Vienna, Austria

Stephen H. Schilling, AB • Department of Pharmacology and Cancer Biology, Duke University Medical Center, Durham, NC, USA

Rolf Schulte-Hermann, MD, PhD • Department of Medicine, Institute of Cancer Research, Unit Toxicology and Prevention, Medical University of Vienna, Vienna, Austria

Joan Seoane, PhD • Medical Oncology Program, Vall d'Hebron University Hospital Research Institute, Barcelona, Spain

Wei Shi, MD, PhD • Developmental Biology Program, Saban Research Institute, Children's Hospital of Los Angeles, Los Angeles, CA, USA

Charles Shuman, MD • Department of Surgery, VA Medical Center, Washington, DC, USA

Patricia J. Sime, MD, FRCP • Division of Pulmonary and Critical Care Medicine, The University of Rochester School of Medicine, Rochester, NY, USA

Serhiy Souchelnytskyi, PhD • Karolinska Biomics Centre, Karolinska University Hospital, Stockholm, Sweden

Michael B. Sporn, MD • Department of Pharmacology, Dartmouth Medical School, Hanover, NH, USA

William G. Stetler-Stevenson, MD, PhD • Cell and Cancer Biology Branch, National Cancer Institute, The National Institutes of Health, Bethesda, MD, USA

Ronald L. Stotish, PhD • MetaMorphix, Inc., Beltsville, MD, USA

Christina H. Stuelten, MD, PhD • Laboratory of Cell Regulation and Carcinogenesis, National Cancer Institute, The National Institutes of Health, Bethesda, MD, USA

Nanjoo Suh, PhD • Department of Chemical Biology, Susan Lehman Cullman Laboratory for Cancer Research, Ernest Mario School of Pharmacy, Rutgers, The State University of New Jersey, Piscataway, NJ, USA

Hiroyuki Suzuki, PhD • Department of Experimental Pathology, Graduate School of Comprehensive Human Sciences, University of Tsukuba, Tsukuba, Ibaraki, Japan

Thomas H. Thatcher, PhD • Division of Pulmonary and Critical Care Medicine, The University of Rochester School of Medicine, Rochester, NY, USA

Sam Thiagalingam, PhD • Department of Medicine (Genetics Program and Cancer Research Center), Boston University School of Medicine, Boston, MA, USA

Sylvie Thuault, PhD • Ludwig Institute for Cancer Research, Biomedical Center, Uppsala, Sweden

Vesna Todorovic, PhD • Department of Cell Biology, New York University School of Medicine, New York, NY, USA

Ulrich Valcourt, PhD • INSERM, Hopital E. Herriot, Lyon, France

N. Gabriela Velasco-Loyden, PhD • Departamento de Biología Celular, Instituto de Fisiología Celular, Universidad Nacional Autónoma de México, México City, México

Kinnimulki Vijayachandra, PhD • Procinth Research Laboratories, Surathkal, India

M. Magdalena Vilchis-Landeros, MD, PhD • Departamento de Biología Celular, Instituto de Fisiología Celular, Universidad Nacional Autónoma de México, México City, México

Xiao-Fan Wang, PhD • Department of Pharmacology and Cancer Biology, Duke University Medical Center, Durham, NC, USA

David Warburton, Dsc, MD, FRCP, FRCS • Developmental Biology Program, Saban Research Institute, Children's Hospital of Los Angeles, Los Angeles, CA, USA

Xiaoli Wu, MD, PhD • Department of Plastic and Reconstructive Surgery, Shanghai Ninth People's Hospital, Shanghai Jiao Tong University School of Medicine, Shanghai, People's Republic of China

Lan Xu, PhD • Program in Molecular Medicine, The University of Massachusetts Medical School, Worcester, MA, USA

Motozo Yamashita, DDS, PhD • Laboratory of Cellular and Molecular Biology, National Cancer Institute, The National Institutes of Health, Bethesda, MD, USA

Motoko Yanagita, MD, PhD • Department of Cardiovascular Medicine, Graduate School of Medicine, Kyoto University, Kyoto, Japan

Ying E. Zhang, PhD • Laboratory of Cellular and Molecular Biology, National Cancer Institute, The National Institutes of Health, Bethesda, MD, USA

Yongliang Zhao, PhD • Center for Radiological Research, Columbia University, New York, NY, USA

Xiangyang Zhu, PhD • MetaMorphix, Inc., Beltsville, MD, USA

I

Basic Concepts of TGF-β Signaling in Normal Physiology and Cancer Pathobiology

1 Component Hardware for Transforming Growth Factor-β Signal Transduction: TGF-β Ligands, TGF-β Receptors, and Smads

Sonia B. Jakowlew

CONTENTS

INTRODUCTION
TGF-β LIGANDS
TGF-β RECEPTORS
SMADS AND TGF-β SIGNAL TRANSDUCTION
INHIBITORY SMADS IN TGF-β SIGNALING
MODULATORS OF SMAD SIGNALING
ACKNOWLEDGMENTS
REFERENCES

Abstract

Transforming growth factor (TGF)-β is a multifunctional regulatory polypeptide that is the prototypical member of a large family of cytokines that controls many aspects of cellular function, including cellular proliferation, differentiation, migration, apoptosis, adhesion, angiogenesis, immune surveillance, and survival. The actions of TGF-β are dependent on several factors including cell type, growth conditions, and the presence of other polypeptide growth factors. The TGF-β signaling pathway has been the subject of several thousand articles that have reinforced the view of the importance of this polypeptide growth factor in normal physiology and pathology. As nearly all cells both produce and respond to TGF-β, understanding the components of the TGF-β signal transduction pathway in normal cells, including the TGF-β ligands, the TGF-β receptors, and the regulatory and inhibitory Smads, and how they may be perturbed in tumor cells, is important. TGF-β exhibits numerous effects that are dependent on the cell type and context, and often appear to be conflicting. Crosstalk with other signaling pathways may provide one explanation that the biological response to TGF-β is often dependent on the extracellular environment of the cell. New therapeutic opportunities may develop from a clearer and more in-depth understanding of the TGF-β signal transduction pathway and the molecular and cellular contexts that permit the tumor suppressor or pro-oncogenic activities of TGF-β.

Key Words: TGF-β ligands; TGF-β receptors; Smads; signal transduction; proliferation; invasion.

From: *Cancer Drug Discovery and Development: Transforming Growth Factor-β in Cancer Therapy, Vol I: Basic and Clinical Biology*
Edited by: Sonia B. Jakowlew © Humana Press Inc., Totowa, NJ

1. INTRODUCTION

Tumor formation in humans is a complicated process that involves multiple events and factors that generally occurs over an extended period of time. Human cancer cells often acquire critical abilities that most normal cells do not ordinarily possess, including resistance to growth inhibition, proliferation without dependence on growth factors, replication without limit, invasion, metastasis, evasion of apoptosis and immune surveillance, and support of angiogenesis through genomic instability. These processes are regulated by complex signal transduction pathways that also control normal cellular homeostasis. Prominent among these signaling pathways is the transforming growth factor-β (TGF-β) pathway. The TGF-β pathway has a complicated role in mediating the ability of cells to participate negatively or positively in growth inhibition, proliferation, replication, invasion, metastasis, apoptosis, immune surveillance, and angiogenesis (Table 1). TGF-β produces context-dependent and cell-specific effects that often appear to be in conflict, including stimulation or inhibition of growth, apoptosis, or differentiation. It is puzzling how such a diverse array of responses can result from binding of TGF-β to a receptor complex that activates a seemingly straightforward signal transduction scheme dependent on shuttling of Smad transducer proteins from the receptor to the nucleus. The genomes of vertebrates encode many TGF-β ligands, fewer types I and II receptors, and only a few Smads. In contrast to the expected simplicity of the signal transduction pathway with few Smads, the cellular responses to TGF-β ligands are complex and context dependent. Despite the prominent role that TGF-β appears to play in regulating multiple cellular processes, use of the TGF-β pathway for molecular targeting has been relatively slow. This is due in part to limited understanding of the molecular mechanisms that are involved in regulating the TGF-β pathway in vivo among many other pathways that could affect outcomes and by the many alterations that can, and do, occur in the TGF-β pathway in different human cancers. This review highlights the findings about the component molecules that compose the system hardware that participates in the TGF-β signaling pathway in human cancer and metastasis that have been made in recent years. The focus will be majorly on the prototypical TGF-β1 ligand and receptors in the TGF-β superfamily and the downstream signaling components.

2. TGF-β LIGANDS

The TGF-β signaling pathway has been the subject of several reviews that have reinforced the view of the importance of this polypeptide growth factor in normal physiology and pathology *(1–5)*. As nearly all cells produce and respond to TGF-β, understanding the roles of TGF-β in tumorigenesis requires insight into the changing response patterns of many interacting cell types, including those of the stoma. TGF-β exhibits numerous effects that are dependent on the cell type and context, and often appear to be conflicting. Crosstalk with other signaling pathways may provide one explanation that the biological response to TGF-β is often dependent on the extracellular environment of the cell *(4–6)*. New therapeutic opportunities may develop from a clearer and more in-depth understanding of the molecular and cellular contexts that permit the tumor suppressor or prooncogenic activities of TGF-β.

More than 60 different TGF-β family members have been identified in multicellular organisms, with a minimum of 29 of these proteins encoded in the human. The high number of ligands may be explained by the need for finely tuned developmental patterns of receptor activation, which can be achieved by differential regulation of ligand expression and activation from latent complexes. Among the 60 proteins that are included in the TGF-β superfamily, there are four TGF-β ligands, five activins, eight bone morphogenetic proteins (BMP), and 15 growth and differentiation factors (GDF). Three TGF-β isoforms have been identified in humans, including TGF-β1, TGF-β2, and TGF-β3. The TGF-β ligands are homodimeric

Table 1
Processes Affected by TGF-β in Tumor Cells

Process	*Effect*
Resistance to growth inhibition	Loss of TGF-β growth inhibition
Cell proliferation	Stimulation of proliferation
Invasion	Promotion of invasiveness
Metastasis	Promotion of metastasis
Replication	Loss of repression of telomere enzymes
Evasion of apoptosis	Loss of apoptosis
Induction of angiogenesis	Induction of angiogenesis
Immune surveillance	Immunosuppression

polypeptides with a molecular weight of 25-kDa. Crystallographic analysis indicates that TGF-β2 consists of two monomers, and each monomer consists of two antiparallel pairs of β-strands that form a flat surface and a separate α–helix *(7)*. Two intrachain disulfide bonds form a ring that is threaded by a third intrachain disulfide bond, and this arrangement has been referred to as the cysteine knot. Each TGF-β isoform is coded for by a unique gene *(8–11)* and is located on a different chromosome *(12–15)* (Table 2). All three TGF-β isoforms are initially synthesized as 55-kDa pro-proteins that consist of an amino-terminal proregion and a carboxy-terminal mature region *(16)*. The proregion facilitates proper dimerization of the proproteins. TGF-β is secreted in a latent, inactive form in which the 12.5-kDa carboxyl-terminal 112 amino acid-long mature form is non-covalently associated with the 80-kDa latency-associated peptide (LAP) amino-terminal remainder *(17,18)*. The LAP forms a complex with the 12.5-kDa TGF-β to keep it inactive, possibly as a reservoir for limiting its bioavailability, as has been attributed to decorin and α_2-macroglobulin *(19,20)*. This complex is often referred to as the small latent TGF-β complex. The small latent TGF-β complex may associate with members of the latent TGF-β-binding protein (LTBP) family to form the large latent TGF-β complex *(21)*. The LTBP family consists of LTBP-1–4 and fibrillins 1–3. A common feature of the LTBP family is the interaction with the LAP during the formation of large latent TGF-β complexes. The liberation of TGF-β from the latent complexes is referred to as activation and may by facilitated by proteases including plasmin and matrix metalloproteinases, by thrombospondin, or by $\alpha_v\beta_6$–integrin *(22)*. The precise steps that are involved in liberation of the bioactive dimer are not completely understood, but may involve cleavage of the LTBP or LAP or both *(23)*. The differences in the sequence of the LAP from the different TGF-βs may translate into differential pathways for activation and may account for the different biological effects of the TGF-β in vivo. The TGF-β dimers are subsequently cleaved by endoproteases at a conserved RXXR amino acid motif located just upstream of the mature TGF-β peptide. Extracellular cleavage is facilitated through furins, which are proprotein convertases that process latent precursor proteins into their biologically active forms *(24,25)*. Exposure of latent TGF-β to nitric oxide results in its nitrosylation and interferes with its ability to neutralize active TGF-β *(26)*. In addition, TGF-β1 can induce its own expression by an autoregulatory mechanism that involves c-Jun N-terminal kinase (JNK) in some cell types *(27–29)*.

3. TGF-β RECEPTORS

Active TGF-β exerts its effects with specific high affinity receptors. In mammals, five TGF-β superfamily type I receptors and seven type II receptors have been identified *(30,31)*. Most types I and II receptors are ubiquitously expressed in adult tissues, although they are

Table 2
Chromosomal Location of TGF-β Signaling Pathway Components in Human

Component	*Gene name*	*Chromosome location*
TGF-β1	TGF-β1	19q13.1–19q13.3
TGF-β2	TGF-β2	1q41
TGF-β3	TGF-β3	14q23–14q24
TGF-β RI	TGF-β RI	9q33–9q34.1
TGF-β RII	TGF-β RII	3q22
TGF-β RIII	TGF-β RIII	1p32–1p33
Smad1	MADH1	4q28
Smad2	MADH2	18q21.1
Smad3	MADH3	15q21–15q22
Smad4	MADH4 (DPC4)	18q21.1
Smad5	MADH5	15q31
Smad6	MADH6	15q21–15q22
Smad7	MADH7	18q21.1
Smad8	MADH9	13q12–13q14

restricted to only certain tissues in embryos. The types I and II receptor are structurally related transmembrane glycoproteins that consist of an extracellular N-terminal ligand-binding domain with more than ten cysteine residues that regulate the dimeric structure, a transmembrane region, and a C-terminal serine/threonine kinase domain. The type I receptors, but not type II receptors, have a highly conserved region that is rich in glycine and serine residues, referred to as the GS domain, in the juxtamembrane domain next to the N-terminus of the kinase domain. The GS domain is a target for the type II receptor kinase, and on its phosphorylation on specific serine and threonine residues, the type I receptor becomes activated *(31,32)*. Being downstream of the type II receptor, the type I receptor plays an important role in determining the specificity of intracellular signals. The divergent L45 loop that is adjacent to the GS region determines specificity among the different type I receptor kinases *(33)*. The types I and II receptor exist as homodimers at the cell surface in the absence of ligands, but have an inherent heteromeric affinity for each other. Only select combinations of types I and II receptor act as ligand-binding signaling complexes. The molecular basis of the selectivity of the type I–type II receptor interactions remains poorly understood, but the structural complement at the interface may help define the selectivity of the receptor combinations. Most of the TGF-β ligands bind with high affinity to the type I receptor, also known as activin receptor-like kinase (ALK), or to the type II receptor, while others bind efficiently only to heteromeric receptor combinations. For example, TGF-β1, TGF-β3, and activins bind efficiently to their respective type II receptors, TGF-β RII, and ActRII/ActRIIB, respectively, without the need for a type I receptor. However, the ligands contact both receptor ectodomains to stabilize the type II-type I receptor complex *(34–36)*. In contrast, BMP-2 and BMP-4 do not bind well to the type II receptor BMPRII, but instead bind efficiently to the type I receptors BMPRIA/ALK3 and BMPRIB/ALK6, and require the heteromeric complex for high affinity binding *(37,38)*. Binding of TGF-β2 or BMP-7 requires both type II and type I receptor ectodomains *(36,39)*.

In addition to binding of related ligands to the same receptor complex, a single ligand often can activate several type II-type I receptor combinations. For example, the BMP-2 homodimer can complex with two BMPRIA ectodomains and shows two receptor-binding epitopes in the ligand that are conserved among BMPs *(37,38)*. Comparison of this complex with that of BMP-7 in complex with ActRII ectodomains shows that the types I and II receptor extracellular domains in the tetrameric receptor complex do not interact with each other, but still allow cooperative ligand binding *(36)*. This cooperativity in receptor binding may be modulated by the flexibility of the ligand *(40)*. The epitopes of TGF-β3 that bind the TGF-β type II receptor are different from the receptor-binding epitopes in BMPs *(35)*. Interestingly, in the complex of TGF-β3 with two TGF-β RII and two TGF-β RI ectodomains, the TGF-β RII and TGF-β RI domains not only contact the ligand, but also interact with each other *(35)*. The differences in the receptor complexes, together with the flexibility in ligand binding to the receptor, provide a structural basis for the versatility of ligand binding to the receptor complexes. Since the signaling responses are defined by the composition of the receptor complex, and particularly by the type I receptor, a ligand can induce different responses depending on the nature of the activated receptor complexes. In addition, the extracellular domain of TGF-β RI binds in vitro with high affinity to complexes of the extracellular domain of TGF-β RII and TGF-β1 or TGF-β3, but not to either ligand or receptor alone *(41)*. Like the ligands, the receptors are also encoded by unique genes *(42–46)*, and are located on different chromosomes *(47,48)* (Table 2). The types I and II receptor are serine/threonine kinases. The promoters for TGF-β1 and its types I and II receptor have been characterized, and have been shown to lack distinct TATA boxes, be highly GC-rich, contain multiple Sp1 sites, and depend on the Sp1 transcription factor for the initiation of transcription *(49–54)*. An active TGF-β receptor complex consists of two molecules each of the types I and II receptor, which are both essential for TGF-β signal transduction and inhibition of cell proliferation.

Accessory cell surface proteins further define the binding efficiency and specificity of the ligand to the receptor complex and add additional complexity. β-glycan and endoglin, a proteoglycan and a glycoprotein, respectively, compose the TGF-β type III receptor (TGF-β RIII) *(55,56)*. β-glycan and endoglin bind TGF-β with high affinity, but have no known role as signaling effectors. Coexpression of β-glycan or endoglin enhances TGF-β responsiveness and binding of TGF-β to the TGF-β RII–TGF-β RI complex. β-glycan also enhances the minimal binding of TGF-β2 to TGF-β RII *(57)*, and endoglin is required for efficient TGF-β signaling through TGF-βRII-ALK1 in endothelial cells *(58)*. Furthermore, β-glycan induces TGF-β signaling in a ligand-independent manner through activation of the p38 pathway *(59)*. The PDZ protein GIPC also regulates β-glycan expression at the cell surface and may further define the response to TGF-β. The cytoplasmic domain of β-glycan also interacts with TGF-β RII, which phosphorylates this domain. This further promotes the interaction of β-glycan with β-arrestin and modulates the internalization of TGF-β receptor complexes *(60)*. Other coreceptors can also act as determinants of ligand binding. For example, Nodal acts through ActRIIB and ActRIB *(61)*. But, efficient binding and signaling by Nodal requires the interaction of Crypto or the related EGF-CFC proteins Cryptic or FRL-1 with ActRIB. Cripto binds Nodal via its EGF domain and ActRIB through its CFC domain. In a similar fashion, the coreceptor DRAGON interacts with ligands and receptors to facilitate BMP signaling as well *(62)*. Use of a series of single amino acid TGF-β type II receptor variants that were generated based on the crystal structure of the TGF-β type II receptor:TGF-β3 complex, showed that TGF-β RII Ile53 and Glu119, which contact TGF-β3 Val92 and Arg25, respectively, together with TGF-β RII Asp32, Glu55, and Glu75, which contact TGF-β3 Arg94, each contribute significantly to ligand binding affinities *(63)*. These contacts likely underlie the lower affinity with which the TGF-β type II receptor binds TGF-β2 as these three ligand residues are unchanged in TGF-β1, but are conservatively substituted in TGF-β2 (Lys25, Ile92, and Lys 94). Several small molecule inhibitors have been developed that block signaling of type I receptors with different abilities that may be useful for inhibiting the signaling activities of TGF-β *(64)*.

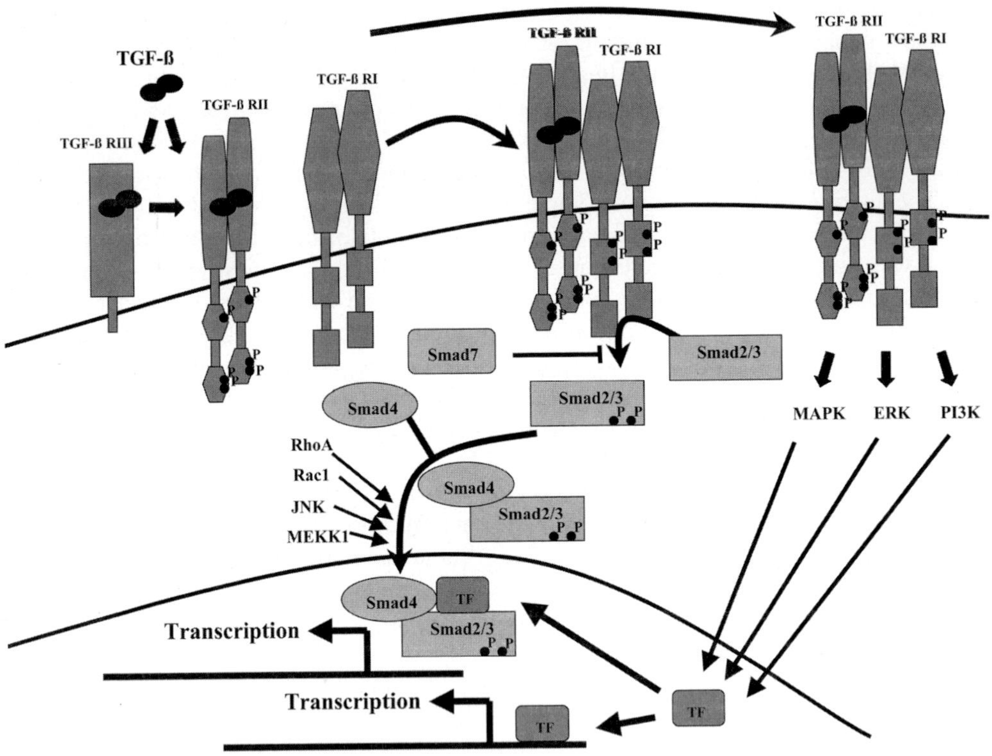

Fig. 1. Mechanism of TGF-β signaling. TGF-β signaling is mediated by Smad-dependent and -independent mechanisms. TGF-β binds the type II TGF-β receptor (TGF-β RII) directly or through the type III TGF-β receptor (TGF-β RIII), inducing association of TGF-β RII with the type I TGF-β receptor (TGF-β RII). TGF-β RII then phosphorylates and activates TGF-β RI, which then phosphorylates Smad2 or Smad3. Phosphorylated Smad2 or Smad3 then associates with Smad4, translocates into the nucleus as a complex, and activates transcription of target genes. Smad7 inhibits TGF-β signaling by preventing activation of Smad2 or Smad3 by TGF-β RI. TGF-β also activates other downstream signaling pathways, which in turn, activate specific downstream transcription factors that may work together with or independent from the Smad proteins to regulate transcriptional responses to TGF-β.

4. SMADS AND TGF-β SIGNAL TRANSDUCTION

The mechanism for TGF-β signaling has been characterized *(1–6)*, and a representation of this mechanism for TGF-β is shown in Fig. 1. TGF-β1 or TGF-β2 or TGF-β3 binds to the TGF-β type II receptor, and this results in the recruitment of the TGF-β type I receptor into the complex *(65,66)*. Competition of retroviral expression of a truncated dominant-negative TGF-β type II receptor was demonstrated in pristane-induced plasma cell tumors for intracellular binding of active TGF-β and restoration of cell surface expression of endogenous TGF-β type II receptor *(67)*. The resulting incapability of the TGF-β receptor complex to participate in signaling was the first demonstration of the formation of an intracellular TGF-β-receptor complex. The TGF-β type III receptor is also capable of binding TGF-β ligands, which subsequently presents the TGF-β ligand to the TGF-β type I receptor *(57)*. Deletion mutagenesis has revealed two regions of the β-glycan ectodomain capable of binding TGF-β, including the endoglin-related region at the amino-terminal half and the uromodulin-related region at the carboxyl-terminal half *(68,69)*. Both TGF-β binding regions of β-glycan bind TGF-β2 with higher affinity than TGF-β1, but only the endoglin-related region increases TGF-β2 labeling of TGF-β RII in vitro *(70)*. Interestingly, expression of β-glycan prevents association between the types I and II receptor in renal epithelial LLC-PK1 cells that lack endogenous β-glycan as a function of the glycosaminoglycan modifications of β-glycan *(71)*. Following TGF-β binding by TGF-β RI, TGF-β RII recruits, binds, and transphosphorylates TGF-β RI on conserved residues in the GS domain, and activates its protein kinase activity *(72,73)*.

The intracellular signal transduction triggered by the kinase activity of TGF-β involves the phosphorylation of Smad family proteins and in turn, complex changes in the transcriptional regulation of various response genes. The Smad family proteins, which include Smads 1–8, are molecules of 42–60-kDa. The Smads are divided into three subclasses depending on their structure and function: the receptor-regulated Smads (R-Smads), common-mediator Smad (Co-Smad), and inhibitory Smads (I-Smads). In general, the R-Smads, Smads 2 and 3, function downstream of the TGF-β/Nodal/Activin ligands, while Smads 1, 5, and 8 are downstream of members of the BMP and GDF subfamilies of ligands. Smads 1–3, 5, and 8 are direct substrates for the TGF-β type I receptor kinase, whereas Co-Smad, Smad 4, participates in Smad complex formation. Smads 6 and 7, the I-Smads, interfere with TGF-β-induced Smad-dependent signal transduction *(74)*. Activation of cell surface receptors by ligands leads to phosphorylation of the R-Smads at two serine residues in a SSXS motif at their extreme C-termini. This phosphorylation allows the R-Smads to form both homomeric and heteromeric complexes with Smad4 that accumulate in the nucleus. There, they are directly involved in transcriptional regulation of target genes in cooperation with other transcription factors.

Activated type I receptor interacts with regulatory Smad proteins, which in epithelial cells, are usually Smad2 or Smad3, that are phosphoserine-binding entities *(75–77)*, and whose genes are encoded on different chromosomes *(78–82)*. Like all of the Smads that have been identified and characterized, Smad2 and Smad3 share a high degree of sequence similarity in two distinct domains. The amino-terminal conserved domain is referred to as the Smad Mad Homology 1 (MH1) domain and the carboxyl-terminal conserved domain is referred to as the Smad Mad Homology 2 (MH2) domain. The MH1 and MH2 domains are separated by a less well-conserved proline-rich linker domain. The β-hairpin loop in the amino-terminal part of the MH1 domain is the direct DNA-binding site of Smad3 and the α-helix 2 plays a role in specific DNA binding and transcriptional activation of Smad3 *(83)*. The MH2 domain can specifically recognize a diverse set of proteins that are unrelated by sequence using a series of hydrophobic patches on the MH2 domain *(84)*. In their inactive configurations, the MH1 and MH2 domains inhibit the function of each other by mutual contact. Upon phosphorylation by the type I receptor, there is a conformational change on the R-Smads enabling them to form a complex with Co-Smad Smad4 *(85)*. In contrast to the considerable flexibility of type II-type I receptor pairing, the combinations of the type I receptors and the R-Smads are highly restricted. This pairing specificity is based on the complementarity of the L45 loop in the type I receptor kinase domain and the L3 loop in the C-terminal of the MH2 domain of R-Smads *(86)*. The L3 loop differs in only two amino acids between Smad2/3 and Smad1/5/8, but this is sufficient to discriminate between the different R-Smads *(87,88)*. TGF-β signaling through ALK5 and activin through ALK4 results in the phosphorylation of Smad2 and Smad3, whereas activated BMP type I receptors, ALK2, ALK3, and ALK6, as well as TGF-β signaling through ALK1 will phosphorylate Smad1 and Smad5 *(2,89–91)*. TGF-β-induced nuclear accumulation of Smad2 is caused exclusively by selective nuclear trapping of phosphorylated, complexed Smad2 *(92)*.

While mutations in the L45 loop region of TGF-β RI do not affect the binding of Smads, mutations in this region of TGF-β RI do affect the binding of other proteins, such as disabled-2 (Dab2) and X-linked inhibitor of apoptosis protein *(93)*. The activities of Smad2 and Smad3 are controlled by post-translational modifications such as phosphorylation and ubiquitination. The levels of phosphorylation of Smad2 and Smad3 induced by TGF-β can be limited by crosstalk between TGF-β and all-trans retinoic acid *(94)*. Phosphorylation events are required for activity of the Smads *(95–97)*. Arginine-462 and cysteine-463 of Smad2, which are in the proximity of the carboxyl-terminal serine residues, contribute to recognition and phosphorylation of the carboxyl-terminus of Smad2 by TGF-β RI *(98)*. Phosphoproteome profiling of MCF-7 human breast epithelial cells treated with TGF-β1 reveals numerous proteins that change their phosphorylation upon addition of TGF-β1, including transcription factor II-I (TFII-I), which can then modulate TGF-β signaling at the transcriptional level *(99)*. Axin, a negative regulator in the Wnt signaling pathway, also regulates the effects of Smad3 on the TGF-β pathway and may function as an adaptor of Smad3 *(100)*. c-Ski and SnoN are transcriptional corepressors that inhibit TGF-β signaling by interaction with Smad2 and Smad3 through the carboxyl-terminal MH2 domain of the Smad proteins and by stabilization of inactive Smad complexes on Smad-binding elements *(101,102)*. Phosphorylated Smad2 or Smad3 can then associate with the common Co-Smad, Smad4, also known as Deleted in pancreatic cancer 4 (DPC4) *(81,103)*. Smad4 is located on the same chromosome as Smad2 (Table 2) *(79,81,103)*. Smad4 is modified by sumoylation *(104)*. Oligo-ubiquitination positively regulates Smad4 function, whereas poly-ubiquitination primarily occurs in unstable cancer cells and leads to protein degradation *(105)*. Unlike R-Smads, Smad4 lacks the C-terminal SSXS motif and neither binds to, nor is phosphorylated by the type I receptor. Smad4 is not a direct substrate of the receptors, but instead, Smad4 interacts with TGF-β-activated R-Smads *(106–108)*. The heteromeric Smad complex then translocates to the nucleus where it regulates transcription through its ability to interact with DNA binding and non-DNA binding transcription factors and cofactors *(91,109–111)*. Since Smads bind DNA either with low affinity via their MH1 domains or not at all, as is the case of Smad2, they require cooperation with other sequence-specific DNA-binding transcription factors to bind efficiently to promoters of target genes. Tissue-specific gene regulation is mediated through the binding of Smads to different sets of tissue-specific transcription factors *(2,77,112)*. The Smad nucleoplasmic shuttling is necessary to regulate transcription of target genes. In unstimulated cells, Smad2 and Smad3 are predominantly cytoplasmic, while Smad4 is distributed throughout the cell *(77)*. The TGF-β-activated phospho-Smad3/Smad4 complex utilizes a mechanism that is independent of importin for nuclear import and engages different nucleoporins for nuclear import compared with the monomeric Smad4 *(113)*. The Smads enter the nucleus with the help of a nuclear localization signal (NLS) expressed upon complex formation of R-Smads and Smad4 *(110)*. Smad4 has a NLS in the MH1 region and a nuclear export signal in the linker region *(114)*, enabling it to shuttle between the cytosol and the nucleus. Mitogen-activated protein (MAP) kinase has been shown to participate in the phosphorylation of Thr276 in the linker region of Smad4, and this phosphorylation can lead to enhanced TGF-β-induced nuclear accumulation, and as a consequence, enhanced transcriptional activity of Smad4 *(115)*. Early adenovirus gene 1 associated with cancer (E1A)-like inhibitor of differentiation-2 can block TGF-β-induced formation of Smad3-Smad4 complexes, and thus suppresses TGF-β signaling *(116)*. Within the heteromeric complex, phospho-Smad3 dominates over Smad4 in nuclear importation and guides the complex to its nuclear destination *(111,113)*. Smad3 interacts with the RING finger protein ROC1 through its MH2 domain with Skp1, Cullin1, and Fbw1a to form an E3 ubiquitin ligase complex that induces ubiquitination of Smad3 and its degradation *(117)*. Smad4 is targeted for degradation by multiple ubiquitin ligases that can simultaneously act on R-Smads and signaling receptors *(118)*.

5. INHIBITORY SMADS IN TGF-β SIGNALING

The inhibitory Smads, I-Smads, Smad6 and Smad7, are natural inhibitors of TGF-β-induced Smad-mediated signaling. I-Smads Smad6 and Smad7, that are located on the same chromosome as Smad2 and Smad4 (Table 2), antagonize TGF-β signaling by binding to the TGF-β type I receptor and interfering with the activation of Smad2 and Smad3 by preventing their interaction with the TGF-β type I receptor and phosphorylation *(119–123)*. The MH2 domain of I-Smads is structurally related to R-Smads and Co-Smad4, whereas their N-terminal region is highly distinct from the other Smads. I-Smads do not possess the C-terminal SSXS domain, which probably enables them to stably bind to the activated type I receptor *(124)*. The I-Smads competitively interfere with R-Smad recruitment and phosphorylation. Smad1 and Smad5 induce Smad6 phosphorylation, whereas Smad3 induces Smad7 expression. BMP signaling induces an inhibitory feedback loop through Smad6 expression, while TGF-β induces an inhibitory feedback loop through Smad7 expression, although BMPs and TGF-β can also induce Smad7 and Smad6 expression, respectively. Smad6 inhibits BMP and TGF-β signaling with similar potency, while Smad7 inhibits TGF-β signaling more efficiently than Smad6 *(125,126)*. Acetylation of I-Smads protects them from ubiquitination and degradation by the E3 ubiquitin ligase Smad ubiquitination regulatory factor-1 (Smurf1) *(127)*. Smurf1 interacts with the type I receptor through the amino-terminal conserved 2 (C2) domain of Smad7, and induces Smad7 ubiquitination and translocation into the cytoplasm, and subsequent turnover and receptor degradation *(128,129)*. Smurf1 also regulates cell polarity and protrusion formation by targeting RhoA for degradation at cellular protrusions as a partner of protein kinase Cxi *(130)*. Neural Precursor Cell Expressed, Developmentally Down-regulated 4-2 (NEDD4-2), structurally similar to Smurfs, also induces ubiquitin-mediated degradation of Smad2 and the type I receptor and negatively regulates TGF-β signaling *(131)*. The nuclear export receptor Chromosomal Region Maintenance 1 (CRM1) is also necessary for nuclear export of Smurf1 and for the negative regulation of TGF-β signaling by Smad7 *(132)*. Arkadia, a protein that enhances signaling activity of Nodal, physically interacts with Smad7, and induces its poly-ubiquitination and degradation *(133)*. Axin is a scaffold protein that activates TGF-β signaling by forming a multimeric complex consisting of Smad7 and Arkadia and promotes degradation of Smad7 *(134)*. Jab1/CSN5, a component of the COP9 signalsome, also associates with Smad7, and when over-expressed, causes translocation of Smad7 from the nucleus to the cytoplasm, and promotes its degradation *(135)*. Smurf2, a C2-WW-HECT domain ubiquitin ligase, associates constitutively with Smad7 and induces export and recruitment to the activated TGF-β receptor where it causes degradation of TGF-β receptors and Smad7 via proteosomal and lysosomal pathways *(136)*. The amino-terminal domain of Smad7 stimulates Smurf activity by recruiting the E2 ubiquitin ligase, UbcH7, to the HECT domain *(137)*. The RING-H2 protein RNF11 is a target of Smurf2 E3 ligase and is over-expressed in breast cancer *(138)*. TGF-β also induces the association of Smurf2 with the transcriptional corepressor SnoN, with Smad2 mediating this interaction for ubiquitin-mediated degradation by the proteosome *(101,139)*. Smad-induced degradation of SnoN can be mediated by the anaphase-promoting complex and the UbcH5 family of ubiquitin-conjugating enzymes *(140)*. SnoN and Ski negatively regulate Smad7 by binding and repressing its promoter in a similar manner *(141)*. Similar to Smurfs, the WW domain-containing protein 1 (WWP1) also associates with Smad7 and induces its export and enhances binding of Smad7 to the type I receptor to cause ubiquitination and degradation of the receptor *(142)*. Using two-hybrid screening, Atrophin I-interacting Protein (AIP4) has been identified as an E3 ubiquitin ligase that specifically targets Smad7 for ubiquitin-dependent degradation without affecting the turnover of activated type I receptor *(143)*. AIP4 can inhibit TGF-β signaling, presumably by enhancing the association of Smad7 with activated type I receptor. Smad7 also interacts with β-catenin and lymphoid enhancer binding factor 1 (LEF1)/T-cell-specific factor (TCF), transcriptional regulators in Wnt signaling, in a TGF-β-dependent manner *(144)*. The physical association of Smad7 and β-catenin is important for TGF-β-induced apoptosis in human prostate cancer cells.

6. MODULATORS OF SMAD SIGNALING

Among a group of potential modulators of Smad signaling are a number of non-Smad proteins, including TGF-β receptor-I interacting protein-1, immunophilin FK506 binding protein 12 (FKBP12), the β-subunit of protein phosphatase 2A, TGF-β receptor I-associated protein (TRAP)-1, Smad anchor for receptor activation, BMP and activin membrane-bound inhibitor (BAMBI), hepatic growth factor-regulated tyrosine kinase substrate (Hgs/Hrs), Smad TGF-β receptor-associated protein (STRAP), activin receptor-interacting protein 1 (ARIP1), sorting nextin 6 (SNX 6), dab2, fas death domain-associated protein (Daxx), eucaryotic translation initiation factor 2α (EIF2α), and TRAP-1-Like Protein (TLP), all of which have been identified as interactors of TGF-β RI and/or TGF-β RII *(145–159)*. Among these, TRAP-1 associates strongly with TGF-β and activin receptor complexes and with Smad4, where it functions as a chaperone for Smad4 *(148,160)*. BAMBI inhibits signaling by BMP and activin as well as by TGF-β *(150)*. The inhibitory effects of BAMBI are mediated by its intracellular domain, which resembles the homodimerization interface of a type I receptor, and by acting as a pseudoreceptor, BAMBI prevents formation of receptor complexes. Smad anchor for receptor activation and Hgs/Hrs have been shown to facilitate Smad2/3 recruitment to activated TGF-β RI *(149,151)* and the adaptor molecule Dab2, which binds TGF-β RI, TGF-β RII, Smad2, and Smad3, has been suggested to bridge the TGF-β receptor complex to the Smad pathway *(156)*. STRAP stabilizes the association between Smad7 and the activated receptor, thus blocking Smad2/3-mediated transcriptional activation *(152)*. A physical association of STRAP with 3-phosphoinositide-dependent protein kinase-1 (PDK1) has been demonstrated that shows that PDK1 facilitates inhibition of TGF-β signaling with STRAP by contributing to the stable association between TGF-β and Smad7 *(161)*. ARIP1 has multiple protein-protein interaction domains, including two WW domains and five PSD-95/Dlg/ZO-1 (PDZ) domains that associate with Smads and activin RII receptors *(153,154)*. SNX 6 interacts with TGF-β RII and inactivated TGF-β RI, and more strongly with activin RIIB *(155)*. EIF2α interacts with TGF-β RII and its over-expression inhibits the TGF-β response *(158)*. TLP inhibits the formation of Smad3/4 complexes in the absence of effects on phosphorylation of Smad3, while it does not affect Smad2 phosphorylation or hetero-oligomerization *(159)*. It has been proposed that TLP might regulate the balance of Smad2 and Smad3 signaling by localizing Smad4 intracellularly.

In addition, CCAAT/enhancer-binding proteins (C/EBP)-β and C/EBP-δ physically interact with Smad3 and Smad4, and Smad3 cooperates with Smad4 and TGF-β signaling to repress the transcriptional activity of C/EBPs *(162)*. Smad3 interacts with Interferon Regulatory Factor-7 (IRF-7), an essential transcription factor for interferon-mediated-α and -β induction, to activate interferon-β transcription *(163)*, and the Tuberous Sclerosis Complex 2 (TSC2) tumor suppressor gene, in which the amino-terminal region of tuberin interacts specifically with the MH2 domain of Smad2 and Smad3 proteins to regulate TGF-β-responsive genes such as $p21^{Cip1}$ *(164)*. FoxO Forkhead transcription factors act as signal transducers at the confluence of Smad, phosphatidylinositol-3 (PI-3) kinase, and telencephalic development factor FoxG1 signaling pathways *(165)*. Smad proteins activated by TGF-β form a complex with FoxO proteins to activate the growth inhibitory gene $p21^{Cip1}$. This process is negatively controlled by the PI-3 kinase pathway and by FoxG1 that binds to FoxO-Smad complexes and blocks $p21^{Cip1}$ expression. Smad3 interactions with the positive regulatory homeobox proteins NKX2.1 and hepatic nuclear factor-3 (HNF-3) underlie the molecular basis for TGF-β-induced repression of Surfactant Protein-B gene transcription in

Ligands
TGF-β1, TGF-β2, TGF-β3

↓

Receptors
TGF-β RI, TGF-β RII, TGF-β RIII

↓

Smad Mediators
Receptor BMP Smads — **Smad1, Smad5, Smad8**
Receptor TGF-β Smads — **Smad2, Smad3**
Inhibitory Smads
Smad6, Smad7

↓

Ubiquitin Ligases
Smurf1, Smurf2

↓

Smad Modulators
TRIP-1, TRAP, FKBP12, SARA, BAMBI, Hgs/Hrs, EIF2a, ARIP1, STRAP, Dab2, Daxx, TLP, SNX6, X-IAP

↓

Smad Degradation
Smurf1, Smurf2, NEDD4-2, CRM1, Arcadia, WWP1, Jab1/CSN5

↓

Smad-dependent Targets
p15Ink4b, p21Cip1, p27Kip1, p57Kip2, c-myc, c-Ski, SnoN, Cdc25A

Fig. 2. Components of the TGF-β signaling pathway. The TGF-β signaling pathway involves the binding of TGF-β ligands to specific menbrane receptors. When the receptors are activated, they interact with and regulate a multitude of intracellular mediators, including Smads that propagate the signal and are modulated and degraded by other proteins to regulate transcription of target genes.

human lung cancer cells *(166)*. Reduced Nkx2.1 expression is detected in lungs of TGF-β1 heterozygous mice compared to wildtype littermates, that is further reduced in carcinogenesis, and that correlates with reduction of Sp1, Sp3, Smad2, Smad3, and Smad4 in lung adenocarcinomas *(167)*. In addition, in vitro data show that regulation of Nkx2.1 by TGF-β1 occurs through TGF-β type II receptor and Smad2, Smad3, and Smad4 signaling and Sp1 and Sp3 in mouse lung cells. In addition, TGF-β signaling may be regulated by receptor internalization *(60,168–170)*. Furthermore, besides Smad-dependent TGF-β signaling, Smad-independent signaling through mitogen-activated protein kinase pathways, PI-3 kinase/AKT, Rho guanine triphosphatases, and protein phosphatase 2A also occurs *(171–176)*. Insulin-like growth factor-I inhibits the ability of TGF-β to regulate expression of several genes by PI-3 kinase/AKT-dependent suppression of the activation of Smad3 in non-tumorigenic rat protate epithelial cells *(177)*. Use of fibroblasts from Smad3 null mice, small molecule inhibitors, and neutralizing antibodies show that matrix-metalloproteinase-9 (MMP-9) expression is induced by tumor cell-derived TGF-β and tumor necrosis factor (TNF)-β, dependent on Smad-, Ras-, and PI-3 kinase-signaling *(178)*. The multitude of components of the TGF-β signaling pathway is shown in Fig. 2. The relative importance and interplay of these various genes and pathways in the changing responses of cells to TGF-β are beginning to be investigated in normal and tumor cells.

ACKNOWLEDGMENTS

It is not possible to include every important contribution that has been made to our understanding of the mechanisms of TGF-β signaling. Apologies are extended to those investigators whose contributions could not be included. This research was supported by the Intramural Research Program of the NIH, National Cancer Institute, Center for Cancer Research.

REFERENCES

1. Attisano L, Wrana JL. Signal transduction by the TGF-β superfamily. Science 2002;296:1646–1647.
2. Derynck R, Zhang YE. Smad-independent and smad-dependent pathways in TGF-β family signaling. Nature 2003;425:577–584.
3. Glick A. TGF-β1, Back to the future. Cancer Biol Ther 2004;3:276–283.
4. Byfield SD, Roberts AB. Lateral signaling enhances TGF-β response complexity. Trends Cell Biol 2004;14:107–111.
5. Massagué J, Seoane J, Wotton D. Smad transcription factors. Genes Dev 2005;19:2783–2810.
6. Kluppel M, Wrana JL. Turning it up a Notch: cross-talk between TGF-β and Notch signaling. Bioessays 2005;27:115–118.
7. Schlunegger MP, Grutter MG. An unusual feature revealed by the crystal structure at 2.2 Å resolution of human transforming growth factor-β2. Nature 1992;358:430–434.
8. Derynck R, Jarrett JA, Chen EY, et al. Human transforming growth factor-β complementary DNA sequence and expression in normal and transformed cells. Nature 1985;316:701–705.
9. Seyedin SM, Thompson AY, Bentz H, et al. Cartilage-inducing factor A: apparent identity to transforming growth factor-β. J Biol Chem 1986;261:5693–5695.
10. Derynck R, Lindquist PB, Lee A, et al. A new type of transforming growth factor-β. EMBO J 1988;7: 3737–3743.
11. ten Dijke P, Hansen P, Iwata KK, Pieler C, Foulkes JG. Identification of another member of the transforming growth factor-β gene family. Proc Natl Acad Sci USA 1988;85:4715–4719.
12. Fujii D, Brissenden JE, Derynck R, Francke U. Transforming growth factor-β maps to human chromosome 19 long arm and to mouse chromosome 7. Somat Cell Mol Genet 1986;12:281–288.
13. Barton DE, Foellmer BE, Du J, Tamm J, Derynck R, Francke U. Chromosomal mapping of genes for transforming growth factors β2 and β3 in man and mouse: dispersion of TGF-β gene family. Oncogene Res 1988;3:323–331.
14. Nishimura DY, Purchio AF, Murray JC. Linkage localization of TGFβ2 and the human homeobox gene HLX1 to chromosome 1q. Genomics 1993;15:357–364.
15. ten Dijke P, Geurts AHM, Foulkes JG, Le Beau MM. Transforming growth factor type β3 maps to human chromosome region q23–q24. Oncogene 1988;3:721–724.
16. Gentry LE, Lioubin MN, Purchio AF, Marquardt H. Molecular events in the processing of recombinant type I pre-pro-transforming growth factor-β to the mature polypeptide. Mol Cell Biol 1988; 8:4162–4168.
17. Barcellos-Hoff MH. Latency and activation in the control of TGF-β. J Mammary Gland Biol Neoplasia 1996;1:353–363.
18. Barcellos-Hoff MH, Ewan KB. Transforming growth factor type-β and breast cancer. Mammary gland development. Breast Cancer Res 2000;2:92–99.
19. Ständer M, Naumann U, Wick W, Weller M. TGF-β and p21: multiple molecular targets of decorin-mediated suppression of neoplastic growth. Cell Tissue Res 1999;296:221–227.
20. Arandjelovic S, Freed TA, Gonias SL. Growth factor-binding sequence in human alpha2-macroglobulin targets the receptor-binding site in transforming growth factor-β. Biochemistry 2003;42:6121–6127.
21. Öklü R, Hesketh R. The latent transforming growth factor-β binding protein (LTBP) family. Biochem J 2000;352:601–610.
22. Annes JP, Munger JS, Rifkin DB. Making sense of latent TGFβ activation. J Cell Sci 2003;116:217–224.
23. Hyytiäinen M, Penttinen C, Keski-Oja J. Latent TGF-β binding proteins: extracellular matrix association and roles in TGF-β activation. Crit Rev Clin Lab Sci 2004;41:233–264.
24. Matthews DJ, Goodman LJ, Gorman CM, Wells JA. A survey of furin substrate specificity using substrate phage display. Protein Sci 1994;3:1197–1205.
25. Leitlein J, Aulwurm S, Waltereit R, et al. Processing of immunosuppressive pro-TGF-β1, 2 by human glioblastoma cells involves cytoplasmic and secreted furin-like proteases. J Immunol 2001;166:7238–7243.

26. Vodovotz Y, Chester L, Chong H, et al. Regulation of transforming growth factor-β1 by nitric oxide. Cancer Res 1999;59:2142–2149.
27. Van Obberghen-Schilling E, Roche NS, Flanders KC, Sporn MB, Roberts AB. Transforming growth factor-β1 positively regulates its own expression in normal and transformed cells. J Biol Chem 1988; 263:7741–7746.
28. Pardoux C, Derynck R. JNK regulates expression and autocrine signaling of TGF-β1. Mol Cell 2004;15:170–171.
29. Ventura JJ, Kennedy NJ, Flavell RA, Davis RJ. JNK regulates autocrine expression of TGF-β1. Mol Cell 2004;15:269–278.
30. Derynck R, Akhurst RJ, Balmain A. TGF-β signaling in cancer: a double-edged sword. Trends Cell Biol 2001;29:117–129.
31. Massagué J. How cells read TGF-β signals. Nat Rev Mol Cell Biol 2000;1:169–178.
32. Heldin C-H, Miyazono K, ten Dijke P. TGF-β signaling from cell membrane to nucleus through SMAD proteins. Nature 1997;390:465–471.
33. Feng XH, Derynck R. A kinase subdomain of transforming growth factor-β (TGF-β) type I receptor determines the TGF-β intracellular signaling specificity. EMBO J 1997;16:3912–3923.
34. Boesen CC, Radaev S, Motyka SA, Patamawenu A, Sun PD. The 1.1 Å crystal structure of human TGF-β type II receptor ligand binding domain. Structure 2002;10:913–919.
35. Hart PJ, Deep S, Taylor AB, Shu Z, Hinck CS, Hinck AP. Crystal structure of the human TβR2 ectodomain-TGF-β3 complex. Nat Struct Biol 2002;9:203–208.
36. Greenwald J, Vega ME, Allendorph GP, Fischer WH, Vale W, Choe S. A flexible activin explains the membrane-dependent cooperative assembly of TGF-β family receptors. Mol Cell 2004;15: 485–489.
37. Kirsch T, Sebald W, Dreyer MK. Crystal structure of the BMP-2-BRIA ectodomian complex. Nat Struct Biol 2000;7:492–496.
38. Keller S, Nickel J, Zhang JL, Sebald W, Mueller TD. Molecular recognition of BMP-2 and BMP receptor IA. Nat Struct Mol Biol 2004;11:481–488.
39. del Re E, Babitt JL, Pirani A, Schneyer AL, Lin HY. In the absence of type III receptor, the transforming growth factor (TGF)-β type II receptor requires the type I receptor to bind TGF-β2. J Biol Chem 2004;279:22,765–22,772.
40. Sebald W, Mueller TD. The interaction of BMP-7 and ActRII implicates a new mode of receptor assembly. Trends Biochem Sci 2003;28:518–521.
41. Zuniga JE, Groppe JC, Cui Y, et al. Assembly of TβRI:TβRII:TGFβ ternary complex in vitro with receptor extracellular domains is cooperative and isoform-dependent. J Mol Biol 2005;354:1052–1068.
42. Attisano L, Carcamo J, Ventura F, Weis FMB, Massagué J, Wrana JL. Identification of human activin and TGF-β type I receptors that form heteromeric kinase complexes with type II receptors. Cell 1993;75:671–680.
43. Ebner R, Chen R-H, Shum L, et al. Cloning of a type I TGF-β receptor and its effect on TGF-β binding to the type II receptor. Science 1993;260:1344–1348.
44. Franzén P, ten Dijke P, Ichijo H, et al. Cloning of TGF-β type I receptor that forms a heteromeric complex with the TGF-β type II receptor. Cell 1993;75: 681–692.
45. Lin HY, Wang X-F, Ng-Eaton E, Weinberg RA, Lodish HF. Expression cloning of the TGF-β type II receptor, a functional transmembrane serine/threonine kinase receptor. Cell 1992;68:775–785.
46. Wang X-F, Lin HY, Ng-Eaton E, Downward HF, Lodish HF, Weinberg RA. Expression cloning and characterization of the TGF-β type III receptor. Cell 1991;67:797–805.
47. Johnson DW, Qumsiyeh M, Benkhalifa M, Marchuk DA. Assignment of human transforming growth factor-β type I and type III receptor genes (TGFBRI and TGFBR3) to 9q33–q34 and 1p32–p33, respectively. Genomics 1995;28:356–357.
48. Mathew S, Murty VVVS, Cheifetz S, George D, Massagué J, Chaganti RSK. Transforming growth factor receptor gene TGFBR2 maps to human chromosome band 3p22. Genomics 1994;20:114–115.
49. Kim S-J, Glick A, Sporn MB, Roberts AB. Characterization of the promoter region of the human transforming growth factor-β1 gene. J Biol Chem 1989;264:402–408.
50. Geiser AG, Kim S-J, Roberts AB, Sporn MB. Characterization of the mouse transforming growth factor-β1 promoter and activation by the Ha-ras oncogene. Mol Cell Biol 1991;11:84–92.
51. Geiser AG, Busam KJ, Kim S-J, et al. Regulation of the transforming growth factor-β1 and -β3 promoters by transcription factor Sp1. Gene 1993;129:223–228.

52. Bloom BB, Humphries DE, Kuang PP, Fine P, Goldstein RH. Structure and expression of the promoter for the R4/ALK5 human type I transforming growth factor-β receptor: Regulation by TGF-β Biochim Biophys Acta 1996;1312:243–248.
53. Bae HW, Geiser AG, Kim DH, et al. Characterization of the promoter region of the human transforming growth factor-β type II receptor gene. J Biol Chem 1995;270:29,460–29,468.
54. Jennings R, Alsarraj M, Wright KL, Muñoz-Antonia T. Regulation of the human transforming growth factor β type II receptor gene promoter by novel Sp1 sites. Oncogene 2001;20:6899–6909.
55. Cheifetz S, Andres JL, Massagué J. The transforming growth factor-β receptor type III is a membrane protoglycan. Domain structure of the receptor. J Biol Chem 1988;263:16,984–16,991.
56. Cheifetz S, Bellon T, Cales C, et al. Endoglin is a component of the transforming growth factor-β receptor system in endothelial cells. J Biol Chem 1992;267: 19,027–19,030.
57. Lopez-Casillas F, Wrana JL, Massagué J. Betaglycan presents ligand to the TGF-β signaling receptor. Cell 1993;73:1435–1444.
58. Lebrin F, Goumans MJ, Jonker L, et al. Endoglin promotes endothelial cell proliferation and TGF-β/ALK1 signal transduction. EMBO J 2004;23:4018–4028.
59. Santander C, Brandan E. Betaglycan induces TGF-β signaling in a ligand-independent manner, through activation of the p38 pathway. Cell Signal 2006;18:1482–1491.
60. Chen W, Kirkbride KC, How T, et al. Beta-arrestin 2 mediates endocytosis of type III TGF-β receptor and down-regulation of its signaling. Science 2003;301:1338–1339.
61. Yeo C, Whitman M. Nodal signals to Smads through Cripto-dependent and Cripto-independent mechanisms. Mol Cell 2001;7:949–957.
62. Samad TA, Rebbapragada A, Bell E, et al. DRAGON, a bone morphogenetic protein co-receptor. J Biol Chem 2005;280:14122–14129.
63. De Crescenzo G, Hinck CS, Shu Z, et al. Three key residues underlie the differential affinity of the TGFβ isoforms for the TGFβ type III receptor. J Mol Biol 2006;355:47–62.
64. Tojo M, Hamashima Y, Hanyu A, et al. The ALK-5 inhibitor A-83-01 inhibits Smad signaling and epithelial-to-mesenchymal transition by transforming growth factor-β. Cancer Sci 2005;96:791–800.
65. Wrana JL, Attisano L, Carcamo J, et al. TGF β signals through a heteromeric protein kinase receptor complex. Cell 1992;71:1003–1014.
66. Moustakas A, Lin HY, Henis YI, Plamondon J, O'Connor-McCourt MD, Lodish HF. The transforming growth factor β receptors types I, II, and III form hetero-oligomeric complexes in the presence of ligand. J Biol Chem 1993;268:22,215–22,218.
67. Fernandez T, Amoroso S, Sharpe S, et al. Disruption of transforming growth factor-β signaling by a novel ligand-dependent mechanism. J Exp Med 2002;195:1247–1255.
68. Lopez-Casillas F, Payne H, Andres JL, Massagué J. Betaglycan can act as a dual modulator of TGF-β access to signaling receptors: mapping of ligand binding and GAG attachment sites. J Cell Biol 1994; 124:557–568.
69. Pepin M-C, Beauchemin M, Plamondon J, O'Connor-McCourt MD. Mapping of the ligand binding domain of the transforming growth factor-β receptor type III by deletion mutagenesis. Proc Natl Acad Sci USA 1994;91:6997–7001.
70. Esparza-Lopez J, Montiel JL, Vilchis-Landeros MM, Okadome T, Miyazono K, Lopez-Casillas F. Ligand binding and functional properties of betaglycan, a co-receptor of the transforming growth factor-β superfamily. Specialized binding regions for transforming growth factor-β and inhibin A. J Biol Chem 2001;276:14,588–14,596.
71. Eickelberg O, Centrella M, Reiss M, Kashgarian M, Wells RG. Betaglycan inhibits TGF-β signaling by preventing type I–type II receptor complex formation. J Biol Chem 2002;277:823–829.
72. Wrana JL, Attisano L, Wieser R, Ventura F, Massagué J. Mechanism of activation of the TGF-β receptor. Nature 1994;370:341–347.
73. Wieser R, Wrana JL, Massagué J. GS domain mutations that constitutively activate TGF-β R-I, the downstream signaling component in the TGF-β receptor complex. EMBO J 1995;14:2199–2208.
74. Park SH. Fine tuning and cross-talking of TGF-β signal by inhibitory Smads. J Biochem Mol Biol 2005;38:9–16.
75. Nakao A, Imamura T, Souchelnytskyi S, et al. TGF-β receptor-mediated signaling through Smad2, Smad3 and Smad4. EMBO J 1997;16:5353–5362.
76. Souchelnytskyi S, Moustakas A, Heldin C-H. TGF-β signaling from a three-dimensional perspective: Insight into selection of partners. Trends Cell Biol 2002;12:304–307.

77. ten Dijke P, Hill CS. New insights into TGF-β-Smad signaling. Trends Biochem Sci 2004;29: 265–273.
78. Eppert K, Scherer SW, Ozcelik H, et al. MADR2 maps to 18q21 and encodes a TGF-β regulated MAD-related protein that is functionally mutated in colorectal carcinoma. Cell 1996;86:543–552.
79. Riggins GJ, Thiagalingam S, Rozenblum E, et al. Mad-related genes in the human. Nature 1996;13: 347–349.
80. Nakao A, Roijer E, Imamura T, et al. Identification of Smad2, a human Mad-related protein in the transforming growth factor β signaling pathway. J Biol Chem 1997;272:2896–2900.
81. Johnson T, Gorunova L, Dawiskiba S, et al. Molecular analyses of the 15q and 18q SMAD genes in pancreatic cancer. Genes Chromo Cancer 1999;24:62–71.
82. Latil A, Pesche S, Valéri A, Fournier G, Cussenot O, Lidereau R. Expression and mutational analysis of the MADR2/Smad2 gene in human prostate cancer. Prostate 1999;40:225–231.
83. Kusanagi K, Kawabata M, Mishima HK, Miyazono K. Alpha-helix 2 in the amino-terminal mad homology 1 domain is responsible for specific DNA binding of Smad3. J Biol Chem 2001;276: 28,155–28,163.
84. Chong PA, Ozdamer B, Wrana JL, Forman-Kay JD. Disorder in a target for the Smad2 Mad homology 2 domain and its implications for binding and specificity. J Biol Chem 2004;279:40,707–40,714.
85. Hata A, Lo RS, Wotton D, Lagna G, Massagué J. Mutations increasing autoinhibition inactivate tumour suppressors Smad2 and Smad4. Nature 1997;388:82–87.
86. Lo RS, Chen YG, Shi Y, Pavletich NP, Massagué J. The L3 loop: a structural motif determining specific interactions between SMAD proteins and TGF-β receptors. EMBO J 1998;17:996–1005.
87. Chen YG, Hata A, Lo RS, et al. Determinants of specificity in TGF-β signal transduction. Genes Dev 1998;12:2144–2152.
88. Persson U, Izumi H, Souchelnytskyi S, et al. The L45 loop in type I receptors for TGF-β family members is a critical determinant in specifying Smad isoform activation. FEBS Lett 1998;434:83–87.
89. Chen YG, Massagué J. Smad1 recognition and activation of transforming growth factor-β family receptors. J Biol Chem 1999;274:3672–3677.
90. Oh SP, Seki T, Goss KA, et al. Activin receptor-like kinase 1 modulates transforming growth factor-β1 signaling in the regulation of angiogenesis. Proc Natl Acad Sci USA 2000;97:2626–2631.
91. Shi Y, Massagué J. Mechanisms of TGF-β signaling from cell membrane to the nucleus. Cell 2003; 113:685–700.
92. Schmierer B, Hill CS. Kinetic analysis of Smad nucleocytoplasmic shuttling reveals a mechanism for transforming growth factor-β-dependent nuclear accumulation of Smads. Mol Cell Biol 2005;25: 9845–9858.
93. Itoh S, Thorikay M, Kowanetz M, et al. Elucidation of Smad requirement in transforming growth factor-β type I receptor-induced responses. J Biol Chem 2003; 278:3751–3761.
94. Cao Z, Flanders KC, Bertolette D, et al. Levels of phospho-Smad2/3 are sensors of the interplay between effects of TGF-β and retinoic acid on monocytic and granulocytic differentiation of HL-60 cells. Blood 2003;101:498–507.
95. Macias-Silva M, Abdollah S, Hoodless PA, Pirone R, Attisano L, Wrana JL. MADR2 is a substrate of the TGF-β receptor and its phosphorylation is required for nuclear accumulation and signaling. Cell 1996;87:1215–1224.
96. Abdollah S, Macias-Silva M, Tsukazaki T, Hayashi H, Attisano L, Wrana JL. TβRI phosphorylation of Smad2 on Ser465 and Ser467 is required for Smad2-Smad4 complex formation and signaling. J Biol Chem 1997;272:27,678–27,685.
97. Souchelnytskyi S, Tamaki K, Engstrom U, Wernstedt C, ten Dijke P, Heldin C-H. Phosphorylation of Ser465 and Ser467 in the C terminus of Smad2 mediates interaction with Smad4 and is required for transforming growth factor β signaling. J Biol Chem 1997;272:107–128.
98. Yakymovych I, Heldin C-H, Souchelnytskyi S. Smad2 phosphorylation by type I receptor: Contribution of arginine 462 and cysteine 463 in the C-terminus of Smad2 for specificity. J Biol Chem 2004;279:35,781–35,787.
99. Stasyk T, Dubrovska A, Lomnytska M, et al. Phosphoproteome profiling of transforming growth factor-β signaling: abrogation of TGF-β1-dependent phosphorylation of TFII-I enhances cooperation of TFII-I and Smad3 in transcription. Mol Biol Cell 2005;16:4765–4780.
100. Furuhashi M, Yagi K, Yanamoto H, et al. Axin facilitates Smad3 activation in the transforming growth factor-β signaling pathway. Mol Cell Biol 2001;21:5132–5141.

101. Mizuide M, Hara T, Furuya T, et al. Two short segments of Smad3 are important for specific interaction of Smad3 with c-Ski and SnoN. J Biol Chem 2003;278:531–536.
102. Suzuki H, Yagi K, Kondo M, Kato M, Miyazono K, Miyazawa K. c-Ski inhibits the TGF-β signaling pathway through stabilization of inactive Smad complexes on Smad-binding elements. Oncogene 2004;23:5068–5076.
103. Hahn SA, Schulte M, Hoque A, et al. DPC4, a candidate tumor suppressor gene at human chromosome 18q21.1. Science 1996;271:350–353.
104. Lee PS, Chang C, Liu D, Derynck R. Sumoylation of Smad4, the common Smad mediator of transforming growth factor-β family signaling. J Biol Chem 2003;278:27,853–27,863.
105. Mören A, Hellman U, Inada Y, Imamura T, Heldin C-H, Moustakas A. Differential ubiquitination defines the functional status of the tumor suppressor Smad4. J Biol Chem 2003;278:33,571–33,582.
106. Langna G, Hata A, Hemmati-Brivanlou A, Massagué J. Partnership between DPC4 and Smad proteins in TGF-β signalling pathways. Nature 1996;383:832–836.
107. Zhang Y, Musci T, Derynck R. The tumor suppressor Smad4/DPC 4 as a central mediator of Smad function. Curr Biol 1997;7:270–276.
108. Yingling JM, Datto MB, Wong C, Frederick JP, Liberati NT, Wang X-F. Tumor suppressor Smad4 is a transforming growth factor-β-inducible DNA binding protein. Mol Cell Biol 1997;17:7019–7028.
109. Inman GJ, Nicolas FJ, Hill CS. Nucleocytoplasmic shuttling of Smads 2, 3, and 4 permits sensing of TGF-β receptor activity. Mol Cell Biol 2002;10:283–294.
110. Reguly T, Wrana J. In or out? The dynamics of Smad nucleocytoplasmic shuttling. Trends Cell Biol 2003;13:216–220.
111. Xu L, Massagué J. Nucleocytoplasmic shuttling of signal transducers. Nat Rev 2004;5:1–11.
112. Feng XH, Derynck R. Specificity and diversity in TGF-β signaling through Smads. Annu Rev Cell Dev Biol 2005;21:659–693.
113. Chen HB, Rud JG, Lin K, Xu L. Nuclear targeting of transforming growth factor-β-activated Smad complexes. J Biol Chem 2005;280:21,329–21,336.
114. Itoh S, Landstrom M, Hermansson A, et al. Transforming growth factor β1 induces nuclear export of inhibitory Smad7. J Biol Chem 1998;273:29,195–29,201.
115. Roelen BA, Cohen OS, Raychowdhury MK, et al. Phosphorylation of threonine 276 in Smad4 is involved in transforming growth factor-β-induced nuclear accumulation. Am J Cell Physiol 2003;285:C823–C830.
116. Lee HJ, Lee JK, Miyake S, Kim S-J. A novel E1A-like inhibitor of differentiation (EID) family member, EID-2, suppresses transforming growth factor-(TGF)-β signaling by blocking TGF-β-induced formation of Smad3–Smad4 complexes. J Biol Chem 2004;279:2666–2672.
117. Fuckuci M, Imamura T, Chiba T, et al. Ligand-dependent degradation of Smad3 by a ubiquitin ligase complex of ROC1 and associated proteins. Mol Biol Cell 2001;12:1431–1443.
118. Mören A, Imamura T, Miyazono K, Heldin C-H, Moustakas A. Degradation of the tumor suppressor by WW and HECT domain ubiquitin ligases. J Biol Chem 2005;280:22,115–22,123.
119. Röijer E, Mören A, ten Dijke P, Stenman G. Assignment of the Smad7 gene (MADH7) to human chromosome 18q21.1 by fluorescence in situ hybridization. Cytogenet Cell Genet 1998;81:189–190.
120. Hayashi H, Abdollah S, Qiu Y, et al. The MAD-related protein Smad7 associates with the TGF-β receptor and functions as an antagonist of TGF-β signaling. Cell 1997;89:1165–1173.
121. Nakao A, Afrakhte M, Mören A, et al. Identification of Smad7, a TGF-β-inducible antagonist of TGF-β signaling. Nature 1997;389:631–635.
122. Hanyu A, Ishidou Y, Ebisawa T, Shimanuki T, Imamura T, Miyazono K. The N domain of Smad7 is essential for specific inhibition of transforming growth factor-β signaling. J Cell Biol 2001;155: 1017–1027.
123. Mochizuki T, Miyazaki H, Hara T, et al. Roles for the MH2 domain of Smad7 in the specific inhibition of transforming growth factor-β superfamily signaling. J Biol Chem 2004;279:31,568–31,574.
124. Afrakhte M, Mören A, Jossen S, et al. Induction of inhibitory Smad6 and Smad7 mRNA by TGF-β family members. Biochem Biophys Res Commun 1998;249:505–511.
125. Miyazono K, Suzuki H, Imamura T. Regulation of TGF-β signaling and its roles in progression of tumors. Cancer Sci 2003;94:230–234.
126. Canalis E, Economides AN, Gazzerro E. Bone morphogenetic proteins, their antagonists, and the skeleton. Endocr Rev 2003;24:218–235.

127. Gronroos E, Hellman U, Heldin C-H, Ericsson J. Control of Smad7 stability by competition between acetylation and ubiquitination. Mol Cell 2002;10:483–493.
128. Ebisawa T, Fukuchi M, Murakami G, et al. Smurf1 interacts with transforming growth factor-β type I receptor through Smad7 and induces receptor degradation. J Biol Chem 2001;276:12,477–12,480.
129. Suzuki C, Murakami G, Fukuchi M, et al. Smurf1 regulates the inhibitory activity of Smad7 by targeting Smad7 to the plasma membrane. J Biol Chem 2002;277:39,919–39,925.
130. Zhang Y, Wang HR, Wrana JL. Smurf1: a link between cell polarity and ubiquitination. Cell Cycle 2004;3:391–392.
131. Kuratomi G, Komuro A, Goto K, et al. NEDD4-2 (neural precursor cell expressed, developmentally down-regulated 4-2) negatively regulates TGF-b (transforming growth factor-β) signaling by inducing ubiquitin-mediated degradation of Smad2 and TGF-β type I receptor. Biochem J 2005;386:461–470.
132. Tajima Y, Goto K, Yoshida M, et al. Chromosomal region maintenance 1 (CRM1)-dependent nuclear export of Smad ubiquitin regulatory factor 1 (Smurf1) is essential for negative regulation of transforming growth factor-β signaling by Smad7. J Biol Chem 2003;278:10716–10721.
133. Koinuma D, Shinozaki M, Komuro A, et al. Arkadia amplifies TGF-β superfamily signaling through degradation of Smad7. EMBO J 2003;22:6458–6470.
134. Liu W, Rui H, Wang J, et al. Axin is a scaffold protein in TGF-β signaling that promotes degradation of Smad7 by Arkadia. EMBO J 2006;25:1646–1658.
135. Kim BC, Lee HJ, Park SH, et al. Jab1/CSN5, a component of the COP9 signalsome, regulates transforming growth factor-β signaling by binding to Smad7 and promoting its degradation. Mol Cell Biol 2004;24:2251–2262.
136. Kavsak P, Rasmussen RK, Causing CG, et al. Smad7 binds to Smurf2 to form an E3 ubiquitin ligase that targets the TGF-β receptor for degradation. Mol Cell 2000;6:1365–1375.
137. Ogunjimi AA, Briant DJ, Pece-Barbara N, et al. Regulation of Smurf2 ubiquitin ligase activity by anchoring the E2 to the HECT domain. Mol Cell 2005;19:297–308.
138. Subramaniam V, Li H, Wong M, et al. The RING-H2 protein RNF11 is overexpressed in breast cancer and is a target of Smurf2 E3 ligase. Br J Cancer 2003;89:1538–1544.
139. Bonni S, Wang HR, Causing CG, Kavsak P, Stroschein SL, Luo K, Wrana JL. TGF-β induces assembly of a Smad2-Smurf2 ubiquitin ligase complex that targets SnoN for degradation. Nat Cell Biol 2001; 3:587–595.
140. Stroschein SL, Bonni S, Wrana JL, Luo K. Smad3 recruits the anaphase-promoting complex for ubiquitination and degradation of SnoN. Genes Dev 2001;15:2822–2836.
141. Briones-Orta MA, Sosa-Garrocho M, Moreno-Alvarez P, Fonseca-Sanchez MA, Macias-Silva M. SnoN co-repressor binds and represses Smad7 gene promoter. Biochem Biophys Res Commun 2006;341:889–894.
142. Komuro A, Imamura T, Saitoh M, et al. Negative regulation of transforming growth factor-β (TGF-β) signaling by WW domain-containing protein 1 (WWP1). Oncogene 2004;23:6914–6923.
143. Lallemand F, Seo SR, Ferrand N, et al. AIP4 restricts transforming growth factor-β signaling through a ubiquitination-independent mechanism. J Biol Chem 2005;280:27,645–27,653.
144. Edlund S, Lee SY, Grimsby S, et al. Interaction between Smad7 and beta-catenin: importance for transforming growth factor β-induced apoptosis. Mol Cell Biol 2005;25:1475–1488.
145. Chen RH, Miettinen PJ, Maruoka EM, Choy L, Derynck R. A WD-domain protein that is associated with and phosphorylated by the type II TGF-β receptor. Nature 1995;377:548–552.
146. Wang T, Li BY, Danielson PD, et al. The immunophilin FKBP12 functions as a common inhibitor of the TGF-β family type I receptors. Cell 1996;86:435–444.
147. Griswold-Prenner J, Kamibayashi C, Maruoka EM, Mumby MC, Derynck R. Physical and functional interactions between type I transforming growth factor-β receptors and B alpha, a WD-40 repeat subunit of phosphatase 2A. Mol Cell Biol 1998;18:6595–6604.
148. Charng MJ, Zhang D, Kinnunen P, Schneider MD. A novel protein distinguishes between quiescent and activated forms of the type I transforming growth factor β receptor. J Biol Chem 1998;273: 9365–9368.
149. Tsukazaki T, Chiang TA, Davison AF, Attisano L, Wrana JL. SARA, a FYVE domain protein that recruits Smad2 to the TGFβ receptor. Cell 1998;95:779–791.
150. Onichtchouk D, Chen Y-G, Dosch R, et al. Silencing of TGF-β signaling by the pseudoreceptor BAMBI. Nature 1999;401:480–485.
151. Miura S, Takeshita T, Asao H, et al. Hgs (Hrs), a FYVE domain protein, is involved in Smad signaling through cooperation with SARA. Mol Cell Biol 2000;20:9346–9355.

152. Datta PK, Moses HL. STRAP and Smad7 synergize in the inhibition of transforming growth factor β signaling. Mol Cell Biol 2000;20:3157–3167.
153. Shoji H, Tsuchida K, Kishi H, et al. Identification and characterization of a PDZ protein that interacts with activin type II receptors. J Biol Chem 2000;275:5485–5492.
154. Tsuchida K, Matsuzaki T, Yamakawa N, Liu ZH, Sugino H. Intracellular and extracellular control of activin function by novel regulatory molecules. Mol Cell Endocrinol 2001;180:25–31.
155. Parks WT, Frank DB, Huff C, et al. Sorting nexin 6, a novel SNX, interacts with the transforming growth factor-β family of receptor serine-threonine kinases. J Biol Chem 2001;276: 19,332–19,339.
156. Hocevar BA, Smine A, Xu X-X, Howe PH. The adaptor molecule Disabled-2 links the transforming growth factor β receptors to the Smad pathway. EMBO J 2001;20:2789–2801.
157. Perlman R, Schiemann WP, Brooks MW, Lodish HF, Weinberg RA. TGF-β-inducd apoptosis is mediated by the adaptor protein Daxx that facillitates JNK activation. Nat Cell Biol 2001;8: 708–714.
158. McGonigle S, Beall MJ, Pearce EJ. Eukaryotic initiation factor 2α subunit associates with TGFβ receptors and 14-3-3ε and acts as a modulator of the TGFβ response. Biochemistry 2002;41: 579–587.
159. Felici A, Wurthner JU, Parks WT, et al. TLP, a novel modulator of TGF-β signaling, has opposite effects on Smad2- and Smad3-dependent signaling. EMBO J 2003;22:4465–4477.
160. Wurthner JU, Frank DB, Felici A, et al. Transforming growth factor-β-associated protein 1 is a Smad4 chaperone. J Biol Chem 2001;276:19,495–19,502.
161. Seong HA, Jung H, Choi HS, Kim KT, Ha H. Regulation of transforming growth factor-β signaling and PDK1 kinase activity by physical association between PDK1 and serine-threonine kinase receptor-associated protein. J Biol Chem 2005;280:42,897–42,908.
162. Choy L, Derynck R. Transforming growth factor-β inhibits adipocyte differentiation by Smad3 interacting with CCAAT/enhancer-binding protein (C/EBP) and repressing C/EBP transactivation function. J Biol Chem 2003;278:9609–9619.
163. Quing J, Liu C, Choy L, Wu RY, Pagano JS, Derynck R. Transforming growth factor β/Smads signaling regulates IRF-7 function and transcriptional activation of the β interferon promoter. Mol Cell Biol 2004;24:1411–1425.
164. Birchenall-Roberts MC, Fu T, Bang OS, et al. Tuberous sclerosis complex 2 gene product interacts with human Smad proteins. A molecular link of two tumor suppressor pathways. J Biol Chem 2004;279:25,605–25,613.
165. Seoane J, Le HV, Shen L, Anderson SA, Massagué J. Integration of Smad and forkhead pathways in the control of neuroepithelial and glioblastoma cell proliferation. Cell 2004;117:211–223.
166. Li C, Zhu NL, Tan RC, Ballard PL, Derynck R, Minoo P. Transforming growth factor-β inhibits pulmonary surfactant protein B gene transcription through Smad3 interactions with NKX2.1 and HNF-3 transcription factors. J Biol Chem 2002;277:38,399–38,408.
167. Kang Y, Hebron H, Ozbun L, Minoo P, Jakowlew SB. Nkx2.1 transcription factor in lung cells and a transforming growth factor-β heterozygous mouse model of lung carcinogenesis. Mol Carcinog 2004;40:212–231.
168. Penheiter SG, Mitchell H, Garamszegi N, Edens M, Dore JJ Jr, Leof EB. Internalization-dependent requirements for transforming growth factor β signaling via the Smad pathway. Mol Cell Biol 2002;22:4750–4759.
169. Hayes S, Chawla A, Corvera S. TGF β receptor internalization into EEA1-enriched early endosomes: Role in signaling to Smad2. J Cell Biol 2002;158:1239–1249.
170. Di Guglielmo GM, Le Roy C, Goodfellow AF, Wrana JL. Distinct endocytic pathways regulate TGF-β receptor signaling and turnover. Nat Cell Biol 2003;5:410–421.
171. Engel ME, McDonnell MA, Law BK, Moses HL. Interdependent Smad and JNK signaling in transforming growth factor-β-mediated transcription. J Biol Chem 1999;274:37,413–37,420.
172. Hocevar BA, Brown TL, Howe PH. TGF-β induces fibronectin synthesis through a c-Jun N-terminal kinase-dependent, Smad4-independent pathway. EMBO J 1999;18:1345–1356.
173. Bakin AV, Tomlinson AK, Bhowmick NA, Moses HL, Arteaga CL. Phosphatidylinositol 3-kinase function is required for transforming growth factor β-mediated epithelial to mesenchymal transition and cell migration. J Biol Chem 2000;275:36,803–36,810.
174. Petritsch C, Beug H, Balmain A, Oft M. TGF-β inhibits p70 S6 kinase via protein phosphatase 2A to induce G(1) arrest. Genes Dev 2000;14:3093–3101.

175. Bhowmick NA, Ghiassi M, Bakin AV, et al. Transforming growth factor-β1 mediates epithelial to mesenchymal transdifferentiation through a Rho-A-dependent mechanism. Mol Biol Cell 2001;12: 27–36.
176. Fink SP, Swinler SE, Lutterbaugh JD, Massagué J, Markowitz S. Transforming growth factor-β-induced growth inhibition in a Smad4 mutant colon carcinoma cell line. Cancer Res 2001;61: 256–260.
177. Song K, Cornelius SC, Reiss M, Danielpour D. Insulin-like growth factor-I inhibits transcriptional responses of transforming growth factor-β by phosphatidylinositol 3-kinase/Akt-dependent suppression of the activation of Smad3 but not Smad2. J Biol Chem 2003;278:38,342–38,351.
178. Stuelten CH, Da Costa Byfield S, Arany PR, Karpova TS, Stetler-Stevenson WG, Roberts AB. Breast cancer cells induce stromal fibroblasts to express MMP-9 via secretion of TNF-α and TGF-β. J Cell Sci 2005;118:2143–2153.

2 TGF-β Signaling in Homeostasis and Cancer

Joan Seoane

CONTENTS

INTRODUCTION
TGF-β SIGNAL TRANSDUCTION
DUAL ROLE OF TGF-β IN CANCER
REGULATION OF CELL PROLIFERATION BY TGF-β
EVADING THE TGF-β ANTIPROLIFERATIVE EFFECT
TGF-β AS AN ONCOGENIC FACTOR
CONCLUSIONS
REFERENCES

Abstract

TGF-β is a member of a large family of cytokines with a crucial role in embryonic development and tissue homeostasis. Disruption of the TGF-β signaling pathway has been implicated in many human diseases including cancer. In normal epithelial cells, TGF-β acts as a tumor suppressor by inhibiting cellular proliferation. During cancer progression, tumor cells escape from the TGF-β antiproliferative response either by acquiring mutations in components of the TGF-β pathway or by selectively inactivating the pathway that leads to cell cycle arrest. In the latter case, TGF-β becomes an oncogenic factor. Over the last years, some of the molecular mechanisms implicated in the TGF-β antiproliferative response have been elucidated and we are beginning to understand how TGF-β is transformed from an anti-tumorigenic factor into an oncogenic factor during cancer progression. This allows a better understanding of cancer biology and helps in the design of better therapeutic protocols against this deadly disease.

Key Words: TGF-β; Smad; proliferation; cancer; signal transduction.

1. INTRODUCTION

Tissue homeostasis in the context of the whole organism requires a tight control over the biology of individual cells. Coordination and regulation of cell behavior is achieved by cell-to-cell communication through a large number of secreted cytokines that signal from the cell surface into the nucleus to regulate gene transcription. TGF-β is a member of a family of cytokines involved in the regulation of embryonic development and tissue homeostasis. TGF-β is secreted by different cell types and, depending on the cellular context, it regulates a diverse array of cellular processes including cell proliferation, morphogenesis, migration, extracellular matrix production, cytokine secretion, apoptosis, and others *(1,2)*.

From: *Cancer Drug Discovery and Development: Transforming Growth Factor-β in Cancer Therapy, Vol I: Basic and Clinical Biology*
Edited by: Sonia B. Jakowlew © Humana Press Inc., Totowa, NJ

Our understanding of the molecular mechanisms implicated in the TGF-β signal transduction pathway has dramatically increased over the last years. One by one, many of the TGF-β molecular pathways that regulate specific gene responses have been elucidated shedding light onto the biology and physiological relevance of TGF-β in normal and disease states. The understanding of the exact mechanisms involved in the control of tissue homeostasis by TGF-β and how those mechanisms are disrupted in cancer is crucial to gain knowledge about cancer biology and might be helpful in order to design new tools for our fight against cancer.

2. TGF-β SIGNAL TRANSDUCTION

TGF-β is a member of a large family of cytokines (42 members in the human genome) characterized by six conserved cysteine residues. The TGF-β family is classified into six subfamilies: the TGF-β, Activin, Nodal, bone morphogenetic protein (BMP), growth and differentiation factor (GDF), and mullerian inhibiting substance (MIS) subfamilies. The active form of the TGF-β ligand is a 25 kDa dimer of two polypeptides stabilized by hydrophobic interactions strengthened by a bisulfide bond. TGF-β is synthesized as the C-terminal domain of a precursor form that is cleaved before secretion from the cell. The N-terminal TGF-β propeptide called latency-associated peptide (LAP) remains noncovalently bound to TGF-β after secretion inhibiting the binding of the active TGF-β forms to its receptors. In most cell types, TGF-β is secreted as a large latent complex composed by the active TGF-β form bound to LAP which in turn is covalently bound to a member of a family of latent TGF-β binding proteins (LTBPs). The latent TGF-β complex interacts with components of the extracellular matrix through the LTBP remaining stored and providing a source of readily accessible ligand. The process of activation of the TGF-β ligand involves the proteolytic cleavage of the LTBP molecule by serine proteases and the release or conformational modification of LAP *(3,4)*.

Once activated, TGF-β signals through two classes of receptors, the TGF-β type I receptor (TβRI) and the TGF-β type II receptor (TβRII). In addition, endoglin and betaglican, also called accessory receptors, bind TGF-β at low affinity and present it to the TGFβ receptors. Type I and II receptors are serine/threonine kinase receptors that form an heterodimeric complex upon TGF-β binding. TGF-β interacts with the ectodomain of the TβRII and allows the subsequent incorporation of the TβRI generating a ligand-receptor complex formed by a ligand dimer, two type I receptors and two type II receptors. The TβRII appears to be a constitutively active kinase that when forms part of the ligand-receptor complex, phosphorylates a characteristic SGSGSG sequence, called the GS domain, present in the type I receptor. Phosphorylation of the type I receptor GS domain leads to the activation of its kinase and turns the GS region into a Smad binding site. Once activated by the TβRII, the TβRI phosphorylates Smads through its catalytic domain *(5)*.

There are eight Smad proteins in humans. Five of them, Smad1, Smad2, Smad3, Smad5 and Smad8 are phosphorylated by the TGF-β family of receptors. These are commonly referred as receptor phosphorylated Smads (RSmads). Phosphorylation of Smad2 and Smad3 is induced by the TGF-β, Activin, and Nodal subfamilies of ligands whereas Smad1, 5 and 8 are phosphorylated in response to the BMP, GDF, and MIS subfamilies of ligands. Smad4, also called co-Smad, is a partner of the RSmads and it is not phosphorylated by the TGF-β receptors. Moreover, two Smad proteins serve as inhibitors of the TGF-β signal, Smad6 and Smad7. They are not phosphorylated by the TGF-β receptors and act as decoys interfering with Smad-receptor or Smad–Smad interaction as well as facilitating the proteasome-mediated degradation of the TGF-β receptors *(6)*.

Smad proteins consist of two globular domains coupled by a linker region. The N-terminal domain, Mad-homology 1 (MH1) domain, contains the DNA binding module formed by a β-hairpin structure conserved in all Smads that recognizes the 5′-AGAC-3′ DNA sequence. The linker region is a flexible segment that contains binding sites for ubiquitin ligases, phosphorylation sites and, in the case of Smad4, a nuclear export signal (NES). The C-terminal domain, Mad-homology 2 (MH2) domain is highly conserved and mediates the interaction of Smads with a large plethora of proteins. It contains the Smad transactivation domain and in the case of RSmads, the MH2 domain has a C-terminal motif, Ser-X-Ser, that is phosphorylated by the activated type I receptor *(5,6)*.

Phosphorylation of the two serine residues in the C-terminal motif generates an acidic tail that binds to a basic pocket in the Smad4 MH2 domain. RSmad bound to Smad4 form oligomers that, in the nucleus interact with DNA-binding partners to form transcriptional complexes. Depending on the DNA-binding cofactor present in the complex or the gene target, RSmad-Smad4 complexes can be heterodimers (RSmad-Smad4) or heterotrimers (two RSmads and one Smad4) *(5–7)*.

In the basal state, Smads are constantly shuttling in and out of the nucleus. Once phosphorylated by the TβRI, RSmads accumulate in the nucleus. Smad intra-cellular localization depends on its interaction with retention factors. The affinity of phosphorylated Smads for cytosolic retention factors is lower than the one of unphosphorylated Smads and concomitantly, phospho-Smads have higher affinity for nuclear retention factors. This results in an accumulation of phosphorylated Smads in the nucleus *(8)*. The Smad anchor for receptor activation (SARA) is the best characterized of the cytosolic retention factors for Smad2 and Smad3. SARA facilitates phoshorylation of Smad by making Smads accessible to the activated receptor. Phosphorylation of Smads induces a decrease in their affinity for SARA and promotes the release of Smads from their cytosolic retention. Other proteins have been described to play a similar adaptor role besides SARA *(9)*. Once in the nucleus, phosphorylated Smads are sequestered by nuclear retention factors. It has been described that the transcription factor FoxH1, a Smad DNA-binding partner, acts as nuclear retention factor. As FoxH1, other nuclear factors might have a similar role preventing the exit of Smads from the nucleus *(5,8)*.

The mechanisms of Smad nuclear import and export have been extensively studied. Smads 2–4 undergo nuclear import by means of direct interactions with nucleoporins in an importin-independent mechanism. These Smads are able to directly interact with different nucleoporins found along the span of the nuclear pore and, due to the asymmetry of the nuclear pore complex, they are thought to move in both directions through the nuclear pore. Smad3 and Smad4 can also interact with importin-β that facilitates their nucleocytoplasmic dynamics *(5,10)*.

In the nucleus, the Smad complex lands on gene promoters to regulate transcription. Smads in isolation have low affinity for DNA and they associate with DNA-binding partners in order to bind gene promoters. Many Smad-binding cofactors have been described. The DNA-binding partner of Smad is usually a transcription factor that interacts on one side with the Smad complex and on the other with their cognate sequences on DNA. Thus, the DNA-binding cofactor facilitates the binding of the Smad complex to a particular promoter region that contains suitable DNA binding elements for Smads and the Smad partner. The presence of a particular cofactor in the Smad complex dictates which gene promoters are targeted. Moreover, many Smad cofactors facilitate the binding of either coactivators or corepressors and determine the transcriptional activity of the Smad complex *(6)*.

DNA-binding cofactors are critical for TGF-β signal transduction and they are responsible for the pleiotropic responses found in such an apparently linear pathway. In many cases, the activity or the expression of the DNA binding partner is regulated by other signaling pathways. In this way, the TGF-β-Smad pathway is modulated and integrated in the signaling context of the cell *(1,6)*.

The TGF-β-Smad pathway is terminated mainly through two mechanisms. Experiments using TGF-β receptor inhibitors have shown that once the signal stops, RSmads get rapidly dephosphorylated and hence inactivated and exported out of the nucleus. Besides dephosphorylation, the RSmad transcriptional activity is terminated via proteasome-mediated degradation. Receptor phosphorylated RSmads are ubiquinated in their linker region and targeted to the 26S proteasome for degradation. Ubiquitination of RSmad is mediated by members of the Smurf family of HECT-domain E3 ubiquitin ligases. It has been shown that ubiquitination does not only target receptor activated Smads but also unphosphorylated Smads can be ubiquitinated in the absence of signal. High levels of Smurf lowers the Smad basal levels and, therefore, decrease the cell response to TGF-β signals. Ubiquitination controls then both the extent and the length of the TGF-β response in a given cell *(6,8)* (Fig. 1).

TGF-β is able to signal in a Smad-independent way *(9)*. Over the last years, several groups have described the TGF-β-mediated activation of several signaling pathways other than the Smad pathway. Several members of the MAPK pathways, ERK-1, ERK-2, p38, and JNK have been shown to be activated in response to TGF-β in various cell types, in certain cases very rapidly, minutes after ligand addition. The exact mechanism of their activation is still unclear although TAK1, a MAPK kinase kinase (MEKK), has been implicated in the process. Moreover, in certain cell types, TGF-β is able to activate the RHO-like GTPases and PI3K and both activations can be linked. The protein phosphatase PP2A, a negative regulator of p70 ribosomal protein S6 kinase, can be activated by TGF-β. The resulting decrease in p70S6K activity contribute to the antiproliferative response mediated by TGF-β.

The Smad independent pathways are not so well characterized as the TGF-β-Smad pathway. TGF-β can signal through different Smad independent pathways depending on the cell type but still, the exact molecular mechanisms linking the TGF-β receptors with the components of those pathways are unknown.

3. DUAL ROLE OF TGF-β IN CANCER

Many of the TGF-β cellular responses are critical for tissue homeostasis and disruption or aberrant regulation of some of them are implicated in several human disorders including cancer *(1,4)*. TGF-β is a potent antiproliferative factor in epithelial and hematopoietic cells *(1)*. Moreover, depending on the cell type, TGF-β can act as an inducer of apoptosis. Due to its antiproliferative and apoptotic functions, TGF-β is considered a tumor suppressor factor. Indeed, during tumor progression, tumor cells tend to lose the tumor-suppressive responses to TGF-β. In a relatively small proportion of cases, tumor cells acquire somatic mutations in components of the TGF-β signal transduction pathway (Smads and TGF-β receptors) in order to evade the TGF-β antiproliferative function. In many other tumors, the TGF-β signal transduction pathway is not affected but cells are specifically resistant to the antiproliferative response to TGF-β. In those tumors, TGF-β becomes an oncogenic factor, inducing proliferation, angiogenesis, invasion and metastasis *(11–13)*. Moreover, TGF-β induces a block on the immune cellular proliferation and differentiation with a particular effect on T cells. Thus, in those tumors that are resistant to the TGF-β antiproliferative response, TGF-β facilitates tumor progression by blocking the antitumoral immune response *(14)*. Usually, tumors where TGF-β acts as a tumorigenic factor express high levels of TGF-β. In many instances, TGF-β is able to induce its own expression generating a malignant autocrine loop.

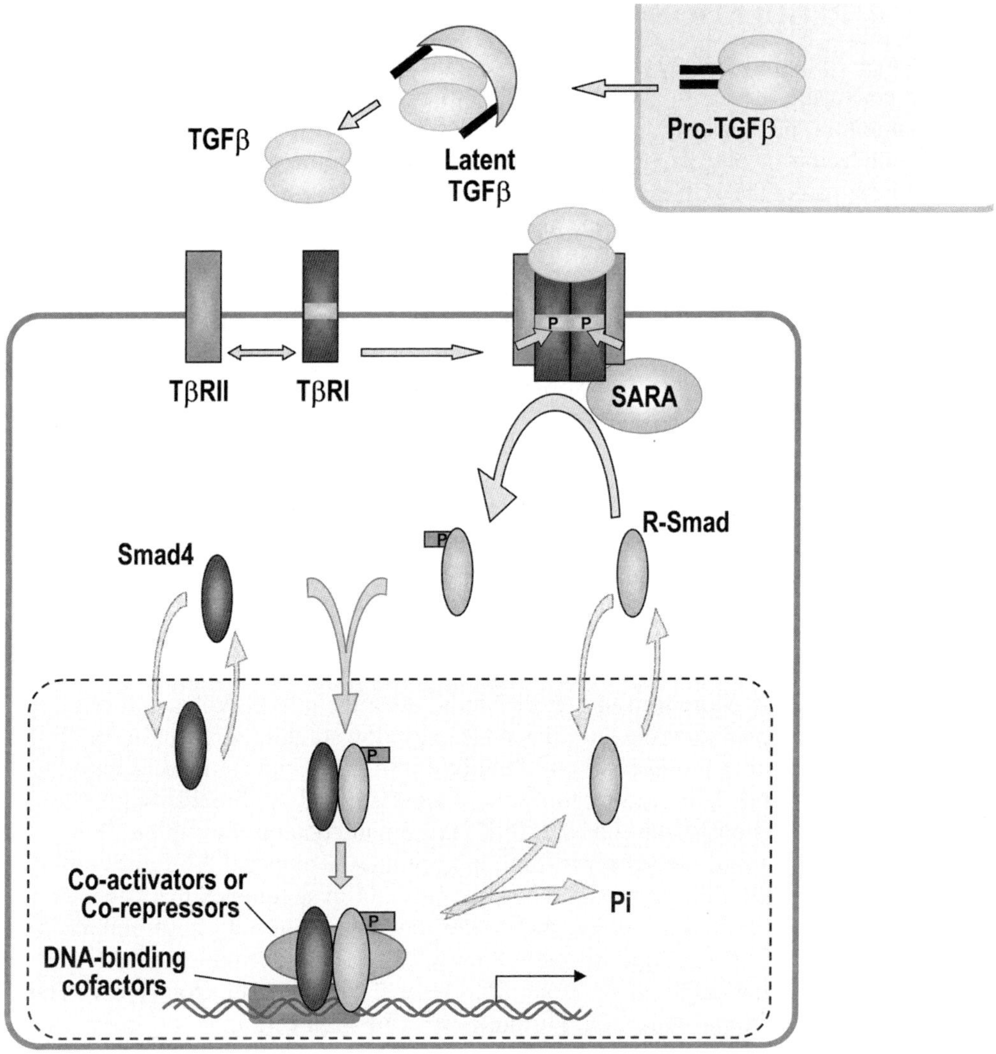

Fig. 1. The TGF-β-Smad pathway. TGF-β is synthesized as a dimeric proprotein (pro-TGF-β). In the Golgi apparatus, the dimeric propeptides are cleaved from the mature TGF-β dimer but still remain attached to the ligand generating a latent TGF-β form. Once TGF-β is liberated from the latent complex, it can bind a TGF-β receptor complex formed by the type I and the type II TGF-β receptors. Upon TGF-β binding, the type II receptor phosphorylates the type I receptor inducing the activity of its serine/threonine kinase. Anchor proteins like SARA capture RSmads for presentation to the activated type I receptor. This facilitates the phosphorylation of RSmads by the type I receptor. RSmads and the co-Smad, Smad4, are continuously shuttling in and out the nucleus. Receptor mediated phosphorylation of RSmad induces their accumulation in the nucleus and their association with the co-Smad, Smad4. In the nucleus, the Smad complex interacts with different DNA-binding cofactors that confers the complex with the affinity to target a specific gene promoter. Recruitment of transcriptional coactivators or corepressors allows the Smad complex to regulate gene transcription. Termination of the TGF-β signal is mediated by RSmad dephosphorylation or proteasome-mediated degradation.

4. REGULATION OF CELL PROLIFERATION BY TGF-β

In normal epithelial cells, TGF-β induce several cytostatic gene responses in order to promote cell cycle arrest. In those cells, TGF-β rapidly induces two cyclin-dependent kinase inhibitors, p21Cip1 and p15Ink4b, and downregulates Myc, Id1 and Id2, all three transcription factors involved in proliferation and inhibitors of differentiation. These five genes are considered part of the TGF-β cytostatic program shared at least by skin, lung and mammary epithelial cells *(1)*. Recently, some of the mechanisms involved in the transcriptional regulation of the cytostatic gene responses by TGF-β have been elucidated. In all cases, except for Id2, an activated Smad complex bound to a DNA-binding cofactor lands on the gene promoter and regulates transcription. In response to TGF-β, phosphorylated Smad3, Smad4 and the transcription factor FoxO interact to form a nuclear complex that binds to a specific region of the *p21Cip1* promoter and activates transcription *(15)*. In addition, activated Smads can interact and synergize with Sp1 trancription factors bound to the proximal region of the *p21Cip1* promoter *(16)*. Transcription of the *p15Ink4b* gene is also induced by a Smad complex that binds to a discrete region of the upstream promoter. Some data indicate that a DNA-binding cofactor facilitates the binding of the Smad complex to the *p15Ink4b* promoter but its identity is still unknown *(17)*. Sp1 transcription factors are also involved in TGF-β-mediated induction of p15Ink4b *(18)*. Myc downregulation by TGF-β is mediated by an E2F4/5-Smad3-Smad4 complex. This complex inhibits Myc expression by recruiting p107, a pocket protein that acts as a transcriptional repressor. This is a very rapid and cell-cycle independent process (19). Id proteins are negative regulators of basic helix-loop-helix transcription factors that involve differentiation. Moreover, Id factors interact with the pocket protein Rb and in this manner induce cell proliferation *(20)*. Therefore, by downregulating Id proteins, TGF-β promotes differentiation and represses proliferation. An ATF3-Smad complex is responsible for Id1 downregulation. ATF3 is a transcriptional repressor and inhibits Id1 expression. In this case, ATF3 is previously induced by TGF-β via a Smad-containing complex. Thus, a two step mechanism is needed for Id1 repression that becomes a so called "self-enabled" gene response where the *Id1* gene repression is mediated by a Smad complex that depends on a prior Smad-mediated response *(21)*. *Id2* is the only gene of the cytostatic program that is not directly regulated by Smads. Myc in cooperation with Max transcriptionally activates *Id2* through an E-box found in the *Id2* promoter *(22)*. Hence, downregulation of Myc by TGF-β prevents *Id2* transcriptional activation. Furthermore, TGF-β in certain cell types is able to induce Mad expression. Mad is a competitor of Myc for Max (the Myc transcriptional partner) forming a Mad-Max complex that binds to the E-box of the *Id2* promoter and represses Id2 expression *(23)* (Fig. 2).

5. EVADING THE TGF-β ANTIPROLIFERATIVE EFFECT

In cancer, the TGF-β cytostatic program is often lost. In most tumor cells, TGF-β is unable to induce p15Ink4b and p21Cip1 as well as downregulate Myc and Id proteins. In certain cases, this is due to the disruption of TGF-β signaling caused by somatic mutations in components of the TGF-β pathway.

The TβRII receptor is inactivated by mutations in gastrointestinal cancers with microsatellite instability. Microsatellite instability is common in many sporadic cancers and it is the result of defects in DNA mismatch repair leading to nucleotide additions or deletions in simple repeated sequences called microsatellites. The TβRII contains one of such repeated sequences, a 10-bp polyadenine repeat, in the extracellular domain, and in most sporadic colon and gastric cancers with microsatellite instability the polyadenine repeat acquire one or two base additions or deletions generating a frameshift mutation and yielding a truncated and inactive TβRII product. Individuals with hereditary nonpolyposis colon cancer, a familial

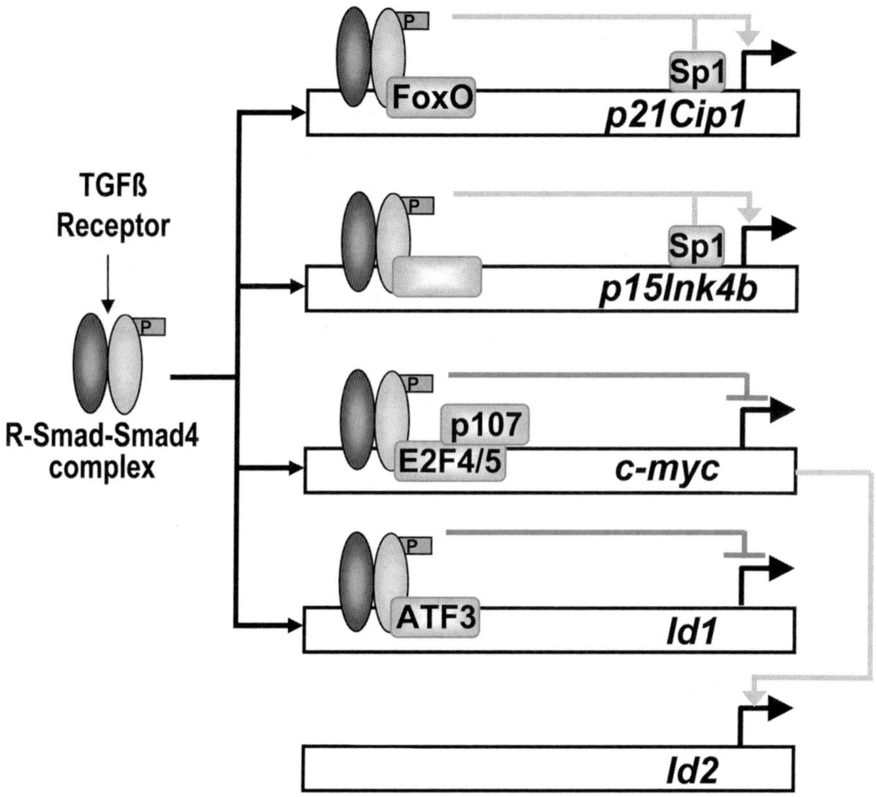

Fig. 2. The TGF-β cytostatic program. In normal epithelial cells, TGF-β regulates the expression of several cytostatic genes in order to induce an arrest in the G1 phase of the cell cycle. TGF-β induces two Cdk inhibitors, p15Ink4b and p21Cip1 and downregulates the expression of Myc, Id1 and Id2. These five gene responses are considered part of the epithelial TGF-β cytostatic program. TGF-β induces p21Cip1 via a Smad-FoxO complex that lands on a specific region of the p21Cip1 promoter. p15Ink4b is induced by a Smad complex that contains a still unknown DNA-binding cofactor. Transcription of Myc and Id1 is repressed by a Smad-E2F4/5-p107 complex and a Smad-ATF3 complex, respectively. Id2 is not regulated directly by Smads. Myc is an inducer of Id2 transcription and TGF-β-mediated downregulation of Myc protein levels prevents Id2 transactivation by Myc.

syndrome characterized by a high incidence of colon, endometrial and gastric cancers, also present mutations in the 10-bp poly-adenine stretch. In most cases, both TβRII alleles have mutations in the polyadenine repeat although, in some cases, the second allele is inactivated by a different mutation in the kinase domain. This indicates that TβRII shares the two-hit inactivation mechanism found in other tumor suppressor genes. Interestingly, mutations in the kinase domain of TβRII have been also found in microsatellite stable colon cancers. Furthermore, other mutations elsewhere in TβRII have been described in T-cell lymphoma, and head and neck carcinomas. Mutations in the TβRI sequence have been described in ovarian, breast, and pancreatic cancers as well as some T-cell lymphomas *(1,4,24)*.

Inactivating mutations in Smad2 and Smad4 are frequently found in some cancers. Missense, nonsense and frameshift mutations as well as small or large deletions are found in both Smads. Most of the missense mutations are found in the Smad MH2 domain and preclude the formation of the Smad2/3-Smad4 heterocomplex *(4,25)*. Tumor derived missense mutations have been also found in the MH1 domain of Smad2 and Smad4. These mutations yield an increase in the affinity of the MH1 domain towards the MH2 domain locking the molecule in an inactive conformation. Smad2 mutations are found in colorectal and lung cancers although with very low frequency. Interestingly, Smad3 has not been found mutated in human cancer to date, however lack of Smad3 protein expression has been described in acute lymphocytic leukemia *(1,4,26)*.

Smad4, also called "deleted in pancreatic carcinoma locus 4", was identified as a candidate tumor suppressor gene in chromosome 18q21 that was deleted or mutated in around 50% of pancreatic carcinomas. Besides pancreatic cancer, mutations in Smad4 have been identified in 10% of all colon cancers and 30% of metastatic colon cancers *(1,4)*. Interestingly, Smad4 is very infrequently mutated in other tumor types. It is important to note that in pancreatic cancer there is a high preponderance of Smad4 mutations, and a very low frequency of mutations in TβRII. Moreover, it has been described that Smad2 and Smad3 can translocate into the nucleus and regulate transcription in response to TGF-β even in the absence of Smad4 in tumor cells from pancreatic cancers *(27)*. Together, these data raises the possibility that in pancreatic tumor cells there is not a selective pressure for blocking the TGF-β pathway but for inactivating Smad4 and acquiring Smad4 independent responses that are most likely different from the ones present in normal cells. Some authors indicate that the TGF-β cytostatic response might be dependent on Smad4 and by inactivating Smad4, tumor cells evade the TGF-β antiproliferative response *(28)*.

Together these results indicate that TGF-β acts as a tumor suppressor gene and that in certain tumor types mutations in the TGF-β pathway that block the TGF-β signal transduction are positively selected in order to avoid the antiproliferative signal of TGF-β. Interestingly, the selective pressure towards mutations of specific components of the TGF-β pathway depends on the tumor type (i.e., TβRII mutations are predominant in gastrointestinal tumors and Smad4 mutations are usually found in pancreatic cancer) suggesting that depending on the cellular context, the cell acquires different selective advantages mutating specific components of the TGF-β pathway.

Only a very small proportion of tumors present mutations in components of the TGF-β pathway, most of them lack the TGF-β antiproliferative response with an intact TGF-β pathway. In those tumors, instead of inhibiting proliferation, TGF-β becomes an oncogenic factor. Work over recent years has begun to shed light onto the mechanisms involved in the selective lose of the antiproliferative pathway. One of the first observations along those lines was that Myc is a repressor of the induction of p21Cip1 and p15Ink4b by TGF-β *(29)*. Miz-1, a zinc finger protein with a POZ domain, binds specifically to a region close to the transcriptional initiator of the *p21Cip1* and *p15Ink4b* genes and recruits Myc to both promoters *(17,30–32)*. The Myc-Miz-1 complex acts as a transcriptional repressor and prevents the induction of p15Ink4b by TGF-β *(17)* and the induction of p21Cip1 in response to TGF-β *(31)* and other inducers as p53 *(31,32)* and Wnt *(33,34)*. Recently, the mechanism of Myc-Miz-1 transcriptional repression has been elucidated. The Myc-Miz-1 complex recruits the DNA methyltransferase, Dnmt3a, to the *p21Cip1* promoter facilitating the methylation and silencing of the gene *(35)*. These results indicated that in order to have full induction of p21Cip1 and p15Ink4b, Myc has to be first downregulated by TGF-β. Hence, tumors with Myc amplification or overexpression might specifically lose the TGF-β antiproliferative response keeping the rest of the TGF-β responses intact.

The induction of p21Cip1 by TGF-β is mediated by a Smad complex that contains the transcription factor FoxO. The transcriptional activity of FoxO factors is regulated by the PI3K-AKT pathway. AKT phosphorylates FoxO in three serine/threonine residues promoting FoxO translocation into the cytosol. Thus, FoxO factors are excluded from the nucleus in tumor cells with high levels of AKT activity. In those cells, a nuclear Smad-FoxO complex cannot be formed and they are unable to induce p21Cip1 in response to TGF-β losing the TGF-β antiproliferative response *(15)*.

A third way through which tumor cells selectively evade the TGF-β antiproliferative response is via the stabilization of TGIF. TGIF is a Smad corepressor that is involved in the regulation of p15Ink4b induction by TGF-β. Hyperactive Ras promotes the phosphorylation of TGIF preventing its ubiquitination and inducing TGIF protein stabilization. TGIF competes with p300, a transcriptional coactivator, for the binding to Smad. The Smad complex that lands on the *p15Ink4b* promoter in tumor cells with high Ras activity contains TGIF instead of p300 and acts as a repressor complex instead of inducing *p15Ink4b* transcription. Therefore, TGF-β is unable to induce p15Ink4b and the TGF-β antiproliferative response is impaired in tumor cells with hyperactive Ras *(36)* (Fig. 3).

6. TGF-β AS AN ONCOGENIC FACTOR

Although the mechanisms involved in the selective lose of the TGF-β antiproliferative response are beginning to be uncovered, not much is known about how TGF-β becomes an oncogenic factor. Lately, some studies have tried to discern the molecular mechanisms implicated in the induction of proliferation, angiogenesis, epithelial-to-mesenchymal transition, and metastasis by TGF-β.

TGF-β has been shown to induce angiogenesis in some tumors. The angiogenic factors, vascular endothelial growth factor (VEGF) and connective-tissue growth factor (CTGF), can be induced by TGF-β in certain tumors *(37,38)*. Moreover, TGF-β can directly induce proliferation of endothelial cells and the expression, secretion and activity of matrix metalloproteases of both endothelial cells and tumor cells *(13)*.

During development and in advanced tumors, epithelial cells transdifferentiate into a mesenchymal-like cell type. This phenomenon is known as epithelial-to-mesenchymal transition (EMT) and allows tumor cells to lose their cell-to-cell contacts and increase their migratory ability. TGF-β is an inducer of EMT in both development and cancer progression and several pathways have been implicated in TGF-β-induced EMT *(13)*. Although EMT might be induced by Smad independent pathways, experiments using TβRI constructs that fail to bind Smads showed that Smads are indispensable for the EMT process *(39,40)*. One of the crucial steps in EMT is the downregulation of e-cadherin, a calcium dependent cell-to-cell receptor. TGF-β induces the expression of Snail and Sip1, which are zinc-finger transcription factors that are well known repressors of e-cadherin *(41,42)*. Therefore, TGF-β-induced EMT is mediated in part by the activation of e-cadherin repressors.

TGF-β can directly foster metastasis to specific tissues. Experiments using transgenic mice that coexpress oncogenic *neu* and the constitutively activated TβRI in the mammary gland, showed that TGF-β signaling enhanced extravasation of breast cells into the lung parenchyma promoting the first step in lung metastasis *(43)*. Moreover, work with the MDA-MB-231 cell line showed that the TGF-β-Smad pathway potentiates metastasis to the bone of athymic mice *(38,44)*. In this case, TGF-β induces IL-11 and the pro-angiogenic factor CTGF and both factors facilitate bone metastasis. IL-11 is known to stimulate osteoclastic bone resorption facilitating tumor cell homing into the bone matrix. The activation of IL-11 and CTGF is mediated by a Smad complex that binds to their promoters. An AP1-Smad complex containing Smad4 is involved in the activation of transcription of IL-11 *(38)*. Smad4 downregulation using RNA interference prevents IL-11 induction by TGF-β in MDA-MB-231 and inhibits bone metastasis *(45)*. In addition, TGF-β stimulates production of the parathyroid-hormone-related peptide (PTHrP) that acts as an osteoclast-activating factor enhancing osteolysis and facilitating metastasis to the bone *(44)*. Both IL-11 and PTHrP establish a vicious cycle by increasing osteolysis and releasing the TGF-β stored in the bone matrix, released TGF-β in turn promotes the induction of more IL-11 and PTHrP from the tumor cells *(46)*.

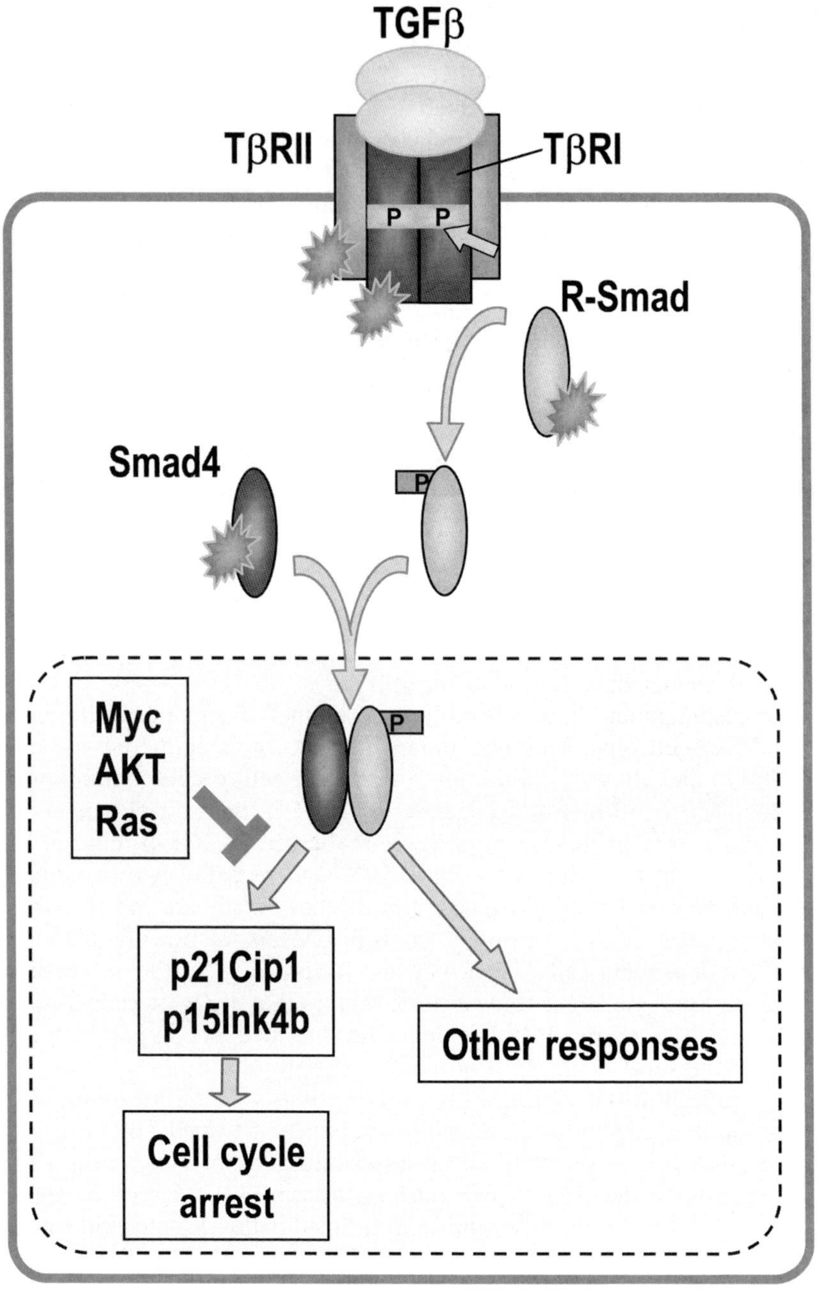

Fig. 3. Escaping the TGF-β antiproliferative response. Tumors escape the TGF-β antiproliferative response via two distinct mechanisms. Some tumor cells acquire mutations in several components of the TGF-β pathway leading to the inactivation of the TGF-β signaling. Mutations in the TβRI, TβRII, RSmad and Smad4 have been found in colon and pancreatic cancers as well as other tumor types. On the other hand, certain tumors evade the TGF-β-mediated cell cycle arrest by selectively avoiding the pathway that leads to the induction of the TGF-β cytostatic program without affecting other TGF-β gene responses. In those tumors, instead of been a tumor suppressor, TGF-β becomes an oncogenic factor. High levels of Myc or high activity of Ras or AKT specifically prevents the induction of p21Cip1 and/or p15Ink4b impairing the ability of TGF-β to inhibit proliferation and facilitating the malignant transformation of TGF-β.

7. CONCLUSIONS

Work over the last years has considerably increased our knowledge about the molecular mechanisms implicated in the TGF-β signal transduction pathway. At a first glance, the TGF-β pathway seems quite linear, but it is turning to be complex, finely regulated and well coordinated with other signaling pathways. Due to its antiproliferative and differentiating effect on epithelial cells, TGF-β is considered a tumor suppressor factor. During oncogenic progression, tumor cells evade TGF-β action either by acquiring mutations in components of the TGF-β pathway and disrupting the TGF-β signal transduction pathway or by selectively inhibiting its antiproliferative response. In the latter case, TGF-β not only loses its antitumorigenic function but becomes an oncogenic factor inducing proliferation, angiogenesis, invasion, metastasis and immune suppression. The mechanisms involved in the malignant transformation of TGF-β remain unclear, although some insights are emerging and are helping to understand how the TGF-β signal transduction pathway is perverted to become an oncogenic pathway.

Advances in the study of the molecular mechanisms that govern tumor progression are providing hopeful and, in some cases, dramatic therapeutic benefits in cancer and have pushed rational molecular targeting to the cutting-edge of cancer therapy. Among several other signal transduction pathways and due to its oncogenic role, the TGF-β pathway is nowadays been evaluated as a potential therapeutic target *(47)*. The dual and complex role of TGF-β in oncogenesis presents a unique challenge that has to be addressed to be able to select the patient population that may benefit from an anti-TGF-β therapy. The understanding of the exact mechanisms involved in the malignant transformation of TGF-β will improve patient stratification and the development of successful therapeutic strategies as well as provide new therapeutic targets to restore the normal TGF-β action.

REFERENCES

1. Siegel PM, Massagué J. Cytostatic and apoptotic actions of TGF-β in homeostasis and cancer. Nat Rev Cancer 2003;3:807–821.
2. Feng XH, Derynck R. Specificity and versatility in TGF-β signaling through Smads. Annu Rev Cell Dev Biol 2005;21:659–693.
3. Annes JP, Munger JS, Rifkin DB. Making sense of latent TGF-β activation. J Cell Sci 2003; 116:217–224.
4. Massagué J. TGF-β signal transduction. Annu Rev Biochem 1998;67:753–791.
5. Shi Y, Massagué J. Mechanisms of TGF-β signaling from cell membrane to the nucleus. Cell 2003;113:685–700.
6. Joan Massagué, David Wotton. Smad transcription factors. Genes Dev 2005;19:2783–2810.
7. Inman GJ, Hill CS. Stoichiometry of active smad-transcription factor complexes on DNA. J Biol Chem 2002;277:51,008–51,016.
8. ten Dijke P, Hill CS. New insights into TGF-β-Smad signalling. Trends Biochem Sci 2004;29: 265–273.
9. Derynck R, Zhang YE. Smad-dependent and Smad-independent pathways in TGF-β family signalling. Nature 2003;425:577–584.
10. Xu L, Massagué J. Nucleocytoplasmic shuttling of signal transducers. Nat Rev Mol Cell Biol 2004; 5:209–219.
11. Roberts AB, Wakefield LM. The two faces of transforming growth factor beta in carcinogenesis. Proc Natl Acad Sci USA 2003;100:8621–8623.
12. Wakefield LM, Roberts AB. TGF-β signaling: positive and negative effects on tumorigenesis. Curr Opin Genet Dev 2002;12:22–29.
13. Derynck R, Akhurst RJ, Balmain A. TGF-β signaling in tumor suppression and cancer progression. Nat Genet. 2001;29:117–129.
14. Li MO, Wan YY, Sanjabi S, Robertson AK, Flavell RA. Transforming growth factor-β regulation of immune responses. Annu Rev Immunol 2006;24:99–146.

15. Seoane J, Le HV, Shen L, Anderson SA, Massagué J. Integration of Smad and forkhead pathways in the control of neuroepithelial and glioblastoma cell proliferation. Cell 2004;117:211–223.
16. Pardali K, Kurisaki A, Moren A, ten Dijke P, Kardassis D, Moustakas A. Role of Smad proteins and transcription factor Sp1 in p21(Waf1/Cip1) regulation by transforming growth factor-β. J Biol Chem 2000;275:29,244–29,256.
17. Seoane J, Pouponnot C, Staller P, Schader M, Eilers M, Massagué J. TGFβ influences Myc, Miz-1 and Smad to control the CDK inhibitor p15INK4b. Nat Cell Biol 2001;3:400–408.
18. Feng XH, Lin X, Derynck R. Smad2, Smad3 and Smad4 cooperate with Sp1 to induce p15(Ink4B) transcription in response to TGF-β. EMBO J 2000;19:5178–5193.
19. Chen CR, Kang Y, Siegel PM, Massagué J. E2F4/5 and p107 as Smad cofactors linking the TGFβ receptor to c-myc repression. Cell 2002;110:19–32.
20. Iavarone A, Garg P, Lasorella A, Hsu J, Israel MA. The helix-loop-helix protein Id-2 enhances cell proliferation and binds to the retinoblastoma protein. Genes Dev 1994;8:1270–1284.
21. Kang Y, Chen CR, Massagué J. A self-enabling TGFβ response coupled to stress signaling: Smad engages stress response factor ATF3 for Id1 repression in epithelial cells. Mol Cell 2003;11:915–926.
22. Lasorella A, Noseda M, Beyna M, Yokota Y, Iavarone A. Id2 is a retinoblastoma protein target and mediates signalling by Myc oncoproteins. Nature 2000;407:592–598.
23. Siegel PM, Shu W, Massagué J. Mad upregulation and Id2 repression accompany transforming growth factor (TGF)-β-mediated epithelial cell growth suppression. J Biol Chem 2003;278: 35,444–35,450.
24. Dumont N, Arteaga CL. Targeting the TGF-β signaling network in human neoplasia. Cancer Cell 2003;3:531–536.
25. Wu JW, Hu M, Chai J, et al. Crystal structure of a phosphorylated Smad2. Recognition of phosphoserine by the MH2 domain and insights on Smad function in TGF-β signaling. Mol Cell 2001;8: 1277–1289.
26. Wolfraim LA, Fernandez TM, Mamura M, et al. Loss of Smad3 in acute T-cell lymphoblastic leukemia. N Engl J Med 2004;351:552–559.
27. Subramanian G, Schwarz RE, Higgins L, et al. Targeting endogenous transforming growth factor beta receptor signaling in SMAD4-deficient human pancreatic carcinoma cells inhibits their invasive phenotype1. Cancer Res 2004;64:5200–5211.
28. Levy L, Hill CS. Smad4 dependency defines two classes of transforming growth factor-β (TGF-β) target genes and distinguishes TGF-β-induced epithelial-mesenchymal transition from its antiproliferative and migratory responses. Mol Cell Biol 2005;25:8108–8125.
29. Warner BJ, Blain SW, Seoane J, Massagué J. Myc downregulation by transforming growth factor beta required for activation of the p15(Ink4b) G(1) arrest pathway. Mol Cell Biol 1999;19:5913–5922.
30. Staller P, Peukert K, Kiermaier A, et al. Repression of p15INK4b expression by Myc through association with Miz-1. Nat Cell Biol 2001;3:392–399.
31. Seoane J, Le HV, Massagué J. Myc suppression of the p21(Cip1) Cdk inhibitor influences the outcome of the p53 response to DNA damage. Nature 2002;419:729–734.
32. Herold S, Wanzel M, Beuger V, et al. Negative regulation of the mammalian UV response by Myc through association with Miz-1. Mol Cell 2002;10:509–521.
33. van de Wetering M, Sancho E, Verweij C, et al. The beta-catenin/TCF-4 complex imposes a crypt progenitor phenotype on colorectal cancer cells. Cell 2002;111:241–250.
34. Wu S, Cetinkaya C, Munoz-Alonso MJ, et al. Myc represses differentiation-induced p21CIP1 expression via Miz-1-dependent interaction with the p21 core promoter. Oncogene 2003;22:351–360.
35. Brenner C, Deplus R, Didelot C, et al. Myc represses transcription through recruitment of DNA methyltransferase corepressor. EMBO J 2005;24:336–346.
36. Lo RS, Wotton D, Massagué J. Epidermal growth factor signaling via Ras controls the Smad transcriptional co-repressor TGIF. EMBO J 2001;20:128–136.
37. Pertovaara L, Kaipainen A, Mustonen T, et al. Vascular endothelial growth factor is induced in response to transforming growth factor-β in fibroblastic and epithelial cells. J Biol Chem 1994;269: 6271–6274.
38. Kang Y, Siegel PM, Shu W, et al. A multigenic program mediating breast cancer metastasis to bone. Cancer Cell 2003;3:537–549.
39. Yu L, Hebert MC, Zhang YE. TGF-β receptor-activated p38 MAP kinase mediates Smad-independent TGF-β responses. EMBO J 2002;21:3749–3759.

40. Itoh S, Thorikay M, Kowanetz M, et al. Elucidation of Smad requirement in transforming growth factor-β type I receptor-induced responses. J Biol Chem 2003;278:3751–3761.
41. Comijn J, Berx G, Vermassen P, et al. The two-handed E box binding zinc finger protein SIP1 downregulates E-cadherin and induces invasion. Mol Cell 2001;7:1267–1278.
42. Peinado H, Quintanilla M, Cano A. Transforming growth factor beta-1 induces snail transcription factor in epithelial cell lines: mechanisms for epithelial mesenchymal transitions. J Biol Chem 2003;278: 21,113–21,123.
43. Siegel PM, Shu W, Cardiff RD, Muller WJ, Massagué J. Transforming growth factor β signaling impairs Neu-induced mammary tumorigenesis while promoting pulmonary metastasis. Proc Natl Acad Sci USA 2003;100:8430–8435.
44. Yin JJ, Selander K, Chirgwin JM, et al. TGF-β signaling blockade inhibits PTHrP secretion by breast cancer cells and bone metastases development. J Clin Invest 1999;103:197–206.
45. Kang Y, He W, Tulley S, et al. Breast cancer bone metastasis mediated by the Smad tumor suppressor pathway. Proc Natl Acad Sci USA 2005;102:13,909–13,914.
46. Mundy GR. Metastasis to bone: causes, consequences and therapeutic opportunities. Nat Rev Cancer 2002;2:584–593.
47. Yingling JM, Blanchard KL, Sawyer JS. Development of TGF-β signalling inhibitors for cancer therapy. Nat Rev Drug Discov 2004;3:1011–1022.

3 TGF-β Availability: Latent TGF-β and Latent TGF-β Binding Proteins

Vesna Todorovic and Daniel B. Rifkin

CONTENTS

INTRODUCTION
TRANSFORMING GROWTH FACTOR-β
LATENT TGF-β BINDING PROTEINS (LTBPS)
ACTIVATION OF LATENT TGF-β
CONCLUSIONS
ACKNOWLEDGMENTS
REFERENCES

Abstract

Transforming growth factor (TGF)-βs are potent regulators of cellular proliferation, apoptosis and differentiation during development and in adult organisms. TGF-βs are secreted as latent complexes, unable to bind their receptors. Latent TGF-β is a dimer of the mature cytokine noncovalently bound to its propeptide, latency associated protein. In addition, latent TGF-βs are normally complexed with another protein, latent TGF-β-binding protein (LTBP). LTBPs are long extracellular molecules that covalently bind the latent form of TGF-β inside the cell, facilitate the cytokine's secretion, direct its distribution and regulate its activation. LTBPs may be covalently linked to the extracellular matrix, thereby storing TGF-β for the future use. There are four distinct LTBP molecules: LTBP1-4, displaying different spatial and temporal distribution and affinity towards TGF-β-binding. Central to regulation of TGF-β activity in the extracellular space is the conversion of latent TGF-β to active TGF-β. TGF-β can be activated by a variety of molecules and environmental conditions. The temporal and spatial gap between TGF-β synthesis and action and the multitude of means to control and influence TGF-β activity, make the biology of this cytokine truly remarkable.

Key Words: Transforming growth factor-β; latency associated protein; latent TGF-β binding protein; extracellular matrix; plasmin; thrombospondin-1; integrin $\alpha_v\beta_6$.

1. INTRODUCTION

The transforming growth factors-βs (TGF-βs) are multipotent cytokines that modulate pivotal cellular processes, such as proliferation, differentiation, and apoptosis, regulate extracellular matrix (ECM) synthesis and proteolysis, and are involved in several developmental processes *(1)*. Consistent with these multiple roles, inappropriate TGF-β activity is associated with a number of pathologies, such as cancer *(2)*, abnormal development and function of many organs *(3,4)*, fibrosis *(5)*, and autoimmune disease *(6)*. The tremendous biological potency

From: *Cancer Drug Discovery and Development: Transforming Growth Factor-β in Cancer Therapy, Vol I: Basic and Clinical Biology*
Edited by: Sonia B. Jakowlew © Humana Press Inc., Totowa, NJ

of active TGF-β is tightly controlled both spatially and temporally at many levels, making the biology of TGF-β one of the most complex and interesting among cytokines.

TGF-β is synthesized as a homodimeric proprotein, and the dimeric propeptide is cleaved intracellularly from the growth factor *(7)*. However, the TGF-β propeptide binds strongly to TGF-β even after the bonds between the propeptide and the mature cytokine are lysed *(8)*. The interaction of the mature TGF-β and its propeptide renders the cytokine inactive; therefore the propeptide is called the latency-associated protein (LAP). The TGF-βs are secreted as tri-molecular complexes, called the large latent complex (LLC), composed of a dimer of the mature TGF-β, a dimer of LAP and a molecule of the latent TGF-β binding protein (LTBP). Upon secretion, the LLC may be covalently integrated into the ECM through the interaction of the LTBP with components of the matrix, thereby storing the cytokine for future use. The extracellular concentration of active TGF-β is primarily regulated by TGF-β activation. A number of different TGF-β activators have been described, including proteases, thrombospondin, and integrins. These seemingly unrelated molecules display different tissue distribution and participate in diverse mechanisms of TGF-β activation in an isoform-specific manner, regulating the process of latent TGF-β activation in a spatially and temporally restricted way.

2. TRANSFORMING GROWTH FACTOR-β

TGF-β is a prototype of a protein superfamily that contains more than 30 members. The large TGF-β superfamily of proteins is classified into several families that include TGF-βs, inhibins and activins, bone morphogenetic proteins (BMPs), and growth and differentiation factors. All of these structurally related cytokines exert a wide range of activities in growth, development and maintenance of cellular and tissue homeostasis.

TGF-β was initially discovered as an "activity" in the conditioned medium from virally transformed cells that stimulated growth of normal rat fibroblasts in soft agar *(9,10)*. This activity was named sarcoma growth factor, but further purification showed that the conditioned medium contained two proteins, named TGF-α and -β, that stimulated cell growth. TGF-β1 was purified from human placenta, blood platelets, and bovine kidneys *(11–13)* and later cloned from a human cDNA library *(14)*. Purification of TGF-β from porcine platelets led to the discovery of a second TGF-β isoform, named TGF-β2 *(15)*, and soon afterwards, a third TGF-β isoform (TGF-β3) was identified from a variety of mammalian cDNA libraries *(16,17)*.

2.1. TGF-β Gene Expression

It is commonly believed that TGF-β is ubiquitously expressed. Indeed, if one disregards specific TGF-β isoforms, TGF-β is omnipresent. However, the different TGF-β isoforms have discrete, dynamic, and sometimes overlapping expression patterns during development and in adult organisms. In situ hybridization studies of TGF-β1-3 mRNA distribution during mouse embryogenesis revealed isoform-specific expression patterns that were developmentally regulated. This type of information provided the first clues for the possible distinct *in vivo* functions for the cytokine. For example, in the developing heart TGF-β2 expression suggested its involvement in the epithelial-mesenchymal transition (EMT) that takes place in the atrio-ventricular (AV) canal and the outflow tract (OFT) during heart septation and remodeling. The septation of the AV canal and the proximal OFT initiates with the formation of specific structures known as endocardial cushions, which are defined by two cellular layers: myocardium and endocardium, lining the ECM called endocardial jelly *(18)*. Cardiac EMT commences when the myocardium underlying the endocardial cushions signals to the endocardium, causing a metamorphosis of certain endocardial cells into motile mesenchymal cells. These mesenchymal cells invade the cardiac jelly, proliferate, condense, and eventually remodel the endocardial cushions into cardiac valves. Two TGF-β isoforms (TGF-β2 and TGF-β1) are expressed in the mouse septating heart *(19)*. TGF-β2 is produced only by the myocardium underlying the endocardial cushions in the AV canal and OFT at the time when the myocardium signals to the endocardium, whereas TGF-β1 expression is limited to the endocardial lining. As EMT advances in the endocardial cushions, TGF-β2 production shifts from the cushion myocardium to the invasive mesenchymal cells, whereas TGF-β1 expression remains restricted to the endocardium. The dynamics of TGF-β2 expression during cardiac septation and remodeling, suggested its involvement in cardiac EMT *(20)*. Indeed, several experiments using endocardial cushion explants in culture confirmed the essential role for TGF-β2, but not other TGF-β isoforms, in cardiac EMT *(21)*. In addition, among the three mouse lines each lacking a specific TGF-β isoform, only *Tgf-β2* knockout mice display defective valvulogenesis in accordance with the disturbed cardiac EMT *(22,23)*.

RNA localization studies can be of great assistance in deciphering the biological function of a particular protein, assuming that the detected mRNA is translated to a biologically active protein. However, in the case of TGF-β, expression studies might not be the most suitable approach for understanding the cytokine's biological function. Understanding of TGF-β synthesis, secretion and, perforce, activation of the latent form are crucial for complete comprehension of TGF-β spatially and temporally tightly controlled action.

2.2. TGF-β Synthesis and Secretion

The TGF-β isoforms are synthesized as ~50-kDa prepro-proteins. The pre region is removed contranslationally and the propeptide dimerizes within the endoplasmic reticulum. These dimeric proproteins are proteolytically cleaved in the Golgi by furine-like proteases to yield a 25 kDa mature growth factor dimer and a 75 kDa propeptide dimer. Despite the cleavage from the mature growth factor, the propeptide and cytokine dimers remain associated noncovalently and their interaction renders the mature TGF-β inactive. In consequence, the propeptide is named the latency associated protein (LAP) and the complex between the LAP and the mature TGFβ is known as the small latent complex (SLC) *(8)* (Fig. 1). The LAP dimer is usually covalently bound to a second gene product, latent TGF-β binding protein, and the trimolecular aggregate is called the large latent complex (LLC) *(24)* (Fig. 1). The LTBP does not confer latency, but facilitates the secretion of TGF-β into the ECM *(25,26)*.

2.3. Conservation of TGF-βs Among Different Species

Given the biological importance of the individual TGF-β isoforms, it is no surprise that the TGF-β isoforms are extraordinarily preserved among species. Bovine, porcine, and simian TGF-β1 cytokines are identical to the human isoform, whereas the mature murine peptide differs only in the substitution of a serine for an alanine at position 75. However, mature chicken TGF-β1 shows only 82% identity to the human homologue. Mature TGF-β2 and –3 peptides are even more conserved across species; the amino acid sequence of chicken, murine, and human TGF-β3 cytokine is completely identical. In addition, a high degree of homology exists between different mature TGF-β isoforms within the same species; for example, human TGF-β1 shares 71% and 76% amino acid sequence identity with the mature TGF-β2 and –3 isoforms, respectively. However, in contrast to the extremely high sequence homology among mature TGF-βs, their respective LAPs are much more divergent (*see* Table 1). Sequence analysis reveals only 34–38% amino acid sequence identity among LAP isoforms (TGF-β1, 2, 3) within a species. However, there is a considerable conservation of individual LAP isoform sequences across species (Table 1). The isoform-specific divergence among LAP amino acid sequences may signify isoform-specific activation

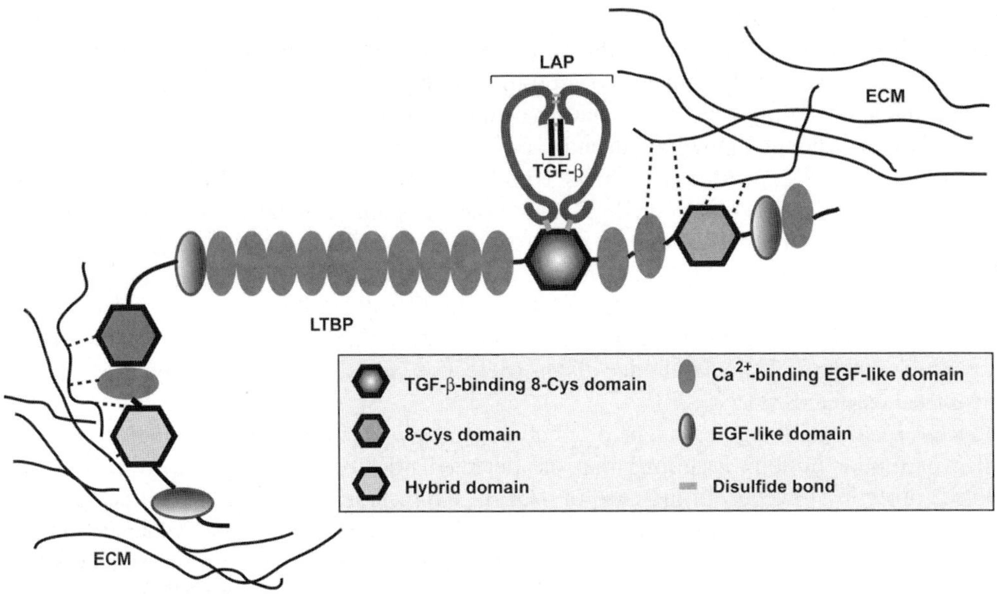

Fig. 1. Schematic representation of the LLC associated with the extracellular matrix (ECM). SLC, consists of dimers of mature TGF-β and its propeptide LAP, covalently binds the third 8-cysteine domain (8-Cys3) of LTBP. Four LTBPs bind to the ECM through their N-terminus; in addition, LTBP-1 and –4 interact with the ECM through their C-terminal modules.

reactions of TGF-β and may explain isoform specific functions in vivo, despite sometimes overlapping expression patterns of the isoforms in vivo and their practically identical functions *in vitro*. For example, latent TGF-β1 and TGF-β3 can be activated by $\alpha_v\beta_6$, whereas TGF-β2 cannot *(5,27)*. This is due to the presence of the integrin-binding sequence RGD in the LAPs of TGF-β1 and –3 but not the TGF-β2 isoform. Additional activation mechanisms for isoforms, especially TGF-β2 remain to be discovered.

2.4. TGF-β LAP

Among the three different mammalian isoforms of LAP, the TGF-β1 LAP (LAP-β1) is the most extensively characterized. LAP-β1 is glycosylated at three glycosylation sites, two of which have mannose-6-phosphate groups that can bind the cell surface mannose-6-phosphate/insulin growth factor II receptor *(28)*. LAP-β1 contains three cysteine residues at amino acid positions 33, 223, and 225 *(29)*. Substitution of Cys223 and 225 in LAP-β1 by serines results in production only of a monomeric precursor form that is easily converted into active TGF-β1, suggesting that Cys223 and 225 are important for dimerization of LAP-β1 by interchain disulfide bonding *(29)*. The third cysteine residue, Cys33, enables disulfide bonding between LAP-β1 and LTBP *(30)*. A mutant form of LAP-β1 with Cys33Ser substitution overexpressed in mouse heart yields atrial fibrosis and delayed myocardial heeling due to a marked increase of active TGF-β1 in the heart *(31)*.

LAP association with the mature TGF-β is critical not only for TGF-β latency but also for the secretion of mature TGF-β, as mutations in LAP-β1 that disrupt its interaction with the mature TGF-β1 prevent cytokine secretion *(32)*. Crosslinking and biological assays indicated that residues 50–85 of LAP-β1 are required for its association with mature TGF-β1. Recently, the work of Young and collaborators further refined LAP-β1/TGF-β1 interacting sequences *(33)*. The authors showed that a conserved sequence LSKL (amino acids 54–57) mediates LAP-β1 binding to the mature TGF-β1. Furthermore, the group identified the

Table 1
The Amino Acid Identities Among TGF-β Latent Proteins Across Humans, Mice and Chickens

	hLAP-β2	*hLAP-β3*	*mLAP-β1*	*mLAP-β2*	*mLAP-β3*	*cLAP-β1*	*cLAP-β2*	*cLAP-β3*
hLAP-β1	35	36	**85**	36	36	**46**	34	36
hLAP-β2	–	46	34	**96**	47	34	**88**	48
hLAP-β3		–	35	47	**97**	37	46	**85**
mLAP-β1			–	35	35	**45**	35	35
mLAP-β2				–	48	35	**87**	48
mLAP-β3					–	37	47	**84**
cLAP-β1						–	34	38
cLAP-β2							–	48

The amino acid sequences of TGF-β latent proteins 1–3 from human (h), mouse (m), and chicken (c) were compared using Blast P BLOSUM62 without a filter. Accession numbers of the compared sequences: human TGF-β1 LAP (*hLAP1;* NP_000651; residues 28-278); human TGF-β2 LAP (*hLAP2;* NP_003229; residues 21-302); human TGF-β3 LAP (*hLAP3;* NP_003230; residues 24-300); mouse TGF-β1 LAP (*mLAP1;* NP_035707; residues 30-278); mouse TGF-β2 LAP (*mLAP2;* AAH11170; residues 21-302); mouse TGF-β3 LAP (*mLAP3;* NP_033394; residues 24-298); chicken TGF-β1 LAP (*cLAP1;* P09531; residues 1-259); chicken TGF-β2 LAP (*cLAP2;* NP_001026216; residues 21-300) and chicken TGF-β3 LAP (*cLAP3;* NP_990785; residues 24-300).

TGF-β1 sequence RKPK (residues 94–87) in the receptor-binding region of mature TGF-β1, as the binding site for LAP.

A human syndrome, called Camurati-Engelmann disease (CED), further emphasizes the importance of LAP-β1 in the proper control of TGF-β latency. This autosomal dominant disease results from mutations in the TGF-β1 LAP sequence and is characterized by hyperosteosis and sclerosis of the diaphysis of the long bones *(34–36)*. Most of the mutations in CED occur at or close to the cysteine residues involved in the interchain bonds of the LAP dimer (Cys 223 and 225). Studies with fibroblasts from three patients with CED mutations at or close to Cys225 indicate that the mutant cells produce substantially more active TGF-β1 than do wild-type cells *(34)*. However, interchain disulfide bonding of LAP-β1 in CED patients occurs efficiently, as judged by LAP-β1 immunoblotting of proteins secreted by CED fibroblasts. One possible explanation for normal level of TGF-β1 synthesis and regular LAP dimerization but considerably higher amounts of active TGF-β in CED patients is that the latent complex of TGF-β1 in CED patients is hyper-responsive to activating proteolytic events *(8)*.

In summary, TGF-β LAP is necessary and sufficient to confer latency to mature TGF-β. However, another protein, LTBP emerges as a crucial regulator of TGF-β activity.

3. LATENT TGF-β BINDING PROTEINS (LTBPs)

LTBPs are large ECM glycoproteins, structurally related to the fibrillins. LTBP-1 was first identified during the characterization of latent TGF-β1 secreted from platelets as a protein that covalently associates with SLC *(37)*. Since the characterization of LTBP-1, three other LTBP isoforms were cloned (LTBP-2, -3, and -4) *(38–40)*. With the exception of LTBP-2, all LTBP isoforms covalently interact with TGF-β LAP. Therefore, it is believed that LTBPs play a dual role: (i) structural, given their association with fibronectin-rich pericellular fibers and fibrillin-containing microfibrils *(7,41)* and (ii) regulatory, as orchestrators of TGF-β availability *(42)*.

3.1. LTBP Structure

The four LTBPs have molecular masses between 120 and 210 kDa. These proteins display modular and highly repetitive structures with the first described LTBP, LTBP-1S, being the prototype for all subsequently described isoforms. LTBP-1S has 17 epidermal growth factor (EGF)-like domains, 14 of which contain consensus Ca^{2+}-binding sequences (Fig.1). In addition, there are four modules, each of approximately 70 amino acids that have eight cysteines that form four disulfide bonds. These eight cysteine (8-Cys) domains are unique to the LTBPs and the fibrillins *(24)*. The N-terminal 8-Cys domain, also called a hybrid domain, has a more divergent sequence and may be derived from an earlier splicing event between an 8-Cys domain and an EGF-like domain. Between the 8-Cys2 and the core of repeating EGF-like motifs there is a proline and basic amino acid-rich region called the hinge domain (*see* below). The general structures of the other LTBPs are similar, with each containing four 8-Cys domains and multiple EGF-like modules (Fig. 2).

Amongst the four LTBPs, the highest degree of sequence diversity is in the hinge domain *(8)*. The hinge sequences of LTBP-1 and -2 are sensitive to proteolytic cleavage *(43,44)*, but the hinge domains of LTBP-3 and -4 are not cleaved *(45)*. The LTBP-1 hinge contains a heparin-binding sequence *(46)*.

The central core of EGF-like domains is reminiscent of the multiple tandem EGF-like repeats in the fibrillins *(47)* (Fig. 2). The repeating EGF-like module core may form a rigid sequence that projects the TGF-β-binding C-terminal region away from the matrix *(48)* facilitating the interaction of latent TGF-β with its activators. However, removal of the EGF-like core from LTBP-1S does not impede latent TGF-β activation by the integrin $\alpha_v\beta_6$ *(49)*. The EGF-like domains flanking the 8-Cys domains may function to restrain 8-Cys domain motion *(30,49)* and facilitate interactions with matrix or fibular proteins as described for fibrillin *(48,50)*.

The 8-Cys3 domains of LTBP-1, -3, and -4 bind the SLC *(30,51)*. The other remaining 8-Cys motifs in LTBP-1, -3, and -4, in addition to all of the 8-Cys domains in LTBP-2 and the fibrillins, do not bind LAP or SLC. The reason for the differential binding of LAP to 8-Cys domains was suggested by Saharinen and Keski-Oja *(52)*, who demonstrated the importance of a dipeptide insertion between cysteines 6 and 7 in the three LAP-binding 8-Cys domains, compared to absences of the insertion in the sequences of the non-LAP-binding 8-Cys. The basis for LLC complex formation was elucidated with the publication of the solution structure of the LTBP-1 8-Cys3 domain *(53)* and comparison of this structure to the structure of the fibrillin-1 non-LAP-binding 8-Cys6. The 8-Cys fold is globular with six β strands and two α-helices. The pairing of the eight cysteines is 1–3, 2–6, 4–7, and 5–8, thereby forming a fairly rigid structure. Although the overall structures of the TGF-β-binding and nonbinding 8-Cys domains are similar, the FP insertion between cysteines 6 and 7 projects a β-strand and displaces the 2–6 cysteine pair into the solvent. The increased accessibility probably enhances the reactivity of the 2–6 disulfide bond permitting disulfide exchange with the pair of cysteine 33 residues in TGF-β1 LAP. The NMR structure also revealed five negatively charged amino acids surrounding the 2–6 disulfide bond suggesting that electrostatic interactions between LAP and the LTBP-1 8-Cys3 domain initiate complex formation. These LAP and 8-Cys3 electrostatic interactions that were also demonstrated by FRET experiments imply that non-covalent protein–protein associations are possible with 8-Cys domains. Covalent bonding example may be only one end of a spectrum of 8-Cys domain binding interactions that might include propeptides of other members of the TGF-β superfamily *(54,55)*.

Different LTBPs have distinct binding specificities for individual TGF-β isoforms. Saharinen and Keski-Oja *(52)* reported that in coexpression experiments, the LTBP-1 and -3 8-Cys3 domains bound all three TGF-β isoforms well, but the LTBP-4 protein 8-Cys3 poorly bound only TGF-β1. However, the in vivo significance of these results is unclear.

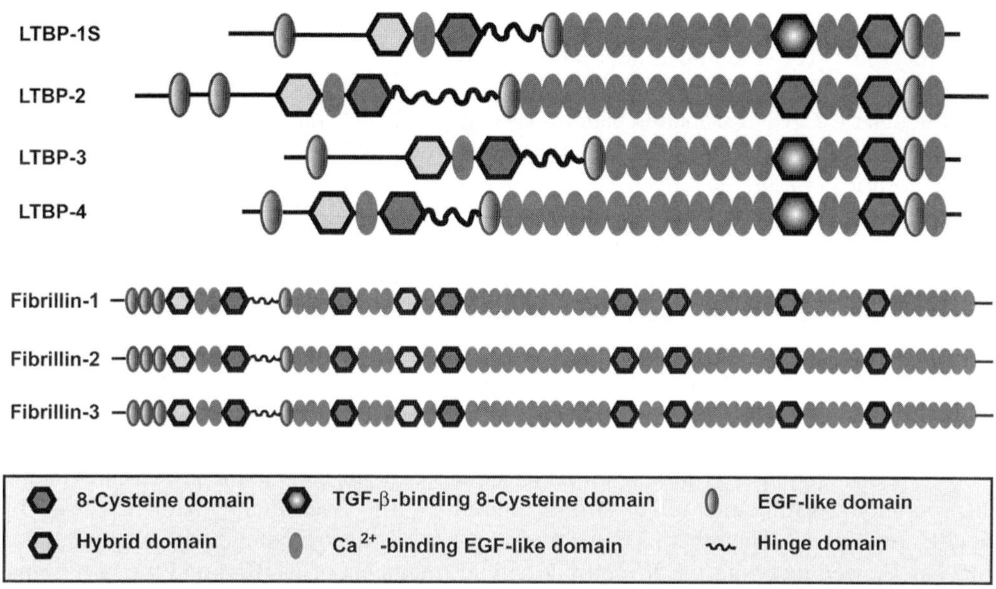

Fig. 2. Schematic representation of the domain modules in LTBPs and fibrillins. The overall domain organization is identical in fibrillins but not amongst LTBPs; with differences only in the numbers of domains that compose the central stretch of calcium-binding epidermal growth factor (EGF)-like repeats and the length of the hinge and spacer sequences between domain modules. The 8-cysteine domain is found only in LTBPs and fibrillins. Multiple Ca^{2+}-binding EGF-like repeats are present in LTBPs and fibrillins.

3.2. LTBP Synthesis and Distribution

Due to usage of different promoters, alternative splicing, and proteolytic processing *(41)*, human LTBPs display structural variability. Northern blotting experiments have revealed LTBP-1 transcripts of 5.2 and 7.0 kb *(56)*. The 5.2 kb transcript corresponds to the initial cDNA isolate, LTBP-1S, whereas the 7.0 kb transcript corresponds to a species of LTBP-1, LTBP-1L, that contains an additional N-terminal 346 amino acids including one EGF-like repeat plus a four-cysteine module, and is quite basic. LTBP-1S and LTBP-1L are generated from the use of separate promoters and alternative splicing between codons 145 and 146 of LTBP-1S *(57)*. Because at least two mRNA species exist for each of the other LTBPs, all of the LTBPs may exist as short and long forms. As the LTBP N-terminal region is important for ECM incorporation, the N-terminal sequence variability might generate LTBPs that have differential affinities and spatial interactions with matrix. Indeed, LTBP-1S and -1L do differ in their association with the ECM, the LTBP-1L being incorporated more efficiently.

Both LTBP-1 and -3 genes produce splice variants that are missing the hinge domain *(41)*. The function of these alternatively spliced forms may relate to the activation of latent TGF-β, as the LTBP-1 hinge region is crucial in $\alpha_v\beta_6$-mediated TGF-β activation *(49)*. The central EGF-like core region of the LTBP molecule is resistant to proteases. Alternative spliced variants within this region are common among all LTBP isoforms. Interestingly, an LTBP-4 splice variant with a deletion of the LAP-binding 8-Cys3 domain exists *(58)*.

Although LTBP-1S is thought to be the most ubiquitous member of the family, the distribution of the other family members is broad. All four LTBPs, although at different levels, are expressed in adult heart, lungs, skeletal muscles, testes, and spleen *(24)*. LTBP-3 is also abundant in brain and prostate, whereas LTBP-4 is expressed in uterus, ovary, and small intestine, and LTBP-2 is expressed particularly strongly in the lung. LTBP-1L has a more restricted distribution than LTBP-1S; it is highly expressed in heart, kidney, lung, testes, and placenta. LTBP-1S, -1L, and -2 are expressed very early during embryonic development, whereas LTBP-3 initiates expression later.

Although the LTBP distribution patterns correspond, in general, to the distribution of tissue phenotypes in mice with mutations in *Ltbp* genes, cultured cells usually produce multiple LTBP isoforms *(45,59)*.

3.3. LTBP Matrix Interactions

LTBP-1 binds to the ECM through both N- and C-terminal regions *(60,61)* (*see* Fig. 1). LTBP-1S peptides containing either the first, second or fourth 8-Cys domains associate with the ECM, with the 8-Cys2 region having the most rapid and extensive incorporation *(61)*. Matrix binding was potentiated by cell-conditioned medium. The association of the N-terminal sequence with the matrix, the requirement for a cell-derived factor, and the presence of a transglutaminase (TGase) site in 8-Cys2 *(61)* suggest that the cell-derived factor is a TGase. LTBP-1 ECM binding may initiate utilizing sequences at the ends of the protein, but matrix incorporation may require initial strong binding by the N-terminal region followed by TGase crosslinking. The other LTBPs are also efficiently deposited in the ECM. Mutational analysis showed that both the N- and C-terminal fragments of LTBP-4 and the N-terminal fragments of LTBP-3 bind readily to fibroblast ECM *(62)*. The C-terminal domains of LTBP-1 and -4 also interact with fibrillin-1 *(63)*. Molecules of LTBP-1 containing only the LAP-binding domain (8-Cys3) fail to target TGF-β to the matrix and yield animals with excess active TGF-β *(64)*.

LTBPs interact with a number of ECM molecules including collagen and fibronectin *(7)*. Isogai et al. *(63)* demonstrated interactions between LTBP-1 and fibrillin-1 and defined the interactive regions of the two proteins; the N-terminal region of fibrillin-1 binds to the C-terminal segment of LTBP-1. The C-terminal segment of LTBP-4 also interacts with the N-terminal sequence of fibrillin-1. The interaction of LTBPs with fibrillin-1 may affect tissue homeostasis, as Neptune et al. *(65)* reported increased active TGF-β within the lungs of fibrillin-1 hypomorphic mice. The decreased amounts of fibrillin-1 in mutant mice may yield excess LLC unable to interact with fibrillin-1 microfibrils. Free LLC then interacts with latent TGF-β activators normally separated from the sequestered LLC. This provocative interpretation of the fibrillin-1 hypomorphic mouse phenotype suggests that defects in structural proteins, such as fibrillin-1, alter growth factor action. Thus, the etiologies of certain connective tissue diseases may reflect an inability to control signaling molecules but not from architectural functions of the mutant proteins.

3.4. In Vitro Approaches to LTBP Function

In vitro studies of LTBP function have demonstrated an association between LTBP and TGF-β activity. LTBP enhances SLC secretion in cultured HEL cells *(25)*. Coexpression of LTBP-1 and SLC also yields complexes with the correct disulfide-binding pattern, as in the absence of LTBP-1, mixed disulfide bonds are formed. These abnormally folded proteins may be recognized and eliminated by ER quality control mechanisms, thereby decreasing the rate and degree of secretion. Bonding to the 8-Cys3 domain ensures correct SLC folding and disulfide bonding, thereby accelerating secretion. In this scene, LTBP is a chaperone for SLC, a concept reinforced by the observation that replacement of the LTBP-binding cysteines in LAP with serines promotes SLC secretion *(49)*.

A number of LTBP-1 functions have been deduced by treating cultured cells with antibodies. Flaumenhaft et al. *(66)* reported that an antibody to LTBP-1 blocked latent TGF-β activation in cocultures of endothelial and smooth muscle cell. This result implied a requirement for LTBP for TGF-β liberation and was consistent with the report that the inhibition of TGase crosslinking of LLC to the ECM blocked latent TGF-β activation *(61)*. Nakajima et al. *(67)* reported an LTBP-1 requirement in endothelial-mesenchymal transformation (EMT) during mouse heart endocardial cushion formation, a process known to require TGF-β *(68)*. Antibodies to LTBP-1 blocked atrio-ventricular endocardial cells EMT cocultured with myocardium, and this effect was neutralized by excess of mature TGF-β2. Gualandris et al. *(69)* provided further evidence that LTBP-1 is part of the regulatory network for TGF-β presentation. When embryonic stem (ES) cells differentiate in the presence of either an antibody to LTBP-1 or LAP, which recombines with and neutralizes free TGF-β, the numbers and organization of cells expressing endothelial cell markers were decreased. Conversely, adding TGF-β to the cultures increased endothelial cell numbers and tube-like structures. These data indicate that interference with LTBP-1 affects endothelial cell maturation in a TGF-β-dependent manner. Dioxin receptor (AhR) is a transcription factor involved in xenobiotic signaling *(70,71)*. However, in the absence of xenobiotics, AhR seems to be related to TGF-β function. Mouse embryonic fibroblasts (MEF) lacking AhR display decreased proliferation and increased apoptosis correlated with augmented TGF-β1 activity, decreased MMP-2 levels and, interestingly, higher production of a specific Ltbp-1 isoform-Ltbp-1L *(72)*. Decrease in Ltbp-1L activity, achieved either at the protein level by using an anti-LTBP-1 antibody or at the RNA level by specifically silencing the *Ltbp-1* mRNA, partially or almost completely restored the normal levels of TGF-β1 and MMP-2 in $AhR^{-/-}$ MEFs, suggesting that Ltbp-1L actively contributes to maintain Tgf-β1 levels *(72,73)*. Thus, all four reports indicate a linkage of LTBP-1 and TGF-β activity.

Not all of the experimental results obtained by treating cells with antibody to LTBP have been interpreted as affecting TGF-β action. Dallas et al. *(74)* reported that antibodies to LTBP-1 blocked development of mineralized foci by rat calvarial cells. The effect of the antibody was not suppressed by addition of TGF-β but was replicated by treatment of the cells with an LTBP-1 antisense oligonucleotide. The description of an LTBP-2 adhesive activity *(75)* for melanoma cells also indicates that LTBPs can have non-TGF-β dependent activities. However, the interpretation of the results of Dallas et al. *(74)* as demonstrating a direct effect of LTBP-1 can be criticized based upon a lack of evidence that the added TGF-β reached the target cells.

3.5. Genetic Approaches to LTBP Function

Null or hypomorphic mutations for three of the LTBPs have been described. Ltbp-$2^{-/-}$ mice die in early gestation (E3.5 and E6.5) *(76)*. The reason for lethality is unknown, but there may be a defect in implantation. The described anti- *(77)* or pro- *(75)* adhesive functions of LTBP-2 may be important during this period of development. Alternatively, Ltbp-2 may bind another signaling molecule essential for development.

The Ltbp-3 null mouse has a number of abnormalities but is fertile and can survive for over two yr *(43)*. Homozygous mutant animals have skeletal and craniofacial abnormalities, developmental emphysema *(78)*, involution of the spleen and thymus *(44)*, and a reduction of CD4/CD8 double positive T cells *(79)*. The skeletal defects are multiple *(80)*; the mice have cranial doming, kyphosis, the long bones are short, and both long bones and vertebrae become osteoarthritic and osteopetrotic as the animals age. Both osteoarthritic and osteopetrotic phenotypes have been described in mice with impaired TGF-β signaling *(81,82)*. Thus, the Ltbp-$3^{-/-}$ phenotypes are consistent with decreased extracellular levels of TGF-β. However, attempts to measure changes in TGF-β in the tissues of these mutant animals were unsuccessful. The craniofacial abnormalities result from the ossification of the synchondroses (growth zones) in the base of the skull. The abnormality in the synchondroses results from a defect in the negative regulatory loop involving Indian hedgehog (IHH) and parathyroid hormone related protein (PTHrP) that mediates chondrocyte differentiation. The IHH expression by prehypertrophic chondrocytes stimulates the production of PTHrP by the perichondrium and resting chondrocytes *(83)*. PTHrP interacts with its receptor, expressed by prehypertrophic chondrocytes, and this slows the rate of chondrocyte differentiation. In *Ltbp-3* null mice, PTHrP expression is decreased, suggesting that TGF-β regulates PTHrP expression. Absence of Ltbp-3 results in lower levels of active TGF-β in response to IHH, decreased TGF-β results in impaired expression of PTHrP, and low PTHrP yields accelerated chondrocyte differentiation.

Ltbp-4 hypomorphic mice were produced as part of a gene trap screen *(84)*. The mutant animals have undetectable levels of Ltbp-4 protein and less than 2% of the normal level of *Ltbp-4* mRNA. The hypomorphic mice develop emphysema at the saccular stage, which is earlier than the emphysema observed in the LTBP-3$^{-/-}$ mice. The lungs are characterized by decreased numbers of septae, larger air spaces, and fragmented elastic fibers. The mice also develop colorectal cancer, a process consistent with both the high level of expression of LTBP-4 in the colon *(1,40)* and the known tumor suppressive effect of TGF-β *(85)*. Immunohistochemical analysis of lung and intestinal tissue from mutant animals revealed a severe decrease in ECM-bound TGF-β1 coincident with normal intracellular staining of the growth factor. Thus, the major phenotypes relate to decreased extracellular levels of TGF-β1.

Koli et al. *(86)*, demonstrated that cells derived from LTBP-4 mutant mice have a defective matrix containing excessive amounts of fibronectin. The mutant cells have decreased levels of active extracellular TGF-β, increased expression of BMP-4, and decreased expression of the BMP-4 inhibitor gremlin. The matrix composition is reversed by over-expressing either gremlin or a dominant negative receptor for BMP-4. Increased BMP-4 and decreased gremlin expression were also observed in tissue sections from mutant mice. These studies demonstrate that the Ltbp-4 modulates extracellular TGF-β levels.

4. ACTIVATION OF LATENT TGF-β

TGF-β activity is regulated at multiple levels. The first level of controlling TGF-β action is the tight spatial and temporal regulation of expression of different TGF-β isoforms. However, only few primary and established cell lines secrete substantial amounts of active TGF-β *(87,88)* and that is because TGF-β is generally secreted in a complex either only with its propeptide-dimer, LAP, or more commonly, in a tertiary complex with LAP and LTBP (LLC). LTBP-1, 3, and 4 bind TGF-β LAP, facilitate the cytokine's secretion and direct its localization in the extracellular milieu. However, although all three LAP-binding LTBPs are detected in the ECM, they display different deposition kinetics *(62)* and may interact with different ECM molecules. Coexpression of TGF-β with a given LTBP isoform may determine the cytokine's ECM sub-localization. Hence, the TGF-β secretion from cells and ECM localization is a second level of control of TGF-β activity. TGF-β deposited in the ECM is latent and is activated in an environmentally controlled manner. Changes in ECM architecture can elicit different mechanisms that activate TGF-β either by liberating the mature cytokine from the latent complex or by exposing only the receptor-interacting domain of the cytokine within the tertiary complex. The molecular targets of these divergent mechanisms of latent TGF-β activation can be LTBPs and/or LAPs. Latent TGF-β activation thus represents a third level of control of TGF-β activity. The soluble mature TGF-β may be sequestered before its ligation to the cell-surface high affinity receptors by TGF-β-binding proteins, including decorin, betaglycan and α2-macroglobulin *(89)*, which introduces a fourth level of control of TGF-β activity. Finally, the differential arrangement of TGF-β receptors and coreceptors on a target cell surface represents another level of control of TGF-β activity.

Despite all these diverse possibilities of regulating TGF-β activity, the most important and interesting process controlling the effects of TGF-β is latent TGF-β activation. Several different mechanisms of latent TGF-β activation have been described, featuring a very divergent group of activators, including proteases, thrombospondin-1 (TSP-1), the integrins $\alpha_v\beta_6$ and $\alpha_v\beta_8$, reactive oxygen species and low or high pH.

4.1. Latent TGF-β Activation by Proteases

Lyons et al. *(90)* who found that plasmin could activate purified latent TGF-β, were the first to suggest the participation of proteases in the activation of latent TGF-β. Later Sato and Rifkin *(91)* showed that in cocultures of endothelial and smooth muscle cells, the constitutively secreted latent TGF-β was activated and this reaction required plasmin. Subsequent experiments indicated that plasmin-mediated latent TGF-β activation required both urokinase and plasmin. Presumably the urokinase activated plasminogen, derived from the serum, to plasmin. Additional experiments indicated that activation of the latent cytokine involved not only the two proteases, but also the participation of the urokinase receptor, the mannose 6–phosphate/ IGF-II cation independent receptor (M6PR), and the enzyme tissue transglutaminase type two (for review *see [61]*). Additional cell types were also shown to activate latent TGF-β by this mechanism *(92)* and the following scheme was suggested to account for the various requirements *(61)*. Upon secretion the LLC was crosslinked to the ECM by tissue transglutaminase. In this form the latent TGF-β was stored for subsequent use. When active TGF-β was required, proteolytic enzyme mobilization was utilized to liberate a soluble but still inactive remnant from the LLC bound in the matrix. This still inactive complex subsequently bound to the M6PR, as the cleavage of the LTBP exposed the previously cryptic mannose 6-phosphate residues within the latent complex. The degraded LLC, once bound to the M6P receptor, was attacked by plasmin, which cleaved a bond in LAP thereby destabilizing the latent complex and releasing free TGF-β. The fact that plasmin associates with the M6P receptor would also focus the reaction to the cell surface and promote the activation reaction *(93)*. Leksa et al. *(94)*, who showed that M6PR/IGF-II and urokinase-type plasminogen activator receptors associate on a membrane of endothelial but not smooth muscle cells, suggested that the formation of this membrane receptor complex is necessary for latent TGF-β activation mediated by plasmin since only the endothelial but not smooth muscle cells were able to activate latent TGF-β.

The cleavage of the LTBP by plasmin is believed to occur within the hinge region because of the sensitivity of the domain to protease cleavage. A number of proteases from both the serine protease and the MMP families are reported to activate latent TGF-β (for review *see [8]*). This promiscuity of activating enzymes may account for the fact that no null mutation in any of the various protease genes reported to activate latent TGF-β yields a phenotype similar to the TGF-β null mutations. Thus, there may be redundancy in protease-mediated activation.

4.2. Latent TGF-β Activation by Thrombospondin-1 (TSP-1)

Thrombospondin-1 (TSP-1) is a large homotrimeric extracellular protein, secreted by many cells (reviewed in *[95]*). Early experiments with human platelet TSP-1 in vitro showed growth inhibition of aortic endothelial cells via the TGF-β pathway, suggesting an interaction between TSP-1 and TGF-β. This observation led to the finding that biologically active TGF-β associates with TSP-1 and that TSP-1 might activate the latent form of TGF-β *(96)*. Indeed, purified TSP-1 converts the latent TGF-β1 secreted by endothelial cells to the biologically active form *(97)*. Crosslinking studies have shown that there is a direct interaction of latent TGF-β1 with TSP-1 *(98)*. Both the large and the SLC can be activated by TSP-1 and antibodies against LTBP-1 do not interfere with TSP-1-mediated

TGF-β1 activation suggesting that LTBP is not involved in this activation process *(99)*. A short amino-acid sequence within the type 1 repeats of TSP-1 binds to a conserved amino-acid sequence (LSKL) near the N-terminus of LAP *(100)*. The LSKL sequence lies within the region of LAP (residues 50–85) previously shown to be necessary for latency. In fact, it has been shown recently that the LSKL residues in LAP-β1 interact with the RKPK sequence in the receptor-binding region of mature TGF-β1 *(33)*. Altogether, these findings suggest the following model for TSP-1-mediated TGF-β activation: TSP-1 binding to LAP-β1 leads to conformational changes of the TGF-β1 prodomain such that the LAP can no longer sustain latency to the mature TGF-β1, but TGF-β1 is not released from the latent complex. In other words, TSP-1, LAP-β1 and TGF-β1 compose a tertiary complex that can elicit TGF-β signaling.

When the TSP-1 binding site for LAP-β1, RFK, is expressed as a peptide, it is able to activate latent TGF-β1 *in vitro* at picomolar concentrations. Conversely, LSKL peptides block TSP-1- or RFK-mediated latent TGF-β1 activation. Since the LSKL sequence is conserved in all three mammalian isoforms of LAP, it is reasonable to believe that TSP-1 can activate all TGF-β isoforms. Indeed, it has been demonstrated that TSP-1 can activate latent TGF-β2 *(100)*. However, the same activity has not been tested yet on TGF-β3.

Tsp-1$^{-/-}$ mice display similar pathology to *Tgf-β1*$^{-/-}$ mice in multiple organ systems, especially lungs and pancreas *(101)*. However, most *Tsp-1* null animals are able to reproduce and no significant inflammation is detected in their organs, whereas *Tgf-β1* KO mice suffer exaggerated inflammation in multiple organs, which ultimately causes their early death. The treatment of three d old *Tsp-1*$^{-/-}$ mice with the KRFK activating peptide induced reversion to the wild type phenotype. A similar treatment of wild-type pups with the LSKL peptide from LAP, which acts as a competitive inhibitor of TSP-1-mediated activation, yielded histopathological changes in lungs similar to those seen either in *Tsp-1*$^{-/-}$ or *TGF-β1*$^{-/-}$ animals. However, the phenotype of *Tsp-1*$^{-/-}$ mice does not resemble those seen in *Tgf-β2* or *Tgf-β3* KO mice, even though TSP-1 can activate TGF-β2 and, by inference, TGF-β3. Therefore, it is possible that TGF-β2 and TGF-β3 are preferentially activated by other mechanisms *in vivo*. In addition, the differences between *Tgf-β1* and *Tsp-1* KO phenotypes indicate that there are other mechanisms for TGF-β1 activation that do not involve TSP-1.

4.3. Latent TGF-β Activation by Integrins

Integrins are heterodimeric transmembrane receptors composed of α and β subunits (reviewed in *[102]*). Several integrin family members can bind TGF-β1 LAP, but only two of them- $\alpha_v\beta_6$ and $\alpha_v\beta_8$, both interact with LAP-β1 and activate latent TGF-β1 *(5,103–105)*. When the cells expressing $\alpha_v\beta_6$ were cocultured with reporter cell line (mink lung epithelial cells that produce luciferase in response to TGF-β), a significant increase in luciferase activity was detected due to TGF-β activation *(5)*. This activity was inhibited by addition of antibodies to either $\alpha_v\beta_6$ or TGF-β. However, when $\alpha_v\beta_6$-expressing cells and TGF-β reporter cells were separated by a permeable filter allowing the flow of the conditioned medium between these two cell populations, little or no TGF-β activity was detected, suggesting that $\alpha_v\beta_6$-mediated TGF-β1 activation does not involve liberation of the mature cytokine from the latent complex *(5)*. Furthermore, when the same cells were transfected with both the $\alpha_v\beta_6$ integrin-expressing construct and the TGF-β driven luciferase gene and plated at different densities, the luciferase activity gradually decreased at lower densities and was eventually lost at densities where cells no longer contacted one another *(102)*. Altogether, these findings suggest that $\alpha_v\beta_6$ binds LAP-β1 and presents the active TGF-β1 only in a paracrine fashion that demands cell-cell contact. Moreover, it appears that $\alpha_v\beta_6$-mediated TGF-β1 activation does not rely simply on the interaction of the $\alpha_v\beta_6$ integrin and

LAP-β1, since the deletion of β6 cytoplasmic domain completely abolishes generation of active TGF-β1 without disrupting $\alpha_v\beta_6$- LAP-β1 interaction. Similarly, the addition of an inhibitor of actin polymerization, cytochalasin D, completely attenuated $\alpha_v\beta_6$ mediated TGF-β1 activation *(5)*. These findings suggest that cytoplasmic proteins, for instance the actin cytoskeleton, might be necessary for $\alpha_v\beta_6$-dependent TGF-β1 activation. However, it appears that not only intracellular proteins are necessary for TGF-β1 activation by $\alpha_v\beta_6$, but also ECM proteins play an important role in this activation mechanism. In a series of experiments, Annes et al. *(49)* demonstrated that the interaction between LAP-β1 and LTBP-1 is essential for $\alpha_v\beta_6$-mediated TGF-β1 activation. Mutational analysis of LTBP-1 revealed that the domains necessary to support $\alpha_v\beta_6$-mediated TGF-β1 activation were the LAP-binding domain (8-Cys3) and, surprisingly, the hinge domain. The LTBP-1 hinge domain has been shown to mediate LTBP-1 interaction with fibronectin *(106)*. Furthermore, the hinge domain of LTBP-3 could not substitute for the LTBP-1 hinge in these activation experiments. Altogether these findings suggest that the principal function of LTBP-1 in $\alpha_v\beta_6$-mediated TGF-β activation is to anchor the latent TGF-β to components of the ECM and that there is a specificity in LTBP family members for performing this activity. Annes et al. suggested a model of $\alpha_v\beta_6$-mediated- TGF-β1 activation by traction. In this model, TGF-β1 is secreted in a complex with LTBP-1 and fixed in the ECM. LAP-β1 binding to the $\alpha_v\beta_6$ integrin on the cell surface creates a retractile force. The magnitude of the force depends on two parameters created by the intracellular interactions of $\alpha_v\beta_6$ integrin (retraction) and extracellular interactions of LTBP-1 (resistance). Once the force generated by the integrin exceeds a threshold, biologically active TGF-β is exposed and can bind its receptors at the cell surface of a nearby cell.

Integrin-LAP-β1 association is mediated by the RGD sequence of LAP-β1, as a mutated form of LAP-β1 in which the RGD sequence is substituted with RGE does not bind any of the integrins. TGF-β3 LAP also contains RGD sequence and indeed, $\alpha_v\beta_6$ can bind and activate latent TGF-β3 *(27)*. Interestingly, TGF-β2 LAP does not contain an RGD sequence and has not been found to associate with any of the integrins. It is possible that latent TGF-β2 binds and is activated by some yet unidentified integrin or that latent TGF-β2 activation is performed by mechanism(s) not mediated by integrins.

The β_6 subunit is present in only a single integrin, $\alpha_v\beta_6$, and the expression of β_6 is limited to the epithelium of healthy adult mammals. Most epithelia generate low levels of β_6 integrin, but upon inflammation or injury, β_6 expression dramatically increases *(107,108)*. Mice lacking the β_6-subunit have persistent lung and skin inflammation but do not develop pulmonary fibrosis even when challenged with anti-cancer and profibrotic drug, bleomicin *(5,109)*. TGF-β1 is a central mediator of tissue fibrosis or healing, and mice lacking TGF-β1 succumb to massive inflammation of multiple organs *(110,111)*. Thus, mice lacking the β_6 integrin partially reproduce the condition caused by complete lack of TGF-β1, displaying lung and skin inflammation. This phenotypic overlap might be a consequence of a vital role for $\alpha_v\beta_6$ integrin in local activation of TGF-β1 by lung and skin epithelia in response to injury and inflammation. However, like the β_6 null mice, the *Tsp-1*$^{-/-}$ mice also partially phenocopy the condition caused by the lack of TGF-β1 during development, suggesting that there might be some overlap of functions for TSP-1 and $\alpha_v\beta_6$ in respect to TGF-β activation. To identify TSP-1 and $\alpha_v\beta_6$ overlapping and nonoverlapping functions, Ludlow and colleagues *(112)* created doubly-null mice for *Tsp-1* and *integrin* β_6 genes. These mice display multiple defects but do not completely reproduce the condition of *Tgf-β1* null mice. In addition, when the phosphorylation of TGF-β signal transducers, Smad2/3, was tested either by immunohistochemistry or Western blots from tissue lysates, the level of pSmad2/3 was actually higher in *Tsp*$^{-/-}$/*integrin* $\beta_6^{-/-}$ animals *(112)*, altogether suggesting that there are mechanisms for TGF-β activation not mediated by TSP-1 nor $\alpha_v\beta_6$.

A second integrin, $\alpha_v\beta_8$, has been recently shown to mediate latent TGF-β1 activation *(104)*. As opposed to $\alpha_v\beta_6$, $\alpha_v\beta_8$ supports TGF-β1 activation even in the absence of the intracellular domain of the β_8-subunit. In addition, $\alpha_v\beta_8$ activation releases mature TGF-β1 into the culture medium of $\alpha_v\beta_8$-expressing cells, and this activation process is completely inhibited by metaloprotease inhibitors. LAP-β1 is a substrate for proteolytic degradation by MT1-MMP. It thus appears that the mechanism of TGF-β1 activation by $\alpha_v\beta_8$ is considerably different than that of $\alpha_v\beta_6$. $\alpha_v\beta_8$ functions more as a tethering factor that binds TGF-β1 LAP and presents it to a cell surface bound metaloprotease that degrades LAP and releases soluble mature TGF-β1 into the extracellular environment.

At present, α_v is the only partner identified for β_8 and therefore ablation of the β_8 gene in mice results in a specific deficiency of $\alpha_v\beta_8$ integrin receptors. Mice lacking the β_8 integrin subunit show defective vasculogenesis during early development, and in later development, defective brain vessel formation yielding brain hemorrhage *(113)*. The capillaries in the yolk sac and brain in $\alpha_v\beta_8$ mutant mice are formed, but are aberrantly patterned with hyper-proliferative endothelial cells. In addition, neuroepithelial cells appear disorganized in the mutant. Integrin β_8 is expressed by neuroepithelial cells and has not been detected in the vasculature. Therefore, it has been proposed that $\alpha_v\beta_8$ integrin provides proper environmental cues for patterning of the embryonic yolk sac and brain vascular network. The defective vascular development seen in β_8 KO mice during midgestation resembles that seen in *Tgf-β1* null embryos, which is consistent with the suggested important role of $\alpha_v\beta_8$ integrin in paracrine TGF-β1 activation.

4.4. Other Mechanisms of Latent TGF-β Activation

Latent TGF-β can be activated *in vitro* by mild protein-denaturing conditions, such as low or high pH, high temperature, detergents, and chaotropic agents, including urea and guanidine hydrochloride. Some of these protein-denaturing conditions, such as mild acid treatment, can be generated *in vivo*. For example, the bone matrix deposited by osteoblasts is rich in latent TGF-β. This ECM is, when necessary, resorbed by osteoclasts who create a mildly acidic environment which can disturb LAP structure yielding active TGF-β *(8)*.

Irradiation leads to latent TGF-β activation, most likely through the local production of reactive oxygen species (ROS) *(114)*. Barcellos-Hoff and coworkers showed that generation of ROS in vitro or in vivo causes side group modifications in LAP and consequently changes in its structure, altogether resulting in the liberation of active TGF-β *(115)*.

5. CONCLUSIONS

TGF-β cytokines perform numerous biological roles highly dependent on the background that is shaped by the temporal, spatial, and environmental conditions. Therefore, the rigorous regulation of TGF-β action is critical for the development and maintenance of cellular and tissue homeostasis. It appears that the activation of latent TGF-β is the most important step in controlling the TGF-β activity. The current literature describes several different mechanisms for TGF-β activation that are isoform- and context-specific, indicating that a complex cytokine like TGF-β demands an equally intricate ways of activation.

ACKNOWLEDGMENTS

This work has been supported by grants from the NIH to DBR.

REFERENCES

1. Taipale J, Saharinen J, Keski-Oja J. Extracellular matrix-associated transforming growth factor-beta: role in cancer cell growth and invasion. Adv Cancer Res 1998;75:87–134.

2. Siegel PM, Massagué J. Cytostatic and apoptotic actions of TGF-beta in homeostasis and cancer. Nat Rev Cancer 2003;3:807–821.
3. Azhar M, Schultz Jel J, Grupp I, et al. Transforming growth factor beta in cardiovascular development and function. Cytokine Growth Factor Rev 2003;14:391–407.
4. Waite KA, Eng C. From developmental disorder to heritable cancer: it's all in the BMP/TGF-beta family. Nat Rev Genet 2003;4:763–773.
5. Munger JS, Huang X, Kawakatsu H, et al. The integrin alpha v beta 6 binds and activates latent TGF beta 1: a mechanism for regulating pulmonary inflammation and fibrosis. Cell 1999;96:319–328.
6. Wahl SM, Chen W. TGF-beta: how tolerant can it be? Immunol Res 2003;28:167–179.
7. Taipale J, Saharinen J, Hedman K, Keski-Oja J. Latent transforming growth factor-beta 1 and its binding protein are components of extracellular matrix microfibrils. J Histochem Cytochem 1996;44:875–889.
8. Annes JP, Munger JS, Rifkin DB. Making sense of latent TGFbeta activation. J Cell Sci 2003;116: 217–224.
9. Roberts AB, Anzano MA, Lamb LC, Smith JM, Sporn MB. New class of transforming growth factors potentiated by epidermal growth factor: isolation from non-neoplastic tissues. Proc Natl Acad Sci USA 1981;78:5339–5343.
10. Moses HL, Branum EL, Proper JA, Robinson RA. Transforming growth factor production by chemically transformed cells. Cancer Res 1981;41:2842–2848.
11. Roberts AB, Anzano MA, Meyers CA, et al. Purification and properties of a type beta transforming growth factor from bovine kidney. Biochemistry 1983;22:5692–5698.
12. Frolik CA, Dart LL, Meyers CA, Smith DM, Sporn MB. Purification and initial characterization of a type beta transforming growth factor from human placenta. Proc Natl Acad Sci USA 1983;80: 3676–3680.
13. Assoian RK, Komoriya A, Meyers CA, Miller DM, Sporn MB. Transforming growth factor-beta in human platelets. Identification of a major storage site, purification, and characterization. J Biol Chem 1983;258:7155–7160.
14. Derynck R, Jarrett JA, Chen, EY, et al. Human transforming growth factor-beta complementary DNA sequence and expression in normal and transformed cells. Nature 1985;316:701–705.
15. Cheifetz S, Weatherbee JA, Tsang ML, et al. The transforming growth factor-beta system, a complex pattern of cross-reactive ligands and receptors. Cell 1987;48:409–415.
16. ten Dijke P, Hansen P, Iwata KK, Pieler C, Foulkes JG. Identification of another member of the transforming growth factor type beta gene family. Proc Natl Acad Sci USA 1988;85:4715–4719.
17. Derynck R, Lindquist PB, Lee A, et al. A new type of transforming growth factor-beta, TGF-beta 3. EMBO J 1988;7:3737–3743.
18. Armstrong EJ, Bischoff J. Heart valve development: endothelial cell signaling and differentiation. Circ Res 2004;95:459–470.
19. Molin DG, Bartram U, Van der Heiden K, et al. Expression patterns of Tgfbeta1-3 associate with myocardialisation of the outflow tract and the development of the epicardium and the fibrous heart skeleton. Dev Dyn 2003;227:431–444.
20. Dickson MC, Slager HG, Duffie E, Mummery CL, Akhurst RJ. RNA and protein localisations of TGF beta 2 in the early mouse embryo suggest an involvement in cardiac development. Development 1993;117:625–639.
21. Camenisch TD, Molin DG, Person A, et al. Temporal and distinct TGFbeta ligand requirements during mouse and avian endocardial cushion morphogenesis. Dev Biol 2002;248:170–181.
22. Bartram U, Molin DG, Wisse LJ, et al. Double-outlet right ventricle and overriding tricuspid valve reflect disturbances of looping, myocardialization, endocardial cushion differentiation, and apoptosis in TGF-beta(2)-knockout mice. Circulation 2001;103:2745–2752.
23. Sanford LP, Ormsby I, Gittenberger-de Groot AC, et al. TGFbeta2 knockout mice have multiple developmental defects that are non-overlapping with other TGFbeta knockout phenotypes. Development 1997;124:2659–2670.
24. Todorovic V, Jurukovski V, Chen Y, Fontana L, Dabovic B, Rifkin DB. Latent TGF-beta binding proteins. Int J Biochem Cell Biol 2005;37:38–41.
25. Miyazono K, Olofsson A, Colosetti P, Helden C-H. A role of the latent TGF-beta 1-binding protein in the assembly and secretion of TGF-beta 1. EMBO J 1991;10:1091–1101.
26. Miyazono K, Ichijo H, Heldin C-H. Transforming growth factor-beta: latent forms, binding proteins and receptors. Growth Factors 1993;8:11–22.
27. Annes JP, Rifkin DB, Munger JS. The integrin alphaVbeta6 binds and activates latent TGFbeta3. FEBS Lett 2002;511:65–68.

28. Purchio AF, Cooper JA, Brunner AM, et al. Identification of mannose 6-phosphate in two asparagine-linked sugar chains of recombinant transforming growth factor-beta 1 precursor. J Biol Chem 1988; 263:14,211–14,215.
29. Brunner AM, Marquardt H, Malacko AR, Lioubin MN, Purchio AF. Site-directed mutagenesis of cysteine residues in the pro region of the transforming growth factor beta 1 precursor. Expression and characterization of mutant proteins. J Biol Chem 1989;264:13,660–13,664.
30. Saharinen J, Taipale J, Keski-Oja J. Association of the small latent transforming growth factor-beta with an eight cysteine repeat of its binding protein LTBP-1. EMBO J 1996;15:245–253.
31. Nakajima H, Nakajima HO, Salcher O, et al. Atrial but not ventricular fibrosis in mice expressing a mutant transforming growth factor-beta(1) transgene in the heart. Circ Res 2000;86:571–579.
32. Sha X, Yang L, Gentry LE. Identification and analysis of discrete functional domains in the pro region of pre-pro-transforming growth factor beta 1. J Cell Biol 1991;114:827–839.
33. Young GD, Murphy-Ullrich JE. Molecular interactions that confer latency to transforming growth factor-beta. J Biol Chem 2004;279:38,032–38,039.
34. Saito T, Kinoshita A, Yoshiura K, et al. Domain-specific mutations of a transforming growth factor (TGF)-beta 1 latency-associated peptide cause Camurati-Engelmann disease because of the formation of a constitutively active form of TGF-beta 1. J Biol Chem 2001;276:11,469–11,472.
35. Kinoshita A, Saito T, Tomita H, et al. Domain-specific mutations in TGFβ1 result in Camurati-Engelmann disease. Nat Genet 2000;26:19–20.
36. Janssens K, ten Dijke P, Ralston SH, Bergmann C, Van Hul W. Transforming growth factor-beta 1 mutations in Camurati-Engelmann disease lead to increased signaling by altering either activation or secretion of the mutant protein. J Biol Chem 2003;278:7718–7724.
37. Kanzaki T, Olofsson A, Moren A, et al. TGF-beta 1 binding protein: a component of the large latent complex of TGF-beta 1 with multiple repeat sequences. Cell 1990;61:1051–1061.
38. Bashir MM, Han MD, Abrams WR, et al. Analysis of the human gene encoding latent transforming growth factor-beta-binding protein-2. Int J Biochem Cell Biol 1996;28:531–542.
39. Moren A, Olofsson A, Stenman G, et al. Identification and characterization of LTBP-2, a novel latent transforming growth factor-beta-binding protein. J Biol Chem 1994;269:32,469–32,478.
40. Saharinen J, Taipale J, Monni O, Keski-Oja J. Identification and characterization of a new latent transforming growth factor-beta-binding protein, LTBP-4. J Biol Chem 1998;273:18,459–18,469.
41. Koli K, Saharinen J, Hyytiainen M, Penttinen C, Keski-Oja J. Latency, activation, and binding proteins of TGF-beta. Microsc Res Tech 2001;52:354–362.
42. Rifkin DB. Latent transforming growth factor-beta (TGF-beta) binding proteins: orchestrators of TGF-beta availability. J Biol Chem 2005;280:7409–7412.
43. Taipale J, Lohi J, Saarinen J, Kovanen PT, Keski-Oja J. Human mast cell chymase and leukocyte elastase release latent transforming growth factor-beta 1 from the extracellular matrix of cultured human epithelial and endothelial cells. J Biol Chem 1995;270:4689–4696.
44. Hyytiäinen M, Taipale J, Heldin C-H, Keski-Oja J. Recombinant latent transforming growth factor beta-binding protein 2 assembles to fibroblast extracellular matrix and is susceptible to proteolytic processing and release. J Biol Chem 1998;273:20,669–20,676.
45. Penttinen C, Saharinen J, Weikkolainen K, Hyytiainen M, Keski-Oja J. Secretion of human latent TGF-beta-binding protein-3 (LTBP-3) is dependent on co-expression of TGF-beta. J Cell Sci 2002; 115:3457–3468.
46. Oklu R, Metcalfe JC, Hesketh TR, Kemp PR. Loss of a consensus heparin binding site by alternative splicing of latent transforming growth factor-beta binding protein-1. FEBS Lett 1998;425:281–285.
47. Downing AK, Knott V, Werner JM, Cardy CM, Campbell ID, Handford PA. Solution structure of a pair of calcium-binding epidermal growth factor-like domains: implications for the Marfan syndrome and other genetic disorders. Cell 1996;85:597–605.
48. Handford PA, Downing AK, Reinhardt DP, Sakai LY. Fibrillin: from domain structure to supramolecular assembly. Matrix Biol 2000;19:457–470.
49. Annes JP, Chen Y, Munger JS, Rifkin DB. Integrin alphaVbeta6-mediated activation of latent TGF-beta requires the latent TGF-beta binding protein-1. J Cell Biol 2004;165:723–734.
50. Lee SS, Knott V, Jovanovic J, et al. Structure of the integrin binding fragment from fibrillin-1 gives new insights into microfibril organization. Structure (Camb) 2004;12:717–729.
51. Gleizes PE, Beavis RC, Mazzieri R, Shen B, Rifkin DB. Identification and characterization of an eight-cysteine repeat of the latent transforming growth factor-beta binding protein-1 that mediates bonding to the latent transforming growth factor-beta1. J Biol Chem 1996;271:29,891–29,896.

52. Saharinen J, Keski-Oja J. Specific sequence motif of 8-Cys repeats of TGF-beta binding proteins, LTBPs, creates a hydrophobic interaction surface for binding of small latent TGF-beta. Mol Biol Cell 2000;11:2691–2704.
53. Lack J, O'Leary JM, Knott V, et al. Solution structure of the third TB domain from LTBP1 provides insight into assembly of the large latent complex that sequesters latent TGF-beta. J Mol Biol 2003; 334:281–291.
54. Altmann CR, Chang C, Munoz-Sanjuan I, et al. The latent-TGFbeta-binding-protein-1 (LTBP-1) is expressed in the organizer and regulates nodal and activin signaling. Dev Biol 2002;248:118–127.
55. Charbonneau NL, Ono RN, Corson GM, Keene DR, Sakai LY. Fine tuning of growth factor signals depends on fibrillin microfibril networks. Birth Defects Res C Embryo Today 2004;72:37–50.
56. Noguera I, Obata H, Gualandris A, Cowin P, Rifkin DB. Molecular cloning of the mouse Ltbp-1 gene reveals tissue specific expression of alternatively spliced forms. Gene 2003;308:31–41.
57. Koski C, Saharinen J, Keski-Oja J. Independent promoters regulate the expression of two amino terminally distinct forms of latent transforming growth factor-beta binding protein-1 (LTBP-1) in a cell type-specific manner. J Biol Chem 1999;274:32,619–32,630.
58. Koli K, Saharinen J, Karkkainen M, Keski-Oja J. Novel non-TGF-beta-binding splice variant of LTBP-4 in human cells and tissues provides means to decrease TGF-beta deposition. J Cell Sci 2001; 114:2869–2878.
59. Weiskirchen R, Moser M, Gunther K, Weiskirchen S, Gressner AM. The murine latent transforming growth factor-beta binding protein (Ltbp-1) is alternatively spliced, and maps to a region syntenic to human chromosome 2p21-22. Gene 2003;308:43–52.
60. Unsold C, Hyytiäinen M, Bruckner-Tuderman L, Keski-Oja J. Latent TGF-beta binding protein LTBP-1 contains three potential extracellular matrix interacting domains. J Cell Sci 2001;114:187–197.
61. Nunes I, Gleizes PE, Metz CN, Rifkin DB. Latent transforming growth factor-beta binding protein domains involved in activation and transglutaminase-dependent cross-linking of latent transforming growth factor-beta. J Cell Biol 1997;136:1151–1163.
62. Koli K, Hyytiäinen M, Ryynanen MJ, Keski-Oja J. Sequential deposition of latent TGF-beta binding proteins (LTBPs) during formation of the extracellular matrix in human lung fibroblasts. Exp Cell Res 2005;310:370–382.
63. Isogai Z, Ono RN, Ushiro S, et al. Latent transforming growth factor beta-binding protein 1 interacts with fibrillin and is a microfibril-associated protein. J Biol Chem 2003;278:2750–2757.
64. Mazzieri R, Jurukovski V, Obata H, et al. Expression of truncated latent TGF-beta-binding protein modulates TGF-beta signaling. J Cell Sci 2005;118:2177–2187.
65. Neptune ER, Frischmeyer PA, Arking DE, et al. Dysregulation of TGF-beta activation contributes to pathogenesis in Marfan syndrome. Nat Genet 2003;33:407–411.
66. Flaumenhaft R, Abe M, Sato Y, et al. Role of the latent TGF-beta binding protein in the activation of latent TGF-beta by co-cultures of endothelial and smooth muscle cells. J Cell Biol 1993;120:995–1002.
67. Nakajima Y, Miyazono K, Kato M, Takase M, Yamagishi T, Nakamura H. Extracellular fibrillar structure of latent TGF beta binding protein-1: role in TGF beta-dependent endothelial-mesenchymal transformation during endocardial cushion tissue formation in mouse embryonic heart. J Cell Biol 1997;136:193–204.
68. Runyan RB, Potts JD, Weeks DL. TGF-beta 3-mediated tissue interaction during embryonic heart development. Mol Reprod Dev 1992;32:152–159.
69. Gualandris A, Annes JP, Arese M, Noguera I, Jurukovski V, Rifkin DB. The latent transforming growth factor-beta-binding protein-1 promotes in vitro differentiation of embryonic stem cells into endothelium. Mol Biol Cell 2000;11:4295–4308.
70. Gonzalez FJ, Fernandez-Salguero P, Ward JM. The role of the aryl hydrocarbon receptor in animal development, physiological homeostasis and toxicity of TCDD. J Toxicol Sci 1996;21:273–277.
71. Fernandez-Salguero PM, Hilbert DM, Rudikoff S, Ward JM, Gonzalez FJ. Aryl-hydrocarbon receptor-deficient mice are resistant to 2,3,7,8-tetrachlorodibenzo-p-dioxin-induced toxicity. Toxicol Appl Pharmacol 1996;140:173–179.
72. Santiago-Josefat B, Mulero-Navarro S, Dallas SL, Fernandez-Salguero PM. Overexpression of latent transforming growth factor-beta binding protein 1 (LTBP-1) in dioxin receptor-null mouse embryo fibroblasts. J Cell Sci 2004;117:849–859.
73. Gomez-Duran A, Mulero-Navarro S, Chang X, Fernandez-Salguero PM. LTBP-1 blockade in dioxin receptor-null mouse embryo fibroblasts decreases TGF-beta activity: Role of extracellular proteases plasmin and elastase. J Cell Biochem 2005.

74. Dallas SL, Miyazono K, Skerry TM, Mundy GR, Bonewald LF. Dual role for the latent transforming growth factor-beta binding protein in storage of latent TGF-beta in the extracellular matrix and as a structural matrix protein. J Cell Biol 1995;131:539–549.
75. Vehvilainen P, Hyytiainen M, Keski-Oja J. Latent transforming growth factor-beta-binding protein 2 is an adhesion protein for melanoma cells. J Biol Chem 2003;278:24,705–24,713.
76. Shipley JM, Mecham RP, Maus E, et al. Developmental expression of latent transforming growth factor beta binding protein 2 and its requirement early in mouse development. Mol Cell Biol 2000; 20:4879–4887.
77. Hyytiäinen M, Keski-Oja J. Latent TGF-beta binding protein LTBP-2 decreases fibroblast adhesion to fibronectin. J Cell Biol 2003;163:1363–1374.
78. Colarossi C, Chen Y, Obata H, et al. Lung alveolar septation defects in Ltbp-3-null mice. Am J Pathol 2005;167:419–428.
79. Chen Y, Dabovic B, Colarossi C, et al. Growth retardation as well as spleen and thymus involution in latent TGF-beta binding protein (Ltbp)-3 null mice. J Cell Physiol 2003;196:319–325.
80. Dabovic B, Chen Y, Colarossi C, et al. Bone abnormalities in latent TGF-[beta] binding protein (Ltbp)-3-null mice indicate a role for Ltbp-3 in modulating TGF-[beta] bioavailability. J Cell Biol 2002;156:227–232.
81. Yang X, Chen L, Xu X, Li C, Huang C, Deng CX. TGF-beta/Smad3 signals repress chondrocyte hypertrophic differentiation and are required for maintaining articular cartilage. J Cell Biol 2001; 153:35–46.
82. Filvaroff E, Erlebacher A, Ye J, et al. Inhibition of TGF-beta receptor signaling in osteoblasts leads to decreased bone remodeling and increased trabecular bone mass. Development 1999;126:4267–4279.
83. Chung UI, Schipani E, McMahon AP, Kronenberg HM. Indian hedgehog couples chondrogenesis to osteogenesis in endochondral bone development. J Clin Invest 2001;107:295–304.
84. Sterner-Kock A, Thorey IS, Koli K, et al. Disruption of the gene encoding the latent transforming growth factor-beta binding protein 4 (LTBP-4) causes abnormal lung development, cardiomyopathy, and colorectal cancer. Genes Dev 2002;16:2264–2273.
85. Zhu Y, Richardson JA, Parada LF, Graff JM. Smad3 mutant mice develop metastatic colorectal cancer. Cell 1998;94:703–714.
86. Koli K, Wempe F, Sterner-Kock A, et al. Disruption of LTBP-4 function reduces TGF-beta activation and enhances BMP-4 signaling in the lung. J Cell Biol 2004;167:123–133.
87. Tucker RF, Shipley GD, Moses HL, Holley RW. Growth inhibitor from BSC-1 cells closely related to platelet type beta transforming growth factor. Science 1984;226:705–707.
88. Holley RW, Armour R, Baldwin JH. Density-dependent regulation of growth of BSC-1 cells in cell culture: growth inhibitors formed by the cells. Proc Natl Acad Sci USA 1978;75:1864–1866.
89. Khalil N. Post translational activation of latent transforming growth factor beta (L-TGF-beta): clinical implications. Histol Histopathol 2001;16:541–551.
90. Lyons RM, Keski-Oja J, Moses HL. Proteolytic activation of latent transforming growth factor-beta from fibroblast-conditioned medium. J Cell Biol 1988;106:1659–1665.
91. Sato Y, Rifkin DB. Inhibition of endothelial cell movement by pericytes and smooth muscle cells: activation of a latent transforming growth factor-beta 1-like molecule by plasmin during co-culture. J Cell Biol 1989;109:309–315.
92. Nunes I, Shapiro RL, Rifkin DB. Characterization of latent TGF-beta activation by murine peritoneal macrophages. J Immunol 1995;155:1450–1459.
93. Godar S, Horejsi V, Weidle UH, Binder BR, Hansmann C, Stockinger H. M6P/IGFII-receptor complexes urokinase receptor and plasminogen for activation of transforming growth factor-beta1. Eur J Immunol 1999;29:1004–1013.
94. Leksa V, Godar S, Schiller HB, et al. TGF-beta-induced apoptosis in endothelial cells mediated by M6P/IGFII-R and mini-plasminogen. J Cell Sci 2005;118:4577–4586.
95. Murphy-Ullrich JE, Poczatek M. Activation of latent TGF-beta by thrombospondin-1: mechanisms and physiology. Cytokine Growth Factor Rev 2000;11:59–69.
96. Murphy-Ullrich JE, Schultz-Cherry S, Hook M. Transforming growth factor-beta complexes with thrombospondin. Mol Biol Cell 1992;3:181–188.
97. Schultz-Cherry S, Murphy-Ullrich JE. Thrombospondin causes activation of latent transforming growth factor-beta secreted by endothelial cells by a novel mechanism. J Cell Biol 1993;122:923–932.
98. Schultz-Cherry S, Lawler J, Murphy-Ullrich JE. The type 1 repeats of thrombospondin 1 activate latent transforming growth factor-beta. J Biol Chem 1994;269:26,783–26,788.

99. Schultz-Cherry S, Ribeiro S, Gentry L, Murphy-Ullrich JE. Thrombospondin binds and activates the small and large forms of latent transforming growth factor-beta in a chemically defined system. J Biol Chem 1994;269:26,775–26,782.
100. Ribeiro SM, Poczatek M, Schultz-Cherry S, Villain M, Murphy-Ullrich JE. The activation sequence of thrombospondin-1 interacts with the latency-associated peptide to regulate activation of latent transforming growth factor-beta. J Biol Chem 1999;274:13,586–13,593.
101. Crawford SE, Stellmach V, Murphy-Ullrich JE, et al. Thrombospondin-1 is a major activator of TGF-beta1 in vivo. Cell 1998;93:1159–1170.
102. Sheppard D. Roles of alphav integrins in vascular biology and pulmonary pathology. Curr Opin Cell Biol 2004;16:552–557.
103. Munger JS, Harpel JG, Giancotti FG, Rifkin DB. Interactions between growth factors and integrins: latent forms of transforming growth factor-beta are ligands for the integrin alphavbeta1. Mol Biol Cell 1998;9:2627–2638.
104. Mu D, Cambier S, Fjellbirkeland L, et al. The integrin alpha(v)beta8 mediates epithelial homeostasis through MT1-MMP-dependent activation of TGF-beta1. J Cell Biol 2002;157:493–507.
105. Lu M, Munger JS, Steadele M, Busald C, Tellier M, Schnapp LM. Integrin alpha8beta1 mediates adhesion to LAP-TGFbeta1. J Cell Sci 2002;115:4641–4648.
106. Fontana L, Chen Y, Prijatelj P, et al. Fibronectin is required for integrin alphavbeta6-mediated activation of latent TGF-beta complexes containing LTBP-1. FASEB J 2005;19:1798–1808.
107. Breuss JM, Gallo J, DeLisser HM, et al. Expression of the beta 6 integrin subunit in development, neoplasia and tissue repair suggests a role in epithelial remodeling. J Cell Sci 1995;108(Pt 6):2241–2251.
108. Breuss JM, Gillett N, Lu L, Sheppard D, Pytela R. Restricted distribution of integrin beta 6 mRNA in primate epithelial tissues. J Histochem Cytochem 1993;41:1521–1527.
109. Huang XZ, Wu JF, Cass D, et al. Inactivation of the integrin beta 6 subunit gene reveals a role of epithelial integrins in regulating inflammation in the lung and skin. J Cell Biol 1996;133:921–928.
110. Kulkarni AB, Huh CG, Becker D, et al. Transforming growth factor beta 1 null mutation in mice causes excessive inflammatory response and early death. Proc Natl Acad Sci USA 1993;90:770–774.
111. Shull MM, Ormsby I, Kier AB, et al. Targeted disruption of the mouse transforming growth factor-beta 1 gene results in multifocal inflammatory disease. Nature 1992;359:693–699.
112. Ludlow A, Yee KO, Lipman R, et al. Characterization of integrin beta6 and thrombospondin-1 double-null mice. J Cell Mol Med 2005;9:421–437.
113. Zhu J, Motejlek K, Wang D, Zang K, Schmidt A, Reichardt LF. beta8 integrins are required for vascular morphogenesis in mouse embryos. Development 2002;129:2891–2903.
114. Barcellos-Hoff MH, Derynck R, Tsang ML, Weatherbee JA. Transforming growth factor-beta activation in irradiated murine mammary gland. J Clin Invest 1994;93:892–899.
115. Barcellos-Hoff MH, Dix TA. Redox-mediated activation of latent transforming growth factor-beta 1. Mol Endocrinol 1996;10:1077–1083.

4 Biological Functions of Latent TGF-β-Binding Proteins and Activation of TGF-β

Marko Hyytiäinen, Katri Koli, and Jorma Keski-Oja

CONTENTS

INTRODUCTION
ACTIVATION OF TGF-β
LATENT TGF-β BINDING PROTEINS
ASSOCIATION OF LTBPS WITH THE ECM
LTBPS AND TGF-β TARGETING IN VIVO
DISEASE ASSOCIATIONS OF LTBPS
LTBPS IN CANCER
CONCLUDING REMARKS
ACKNOWLEDGMENTS
REFERENCES

Abstract

TGF-βs 1–3 are secreted from the cells as latent complexes composed of dimeric mature growth factors, the respective N-terminal propeptides latency-associated peptide (LAP) and one of the four latent TGF-β binding proteins (LTBPs). The non-covalently associated LAP renders each mature TGF-β latent; they cannot bind to the cell surface receptors and mediate their effects. LTBPs augment the secretion of each small latent complex and directs them to the extracellular matrix. The activation of TGF-β (release of mature TGF-β from the latent complex) is a central way to control the activity of TGF-β. The activation can be achieved by several proteinases, by the extracellular matrix (ECM) glycoprotein thrombospondin-1 and via a subset of cell surface integrins. Similarly, the association of the propeptides TGF-βs with different LTBPs, and subsequent different localization to the extracellular space, and potentially differential susceptibility to the activation events regulate the availability of each TGF-β; in an efficient and specific manner.

Key Words: LTBP; TGF-β; extracellular matrix; latency; activation.

1. INTRODUCTION

Most studies on the effects and significance of transforming growth factor (TGF-β) have been carried out with purified, active TGF-β. TGF-β is, however, secreted from the cells as a latent complex. The latent form is not able to bind to its signaling receptors and mediate

From: *Cancer Drug Discovery and Development: Transforming Growth Factor-β in Cancer Therapy, Vol I: Basic and Clinical Biology*
Edited by: Sonia B. Jakowlew © Humana Press Inc., Totowa, NJ

its effects. Latency is an important step in the modulation of TGF-β activity, and is related to the ability of latent TGF-β to bind to the extracellular matrix, ECM via latent TGF-β-binding proteins, LTBPs. Mature TGF-βs are processed from two 55-kDa polypeptides. The 55-kDa proteins dimerize and get proteolytically processed by furin like proteases during secretion from the cells to yield the mature 25-kDa growth factor, which remain in the complex with the N-terminal propeptide parts (latency associated peptide; LAP) in the extracellular space (Fig. 1). This complex is called the small latent TGF-β (SL-TGF-β), which in turn, associates covalently with an additional protein, latent TGF-β binding protein (LTBP) *(1,2)*. These two form large latent TGF-β (LL-TGF-β) complexes, and represent the forms of TGF-β secreted by most cells.

2. ACTIVATION OF TGF-β

2.1. In vitro Methods

The LAP part of the TGF-β complex is sensitive to conformational changes caused by changes in the environment *(3)*. Conformationally affected LAP can no longer mask the mature TGF-β, resulting in its activation. For this reason, TGF-β can be activated in vitro by the buffer conditions of solutions, such as changes in pH (low or high pH), by chaotropic agents like urea and guanidine hydrochloride, or detergents, as well as by increasing the temperature *(4)*. Removal of carbohydrate structures from LAP can result in activation of TGF-β *(5,6)*. Cleavage of LAP by proteases such as plasmin leads to activation of TGF-β *(7)*.

Interestingly, the interaction/release of mature TGF-β from the intact SL-complex is reversible. Recombinant LAP peptides without the mature growth factor can bind and neutralize active TGF-β in vitro *(8,9)*, and application of recombinant LAP in mice decreases the activity of TGF-β in liver *(10)*. Whether LAP binds active TGF-β reversibly in vivo is an open question.

TGF-β can be activated by radiation *(11)*, and more generally by events causing increase of reactive oxygen species (ROS) in the extracellular space *(12)*. The activation occurs palusibly by direct modification of LAP by ROS, which modifies the structure of latent TGF-β resulting in its activation *(12)*.

Low pH, radiation and the protease plasmin (*see* Section 2.2.) are likely to be involved in TGF-β activation in vivo as well *(11,13)*.

2.2. Roles of Proteases

Platelet derived latent TGF-β gets activated when blood clots are dissolved with plasmin *(14)*. Plasminogen and its activator uPA are involved in TGF-β activation in cocultures of endothelial and smooth muscle cells as well *(15,16)*. In this activation cascade, LL-TGF-β gets initially incorporated to the ECM via LTBP *(17,18)*. The complex is then proteolytically cleaved from the matrix (Fig. 2), and associates with the cell surface via mannose-6-phosphate of LAP to mannose-6-phosphate/IGF-II receptor *(16)*. LTBP-1 may also be involved in the cell surface localization *(19)*. Latent TGF-β located at the cell surface is subsequently cleaved by plasmin *(20)* so that the mature TGF-β gets released from the complex. Retinoids activate TGF-β in cell cultures partially via similar mechanisms *(21)*. Plasmin and plasminogen activator activity are needed for the generation of active TGF-β, but mannose-6-phosphate has no effect in this system *(21)*. Interestingly, plasminogen deficient mice do not have any phenotypic similarities with mice deficient in any of the three TGF-βs *(22)*, suggesting that other, redundant enzymes or activation mechanism can replace the missing plasmin.

Certain ECM proteinases can directly cleave the LAP of TGF-β1, leading to TGF-β activation. Matrix metalloproteinases (MMP)-9 and MMP-2 can also effectively activate

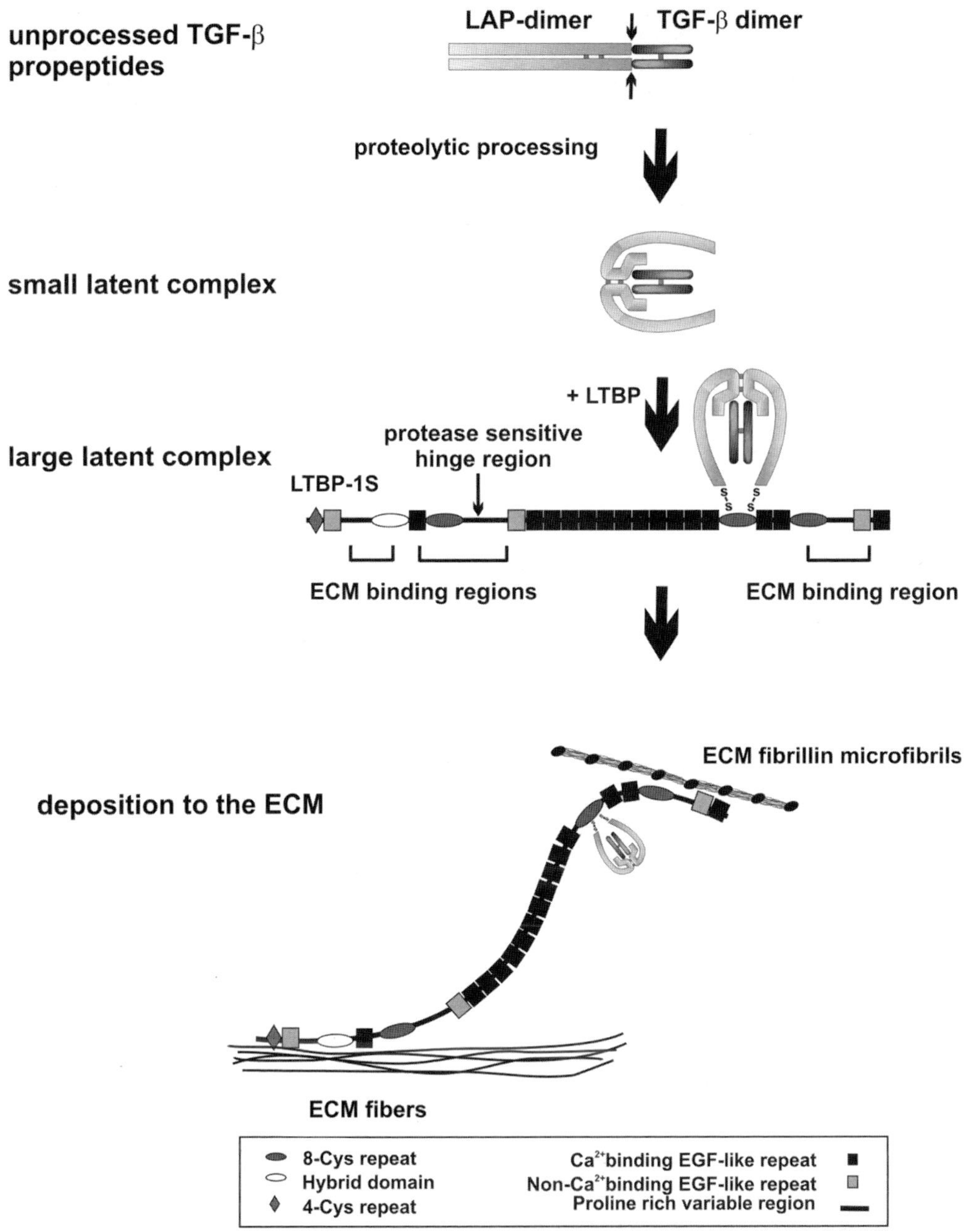

Fig. 1. Assembly of latent TGF-β complexes. TGF-β is synthesized in the cells as a proform. The proform is proteolytically processed, dimerized and disulfide-linked to an LTBP to yield large latent TGF-β (LL-TGF-β), which is secreted from the cells. LL-TGF-β binds to the ECM fibers via the N- and C-terminal domains of LTBPs.

LL-TGF-β *(23)*. Especially MMP-9 is an efficient activator of TGF-βs when complexed with CD44 at the cell surface. At least these two MMPs have the ability to digest the LAPs of TGF-βs. The cleavage and activation of the LAP of TGF-β2 was the most efficient one *(23)*. In addition, latent TGF-β1 can be activated by matrix metalloproteinases derived from extracellular microvesicles *(24–27)*. This process is implicated especially in chondrocytes

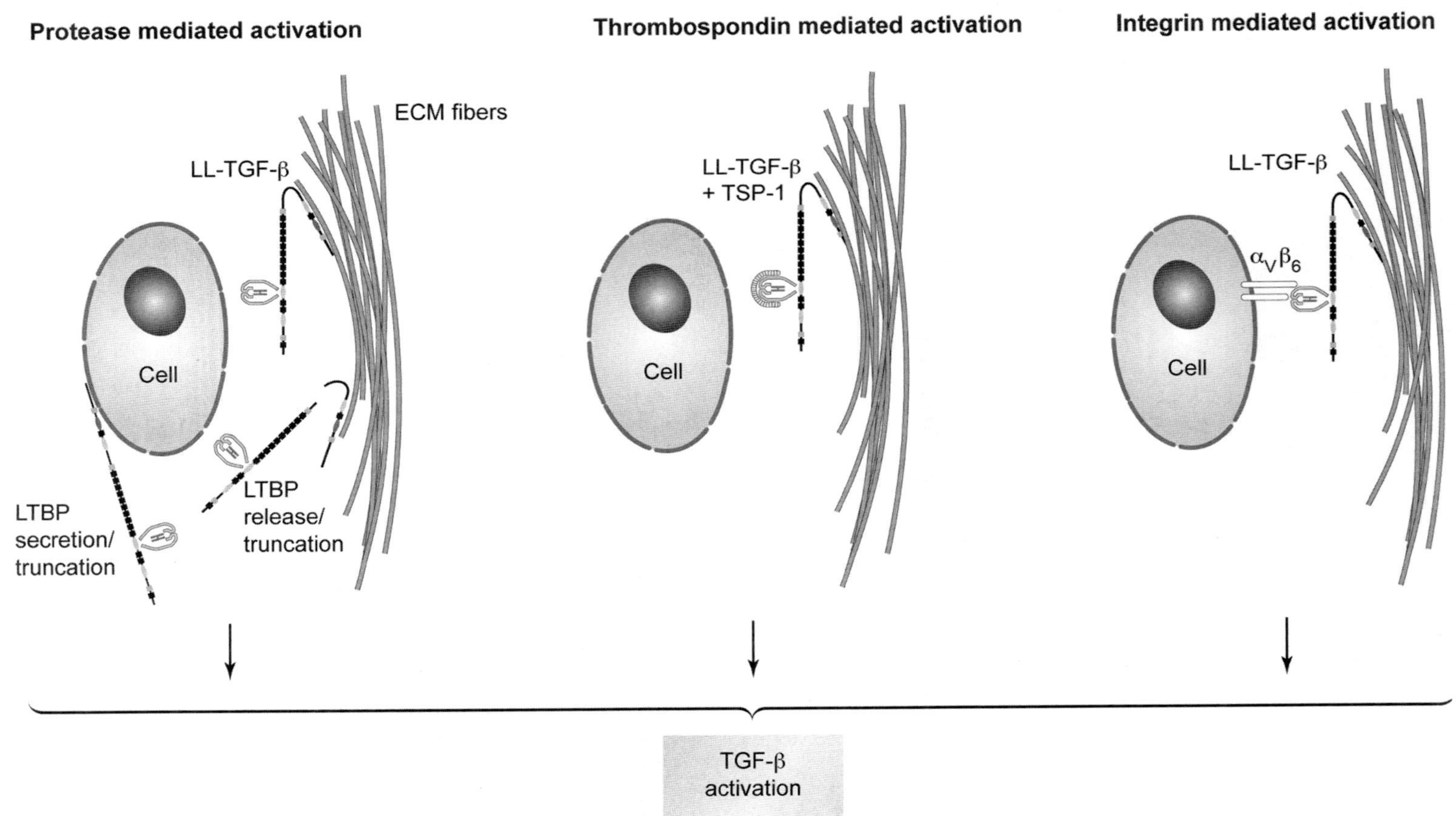

Fig. 2. Different modes of TGF-β activation. TGF-β can be activated via several different mechanisms, of which protease-mediated, thrombospondin-mediated and integrin-mediated are illustrated here. In the protease-mediated activation, LL-TGF-β associates with the ECM, and is released proteolytically. Thereafter, the LAP part of the complex is proteolytically cleaved, which releases mature, active TGF-β from the complex. In thrombospondin-mediated activation TSP-1 binds to the LAP and active TGF-β present in the matrix, and changes the conformation of the complex so that TGF-β receptors can recognize the activated growth factor. In the integrin-mediated activation, integrin $\alpha_v\beta_6$ binds to RGD sequence of LAP of LL-TGF-β, which is anchored to the matrix. This leads again to the changes in the conformation of the complex, and the availability of active TGF-β for the TGF-β receptors.

undergoing endochondral ossification. Calpain, a cysteine protease implicated in numerous physiological processes *(28)* can activate SL-TGF-β1 at least in bovine capillary endothelial cells *(29)*. Calpain cleaves TGF-β1 LAP in vitro, and this means of activation takes plausibly place at the cell surface. Stimulation of the rat mast cells to degranulate results in the release of chymase together with latent TGF-β, which is then activated by chymase *(30)*.

2.3. Thrombospondin Mediated Activation

The multifunctional extracellular matrix protein thrombospondin-1 is able to activate both small and large latent complexes of TGF-β *(31)*. However, contrasting reports have also been published *(32,33)*. In this activation cascade, a defined sequence in thrombospondin interacts with the LAP propeptide *(34)*. Another site of thrombospondin interacts simultaneously with the mature TGF-β. This changes the conformation and affinity of LAP to the mature TGF-β. The bound TGF-β is thus rendered into the active form (Fig. 2). Activation of TGF-β is specific for thrombospondin-1, and it has been speculated that thrombospondin-2 might act as a competitive inhibitor in TGF-β activation *(35)*. However, in vivo evidence suggest that thrombospondin-2 has no effect on TGF-β activation in healing wounds *(36)*. Thrombospondin-1 is likely to act as an in vivo activator of TGF-β *(37)*. Mice deficient for thrombospondin-1 share many phenotypic similarities with TGF-β knockout mice. The lung phenotypes, for example, are similar. In addition, thrombospondin-1 participates in the activation of TGF-β in glomerulonephritis in vivo *(38)*. Thrombospondin-1 has also been observed to cooperate with plasmin-mediated activation of TGF-β in bleomycin-induced pulmonary inflammation *(39)*.

Activated platelets secrete large amounts of both thrombospondin-1 and latent TGF-β during degranulation. Interestingly, platelet derived TGF-β gets activated similarly in normal and thrombospondin-1 null mice. This suggests that other TGF-β activation mechanisms take place in certain tissues *(37)*.

2.4. Activation by Integrins

The amino acid sequence RGD is a recognition site for integrins, a family of heterodimeric cell surface adhesion receptors. This sequence is part of several ECM proteins. The LAP propeptides of TGF-β1 and TGF-β3 contain the integrin recognition sequence RGD *(40,41)*. The RGD sequence in LAP can be recognized by integrins $\alpha_v\beta_1$, $\alpha_v\beta_6$, $\alpha_v\beta_8$, $\alpha_v\beta_3$, $\alpha_v\beta_5$, and $\alpha_8\beta_1$ *(42)*. However, the binding of an integrin does not necessarily lead to TGF-β activation. The binding of cells to TGF-β1·LAP via $\alpha_v\beta_6$, but not via $\alpha_v\beta_1$ is known to lead to TGF-β activation *(43)*.

In the activation of TGF-β by integrin $\alpha_v\beta_6$, the integrin associates with the RGD sequence of TGF-β1·LAP in the latent TGF-β complex (Fig. 2). The binding changes the conformation of LAP, resulting the growth factor to be available to the cell surface TGF-β receptors without dissociation of the growth factor *(42)*. Organized actin cytoskeleton of the activating cells and subsequent relocalization of $\alpha_v\beta_6$ integrins to the focal adhesions are needed, suggestive of the integrin clustering for TGF-β activation to occur *(43)*. In addition, several molecules with an ability to inhibit protease mediated TGF-β activation do not inhibit integrin-mediated activation *(43)*. Mice deficient in either β_6–integrin or TGF-β1 share some phenotypic properties, such as inflammation in the lungs and skin. In addition, β_6 integrin deficient mice have resistance against bleomycin induced lung fibrosis, suggesting that TGF-β activation is impaired in these animals. TGF-β3 is known to be activated by mechanisms involving integrin binding as well *(44)*, whereas TGF-β2 has no RGD sequence in its LAP part, and is consequently not activated by integrins $\alpha_v\beta_6$ or $\alpha_v\beta_8$ *(44,45)*.

Binding of integrin $\alpha_v\beta_8$ to SL-TGF-β1 is also able to bring about the activation *(45)*. The binding of integrin to TGF-β1 LAP is not sufficient for the activation of TGF-β1, but needs a protease-mediated event. In integrin transfected mink lung epithelial cells $\alpha_v\beta_8$ integrin binds to the LAP of TGF-β1, which recruits membrane type 1 metalloproteinase (MT1-MMP) to the complex at the cell surface. MT1-MMP, in turn, cleaves LAP releasing active TGF-β from the latent complex *(45)*. This means of activation is relevant in vivo, as β_8 integrin expressing cancer cells are less tumorigenic in nude mice, and the decreased growth rate is mediated by increased TGF-β activation *(45)*. Recently, $\alpha_v\beta_5$ integrin was implicated in TGF-β activation in fibroblasts from systemic sclerosis patients *(46)*.

Integrin $\alpha_5\beta_1$ recognizes RGD sequences on its target ECM substrates. This integrin seems to have a positive effect on TGF-β activation as well *(47)*. However, it is likely that $\alpha_5\beta_1$ does not bind to LAP of TGF-β, but the decreased TGF-β activation in integrin α_5 null cells is caused by the decreased deposition of fibronectin to the ECM, and subsequent loss of LTBP·TGF-β complexes from the ECM *(47)*.

3. LATENT TGF-β BINDING PROTEINS

LTBPs 1–4 *(48–55)* are 125–240 kDa extracellular matrix proteins, which belong to the fibrillin/LTBP -gene family (illustrated in Fig. 3). They consist mainly of two types of repeating cysteine-rich domains, epidermal growth factor (EGF) like repeats *(56)* and so called eight-cysteine (8-Cys) domains. 8-Cys domains have been detected only in LTBPs and fibrillins (*see* below), and the presence of this domain defines the family. LTBPs contain, in general, 15–20 EGF-like repeats, three 8-Cys repeats, and a hybrid domain, which resembles both EGF and 8-Cys repeats.

Fibrillins are components of elastic tissue 10-nm microfibrils *(57–59)*, and their defects are responsible for matrix fragility observed in patients with the Marfan syndrome and congenital contractural arachnodactyly disorders *(60,61)*. LTBPs and fibrillins share a similar protein domain structure. Fibrillins are considerably larger glycoproteins than the LTBPs (Fig. 3). Fibrillins consist of >40 EGF-like repeats, seven 8-Cys repeats and of two hybrid domains each *(62)*.

LTBPs have varying degrees of functional and structural similarities and differences. LTBPs -1 and -3 form effectively covalent bonds with all three TGF-βs, whereas LTBP-4 can form complexes with TGF-β1 only. On the contrary, LTBP-2 is unable to form complexes with any of the TGF-βs *(63)* (*see* Table 1). The specific roles of the LTBPs in mediating and targeting the effects of TGF-β are under intensive research.

3.1. LTBP-1, the Prototype LTBP

LTBP-1 was originally identified and purified from platelets as a binding protein for TGF-β1 *(48,64)*. LTBP-1 is secreted from cultured cells in a form devoid of TGF-β in about tenfold excess in lung fibroblasts *(18)*. This suggested that LTBP-1 has additional roles not related to TGF-β. There are two distinct mRNA species for LTBP-1 differing substantially in size *(49,65)*, designated as LTBP-1L and LTBP-1S, (*see* Fig. 3). Mouse LTBP-1 has similar alternative splice variants *(66,67)*. The larger form LTBP-1L is a splice variant, which has an N-terminal extension of 346 amino acids not found in the smaller LTBP-1S isoform *(68)*. LTBP-1 has several other splice variants, which lack specific domains of LTBP-1 *(1)*, (Fig. 3). These splice variants have been suggested to have different characteristics, e.g., glycosylation or sensitivity for proteases *(69–71)*.

LTBP-1 is ubiquitously expressed. In multitissue Northern hybridization analysis LTBP-1L is mainly expressed in the heart, placenta, kidney and prostate, whereas LTBP-1S has a wider expression pattern and appears also in the lung, skeletal muscle, testis and ovary *(68)*. *Xenopus* LTBP-1 seems to be expressed in the most dorsal tissues of the developing embryo *(72,73)*. The expression of LTBP-1 is induced by retinoid acid and

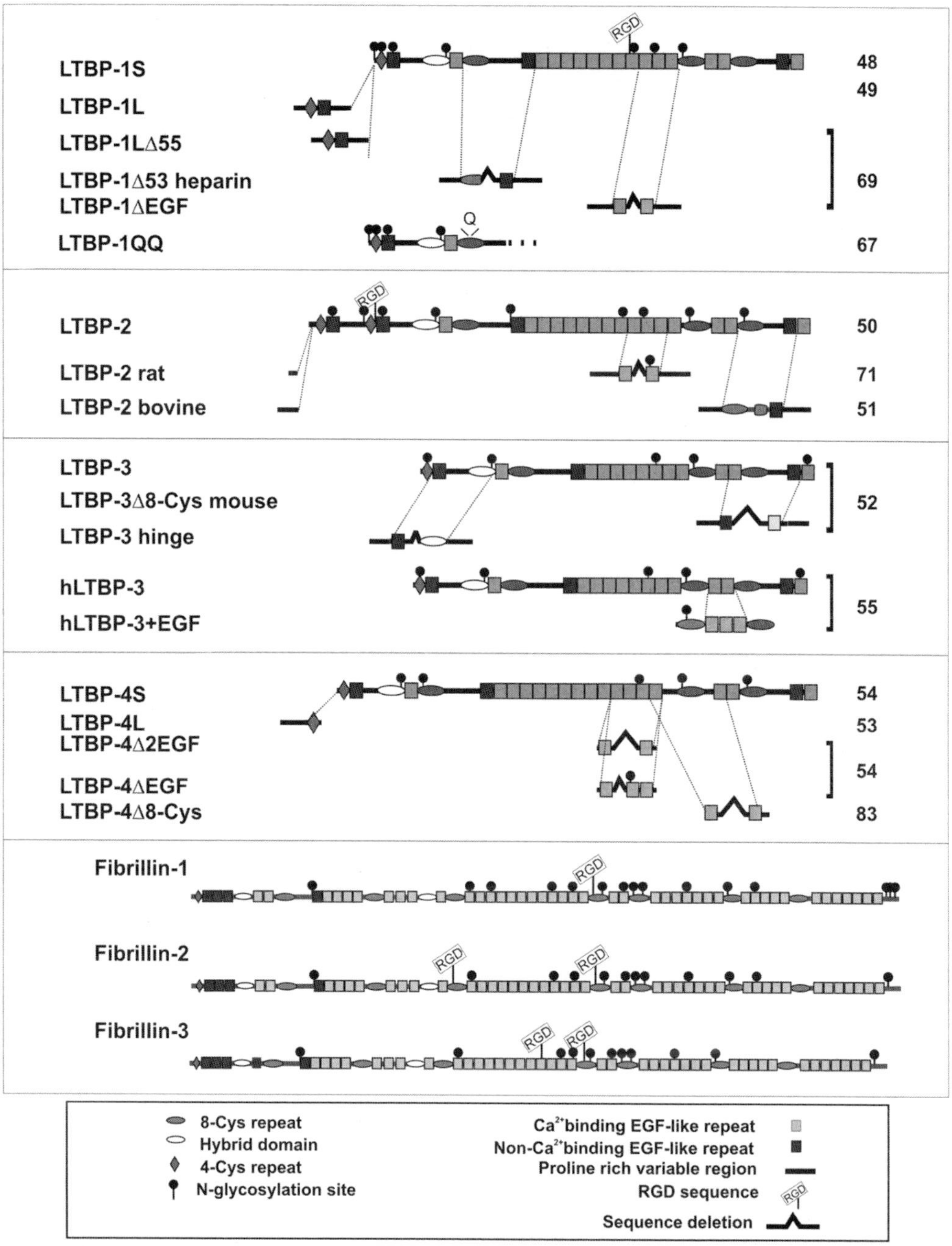

Fig. 3. Domain structures and splice variants of LTBP/fibrillin family members. LTBPs are composed mainly of cysteine-rich domains interrupted by variable proline-rich domains. There are numerous splice variants, which are formed by deleting or adding variable domains to the LTBPs (references on the right). Splice variants of fibrillins are not shown here.

dexamethasone in cultured cells *(74)*, as well as with parathyroid hormone *(75)*. In rat cirrhotic fat storing cells, platelet derived growth factor BB enhances the expression of LTBP-1 *(76)*. Interestingly, LTBP-1 expression is induced by TGF-β *(77)*, suggesting a positive feedback loop for TGF-β system.

Table 1
Capability of Different LTBPs to form Covalent Complexes with TGF-βs

	TGF-β1	*TGF-β2*	*TGF-β3*
LTBP-1	+++	+++	+++
LTBP-2	–	–	–
LTBP-3	+++	+++	+++
LTBP-4	+	–	–

Adapted from ref. *63*.

3.2. LTBP-2, the Non-TGF-β Binding LTBP

LTBP-2 deficient mice die very early during embryogenesis *(78)*. The phenotypes of knockout mice of the three TGF-βs differ from this. In our coexpression analyses we unexpectedly found that LTBP-2 is unable to form complexes with any of the three TGF-βs *(63)*. This indicates that LTBP-2 has roles not related to TGF-β. LTBP-2 is predominantly expressed in the lung, and also in the heart, placenta, liver, and skeletal muscle *(50)*. During rodent development LTBP-2 is expressed in the lung, dermis, perichondrium, arterial vessels, epicardium, pericardium, and heart valves *(78)*.

LTBP-2 has numerous cell adhesive functions. It mediates the adhesion of melanoma cells via integrins $\alpha_3\beta_1$ and $\alpha_6\beta_1$ *(79)*. On the other hand, LTBP-2 has effects on the morphology of fibroblasts binding to fibronectin. They turn to a less adherent form, and LTBP-2 has an unexpected antiadhesive property on these cells *(80)*. LTBP-2 thus seems to display complex roles by regulating cell adhesion and migration, instead of playing roles in the biology of TGF-βs.

3.3. LTBP-3, Functions in the Bone

LTBP-3 was identified on the basis of its sequence similarity with LTBP-1. It is expressed for example in the central nervous system, somites and cardiovascular tissue, and in the developing bone during mouse development *(52)*. The highest expression of LTBP-3 in human has been detected in skeletal muscle, heart, prostate, and ovaries *(55)*. LTBP-3 expression is induced by TGF-β in a manner similar to LTBP-1 *(81)*. Tissue in situ hybridization analysis of mouse embryos revealed wide expression pattern of LTBP-3 in mesenchymal cells *(52)*. LTBP-3 has important functions in the development and functions of the bone, as indicated by the careful investigation of LTBP-3 null mice *(82)* (*see* Section 5).

3.4. LTBP-4, Lung Development and Colon Cancer

LTBP-4 was identified from sequence databases based on its similarity with LTBP-1 *(53,54)*. The overall domain structure of LTBP-4 is similar to the other LTBPs. However, some additional proline-rich areas interrupt the cysteine rich domains in LTBP-4. LTBP-4 is predominantly expressed in the aorta, heart, small intestine, and ovaries *(53,54)*.

Two amino terminal variants similar to LTBP-1L and -1S have been identified for LTBP-4 *(54)*. In addition, LTBP-4 has a potentially important splice variant LTBP-4ΔCys-83rd, which lacks the third 8-Cys repeat containing the SL-TGF-β1 binding site. The existence and variable expression pattern *(83)* of splice variants unable to bind SL-TGF-β provides another way of regulating and targeting TGF-β activity.

LTBP-4 has functions in the formation of lungs, and deficiency leads to colon cancer *(84)*. Unexpectedly, cultured LTBP-4 null cells display a fibrotic phenotype *(85)*.

3.5. Significance of LTBPs

The existence of four LTBPs, their various alternatively spliced forms and diverse mRNA expression patterns in different tissues and developmental stages *(86)* suggest a substantial variety of functional properties for LTBPs. Switching of isoform expression during development may be typical of LTBPs and important for the spatial and temporal activation of TGF-β. Detailed analyses of the expression patterns of different forms of LTBPs in various stages of development will open new paths for the understanding of the functions of the four different LTBPs.

4. ASSOCIATION OF LTBPs WITH THE ECM

While LTBPs were identified as binding proteins and carriers for latent TGF-β, the fibrillins were identified as structural ECM components. This finding led to studies, where the LTBPs were identified as components of the ECM as well (*see* Fig. 1). Thereafter, LTBPs have been found to have roles in targeting TGF-β activity in the extracellular matrix. In spite of the fact that three of the LTBPs have important roles in the biology of TGF-β, they may have structural roles as well.

LTBPs-1 and -4 colocalize with the small latent TGF-β1 in the fibrillar ECM structures *(84,87)*. Matrix bound LTBPs do not, however, always associate with TGF-β *(18,87)*. LTBP-1 codistributes initially with fibronectin, as well as with fibrillin-1 in the ECM of cultured cells *(87)*. Colocalization status appears to depend on the age of the cell culture. Namely, a few days after plating of the cells, LTBP-1 colocalizes with fibronectin. In fact, the assembly of fibronectin into the ECM is a prerequisite for LTBP-1 association with the ECM *(88)*. Interestingly, in the course of time, usually after a week of culture, the distribution changes, and LTBP-1 codistributes with fibrillin-1 *(87,89)*. This indicates that LTBP-1 containing matrix undergoes remodeling during maturation. LTBP-1 has been found to colocalize with fibrillin-1 in tissues as well *(90,91)*.

At least the C-terminal domains of LTBPs -1 and -4 interact directly with fibrillin-1 (*see* Figs. 1 and 4), and LTBP-1 colocalizes with fibrillin in cell cultures as well as in vivo *(87,91)*. LTBP-1 has, however, not been detected in microfibril extracts purified from tissues *(91)*. The major ECM interacting sites of LTBPs are located at the N-terminus of the proteins *(92,93)*, whereas the LTBP-1–fibrillin interaction site is in the C-terminus. These results suggest that LTBP-1 is not an integral component of the microfibrils, but acts more like a microfibril- associated protein. LTBP-2 colocalizes with elastic fiber microfibrils in bovine tendons *(51)*, and in fibrillar structures in the ECM of cultured cells *(94)*.

LTBP-3 has slightly lower molecular mass than the other LTBPs, and it is most prominently expressed in the bone. LTBPs–3 and -4 are deposited in the ECM of human fibroblasts considerably later than LTBP-1 *(93)*, and these LTBPs do not colocalize so clearly with fibronectin in the ECM. Interestingly, the C-terminal domain of LTBP-3 does not associate with the ECM as corresponding fragments of LTBPs-1 and -4 *(93)*. This has implications for the assembly of LTBP/TGF-β complexes as well; different spatial and temporal ECM association of LTBP/TGF-β complexes and additional differential LTBP complex formation with the different TGF-β isoforms (Table 1) may regulate TGF-β activation in tissues.

LTBPs-1 and -4 are secreted from the cells at least to some extent in excess of SL-TGF-β. Indicative of the tight control of the functions of LTBP-3, it does not appreciably get secreted from the cells without TGF-β in culture. This finding has interesting implications both in the biology of LTBP-3, and in its role in the targeting of the effects of TGF-β *(55,95)*. This may also be indicative of a chaperone function of LTBP-3 for TGF-β.

The main ECM interacting domains are located at the N-terminus of LTBPs *(92,93,96)* (Fig. 4). The long form LTBP-1L has been reported to associate more efficiently with the

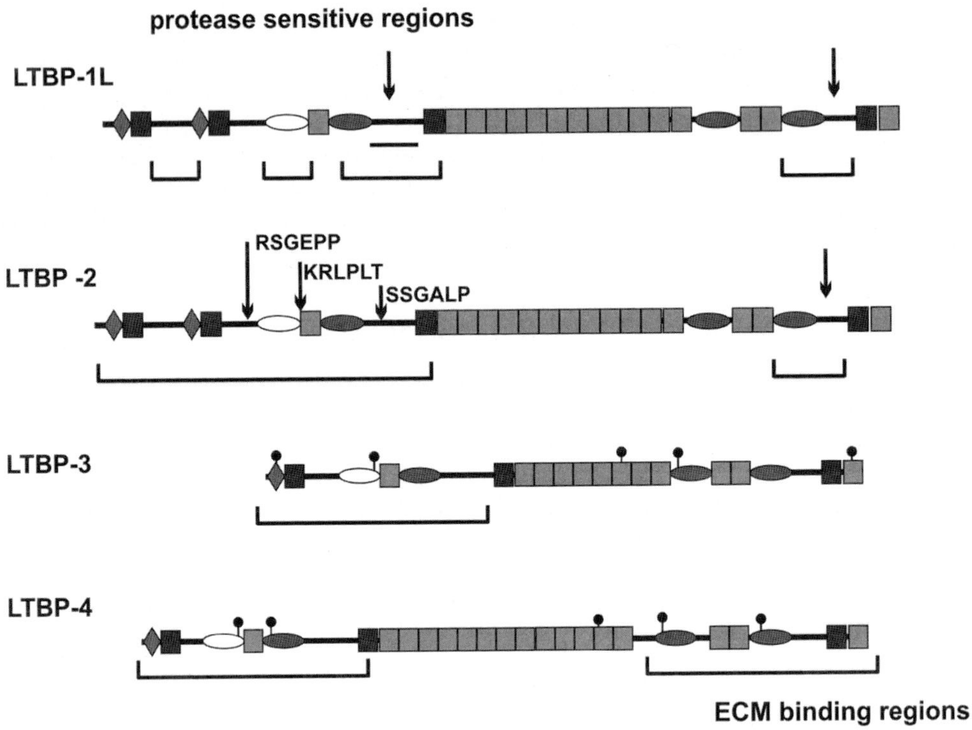

Fig. 4. Extracellalur matrix binding sites and proteolytically sensitive sites in LTBPs. LTBPs associate with the ECM via their N- and C-terminal domains, with the exception that the C-terminus of LTBP-3 does not associate with the ECM. The figure illustrates the so far identified ECM-interacting domains (square brackets). It is likely that the ECM-binding domains contain several independent sites interacting with different ECM molecules. Numerous proteases are capable of digesting LTBPs. The protease sensitive sites have been partially mapped from LTBPs-1 and -2 (arrows). The N-terminal amino acid sequences of proteolytically processed LTBP-2 fragments are illustrated. The proteolytically sensitive sites seem to localize to proline-rich regions of the molecules, and the other proline-rich regions of LTBPs are likely to be protease sensitive as well.

extracellular matrix than LTBP-1S *(68)*. This ability may be related to the additional splice variant LTBP-1QQ codistributing with LTBP-1L. In this splice variant, one glutamine residue is added to LTBP-1 *(67)*, and this glutamine may act as a transglutaminase target for more efficient ECM association.

These observations emphasize the role of the N-terminus in mediating the interactions of LTBPs with different ECM components. The diverse mRNA expression patterns of LTBP-1L and LTBP-1S forms in different human and mouse tissues *(67,68)* and the fact that these forms have independent promoters *(77)* suggest that they are regulated in a tissue-specific manner.

Proteolytic cleavage of the protein at its N-terminus by plasmin results in the release of the central part of LTBP from the matrix *(18,94)*. In addition, the C-terminus of LTBPs contain a potential cleavage site facilitating their proteolytic release from the ECM *(92,94)*. The central part of LTBP-1 contains the domain for the SL-TGF-β binding (Fig. 1). The cleavage of LTBP-1 thus results in the release of the truncated LL-TGF-β from the ECM. Some proteases of the serine protease family, such as plasmin, can release latent TGF-β complex from the ECM *(18,97)*. MMPs have also been implicated in this process *(27)*. The release from the ECM is plausibly the first step of TGF-β activation in certain cascades *(98)*.

This is seen in transgenic mice expressing modified LTBP-1 unable to associate with the ECM. The expression of this form leads to excessive TGF-β activation in the skin *(99)*.

However, since TGF-β can get activated by a number of alternative mechanisms, the presence of latent TGF-β complexes in the ECM may even be a prerequisite for the activation of latent TGF-β. For example, as integrin $\alpha_v\beta_6$ recognizes and activates ECM bound LL-TGF-β complex *(42,43,100)*, the shedding of LTBP from the matrix is likely to negatively regulate the activation of TGF-β. Indeed, the loss of fibronectin ECM, and subsequent loss of ECM association of LTBP-1·TGF-β leads to decreased integrin mediated TGF-β activation *(47)*.

At least LTBP-1 seems to direct TGF-β for activation under certain conditions *(101)*. Loss of LTBP-4 leads, in turn, to decreased activation of TGF-β in the lung and colon, but not in the kidney *(84,85)*. This indicates that different LTBPs drive the secretion and association with the ECM, and even more importantly, the activation of TGF-β in different tissues.

The ECM association of growth factors of TGF-β family via LTBPs may not be limited to TGF-βs only. Recently, the prodomain of BMP-7, a TGF-β family member, was reported to interact with fibrillin-1, suggesting that this growth factor is targeted to the ECM via mechanisms similar to that of TGF-βs *(102)*.

5. LTBPs AND TGF-β TARGETING IN VIVO

In vitro studies on murine embryonic cell differentiation indicate that anti-LTBP-1 antibodies are able to prevent endothelial cell differentiation. The same effect was observed by neutralizing antibodies against TGF-β1 indicating a TGF-β dependent mechanism in this process *(103)*. *Xenopus* LTBP-1 modulates the activity of activin and nodal in the extracellular space during embryonic development *(72)*.

A number of recent reports have addressed the roles of LTBPs by gene-targeted mice (Table 2). These studies have provided evidence for the specific roles of these extracellular matrix proteins in the biology of TGF-β.

Mice deficient in LTBP-3 display abnormalities in their bones, pinpointing the importance of this carrier protein for correct TGF-β deposition in the bone *(82,104,105)*. Shortly after birth these mice develop cranio-facial defects, and at the age of several months, osteosclerosis and osteoarthritis *(104)*. In addition, LTBP-3 –/– mice have decreased growth rate and involution of spleen and thymus *(106)*, and lung alveolar septation defects *(107)*. These changes are consistent with the phenotypes of mice with defective TGF-β signaling *(108–110)* suggesting a role for LTBP-3 in mediating the effects of TGF-β in bone.

Unexpectedly, LTBP-4 has the ability to form complexes with TGF-β1 only *(63)*. Mice deficient in LTBP-4 develop cardiomyopathy and pulmonary emphysema. These mice develop also colorectal cancer *(84)* indicating the important role of LTBP-4 targeted TGF-β1 in negative regulation of cell growth. TGF-β signaling decreases in affected tissues, showing direct dependence of TGF-β signaling on LTBP-4.

As observed before, TGF-β1 has a suppressive function on the expression of TGF-β2 and -3. Removal of LTBP-4 mediated TGF-β1 suppression caused dramatic upregulation of other two TGF-βs as well as a number of other genes *(85)*. This is of particular importance, since the other LTBPs could not rescue this phenotype.

6. DISEASE ASSOCIATIONS OF LTBPs

Chromosomal locations of all the human LTBP genes have been determined by fluorescence in situ hybridization. However, genetic linkages of LTBPs with any disease locuses have not been found yet. The expression of LTBP-1 is elevated in several fibrotic conditions. Elevated levels of TGF-βs and LTBP-1 have been observed during chronic rejection of

Table 2
Phenotypes of Different LTBP Knockout Mice

LTBP-1L	Heart defect (V. Todorovic and D. Rifkin, submited 2007)
LTBP-2	E3.5–E6.5 embryonic lethal *(78)*
LTBP-3	Defects in bone and lung formation *(82,107)*
LTBP-4	Colorectal cancer, lung emphysema, cardiomegaly *(84)*

cardiac allografts in rats *(111)* and tuberculous pleurisy, where LTBP-1 was localized in immature fibrotic areas of the pleura by immunohistochemistry *(112)*. In human dermis LTBP-1 and TGF-β1 colocalized with elastic fibers supporting the repository role for LTBP-1 also in vivo *(113)*. Accordingly, the immunoreactivity for both LTBP-1 and TGF-β1 were lost in anetoderma, which is characterized by the absence of elastic fibers in the dermis.

TGF-β is implicated in virtually all fibrotic conditions *(114)*. LTBP-1 expression is affected in fibrotic conditions as well. Pseudoexfoliation syndrome is characterized by pathologic production and accumulation of fibrillar extracellular material in ocular tissues. Increased levels of TGF-β1, LTBP-1, and LTBP-2 have been detected in aqueous humor of pseudo-exfoliation patients suggesting their involvement in the accumulation of abnormal extracellular elastic material *(115)*. All LTBP isoforms are expressed in normal and fibrotic liver *(116, 117)*, and the expression of LTBP-1 is increased together with TGF-β in fibrosis of liver *(118)*. LTBP-1 expression is also increased in adriamycin-induced nephropathy *(119)* and chronic pancreatitis *(120)*. Another example of increased LTBP-1 expression in fibrotic liver is dioxin receptor null mice. Dioxin receptor negatively regulates LTBP-1 expression *(121)*, and mice deficient in dioxin receptor develop portal liver fibrosis. LTBP-1 expression is increased together with TGF-β, especially in fibrotic parts of the liver in those animals *(122)*.

LTBPs are involved in the Marfan syndrome as well. Fibrillin-1 deficiency in mutant mice having Marfan syndrome like phenotype causes the excessive activation of TGF-β, resulting in apoptosis in the developing lung *(123)*. Patients suffering from the Marfan syndrome have similar lung abnormalities. Interestingly, TGF-β activity increases in these mice, leading to pulmonary emphysema. Since at least LTBPs-1 and -4 interact with fibrillin-1 *(91)*, latent TGF-β complexes may not be able to associate with the ECM of these mice. In contrast, TGF-β is secreted from these cells resulting in immediate, uncontrolled TGF-β activation. Therefore, fibrillin-1 has plausibly a key role in regulating TGF-β activation in the lung. Fibrillin-1 is likely to sequester LL-TGF-β complexes and limit its activation *(124)*.

An apparent contrast exists between the results obtained using fibrillin-1 and LTBP-4 deficient mice *(84,123)*. In the latter, TGF-β activity is decreased, whereas in the former it is increased. In spite of this, both mice develop emphysema. Therefore it appears that an optimal level or properly targeted TGF-β activity is needed for the normal development. β6-integrin is able to activate SL-TGF-β1 *(43,100)* and lack of β6-integrin in mice leads to emphysema *(125)* again implicating TGF-β in the development of this lung phenotype. Emphysema caused by defective TGF-β activation is dependent on increased expression of matrix MMP-12. This implicates complex relationships between TGF-β, its activation and subsequent effects at the tissue level.

7. LTBPs IN CANCER

TGF-β has a biphasic role in cancer. In some forms, TGF-β expression is decreased, whereas in certain other ones its expression is increased. In addition, the cancer cell responses to TGF-β vary depending on the stage and type of cancer. Usually, inaggressive cancer types respond to TGF-β by decreasing their growth rate, whereas malignant advanced tumors have lost their ability to respond to TGF-β growth inhibition *(126)*.

LTBP-1 expression is increased together with TGF-β in pancreatic carcinoma *(127)* and in colorectal adenomas *(128)*. In contrast, its expression decreases in liver cancers *(129)*, as well as in ovarian neoplasms *(130)*. In neuroendocrine tumor cells TGF-β expression is increased, while those cells express only low levels of LTBP-1. However, stromal cells express LTBP-1 *(131)*.

TGF-β is expressed in normal, benign and malignant prostate tissue. However, LTBP-1 expression is lost in prostate carcinoma, suggesting an important role for LTBP-1 in maintaining this tissue in a quiescent state *(132)*. In gastrointestinal carcinoma, LTBP-1 is expressed in stromal cells, but not in carcinoma cells, whereas TGF-β is expressed in both cell types *(133)*.

The expression patterns of LTBP-1 are very much dependent on the stage and type of cancer, like those of TGF-β. However, the expression patterns of LTBP-1 and TGF-βs are only partially overlapping, suggesting independent roles for LTBP-1, as well as the participation of other LTBPs in the regulation of TGF-β activity.

LTBP-4 has been implicated in colorectal cancer *(84)*. LTBP-4 null mice develop colorectal cancer, and TGF-β activity is decreased in a manner similar to the lungs of those animals.

8. CONCLUDING REMARKS

Increasing evidence indicates the crucial role for LTBPs in mediating the effects of TGF-β, an important growth regulator in development and pathological situations. As TGF-βs associate with LTBPs, they potentially regulate TGF-β activity via different spatial and temporal localization, different susceptibility of TGF-β complexes for activation, and different expression patterns during development and tumorigenesis. The expression analyses of LTBPs in various tumors have not, however, been extensively carried out. Therefore, more thorough investigation of expression patterns of LTBPs, and more careful evaluation of the role of LTBPs in the activation of TGF-β in tissues are likely to reveal how LTBPs contribute to the activity of TGF-βs, and pathogenicity of the tumors.

ACKNOWLEDGMENTS

Our original research has been supported by grants from the Academy of Finland, the Finnish Cancer Foundation, the Sigrid Juselius Foundation, Biocentrum Helsinki, National Technology Agency (TEKES), Maud Kuistila Foundation, Foundation for the Finnish Cancer Institute, Novo Nordisk Foundation, Biomedicum Helsinki Foundation, Helsinki University Hospital Fund and the University of Helsinki. Sami Starast and Anne Remes are acknowledged for excellent technical assistance.

REFERENCES

1. Hyytiäinen M, Penttinen C, Keski-Oja J. Latent TGF-beta binding proteins: extracellular matrix association and roles in TGF-beta activation. Crit Rev Clin Lab Sci 2004;41:233–264.
2. Rifkin DB. Latent transforming growth factor-beta (TGF-beta) binding proteins: orchestrators of TGF-β availability. J Biol Chem 2005;280:7409–7412.
3. McMahon GA, Dignam JD, Gentry LE. Structural characterization of the latent complex between transforming growth factor beta 1 and beta 1-latency-associated peptide. Biochem J 1996;313:343–351.
4. Brown PD, Wakefield LM, Levinson AD, Sporn MB. Physicochemical activation of recombinant latent transforming growth factor-beta's 1, 2, and 3. Growth Factors 1990;3:35–43.
5. Schultz-Cherry S, Hinshaw VS. Influenza virus neuraminidase activates latent transforming growth factor-beta. J Virol 1996;70:8624–8629.
6. Miyazono K, Heldin C-H. Role for carbohydrate structures in TGF-beta1 latency. Nature 1989;338: 158–160.
7. Lyons RM, Keski-Oja J, Moses HL. Proteolytic activation of latent transforming growth factor-beta from fibroblast-conditioned medium. J Cell Biol 1988;106:1659–1665.

8. Gentry LE, Nash BW. The pro domain of pre-pro-transforming growth factor-beta1 when independently expressed is a functional binding protein for the mature growth factor. Biochemistry 1990;29: 6851–6857.
9. Yang Y, Dignam JD, Gentry LE. Role of carbohydrate structures in the binding of beta1-latency-associated peptide to ligands. Biochemistry 1997;36:11,923–11,932.
10. Bottinger EP, Factor VM, Tsang ML, et al. The recombinant proregion of transforming growth factor-beta1 (latency- associated peptide) inhibits active transforming growth factor-beta1 in transgenic mice. Proc Natl Acad Sci USA 1996;93:5877–5882.
11. Barcellos-Hoff MH, Derynck R, Tsang ML, Weatherbee JA. Transforming growth factor-beta activation in irradiated murine mammary gland. J Clin Invest 1994;93:892–899.
12. Barcellos-Hoff MH, Dix TA. Redox-mediated activation of latent transforming growth factor-beta1. Mol Endocrinol 1996;10:1077–1083.
13. Oursler MJ. Osteoclast synthesis and secretion and activation of latent transforming growth factor-beta. J Bone Miner Res 1994;9:443–452.
14. Grainger DJ, Wakefield L, Bethell HW, Farndale RW, Metcalfe JC. Release and activation of platelet latent TGF-beta in blood clots during dissolution with plasmin. Nat Med 1995;1:932–937.
15. Antonelli-Orlidge A, Saunders KB, Smith SR, D'Amore PA. An activated form of transforming growth factor beta is produced by cocultures of endothelial cells and pericytes. Proc Natl Acad Sci USA 1989;86:4544–4548.
16. Dennis PA, Rifkin DB. Cellular activation of latent transforming growth factor beta requires binding to the cation-independent mannose 6-phosphate/insulin-like growth factor type II receptor. Proc Natl Acad Sci USA 1991;88:580–584.
17. Nunes I, Gleizes PE, Metz CN, Rifkin DB. Latent transforming growth factor-beta binding protein domains involved in activation and transglutaminase-dependent cross-linking of latent transforming growth factor-beta. J Cell Biol 1997;136:1151–1163.
18. Taipale J, Miyazono K, Heldin CH, Keski-Oja J. Latent transforming growth factor-β1 associates to fibroblast extracellular matrix via latent TGF-beta binding protein. J Cell Biol 1994;124:171–181.
19. Flaumenhaft R, Abe M, Sato Y, et al. Role of the latent TGF-beta binding protein in the activation of latent TGF-beta by co-cultures of endothelial and smooth muscle cells. J Cell Biol 1993;120:995–1002.
20. Flaumenhaft R, Kojima S, Abe M, Rifkin DB. Activation of latent transforming growth factor-beta. Adv Pharmacol 1993;24:51–76.
21. Kojima S, Rifkin DB. Mechanism of retinoid-induced activation of latent transforming growth factor-beta in bovine endothelial cells. J Cell Physiol 1993;155:323–332.
22. Romer J, Bugge TH, Pyke C, et al. Impaired wound healing in mice with a disrupted plasminogen gene. Nat Med 1996;2:287–292.
23. Yu Q, Stamenkovic I. Cell surface-localized matrix metalloproteinase-9 proteolytically activates TGF-beta and promotes tumor invasion and angiogenesis. Genes Dev 2000;14:163–176.
24. Dangelo M, Sarment DP, Billings PC, Pacifici M. Activation of transforming growth factor beta in chondrocytes undergoing endochondral ossification. J Bone Miner Res 2001;16:2339–2347.
25. D'Angelo M, Billings PC, Pacifici M, Leboy PS, Kirsch T. Authentic matrix vesicles contain active metalloproteases (MMP). A role for matrix vesicle-associated MMP-13 in activation of transforming growth factor-β. J Biol Chem 2001;276:11,347–11,353.
26. Maeda S, Dean DD, Gay I, Schwartz Z, Boyan BD. Activation of latent transforming growth factor beta1 by stromelysin 1 in extracts of growth plate chondrocyte-derived matrix vesicles. J Bone Miner Res 2001;16:1281–1290.
27. Maeda S, Dean DD, Gomez R, Schwartz Z, Boyan BD. The first stage of transforming growth factor beta1 activation is release of the large latent complex from the extracellular matrix of growth plate chondrocytes by matrix vesicle stromelysin-1 (MMP-3). Calcif Tissue Int 2002;70:54–65.
28. Perrin BJ, Huttenlocher A. Calpain. Int J Biochem Cell Biol 2002;34:722–725.
29. Abe M, Oda N, Sato Y. Cell-associated activation of latent transforming growth factor-beta by calpain. J Cell Physiol 1998;174:186–193.
30. Lindstedt KA, Wang Y, Shiota N, et al. Activation of paracrine TGF-beta1 signaling upon stimulation and degranulation of rat serosal mast cells: a novel function for chymase. FASEB J 2001;15: 1377–1388.
31. Murphy-Ullrich JE, Poczatek M. Activation of latent TGF-beta by thrombospondin-1: mechanisms and physiology. Cytokine Growth Factor Rev 2000;11:59–69.

32. Bailly S, Brand C, Chambaz EM, Feige JJ. Analysis of small latent transforming growth factor-beta complex formation and dissociation by surface plasmon resonance. Absence of direct interaction with thrombospondins. J Biol Chem 1997;272:16,329–16,334.
33. Grainger DJ, Frow EK. Thrombospondin 1 does not activate transforming growth factor-beta1 in a chemically defined system or in smooth-muscle-cell cultures. Biochem J 2000;350 (Pt 1):291–298.
34. Ribeiro SM, Poczatek M, Schultz-Cherry S, Villain M, Murphy-Ullrich JE. The activation sequence of thrombospondin-1 interacts with the latency-associated peptide to regulate activation of latent transforming growth factor-beta. J Biol Chem 1999;274:13,586–13,593.
35. Schultz-Cherry S, Chen H, Mosher DF, et al. Regulation of transforming growth factor-beta activation by discrete sequences of thrombospondin 1. J Biol Chem 1995;270:7304–7310.
36. Agah A, Kyriakides TR, Lawler J, Bornstein P. The lack of thrombospondin-1 (TSP1) dictates the course of wound healing in double-TSP1/TSP2-null mice. Am J Pathol 2002;161:831–839.
37. Crawford SE, Stellmach V, Murphy-Ullrich JE, et al. Thrombospondin-1 is a major activator of TGF-beta1 in vivo. Cell 1998;93:1159–1170.
38. Daniel C, Wiede J, Krutzsch HC, et al. Thrombospondin-1 is a major activator of TGF-beta in fibrotic renal disease in the rat in vivo. Kidney Int 2004;65:459–468.
39. Yehualaeshet T, O'Connor R, Begleiter A, Murphy-Ullrich JE, Silverstein R, Khalil N. A CD36 synthetic peptide inhibits bleomycin-induced pulmonary inflammation and connective tissue synthesis in the rat. Am J Respir Cell Mol Biol 2000;23:204–212.
40. Derynck R, Jarrett JA, Chen EY, et al. Human transforming growth factor-beta complementary DNA sequence and expression in normal and transformed cells. Nature 1985;316:701–705.
41. Derynck R, Lindquist PB, Lee A, et al. A new type of transforming growth factor-beta, TGF-beta3. EMBO J 1988;7:3737–3743.
42. Sheppard D. Integrin-mediated activation of latent transforming growth factor-beta. Cancer Metastasis Rev 2005;24:395–402.
43. Munger JS, Huang X, Kawakatsu H, et al. The integrin alpha vbeta6 binds and activates latent TGF-beta1: a mechanism for regulating pulmonary inflammation and fibrosis. Cell 1999;96:319–328.
44. Annes JP, Rifkin DB, Munger JS. The integrin alphaVbeta6 binds and activates latent TGF-beta3. FEBS Lett 2002;511:65–68.
45. Mu D, Cambier S, Fjellbirkeland L, et al. The integrin alpha(v)beta8 mediates epithelial homeostasis through MT1-MMP-dependent activation of TGF-beta1. J Cell Biol 2002;157:493–507.
46. Asano Y, Ihn H, Yamane K, Jinnin M, Mimura Y, Tamaki K. Involvement of alphavbeta5 integrin-mediated activation of latent transforming growth factor beta1 in autocrine transforming growth factor-beta signaling in systemic sclerosis fibroblasts. Arthritis Rheum 2005;52:2897–2905.
47. Fontana L, Chen Y, Prijatelj P, et al. Fibronectin is required for integrin alphavbeta6-mediated activation of latent TGF-beta complexes containing LTBP-1. FASEB J 2005;19:1798–1808.
48. Kanzaki T, Olofsson A, Moren A, et al. TGF-beta1 binding protein: a component of the large latent complex of TGF-beta1 with multiple repeat sequences. Cell 1990;61:1051–1061.
49. Tsuji T, Okada F, Yamaguchi K, Nakamura T. Molecular cloning of the large subunit of transforming growth factor type beta masking protein and expression of the mRNA in various rat tissues. Proc Natl Acad Sci USA 1990;87:8835–8839.
50. Moren A, Olofsson A, Stenman G, et al. Identification and characterization of LTBP-2, a novel latent transforming growth factor-beta-binding protein. J Biol Chem 1994;269:32,469–32,478.
51. Gibson MA, Hatzinikolas G, Davis EC, Baker E, Sutherland GR, Mecham RP. Bovine latent transforming growth factor beta1-binding protein 2: molecular cloning, identification of tissue isoforms, and immunolocalization to elastin-associated microfibrils. 1995.
52. Yin W, Smiley E, Germiller J, et al. Isolation of a novel latent transforming growth factor-beta binding protein gene (LTBP-3). J Biol Chem 1995;270:10,147–10,160.
53. Giltay R, Kostka G, Timpl R. Sequence and expression of a novel member (LTBP-4) of the family of latent transforming growth factor beta-binding proteins. FEBS Lett 1997;411:164–168.
54. Saharinen J, Taipale J, Monni O, Keski-Oja J. Identification and characterization of a new latent transforming growth factor-beta-binding protein, LTBP-4. J Biol Chem 1998;273:18,459–18,469.
55. Penttinen C, Saharinen J, Weikkolainen K, Hyytiäinen M, Keski-Oja J. Secretion of human latent TGF-beta-binding protein-3 (LTBP-3) is dependent on co-expression of TGF-beta. J Cell Sci 2002;115:3457–3468.
56. Davis CG. The many faces of epidermal growth factor repeats. New Biol 1990;2:410–419.
57. Handford PA, Downing AK, Reinhardt DP, Sakai LY. Fibrillin: from domain structure to supramolecular assembly. Matrix Biol 2000;19:457–470.

58. Rosenbloom J, Abrams WR, Mecham R. Extracellular matrix 4: the elastic fiber. FASEB J 1993;7: 1208–1218.
59. Sakai LY, Keene DR. Fibrillin: monomers and microfibrils. Methods Enzymol 1994;245:29–52.
60. Lee B, Godfrey M, Vitale E, et al. Linkage of Marfan syndrome and a phenotypically related disorder to two different fibrillin genes. Nature 1991;352:330–334.
61. Ramirez F. Fibrillin mutations in Marfan syndrome and related phenotypes. Curr Opin Genet Dev 1996;6:309–315.
62. Ramirez F, Pereira L. The fibrillins. Int J Biochem Cell Biol 1999;31:255–259.
63. Saharinen J, Keski-Oja J. Specific sequence motif of 8-Cys repeats of TGF-beta binding proteins, LTBPs, creates a hydrophobic interaction surface for binding of small latent TGF-beta. Mol Biol Cell 2000;11:2691–2704.
64. Miyazono K, Hellman U, Wernstedt C, Heldin C-H. Latent high molecular weight complex of transforming growth factor beta1. Purification from human platelets and structural characterization. J Biol Chem 1988;263:6407–6415.
65. Miyazono K, Olofsson A, Colosetti P, Heldin C-H. A role of the latent TGF-beta1-binding protein in the assembly and secretion of TGF-beta1. EMBO J 1991;10:1091–1101.
66. Weiskirchen R, Moser M, Gunther K, Weiskirchen S, Gressner AM. The murine latent transforming growth factor-beta binding protein (Ltbp-1) is alternatively spliced, and maps to a region syntenic to human chromosome 2p21-22. Gene 2003;308:43–52.
67. Noguera I, Obata H, Gualandris A, Cowin P, Rifkin DB. Molecular cloning of the mouse Ltbp-1 gene reveals tissue specific expression of alternatively spliced forms. Gene 2003;308:31–41.
68. Olofsson A, Ichijo H, Moren A, ten Dijke P, Miyazono K, Heldin CH. Efficient association of an amino-terminally extended form of human latent transforming growth factor-beta binding protein with the extracellular matrix. J Biol Chem 1995;270:31,294–31,297.
69. Öklü R, Metcalfe JC, Hesketh TR, Kemp PR. Loss of a consensus heparin binding site by alternative splicing of latent transforming growth factor beta binding protein-1. FEBS Lett 1998;425: 281–285.
70. Michel K, Roth S, Trautwein C, Gong W, Flemming P, Gressner AM. Analysis of the expression pattern of the latent transforming growth factor beta binding protein isoforms in normal and diseased human liver reveals a new splice variant missing the proteinase-sensitive hinge region. Hepatology 1998;27: 1592–1599.
71. Krohn K. TGF-beta1-dependent differential expression of a rat homolog for latent TGF-beta binding protein in astrocytes and C6 glioma cells. Glia 1999;25:332–342.
72. Altmann CR, Chang C, Munoz-Sanjuan I, et al. The latent-TGFbeta-binding-protein-1 (LTBP-1) is expressed in the organizer and regulates nodal and activin signaling. Dev Biol 2002;248:118–127.
73. Quarto N. Molecular cloning and expression study of Xenopus latent TGF-beta binding protein-1 (LTBP-1). Gene 2002;290:53–61.
74. Weikkolainen K, Keski-Oja J, Koli K. Expression of latent TGF-beta binding protein LTBP-1 is hormonally regulated in normal and transformed human lung fibroblasts. Growth Factors 2003;21:51–60.
75. Kwok S, Qin L, Partridge NC, Selvamurugan N. Parathyroid hormone stimulation and PKA signaling of latent transforming growth factor-beta binding protein-1 (LTBP-1) mRNA expression in osteoblastic cells. J Cell Biochem 2005;95:1002–1011.
76. Westhoff JH, Sawitza I, Keski-Oja J, Gressner AM, Breitkopf K. PDGF-BB induces expression of LTBP-1 but not TGF-beta1 in a rat cirrhotic fat storing cell line. Growth Factors 2003;21:121–130.
77. Koski C, Saharinen J, Keski-Oja J. Independent promoters regulate the expression of two amino terminally distinct forms of latent transforming growth factor-beta binding protein-1 (LTBP-1) in a cell type-specific manner. J Biol Chem 1999;274:32,619–32,630.
78. Shipley JM, Mecham RP, Maus E, et al. Developmental expression of latent transforming growth factor-β binding protein 2 and its requirement early in mouse development. Mol Cell Biol 2000;20: 4879–4887.
79. Vehviläinen P, Hyytiäinen M, Keski-Oja J. Latent TGF-beta binding protein LTBP-2 mediates melanoma cell adhesion. J Biol Chem 2003;25:25.
80. Hyytiäinen M, Keski-Oja J. Latent TGF-beta binding protein LTBP-2 decreases fibroblast adhesion to fibronectin. J Cell Biol 2003;163:1363–1374.
81. Kantola AK, Keski-Oja J, Koli K. Induction of human LTBP-3 promoter activity by TGF-beta1 is mediated by Smad3/4 and AP-1 binding elements. Gene 2005;363:142–150.

82. Dabovic B, Chen Y, Colarossi C, et al. Bone abnormalities in latent TGF-[beta] binding protein (Ltbp)-3-null mice indicate a role for Ltbp-3 in modulating TGF-[beta] bioavailability. J Cell Biol 2002;156: 227–232.
83. Koli K, Saharinen J, Karkkainen M, Keski-Oja J. Novel non-TGF-beta-binding splice variant of LTBP-4 in human cells and tissues provides means to decrease TGF-beta deposition. J Cell Sci 2001;114: 2869–2878.
84. Sterner-Kock A, Thorey IS, Koli K, et al. Disruption of the gene encoding the latent transforming growth factor-beta binding protein 4 (LTBP-4) causes abnormal lung development, cardiomyopathy, and colorectal cancer. Genes Dev 2002;16:2264–2273.
85. Koli K, Wempe F, Sterner-Kock A, et al. Disruption of LTBP-4 function reduces TGF-beta activation and enhances BMP-4 signaling in the lung. J Cell Biol 2004;167:123–133.
86. Saharinen J, Hyytiäinen M, Taipale J, Keski-Oja J. Latent transforming growth factor-beta binding proteins (LTBPs)—structural extracellular matrix proteins for targeting TGF-beta action. Cytokine Growth Factor Rev 1999;10:99–117.
87. Taipale J, Saharinen J, Hedman K, Keski-Oja J. Latent transforming growth factor-beta1 and its binding protein are components of extracellular matrix microfibrils. J Histochem Cytochem 1996;44: 875–889.
88. Dallas SL, Sivakumar P, Jones CJ, et al. Fibronectin regulates latent transforming growth factor-beta (TGF beta) by controlling matrix assembly of latent TGF beta-binding protein-1. J Biol Chem 2005;280: 18,871–18,880.
89. Dallas SL, Keene DR, Bruder SP, et al. Role of the latent transforming growth factor beta binding protein 1 in fibrillin-containing microfibrils in bone cells in vitro and in vivo. J Bone Miner Res 2000; 15:68–81.
90. Sinha S, Heagerty AM, Shuttleworth CA, Kielty CM. Expression of latent TGF-beta binding proteins and association with TGF-beta1 and fibrillin-1 following arterial injury. Cardiovasc Res 2002;53: 971–983.
91. Isogai Z, Ono RN, Ushiro S, et al. Latent transforming growth factor beta-binding protein 1 interacts with fibrillin and is a microfibril-associated protein. J Biol Chem 2003;278:2750–2757.
92. Unsöld C, Hyytiainen M, Bruckner-Tuderman L, Keski-Oja J. Latent TGF-beta binding protein LTBP-1 contains three potential extracellular matrix interacting domains. J Cell Sci 2001;114: 187–197.
93. Koli K, Hyytiäinen M, Ryynänen MJ, Keski-Oja J. Sequential deposition of latent TGF-beta binding proteins (LTBPs) during formation of the extracellular matrix in human lung fibroblasts. Exp Cell Res 2005;310:370–382.
94. Hyytiäinen M, Taipale J, Heldin C-H, Keski-Oja J. Recombinant latent transforming growth factor beta-binding protein 2 assembles to fibroblast extracellular matrix and is susceptible to proteolytic processing and release. J Biol Chem 1998;273:20,669–20,676.
95. Chen Y, Dabovic B, Annes JP, Rifkin DB. Latent TGF-beta binding protein-3 (LTBP-3) requires binding to TGF-beta for secretion. FEBS Lett 2002;517:277–280.
96. Saharinen J, Taipale J, Keski-Oja J. Association of the small latent transforming growth factor-beta with an eight cysteine repeat of its binding protein LTBP-1. EMBO J 1996;15:245–253.
97. Taipale J, Lohi J, Saarinen J, Kovanen PT, Keski-Oja J. Human mast cell chymase and leukocyte elastase release latent transforming growth factor-beta1 from the extracellular matrix of cultured human epithelial and endothelial cells. J Biol Chem 1995;270:4689–4696.
98. Taipale J, Keski-Oja J. Growth factors in the extracellular matrix. FASEB J 1997;11:51–59.
99. Mazzieri R, Jurukovski V, Obata H, et al. Expression of truncated latent TGF-beta-binding protein modulates TGF-beta signaling. J Cell Sci 2005;118:2177–2187.
100. Annes JP, Chen Y, Munger JS, Rifkin DB. Integrin alphaVbeta6-mediated activation of latent TGF-beta requires the latent TGF-beta binding protein-1. J Cell Biol 2004;165:723–734.
101. Gomez-Duran A, Mulero-Navarro S, Chang X, Fernandez-Salguero PM. LTBP-1 blockade in dioxin receptor-null mouse embryo fibroblasts decreases TGF-beta activity: Role of extracellular proteases plasmin and elastase. J Cell Biochem 2006;97:380–392.
102. Gregory KE, Ono RN, Charbonneau NL, et al. The prodomain of BMP-7 targets the BMP-7 complex to the extracellular matrix. J Biol Chem 2005;280:27,970–27,980.
103. Gualandris A, Annes JP, Arese M, Noguera I, Jurukovski V, Rifkin DB. The latent transforming growth factor-beta-binding protein-1 promotes in vitro differentiation of embryonic stem cells into endothelium. Mol Biol Cell 2000;11:4295–4308.

104. Dabovic B, Chen Y, Colarossi C, Zambuto L, Obata H, Rifkin DB. Bone defects in latent TGF-beta binding protein (Ltbp)-3 null mice; a role for Ltbp in TGF-beta presentation. J Endocrinol 2002;175: 129–141.
105. Dabovic B, Levasseur R, Zambuto L, Chen Y, Karsenty G, Rifkin DB. Osteopetrosis-like phenotype in latent TGF-beta binding protein 3 deficient mice. Bone 2005;37:25–31.
106. Chen Y, Dabovic B, Colarossi C, et al. Growth retardation as well as spleen and thymus involution in latent TGF-beta binding protein (Ltbp)-3 null mice. J Cell Physiol 2003;196:319–325.
107. Colarossi C, Chen Y, Obata H, et al. Lung alveolar septation defects in Ltbp-3-null mice. Am J Pathol 2005;167:419–428.
108. Filvaroff E, Erlebacher A, Ye J, et al. Inhibition of TGF-beta receptor signaling in osteoblasts leads to decreased bone remodeling and increased trabecular bone mass. Development 1999;126: 4267–4279.
109. Janssens K, Gershoni-Baruch R, Guanabens N, et al. Mutations in the gene encoding the latency-associated peptide of TGF-beta1 cause Camurati-Engelmann disease. Nat Genet 2000;26:273–275.
110. Erlebacher A, Derynck R. Increased expression of TGF-beta2 in osteoblasts results in an osteoporosis-like phenotype. J Cell Biol 1996;132:195–210.
111. Waltenberger J, Wanders A, Fellstrom B, Miyazono K, Heldin CH, Funa K. Induction of transforming growth factor-beta during cardiac allograft rejection. J Immunol 1993;151:1147–1157.
112. Maeda J, Ueki N, Ohkawa T, et al. Local production and localization of transforming growth factor-beta in tuberculous pleurisy. Clin Exp Immunol 1993;92:32–38.
113. Karonen T, Jeskanen L, Keski-Oja J. Transforming growth factor beta1 and its latent form binding protein-1 associate with elastic fibres in human dermis: accumulation in actinic damage and absence in anetoderma. Br J Dermatol 1997;137:51–58.
114. Border WA, Noble NA. Transforming growth factor-beta in tissue fibrosis. N Engl J Med 1994;331: 1286–1292.
115. Schlotzer-Schrehardt U, Zenkel M, Kuchle M, Sakai LY, Naumann GO. Role of transforming growth factor-beta1 and its latent form binding protein in pseudoexfoliation syndrome. Exp Eye Res 2001;73: 765–780.
116. Gong W, Roth S, Michel K, Gressner AM. Isoforms and splice variant of transforming growth factor beta binding protein in rat hepatic stellate cells. Gastroenterology 1998;114:352–363.
117. Mangasser-Stephan K, Gartung C, Lahme B, Gressner AM. Expression of isoforms and splice variants of the latent transforming growth factor beta binding protein (LTBP) in cultured human liver myofibroblasts. Liver 2001;21:105–113.
118. Kinnman N, Andersson U, Hultcrantz R. In situ expression of transforming growth factor-beta1-3, latent transforming growth factor-beta-binding protein and tumor necrosis factor-alpha in liver tissue from patients with chronic hepatitis C. Scand J Gastroenterol 2000;35:1294–1300.
119. Tamaki K, Okuda S, Miyazono K, Nakayama M, Fujishima M. Matrix-associated latent TGF-beta with latent TGF-beta binding protein in the progressive process in adriamycin-induced nephropathy. Lab Invest 1995;73:81–89.
120. van Laethem JL, Deviere J, Resibois A, et al. Localization of transforming growth factor beta1 and its latent binding protein in human chronic pancreatitis. Gastroenterology 1995;108:1873–1881.
121. Santiago-Josefat B, Mulero-Navarro S, Dallas SL, Fernandez-Salguero PM. Overexpression of latent transforming growth factor-beta-binding protein 1 (LTBP-1) in dioxin receptor-null mouse embryo fibroblasts. J Cell Sci 2004;117:849–859.
122. Corchero J, Martin-Partido G, Dallas SL, Fernandez-Salguero PM. Liver portal fibrosis in dioxin receptor-null mice that overexpress the latent transforming growth factor-beta-binding protein-1. Int J Exp Pathol 2004;85:295–302.
123. Neptune ER, Frischmeyer PA, Arking DE, et al. Dysregulation of TGF-beta activation contributes to pathogenesis in Marfan syndrome. Nat Genet 2003;33:407–411.
124. Kaartinen V, Warburton D. Fibrillin controls TGF-beta activation. Nat Genet 2003;33:331–332.
125. Morris DG, Huang X, Kaminski N, et al. Loss of integrin alpha(v)beta6-mediated TGF-beta activation causes Mmp12-dependent emphysema. Nature 2003;422:169–173.
126. Pasche B. Role of transforming growth factor beta in cancer. J Cell Physiol 2001;186:153–168.
127. Van Laethem JL, Resibois A, Rickaert F, et al. Different expression of transforming growth factor-beta1 in pancreatic ductal adenocarcinoma and cystic neoplasms. Pancreas 1997;15:41–47.
128. Van Laethem JL, Resibois A, Adler M, Peny MO, Ohtani H, Robberecht P. Localization of transforming growth factor-beta 1 precursor and latent TGF-beta1 binding protein in colorectal adenomas. Dis Sci 1996; 41:1741–1748.

129. Roth-Eichhorn S, Heitmann B, Flemming P, Kubicka S, Trautwein C. Evidence for the decreased expression of the latent TGF-beta binding protein and its splice form in human liver tumours. Scand J Gastroenterol 2001;36:1204–1210.
130. Henriksen R, Gobl A, Wilander E, Oberg K, Miyazono K, Funa K. Expression and prognostic significance of TGF-beta isotypes, latent TGF-beta1 binding protein, TGF-beta type I and type II receptors, and endoglin in normal ovary and ovarian neoplasms. Lab Invest 1995;73:213–220.
131. Chaudhry A, Oberg K, Gobl A, Heldin CH, Funa K. Expression of transforming growth factors beta1, beta2, beta3 in neuroendocrine tumors of the digestive system. Anticancer Res 1994;14: 2085–2091.
132. Eklöv S, Funa K, Nordgren H, et al. Lack of the latent transforming growth factor-beta binding protein in malignant, but not benign prostatic tissue. Cancer Res 1993;53:3193–3197.
133. Mizoi T, Ohtani H, Matsuno S, Nagura H. Immunohistochemical identification of transforming growth factor-beta and its binding protein in human gastrointestinal carcinoma. Tohoku J Exp Med 1992;168:271–273.

5 Regulators of Smad2/3 Transcription and Phosphorylation

Azeddine Atfi and Marie-Francoise Bourgeade

CONTENTS

INTRODUCTION
REGULATION OF RECEPTOR-DEPENDENT SMAD PHOSPHORYLATION
TRANSLOCATION OF SMADS TO THE NUCLEUS
BASIS OF TRANSCRIPTIONAL ACTIVATION BY SMAD PROTEINS
COACTIVATORS OF SMAD PROTEINS
SMAD COREPRESSORS
CONCLUDING REMARKS
REFERENCES

Abstract

Transforming growth factor beta (TGF-β) is a multifunctional cytokine that controls a wide variety of cellular processes, such as cell migration, adhesion, differentiation, proliferation, and cell death. Disruption of components of the TGF-β signaling pathway is associated with human diseases, including cancer. TGF-β initiates signaling from the cell surface by contacting two distantly related transmembrane serine/threonine kinases called receptors I (TβRI) and II (TβRII). Upon ligand binding, the constitutively active TβRII phosphorylates and activates TβRI, which in turn phosphorylates Smad2 or Smad3 on two serines at the carboxyl terminus, within a highly conserved SSXS motif. Following phosphorylation, Smad2 and Smad3 associate with the shared partner Smad4 and translocate to the nucleus where Smad complexes, in cooperation with coactivators and corepressors, participate in transcriptional regulation of TGF-β-responsive genes. Smad phosphorylation, cellular distribution, and activation are tightly regulated via crosstalk with other signaling pathways or through functional interactions with Smad partners and modulators that determine the specific target genes. The present knowledge of the mechanisms controlling phosphorylation and activation of Smad proteins is reviewed here.

Key Words: Smad phosphorylation; nuclear translocation of Smads; Smad transcriptional activity; Smads coactivators; Smad corepressors.

1. INTRODUCTION

The switch from normal to tumor cells that results in cancer can arise from a variety of alterations in normal cell function. In many cases, tumor cells develop when normal progenitor cells lose control of signaling pathways that regulate responses to negative growth-regulatory factors. Because many cancers of epithelial and lymphoid origin develop resistance to the neg-

From: *Cancer Drug Discovery and Development: Transforming Growth Factor-β in Cancer Therapy, Vol I: Basic and Clinical Biology*
Edited by: Sonia B. Jakowlew © Humana Press Inc., Totowa, NJ

ative growth-regulatory effects of TGF-β, it has been postulated that one of the mechanisms whereby cells undergo neoplastic transformation and escape from normal growth control involves an altered response to TGF-β. TGF-β initiates responses by contacting two types of transmembrane serine/threonine kinases called receptors type I and type II, promoting activation of the type I by the type II kinase. The activated type I receptor then propagates the signal to the nucleus by phosphorylating the specific intracellular mediators Smad2 and Smad3. Once phosphorylated, Smad2 and Smad3 associate with the shared partner Smad4 and the complexes accumulate in the nucleus where they regulate the expression of TGF-β target genes through cooperative interactions with transcriptional partners. The disruption of TGF-β signaling, either via mutational inactivation of components of the signaling pathway, or by modulation of their expression or function, is now known to play an important role in tumor progression. In this chapter, we will focus on current research efforts investigating the molecular mechanisms of Smad activation and regulation that could contribute to our understanding on the role of TGF-β signaling in malignant transformation.

2. REGULATION OF RECEPTOR-DEPENDENT SMAD PHOSPHORYLATION

Genetic and biochemical studies have described a family of intracellular effectors known as the Smad proteins, acting downstream of TGF-β receptors *(1–3)*. Three different classes of Smads have been defined: the receptor-regulated Smad (R-Smads), the common mediator Smads (Co-Smads), and the antagonistic or inhibitory Smads (I-Smads). The R-Smads interact with the activated type I receptors and become activated through the phosphorylation on a C-terminal SXS motif *(4–7)*. Smad2 and Smad3 are substrates for TGF-β and activin receptors, whereas Smad1, 5 and 8 are substrates for the type I receptor for bone morphogenic protein (BMP), another member of the TGF-β superfamily *(2)*. Phosphorylated R-Smads then form a complex with the common partner Smad4, and translocate into the nucleus where they regulate the expression of TGF-β-responsive genes. In contrast to R-Smads and Smad4, the antagonistic Smads, Smad6 and Smad7, appear to block signal transduction by functioning as interfering decoys in receptor-R-Smads or R-Smads/Smad4 interactions *(2,3,8)*.

The R-Smad and Co-Smad proteins contain two conserved polypeptide segments, the N-terminal MH1 (MAD-homology 1) domain and the C-terminal MH2 domain linked by a less conserved linker region *(3,9)*. The MH1 domains of Smad3 and Smad4 exhibit sequence-specific DNA binding activity and associate with DNA binding proteins. In contrast to Smad3 and Smad4, Smad2 lacks DNA-binding activity because of an insert of 30 amino acids located in its MH1 domain, but a splicing variant of Smad2 with a deletion of this 30 amino-acid insertion has similar DNA-binding properties as Smad3 *(10)*. In addition to mediating the binding of Smads to DNA, the MH1 domain can negatively regulate the function of the MH2 domain through an autoinhibition mechanism, and receptor-mediated phosphorylation appears to relieve these two domains from their mutually inhibitory interaction *(11)*. Interestingly, removal of the MH1 domains of R-Smads yields truncated proteins that constitutively activate transcription *(12,13)*. The N-terminal domains of I-Smads display weak sequence homology to the MH1 domains of R-Smads and lack DNA binding activity.

The MH2 domain is highly conserved among all Smad proteins and is responsible for a large number of distinct protein-protein interactions that include: R-Smads–receptor interaction, binding of Smad2 to the anchor protein SARA (Smad anchor for receptor activation), formation of R-Smads-Smad4 complexes, and association of Smads with a large variety of DNA binding factors and transcriptional coactivators or corepressors *(3,8,9)*.

In the case of R-Smad/receptor interactions, functional and structural analyses of R-Smads have identified the α-helix 1 (H1) and three residues within the L3 loop of the MH2 domain as determinants that control specificity of the interaction with the receptor *(14,15)*. On the receptor side, the nine-amino acid L45 loop, which connects β-strands 4 and 5 in the type I receptor kinase domain, specifies interactions with Smads *(16,17)*. Indeed, mutation of these residues in R-Smads or receptor is sufficient to abolish the interaction and activation of Smad signaling pathway by the ligand-activated receptor *(17)*.

The identification of various proteins that participate in the interaction of R-Smads with the receptors has suggested that the subcellular localization of these signaling mediators plays an important role in the pathway. For instance, R-Smad binding to the receptors is facilitated by SARA, a protein that contains a FYVE (Fab1/YOTP/Vac1p/EEA1) domain and is predominantly concentrated in PtdIn3P-enriched early endosomes *(18,19)*. The interaction of Smad2 and SARA occurs between the MH2 domain of Smad2 and the Smad-binding domain (SBD) of SARA that is located between residues 665 and 750 *(18)*. Structure based investigations have revealed that a coiled region, an α-helix, and a β-strand on the SARA SBD are important determinants that mediate the interaction of SARA with the MH2 domain of unphosphorylated Smad2 *(20)*. At steady state, SARA-bound Smad2 is localized in early endosomes to which the activated receptor complex is internalized via clathrin-coated pits *(19)*. Therefore, clathrin-mediated endocytosis seems to play a crucial role in TGF-β signaling by controlling the colocalization of receptors with the SARA/Smad2 complexes in early endosomes. The importance of clathrin-mediated endocytic pathway in TGF-β signaling is also manifested by the recent finding that cPML, a cytoplasmic form of promyelocytic leukemia protein, is required for the initiation of TGF-β signaling by promoting Smad2 binding to SARA and facilitating the recruitment of SARA/Smad2 and TGF-β receptors to the early endosomes *(21)*. Furthermore, other adaptor proteins, including Disabled-2 (DAB2) and Hgs, have been shown to participate in TGF-β signaling probably by controlling the accumulation of Smad2/Smad3 and receptors in the early endosomes. In fact, DAB2, which is enriched in clathrin-coated vesicles, binds both the receptor and Smad2 or Smad3 and promotes receptor-mediated phosphorylation of these R-Smads and TGF-β signaling *(22)*. Hgs contains a FYVE-domain, suggesting that it plays a role similar to that of SARA by bringing R-Smads to the early endosomes where the activated receptors are enriched *(23)*.

In addition to clathrin, TGF-β receptors can also associate with caveolin, a protein present in plasma membrane invaginations called caveolae, which leads to their internalization into caveolin1-positive vesicles with subsequent degradation through the proteasome pathway *(19)*. Consistent with this notion, the caveolin1-positive vesicles were found to associate with the inhibitory Smad, Smad7, which is known to mediate the association of the ubiquitin ligases Smurf1 and Smurf2 to receptors, leading to their degradation *(19)*. In the basal state, Smad7 and Smurf1/2 are localized to the nucleus and activation of TGF-β signaling induces the mobilization of Smad7/Smurf complexes from the nucleus to the plasma membrane, where Smad7 directly binds the activated type I TGF-β receptor and mediates its association with Smurf1/2 *(24,25)*. In this complex, Smad7 itself undergoes ubiquitination and degradation through the proteasome pathway. More recently, other ubiquitin ligases, including Tiul1 and AIP4, have been reported to cooperate with Smad7 in the negative regulation of TGF-β signaling *(26,27)*. Like, Smurf1/2, AIP4 and Tiul1 contain the distinctive structural features of the HECT (homologous to E6-AP carboxyl terminus) subclass of E3 ubiquitin ligases. These features include a phospholipid/calcium-binding C2 domain, WW domains, which mediate protein-protein interactions by binding to PPXY (PY) motifs on partner proteins, and a carboxyl-terminal HECT catalytic domain *(28)*. By interacting with the PY motif of Smad7, Smurf1, Smurf2, and Tiul1 can be recruited to TGF-β receptors to induce their degradation. Structure-function analyses have indicated that the ubiquitin ligase activity of Smurf1, Smurf2, and Tiul1 is essential for ubiquitin-dependent degradation of TGF-β receptors/Smad7 complexes *(24–26)*. In marked contrast to Smurf1, Smurf2, and Tiul1, expression of AIP4 has no effect on the turnover of the activated TGF-β type I receptor but enhances the association of Smad7 with the activated type I receptor independent of its ubiquitin ligase activity *(27)*. This function of AIP4 seems to play an important role in the negative regulation of TGF-β signaling since Smad7 can also act in another mechanistic manner by competitively interfering with Smad2 and Smad3 from binding to and being phosphorylated by the activated type I receptor *(29)*.

The expression of Smad7 is increased in response to TGF-β in several cell types, which provides a negative feedback loop to control TGF-β activity. The expression of Smad7 can also be induced by a variety of antagonistic signaling pathways, which negatively TGF-β signaling. For example, the proinflammatory cytokines IFN-γ (interferon-γ) and TNF-α (tumor necrosis factor-α), signaling via STAT1 (signal transducer and activator of transcription) and NF-κB respectively, activate Smad7 expression, leading to inhibition of R-Smads phosphorylation and the associated downregulation of TGF-β signaling *(2)*.

3. TRANSLOCATION OF SMADS TO THE NUCLEUS

In unstimulated cells, R-Smads are predominantly localized in the cytoplasm, whereas Smad4 is distributed in both the cytoplasm and the nucleus. Nuclear translocation of R-Smads does not require Smad4, although Smad4 translocates with R-Smads *(3,8)*. The retention of R-Smads in the cytoplasm is primilary mediated by interaction with SARA, which binds through its SBD to a region in the MH2 domain of Smad2, referred to as the "hydrophobic corridor", and blocks R-Smads nuclear import *(30)*. Receptor-mediated phosphorylation decreases the affinity of R-Smads for SARA, releasing Smad2 and exposing its nuclear import function. The Smad nuclear import function inhibited by SARA occurs in a manner independent of the nuclear localization signal (NLS) import pathway, which mediates the import of proteins containing NLS, a lysine- and arginin-rich sequence. Several studies have indicated that the Smad MH1 domain contains a conserved lysine-rich domain that contributes to the nuclear import of Smad3 and Smad4 through binding to importin-α and importin-β1, respectively *(31–34)*. The nuclear import function of Smad2 does not involve importins, suggesting alternative mechanisms for its nuclear import and export. Indeed, the MH2 domain of Smad2 binds directly to components of the nuclear pore complex, the nucleoporins CAN/Nup1 and Nup153 *(35)*. Structural studies have indicated that the hydrophobic corridor in the MH2 domain of Smad2 overlaps the CAN/Nup1 and Nup153 binding site. C-terminal phosphorylation of the MH2 domain and consequent conformational changes decreases its binding for SARA, freeing the hydrophobic corridor of Smad2 for association with the nucleoporins CAN/Nup1 and Nup153 *(20)*. As the hydrophobic corridor in the MH2 domain of Smad2 also serves as a binding site for several transcription partners in the nucleus *(36)*, this model predicts that SARA and nuclear Smad partners may compete with CAN/Nup1 and Nup153, allowing constant shuttling of Smad2 between the cytoplasm and the nucleus. Consistent with these observations, dephosphorylation of R-Smads in the nucleus results in the dissociation of Smad complexes and the export of inactive Smads to the cytoplasm *(37)*. Smad4, which accumulates in the nucleus by association with activated R-Smads, also undergoes continuous nucleocytoplasmic shuttling owing to the combined activities of a constitutively active NLS in the MH1 domain and a nuclear export signal (NES) in the linker region *(38)*. Inhibition of the nuclear transport receptor CRM1 (required for chromosome region maintenance) by the drug leptomycin B or mutations in the NES of Smad4 abolish the nuclear export of Smad4, which lends support to the idea that NES may function via a CRM1-dependent pathway *(38)*. Upon activation of TGF-β signaling, the NES of Smad4 may be masked by the formation of R-Smad-Smad4 complexes, facilitating the nuclear retention of Smad4.

The molecular mechanisms governing the nuclear accumulation of activated Smads are tightly regulated and subjected to many levels of positive and negative regulations. For example, mitogenic growth factors, such as EGF (epidermal growth factor) and HGF (hepatocyte growth factor) that operate through activation of the Ras signaling pathway can directly interfere with Smad-dependent responses by attenuating ligand-induced accumulation of R-Smads in the nucleus. These effects are mediated by the member of the MAPK (mitogen activated protein kinase) family, ERK (extracellular-regulated kinase), which phosphorylates three SP sites in the linker regions of Smad2 and Smad3 *(39)*. Mutating these residues restored Smad nuclear accumulation and Smad-dependent responses. Currently, the mechanism by which the ERK-dependent phosphorylation induces nuclear exclusion of R-Smad is not known, but it could function to modulate the balance of Smad shuttling by affecting the Smad interaction with functional partners known to be involved in the subcellular distribution of Smads, such as SARA and nuclear transcriptional partners. Similar to ERK, activation of CaMKII (calcium-calmodulin-dependent kinase II), which has been implicated in different cellular processes, restricts TGF-β signaling by inhibiting the Smad nuclear translocation. This inhibitory effect of CaMKII occurs as a result of Smad2 phosphorylation by CaMKII in the linker region at several residues that are different from the ERK sites *(40,41)*. In addition to phosphorylation by ERK and CaMKII, R-Smads can also be phosphorylated by another member of the MAPK family, JNK (c-Jun N-terminal kinase), although the phosphorylation sites involved in this process remain unknown. In contrast to ERK and CaMKII, phosphorylation of R-Smads by JNK seems to enhance their nuclear translocation *(42)*. Finally, the nuclear accumulation of Smad3 can be limited by the protein kinase AKT in a Smad phosphorylation-independent manner *(43,44)*. The mechanism that accounts for the sequestration of Smad3 by AKT appears to involve a direct association of AKT with unphosphorylated Smad3. Receptor-mediated phosphorylation of Smad3 decreases the partnering of AKT with Smad3, suggesting that the balance of the ratio between AKT and Smad3 may play a role in the AKT-dependent suppression of TGF-β signaling. All together, these observations suggest that regulation of Smad nuclear translocation by MAPK and other kinases might allow for modulation of the magnitude of TGF-β signals depending on cell type or in response to other signal inputs.

4. BASIS OF TRANSCRIPTIONAL ACTIVATION BY SMAD PROTEINS

Once in the nucleus, Smads function to target specific gene promoters through their ability to associate directly with DNA or through cooperative interactions with DNA-binding cofactors. The ability of Smads to bind directly with DNA was first demonstrated by the identification of CAGA boxes within the promoter of the *PAI-1* (plasminogen activator inhibitor 1) gene *(45,46)*. In fact, Smad3 MH1 domain interacts through a hairpin motif with the DNA sequence CAGA (also called Smad Binding Element or SBE) and its reverse complement, GTCT, and this sequence has been identified within the promoter regions of other TGF-β-regulated genes, including *JunB*, *c-Jun*, or *Smad7 (8,9)*. In many cases, mutation in the Smad binding site is sufficient to reduce TGF-β-mediated transcriptional activation, which supports the idea that Smad DNA binding is important for the activation of TGF-β target genes. However, Smad-DNA binding alone does not contribute to transcriptional activation of all TGF-β-responsive genes since Smad2 lacks DNA-binding activity because of an insert of 30 amino acids located in the hairpin of its MH1 domain.

Because the SBE sequence is frequently present throughout the genome and because the affinity of DNA binding by Smad proteins is very weak, it is likely that additional DNA contacts as well as cooperation with other DNA binding proteins are necessary to elicit specific, high-affinity binding of a Smad complex to TGF-β-responsive genes. There are now a growing number of examples in which Smad proteins achieve transcriptional activation of target genes by binding to specific promoter elements through physical interaction with other DNA-binding cofactors. These interactions, which stabilize the higher order DNA binding complex, occur through either the Smads' MH1 or MH2 domains, depending on the DNA-binding cofactor. For example, several TGF-β-responsive promoters contain AP-1 sites, which are activated by heterodimers of c-Jun and c-Fos. AP-1 and Smad3/Smad4 complexes bind to DNA on separate sites located close from each other and this allows Smad3 to directly bind c-Jun via its MH1 domain plus linker domain and indirectly with c-Fos through its MH2 domain. Overexpression of Smad3 can potentiate transcription from AP-1-containing reporter genes, which may explain why c-Jun, when interacting with Smad3, binds the AP-1 site with higher affinity that in the absence of Smad3 *(47)*. In addition to AP-1 complexes, Smad3 can also form higher order DNA binding complex by binding to the basic helix-loop-helix (bHLH) leucine zipper protein ATF2 that targets cAMP response elements (CRE) either as a homodimer or as heterodimer with c-Jun. Smad3 or Smad4 binds ATF2 through their MH1 domain and cooperate with ATF2 to activate CRE-containing elements *(48,49)*. Other examples of functional cooperativity between Smad and DNA-binding partners include TFE3 and some members of the forkhead family of transcription factors. TFE3 is a bHLH transcription factor that binds to the DNA sequence CACGTG (known as the E-box). TEF3 binding to this site allows recruitment of Smad3/Smad4 complexes to SBE that lies 3 base pairs downstream from the TFE3 binding site in the *PAI-1* promoter, and any alteration of the spacing between SBE and E-Box abrogates TGF-β-induced transcription *(50)*. In the case of members of the forkhead family of transcription factors, FoxO1, FoxO3a, and FoxO4 bind specifically and directly to Smad3 and Smad4 and transactivate the promoter of *p21Cip1* (a CDK inhibitor) by targeting a region containing a consensus forkhead-binding element (FHBE) immediately upstream of SBE *(51)*. Interestingly, loss of function of FoxO3 after its inactivation by Bcr-Abl through AKT is linked directly to the loss of TGF-β-mediated cytostatic effects *(52)*. Taken together, these observations indicate that Smads must cooperate among each other and with other DNA binding proteins to elicit specific transcriptional responses. There is now a growing body of evidence that these mechanisms of cooperativity take place between Smad and many different families of DNA binding factors some of which may be ubiquitous and others may function in a cell-type or a cell-context specific manner *(2,3,8)*.

Another group of proteins facilitating Smad binding to target promoters seems to function purely as DNA-binding adaptors. The first and most extensively characterized member of this group is the Xenopus xFAST1/FoxH1 (forkhead activin signal transducer 1/forkhead box protein H1), a DNA-binding protein that has no intrinsic DNA transactivity *(53)*. xFAST1/FoxH1 is a winged-helix/forkhead DNA binding factor that binds to an activin responsive element (ARE) in the promoter region of the *mix2* gene. The interaction of xFast/FoxH1 with ligand-activated Smad2 is mediated through the Smad interaction domain (SID) of xFast/FoxH1 and a proturing element (α-helix 2) on the MH2 domain of Smad2 *(53)*. Analysis of DNA/protein complexes showed that xFast/FoxH1 constitutively binds ARE elements and that activated Smad2 interacts with xFast/FoxH1 to recruit Smad4 into the nuclear complex. In this complex, the Smad4 MH1 domain binds DNA on a Smad binding site adjacent to ARE, and this contact stabilizes the DNA binding by the Smad2/Fast1 complexes *(54)*. Similar mechanisms have been described for the human and mouse FASTs (FAST1 and FAST2) that mediate activation of the homeobox gene *goosecoid* following activin and TGF-β stimulation *(54)*. Extensive analysis of the *goosecoid* promoter revealed that Smad2 activates transcription in cooperation with Smad4 and Fast1, whereas Smad3 blocks Smad2-dependent activation probably by competing with Smad4 for binding to the Smad

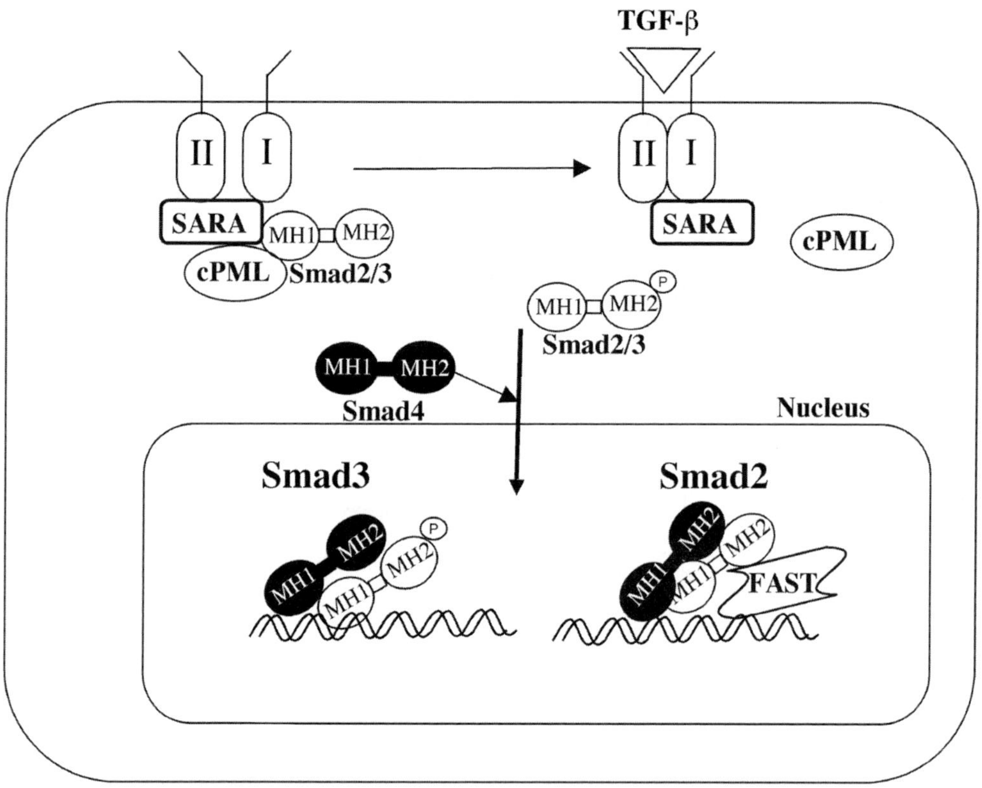

Fig. 1. General mechanism of Smad phosphorylation and activation.

site that lies adjacent to ARE. This scenario seems to be specific for the *goosecoid* gene because the ARE of the Xenopus *mix2* gene appears to be activated by either Smad2 or Smad3 *(54)* (Fig. 1).

5. COACTIVATORS OF SMAD PROTEINS

When fused to the GAL4-DNA binding domain, the MH2 domains of Smads can drive transcriptional activation of the heterologous promoter containing GAL4-binding sites, raising the interesting possibility that Smads may directly associate with the basal transcriptional machinery *(55)*. In fact, it is now established that Smads modulate transcription in response to ligand by interacting with the general transcription coactivators CBP (CREB binding protein) and p300 *(3,8,9)* (Fig. 2). Mapping the domains mediating these interactions has revealed that Smad2 and Smad3 form a complex with CBP through a physical contact between the MH2 domain of R-Smads and a region that corresponds to residues 1981 to 2175 in CBP, and efficient interaction requires receptor-mediated phosphorylation of the SXS motif in R-Smads. Smad4 also associates with CBP/p300 but this interaction is most likely indirect, since only Smad2 and Smad3 exhibit efficient binding in yeast two-hybrid assays. This indirect interaction of CBP with Smad4 is likely of physiological significance, as it stabilizes the interaction of R-Smads with CBP/p300. In addition to CBP/p300, R-Smads binds two other proteins pCAF and Gcn5 that play an important role in the TGF-β signaling pathway by a mechanism resembling the one engaged by CBP/p300 *(56,57)*. CBP and p300, which possess acetyl transferase activity, act as coactivators by bringing Smad2 or Smad3 within the proximity of the general transcription machinery and by modifying the chromatin

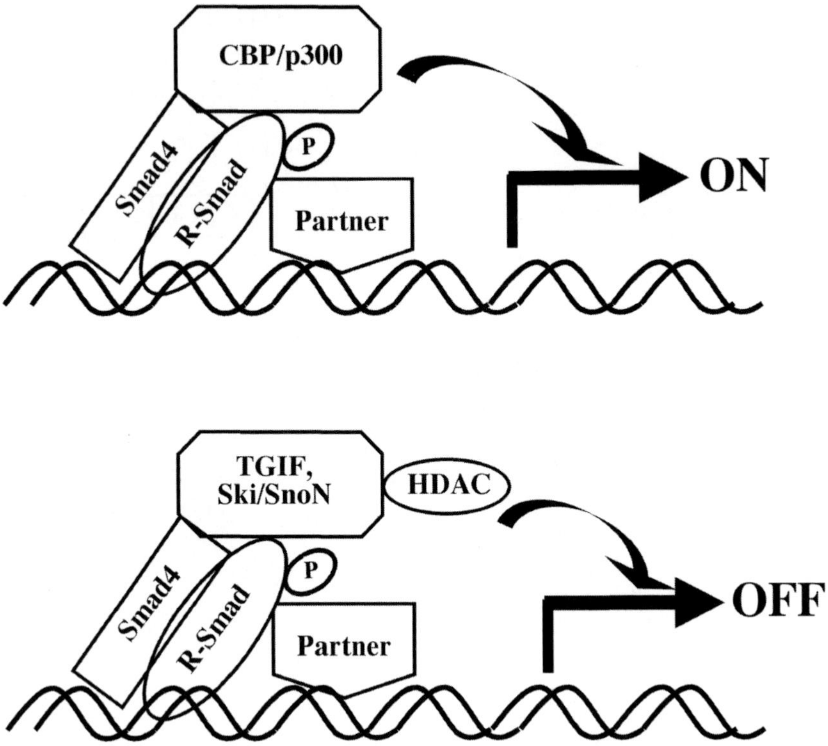

Fig. 2. Functional cooperativity of Smads with coactivators and corepressors.

structure through histone acetylation. The importance of chromatin modification in the TGF-β signaling is underscored by the finding that overexpression of E1A, a viral protein known to sequester CBP/p300, can block the transcriptional activity of Smads in a TGF-β-dependent manner. Conversely, the ability of Smads to activate transcription was enhanced by overexpression of MGS1, a coactivator that stabilizes the association of CBP with Smad4. Similarly, ZEB1 also enhances TGF-β signaling by promoting the Smad3/CBP interaction *(3,8,9)*. However, the mechanism of transcriptional activation through chromatin modulation is not restricted to the histone acetyl transferases described above since other proteins devoided of histone acetyl transferase activity were described as being coactivators partners for Smad proteins. Such partners include SMIF (Smad4-interacting transcriptional coactivator), a protein with transcriptional coactivation potential that functions in TGF-β signaling by specifically associating with Smad4 irrespective of the association of Smad4 with CBP *(58)*. The roles of SMIF in TGF-β signaling remain to be determined.

6. SMAD COREPRESSORS

Within the nucleus, the transcriptional function of Smad proteins can be limited by a variety of corepressors, the most characterized of which being TGIF (TG-interacting factor), Evi1(Ecotropic virus integration site-1), and the two related protooncongenes c-Ski and SnoN.

TGIF is a member of the three-amino acid extension loop (TALE) superfamily of homeodomain proteins that directly interacts with the MH2 domains of Smad2 and Smad3 following activation of TGF-β signaling *(59)*. The mechanism of TGIF-mediated repression of Smad2 transcriptional activity has been primarily attributed to the ability of TGIF to compete with CBP/p300 for R-Smads interaction or to recruit to Smad2 a transcriptional corepressor containing HDAC (histone deacetylases) via association with the general corepressors mSin3 and CtBP (C-terminal binding protein) *(59–61)*. However, the specific role(s) of the HDAC corepressor complex in this process is unclear, since a recent work has shown that expression of a holoprosencephaly mutant form of TGIF that still associates with both Smad2 and the HDAC repressor complex, failed to repress TGF-β-induced transcription *(62)*. The search for TGIF-interacting proteins has recently led to the identification of the ubiquitin ligase Tiul1 that functions in conjunction with TGIF to turn off TGF-β signaling *(26)*. TGIF does not seem to be targeted for degradation by Tiul1 but rather functions to form with Smad2 and Tiul1 a stable complex from which Tiul1 then induces ubiquitin-dependent degradation of Smad2. However, Tiul1 cannot be the only TGIF target, as repression by TGIF requires the presence of the transcription factor c-Jun, which is involved in the assembly of the TGIF/Smad2 complex as well as interference of Smad binding to the activators CBP/p300 *(63)*. Since activation of the JNK pathway can stabilize the TGIF/c-Jun complex, this model also provides new insights into the negative regulation of Smad2 by the JNK pathway *(45,63)*. Although the role of TGIF in negative regulation of TGF-β signaling during the course of tumorigenesis remains to be determined, the fact that TGIF is an established inhibitor of the tumor suppressor Smad2 suggests that TGIF might interfere with a complex tumor suppressor network that integrates TGF-β signals.

c-Ski is the cellular homolog of v-Ski, originally identified as the oncogene present in avian Sloan-Kettering viruses that induces transformation of chicken embryo fibroblasts. The v-Ski oncoprotein is truncated at the carboxyl terminus by 312 amino acids relative to the c-Ski protein, but this truncation does not play a role in the activation of *ski* as an oncogene. Overexpression of either c-Ski or v-Ski induces morphological transformation and anchorage-independent growth in chicken and quail embryo fibroblasts, indicating that the transforming activity is attributable to overexpression, not truncation, of the c-Ski protein *(64)*. Recombinant c-Ski protein purified from bacteria cannot directly bind to DNA, but c-Ski in nuclear extracts from mammalian cell cultures does bind to DNA suggesting that c-Ski lacks a DNA binding ability in its own and instead regulates transcription via its association with other proteins. Recently, diverse types of studies converged on the conclusions that Ski binding to DNA is mediated in part through its association with the R-Smad/Smad4 heteromeric complexes *(64,65)*. Because c-Ski was found to be a component of the HDAC complex through binding to N-CoR and mSin3 *(64,65)*, it has been postulated that the mechanism used by c-Ski to repress Smad signaling involves recruitment of a transcriptional repressor complex to Smads. It is interesting to note that this function of c-Ski is modulated by activation of the JNK pathway, which enhances the association of c-Ski and c-Jun, thereby allowing the stabilization of the transcriptional repressor complex *(66)*. However, alternative mechanisms of repression of Smad signaling may also exist since c-Ski has been shown to prevent the formation of a functional R-Smad/Smad4 heteromeric complex or the phosphorylation of R-Smads by the activated type I receptor, thereby inactivating their ability to activate transcription *(67,68)*. Furthermore, the association of c-Ski with R-Smads interferes with binding of Smads to the transcriptional activators CBP/p300 *(64)*.

Like Ski, SnoN interacts with Smads and represses their ability to activate target genes by assembling a repressor complex containing N-CoR, mSin3, and a HDAC *(69)*. SnoN is degraded soon after cell exposure to TGF-β, but accumulates again later, suggesting that SnoN may keep Smads from activating transcription in the basal state and may help to terminate TGF-β action later *(70)*. At least two related mechanisms have been described to explain the inhibitory effect of TGF-β on the accumulation of SnoN. In the presence of ligand, phosphorylated Smad2 recruits Smurf2 to SnoN, resulting in the ubiquitin-dependent degradation of SnoN by the proteasome system *(71)*. For the second mechanism, instead of recruiting Smurf2, Smad2 mediates the interaction of SnoN with the E3 ubiquitin ligase APC (Anaphase promoting complex), with an outcome resembling the one imposed by Smurf2 *(72)*.

Evi1 is a protooncogene that has been reported to interact with Smad3, but not Smad2, and is able to inhibit the activation of TGF-β-responsive promoters by disrupting the binding of the Smad3/Smad4 complex to DNA (*73*). More recent studies revealed that two additional mechanisms are likely to be responsible for the inhibitory effect of Evi1 on TGF-β responses: first Evi1 can interact with CtBP, and this interaction is required for the Evi1-dependent inhibitory effect on TGF-β signaling, and second, Evi1 can directly bind HDAC1 and the repression of TGF-β signaling by Evi1 is relieved by the HDAC inhibitor trichostatin A (TSA) *(74,75)*.

Additional inhibitors of the transcriptional functions of Smad proteins include SNIP1 and SIP1, which interact with Smads on TGF-β-responsive promoters and repress their ability to activate TGF-β target genes by competing with CBP/p300 for Smad interaction or recruiting the general corepressor CtBP, respectively *(76,77)*. Clearly, when Smads associate with corepressors that have their own co-repressor-recruiting ability, Smads repress transcription by modifying the composition of coactivator complexes or the kinetics of their assembly. Recently, it turns out that Smads can act themselves as transcriptional repressors depending on the target genes. Analysis of the TGF-β-dependent cell cycle arrest in epithelial cells, which have been largely investigated because of its relevance to cancer, has provided an example of the direct repression by Smads. In the absence of TGF-β signaling, a complex containing Smad3, the transcription factors E2F4/5 and DP1, and the corepressor p107 preexists in the cytoplasm *(78)*. Receptor-mediated phosphorylation of Smad3 induces the nuclear translocation of this Smad repressor complex, which in turn associates with a composite Smad-E2F site on the *c-myc* promoter, leading to its repression. Different mechanisms of Smad-mediated repression operate in the inhibition of osteoblast or myogenic differentiation. In the first case, Smad3 interacts with Runx2 and inhibits Runx2 transcriptional activity by a mechanism that does not involve binding of Smad3 to the promoter or decreased binding of Runx2 to its cognate DNA sequence *(79)*. Subsequent studies have demonstrated that the repressive function of Smad3 occurs through the action of the class IIa HDAC4 and HDAC5, which are recruited through interaction with Smad3 to the Smad3/Runx2 complex at the Runx2-binding DNA sequence *(80)*. For the TGF-β-dependent inhibition of myogenic differentiation, Smad3 interacts with MyoD and myogenin and interferes with their heterodimerization or oligomerization with the partner E12/47, thereby decreasing the DNA binding of MyoD and Myogenin. Furthermore, Smad3 also associates with MEF2C and this interaction interferes with the assembly of a complex containing MyoD and CRIP1, a protein that functions as a coactivator for MEF2C *(81,82)*. In addition to repressing DNA either directly or by associating with transcriptional repressors, several studies indicate that Smads may function in different ways to repress the transcriptional activity of a large variety of transcriptional factors, but the mechanism for these cases of repression have not been characterized *(1–3)*.

7. CONCLUDING REMARKS

The phosphorylation of R-Smads by the activated type I TGF-β receptor and their subsequent heterodimerization with Smad4 and translocation to the nucleus where they bind the promoters of target genes form the basis for a model how Smad proteins work to transmit TGF-βsignals from the plasma membrane to the nucleus. An important choice that must be made in the nucleus is whether an activated Smad complex will activate or repress gene expression. It is likely that the responses depend as much on the activity of partners and modulators that exert control and determine the specific target genes. By focusing here on the central component of the TGF-β signaling pathway, we have largely described step by step its main defining features without providing much details on its link with other signaling pathways. Despite much progress over the past decade, many unanswered questions remain about the noncanonical TGF-β signaling pathways and much work is needed for the delineation of the mechanisms of crosstalk between these separate pathways. Once these pathways are defined and their contribution to specific cellular and context-dependent effects of TGF-β are established, it will be possible to develop new molecular targets for therapy trials.

REFERENCES

1. Heldin C-H, Miyazono K, ten Dijke P. TGF-β signalling from cell membrane to nucleus through SMAD proteins. Nature 1997;390:465–471.
2. Massagué J, Chen YG. Controlling TGF-β signaling. Genes Dev 2000;14:627–644.
3. Shi Y, Massagué J. Mechanisms of TGF-β signaling from cell membrane to the nucleus. Cell 2003; 113:685–700.
4. Kretzschmar M, Liu F, Hata A, Doody J, Massagué J. The TGF-β family mediator Smad1 is phosphorylated directly and activated functionally by the BMP receptor kinase. Genes Dev 1997;11:984–995.
5. Macias-Silva M, Abdollah S, Hoodless PA, Pirone R, Attisano L, Wrana JL. MADR2 is a substrate of the TGF-β receptor and its phosphorylation is required for nuclear accumulation and signaling. Cell 1996;87:1215–1224.
6. Abdollah S, Macias-Silva M, Tsukazaki T, Hayashi H, Attisano L, Wrana JL. TbetaRI phosphorylation of Smad2 on Ser465 and Ser467 is required for Smad2-Smad4 complex formation and signaling. J Biol Chem 1997;272:27,678–27,685.
7. Souchelnytskyi S, Tamaki K, Engstrom U, Wernstedt C, ten Dijke P, Heldin C-H. Phosphorylation of Ser465 and Ser467 in the C terminus of Smad2 mediates interaction with Smad4 and is required for transforming growth factor-beta signaling. J Biol Chem 1997;272:28,107–28,115.
8. Feng XH, Derynck R. Specificity and Versatility in TGF-β signaling through Smads. Annu Rev Cell Dev Biol 2005;21:659–693.
9. Attisano L, Wrana JL. Smads as transcriptional co-modulators. Curr Opin Cell Biol 2000;12:235–243.
10. Dennler S, Huet S, Gauthier JM. A short amino-acid sequence in MH1 domain is responsible for functional differences between Smad2 and Smad3. Oncogene 1999;18:1643–1648.
11. Hata A, Lo RS, Wotton D, Lagna G, Massagué J. Mutations increasing autoinhibition inactivate tumour suppressors Smad2 and Smad4. Nature 1997;388:82–87.
12. LeSueur JA, Graff JM. Spemann organizer activity of Smad10. Development 1999;126:137–146.
13. Howell M, Itoh F, Pierreux CE, Valgeirsdottir S, Itoh S, ten Dijke P, Hill CS. Xenopus Smad4beta is the co-Smad component of developmentally regulated transcription factor complexes responsible for induction of early mesodermal genes. Dev Biol 1999;214:354–369.
14. Lo RS, Chen YG, Shi Y, Pavletich NP, Massagué J. The L3 loop: a structural motif determining specific interactions between SMAD proteins and TGF-β receptors. EMBO J 1998;17:996–1005.
15. Chen YG, Massagué J. Smad1 recognition and activation by the ALK1 group of transforming growth factor-beta family receptors. J Biol Chem 1999;274:3672–3677.
16. Feng XH, Derynck R. A kinase subdomain of transforming growth factor-beta (TGF-β) type I receptor determines the TGF-β intracellular signaling specificity. EMBO J 1997;16:3912–3923.
17. Chen YG, Hata A, Lo RS, Wotton D, Shi Y, Pavletich N, Massagué J. Determinants of specificity in TGF-β signal transduction. Genes Dev 1998;12:2144–2152.
18. Tsukazaki T, Chiang TA, Davison AF, Attisano L, Wrana JL. SARA, a FYVE domain protein that recruits Smad2 to the TGF-β receptor. Cell 1998;95:779–791.
19. Di Guglielmo GM, Le Roy C, Goodfellow AF, Wrana JL. Distinct endocytic pathways regulate TGF-β receptor signalling and turnover. Nat Cell Biol 2003;5:410–421.
20. Wu G, Chen YG, Ozdamar B, et al. Structural basis of Smad2 recognition by the Smad anchor for receptor activation. Science 2000;287:92–97.
21. Lin HK, Bergmann S, Pandolfi PP. Cytoplasmic PML function in TGF-β signalling. Nature 2004;431:205–211.
22. Hocevar BA, Smine A, Xu XX, Howe PH. The adaptor molecule Disabled-2 links the transforming growth factor beta receptors to the Smad pathway. EMBO J 2001;20:2789–2801.

23. Miura S, Takeshita T, Asao H, et al. Hgs (Hrs), a FYVE domain protein, is involved in Smad signaling through cooperation with SARA. Mol Cell Biol 2000;20:9346–9355.
24. Kavsak P, Rasmussen RK, Causing CG, et al. Smad7 binds to Smurf2 to form an E3 ubiquitin ligase that targets the TGF-β receptor for degradation. Mol Cell 2000;6:1365–1375.
25. Ebisawa T, Fukuchi M, Murakami G, et al. Smurf1 interacts with transforming growth factor-beta type I receptor through Smad7 and induces receptor degradation. J Biol Chem 2001;276: 12,477–12,480.
26. Seo SR, Lallemand F, Ferrand N, et al. The novel E3 ubiquitin ligase Tiul1 associates with TGIF to target Smad2 for degradation. EMBO J 2004;23:3780–3792.
27. Lallemand F, Seo SR, Ferrand N, et al. AIP4 restricts transforming growth factor-beta signaling through a ubiquitination-independent mechanism. J Biol Chem 2005;280:27,645–27,653.
28. Ciechanover A, Orian A, Schwartz AL. Ubiquitin-mediated proteolysis: biological regulation via destruction. Bioessays 2000;22:442–451.
29. Hayashi H, Abdollah S, Qiu Y, et al. The MAD-related protein Smad7 associates with the TGF-β receptor and functions as an antagonist of TGF-β signaling. Cell 1997;89:1165–1173.
30. Xu L, Chen YG, Massagué J. The nuclear import function of Smad2 is masked by SARA and unmasked by TGF-β-dependent phosphorylation. Nat Cell Biol 2000;2:559–562.
31. Xiao Z, Liu X, Lodish HF. Importin beta mediates nuclear translocation of Smad 3. J Biol Chem 2000; 275:23,425–23,428.
32. Xiao Z, Liu X, Henis YI, Lodish HF. A distinct nuclear localization signal in the N terminus of Smad 3 determines its ligand-induced nuclear translocation. Proc Natl Acad Sci USA 2000;97:7853–7858.
33. Xiao Z, Watson N, Rodriguez C, Lodish HF. Nucleocytoplasmic shuttling of Smad1 conferred by its nuclear localization and nuclear export signals. J Biol Chem 2001;276:39,404–39,410.
34. Kurisaki A, Kose S, Yoneda Y, Heldin C-H, Moustakas A. Transforming growth factor-beta induces nuclear import of Smad3 in an importin-beta1 and Ran-dependent manner. Mol Biol Cell 2001;12: 1079–1091.
35. Xu L, Kang Y, Col S, Massagué J. Smad2 nucleocytoplasmic shuttling by nucleoporins CAN/Nup214 and Nup153 feeds TGF-β signaling complexes in the cytoplasm and nucleus. Mol Cell 2002;10: 271–282.
36. Randall RA, Germain S, Inman GJ, Bates PA, Hill CS. Different Smad2 partners bind a common hydrophobic pocket in Smad2 via a defined proline-rich motif. EMBO J 2002;21:145–156.
37. Inman GJ, Nicolas FJ, Hill CS. Nucleocytoplasmic shuttling of Smads 2, 3, and 4 permits sensing of TGF-β receptor activity. Mol Cell 2002;10:283–294.
38. Pierreux CE, Nicolas FJ, Hill CS. Transforming growth factor β-independent shuttling of Smad4 between the cytoplasm and nucleus. Mol Cell Biol 2000;20:9041–9054.
39. Kretzschmar M, Doody J, Timokhina I, Massagué J. A mechanism of repression of TGF-β/Smad signaling by oncogenic Ras. Genes Dev 1999;13:804–816.
40. Abdel-Wahab N, Wicks SJ, Mason RM, Chantry A. Decorin suppresses transforming growth factor-beta-induced expression of plasminogen activator inhibitor-1 in human mesangial cells through a mechanism that involves Ca2+-dependent phosphorylation of Smad2 at serine-240. Biochem J 2002; 362:643–649.
41. Wicks SJ, Lui S, Abdel-Wahab N, Mason RM, Chantry A. Inactivation of smad-transforming growth factor beta signaling by Ca(2+)-calmodulin-dependent protein kinase II. Mol Cell Biol 2000;20: 8103–8111.
42. Engel ME, McDonnell MA, Law BK, Moses HL. Interdependent SMAD and JNK signaling in transforming growth factor-beta-mediated transcription. J Biol Chem 1999;274:37,413–37,420.
43. Conery AR, Cao Y, Thompson EA, Townsend CM, Jr., Ko TC, Luo K. Akt interacts directly with Smad3 to regulate the sensitivity to TGF-β induced apoptosis. Nat Cell Biol 2004;6:366–372.
44. Remy I, Montmarquette A, Michnick SW. PKB/Akt modulates TGF-β signalling through a direct interaction with Smad3. Nat Cell Biol 2004;6:358–365.
45. Dennler S, Prunier C, Ferrand N, Gauthier JM, Atfi A. c-Jun inhibits transforming growth factor beta-mediated transcription by repressing Smad3 transcriptional activity. J Biol Chem 2000;275: 28,858–28,865.
46. Zawel L, Dai JL, Buckhaults P, et al. Human Smad3 and Smad4 are sequence-specific transcription activators. Mol Cell 1998;1:611–617.
47. Zhang Y, Feng XH, Derynck R. Smad3 and Smad4 cooperate with c-Jun/c-Fos to mediate TGF-β-induced transcription. Nature 1998;394:909–913.

48. Hanafusa H, Ninomiya-Tsuji J, Masuyama N, et al. Involvement of the p38 mitogen-activated protein kinase pathway in transforming growth factor-beta-induced gene expression. J Biol Chem 1999;274:27,161–27,167.
49. Sano Y, Harada J, Tashiro S, Gotoh-Mandeville R, Maekawa T, Ishii S. ATF-2 is a common nuclear target of Smad and TAK1 pathways in transforming growth factor-β signaling. J Biol Chem 1999;274: 8949–8957.
50. Hua X, Liu X, Ansari DO, Lodish HF. Synergistic cooperation of TFE3 and smad proteins in TGF-β-induced transcription of the plasminogen activator inhibitor-1 gene. Genes Dev 1998;12: 3084–3095.
51. Seoane J, Le HV, Shen L, Anderson SA, Massagué J. Integration of Smad and forkhead pathways in the control of neuroepithelial and glioblastoma cell proliferation. Cell 2004;117:211–223.
52. Atfi A, Abecassis L, Bourgeade MF. Bcr-Abl activates the AKT/Fox O3 signalling pathway to restrict transforming growth factor-beta-mediated cytostatic signals. EMBO Rep 2005;6:985–991.
53. Chen X, Weisberg E, Fridmacher V, Watanabe M, Naco G, Whitman M. Smad4 and FAST-1 in the assembly of activin-responsive factor. Nature 1997;389:85–89.
54. Labbe E, Silvestri C, Hoodless PA, Wrana JL, Attisano L. Smad2 and Smad3 positively and negatively regulate TGF-β-dependent transcription through the forkhead DNA-binding protein FAST2. Mol Cell 1998;2:109–120.
55. Lagna G, Hata A, Hemmati-Brivanlou A, Massagué J. Partnership between DPC4 and SMAD proteins in TGF-β signalling pathways. Nature 1996;383:832–836.
56. Itoh S, Ericsson J, Nishikawa J, Heldin C-H, ten Dijke P. The transcriptional co-activator P/CAF potentiates TGF-β/Smad signaling. Nucleic Acids Res 2000;28:4291–4298.
57. Kahata K, Hayashi M, Asaka M, et al. Regulation of transforming growth factor-beta and bone morphogenetic protein signalling by transcriptional coactivator GCN5. Genes Cells 2004;9:143–151.
58. Bai RY, Koester C, Ouyang T, et al. SMIF a Smad4-interacting protein that functions as a co-activator in TGFbeta signalling. Nat Cell Biol 2002;4:181–190.
59. Wotton D, Lo RS, Lee S, Massagué J. A Smad transcriptional corepressor. Cell 1999;97:29–39.
60. Melhuish TA, Wotton D. The interaction of the carboxyl terminus-binding protein with the Smad corepressor TGIF is disrupted by a holoprosencephaly mutation in TGIF. J Biol Chem 2000;275: 39,762–39,766.
61. Wotton D, Knoepfler PS, Laherty CD, Eisenman RN, Massagué J. The Smad transcriptional corepressor TGIF recruits mSin3. Cell Growth Differ 2001;12:457–463.
62. Gripp KW, Wotton D, Edwards MC, et al. Mutations in TGIF cause holoprosencephaly and link NODAL signalling to human neural axis determination. Nat Genet 2000;25:205–208.
63. Pessah M, Prunier C, Marais J, et al. c-Jun interacts with the corepressor TG-interacting factor (TGIF) to suppress Smad2 transcriptional activity. Proc Natl Acad Sci USA 2001;98:6198–6203.
64. Luo K. Ski and SnoN: negative regulators of TGF-β signaling. Curr Opin Genet Dev 2004;14:65–70.
65. Sun Y, Liu X, Eaton EN, Lane WS, Lodish HF, Weinberg RA. Interaction of the Ski oncoprotein with Smad3 regulates TGF-β signaling. Mol Cell 1999;4:499–509.
66. Pessah M, Marais J, Prunier C, et al. c-Jun associates with the oncoprotein Ski and suppresses Smad2 transcriptional activity. J Biol Chem 2002;277:29,094–29,100.
67. Wu JW, Krawitz AR, Chai J, et al. Structural mechanism of Smad4 recognition by the nuclear oncoprotein Ski: insights on Ski-mediated repression of TGF-β signaling. Cell 2002;111:357–367.
68. Prunier C, Pessah M, Ferrand N, Seo SR, Howe P, Atfi A. The oncoprotein Ski acts as an antagonist of transforming growth factor-beta signaling by suppressing Smad2 phosphorylation. J Biol Chem 2003;278:26,249–26,257.
69. Sun Y, Liu X, Ng-Eaton E, Lodish HF, Weinberg RA. SnoN and Ski protooncoproteins are rapidly degraded in response to transforming growth factor beta signaling. Proc Natl Acad Sci USA 1999;96: 12,442–12,447.
70. Stroschein SL, Wang W, Zhou S, Zhou Q, Luo K. Negative feedback regulation of TGF-β signaling by the SnoN oncoprotein. Science 1999;286:771–774.
71. Bonni S, Wang HR, Causing CG, et al. TGF-β induces assembly of a Smad2-Smurf2 ubiquitin ligase complex that targets SnoN for degradation. Nat Cell Biol 2001;3:587–595.
72. Stroschein SL, Bonni S, Wrana JL, Luo K. Smad3 recruits the anaphase-promoting complex for ubiquitination and degradation of SnoN. Genes Dev 2001;15:2822–2836.
73. Kurokawa M, Mitani K, Irie K, et al. The oncoprotein Evi-1 represses TGF-β signalling by inhibiting Smad3. Nature 1998;394:92–96.

74. Izutsu K, Kurokawa M, Imai Y, Maki K, Mitani K, Hirai H. The corepressor CtBP interacts with Evi-1 to repress transforming growth factor beta signaling. Blood 2001;97:2815–2822.
75. Alliston T, Ko TC, Cao Y, Liang YY, Feng XH, Chang C, Derynck R. Repression of bone morphogenetic protein and activin-inducible transcription by Evi-1. J Biol Chem 2005;280:24,227–24,237.
76. Kim RH, Wang D, Tsang M, et al. A novel smad nuclear interacting protein, SNIP1, suppresses p300-dependent TGF-β signal transduction. Genes Dev 2000;14:1605–1616.
77. Watanabe M, Whitman M. The role of transcription factors involved in TGF-β superfamily signaling during development. Cell Mol Biol (Noisy-le-grand) 1999;45:537–543.
78. Chen CR, Kang Y, Siegel PM, Massagué J. E2F4/5 and p107 as Smad cofactors linking the TGF-β receptor to c-myc repression. Cell 2002;110:19–32.
79. Alliston T, Choy L, Ducy P, Karsenty G, Derynck R. TGF-β-induced repression of CBFA1 by Smad3 decreases cbfa1 and osteocalcin expression and inhibits osteoblast differentiation. EMBO J 2001;20: 2254–2272.
80. Kang JS, Alliston T, Delston R, Derynck R. Repression of Runx2 function by TGF-β through recruitment of class II histone deacetylases by Smad3. EMBO J 2005;24:2543–2555.
81. Liu D, Black BL, Derynck R. TGF-β inhibits muscle differentiation through functional repression of myogenic transcription factors by Smad3. Genes Dev 2001;15:2950–2966.
82. Liu D, Kang JS, Derynck R. TGF-β-activated Smad3 represses MEF2-dependent transcription in myogenic differentiation. EMBO J 2004;23:1557–1566.

6 Interaction of Smad4 and Embryonic Liver Fodrin-β-spectrin in Hyperplasia, Neoplasia, and Tumor Suppression

Sang Soo Kim, Charles Shuman, and Lopa Mishra

CONTENTS

INTRODUCTION
SMAD4
EMBRYONIC LIVER FODRIN
ANIMAL MODEL
CONCLUSIONS
ACKNOWLEDGMENTS
REFERENCES

Abstract

Transforming growth factor (TGF)-β is both a suppressor and promoter of tumorigenesis; however, its contribution to early tumor suppression and staging remains largely unknown. In search of the mechanism of early tumor suppression, we focus on Smad4, a transducer of TGF-β signaling, and embryonic liver fodrin (ELF), a β-spectrin, in tumorigenesis by linking a major, dynamic scaffolding protein and a key signaling protein. ELF activates and modulates Smad4 activation of the TGF-β response to confer cell polarity, to maintain cell architecture, and to inhibit epithelial-to-mesenchymal transition. In human gastric tumor samples, a significant loss of ELF and reduction of Smad4 expression was recently found. An examination of *elf*$^{+/-}$, *Smad4*$^{+/-}$, and *elf*$^{+/-}$/*Smad4*$^{+/-}$ mice revealed that disruption of ELF and Smad4 displayed a synergistic effect on tumor formation in terms of incidence and duration, suggesting a cooperative interaction between ELF and Smad4 leading to enhanced tumor suppression. This review examines our understanding of the significance of the interactions between ELF and Smad4, through regulation of the TGF-β signaling pathway and repression of tumor formation.

Key Words: TGF-β; ELF; Smad4; knockout mice; tumorigenesis.

1. INTRODUCTION

Transforming growth factor (TGF)-β is an important regulator of cell differentiation, proliferation, migration, and apoptosis in several types of cells including epithelial, endothelial,

From: *Cancer Drug Discovery and Development: Transforming Growth Factor-β in Cancer Therapy, Vol I: Basic and Clinical Biology*
Edited by: Sonia B. Jakowlew © Humana Press Inc., Totowa, NJ

Table 1
The Mutations of TGF-β Signaling Pathway in Human Cancer

Name	*Function*	*Major type of cancer*	*Reference*
TGFB1	Ligand	Upregulated in tumor cells	*(5)*
TGFBRII	TGF-β receptor type II	Colorectal, Gastric	*(6–9)*
Smad2	Downstream transducers R-Smad	Colorectal	*(10,11)*
Smad4	Downstream transducers Co-Smad	Pancreas, Colorectal, Juvenile polyposis	*(12–14)*
ELF	β-spectrin	Colorectal, Gastric	*(15,16)*

hematopoietic, neural, and certain types of mesenchymal cells. Their signaling mechanism is mediated through types I and II receptors, transmembrane serine/threonine kinases, and transmitted to Smads, specific intracellular mediators. Activation of Smads results in nuclear translocation and activation of gene expression *(1–4)*. Vertebrates possess at least nine Smad proteins falling into three functional classes: (i) receptor-activated Smads (R-Smads): Smad1, Smad2, Smad3, Smad5, and Smad8; (ii) co-mediator Smads: Smad4 and Smad10; and (iii) inhibitory Smads: Smad6 and Smad7. Inhibition of cell proliferation is central to the TGF-β response and escape from this response is a hallmark of many cancer cells. Many aspects of cancer involve inactivation of the TGF-β pathway as summarized in Table 1. Especially Smad4, also known as DPC4 in humans, is commonly inactivated in human pancreatic and gastrointestinal tumors *(17–19)*.

The activities of Smads can be modulated by several adaptors such as embryonic liver fodrin (ELF), filamin or Smad anchor for receptor activation (SARA) as well as functional interactions with multiple signal transduction pathways *(20,21)*. These adaptor proteins play a critical role in localizing Smad and facilitating the functions of TGF-β. ELF is a key protein involved in endodermal stem/progenitor cell commitment to foregut lineage *(21)*. It is also crucial for protein sorting, cell adhesion, and the development of a polarized differentiated epithelial cell. In prior studies, homozygous loss of *elf* disrupts the TGF-β signaling pathway and displays similar phenotypes with Smad-null mice. Moreover, intercrossing *elf* mutants and *Smad4* mutants thereby generating double heterozygotes revealed a marked increase in tumorigenesis.

Increases or decreases in the effects of the TGF-β pathway have been linked to numerous diseases, including cancers of the colon, gastrointestinal tract, and pancreas. This review will discuss the interactions of ELF and Smad4 by which TGF-β mediates its cellular functions, focusing on its role in disease, particularly tumor suppression.

2. SMAD4

Smads, the only substrates for type I receptor kinases known to have a signaling function, were first identified as the products of the *Drosophila Mad* and *C. elegans Sma* genes, which lie downstream of the BMP-analogous ligand-receptor systems in these organisms *(22,23)*. The name of Smad comes from "The Mothers Against Decapentaplegic" (Mad) gene in *Drosophila* and the related Sma genes in *C. elegans* which are implicated in signal transduction. The proposed root symbol was Smad, a merger of Sma and Mad, which serves to differentiate these proteins from unrelated gene products previously called Mad *(24)*.

The human genome encodes eight Smad family members, and related proteins are known in the several species including rat, mouse, *Xenopus*, zebrafish, *Drosophila* and *C. elegans*.

Smads are ubiquitously expressed throughout development and in all adult tissues *(25,26)*. Functionally, Smads fall into three subfamilies: receptor-activated R-Smads (Smad1, Smad2, Smad3, Smad5, and Smad8), which become phosphorylated by the type I receptors; common mediator Smads (Smad4), which oligomerize with activated R-Smads; and inhibitory Smads (Smad6 and Smad7), which are induced by TGF-β family members. The latter exert a negative feedback effect by competing with R-Smads for receptor interaction and by marking the receptors for degradation. Smads have two conserved domains, the N-terminal Mad homology 1 (MH1) and C-terminal Mad homology 2 (MH2) domains. The MH1 domain is highly conserved among R-Smads and Co-Smads; however, the N-terminal parts of I-Smads have only weak sequence similarity to MH1 domains. Sequence and structural analyses indicate that the MH1 domain is homologous to the diverse His-Me (histidine-metal-ion) finger family of endonucleases, and it may have evolved from an ancient enzymatic domain that had lost its catalytic activity but retained its DNA-binding properties *(27)*. The MH1 domain regulates nuclear import and transcription by binding to DNA and interacting with several nuclear proteins. The MH2 domain is highly conserved among all Smads. Its structure contains several α-helices and loops, which surround a ß-sandwich *(28)*, and it resembles the forkhead-associated domain, a phosphopeptide-binding domain common in transcription and signaling factors *(29)*. The MH2 domain regulates Smad oligomerization, recognition by type I receptors and interacts with cytoplasmic adaptors and several transcription factors.

Smad3 is a direct mediator of transcriptional activation by the TGF-β receptor. Its target genes in epithelial cells include cyclin-dependent kinase inhibitors that generate a cytostatic response *(30)*. Smad3 can also mediate transcriptional repression of the growth-promoting gene, MYC, by forming a complex containing Smad3, the transcription factors E2F4, E2F5, DP1, and p107 which preexist in the cytoplasm. In response to TGF-β, this complex moves into the nucleus and associates with Smad4, recognizing a composite Smad-E2F site on MYC for repression. Previously known as the ultimate recipients of CDK regulatory signals, E2F4/E2F5 and p107 act as transducers of TGF-β receptor signals upstream of CDK. Smad proteins therefore mediate transcriptional activation or repression depending on their associated partners.

TGF-β stimulation leads to phosphorylation and activation of R-Smads including Smad2 and Smad3, which form complexes with Smad4 which accumulate in the nucleus and regulate transcription of target genes *(31)*. In TGF-β stimulation of epithelial cells, receptors remain active for at least 3–4 h, and continuous receptor activity is required to maintain active Smads in the nucleus and for TGF-β induced transcription. Continuous nucleocytoplasmic shuttling of the Smads during active TGF-β signaling provides the mechanism, whereby the intracellular transducers of the signal continuously monitor receptor activity. These data explain how, at all times, the concentration of active Smads in the nucleus is directly dictated by the levels of activated receptors in the cytoplasm.

Smad4 plays a pivotal role in signal transduction of the TGF-β superfamily by hetero-oligomerization to mediating transcriptional activation of target genes *(32)*. In transcriptional activators of human TGF-β signaling, Smad3 and Smad4 proteins could specifically recognize an identical 8-bp palindromic sequence (GTCTAGAC). Tandem repeats of this palindrome conferred striking TGF-β responsiveness to a minimal promoter. This responsiveness was abrogated by targeted deletion of the cellular Smad4 gene *(33)*.

To directly test the hypothesis that the Smad4 gene is a tumor suppressor that is critical for transmitting signals from TGF-β family ligands, Smad4 was deleted through homologous recombination in human colorectal cancer cells *(34)*. This deletion abrogated signaling from TGF-β, as well as from the TGF-β family member, activin. These results indicate that mutational inactivation of Smad4 results TGF-β unresponsiveness and give a basis for understanding the physiologic role of this gene in tumorigenesis.

About 90% of human pancreatic carcinomas show allelic loss at chromosome 18q. To identify candidate tumor suppressor genes on 18q, a series of studies revealed homozygous deletions of 18q21.1 in 25 of 84 pancreatic tumors. DPC4, was identified to be the candidate tumor suppressor, and was found to be similar in sequence to a *Drosophila melanogaster* gene (Mad) implicated in a TGF-β like signaling pathway. Interestingly, inactivating mutations in DPC4 were identified in six of 27 pancreatic carcinomas that did not have homozygous deletions at 18q21.1 *(12)*. Tumors other than pancreatic also developed an alteration of the DPCA4 locus but not in high incidence in breast, ovarian, and head and neck squamous cell carcinomas *(35,36)*. Another dramatic contribution of Smad4 comes from gastrointestinal tract. Juvenile polyposis syndrome (JPS) is an autosomal dominant condition that predisposes gene carriers to various types of tumors. The diagnosis is based on the occurrence of hamartomatous gastrointestinal polyps that turn into malignant lesions in approximately 20% of cases *(37)*. In the case of JPS, several reports demonstrated that germline mutations of Smad4 contribute to JPS. In a large Iowa kindred with generalized JPS and gastrointestinal cancer, a Smad4 gene within a region on 18q21.1 was defined as the candidate tumor suppression by linkage analysis in a kindred with JPS *(38–41)*. This interval contains two putative tumor suppressor genes, DCC and Smad4. After sequencing 14 DCC exons and all 11 Smad4 exons, a deletion was detected which created a frameshift and a new stop codon. The mutant Smad4 proteins were predicted to be truncated at the carboxyl terminus and to lack sequences required for normal function.

In pancreatic and colorectal cancer, inactivation of Smad4 through homozygous deletion or intragenic mutation occurs frequently in association with malignant progression. However, mutation of this gene is seen only occasionally in the rest of human cancers. The majority of Smad4 gene mutations in human cancers are missense, nonsense, and frameshift mutations at the mad homology 2 region (MH2), which interfere with the homo-oligomer formation of Smad4 protein and the hetero-oligomer formation between Smad4 and Smad2 proteins, resulting in disruption of TGF-β signaling *(13)*.

3. EMBRYONIC LIVER FODRIN

ELF is a member of the spectrin family. Spectrins are tetrameric, cytoskeletal proteins essential for determination of cell shape, resilience of membranes to mechanical stress, positioning of transmembrane proteins, and organization of organelles and for molecular trafficking. α- and β-spectrin subunits form antiparallel dimers that self associate to give the spectrin tetramer. β-subunits, such as ELF, contain most of the spectrin binding activity *(42)*. Various isoforms of spectrin and its key adaptor protein, ankyrin, are present on the Golgi, on transport intermediates, and on the membranes of the endolysosomal pathway. Disruption of spectrin function either by dominant negative spectrin mutants or by antispectrin antibodies blocks anterograde transport in the secretory pathway, and directs regulation of the interaction between spectrin and Golgi membranes by the small G protein ADP-ribosylation factor (ARF) *(43,44)*.

The general primary structure of human β-spectrin has been deduced from human brain cDNA. β-Spectrin is encoded by a gene located on human chromosome 2p21. β-Spectrins have closely related N-terminal domains implicated in binding to actin, and 17 copies of a 106-residue repeat motif with consensus residues that are highly conserved between β-spectrins as well as α-spectrins. C-terminal domains of β-spectrins are candidate regions that associate with α-spectrin. Northern blot analysis of several rat tissues detected highest expression in lung, followed by kidney, brain, thymus, heart, and liver. Western blot analysis of cytosolic and membrane fractions of rat tissues revealed β-spectrins were enriched in membrane fractions of brain, kidney, and lung *(45–47)*.

Recently three isoforms of mouse β-spectrin were cloned and called "ELF" *(21)*. The longest isoform, elf3, consists of 2154 residues and is characterized by an actin-binding domain, a long repeat domain, and a short regulatory domain remarkable for the absence of a PH domain. Northern blot analysis demonstrated an abundant 9.0-kb elf3 transcript in brain, liver, and heart tissues. Immunohistochemical studies demonstrated ELF labeling of the basolateral or sinusoidal membrane surface, as well as a granular cytoplasmic pattern in hepatocytes. Antisense studies utilizing cultured liver explants show a vital role of *elf3* in hepatocyte differentiation and intrahepatic bile duct formation. The differential expression, tissue localization, and functional studies demonstrate the importance of *elf3* in modulating interactions between various components of the cytoskeleton proteins controlling liver and bile duct development.

TGF-β signals are conveyed through serine-threonine kinase receptors at the cell surface to specific intracellular mediators, the Smad proteins. Activation of Smad proteins results in their translocation to the nucleus and subsequent activation of gene expression. Activation of Smads can be modulated by adaptor proteins in the cytosol such as filamin and Smad anchor for receptor activation. Because such adaptors can control Smad access to TGF-β receptors, which activate Smad at the cell surface membrane, they play a critical role in facilitating TGF-β functions such as growth, differentiation, vascular remodeling, and cell fate specification *(48–51)*.

Cytoskeletal proteins belonging to the β-spectrin family are thought to regulate signal transduction by working as adaptor molecules *(52–54)*. When expression of *elf* was blocked by antisense, liver formation was inhibited *(21)*, and a phenotype similar to mice with compound haploinsufficiency at *Smad2* and *Smad3* loci was seen *(55,56)*. Moreover, ELF-deficient mice (determined by gene targeting) display a similar phenotype as other Smad-null mice including abnormal anatomy of primary brain vesicles, craniofacial abnormalities, aberrant gut formation, and distorted liver architecture *(20)*. Yolk sac blood vessel dilatation observed in some of the $elf^{-/-}$ mutants is reminiscent of the TGFBI, TGFBII, and Smad5 mutants, suggesting a role for ELF in TGF-β signaling *(57,58)*. Analysis of mouse embryonic fibroblasts (MEFs) derived from wild-type and $elf^{-/-}$ mutants showed that the $elf^{-/-}$ mouse embryonic fibroblasts did not respond to TGF-β stimulation. It has been also shown that TGF-β triggers phosphorylation and association of ELF with Smad3 and Smad4, followed by nuclear translocation. ELF deficiency results in mislocalization of Smad3 and Smad4 and loss of the TGF-β-dependent transcriptional response, which could be rescued by overexpression of the carboxyl terminal region of ELF.

Thus, disruption of ELF results in the loss of Smad3/Smad4 activation and a disruption of the TGF-β pathway. Therefore it has been of great interest to determine whether the loss of ELF is involved the tumorigenesis of tissue that has already shown TGF-β-dependency. Recent publications show that a significant loss of ELF and reduced Smad4 expression result in mouse and in human gastric cancer *(15)*. Furthermore, the rescue of E-cadherin-dependent homophilic cell-cell adhesion by ectopic expression of full-length ELF suggest that ELF has an essential role in tumor suppression in gastrointestinal cancers.

4. ANIMAL MODEL

To date, three independent groups have generated mice with targeted deletion of Smad4 (*59–61*; Table 2). Even though each group employed a different targeting strategy, there were similarities among the phenotypes of the homozygous mutants. Homozygous Smad4 mutant mice die around day 7 of embryogenesis. Mutant embryos showed reduced size, failure to gastrulate or express a mesodermal marker, and show abnormal visceral endoderm development. Growth retardation of the Smad4-deficient embryos results in reduced cell proliferation rather than increased apoptosis. These findings indicate that Smad4 is required

Table 2
Tumor Related Mice Phenotypes of Smad4 and Elf

Mice	*Phenotypes*	*Tissues*	*Ref*
$Smad4^{Mmt}/Smad4^{Mmt}$	Embryonic lethality (E7)		*(59)*
$Smad4^{Mmt}/Smad4^{+}$	Juvenile Intestinal Polyposis	Digestive/alimentary	*(59,62–65)*
$Smad4^{Mmt}/Smad4^{+}$, Apc^{-}/Apc^{+}	Intestinal adenocarcinoma/ epidermal cyst	Digestive/skin	*(59)*
$Smad4^{Mak}/Smad4^{Mak}$	Embryonic lethality (E7.5)/ reduced embryo size/ no development of demarcation/ disorganized visceral endoderm and extraembryonic region/ arrested development of gastrulation/failure to form mesoderm		*(60)*
$Smad4^{Cxd}/Smad4^{Cxd}$	Embryonic lethality (E6.5–E8.5)/ reduced embryo size/ abnormal extraembryonic portion of the egg cylinder/arrested development of gastrulation/ failure to form mesoderm		*(61)*
$Smad4^{Cxd}/Smad4^{+}$	Juvenile Intestinal Polyposis	Digestive/alimentary	*(66)*
$Smad4^{Co}/Smad4^{Co}$/ MMTV-cre	Mammary abscess/Squamous carcinoma	Breast/skin	*(61,67,68)*
$Elf^{-/-}$	Embryonic lethality (E8.5–E16.5)/ reduced embryo size/ abnormal morphology or development (cardiovascular, craniofacial, digestive, liver, muscle, brain, yolk sac)		*(20)*
$Elf^{+/-}$	Carcinoma	Liver	
$Elf^{+/-}$, $Smad4^{Cxd}/Smad4^{+}$	Carcinoma/Hyperplasia	Digestive/pancreas	*(15,16)*

for the differentiation of the visceral endoderm and pattern formation of embryogenesis. Although homozygous *Smad4* mutant mice were embryonically lethal, the heterozygotes were fertile and appeared normal up to the age of 1 yr. Upon further investigation, however, *Smad4*$^{+/-}$ mice began to develop polyposis in the fundus and antrum when they were older, and in the duodenum and cecum in older animals at a lower frequency. With increasing age, polyps in the antrum show sequential changes from hyperplasia, to dysplasia, in-situ carcinoma, and finally invasion *(62,66)*. These alterations are initiated by a dramatic expansion of the gastric epithelium where Smad4 is expressed. Moreover, loss of the remaining *Smad4* wild-type allele was detected only in later stages of tumor progression, suggesting that haploinsufficiency of Smad4 is sufficient for tumor initiation and leads loss of heterozygosity. These findings were consistent that germline *Smad4* mutations are found in a subset of familial juvenile polyposis. Moreover, when the *Smad4*$^{+/-}$ mice were introduced the APC heterozygous mutation, these compound heterozygotes carrying both mutations developed more malignant tumors in intestinal polyps than those in the simple heterozygotes. In order to overcome embryonic lethality and to determine whether the tissue specific contribution

Table 3
Classification of Tumors in Smad $4^{+/-}$, Elf $^{+/-}$, and Smad$^{+/-}$/elf $4^{+/-}$ Mice

Mice	*Pathology*	*Tissue*
Smad4$^{+/-}$	Squamous cell carcinoma	Stomach, Mandible
	Hyperplastic	Stomach
	Tubular Adenoma	Duodenum, Gastric mucosa
	Scrotal abscess	Testis
	Fibromyxolipoma	Foreleg
	Hyperplastic Brunners gland	Duodenum
elf $^{+/-}$	Tumor	Kidney
	Hepatocellular Carcinoma	Liver
elf $^{+/-}$/Smad4$^{+/-}$	Adenocarcinoma	Stomach
	Hyperkeratosis	Forestomach
	Hyperplastic	Stomach, Duodenum
	Adenoma	Colon
	Hyperplasia	Mandible, Skin
	Hypertrophic	Submandibular salivary glands
	Periductal lymphocytes	Exorbital lacrimal gland
	Peri-islet lymphocytes	Pancreas

of Smad4 for tumorigenesis, Yang et al. (2002) used a conditional knockout approach to generate a Smad4 mutation in specific tissues or cells. The disruption of Smad4 using a Cre-loxP approach displayed mammary gland hyperplasia and skin squamous cell carcinoma formation, implicating the role of Smad4 as a general tumor suppressor in several tissues *(67,68)*.

While disruption of ELF, a β-spectrin, leads to disruption of TGF-β signaling through Smad proteins in mice *(20)*, *elf$^{-/-}$* mice exhibit a phenotype similar to *Smad2$^{+/-}$/Smad3$^{+/-}$* compound mutant mice of midgestational death due to gastrointestinal, liver, neural, and heart defects. ELF deficiency results in mislocalization of Smad3 and Smad4 and loss of the TGF-β-dependent transcriptional response, which could be rescued by overexpression of the carboxyl terminal region of ELF. The *elf* gene expression is therefore an important modulator of TGF-β activation.

An examination of *elf$^{+/-}$*, *Smad4$^{+/-}$*, and *elf$^{+/-}$/Smad4$^{+/-}$* mice for tumor development revealed that 40% of *elf$^{+/-}$* mice (8/20) developed tumors of varying etiology (Table 3). This tumor incidence was comparable to that seen in *Smad4$^{+/-}$* mice (45%, 9/20). Interestingly, 90% (18/20) of *elf$^{+/-}$/Smad4$^{+/-}$* double heterozygous mutants developed tumors. The major tumorigenic alteration associated with the *elf$^{+/-}$/Smad4$^{+/-}$* mice, by 7–9 mo of age is development of gastric hyperplasia and cancer. *Elf$^{+/-}$/Smad4$^{+/-}$* mice develop spontaneous tumors as a function of age. Tumors in these mice were usually seen at an age ranging between 7 and 10 mo and histopathological examination of these tumors by veterinary pathologists indicated that these tumors were of varying cellular origin (Table 3). On the other hand, development of spontaneous tumors in *elf* or *Smad4* heterozygous animals occurred with a delayed onset later than a year, suggesting a cooperative interaction between ELF and Smad4 leading to enhanced tumorigenesis.

In the gastric cancers, further analysis has revealed that E-cadherin accumulation at cell–cell contacts and E-cadherin/β-catenin-dependent epithelial cell–cell adhesion is disrupted in *elf$^{+/-}$/Smad4$^{+/-}$* mutant gastric epithelial cells. Subcellular fractionation revealed that E-cadherin is expressed mainly at the cell membrane after TGF-β stimulation. In contrast, *elf$^{+/-}$/Smad4$^{+/-}$* mutant tissues showed abnormal distribution of E-cadherin which could be rescued by overexpression of ELF but not Smad3 or Smad4 *(15)*.

The presence of such a dramatic phenotype as gastric and colon cancer in nearly all of the mice highlights the critical importance of gene dosage in this family. Haploinsufficiency phenotypes result in holoprosencephaly, liver hypoplasia as well as gastric and colon carcinoma. Aberrant expression of ELF has been described in multiple tumors *(15,16)*. ELF as a member of spectrin family may be involved in the generation of cell polarity and protein sorting, it is conceivable that abnormalities in ELF function could result in the partial or complete loss of cellular polarity seen in tumor cells. Actually, ELF by virtue of its involvement in Smad4 localization and subsequent activation of Smad4 may present as a tumor suppressor: aberrations in this interaction thus resulting in tumorigenesis.

5. CONCLUSIONS

In human tumor formation, TGF-β activity is also associated with increased oncogenicity in promoting invasion and motility, as well as indirect effects on angiogenesis and immune surveillance *(70,71)*. For instance, TGF-β mediated repression of E-cadherin with loss of E-cadherin expression results in the translocation of β-catenin from cell–cell contacts to the cytoplasm and the induction of epithelial–mesenchymal transitions, leading to an invasive phenotype *(72)*. On the other hand, Smad4 induces E-cadherin with recruitment of catenins to the plasma membrane *(73)*. Therefore it is important to link ELF in cell–cell adhesion mediated by TGF-β signaling with modulation of E-cadherin and β-catenin in the *elf*$^{+/-}$/*Smad4*$^{+/-}$ mutants which is particularly interesting in view of the tumor suppressor role of E-cadherin *(74)*. Lack of E-cadherin accumulation at cell–cell contacts results in loss of β-catenin localization to cell–cell contacts, leading to an epithelioid morphology, and increased cell motility. This abnormality could induce the scattered morphology and the invasive nature of diffuse cancers *(73,75)*. In diffuse-type carcinoma, an important molecular mechanism determining its histologic phenotype is thought to be loss of E-cadherin-mediated cell–cell adhesion. Several means of silencing E-cadherin gene expression in carcinoma have been observed, including methylation of the E-cadherin promoter, E-cadherin gene truncation mutations and loss of heterozygosity of the E-cadherin locus *(76–79)*. However loss of ELF expression is associated with invasive gastric carcinoma and provides the first indication that induction of E-cadherin localization by such cytoskeletal/adaptor factors plays a role in suppression of these tumors.

The results from these studies shed light on previous studies which demonstrated the accumulation of E-cadherin-mediated adherens junctions, the membrane cytoskeleton, and the Na^+/K^+-ATPase and their regulation through sequential protein expression *(80,81)*. The hierarchy of events leading to polarized protein stabilization and accumulation is, however, unclear. For instance, the establishment of spatial coordinates during the differentiation of polarized cells involves a positional cue from cadherins that results in the targeting of α- and β-Spectrin to a discrete plasma membrane domain. Spectrin tetramers become linked directly or indirectly to a number of integral proteins, such as members of the L1 cell adhesion protein family, to specific points on cell membranes. The spectrin tetramer is then able to capture and stabilize additional membrane interacting proteins, forming the characteristic profile of a polarized membrane domain *(52,56)*.

In the ureteric bud, for example, the distribution of Spectrin undergoes dramatic changes during the development of a continuous monolayer *(82)*. The model for the assembly and localization of the membrane cytoskeleton and Na^+/K^+-ATPase was considered to be as follows: before induction of the ureteric bud, mesenchymal cells express abundant α- and β-Spectrin. Induction then results in E-cadherin expression. Newly synthesized E-cadherin immediately binds to cytosolic α- and β-Spectrin and is transported to the cell surface where it forms homotypic contacts with E-cadherin molecules on adjacent cells, whereby the future lateral plasma membrane is defined. At slightly later times, ankyrin expression begins, allowing Na^+/K^+-ATPase to be recruited to the membrane in these studies. Genetic methods now provide an opportunity to test this model. In our mouse model, where ELF, a non-PH domain β-spectrin with unique 5′ and 3′ exons is specifically disrupted, we have been able to confirm loss of ELF by a number of approaches as described previously and in this study *(16,20)*. In addition, the *elf* knockout phenotype reinforces analyses demonstrating the role of β-spectrins, and more recently ELF, in Na^+/K^+-ATPase localization *(16,80)*. Indeed, the collective studies examining normal ELF function (utilizing peptide-specific ELF antibodies) and rescue studies further corroborate the remarkable role of this β-spectrin in linking TGF-β signaling with E-cadherin induced tumor suppression.

These studies also indicate that ELF is significantly involved in the targeting of proteins such as E-cadherin to the plasma membrane of polarized epithelial cells. Indeed, a dramatic loss of E-cadherin-induced cell–cell adhesion is seen in the *elf* mutant cells. Thus, while initial spatial cues are most probably determined by the cadherins, β-Spectrins, such as ELF, play a major role in further recruiting E-cadherin molecules to the lateral membrane of the cell surface and, potentially, the development of a continuous monolayer. Taken together, these studies are consistent with the role for the spectrin skeleton in stabilizing membrane proteins and preventing their endocytosis after delivery to the membrane. It is possible that spectrin also interacts within the secretory pathway to sort E-cadherin and other proteins before its arrival at the cell membrane, and it will be important to address these issues in future studies *(80,83,84)*.

Indeed, β-spectrins and ELF bind to E-cadherin via β-catenin *(85)*, an important mediator of the *Wnt* signaling pathway, which has a central role in colorectal carcinoma, controlling the switch between proliferation and differentiation in intestinal epithelial cells *(86,87)*. Disruption of the G1 phase of the cell cycle blocks a genetic program that is physiologically active in the proliferative compartment of colon crypts. Regulation of intracellular β-catenin signaling through APC, originally identified from familial adenomatous polyposis patient studies, and by p53 potentially closely links the *Wnt* signaling to TGF-β signaling pathways *(88,89)*. With this scenario, it will be interesting to delineate the crosstalk between signaling pathways that involves ELF and the key players in colorectal carcinogenesis, such as *Wnt* signaling *(90)*. In view of the above findings, loss of ELF potentially contributes to the events that lead to onset of cancer and it is possible that ELF plays an important role in tumor suppressor mechanisms that can be recognized as a potential early marker in carcinoma.

ACKNOWLEDGMENTS

This work was supported by Sally Funderberg award (L. M.), NIH grants RO1 DK56111 (L. M.), RO1 CA106614 (L. M.), Veterans Affairs Merit Award (L. M.) and the National Cancer Center, Korea #0610660 (S. S. K.). The authors wish to thank Tiffany Blake for critical review of the manuscript.

REFERENCES

1. Attisano L, Wrana JL. Mads and Smads in TGF-β signaling. Curr Opin Cell Biol 1998;10:188–194.
2. Derynck R, Zhang YE. Smad-dependent and Smad-independent pathways in TGF-β family signalling. Nature 2003;425:577–584.
3. Heldin C-H, Miyazono K, ten Dijke P. TGF-β signaling from cell membrane to nucleus through Smad proteins. Nature 1997;390:465–467.
4. Shi Y, Massagué J. Mechanisms of TGF-β signaling from cell membrane to nucleus. Cell 2003; 113: 685–700.
5. Blobe GC, Schiemann WP, Lodish HF. Role of transforming growth factor β in human disease. New Eng J Med 2000;342:1350–1358.

6. Lu SL, Kawabata M, Imamura T, et al. HNPCC associated with germline mutation in the TGF-β type II receptor gene. Nature Genet 1998;19:17–18.
7. Markowitz S, Wang J, Myeroff L, et al. Inactivation of the type II TGF-β receptor in colon cancer cells with microsatellite instability. Science 1995;268:1336–1338.
8. Myeroff LL, Parsons R, Kim SJ, et al. A transforming growth factor β receptor type II gene mutation common in colon and gastric but rare in endometrial cancers with microsatellite instability. Cancer Res 1995;55:5545–5547.
9. Tannergard P, Liu T, Weger A, Nordenskjold M, Lindbloom A. Tumorigenesis in colorectal tumors from patients with hereditary non-polyposis colorectal cancer. Hum Genet 1997;101:51–55.
10. Eppert K, Scherer SW, Ozcelik H, et al. MADR2 maps to 18q21 and encodes a TGF-β-regulated MAD-related protein that is functionally mutated in colorectal carcinoma. Cell 1996;86: 543–552.
11. Riggins GJ, Thiagalingam S, Rozenblum E, et al. Mad-related genes in the human. Nature Genet 1996;13:347–349.
12. Hahn SA, Schutte M, Hoque TMS, et al. DPC4, a candidate tumor suppressor gene at human chromosome 18q21.1. Science 1996;271:350–354.
13. Miyaki M, Kuroki T. Role of Smad4 (DPC4) inactivation in human cancer. Biochem Biophys Res Commun 2003;306:799–804.
14. Thiagalingam S, Lengauer C, Leach FS, et al. Evaluation of candidate tumour suppressor genes on chromosome 18 in colorectal cancers. Nature Genet 1996;13:343–346.
15. Katuri V, Tang Y, Marshall B, Rashid A, et al. Inactivation of ELF/TGF-β signaling in human gastrointestinal cancer. Oncogene 2005;24:8012–8024.
16. Tang Y, Katuri V, Srinivasan R, et al. Transforming growth factor-β suppresses nonmetastatic colon cancer through Smad4 and adaptor protein ELF at an early stage of tumorigenesis. Cancer Res 2005;65:4228–4237.
17. Powell SM, Harper JC, Hamilton SR, Robinson CR, Cummings OW. Inactivation of Smad4 in gastric carcinomas. Cancer Res 1997;57:4221–4224.
18. Weinstein M, Yang X, Deng C. Functions of mammalian Smad genes as revealed by targeted gene disruption in mice. Cytokine Growth Factor Rev 2000;11:49–58.
19. Xu J, Attisano L. Mutations in the tumor suppressors Smad2 and Smad4 inactivate TGF-β signaling by targeting Smads to the ubiquitin-proteasome pathway. Proc Natl Acad Sci USA 2000;97: 4820–4825.
20. Tang Y, Katuri V, Dillner A, Mishra B, Deng CX, Mishra L. Disruption of transforming growth factor-β signaling in ELF β-spectrin-deficient mice. Science 2003;299:574–577.
21. Mishra L, Cia T, Yu P, Monger P, Mishra B. elf3 encodes a novel 200 KD β-spectrin: Role in liver development. Oncogene 1999;18:353–364.
22. Patterson GI, Padgett RW. TGF β-related pathways. Roles in Caenorhabditis elegans development. Trends Genet 2000;16:27–33.
23. Whitman M. Smads and early developmental signaling by the TGFβ superfamily. Genes Dev 1998;12: 2445–2462.
24. Derynck R, Gelbart WM, Harland RM, et al. Nomenclature: vertebrate mediators of TGF-β family signals. Cell 1996;87:173.
25. Flanders KC, Kim ES, Roberts AB. Immunohistochemical expression of Smads 1–6 in the 15-day gestation mouse embryo: signaling by BMPs and TGF-βs. Dev Dyn 2001;220:141–154.
26. Luukko K, Ylikorkala A, Makela TP. Developmentally regulated expression of Smad3, Smad4, Smad6, and Smad7 involved in TGF-β signaling. Mech Dev 2001;101:209–212.
27. Grishin NV. Mh1 domain of Smad is a degraded homing endonuclease. J Mol Biol 2001;307:31–37.
28. Shi Y. Structural insights on Smad function in TGFß signaling. Bio essays 2001;23:223–232.
29. Li J, Lee GI, Van Doren SR, Walker JC. The FHA domain mediates phosphoprotein interactions. J Cell Sci 2000;113:4143–4149.
30. Chen CR, Kang Y, Siegel PM, Massagué J. E2F4/5 and p107 as Smad cofactors linking the TGFβ receptor to c-myc repression. Cell 2002;110:19–32.
31. Inman GJ, Nicolas FJ, Hill CS. Nucleocytoplasmic shuttling of Smads 2, 3, and 4 permits sensing of TGF-β receptor activity. Mol Cell 2002;10:283–294.
32. Shioda T, Lechleider RJ, Dunwoodie SL, et al. Transcriptional activating activity of Smad4: roles of SMAD hetero-oligomerization and enhancement by an associating transactivator. Proc Nat Acad Sci USA 1998;95:9785–9790.
33. Zawel L, Dai JL, Buckhaults P, et al. Human Smad3 and Smad4 are sequence-specific transcription activators. Mol Cell 1998;1:611–617.

34. Zhou S, Buckhaults P, Zawel L, et al. Targeted deletion of Smad4 shows it is required for transforming growth factor β and activin signaling in colorectal cancer cells. Proc Nat Acad Sci USA 1998;95:2412–2416.
35. Kim SK, Fan Y, Papadimitrakopoulou V, et al. DPC4, a candidate tumor suppressor gene, is altered infrequently in head and neck squamous cell carcinoma. Cancer Res 1996;56:2519–2521.
36. Schutte M, Hruban RH, Hedrick L, et al. DPC4 gene in various tumor types. Cancer Res 1996;56: 2527–2530.
37. Handra-Luca A, Condroyer C, de Moncuit C, et al. Vessels' morphology in SMAD4 and BMPR1A-related juvenile polyposis. Am J Med Genet 2005;138A: 113–117.
38. Friedl W, Uhlhaas S, Schulmann K, et al. Juvenile polyposis: massive gastric polyposis is more common in MADH4 mutation carriers than in BMPR1A mutation carriers. Hum Genet 2002;111:108–111.
39. Houlston R, Bevan S, Williams A, et al. Mutations in DPC4 (SMAD4) cause juvenile polyposis syndrome, but only account for a minority of cases. Hum Mol Genet 1998;7:1907–1912.
40. Howe JR, Roth S, Ringold JC, et al. Mutations in the SMAD4/DPC4 gene in juvenile polyposis. Science 1998;280:1086–1088.
41. Howe JR, Sayed MG, Ahmed AF, et al. The prevalence of MADH4 and BMPR1A mutations in juvenile polyposis and absence of BMPR2, BMPR1B, and ACVR1 mutations. J Med Genet 2004;41: 484–491.
42. Hayes NV, Phillips GW, Carden MJ, Baines AJ. Definition of a sequence unique in β II spectrin required for its axonspecific interaction with fodaxin. J Neurochem 1997;68:1686–1695.
43. Fucini RV, Navarrete A, Vadakkan C, et al. Activated ADP-ribosylation factor assembles distinct pools of actin on golgi membranes. J Biol Chem. 2000;275:18,824–18,829.
44. Godi A, Santone I, Pertile P, et al. ADP ribosylation factor regulates spectrin binding to the Golgi complex. Proc Natl Acad Sci USA 1998;95:8607–8612.
45. Chang JG, Scarpa A, Eddy RL, et al. Cloning of a portion of the chromosomal gene and cDNA for human β-fodrin, the nonerythroid form of β-spectrin. Genomics 1993;17:287–293.
46. Hu RJ, Watanabe M, Bennett V. Characterization of human brain cDNA encoding the general isoform of β-spectrin. J Biol Chem 1992;267:18,715–18,722.
47. Watkins PC, Eddy R, Forget BG, Chang JG, Rochelle R, Shows TB. Assignment of a non-erythroid spectrin gene to human chromosome 2. Am J Hum Genet 1998;43:A161.
48. Golestaneh N, Mishra B. TGF-β, neuronal stem cells and glioblastoma. Oncogene 2005;24: 5722–5730.
49. Mishra L, Katuri V, Evans S. The Role of PRAJA and ELF in TGF-β Signaling and Gastric Cancer. Cancer Biol Ther 2005;4:694–699.
50. Moustakas A, Souchelnytskyi S, Heldin CH. Smad regulation in TGF-β signal transduction J Cell Sci 2001;114:4359–4369.
51. Wrana JL, Attisano L. The Smad pathway. Cytokine Growth Factor Rev 2000;11:5–13.
52. Bennett V, Baines AJ. Spectrin and Ankyrin-Based Pathways: Metazoan Inventions for Integrating Cells Into Tissues. Physiol Rev 2001;81:1353–1392.
53. Nelson WJ, Veshnock PJ. Dynamics of membrane-skeleton (fodrin) organization during development of polarity in Madin-Darby canine kidney epithelial cells. J Cell Biol 1986;103:1751–1765.
54. De Matteis MA, Morrow JS. Spectrin tethers and mesh in the biosynthetic pathway. J Cell Sci 2000;113:2331–2343.
55. Weinstein M, Monga SP, Liu Y, et al. Smad proteins and hepatocyte growth factor control parallel regulatory pathways that converge on β1-integrin to promote normal liver development. Mol Cell Biol 2001;21:5122–5131.
56. Mishra L, Cai T, Levine A, et al. Identification of elf1, a β-spectrin, in early mouse liver development. Int J Dev Biol 1998;42:221–224.
57. Goumans MJ, Mummery C. Functional analysis of the TGFβ receptor/Smad pathway through gene ablation in mice. Int J Dev Biol 2000;44:253–265.
58. Larsson J, Goumans MJ, Sjostrand LJ, et al. Abnormal angiogenesis but intact hematopoietic potential in TGF-β type I receptor-deficient mice. EMBO J 2001;20:1663–1673.
59. Takaku K, Oshima M, Miyoshi H, Matsui M, Seldin MF, Taketo MM. Intestinal tumorigenesis in compound mutant mice of both Dpc4 (Smad4) and Apc genes. Cell 1998;92:645–656.
60. Sirard C, de la Pompa JL, Elia A, et al. The tumor suppressor gene Smad4/Dpc4 is required for gastrulation and later for anterior development of the mouse embryo. Genes Dev 1998;12: 107–119.

61. Yang X, Li C, Xu X, Deng C. The tumor suppressor SMAD4/DPC4 is essential for epiblast proliferation and mesoderm induction in mice. Proc Natl Acad Sci USA 1998;95:3667–3672.
62. Takaku K, Miyoshi H, Matsunaga A, Oshima M, Sasaki N, Taketo MM. Gastric and duodenal polyps in Smad4 (Dpc4) knockout mice. Cancer Res 1999;59:6113–6117.
63. Seno H, Oshima M, Ishikawa TO, et al. Cyclooxygenase 2- and prostaglandin E(2) receptor EP(2)-dependent angiogenesis in Apc(Delta716) mouse intestinal polyps. Cancer Res 2002;62: 506–511.
64. Takeda H, Miyoshi H, Tamai Y, Oshima M, Taketo MM. Simultaneous expression of COX-2 and mPGES-1 in mouse gastrointestinal hamartomas. Br J Cancer 2004;90:701–704.
65. Taketo MM, Takaku K. Gastrointestinal tumorigenesis in Smad4 (Dpc4) mutant mice. Hum Cell 2000;13:85–95.
66. Xu X, Brodie SG, Yang X, et al. Haploid loss of the tumor suppressor Smad4/Dpc4 initiates gastric polyposis and cancer in mice. Oncogene 2000;19:1868–1874.
67. Li W, Qiao W, Chen L, et al. Squamous cell carcinoma and mammary abscess formation through squamous metaplasia in Smad4/Dpc4 conditional knockout mice. Development 2003;130: 6143–6153.
68. Qiao W, Li AG, Owens P, Xu X, Wang XJ, Deng CX. Hair follicle defects and squamous cell carcinoma formation in Smad4 conditional knockout mouse skin. Oncogene 2006;25:207–217.
69. Yang X, Li C, Herrera PL, Deng CX. Generation of Smad4/Dpc4 conditional knockout mice. Genesis 2002;32:80–81.
70. Kretzschmar M, Doody J, Timokhina I, Massagué J. A mechanism of repression of TGFβ/Smad signaling by oncogenic Ras. Genes Dev 1999;13:804–816.
71. Roberts AB, Sporn MB. The transforming growth factor-βs. In: Peptide Growth Factors and Their Receptors, Handbook of Experimental Pharmacology. (Sporn MB, Roberts AB, eds.), Springer-Verlag, Berlin, Germany: 1990:419–472.
72. Peinado H, Quintanilla M, Cano A. Transforming growth factor β-1 induces snail transcription factor in epithelial cell lines: mechanisms for epithelial mesenchymal transitions. J Biol Chem 2003;278: 21,113–21,123.
73. Muller N, Reinacher-Schick A, Baldus S, et al. Smad4 induces the tumor suppressor E-cadherin and P-cadherin in colon carcinoma cells. Oncogene 2002;21:6049–6058.
74. Li QL, Ito K, Sakakura C, et al. Causal relationship between the loss of RUNX3 expression and gastric cancer. Cell 2002;109:113–124.
75. Sormunen RT, Leong AS, Vaaraniemi JP, Fernando SS, Eskelinen SM. Immunolocalization of the fodrin, E-cadherin, and β-catenin adhesion complex in infiltrating ductal carcinoma of the breast-comparison with an in vitro model. J Pathol 1999;187:416–423.
76. Brooks-Wilson AR, Kaurah P, Suriano G, et al. Germline E-cadherin mutations in hereditary diffuse gastric cancer: assessment of 42 new families and review of genetic screening criteria. J Med Genet 2004;41:508–517.
77. Rosivatz E, Becker I, Specht K, et al. Differential expression of the epithelial-mesenchymal transition regulators snail, SIP1, and twist in gastric cancer. Am J Pathol 2002;161:1881–1891.
78. Scartozzi M, Galizia E, Freddari F, Berardi R, Cellerino R, Cascinu S. Molecular biology of sporadic gastric cancer: prognostic indicators and novel therapeutic approaches. Cancer Treat Rev 2004;30: 451–459.
79. Yang J, Mani SA, Donaher JL, et al. Twist, a master regulator of morphogenesis, plays an essential role in tumor metastasis. Cell 2004;117:927–939.
80. Nelson WJ, Hammerton RW. A membrane-cytoskeletal complex containing Na^+, K^+-ATPase, ankyrin, and fodrin in Madin-Darby canine kidney (MDCK) cells: implications for the biogenesis of epithelial cell polarity. J Cell Biol 1989;108:893–902.
81. Piepenhagen PA, Nelson WJ. Biogenesis of polarized epithelial cells during kidney development in situ: roles of E-cadherin-mediated cell-cell adhesion and membrane cytoskeleton organization. Mol Biol Cell 1998;9:3161–3177.
82. Pinder JC, Baines AJ. A protein accumulator. Nature 2000;406:253–254.
83. Fath KR, Trimbur GM, Burgess DR. Molecular motors and a spectrin matrix associate with Golgi membranes in vitro. J Cell Biol 1997;139:1169–1181.
84. Nelson WJ, Nusse R. Convergence of Wnt, β-catenin, and cadherin pathways. Science 2004;303: 1483–1487.
85. Pradhan D, Lombardo CR, Roe S, Rimm DL, Morrow JS. α-Catenin binds directly to spectrin and facilitates spectrin-membrane assembly in vivo. J Biol Chem 2001;276:4175–4181.

86. Bienz M, Clevers H. Linking colorectal cancer to Wnt signaling. Cell 2000;103:311–320.
87. Van de Wetering M, Sancho E, Verweij C, et al. The β-catenin/TCF-4 complex imposes a crypt progenitor phenotype on colorectal cancer cells. Cell 2002;111:241–250.
88. Abraham SC, Wu TT, Klimstra DS, et al. Distinctive molecular genetic alterations in sporadic and familial adenomatous polyposis-associated pancreatoblastomas: frequent alterations in the APC/ß-catenin pathway and chromosome 11p. Am J Pathol 2001;159:1619–1627.
89. Cagatay T, Ozturk M. P53 mutation as a source of aberrant ß-catenin accumulation in cancer cells. Oncogene 2002;21:7971–7980.
90. Mishra L, Shetty K, Tang Y, Stuart A, Byers SW. The role of TGF-β and Wnt signaling in gastrointestinal stem cells and cancer. Oncogene 2005;24:5775–5789.

7 Regulation of Smad Activity by Phosphorylation

Fang Liu

CONTENTS

INTRODUCTION
RECEPTOR-MEDIATED PHOSPHORYLATION
SMAD PHOSPHATASES
SMAD PHOSPHORYLATION BY MAP KINASE FAMILY MEMBERS
SMAD PHOSPHORYLATION BY CDK
SMAD PHOSPHORYLATION BY OTHER KINASES
CONCLUDING REMARKS
ACKNOWLEDGMENTS
REFERENCES

Abstract

Transforming growth factor-β (TGF-β) regulates a wide variety of biological activities. Smad proteins play a major role in transducing the TGF-β family signal at the cell surface into gene regulation in the nucleus. In response to TGF-β, Smad2 and Smad3 are activated through phosphorylation by the TGF-β receptor at the C-tail SSXS motif. The activated Smad2 and Smad3 then form heteromeric complexes with Smad4, and together accumulate in the nucleus to regulate transcription of target genes. Similarly, Smad1 is activated through phosphorylation by the BMP receptor at the C-tail SSXS motif. Smad phosphatases that dephosphorylate the SSXS motifs have been identified as key regulators in the termination of the TGF-β family signals. Smad proteins are also phosphorylated by other kinases including the MAP kinase family members, cyclin-dependent kinases, protein kinase C, and Ca^{2+}-calmodulin-dependent kinase II. Phosphorylation by these kinases regulates Smad activity in several modes, such as nuclear accumulation, DNA binding, and transcriptional activity, thus affecting proliferation, apoptosis, and other activities. Identification of the various kinases and phosphatases that phosphorylate/dephosphorylate Smads and elucidation of the mechanisms by which they regulate Smads activities help develop more effective therapies against cancer.

Key Words: TGF-β; Smad; phosphorylation; phosphatase; receptor kinase; MAP kinase; CDK; cancer.

1. INTRODUCTION

Transforming growth factor-β (TGF-β) regulates essentially all aspects of biological activities, such as cell proliferation, differentiation, adhesion, motility, apoptosis, angiogenesis, and immune function *(1,2)*. Smad proteins play critical roles in transmitting the TGF-β

From: *Cancer Drug Discovery and Development: Transforming Growth Factor-β in Cancer Therapy, Vol I: Basic and Clinical Biology*
Edited by: Sonia B. Jakowlew © Humana Press Inc., Totowa, NJ

signal *(3–8)*. TGF-β signals through two types of transmembrane serine/threonine kinase receptors, the type I and type II receptors *(3–8)*. Upon TGF-β binding to its type II receptor, the type I receptor is recruited to the complex. In the ligand-receptor complex, the type II receptor has a constitutive kinase activity. It trans-phosphorylates the type I receptor and consequently activates it. The activated type I receptor plays a major role in specifying downstream signaling events *(3–8)*. It activates Smad proteins that transmit the TGF-β signal at the cell surface into gene regulation in the nucleus, leading to various biological effects *(3–8)*.

The Smad family can be divided into three groups *(3–8)*. One group includes the receptor regulated Smads (R-Smads). For the TGF-β and activin signaling pathways, the TGF-β type I receptor and the analogous activin type I receptor phosphorylate the SSXS motif in the C-tail of Smad2 and Smad3 *(9–12)*. Similarly, for the bone morphogenetic protein (BMP) pathways, the BMP type I receptors phosphorylate the SSXS motif in the C-tail of Smad1, Smad5, and Smad8 *(13,14)*. The second group includes the common Smads (Co-Smad) that participate in the signaling pathways of various TGF-β family members. Smad4 is the only Co-Smad in mammalian cells. While Smad4 is not phosphorylated by receptors, Smad4 can form heteromeric complexes with receptor phosphorylated Smads, and can participate in transcriptional regulation of TGF-β, activin, and BMP responsive genes *(15–17)*. The third group includes inhibitory Smads (I-Smads) that antagonize the function of receptor-regulated Smads. Smad7 inhibits TGF-β family signaling, and Smad6 inhibits BMP signaling *(18–21)*. Smad7 and Smad6 are direct target genes of activated Smad complexes. Their transcription is induced by ligand treatment, therefore providing a negative feedback loop control of TGF-β family signaling *(22–26)*.

Smads contain conserved N-terminal and C-terminal domains, connected by a divergent proline-rich linker region *(3–8)*. The Smad N-terminal domains are responsible for binding to DNA and Smads usually bind to DNA in cooperation with cellular factors *(3–8,27–30)*. The transcriptional activities reside in the C-terminal domains of R-Smads and Smad4 *(31–39)*. In addition, the linker regions of Smad3 and Smad4 have been shown to contain transcriptional activation domains *(40–43)*. Smads activate transcription by recruitment of histone acetyl transferases, such as p300/CBP, P/CAF and GCN5, and by interaction with components in the ARC/mediator complex *(32–44)*.

The TGF-β family-induced receptor phosphorylation of R-Smads at the SSXS motif in their C-tails is necessary for formation of heteromeric complexes with Smad4 and accumulation in the nucleus *(9,11,12)* (Fig. 1). Smad phosphatases dephosphorylate the SSXS motifs to turn off the TGF-β family signal and to recycle Smads *(45–48)* (Fig. 1). In addition to the TGF-β receptor, several other kinases have been shown to phosphorylate Smad proteins, which include members of the mitogen-activated protein kinase (MAPK) superfamily, cyclin-dependent kinases (CDKs), Ca^{2+}-calmodulin-dependent kinase II (CaMKII), and protein kinase C (PKC) *(7,49)*. Thus, Smads integrate signaling pathways activated by a variety of stimuli. This chapter is centered on regulation of Smad activity through phosphorylation by the various kinases and dephosphorylation by phosphatases as well as their implications in cancer.

2. RECEPTOR-MEDIATED PHOSPHORYLATION

TGF-β receptor mediated phosphorylation of Smad2 is mapped to the two serine residues, S465 and S467 in the C-tail SSXS motif (Fig. 2), identified by phosphopeptide mapping *(9,11,12)*. The phosphorylation occurs in an obligate order: phosphorylation of S465 requires the prior phosphorylation of S467 *(12)*. The TGF-β type I receptor kinase apparently initially recognizes only S467, which upon phosphorylation then provides a new recognition sequence, allowing phosphorylation of S465. The requirement of S467

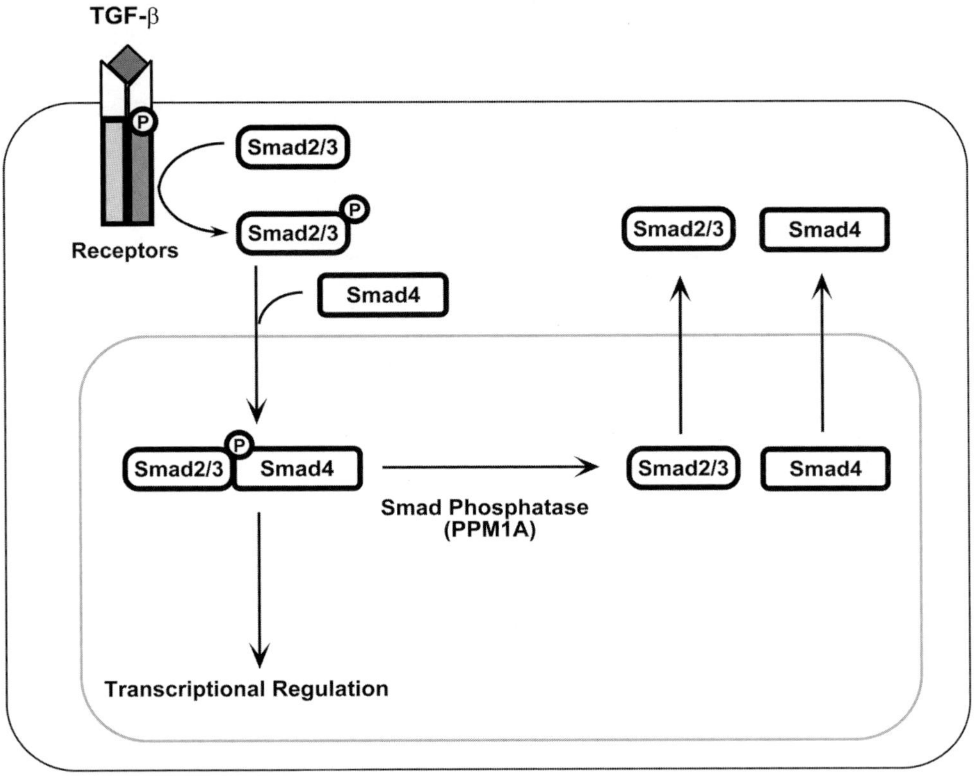

Fig. 1. A model for Smad2/3 phosphorylation by the TGF-β receptor and dephosphorylation by a Smad phosphatase. In response to TGF-β, Smad2 and Smad3 are phosphorylated by the receptor. The activated Smad2/3 dissociate from the receptor, form heteromeric complexes with Smad4 and together accumulate in the nucleus to regulate transcription. PPM1A is identified as a Smad phosphatase. PPM1A binds to phosphorylated Smad2/3 and promotes their dissociation with Smad4, presumably through dephosphorylation of Smad2/3. This leads to nuclear export of Smads and turning off the TGF-β signal.

phosphorylation could be bypassed by replacement of S467 with an aspartic acid residue *(12)*. Thus, the appearance of a negative charge at the place of S467 is sufficient for recognition of S465 as a phosphate acceptor. The necessity for sequential phosphorylation of Smad2 might increase the specificity of phosphorylation, thereby providing a safeguard against inappropriate activation by certain other kinases. It may also create a certain threshold level for Smad2 activation, in which activation occurs only after both serine residues have become phosphorylated *(12)*.

In the SSXS motif, although S464 in Smad2 is not phosphorylated, it is important for efficient phosphorylation of Smad2 at S467 and S465. Mutation of S464 to alanine reduces S467 and S465 phosphorylation in vivo *(11)*. One explanation is that S464 mutation to alanine interferes, in part, with phosphorylation at S467 and S465. The S464A mutant also functions in a dominant negative fashion, albeit somewhat less efficiently than the dominant effects of S465A and S467A mutants *(11)*.

Smad2 and Smad3 interact with the TGF-β type I receptor transiently *(9,11,12)*. The interactions between Smad2/3 and the receptor are mediated by the L3 loop in the C-terminal domain of Smad2/3 and L45 loop in the TGF-β type I receptor *(50–53)*. The conserved L3 loop differs by only two amino acids in the TGF-β receptor regulated Smad2 and Smad3

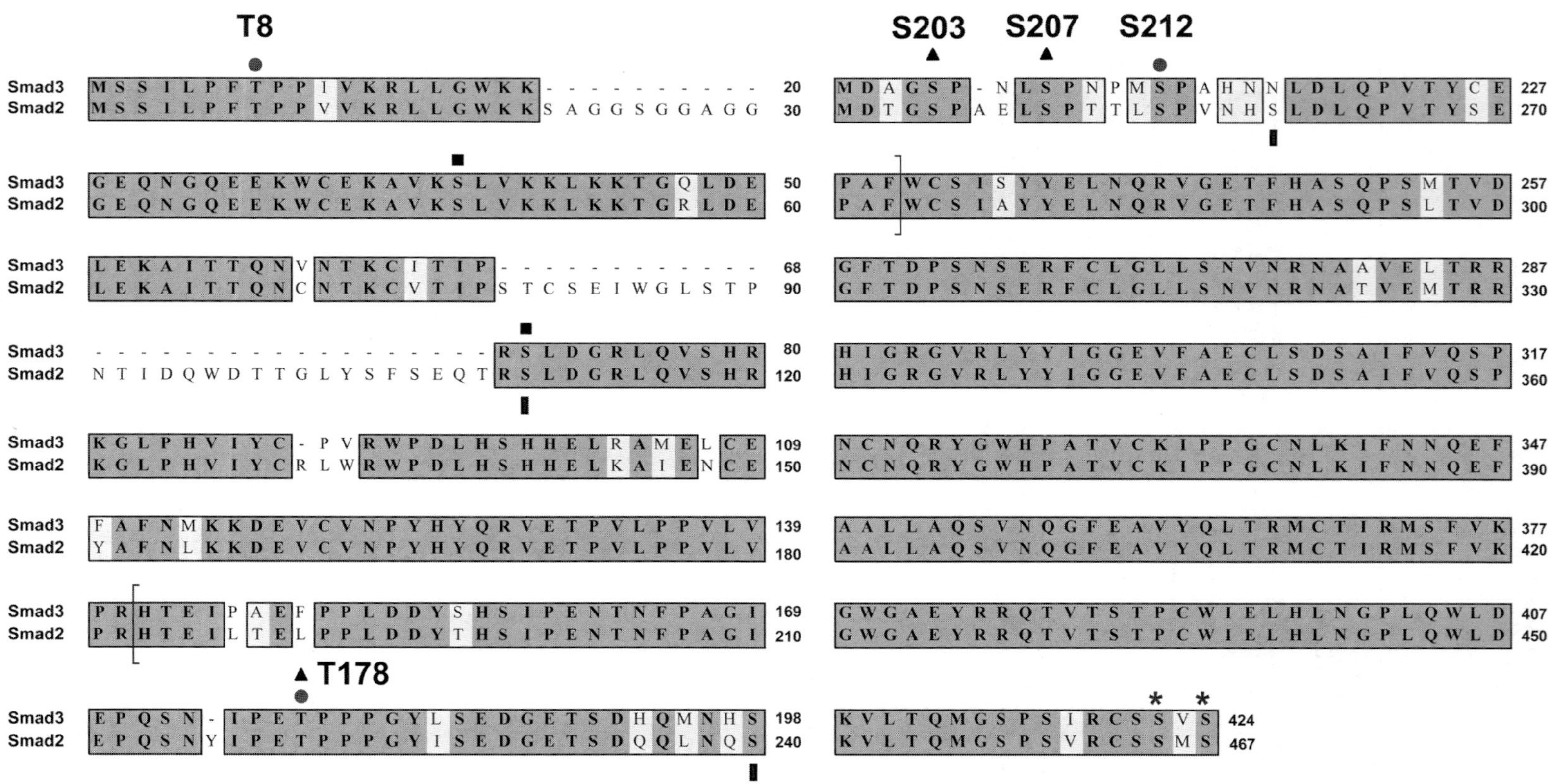

Fig. 2. Mapped phosphorylation sites in Smad3 and Smad2. The linker regions of Smad3 and Smad2 are marked by brackets. For Smad3, it is amino acids 142-230. For Smad2, it is amino acids 183-273. TGF-β receptor kinase phosphorylates the SSXS motif in Smad2 and Smad3 (asterisks). In Smad3, T8, T178, and S212 are phosphorylated by CDK4 and CDK2 (dots); S207, S203 and T178 are phosphorylated by ERK MAP kinase in response to EGF (triangles); S37 and S70 are phosphorylated by PKC (squares). In Smad2, S47 and S110 in the analogous positions are also phosphorylated by PKC; CaMKII phosphorylates S240, S260 and S110 (long bars).

versus the BMP receptor regulated Smad1, Smad5, and Smad8 *(52)*. Interestingly, the difference in these two amino acids is usually sufficient for discrimination by the TGF-β receptor versus the BMP receptor *(52)*. Under certain circumstances, Smad1 can be phosphorylated in response to TGF-β treatment, presumably by the TGF-β type I receptor *(54)*. The crystal structure of the C-terminal domain of Smad2 is similar to that of Smad4, but the C-terminal domain of Smad2 contains an extended basic pocket near the L3 loop *(55–57)*. This basic pocket in Smad2 has been proposed to serve as a docking site for the phosphorylated GS domain of the activated TGF-β type I receptor *(56)*.

The transient interactions between receptors and Smad2/3 were shown by affinity labeling the receptors with [^{125}I]-TGF-β, the most sensitive detection method *(9,11,12)*. The interaction requires the kinase activity of the type II receptor, which phosphorylates and activates the type I receptor. In the presence of a kinase defective type II receptor, no interaction between receptors and Smad2, and presumably Smad3 as well, can be detected *(9)*. In the presence of both wild type TGF-β types II and I receptors, little or none of Smad2 can be detected for interaction with the [^{125}I]-TGF-β labeled receptors due to the nature of transient interactions. In the presence of wild type TGF-β type II receptor and a kinase defective type I receptor, however, Smad2 can be readily detected for interaction with the [^{125}I]-TGF-β labeled receptors *(9)*. When the C-tail phosphorylation sites are mutated, Smad2 can be readily detected for interaction with wild type TGF-β types II and I receptors *(11,12)*. As described above, the C-tail phosphorylation mutants function as dominant negative mutants. The dominant negative effects could be explained in part by their strong interaction with the wild-type TGF-β receptor, which may prevent interaction with and phosphorylation of Smad2 and Smad3, thereby blocking signal transduction.

Smad2 and Smad3 are unable to interact with the receptors on their own. Instead, they require other proteins to assist in this process. SARA (Smad anchor for receptor activation) functions to recruit Smad2 and Smad3 to the TGF-β receptor *(58,59)*. SARA contains a FYVE domain for membrane localization, a Smad binding domain (SBD) for binding to Smad2/3, and a carboxyl-terminal domain for interacting with the receptor kinase *(58)*. At the basal state, Smad2 and Smad3 are bound by SARA. Structural studies indicate that SARA binds and stabilizes monomeric forms of Smad2 and Smad3, and inhibits Smad2 and Smad3 trimerization *(60,61)*. The cytoplasmic promyelocytic leukaemia protein (cPML) physically interacts with both Smad2/3 and SARA and is required for association of Smad2/3 with SARA *(62)*. cPML can also bind to the receptors. Upon TGF-β binding and activation of the receptor, cPML can promote the transfer of the complex containing TGF-β receptors, SARA, Smad2/3, and probably cPML itself into early endosome *(62)*. cPML then dissociates from the complex, followed by SARA presenting Smad2 and Smad3 for receptor recognition, and precisely positioning the phosphorylation sites of Smad2 and Smad3 in the kinase catalytic center *(59–62)*. The phosphorylated Smad2/3 then dissociate from SARA and the receptor, ready to propagate the signal to downstream events *(58–62)*.

Dissociation of the phosphorylated Smad2/3 is necessary for the subsequent signaling events, and the dissociation from the receptor requires both S467 and S465 being phosphorylated *(11,12)*. As described above, a Smad2 mutant with S467 replaced by aspartic acid, in which phosphorylation of S465 still occurs, also had a dominant negative effect *(12)*. Thus, although the presence of a negative charge at the position of S467 allows phosphorylation of S465 by TGF-β type I receptor, this is not sufficient for efficient dissociation of phosphorylated Smad2 from the receptor kinase. Similarly, when both S467 and S465 were replaced with aspartic acid residues, the mutant Smad2 showed a strong interaction with TGF-β type I receptor and a dominant negative effect on TGF-β signaling *(12)*. These observations highly suggest the necessity for simultaneous phosphorylation of S467 and S465 for efficient dissociation of activated Smad2 from the receptor kinase.

After Smad2/3 is phosphorylated by the receptor and dissociated from the SARA and the receptor, they form complexes with Smad4 and together accumulate in the nucleus. Smad2 is predominantly present in the cytoplasm at basal state, whereas a significant proportion of Smad3 is present in the nucleus even at basal state *(45)*. Thus, most experiments on TGF-β-induced nuclear accumulation have been analyzed using Smad2 as a model system. Smad2 phosphorylation by the receptor is necessary for ligand-induced nuclear accumulation. When the S467 and S465 are mutated, TGF-β induced nuclear accumulation of Smad2 is abrogated *(11,12)*. Importantly, mutation of the S467 and S465 also disrupts TGF-β-induced nuclear accumulation of Smad4 *(12)*. Thus, receptor mediated phosphorylation of Smad2 and Smad3 at the C-tail SSXS motif results in dissociation from SARA and the receptors, and is important for subsequent complex formation with Smad4 and together accumulation in the nucleus *(11,12)*.

3. SMAD PHOSPHATASES

Since the C-tail SSXS motif phosphorylation plays such a pivotal role in activation of the Smad signaling pathways, it is imperative to identify the phosphatases that turn off the Smad signaling. Recently, two Smad phosphatases that dephosphorylates the C-tail SSXS motif have been identified *(45–48)*. Through overexpression of catalytic subunits for 39 different human phosphatases, PPM1A/PP2C-α is found to be the only one that reduces Smad2/3 C-tail phosphorylation induced by the constitutively active TGF-β type I receptor *(45,46)*. In an in vitro assay, PPM1A can dephosphorylate the semi-synthetic phosphorylated Smad2 at the C-tail. PPM1A is localized in the nucleus in the absence as well as in the presence of TGF-β. Both wild type and catalytic inactive PPM1A can interact with Smad2. However, only the C-tail phosphorylated Smad2, but not the unphosphorylated Smad2, can interact with PPM1A. Knockdown of PPM1A by shRNA enhances nuclear levels of phosphorylated Smad2/3, and consequently the nuclear levels of Smad2-Smad4 complex. Conversely, overexpression of wild type PPM1A, but not a catalytically inactive PPM1A, abolishes the formation of TGF-β-induced Smad2/3-Smad4 complexes. Moreover, PPM1A promotes Smad2 nuclear export *(45,46)*. These observations suggest that PPM1A dephosphorylates Smad2 and Smad3 in the nucleus, leading to the dissociation of Smad2/3-Smad4 complexes and subsequent nuclear export (Fig. 1). Previous studies proposed that Smad phosphatases are present in the nucleus to dephosphorylate Smads, allowing them to recycle back to the cytoplasm *(63,64)*. PPM1A apparently fits these criteria.

PPM1A knockdown cells have an increased sensitivity to the growth-inhibitory effects of TGF-β. Accordingly, depletion of PPM1A increased TGF-β-induced upregulation of p15 and p21 and downregulation of c-myc *(45,46)*. Interestingly, depletion of PPM1A also increased the basal levels of p15, p21, and c-myc. The underlying mechanisms for the increased basal levels of p15, p21 and c-myc remain to be addressed. Depletion of PPM1A appears to enhance global TGF-β signaling, as the TGF-β responses for plasminogen activator inhibitor-1 and fibronectin are also increased *(45,46)*. Conversely, conditional overexpression of PPM1A, even just at a level comparable to that of the endogenous PPM1A, reduces TGF-β responses *(45,46)*. These observations further support the role of PPM1A in turning off the TGF-β signal.

The role of PPM1A is also supported by study of embryonic development. The zebrafish PPM1A, which shares 75% amino acid sequence identity to the human PPM1A, displays strong activity towards dephosphorylating both Smad2 and Smad3. Injection of PPM1A mRNA into zebrafish embryos antagonizes the dorsalizing activity of Nodal/Smad2/Smad3 *(45,46)*. Thus, these observations provide strong evidence for the physiological role of PPM1A in the dephosphorylation of the activated Smad2 and Smad3.

PPM1A has been previously shown to dephosphorylate other substrates. PPM1A specifically interacts with and dephosphorylates the phosphorylated substrates, including p38 MAPK, CDK2, phosphatidylinositol 3-kinase (PI3K), and Axin *(65–68)*. PPM1A seems not to influence TGF-β-mediated activation of p38 MAPK and PI3K. In addition, these kinases do not mediate the inhibitory effect of PPM1A on Smad signaling *(45)*. PPM1A has also been reported to inhibit cell proliferation and activation of apoptosis *(69)*. Conditional overexpression of PPM1A in 293 cells can lead to G2-M cell cycle arrest and apoptosis *(69)*. Currently we do not know how these functions fit with its role to turn off the TGF-β signal. TGF-β seems to have no effect on the subcellular localization, expression level and activity of PPM1A *(45)*. It will be extremely important to determine how PPM1A is regulated to control TGF-β signaling.

Interestingly, another very different phosphatase has been identified for the DPP (decapentaplegic), a BMP-like pathway in *Drosophila (47,48)*. Through RNA interference (RNAi)-mediated screen by knocking down all 44 serine/threonine phosphatases in *Drosophila* S2 cells, the pyruvate dehydrogenase phosphatase (PDP) has been found to be necessary for dephosphorylation of Mothers Against DPP (MAD), a *Drosophila* Smad *(47,48)*. When the PDP is knocked down, it resulted in sustained phosphorylation of MAD. Using the powerful genetics in *Drosophila*, it has been shown that the level of the phosphorylated MAD is increased in two *Drosophila* strains that are defective in the PDP gene or its expression *(47,48)*. One strain harbors a piggyback transposon just upstream of the first exon of the PDP gene, and the other strain contains a chromosome deletion that includes the PDP gene. Recombinant GST-PDP can specifically dephosphorylate phosphorylated MAD in vitro. A direct interaction between phosphorylated MAD and PDP can be detected by coimmunoprecipitation assay. Interestingly, although the C-tail phosphorylation is not required for MAD and PDP interaction, the C-tail phosphorylation substantially increases the interaction *(47,48)*. The two mammalian homologues of the *Drosophila* PDP are important in dephosphorylating BMP-activated Smad1 but not TGF-β-activated Smad2 or Smad3 *(45,47)*. Thus, the PDPs specifically inactivate Smads in the BMP/DPP pathway.

The subcellular localization of PDP attracts much attention. In live *Drosophila* S2 cells, PDP fused to the green fluorescent protein is distributed throughout the cell. Moreover, it exhibits a punctate pattern that overlaps with mitochondria. Such distribution of PDP is not changed in the presence of DPP *(47,48)*. Overexpression of PDP-green fluorescent protein seems not to prevent nuclear accumulation of MAD, possibly due to the receptor activation being far more potent than PDP activity under the experimental condition or owing to the necessity to use more sensitive and quantitative methods to measure the changes in the kinetics of MAD nuclear import and export *(47,48)*. Thus, PDP appears to be broadly distributed and can therefore gain access to substrates in mitochondria, cytoplasm, and nucleus. Indeed, in both the cytosolic and nuclear fractions, coimmunoprecipitation of phospho-MAD and PDP could be detected *(47)*. Less phospho-MAD was coimmunoprecipitated with PDP in the nuclear fraction, which could be due to the fact that less PDP was present in the nucleus and that the high salt buffer for nuclear extraction may not be friendly for the phospho-MAD and PDP interaction *(47)*. Since Smad phosphatases were proposed to function in the nucleus *(63,64)*, it will be important to determine in the future where dephosphorylation of MAD and Smad1 occurs, in the cytoplasm, the nucleus, or both?

What is the functional relevance of PDP being a Smad1/MAD phosphatase? PDP is known to dephosphorylate the pyruvate dehydrogenase complex in mitochondria, which catalyzes the conversion of pyruvate to acetyl CoA. Dephosphorylation counteracts the inhibitory activity of pyruvate dehydrogenase kinase and thus reactivates the pyruvate dehydrogenase complex *(48)*. It is possible that the BMP signaling pathways are connected with the metabolic pathways *(48)*.

Since MAD is still dephosphorylated to certain extent in the PDP RNAi knockdown cells *(47,48)*, it is possible that additional Smad1 phosphatases for the C-tail SSXS motifs are present. Interestingly, PPM1A also mediates Smad1 C-tail dephosphorylation *(45)*. In addition, other phosphatases for Smad1 also appear to exist *(45)*. Thus, several phosphatases seem to participate in the dephosphorylation of Smad1 SSXS motif. Future studies are necessary to determine how these phosphatases are regulated to coordinately turn off the BMP signal.

4. SMAD PHOSPHORYLATION BY MAP KINASE FAMILY MEMBERS

Except for the receptor-mediated phosphorylation, ERK was the first kinase shown to phosphorylate Smads. It was shown that ERK phosphorylation of Smad1 inhibits its nuclear accumulation *(70)*. ERK phosphorylation of Smad1 occurs within the four putative ERK phosphorylation sites in the linker region. When the four putative phosphorylation sites are mutated, ERK-mediated phosphorylation is abolished *(70)*. Since Smad1 seems to be a good substrate for ERK and each of the four putative phosphorylation sites in Smad1 fits the ERK consensus phosphorylation site PXS/TP, each of them is likely phosphorylated by ERK. Mutation of the four ERK phosphorylation sites leads to Smad1 predominantly localized in the nucleus *(70)*. ERK phosphorylation of Smad1 results in the inhibition of Smad1 transcriptional activity, contributing to neural induction in *Xenopus (71,72)*. When the four ERK phosphorylation motifs are mutated to nonphosphorylatable alanine, the mutant has a higher activity than the wild type Smad1 to induce ventralizing effect in *Xenopus (73)*. Mutant mice carrying mutation of the four ERK phosphorylation motifs and two other conserved serine residues in the linker region of Smad1 have been generated *(74)*. The mutant mice survive embryogenesis but exhibit defects in gastric epithelial homeostasis correlated with changes in cell contacts, actin cytoskeleton remodeling, and nuclear β-catenin accumulation. In addition, formation of primordial germ cells is impaired in the mutant mice *(74)*. These observations highlight the importance of ERK phosphorylation in control of Smad1 activity *(72,75)*.

For the activin-mediated mesoderm induction in *Xenopus*, wild type Smad2 loses its competence by its inability to localize in the nucleus in response to activin at stage 11. When three putative ERK phosphorylation sites in the Smad2 linker region are mutated, the mutant Smad2 responds to activin to accumulate in the nucleus. Thus, nuclear exclusion of Smad2 provides a mechanism for loss of competence in *Xenopus (76)*. In mammalian cells, the effects of ERK phosphorylation of Smad2/3 on their subcellular localization have been investigated in a number of studies. The initial study showed that activation of Ras, and to a less extent, activation of ERK, inhibits TGF-β-induced Smad2/3 nuclear accumulation *(77)*. Subsequent studies have reported that ERK phosphorylation of Smad2/3 can inhibit, promote, or no effect on their localization and transcriptional activities *(78–87)*. A recent study showed that the ERK mediated inhibition of Smad2/3 subcellular localization depends on the expression level of an activated Ras *(87)*. When the activated Ras is expressed at high levels, which is more than five fold than the endogenous activated Ras levels, it leads to the inhibition of TGF-β-induced nuclear accumulation of Smad2/3. When the activated Ras is expressed at low levels, which is only two fold higher than the endogenous levels of activated Ras, it has no effect on the TGF-β-induced nuclear accumulation of Smad2/3 *(87)*. This suggests that in normal cells, Ras activation by other growth factors or by TGF-β does not affect Smad localization. Since high levels of activated Ras are present in a significant proportion of cancer cells *(88)*, it can inhibit Smad2/3 activity by sequestrating them in the cytoplasm.

The ERK phosphorylation sites have been mapped to S207, S203 and T178 in the linker region of Smad3 (Fig. 1), and S207 is the best ERK phosphorylation site in Smad3 *(86)*. It is the only site that is recognized to be an ERK phosphorylation site under high stringency screen by the Scansite, a computer program that recognizes the various phosphorylation sites and other signature motifs. S207 is a very sensitive site, significantly responding to EGF even at low doses (1 ng/ml). It shows much higher induction of phosphorylation by EGF than the S203 and T178 sites. S207 also shows the quickest response to EGF treatment. Its phosphorylation is markedly increased after treatment with EGF for only 5 min, whereas significant phosphorylation on S203 or T178 occurred only after EGF treatment for 15 min *(86)*. EGF-induced phosphorylation of Smad3 peaks at roughly 15–30 min at the three sites. Their phosphorylation is sustained, significant phosphorylation still remaining even after 4–8 h *(86)*, an important feature that allows to impact on gene regulation *(89,90)*. When the S207, S203 and T178 are mutated to nonphosphorylatable alanine or valine, the mutant Smad3 has an increased capacity to activate a Smad target gene, suggesting that ERK phosphorylation inhibits Smad3 activity *(86)*.

The effects of ERK phosphorylation appear to be context-dependent based on a number of studies so far, which help to explain that some disparate effects of ERK phosphorylation on Smad2/3 subcellular localization and transcriptional activities have been reported *(76–87)*. EGF can also inhibit the antiproliferative effects of TGF-β in a Smad-independent manner as found in primary ovary cells *(91)*. Many studies on TGF-β and EGF are based on tissue culture experiments, in which TGF-β and EGF can be added separately. In living organisms, it is much more complex, as TGF-β and EGF coexist. It is fully expected that TGF-β and EGF can have opposing as well as synergistic effects in a context dependent manner.

JNK (c-Jun N-terminal kinase), which belongs to the MAP kinase superfamily, can be activated by TGF-β treatment *(92,93)*. It has been reported that TGF-β activated JNK can phosphorylate Smad2 or Smad3, and such phosphorylation can facilitate their activation by the TGF-β receptor complex and their nuclear accumulation *(92,93)*. Based on the chemical inhibitor SP600125, it has been shown that S207 and/or S212 in the Smad3 linker region can be phosphorylated by JNK *(94)*. Because SP600125 cross inhibits a variety of kinases, as indicated by the observation that 13 out of total 28 kinases analyzed are inhibited with similar or even greater potency than inhibition of JNK *(95)*, the JNK phosphorylation sites in Smad3 remain to be further examined. It has been proposed that JNK phosphorylation of Smad3 is involved in the invasion by activated mesenchymal cells and contributes to colon cancer progression *(94,96–99)*. It is necessary to point out that TGF-β-induced JNK activation is usually rapid and transient, peaking at about 10 min and declining to the basal level at ~20 min (*[94]*, and our unpublished results). In addition, JNK activation in response to TGF-β cannot be detected in some cell lines (our unpublished results). Thus, the primary role of JNK on Smad phosphorylation may be to initiate a phosphorylation event. The sustained phosphorylation may be carried out by other kinases. JNK also plays another role in TGF-β signaling. Through the use of $JNK1^{-/-}JNK2^{-/-}$ mouse embryonic fibroblasts, it has been shown that JNK inhibits expression and autocrine signaling of TGF-β1 *(100,101)*.

p38 MAP kinase is also activated in response to TGF-β. The activation is dependent on TAK1 (TGF-β activated kinase 1) *(102)*. The early activation of p38 is independent of Smad proteins *(103)*, whereas the delayed activation of p38 relies on Smad-dependent GADD45β expression *(104)*. An initial study indicated that TGF-β mediated activation of p38 MAP kinase is required for TGF-β-induced apoptosis, epithelial-to-mesenchymal transition (EMT), but not growth arrest, based on results in the normal murine mammary gland NMuMg cell line *(103)*. A subsequent study showed that TGF-β induced activation of p38 MAP kinase contributes to the TGF-β growth inhibition in the MCF10A1h human breast cancer cell line *(105)*. It is possible that TGF-β-activated p38 may have distinct roles for growth inhibition in a context-dependent manner. Notably, p38 plays an important role in EMT and apoptosis *(106–108)*. Smad3 is essential for TGF-β-induced EMT as well as apoptosis. For instance, TGF-β-induced EMT is lost in Smad3 knockout kidney epithelial cells and lens epithelial cells *(109–111)*. The TGF-β-induced EMT and apoptosis are lost in the conditional Smad3 knockout hepatocytes *(112)*. Smad3 markedly reduces carcinogen-induced hepatocarcinoma in transgenic mice by sensitizing hepatocytes to undergo apoptosis through downregulation of Bcl-2, a major inhibitor for apoptosis *(108)*. TGF-β-induced activation of p38 is involved in this process *(108)*. Since p38 can phosphorylate Smad3 in the linker region *(94,113)*, it will be important to address whether p38 phosphorylation of Smad3 plays an important role in the regulation of tumor initiation and progression as well as metastasis.

5. SMAD PHOSPHORYLATION BY CDK

TGF-β potently inhibits cell proliferation by causing cell cycle arrest at the G1 phase *(49,114–116)*. The CDKs that function during the G1 phase include the homologous CDK4 and CDK6 as well as CDK2 *(117)*. CDK4/6 are activated by D type cyclins in early-mid G1 phase, whereas CDK2 complexes with E- and A-type cyclins during late G1 and S phase, respectively. G1 CDKs are negatively controlled by two classes of CDK inhibitors. The first class includes four members of the INK4 family (p16, p15, p18, and p19), which specifically inhibit cyclin D associated CDK4 or CDK6 activities. The second class includes three members of the Cip/Kip family (p21, p27, and p57), which inhibit cyclin E or cyclin A associated CDK2 activity *(117)*. TGF-β inhibits G1 cell cycle progression by blocking the CDK activities *(49,114–116)*. TGF-β downregulates the expression of c-myc, inhibitor of differentiation (Id), and upregulates the expression of the CDK inhibitors p15 and p21 *(49,114–116,118–122)*. The expression level as well as the activity of CDC25A is also downregulated by TGF-β *(123–125)*. Smad3 plays a critical role in mediating the TGF-β growth inhibitory responses. For example, the TGF-β antiproliferative responses are greatly diminished in a variety of Smad3$^{-/-}$ primary cells, such as T cells, keratinocytes, astrocytes, embryonic fibroblasts, and hepatocytes *(112,126–131)*. For TGF-β downregulation of c-myc gene expression, a preassembled complex containing Smad3, E2F4 (or E2F5) and p107 moves into the nucleus in response to TGF-β, associates with Smad4 and binds to a composite site for E2F and Smad3, leading to the inhibition of c-myc transcription *(132–135)*. c-myc downregulation is also important for subsequent transcriptional activation of the p15 and p21 genes *(136–140)*. Smad2, Smad3, and Smad4 together with cellular factors, such as Sp1, FoxO, and Runx3, activate the expression of p15 and p21 *(141–145)*. The induced p15 binds to CDK4 and CDK6, thus blocking their interaction with cyclin D and inhibiting their activities. As a result, p27 that is bound to the cyclin D-CDK4/6 is released, and then binds to the cyclin E-CDK2 complex, inhibiting its activity. p21 also binds to cyclin E-CDK2 and blocks its activity. Thus, the coordinated inhibition of CDK4/6 and CDK2 activities leads to cell cycle arrest induced by TGF-β *(49,114–116)*.

Smad3, which plays an important role in mediating the TGF-β antiproliferative responses, is a physiological substrate for both CDK4 and CDK2 *(49,130,131)*. Except for the Rb family members, Smad3 is the only substrate demonstrated for CDK4. Endogenous Smad3 is phosphorylated in a cell cycle-dependent manner, peaking close to the G1-S transition, consistent with G1 CDKs phosphorylation of Smad3 *(130)*. The peak of Smad3 phosphorylation is a couple of hours ahead of that of Rb, suggesting that Smad3 is more phosphorylated by CDK4/6 than by CDK2 compared with Rb *(130)*. Overexpression of cyclins D, E, or A, which activates endogenous CDK4/6 or CDK2, markedly increases Smad3 phosphorylation. p16 and p21, which inhibit CDK4/6 and CDK2 activities, respectively, block cyclin D- and cyclin E-mediated Smad3 phosphorylation, respectively *(130)*. These observations further support Smad3 as a CDK substrate.

The Smad3 phosphorylation sites by CDK4 and CDK2 have been mapped both in vivo and in vitro *(130)*. When the four putative sites in the linker region (T178, S203, S207, S212) and one putative site (T8) in the N-terminal domain are simultaneously mutated, the phosphorylation by both CDK4 and CDK2 are dramatically reduced in vivo and in vitro *(130)*. To identify the exact phosphorylation sites, specific phosphopeptide antibodies against each of the five sites were generated. All the five sites can be phosphorylated by both CDK4 and CDK2 in vitro, whereas only T8, T178 and S212 are phosphorylated in vivo by CDK4 and CDK2 *(130)* (Fig. 2). Their phosphorylation displays a cell cycle-dependent manner, in parallel with G1 CDK activities. Their phosphorylation is reduced in cells expressing shRNA to target CDK4 or CDK2 degradation. Their phosphorylation levels are also in parallel with the status of the CDK4 genotype: reduced phosphorylation in $CDK4^{neo/neo}$ mouse embryonic fibroblasts that express very little CDK4, and in contrast, increased phosphorylation in mouse embryonic fibroblasts that contain a knockin $CDK4^{R24C}$, which confers increased CDK4 activity through its ability to prevent p16 binding. All these lines of evidence support the conclusion that these three sites are phosphorylated by CDK4 and CDK2 in vivo *(49,130)*.

Mutation of the CDK phosphorylation sites in Smad3, serine to alanine and threonine to valine, increases its ability to activate the expression of p15 in both a reporter gene assay as well as by Northern blot analysis for endogenous gene in the absence as well as in the presence of TGF-β *(130)*. When the CDK phosphorylation mutant is fused to the GAL4 DNA binding domain, it has a higher transcriptional activity, suggesting that CDK phosphorylation of Smad3 inhibits its activity *(130)*. The CDK phosphorylation mutant also has an increased capacity to downregulate the expression of c-myc, which raises the possibility that an active repression mechanism may be involved in the downregulation of the c-myc gene. Smad3 inhibits cell cycle progression from the G1 to S phase *(49,130,131)*. Accordingly, $Smad3^{-/-}$ fibroblasts proliferate faster than the littermate wild type control cells *(126,130)*. Introduction of wild type Smad3 into $Smad3^{-/-}$ fibroblasts inhibits cell cycle progression from G1 to S phase. Mutation of the CDK phosphorylation sites increases this ability *(130)*. In addition, introduction of wild type Smad3, and to a greater extent, the CDK phosphorylation mutant Smad3, restores the TGF-β growth-inhibitory responses to the $Smad3^{-/-}$ fibroblasts *(130)*. Taken together, these observations indicate that CDK phosphorylation of Smad3 facilitates cell cycle progression.

The Rb pathway, in which p16 inhibits CDK4/6 and prevents them from phosphorylating Rb, is inactivated in almost all human cancers *(146–149)*. p16 inactivation is one of the most frequent events in human cancers. p16 is inactivated through deletion, mutation, or methylation of its promoter in a wide variety of cancers *(146–149)*. Cyclin D1 is amplified, translocated, or overexpressed in many cancers. Cyclin D2 and cyclin D3 amplification or overexpression also occurs in several types of cancers *(146–149)*. CDK4 is amplified and overexpressed in a variety of tumors and tumor cell lines. Some of them, mainly gliomas, sarcomas, breast tumors and carcinomas of the uterine cervix display co-amplification with the MDM2 locus as part of the amplicon *(146–149)*. The CDK6 gene is also amplified or translocated in certain types of cancers. Point mutations in CDK4 and CDK6 also occur in certain cancers or tumor cell lines. For example, $CDK4^{R24C}$, which prevents p16 binding, occurs in melanoma patients *(149)*. Frequent mutation of Rb itself, however, is restricted to only a subset of human cancers, including retinoblastomas, small cell lung cancers, osteosarcomas, and bladder carcinomas *(146,148,150)*. Thus, most cancers may inactivate both Rb and Smad3. Inhibition of the activities of Smad3, and presumably Smad2 as well by CDK phosphorylation is likely to contribute to tumorigenesis and resistance to the TGF-β growth-inhibitory effects in cancer.

TGF-β is a double-edged sword in cancer, being a potent tumor suppressor during early stages of cancer but often promotes tumor progression and metastasis at late stages of cancer *(151–153)*. Whereas it is essentially established that Smad3 inhibits tumorigenesis in early stages of cancer, it seems that its role in the later stages of cancer can be either inhibitory or promoting in a context-dependent manner (*[108,154]* as examples). Future studies are necessary to investigate the role of CDK phosphorylation of Smad3 and Smad2 not only in tumor initiation but also in tumor progression and metastasis.

6. SMAD PHOSPHORYLATION BY OTHER KINASES

CaMKII has been shown to induce in vivo phosphorylation of Smad2 and Smad4, and to a much less extent, Smad3. The CaMKII phosphorylation of Smad2 occurs in the linker (S240 and S260) and in the N-terminal domain (S110) (Fig. 2), and the phosphorylation results in the inhibition of TGF-β-inducible nuclear accumulation of Smad2 *(155)*. TGF-β-dependent transcriptional responses can be prevented by expression of a constitutively activated CaMKII *(155)*. Thus, CaMKII antagonizes TGF-β responses.

Smad signaling is also modulated by PKC *(156)*. PKC directly phosphorylates Smad3 and Smad2 both in vivo and in vitro. Through phospho amino acid analysis and Edman degradation of ^{32}P-labeled tryptic peptides, the phosphorylation has been mapped to the S37 and S70 in the N-terminal domain of Smad3 and S47 and S110 in the N-terminal domain of Smad2 (Fig. 2). Furthermore, phosphopeptide mapping of Smad3 and Smad2, with the respective serines mutated to alanine residues, confirmed that these sites are indeed phosphorylated by activated PKC. Based on the crystal structure of the MH1 domain of Smad3 *(30)*, phosphorylation of the two PKC sites is predicted to affect the ability of Smad3 to bind to DNA. Indeed, it has been shown that PKC phosphorylation disrupts the DNA binding activity of Smad3, thus inhibiting TGF-β/Smad dependent transcriptional responses *(156)*. The tumor promoter phorbol 12-myristate 13-acetate (PMA) activates PKC. Inactivation of Smad3 DNA binding activity by PKC phosphorylation renders cells more sensitive to transformation by PMA. PKC phosphorylation of Smad3 also plays a critical role for PMA to inactivate TGF-β-induced cell death. Thus, PKC phosphorylation of Smad3 inhibits both the growth inhibitory and apoptotic action of TGF-β *(156)*.

In addition to R-Smads, activities of Smad4 and I-Smads are also regulated by phosphorylation. For example, phosphorylation of Smad4 at T276 regulates its TGF-β-induced nuclear accumulation *(157)*. Phosphorylation of Smad7 at S249 affects its TGF-β-independent transcriptional activity *(158)*. The identities of the responsible kinases remain to be fully investigated.

7. CONCLUDING REMARKS

Since the discovery of Smads 10 yr ago, great progresses have been made in the identification of the kinases and phosphatases that regulate the activities of Smads. In the mean time, many interesting questions still remain to be addressed. Receptor-mediated phosphorylation of R-Smads at the C-tail SSXS motifs enables R-Smads to dissociate from the receptor, associate with Smad4, and accumulate together in the nucleus to regulate transcription. In addition to the C-tail phosphorylation, it will be essential to thoroughly and systematically identify additional sites in R-Smads that are rapidly phosphorylated in response to TGF-β, to determine which kinases are responsible for such phosphorylation, and to analyze what are the functional significance for the phosphorylation.

Smad phosphatases play a critical role in the termination of the TGF-β family signals. PPM1A dephosphorylates the SSXS motif in Smad2 and Smad3, whereas PDP, PPM1A, and potentially additional phosphatases dephosphorylate the SSXS motif in Smad1. Obviously, it will be important to examine the correlations between these phosphatases with cancer and other diseases. It will also be important to determine whether a TGF-β pathway-specific phosphatase is present, and how different phosphatases for Smad1 SSXS motif participate in turning off the BMP signal. In addition, it will be necessary to identify the phosphatases that dephosphorylate the linker regions of R-Smads.

Smads are also phosphorylated by MAP kinase superfamily members and by CDKs. Phosphorylation of Smad1, Smad2 or Smad3 by ERK can lead to inhibition of nuclear accumulation and/or inhibition of their transcriptional activity in some studies. It is necessary to emphasize that the effects of Smad phosphorylation by MAP kinase superfamily members are context-dependent. CDK phosphorylation of Smad3 inhibits its transcriptional activity and antiproliferative activity. Since part of the CDK phosphorylation and all MAP kinase-mediated phosphorylation occur in the Smad3 linker region, which contains a transcriptional activation domain, it will be of great interest to determine the exact mechanisms by which the linker phosphorylation regulates Smad3 transcriptional activity. Cancer cells often contain high levels of ERK and CDK activities. ERK and CDK may synergize to inhibit Smad activity. Furthermore, the interplay involving TGF-β, EGF, MAP kinase family members, CDK, and Smads likely play a critical role in tumorigenesis, tumor progression and metastasis. Understanding of these events may lead to advances in cancer therapy.

ACKNOWLEDGMENTS

I thank Drs. X.-H. Feng, A. K. Kamaraju, K. Matsuzaki, M. Reiss, A. B. Roberts, W. Xie, and Y. E. Zhang and members of my laboratory for helpful discussions. I also thank the American Association for Cancer Research-National Foundation for Cancer Research, the Burroughs Wellcome Fund, the Sidney Kimmel Foundation for Cancer Research, the Pharmaceutical Research and Manufacturer of America Foundation, the Emerald Foundation, the New Jersey Commission on Cancer Research, the Department of Defense Breast Cancer Research Program, and the National Institutes of Health for support of my research.

REFERENCES

1. Roberts AB, Sporn MB. The transforming growth factor-betas. In: Peptide Growth Factors and Their Receptors, Eds Sporn MB, Roberts AB. (Springer-Verlag, Heidelberg) 1990;pp.419–472.
2. Massagué J. The transforming growth factor-beta family. Annu Rev Cell Biol 1990;6:597–641.
3. Heldin C-H, Miyazono K, ten Dijke P. TGF-beta signalling from cell membrane to nucleus through SMAD proteins. Nature 1997;390:465–471.
4. Massagué J, Chen YG. Controlling TGF-beta signaling. Genes Dev 2000;14:627–644.
5. Shi Y, Massagué J. Mechanisms of TGF-beta signaling from cell membrane to the nucleus. Cell 2003;113:685–700.
6. Liu F. Receptor-regulated Smads in TGF-beta signaling. Front Biosci 2003;8:S1280–S1303.
7. Derynck R, Zhang YE. Smad-dependent and Smad-independent pathways in TGF-β family signalling. Nature 2003;425:577–584.
8. ten Dijke P, Hill CS. New insights into TGF-beta-Smad signalling. Trends Biochem Sci 2004;29: 265–273.
9. Macias-Silva M, Abdollah S, Hoodless PA, Pirone R, Attisano L, Wrana JL. MADR2 is a substrate of the TGFbeta receptor and its phosphorylation is required for nuclear accumulation and signaling. Cell 1996;87:1215–1224.
10. Zhang Y, Feng X, We R, Derynck R. Receptor-associated Mad homologues synergize as effectors of the TGF-beta response. Nature 1996;383:168–172.
11. Abdollah S, Macias-Silva M, Tsukazaki T, Hayashi H, Attisano L, Wrana JL. TbetaRI phosphorylation of Smad2 on Ser465 and Ser467 is required for Smad2-Smad4 complex formation and signaling. J Biol Chem 1997;272:27,678–27,685.
12. Souchelnytskyi S, Tamaki K, Engstrom U, Wernstedt C, ten Dijke P, Heldin C-H. Phosphorylation of Ser465 and Ser467 in the C terminus of Smad2 mediates interaction with Smad4 and is required for transforming growth factor-beta signaling. J Biol Chem 1997;272:28,107–28,115.

13. Hoodless PA, Haerry T, Abdollah S, et al. MADR1, a MAD-related protein that functions in BMP2 signaling pathways. Cell 1996;85:489–500.
14. Kretzschmar M, Liu F, Hata A, Doody J, Massagué J. The TGF-beta family mediator Smad1 is phosphorylated directly and activated functionally by the BMP receptor kinase. Genes Dev 1997;11: 984–995.
15. Lagna G, Hata A, Hemmati-Brivanlou A, Massagué J. Partnership between DPC4 and SMAD proteins in TGF-beta signalling pathways. Nature 1996;383:832–836.
16. Liu F, Pouponnot C, Massagué J. Dual role of the Smad4/DPC4 tumor suppressor in TGFβ-inducible transcriptional complexes. Genes Dev 1997;11:3157–3167.
17. Zhou S, Buckhaults P, Zawel L, et al. Targeted deletion of Smad4 shows it is required for transforming growth factor beta and activin signaling in colorectal cancer cells. Proc Natl Acad Sci USA 1998;95:2412–2416.
18. Hayashi H, Abdollah S, Qiu Y, et al. The MAD-related protein Smad7 associates with the TGFbeta receptor and functions as an antagonist of TGFbeta signaling. Cell 1997;89:1165–1173.
19. Nakao A, Afrakhte M, Moren A, et al. Identification of Smad7, a TGFbeta-inducible antagonist of TGF-beta signalling. Nature 1997;389:631–635.
20. Imamura T, Takase M, Nishihara A, et al. Smad6 inhibits signalling by the TGF-beta superfamily. Nature 1997;389:622–626.
21. Hata A, Lagna G, Massagué J, Hemmati-Brivanlou A. Smad6 inhibits BMP/Smad1 signaling by specifically competing with the Smad4 tumor suppressor. Genes Dev 1998;12:186–197.
22. Brodin G, Ahgren A, ten Dijke P, Heldin C-H, Heuchel R. Efficient TGF-beta induction of the Smad7 gene requires cooperation between AP-1, Sp1, and Smad proteins on the mouse Smad7 promoter. J Biol Chem 2000;275:29,023–29,030.
23. Denissova NG, Pouponnot C, Long J, He D, Liu F. Transforming growth factor beta-inducible independent binding of SMAD to the Smad7 promoter. Proc Natl Acad Sci USA 2000;97:6397–6402.
24. Ishida W, Hamamoto T, Kusanagi K, et al. Smad6 is a Smad1/5-induced smad inhibitor. Characterization of bone morphogenetic protein-responsive element in the mouse Smad6 promoter. J Biol Chem 2000;275:6075–6079.
25. Nagarajan RP, Zhang J, Li W, Chen Y. Regulation of Smad7 promoter by direct association with Smad3 and Smad4. J Biol Chem 1999;274:33,412–33,418.
26. von Gersdorff G, Susztak K, Rezvani F, Bitzer M, Liang D, Bottinger EP. Smad3 and Smad4 mediate transcriptional activation of the human Smad7 promoter by transforming growth factor beta. J Biol Chem 2000;275:11,320–11,326.
27. Yingling JM, Datto MB, Wong C, Frederick JP, Liberati NT, Wang XF. Tumor suppressor Smad4 is a transforming growth factor beta-inducible DNA binding protein. Mol Cell Biol 1997;17: 7019–7028.
28. Dennler S, Itoh S, Vivien D, ten Dijke P, Huet S, Gauthier JM. Direct binding of Smad3 and Smad4 to critical TGF beta-inducible elements in the promoter of human plasminogen activator inhibitor-type 1 gene. EMBO J 1998;17:3091–3100.
29. Zawel L, Dai JL, Buckhaults P, et al. Human Smad3 and Smad4 are sequence-specific transcription activators. Mol Cell 1998;1:611–617.
30. Shi Y, Wang YF, Jayaraman L, Yang H, Massagué J, Pavletich NP. Crystal structure of a Smad MH1 domain bound to DNA: insights on DNA binding in TGF-β signaling. Cell 1998;94:585–594.
31. Liu F, Hata A, Baker JC, et al. A human Mad protein acting as a BMP-regulated transcriptional activator. Nature 1996;381:620–623.
32. Feng XH, Zhang Y, Wu RY, Derynck R. The tumor suppressor Smad4/DPC4 and transcriptional adaptor CBP/p300 are coactivators for smad3 in TGF-beta-induced transcriptional activation. Genes Dev 1998;12:2153–2163.
33. Janknecht R, Wells NJ, Hunter T. TGF-beta-stimulated cooperation of smad proteins with the coactivators CBP/p300. Genes Dev 1998;12:2114–2119.
34. Nishihara A, Hanai JI, Okamoto N, et al. Role of p300, a transcriptional coactivator, in signalling of TGF-beta. Genes Cells 1998;3:613–623.
35. Pouponnot C, Jayaraman L, Massagué J. Physical and functional interaction of SMADs and p300/CBP. J Biol Chem 1998;273:22,865–22,868.
36. Shen X, Hu PP, Liberati NT, Datto MB, Frederick JP, Wang X-F. TGF-β-induced phosphorylation of Smad3 regulates its interaction with coactivator p300/CREB-binding protein. Mol Biol Cell 1998;9: 3309–3319.

37. Topper JN, DiChiara MR, Brown JD, et al. CREB binding protein is a required coactivator for Smad-dependent, transforming growth factor beta transcriptional responses in endothelial cells. Proc Natl Acad Sci USA 1998;95:9506–9511.
38. Itoh S, Ericsson J, Nishikawa J, Heldin C-H, ten Dijke P. The transcriptional co-activator P/CAF potentiates TGF-beta/Smad signaling. Nucleic Acids Res 2000;28:4291–4298.
39. Kahata K, Hayashi M, Asaka M, et al. Regulation of transforming growth factor-beta and bone morphogenetic protein signalling by transcriptional coactivator GCN5. Genes Cells 2004;9:143–151.
40. de Caestecker MP, Yahata T, Wang D, et al. The Smad4 activation domain (SAD) is a proline-rich, p300-dependent transcriptional activation domain. J Biol Chem 2000;275:2115–2122.
41. Qin B, Lam SS, Lin K. Crystal structure of a transcriptionally active Smad4 fragment. Structure 1999;7:1493–1503.
42. Wang G, Long J, Matsuura I, He D, Liu F. The Smad3 linker region contains a transcriptional activation domain. Biochem J 2005;386:29–34.
43. Prokova V, Mavridou S, Papakosta P, Kardassis D. Characterization of a novel transcriptionally active domain in the transforming growth factor β-regulated Smad3 protein. Nucleic Acids Res 2005;33:3708–3721.
44. Kato Y, Habas R, Katsuyama Y, Naar AM, He X. A component of the ARC/Mediator complex required for TGF beta/Nodal signalling. Nature 2002;418:641–646.
45. Lin X, Duan X, Liang YY, et al. PPM1A functions as a Smad phosphatase to terminate TGFbeta signaling. Cell 2006;125:915–928.
46. Schilling SH, Datto MB, Wang XF. A phosphatase controls the fate of receptor-regulated Smads. Cell 2006;125:838–840.
47. Chen HB, Shen J, Ip YT, Xu L. Identification of phosphatases for Smad in the BMP/DPP pathway. Genes Dev 2006;20:648–653.
48. Hill CS. Turning off Smads: identification of a Smad phosphatase. Dev Cell 2006;10:412–413.
49. Liu F. Smad3 phosphorylation by cyclin-dependent kinases. Cytokine Growth Factor Rev 2006;17:9–17.
50. Feng XH, Derynck R. A kinase subdomain of transforming growth factor-beta (TGF-beta) type I receptor determines the TGF-beta intracellular signaling specificity. EMBO J 1997;16:3912–3923.
51. Lo RS, Chen YG, Shi Y, Pavletich NP, Massagué J. The L3 loop: a structural motif determining specific interactions between SMAD proteins and TGF-beta receptors. EMBO J 1998;17:996–1005.
52. Chen YG, Hata A, Lo RS, et al. Determinants of specificity in TGF-β signal transduction. Genes Dev 1998;12:2144–2152.
53. Huse M, Muir TW, Xu L, Chen YG, Kuriyan J, Massagué J. The TGF beta receptor activation process: an inhibitor- to substrate-binding switch. Mol Cell 2001;8:671–682.
54. Liu X, Yue J, Frey RS, Zhu Q, Mulder KM. Transforming growth factor beta signaling through Smad1 in human breast cancer cells. Cancer Res 1998;58:4752–4757.
55. Shi Y, Hata A, Lo RS, Massagué J, Pavletich NP. A structural basis for mutational inactivation of the tumour suppressor Smad4. Nature 1997;388:87–93.
56. Wu G, Chen YG, Ozdamar B, et al. Structural basis of Smad2 recognition by the Smad anchor for receptor activation. Science 2000;287:92–97.
57. Wu JW, Hu M, Chai J, et al. Crystal structure of a phosphorylated Smad2. Recognition of phosphoserine by the MH2 domain and insights on Smad function in TGF-beta signaling. Mol Cell 2001;8:1277–1289.
58. Tsukazaki T, Chiang TA, Davison AF, Attisano L, Wrana JL. SARA, a FYVE domain protein that recruits Smad2 to the TGFbeta receptor. Cell 1998;95:779–791.
59. Wu G, Chen YG, Ozdamar B, et al. Structural basis of Smad2 recognition by the Smad anchor for receptor activation. Science 2000;287:92–97.
60. Qin BY, Lam SS, Correia JJ, Lin K. Smad3 allostery links TGF-beta receptor kinase activation to transcriptional control. Genes Dev 2002;16:1950–1963.
61. Moustakas A, Heldin C-H. From mono- to oligo-Smads: the heart of the matter in TGF-beta signal transduction. Genes Dev 2002;16:1867–1871.
62. Lin HK, Bergmann S, Pandolfi PP. Cytoplasmic PML function in TGF-beta signalling. Nature 2004;431:205–211.
63. Inman GJ, Nicolas FJ, Hill CS. Nucleocytoplasmic shuttling of Smads 2, 3, and 4 permits sensing of TGF-beta receptor activity. Mol Cell 2002;10:283–294.
64. Xu L, Kang Y, Col S, Massagué J. Smad2 nucleocytoplasmic shuttling by nucleoporins CAN/Nup214 and Nup153 feeds TGFbeta signaling complexes in the cytoplasm and nucleus. Mol Cell 2002;10:271–282.

65. Takekawa M, Maeda T, Saito H. Protein phosphatase 2Calpha inhibits the human stress-responsive p38 and JNK MAPK pathways. EMBO J 1998;17:4744–4752.
66. Cheng A, Ross KE, Kaldis P, Solomon MJ. Dephosphorylation of cyclin-dependent kinases by type 2C protein phosphatases. Genes Dev 1999;13:2946–2957.
67. Yoshizaki T, Maegawa H, Egawa K, et al. Protein phosphatase-2C alpha as a positive regulator of insulin sensitivity through direct activation of phosphatidylinositol 3-kinase in 3T3-L1 adipocytes. J Biol Chem 2004;279:22,715–22,726.
68. Strovel ET, Wu D, Sussman DJ. Protein phosphatase 2C alpha dephosphorylates axin and activates LEF-1-dependent transcription. J Biol Chem 2000;275:2399–2403.
69. Ofek P, Ben-Meir D, Kariv-Inbal Z, Oren M, Lavi S. Cell cycle regulation and p53 activation by protein phosphatase 2C alpha. J Biol Chem 2003;278:14,299–14,305.
70. Kretzschmar M, Doody J, Massagué J. Opposing BMP and EGF signalling pathways converge on the TGF-β family mediator Smad1. Nature 1997;389:618–622.
71. Pera EM, Ikeda A, Eivers E, De Robertis EM. Integration of IGF, FGF, and anti-BMP signals via Smad1 phosphorylation in neural induction. Genes Dev 2003;17:3023–3028.
72. De Robertis EM, Kuroda H. Dorsal-ventral patterning and neural induction in Xenopus embryos. Annu Rev Cell Dev Biol 2004;20:285–308.
73. Sater AK, El-Hodiri HM, Goswami M, et al. Evidence for antagonism of BMP-4 signals by MAP kinase during Xenopus axis determination and neural specification. Differentiation 2003;71:434–444.
74. Aubin J, Davy A, Soriano P. In vivo convergence of BMP and MAPK signaling pathways: impact of differential Smad1 phosphorylation on development and homeostasis. Genes Dev 2004;18: 1482–1494.
75. Massagué J. Integration of Smad and MAPK pathways: a link and a linker revisited. Genes Dev 2003;17:2993–2997.
76. Grimm OH, Gurdon JB. Nuclear exclusion of Smad2 is a mechanism leading to loss of competence. Nat Cell Biol 2002;4:519–522.
77. Kretzschmar M, Doody J, Timokhina I, Massagué J. A mechanism of repression of TGF-β/Smad signaling by oncogenic Ras. Genes Dev 1999;13:804–816.
78. de Caestecker MP, Parks WT, Frank CJ, et al. Smad2 transduces common signals from receptor serine-threonine and tyrosine kinases. Genes Dev 1998;12:1587–1592.
79. Blanchette F, Rivard N, Rudd P, Grondin F, Attisano L, Dubois CM. Cross-talk between the p42/p44 MAP kinase and Smad pathways in transforming growth factor β1-induced furin gene transactivation. J Biol Chem 2001;276:33,986–33,994.
80. Hayashida T, Decaestecker M, Schnaper HW. Cross-talk between ERK MAP kinase and Smad signaling pathways enhances TGF-β-dependent responses in human mesangial cells. FASEB J 2003;17: 1576–1578.
81. Funaba M, Zimmerman CM, Mathews LS. Modulation of Smad2-mediated signaling by extracellular signal-regulated kinase. J Biol Chem 2002;277:41,361–41,368.
82. Hu PP, Shen X, Huang D, Liu Y, Counter C, Wang XF. The MEK pathway is required for stimulation of p21(WAF1/CIP1) by transforming growth factor-β. J Biol Chem 1999;274: 35,381–35,387.
83. Lehmann K, Janda E, Pierreux CE, et al. Raf induces TGFβ production while blocking its apoptotic but not invasive responses: a mechanism leading to increased malignancy in epithelial cells. Genes Dev 2000;14:2610–2622.
84. Yue J, Mulder KM. Requirement of Ras/MAPK pathway activation by transforming growth factor β for transforming growth factor β1 production in a Smad-dependent pathway. J Biol Chem 2000;275:30,765–30,773.
85. Mulder KM. Role of Ras and Mapks in TGFbeta signaling. Cytokine Growth Factor Rev 2000;11: 23–35.
86. Matsuura I, Wang G, He D, Liu F. Identification and characterization of ERK MAP kinase phosphorylation sites in Smad3. Biochemistry 2005;44:12,546–12,553.
87. Kfir S, Ehrlich M, Goldshmid A, Liu X, Kloog Y, Henis YI. Pathway- and expression level-dependent effects of oncogenic N-Ras: p27(Kip1) mislocalization by the Ral-GEF pathway and Erk-mediated interference with Smad signaling. Mol Cell Biol 2005;25:8239–8250.
88. McCormick F. Signalling networks that cause cancer. Trends Cell Biol 1999;9:M53–M56.
89. Murphy LO, Smith S, Chen RH, Fingar DC, Blenis J. Molecular interpretation of ERK signal duration by immediate early gene products. Nat Cell Biol 2002;4:556–564.

90. Murphy LO, Blenis J. MAPK signal specificity: the right place at the right time. Trends Biochem Sci 2006;31:268–275.
91. Dunfield LD, Nachtigal MW. Inhibition of the antiproliferative effect of TGFβ by EGF in primary human ovarian cancer cells. Oncogene 2003;22:4745–4751.
92. Brown JD, DiChiara MR, Anderson KR, Gimbrone MA Jr, Topper JN. MEKK-1, a component of the stress (stress-activated protein kinase/c-Jun N-terminal kinase) pathway, can selectively activate Smad2-mediated transcriptional activation in endothelial cells. J Biol Chem 1999;274:8797–8805.
93. Engel ME, McDonnell MA, Law BK, Moses HL. Interdependent SMAD and JNK signaling in transforming growth factor β-mediated transcription. J Biol Chem 1999;274:37,413–37,420.
94. Mori S, Matsuzaki K, Yoshida K, et al. TGF-β and HGF transmit the signals through JNK-dependent Smad2/3 phosphorylation at the linker regions. Oncogene 2004;23:7416–7429.
95. Bain J, McLauchlan H, Elliott M, Cohen P. The specificities of protein kinase inhibitors: an update. Biochem J 2003;371:199–204.
96. Yamagata H, Matsuzaki K, Mori S, et al. Acceleration of Smad2 and Smad3 phosphorylation via c-Jun NH(2)-terminal kinase during human colorectal carcinogenesis. Cancer Res 2005;65:157–165.
97. Yoshida K, Matsuzaki K, Mori S, et al. Transforming growth factor-beta and platelet-derived growth factor signal via c-Jun N-terminal kinase-dependent Smad2/3 phosphorylation in rat hepatic stellate cells after acute liver injury. Am J Pathol 2005;166:1029–1039.
98. Matsuzaki K, Okazaki K. Transforming growth factor-beta during carcinogenesis: the shift from epithelial to mesenchymal signaling. J Gastroenterol 2006;41:295–303.
99. Matsuzaki K. Smad3 phosphoisoform-mediated signaling during sporadic human colorectal carcinogenesis. Histol Histopathol 2006;21:645–662.
100. Ventura JJ, Kennedy NJ, Flavell RA, Davis RJ. JNK regulates autocrine expression of TGF-beta1. Mol Cell 2004;15:269–278.
101. Pardoux C, Derynck R. JNK regulates expression and autocrine signaling of TGF-beta1. Mol Cell 2004;15:170–171.
102. Yamaguchi K, Shirakabe K, Shibuya H, et al. Identification of a member of the MAPKKK family as a potential mediator of TGF-beta signal transduction. Science 1995;270:2008–2011.
103. Yu L, Hebert MC, Zhang YE. TGF-beta receptor-activated p38 MAP kinase mediates Smad-independent TGF-beta responses. EMBO J 2002;21:3749–3759.
104. Takekawa M, Tatebayashi K, Itoh F, Adachi M, Imai K, Saito H. Smad-dependent GADD45beta expression mediates delayed activation of p38 MAP kinase by TGF-beta. EMBO J 2002;21:6473–6482.
105. Kamaraju AK, Roberts AB. Role of Rho/ROCK and p38 MAP kinase pathways in TGF-β-mediated Smad-dependent growth inhibition of human breast carcinoma cells in vivo. J Biol Chem 2004;280: 1024–1036.
106. Zarubin T, Han J. Activation and signaling of the p38 MAP kinase pathway. Cell Res 2005;15:11–18.
107. Zohn IE, Li Y, Skolnik EY, Anderson KV, Han J, Niswander L. p38 and a p38-interacting protein are critical for downregulation of E-cadherin during mouse gastrulation. Cell 2006;125:957–969.
108. Yang YA, Zhang GM, Feigenbaum L, Zhang YE. Smad3 reduces susceptibility to hepatocarcinoma by sensitizing hepatocytes to apoptosis through downregulation of Bcl-2. Cancer Cell 2006;9: 445–457.
109. Sato M, Muragaki Y, Saika S, Roberts AB, Ooshima A. Targeted disruption of TGF-beta1/Smad3 signaling protects against renal tubulointerstitial fibrosis induced by unilateral ureteral obstruction. J Clin Invest 2003;112:1486–1494.
110. Saika S, Kono-Saika S, Ohnishi Y, et al. Smad3 signaling is required for epithelial-mesenchymal transition of lens epithelium after injury. Am J Pathol 2004;164:651–663.
111. Roberts AB, Tian F, Byfield SD, et al. Smad3 is key to TGF-β-mediated epithelial-to-mesenchymal transition, fibrosis, tumor suppression and metastasis. Cytokine Growth Factor Rev 2006;17:19–27.
112. Ju W, Ogawa A, Heyer J, et al. Deletion of Smad2 in mouse liver reveals novel functions in hepatocyte growth and differentiation. Mol Cell Biol 2006;26:654–667.
113. Furukawa F, Matsuzaki K, Mori S, et al. p38 MAPK mediates fibrogenic signal through Smad3 phosphorylation in rat myofibroblasts. Hepatology 2003;38:879–889.
114. Massagué J, Blain SW, Lo RS. TGF-β signaling in growth control, cancer, and heritable disorders. Cell 2000;103:295–309.
115. Siegel PM, Massagué J. Cytostatic and apoptotic actions of TGF-beta in homeostasis and cancer. Nat Rev Cancer 2003;3:807–821.
116. Massagué J. G1 cell-cycle control and cancer. Nature 2004;432:298–306.

117. Sherr CJ, Roberts JM. CDK inhibitors: positive and negative regulators of G1-phase progression. Genes Dev 1999;13:1501–1512.
118. Pietenpol JA, Stein RW, Moran E, et al. TGF-β1 inhibition of c-myc transcription and growth in keratinocytes is abrogated by viral transforming proteins with pRB binding domains. Cell 1990;61: 777–785.
119. Hannon GJ, Beach D. p15INK4B is a potential effector of TGF-β-induced cell cycle arrest. Nature 1994;371:257–261.
120. Li JM, Nichols MA, Chandrasekharan S, Xiong Y, Wang X-F. Transforming growth factor β activates the promoter of cyclin-dependent kinase inhibitor p15INK4B through an Sp1 consensus site. J Biol Chem 1995;270:26,750–26,753.
121. Datto MB, Li Y, Panus JF, Howe DJ, Xiong Y, Wang X-F. Transforming growth factor β induces the cyclin-dependent kinase inhibitor p21 through a p53-independent mechanism. Proc Natl Acad Sci USA 1995;92:5545–5549.
122. Kang Y, Chen CR, Massagué J. A self-enabling TGF-β response coupled to stress signaling: Smad engages stress response factor ATF3 for Id1 repression in epithelial cells. Mol Cell 2003;11: 915–926.
123. Iavarone A, Massagué J. E2F and histone deacetylase mediate transforming growth factor β repression of cdc25A during keratinocyte cell cycle arrest. Mol Cell Biol 1999;19:916–922.
124. Bhowmick NA, Ghiassi M, Aakre M, Brown K, Singh V, Moses HL. TGF-β-induced RhoA and p160ROCK activation is involved in the inhibition of Cdc25A with resultant cell-cycle arrest. Proc Natl Acad Sci USA 2003;100:15,548–15,553.
125. Ray D, Terao Y, Nimbalkar D, et al. Transforming growth factor β facilitates β-TrCP-mediated degradation of Cdc25A in a Smad3-dependent manner. Mol Cell Biol 2005;25:3338–3347.
126. Datto MB, Frederick JP, Pan L, Borton AJ, Zhuang Y, Wang XF. Targeted disruption of Smad3 reveals an essential role in transforming growth factor β-mediated signal transduction. Mol Cell Biol 1999;19:2495–2504.
127. Yang X, Letterio JJ, Lechleider RJ, et al. Targeted disruption of SMAD3 results in impaired mucosal immunity and diminished T cell responsiveness to TGF-β. EMBO J 1999;18:1280–1291.
128. Ashcroft GS, Yang X, Glick AB, et al. Mice lacking Smad3 show accelerated wound healing and an impaired local inflammatory response. Nature Cell Biol 1999;1:260–266.
129. Rich JN, Zhang M, Datto MB, Bigner DD, Wang X-F. Transforming growth factor-β-mediated p15INK4B induction and growth inhibition in astrocytes is Smad3-dependent and a pathway prominently altered in human glioma cell lines. J Biol Chem 1999;274:35,053–35,058.
130. Matsuura I, Denissova NG, Wang G, He D, Long J, Liu F. Cyclin-dependent kinases regulate the antiproliferative function of Smads. Nat 2004;430:226–231.
131. Liu F, Matsuura I. Inhibition of Smad antiproliferative function by CDK phosphorylation. Cell Cycle 2005;4:63–66.
132. Chen CR, Kang Y, Massagué J. Defective repression of c-myc in breast cancer cells: A loss at the core of the transforming growth factor growth β arrest program. Proc Natl Acad Sci USA 2001;98: 992–999.
133. Yagi K, Furuhashi M, Aoki H, et a. c-myc is a downstream target of the Smad pathway. J Biol Chem 2002;277:854–861.
134. Chen C-R, Kang Y, Siegel PM, Massagué J. E2F4/5 and p107 as Smad cofactors linking the TGF-ß receptor to c-myc repression. Cell 2002;110:19–32.
135. Frederick JP, Liberati NT, Waddell DS, Shi Y, Wang XF. Transforming growth factor β-mediated transcriptional repression of c-myc is dependent on direct binding of Smad3 to a novel repressive Smad binding element. Mol Cell Biol 2004;24:2546–2559.
136. Staller P, Peukert K, Kiermaier A, et al. Repression of p15INK4b expression by Myc through association with Miz-1. Nat Cell Biol 2001;3:392–399.
137. Seoane J, Pouponnot C, Staller P, Schader M, Eilers M, Massagué J. TGF-β influences Myc, Miz-1 and Smad to control the CDK inhibitor p15INK4b. Nat Cell Biol 2001;3:400–408.
138. Claassen GF, Hann SR. A role for transcriptional repression of p21CIP1 by c-Myc in overcoming transforming growth factor β-induced cell-cycle arrest. Proc Natl Acad Sci USA 2000;97:9498–9503.
139. Seoane J, Le HV, Massagué J. Myc suppression of the p21(Cip1) Cdk inhibitor influences the outcome of the p53 response to DNA damage. Nature 2002;419:729–734.
140. Feng XH, Liang YY, Liang M, Zhai W, Lin X. Direct interaction of c-Myc with Smad2 and Smad3 to inhibit TGF-β-mediated induction of the CDK inhibitor p15(Ink4B). Mol Cell 2002;9:133–143.

141. Feng X-H, Lin X, Derynck R. Smad2, Smad3 and Smad4 cooperate with Sp1 to induce p15Ink4B transcription in response to TGF-β. EMBO J 2000;19:5178–5193.
142. Moustakas A, Kardassis D. Regulation of the human p21/WAF1/Cip1 promoter in hepatic cells by functional interaction between Sp1 and Smad family members. Proc Natl Acad Sci USA 1998;95:6733–6738.
143. Pardali K, Kurisaki A, Moren A, ten Dijke P, Kardassis D, Moustakas A. Role of Smad proteins and transcription factor Sp1 in p21Waf1/Cip1 regulation by transforming growth factor-β. J Biol Chem 2000;275:29,244–29,256.
144. Seoane J, Le HV, Shen L, Anderson SA, Massagué J. Integration of Smad and forkhead pathways in the control of neuroepithelial and glioblastoma cell proliferation. Cell 2004;117:211–223.
145. Chi XZ, Yang JO, Lee KY, et al. RUNX3 suppresses gastric epithelial cell growth by inducing p21(WAF1/Cip1) expression in cooperation with transforming growth factor β-activated SMAD. Mol Cell Biol 2005;25:8097–8107.
146. Sherr CJ. Cancer cell cycles. Science 1996;274:1672–1677.
147. Hall M, Peters G. Genetic alterations of cyclins, cyclin-dependent kinases, and Cdk inhibitors in human cancer. Adv Cancer Res 1996;68:67–108.
148. Sherr CJ, McCormick F. The RB and p53 pathways in cancer. Cancer Cell 2002;2:103–112.
149. Ortega S, Malumbres M, Barbacid M. Cyclin D-dependent kinases, INK4 inhibitors and cancer. Biochim Biophys Acta 2002;1602:73–87.
150. Weinberg RA. The retinoblastoma protein and cell cycle control. Cell 1995;81:323–330.
151. Derynck R, Akhurst RJ, Balmain A. TGF-β signaling in tumor suppression and cancer progression. Nat Genet 2001;29:117–129.
152. Roberts AB, Wakefield LM. The two faces of transforming growth factor β in carcinogenesis. Proc Natl Acad Sci USA 2003;100:8621–8623.
153. Miyazono K, Suzuki H, Imamura T. Regulation of TGF-beta signaling and its roles in progression of tumors. Cancer Sci 2003;94:230–234.
154. Tian F, DaCosta Byfield S, Parks WT, et al. Reduction in Smad2/3 signaling enhances tumorigenesis but suppresses metastasis of breast cancer cell lines. Cancer Res 2003;63:8284–8292.
155. Wicks SJ, Lui S, Abdel-Wahab N, Mason RM, Chantry A. Inactivation of Smad-transforming growth factor β signaling by Ca(2+)-calmodulin-dependent protein kinase II. Mol Cell Biol 2000;20: 8103–8111.
156. Yakymovych I, ten Dijke P, Heldin C-H, Souchelnytskyi S. Regulation of Smad signaling by protein kinase C. FASEB J 2001;15:553–555.
157. Roelen BA, Cohen OS, Raychowdhury MK, et al. Phosphorylation of threonine 276 in Smad4 is involved in transforming growth factor-beta-induced nuclear accumulation. Am J Physiol Cell Physiol 2003;285:C823–C830.
158. Pulaski L, Landstrom M, Heldin C-H, Souchelnytskyi S. Phosphorylation of Smad7 at Ser-249 does not interfere with its inhibitory role in transforming growth factor-beta-dependent signaling but affects Smad7-dependent transcriptional activation. J Biol Chem 2001;276:14,344–14,349.

8 Nuclear Targeting of TGF-β-Activated Smads in Normal and Tumor Biology

Lan Xu

CONTENTS

INTRODUCTION
THE MECHANISM OF NUCLEAR IMPORT AND EXPORT OF R-SMADS
NUCLEAR IMPORT AND EXPORT OF SMAD4
NUCLEAR TARGETING OF TGF-β ACTIVATED SMAD COMPLEXES
DYSREGULATION OF NUCLEAR TARGETING OF SMADS
CONCLUDING REMARKS
REFERENCES

Abstract

Smad proteins spontaneously shuttle between cytoplasm and nucleus. Such intracellular dynamics is essential for Smads to link transforming growth factor-β (TGF-β) signaling at the cell surface and the ensuing transcriptional regulation in the nucleus. The subcellular distribution of Smads is controlled so that they accumulate in the nucleus in response to TGF-β stimulation, which confers inducibility to the expression of TGF-β target genes. Oncogenic factors can impede nuclear translocation of Smads and hence interfere with the growth inhibitory function of TGF-β. Recent studies have unraveled molecular mechanisms underlying the transnuclear pore complex movements of Smads as well as factors that regulate these processes. With this information, we are beginning to understand how nuclear targeting of Smads is controlled by TGF-β and other signaling pathways.

Key Words: Nuclear import; nuclear export; nucleocytoplasmic shuttling; Smad; TGF-β.

1. INTRODUCTION

Nucleocytoplasmic trafficking of the Smad family proteins is an integral component of the TGF-β signal transduction pathway. TGF-β drives nuclear accumulation of TGF-β-receptor-activated Smads and Smad4, whereas termination of the signal is accompanied by nuclear export of these Smads. Thus the concentration of Smads in the nucleus directly reflects the magnitude of the extracellular TGF-β signal. The duration of Smads accumulation in the nucleus is also an important determinant of TGF-β-induced gene expression pattern and the nature of cellular responses to TGF-β. The ability to translocate across the nuclear envelop in either direction enables Smads to serve the dual functions of sensing TGF-β signal in the cytoplasm and regulating gene expression in the nucleus. Recent studies on intracellular movement of Smads have yielded much information on the machineries responsible for nuclear import and export of Smads, and how these processes are regulated

From: *Cancer Drug Discovery and Development: Transforming Growth Factor-β in Cancer Therapy, Vol I: Basic and Clinical Biology*
Edited by: Sonia B. Jakowlew © Humana Press Inc., Totowa, NJ

by TGF-β signaling. Of particular interest are examples of mitogenic signals or oncogenic mutations that interfere with nuclear accumulation of Smads, leading to impaired anti-proliferation functions of TGF-β.

1.1. TGF-β-Regulated Subcellular Distribution of Smads

First through genetic analyses in *Drosophila melanogaster*, the family of Smad proteins were discovered as the essential mediators of TGF-β signal transduction *(1–5)*. Based on their functions Smads are grouped into receptor-activated Smad (R-Smads), common Smad (Co-Smad), and inhibitor Smad (I-Smad) *(2–5)*. Immunocytochemistry studies revealed compartmentalized localization of these Smad proteins in cells and more importantly that Smads undergo nuclear translocation in response to TGF-β stimulation *(6–8)*. Such intracellular trafficking of various Smads is crucial for the Smads to either activate or dampen TGF-β signaling.

The R-Smads include Smad2/3 downstream of activin or TGF-β, and Smad1/5/8 in the bone morphogenetic protein (BMP) pathways. At the basal state, R-Smads mostly reside in the cytoplasm or distribute evenly throughout the cell, and upon stimulation by TGF-β cytokines, R-Smads rapidly accumulate in the nucleus *(6,7)*. R-Smads are phosphorylated at the C-terminal SSXS motif by their cognate transmembrane TGF-β receptor kinases upon ligand stimulation, and R-Smads are transcriptional regulators once they reside in the nucleus *(9–12)*. Therefore the spatial movement of Smads, first to the membrane-embedded TGF-β receptors and then into the nucleus, allows the cell to sense extracellular TGF-β and translate that signal accordingly into transcriptional responses. In contrast, when TGF-β signaling is blocked by kinase inhibitors against the TGF-β receptor, R-Smads are quickly dephosphorylated and promptly exported out the nucleus *(13,14)*. Thus, nuclear export of R-Smads leads to an end of the transcriptional responses and replenishes the cytoplasm with R-Smads for sensing TGF-β signal again.

The Co-Smad, Smad4, also undergoes nuclear accumulation once the cells receive TGF-β signal *(8)*. Smad4 is not phosphorylated by the TGF-β receptor kinases, but it tightly associates with C-terminally phosphorylated-R-Smads to form heterotrimer complexes *(15,16)*. Therefore, the phospho-R-Smads and Smad4 likely enter the nucleus as a complex. However, in cells carrying genetic deletion of Smad4, TGF-β still induces nuclear accumulation of R-Smads, indicating that Smad4 is not required for nuclear translocation of activated R-Smads *(17)*.

Smad7, one of the I-Smads, moves in the opposite direction of the R-Smads and Smad4 upon TGF-β stimulation. Several studies have shown that Smad7 is mostly present in the nucleus at the basal state *(18–20)*. Yet, the primary site of Smad7 action, namely binding to and degradation of the TGF-β receptors, is the cell membrane *(21–23)*. Indeed, Smad7 has been reported to exit the nucleus in response to TGF-β *(18,19)*. Therefore, in a typical feedback manner, Smad7 does not start to mitigate TGF-β signaling until the signal has reached into the nucleus which prompts nuclear export of Smad7.

The above conclusions on subcellular localization of Smads at basal and TGF-β-stimulated states were mostly based on immunocytochemistry analyses, which only give static pictures of the distribution of Smads in the cells. With assays that are more revealing of dynamic movement of proteins, such as heterokaryon assays, live cell fluorescence recovery after photobleaching and fluorescence loss in photobleaching, R-Smads and Smad4 turned out to actively undergo bidirectional shuttling across the nuclear envelop *(14,24–26)*. So instead of mobilizing R-Smads and Smad4 to move across the nuclear envelop, the role of TGF-β signaling is rather to restrict and target Smads to the proper locations where they exert their functions.

1.2. The General Nuclear Transport Machinery

In eukaryotes, the nuclear envelope acts as a barrier and conduit that controls exchange of molecules between the cytoplasm and nucleus. Macromolecules such as proteins and RNA enter and exit the nucleus via the nuclear pore complexes (NPCs). NPCs are multiprotein structures that span the lipid bilayer of the nuclear envelop *(27,28)*. The building blocks of NPCs are about 30 nucleoporins, many of which contain characteristic repeats of phenylalanine-glycine (FG) residues *(29)*. Multiple copies of each of the nucleoporins assemble into NPCs that approach 125 MDa in size. NPCs are highly structured, with each nucleoporin positioned at specific loci. Some nucleoporins are symmetrically distributed at both entrances while the others are located toward either cytoplasm or nucleoplasm. The NPCs are dynamic and mobile, rather than being the inert structural element of the NPCs, the nucleoporins actively participate in both nuclear import and export by recruiting the transport cargos to the NPC and releasing them at the conclusion of the transport processe *(30–32)*.

Most of the studied nuclear import and export events are mediated by importins and exportins, respectively *(33–35)*. Importins and exportins are structurally similar and belong to the importin-β superfamily *(36)*. One common property of these transport factors is their direct interaction with the phenylalanine/glycine-(FG-) repeats that are commonly found in many nucleoporins. On the other hand, importins and exportins recognize and bind to unique structural elements in their cargos. Therefore, importins and exportins are essentially adaptor molecules that bridge the interaction between their substrates and nucleoporins. To date, there are over 20 importins and exportins found in human genome, presumably each responsible for importing or exporting a number of cargos *(36)*.

Owing to their ability to interact with nucleoporins, both importins and exportins can spontaneously shuttle across the NPCs in both directions *(37–39)*. This raised a critical issue of what determines the directionality in nuclear transport. It turns out that the small GTPase Ran in its GTP-bound form (RanGTP) is a crucial regulator of association of importins and exportins with their cargos: RanGTP disrupts importins interaction with their substrates while it facilitates exportins binding to their cargos *(33–35)*. RanGEF, which catalyzes loading of GTP onto Ran, is present exclusively in the nucleus, whereas RanGAP (RanGTPase activating proteins) which catalyzes RanGTP hydrolysis into RanGDP, mostly resides in the cytoplasm. In this way an asymmetry of RanGTP is established across the NPCs. Therefore upon entering the RanGTP rich nucleus, the cargo molecule is unloaded from importin, and the empty importin can shuttle back to the cytoplasm to engage another cargo molecule. On the other hand, the high concentration of RanGTP in the nucleus promotes exportins interaction with their cargos, and the complex comes apart once it reaches into cytoplasm. Therefore, the RanGTP gradient sets up the directionality for both nuclear import and export by regulating the interaction between importins and exportins with their respective cargos.

2. THE MECHANISM OF NUCLEAR IMPORT AND EXPORT OF R-SMADS

Most of the importin/exportin-mediated nuclear transport is responsible for transporting cargoes constitutively, such as transporting nuclear localization signal (NLS)-containing proteins into the nucleus or RNA into the cytoplasm. It is conceivable that for signaling molecules such as Smads, more fluidity may be required, as they need to move in either direction in response to the signals. Indeed, studies on Smad transport pathways revealed unique properties in their nuclear transport mechanisms.

2.1. Mechanisms of Nuclear Import

A reconstituted in vitro nuclear import assay, which has been instrumental in elucidation of importin-mediated nuclear transport mechanism, was utilized to dissect the import pathway for Smads *(40,41)*. The behavior of Smad2 and Smad3 in the nuclear import assay revealed unique features of nuclear import of these Smads *(42,43)*. First, Smad2 and Smad3 could enter the nucleus when they are not phosphorylated at the C-terminus by TGF-β *(42,43)*. Second, efficient nuclear import of Smad2 and Smad3 could take place without the supply of importins. Even when importin-cargo association in general was inhibited by excessive RanGTP, nuclear import of unphosphorylated Smad2 and Smad3 was unaffected *(42,43)*. Third, when the NPCs are blocked, either by wheat germ agglutanin or by a dominant negative importin-β fragment that irreversibly occupies nucleoporins, Smad2 and Smad3 failed to enter the nucleus *(42,43)*. Fourth, when Smad2 or Smad3 was bound by a small fragment of the protein, Smad adaptor for receptor activation (SARA), their nuclear import was largely inhibited *(42,43)*. These observations led to the conclusion that Smad2 and Smad3 are spontaneously and intrinsically capable of entering the nucleus without being activated by TGF-β receptor kinases *(42,43)*. This is consistent with the observation that at the basal state Smad2 and Smad3 shuttle across the nuclear envelop. Moreover, although nuclear entrance of Smad2 and Smad3 requires passage through the NPCs, they can apparently dock onto and migrate through the NPCs without the involvement of importins. This is owing to the ability of Smads to directly interact with nucleoporins such as Nup214/CAN and Nup153 *(24,42)*. Such binding to nucleoporins is mediated by a few "hydrophobic patches" in the C-terminal MH2 domain of Smad2 and Smad3 *(24)*. Interestingly this same region is also the interaction interface between Smad2/3 and SARA *(44,45)*, which explains how SARA could block the spontaneous nuclear import of Smad2/3 *(24,42,43)*.

A separate nuclear import pathway have also been proposed for Smad3 which involves the participation of importin-β *(46–48)*. A lysine rich region resembling NLS in the N-terminal MH1 domain of Smad3 was found to contain nuclear import activity and bind importin-β1 *(47–49)*. This suggested a possible alternative pathway for importing Smad3 into the nucleus. However, in the reconstituted nuclear transport assay, this importin-β1-mediated nuclear import through the MH1 domain was much weaker than that mediated by the MH2 domain which is independent of importins *(42)*. Moreover, although similar NLS-like sequence exists in Smad2, the difference in flanking regions prevents Smad2 from binding importin-β1 directly, so this importin-β1-dependent mechanism does not apply to Smad2 *(48)*.

The mechanisms described above may well explain how R-Smads enter the nucleus during the nucleocytoplasmic shuttling at the basal state when they are monomers *(50)*. But once the cells are stimulated with TGF-β, R-Smads are tightly associated with Smad4 to form high molecular weight complexes. Does the R-Smad/Smad4 complex utilize the same machinery? Using highly purified and homogenously phosphorylated Smad3 complexed with Smad4, Chen et al. reached the conclusion that the TGF-β activated Smad3/Smad4 complex can also enter the nucleus independent of importins, and the complex also interacts with nucleoporins *(51)*. Compared to unphosphorylated Smad3, the complex did not exhibit accelerated nuclear import *(51)*. This conclusion agrees well with recent studies in living cells which also suggest that the rate of Smad3 nuclear import is not increased when cells are stimulated by TGF-β *(26)*. All these data seem to support the hypothesis that enhanced nuclear import efficiency is not the primary mechanism through which TGF-β drives R-Smads into the nucleus.

2.2. Nuclear Export

In cells exposed to TGF-β, Smad2/3 become exclusively localized in the nucleus. But when the receptor kinase activity was inhibited by the compound SB431542, the same pool of Smad2/3 rapidly returned to the cytoplasm *(14)*. Such nuclear export correlates directly with dephosphorylation of Smad2/3 *(14,24)*. Moreover, even when Smad2 was fused with an NLS and targeted into the nucleus, the NLS-Smad2 exhibited nuclear export activity in a heterokaryon assay *(24)*. These observations provide definitive evidence that Smad2/3 harbor intrinsic nuclear export activity. Nuclear export of Smad2/3 is insensitive to inhibition by the drug Leptomycin B (LMB), thus excluding the involvement of the exportin CRM-1 *(14,24)*. Interestingly, when the nucleoporin-binding domain in Smad2 is occupied by transcription factors FoxH1 or ATF3, Smad2 becomes sequestered in the nucleus in the absence of TGF-β signaling *(24,52)*. This suggests that direct nucleoporins-interaction by Smad2 may be important for Smad2 nuclear export, although the participation of exportins other than CRM-1 cannot be ruled out.

In contrast, subcellular localization of Smad1 appeared to be sensitive to LMB treatment, indicating that it is exported by CRM-1 *(53)*. Indeed there is an NES-like element in Smad1, the mutation of which resulted in BMP-independent nuclear accumulation of Smad1 *(53)*. It is peculiar that this NES-like sequence is highly conserved in other Smads, and yet clearly Smad2 and Smad3 are not subject to CRM-1-mediated nuclear export *(14,24,54)*. Smad1 has not been shown to directly interact with CRM-1, thus it remains an issue if Smad1 is directly exported by CRM-1 or by a piggyback mechanism.

2.3. Nucleocytoplasmic Shuttling vs Retention

At the basal state, R-Smads shuttle continuously between the cytoplasm and nucleus, consistent with their ability of spontaneous nuclear import and export without TGF-β stimulation *(14,24–26)*. R-Smads can also be sequestered in nucleus or cytoplasm by retention factors. The best examples so far include SARA, FoxH1 and ATF3 *(24,43,52,55)*. These factors share the same property of binding to the hydrophobic regions in the MH2 domain of Smad2/3, and their overexpression in cells is sufficient to restrict endogenous Smad2/3 in the endosome (in the case of SARA) or nucleus (in the cases of FoxH1 and ATF3), completely independent of TGF-β signal. Binding of SARA or FoxH1 interfere with Smad2 interaction with nucleoporins such as CAN/Nup214 and Nup153, which suggests the mechanism of retention *(24)*. It is conceivable that other Smad2/3 interacting proteins, especially those binding through the hydrophobic pockets, also have the potential to act as retention factors. Therefore, in addition to the nuclear import and export machinery, the retention factors are another important element in controlling subcellular localization of R-Smads.

3. NUCLEAR IMPORT AND EXPORT OF SMAD4

Smad4 undergoes nuclear accumulation in response to TGF-β, but nuclear translocation of R-Smads does not rely on Smad4 *(8)*. In cells with genetic deletion of Smad4, the efficiency and extent of nuclear concentration of R-Smads in response to TGF-β is similar to that in cells expressing Smad4 *(17)*. On the other hand, when nuclear export of Smad4 was specifically inactivated, Smad4 becomes exclusively in the nucleus in the absence of TGF-β signal while R-Smads remain in the cytoplasm, indicating that nuclear import and export mechanisms for Smad4 and R-Smads are separate in basal state cells when they are not associated with each other in a heterotrimeric complex *(14,24,54,56)*.

3.1. Nuclear Import

The behavior of Smad4 in the in vitro reconstituted nuclear transport assay is similar to that of Smad2 and Smad3: importin appears to be dispensible and the dominant negative RanQ69L-GTP does not inhibit nuclear import of Smad4 *(42)*. Protein–protein interaction experiments also revealed direct contact between Smad4 and CAN/Nup214 *(42)*. Therefore, like the R-Smads, Smad4 has the ability to directly engage nucleoporins for nuclear import. However, unlike Smad2 and Smad3, the C-terminal part of Smad4 is not sufficient for interaction with nucleoporins owing to differences in structure. In fact nucleoporin interaction appears to require intact full-length Smad4 *(42)*.

The N-terminal MH1 domain of Smad4 contains a cluster of Lys-Arg residues that resemble NLS. Mutagenesis analysis suggest that these residues are important for Smad4 nuclear import *(47,54)*. This Lys-Arg cluster in Smad3 was shown to bind importin-β1 *(48)*, so it is surprising that this same sequence in Smad4 interacted with importin-α instead *(47)*, because importin-α and importin-β are very different in primary sequence and structure. In reconstituted nuclear import assay, importin-α very weakly enhanced nuclear import of the Smad4 MH1 domain, at a concentration much higher than what is needed for a typical NLS-containing protein *(42)*. Therefore, similar to the case in Smad3, importins may provide an alternative or complementary mechanism for nuclear import in addition to the nonconventional importin-independent pathway.

3.2. Nuclear Export

The nuclear export mechanism for Smad4 is well understood. LMB, a specific inhibitor of CRM-1 activity, induces nuclear concentration of Smad4 in the absence of TGF-β signaling *(54,56)*. This not only confirms the spontaneous, TGF-β-independent nuclear import ability of Smad4, but also indicated that inhibition of CRM-1-mediated export alone is sufficient in driving Smad4 into the nucleus. Although all Smads are quite similar in sequence, only Smad4 contains a canonical Ile/Leu rich nuclear export signal (NES) motif *(54,56)*. Smad4 interacts with CRM-1 through this NES, and more importantly, like all other typical CRM-1 cargos, such interaction is strictly dependent on the presence of RanGTP *(51,54)*. Therefore it is clear that CRM-1 directly exports Smad4 out of the nucleus.

4. NUCLEAR TARGETING OF TGF-β ACTIVATED SMAD COMPLEXES

In the absence of TGF-β, nuclear import and export of R-Smads is separate from that of Smad4. But when the cells are stimulated with TGF-β, R-Smads and Smad4 move as a stable complex. Recent studies have begun to unravel how nuclear import and export properties of Smads change as a consequence of R-Smads-Smad4 association and how through regulating the interaction between Smads and their transport and retention factors that TGF-β drives R-Smads and Smad4 into the nucleus *(51)*.

4.1. Smad4

The mechanism of TGF-β-induced nuclear accumulation is best understood in the case of Smad4. Either TGF-β stimulation or LMB treatment alone could target Smad4 into the nucleus, suggesting that TGF-β could overcome CRM-1-mediated nuclear export, either by enhancing the nuclear import rate or disabling the CRM-1 export pathway.

Smad4 is mostly a monomer in basal state cells while it tightly associates with phosphorylated Smad2 or Smad3 in a heterotrimer (i.e., 2 Smad2 or Smad3: 1 Smad4) configuration upon TGF-β stimulation *(15,16,50,57)*. In reconstituted nuclear import assays, the kinetics of nuclear import of the monomeric or complexed Smad4 are very similar *(51)*. The only difference is that the two forms of Smad4 could employ different nucleoporins in their migration through the NPCs, as evidenced by the observation that blocking a subset of nucleoporins by the importin-β (45-462). On the other hand, Smad4 in the nucleus was readily exported by recombinant CRM-1, but the heterotrimer complex still resided in the nucleus under the same assay condition *(51)*. Indeed, once Smad4 is complexed with phospho-Smad3, it is no longer able to bind CRM-1 *(51)*. Although Smad4 is also bound to its target DNA site through a motif close to the NES, the binding of DNA element did not seem to affect Smad4 engagement to CRM-1 *(51)*.

Therefore, inactivation of CRM-1-mediated nuclear export is the primary mechanism by which TGF-β targets Smad4 into the nucleus. TGF-β-induced association of phospho-Smad2/3 with Smad4 is required and sufficient to retain Smad4 in the nucleus.

4.2. R-Smads

Nuclear import properties of the monomeric and trimeric forms of Smad2 and Smad3 appear to be similar in the reconstituted nuclear import assay, and there was no significant difference in the rate of nuclear import as well *(43,51)*. This is further validated by observation in live cells that the rate of Smad2 nuclear import was unaffected by TGF-β-induced phosphorylation of Smad2 *(26)*. On the other hand, the mobility of phospho-Smad2 in the nucleus is considerably slower than that of unphosphorylated Smad2 *(26)*. Evidence from a quantitative nuclear export assay also supports the view that TGF-β stimulation impedes nuclear export of Smad2 *(24)*. All these observations seem to suggest that reduced nuclear export rate is the main reason that TGF-β-phosphorylated Smad2 accumulates in the nucleus, a scenario similar to that of Smad4 as described in Section 4.1. Because CRM-1 is clearly not involved in nuclear export of Smad2 or Smad3, validation of such hypothesis must await identification of the export factor(s) for R-Smads.

Change in the rate of nuclear import or export is not the only conceivable way by which TGF-β can drive R-Smads into the nucleus. In living cells, the mobility of Smad2 is much lower than expected for a freely diffusible protein, indicating that it is being retained in the cytoplasm *(25)*. Indeed, the endosome-associated protein SARA binds to Smad2 and Smad3, and the binding prevents these Smad to interact with nucleoporins and inhibits nuclear import of Smad2 and Smad3 *(24,43)*. More importantly, phosphorylation of Smad2 and Smad3 induces conformational changes that lead to their dissociation from SARA *(43–45)*. Therefore, release from cytoplasmic retention is one potential mechanism through which TGF-β enhances nuclear accumulation of Smad2 and Smad3 in the nucleus. However, because SARA exclusively localizes within endosomes and yet Smad2 and Smad3 are evenly distributed throughout the cytoplasm, retention factors other than SARA are likely to be involved and remain to be discovered.

Overexpression of nuclear factors such as FoxH1 and ATF3 results in nuclear accumulation of Smad2 and Smad3 in the absence of TGF-β, demonstrating that nuclear retention factors can provide sufficient impetus to confine Smad2/3 into the nucleus *(24,52)*. Hence, nuclear factors whose association with Smad2/3 is enhanced by TGF-β could be another major determinant of nuclear accumulation of R-Smads.

These recent developments in the study of intracellular Smad movement support an emerging paradigm regarding the trafficking and subcellular localization of R-Smads and Smad4 (Fig. 1). These Smads are constantly cycling through the NPCs and their subcellular distribution at basal state reflects an equilibrium among import/export and retention forces. The TGF-β signal, by inducing phosphorylation of R-Smads and the subsequent hetero-trimerization with Smad4, affects interaction of these Smads with nuclear export factors and their nuclear or cytoplasmic retention factors. Such changes in affinity between TGF-β-activated Smads and their partners impede the movement of Smads once they enter the nucleus and eventually restrict R-Smads and Smad4 in the nucleus *(58)*. In this way, TGF-β limits Smads location to the nucleus.

5. DYSREGULATION OF NUCLEAR TARGETING OF SMADS

TGF-β exerts potent antiproliferative effects on a number of cell types particularly epithelial cells *(59–62)*. Cancer cells originated from epithelium have often adopt various

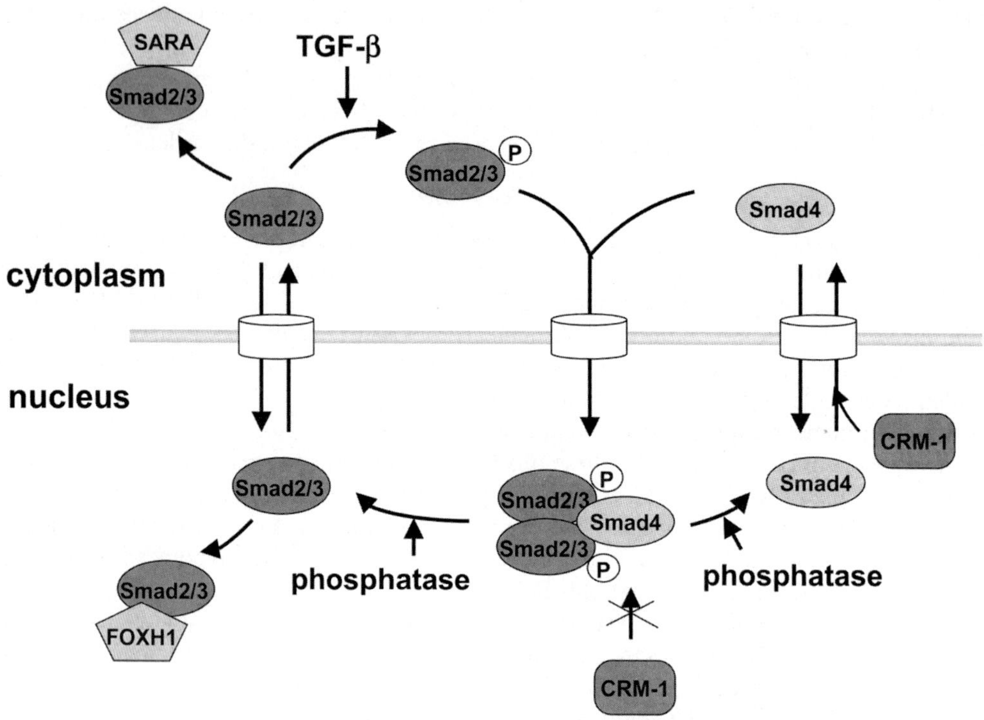

Fig. 1. Control of nucleocytoplasmic distribution of Smad2/3 and Smad4 by TGF-β. Smad2/3 spontaneously shuttles across the nuclear pore complexes through direct interaction with nucleoporins. Smad2/3 can be sequestered by cytoplasmic or nuclear retention factors such as SARA and FoxH1. Upon TGF-β-induced phosphorylation, Smad2/3 forms heterotrimer complexes with Smad4 and becomes restricted in the nucleus. Yet unknown phosphatase activities are involved in dephosphorylation of Smad2/3 which leads to dissociation of the complex and return of Smad2/3 and Smad4 to the nucleocytoplasmic shuttling. Smad4 is intrinsically capable of translocating into the nucleus but is constantly exported by CRM-1. The heterotrimer formation with phosphorylated Smad2/3 shields Smad4 from CRM-1 and allows Smad4 to concentrate in the nucleus.

mechanisms to counter such cytostatic function of TGF-β, including disruption of nuclear accumulation of R-Smads and Smad4 *(59–62)*. This could be achieved either by somatic mutations in Smads and/or crosstalk from mitogenic signals. Reduction in the amount and duration of Smads in the nucleus can lead to failure of TGF-β to regulate its target gene expression and renders the cells insensitive to growth inhibition by TGF-β.

5.1. Cancer-Causing Mutations

Smad4 is a tumor suppressor that is mutated in pancreatic, colon and other cancers *(63,64)*. Many of the mutations were missense point mutations and are clustered in the C-terminal MH2 domain of Smad4. Based on X-ray crystal structural analysis, some of these point mutations are expected to affect the interaction of Smad4 with phospho-Smad2/3 *(16,65)*. Because binding of phospho-Smad2/3 is key to retain Smad4 in the nucleus, such Smad4 mutants are conceivably also defective in nuclear accumulation upon TGF-β stimulation. This idea was tested in two of the mutants, R361C and D351H. Indeed, both mutants are unable to bind phospho-Smad2, and both have much weakened nuclear accumulation in response to TGF-β *(51)*. Thus, the inability to accumulate in the nucleus in response to TGF-β is the primary molecular defect in these cancer-related Smad4 mutants. The diminished nuclear concentration of these Smad4 mutants is not due to defects in nuclear import, as these mutants readily accumulate in the nucleus when cells are treated with LMB, just as efficiently as the wild type Smad4 *(51)*. These observations reinforced the direct link between Smad4 binding to phospho-Smad2/3 and Smad4 targeting to the nucleus by TGF-β. It is likely that other Smad4 mutants with reduced affinity for phospho-Smad2/3 are also defective in nuclear accumulation, which would account for the inability of TGF-β to inhibit cell proliferation in these cells *(66)*.

5.2. Crosstalk from Other Signals

TGF-β signaling could be interfered by various other signals, which may mitigate the physiological response to TGF-β *(67–73)*. Several examples of crosstalk involve phosphorylation of Smads at Ser/Thr residues different from the C-terminus SXS motif, by kinases other than the TGF-β receptor kinase *(68,69,71,72)*. MAP kinases downstream of mitogenic signals can phosphorylate several residues in the linker region of Smad1, Smad2 and Smad3 *(68,69)*. In mammalian cells and during Xenopus embryo development, phosphorylation of these residues in R-Smads have been shown to attenuate nuclear accumulation of these Smads in response to TGF-β or activin *(68,69,74)*. The calcium and calmodulin dependent kinase, CaMK II, also phosphorylates Smad3 on a number of residues in the linker region, which are different from the MAP kinase sites *(71)*. These phosphorylation events also seem to result in reduced nuclear concentration of Smad3 *(71)*. More recently, G protein coupled receptor kinase 2 (GRK2) was also found to be capable of phosphorylating the linker region of Smad2, which also leads to decreased nuclear accumulation of Smad2 *(72)*. These kinases, which are activated by various mitogenic signals, provide a mechanism by which the pro-growth factors can antagonize the antiproliferation function exerted by TGF-β.

How do these phosphorylation events impede nuclear accumulation of Smads? It was shown that GRK2-mediated phosphorylation in the linker region inhibited phosphorylation of the C-terminal SXS motif by TGF-β, providing a likely mechanism by which GRK2 affects TGF-β-induced nuclear accumulation of Smad2 *(72)*. However, linker phosphorylation by MAP kinases or CaMK II was not known to interfere with the C-terminal phosphorylation, and how they perturb nuclear accumulation of Smad2 or Smad3 remains a challenging issue *(68,69,71)*.

Crosstalk from other signaling pathways can also inhibit nuclear targeting of R-Smads through mechanisms that do not involve phosphorylation of R-Smads. The Akt kinase has been shown to sequester Smad3 in the cytoplasm, inhibiting phosphorylation of Smad3 and its nuclear translocation in response to TGF-β *(75,76)*. These functions of Akt are not dependent on its kinase activity, and are most likely due to competition with other factors for binding to Smad3. This provides a mechanism through which the progrowth or survival signals antagonize the apoptotic function of TGF-β *(75,76)*.

5.3. TGF-β Signal Strength and Duration of Smad Accumulation in the Nucleus

The inability of TGF-β to induce cell cycle arrest is a hallmark of many epithelial-derived cancer cells. In many cases, such defects in TGF-β signaling are not due to loss-of-function mutations in the core components of the signaling pathway. It was recently shown that in the case of pancreatic cancer cell lines Pac-1 and PT45, the duration of Smad2/3 accumulation in the nucleus in response to TGF-β was much shortened, which directly correlated to failure of TGF-β to increase p21 level and inhibit cell proliferation *(77)*. Interestingly, not all of TGF-β target genes are as sensitive as p21 is to the duration of Smad2/3 in the nucleus. In the case of p21, the high turnover rate of p21 mRNA demands persistent activation of its transcription by Smads in order to reach an elevated level sufficient to arrest cell cycle progression *(77)*. Thus these observations revealed an unique mechanism through which the antiproliferation function of TGF-β is selectively disarmed in cancer cells *(77)*.

This study also demonstrated that in order to turn on the cytostatic program in cells, both the quantity and duration of Smad in the nucleus are critical *(77)*. Given the crucial requirement of C-terminal phosphorylation for targeting R-Smads to the nucleus, it is conceivable that factors impinging on this phosphorylation event can all be important regulatory inputs. These would include the concentration of TGF-β the cells are exposed to, the efficacy of the type I and type II TGF-β receptor kinases, and phosphatases of R-Smads. These factors could all be potential targets of intervention by oncogenic forces that inhibit the cytostatic program induced by TGF-β.

6. CONCLUDING REMARKS

Subcellular distribution of Smads is critical for transduction and modulation of TGF-β signals. We have begun to understand how R-Smads and Smad4 translocate across the NPCs, and how the TGF-β-regulated interplay between the nuclear transport machinery and retention factors dictates the subcellular location of these Smads. There are still notable gaps in our knowledge on this subject. We know little about trafficking of R-Smads to the receptor kinases and their subsequent movement to the NPCs: are these free-diffusion events or are they facilitated and directed by certain factors? We do not know if other posttranslational modification of Smads, such as acetylation and sumoylation may affect nucleocytoplasmic trafficking of Smads *(78–80)*. The control of distribution of Smad7 remains unknown. Information on these subjects will further advance our understanding of regulatory inputs that impinge on intracellular movement of Smads and modulate the complex biological functions of TGF-β in normal and tumor cells.

REFERENCES

1. Raftery LA, Twombly V, Wharton K, Gelbart WM. Genetic screens to identify elements of the decapentaplegic signaling pathway in Drosophila. Genetics 1995;139:241–254.
2. Derynck R, Zhang Y, Feng XH. Smads: transcriptional activators of TGF-beta responses. Cell 1998;95:737–740.
3. Attisano L, Wrana JL. Signal transduction by the TGF-beta superfamily. Science 2002;296:1646–1647.
4. Massagué J, Wotton D. Transcriptional control by the TGF-beta/Smad signaling system. EMBO J 2000;19:1745–1754.
5. Heldin C-H, Miyazono K, ten Dijke P. TGF-β signalling from cell membrane to nucleus through SMAD proteins. Nature 1997;390:465–471.
6. Baker J, Harland RM. A novel mesoderm inducer, mMadr-2, functions in the activin signal transduction pathway. Genes Dev 1996;10:1880–1889.
7. Macias-Silva M, Abdollah S, Hoodless PA, Pirone R, Attisano L, Wrana JL. MADR2 is a substrate of the TGFβ receptor and phosphorylation is required for nuclear accumulation and signaling. Cell 1996;87:1215–1224.
8. Liu F, Pouponnot C, Massagué J. Dual role of the Smad4/DPC4 tumor suppressor in TGFβ-inducible transcriptional responses. Genes Dev 1997;11:3157–3167.
9. Abdollah S, Macias-Silva M, Tsukazaki T, Hayashi H, Attisano L, Wrana JL. TbetaRI phosphorylation of Smad2 on Ser465 and Ser467 is required for Smad2-Smad4 complex formation and signaling. J Biol Chem 1997;272:27,678–27,685.
10. Liu F, Hata A, Baker J, et al. A human Mad protein acting as a BMP-regulated transcriptional activator. Nature 1996;381:620–623.
11. Souchelnytskyi S, Tamaki K, Engstrom U, Wernstedt C, ten Dijke P, Heldin C-H. Phosphorylation of Ser465 and Ser467 in the C terminus of Smad2 mediates interaction with Smad4 and is required for transforming growth factor-beta signaling. J Biol Chem 1997;272:28,107–28,115.
12. Kretzschmar M, Liu F, Hata A, Doody J, Massagué J. The TGF-β mediator Smad1 is directly phosphorylated and functionally activated by the BMP receptor kinase. Genes Dev 1997;11:984–995.

13. Inman GJ, Nicolas FJ, Callahan JF, et al. SB-431542 is a potent and specific inhibitor of transforming growth factor-beta superfamily type I activin receptor-like kinase (ALK) receptors ALK4, ALK5, and ALK7. Mol Pharmacol 2002;62:65–74.
14. Inman GJ, Nicolas FJ, Hill CS. Nucleocytoplasmic shuttling of Smads 2, 3, and 4 permits sensing of TGF-beta receptor activity. Mol Cell 2002;10:283–294.
15. Lagna G, Hata A, Hemmati-Brivanlou A, Massagué J. Partnership between DPC4 and SMAD proteins in TGFβ signalling pathways. Nature 1996;383:832–836.
16. Chacko BM, Qin BY, Tiwari A, et al. Structural basis of heteromeric smad protein assembly in TGF-Beta signaling. Mol Cell 2004;15:813–823.
17. Calonge MJ, Massagué J. Smad4/DPC4 silencing and hyperactive Ras jointly disrupt transforming growth factor-beta antiproliferative responses in colon cancer cells. J Biol Chem 1999;274: 33,637–33,643.
18. Zhu HJ, Iaria J, Sizeland AM. Smad7 differentially regulates transforming growth factor beta-mediated signaling pathways. J Biol Chem 1999;274:32,258–32,264.
19. Itoh S, Landstrom M, Hermansson A, et al. Transforming growth factor beta1 induces nuclear export of inhibitory Smad7. J Biol Chem 1998;273:29,195–29,201.
20. Bai S, Cao X. A nuclear antagonistic mechanism of inhibitory Smads in transforming growth factor-beta signaling. J Biol Chem 2002;277:4176–4182.
21. Nakao A, Afrakhte M, Morén A, et al. Identification of Smad7, a TGFβ-inducible antagonist of TGF-β signaling. Nature 1997;389:631–635.
22. Suzuki C, Murakami G, Fukuchi M, et al. Smurf1 regulates the inhibitory activity of Smad7 by targeting Smad7 to the plasma membrane. J Biol Chem 2002;277:39,919–39,925.
23. Kavsak P, Rasmussen RK, Causing CG, et al. Smad7 binds to Smurf2 to form an E3 ubiquitin ligase that targets the TGF beta receptor for degradation. Mol Cell 2000;6:1365–1375.
24. Xu L, Kang Y, Col S, Massagué J. Smad2 nucleocytoplasmic shuttling by nucleoporins CAN/Nup214 and Nup153 feeds TGFbeta signaling complexes in the cytoplasm and nucleus. Mol Cell 2002;10: 271–282.
25. Nicolas FJ, De Bosscher K, Schmierer B, Hill CS. Analysis of Smad nucleocytoplasmic shuttling in living cells. J Cell Sci 2004;117:4113–4125.
26. Schmierer B, Hill CS. Kinetic analysis of Smad nucleocytoplasmic shuttling reveals a mechanism for transforming growth factor β-dependent nuclear accumulation of Smads. Mol Cell Biol 2005;25: 9845–9858.
27. Fahrenkrog B, Aebi U. The nuclear pore complex: nucleocytoplasmic transport and beyond. Nat Rev Mol Cell Biol 2003;4:757–766.
28. Suntharalingam M, Wente SR. Peering through the pore: nuclear pore complex structure, assembly, and function. Dev Cell 2003;4:775–789.
29. Cronshaw JM, Krutchinsky AN, Zhang W, Chait BT, Matunis MJ. Proteomic analysis of the mammalian nuclear pore complex. J Cell Biol 2002;158:915–927.
30. Lyman SK, Gerace L. Nuclear pore complexes: dynamics in unexpected places. J Cell Biol 2001;154: 17–20.
31. Ryan KJ, Wente SR. The nuclear pore complex: a protein machine bridging the nucleus and cytoplasm. Curr Opin Cell Biol 2000;12:361–371.
32. Rout MP, Aitchison JD. The nuclear pore complex as a transport machine. J Biol Chem 2001;276: 16,593–16,596.
33. Weis K. Regulating Access to the Genome. Nucleocytoplasmic transport throughout the cell cycle. Cell 2003;112:441–451.
34. Gorlich D, Kutay U. Transport between the cell nucleus and the cytoplasm. Annu Rev Cell Dev Biol 1999;15:607–660.
35. Macara IG. Transport into and out of the nucleus. Microbiol Mol Biol Rev 2001;65:570–594, table of contents.
36. Strom AC, Weis K. Importin-beta-like nuclear transport receptors. Genome Biol 2001;2:reviews 3008. doi:10.1186/gb-2001.
37. Ribbeck K, Kutay U, Paraskeva E, Gorlich D. The translocation of transportin-cargo complexes through nuclear pores is independent of both Ran and energy. Curr Biol 1999;9:47–50.
38. Englmeier L, Olivo JC, Mattaj IW. Receptor-mediated substrate translocation through the nuclear pore complex without nucleotide triphosphate hydrolysis. Curr Biol 1999;9:30–41.
39. Nakielny S, Dreyfuss G. Import and export of the nuclear protein import receptor transportin by a mechanism independent of GTP hydrolysis. Curr Biol 1998;8:89–95.

40. Adam SA, Marr RS, Gerace L. Nuclear protein import in permeabilized mammalian cells requires soluble cytoplasmic factors. J Cell Biol 1990;111:807–816.
41. Adam EJ, Adam SA. Identification of cytosolic factors required for nuclear location sequence-mediated binding to the nuclear envelope. J Cell Biol 1994;125:547–555.
42. Xu L, Alarcon C, Col S, Massagué J. Distinct domain utilization by Smad3 and Smad4 for nucleoporin interaction and nuclear import. J Biol Chem 2003;278:42,569–42,577.
43. Xu L, Chen YG, Massagué J. The nuclear import function of Smad2 is masked by SARA and unmasked by TGFb-dependent phosphorylation. Nat Cell Biol 2000;2:559–562.
44. Tsukazaki T, Chiang TA, Davison AF, Attisano L, Wrana JL. SARA, a FYVE domain protein that recruits Smad2 to the TGFbeta receptor. Cell 1998;95:779–791.
45. Wu G, Chen YG, Ozdamar B, et al. Structural basis of Smad2 recognition by the Smad anchor for receptor activation. Science 2000;287:92–97.
46. Xiao Z, Liu X, Lodish HF. Importin beta mediates nuclear translocation of Smad 3. J Biol Chem 2000;275:23,425–23,428.
47. Xiao Z, Latek R, Lodish HF. An extended bipartite nuclear localization signal in Smad4 is required for its nuclear import and transcriptional activity. Oncogene 2003;22:1057–1069.
48. Kurisaki A, Kose S, Yoneda Y, Heldin C-H, Moustakas A. Transforming growth factor-beta induces nuclear import of Smad3 in an importin-beta1 and Ran-dependent manner. Mol Biol Cell 2001;12: 1079–1091.
49. Xiao Z, Liu X, Henis YI, Lodish HF. A distinct nuclear localization signal in the N terminus of Smad 3 determines its ligand-induced nuclear translocation. Proc Natl Acad Sci USA 2000;97:7853–7858.
50. Jayaraman L, Massagué J. Distinct oligomeric state of SMAD proteins in the transforming growth factor-beta pathway. J Biol Chem 2000;275:40,710–40,717.
51. Chen HB, Rud JG, Lin K, Xu L. Nuclear targeting of transforming growth factor-beta-activated Smad complexes. J Biol Chem 2005;280:21,329–21,336.
52. Kang Y, Chen CR, Massagué J. A Self-enabling TGFbeta response coupled to stress signaling: Smad engages stress response factor ATF3 for Id1 repression in epithelial cells. Mol Cell 2003;11:915–926.
53. Xiao Z, Brownawell AM, Macara IG, Lodish HF. A novel nuclear export signal in Smad1 is essential for its signaling activity. J Biol Chem 2003;278:34,245–34,252.
54. Pierreux CE, Nicolas FJ, Hill CS. Transforming growth factor beta-independent shuttling of Smad4 between the cytoplasm and nucleus. Mol Cell Biol 2000;20:9041–9054.
55. Hoodless PA, Tsukazaki T, Nishimatsu S, Attisano L, Wrana JL, Thomsen GH. Dominant-negative Smad2 mutants inhibit activin/Vg1 signaling and disrupt axis formation in Xenopus. Dev Biol 1999; 207:364–379.
56. Watanabe M, Masuyama N, Fukuda M, Nishida E. Regulation of intracellular dynamics of Smad4 by its leucine-rich nuclear export signal. EMBO Rep 2000;1:176–182.
57. Kawabata M, Inoue H, Hanyu A, Imamura T, Miyazono K. Smad proteins exist as monomers in vivo and undergo homo- and hetero-oligomerization upon activation by serine/threonine kinase receptors. EMBO J 1998;17:4056–4065.
58. Xu L, Massagué J. Nucleocytoplasmic shuttling of signal transducers. Nat Rev Mol Cell Biol 2004; 5:209–219.
59. Massagué J, Blain SW, Lo RS. TGFbeta signaling in growth control, cancer, and heritable disorders. Cell 2000;103:295–309.
60. Derynck R, Akhurst RJ, Balmain A. TGF-beta signaling in tumor suppression and cancer progression. Nat Genet 2001;29:117–129.
61. Siegel PM, Massagué J. Cytostatic and apoptotic actions of TGF-beta in homeostasis and cancer. Nat Rev Cancer 2003;3:807–821.
62. Akhurst RJ, Derynck R. TGF-beta signaling in cancer: a double-edged sword. Trends Cell Biol 2001;11:S44–S51.
63. Hahn SA, Schutte M, Hoque ATMS, et al. DPC4, a candidate tumor suppressor gene at human chromosome 18q21.1. Science 1996;271:350–353.
64. Schutte M, Hruban RH, Hedrik L, et al. DPC4 gene in various tumor types. Cancer Res. 1996; 56:2527–2530.
65. Wu JW, Hu M, Chai J, et al. Crystal structure of a phosphorylated Smad2. Recognition of phosphoserine by the MH2 domain and insights on Smad function in TGF-beta signaling. Mol Cell 2001;8: 1277–1289.

66. Iacobuzio-Donahue CA, Song J, Parmiagiani G, Yeo CJ, Hruban RH, Kern SE. Missense mutations of MADH4: characterization of the mutational hot spot and functional consequences in human tumors. Clin Cancer Res 2004;10:1597–1604.
67. Bitzer M, von Gersdorff G, Liang D, et al. A mechanism of supression of TGF-β/SMAD signaling by NFκB/RelA. Genes Dev 2000;14:187–197.
68. Kretzschmar M, Doody J, Massagué J. Opposing BMP and EGF signalling pathways converge on the TGFβ family mediator Smad1. Nature 1997;389:618–622.
69. Kretzschmar M, Doody J, Timokhina I, Massagué J. A mechanism of repression of TGFβ/Smad signaling by ongenic ras. Genes Dev 1999;13:804–816.
70. Ulloa L, Doody J, Massagué J. Inhibition of transforming growth factor-beta/SMAD signalling by the interferon-gamma/STAT pathway. Nature 1999;397:710–713.
71. Wicks SJ, Lui S, Abdel-Wahab N, Mason RM, Chantry A. Inactivation of smad-transforming growth factor beta signaling by Ca(2+)-calmodulin-dependent protein kinase II. Mol Cell Biol 2000;20: 8103–8111.
72. Ho J, Cocolakis E, Dumas VM, Posner BI, Laporte SA, Lebrun JJ. The G protein-coupled receptor kinase-2 is a TGFbeta-inducible antagonist of TGFbeta signal transduction. EMBO J 2005;24: 3247–3258.
73. Matsuura I, Denissova NG, Wang G, He D, Long J, Liu F. Cyclin-dependent kinases regulate the antiproliferative function of Smads. Nature 2004;430:226–231.
74. Grimm OH, Gurdon JB. Nuclear exclusion of Smad2 is a mechanism leading to loss of competence. Nat Cell Biol 2002;4:519–522.
75. Conery AR, Cao Y, Thompson EA, Townsend CM, Jr., Ko TC, Luo K. Akt interacts directly with Smad3 to regulate the sensitivity to TGF-beta induced apoptosis. Nat Cell Biol 2004;6:366–372.
76. Remy I, Montmarquette A, Michnick SW. PKB/Akt modulates TGF-beta signalling through a direct interaction with Smad3. Nat Cell Biol 2004;6:358–365.
77. Nicolas FJ, Hill CS. Attenuation of the TGF-beta-Smad signaling pathway in pancreatic tumor cells confers resistance to TGF-beta-induced growth arrest. Oncogene 2003;22:3698–3711.
78. Lee PS, Chang C, Liu D, Derynck R. Sumoylation of Smad4, the common Smad mediator of transforming growth factor-beta family signaling. J Biol Chem 2003;278:27,853–27,863.
79. Lin X, Liang M, Liang YY, Brunicardi FC, Feng XH. SUMO-1/Ubc9 promotes nuclear accumulation and metabolic stability of tumor suppressor Smad4. J Biol Chem 2003;278:31,043–31,048.
80. Gronroos E, Hellman U, Heldin C-H, Ericsson J. Control of Smad7 stability by competition between acetylation and ubiquitination. Mol Cell 2002;10:483–493.

9 Ski, SnoN, and Akt as Negative Regulators of Smad Activity: Balancing Cell Death and Cell Survival

Erwan Le Scolan and Kunxin Luo

Contents

Introduction
The TGF-β Signaling Pathway
Ski/SnoN as Repressors of the TGF-β Signaling Pathway
Crosstalk of Akt with TGF-β Signaling Pathway: Akt as a Repressor of the Smad Proteins
Conclusions
Acknowledgments
References

Abstract

Cytokines of the transforming growth factor-β (TGF-β) superfamily exert many of their diverse effects via the recruitment of the Smad proteins to activate transcription of TGF-β target genes. Activities of the Smad proteins have been shown to be regulated through interaction with cellular partners. Among these, the protooncogenes of the Ski family, Ski, and SnoN, and the protein kinase Akt can negatively modulate the Smad signaling via differents mechanisms to affect various downstream TGF-β responses. Here, we review how Ski, SnoN, and Akt act as repressors of Smad activity to antagonizes TGF-β signaling.

Key Words: TGF-β; Smad; Ski; SnoN; Akt; tumorigenesis.

1. INTRODUCTION

The transforming growth factor-β (TGF-β) superfamily of cytokines regulates many cellular processes including cell growth, cell survival and differentiation *(1–3)*. By balancing cell proliferation and cell death, TGF-β functions as an important regulator of cell homeostasis, and alterations of the TGF-β signaling pathway have been implicated in the development of human cancer and other diseases *(4,5)*. During cancer development, TGF-β signaling acts as a tumor suppressor at early stages of tumorigenesis, but as a promoter of tumor invasiveness and metastasis at later stages *(2)*. A better understanding of the mechanism of TGF-β receptor signal transduction is therefore a crucial first step toward a better understanding and treatment of human cancer.

From: *Cancer Drug Discovery and Development: Transforming Growth Factor-β in Cancer Therapy, Vol I: Basic and Clinical Biology*
Edited by: Sonia B. Jakowlew © Humana Press Inc., Totowa, NJ

Many biological effects of TGF-β are mediated by the Smad proteins *(6–8)*. The Smad proteins activate expression of TGF-β responsive genes by interacting with transcriptional coactivators and are often subjected to positive and negative regulation by other cellular cofactors and pathways. Here, we review how the Ski family of proteins antagonizes TGF-β signaling by repressing activity of the Smad proteins. Following several recent studies identifying a new crosstalk between Akt and TGF-β signaling pathway, we will also discuss how Akt can antagonize TGF-β-induced apoptosis through interaction with Smad3.

2. THE TGF-β SIGNALING PATHWAY

Cytokines of the TGF-β superfamily, including TGF-β, bone morphogenetic protein (BMP), and activin, exert their diverse effects through a tetrameric complex of two serine-threonine kinase receptors, called type I and type II receptors *(8)*. Upon ligand binding, the type II receptor phosphorylates and activates the type I receptor. Activated type I receptor phosphorylates the receptor-activated Smad (R-Smad) proteins, including Smad2 and Smad3 for TGF-β and activin signaling, and Smad1, Smad5, and Smad8 for BMP signaling. The phosphorylated R-Smads then form heteromeric complexes with a common-mediator Smad (Co-Smad), Smad4, and translocate into the nucleus. In collaboration with other transcription factors, the heteromeric Smad complexes bind to promoter DNA and activate or repress transcription of TGF-β target genes *(9–11)*. The activities of the Smad proteins are additionally regulated through interaction with cellular partners including proteins that regulate the stability, intracellular localization, post-translational modification and transcriptional activity of the Smad proteins *(12)*. Among the corepressors of Smad proteins are the Ski family of protooncoproteins that play important roles in regulation of oncogenic transformation as well as embryonic development.

3. SKI/SNON AS REPRESSORS OF THE TGF-β SIGNALING PATHWAY

3.1. The Ski/SnoN Protooncogenes

Ski was initially identified as the oncogene (v-*ski*) of the avian Sloan-Kettering retrovirus *(13)*. The cellular homolog of v-*ski*, c-*ski*, and the closely related *snoN* were cloned later based on sequence homology *(14)*. Human c-*Ski* and SnoN are nuclear proteins of 728 and 684 amino acids, respectively *(14,15)*. Structurally, proteins of the Ski family share a highly conserved amino (N)-terminal region (Ski homology region) that is both necessary and sufficient for the transforming and differentiation activities of Ski and SnoN (Fig. 1A) *(16,17)*. Within the Ski homologous region, four residues, Cys247, Cys250, His262 and His264 coordinate a bound zinc atom to form a novel class of Cys_2His_2 type zinc finger motif. These residues are conserved in all Ski and Sno proteins as well as the related *Fussel-18* and *Corl1* and are essential in stabilizing the three-dimensional fold of this domain. This domain is highly homologous to the SAND domain (named after Sp100, AIRE-1, NucP41/75 and DEAF-1), an evolutionarily conserved structural motif found in several nuclear proteins that are involved in chromatin dependant transcriptional repression. The SAND domain in these repressor proteins use an I-loop to bind DNA, while Ski uses the same I-loop in its SAND domain to interact with the L3 loop region of Smad4. The carboxyl-terminal portion of Ski is less conserved among members of the Ski family. Toward the carboxyl-termini of Ski and SnoN, a short sequence has been reported to mediate homo- and heterodimerization of Ski and SnoN *(18,19)*. Although the physiological significance of this dimerization is not clear, Cohen et al., reported that heterodimers of Ski and SnoN are more potent transforming agents than homodimers of Ski or SnoN *(20)*. When overexpressed, Ski induces oncogenic transformation of chicken embryo fibroblasts and muscle differentiation of quail embryonic cells *(21)*. Ski homologs have been identified in human, mouse, chicken, *Xenopus* and zebra

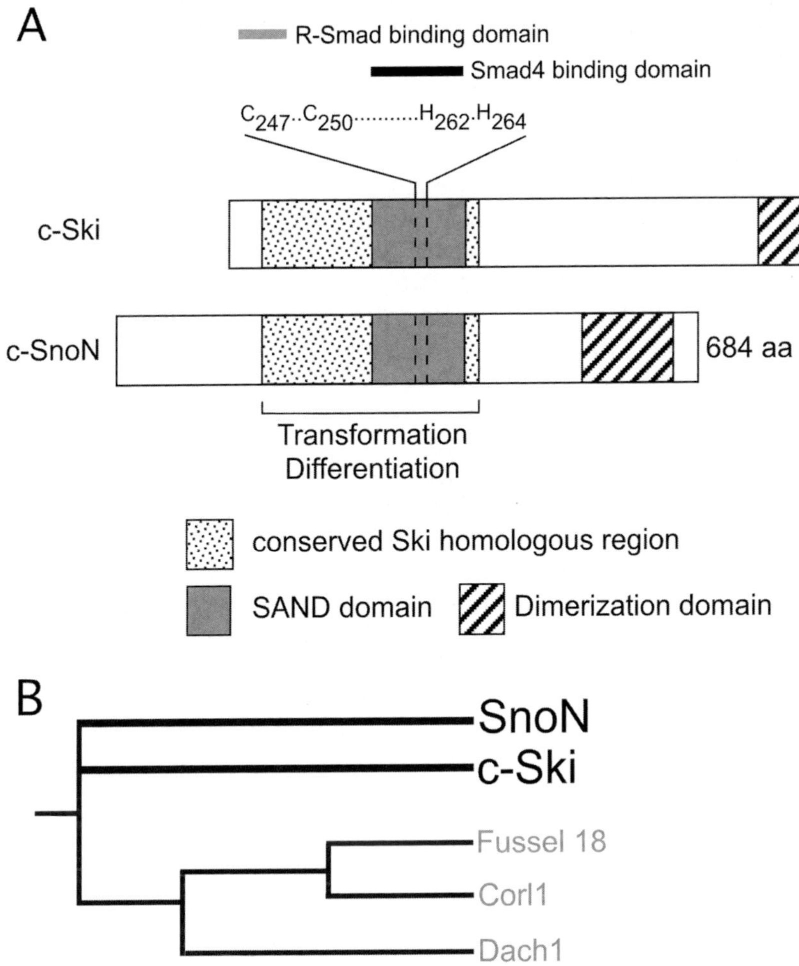

Fig. 1. The Ski family. (**A**) Structural organization of Ski and SnoN. The number of amino acid residues for each protein is indicated to the right. The Smad4 binding domain and R-Smad binding domain are indicated on the top. Dash lines in the SAND domain indicate the conserved Cys2His2 zinc finger motif. (**B**) Phylogenic analysis of the Ski homologous region of c-*Ski*, SnoN, Fussel 18, Corl1 and Dach1. Phylogenic analysis was performed using ClustalW method.

fish *(16)*. In adult mice, *ski* is expressed in all tissues but at a low level *(22)*. Its expression is elevated during certain stages of embryonic development especially in the neural tube, the migrating neural crest cells *(22)* and skeletal muscle *(23)*. In vivo studies in *Xenopus*, zebra fish and mice have shown essential and conserved functions of Ski in the development of neuronal, craniofacial and muscle lineages. Overexpression of Ski induces autonomous neural axis formation in *Xenopus (24,25)*. Ectopic expression of Ski in zebra fish disrupts gastrulation and neural patterning *(26)* and transgenic mice overexpressing a Ski fragment exhibit hypertrophy of type II skeletal muscle fibers *(27)*. Consistent with the function of Ski in the differentiation of neuronal and muscle lineages, *ski*-null pups suffer from exencephaly owing to a defect in cranial neural tube closure, various craniofacial defects and a marked decreased in skeletal muscle mass *(28)*. Interestingly, some of these phenotypes are reminiscent of some of the features observed in individuals diagnosed with the 1p36 deletion syndrome, a disorder caused by monosomy of the short arm of human chromosome 1p *(29)*. The human c-*ski* gene has been mapped to the 1p36 region subsequently and is found to be deleted in all individuals with this syndrome. Thus loss of Ski may be responsible for some of the craniofacial defects observed in these individuals.

SnoN (for ski-related novel) is a closely related member of the Ski family of oncoproteins (Fig. 1B). Two alternatively spliced forms of SnoN, SnoA and SnoI, have been identified in human and another isoform, SnoN2, have been detected in mouse *(14,30)*. In human, the expression of SnoN and SnoA appears to be ubiquitous while that of SnoI is restricted to malignant skeletal muscle *(30)*. When overexpressed, SnoN also induces transformation of chicken embryo fibroblast and muscle differentiation of quail embryo cells *(31)*. The role of SnoN in mouse development is not resolved. Two groups reported three mouse strains lacking the *sno* gene by targeted deletion or with a reduced *sno* expression by deletion of the *snoN* promoter region *(32,33)*. One mouse strain showed embryonic lethality before E3.5 *(33)*, while two other strains are viable and display only a defect in T-cell activation associated with an increased sensitivity to TGF-β *(32)*. The reason for this difference in phenotype between these mouse strains is still not understood.

Recently, two distantly *ski*-related genes, *Fussel-18 (34)* and *Corl1 (35)* have been identified in mouse and human (Fig. 1B). Fussel-18 encodes a protein of 297 amino acids and is mainly consisted of the Ski homologous region with a 38% sequence homology to Ski *(34)*. In contrast to the ubiquitous expression pattern of Ski and SnoN, Fussel-18 expression is restricted to neuronal tissues. It can bind Smad2 and Smad3 and repress TGF-β-induced transcription in a luciferase reporter assay. However, the biological activity of this protein still remains to be determined. Corl1 (corepressor for Lbx1) has been identified as a transcription factor that regulates neuronal differentiation in the developing central nervous system. Corl1 encodes a nuclear protein of 936 amino acids with homology to Ski in its amino and carboxyl terminal regions. However, interaction between Corl1 and Smad proteins has not yet been reported.

3.2. Mechanism of Ski and SnoN Function

Because Ski and SnoN do not contain any intrinsic catalytic activity, they must carry out biological functions through interaction with other cellular partners. Using multiple strategies, Ski and SnoN have been found to interact with the Smad proteins and antagonize TGF-β signaling *(36–40)* (Fig. 2). Other Ski interacting proteins have been identified since. The mechanism and physiological significance of these interactions are discussed in Section 3.2.1. and 3.2.2.

3.2.1. Ski/SnoN Interact with Smad Proteins to Antagonize TGF-β Signaling

We and others have found that Ski and SnoN interact with Smad2, Smad3, and Smad4, and repress their ability to activate TGF-β responsive genes. Overexpression of Ski or SnoN blocks the ability of cells to undergo cell cycle arrest in response to TGF-β. This ability to antagonize TGF-β-induced growth arrest may contribute to the transforming activity of Ski and SnoN, at least in chicken embryo fibroblasts *(41)*. Mutations in Ski and SnoN that disrupt their interactions with the Smad proteins markedly impair their ability to induce anchorage independent growth in chicken embryo fibroblasts. In addition to repressing TGF-β signaling, Ski, but not SnoN, interacts with the BMP-specific Smad (Smad1 and Smad5) in a ligand-dependant manner and blocks BMP signaling in both *Xenopus* oocytes and mammalian cells *(24)*. This ability to antagonize BMP signaling results in neuralization by Ski in the *Xenopus* embryo and blocking of osteoblast differentiation of murine bone marrow stroma cells *(24)*.

Subsequent studies have investigated the molecular mechanism of repression of the Smad proteins by Ski and SnoN. The first hint of how Ski and SnoN may repress Smad activity came from reports that Ski and SnoN physically associate with transcriptional corepressors,

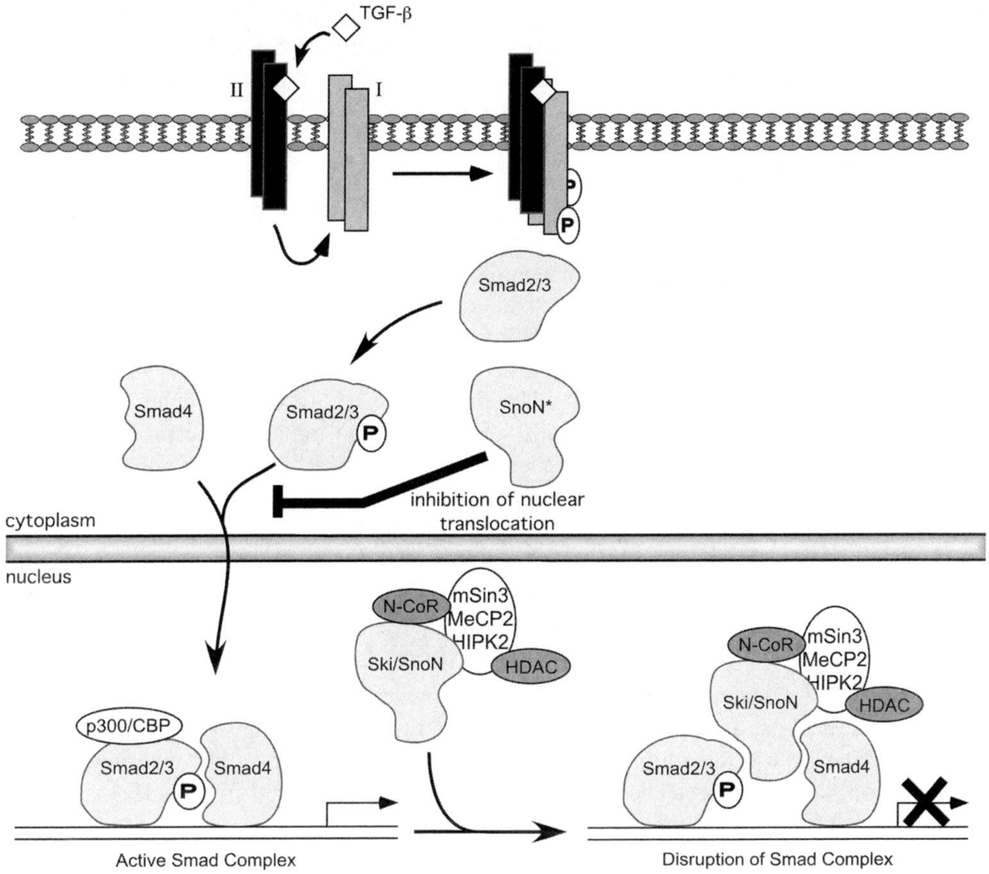

Fig. 2. Mechanism of repression of the Smad proteins by Ski and SnoN. SnoN*: cytoplasmic SnoN in normal mammary epithelial cells and keratinocytes.

including the nuclear hormone corepressor N-CoR, histone deacetylase, mSin3A, the methyl CpG-binding protein MeCP2 and HIPK2 *(42–45)*. Through simultaneous interaction with the Smad complex, Ski, and SnoN recruit the transcriptional corepressors to the TGF-β responsive promoters to inhibit transcription of TGF-β target genes. In addition, Ski and SnoN can also block the binding of transcriptional coactivator p300/CBP to Smad3 *(40,46)*. Finally, the crystal structure of Ski SAND domain in complex with the MH2 domain of Smad4 provides a better understanding of Ski-mediated repression of the Smads *(46)*. The Smad4-Ski interaction is essential for the repressive activity of Ski as in cells lacking Smad4, Ski fails to repress TGF-β signaling *(36,41)*, and a Ski mutant which cannot bind Smad4 was unable to repress BMP signaling *(24,47)*. Analysis of the crystal structure reveals that the Ski-binding surface on Smad4 overlaps significantly with that required for binding of the phosphorylated R-Smads. Consequently, interaction of Ski with Smad4 disrupts the formation of functional heterocomplexes between the R-Smads and Smad4. Thus, Ski and SnoN represses Smad-mediated transcription through at least three major mechanisms, by disrupting a functional heteromeric Smad complex, by blocking the interaction of the R-Smads to p300/CBP and by recruiting a transcription corepressor complex.

Ski oncoprotein has also been shown to regulate the transcriptional activity of Smad2 through its association with c-*Jun* *(48)*. Activated c-*Jun* has been shown to inhibit TGF-β signaling in cooperation with TGF-β by blocking the recruitment of p300/CBP to activated

Smad2 *(49)*. In the absence of TGF-β, activated c-Jun interacts with Ski to stabilize the Ski/Smad2 complex and inhibits the transcriptional activity of Smad2. The activation of TGF-β signaling results in dissociation of c-*Jun* from Ski. Thus, this interaction may function to maintain the repressed state of Smad2 in the absence of TGF-β.

3.2.2. Other Ski/SnoN Interacting Proteins

Ski and SnoN are distantly related to the retinal determination protein Dachshund. The region of highest homology is located within the amino-terminal portion of the Ski homologous region, known as the Dachshund homologous domain (DHD). The crystal structure of the Ski-DHD *(50)* reveals that it is consisted of a mixed αβ structure characteristic of the winged-helix/forkhead class of DNA binding proteins. However, the Ski-DHD does not appear to have the DNA binding activity. Evidence suggests that Ski possibly uses this domain to bind to protein partners. This domain may be involved in the interaction of Ski with N-CoR, Skip (Ski interacting protein) *(51,52)* and the transforming protein of acute promyelocytic leukemia-retinoic acid receptor β (PML-RARβ) *(53)*. Ski has also been reported to interact with other proteins including RAR, Gli3, RB, c-*jun*, and C184M *(48,54–57)*. The biological significance of most of these interactions has not been fully characterized. Some of these interactions may conceivably contribute to the role of Ski in repression of TGF-β signaling.

3.3. Regulation of Ski and SnoN

Ski and SnoN are expressed in almost all normal cells and tissues but at low levels *(14,15)*. Upregulation of Ski and SnoN expression occurs during certain stages of embryonic development and has been detected in some human cancers *(22,23,58–61)*. Here, we review how expression of Ski and SnoN can be regulated at the levels of gene amplification, transcriptional activation, posttranslational modification and protein stability.

Expression of SnoN, but not Ski, is tightly controlled by TGF-β *(38)*. Within 30 min of TGF-β stimulation, SnoN is rapidly degraded, allowing activation of TGF-β target genes. The degradation of SnoN requires the ubiquitin-dependent proteasome and can be mediated by the recruitment of two ubiquitin ligases, the anaphase promoting complex (APC) or Smurf2 *(62–64)*. Two hours after stimulation, TGF-β induces a marked increased in SnoN transcription. This increase in SnoN transcription is mediated by the direct binding of the Smad2/Smad4 complex to the TGF-β-responsive element in the *snoN* promoter and can be inhibited by the Smad3/Smad4 complex through a novel Smad inhibitory site *(65)*. In addition to turning off TGF-β signaling in a negative feedback manner, this later increase in SnoN expression may affect cellular growth and transformation in a manner independent of TGF-β signaling. In fibroblast cells that are transformed by TGF-β, this later increase in *snoN* transcription in response to TGF-β is more pronounced and of a longer duration. Using siRNA specific for SnoN and specific inhibitors of TGF-β receptor kinase, this prolonged induction of SnoN expression by TGF-β has been shown to be required for the transformation of these fibroblasts *(65)*.

Upregulation of SnoN expression in human cancer has been shown to occur through gene amplification, transcriptional activation or increased stability *(58,60,66)*. SnoN is located on chromosome 3q26, a locus that is frequently amplified in many human cancers. In particular, amplification of the *snoN* gene has been detected in esophageal cancer *(60)* as well as in ovarian cancer (Joe Gray, unpublished observation), and this may partially account for the increased expression of SnoN in these cancer cells. In many lung cancer cells, the increase in *snoN* expression occurs also at the level of transcription *(124)*. Recently, it has been reported that in some esophageal cancer cells, SnoN is resistant to TGF-β-induced degradation and as a result is present at a high level in these cells, leading to the loss of TGF-β responsiveness *(66)*.

SnoN and Ski may also be regulated by alterations in intracellular localization. SnoN and Ski have always been thought as nuclear proteins. Indeed, in most tissue culture cell lines and cancer cell lines, they are nuclear *(14,30,31)*. However, recent studies suggest that SnoN and Ski can be present in the cytoplasm under certain conditions *(59,67)*. In melanoma, Ski subcellular localization has been reported to change from nuclear in preinvasive melanomas to cytoplasmic in metastatic melanomas *(59)*. We recently reported that while SnoN is localized exclusively in the nucleus in cancer tissues or cells, in normal mammary and skin tissues as well as nontumorigenic mammary epithelial cell lines and primary keratinocytes, SnoN is predominantly cytoplasmic *(67)*. Upon morphological differentiation or cell cycle arrest, SnoN translocates into the nucleus. The cytoplasmic SnoN also represses TGF-β signaling. However, unlike nuclear SnoN that inhibits the transcriptional activity of the Smad complexes, cytoplasmic SnoN antagonizes Smad function by sequestering the Smad proteins in the cytoplasm. In addition, cytoplasmic SnoN is resistant to TGF-β-induced degradation and therefore is more potent than nuclear SnoN in repressing TGF-β signaling. Although the exact roles of cytoplasmic vs nuclear SnoN in regulation of epithelial cell biology are yet to be determined, it is likely that cytoplasmic and nuclear SnoN may produce different patterns of downstream TGF-β responses, and this may influence the proliferation or differentiation states of epithelial cells.

Finally, activities of SnoN and Ski could be regulated by posttranslational modification. Ski has been shown to be a phosphoprotein *(68,69)*. Phosphorylation of Ski occurs during mitosis, suggesting that its activity may be regulated in a cell cycle dependent manner *(68)*. Consistent with this, one report suggests that the stability of Ski is regulated by the ubiquitin-conjugating enzyme Cdc34 in a cell-cycle dependant manner *(70)*. The steady-state level of SnoN has also been shown to vary throughout cell cycle, being low at G1 owing to ubiquitination by the APC *(62)*. SnoN can also be regulated by sumoylation, and this may promote the myogenic activity of SnoN *(125)*.

3.4. Ski and SnoN in Tumorigenesis

Although Ski and SnoN have been identified initially as oncogenes, the precise roles of Ski and SnoN in carcinogenesis is still controversial. SnoN clearly has a prooncogenic activity. When overexpressed, SnoN induces transformation of chicken embryo fibroblast *(31)*. Elevated expression of SnoN has been reported in many human cancer cell lines including those derived from breast cancer, esophageal cancer, lung cancer, and rhabdomyosarcoma *(14,30,58,60)*. In addition, the *snoN* gene is located at chromosome 3q26, a known oncogene locus that is frequently amplified in some human cancers including squamous-cell carcinomas of the esophagus and ovarian cancer *(60)*. Moreover, reducing SnoN expression in human breast and lung cancer cells by a siRNA approach markedly decreases the mitogenic transformation in vitro and tumor growth in vivo *(124)*. Thus, high levels of SnoN appear to promote tumor cell growth. However, Shinagawa et al. reported that SnoN+/– mice are more sensitive to chemical-induced tumorigenesis, suggesting that SnoN may also contain an antitumorigenic activity *(33)*.

The role of Ski in human carcinogenesis is even less understood than that of SnoN. Evidence supporting a prooncogenic activity of ski includes its ability to promote anchorage-independent growth of chicken and quail embryo fibroblasts when overexpressed as well as the report that increased *ski* expression may correlate with malignant progression of esophageal and melanoma cancer *(59,71)*. In a recent study of 179 colorectal tumor biopsies, amplification of Ski has been found in 10.1% of these tumor samples, and this amplification appears to serve as a negative prognostic marker in early-stage colorectal cancer *(72)*. However, there are also reports suggesting an antitumorigenic role of Ski. Similar to that observed in SnoN+/– mice, Ski+/– mice also display an increased susceptibility to chemical induced tumorigenesis *(73)*. Interestingly, the human *ski* gene is located at chromosome 1p36, a well-known tumor suppressor locus frequently deleted in familial melanoma and neuroblastoma *(74,75)*. In contradiction to previous report that Ski is overexpressed in human melanomas, a more recent study *(76)* suggests that Ski expression is actually reduced in some melanoma cell lines.

The complexity of the effects of Ski and SnoN on tumorigenesis may partially reflect the dual roles of TGF-β in malignant progression. TGF-β, through activation of the Smads, acts as a growth factor inhibitor in nontransformed cells and at early stages of tumorigenesis. This growth inhibitory pathway is often inactivated during malignant progression of many human tumors *(77)*. At later stages of tumorigenesis, TGF-β, through both Smad-dependent and Smad-independent mechanisms *(78)*, functions as a promoter of tumor invasiveness and metastasis through its ability to induce epithelial to mesenchymal transdifferentiation (EMT) and modulate extracellular microenvironment *(2)*. As negative regulators of TGF-β signaling, it is conceivable that Ski and SnoN may also possess both tumor promoting and tumor suppressive activities at different stages of tumorigenesis. Indeed, we found recently that SnoN promotes tumor growth but inhibits EMT and tumor metastasis *(124)*.

3.5. Summary

Clearly, more studies are needed in order to understand the functions of Ski and SnoN in tumorigenesis. Past studies have mostly focused on the roles of these proteins in mitogenic transformation in tissue culture cell lines. Their functions in regulation of EMT and tumor metastasis are not understood. Particularly, it is not clear how the prooncogenic activity and antitumorigenic activity of Ski and SnoN are regulated during malignant progression and under what cellular contexts they are manifested. Genetic engineered mouse models allowing inducible or tissue specific expression (or knockdown) of wild-type (or mutant) Ski and SnoN protein should help us to resolve this question. It is also conceivable that Ski and SnoN may regulate other intracellular signaling pathways in a Smad-independent manner, and these activities may contribute to some of the effects of Ski and SnoN on tumorigenesis. As we know more about these proteins, it becomes apparent that Ski and SnoN are not simply abundant proteins with similar functions. Although they contain sequence homology and share some cellular partners, they have distinct functions and are regulated differently. Future studies will no doubt give additional information about these interesting aspects of Ski and SnoN function.

4. CROSSTALK OF AKT WITH TGF-β SIGNALING PATHWAY: AKT AS A REPRESSOR OF THE SMAD PROTEINS

The Smad pathway is the most characterized signaling pathway activated by the TGF-β superfamily. However, TGF-β also induces many Smad-independent signaling pathways including mitogen-activated protein kinase (MAPK)/p38/JNK, Rho small GTPase and PI-3 kinase/Akt signaling pathways *(79,80)*. PI-3kinase/Akt has been shown to be involved in regulation of TGF-β-induced apoptosis as well as EMT via different mechanisms. In mesenchymal cells and in certain mammary epithelial cells undergoing EMT, TGF-β can activate the PI3K/Akt pathway to promote cell survival *(78,81,82)*. The hyperactivation of Akt in these cells may cooperate with TGF-β signaling to promote tumor development. In addition to mediating the TGF-β effects, PI3K/Akt also directly modulates the activity of Smad proteins to antagonize TGF-β-induced apoptosis. In this part of the review, we will focus on the crosstalk between the PI3K/Akt pathway and the Smad proteins and discuss how negative regulation of the Smad activity by Akt may contribute to tumor development.

4.1. Akt/PKB

Akt, also known as protein kinase B, is a serine/threonine kinase of the AGC (cAMP-dependent protein kinase A/protein kinase G/protein kinase C) superfamily. Akt is implicated in the regulation of a wide variety of cellular processes such as cell metabolism, proliferation and survival *(83–85)*. Three isoforms of Akt have been identified in mammals: Akt 1, Akt2, and Akt3 *(89,90,124–127)*. Akt 1 is ubiquitously expressed whereas Akt 2 expression is expressed at a lower level and restricted to skeletal muscle, heart, kidney and liver *(86–88)*. Akt 3 expression is also low in most of tissues except testis and brain *(89,90)*. These Akt isoforms are highly homologous and all contain a conserved amino-terminal pleckstrin homology (PH) domain, a central kinase domain containing the Thr308 phosphorylation site and a carboxyl-terminal regulatory domain containing the Ser473 phosphorylation site *(91)*.

Optimal activation of Akt requires both localization to the plasma membrane as well as phosphorylation of T308 and S473 *(83,91,92)*. Growth factor stimulation results in the activation of the PI3-kinase and a subsequent increase in expression of PI3,4P and PI3,4,5P at the plasma membrane. Binding of these lipids to the Akt PH domain leads to translocation of Akt to the plasma membrane *(93–98)*. This brings Akt to a close proximity to regulatory kinases such as PDK-1. In addition, the interaction of lipids with Akt induces a conformational change in Akt that renders it accessible to phosphorylation at T308 and S473, leading to the activation of Akt *(99–103)*. T308 is phosphorylated by PDK1 *(104)*, while the mechanism mediating S473 phosphorylation is not completely understood. Some evidences suggest that S473 could be phosphorylated by PDK1 *(105)*, the integrin-liked kinase *(106,107)* or that Akt could also autophosphorylate S473 *(108,109)*. Once activated, Akt mediates its cellular effect via the phosphorylation of different targets containing a consensus RXRXXS/T-bulky hydrophobic motif. Among them are components of apoptotic machinery and other cellular pathways including Bad, Caspase-9, GSK-3β, mTOR, NFκB, CREB, and transcription factor of the forkhead family FOXO3a (FKHRL1) *(84,110)*.

4.2. Akt as a Repressor of the TGF-β Signaling Pathway

A possible link between the PI3 kinase/Akt and TGF-β-induced apoptosis was first hinted in experimental models of liver regeneration. In this system, liver injury usually initiates a rapid cell proliferation and liver regeneration, a process that is effectively inhibited by TGF-β through its ability to induce apoptosis and cell cycle arrest of hepatocytes. This negative effect of TGF-β on hepatocyte growth can be antagonized by insulin, insulin-like growth factor-1 (IGF-1), interleukin-6, hepatocyte growth factor as well as insulin receptor substrate-1 (IRS-1) both in vivo and in vitro *(111,112)*. Moreover, overexpression of an activated PI3 kinase, IRS-1 or Akt in Hep3B hepatoma cells renders cells resistant to TGF-β induced apoptosis *(111–113)*. Thus, activation of Akt seems to be necessary to protect liver cells from TGF-β induced apoptosis.

Different models have been proposed to explain the mechanism by which Akt interferes with TGF-β signaling and blocks apoptosis. It was first described that in human Hep3B cells, the inhibitory effect of Akt was mediated through the inactivation of caspase-3 *(112)*. However, the mechanism by which Akt inhibits caspase activation has not been described. Two recent studies have uncovered a new mechanism by which Akt can prevent TGF-β-induced apoptosis independently of its kinase activity through a physical interaction between Akt and Smad3 *(114,115)* Akt interacts directly with unphosphorylated Smad3, but not with the highly homologous Smad2, to sequester it outside the nucleus, preventing its phosphorylation and nuclear translocation. In Hep3B cells, formation of the Akt/Smad3 complex is stimulated by insulin and inhibited by TGF-β. Because Smad3, but not Smad2, mediates the apoptotic signaling of TGF-β, this interaction of Akt with Smad3 specifically affects the apoptotic response of TGF-β without affecting the ability of cells to undergo cell cycle arrest. Moreover, the ratio of Smad3 to Akt in a given cell determines the sensitivity to TGF-β-induced apoptosis. This model thus explains why some cells undergo apoptosis in response to TGF-β while others exhibit only cell cycle arrest. Also consistent with this model, IGF-1 was found to inhibit TGF-β-induced transcription in a nontumorigenic rat prostate epithelial cell line through the inhibition of Smad3 phosphorylation but not that of Smad2 via Akt *(116)*. Although the kinase activity of Akt was initially reported to be required for this inhibition of TGF-β-induced transcription, it was later found to be dispensable *(117)*.

Akt could also repress Smad activity indirectly through phosphorylating the FOXO transcription factors of the fork-head family. The FOXO proteins regulate multiple cellular processes including cell metabolism, cell cycle progression and DNA repair *(118–121)*. In response to growth factor stimulation, the FOXO proteins are phosphorylated by Akt *(122)*, resulting in nuclear exclusion and inhibition of the transcriptional activity. In telencephalic neuroepithelium and in glioblastoma brain tumor cells, some of the FOXO proteins including FOXO1, FOXO3, or FOXO4 form a complex with Smad4 and Smad3 to activate $p21^{CIP1}$ expression in response to TGF-β *(123)*. Phosphorylation of these FOXO proteins by Akt inhibits the formation of this complex and leads to resistance of gliobastoma cells to TGF-β-induced cytostasis. These results illustrate another way by which Akt can indirectly repress Smad activity through inactivating its downstream targets.

5. CONCLUSIONS

TGF-β signaling pathways play critical and complex roles in mammalian tumorigenesis and are therefore subjected to positive and negative regulations through multiple mechanisms from plasma membrane to the nucleus. The above described pathways are just two examples of how the cytostatic and apoptotic programs of the Smad proteins are modulated through interaction with cellular partners and crosstalk with other signaling pathways. Analyses of these interactions and crosstalks have helped to establish new paradigms involved in the regulation of the Smad proteins. Future work will no doubt provide a better understanding of how the complex Smad signaling network functions under physiological conditions and how deregulation of the network contributes to malignant progression.

ACKNOWLEDGMENTS

Research on Smad signaling and regulation by Ski, SnoN, and Akt in our laboratory is supported by grants from the NIH R01s and Philip Morris External Research Grant to K.L.

REFERENCES

1. Roberts AB, Flanders KC, Heine UI, et al. Transforming growth factor-beta: multifunctional regulator of differentiation and development. Philos Trans R Soc Lond B Biol Sci 1990;327:145–154.
2. Derynck R, Akhurst RJ, Balmain A. TGF-beta signaling in tumor suppression and cancer progression. Nat Genet 2001;29:117–129.
3. Moses HL, Yang EY, Pietenpol JA. TGF-beta stimulation and inhibition of cell proliferation: new mechanistic insights. Cell 1990;63:245–247.
4. Blobe GC, Schiemann WP, Lodish HF. Role of transforming growth factor beta in human disease. N Engl J Med 2000;342:1350–1358.
5. Massagué J, Blain SW, Lo RS. TGFbeta signaling in growth control, cancer, and heritable disorders. Cell 2000;103:295–309.
6. ten Dijke P, Hill CS. New insights into TGF-beta-Smad signalling. Trends Biochem Sci 2004;29: 265–273.
7. Heldin C-H, Miyazono K, ten Dijke P. TGF-beta signalling from cell membrane to nucleus through SMAD proteins. Nature 1997;390:465–471.

8. Shi Y, Massagué J. Mechanisms of TGF-beta signaling from cell membrane to the nucleus. Cell 2003; 113:685–700.
9. Attisano L, Wrana JL. Smads as transcriptional co-modulators. Curr Opin Cell Biol 2000;12: 235–243.
10. Miyazono K, ten Dijke P, Heldin C-H. TGF-beta signaling by Smad proteins. Adv Immunol 2000;75: 115–157.
11. Massagué J, Wotton D. Transcriptional control by the TGF-beta/Smad signaling system. EMBO J 2000;19:1745–1754.
12. Zhang Y, Derynck R. Regulation of Smad signalling by protein associations and signalling crosstalk. Trends Cell Biol 1999;9:274–279.
13. Li Y, Turck CM, Teumer JK, Stavnezer E. Unique sequence, ski, in Sloan-Kettering avian retroviruses with properties of a new cell-derived oncogene. J Virol 1986;57:1065–1072.
14. Nomura N, Sasamoto S, Ishii S, Date T, Matsui M, Ishizaki R. Isolation of human cDNA clones of ski and the ski-related gene, sno. Nucleic Acids Res 1989;17:5489–5500.
15. Pearson-White S, Crittenden R. Proto-oncogene Sno expression, alternative isoforms and immediate early serum response. Nucleic Acids Res 1997;25:2930–2937.
16. Liu X, Sun Y, Weinberg RA, Lodish HF. Ski/Sno and TGF-beta signaling. Cytokine Growth Factor Rev 2001;12:1–8.
17. Luo K. Ski and SnoN: negative regulators of TGF-beta signaling. Curr Opin Genet Dev 2004;14: 65–70.
18. Heyman HC, Stavnezer E. A carboxyl-terminal region of the ski oncoprotein mediates homodimerization as well as heterodimerization with the related protein SnoN. J Biol Chem 1994;269: 26,996–27,003.
19. Nagase T, Nomura N, Ishii S. Complex formation between proteins encoded by the ski gene family. J Biol Chem 1993;268:13,710–13,716.
20. Cohen SB, Zheng G, Heyman HC, Stavnezer E. Heterodimers of the SnoN and Ski oncoproteins form preferentially over homodimers and are more potent transforming agents. Nucleic Acids Res 1999;27:1006–1014.
21. Colmenares C, Stavnezer E. The ski oncogene induces muscle differentiation in quail embryo cells. Cell 1989;59:293–303.
22. Lyons GE, Micales BK, Herr MJ, et al. Protooncogene c-ski is expressed in both proliferating and postmitotic neuronal populations. Dev Dyn 1994;201:354–365.
23. Namciu S, Lyons GE, Micales BK, Heyman HC, Colmenares C, Stavnezer E. Enhanced expression of mouse c-ski accompanies terminal skeletal muscle differentiation in vivo and in vitro. Dev Dyn 1995;204:291–300.
24. Wang W, Mariani FV, Harland RM, Luo K. Ski represses bone morphogenic protein signaling in Xenopus and mammalian cells. Proc Natl Acad Sci USA 2000;97:14,394–14,399.
25. Amaravadi LS, Neff AW, Sleeman JP, Smith RC. Autonomous neural axis formation by ectopic expression of the protooncogene c-ski. Dev Biol 1997;192:392–404.
26. Kaufman CD, Martinez-Rodriguez G, Hackett PB, Jr. Ectopic expression of c-ski disrupts gastrulation and neural patterning in zebrafish. Mech Dev 2000;95:147–162.
27. Sutrave P, Kelly AM, Hughes SH. Ski can cause selective growth of skeletal muscle in transgenic mice. Genes Dev 1990;4:1462–1472.
28. Berk M, Desai SY, Heyman HC, Colmenares C. Mice lacking the ski proto-oncogene have defects in neurulation, craniofacial, patterning, and skeletal muscle development. Genes Dev 1997;11:2029–2039.
29. Colmenares C, Heilstedt HA, Shaffer LG, et al. Loss of the SKI proto-oncogene in individuals affected with 1p36 deletion syndrome is predicted by strain-dependent defects in Ski–/– mice. Nat Genet 2002;30:106–109.
30. Pearson-White S. SnoI, a novel alternatively spliced isoform of the ski protooncogene homolog, sno. Nucleic Acids Res 1993;21:4632–4638.
31. Boyer PL, Colmenares C, Stavnezer E, Hughes SH. Sequence and biological activity of chicken snoN cDNA clones. Oncogene 1993;8:457–466.
32. Pearson-White S, McDuffie M. Defective T-cell activation is associated with augmented transforming growth factor Beta sensitivity in mice with mutations in the Sno gene. Mol Cell Biol 2003;23: 5446–5459.
33. Shinagawa T, Dong HD, Xu M, Maekawa T, Ishii S. The sno gene, which encodes a component of the histone deacetylase complex, acts as a tumor suppressor in mice. EMBO J 2000;19:2280–2291.

34. Arndt S, Poser I, Schubert T, Moser M, Bosserhoff AK. Cloning and functional characterization of a new Ski homolog, Fussel-18, specifically expressed in neuronal tissues. Lab Invest 2005;85: 1330–1341.
35. Mizuhara E, Nakatani T, Minaki Y, Sakamoto Y, Ono Y. Corl1, a novel neuronal lineage-specific transcriptional corepressor for the homeodomain transcription factor Lbx1. J Biol Chem 2005;280: 3645–3655.
36. Xu W, Angelis K, Danielpour D, et al. Ski acts as a co-repressor with Smad2 and Smad3 to regulate the response to type beta transforming growth factor. Proc Natl Acad Sci USA 2000;97:5924–5929.
37. Sun Y, Liu X, Eaton EN, Lane WS, Lodish HF, Weinberg RA. Interaction of the Ski oncoprotein with Smad3 regulates TGF-beta signaling. Mol Cell 1999;4:499–509.
38. Stroschein SL, Wang W, Zhou S, Zhou Q, Luo K. Negative feedback regulation of TGF-beta signaling by the SnoN oncoprotein. Science 1999;286:771–774.
39. Luo K, Stroschein SL, Wang W, et al. The Ski oncoprotein interacts with the Smad proteins to repress TGFbeta signaling. Genes Dev 1999;13:2196–2206.
40. Akiyoshi S, Inoue H, Hanai J, et al. c-Ski acts as a transcriptional co-repressor in transforming growth factor-beta signaling through interaction with smads. J Biol Chem 1999;274:35,269–35,277.
41. He J, Tegen SB, Krawitz AR, Martin GS, Luo K. The transforming activity of Ski and SnoN is dependent on their ability to repress the activity of Smad proteins. J Biol Chem 2003;278:30,540–30,547.
42. Tokitou F, Nomura T, Khan MM, et al. Viral ski inhibits retinoblastoma protein (Rb)-mediated transcriptional repression in a dominant negative fashion. J Biol Chem 1999;274:4485–4488.
43. Nomura T, Khan MM, Kaul SC, et al. Ski is a component of the histone deacetylase complex required for transcriptional repression by Mad and thyroid hormone receptor. Genes Dev 1999;13: 412–423.
44. Kokura K, Kaul SC, Wadhwa R, et al. The Ski protein family is required for MeCP2-mediated transcriptional repression. J Biol Chem 2001;276:34,115–34,121.
45. Harada J, Kokura K, Kanei-Ishii C, et al. Requirement of the co-repressor homeodomain-interacting protein kinase 2 for ski-mediated inhibition of bone morphogenetic protein-induced transcriptional activation. J Biol Chem 2003;278:38,998–39,005.
46. Wu JW, Krawitz AR, Chai J, et al. Structural mechanism of Smad4 recognition by the nuclear oncoprotein Ski: insights on Ski-mediated repression of TGF-beta signaling. Cell 2002;111:357–367.
47. Takeda M, Mizuide M, Oka M, et al. Interaction with Smad4 is indispensable for suppression of BMP signaling by c-Ski. Mol Biol Cell 2004;15:963–972.
48. Pessah M, Marais J, Prunier C, et al. c-Jun associates with the oncoprotein Ski and suppresses Smad2 transcriptional activity. J Biol Chem 2002;277:29,094–29,100.
49. Pessah M, Prunier C, Marais J, et al. c-Jun interacts with the corepressor TG-interacting factor (TGIF) to suppress Smad2 transcriptional activity. Proc Natl Acad Sci USA 2001;98:6198–6203.
50. Wilson JJ, Malakhova M, Zhang R, Joachimiak A, Hegde RS. Crystal structure of the dachshund homology domain of human SKI. Structure (Camb) 2004;12:785–792.
51. Dahl R, Wani B, Hayman MJ. The Ski oncoprotein interacts with Skip, the human homolog of Drosophila Bx42. Oncogene 1998;16:1579–1586.
52. Prathapam T, Kuhne C, Hayman M, Banks L. Ski interacts with the evolutionarily conserved SNW domain of Skip. Nucleic Acids Res 2001;29:3469–3476.
53. Khan MM, Nomura T, Kim H, et al. Role of PML and PML-RARalpha in Mad-mediated transcriptional repression. Mol Cell 2001;7:1233–1243.
54. Dahl R, Kieslinger M, Beug H, Hayman MJ. Transformation of hematopoietic cells by the Ski oncoprotein involves repression of retinoic acid receptor signaling. Proc Natl Acad Sci USA 1998;95: 11,187–11,192.
55. Medrano EE. Repression of TGF-beta signaling by the oncogenic protein SKI in human melanomas: consequences for proliferation, survival, and metastasis. Oncogene 2003;22:3123–3129.
56. Kokura K, Kim H, Shinagawa T, Khan MM, Nomura T, Ishii S. The Ski-binding protein C184M negatively regulates tumor growth factor-beta signaling by sequestering the Smad proteins in the cytoplasm. J Biol Chem 2003;278:20,133–20,139.
57. Dai P, Shinagawa T, Nomura T, et al. Ski is involved in transcriptional regulation by the repressor and full-length forms of Gli3. Genes Dev 2002;16:2843–2848.
58. Zhang F, Lundin M, Ristimaki A, et al. Ski-related novel protein N (SnoN), a negative controller of transforming growth factor-beta signaling, is a prognostic marker in estrogen receptor-positive breast carcinomas. Cancer Res 2003;63:5005–5010.

59. Reed JA, Bales E, Xu W, Okan NA, Bandyopadhyay D, Medrano EE. Cytoplasmic localization of the oncogenic protein Ski in human cutaneous melanomas in vivo: functional implications for transforming growth factor beta signaling. Cancer Res 2001;61:8074–8078.
60. Imoto I, Pimkhaokham A, Fukuda Y, et al. SNO is a probable target for gene amplification at 3q26 in squamous-cell carcinomas of the esophagus. Biochem Biophys Res Commun 2001;286:559–565.
61. Fumagalli S, Doneda L, Nomura N, Larizza L. Expression of the c-ski proto-oncogene in human melanoma cell lines. Melanoma Res 1993;3:23–27.
62. Wan Y, Liu X, Kirschner MW. The anaphase-promoting complex mediates TGF-beta signaling by targeting SnoN for destruction. Mol Cell 2001;8:1027–1039.
63. Bonni S, Wang HR, Causing CG, et al. TGF-beta induces assembly of a Smad2-Smurf2 ubiquitin ligase complex that targets SnoN for degradation. Nat Cell Biol 2001;3:587–595.
64. Stroschein SL, Bonni S, Wrana JL, Luo K. Smad3 recruits the anaphase-promoting complex for ubiquitination and degradation of SnoN. Genes Dev 2001;15:2822–2836.
65. Zhu Q, Pearson-White S, Luo K. Requirement for the SnoN oncoprotein in transforming growth factor beta-induced oncogenic transformation of fibroblast cells. Mol Cell Biol 2005;25:10,731–10,744.
66. Edmiston JS, Yeudall WA, Chung TD, Lebman DA. Inability of transforming growth factor-beta to cause SnoN degradation leads to resistance to transforming growth factor-beta-induced growth arrest in esophageal cancer cells. Cancer Res 2005;65:4782–4788.
67. Krakowski AR, Laboureau J, Mauviel A, Bissell MJ, Luo K. Cytoplasmic SnoN in normal tissues and nonmalignant cells antagonizes TGF-beta signaling by sequestration of the Smad proteins. Proc Natl Acad Sci USA 2005;102:12,437–12,442.
68. Marcelain K, Hayman MJ. The Ski oncoprotein is upregulated and localized at the centrosomes and mitotic spindle during mitosis. Oncogene 2005;24:4321–4329.
69. Sutrave P, Copeland TD, Showalter SD, Hughes SH. Characterization of chicken c-ski oncogene products expressed by retrovirus vectors. Mol Cell Biol 1990;10:3137–3144.
70. Macdonald M, Wan Y, Wang W, et al. Control of cell cycle-dependent degradation of c-Ski proto-oncoprotein by Cdc34. Oncogene 2004;23:5643–5653.
71. Fukuchi M, Nakajima M, Fukai Y, et al. Increased expression of c-Ski as a co-repressor in transforming growth factor-beta signaling correlates with progression of esophageal squamous cell carcinoma. Int J Cancer 2004;108:818–824.
72. Buess M, Terracciano L, Reuter J, et al. Amplification of SKI is a prognostic marker in early colorectal cancer. Neoplasia 2004;6:207–212.
73. Shinagawa T, Nomura T, Colmenares C, Ohira M, Nakagawara A, Ishii S. Increased susceptibility to tumorigenesis of ski-deficient heterozygous mice. Oncogene 2001;20:8100–8108.
74. Brodeur GM. Neuroblastoma: biological insights into a clinical enigma. Nat Rev Cancer 2003;3: 203–216.
75. Smedley D, Sidhar S, Birdsall S, et al. Characterization of chromosome 1 abnormalities in malignant melanomas. Genes Chromosomes Cancer 2000;28:121–125.
76. Poser I, Rothhammer T, Dooley S, Weiskirchen R, Bosserhoff AK. Characterization of Sno expression in malignant melanoma. Int J Oncol 2005;26:1411–1417.
77. Levy L, Hill CS. Alterations in components of the TGF-beta superfamily signaling pathways in human cancer. Cytokine Growth Factor Rev 2005.
78. Bhowmick NA, Ghiassi M, Bakin A, et al. Transforming growth factor-beta1 mediates epithelial to mesenchymal transdifferentiation through a RhoA-dependent mechanism. Mol Biol Cell 2001;12: 27–36.
79. Moustakas A, Heldin C-H. Non-Smad TGF-beta signals. J Cell Sci 2005;118:3573–3584.
80. Derynck R, Zhang YE. Smad-dependent and Smad-independent pathways in TGF-beta family signalling. Nature 2003;425:577–584.
81. Shin I, Bakin AV, Rodeck U, Brunet A, Arteaga CL. Transforming growth factor beta enhances epithelial cell survival via Akt-dependent regulation of FKHRL1. Mol Biol Cell 2001;12:3328–3339.
82. Bakin AV, Tomlinson AK, Bhowmick NA, Moses HL, Arteaga CL. Phosphatidylinositol 3-kinase function is required for transforming growth factor beta-mediated epithelial to mesenchymal transition and cell migration. J Biol Chem 2000;275:36,803–36,810.
83. Brazil DP, Hemmings BA. Ten years of protein kinase B signalling: a hard Akt to follow. Trends Biochem Sci 2001;26:657–664.
84. Datta SR, Brunet A, Greenberg ME. Cellular survival: a play in three Akts. Genes Dev 1999;13: 2905–2927.

85. Yang ZZ, Tschopp O, Baudry A, Dummler B, Hynx D, Hemmings BA. Physiological functions of protein kinase B/Akt. Biochem Soc Trans 2004;32:350–354.
86. Bellacosa A, Franke TF, Gonzalez-Portal ME, et al. Structure, expression and chromosomal mapping of c-akt: relationship to v-akt and its implications. Oncogene 1993;8:745–754.
87. Altomare DA, Lyons GE, Mitsuuchi Y, Cheng JQ, Testa JR. Akt2 mRNA is highly expressed in embryonic brown fat and the AKT2 kinase is activated by insulin. Oncogene 1998;16:2407–2411.
88. Altomare DA, Guo K, Cheng JQ, Sonoda G, Walsh K, Testa JR. Cloning, chromosomal localization and expression analysis of the mouse Akt2 oncogene. Oncogene 1995;11:1055–1160.
89. Brodbeck D, Cron P, Hemmings BA. A human protein kinase Bgamma with regulatory phosphorylation sites in the activation loop and in the C-terminal hydrophobic domain. J Biol Chem 1999;274: 9133–9136.
90. Nakatani K, Sakaue H, Thompson DA, Weigel RJ, Roth RA. Identification of a human Akt3 (protein kinase B gamma) which contains the regulatory serine phosphorylation site. Biochem Biophys Res Commun 1999;257:906–910.
91. Nicholson KM, Anderson NG. The protein kinase B/Akt signalling pathway in human malignancy. Cell Signal 2002;14:381–395.
92. Galetic I, Andjelkovic M, Meier R, Brodbeck D, Park J, Hemmings BA. Mechanism of protein kinase B activation by insulin/insulin-like growth factor-1 revealed by specific inhibitors of phosphoinositide 3-kinase–significance for diabetes and cancer. Pharmacol Ther 1999;82:409–425.
93. Currie RA, Walker KS, Gray A, et al. Role of phosphatidylinositol 3,4,5-trisphosphate in regulating the activity and localization of 3-phosphoinositide-dependent protein kinase-1. Biochem J 1999;337 (Pt 3):575–583.
94. Sable CL, Filippa N, Filloux C, Hemmings BA, Van Obberghen E. Involvement of the pleckstrin homology domain in the insulin-stimulated activation of protein kinase B. J Biol Chem 1998;273:29,600–29,606.
95. Zhang X, Vik TA. Growth factor stimulation of hematopoietic cells leads to membrane translocation of AKT1 protein kinase. Leuk Res 1997;21:849–856.
96. Wijkander J, Holst LS, Rahn T, et al. Regulation of protein kinase B in rat adipocytes by insulin, vanadate, and peroxovanadate. Membrane translocation in response to peroxovanadate. J Biol Chem 1997;272:21,520–21,526.
97. Meier R, Alessi DR, Cron P, Andjelkovic M, Hemmings BA. Mitogenic activation, phosphorylation, and nuclear translocation of protein kinase Bbeta. J Biol Chem 1997;272:30,491–30,497.
98. Andjelkovic M, Alessi DR, Meier R, et al. Role of translocation in the activation and function of protein kinase B. J Biol Chem 1997;272:31,515–31,524.
99. Klippel A, Kavanaugh WM, Pot D, Williams LT. A specific product of phosphatidylinositol 3-kinase directly activates the protein kinase Akt through its pleckstrin homology domain. Mol Cell Biol 1997;17:338–344.
100. Frech M, Andjelkovic M, Ingley E, Reddy KK, Falck JR, Hemmings BA. High affinity binding of inositol phosphates and phosphoinositides to the pleckstrin homology domain of RAC/protein kinase B and their influence on kinase activity. J Biol Chem 1997;272:8474–8481.
101. Franke TF, Kaplan DR, Cantley LC, Toker A. Direct regulation of the Akt proto-oncogene product by phosphatidylinositol-3,4-bisphosphate. Science 1997;275:665–668.
102. James SR, Downes CP, Gigg R, Grove SJ, Holmes AB, Alessi DR. Specific binding of the Akt-1 protein kinase to phosphatidylinositol 3,4,5-trisphosphate without subsequent activation. Biochem J 1996; 315(Pt 3):709–713.
103. Datta K, Bellacosa A, Chan TO, Tsichlis PN. Akt is a direct target of the phosphatidylinositol 3-kinase. Activation by growth factors, v-src and v-Ha-ras, in Sf9 and mammalian cells. J Biol Chem 1996;271:30,835–30,839.
104. Alessi DR, Deak M, Casamayor A, et al. 3-Phosphoinositide-dependent protein kinase-1 (PDK1): structural and functional homology with the Drosophila DSTPK61 kinase. Curr Biol 1997;7:776–789.
105. Balendran A, Casamayor A, Deak M, et al. PDK1 acquires PDK2 activity in the presence of a synthetic peptide derived from the carboxyl terminus of PRK2. Curr Biol 1999;9:393–404.
106. Delcommenne M, Tan C, Gray V, Rue L, Woodgett J, Dedhar S. Phosphoinositide-3-OH kinase-dependent regulation of glycogen synthase kinase 3 and protein kinase B/AKT by the integrin-linked kinase. Proc Natl Acad Sci USA 1998;95:11,211–11,216.
107. Persad S, Attwell S, Gray V, et al. Regulation of protein kinase B/Akt-serine 473 phosphorylation by integrin-linked kinase: critical roles for kinase activity and amino acids arginine 211 and serine 343. J Biol Chem 2001;276:27,462–27,469.

108. Laine J, Kunstle G, Obata T, Sha M, Noguchi M. The protooncogene TCL1 is an Akt kinase coactivator. Mol Cell 2000;6:395–407.
109. Toker A, Newton AC. Akt/protein kinase B is regulated by autophosphorylation at the hypothetical PDK-2 site. J Biol Chem 2000;275:8271–8274.
110. Song G, Ouyang G, Bao S. The activation of Akt/PKB signaling pathway and cell survival. J Cell Mol Med 2005;9:59–71.
111. Hong F, Nguyen VA, Shen X, Kunos G, Gao B. Rapid activation of protein kinase B/Akt has a key role in antiapoptotic signaling during liver regeneration. Biochem Biophys Res Commun 2000;279: 974–979.
112. Chen RH, Su YH, Chuang RL, Chang TY. Suppression of transforming growth factor-beta-induced apoptosis through a phosphatidylinositol 3-kinase/Akt-dependent pathway. Oncogene 1998;17: 1959–1968.
113. Tanaka S, Wands JR. Insulin receptor substrate 1 overexpression in human hepatocellular carcinoma cells prevents transforming growth factor beta1-induced apoptosis. Cancer Res 1996;56:3391–3394.
114. Remy I, Montmarquette A, Michnick SW. PKB/Akt modulates TGF-beta signalling through a direct interaction with Smad3. Nat Cell Biol 2004;6:358–365.
115. Conery AR, Cao Y, Thompson EA, Townsend CM, Jr., Ko TC, Luo K. Akt interacts directly with Smad3 to regulate the sensitivity to TGF-beta induced apoptosis. Nat Cell Biol 2004;6:366–372.
116. Song K, Cornelius SC, Reiss M, Danielpour D. Insulin-like growth factor-I inhibits transcriptional responses of transforming growth factor-beta by phosphatidylinositol 3-kinase/Akt-dependent suppression of the activation of Smad3 but not Smad2. J Biol Chem 2003;278:38,342–38,351.
117. Danielpour D, Song K. Cross-talk between IGF-I and TGF-beta signaling pathways. Cytokine Growth Factor Rev 2006;17(1–2):59–74.
118. Tran H, Brunet A, Grenier JM, et al. DNA repair pathway stimulated by the forkhead transcription factor FOXO3a through the Gadd45 protein. Science 2002;296:530–534.
119. Nakae J, Kitamura T, Silver DL, Accili D. The forkhead transcription factor Foxo1 (Fkhr) confers insulin sensitivity onto glucose-6-phosphatase expression. J Clin Invest 2001;108:1359–167.
120. Medema RH, Kops GJ, Bos JL, Burgering BM. AFX-like Forkhead transcription factors mediate cell-cycle regulation by Ras and PKB through p27kip1. Nature 2000;404:782–787.
121. Brunet A, Park J, Tran H, Hu LS, Hemmings BA, Greenberg ME. Protein kinase SGK mediates survival signals by phosphorylating the forkhead transcription factor FKHRL1 (FOXO3a). Mol Cell Biol 2001;21:952–965.
122. Brunet A, Bonni A, Zigmond MJ, et al. Akt promotes cell survival by phosphorylating and inhibiting a Forkhead transcription factor. Cell 1999;96:857–868.
123. Seoane J, Le HV, Shen L, Anderson SA, Massagué J. Integration of Smad and forkhead pathways in the control of neuroepithelial and glioblastoma cell proliferation. Cell 2004;117:211–223.
124. Zhu Q, Krakowski AR, Dunham EE, et al. Dual role of SnoN in mammalian tumorigenesis. Mol Cell Biol 2007;27(1):324–339.
125. Wrighton KH, Liang M, Bryan B, et al. TGF-β independent regulation of myogenesis by SnoN sumoylation. J Biol Chem 2007;282:6517–6524.

10 Transforming Growth Factor-β in Cancer Therapy: Smurfs in TGF-β Signaling and Regulation of Bone Homeostasis and Cancer

Ying E. Zhang and Motozo Yamashita

CONTENTS

INTRODUCTION
STRUCTURAL FEATURES OF SMURF1 AND SMURF2
REGULATION OF SMAD-DEPENDENT TGF-β/BMP SIGNALING BY SMURFS
REGULATION OF SMAD-INDEPENDENT TGF-β/BMP SIGNALING BY SMURFS
ROLE OF SMURFS IN BONE HOMEOSTASIS, CELL POLARITY AND CANCER
CONCLUDING REMARKS
ACKNOWLEDGMENTS
REFERENCES

Abstract

Controlled proteolysis mediated by Smad ubiquitination regulatory factors (Smurfs) plays a crucial role in modulating cellular responses to ligands of the TGF-β superfamily. Accumulating biochemical evidence has demonstrated that Smurfs regulate TGF-β or BMP signaling either by targeting the receptor-regulated Smads for degradation or by using Smad7 as an adaptor to degrade the ligand-activated receptors, Smad4 or other associated proteins. Recent *in vivo* biological assays indicate that Smurfs can also target other proteins outside the canonical Smad-pathway. Smurf1 controls osteoblast function by modulating the activity of the MEKK2-JNK pathway, and thereby regulates bone homeostasis. Smurf1 also regulates cell polarity and tight junctions by fine-tuning the RhoA level through Par6 polarity complex at cellular protrusions. This chapter summarizes the current finding of molecular mechanisms of Smurfs action and their biological activities, and discusses the potential involvement of Smurfs in tumorigenesis.

Key Words: Smurf; Smad; TGF-β; ubiquitin; E3 ligase; HECT domain.

1. INTRODUCTION

Transforming growth factor-β (TGF-β) and its related growth factors such as activin and bone morphogenetic protein (BMP), exert diverse biological effects in a wide range of cellular

From: *Cancer Drug Discovery and Development: Transforming Growth Factor-β in Cancer Therapy, Vol I: Basic and Clinical Biology*
Edited by: Sonia B. Jakowlew © Humana Press Inc., Totowa, NJ

processes ranging from growth and differentiation to apoptosis *(1)*. TGF-β itself is well known as a growth inhibitor of cells descendant of the epithelium lineage, and escape from this growth inhibition is a characteristic feature of cancer cells. As such, the molecular mechanism that controls TGF-β signaling often becomes the target of oncogenic mutation that enables neoplastic transformation *(2,3)*. At the cell surface, signaling by TGF-β family of polypeptide growth factors is mediated by a heteromeric complex of two types of transmembrane receptors. Upon ligand activation they phosphorylate a class of downstream signal transducers, known as Smads *(4,5)*. Once phosphorylated, these receptor-activated Smads (R-Smads) bind to a common Smad (i.e., Smad4), and translocate into the nucleus where the R-Smads/Smad4 complexes associate with other transcription factors to either activate or repress transcription of a selected set of target genes. A third class of Smads, known as inhibitory Smads (I-Smads) and comprising Smad6 and Smad7, antagonizes the R-Smad-mediated signaling responses by binding to ligand-activated receptors. Recently, increasing experimental evidence indicates that the TGF-β receptor complex can activate a number of non-Smad downstream pathways, in particular, those mediated by MAP kinases and Rho GTPases *(4,6)*. As genetic or epigenetic alterations of different components of the TGF-β signaling pathway have been reported in a spectrum of human developmental or hyperproliferative disorders and various forms of cancers *(2,3)*, a thorough molecular depiction of TGF-β signaling and its control mechanisms will not only serve as a paradigm for understanding the physiological roles of this large family of ligands but will also generate insight for devising novel therapeutic strategies for relevant clinical conditions.

This work concerns the recently identified Smad ubiquitin regulatory factors (Smurfs), which are members of the HECT (homology to E6AP carboxyl terminus) family of ubiquitin ligases. HECT domain proteins constitute a major subclass of E3 ubiquitin ligases *(7)*. Covalent attachment of a chain of the highly conserved 76 amino acid ubiquitin to selected protein targets signals their subsequent destruction in the 26S proteasomes. Protein degradation by the ubiquitin/proteasome system has emerged as a vital control mechanism in a variety of cellular processes, such as heat shock response, cell cycle progression, DNA repair, signal transduction, and transcription *(8)*. This system is comprised of three sequential acting enzymes: ubiquitin activating enzyme (E1), ubiquitin-conjugase (E2), and ubiquitin ligase (E3). The E3 ubiquitin ligase usually comes in direct contact with proteins destined to be degraded in the proteasomes, thus playing a crucial role in defining substrate specificity. Smurfs have been shown to modulate TGF-β/BMP signaling by directly ubiquitinating R-Smads for degradation in proteasomes or by using Smads as adaptors for marking receptors or associated transcription partners for degradation. Recent studies with knock-out mice indicate that Smurf1 also controls Smad-independent signaling responses to BMP in osteoblasts. Here, the mechanistic action of Smurfs and their biological function will be considered in detail and their potential involvement in tumorigenesis will be discussed.

2. STRUCTURAL FEATURES OF SMURF1 AND SMURF2

Two Smurf genes, Smurf1 and Smurf2 are present in vertebrate genomes. The first of these two, HECT-domain containing E3 ubiquitin ligases, Smurf1, was identified in *Xenopus laevis* as a binding partner of R-Smads specific for the BMP pathway in a yeast two-hybrid screen and was so characterized based on its ability to ubiquitinate Smad1 and attenuate BMP signaling during amphibian embryonic development *(9)*. Subsequent searches of expressed sequence tag databases and biochemical characterization turned up human ortholog of Smurf1 *(9)* as well as the paralogous gene Smurf2 *(10–12)*. Smurf1 and Smurf2 share 80% of amino acid sequence identity and several distinctive structural features including an amino-terminal phospholipid/calcium-binding C2 domain, two to three copies of WW repeats that mediate protein–protein interaction with a PPXY (PY) motif frequently present in target proteins, and a carboxyl-terminal HECT domain (Fig. 1). HECT domains

A

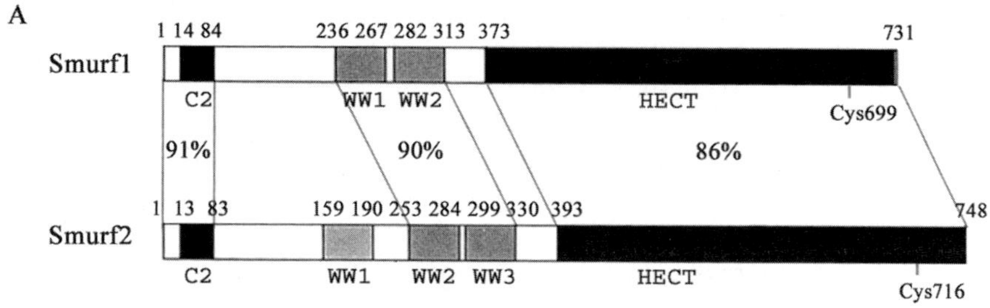

B WW domains

```
Smurf1 WW1   236 LPEGYEQRTTVQGQVYFLHTQTGVSTWHDPRI 267
       WW2   282 LPPGWEVRSTVSGRIYFVDHNNRTTQFTDPRL 313

Smurf2 WW1   159 LPDGWEERRTASGRIQYLNHITRTTQWERPTR 190
       WW2   253 LPEGYEQRTTQQGQVYFLHTQTGVSTWHDPRV 284
       WW3   299 LPPGWEIRNTATGRVYFVDHNNRTTQFTDPRL 330
```

Fig. 1. Structural features of Smurfs. (**A**) Schematic representation of Smurf1 and Smurf2. C2, WW and HECT domains are defined by NCBI conserved domain search. Percentage of amino acid residue identity within these domains and the conserved cysteine residue that is required for HECT domain E3 ligase activity are indicated. (**B**) Amino acid sequence of WW domains of Smurf1 and Smurf2.

are directly responsible for catalyzing the transfer of ubiquitin moieties to target proteins that are bound either directly by the E3 ligase or indirectly through adaptor proteins. The recently determined crystal structure of the Smurf2 HECT domain revealed a larger N-terminal lobe and a smaller C-terminal lobe connected via a short flexible linker *(13)*. The larger N-terminal lobe contains the E2 binding site, whereas the smaller C-terminal lobe houses the catalytic ubiquitin acceptor cysteine. E2 catalyzes a thioester covalent attachment of ubiquitin to this conserved cysteine residue. This topographic architecture closely resembles that of the HECT domains of E6AP and WWP1 *(14,15)*. However, using the co-crystal structure of E6AP and its cognate E2 partner UbcH7 as a guide, the Smurf2 HECT domain was found to adopt a more open structure owing to difference in subdomains of the N-terminal lobe, relative orientation of the N- and C-terminal lobes and the flexibility of interconnecting linkers *(13)*. Rationalized functional analysis indicated that the Smurf2-HECT domain has a suboptimal E2 binding pocket, providing a structural basis for regulation of the ubiquitin ligase activity through the action of auxiliary proteins *(13)*.

Smurf1 and Smurf2 contain two and three WW domains, respectively (Fig. 1). The WW domain derives its name from the presence of two highly conserved tryptophans and a conserved proline in an approximately 30 amino acid stretch *(16)*. WW domains have a preference for binding to small proline-rich sequences, most common of which is the PPXY (PY) motif, with varying degrees of substrate specificity. Each of the two WW domains in Smurf1 and the last two WW domains of Smurf2 is required for interaction with the PY motifs present in the linker regions of Smads; removal of either one of these WW domains abolishes Smurf-Smad interaction *(9–12)*.

The N-terminal phospholipid binding C2 domain was first identified as a Ca^{2+} binding site in protein kinase C but has subsequently been found in various proteins including those involved in signal transduction and membrane trafficking *(17,18)*. The C2 domain of NEDD4, which is structurally similar to that of Smurfs, was found to bind phospholipids and to be responsible for targeting NEDD4 to plasma membrane *(19,20)*. A similar membrane-targeting function has been shown for Smurf1 using a mutant that lacks the C2 domain (Smurf1ΔC2) *(21)*. Both wild-type Smurf1 and Smurf1ΔC2 are able to bind Smad7 and induce its nuclear export. Once in the cytoplasm, however, Smurf1ΔC2 became stationary, losing its ability to move to the plasma membrane or to recruit Smad7 to the cell surface TGF-β receptor complex. Moreover, Smurf1ΔC2 also lost ability to induce ubiquitination and degradation of the TGF-β receptors and failed to enhance the inhibitory activity of Smad7 *(21)*, although it still retained the ability to ubiquitinate Smad7. These data demonstrate the significant role of the C2 domain of Smurf1 in membrane-targeting.

3. REGULATION OF SMAD-DEPENDENT TGF-β/BMP SIGNALING BY SMURFS

3.1. Regulation of R-Smad Protein Levels by Smurfs

The PY motif recognized by WW domains is universally present in the linker region of most R-Smads and inhibitor Smads, except Smad8 *(22)*. Smurf1 interacts with the PY motifs of Smad1 and Smad5 via the WW-domains, but does not recognize Smad4 because the later lacks a discernible PY motif *(9)*. Deletion of the PY motif in Smad1 abolishes its interaction with Smurf1 *(9)*. Subsequently, a Smurf1 fragment, containing the two WW domains, was found to interact with Smad2 and Smad3 in a high-throughput yeast two-hybrid screen of a human placenta cDNA library *(23)*. Likewise, Smurf2 also interacts with the PY motifs of Smad1, 2, 3, and 5 via the last two of its three WW domains *(11)*, albeit its interaction with Smad7 is the strongest (discussed on page 160).

In developing *Xenopus* embryos, forced expression of either Smurf1 or Smurf2 was found to cause dorsalization of the ventral mesoderm and promote ectopic neuronal growth in the ectoderm, two features that are characteristic of diminishing BMP signaling *(9,11)*. However, the effect of Smurf2 on dorsal induction appeared to be relatively weaker than that of Smurf1 *(11)*. In cultured mammalian cell systems, association with either Smurf leads to ubiquitination and subsequent proteasomal degradation of Smad1 and Smad5. Initially, the interaction with Smurfs and the reduction of steady state protein levels of Smad1 and Smad5 was found to be independent of BMP receptor activation *(9,11)*. This indicates that Smurf1 and Smurf2 probably do not function downstream of activated Smad1 or Smad5, but rather regulate the available pool of specific R-Smads to prevent spurious activation of the BMP pathway (Fig. 2). During development, the size of R-Smad pool may ultimately affect the cellular responsiveness in a morphogenic BMP signaling gradient that controls tissue patterning and differentiation. For example in early *Xenopus* embryos, a complete inhibition of BMP signaling leads to neural and dorsal mesoderm development, whereas a partial interference with BMP signaling results in the formation of a cement gland, neural crest and lateral mesoderm *(24)*. Unrestrained BMP signaling and/or Smad1 activation induce the formation of epidermis and ventral mesoderm *(25,26)*.

In addition to regulating the available pools of constitutive Smad1 and 5 protein levels, Smurf2 was shown to bind receptor activated Smad2 *(10,27)*. In human keratinocytes, endogenous Smad2 is targeted for ubiquitination and proteasomal degradation in response to TGF-β stimulation *(10)*. In this case, destruction of activated Smad2 was shown to occur in the nucleus. As Smurf2 is primarily a nuclear protein, it is likely that TGF-β signaling is required to mobilize Smad2 into the nucleus, where it is targeted by Smurf2 mediated ubiquitination, thus forming a negative feedback control loop.

Of particular note, the steady state levels of Smad1, Smad5, Smad2 and Smad3 were shown to be unchanged in the recently published Smurf1-deficient mice *(28)*. It is possible that Smurf1 normally serves a very restricted role *in vivo* regardless of its other potentials. Alternatively, Smurf1 could have overlapping functions with Smurf2 that also has the ability to degrade R-Smads. Therefore, final resolution of this issue awaits the detailed analysis of Smurf1 and Smurf2 doubly deficient mice.

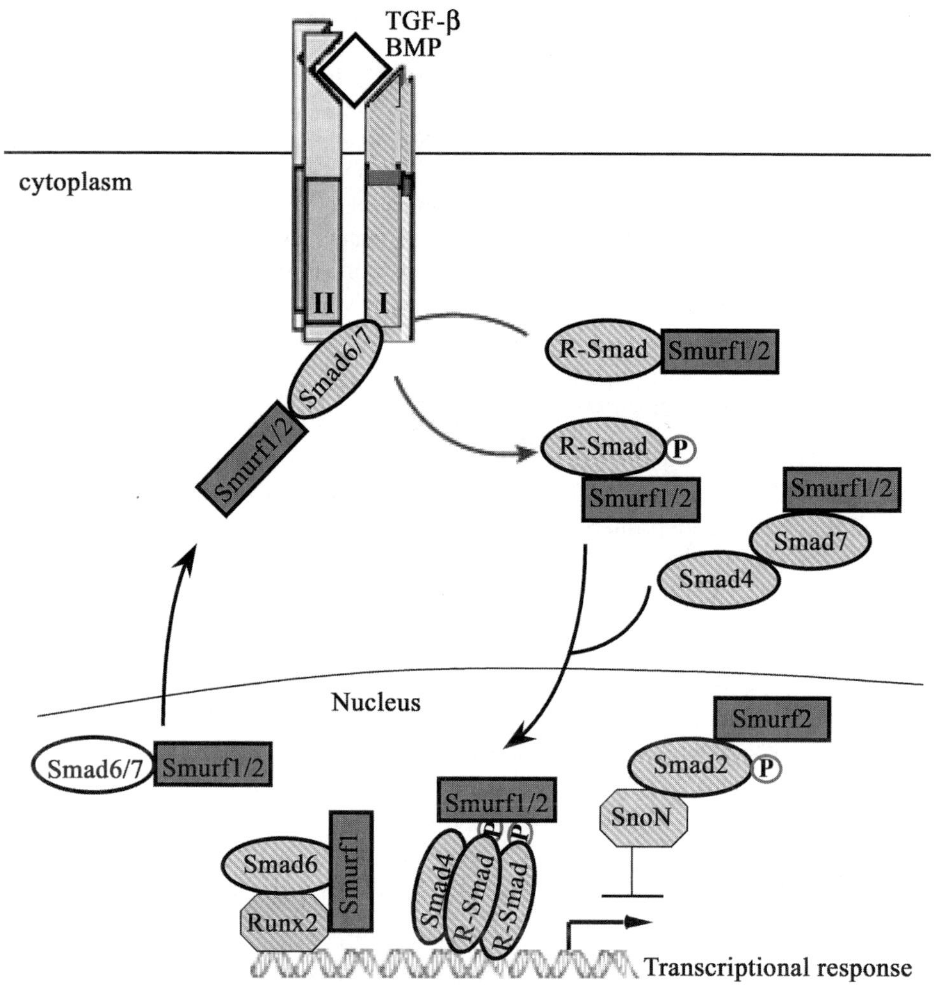

Fig. 2. Regulation of Smad-dependent TGF-β/BMP signaling by Smurfs. Smurfs regulate both basal and activated R-Smads and use I-Smads as well as R-Smads to target receptors, Smad4, SnoN and Runx2. Targets of Smurf-mediated degradation are indicated by stripped pattern.

3.2. R-Smads and I-Smads as Adaptors for Smurfs

In addition to Smurf-mediated ubiquitination and degradation, several laboratories have reported that R-Smads, i.e., Smad2 and Smad3, can function as adaptors to recruit Smurfs to target R-Smad associated proteins for degradation (Fig. 2). Ski and SnoN are two closely related nuclear proto-oncogenes that bind Smads and act as Smad-associated transcription repressors of TGF-β target genes *(29)*. It has been shown that TGF-β stimulation results in the formation of a Smurf2/Smad2/SnoN complex and that affects Smurf2-mediated ubiquitination and proteasomal degradation of SnoN *(27)*. Like Smad4, SnoN itself lacks a PY motif and cannot bind directly to Smurfs. Because SnoN is a negative regulator of TGF-β-mediated transcriptional response, its degradation may thereby restore TGF-β signaling. On the other hand, Smurfs can also form a ternary complex with Smad2 and Smad4 *(30)*. In this case, Smad2 acts as an adaptor to mediate degradation of Smad4, thus attenuating TGF-β signaling. The reason why the same Smad2/Smurf complex is involved in both down regulation and restoration of TGF-β signaling remains unclear, but it may reflect a difference in cell types or perhaps the timing of the signal to properly regulate specific physiological response within the pathway.

Similar to R-Smads, two I-Smads, Smad6 and Smad7, that antagonize TGF-β superfamily signaling by firmly interacting with the type I receptor, can also function as adaptors to engage the Smurf-mediated degradation (Fig. 2). Normally, Smad7 resides in the nucleus in unstimulated cells and can associate with either Smurf1 or Smurf2 through its PY motif *(12,22)*. In the case of interaction between Smurf1 and Smad7, chromosomal region maintenance 1 (CRM1) recognizes an importin β-related nuclear export signal present within the Smurf1 HECT domain to affect nuclear export of Smurf1/Smad7 complex *(31)*. In the cytoplasm, either Smurf1 or Smurf2, can recruit Smad7 to the plasma membrane possibly via the lipid binding C2 domain. At the cell surface, the Smurfs/Smad7 complex associate with the receptors and cause their ubiquitination and degradation *(12,22)*. The Smad7/Smurf2 mediated ubiquitination-proteasome pathway appears to play an important role in controlling the membrane trafficking and internalization of TGF-β receptors, which can be internalized via either a clathrin- or a raft/caveolar-dependent pathway *(32)*. While the Smad7/Smurf2 complex is found to associate with the receptor complex in the lipid-rich rafts on plasma membrane, primarily the internalized receptors are present in the caveolin-positive compartments. Like Smad7, Smad6 has also been shown to function as an adaptor for recruiting Smurf1 to the BMP receptors and mediate receptor turn over *(33)*. In all the above cases, ligand-induced activation of the receptor complex is required. Moreover, both of these I-Smads act as adaptors to mediate ubiquitination and proteasomal degradation of non-receptor proteins such as Smad4 and Runx2 in vitro *(30,34)*.

Recent biochemical analysis revealed that in addition to the PY motif, Smad7 forms a second contact with Smurf2 via its N-terminal domain (NTD), which recognizes the catalytic HECT domain *(13)*. As indicated by the crystal structural model, the Smurf2 HECT domain adopts a relatively open configuration that constitutes a suboptimal binding pocket for its cognate E2 enzyme, UbcH7. Thus, the NTD of Smad7 may play a crucial role in facilitating of a stable E2 recruitment. Consistent with this prediction, Smad7-NTD was found to enhance auto-ubiquitination of Smurf2, and point mutations within the interface between the NTD of Smad7 and the Smurf2 HECT domain can either augment or diminish the binding of UbcH7 and influence auto-ubiquitination of Smurf2 accordingly *(13)*. Therefore, Smad7 links substrate recognition and E2 recruitment of Smurf2, and acts to direct Smurf2 activity in a spatial and temporal manner. For example, in the absence of receptor activation, the Smad7-dependent enhancement of auto-ubiquitination and degradation of Smurf2 might function as a feedback mechanism to control basal activity of Smurf2; in the presence of receptor activation, Smad7 recruits Smurf2 to cause ubiquitination and turn over of Smad7-receptor complex and leads to inhibition of TGF-β signaling. Interestingly, Smad7 is acetylated at two lysine residues in its NTD by the transcriptional activator p300, which serves to protect Smad7 from Smurf1-mediated ubiquitination *(35)*. It will be interesting to determine whether this will protect Smad7-associated proteins from degradation and whether this affects regulation of the Smurf2-HECT domain by the NTD.

4. REGULATION OF SMAD-INDEPENDENT TGF-β/BMP SIGNALING BY SMURFS

4.1. Controlling Basal Activity of MEKK2-JNK Cascade

Now, it is established that TGF-β can signal independently of Smads through mitogen-activated protein kinases (MAPKs) or Rho-like GTPases *(4,6)*. Mounting evidence has indicated that these noncanonical pathways are directly activated by the ligand-stimulated receptors and regulate cellular response to TGF-β either in parallel to or in conjunction with

Smad signaling, or converge onto Smads to control Smad activity. The multiplexity of the cytoplasmic signaling conduit is consistent with the functional cooperation between Smads and c-Jun, JunB, ATF-2, or other co-activator/repressors in controlling nuclear transcription. In addition, MAPKs, which include Erk, JNK and p38, were found to phosphorylate Smads at various sites in the linker region connecting the two conserved MH1 and MH2 domains. Phosphorylating at these sites either inhibit or enhance Smad activity. The role of Smurf1 in regulating Smad-independent TGF-β/BMP signaling was first revealed in studies of Smurf1-deficient mice *(28)*. In Smurf1-deficient cells obtained from Smurf1-knockout mice, phosphorylated MEKK2 accumulated in osteoblasts. Like many Smad proteins, MEKK2 contains a PY motif that affects binding of the WW-domains. Smurf1 was shown to specifically bind MEKK2 and target it for ubiquitination and degradation *in vivo (28)*. Through MEKK2, Smurf1 controls the basal activity of JNK, a MAP kinase downstream of MEKKs and upstream of the Jun and ATF families of transcription factors (Fig. 3). Consequentially, certain BMP and TGF-β-mediated transcriptional responses are enhanced in *Smurf1*-deficient cells *(28)*. The ascribed effect of Smurf1 deficiency does not result from up-regulation of R-Smads and the receptors, because both the basal protein levels of Smad1, Smad2, Smad3, Smad5 and the levels of their ligand-activated forms are normal in the absence of Smurf1 *(28)*. Therefore, it is most likely due to a compensatory up-regulation of Smurf2 in *Smurf1*-deficient cells *(28)*. Thus by regulating MEKK2-JNK activity, Smurf1 ultimately determines the outcomes of biological response of the TGF-β/BMP signaling pathway.

4.2. Regulation of RhoA at Cellular Protrusions

Members of the Rho family of small guanosine triphosphatases are known for their important functions in the dynamic regulation of the actin cytoskeleton that powers cell shape changes, migration, and polarity *(36,37)*. TGF-β stimulation induces accumulation of receptor complexes in tight junctions and phosphorylation of Par6 at serine residue 345 *(38)*. Par6, a scaffold protein regulating epithelial cell polarity *(39)*, binds and recruits Smurf1 to the activated receptor complex at tight junctions in polarized epithelial cell sheets. The Par6-Smurf1 complex then mediates localized ubiquitination and turn over of RhoA at cellular protrusions, which enables a TGF-β-dependent dissolution of tight junctions, a prerequisite for epithelial to mesenchymal transition (EMT) *(38)*. Mutation of serine 345 of Par6 to alanine blocks Smurf1-mediated degradation of RhoA in response to TGF-β. The Smurf1-mediated degradation of RhoA is a localized event and requires presence of Smurf1 at lamellipodial- and filopodial-like protrusions. PKCζ, an effector of Cdc42/Rac1-Par6 polarity complex, can bind directly to Smurf1 to regulate Smurf1 localization and RhoA degradation at cellular protrusions *(38,40)*, indicating that the activity of Smurf1 toward RhoA is locally restricted to sites of active protrusion. In support of this, silencing Smurf1 expression by siRNA-mediated RNA interference did not lead to marked changes in total RhoA protein levels, but instead to RhoA and associated F-actin accumulated in cellular protrusions *(40)*. Sites of RhoA ubiquitination by Smurf1 mapped to two residues, lysine 6 and 7. Mutation of these two lysines inhibits dissolution of tight junctions and EMT in response to TGF-β *(38)*. Thus, through regulating RhoA degradation, Smurf1 controls TGF-β-dependent dissolution of tight junctions and rearrangement of the actin cytoskeleton (Fig. 3).

5. ROLE OF SMURFS IN BONE HOMEOSTASIS, CELL POLARITY AND CANCER

5.1. Function of Smurf1 in Skeletal and Bone Homeostasis

The role of Smurf1 in osteoblasts was first implicated in studies employing in vitro differentiation of pluripotent mouse C2C12 myoblasts *(41)*. In this system, specifically,

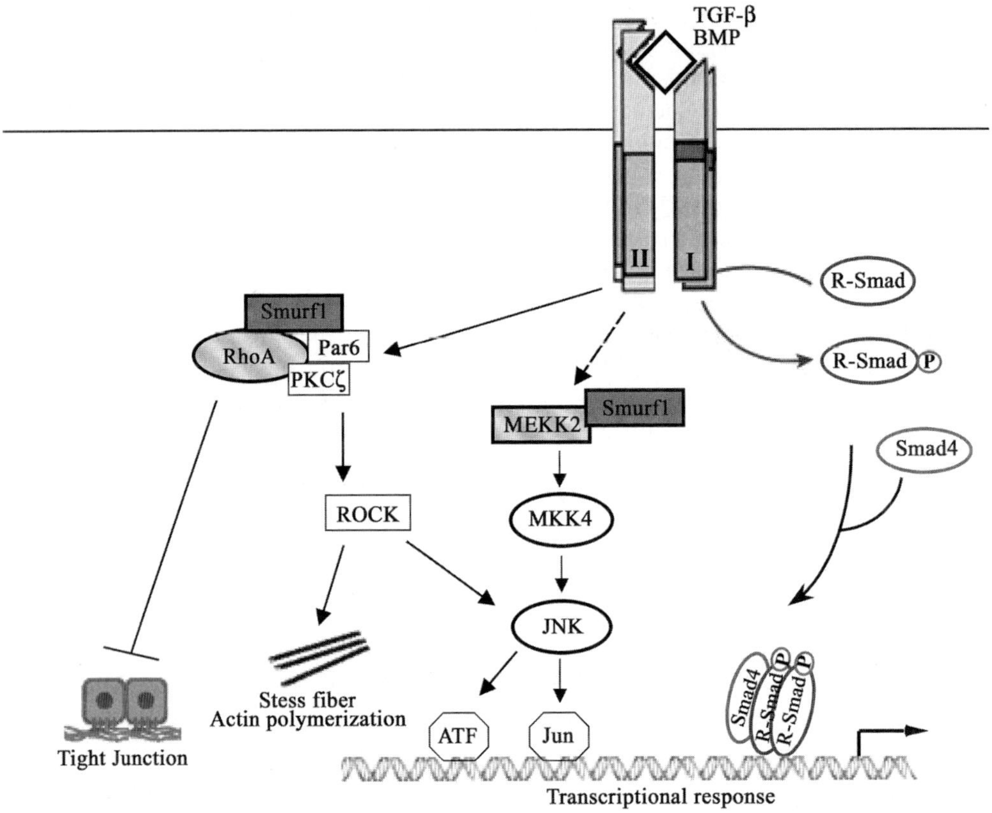

Fig. 3. Regulation of Smad-independent TGF-β/BMP signaling by Smurfs. Smurf1 regulate MEKK2 level through interaction with MEKK2; Smurf1 regulate RhoA levels through a TGF-β-dependent pathway that involves Par6 and PKCζ-dependent recruitment of Smurf1 to RhoA at cellular protrusion.

Smurf1 blocks the BMP-induced osteoblast conversion of C2C12 cells by inducing degradation of Smad5, while leaving the ability of TGF-β to inhibit myoblast differentiation unaffected. Smurf1 has also been reported to induce the degradation Runx2, an osteoblast-specific transcription factor *(42)*. Overexpression of Smurf1 in 2T3 osteoblast progenitor cells inhibits BMP signaling and osteoblast differentiation by targeting Smad1 and Runx2 for degradation. Furthermore, transgenic mice carrying a Smurf1 transgene powered by an osteoblast-specific type I collagen promoter prevents osteoblast differentiation and reduces bone formation *(43)*.

Consistent with the role of Smurf1 in controlling osteoblast function, *Smurf1*-deficient mice are perinatally normal but exhibit an age-dependent increase of bone mass due to enhanced osteoblast activity *(28)*. During bone remodeling, biochemical studies in *Smurf1*-deficient cells indicate that Smurf1 is required for limiting cellular response to BMP. However, Smurf1 neither exerts its control of osteoblast function through Smad-dependent canonical BMP signaling, nor affect Runx2 protein level. Instead, Smurf1 promotes ubiquitination and destruction of MEKK2 upstream in the JNK signaling cascade, which has the tenacity to mobilize the Jun/ATF family of transcription factors to regulate osteoblast activity, as *Smurf1*$^{-/-}$ osteoblast cells have an elevated MEKK2-JNK activity and sensitized response to BMP-2 *(28)*. Thus, the facts that activated MEKK2-JNK is capable of stimulating wild-type osteoblasts and JNK inhibitor can block the enhanced osteogenic activity associated with *Smurf1*$^{-/-}$ osteoblasts indicate that the elevated JNK activity, and hence the activated Jun/ATF, is cooperating with Smads or Runx2 to activate transcription.

Detailed histological and molecular characterization indicates that the bone phenotype displayed in Smurf1 loss-of-function studies was quite different from that in gain-of-function perturbations. When ectopically expressed, Smurf1 interacts, directs with ubiquitin/proteasome-mediated degradation of BMP pathway specific Smads and/or Runx2 in osteoblast progenitor cells and reduces both the number and proliferative potential of mature osteoblasts in the transgenic mice *(43)*. As such it affects the commitment of osteoblast precursors during the early stage of osteoblast differentiation, rather than the bone forming activities of mature osteoblasts. These mechanisms can operate in cells upon overexpression of Smurf1, but they may not be the obligatory physiological functions of Smurf1. Analysis of *Smurf1*-deficient mice indicates that *Smurf1*$^{-/-}$ mice are born with normal stature and without discernible skeletal patterning defects, indicating that Smurf1 neither essential for embryonic skeletal development nor necessary for postnatal endochondral ossification of cartilages near endochondral growth plates that is required for longitudinal bone elongation *(28)*. At endochondral growth plates chondrocytes synthesize and deposit cartilage specific extracellular matrix, hyperproliferate and eventually die. At the same time, osteoblasts and blood vessels from surrounding mesenchyme invade the hypertrophic zone of chondrocytes and eventually replace the cartilages with mineralized bones. It has been suggested that chondrocytes and osteoblasts are derived from a common origin as descendants of mesenchymal stem cells *(44)*. It is unlikely that Smurf1 plays an indispensable role in the early commitment of this cell lineage because the number of mature osteoblasts remains the same in *Smurf1*$^{-/-}$ mice as in wild type littermates *(28)*. Studies of molecular markers also seem to corroborate this notion as no change was detected in the expression level of Runx2, the early marker of osteoblast differentiation. Whereas expression of α1 collagen type 1, α2 collagen type 1, osteocalcin, and bone sialoprotein, all of which are markers of later stages of osteoblast maturation and osteoblast function, were increased *(28)*. Thus, it appears that at physiological levels, Smurf1 acts as a negative regulator of osteoblast function rather than osteoblast differentiation to maintain normal osteoblast physiology and activity.

Besides osteoblasts, Smurfs may also have a physiological role in chondrocytes and articular cartilage as both Smurf1 and Smurf2 are expressed there *(28)*. In one experiment with transgenic mice overexpressing Smurf1 in chondrocytes under the control of α2 (XI) collagen chain gene promoter (Col11α2), no abnormality in chondrocyte or endochondral ossification was observed. However, in another experiment with Smad6/Smurf1 doubly transgenic mice, Smurf1 transgene was found to enhance the inhibitory role of Smad6 that caused a delay in chondrocyte hypertrophy and endochondral ossification *(45)*. In a 5-azacytidine (Aza) induce mouse model designed to examine articular cartilage maturation, which depends on the maturational arrest of chondrocytes before terminal hypertrophic differentiation occurs, Aza was found to induce expression of maturation markers such as type X collagen, Indian hedgehog, and alkaline phosphatase *(46)*. A shift from TGF-β-type signaling response that suppresses maturation to BMP-type response that accelerates maturation was observed in the Aza-treated cells *(46)*. The down-regulation of TGF-β signaling in Aza-treated articular chondrocytes correlates with increased Smurf2 protein, suggesting that Smurf2 may play a role in articular chondrocyte maturation through inhibiting the Smad2 and/or Smad3 mediated TGF-β signaling.

5.2. Regulation of Cell Polarity Network and Transformed Phenotype of Cancer Cells

The notion that Smurf1 regulates cell polarity and protrusion formation comes from the finding that RhoA is a target for Smurf1-mediated degradation *(38,40)*. RhoA functions to maintain apical-basal polarity and cell junctions in colonic epithelia and keratinocytes *(47,48)*. RhoA is also a key player in the TGF-β-induced EMT response *(38,49)*. EMT is a transient change in cell structure often associated with weaker cell–cell interactions and acquisition of motile and invasive properties of the cells that are prerequisite for progressing to advanced metastatic tumors *(50)*. In later stages of tumorigenesis, TGF-β promotes tumor growth by inducing EMT *(51, 52)*. The finding that the Par6-Smurf1-RhoA pathway is required for TGF-β-mediated EMT suggests that Smurf1 may have a function in cancer progression. In HEK293T cells, a discernible, well-organized stress fiber network is absent, typical of transformed phenotype. Knockdown Smurf1 expression using siRNA caused these cells to lose their protrusions, rearrange the actin cytoskeleton to a cortical pattern and assume a cuboidal morphology *(40)*. In addition, motility of HEK293T cells assayed through a modified boyden chamber was also reduced *(40)*. Thus, knockdown Smurf1 expression suppresses the tumorigenic morphology and motility of HEK293T cells. Therefore, it is possible that Smurf1 plays a key role in maintaining the protrusive activity and transformed phenotype of cancer cells.

5.3. Implications of Smurf2 Function in Tumorigenesis

In epithelial cells and during the early stages of tumorigenesis, TGF-β acts as a tumor suppressor by maintaining tissue homeostatsis *(2,3)*. Because Smurfs regulate many components of TGF-β signaling, dysregulated expression of Smurfs may profoundly affect the proper signaling of TGF-β and related factors, thus contributing to cancer development. Consistent with this hypothesis, high-level expression of Smurf2 correlates with poor prognosis in patients with esophageal squamous cell carcinoma *(53)*. In addition, the elevated Smurf2 expression correlates with decreased phospho-Smad2, suggesting that Smurf2 might alter TGF-β signaling by targeting phospho-Smad2 for degradation in esophageal squamous cell carcinoma *(53)*. In invasive breast cancer, RNF11, a RING-H2 protein that is highly expressed, was also shown to interact with HECT family of E3 ligases, including Smurf2 and AIP4 *(54)*. RNF11 also binds AMSH (associated molecule with the SH3 domain of STAM) *(55)*, a protein that positively regulates TGF-β/BMP signaling through interaction with Smad6 and Smad7 and inhibits the inhibitory role of I-Smads on TGF-β/BMP signaling *(56)*. It was shown that RNF11 could recruit AMSH to Smurf2 for ubiquitination and degradation, thus inhibit AMSH-enhanced TGF-β response *(55)*. These observations imply that Smurf2 could have a role during tumor development and might be a novel target for cancer therapeutics.

In contrast, it has been reported that Smurf2 is upregulated in human fibroblasts undergoing replicative senescence in response to telomere attrition and that telomerase expression in such cells prevents Smurf2 up-regulation *(57)*. In addition, retroviral-mediated expression of Smurf2 is sufficient to produce the senescence phenotype in both early passage fibroblasts and telomerase-immortalized cells *(57)*. Interestingly, the senescence-inducing actions of Smurf2 require its three WW domains and HECT domains, yet it is independent of its ubiquitin ligase activity. Instead, Smurf2 recruits the Rb and p53 pathways for senescence induction although the underlying molecular mechanism is not understood. These results suggest that Smurf2 might have a tumor-suppressor role through induction of replicative senescence that impedes the accumulation of mutations required for neoplastic transformation.

6. CONCLUDING REMARKS

In the past few years, considerable progress has been made towards understanding the signaling networks and biological function of Smurfs. The context-dependent, diverse array of signaling mechanisms regulated by Smurfs described in this review play important roles in a variety of cellular functions. However, more questions must be answered before the complete and accurate picture of the physiological and pathological function of Smurfs can be understood. It remains unclear, for example, how Smads are physiological targets of Smurf function. More importantly, a major challenge now is to determine how the specificity in Smurf function is achieved, and what makes Smurf1 different from Smurf2. The combination of gene disruption analysis using RNA interference and/or mouse gene targeting approaches will be powerful tools in answering these questions.

Finally, as we continue to advance our understanding of the function of Smurfs and how subtle perturbation of Smurf function can result in pathological situations, considerations should be paid to the development of practical and clinical approaches for pharmacological interventions that target at Smurfs for treatment of a variety of diseases, including cancer.

ACKNOWLEDGMENTS

We thank editorial assistance of the NCI Fellows Editorial Board. Research in Y. E. Zhang's group is supported by the Intramural Research Program of the NIH, National Cancer Institute, Center for Cancer Research. M.Y. is a fellow of Japan Society for the Promotion of Science.

REFERENCES

1. Massagué J. TGF-beta signal transduction. Annu Rev Biochem 1998;67:753–791.
2. Derynck R, Akhurst RJ, Balmain A. TGF-beta signaling in tumor suppression and cancer progression. Nat Genet 2001;29(2):117–129.
3. Siegel PM, Massagué J. Cytostatic and apoptotic actions of TGF-beta in homeostasis and cancer. Nat Rev Cancer 2003;3(11):807–821.
4. Derynck R, Zhang YE. Smad-dependent and Smad-independent pathways in TGF-beta family signalling. Nature 2003;425(6958):577–584.
5. Shi Y, Massagué J. Mechanisms of TGF-beta signaling from cell membrane to the nucleus. Cell 2003;113(6):685–700.
6. Moustakas A, Heldin C-H. Non-Smad TGF-beta signals. J Cell Sci 2005;118(Pt 16):3573–3584.
7. Pickart CM. Mechanisms underlying ubiquitination. Annu Rev Biochem 2001;70:503–533.
8. Hershko A, Ciechanover A. The ubiquitin system. Annu Rev Biochem 1998;67:425–479.
9. Zhu H, Kavsak P, Abdollah S, Wrana JL, Thomsen GH. A SMAD ubiquitin ligase targets the BMP pathway and affects embryonic pattern formation. Nature 1999;400(6745):687–693.
10. Lin X, Liang M, Feng XH. Smurf2 is a ubiquitin E3 ligase mediating proteasome-dependent degradation of Smad2 in transforming growth factor-beta signaling. J Biol Chem 2000;275(47):36,818–36,822.
11. Zhang Y, Chang C, Gehling DJ, Hemmati-Brivanlou A, Derynck R. Regulation of Smad degradation and activity by Smurf2, an E3 ubiquitin ligase. Proc Natl Acad Sci USA 2001;98(3):974–979.
12. Kavsak P, Rasmussen RK, Causing CG, et al. Smad7 binds to Smurf2 to form an E3 ubiquitin ligase that targets the TGF beta receptor for degradation. Mol Cell 2000;6(6):1365–1375.
13. Ogunjimi AA, Briant DJ, Pece Barbara N, et al. Regulation of Smurf2 ubiquitin ligase activity by anchoring the E2 to the HECT domain. Mol Cell 2005;19(3):297–308.
14. Huang L, Kinnucan E, Wang G, et al. Structure of an E6AP-UbcH7 complex: insights into ubiquitination by the E2-E3 enzyme cascade. Science 1999;286(5443):1321–1326.
15. Verdecia MA, Joazeiro CA, Wells NJ, et al. Conformational flexibility underlies ubiquitin ligation mediated by the WWP1 HECT domain E3 ligase. Mol Cell 2003;11(1):249–259.
16. Zarrinpar A, Lim WA. Converging on proline: the mechanism of WW domain peptide recognition. Nat Struct Biol 2000;7(8):611–613.
17. Nalefski EA, Falke JJ. The C2 domain calcium-binding motif: structural and functional diversity. Protein Sci 1996;5(12):2375–2390.
18. Ponting CP, Parker PJ. Extending the C2 domain family: C2s in PKCs delta, epsilon, eta, theta, phospholipases, GAPs, and perforin. Protein Sci 1996;5(1):162–166.
19. Plant PJ, Yeger H, Staub O, Howard P, Rotin D. The C2 domain of the ubiquitin protein ligase Nedd4 mediates Ca^{2+}-dependent plasma membrane localization. J Biol Chem 1997;272(51):32,329–32,336.
20. Plant PJ, Lafont F, Lecat S, Verkade P, Simons K, Rotin D. Apical membrane targeting of Nedd4 is mediated by an association of its C2 domain with annexin XIIIb. J Cell Biol 2000;149(7):1473–1484.

21. Suzuki C, Murakami G, Fukuchi M, et al. Smurf1 regulates the inhibitory activity of Smad7 by targeting Smad7 to the plasma membrane. J Biol Chem 2002;277(42):39,919–39,925.
22. Ebisawa T, Fukuchi M, Murakami G, et al. Smurf1 interacts with transforming growth factor-beta type I receptor through Smad7 and induces receptor degradation. J Biol Chem 2001;276(16):12,477–12,480.
23. Colland F, Jacq X, Trouplin V, et al. Functional proteomics mapping of a human signaling pathway. Genome Res 2004;14(7):1324–1332.
24. Hemmati-Brivanlou A, Melton D. Vertebrate embryonic cells will become nerve cells unless told otherwise. Cell 1997;88(1):13–17.
25. Thomsen GH. Xenopus mothers against decapentaplegic is an embryonic ventralizing agent that acts downstream of the BMP-2/4 receptor. Development 1996;122(8):2359–2366.
26. Wilson PA, Lagna G, Suzuki A, Hemmati-Brivanlou A. Concentration-dependent patterning of the Xenopus ectoderm by BMP4 and its signal transducer Smad1. Development 1997;124(16): 3177–3184.
27. Bonni S, Wang HR, Causing CG, et al. TGF-beta induces assembly of a Smad2-Smurf2 ubiquitin ligase complex that targets SnoN for degradation. Nat Cell Biol 2001;3(6):587–595.
28. Yamashita M, Ying SX, Zhang GM, et al. Ubiquitin ligase Smurf1 controls osteoblast activity and bone homeostasis by targeting MEKK2 for degradation. Cell 2005;121(1):101–113.
29. Luo K. Ski and SnoN: negative regulators of TGF-beta signaling. Curr Opin Genet Dev 2004;14(1): 65–70.
30. Moren A, Imamura T, Miyazono K, Heldin C-H, Moustakas A. Degradation of the tumor suppressor Smad4 by WW and HECT domain ubiquitin ligases. J Biol Chem 2005;280(23):22,115–22,123.
31. Tajima Y, Goto K, Yoshida M, et al. Chromosomal region maintenance 1 (CRM1)-dependent nuclear export of Smad ubiquitin regulatory factor 1 (Smurf1) is essential for negative regulation of transforming growth factor-beta signaling by Smad7. J Biol Chem 2003;278(12):10,716–10,721.
32. Di Guglielmo GM, Le Roy C, Goodfellow AF, Wrana JL. Distinct endocytic pathways regulate TGF-beta receptor signalling and turnover. Nat Cell Biol 2003;5(5):410–421.
33. Murakami G, Watabe T, Takaoka K, Miyazono K, Imamura T. Cooperative inhibition of bone morphogenetic protein signaling by Smurf1 and inhibitory Smads. Mol Biol Cell 2003;14(7): 2809–2817.
34. Shen R, Chen M, Wang YJ, et al. Smad6 interacts with Runx2 and mediates Smurf1-induced Runx2 degradation. J Biol Chem 2005;281(6):3569–3576.
35. Gronroos E, Hellman U, Heldin C-H, Ericsson J. Control of Smad7 stability by competition between acetylation and ubiquitination. Mol Cell 2002;10(3):483–493.
36. Perez-Moreno M, Jamora C, Fuchs E. Sticky business: orchestrating cellular signals at adherens junctions. Cell 2003;112(4):535–548.
37. Burridge K, Wennerberg K. Rho and Rac take center stage. Cell 2004;116:167–179.
38. Ozdamar B, Bose R, Barrios-Rodiles M, Wang HR, Zhang Y, Wrana JL. Regulation of the polarity protein Par6 by TGF beta receptors controls epithelial cell plasticity. Science 2005;307(5715): 1603–1609.
39. Gao L, Joberty G, Macara IG. Assembly of epithelial tight junctions is negatively regulated by Par6. Curr Biol 2002;12(3):221–225.
40. Wang HR, Zhang Y, Ozdamar B, et al. Regulation of cell polarity and protrusion formation by targeting RhoA for degradation. Science 2003;302(5651):1775–1779.
41. Ying SX, Hussain ZJ, Zhang YE. Smurf1 facilitates myogenic differentiation and antagonizes the bone morphogenetic protein-2-induced osteoblast conversion by targeting Smad5 for degradation. J Biol Chem 2003;278(40):39,029–39,036.
42. Zhao M, Qiao M, Oyajobi BO, Mundy GR, Chen D. E3 ubiquitin ligase Smurf1 mediates core-binding factor alpha1/Runx2 degradation and plays a specific role in osteoblast differentiation. J Biol Chem 2003;278(30):27,939–27,944.
43. Zhao M, Qiao M, Harris SE, Oyajobi BO, Mundy GR, Chen D. Smurf1 inhibits osteoblast differentiation and bone formation in vitro and in vivo. J Biol Chem 2004;279(13):12,854–12,859.
44. Ducy P, Zhang R, Geoffroy V, Ridall AL, Karsenty G. Osf2/Cbfa1: a transcriptional activator of osteoblast differentiation. Cell 1997;89(5):747–754.
45. Horiki M, Imamura T, Okamoto M, et al. Smad6/Smurf1 overexpression in cartilage delays chondrocyte hypertrophy and causes dwarfism with osteopenia. J Cell Biol 2004;165(3):433–445.
46. Zuscik MJ, Baden JF, Wu Q, et al. 5-azacytidine alters TGF-beta and BMP signaling and induces maturation in articular chondrocytes. J Cell Biochem 2004;92(2):316–331.

47. Vaezi A, Bauer C, Vasioukhin V, Fuchs E. Actin cable dynamics and Rho/Rock orchestrate a polarized cytoskeletal architecture in the early steps of assembling a stratified epithelium. Dev Cell 2002;3(3): 367–381.
48. Fukata M, Kaibuchi K. Rho-family GTPases in cadherin-mediated cell-cell adhesion. Nat Rev Mol Cell Biol 2001;2(12):887–897.
49. Bhowmick NA, Ghiassi M, Bakin A, et al. Transforming growth factor-beta1 mediates epithelial to mesenchymal transdifferentiation through a RhoA-dependent mechanism. Mol Biol Cell 2001;12(1): 27–36.
50. Radisky DC. Epithelial-mesenchymal transition. J Cell Sci 2005;118(Pt 19):4325–4326.
51. Zavadil J, Bottinger EP. TGF-beta and epithelial-to-mesenchymal transitions. Oncogene 2005;24(37): 5764–5774.
52. Janda E, Lehmann K, Killisch I, et al. Ras and TGF beta cooperatively regulate epithelial cell plasticity and metastasis: dissection of Ras signaling pathways. J Cell Biol 2002;156(2):299–313.
53. Fukuchi M, Fukai Y, Masuda N, et al. High-level expression of the Smad ubiquitin ligase Smurf2 correlates with poor prognosis in patients with esophageal squamous cell carcinoma. Cancer Res 2002;62(24):7162–7165.
54. Subramaniam V, Li H, Wong M, et al. The RING-H2 protein RNF11 is overexpressed in breast cancer and is a target of Smurf2 E3 ligase. Br J Cancer 2003;89(8):1538–1544.
55. Li H, Seth A. An RNF11: Smurf2 complex mediates ubiquitination of the AMSH protein. Oncogene 2004;23(10):1801–1808.
56. Itoh F, Asao H, Sugamura K, Heldin C-H, ten Dijke P, Itoh S. Promoting bone morphogenetic protein signaling through negative regulation of inhibitory Smads. EMBO J 2001;20(15):4132–4142.
57. Zhang H, Cohen SN. Smurf2 up-regulation activates telomere-dependent senescence. Genes Dev 2004;18(24):3028–3040.

11 Involvement of km23 Dynein Light Chains in TGF β Signaling

Qunyan Jin, Guofeng Gao, and Kathleen M. Mulder

CONTENTS

INTRODUCTION
THE km23/LC7/ROBL DLC FAMILY
km23-1
km23-2
A MODEL FOR km23-1 AND km23-2 IN RECRUITMENT OF TGF-B SIGNALING ENDOSOMES FOR INTRACELLULAR TRANSPORT ALONG MTs
FUTURE DIRECTIONS
ACKNOWLEDGMENTS
REFERENCES

Abstract

We have identified km23-1 and km23-2 as TGF-β signaling intermediates that are also light chains of the minus-end directed motor protein dynein. The km23/LC7/robl dynein light chains (DLCs) belong to an ancient superfamily of proteins, known as the MglB superfamily. Our NMR structural studies demonstrate that km23-1 adopts the structure of a stable homodimer, similar to that of the p14/MAPK scaffolding protein (MP1) heterodimer. Furthermore, our results suggest that km23-1/km23-2 may function as motor receptors to recruit Smads2/3 to the dynein motor complex for endosomal transport, prior to Smad nuclear translocation. Thus, TGF-β signaling endosomes may be natural cargo of the dynein complex. Here, we present a model for how km23 family members may function as motor receptors, and describe the evidence that we have accumulated thus far to support such a model. For example, we have demonstrated that the kinase activity of TGF-β receptor RII directly phosphorylates km23-1 on specific serine residues, resulting in attachment of the km23 DLCs to the rest of the multi-subunit dynein complex. This dynein binding function of km23 appears to be required for the endosomal transport of TGF-β signaling components, because disruption of the motor complex blocks transcriptional effects of TGF-β downstream. Alterations in km23-1 found in the ovarian cancer patients appear to modify km23-1 functions in both binding to the dynein motor complex and TGF-β signaling. In addition, forced expression of km23-1 in ovarian cancer cells suppresses their growth in vitro, and tumorgenesis in vivo, by increasing the number of cells in the G2/M phase of the cell cycle. Our results implicate km23-1 as a novel, TGF β-related anti-cancer target for the development of diagnostics and therapeutics.

From: *Cancer Drug Discovery and Development: Transforming Growth Factor-β in Cancer Therapy, Vol I: Basic and Clinical Biology*
Edited by: Sonia B. Jakowlew © Humana Press Inc., Totowa, NJ

Key Words: km23; dynein; TGF-β; signal transduction; cancer; growth factor; motor protein; DYNLRB; Smad; trafficking.

1. INTRODUCTION

Transforming growth factor β (TGF-β) is the prototype for a family of highly conserved ubiquitous peptides that show remarkable diversity in the biological actions they mediate, which include effects on cell growth, cell death, cell differentiation, and the extracellular matrix *(1–4)*.

TGF-β has been shown to initiate at least two major signaling pathways, the Smad and Ras/mitogen-activated protein (MARK) pathways, to regulate cellular and tissue activities *(1–4)*. In the Smad pathway, TGF-β initiates its response by binding and bringing together two pairs of structurally similar, single-pass transmembrane receptors, classified as types I (TβRI) and types II (TβRII). The ligand-mediated assembly of the receptor complex triggers an intracellular phosphorylation cascade initiated by the kinase activity of TβRII, which transphosphorylates the adjacent TβRI. This event, in turn, results in phosphorylation of receptor-regulated Smads (RSmads), which together with various transcriptional coactivators and corepressors, regulate the transcription of target genes *(1–5)*.

The Ras/MAPK pathways are other major pathways that transduce signals initiated by TGF-β. Three distinct groups of MAPKs have been identified in mammalian cells, including extracellular signal-regulated kinases (ERKs), Jun N-terminal kinases (JNKs)/stress-activated protein kinases (SAPKs), and p38 *(2)*. We have demonstrated that TGF-β can activate ERKs within 5–30 min in TGF-β-sensitive, but not in TGF-β-resistant untransformed epithelial cells *(6–8)*.

Moreover, we demonstrated that overexpression of a dominant-negtive mutant of Ras (RasN 17), or addition of the specific MAP/ERK kinase (MEK1) inhibitor PD98059, blocked the ability of TGF-β to activate ERK1 in epithelial cells *(9,10)*. In addition, we have observed that overexpression of RasN 17 abrogated the ability of TGF-β to up-regulate the cell cycle inhibitors $p27^{Kip1}$ and $p21^{Cip1}$. We and other groups have demonstrated that TGF-β can also activate the SAPKs/JNKs in a wide variety of cell types *(7,8,11–13)*. In addition to the ERKs and JNKs, TGF-β has been shown to activate p38 in several cell types *(14–16)*. Since our initial demonstration that TGF-β could activate the Ras/MAPK pathways in untransformed epithelial cells, additional data have accumulated to indicate that TGF-β activation of the Ras/MAPK pathways plays a critical role in TGF-β signaling *(2,12,17,18)*.

In addition to Smads and MAPKs, several TGF-β receptor-interacting factors and Smad-interacting factors have been reported *(4)*. Figure 1 depicts some of these TGF-β signaling pathways for epithelial cells. For example, Smad anchor for receptor activation (SARA) *(19)* and Disabled-2 (Dab2) *(20)* have been shown to bind both the TβRs and the RSmads to promote Smad phosphorylation and TGF-β signaling. TβRI-associated protein-1 (TRAP-1) has been shown to bind selectively to inactive TβR and activin receptor complexes, forming a transient complex with Smad4 upon signal activation, and potentially facilitating transfer of Smad4 to the acceptor Smad2/3 *(21)*. More recently, TRAP-1-like protein (TLP) has been shown to function as an adaptor protein, coupling the TβR complex to the Smad pathway, with the unique role of regulating the balance between Smad3 and Smad2 signaling *(22)*. Other proteins that have been described in the literature, such as Daxx *(23)*, the immunophilin $FKBP_{12}$ *(24)*, TβR-interacting protein 1 (TRIP-1) *(25)*, the β subunit of phosphatase 2A (PP2A) *(26)*, and others, have all been identified as interactors of TβRI and/or TβRII. These TβR-interacting proteins often regulate receptor function through non-Smad pathways. In summary, TGF-β can initiate at least two prominent signaling cascades, the Smad and the Ras/MAPK pathways, in order to regulate a wide variety of biological responses.

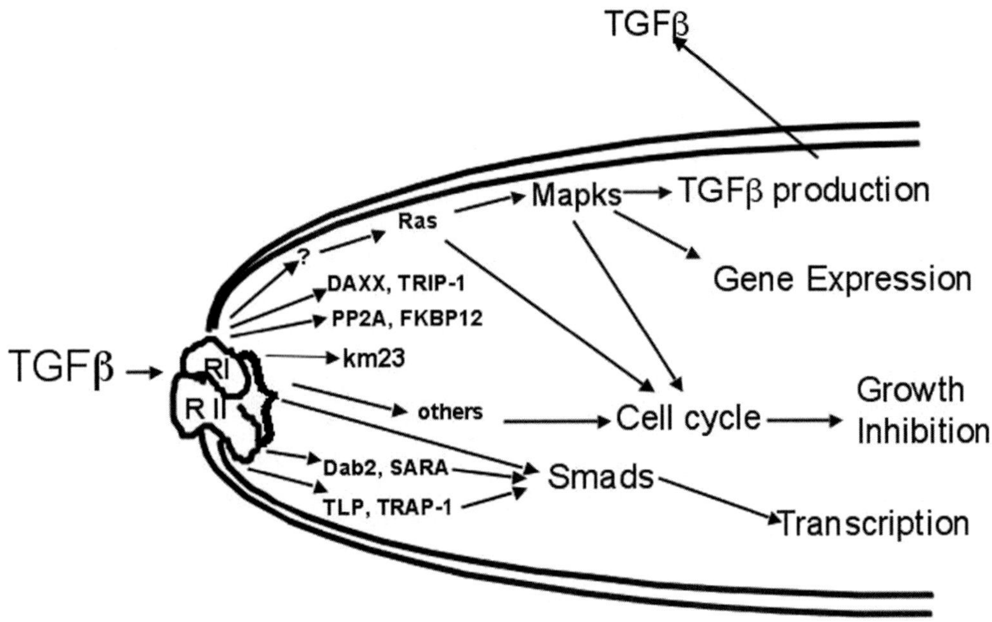

Fig. 1. Summary of TGF-β signaling pathways in epithelial cells. DAXX, death-associated protein.

With the increasing numbers of laboratories studying TGF-β signaling, additional TGF-β signaling components and pathways are likely to be discovered to mediate the diverse biological responses of this polypeptide factor. We have developed a novel screen to identify TβR-interacting proteins that should not include many of these other TβR-interacting factors *(27)*. Using this system, we have identified km23-1 and km23-2 as TGF-β signaling intermediates and that are also light chains of the minus-end directed motor protein dynein *(27,28)*. Each of these will be discussed in detail in the remaining section of this chapter.

Cytoplasmic dynein is a motor complex that transports membrane vesicles and diverse motor cargoes along microtubules (MTs) in a retrograde manner *(29–32)*. It plays a wide variety of functions in mitotic spindle assembly and orientation, positioning of the Golgi apparatus, and transport of various intracellular organelles, including endosomes and lysosomes *(29–34)*. The dynein motor complex is a large multimeric complex, generally consisting of two heavy chains (HC), two intermediate chains (IC), several light-intermediate chains (LIC), several light chains (LC), and other adaptor or accessory proteins (Fig. 2). One such adaptor complex is dynactin, which is important for cargo binding and consists of several subunits itself *(29–32)*.

Three distinct classes of DLCs have been identified in mammals, including the LC8, Tctex-1/rp3, and km23/LC7/robl class *(31,35,36)*. In addition to binding the DIC at distinct regions *(37)*, light chains have also been shown to interact with a number of cargoes to exert diverse functions *(29–32)*. LC8 binds to neuronal nitric oxide synthase (nNOS) *(38,39)*, and to the proapoptotic Bcl-2 family member (Bim) *(40)*. LC8 has also been reported to be a physiologically interacting substrate of p21-activated kinase 1 (Pak1) *(41)*. Tctex-1 has been shown to interact with various cellular components, including the cytoplasmic tails of the rhodopsin receptor *(42)*, the bone morphogenetic protein receptor type II (BMPRII) *(43)*, the parathyroid hormone receptor *(44)*, the neurotrophin receptor Trk *(45)*, and the polio virus receptor CD155 *(46)*. rp3 has also been shown to interact with the herpes simplex virus type 1 capsid protein VP26 and proposed to mediate retrograde cellular transport of viral

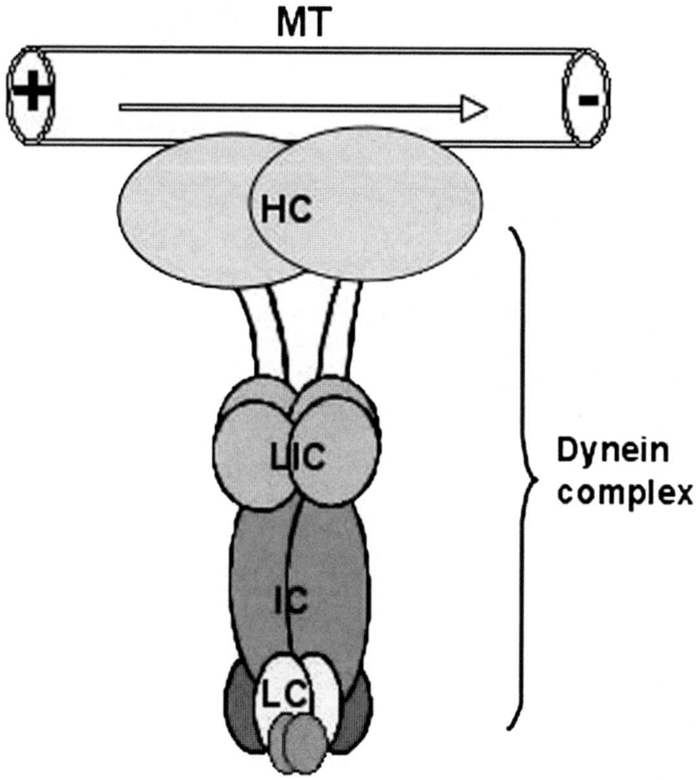

Fig. 2. Organization of dynein, and its interaction with microtubules. Arrow indicates the direction of dynein movement. The minus (–) and plus (+) ends of the microtubule are indicated.

capsids *(47)*. However, little is known about the binding partners of the km23/LC7/robl class. As mentioned above, our laboratory first discovered km23 as a novel TβR-interacting protein *(28)*. Since another mammalian form of km23 has been identified, termed km23-2 *(28,48,49)*, we now refer to km23 as km23-1.

In this review we will summarize the structure, activation, and cellular functions of the km23 DLCs. A major part of the review will focus on the function of km23 DLCs in TGF β signaling pathways.

2. THE km23/LC7/robl DLC FAMILY

km23/LC7/robl DLCs belong to an ancient superfamily of proteins, known as the MglB superfamily, with representative members in all three kingdoms of life *(28,50,51)*. Table 1 lists the percent homologies, identities, and similarities of the DLCs of the km23/LC7/roadblock family. Members of the MglB superfamily have been implicated in NTPase regulation in bacteria and archae, and in regulating the MAPK and other signaling pathways in eukaryotes *(51)*. Recently, our NMR results have demonstrated that km23-1 adopts the structure of a stable homodimer *(48)*, similar to that of the p14/ MAPK scaffolding protein (MP1) heterodimer. MP1 and p14 are two distantly related members of the km23/robl/MglB superfamily, distinct from the LC8 and Tctex-1 classes of DLCs, which also adopt a homodimeric conformation *(52,53)*. In addition, our results and those of another group have shown that km23-1 can form homodimers or heterodimers with km23-2 in vitro and in vivo

Table 1
Comparision of km23-1 to Some Other km23/LC7/robl Family Members

Homologue	*Species*	*% Homology*	*% Identity*	*% Similarity*	*Amino acids*
ZFIN	Danio rerio	75	80	93	96
km23-2	Homo sapiens	70	77	91	96
robl	Drosophila melanogaster	67	71	81	97
chiLC7	Chlamydomonas	59	55	74	105
B15	Spermatozopsis	58	56	69	98
T24H10.6	Caenorhabditis elegans	56	47	76	95
bxd	Drosophila melanogaster	42	23	51	101
LMAJFV1	Leishmania	33	33	32	103

(48,54). Moreover, NMR structural studies have identified a specific region of km23-1 that may be important for DIC binding and cargo attachment *(48)*.

3. km23-1

3.1. km23-1 Introduction

Human km23-1 is a 96-amino acid protein encoded by a 291-base pair open reading frame. km23-1 is localized on human chromosome 20q11.21 *(27)*, and is ubiquitously expressed in human tissues and cell lines *(55)*. Cell fractionation and Western blot analyses indicated that km23-1 is an 11 kDa cytoplasmic protein *(55)*. NMR structural data have shown that the conserved surface residues of km23-1 are located predominantly on a single face of the molecule, where the three conserved serines in km23-1 exist. Adjacent to this surface is a large of cleft formed by the incomplete overlap of the loops from the two monomers *(48)*.

3.2. km23-1: A TGF-β Signaling Intermediate That is Also a DLC

Using a novel screening method, km23-1 was identified by its ability to interact with the phosphorylated cytoplasmic regions of TβRII and TβRI *(28)*. TGF-β was shown to induce the interaction of TβRII with km23-1 in two different TGF-β-responsive epithelial cell lines at early times (within 5 min) after TGF-β addition *(28)*. The phosphorylation of km23-1 is TGF-β-dependent, in that epidermal growth factor (EGF) is unable to phosphorylate km23-1 *(55)*. In addition, no interaction between the full-length BMPRII and km23-1 was observed in a GST pull-down assay (R. Machado and R. C. Trembath, personal communication), further suggesting that km23-1 is more specifically involved in TGF-β signaling. Phosphoamino acid analyses indicated that km23-1 is phosphorylated primarily on serine residues in response to TGF-β receptor activation *(28)*. We have shown that the conserved serine residues, particularly S32 and S73, are critical sites for phosphorylation by the TβRs *(48)*. More importantly, TβRII is absolutely required for km23-1 phosphorylation and recruitment of km23-1 to the rest of dynein motor through the DIC *(28)*. Since km23-1 is a light chain of dynein *(28)*, it was of interest to determine the site at which km23-1 binds to DIC and/or to other partners. NMR structural results have indicated that the surface that includes the two N-terminal-helices may be involved in DIC binding and that phosphorylation of S32, S73, or both, by TβR activation may regulate this process *(48)*. Future studies will address this in more detail.

Our results thus far indicate that km23-1 may function as a signaling intermediate for TGF-β. If this were the case, we would expect that km23-1 could mediate some of the known TGF-β signaling events. For example, previous results have indicated that JNK activation by TGF-β may play a role in TGF-β-mediated growth inhibition, either through amplification of TGF-β 1 production, via crosstalk with the Smads, and/or by regulation of cell cycle inhibitors *(56,57)*. In Mv1Lu cells firmly expressing km23-1, we have shown that overexpression of km23-1 induces specific TGF-β responses, including JNK activation, c-*Jun* phosphorylation, and cell growth inhibition *(28)*. Moreover, consistent with a role of km23-1 in TGF-β signaling, siRNA blockade of km23-1 expression resulted in a decrease in specific cellular responses to TGF-β, including an induction of fibronectin expression and an inhibition of cell cycle progression *(48)*. This is the first demonstration of a link between cytoplasmic dynein and a natural growth inhibitory cytokine, suggesting a novel function for km23-1 in TGF-β signaling.

3.3. km23-1: A Signaling Intermediate in Smad2-Dependent TGF-β Signaling

Our previous results suggest that km23-1 may function as a *motor receptor* to recruit TGF-β signaling components to the dynein motor for intracellular transport along MTs *(28)*. Because Smad2/3 are critical intermediates in TGF-β signaling *(58,59)*, it was of interest to determine whether Smad2/3 is one of the cargos that km23-1 links to the dynein motor. We have demonstrated that km23-1 co-localizes with Smad2 at very early times after TGF-β addition to cells, prior to Smad2 translocation to the nucleus *(27)*. Further, km23-1 interacts with Smad2, but not Smad3 in vitro. TGF-β regulates the interaction of endogenous km23-1 and Smad2 in vivo. More significantly, blockade of km23-1 resulted in a reduction in Smad2-dependent activin-responsive element (ARE)-Lux transcriptional activity, but not in Smad3-dependent Smad-binding element $(SBE)_2$-Luc transcriptional regulation (Fig. 3). In addition, our results showed that in cells transfected with dynamitin, Smad2 was expressed diffusely throughout the cell after TGF-β treatment, indicating that Smad2 had not translocated to the nucleus.

Overexpression of dynamitin has previously been shown to block dynactin function and disrupt the dynactin/dynein complex *(60)*. It is used frequently in the dynein field as a method for disrupting dynein transport of cargo *(32)*. Thus, our results demonstrate that dynein-mediated transport of Smad2 is required for Smad2 to translocate to the nucleus. This is not meant to imply that other mechanisms are not also involved in the many events that must occur before Smad2 arrives in the nucleus. It just emphasizes the importance of km23-1, and the rest of the motor, in the intracellular transport of Smad2, because the DLCs are involved in linking the cargo to the rest of the motor. Collectively, our results indicate that km23-1 recruits Smad2 to the rest of dynein motor for intracellular transport along MTs.

3.4. km23-1: A Light Chain of the Motor Protein Dynein Frequently Altered in Ovarian Cancer

Alterations in TGF-β-signaling components and pathways have been reported in several types of human cancers *(1,2)*, including epithelial ovarian cancer *(61,62)*. Because, we had identified km23-1 as a TβR-interacting protein and TGF-β signaling components are known to be altered in cancers, it was of interest to determine whether km23-1 alterations would exist in human cancers. We developed a highly sensitive method to detect alterations in human cancer tissues *(63)*. Using two microscale technologies, laser-capture microdissection and nested reverse-transcription-PCR, we demonstrated that km23-1 was altered at a high frequency in human ovarian cancer patients *(27,63)*. We identified various mutations

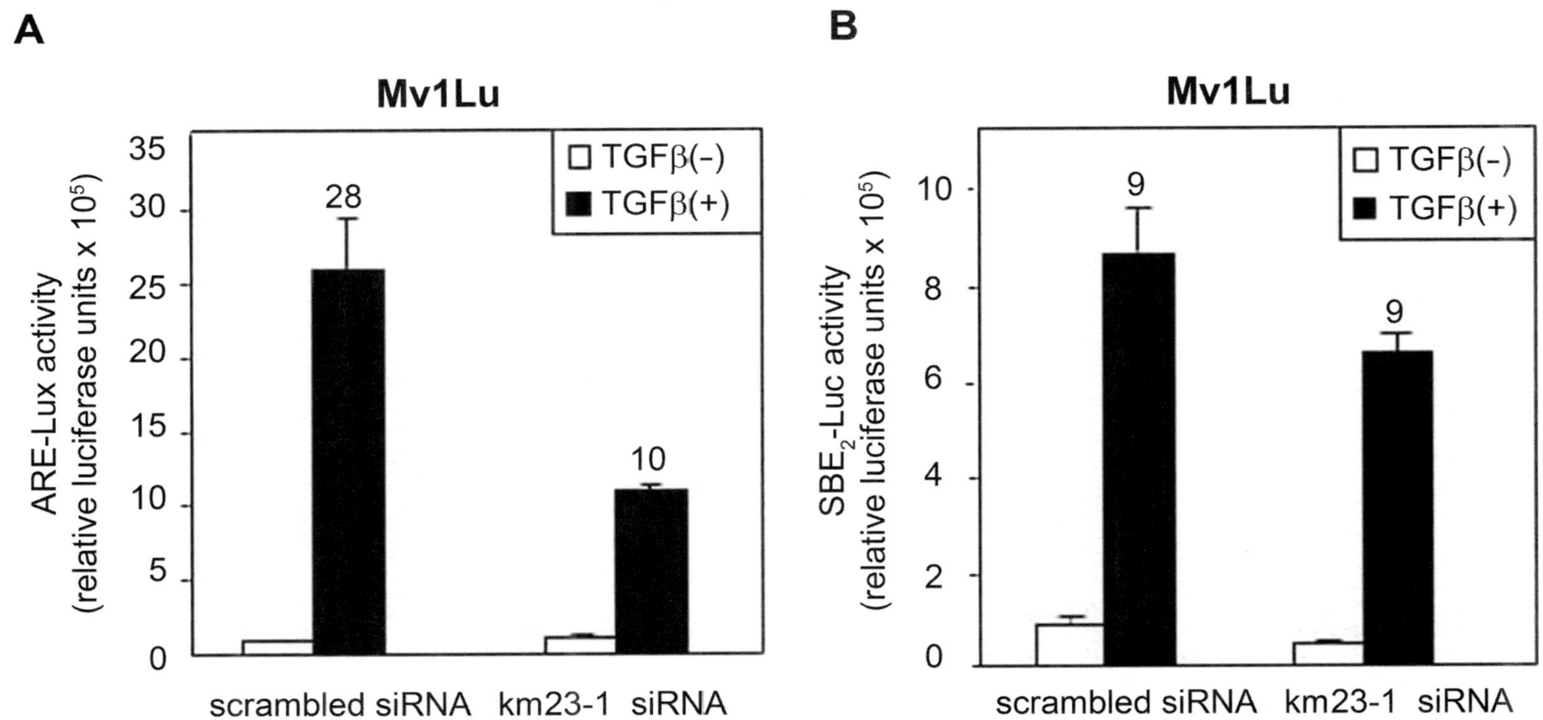

Fig. 3. siRNA blockade of endogenous km23-1 inhibits Smad2/4-dependent transcription in TGF-β/activin reporter assays, but has no effect on Smad3/4-dependent SBE transcriptional activation. (**A**) Mv1Lu cells were transfected with increasing amounts of either km23-1 siRNA or scrambled siRNA along with ARE-lux and forkhead activin signal transducer (FAST)-1. To normalize transfection efficiencies, renilla was co-transfected as an internal control. Twenty-four hours after transfection, the medium was replaced with serum-free medium for 1 h, followed by incubation of cells in the absence (open bar) and presence (black bar) of TGF-β (5 ng/ml) for an additional 18 h. Luciferase activity was measured using the Dual Luciferase Reporter Assay System. All reporter assays were performed in triplicate. The results shown are representative of two similar experiments. (**B**) Mv1Lu cells were transfected with increasing amounts of either km23-1 siRNA or scrambled siRNA along with SBE_2-Luc, and then analyzed as for **A**. The results shown are representative of two similar experiments.

in km23-1 in 8 out of 19 ovarian cancer patients, and calculated a total km23-1 alteration rate of 42%. These mutations were not present in tumor-free tissues *(27–63)*. Such a high alteration rate of km23-1 in ovarian cancer suggests that km23-1 may play an important role in ovarian cancer development and/or progression. More importantly, novel altered forms of km23-1 found in the ovarian cancer patients displayed a disruption in the DIC binding function of km23-1, as well as an inhibition of TGF-β-dependent transcriptional activation of both the 3TP-lux and the ARE-Lux transcriptional reporters *(63)*. Our data suggest that the alterations in km23-1 found in the ovarian cancer patients may modify km23-1 functions in TGF-β signaling and tumorgenesis.

3.5. km23-1: A Potential Tumor Suppressor

km23-1 is frequently mutated in human epithelial ovarian cancer tissues and cell lines, and its overexpression in untransformed epithelial cells suppresses growth *(27,28,63)*. To further investigate the tumor suppressive effect of wild-type (wt) km23-1 specifically in ovarian cancer cells, we stably expressed wt km23-1 in an inducible fashion in the SK-OV-3 human ovarian cancer cell line, which expresses a truncated formed km23-1, β exon-3-km23-1. In the Tet-Off inducible system we employed, removal of doxycycline (dox) resulted in an induction of the stably expressed wt km23-1 within 4–5 d in both a km23-1 pool and a representative clone that had been selected (#14). In contrast, stable expression of the empty vector construct in the same SK-OV-3-pBig2r system (EV pool) did not result in induction of km23-1 upon dox removal. Induction of km23-1 resulted in a significant decline in viable cell number in monolayer culture compared to the EV pool. Similarly, expression of km23-1 by removal of dox in both the km23-1 pool and clone #14 resulted in a 5- to 10-fold decrease in the number of colonies formed in soft agar, compared to the number of colonies formed by withdrawal of dox from the EV pool cells. To investigate the effect of wt km23-1 on in vivo tumor growth, we injected athymic nude mice with SK-OV-3-pBig2r cells. Consistent with the growth suppressive effect in vitro, induction of wt km23-1 led to a 3- to 5-fold decline in xenograft tumor size and number in in vivo. Fluorescence-activated cell sorting (FACS) analysis of the cell cycle distribution revealed that wt km23-1 overexpression led to a 2-fold increase in G2/M DNA content in the km23-1 pool upon dox removal. Overall, our results suggest that km23-1 suppresses ovarian cancer cell growth by increasing the number of cells in the G2/M phase of the cell cycle.

3.6. km23-1 is Required for TGF-β1 Production in Human Colon Carcinoma Cells

In tumor cells that have lost responsiveness to the growth inhibitory effects of TGF-β, increased TGF-β production by the tumor cells often contributes to cancer progression primarily through paracrine mechanisms *(2)*. We have previously identified the signaling pathways mediating TGF-β1 production *(56)*. Since MAPKs were required for TGF-β1 production, and we have shown that forced expression of km23-1 activates JNK *(28)*, it was of interest to determine whether km23-1 was required for TGF-β1 production in human cancer cells. To assess this, we stably transfected RKO human colon carcinoma cells with km23-1 siRNA, and demonstrated that km23-1 siRNA significantly suppressed both AP-1 binding and luciferase reporter activity at the proximal AP-1 site in the TGF-β1 promoter, previously shown to be required for TGF-β1 expression *(56)*. Further, km23-1 siRNA-transfected RKO cells displayed a reduction in the amount of TGF-β1 secreted into the conditioned medium (CM) of the cells, and in the biological activity of the secreted TGF-β1. More significantly, CM from km23-1 siRNA-transfected RKO cell cultures reduced the cellular migration of NIH 3T3 cells. Co-culture of km23-1 siRNA-transfected RKO cells with NIH 3T3 cells suppressed NIH 3T3 cell mitogenesis. Thus, our results implicate km23-1 as one of the targets that could be exploited to suppress TGF-β1 production by tumor cells, thereby suppressing the migratory behavior associated with the malignant phenotype.

4. km23-2

4.1. km23-2 Introduction

Human km23-2 is also a 96-amino acid protein encoded by a 291-base pair open reading frame. Human km23-2 localizes to chromosome 16 (16q23.3). Previous reports have shown that chromosome 16q23, a chromosome region frequently lost in many tumors *(64)*, including hepatocellular carcinoma *(65)* and prostate cancer *(66)*. Western blot analyses indicated that human km23-2 is also an 11 kDa cytoplasmic protein. km23-2 displays 70% homology and 77% identity with the human km23-1, we have identified *(27,28)*. The three-dimensional (3D) structure of km23-2 has suggested that there are differences in potential binding properties of km23-1 and km23-2. For example, among the 27 interfacial residues in km23-1, three differ from those in km23-2. These include M46, F49, and N76 in km23-1, which are substituted to L46, L49, and H76 in km23-2 *(48)*. To identify differences in potential binding properties of km23-1 and km23-2, we have analyzed the location of the residues that differ between these isoforms. Our results suggest that the residues that differ between km23-1 and km23-2 may not alter the DIC binding function of the DLC, but they may play some role in the regulation of specific transport and/or motor regulatory activities *(48)*.

4.2. km23-2: A New DLC That Regulates TGF-β Signaling

As mentioned earlier, km23-2 is a member of the km23/LC7/robl family of LCs of the motor protein dynein in mammalian cells *(27,28)*. Because we have demonstrated that km23-1 is required for Smad2-dependent TGF-β signaling, it was of interest to determine the function of km23-2 in TGF-β signaling. We have found that in the presence of TGF-β, but not in the absence, km23-2 was present in early endosomes with the TβRs and Smad3. Furthermore, km23-2 interacted with TβRII in the presence of TGF-β. Expression of both TβRs with km23-2 resulted in a modest level of km23-2 phosphorylation. However, TGF-β treatment for 15 min resulted in a significant increase km23-2 phosphorylation. More significantly, this phosphorylation of km23-2 was completely blocked upon expression of the dominant-negative kinase deficient RII (KNRII). To further confirm whether the kinase activity of TβRII and/or TβRI was required for km23-2 phosphorylation, we performed similar in vivo phosphorylation assays *(28)* in Mv1Lu cells expressing endogenous TβRs and in various receptor mutant lines derived from Mv1Lu cells *(67–69)*. Upon TGF-β treatment, km23-2 was highly phosphorylated in the parental Mv1Lu cells, and similarly phosphorylated in R1B cells lacking TβRI. However, the signal was weaker in the R1B cells than in the parental Mv1Lu cells, suggesting that the presence of TβRI may enhance the phosphorylation of km23-2 through TβRII, although the kinase activity of TβRI may not be required for km23-2 phosphorylation. km23-2 was not phosphorylated in DR26 cells lacking TβRII. These results provide additional evidence that the kinase activity of TβRII is required for km23-2 phosphorylation.

As mentioned earlier, km23-2 is a member of the km23/LC7/robl family of LCs of the motor protein dynein in mammalian cells *(27,28,49)*. It is thought that DLCs may be important for specifying the nature of the cargo that will be transported by the motor *(33,34)*. Therefore, it is likely that extracellular factors (such as growth factors, cytokines, and so on) might be able to select the particular DLCs to be recruited to the motor in specific cellular contexts to specify and carry the cargo. Our results have demonstrated that TGF-β rapidly induced the recruitment of km23-2 to DIC. To provide definitive evidence that TβR activation is required for the km23-2-DIC interaction, we examined the interaction between km23-2 and DIC in the absence and presence of KNRII. Our data have demonstrated that expression of KNRII completely blocked this interaction between km23-2 and DIC in the presence of TGF-β. Collectively, our results strongly suggest that km23-2 is a new DLC that regulates in TGF-β signaling.

4.3. km23-2 is Required for Smad3-Dependent TGF-β Signaling

Differential binding of Smad2 and Smad3 to signaling components has been reported previously *(70)*. Because km23-1 interacts with Smad2, but not with Smad3 *(71)*, suggesting that Smad2 may be a specific cargo for intracellular transport by km23-1, it was of interest to determine whether there was any differential regulation of the RSmads by km23-2. Immunoprecipitation (IP)/blot and luminescence-based mammalian interaction mapping (LUMIER) analyses showed that TGF-β stimulated the interaction of km23-2 preferentially with Smad3, relative to that with Smad2. Furthermore, siRNA blockade of km23-2 resulted in a reduction of TGF-β stimulation of SBE_2-Luc transcriptional activity, while a reduction in TGF-β stimulation of Smad2-dependent ARE-Lux transcriptional activity did not occur under similar conditions (Fig. 4). These results suggested that km23-2′s role in mediating transcriptional activation of Smad3 might be somewhat specific for this RSmad. In addition, blockade of km23-2 decreased TGF-β activation of the human Smad7 promoter Smad7-Luc, an endogenous Smad3-target promoter. More significantly, we have found that blockade of km23-2 decreased TGF-β induction of plasminogen activator inhibitor-1 (PAI-1) gene expression. Smad7 and PAI-1 have been shown to be Smad3-specific target genes *(72,73)*. Thus, our data indicate that km23-2′s role in mediating transcriptional activation of Smad3 may be relatively specific for this RSmad.

5. A MODEL FOR km23-1 AND km23-2 IN RECRUITMENT OF TGF-β SIGNALING ENDOSOMES FOR INTRACELLULAR TRANSPORT ALONG MTs

Based upon our results and those of others, we propose a model for km23-1/km23-2 action in the recruitment of TGF-β signaling endosomes for intracellular transport along MTs (Fig. 5). km23-1 is shown as an example. According to this model, in the absence of ligand, km23-1 and Smad2 are present in complexes unrelated to TGF-β activation. Within minutes of ligand binding, activated TβRs are internalized into EEA1/SARA-enriched endosomes, where Smad2 is recruited by SARA *(74–77)*. However, little is known about the precise mechanisms directing intracellular localization of Smad3. Recent reports have indicated that Smad3 may also be present with SARA in early endosomes *(78)*, but that the RSmads may be phosphorylated in different compartment locales *(79)*. Our sucrose gradient analyses indicated that Smad3 was also present in EEA1-enriched fractions, but these compartments may be distinct from those in which Smad2 resides *(71)*. We have also demonstrated that once km23-1 was phosphorylated by TβRII, during the same time period that Smad2 is phosphorylated by TβRI *(4)*, km23-1 selectively interacted with the TβR/Smad2 complex, and recruited the TGF-β signaling endosomes to the dynein motor through the DIC-km23-1 interaction *(71)*. In support of our results and this model, dynein is known to mediate the association of endosomal membranes with MTs *(80,81)*. After attachment of the TGF-β signaling endosomes to the rest of the motor, we propose that km23-1/dynein transports the TGF-β signaling endosomes along the MTs to the next endosomal compartments. Our results also revealed that overexpression of dynamitin inhibited Smad2 intracellular transport, indicating that the dynein complex is required for Smad2 intracellular transport *(71)*. Eventually, Smad2 translocates to the nucleus for transcriptional

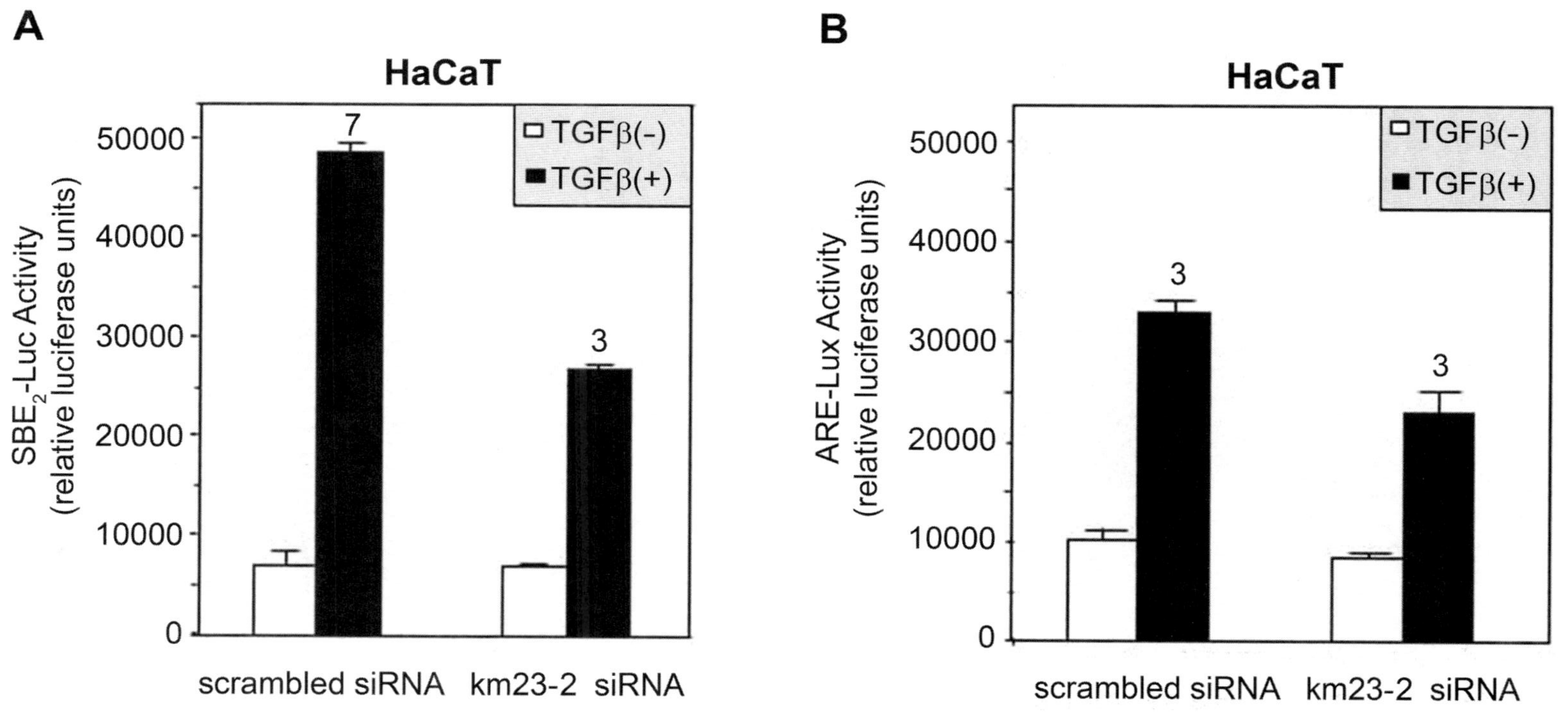

Fig. 4. siRNA blockade of endogenous km23-2 expression results in significant inhibition of Smad3-dependent SBE_2-Luc activation, but has no effect on inhibition of Smad2-dependent ARE-Lux activation in HaCaT cells. (**A**) siRNA blockade of endogenous km23-2 expression results in significant inhibition of Smad3-dependent SBE_2-Luc activation in HaCaT cells. HaCaT cells were transfected with increasing amounts of either km23-2 Stealth siRNA or scrambled Stealth siRNA control along with SBE_2-Luc. Luciferase assays were performed as for Figure 3. (**B**) siRNA blockade of endogenous km23-2 expression has no effect on inhibition of Smad2-dependent ARE-Lux activation in HaCaT cells. Experiments were performed as for (A), except that ARE-lux and FAST-1 were used instead of SBE_2-Luc.

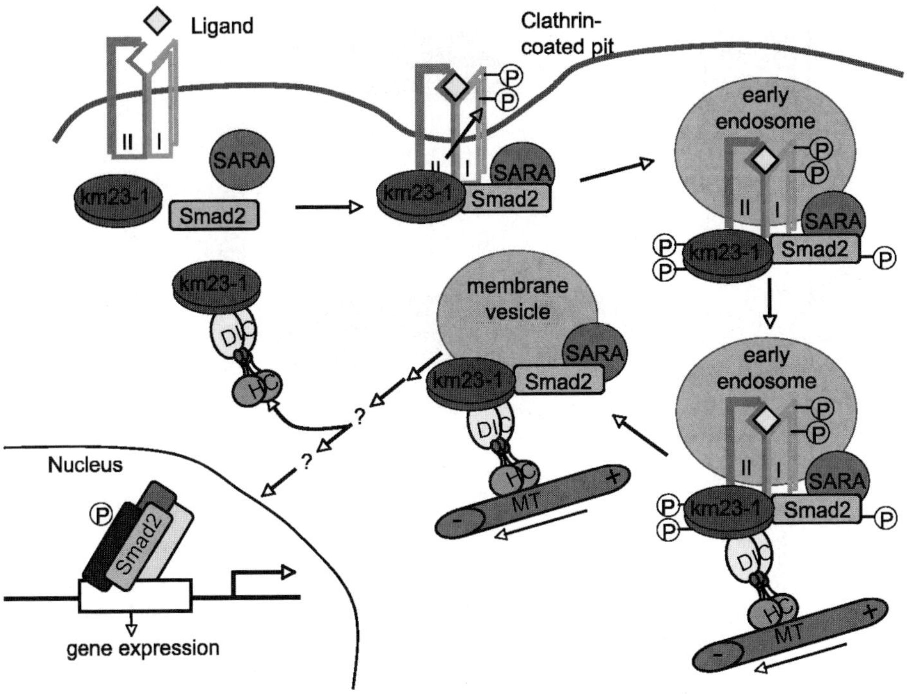

Fig. 5. Model for km23-1 recruitment of TGF-β signaling endosomes containing Smad2 for intracellular transport along MTs. In the absence of ligand, km23-1 and Smad2 may be in an inactive complex. Upon TGF-β stimulation, activated TβRs internalize into EEA1/SARA-enriched endosomes, where Smad2 is recruited by SARA. Upon phosphorylation of km23-1 and Smad2 by TβRs, DIC is recruited to form a motor complex, and then tethers endosomal membranes to MTs. km23-1 and the rest of dynein complex transports TGF-β signaling endosomes along MTs to the next endosomal compartment. Eventually, Smad2 and km23-1 are segregated to different compartments, and Smad2 translocates to the nucleus for transcriptional regulation of target genes.

regulation of specific target genes. Similar events may occur for Smad3 through its regulation by km23-2 (not shown).

6. FUTURE DIRECTIONS

6.1. km23: A Novel Anti-Cancer Target for the Development of Diagnostics and Therapeutics

TGF-β is growth inhibitory for normal cells of endothelial, hematopoietic, neuronal, and epithelial origin *(2,4)*. However, cancers are often refractory to this growth inhibitory effect, owing to genetic loss of TGF-β signaling components or, more commonly, perturbation of TGF-β signaling pathways *(2,4)*. We have shown that km23-1 is altered at high frequency (42%) in epithelial ovarian cancer *(27,63)*. The alterations in km23-1 appear to disrupt the ability of km23-1 to bind to the DIC and regulate TGF-β-dependent transcriptional events. Thus, km23-1 may play a role in ovarian cancer formation or progression through a mechanism involving a loss of TGF-β signaling *(63)*. Our results have suggested that km23-1 may function as a motor receptor to recruit TGF-β signaling components (such as Smad2, JNK, and ERK) to the rest of the dynein motor for intracellular transport. However, when km23-1 is altered, the cargo is unable to reach the correct destination in the cell (Fig. 6). As a result, a traffic jam occurs and causes chaos in the cell. Our future studies will test drugs

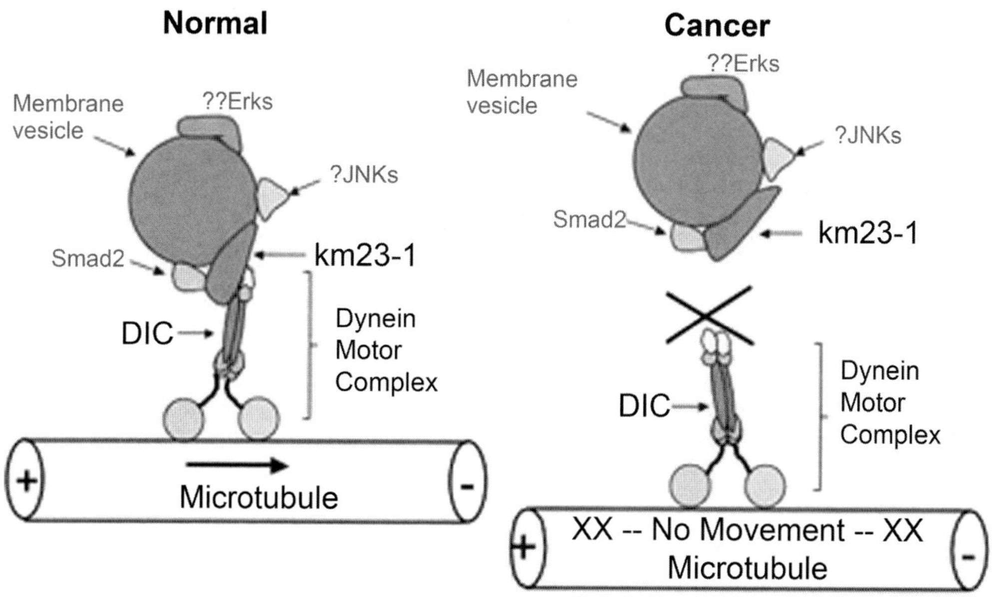

Fig. 6. Model for the effect of km23 alterations on intracellular transport and TGF-β signaling in cancer. In non-cancer cells, km23-1 may function as a motor receptor to recruit TGF-β signaling components (such as Smad2, JNK, and ERK) to the rest of the dynein motor for intracellular transport. However, in cancer cells, when km23-1 is altered, the cargo is unable to reach the motor complex, and TGF-β signaling is disrupted. A traffic jam may occur to cause chaos in the cell.

that would target km23-1 to override the defects caused by the km23-1 alterations in the cancer cells.

As mentioned earlier, we have determined the precise 3D structure of km23-1 and km23-2 *(48)*. Future studies are directed at determining the structural alterations resulting from the mutations in km23-1 that we have identified in the ovarian cancer patients. 3D structural data would help to determine how km23-1 mutations modify km23-1 functions, and to assist in the development of novel km23-1-based therapeutics for ovarian cancer.

ACKNOWLEDGMENTS

We thank Dr. Scott E. Kern (John Hopkins Oncology Center, Baltimore, MD) for SBE_2-Luc; Dr. Malcolm Whitman (Harvard Medical School, Boston, MA) for the ARE-Lux and FAST-1 constructs; Dr. Joan Massagué (Memorial Sloan-Kettering Cancer Center, New York, NY) for the R1B and DR26 cells; Dr. Richard B. Vallee (Columbia University) for the dynamitin-myc; Dr. Erwin Botinger (Mount Sinai School of Medicine, New York, NY) for the Smad7-Luc; Dr. Jeffrey L. Wrana (Samuel Lundenfeld Res. Institute, Toronto, Canada) for the pCMV5-HA-RII, hRL-Smad2, hRL-Smad3 constructs. The work was supported by National Institutes of Health Grants CA90765, CA92889, and CA100239, and Dept of Defense # DAMD17-03-01-0287 to K.M.M.

REFERENCES

1. Massagué J, Blain SW, Lo RS. TGF-β signaling in growth control, cancer, and heritable disorders. Cell 2000;103:295–309.
2. Yue J, Mulder KM. Transforming growth factor-β signal transduction in epithelial cells. Pharmacol Ther 2001;91:1–34.

3. Shi YG, Massagué J. Mechanism of TGF-β signaling from cell membrane to the nucleus. Cell 2003; 113:685–700.
4. Derynck R, Zhang YE. Smad-dependent and Smad-independent pathways in TGF β-family signaling. Nature 2003;425:577–584.
5. Attisano L, Wrana JL. Signal transduction by the TGF-β superfamily. Science 2002;296:1646–1647.
6. Hartsough MT, Mulder KM. Transforming growth factor β activation of p44mapk in proliferating cultures of epithelial cells. J Biol Chem 1995;270:7117–7124.
7. Frey RS, Mulder KM. TGF β regulation of mitogen-activated protein kinases in human breast cancer cells. Cancer Lett 1997;117:41–50.
8. Frey RS, Mulder KM. Involvement of extracellular signal-regulated kinase 2 and stress-activated protein kinase/Jun N-terminal kinase activation by transforming growth factor β in the negative growth control of breast cancer cells. Cancer Res 1997;57:628–633.
9. Hartsough MT, Frey RS, Zipfel PA, et al. Altered transforming growth factor signaling in epithelial cells when ras activation is blocked. J Biol Chem 1996;271:22,368–22,375.
10. Yue J, Buard A, Mulder KM. Blockade of TGF β3 up-regulation of p27Kip1 and p21Cip1 by expression of RasN17 in epithelial cells. Oncogene 1998;17:47–55.
11. Atfi A, Djelloul S, Chastre E, Davis R, Gespach C. Evidence for a role of Rho-like GTPases and stress-activated protein kinase/c-Jun N-terminal kinase (SAPK/JNK) in transforming growth factor β-mediated signaling. J Biol Chem 1997;272:1429–1432.
12. Hartsough MT, Mulder KM. Transforming growth factor-β signaling in epithelial cells. Pharmacol Ther 1997;75:21–41.
13. Hocevar BA, Brown TL, Howe PH. TGF-β induces fibronectin synthesis through a c-Jun N-terminal kinase-dependent, Smad4-independent pathway. EMBO J 1999;18:1345–1356.
14. Ravanti L, Hakkinen L, Larjava H, et al. Transforming growth factor-β induces collagenase-3 expression by human gingival fibroblasts via p38 mitogen-activated protein kinase. J Biol Chem 1999;274:37,292–37,300.
15. Adachi-Yamada T, Nakamura M, Irie K, et al. p38 mitogen-activated protein kinase can be involved in transforming growth factor β superfamily signal transduction in Drosophila wing morphogenesis. Mol Cell Biol 1999;19:2322–2329.
16. Hanafusa H, Ninomiya-Tsuji J, Masuyama N, et al. Involvement of the p38 mitogen-activated protein kinase pathway in transforming growth factor-β-induced gene expression. J Biol Chem 1999;274:27,161–27,167.
17. Mulder KM, Morris SL. Activation of p21ras by transforming growth factor β in epithelial cells. J Biol Chem 1992;267:5029–5031.
18. Mulder KM. Role of Ras and Mapks in TGF β signaling. Cytokine Growth Factor Rev 2000;11:23–35.
19. Tsukazaki T, Chiang TA, Davison AF, Attisano L, Wrana JL. SARA, a FYVE domain protein that recruits Smad2 to the TGF β receptor. Cell 1998;95:779–791.
20. Hocevar BA, Smine A, Xu X-X, Howe PH. The adaptor molecule Disabled-2 links the transforming growth factor β receptors to the Smad pathway. EMBO J 2001;20:2789–2801.
21. Wurthner JU, Frank DB, Felici A, et al. Transforming growth factor-β receptor-associated protein 1 is a Smad4 chaperone. J Biol Chem 2001;276:19,495–19,502.
22. Felici A, Wurthner JU, Parks WT, et al. TLP, a novel modulator of TGF-β signaling, has opposite effects on Smad2- and Smad3-dependent signaling. EMBO J 2003;22:4465–4477.
23. Perlman R, Schiemann WP, Brooks MW, Lodish HF, Weinberg RA. TGF-β-induced apoptosis is mediated by the adaptor protein Daxx that facilitates JNK activation. Nat Cell Biol 2001;8:708–714.
24. Wang D, Kanuma T, Mizunuma H, et al. Analysis of specific gene mutations in the transforming growth factor-β signal transduction pathway in human ovarian cancer. Cancer Res 2000;60:4507–4512.
25. Chen RH, Miettinen PJ, Maruoka EM, Choy L, Derynck R. A WD-domain protein that is associated with and phosphorylated by the type II TGF-β receptor. Nature 1995;377:548–552.
26. Griswold-Prenner I, Kamibayashi C, Maruoka EM, Mumby MC, Derynck R. Physical and functional interactions between type I transforming growth factor β receptors and Balpha, a WD-40 repeat subunit of phosphatase 2A. Mol Cell Biol 1998;18:6595–6604.
27. Ding W, Mulder KM. km23: A novel TGF β signaling target altered in ovarian cancer. In: Molecular Targeting and Signal Transduction (Kumar R. ed.,), pp.315–327. Kluwer Academic Publishers, Norwell, MA 2004.
28. Tang Q, Staub CM, Gao G, et al. A novel transforming growth factor-β receptor-interacting protein that is also a light chain of the motor protein dynein. Mol Biol Cell 2002;13:4484–4496.
29. Hirokawa N. Kinesin and dynein superfamily proteins and the mechanism of organelle transport. Science 1998;279:519–526.
30. King SJ, Bonilla M, Rodgers ME, Schroer TA. Subunit organization in cytoplasmic dynein subcomplexes. Protein Sci 2002;11:1239–1250.

31. Vale RD. The molecular motor toolbox for intracellular transport. Cell 2003;112:467–480.
32. Vallee RB, Williams JC, Varma D, Barnhart LE. Dynein: An ancient motor protein involved in multiple modes of transport. J Neurobiol 2004;58:189–200.
33. Kamal A, Goldstein LS. Principles of cargo attachment to cytoplasmic motor proteins. Curr Opin Cell Biol 2002;14:63–68.
34. Mallik R, Gross SP. Molecular motors: strategies to get along. Curr Biol 2004;14:R971–R982.
35. Williams JC, Xie H, Hendrickson WA. Crystal structure of dynein light chain TcTex-1. J Biol Chem 2005;280:21,981–21,986.
36. Wu H, Maciejewski MW, Takebe S, King SM. Solution structure of the Tctex1 dimer reveals a mechanism for dynein-cargo interactions. Structure (Camb) 2005;13:213–223.
37. Susalka SJ, Nikulina K, Salata MW, et al. The roadblock light chain binds a novel region of the cytoplasmic Dynein intermediate chain. J Biol Chem 2002;277:32,939–32,946.
38. Jaffrey SR, Snyder SH. PIN: an associated protein inhibitor of neuronal nitric oxide synthase. Science 1996;274:774–777.
39. Fan JS, Zhang Q, Li M, et al. Protein inhibitor of neuronal nitric-oxide synthase, PIN, binds to a 17-amino acid residue fragment of the enzyme. J Biol Chem 1998;273:33,472–33,481.
40. Puthalakath H, Huang DC, O'Reilly LA, King SM, Strasser A. The proapoptotic activity of the Bcl-2 family member Bim is regulated by interaction with the dynein motor complex. Mol Cell 1999;3:287–296.
41. Vadlamudi RK, Bagheri-Yarmand R, Yang Z, et al. Dynein light chain 1, a p21-activated kinase 1-interacting substrate, promotes cancerous phenotypes. Cancer Cell 2004;5:575–585.
42. Tai AW, Chuang JZ, Bode C, Wolfrum U, Sung CH. Rhodopsin's carboxy-terminal cytoplasmic tail acts as a membrane receptor for cytoplasmic dynein by binding to the dynein light chain Tctex-1. Cell 1999;97:877–887.
43. Machado RD, Rudarakanchana N, Atkinson C, et al. Functional interaction between BMPR-II and Tctex-1, a light chain of Dynein, is isoform-specific and disrupted by mutations underlying primary pulmonary hypertension. Hum Mol Genet 2003;12:3277–3286.
44. Sugai M, Saito M, Sukegawa I, et al. PTH/PTH-related protein receptor interacts directly with Tctex-1 through its COOH terminus. Biochem Biophys Res Commun 2003;311:24–31.
45. Yano H, Lee FS, Kong H, et al. Association of Trk neurotrophin receptors with components of the cytoplasmic dynein motor. J Neurosci 2001;21:1–7.
46. Mueller S, Cao XM, Welker R, Wimmer E. Interaction of the poliovirus receptor CD155 with the dynein light chain Tctex-1 and its implication for poliovirus pathogenesis. J Biol Chem 2002;277:7897–7904.
47. Douglas MW, Diefenbach RJ, Homa FL, et al. Herpes simplex virus type 1 capsid protein VP26 interacts with dynein light chains RP3 and Tctex1 and plays a role in retrograde cellular transport. J Biol Chem 2004;279:28,522–28,530.
48. Ilangovan U, Ding W, Zhong Y, et al. Structure and dynamics of the homodimeric dynein Light Chain km23. J Mol Biol 2005;352:338–354.
49. Jin Q, Gao G, Mulder KM. Requirement of a new dynein light chain in Smad3-dependent TGF β signaling. Submitted 2006.
50. Bowman AB, Patel-King RS, Benashski SE, McCaffery JM, Goldstein LS, King SM. Drosophila roadblock and Chlamydomonas LC7: a conserved family of dynein-associated proteins involved in axonal transport, flagellar motility, and mitosis. J Cell Biol 1999;146:165–180.
51. Koonin EV, Aravind L. Dynein light chains of the Roadblock/LC7 group belong to an ancient protein superfamily implicated in NTPase regulation. Curr Biol 2000;10:R774–R776.
52. Lunin VV, Munger C, Wagner J, Ye Z, Cygler M, Sacher M. The structure of the MAPK scaffold, MP1, bound to its partner, p14. A complex with a critical role in endosomal MAP kinase signaling. J Biol Chem 2004;279:23,422–23,430.
53. Kurzbauer R, Teis D, de Araujo ME, et al. Crystal structure of the p14/MP1 scaffolding complex: how a twin couple attaches mitogen-activated protein kinase signaling to late endosomes. Proc Natl Acad Sci USA 2004;101:10,984–10,989.
54. Nikulina K, Patel-King RS, Takebe S, Pfister KK, King SM. The Roadblock light chains are ubiquitous components of cytoplasmic dynein that form homo- and heterodimers. Cell Motil Cytoskeleton 2004; 57:233–245.
55. Jin Q, Ding W, Staub CM, Gao G, Tang Q, Mulder KM. Requirement of km23 for TGF β-mediated growth inhibition and induction of fibronectin expression. Cell Signal 2005;17:1363–1372.
56. Yue J, Mulder KM. Requirement of Ras/MAPK pathway activation by transforming growth factor β for transforming growth factor β1 production in a Smad-dependent pathway. J Biol Chem 2000;275: 30,765–30,773.

57. Derynck R, Akhurst RJ, Balmain A. TGF-β signaling in tumor suppression and cancer progression. Nat Genet 2001;29:117–129.
58. Attisano L, Wrana JL. Mads and Smads in TGF β signalling. Curr Opin in Cell Biol 1998;10:188–194.
59. Massagué J. TGF-β signal transduction. Annu Rev Biochem 1998;67:753–791.
60. Burkhardt JK, Echeverri CJ, Nilsson T, Vallee RB. Overexpression of the dynamitin (p50) subunit of the dynactin complex disrupts dynein-dependent maintenance of membrane organelle distribution. J Cell Biol 1997;139:469–484.
61. Lynch MA, Nakashima R, Song H, et al. Mutation analysis of the transformation growth factor β receptor type II gene in human ovarian carcinoma. Cancer Res 1998;58:4227–4232.
62. Chen T, Triplett J, Dehner B, et al. Transforming growth factor-β receptor type I gene is frequently mutated in ovarian carcinomas. Cancer Res 2001;61:4679–4682.
63. Ding W, Tang Q, Espina V, Liotta LA, Mauger DT, Mulder KM. A transforming growth factor-β receptor-interacting protein frequently mutated in human ovarian cancer. Cancer Res 2005;65:6526–6533.
64. Paige AJW, Taylor KJ, Stewart A, et al. A 700 kb physical map of a region of 16q23.2 homozygously deleted in multiple cancers and spanning the common fragile site FRA16D. Cancer Res 2000;60:1690–1696.
65. Balsara BR, Pei J, De Rienzo A, et al. Human hepatocellular carcinoma is characterized by a highly consistent pattern of genomic imbalances, including frequent loss of 16q23.1-24.1. Genes Chromosomes Cancer 2001;30:245–253.
66. Yakicier MC, Legoix P, Vaury C, et al. Identification of homozygous deletions at chromosome 16q23 in aflatoxin B1 exposed hepatocellular carcinoma. Oncogene 2001;20:5232–5238.
67. Boyd FT, Massagué J. Transforming growth factor-β inhibition of epithelial cell proliferation linked to the expression of a 53-kDa membrane receptor. J Biol Chem 1989;264:2272–2278.
68. Laiho M, Weis FM, Boyd FT, Ignotz RA, Massagué J. Responsiveness to transforming growth factor-β (TGF-β) restored by genetic complementation between cells defective in TGF-β receptors I and II. J Biol Chem 1991;266:9108–9112.
69. Laiho M, Weis MB, Massagué J. Concomitant loss of transforming growth factor (TGF)-β receptor types I and II in TGF-β-resistant cell mutants implicates both receptor types in signal transduction. J Biol Chem 1990;265:18,518–18,524.
70. Remy I, Montmarquette A, Michnick SW. PKB/Akt modulates TGF-β signalling through a direct interaction with Smad3. Nat Cell Biol 2004;6:358–365.
71. Jin Q, Ding W, Mulder KM. Requirement for DYNLRB1/km23 in a TGF β-dependent Smad2 signaling pathway. J Biol Chem In revision. (2005).
72. von Gersdorff G, Susztak K, Rezvani F, Bitzer M, Liang D, Bottinger EP. Smad3 and Smad4 mediate transcriptional activation of the human Smad7 promoter by transforming growth factor β. J Biol Chem 2000;275:11,320–11,326.
73. Dennler S, Itoh S, Vivien D, ten Dijke P, Huet S, Gauthier JM. Direct binding of Smad3 and Smad4 to critical TGFβ-inducible elements in the promoter of human plasminogen activator inhibitor-type 1 gene. EMBO J 1998;17:3091–3100.
74. Anders RA, Arline SL, Dore JJE, Jr., Leof EB. Distinct endocytic responses of heteromeric and homomeric transforming growth factor β receptors. Mol. Biol. Cell 1997;11:2133–2143.
75. Anders RA, Dore JJE, Jr., Arline SL, Garamszegi N, Leof EB. Differential requirement for type I and type II transforming growth factor β receptor kinase activity in ligand-mediated receptor endocytosis. J Biol Chem 1998;273:23,118–23,125.
76. Di Guglielmo GM, Le Roy C, Goodfellow AF, Wrana JL. Distinct endocytic pathways regulate TGF-β receptor signalling and turnover. Nat Cell Biol 2003;5:382–384.
77. Hayes S, Chawla A, Corvera S. TGF β receptor internalization into EEA1-enriched early endosomes: role in signaling to Smad2. J Cell Biol 2002;158:1239–1249.
78. Runyan CE, Schnaper HW, Poncelet AC. The role of internalization in transforming growth factor β1-induced Smad2 association with Smad anchor for receptor activation (SARA) and Smad2-dependent signaling in human mesangial cells. J Biol Chem 2005;280:8300–8308.
79. Felici A, Wurthner JU, Parks WT, et al. TLP, a novel modulator of TGF-β signaling, has opposite effects on Smad2- and Smad3-dependent signaling. EMBO J 2003;22:4465–4477.
80. Goltz JS, Wolkoff AW, Novikoff PM, Stockert RJ, Satir P. A role for microtubules in sorting endocytic vesicles in rat hepatocytes. Proc Natl Acad Sci USA 1992;89:7026–7030.
81. Oda H, Stockert RJ, Collins C, et al. Interaction of the microtubule cytoskeleton with endocytic vesicles and cytoplasmic dynein in cultured rat hepatocytes. J Biol Chem 1995;270:15,242–15,249.

12 TGF-β Signaling in Fibroblastic Cells and Oncogenesis

Neil A. Bhowmick

CONTENTS

INTRODUCTION
TGF-β SIGNALING AND THE TISSUE STROMA
TGF-β RESPONSIVITY OF STROMAL FIBROBLASTS CAN REGULATE CANCER INITIATION
TGF-β RESPONSIVITY OF STROMAL FIBROBLASTS CAN REGULATE CANCER PROGRESSION
TGF-β RESPONSIVITY OF THE HOST CAN REGULATE CANCER METASTASIS
CONCLUDING REMARKS
REFERENCES

Abstract

The growth factor and cytokine, transforming growth factor-beta (TGF-β), is known for both its tumor suppressive and inductive properties. Its understanding has recently been expanded from epithelial to stromal fibroblastic cells. The observation that the loss of TGF-β responsivity in the stromal fibroblastic cells can initiate neoplastic progression in adjacent epithelial cells from the prostate and the forestomach indicated that TGF-β has a tumor suppressive role not only in responsivity of epithelia but through paracrine signals from the stroma. Interestingly, the transgenic mouse model with the loss of TGF-β receptor type II in the stromal fibroblasts had little overt neoplastic effect on the other tissues. In particular, the stromal cells of the mammary gland expressed some of the same oncogenic growth factors as that of the prostate and the forestomach yet only ductal development was altered. However, when the same mammary stromal cells devoid of TGF-β responsivity was recombined with mammary epithelial cells expressing the polyoma middle T antigen, the mammary tumors that formed had greater infiltration in the circulation. These reports suggest stromal TGF-β responsivity is involved in the initiation, progression and metastasis of cancer in a tissue specific manner.

Key Words: Stromal cells; mesenchyme; TGF-β; forestomach; prostate; tumorigenesis.

1. INTRODUCTION

The transforming growth factor-beta (TGF-β) family of TGF-β1, TGF-β2, and TGF-β3 are multifunctional cytokines responsible for context-specific inhibition or stimulation

From: *Cancer Drug Discovery and Development: Transforming Growth Factor-β in Cancer Therapy, Vol I: Basic and Clinical Biology*
Edited by: Sonia B. Jakowlew © Humana Press Inc., Totowa, NJ

of cell proliferation, modulation of cell differentiation, control of extracellular matrix synthesis and degradation, control of mesenchymal-epithelial interactions, control of carcinogenesis, and mediation of wound healing *(1)*. TGF-β is found highly abundant in solid tumors and even detected in circulation of cancer patients *(2–6)*. In particular, the role of TGF-β on cell proliferation has been a point of focus in cancer research. TGF-β was originally purified through its observed capacity to transform fibroblastic cells in culture as determined by anchorage-independent growth in soft agar *(3)*. Given that TGF-β and its cognate receptors are expressed by nearly all cells, transformed and normal, and could be found in all tissues and platelets, it was evident that the molecule must have physiologic functions distinct from its transforming activity. The paradigm that has emerged indicate TGF-β could have a growth inhibitory or tumor suppressor role in cells prior to transformation, however, following cancer progression TGF-β could support further tumor growth and even metastasis *(7)*. However, this observation is based on the role of TGF-β on epithelial or tumor cells, not the tumor microenvironment or the host as a whole. Globally, the various roles of TGF-β can be attributed to the multiple parallel downstream signaling pathways activated by the ligand-receptor complex on the cell surface and the crosstalk with other signals unique to a particular cell type or environment.

It is well established that TGF-β signaling plays a significant role in epithelial-mesenchymal interactions in embryonic development *(8)*. In development the mesenchyme is not just a passive substratum for the epithelium, but takes an active role by generating signals necessary for organ development *(9)*. However, the molecular basis for TGF-β effects in mature tissues and tumors have overwhelmingly provided by studies on epithelial cells in culture. While these studies have indeed been informative for understanding cytoplasmic and nuclear signal transduction of TGF-β, much more needs to be done to understand its role in complex tissues within a host. Transgenic mice overexpressing activated-TGF-β1 in the mammary gland showed resistance to tumorigenesis by TGF-α overexpression or 7,12-dimethylbenz (α) anthracene (DMBA) treatment *(10)*. These studies helped support the concept of the tumor suppressive properties of TGF-β in nontransformed tissues. The subsequent tumor promoting capacity of TGF-β in vivo have been supported by a number of studies, mainly with preestablished tumors in mice in combination with the neutralization of TGF-β by various means including soluble TGF-β receptor fragments, monoclonal antibodies, and more recently specific TGF-β receptor kinase inhibitors *(11–15)*. These latter results have instigated the development of TGF-β neutralizing therapeutics for cancer treatment currently in early clinical trials. However, in many of these studies supporting both the tumor suppressive and tumor promoting roles of TGF-β, little has been done to establish the role of the stroma in the tumor response. Early on, the ability of TGF-β to stimulate a stromal response that included angiogenesis and formation of extracellular matrix was observed with the use of wire mesh chambers implanted into the backs of rats and injected with TGF-β *(16)*. These observations were further verified by injection of TGF-β under the neck skin of newborn mice to show accumulation of granulation tissue with chemotactic properties for inflammatory cells and stimulation of extracellular matrix production by fibroblasts *(17,18)*. Thus both cancer cells and its host are targets for TGF-β signaling. The potent immunoregulatory properties of TGF-β, particularly the dampening of T-cell activation and cytotoxicity program *(19)*, are discussed briefly in this review. The remainder of this work will be devoted to possible mechanisms TGF-β affects the stromal environment of neoplastic tissues with primary focus on a major cellular component, the cells derived from the mesenchymal fibroblast lineage (excluding adipocytes and bone). Generally speaking the role of TGF-β responsivity in smooth muscle, myofibroblasts, and fibroblastic cells will be discussed, depending on the tissue and the state of neoplastic conversion.

2. TGF-β SIGNALING AND THE TISSUE STROMA

Studies in the late 1980's and early 1990's also identified roles for each of the TGF-β isoforms in embryogenesis. Both in situ hybridization and use of isoform-specific antibodies demonstrated that the TGF-βs are localized in specific patterns *(20–22)*, both spatially and temporally during embryogenesis and suggested specific roles in processes such as cardiogenesis, palate formation, and osteogenesis *(23–27)*. Each of these has now been confirmed by isoform-specific gene targeting and has been studied in detail mechanistically *(28–30)*.

TGF-βs are secreted by both stromal and epithelial compartments as inactive precursors, which have to be proteolytically cleaved in order to bind to their receptors *(31)*. It exerts its effects through binding to the TGF-β type II receptor (TβRII) and subsequent recruitment of the type I receptor (TβRI) for downstream cytoplasmic signaling through multiple parallel signaling pathways *(32)*. The activated TGF-β ligand-receptor complex conveys the signal to the nucleus through the phosphorylation and activation of the SMAD signaling pathway *(33)* as well as other parallel downstream signaling pathways involving RhoA *(34)*, stress kinases (i.e., JNK and p38MAPK) *(35–37)*, phosphatase2A *(38)*, and PI3-kinase/AKT *(39)*. The established paradigm of TβRI phosphorylation of Smad2 and Smad3 in conjunction with Smad4 recruitment is important in TGF-β-mediated G1 cell cycle arrest *(40)*. The SMAD signaling pathway up-regulates the expression of cyclin-dependent kinase (cdk) inhibitors, p15INK4A, p16INK4B, p27Kip1, and p21Cip1, thereby suppressing cyclin D and/or cyclin E-associated hyperphosphorylation of Rb and S-phase cell cycle progression (reviewed in *[41]*). SMAD signaling is also associated with the transcriptional and post-translational downregulation of the mitogenic proto-oncogene, c-*myc (42–44)*. Smad7, also stimulated by TGF-β, can block Smad2/3 interaction with TβRI and has been shown to promote apoptosis in hepatocytes and prostate cells. The RhoA, PI3-kinase/AKT, and p38MAPK signaling downstream of the TGF-β ligand-receptor complex can mediate epithelial to mesenchymal transdifferentiation. Approximately 36% of the stromal fibroblasts in a fibrotic kidney have been attributed to such renal epithelial transdifferentiation *(45)*. The contribution of epithelial transdifferntiation of tumor cells to the stroma or its relevance to metastatic potential is not clear. However, TGF-β stimulation of RhoA and downstream p160ROCK activation is not only involved in changes in cell morphology and motility but also cooperate with SMAD signaling to inhibit epithelial cell proliferation *(34,46)*. Figure 1A summarizes the TGF-β downstream signals in a simplified cartoon. Although much is known about TGF-β signaling in epithelial cells, the role of stromal-TGF-β signaling on epithelial development has had little attention.

The signaling pathways depicted in Figure 1A are found to differ in cancer cells through mutations or epigenetic mechanisms. Known mutations affecting the TGF-β signaling pathway in human cancers *(47)* include: the inactivation of TβRII through microsatellite instability in gastric and colon cancer *(48–50)*, the inactivation of TβRI in lymphomas and breast cancers *(51,52)*, Smad4 loss in colon, gastric, and pancreatic cancers *(53–56)*, Smad2 inactivation in colon cancer, leukemia, and lymphoma *(57–59)*, Evi-1 overexpressed during blast crisis of chronic myelogenous leukemia (CML) that results in an inhibition of Smad3 *(60–62)*. Other downstream effectors that are impacted by other signaling pathways including the TGF-β pathway, myc, Ras, Rb, and the family of cyclin dependent kinase inhibitors (CDKIs). However, nontransformed epithelia and fibroblasts in culture can also have differing downstream transduction of the same TGF-β ligand-receptor complex. One such example was observed with mouse embryonic fibroblast, NIH3T3, and a nontransformed mouse mammary cell line, NMuMG, where TGF-β treatment resulted in Smad2 activation by both NMuMG and NIH3T3 cells, yet RhoA activation was found by NMuMG cells, not NIH3T3 cells (Fig. 1B) *(46)*. While by no means can NIH3T3 cells or NMuMG be representative

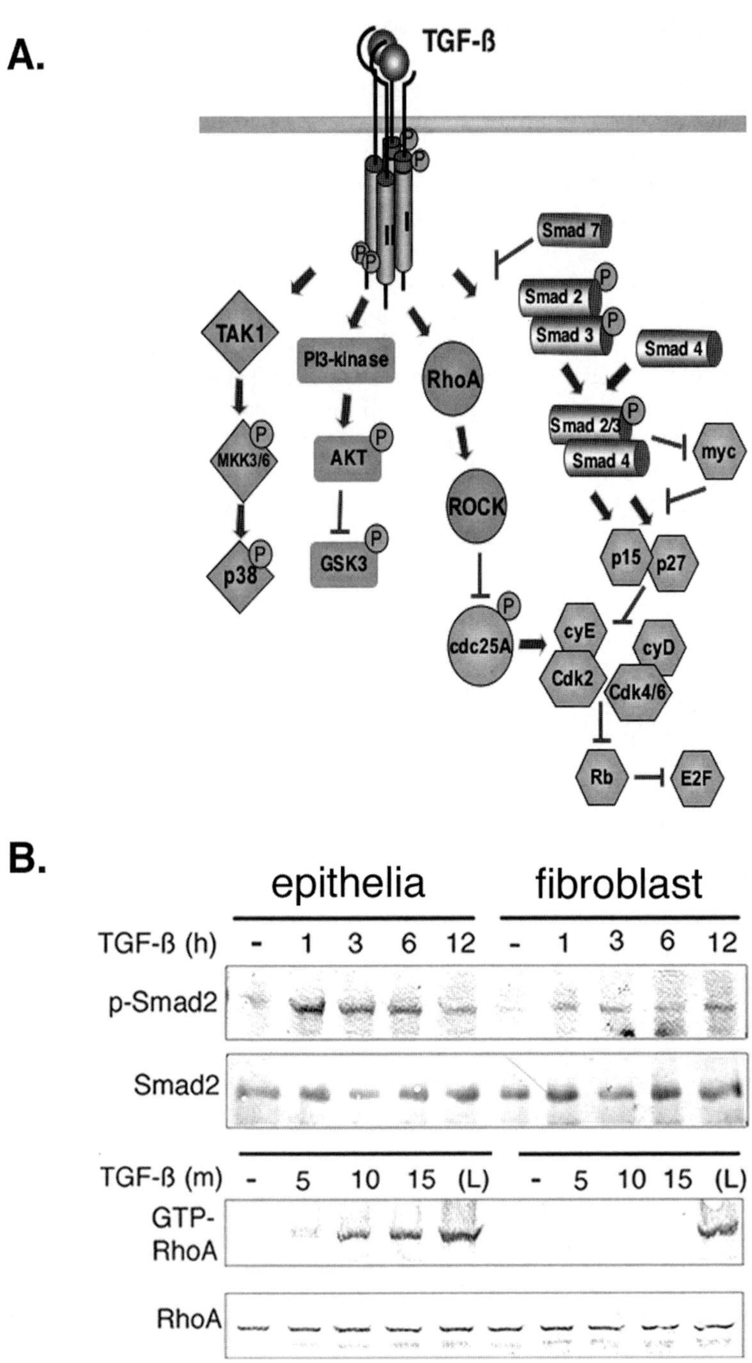

Fig. 1. (A) The activated TGF-β-receptor complex simulates the p38MAPK, PI3kinas/AKT, RhoA, and SMAD signaling pathways. In epithelial cells RhoA activation cooperates with the p38MAPK and PI3kinase pathways to initiate cytoskeletal changes as well as interacts with the Smad signaling pathway to maintain Rb in a hypo-phosphorylated state for G1 arrest of the cell cycle. **(B)** However differential TGF-β-mediated RhoA responsiveness is observed between NIH3T3 fibroblasts and NMuMG mammary epithelial cells. The two cell lines were incubated with TGF-β for the indicated times, lysed, and analyzed for phosphorylated and total Smad2 by Western blotting. RhoA activation

of a generalized phenomena in fibroblasts or epithelia in vivo, it suggests further differences between the two cells types and their differing responses to TGF-β.

TGF-β contributes to the stromal framework through stimulating the synthesis of extra cellular matrix (ECM) components, including fibronectin, certain collagens, chondroitin/dermatan sulfate, biglycan, decorin, osteopontin, osteonectin, tenascin, and thrombospondin (reviewed in *[63]*). In addition, TGF-β stimulates the synthesis of inhibitors of matrix degrading enzymes, including plasminogen activator inhibitor-1 (PAI-1) and tissue inhibitor of matrix metalloproteinase (TIMP) *(64,65)*. Conversely, TGF-β was shown to inhibit expression of matrix degrading proteases such as collagenase, stromelysin, and plasminogen activator *(64,66,67)*. Thus, the increase in inhibitors and decrease in matrix proteases causes increased matrix accumulation. TGF-β treatment induced increase in integrin expression resulting in modulation of cell binding to matrix components *(68–70)*. Other studies have suggested that a cooperative requirement of integrin β1 for TGF-β-mediated p38MAPK activation in mammary epithelial cells. In these studies, the neutralization of integrin β1 binding in NMuMG cells was accompanied by the loss TGF-β-mediated p38MAPK activation *(71)*. A transgenic mouse model with the concomitant conditional knockout of integrin-β1 and expression of MMTV-polyoma-middle T in mammary epithelial cells suppressed mammary tumorigenesis where in the control MMTV-polyoma middle T expressing mice developed tumors *(72)*. Together these studies not only suggest a role for TGF-β in the production of ECM components, but possibly a mechanism by which the epithelial response can differ in response to the stromally-derived ECM. In view of the unsuccessful therapeutic attempts to suppress ECM degradation through matrix metaloproteinase inhibition as a method for stopping metastatic progression of cancer, techniques to alter tumor responsivity to the cellular stromal environment may prove to be more beneficial.

3. TGF-β RESPONSIVITY OF STROMAL FIBROBLASTS CAN REGULATE CANCER INITIATION

It has long been observed that solid tumors are associated with stromal cells that differ phenotypically from that of the normal tissue architecture. Even precancerous lesions in the human prostate gland are associated with phenotypic changes in the immediate stroma *(42)*. These stromal changes are typically characterized by the stimulation of matrix, increased microvessel density, and general desmoplastic appearance of the stromal cells. Conversely, various studies have examined the role of the stroma in promoting tumorigenesis using xenografted carcinoma models. Both the tumor incidence and latency of subcutaneously injecting transformed mammary epithelium is nearly half of that when the same transformed mammary epithelial cells were coinjected with normal human mammary fibroblasts *(73)*. Similarly, another study showed that xenografting nontumorigenic prostatic epithelium, immortalized by the expression of the large T-antigen, with carcinoma-associated fibroblasts resulted in increased tumor incidence and growth compared to those combined with normal prostatic stromal cells *(74)*. Under conditions of suboptimal vasculature availability (e.g., subcutaneous injections), tumorigenesis can be promoted by the addition of a matrix, such as matrigel in combination with the stromal and epithelial cells *(75)*. Thus, the uniqueness of the tumor-associated stroma has multiple roles in supporting the tumor through

was measured by adsorbing cell lysates to GST-RBD (Rhotekin Rho binding domain) beads *(34,106)*. GTP-bound and total RhoA expression was determined by Western blotting. As a positive control, cells were treated for 5 min with lysophosphatidic acid (L). Despite similar TGF-β stimulation of Smad2 activation, RhoA activation is not observed in the NIH3T3 cells.

providing a source of growth factors, blood supply, and ECM framework. TGF-β plays a role in each of these components critical to the initiation of cancer *(76,77)*.

Epithelial, endothelial, inflammatory, and fibroblastic cells in a normal tissue or tumor often express and have the capacity to respond to the TGF-β ligands through autocrine and paracrine mechanisms. A method of determining the specific responsiveness of a cellular component for TGF-β without affecting ligand availability is through conditionally knocking out the TβRII. A mechanism for ablating TβRII expression in fibroblasts is by employing the Cre-lox methodology *(78)*. To accomplish complete loss of TGF-β signaling in the stroma, transgenic mouse lines were generated having *loxP* sites in introns 1 and 2 (*Tgfbr2*$^{floxE2/floxE2}$ *[79]*). To conditionally knockout TβRII in the fibroblastic compartment, the Flox mouse line was crossed with a mouse having the *Cre* transgene under the control of the FSP1 (fibroblast specific protein/S100A4) promoter to ablate TGF-β function in these cells. FSP1 is expressed in fibroblasts in adult mice *(80,81)* and is not expressed during embryologic development until after E8.5. At E12.5 and afterwards, FSP1 expression is detected in cells of mesenchymal origin and of a fibroblastic phenotype *(82)*. Mice with the green fluorescent protein (GFP) gene driven by the FSP1 promoter suggest its expression in various tissues including the skin, pancreas, prostate, mammary gland, and kidney *(83)*. As a result of the widespread stromal recombination, the Tgfbr2^{fspko} mice have a striking set of phenotypic changes, including stromal hypercellularity in the forestomach, prostate, bladder, and early mortality. Intriguingly, tissues reported to be developmentally regulated by TGF-β such, as the kidney, lung, and skin were not adversely affected by the loss of stromal TGF-β signaling *(83)*. However, the Tgfbr2^{fspko} mice consistently die at approximately 8 wk of age, by a mechanism that is not completely clear. The ability to inhibit TGF-β signaling in a specific tissue compartment has allowed for the better understanding of TGF-β signaling in vivo in many tissues.

The composition of the stroma was altered in the prostate and forestomach of the Tgfbr2^{fspKO} mice prior to observable changes in the epithelial compartment. The sequence of phenotypic events in the two tissues was similar with respect to initial stromal expansion and increased desmoplastic appearance. This was followed by the hyperplasia of the adjacent epithelial cells. Subsequently, the Tgfbr2^{fspko} mice developed preneoplastic lesions in the prostate (prostatic intraepithelial neoplasia [PIN]) and squamous cell carcinoma in the forestomachs by 5 wk of age (Fig. 2) *(83)*. This mouse model has helped establish the importance of TGF-β responsivity of the stroma in tumor initiation. This has also become the basis for a wide-reaching hypothesis on the role of stroma in the initiation and progression of cancer in the absence of preexisting genetic lesions in the epithelia.

The understanding of stromally mediated initiation of tumorigenesis per say, is in its infancy compared to cancers resulting from direct genetic or epigenetic changes in the epithelia. However, the concept of inflammation as an initiating factor in carcinogenesis has been a focus of cancer preventative research for some time; examples: (1) chronic pancreatitis is the leading cause of pancreatic cancer *(84,85)*, (2) ulcerative colitis can lead to colon cancer *(86)*, and (3) Barrett's esophagus can predispose individuals to esophageal cancer *(87)*. These observations have lead to the clinical examination of antiinflammatory drugs for cancer prevention and amelioration. Most visibly, the class of cox-2 inhibitors have been used in various clinical trials. The inflammation phenotype involves the stromal compartments long before the epithelia are affected. The epithelia of specific tissues develop mutations as a result of chronic exposure to various chemical carcinogens (e.g., liver cirrhosis from alcohol abuse leading to hepatocellular carcinoma *[88,89]*). However, the inflammation of the tissue is evidence for stromal involvement prior to the carcinogenic process. Even in the absence of inflammatory immune cells, the mesenchymal stromal cells acquire the phenotype analogous to that found in tumor-associates stromal cells that include the elevated expression of TGF-β as well as angiogenic and growth promoting factors.

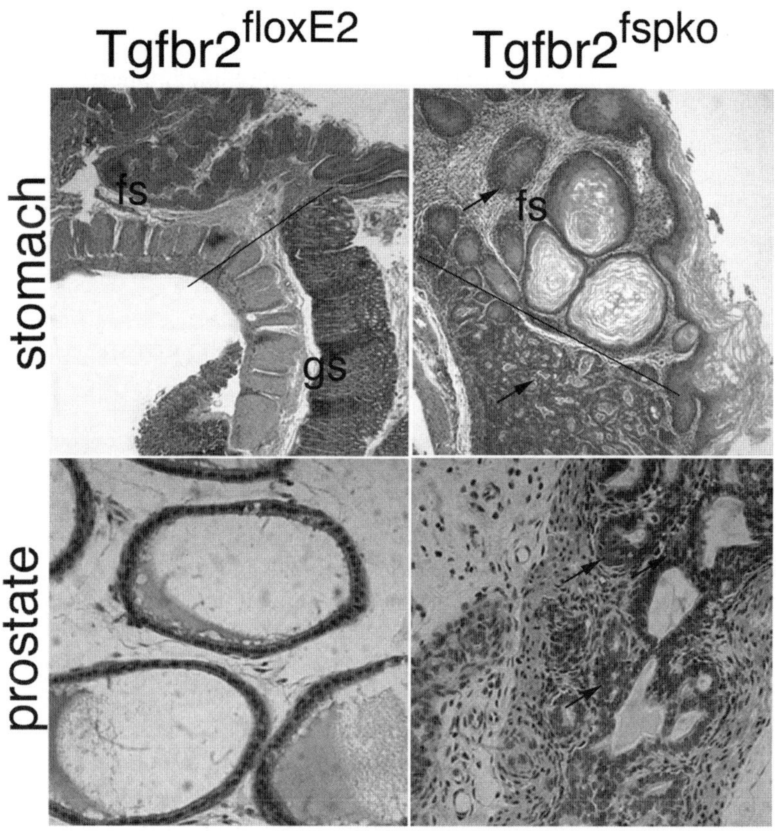

Fig. 2. The conditional deletion of Tgfbr2 gene expression in fibroblastic cells results in neoplastic conversion of epithelial cells of the forestomach and prostate tissues in 6-wk old mice *(84)*. The stomachs of Tgfbr2$^{floxE2/floxE2}$ (control) and Tgfbr2^{fspKO} were compared by hematoxylin and eosin staining of tissue sections. The upper portion of the stomach panels show forestomach (fs) and the lower portion of the panel is the glandular stomach (gs). The normal forestomach is characterized by stratified squamous epithelia with sparse fibroblast in the submucosa as seen in Tgfbr2$^{floxE2/floxE2}$ mice. The Tgfbr2^{fspKO} mice develop squamous cell carcinoma (arrow) with abundant stromal cells and hyperkeratinization invading into the glandular stomach space. The lower panel shows dorso-lateral prostate glands of Tgfbr2$^{floxE2/floxE2}$ and Tgfbr2^{fspKO} mice. Hematoxylin and eosin stained sections show histology of a single layer of epithelia with a minor stromal component found in the Tgfbr2$^{floxE2/floxE2}$ prostate. In contrast, the Tgfbr2^{fspKO} develop prostatic intraepithelial neoplasia (arrow), where hyperplastic epithelia with nuclear atypia accompany abundant stroma.

There is evidence to suggest the effects of the ablation of TGF-β responsivity in stromal fibroblasts is owing to differences in stromal secreted factors. The first is the observation that only approximately 30–50% of the prostatic and forestomach stromal cells show deficiency in TGF-β signaling in the Tgfbr2^{fspko} mice *(83)*. Further, unpublished data in our laboratory show that conditioned media from prostatic stromal cells can stimulate proliferation of primary prostatic epithelial cells. Thus, ECM expression differences are unlikely cause of the oncogenesis. The secreted factors found in the Tgfbr2^{fspko} mouse prostates and forestomach stroma include hepatocyte growth factor (HGF), transforming growth factor-α (TGF-α), and macrophage-stimulating protein (MSP) go on to stimulate the cognate receptors c-*Met*, Egfr1, and Ron, respectively in the adjacent epithelia *(83,90)*. More interestingly, many of the tissues that did not exhibit neoplastic conversion like the esophagus, stomach body, and mammary gland in the Tgfbr2^{fspko} mice also showed overexpression of these same oncogenic factors *(90,91)*. These results can have multiple implications including the presence of other tissue-specific oncogenic factors not yet identified as well as the potential of different tolerances for oncogenic signals by different tissues. Presumably, tissues like the skin, bladder, and stomach must have higher tolerances for toxins in the environment than other tissues. So, both the explanations are feasible. However, the tissue-specific differences in the stromal cells themselves also need to be considered in their ability to support and even initiate the neoplastic conversion of the adjacent epithelia.

4. TGF-β RESPONSIVITY OF STROMAL FIBROBLASTS CAN REGULATE CANCER PROGRESSION

It is established that the stroma has a supportive role in solid tumors. As such it is specifically termed a *reactive stroma*. The specialized terminology for the carcinoma-associated stroma is owing to its altered phenotype and gene expression, versus stromal cells associated with the normal tissue. The analogies of the cancer-associated stroma and that found in wound repair or other inflammatory responses are based on various similarities including extracellular matrix deposition, stromal cell phenotype, and pro-angiogenic properties *(92,93)*. These stromal changes are a direct result of the increased level of growth factor and cytokine expression, such as TGF-β, vascular endothelial growth factor (VEGF), HGF, and TGF-α found in many tumor types characterized to promote tumor growth. Because, TGF-β is abundant in both inflammation and cancer, often it is described to instigate these functions of the reactive stroma *(94)*. While these are documented stromal changes that occur in preneoplastic and tumor tissues, in apparent contradiction, the loss of TGF-β responsivity in the Tgfbr2^{fspko} mouse model resulted in these changes *prior* to observable epithelial changes *(77,83)*. The fact that the loss of TGF-β responsivity of the stromal fibroblast-derived cells results in the tumorigenesis, suggest that under normal conditions these stromal cells can provide a generalized tumor suppressive function. TGF-β downstream signals are reported to inhibit the expression of HGF *(95)*. Thus TGF-β can have tumor suppressive role via the specific stromal cells even during the stages of tumor initiation (as described above) as well as under further progression of the tumor. The tumor promoting activities of TGF-β can be a result of its responsivity in epithelial, endothelial, and immune cells. An apparent synergistic cooperation of the TGF-β and TNF-α signaling at the level of p38MAPK activation is found in colonic epithelial cells *(96)*. This suggests a further required qualification of the dual role paradigm of TGF-β to include its effect on the individual host components and the epithelia.

As opposed to the reactive stroma the previous section discussed a mechanism by which the epithelia can have a reactive characteristic. Not surprisingly, both cellular components respond to each other. More specifically, many ligands that are primarily expressed by the stromal cells have their cognate receptors in epithelial cells. While this is true for HGF ligand and c-*Met* receptor in normal tissues, glioblastoma, esophageal squamous cell carcinomas, and gastric cancers have reports of both HGF and c-*Met* expressed by the epithelial tumor tissue *(97–99)*. Interestingly, VEGF normally act in a paracrine fashion to stimulate neovasculature, but coexpression of VEGF and its cognate receptor (Flt-1) are frequently found in lymphomas and myelomas *(100)* as well as breast cancer stimulating proliferation *(101,102)*. Such a switch from paracrine to autocrine signaling can conceivably make the tumor cells independent of its stromal requirements. Thus, the tumor can be refractive to tumor suppressor effects of the stroma and potentially capable of further metastasis progression.

5. TGF-β RESPONSIVITY OF THE HOST CAN REGULATE CANCER METASTASIS

In the paradigm of the dual tumor suppressor and tumor promoter roles of TGF-β, one of the primary concerns is the prometastasis role of TGF-β. This is supported by in vitro signaling data suggesting the TGF-β-mediated activation of epithelial to mesenchymal transdifferentiation through multiple downstream pathways associated with increased motility. There are also multiple supporting studies involving mouse models especially examining the metastatic progression of breast cancer. The overexpression of activated TGF-β type I receptor and c-*Neu* in the mammary epithelia by the mouse mammary tumor virus (MMTV) promoter resulted in elevated metastasis, compared to that observed in the cancer initiating MMTV-c-*Neu* mice alone. The overexpression of active-TGFβ1 in the mammary gland in the same MMTV-c-*Neu* mouse background also showed elevated metastatic potential compared to the MMTV-c-*Neu* mice *(103)*. These observations have resulted in the development of pharmacologic inhibitors for TGF-β signaling at the level of the ligand and kinase inhibition of the TβRI. Pre-clinical studies in mice that normally develop lung metastasis through the expression of the polyoma middle T antigen (PyMT) by the MMTV promoter, showed appreciable inhibition of metastasis *(12)*. But these results are confounded by two other studies utilizing the Tgfbr2$^{floxE/floxE2}$ mouse model to ablate TGF-β signaling specifically in the mammary epithelia (by MMTV-Cre) and the stromal fibroblasts (by FSP-Cre) *(83,90)*. Interestingly, the ablation of TGF-β signaling in the fibroblasts with PyMT expression in the mammary epithelia resulted in larger tumors and a significant number of mammary cancer cells in the circulation compared to PyMT expression in the mammary epithelia combined with normal stromal cells *(90)*. No elevation in metastatic lesions to the lung was found with TβRII loss in the stroma *(90)*. However, the ablation of TGF-β signaling in the mammary epithelia resulted in a significant elevation of lung metastasis number and volume over the MMTV-PyMT mice *(104)*. Moreover, the cells that metastasized to the lung showed no activation of Smad2, as evidence that TGF-β signaling in the tumor cell is not necessary for metastasis *(104)*. These results overwhelmingly suggest the use of caution for pharmacologic inhibitors for TGF-β in the clinic. However, the question remained how the pharmacologic inhibitors suppressed mammary cancer metastasis. Because, the conditional knockout of Tgfbr2 in the epithelia and the stromal fibroblasts, both resulted in increased extravasation of the tumor cells that only left systemic host suppression of TGF-β signaling as a possibility *(91,104)*. Most recently work by Drs. Dori Thomas and Joan Massagué (Sloan-Kettering Cancer Center, New York) described how TGF-β inhibits cytotoxic T-cell activity at a systemic level through the inhibition of expression of cytolytic genes: perforin, granzyme A, granzyme B, Fas ligand, and interferonγ *(19)*. Activated cytotoxic T-cell have the capacity to home and kill cancer cells, but TGF-β can suppress at least the lytic capacity of these cells. Thus presumably the neutralization of systemic TGF-β restored tumor-specific cytotoxic T-cell activity. Together, this suggests the need for (cytotoxic T-cell) specific inhibition of TGF-β activity in future pharmaceuticals.

6. CONCLUDING REMARKS

Traditionally, mesenchymally derived stromal cells are characterized as fibroblasts, myofibroblasts, of smooth muscle cells depending of the tissue being described. These characterizations are based on phenotypic architecture and differential protein expression that serve as markers. For example, fibroblasts throughout an organism express vimentin and/or FSP-1 (S100A4). However, even tissues of close proximity as the esophagus and the forestomach (both lined by a continuous squamous epithelial layer) develop very differently in the Tgfbr2^{fspko} mouse model *(83)*. Despite the common loss of stromal TGF-β responsivity, only the forestomach develop squamous cell carcinoma in Tgfbr2^{fspko} mice *(83)*. This illustrates the potential of multiple types of stromal fibroblasts, comparable to the epithelial types within any complex tissue architecture. The differences in the stromal compartment in the various tissues are further complicated when the epithelia become neoplastic. There is a well-documented change in the adjacent stromal cells that can potentiate tumor progression and metastasis. Although, the phenotypic changes observed in various solid tumor stromal cells have been surprisingly similar, there is very likely to be tissue specific differences we are not yet aware. Much of our ignorance in the understanding of this compartment of the cancer tissue can be attributed to the deficiency of cell-specific tools, until recently. Now that the stromal compartment has been shown to have the potential to mediate cancer initiation, progression, and metastasis, it has emerged as a viable target for prognostic markers and therapeutic development.

The epithelial and stromal tissue compartments communicate through soluble factors and matrix components. Because, these make up a large portion of the expressed genome, it is difficult to know what is a good target specifically for any one cancer for any one individual. Drugs developed by rational design include those targeting toward MMP, EGF/TGFα, VEGF, TNFα, COX2, and TGF-β. While, the specific pharmaceuticals have had some level of success over more arbitrary chemotherapeutic drugs that kill proliferative cells, none are immune to unintended side effects or resistance development. This chapter only addresses the risk of therapy with TGF-β antagonists based on potential for prostate cancer initiation, acceleration of preneoplastic lesions, or induction of metastasis of cancers. However, it can be a lesson for all such therapeutics that affects the activity of factors that are important for healthy and neoplastic cells. Because, the stromal cells are not generally neoplastic, it is not thought to have the genetic instability common to the epithelial cancer cells. This trait may delay drug resistance development. With regard to the unintended side effects, this may always be problem without further research toward highly individualized molecular-targeted therapy. However, for more attainable goal, targeting specific downstream signaling proteins associated with tumorigenesis will likely yield fewer effects on other pathways. This is especially important for multifunctional cytokines like TGF-β.

REFERENCES

1. Blobe GC, Schiemann WP, Lodish HF. Role of transforming growth factor beta in human disease. N Engl J Med 2000;342:1350–1358.
2. Moses HL, Branum EL, Proper JA, et al. Transforming growth factor production by chemically transformed cells. Cancer Res 1981;41:2842–2848.
3. Childs CB, Proper JA, Tucker RF, et al. Serum contains a platelet-derived transforming growth factor. Proc Natl Acad Sci USA 1982;79:5312–5316.
4. Coffey RJ, Jr., Shipley GD, Moses HL. Production of transforming growth factors by human colon cancer lines. Cancer Res 1986;46:1164–1169.
5. Krasagakis K, Tholke D, Farthmann B, et al. Elevated plasma levels of transforming growth factor (TGF)-beta1 and TGF-beta2 in patients with disseminated malignant melanoma. Br J Cancer 1998; 77:1492–1494.
6. Huang A, Gilmour JW, Imami N, et al. Increased serum transforming growth factor-beta1 in human colorectal cancer correlates with reduced circulating dendritic cells and increased colonic Langerhans cell infiltration. Clin Exp Immunol 2003;134:270–278.
7. Roberts AB, Wakefield LM. The two faces of transforming growth factor beta in carcinogenesis. Proc Natl Acad Sci USA 2003;100:8621–8623.
8. Roberts AB, Flanders KC, Heine UI, et al. Transforming growth factor-beta: multifunctional regulator of differentiation and development. Philos Trans R Soc Lond B Biol Sci 1990;327:145–154.
9. Baskin LS, Hayward SW, Young P, et al. Role of mesenchymal-epithelial interactions in normal bladder development. J Urol 1996;156:1820–1827.

10. Pierce DF, Jr., Gorska AE, Chytil A, et al. Mammary tumor suppression by transforming growth factor beta 1 transgene expression. Proc Natl Acad Sci USA 1995;92:4254–4258.
11. Yang YA, Dukhanina O, Tang B, et al. Lifetime exposure to a soluble TGF-beta antagonist protects mice against metastasis without adverse side effects. J Clin Invest 2002;109:1607–1615.
12. Muraoka RS, Dumont N, Ritter CA, et al. Blockade of TGF-beta inhibits mammary tumor cell viability, migration, and metastases. J Clin Invest 2002;109:1551–1559.
13. Sawyer JS, Anderson BD, Beight DW, et al. Synthesis and activity of new aryl- and heteroaryl-substituted pyrazole inhibitors of the transforming growth factor-beta type I receptor kinase domain. J Med Chem 2003;46:3953–3956.
14. Laping NJ, Grygielko E, Mathur A, et al. Inhibition of transforming growth factor (TGF)-beta1-induced extracellular matrix with a novel inhibitor of the TGF-beta type I receptor kinase activity: SB-431542. Mol Pharmacol 2002;62:58–64.
15. Ge R, Rajeev V, Subramanian G, et al. Selective inhibitors of type I receptor kinase block cellular transforming growth factor-beta signaling. Biochem Pharmacol 2004;68:41–50.
16. Sporn MB, Roberts AB, Shull JH, et al. Polypeptide transforming growth factors isolated from bovine sources and used for wound healing in vivo. Science 1983;219:1329–1331.
17. Roberts AB, Sporn MB, Assoian RK, et al. Transforming growth factor type beta: rapid induction of fibrosis and angiogenesis in vivo and stimulation of collagen formation in vitro. Proc Natl Acad Sci USA 1986;83:4167–4171.
18. Roberts CJ, Birkenmeier TM, McQuillan JJ, et al. Transforming growth factor-beta stimulates the expression of fibronectin and of both subunits of the human fibronectin receptor by cultured human lung fibroblasts. J Biol Chem 1988;263:4586–4592.
19. Thomas DA, Massagué J. TGF-beta directly targets cytotoxic T cell functions during tumor evasion of immune surveillance. Cancer Cell 2005;8:369–380.
20. Pelton RW, Saxena B, Jones M, et al. Immunohistochemical localization of TGF beta 1, TGF beta 2, and TGF beta 3 in the mouse embryo: expression patterns suggest multiple roles during embryonic development. J Cell Biol 1991;115:1091–1105.
21. Pelton RW, Nomura S, Moses HL, et al. Expression of transforming growth factor beta 2 RNA during murine embryogenesis. Development 1989;106:759–767.
22. Pelton RW, Johnson MD, Perkett EA, et al. Expression of transforming growth factor-beta 1, -beta 2, and beta 3 mRNA and protein in the murine lung. Am J Respir Cell Mol Biol 1991;5:522–530.
23. Akhurst RJ, Lehnert SA, Faissner A, et al. TGF beta in murine morphogenetic processes: the early embryo and cardiogenesis. Development 1990;108:645–656.
24. Abbott BD, Birnbaum LS. Retinoic acid-induced alterations in the expression of growth factors in embryonic mouse palatal shelves. Teratology 1990;42:597–610.
25. Pelton RW, Hogan BL, Miller DA, et al. Differential expression of genes encoding TGFs beta 1, beta 2, and beta 3 during murine palate formation. Dev Biol 1990;141:456–460.
26. Heine U, Munoz EF, Flanders KC, et al. Role of transforming growth factor-beta in the development of the mouse embryo. J Cell Biol 1987;105:2861–2876.
27. Gerstenfeld LC, Cullinane DM, Barnes GL, et al. Fracture healing as a post-natal developmental process: molecular, spatial, and temporal aspects of its regulation. J Cell Biochem 2003;88:873–884.
28. Proetzel G, Pawlowski SA, Wiles MV, et al. Transforming growth factor-beta 3 is required for secondary palate fusion. Nat Genet 1995;11:409–414.
29. Kulkarni AB, Huh CG, Becker D, et al. Transforming growth factor beta 1 null mutation in mice causes excessive inflammatory response and early death. Proc Natl Acad Sci USA 1993;90:770–774.
30. Sanford LP, Ormsby I, Gittenberger-de Groot AC, et al. TGFbeta2 knockout mice have multiple developmental defects that are non-overlapping with other TGFbeta knockout phenotypes. Development 1997; 124:2659–2670.
31. Sporn MB, Roberts AB. Transforming growth factor-beta: recent progress and new challenges. J Cell Biol 1992;119:1017–1021.
32. Attisano L, Wrana JL. Signal transduction by the TGF-beta superfamily. Science 2002;296: 1646–1647.
33. Zhang Y, Feng X, We R, et al. Receptor-associated Mad homologues synergize as effectors of the TGF-beta response. Nature 1996;383:168–172.
34. Bhowmick NA, Ghiassi M, Bakin A, et al. Transforming growth factor-beta1 mediates epithelial to mesenchymal transdifferentiation through a RhoA-dependent mechanism. Mol Biol Cell 2001; 12:27–36.

35. Atfi A, Djelloul S, Chastre E, et al. Evidence for a role of Rho-like GTPases and stress-activated protein kinase/c-Jun N-terminal kinase (SAPK/JNK) in transforming growth factor beta-mediated signaling. J Biol Chem 1997;272:1429–1432.
36. Engel ME, McDonnell MA, Law BK, et al. Interdependent SMAD and JNK signaling in transforming growth factor- beta-mediated transcription. J Biol Chem 1999;274:37,413–37,420.
37. Wang W, Zhou G, Hu MC, et al. Activation of the hematopoietic progenitor kinase-1 (HPK1)-dependent, stress-activated c-Jun N-terminal kinase (JNK) pathway by transforming growth factor beta (TGF-beta)-activated kinase (TAK1), a kinase mediator of TGF beta signal transduction. J Biol Chem 1997;272:22,771–22,775.
38. Griswold-Prenner I, Kamibayashi C, Maruoka EM, et al. Physical and functional interactions between type I transforming growth factor beta receptors and Balpha, a WD-40 repeat subunit of phosphatase 2A. Mol Cell Biol 1998;18:6595–6604.
39. Krymskaya VP, Hoffman R, Eszterhas A, et al. TGF-beta 1 modulates EGF-stimulated phosphatidylinositol 3-kinase activity in human airway smooth muscle cells. Am J Physiol 1997;273: L1220–L1227.
40. Kretzschmar M, Massagué J. SMADs: mediators and regulators of TGF-beta signaling. Curr Opin Genet Dev 1998;8:103–111.
41. Massagué J, Blain SW, Lo RS. TGFbeta signaling in growth control, cancer, and heritable disorders. Cell 2000;103:295–309.
42. Alexandrow MG, Moses HL. Transforming growth factor beta and cell cycle regulation. Cancer Res 1995;55:1452–1457.
43. Feng XH, Liang YY, Liang M, et al. Direct interaction of c-Myc with Smad2 and Smad3 to inhibit TGF-beta-mediated induction of the CDK inhibitor p15(Ink4B). Mol Cell 2002;9:133–143.
44. Yagi K, Furuhashi M, Aoki H, et al. c-myc is a downstream target of the Smad pathway. J Biol Chem 2002;277:854–861.
45. Kalluri R, Neilson EG. Epithelial-mesenchymal transition and its implications for fibrosis. J Clin Invest 2003;112:1776–1784.
46. Bhowmick NA, Ghiassi M, Aakre M, et al. TGF-β-induced RhoA and p160ROCK activation is involved in the inhibition of Cdc25A with resultant cell-cycle arrest. Proc Natl Acad Sci USA 2003; 100:15,548–15,553.
47. Roberts AB. TGF-beta signaling from receptors to the nucleus. Microbes Infect 1999;1:1265–1273.
48. Myeroff LL, Parsons R, Kim SJ, et al. A transforming growth factor beta receptor type II gene mutation common in colon and gastric but rare in endometrial cancers with microsatellite instability. Cancer Res 1995;55:5545–5547.
49. Chung YJ, Song JM, Lee JY, et al. Microsatellite instability-associated mutations associate preferentially with the intestinal type of primary gastric carcinomas in a high-risk population. Cancer Res 1996;56:4662–4665.
50. Grady WM, Rajput A, Myeroff L, et al. Mutation of the type II transforming growth factor-beta receptor is coincident with the transformation of human colon adenomas to malignant carcinomas. Cancer Res 1998;58:3101–3104.
51. Pasche B, Kolachana P, Nafa K, et al. TbetaR-I(6A) is a candidate tumor susceptibility allele. Cancer Res 1999;59:5678–5682.
52. Woszczyk D, Gola J, Jurzak M, et al. Expression of TGF beta1 genes and their receptor types I, II, and III in low- and high-grade malignancy non-Hodgkin's lymphomas. Med Sci Monit 2004;10: CR33–CR37.
53. de Winter JP, Roelen BA, ten Dijke P, et al. DPC4 (SMAD4) mediates transforming growth factor-beta1 (TGF-beta1) induced growth inhibition and transcriptional response in breast tumour cells. Oncogene 1997;14:1891–1899.
54. Schutte M, Hruban RH, Hedrick L, et al. DPC4 gene in various tumor types. Cancer Res 1996; 56:2527–2530.
55. Hahn SA, Schutte M, Hoque AT, et al. DPC4, a candidate tumor suppressor gene at human chromosome 18q21.1. Science 1996;271:350–353.
56. O'Brien C. New tumor suppressor found in pancreatic cancer. Science 1996;271:294.
57. Ikezoe T, Takeuchi S, Kamioka M, et al. Analysis of the Smad2 gene in hematological malignancies. Leukemia 1998;12:94–95.
58. Wieser R, Gruber B, Rieder H, et al. Mutational analysis of the tumor suppressor Smad2 in acute lymphoid and myeloid leukemia. Leukemia 1998;12:1114–1118.

59. Nakao A, Roijer E, Imamura T, et al. Identification of Smad2, a human Mad-related protein in the transforming growth factor beta signaling pathway. J Biol Chem 1997;272:2896–2900.
60. Mitani K, Ogawa S, Tanaka T, et al. Generation of the AML1-EVI-1 fusion gene in the t(3;21)(q26;q22) causes blastic crisis in chronic myelocytic leukemia. EMBO J 1994;13:504–510.
61. Ogawa S, Kurokawa M, Tanaka T, et al. Increased Evi-1 expression is frequently observed in blastic crisis of chronic myelocytic leukemia. Leukemia 1996;10:788–794.
62. Mitani K. Molecular mechanisms of leukemogenesis by AML1/EVI-1. Oncogene 2004;23:4263–4269.
63. Massagué J, Cheifetz S, Boyd FT, et al. TGF-beta receptors and TGF-beta binding proteoglycans: recent progress in identifying their functional properties. Ann N Y Acad Sci 1990;593:59–72.
64. Laiho M, Saksela O, Keski-Oja J. Transforming growth factor beta alters plasminogen activator activity in human skin fibroblasts. Exp Cell Res 1986;164:399–407.
65. Overall CM, Wrana JL, Sodek J. Independent regulation of collagenase, 72-kDa progelatinase, and metalloendoproteinase inhibitor expression in human fibroblasts by transforming growth factor-beta. J Biol Chem 1989;264:1860–1869.
66. Kerr LD, Olashaw NE, Matrisian LM. Transforming growth factor beta 1 and cAMP inhibit transcription of epidermal growth factor- and oncogene-induced transin RNA. J Biol Chem 1988;263: 16,999–17,005.
67. Edwards DR, Murphy G, Reynolds JJ, et al. Transforming growth factor beta modulates the expression of collagenase and metalloproteinase inhibitor. EMBO J 1987;6:1899–1904.
68. Roberts CJ, Birkenmeier TM, McQuillan JJ, et al. Transforming growth factor beta stimulates the expression of fibronectin and of both subunits of the human fibronectin receptor by cultured human lung fibroblasts. J Biol Chem 1988;263:4586–4592.
69. Heino J, Ignotz RA, Hemler ME, et al. Regulation of cell adhesion receptors by transforming growth factor-beta. Concomitant regulation of integrins that share a common beta 1 subunit. J Biol Chem 1989;264:380–388.
70. Ignotz RA, Heino J, Massagué J. Regulation of cell adhesion receptors by transforming growth factor-beta. Regulation of vitronectin receptor and LFA-1. J Biol Chem 1989;264:389–392.
71. Bhowmick NA, Zent R, Ghiassi M, et al. Integrin beta 1 signaling is necessary for transforming growth factor-beta activation of p38MAPK and epithelial plasticity. J Biol Chem 2001;276:46,707–46,713.
72. White DE, Kurpios NA, Zuo D, et al. Targeted disruption of beta1-integrin in a transgenic mouse model of human breast cancer reveals an essential role in mammary tumor induction. Cancer Cell 2004;6:159–170.
73. Elenbaas B, Spirio L, Koerner F, et al. Human breast cancer cells generated by oncogenic transformation of primary mammary epithelial cells. Genes Dev 2001;15:50–65.
74. Olumi AF, Grossfeld GD, Hayward SW, et al. Carcinoma-associated fibroblasts direct tumor progression of initiated human prostatic epithelium. Cancer Res 1999;59:5002–5011.
75. Tuxhorn JA, McAlhany SJ, Dang TD, et al. Stromal cells promote angiogenesis and growth of human prostate tumors in a differential reactive stroma (DRS) xenograft model. Cancer Res 2002; 62:3298–3307.
76. Hanahan D, Folkman J. Patterns and emerging mechanisms of the angiogenic switch during tumorigenesis. Cell 1996;86:353–364.
77. Bhowmick NA, Moses HL. Tumor-stroma interactions. Curr Opin Genet Dev 2005;15:97–101.
78. Sternberg N, Smoller D, Braden T. Three new developments in P1 cloning. Increased cloning efficiency, improved clone recovery, and a new P1 mouse library. Genet Anal Tech Appl 1994;11: 171–180.
79. Chytil A, Magnuson MA, Wright CV, et al. Conditional inactivation of the TGF-beta type II receptor using Cre:Lox. Genesis 2002;32:73–75.
80. Iwano M, Plieth D, Danoff TM, et al. Evidence that fibroblasts derive from epithelium during tissue fibrosis. J Clin Invest 2002;110:341–350.
81. Iwano M, Fischer A, Okada H, et al. Conditional abatement of tissue fibrosis using nucleoside analogs to selectively corrupt DNA replication in transgenic fibroblasts. Mol Ther 2001;3:149–159.
82. Strutz F, Okada H, Lo CW, et al. Identification and characterization of a fibroblast marker: FSP1. J Cell Biol 1995;130:393–405.
83. Bhowmick NA, Chytil A, Plieth D, et al. TGF-β signaling in fibroblasts modulates the oncogenic potential of adjacent epithelia. Science 2004;303:848–851.
84. Jura N, Archer H, Bar-Sagi D. Chronic pancreatitis, pancreatic adenocarcinoma and the black box in-between. Cell Res 2005;15:72–77.

85. Carey LC. Acute and chronic pancreatitis. Surg Clin North Am 1975;55:325–338.
86. Farrell RJ, Peppercorn MA. Ulcerative colitis. Lancet 2002;359:331–340.
87. Tannapfel A. Molecular findings in Barrett's epithelium. Dig Dis 2004;22:126–133.
88. McGlynn KA, London WT. Epidemiology and natural history of hepatocellular carcinoma. Best Pract Res Clin Gastroenterol 2005;19:3–23.
89. London WT, Evans AA, Buetow K, et al. Molecular and genetic epidemiology of hepatocellular carcinoma: studies in China and Senegal. Princess Takamatsu Symp 1995;25:51–60.
90. Cheng N, Bhowmick NA, Chytil A, et al. Loss of TGF-beta type II receptor in fibroblasts promotes mammary carcinoma growth and invasion through upregulation of TGF-alpha-, MSP- and HGF-mediated signaling networks. Oncogene 2005;24:5053–5068.
91. Bhowmick NA, Neilson EG, Moses HL. Stromal fibroblasts in cancer initiation and progression. Nature 2004;432:332–337.
92. Ronnov-Jessen L, Petersen OW, Bissell MJ. Cellular changes involved in conversion of normal to malignant breast: importance of the stromal reaction. Physiol Rev 1996;76:69–125.
93. Noel A, Foidart JM. The role of stroma in breast carcinoma growth in vivo. J Mammary Gland Biol Neoplasia 1998;3:215–225.
94. Tuxhorn JA, Ayala GE, Smith MJ, et al. Reactive stroma in human prostate cancer: induction of myofibroblast phenotype and extracellular matrix remodeling. Clin Cancer Res 2002;8:2912–2923.
95. Joseph H, Gorska AE, Sohn P, et al. Overexpression of a kinase-deficient transforming growth factor-beta type II receptor in mouse mammary stroma results in increased epithelial branching. Mol Biol Cell 1999;10:1221–1234.
96. Bates RC, Mercurio AM. Tumor necrosis factor-alpha stimulates the epithelial-to-mesenchymal transition of human colonic organoids. Mol Biol Cell 2003;14:1790–1800.
97. Thiery JP, Chopin D. Epithelial cell plasticity in development and tumor progression. Cancer Metastasis Rev 1999;18:31–42.
98. Moriyama T, Kataoka H, Koono M, et al. Expression of hepatocyte growth factor/scatter factor and its receptor c-Met in brain tumors: evidence for a role in progression of astrocytic tumors (Review). Int J Mol Med 1999;3:531–536.
99. Tahara E. Cancer-stromal interaction through growth factor/cytokine networks implicated in growth of stomach cancer. Princess Takamatsu Symp 1994;24:187–194.
100. Gerber HP, Ferrara N. The role of VEGF in normal and neoplastic hematopoiesis. J Mol Med 2003;81:20–31.
101. Weigand M, Hantel P, Kreienberg R, et al. Autocrine vascular endothelial growth factor signalling in breast cancer. Evidence from cell lines and primary breast cancer cultures in vitro. Angiogenesis 2005;8:197–204.
102. Guo P, Fang Q, Tao HQ, et al. Overexpression of vascular endothelial growth factor by MCF-7 breast cancer cells promotes estrogen-independent tumor growth in vivo. Cancer Res 2003;63:4684–4691.
103. Muraoka RS, Koh Y, Roebuck LR, et al. Increased malignancy of Neu-induced mammary tumors overexpressing active transforming growth factor beta1. Mol Cell Biol 2003;23:8691–8703.
104. Forrester E, Chytil A, Bierie B, et al. Effect of conditional knockout of the type II TGF-beta receptor gene in mammary epithelia on mammary gland development and polyomavirus middle T antigen induced tumor formation and metastasis. Cancer Res 2005;65:2296–2302.

13 Mechanisms of TGF-β-Induced Apoptosis in Cancer Cells

Tae-Aug Kim and Seong-Jin Kim

CONTENTS

SUMMARY
INTRODUCTION
MOLECULAR MECHANISMS OF TGF-β-INDUCED APOPTOSIS
CONCLUSION
ACKNOWLEDGMENTS
REFERENCES

Abstract

Apoptosis is a common regulatory process of multicellular organisms. Transforming growth factor beta (TGF-β) has essential roles in a variety of apoptotic pathways including the mitochondrial apoptotic, death receptor, and other intracellular signaling pathways. The TGF-β-mediated apoptotic process involves not only intracellular proapoptotic responses but also anti-apoptotic signals. Resistance to TGF-β regulatory signals is the most indicative characteristic of many cancer cells during tumorigenesis. Therefore, controlling the homeostatic balance of these regulatory signals is critical for the prevention of tumorigenesis. Understanding the mechanisms of TGF-β-induced apoptosis in cancer cells will provide new insight of anticancer therapy.

Key Words: TGF-β; apoptosis; hepatoma; gastric cancer.

1. INTRODUCTION

Apoptosis, a particular morphological manifestation of programmed cell death, is a common regulatory process of multicellular organisms. The term apoptosis has been used as an active process that depends on the execution of a defined sequence of signaling events that lead to cell demise. It is a precisely regulated phenomenon essential for many biological processes, such as embryonic development, regulation of the immune system, and normal homeostasis of multicellular organisms *(1–3)*. Apoptosis also operates in adult organisms to maintain normal cellular homeostasis, which is particularly important with respect to the development of disease in human beings. During the past two decades, hundreds of genes that control the initiation, execution, and regulation of apoptosis have been explored. Among them, TGF-β plays an important role in apoptosis in vivo and in vitro. TGF-β-mediated apoptosis is involved in the elimination of damaged or abnormal cells from normal tissue in vivo, and these apoptotic effects are shown in various cell types as described in Table 1. The TGF-β-mediated apoptotic process involves not only intracellular proapoptotic

From: *Cancer Drug Discovery and Development: Transforming Growth Factor-β in Cancer Therapy, Vol I: Basic and Clinical Biology*
Edited by: Sonia B. Jakowlew © Humana Press Inc., Totowa, NJ

Table 1
Mechanistic Events of TGF-β-Induced Apoptosis in Cancer Cells

Cell type	*Intracellular modulators*	*Activated caspases*	*Reference*
Hepatoma	BAD, Smad2/3, cIAP	3	*(10)*
	Bax, Bcl-xL	3, 8, 9	*(104–106)*
	ERK, MKK6, SEK1, JNK		*(13)*
	p38, MLK3		*(14)*
	TRAIL		*(28)*
	TIEG	3	*(104,105)*
	Cytochrome C, Apaf-1		*(111,113)*
	p53, p21 WAF1, Bcl, NFkB		*(112)*
Gastric Cancer	Bid, FADD, Smad3/7	8	*(48)*
	RUNX3, p21CIP1		*(59,65,66,69)*
	Smad7		*(60)*
	p53		*(107)*
Prostate Cancer	bcl-2	1	*(39,43)*
	Smad7		*(108)*
	Smad2/3/4/6/7		*(40)*
Endothelial Cells	CD222, CD87		*(85)*
	RUNX2/3, p21 CIP1		*(88)*
Nervous System	TNF-α		*(109)*
	Smad/MAPK	3, 8, 9	*(101)*
	TNF-α, TIEG1,bcl-X(L)		*(102,103)*
	Bad		*(114)*
	p21waf1, p27kip1		*(110)*

responses but also antiapoptotic signals. TGF-β provides signals for both cell survival and apoptosis depending upon the cell type and physiological context. Cancer cells become refractory to this regulatory signal and thus do not undergo apoptosis under appropriate conditions. In most cases, the TGF-β signaling pathway is proapoptotic, but the molecular mechanism that allows TGF-β to induce apoptosis is still controversial. However, the homeostatic balance of these regulatory signals is critical for the prevention of tumorigenesis. In this review, we summarize the molecular mechanism of TGF-β-induced apoptosis in a variety of human cancers. Understanding the mechanisms of TGF-β-induced apoptosis in normal and cancer cells will provide new insight of anticancer therapy because increasing the sensitivity of tumor cells to anticancer therapy is tightly correlated with the induction of apoptosis by anticancer drugs.

2. MOLECULAR MECHANISMS OF TGF-β-INDUCED APOPTOSIS

2.1. Hepatoma

The growth and mass of the liver can be regulated during development and adult life. Liver cells proliferate during the postnatal period, whereas adult hepatocytes no longer actively proliferate. After injury to the liver by partial hepatectomy, the liver can regenerate. This regeneration process involves a wide variety of growth factors controlling proliferation, apoptosis, and differentiation. Many observations indicate that TGF-β plays an important role in hepatocyte growth inhibition and is a potent inducer of apoptosis in the liver *(4,5)*. The molecular mechanism of TGF-β-induced apoptosis has been studied in the hepatoma

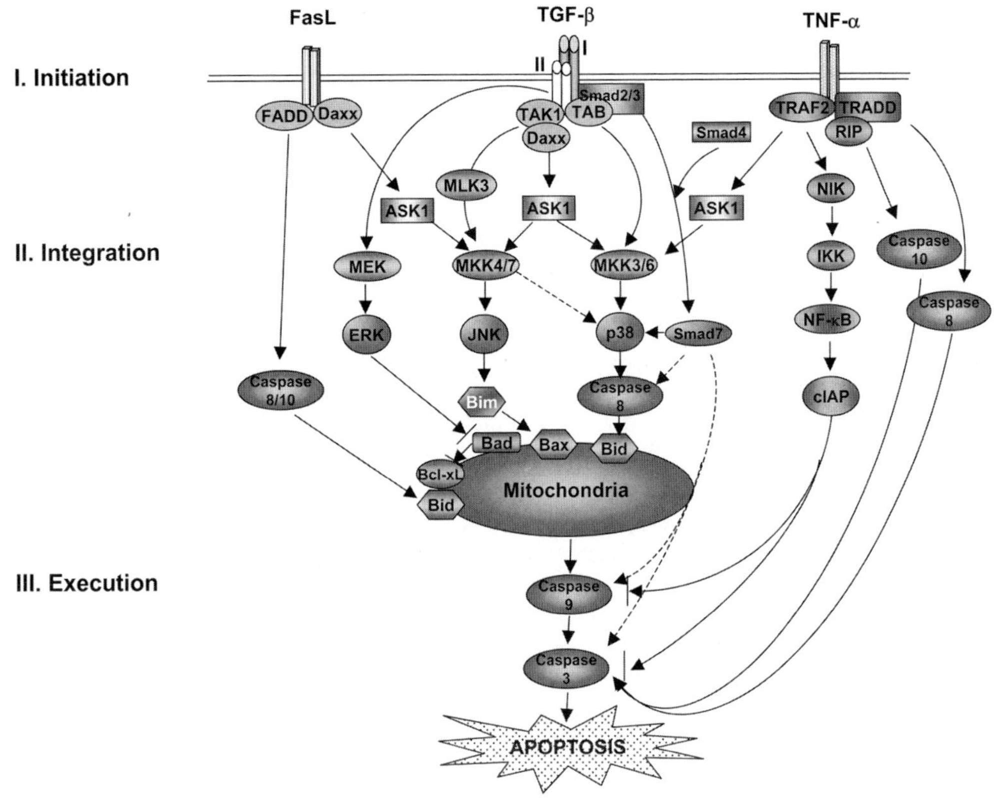

Fig. 1. Schematic diagram of TGF-β-induced apoptosis. Apoptotic events are divided into three stages. The initiation step involves death ligands binding to their respective receptors. Intracellular signaling pathways are operated in integration stages. Finally, execution of apoptosis can be processed through the mitochondrial pathway. TGF-β (transforming growth factor β1), TNF-α (tumor necrosis factor-α), Daxx (death-associated protein), ASK1 (apoptosis signal-regulating kinase-1), NF-κB (nuclear factor-κB), IKK (IκB kinase) JNK (c-*Jun*-N-terminal kinase), MLK3 (mixed lineage kinase3), ERK (extracellular signal-regulated kinase), MKK (mitogen-activated protein kinase kinase), Bid (BH3-Bcl-2-homology domain 3), Bim (Bcl-2-interacting mediator of cell death), Bax (B-cell lymphoma 2-associated protein X), TAB (TGF-β-activated kinase 1), and TAB1 (TAK1-binding protein).

cell lines, FaO and Hep3B, and in primary hepatocytes. Three different apoptotic pathways are involved in this mechanism: (i) the mitochondrial apoptotic pathway, (ii) involvement of intracellular regulators of apoptosis, and (iii) the death receptor pathway (Fig. 1). First, the mitochondrial apoptotic pathway indicates the critical role of Bcl-family proteins, acting to either inhibit or promote cell death *(6,7)*. Posttranscriptional modifications, together with the modulation of the levels of expression and subcellular localization of inhibitors (Bcl-2 and Bcl-X_L) and promoters (BAX, BAD, BID, and BIK), determine how cells respond to apoptotic stimuli *(6–8)*. The 15-kDa truncated BAD, a proapoptotic member of the Bcl-2 family of proteins, is a more potent inducer of apoptosis than the wild-type (WT) protein, whereas a mutant BAD resistant to caspase 3 cleavage is a weaker apoptosis inducer *(9)*. We have reported that TGF-β enhances cleavage of full-length BAD during TGF-β-mediated apoptosis of FaO cells in a Smad3-dependent manner *(10)*. A caspase 3-resistant mutant BAD showed less proapoptotic activity than the WT protein, whereas overexpression of truncated BAD accelerates TGF-β-dependent apoptosis. Furthermore, TGF-β-dependent apoptosis in rat FaO cells is potently enhanced by expression of Smad3 and is less efficiently enhanced by Smad2; this apoptosis can be blocked by antisense Smad3, suggesting that Smad3 may be an important mediator of TGF-β-induced apoptosis in hepatoma cells. It has been reported that regenerating hepatocytes are more resistant to the apoptosis induced by TGF-β than fetal hepatocytes. Regenerating cells show increased levels of some antiapoptotic proteins, such as Bcl-x_L or cIAP-1, and have higher intracellular glutathione content, which could confer resistance to apoptosis induced by TGF-β during liver regeneration *(11)*.

The second mechanism is the involvement of intracellular signaling molecules of apoptosis such as JNK, p38MAP kinase, Akt, NF-kB, and Smad7. Mitogen-activated protein (MAP) kinases are a group of protein serine/threonine kinases that are activated in response to a wide variety of external signals and mediate signal transduction cascades that play an important regulatory role in cell growth, differentiation, and apoptosis *(12)*. During TGF-β-induced apoptosis, three MAP kinases (ERK, JNK, and p38 kinase) showed simultaneous sustained activation in the rat hepatoma cell line FaO *(13)*. TGF-β-induced apoptosis was markedly enhanced when ERK activation was selectively inhibited by the MAPK/ERKK inhibitor PD98059. In contrast, TGF-β-induced apoptosis can be suppressed by both interference of p38 activity by overexpression of the dominant-negative (DN) MKK6 mutant and by inhibition of the JNK pathway by overexpression of the DN SEK1 mutant. Recently, it has been shown that ectopic expression of GADD45b in AML12 murine hepatocytes activates p38 and triggers apoptotic cell death, suggesting that GADD45b participates in TGF-β-induced apoptosis by acting upstream of p38 activation. We have also reported that mixed lineage kinase 3 (MLK3), an upstream activator of the MAP kinase pathway mediates TGF-β-induced apoptosis signaling in hepatoma cells *(14)*. MLKs are a family of serine/threonine protein kinases that function in a phosphorelay module to control the activity of specific MAPKs *(15–18)*. MLKs act as MKK kinases and lead to the activation of JNKs via the activation of MKKs *(19–22)*. Studies on neuronal and other cell types indicate that the MLK family members are likely to act between Rac1/Cdc42 and MKK4 and -7 in death signaling.

Third, the role of different death receptor/ligand systems has been studied in TGF-β-induced apoptosis. Both death receptors and TGF-β receptors use a common signaling pathway via caspases for apoptosis induction. These receptor-signaling mechanisms may be linked. TNF-related apoptosis-inducing ligand (TRAIL) has the ability to preferentially kill a wide variety of tumor cell lines *(23,24)*, whereas most normal cells are resistant to TRAIL, both in vitro and in vivo *(25,26)*. The expression of TRAIL is upregulated upon exposure of liver cell lines to TGF-β, and TRAIL is a major contributor to apoptosis mediated by TGF-β in hepatoma cells *(27)*. TGF-β did not sensitize TRAIL-resistant cells to TRAIL-induced apoptosis. Hence, TGF-β-induced TRAIL expression in tumor cells that have acquired TRAIL resistance may even support tumor progression. However, when expressed in normal tissue, TRAIL has been shown to inhibit tumor development and metastasis, because the absence of TRAIL results in an increase in metastasis in the liver *(28)*. Thus, TRAIL upregulation by TGF-β may serve diverse purposes, depending on the environment or the TRAIL sensitivity status of the tumor, respectively. TGF-β-induced apoptosis in primary human liver cells has not been convincingly shown thus far, whereas TGF-β sensitivity in human hepatoma cell lines as well as rodent liver cells is well established *(29,30)*. Following TGF-β treatment, TRAIL expression was clearly induced in primary human hepatocytes on the mRNA level as well as the protein level. The development of a TRAIL-resistant phenotype in tumor cells may be responsible for the development of resistance toward TGF-β-induced growth inhibition.

2.2. Prostate Cancer

The apoptotic signals of TGF-β may also involve oncogene and tumor suppressor-gene products. TGF-β can suppress transcription of the c-*myc* protooncogene or it can retain the Rb protein in its underphosphorylated, growth-suppressive state *(31,32)*. TGF-β has been established as a physiological regulator of prostate growth because of its ability to inhibit cell proliferation and induce apoptosis *(33,34)*. TGF-β can simultaneously inhibit normal prostate epithelial cell proliferation and directly activate prostate apoptosis in the presence of physiological levels of androgens *(33,35,36)*. While serving a growth inhibitory/apoptotic role in the normal prostate, overproduction of TGF-β in prostate cancer may contribute to tumor progression. Earlier studies demonstrated that a marked decrease in the expression of TGF-β type I receptor (TβRI) and type II receptor (TβRII) in prostate tumors correlated with tumor grade *(37,38)*. Enforced expression of TβRII in the LNCaP prostate carcinoma cell line restores sensitivity to TGF-β, which results in growth arrest and apoptosis induction *(39)*. More significantly, LNCaP TβRII cells exhibit decreased tumorigenicity in vivo through downregulation of bcl-2 expression and induction of caspase-1 expression *(39)*. Prostate tumors with functional defects impair the intercellular transduction of the TGF-β signal.

In the prostate, TGF-β-induced apoptosis is mediated by the Smad-dependent pathway. The levels of Smad expression and activation correlate with apoptosis induction following castration and are reduced in the prostate tumor samples compared with normal prostate samples, suggesting that lower Smad expression resulted in decreased apoptotic sensitivity to androgen withdrawal *(40)*. Evidence emerging from several studies suggests that the caspase cascade is involved in the execution of apoptosis in prostate cancer cells in response to diverse stimuli *(39–42)*. It has been reported that TGF-β induces prostate cancer cell apoptosis via upregulation of caspase-1 expression *(43)*. The apoptosis regulatory role of TGF-β in the prostate and the significance of its key receptor TβRII as a potential tumor suppressor have been well documented *(44,45)*. Together with current studies, these findings suggest that timing the coordination of the administration of chemotherapeutic agents that inhibit transcription or translation with those that induce apoptosis could enhance the effectiveness of therapy.

2.3. Gastric Cancer

Gastric epithelial cells exhibit a rapid rate of turnover, which requires an appropriate balance between the proliferation of progenitor cells and the loss of mature cells *(46)*. Loss of mature epithelial cells at the gastric surface can be mediated mainly by physiological cell death, which always occurs at the surface of the gastrointestinal tract. In the gastric mucosa, 1% to 3% of the epithelial cells were reported to show morphological features of apoptosis at any given time under physiological conditions *(47)*. The physiological loss of gastric epithelial cells is necessary for the maintenance of tissue homeostasis associated with the exchange of mature epithelial cells to fresh proliferating cells, and any defects in this epithelial cell death pathway may be a contributing factor to disease development.

Two major apoptotic pathways, the death receptor and the mitochondrial apoptotic pathway, have been documented in human gastric cells. TGF-β-induced apoptosis of human gastric SNU-620 carcinoma cells is caused by the Fas death pathway in a Fas ligand-independent manner. This Fas death pathway activated by TGF-β is linked to the mitochondrial apoptotic pathway via caspase-8-mediated Bid cleavage *(48)*. TGF-β induced expression and activation of Fas and subsequent caspase-8-mediated Bid cleavage. Interestingly, expression of dominant negative FADD and treatment with caspase-8 inhibitor efficiently prevented TGF-β-induced apoptosis. The TGF-β-mediated apoptotic process was effectively inhibited by Smad3 knockdown and completely abrogated by Smad7 expression, suggesting that the Smad3 pathway acts upstream of the Fas-FADD pathway during TGF-β-induced apoptosis in SNU-620 cells.

Involvement of intracellular signaling molecules of apoptosis has been described in gastric cancer. Among them, Smad proteins are involved in TGF-β-induced apoptosis *(49,50)*.

However, their roles in apoptosis are still ambiguous and controversial. It has been reported that distinct Smads play a role in TGF-β-induced apoptosis *(10,49–51)*. Moreover, the proapoptotic or antiapoptotic roles of Smad7 have been reported *(52–54)*. Smad3 knockdown potently suppressed Fas induction and subsequent apoptotic events induced by TGF-β, which were also completely inhibited by Smad7 overexpression. Accordingly, these findings indicate the proapoptotic role of Smad3 and the antiapoptotic role of Smad7 in gastric carcinoma cells. Furthermore, abnormal expression of TGF-β receptors and inactive mutations of *Smad* family genes are frequently detected in human gastric carcinomas *(55–57)*. Levels of physiological apoptosis are abnormally low in the epithelium of glandular stomachs and forestomachs in *Runx3*-deficient mice *(58)* and *Smad7*-transgenic mice *(59)*, respectively, both of which exhibit defects in TGF-β-induced signaling. Hyperplasia in the gastric epithelium of these mutant mice might be caused not only by loss of cell proliferation-inhibitory activity but also by reduced apoptotic loss of epithelial cells from the surface of the stomach. TGF-β may be involved in the regulation of turnover of gastric epithelial cells through its proapoptotic activity.

The *RUNX* family of transcription factors plays a pivotal role in normal development and neoplasias. In mammals, *RUNX3* is involved in neurogenesis *(60,61)* and thymopoiesis *(62,63)* and functions as a tumor suppressor of gastric cancer *(59,64)*. About 60% of primary gastric cancer specimens do not express *RUNX3* owing to a combination of hemizygous deletion and DNA hypermethylation of the *RUNX3* promoter region *(59,64–66)*. *RUNX3* activity is closely associated with TGF-β signaling because the gastric mucosa of the *Runx3* knockout mouse is less sensitive to TGF-β, which induces both cell cycle arrest and apoptosis. *RUNX3* is required for TGF-β-dependent induction of $p21^{cip1}$ expression in stomach epithelial cells. Overexpression of *RUNX3* potentiates TGF-β-dependent endogenous $p21^{cip1}$ induction. In cooperation with Smads, RUNX3 synergistically activates the $p21^{cip1}$ promoter. In contrast, *RUNX3-R122C*, a mutation identified in a gastric cancer patient, abolished the ability to activate the $p21^{cip1}$promoter or cooperate with Smads *(67)*.

2.4. Endothelial Cells

The balance between cell death and survival within blood vessels is decisive for the physiology of the vasculature. It also contributes to the pathogenesis of many diseases of the vascular system when dysregulated *(68)*. Endothelial cell apoptosis plays a role in angiogenesis, a critical process in wound healing, inflammation, cancer development, and embryogenesis *(69)*. While the apoptotic executioner machinery of endothelial cells is the same as in other models of apoptosis, the dependence of endothelial cells on a variety of growth factors implies an important role for cytokines in the regulation of endothelial apoptotic response. A network of cytokines, proteases, matrix proteins, and other factors *(70–73)* tightly coordinates this control of vasculogenesis and angiogenesis. During vascular remodeling, TGF-β plays a pivotal role by acting pleiotropically on the major cell types of the vasculature: endothelial cells (ECs), smooth-muscle cells (SMCs), and pericytes *(74–78)*. TGF-β led to both proapoptotic effects on ECs and antiapoptotic effects on SMCs *(79,80)*. Genetic knockout of TGF-β, its receptor and downstream signaling molecules have demonstrated the essential role of TGF-β signaling in vasculogenesis *(81)*. TGF-β must be activated from the latent form (LTGF-β) to induce biological responses. The proper regulation of the amount of active TGF-β is essential for the function of the microvascular unit. Apoptosis of HUVECs is associated with the activation of TGF-β, which also accompanies apoptosis in Mv1Lu cells, implying that the activation of TGF-β is a common feature of the apoptotic response in different cell types *(82)*. These studies are complementary to reports on the involvement of endogenous TGF-β in radiation-induced apoptosis of embryonic epithelial cells, and on the induction of TGF-β synthesis during apoptosis and secretion of both latent and active

TGF-β by apoptotic T-cells *(83,84)*. Leksa et al. *(85)* reported that mini-plasminogen (mini-Plg) is a high-affinity ligand for the mannose-6-phosphate/insulin-like-growth-factor-II receptor (CD222) and is essential for the activation of TGF-β by the (urokinase-type plasminogen-activator receptor) CD87–CD222 complex to induce apoptosis in endothelial cells. The complex between CD222 and CD87 is important for the release of active TGF-β that, in turn, induces EC apoptosis in a caspase-dependent manner. A specific binding of mini-Plg to CD222 and proteolytic activation of TGF-β by mini-Plm generated on the complex trigger the process.

Another molecular mechanism of TGF-β-mediated apoptosis in endothelial cells involves the RUNX gene family. As discussed above, the normal function of RUNX3 in gastric epithelial cells has been found to be the suppression of uncontrolled growth during development *(59,86)*. Cells from mice in which the *Runx3* gene was inactivated failed to exhibit growth suppression or apoptosis in response to TGF-β. The *RUNX2* gene encodes the largest protein of the Runx transcription factor family *(87)*. RUNX2 contains a QA domain at the N-terminus and exon 8 near the C-terminus sequences that are missing in the RUNX1 and RUNX3 proteins. Recently, Sun et al. *(88)* showed that transfection of the *RUNX2Δ8* gene in endothelial cells resulted in the expression of a protein similar to RUNX3; both RUNX2Δ8 and RUNX3 (but not RUNX2) lack the exon 8 domain. RUNX2Δ8-transfected cells were more sensitive to TGF-β-induced apoptosis than RUNX2 transfectants. The RUNX2 gene was a strong repressor of the promoter of the cyclin-dependent kinase inhibitor, $p21^{CIP1}$, while RUNX2Δ8 was a competitive inhibitor of RUNX2 and exhibited weak repressional activity. These studies indicate that endothelial cells regulate growth and apoptosis in part by alternative splicing events in the RUNX2 transcription factor to affect the TGF-β signaling pathway. The exon 8 domain of RUNX2 may contribute to the strong repressional activity of RUNX2 for some target gene promoters. This specific mechanism of TGF-β-mediated apoptosis in endothelial cells is thus a potential novel target to be considered for the treatment of pathological vascular disorders.

2.5. Nervous System

The regulation of the balance of neuronal survival and death is a permanent feature in nervous system development, maintenance, degeneration, and repair. Accordingly, TGF-β, which is normally present at low levels in healthy adult central nerve system (CNS) cells, is rapidly upregulated after injury and induces expression of many injury response genes *(89)*. The main sources of TGF-β1 in the injured brain are astrocytes and microglia, but neurons can produce it as well *(90)*. TGF-β2 and TGF-β3 are regulated mainly by hormonal and developmental signals, and it has been speculated that they play a role in CNS development. Numerous studies used cultured neurons to show a protective effect of TGF-β against various toxins and injurious agents *(90,91)*. Similar results were obtained when TGF-βs were delivered intracerebrally or by using viral vectors adult rodent brains after ischemic injury *(91)*. Moreover, lack of TGF-β expression in neonatal $Tgfb1^{-/-}$ mice results in a widespread increase in degenerating neurons, accompanied by reduced expression of synaptophysin and laminin and prominent microgliosis *(92)*. Lack of TGF-β1 also strongly reduces survival of primary neurons cultured from $Tgfb1^{-/-}$ mice. TGF-β1 deficiency in adult $Tgfb1^{-/+}$ mice results in increased neuronal susceptibility to excitotoxic injury, whereas astroglial overexpression of TGF-β1 protects adult mice against neurodegeneration in acute, excitotoxic, and chronic injury paradigms. In addition, antibody neutralization of TGF-β2 inhibits proliferation and activates apoptosis of cerebellar cell precursors in the developing cerebellum *(93)*. These studies demonstrate that TGF-β is a survival factor for mouse CNS neurons in vivo and in cell culture, suggesting a nonredundant function for TGF-β in maintaining neuronal integrity, survival of CNS neurons, and in regulating microglial activation. In contrast, in vivo neutralization of TGF-βs with antibodies during early CNS development prevented ontogenetic neuronal death in the chick embryo cilliary ganglion and retina *(94,95)*. TGF-β is essentially required to regulate programmed cell death in the central retina. Application of TGF-β neutralizing antibodies resulted in a decrease of TUNEL-positive cells in the central retina. De Luca et al. *(96)* demonstrated that TGF-βs can provoke apoptosis in cultured immature cerebella granule cells unless they were cultured with high concentrations of potassium, which led to membrane depolarization and resembled in vitro the neuronal activity that results from the first afferent synaptic connection established in vivo. This result suggested that TGF-βs act as an apoptotic signal involved in the deletion of cells that do not reach their targets during cerebella development. These findings suggested that TGF-β limits the expansion of postmitotic neuronal precursor populations on the one hand by promoting cell death and, on the other hand, by supporting the survival of those neurons that have successfully reached their target area and gained supportive synaptic connectivity *(96)*. Indeed, cerebella granule cells that establish functional synaptic connection with mossy fibers become resistant to TGF-β-mediated apoptosis. In primary rat oligodendrocyte progenitor cultures, TGF-β induces both growth arrest and apoptosis. Cell death of the oligodendrocyte lineage occurs also as a normal process during development *(97)*. This process has been understood as a mechanism to adjust the number of myelinating cells to the number of axons formed by neuronal cells *(98)*. Like neurons, oligodendrocytes are generated in excess and eliminated by apoptosis. It has been shown that oligodendrocyte progenitors and newly formed oligodendrocytes are very susceptible to apoptosis *(98,99)*. The proliferation of oligodendrocyte progenitors can be inhibited by TGF-β, while differentiation into mature oligodendrocytes is promoted *(100)*. Schuster et al. *(101)* used the oligodendroglial cell line OLI-neu as a model system to investigate the role of TGF-β in oligodendroglial cell death and proliferation. TGF-β-mediated cell death is accompanied by caspase 3 activation and activation of the Smad pathway. Cell death can be blocked by a pan-caspase inhibitor as well as by inhibitors of caspase 8 and 9. The investigation of pro- and antiapoptotic signaling pathways showed downregulation of the p42/p44 MAPK pathway, which normally contributes to cellular survival and differentiation. Recent studies showed that TNF-α and TIEG1, a Kruppel-like Zn-finger transcription factor, facilitated TGF-β-mediated apoptosis in OLI-neu cells *(102,103)*. These studies indicated that TGF-β induced apoptotic cell death in cells of oligodendroglial origin, whereby the signaling cascade involved not only the downregulation of antiapoptotic signaling such as bcl-X(L), leading to the activation of caspases, but also involved Smad-dependent pathways.

3. CONCLUSION

Taken together, TGF-β has complex and even opposing effects on its target cellular components in a physiological context and cell type dependent manner. In many cases, human cancers become resistant to the antiproliferative effects of TGF-β and do not undergo the apoptotic process under appropriate conditions. Although Smad proteins are the predominant signaling molecules in the TGF-β signaling pathway, other apoptosis pathways such as the mitochondrial apoptotic pathway, intracellular regulators of apoptosis, and the death receptor pathway may significantly contribute to TGF-β-induced apoptosis through cross-talks with each other. Controlling the homeostatic balance of TGF-β-mediated apoptotic and non-apoptotic signals may play an important role in the regulation of cell viability under normal physiological conditions, because complex networks between different intercellular signaling pathways determine the cellular responses to intracellular cell fate and to the extracellular environment. Therefore, defining the precise molecular elements that are involved in the intrinsic or extrinsic apoptosis pathway in tumors will provide new insight to develop targeted therapies for a less toxic and more effective treatment in individualized cancer.

ACKNOWLEDGMENTS

The authors would like to thank Amy Hobbie for her critical reading in manuscript preparation.

REFERENCES

1. Vaux DL, Korsmeyer SJ. Cell death in development. Cell 1999;96:245–254.
2. Fas SC, Fritzsching B, Suri-Payer E, Krammer PH. Death receptor signaling and its function in the immune system. Curr Dir Autoimmun 2006;9:1–17.
3. Twomey C, McCarthy JV. Pathway of apoptosis and importance in development. J Cell Mol Med 2005;9:345–359.
4. Bissell DM, Roulot D, George J. Transforming growth factor β and the liver. Hepatology 2001; 34:859–867.
5. Schuster N, Krieglstein K. Mechanisms of TGF-β-mediated apoptosis. Cell Tissue Res 2002; 307:1–14.
6. Korsmeyer SJ, Gross A, Harada H, et al. Death and survival signals determine active/inactive conformations of pro-apoptotic BAX, BAD, and BID molecules. Cold Spring Harbor Symp Quant Biol 1999;64:343–350.
7. Harada HB, Becknell M, Wilm M, et al. Phosphorylation and inactivation of BAD by mitochondria-anchored protein kinase A Mol Cell 1999;3:413–422.
8. Oltvai ZN, Korsmeyer S. Checkpoints of dueling dimers foil death wishes. Cell 1994;79:189–192.
9. Condorelli F, Salomoni P, Cotteret S, et al. Caspase cleavage enhances the apoptotis-inducing effects of BAD. Mol. Cell. Biol. 2001;21:3025–3036.
10. Kim BC, Mamura M, Choi KS, et al. Transforming growth factor β1 induces apoptosis through cleavage of BAD in a Smad3-dependent mechanism in FaO hepatoma cells. Mol Cell Biol 2002;22:1369–1378.
11. Herrera B, Alvarez AM, Beltran J, et al. Resistance to TGF-β-induced apoptosis in regenerating hepatocytes. J Cell Physiol 2004;201:385–392.
12. Robinson MJ, Cobb MH. Mitogen-activated protein kinase pathways. Curr Opin Cell Biol 1997;9: 180–186.
13. Park HJ, Kim BC, Kim SJ, et al. Role of MAP kinases and their cross-talk in TGF-β1-induced apoptosis in FaO rat hepatoma cell line. Hepatology 2002;35:1360–1371.
14. Kim KY, Kim BC, Xu Z, et al. Mixed Lineage Kinase 3 (MLK3)-activated p38 MAP kinase mediates transforming growth factor-β-induced apoptosis in hepatoma cells. J Biol Chem 2004; 279:29,478–29,484.
15. Dorow DS, Devereux L, Dietzsch E, et al. Identification of a new family of human epithelial protein kinases containing two leucine/isoleucine-zipper domains. Eur J Biochem 1993;213:701–710.
16. Ing YL, Leung IW, Heng HH, et al. MLK-3: identification of a widely-expressed protein kinase bearing an SH3 domain and a leucine zipper-basic region domain. Oncogene 1994;9:1745–1750.
17. Gallo KA, Mark MR, Scadden DT, et al. Identification and characterization of SPRK, a novel src-homology 3 domain-containing proline-rich kinase with serine/threonine kinase activity. J Biol Chem 1994;269:15,092–15,100.
18. Gallo KA, Johnson GL. Mixed-lineage kinase control of JNK and p38 MAPK pathways. Nat Rev Mol Cell Biol 2002;3:663–672.
19. Rana A, Gallo K, Godowski P, et al. The mixed lineage kinase SPRK phosphorylates and activates the stress-activated protein kinase activator, SEK-1. J Biol Chem 1996;271:19,025–19,028.
20. Tibbles LA, Ing YL, Kiefer F, et al. MLK-3 activates the SAPK/JNK and p38/RK pathways via SEK1 and MKK3/6. EMBO J 1996;15:7026–7035.
21. Teramoto H, Coso OA, Miyata H, et al. Signaling from the small GTP-binding proteins Rac1 and Cdc42 to the c-Jun N-terminal kinase/stress-activated protein kinase pathway. J Biol Chem 1996; 271:27,225–27,228.
22. Hirai S, Katoh M, Terada M, et al. MST/MLK2, a member of the mixed lineage kinase family, directly phosphorylates and activates SEK1, an activator of c-Jun N-terminal kinase/stress-activated protein kinase. J Biol Chem 1997;272:15,167–15,173.
23. Wiley SR, Schooley K, Smolak PJ, et al. Identification and characterization of a new member of the TNF family that induces apoptosis. Immunity 1995;3:673–682.

24. Walczak H, Degli-Esposti MA, Johnson RS, et al. TRAIL-R2: a novel apoptosis-mediating receptor for TRAIL. EMBO J 1997;16:5386–5397.
25. Ashkenazi A, Pai RC, Fong S, et al. Safety and antitumor activity of recombinant soluble Apo2 ligand. J Clin Invest 1999;104:155–162.
26. Walczak H, Miller RE, Ariail K, et al. Tumoricidal activity of tumor necrosis factor-related apoptosis-inducing ligand in vivo. Nat Med 1999;5:157–163.
27. Herzer K, Ganten TM, Schulze-Bergkamen H, et al. Transforming growth factor beta can mediate apoptosis via the expression of TRAIL in human hepatoma cells. Hepatology 2005;42:183–92.
28. Cretney E, Takeda K, Yagita H, et al. Increased susceptibility to tumor initiation and metastasis in TNF-related apoptosis-inducing ligand-deficient mice. J Immunol 2002;168:1356–1361.
29. Lin JK, Chou CK. In vitro apoptosis in the human hepatoma cell line induced by transforming growth factor beta1. Cancer Res 1992;52:385–388.
30. Oberhammer F, Bursch W, Parzefall W, et al. Effect of transforming growth factor beta on cell death of cultured rat hepatocytes. Cancer Res 1991;51:2478–2485.
31. Laiho M, DeCaprio JA, Ludlow JW, et al. Growth inhibition by TGF-beta linked to suppression of retinoblastoma protein phosphorylation. Cell 1990;62:175–185.
32. Geiser AG, Burmester JK, Webbink R, et al. Inhibition of growth by transforming growth factor-following fusion of two non-responsive human carcinoma cell lines. J Biol Chem 1992;267:2588–2593.
33. Kyprianou N, Isaacs JT. Activation of programmed cell death in the rat ventral prostate after castration. Endocrinology 1988;122:552–562.
34. Sutkowski DM, Fong CJ, Sensibar JA, et al. Interaction of epidermal growth factor and transforming growth factor-beta in human prostatic epithelial cells in culture. Prostate 1992;21:133–143.
35. Kyprianou N, Isaacs JT. Expression of transforming growth factor-beta in the rat ventral prostate during castration-induced programmed cell death. Mol Endocrinol 1989;3:1515–1522.
36. Sgambato A, Doki Y, Schieren I, Weinstein IB. Effects of cyclin E overexpression on cell growth and response to transforming growth factor beta depend on cell context and p27Kip1 expression. Cell Growth Differ 1997;8:393–405.
37. Kim IY, Ahn HJ, Zelner DJ, et al. Loss of expression of transforming growth factor beta type I and type II receptors correlates with tumor grade in human prostate cancer tissues. Clin Cancer Res. 1996;2:1255–1261.
38. Williams RH, Stapleton AM, Yang G, et al. Reduced levels of transforming growth factor β receptor type II in human prostate cancer: an immunohistochemical study. Clin Cancer Res 1996;2:635–640
39. Guo Y, Kyprianou N. Restoration of transforming growth factor-β signaling pathway in human prostate cancer cells suppresses tumorigenicity via induction of caspase-1-mediated apoptosis. Cancer Res 1999;59:1366–1371.
40. Brodin G, ten Dijke PT, Funa K, et al. Increased Smad expression and activation are associated with apoptotsis in normal and malignant prostate after castration. Cancer Res 1999;59:2731–2738.
41. Marcelli M, Cunningham GR, Walkup M, et al. Signaling pathway activated during apoptosis of the prostate cancer cell line LNCaP: overexpression of caspase-7 as a new gene therapy strategy for prostate cancer. Cancer Res 1999;59:382–390.
42. Bowen C, Voeller HJ, Kikly K, Gelmann EP. Synthesis of pro-caspase-3 and -7 during apoptosis in prostate cancer cells. Cell Death Differ 1999;6:394–401.
43. Zhu B, Kyprianou N. Transforming growth factor β and prostate cancer. Cancer Treat Res. 2005;126: 157–173.
44. Bruckheimer EM, Kyprianou N. Apoptosis in prostate carcinogenesis; A growth regulator and a therapeutic target. Cell Tissue Res. 2000;301:153–162.
45. Bello-DeOcampo D, Tindall DJ. TGF-β/Smad signaling in prostate cancer. Current Drug Targets 2003;4:197–207.
46. Hall PA, Coates PJ, Ansari B, Hopwood D. Regulation of cell number in the mammalian gastrointestinal tract: the importance of apoptosis. J Cell Sci 1994;107:3569–3577.
47. von Herbay A, Rudi J. Role of apoptosis in gastric epithelial turnover. Microsc Res Tech 2000;48: 303–311.
48. Kim SG, Jong HS, Kim TY, et al.Transforming growth factor-beta1 induces apoptosis through Fas ligand-independent activation of the Fas death pathway in human gastric SNU-620 carcinoma cells. Mol Biol Cell 2004;15:420–434.
49. Yamamura Y, Hua X, Bergelson S, Lodish HF. Critical role of Smads and AP-1 complex in transforming growth factor-β-dependent apoptosis. J Biol Chem 2000;275:36,295–36,302.

50. Yanagisawa K, Osada H, Masuda A, et al. Induction of apoptosis by Smad3 and down-regulation of Smad3 expression in response to TGF-β in human normal lung epithelial cells. Oncogene 1998;17:1743–1747.
51. Dai JL, Bansal RK, Kern SE. G1 cell cycle arrest and apoptosis induction by nuclear Smad4/Dpc 4, phenotypes reversed by a tumorigenic mutation. Proc Natl Acad Sci USA 1999;96:1427–1432.
52. Park JG, et al. Establishment and characterization of human gastric carcinoma cell lines. Int J Cancer 1997;70:443–449.
53. Lallemand F, Mazars A, Prunier C, et al. Smad7 inhibits the survival nuclear factor kappaB and potentiates apoptosis in epithelial cells. Oncogene 2001;20:879–884.
54. Edlund S, Bu S, Schuster N, et al. Transforming growth factor-β1 (TGF-β)-induced apoptosis of prostate cancer cells involves Smad7-dependent activation of p38 by TGF-β-activated kinase 1 and mitogen-activated protein kinase kinase 3. Mol Bio. Cell 2003;14:529–544.
55. Ju HR, Jung U, Sonn CH, et al. Aberrant signaling of TGF-β1 by the mutant Smad4 in gastric cancer cells. Cancer Lett 2003;196:197–206.
56. Kang SH, Bang YJ, Im YH, et al. Transcriptional repression of the transforming growth factor-β type I receptor gene by DNA methylation results in the development of TGF-β resistance in human gastric cancer. Oncogene 1999;18:7280–7286.
57. Park K, Kim SJ, Bang YJ, et al. Genetic changes in the transforming growth factor beta (TGF-β) type II receptor gene in human gastric cancer cells: correlation with sensitivity to growth inhibition by TGF-β. Proc Natl Acad Sci USA 1994;91:8772–8776.
58. He W, Li AG, Wang D, et al. Overexpression of Smad7 results in severe pathological alterations in multiple epithelial tissues. EMBO J. 2002;21:2580–2590.
59. Li QL, Ito K, Sakakura C, et al. Causal relationship between the loss of RUNX3 expression and gastric cancer. Cell 2002;109:113–124.
60. Inoue K, Ozaki S, Shiga T, et al. Runx3 controls the axonal projection of proprioceptive dorsal root ganglion neurons. Nat Neurosci 2002;5:946–954.
61. Levanon D, Bettoun D, Harris-Cerruti C, et al. The Runx3 transcription factor regulates development and survival of TrkC dorsal root ganglia neurons. EMBO J 2002;21:3454–3463.
62. Taniuchi I, Osato M, Egawa T, et al. Differential requirements for Runx proteins in CD4 repression and epigenetic silencing during T lymphocyte development. Cell 2002;111:621–633.
63. Woolf E, Xiao C, Fainaru O, et al. Runx3 and Runx1 are required for CD8 T cell development during thymopoiesis. Proc Natl Acad Sci USA 2003;100:7731–7736.
64. Guo WH, Weng LQ, Ito K, et al. Inhibition of growth of mouse gastric cancer cells by Runx3, a novel tumor suppressor. Oncogene 2002;21:8351–8355.
65. Kim TY, Lee HJ, Hwang KS, et al. Methylation of RUNX3 in various types of human cancers and premalignant stages of gastric carcinoma. Lab Invest 2004;84:479–484.
66. Oshimo Y, Oue N, Mitani Y, et al. Frequent loss of RUNX3 expression by promoter hypermethylation in gastric carcinoma. Pathobiology 2004;71:137–143.
67. Chi XZ, Yang JO, Lee KY, Ito K, RUNX3 suppresses gastric epithelial cell growth by inducing p21(WAF1/Cip1) expression in cooperation with transforming growth factor β-activated SMAD. Mol Cell Biol 2005;25:8097–8107.
68. Carmeliet P. Angiogenesis in health and disease. Nat Med 2003;9:653–660.
69. Chavakis E, Dimmeler S. Regulation of endothelial cell survival and apoptosis during angiogenesis. Arterioscler Thromb Vasc Biol 2002;22:887–893.
70. Folkman J, D'Amore PA. Blood vessel formation: what is its molecular basis? Cell 1996;87:1153–1155.
71. Jackson C. Matrix metalloproteinases and angiogenesis. Curr Opin Nephrol Hypertens 2002;11: 295–299.
72. Pepper MS. Extracellular proteolysis and angiogenesis. Thromb Haemost 2001;86:346–355.
73. Risau W. Mechanisms of angiogenesis. Nature 1997;386:671–674.
74. Fajardo LF, Prionas SD, Kwan HH, et al.Transforming growth factor beta1 induces angiogenesis in vivo with a threshold pattern. Lab Invest 1996;74:600–608.
75. Hirschi KK, Rohovsky SA, D'Amore PA. PDGF, TGF-β, and heterotypic cell-cell interactions mediate endothelial cell-induced recruitment of 10T1/2 cells and their differentiation to a smooth muscle fate. J Cell Biol 2003;141;805–814.
76. Koh GY, Kim SJ, Klug MG, et al. Targeted expression of transforming growth factor-β1 in intracardiac grafts promotes vascular endothelial cell DNA synthesis. J Clin Invest 1995;95:114–121.
77. Madri JA, Bell L, Merwin JR. Modulation of vascular cell behavior by transforming growth factors beta. Mol Reprod Dev 1992;32:121–126.

78. Vernon RB, Sage EH. A novel, quantitative model for study of endothelial cell migration and sprout formation within three-dimensional collagen matrices. Microvasc Res 1999;57:118–133.
79. Pollman MJ, Naumovski L, Gibbons GH. Vascular cell apoptosis: cell type-specific modulation by transforming growth factor-β1 in endothelial cells versus smooth muscle cells. Circulation 1999; 99:2019–2026.
80. Schulick AH, Taylor AJ, Zuo W, et al. Overexpression of transforming growth factor beta1 in arterial endothelium causes hyperplasia, apoptosis, and cartilaginous metaplasia. Proc Natl Acad Sci USA 1998;95:6983–6988.
81. Lebrin F. Deckers M, Bertolino P, et al. TGF-β receptor function in the endothelium. Cardiovascular Res 2005;65:599–608.
82. Solovyan VT, Keski-Oja J. Apoptosis of human endothelial cells is accompanied by proteolytic processing of latent TGF-β binding proteins and activation of TGF-β̃. Cell Death Differ 2005;12:815–826.
83. Hodge S, Hodge G, Flower R, et al. Up-regulation of production of TGF-β and IL-4 and down-regulation of IL-6 by apoptotic human bronchial epithelial cells. Immunol Cell Biol 2002;80:537–543.
84. Chen W, Jin W, Tian H, et al.Requirement for transforming growth factor-1 in controlling T cell apoptosis. J Exp Med 2001;194:439–453.
85. Leksa V, Godar S, Schiller HB, et al. TGF-β-induced apoptosis in endothelial cells mediated by M6P/IGFII-R and mini-plasminogen. J Cell Sci 2005;118:4577–4586.
86. Ito Y, Miyazono K. RUNX transcription factors as key targets of TGF-β superfamily signaling Curr Opin Genet Dev 2003;13:43–47.
87. Westendorf JJ, Hiebert SW. Mammalian runt-domain proteins and their roles in hematopoiesis, osteogenesis, and leukemia J Cell Biochem 2003;32/33:51–58.
88. Sun L, Vitolo MI, Qiao M, et al. Regulation of TGF-β-mediated growth inhibition and apoptosis by RUNX2 isoforms in endothelial cells. Oncogene 2004;23:4722–4734.
89. Finch CE, Laping NJ, Morgan TE, et al. TGF-β1 is an organizer of responses to neurodegeneration. J Cell Biochem 1993;53:314–322.
90. Flanders KC, Ren RF, Lippa CF. Transforming growth factor-βs in neurodegenerative disease. Prog Neurobiol 1998;54:71–85.
91. Unsicker K, Krieglstein K. TGF-betas and their roles in the regulation of neuron survival. Adv Exp Med Biol 2002;513:353–374.
92. Brionne TC, Tesseur I, Masliah E, Wyss-Coray T. Loss of TGF-β1 leads to increased neuronal cell death and microgliosis in mouse brain. Neuron 2003;40:1133–1145.
93. Elvers M, Pfeiffer J, Kaltschmidt C, Kaltschmidt B. TGF-β2 neutralization inhibits proliferation and activates apoptosis of cerebellar granule cell precurors in the developing cerebellum. Mech Dev 2005;122:587–602.
94. Dünker N, Schuster N, Krieglstein K. TGF-β modulates programmed cell death in the retina of the developing chick embryo. Development 2001;128:1933–1942.
95. Krieglstein K, Richter S, Farkas L, et al. Reduction of endogenous transforming growth factors beta prevents ontogenetic neuron death. Nat Neurosci 2000;3:1085–1090.
96. de Luca MW, Fontana A. TGF-β-induced apoptosis of cerebellar granule neurons is prevented by depolarization, J. Neurosci. 1996;16:4174–4185.
97. Casaccia-Bonnefil P. Cell death in the oligodendrocyte lineage: a molecular perspective of life/death decicisions in development and disease. Glia 2000;29:124–135.
98. Barres BA, Hart IK, Coles HS, et al. Cell death in the oligodendrocyte lineage. J Neurobiol 1992;23: 1221–1230.
99. Scurlock B, Dawson G. Differential responses of oligodendrocytes to tumor necrosis factor and other pro-apoptotic agents: role of ceramide in apoptosis. J Neurosci Res 1999;55:514–522.
100. McKinnon RD, Piras G, Ida JA Jr, Dubois-Dalcq M. A role for TGF-β in oligodendrocyte differentiation. J Cell Biol 1993;121:1397–1407.
101. Schuster N, Bender H, Philippi A, et al. TGF-β induces cell death in the oligodendroglial cell line OLI-neu. Glia 2002;40:95–108.
102. Schuster N, Bender H, Rossler O, et al. Transforming growth factor-beta and tumor necrosis factor-a cooperate to induce apoptosis in the oligodendroglial cell line OLI-new. J Neurosci Res 2003;73: 324–333.
103. Bender H, Wang Z, Schuster N, et al. TIEG1 facilitates transforming growth factor-β-mediated apoptosis in the oligodendroglial cell line OLI-neu. J Neurosci Res 2004;75:344–352.

104. Ribeiro A, Bronk SF, Roberts PJ, et al. The transforming growth factor beta(1)-inducible transcription factor TIEG1 mediates apoptosis through oxidative stress. Hepatology 1999;30:1490–1497.
105. Buenemann CL, Willy C, Buchmann A, et al. Transforming growth factor-β1-induced Smad signaling, cell-cycle arrest and apoptosis in hepatoma cells. Carcinogenesis 2001;22:447–452.
106. Shima Y, Nakao K, Nakashima T, et al. Activation of caspase-8 in transforming growth factor-β-induced apoptosis of human hepatoma cells. Hepatology 1999;30:1215–1222.
107. Yamamura Y, Hua X, Bergelson S, Lodish HF. Critical role of SMADS and AP-1 complex in transforming growth factor-β-dependent apoptosis. J Biol Chem 2000;275:36,295–36,302.
108. Landström M, Heldin NE, Bu S, et al. Smad7 mediates apoptosis induced by transforming growth factor β in prostatic carcinoma cells. Curr Biol 2000;10:535–538.
109. Skoff AM, Lisak RP, Bealmear B, Benjamins JA. TNF-alpha and TGF-beta act synergistically to kill Schwann cells. J Neurosci Res 1998;53:747–756.
110. Yu C, Takeda M, Soliven B. Regulation of cell cycle proteins by TNF-alpha and TGF-beta in cells of oligodendroglial lineage. J Neuroimmunol 2000;108:2–10.
111. Chipuk JE, Bhat M, Hsing AY, et al. Bcl-xL blocks transforming growth factor-β1-induced apoptosis by inhibiting cytochrome c release and not by directly antagonizing Apaf-1 dependent caspase activation in prostate epithelial cells. J Biol Chem 2001;276:2614–2621.
112. Saile B, Matthes N, El Armouche H, et al. The bcl, NFkappaB and p53/p21WAF1 systems are involved in spontaneous apoptosis and in the anti-apoptotic effect of TGF-β or TNF-α on activated hepatic stellate cells. Eur J Cell Biol 2001;80:554–561.
113. Freathy C, Brown DG, Roberts RA, Cain K. Transforming growth factor-β1 induces apoptosis in rat FaO hepatoma cells via cytochrome c release and oligomerization of Apaf-1 to form a 700-kd apoptosome caspase-processing complex. Hepatology 2000;32:750–760.
114. Zhu Y, Yang GY, Ahlemeyer B, et al. Transforming growth factor-β1 increases Bad phosphorylation and protects neurons against damage. J Neurosci 2002;22:3898–3909.

14 Mechanisms of Cell Cycle Regulation by TGF-β Disabled in Cancer

Leslie I. Gold and Jon Lecanda

CONTENTS

INTRODUCTION
TGF-β EFFECTS ON CELL CYCLE PROTEINS OBVIATED IN HUMAN CANCERS
THE REGULATION OF THE CDKI, P27KIP1, BY TGF-β
REFERENCES

Abstract

Disabling the TGF-β signaling pathway is a widespread means by which carcinoma cells subvert the normally potent growth arresting effect of TGF-β to enable cancer cell growth. We have shown disrupted TGF-β signaling concomitant with loss of growth inhibition by TGF-β occurs early in endometrial carcinogenesis. As an effector of cell cycle arrest, TGF-β inhibits synthesis of certain cyclin/cdks and directly induces cdk inhibitors (CDKI), such as $p15^{Ink4b}$(p15) and $p21^{Waf1/Cip1}$ (p21), to inhibit phosphorylation of Rb, thereby preventing progression to S phase. Although the level of the CDKI, $p27^{Kip1}$ (p27) is increased in a variety of TGF-β treated cells, evidence for the regulation of p27 by TGF-β in epithelial cells has largely been indirect. Now, we show that TGF-β dose-dependently increases both nuclear and cytoplasmic levels of p27 in primary cultures of normal endometrial epithelial cells (EECs) by preventing its degradation through the downregulation of Skp2 and cks1, critical components of the SCF^{skp2} complex, that target p27 for ubiquitin-mediated degradation. Thus, the levels of p27 and Cks1/Skp2 are inversely regulated by TGF-β to achieve normal cell cycle arrest in late G1. In contrast, the lack of TGF-β signaling in primary endometrial carcinoma (ECA) cells promotes rapid proteasomal degradation of p27 for uncontrolled cell growth. The increase in p27 and decrease in Cks1/Skp2 is recapitulated in vivo in normal secretory phase endometrium (SE) compared to all grades of ECA, which show the reverse. Transient transfection of Smad2 into a TGF-β-responsive ECA cell line, HEC-1A, decreased the levels of cks1 and a small molecule inhibitor of TGF-β receptor I kinase (SD-208) blocked the ability of TGF-β to decrease cks1 and increase p27 indicating that these effects result from direct classic TGF-β signaling. Importantly, knocking down p27 with a siRNA approach blocked TGF-β mediated growth inhibition in HEC-1A, further substantiating that TGF-β mediates growth inhibition through p27 in these cells. In addition, we further show that estrogen causes MAPK-driven ubiquitin-mediated degradation of p27 for proliferation and that progesterone induces marked accumulation of p27 to block cell growth. Thus, our results suggest that TGF-β and ovarian hormones converge on the regulation of intra-cellular p27 levels as a major target for endometrial growth. Because ECA is an estrogen-induced cancer, related to unopposed estrogen, and progesterone is therapeutic for this disease, we propose that the pathogenesis of ECA may be related to both the lack of TGF-β signaling and the absence of progesterone (as in menopause and anovulation), causing continuous estrogen-driven proteasomal

From: *Cancer Drug Discovery and Development: Transforming Growth Factor-β in Cancer Therapy, Vol I: Basic and Clinical Biology*
Edited by: Sonia B. Jakowlew © Humana Press Inc., Totowa, NJ

degradation of p27. Finally, soluble mediators derived from stromal cells in the ECA tumor microenvironment, but not from normal stromal cells, decrease the levels of p27 and increase the levels of cks1, while increasing growth of HEC-1A cells, implicating a role for stromal cells in endometrial carcinogenesis.

Temporally regulated ubiquitin-mediated destruction of cell cycle proteins precisely control the levels of proteins to block or induce cell cycle progression. Our results suggest that TGF-β negatively regulates destruction of p27 protein to maintain sufficient levels of this CDKI for cell cycle arrest. As such, TGF-β facilitates a counter-destruction mechanism of cell cycle regulation. As Skp2 and cks1 are markedly increased in ECA (and other human cancers) perpetuating destruction of p27 thereby permitting uncontrolled growth, these proteins may be novel rational targets for the prevention and therapy of this cancer.

Key Words: p27kip1; cks1; skp2; endometrial cancer; ubiquitin–proteasome pathway.

1. INTRODUCTION

The tumor suppressor function of TGF-β and members of its signaling cascade more recently, have been challenged by a role for TGF-β in epithelial to mesenchymal transition (EMT) and metastasis *(1–4)*. However, although TGF-β is prooncogenic in the later stages of malignant progression, its role in preventing the initial escape of cancer cells from growth regulation remains important to define on a molecular level for the development of new concepts that can translate into cancer prevention. Moreover, it is important to delineate the intricacies of how TGF-β normally regulates the cell cycle for a better understanding of mechanisms and intracellular pathways of TGF-β action that have become awry in cancer cells as they are potential targets for therapeutic reconstitution. As the complexities for the dual role of TGF-β as a tumor suppressor and perpetuator of malignant progression unravel, it is most important to take into account both the context-dependent and cell-type dependent nature of the responses observed for both normal and cancer cells.

Generally, TGF-β inhibits the mRNA and protein synthesis of certain cyclin/cdks, and induces inhibitors of cdks (CDKI) to block kinase activity needed for cell cycle progression *(5)*. Whereas the levels of proteins belonging to the cell cycle machinery are regulated by cues and rate of gene transactivation and protein translation, the precise timing of degradation of these critical proteins has become equally important in the regulation of the cell cycle, including mitosis. To this point, specific ubiquitin E3 ligases peak and plummet throughout the phases of the cell cycle targeting cyclin, cdks, CDKIs, and other components of the cell cycle for ubiquitin-mediated destruction by the 26S proteasome, in perfect time for normal cell cycle transition *(6)*. Thus, in addition to the underlying gene mutation-based oncogenesis, aberrant levels of ligases at inappropriate times in the cell cycle, that degrade CDKIs important for blocking G1 to S transition, would arm cancer cells with the ability to uncontrollably proliferate. Indeed, high levels of the E3 ligase, Skp2, target the CDKI, $p27^{Kip1}$, for proteasomal degradation in a number of human cancers *(7–11)*. Thus, although p27 can be regulated at the levels of transcription and translation, rapidly growing evidence has revealed an exquisite complex multitude of post-translational mechanisms that regulate the fate of p27 function, thereby providing clues for mechanisms involved in malignant growth *(8)*. Phosphorylation of p27 by specific kinases at different sites on the molecule determine both the specific systems involved in degradation and nuclear cytoplasmic shuttling. In fact, expulsion of p27 and/or prevention of its importation into the nucleus, where its major function is to inhibit cdk2 activity to block progression to S phase, is presently recognized as a means of fostering cancer growth.

The contents of this chapter will provide an overview of how TGF-β achieves its cytostatic effect and how the loss of this TGF-β tumor suppressor function in human cancers aberrantly effects cell cycle regulation, with particular emphasis on the current status of the function of the CDKI, p27 in human cancers *(5,8,12–15)*. The goal of our studies has been to reveal mechanisms involved in endometrial carcinogenesis. For this purpose, we isolated cells from both malignant and normal endometrium for primary cultures in order to compare normal and malignant cell growth and to aim at ensuring optimal physiological relevance of our results. These studies suggest that TGF-β prevents degradation of p27 by downregulating Skp2 and Cks1 that target its ubiquitin-mediated degradation, as a means of normally inhibiting EEC growth and that this normal function is obviated in endometrial cancer cells that lack TGF-β signaling. This mechanism, downstream from TGF-β signaling may play a significant role in the pathogenesis of ECA and potentially, other human cancers. The negative regulation of SCF^{skp2} ubiquitin-mediated degradation of p27 provides a novel perspective on a role for TGF-β in controlling cell growth and a significant portion of this chapter will be devoted to our experimental support for this effect. Further, we will seize the opportunity to report on newly discovered mechanisms that have been shown to regulate p27 in normal and cancer cell growth. In the final part of this chapter, we will briefly draw provocative parallels between TGF-β and p27 in their paradoxical functions as tumor suppressors and as mediators of EMT migration/metastasis through cytoskeletal effects. Identifying mechanisms and molecular targets associated with TGF-β separately, as a tumor suppressor and as a proonco-genic protein, as well as the internal intracellular differences that enable this switch of TGF-β function during progression should be described for each type of human cancer. Along with the individual genetic and proteomic signatures from each patient, we may define and assign appropriate preventative and therapeutic strategies.

2. TGF-β EFFECTS ON CELL CYCLE PROTEINS OBVIATED IN HUMAN CANCERS

2.1. TGF-β and the Cell Cycle

2.1.1. Aberrant TGF-β Signaling in Human Cancers

As TGF-β signaling in cancer is described in this book, this chapter will focus on defects in TGF-β signaling as they relate to aberrant cell cycle control. Briefly, as mutations disrupting the growth inhibitory function of TGF-β have been found in nearly all members of the signaling cascade *(16–18)*, a tumor suppressor role has been assigned to all components of the classic TGF-β signaling machinery. Aside from the integrity of TGF-β signaling, another level of regulation of TGF-β function is the ability of cells to locally activate the latent precursor molecule generally bound to the extracellular matrix to release the 25 kDa bioactive dimer *(19,20)*. TGF-β receptors are expressed on nearly all cell types and both a simplistic TGF-β-specific signal transduction pathway and a complex network of additional interacting intracellular pathways that lend versatility to the TGF-β response have been described *(21–23)*. There are two cell surface receptors, TβRI and TβRII, that possess serine/threonine kinase activity. Following ligand binding of any of the three (mammalian) TGF-β isoforms to TβRII, TβRI is recruited and phosphorylated. Then, conformationally activated, TGF-βRI phosphorylates the downstream transcription factors, receptor-specific Smad2 or Smad3 (R-Smads) in their C-terminal serine-any amino acid (X)-serine (SXS) motifs. Subsequently, R-Smads collaborate with Co-Smad4, which translocates to the nucleus to bind to promoters in conjunction with specific DNA-binding transcription factors, coactivators, or corepressors that specify TGF-β-driven responses. Nucleocytoplasmic shuttling and sequestration of receptors into endocytic vesicles influence TGF-β-responses *(24–29)*. Importantly, TGF-β receptors can both signal through other signaling pathways and through Smads by cross-talk with mitogen-associated kinase (MAPK) and numerous other signaling pathways *(22)*. For example, kinases from other pathways have been shown to signal by phosphorylation of Smads within their middle linker region, rather than the TbRI-specific SXS sites (e.g., PKC, CaMKII, Akt, Ras), causing inhibition of Smad function *(22,30,31)*. An inhibitory, I-Smad7, can be upregulated to block Smad2/3 signaling *(32,33)*. Some of these signaling routes, particularly MAPK- and Akt-related, have been shown to occur in nearly all human cancers, cancer cell lines, and in various context-dependent situations *(2,14,23,34)*.

2.1.2. Targets of TGF-β Regulation: The Cyclins and Cyclin-Dependent Kinase Inhibitors

The diverse pathways for TGF-β signaling culminate in a multitude of effects on various components of the cell cycle for normal growth inhibition of epithelial-derived cells in a context-dependent and cell-type specific manner. This provides genetically altered or dysplastic cells with a plethora of approaches to disengage for malignant growth. In addition, endothelial cells, hematopoietic cells, and lymphocytes are growth inhibited by TGF-β *(35)*. In contrast, TGF-β generally stimulates the growth of mesenchymally-derived cells *(36)*. The growth of a tumor epitomizes both the loss of normal growth control and decreased apoptosis such that tumor volume is increased. TGF-β effects both functions to maintain homeostasis of growth and programmed cell death in different cells and under specific conditions *(5,13,14)*. The major effect of TGF-β on the cell cycle is to regulate G1 to S transition. The cell cycle is driven by specific cyclins and their respective catalytic partners, cyclin-dependent kinases (cdks), that phosphorylate the pocket proteins, Rb, p130, and p107, the masters of the cell cycle (Fig. 1). As the pocket proteins are stochastically phosphorylated with cell cycle progression, they release E2F transcription factors to activate genes involved in proliferation such as c-*myc* and cyclinE1. An E2F-1-Rb complex has been shown to act as a transcriptional repressor to mediate cell cycle arrest by TGF-β (and contact inhibition) *(37)*.

Prompted by unexpected differential phenotypes obtained with specific E2F isoform null mice and further pursued using a ChIP-on-chip approach, the regulation of proliferating/ cycling vs quiescent cells with respect to the various combinations of eight E2F and three pocket proteins with their respective target genes, have revealed a new emerging complexity of cell cycle regulation by these proteins *(38–40)*. These studies show that multiple distinct networks of pocket protein interactions on specific target genes at specific phases of the cell cycle temporally regulate proliferating cells. E2F1-3 interact with Rb as transcriptional activators in continuously cycling cells while E2F4,7,8 act with p130 and p107 as transcriptional repressors, as shown in quiescent cells *(41)*. For example, in mitogenically-stimulated quiescent cells, both Rb and p130 are phosphorylated (by cyclinD-dependent kinases) and in cells that have exited the cell cycle, p130/E2F4 complexes operate to maintain quiescence.

The simplistic view of the dogma for cell cycle regulation/progression has been under scrutiny from additional studies showing that disruption of genes encoding the three D cyclins, two E cyclins, and cdk2 has little effect on fetal mouse development *(5,42,43)*. Nonetheless, until further information is revealed, the classic view is as follows (Figure 1): The D-type cyclins (D1, D2, and D3) couple with Cdk4 and cdk6 and the E-type cyclins (E1 and E2), partner with cdk2 for entry and progression through G1 and on to S phase for the synthesis of DNA. Cyclin A binds to cdk2 and then cdk1 continues to phosphorylate Rb through S phase followed by cyclin B/cdk1, which subsequently drives the cell cycle through G2 to M (prophase through anaphase) *(44)*. G1 and G2 are the gaps between the major functional phases of DNA synthesis and mitosis; TGF-β acts on both G phases. Quiescent cells exist in G0 and can be recruited into the cell cycle at the G1 point (G0/G1) by extracellular signals including cell adhesion and mitogens *(45–47)*. The point in G1, when a cell becomes committed to progressing through the cell cycle is referred to as the restriction point (R); this point divides G1 into early and late phases and cyclin D/cdk4,6 complexes phosphorylate Rb to move the cell cycle through to late G1. TGF-β blocks G1 to S progression in the late

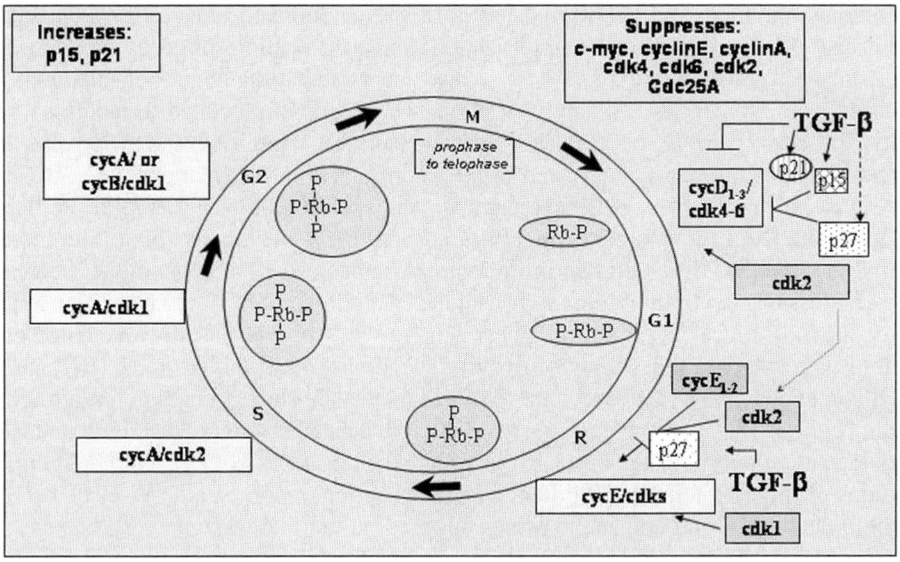

Fig. 1. Simplistic view of TGF-β actions on the cell cycle. The cyclins (cyc) and cyclin-dependent kinases (cdks) form a complex that incrementally phosphorylate (P) the retinoblastoma protein (Rb) for progression through G1 onto S phase. $p15^{ink4b}$ (ink family), $p27^{kip1}$, and $p21^{Cip1}$ (cip/kip family) are cyclin-dependent kinase inhibitors (CDKI) that inhibit cdk kinase activity thereby blocking Rb phosphorylation and preventing progression to S phase (block in G1). $p15^{ink4b}$ is specific for cycD1-3 and inhibits cdk4,6, activity. $p27^{kip1}$ (cip/kip family) interacts with most cyclins but specifically inhibits cyclinE/cdk2 in late G1. TGF-β directly suppresses cycA, cycE, and the cdks and directly upregulates $p21^{Cip}$ and $p15^{ink4b}$. The action of TGF-β on $p27^{kip1}$ ensues following an increase in the concentration of $p15^{ink4b}$ that displaces p27kip1 from cycD/cdk4,6 to cycE/cdk2 to block cdk2 activity, leaving Rb in a hypophosphorylated state for G1 block (*see* Section 2.1). Cells are committed to cell cycle progression at the restriction point (R).

G1 phase. However, just prior to entering S phase (approximately 2 h), TGF-β can no longer effect G1 arrest *(48,49)*.

TGF-β decreases the synthesis of cyclin E and cyclin A as well as cdk4, cdk6, and cdk2 that promote cell cycle progression. However, cdk4 and cdk2 have been shown to phosphorylate specific sites in the middle linker region of Smad2 and Smad3 (different than MAPK sites), interfering with TGF-β signaling, as a negative regulatory loop to inhibit Smad-dependent transactivation of genes involved in cell-cycle arrest *(50)*. As increased cdk activity is found in cancer cells *(44,51)*, this is an additional route that would lead to enhanced tumorigenesis with the loss of TGF-β responsiveness. In addition, TGF-β inhibits the phosphatase Cdc25A, which is important in activating both cdk2 and cdk4,6 kinase activity thereby reducing Rb and p130 phosphorylation for G1 arrest *(52–54)*. Furthermore, to inhibit Cdc25A, TGF-β induces activation of RhoA, a protein involved in cytoskeletal organization and also, shown to mediate cell cycle progression *(55)*. Subsequently, the downstream effector of RhoA, $p160^{ROCK}$, translocates to the nucleus to inhibit Cdc25A by a posttranslational modification, possibly phosphorylation *(56,57)*. Thus, RhoA is required for cell cycle arrest in certain cells. Interestingly, whereas the Smad3 linker region contains phosphorylation sites for p38MAPK and Rho/ROCK (Ser203 and Ser207), shown to be required to mediate growth inhibition in human breast cancer cells in vivo *(58)*, these identical signaling pathways, including the PI3K pathway mediate EMT *(59)*. TGF-β also inhibits CAK (cdk activating kinase) to block cdk2 activity *(60)*.

Deletions, mutations, and abnormal levels of the cyclins and cdks have been found in human cancers *(61)*. Notably, gene amplifications and translocations of genes encoding cyclin D1 and cdk4 cause their overexpression in many cancers, including in 43% of squamous, 34% esophageal, and 50% breast carcinomas *(51,62,63)*. Specifically, related to increased kinase activity, the cdk4 gene is mutated in 4% of melanomas, amplified and overexpressed in glioblastomas (50%), breast cancer and osteosarcomas (16%), and overexpressed in 73% of uterine/cervical cancers *(61)*. Similarly, cdk6 is overexpressed in over 90% of acute lymphocytic leukemia (ALL) and other lymphomas ad amplified in gliomas (5%). The increased expression of cyclin D1 for growth promotion was enhanced by a concomitant decrease in TβRII and loss of growth inhibition in an esophageal cell line *(64)*. Aside from increased cdk activity, the most prevalent target for cancer cells to block or divert TGF-β-mediated growth inhibition is to obviate the effects of cyclin-dependent kinase inhibitors (CDKI), that are normally upregulated by TGF-β to achieve block in late G1. Thus, a major contributing factor to carcinogenesis (80% of human cancers) is the imbalance between cdk activity and CDKIs for misregulation/phosphorylation of Rb or p130 resulting from both genetic and epigenetic alterations of these proteins and further, epigenetic dysregulation of the cell cycle can result from both the genetic and epigenetic events imposed on the TGF-β signaling pathways.

2.1.3. Targets of TGF-β Upregulation: Cyclin-Dependent Kinase Inhibitors

The most significant and direct means for TGF-β to mediate cell cycle arrest is to inhibit the kinase activity of cdks achieved by inducing the synthesis $p15^{Ink4b}$ and $p21^{Cip1}$ genes, members of the INK4 and CIP/KIP families of CDKIs, respectively *(13,65–68)* and to downregulate c-*myc* (as described in Section 2.1.4.). The CDKIs are considered to be tumor suppressors, because they directly and effectively block cell cycle progression. Other members of the CIP/KIP family of tumor suppressors are $p27^{Kip1}$ and $p57^{Kip2}$, and of the INK family, $p16^{INK4a}$, $p18^{INK4c}$, and $p19^{INK4d}$ [61]. Interestingly, the $p16^{INK4a}$ gene locus encodes another protein, p14ARF (p19ARF [alternate reading frame] in mice), which contains a different promoter, in which only certain exons are translated using different open reading frames *(69)*. Whereas $p15^{Ink4b}$ blocks cdk4,6 activity early in G_1 phase, $p21^{Cip1}$ blocks cdk2 latter in G_1 just prior to S phase entry. In addition, the CIP/KIP CDKIs block the cell cycle in G_2 phase as well *(70)*. Whereas TGF-β directly, rapidly, and specifically upregulates p15 downstream from Smad signaling *(5,13)*, it can still induce growth arrest in p15 null cells most likely through inhibiting Cdc25A phophatase activity *(54)*. Although, the p15 promoter does not contain canonical Smad binding sites *(71)*, the gene is transactivated by the interaction of Smad2, Smad3, and Smad4 with the zinc finger transcription factor, Sp1; the p21 promoter is similarly transactivated in this way. The activation of p21 by TGF-β is independent of p53 (in p53 mutant pancreatic cells) *(72)*. In addition, p21 is a downstream facilitator of the p53 pathway arresting cell growth and causing apoptosis following DNA damage. Notably, p53 is the most commonly mutated gene in human cancer *(69,73)*. Both the p15 and p21 genes as well as other genes involved in cytostasis are activated in concert by Smads and the Forkhead box transcription factors, FoxO1, FoxO3, and FoxO4. The FoxO factors bind directly to Smad3 and Smad4 and simultaneously to the FHBE and SBE sites, respectively, contiguous sites on the promoters *(74,75)*. The activation of PI3K/Akt negatively regulates p21 by phosphorylating FoxO causing its exclusion from the nucleus in human keratinocytes and glioblastoma cells lines *(75)*. Additionally, FoxG1 blocks TGF-β-induced p21 expression by binding to FoxO, thereby causing TGF-β resistance and enabling proliferation of glioblastoma cells. Therefore, constitutive activation of the PI3K/Akt pathway, found associated with cancer *(76,77)*, may be one means of promoting TGF-β resistance by inhibition of the TGF-β-induced p21 response. TGF-β was able to induce p21 in a Smad4 null pancreatic cell line, BxPC-3, via Smad2 and Smad3 binding to the p21 promoter, and mediate growth inhibition *(78)*. In contrast, in other studies, using the pancreatic tumor cell line COLO-357 and HaCaT cells (human keratinocytes), Smad4 was essential for the function of TGF-β-mediated growth inhibition *(79)* and for upregulation of p21 in MDA-MB-468 breast carcinoma cells and HaCaT cells *(80)*. Growth arrest with concomitant increase in p21 and p27 in MCF-7 cells was achieved by solely restoring expression of TβRII *(81)*. Genetic deletions and hypermethylation of the p15 and p16 genes are frequently associated with many human cancers *(44,82,83)*. For example, alterations of the INK4b locus (at 9p21) encoding p15 occurs at a rate of 90% in leukemia, 55% in lymphoma, and 30–50% in gliomas *(61)*.

Of the CIP/KIP family of CDKIs, $p27^{kip1}$ (p27) accumulates in response to TGF-β to purportedly inhibit cdk2 activity (discussed below in Part 3.1) *(8,15,84,85)*. As shown by gene arrays, $p57^{KIP2}$ is the CDKI that is directly induced by TGF-β for cell cycle arrest in G_1 in human hematopoietic cells by Smad3/Smad4 binding in the proximal promoter region *(86)*. p57 appears to be a critical tumor suppressor in hematologic malignancies, which is owing to silencing of the gene by hypermethylation, as shown in 30–55% of acute lymphoblastic and myelogenous leukemias and B-cell lymphomas *(87–89)*. Conversely, TGF-β induces a decrease in p57, but not p27, in osteoblasts as an effector of growth stimulation by increasing cdk2 and preventing cell differentiation and it was further shown that TGF-β lowered p57 levels, through Smad2, Smad3, and Smad4, by inducing ubiquitin-mediated p57 degradation, which was inhibited by the inhibitory, Smad7 *(90,91)*.

2.1.4. Target of TGF-β/Smad Signaling: The c-*myc* Transcription Factor

The downregulation of the protooncogene and transcription factor, c-*myc*, by TGF-β is essential for TGF-β to exert cytostasis and particularly for transactivation of the CDKIs, p15 and p21 *(92–95)*. In fact, enforced expression of c-*myc* overrides the ability of TGF-β to induce growth arrest *(95–97)*, at least in part, by inhibiting the ability of TGF-β to separately induce both p15 and p21 *(92,93,95)*. Whereas c-*myc* in association with Miz1 (the c-*myc* interacting zinc finger protein) directly represses both p15 and p21 genes, Smad3 downregulates c-*myc* *(98)*. Additionally, c-*myc* directly interacts with Smad2 and Smad3 to block their interaction with Sp1, forming a c-*myc*/Smad-Sp1 nucleoprotein complex to block Sp1-dependent transcription of p15 and p21 by TGF-β *(98,99)*. Aside from the classic Smad binding element (SBE), a novel Smad3 binding site, repressive Smad binding site (RSBE), within the TGF-β inhibitory element (TIE) of the c-*myc* promoter, also containing sites for E2F-4 and p107, was identified which functions both to repress and activate the c-*myc* promoter *(100)*. Moreover, a corepressor complex composed of Smad3 directly bound to E2F4 or E2F5 was shown to be recruited to the TIE on the c-*myc* promoter in response to TGF-β further requiring p107 for c-*myc* repression by TGF-β *(101)*. The SBE site adjacent to the TIE on the c-*myc* promoter was not involved in TGF-β mediated c-*myc* repression *(100)*. These studies show that both the SBE and the RSBE in conjunction with specific coactivators or corepressors that are cell and context dependent determine the fate of whether Smad3 will engage in repression or activation of c-*myc*. Taken together, the interaction of Smad and *myc* are opposing as Smad inhibits c-*myc* downstream of TGF-β cytostatic function and in turn, c-*myc* blocks Smad, to block TGF-β function *(5)*. Thus, the central cause for resistance to TGF-β growth arrest converges on c-*myc* and p15. This can be achieved by disabling the Smad pathway of TGF-β signaling, by c-*myc* directly interacting with Smad2 and Smad3 to block Smad-dependent transcription of p15, and by c-*myc* directly repressing basal transcription of the p15 gene *(95,102)*.

The action of c-*Myc*, l-*myc*, and n-*Myc* induce D-type cyclins in association with the transcription factor, Max which as discussed in Section 2.1.2., are increased in many human cancers *(103)*. Amplification of the c-*myc* gene and overexpression of the protein, leading to TGF-β resistance, has been shown in about 30% of human cancers, including colon and breast carcinomas, and in the appropriate context, even Smad3 itself can transactivate this protooncogene *(104–111)*. The inhibition of c-*myc* appears to be a reasonable target for cancer therapy and in fact, c-*myc* antisense has been shown to arrest growth in certain cancer cells *(112–114)*. The Smad proteins interact with close to 40 DNA-binding transcription factors to enable diverse responses thoroughly reviewed elsewhere *(22)*. Prominent among them are Id1–Id3, that act like c-*myc* as they suppress CDKIs and are downregulated by TGF-β to effectuate growth arrest *(5,13,115–117)*, the RUNX proteins *(118)*, coactivators, such as CBP/p300 and the corepressors, c-*ski*, snoN, TGIF, all which can be perturbed from normal function, in cancer *(5,22)*.

3. THE REGULATION OF THE CDKI, $p27^{KIP1}$, BY TGF-β

3.1. $p27^{kip1}$ *Effects on the Cell Cycle*

The cyclin-dependent kinase inhibitor (CKI), $p27^{kip1}$ (p27 hereafter), regulates cell cycle progression from G0 through G1 and onto S phase and is considered to be a focal point for loss of tumor suppressor activity in a large number of human cancers *(8,12,15)*. Unique to CDKIs, p27 has a dual role in regulating the cell cycle *(119,120)*. It acts early in G1 to aid in assembly, activation, and localization of cyclinD/cdk2 complexes for phosphorylation of Rb and thus, progression to S phase. Antithetically, however, p27 prevents further phosphorylation of Rb by binding cyclinE/cdk2 and cyclin A/cdk2 complexes to enforce G1 arrest. p27 was discovered associated with cyclinE/cdk2 and cyclinA/cdk2 in which the kinase activity of cdk2 was blocked in TGF-β arrested *(121–123)* and contact inhibited epithelial cells *(124)*. An increase in p27 level was also observed in cells during growth arrest induced by low serum *(125)*, growth in suspension *(126)*, and in cells arrested in G1 by lovastatin *(123)*. A direct effect of TGF-β on increasing nuclear levels of p27 for inhibition of G1 cyclin-cdks has previously not been observed in epithelial cells. Ostensibly, the mechanism for coordinate increase in p15 and p27 by TGF-β results from a direct induction of p15 mRNA, causing a concomitant release of p27 from the cyclin D–cdk complex and its subsequent association with cyclin E–cdk2 to inhibit cdk2 activity and block progression to S phase *(65–67,121,127,128)* (Fig. 1). In this way, the increase in p15 destabilizes and displaces p27 binding to cyclinD/cdk4. However, because cells containing nonfunctional p15 are able to accumulate p27 and inhibit cyclin E–cdk2 in response to TGF-β and in addition, both p15 and p27 null cells can still undergo TGF-β-mediated growth inhibition *(54,129)*, either compensatory or alternative mechanisms occur. Indeed, TGF-β induced an increase in p21 and p130 binding to cyclin E–cdk2, causing growth inhibition in one study *(130)*. Another study showed that TGF-β effects were related to modulation of the phosphorylation of p27 *(131)*.

3.1.1. Posttranslational Mechanisms Regulate p27 Levels and Subcellular Distribution

Recently, there has been strong interest in conceptually viewing the temporal posttranslational regulation of the levels of the masters of the cell cycle: the cyclins-cdks and their CDK inhibitors. The levels of these proteins are exquisitely coordinated with the phases of the cell cycle by ubiquitin-mediated proteasomal degradation pervading during both S phase progression and G1 arrest (in addition to and equally important to activation of transcription and translation). Accordingly, the intracellular levels and subcellular localization of p27 are tightly regulated by phosphorylation of the molecule at specific sites and through different protein interactions throughout the cell cycle *(8,12,132–139)*. In addition, the concentration of p27 in the nucleus is coordinately controlled in conjunction with phases of the cell cycle.

Maximal levels occur at G1 and G0, when p27 inhibition of cyclinE-cdk2 is highest *(65,123,140)*. Subsequently, p27 levels dissipate with progression from G1 to S and are lowest during S phase when phosphorylation of p27 at T187 by cyclin E-cdk2 signals ubiquitin-mediated degradation for cycle progression *(141–144)*. As such, mutant p27 (T187A) is not phosphorylated and therefore, resistant to ubiquitin-mediated degradation by proteasomes preventing S phase progression *(11)*. The dual nature of p27 to inhibit cyclinE–cdk2 activity and arrest growth while this same cdk complex phosphorylates p27 causing its destruction for S phase progression, is purportedly dictated by relative binding affinities *(141)*. Specifically, at low affinity and equilibrium binding to the cyclinE–cdk2 complex, p27 is a substrate for cdk2 activity, while at high-affinity binding it inhibits the activity of this complex. It is envisioned that this reciprocal interaction ensures forward transition from G1 to S by providing a negative regulatory feedback loop *(141,142,145)*. In addition, phosphorylation of p27 at T187 by cyclin A–cdk2 and cyclin B–cdk1 regulates p27 turnover *(146)*.

Ubiquitin-mediated degradation of proteins by the 26S proteasome proceeds sequentially through an ATP-dependent enzymatic cascade involving the E1, ubiquitin-activating enzyme, an E2 ubiquitin-conjugating enzyme (Ubc), and an E3 ubiquitin ligase. The specificity for substrate is dictated by the last two enzymes with the E3 ligase as the final enzyme that attaches the ubiquitin molecules, covalently bound to E2, to lysines on the protein targeted for destruction *(147)*. Ubiquitylation of p27, purported to be on lysines 134, 153, and 165 requires phosphorylation at T187 thereby permitting its recognition and subsequent binding by S-phase kinase interacting protein-2 (Skp2), the E3 ligase F-box type protein specific for p27 *(8,148–150)*. Skp2 is part of the SCF (Skp-Cullin-F-box) scaffold ubiquitin protein ligase complex (Ubls), which has dual binding specificity for p27 and for being tethered to the SCF core containing cul1, skp1, roc1, and rbx1, through skp1 *(8,151–153)*. Cks1 binds to Skp2 to induce the requisite conformational change for interfacing with phosphorylated p27 *(154,155)*. This route of p27 degradation is reported to occur in the nucleus during S-G_2, thereby permitting high-nuclear cdk2 activity and thus, cycle progression *(133,156)*. However, likely cell-type specific differences account for a study showing that p27 is escorted to the cytoplasm by Skp2 for degradation in a lymphoblastoid B-cell line *(157)*. Thus, both Cks1 and Skp2 are rate limiting in the destruction of p27 for the transition from the quiescent to proliferative state. Accordingly, Skp2 and Cks1 null mice have elevated levels of p27 and reduced growth rates with *(158)* associated functional consequences *(151,152, 159,160)*. In addition, MAPK, constitutively active in many human cancers, has been shown to phosphorylate p27 at T187 in breast tumor cells *(161,162)*. Chemical blocking of EGF receptor kinase and thus, MAPK, increased p27 stability and delayed tumor formation in a mouse mammary model of tumorigenesis. Indirectly, oncogenic Ras signaling induced by growth factors (e.g., EGF and PDGF) led to p27 degradation via the GTP-binding protein RhoA via activation of cyclinE–cdk2 *(163,164)*. However, Ras activation of MAPK may exert kinase activity directly on p27 *(165)*. Recently, it has been shown that tyrosine phosphorylation of p27 on Y88 hastens the kinetics of phosphorylation at T187 by cdk2 causing a conformational change in p27 *(166)*. The implication of these studies to cancer is the high rate of tyrosine kinase activity found in malignancies, which would facilitate p27 degradation for S phase progression.

In addition to the SCF^{skp2} complex, the anaphase promoting complex (APC^{cdh1}) targets specific cell cycle proteins for degradation *(6)*. The APC complex regulates entry (late G2/Anaphase) and exit from mitosis (late mitosis/early G1) with substrate specificity for cyclins A and B, cdc20, cdc5, and interestingly Skp2 and Cks1 *(167,168)*. The Skp2 complex controls G1 to S transition and displays substrate specificity for the cell cycle proteins, cyclin E, cyclin D1, p27, p21, c-*myc*, FOXO1 and the pocket protein, p130 *(8,159,169,170)*.

Because the APC causes destruction of Skp2, higher levels of p27 would be maintained (to block cdk2 activity). Hence, the APC complex can control the span of time that the cell cycle will remain in G1. In addition, TGF-β-induced FOXO1-mediated transactivation of p27 mRNA was blocked by Akt-driven phosphorylation-dependent ubiquitin-mediated degradation of FOXO1 following Skp2 ubiquitylation implicating another means of regulation of p27 levels by Skp2 (preventing its transcription) *(171)*. Taken together, one can envision that cancer cells are also proficient at harnessing the ubiquitin system for the benefit of unchecked proliferation and as such, proteins involved in destruction of p27, such as Skp2, are elevated in human cancers and correlate with high grade malignancy *(7,172–180)*.

Prompted by studies showing that proteasome-dependent degradation of p27 at G0–G1 transition proceeds normally in Skp2 null mice, a mitogen-induced Skp2-independent pathway of p27 degradation was identified requiring phosphorylation of p27 at S10, by hKis, and subsequent localization of p27 to the cytoplasm *(133,181–183)*. The E3 ligase identified as the mediator of this Skp2-independent proteasomal degradation of p27 is the kip1 ubiquitylation-promoting complex 2 (KPC2), which together with KPC1 mediates p27 degradation during G1 *(132)*. MAPK, PI3K/Akt, and ERK2 also phosphorylate p27 on S10 *(136,182)*, which has been shown to be hepatocyte growth factor (HGF)-driven *(184)*. Nuclear export of p27 is mediated by the nuclear exportin, CRM1 and a RanGTP gradient causing degradation of p27 in the cytoplasm early in G1 *(185)*. This allows a step-wise activation of cyclinE–cdk2 followed by phosphorylation of p27 at T187 for degradation in late G1 (for S phase progression). Cell cycle dependent phosphorylation of p27 on S10 occurs at a rate of 70% of the total *(133,182)*. Interestingly, oncogenic Ras was shown to induce loss of TGF-β growth inhibitory activity by mislocalizing p27 along with cdk6 to the cytoplasm *(186)*. Recent mutational analysis studies show that S10 phosphorylation of p27 regulates both export to the cytoplasm and its association with cyclin/cdk complexes via Arg30, Leu32, Phe 62, and Phe64 within the p27 molecule *(12)*. Importantly, lung tumorigenesis was inhibited by blocking Ras-induced cytoplasmic translocation with a p27 mutant S10A. Conversely, phosphorylation of p27 on S10 by retinoic acid (RA) treatment of the Lan-5 brain tumor cell line increased p27 stability and nuclear accumulation for growth inhibition *(187)*.

In addition, p27 can be phosphorylated on T157 by Akt, the downstream substrate of the PI3K[PKB]/Akt intracellular signaling pathway, again causing cytoplasmic localization of p27 *(138,188)*. The correlation of high levels of Akt and p27 in the cytoplasm of breast cancer tissue may account for unrestricted cdk2 activity in the nucleus to enable continuous progression through the cell cycle *(189)*. PI3K/Akt-dependent phosphorylation of T157 and T198, but not T187 or S10, of p27, along with its mislocalization to the cytoplasm, was shown to be involved in thryroid carcinogenesis *(190)*. Cytoplasmic sequestration of p27 has been shown in esophageal *(191)*, colon *(192)*, ovarian *(193,194)*, and breast carcinomas *(188,189)*, and the presence of p27 in the cytoplasm has been associated with decreased survival *(195,196)*. It has been shown that p27 binds nuclear importins for transport into the nucleus and phosphorylation of p27 at T157 is proposed to suppress this activity by competitive binding of 14-3-3 to the nuclear localization signal region (NLS, residues 153–170) on p27 that binds importin α5 *(136,197)*. Thus, activated Akt in cancer cells may indeed cause retention of p27 in the cytoplasm by mediating the binding of 14-3-3 to p27. Different from human, murine p27 lacks the T157 site and thus, cannot be directed to the cytoplasm by PI3K/Akt phosphorylation *(139)*. In the mouse, constitutive activation of N-Ras caused cytoplasmic mis-localization of p27 via the Ral-guanine nucleotide exchange factor pathway (RAL-GEF) which also blocked Smad2/3 from entering the nucleus for TGF-β-mediated growth inhibition; human p27 is redirected by both mechanisms. PI3K also induces c-*myc* translation and both c-*myc* and Ras regulate the levels of p27, as well as PI3K, directly causing p27 expulsion from the nucleus. Thus, the increase in cyclinE-cdk2 activity causing S phase progression, by coexpression of Ras and c-*myc*, converge on the loss of nuclear p27 regulated by these phosphorylation events *(198,199)*.

A major protein involved in nuclear export of p27 and its degradation in the cytoplasm is JAB1 *(200,201)*. Thus, another level of the regulation of p27 levels is in the phosphorylation independent transport of p27 from the nucleus. However, the subsequent degradation step requires p27 phosphorylation presumably at T187, indicating that T187-dependent degradation of p27 can occur in both the nucleus and cytoplasm. Conversely, the vitamin D3 upregulated protein 1(VDUP1) binds to JAB1 and blocks JAB-1-mediated translocation of p27 from the nucleus to the cytoplasm thereby restoring p27 stability; VDUP1 itself is decreased in cancer *(202)*. Whereas there was an inverse relationship between high Jab1 and low p27 expression in node negative breast cancers, neither could be correlated with disease-free survival *(203)*. Finally, the nuclear-pore associated protein N-PAP60 has recently been shown to be required for cyclin-E mediated degradation of p27 by mediating its expulsion into the nuclear envelope connected to the ER where proteasomes are shown to be concentrated *(204,205)*. It is proposed that the interaction of p27 with PAP60 may be involved in the presentation of ubiquitinated p27 to the proteasomes *(15)*.

3.2. TGF-β/Smad Signaling Rescues p27 from Proteasomal Degradation

3.2.1. Disrupted TGF-β Signaling Targets p27 in Endometrial Carcinoma

Endometrial carcinoma (ECA) is the most common gynecological malignancy, which occurs at a rate of approximately 40,000 cases per year *(206–208)*. The endometrial gland and surrounding stromal cells are growth regulated by growth factors and ovarian hormones by autocrine and paracrine mechanisms. Estrogens, in the proliferative phase of the menstrual cycle, and progesterone from the corpus luteum, in the secretory phase, stimulate and inhibit endometrial cell growth, respectively, in a cyclical manner. While the mechanisms involved in the pathogenesis of ECA are poorly understood, endometriod type I carcinoma, which occurs in 85% of all ECAs, is estrogen-induced (e.g., unopposed estrogen as in anovulation and estrogen-replacement therapy). In addition, Tamoxifen (selective estrogen receptor modulator [SERM]) used for breast cancer therapy is associated with an approximate 66% risk for development of ECA *(209–211)*. Importantly, because glandular hyperplasia precedes carcinoma in 82% of cases *(206–208)*, the endometrium provides a means to study early molecular events that might occur in neoplastic development of this tumor. Progestins, because of their growth inhibitory activity on the endometrium, are used to treat hyperplasia and carcinoma with 62% success *(212–216)*.

Critical to understanding mechanisms of loss of growth regulation disabled in cancer, it is important to compare normal cells with their malignant cell counterparts. To this aim, using primary cell cultures of normal endometrial epithelial cells (EECs) from proliferative (PE) and secretory phase (SE) endometrium and carcinoma cells, that all retained their respective phenotypes, we showed that whereas normal EECs were growth inhibited (55% of controls) by TGF-β (10 pM), ECA cells were unresponsive owing to lack of TGF-β signaling *(217)*. The overall objective of our studies is to reveal downstream targets of TGF-β signaling that confer normal growth inhibitory activity that malfunction in human cancers owing to loss of TGF-β signaling. Although TGF-β and p27 act similarly as the master switches that halt cells at the G1/S checkpoint, the effect of TGF-β on p27 has mainly been indirect. In an effort to identify the downstream effector(s) of TGF-β-mediated growth inhibition obviated in ECA, we compared the effect of TGF-β on p27 mRNA and protein levels in normal primary cultures of EECs and ECA cells *(292)*. TGF-β did not induce p27 mRNA in EECs or ECA cells, which contained equal basal levels of p27. In contrast, TGF-β strongly induced a dose-dependent increase in p27 protein in the normal EEC cultures with greater distribution in the nuclear and cytoplasmic fractions (Fig. 2). However, ECA cells did not contain any p27 and, as expected, no response

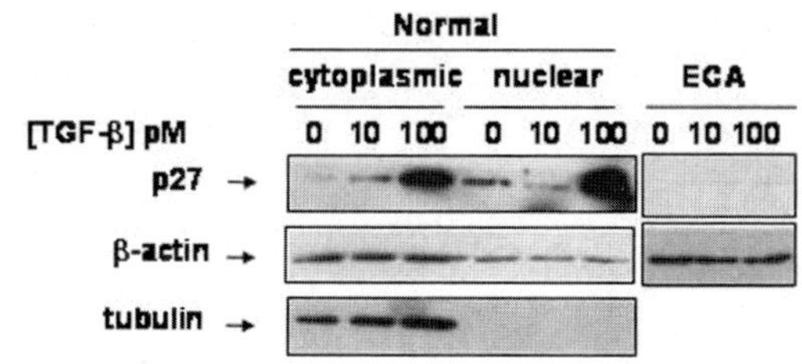

Fig. 2. TGF-β induces a dose-dependent increase in p27[kip1] in primary cultures of endometrial epithelial cells (EECs) but not primary endometrial carcinoma (ECA) cells. EECs and ECA cultures derived from human uteri were isolated and seeded onto tissue culture dishes, as described (Parekh et al., 2002). After 24 h, the cells were switched to serum-free conditions, increasing doses of TGF-β were added, and the cells incubated for 48 h. As shown in the immunoblot using antibodies to p27[kip1], TGF-β induced an increase in p27[kip1] in the cytoplasmic and nuclear fractions in normal EECs but not ECA cells, presumably, as previously shown, owing to dysregulated TGF-β signaling. Tubulin was used as a marker for the cytoplasmic fraction (without subcellular fraction in total cell lysates, normal, $n = 9$; endometrial carcinoma, $n = 3$; with subcellular fractionation, normal, $n = 2$).

to TGF-β was obtained, ostensibly because these cells lack TGF-β signaling *(217)*. The absence of p27 in primary ECA cells by immunoblotting was demonstrated in vivo in the glands of patient's ECA tissue samples (Fig. 3, Panel D) irrespective of grade (grades I–III); a statistically significant decrease in p27 immunoreactivity compared to normal differentiated SE (Panel B) ($p \leq 0.001$) was obtained. Hyperplastic endometrium had moderate cytoplasmic immunostaining (Panel C). Interestingly, there was a statistically significant correlation of a decrease of Smad2P (signifies TGF-β signaling) and p27 in the glands of the tissue specimens of the same patients (Spearman rank correlation value = 0.54; $p < 0.0002$). As p27 is absent in grade I ECAs and localizes to the cytoplasm in hyperplasia, as shown for the decrease in TGF-β signaling, the decrease in p27 expression appears to be an early event in the pathogenesis of ECA as well. Moreover, the cytoplasmic localization of p27 in hyperplasia may represent an initial exclusion of p27 from the nucleus where it functions to block cdk2 activity, prior to its complete loss, to ensure that proliferation is enabled in ECA. Similar to our studies, it has been shown that the increased expression of TGF-β ligands in endocervical cancer tissue compared to normal is consistent with a decrease in p27 *(218)*. However, unlike our studies, TGF-β receptors were upregulated but their functional status was not assessed.

Reduced expression of p27 protein is frequent in many human carcinomas including breast *(219–221)*, non-small cell lung cancer *(222)*, ovary *(193)*, endometrium *(223–226)*, and in brain and lymphomas *(195,227)*. Furthermore, generally, reduced levels or complete loss of p27 correlate with higher tumor grade, poor survival *(195,228,229)*, and disease recurrence *(193)*. Therefore, unlike ECA showing decreased p27 irrespective of grade, decreased expression of p27 protein has been considered as an independent prognostic marker in certain human cancers, such as breast and gastric cancers *(220,222)*. Most tumor suppressors, such as p15 and p16 as well as components of the TGF-β signaling pathway are notably direct targets for gene mutations in human cancers and therefore conform to the classic Knudsonian model of tumorigenesis *(44)*. Although, p27 is the critical CDKI that blocks cell cycle progression, loss of a single allele of the human p27 gene at 12p13 occurs at a very low rate and is mainly found in lymphoblastic leukemias; inactivation of a second allele is even more rare *(195,230–234)*. Moreover, in murine models of oncogenesis, p27 has been shown to be haploinsufficient and thus, both normal alleles are necessary for complete tumor suppression *(15,235–237)*. However, mice with homozygous deletion of p27 develop pituitary hyperplasia and are rendered predisposed to tumor formation by DNA damaging agents *(70,235,237,238)*.

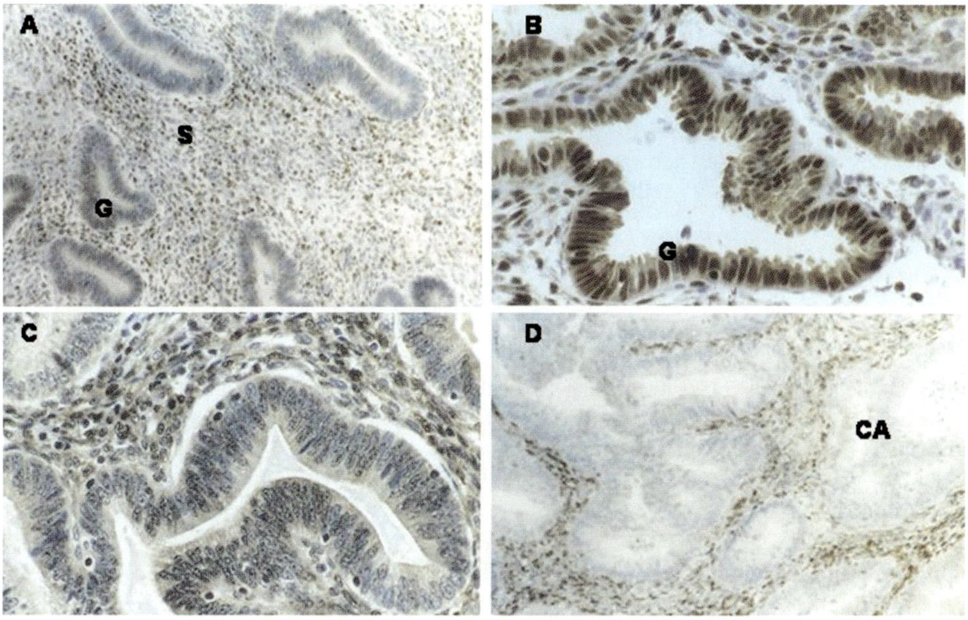

Fig. 3. Marked decrease in p27kip1 expression in endometrial carcinoma by immunohistochemical analysis. Whereas intense nuclear expression of $p27^{kip1}$ was shown in secretory endometrial glands (Panel **B**, $n = 6$), the proliferative glands (Panel **A**, $n = 10$) showed faint cytoplasmic immunostaining. The absence of $p27^{kip1}$ in the malignant glands (loss of glandular architecture) of endometrial carcinoma tissue, grade I/III (nuclear/cytoplasmic grade by the WHO system) (Panel **D**, $n = 19$) was statistically significant compared to secretory endometrium ($p \leq 0.001$). Endometrial hyperplasia (Panel **C**, $n = 8$) expressed moderate levels of $p27^{kip1}$ in the cytoplasm. Certain stromal cells express $p27^{kip1}$ in both normal and malignant tissue. G = glands; S = stroma; CA = cancer. Magnification: Panels **A**, **D**, ×100; **B**, **C**, ×200.

3.2.2. TGF-β Prevents Degradation of p27 for Normal Growth Inhibition: Inverse Relationship Between Cks1/Skp2 and p27

p27 mRNA levels are constant throughout the cell cycle and although under translational control *(239,240)*, intracellular levels are predominantly controlled posttranslationally by degradation through the ubiquitin-proteasome pathway *(147,148,241)*. Accordingly, the decrease of p27 in human malignancies is related to increased and uncontrolled proteasomal degradation. Because of the high impact of the loss of p27 that has already been shown in human cancers *(8,15,242)*, this has been an area of intensive study. Similarly, we obtained a very high rate of p27 degradation in ECA. In an in vitro proteasome degradation assay *(241)*, ECA tissue lysates ubiquitylated and degraded 50% of exogenously added recombinant p27 (rp27) in 30 min with complete obliteration of rp27 by 3 h compared to SE tissue lysates in which p27 was not ubiquitylated and still remained intact at 20 h *(292)*. Consistent with the low p27 in the glands of normal PE (Fig. 3, Panel A), these tissue lysates similarly ubiquitylated and degraded rp27. Importantly, there was 100% correlation between the presence of p27 in the glands in vivo (Fig. 3) with the ability of the tissue lysates from the same patients to degrade p27 in vitro, strongly suggesting that the absence of p27 from the carcinomas that we observed by immunohistochemistry (IHC) is owing to a high rate of proteasomal degradation in ECA. Proteasome inhibitors, such as PS-341 (Velcade) are currently being used in clinical trials as cancer therapeutics to restore intracellular p27 levels. Whereas these trials are encouraging *(243–246)*, it is notable that cell cycle proteins promoting growth are similarly

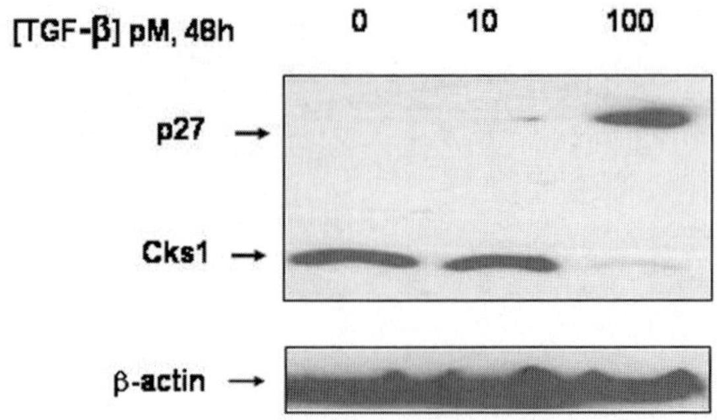

Fig. 4. TGF-β has an inverse effect on the levels of p27^{kip1} and Cks1 in primary cultures of endometrial epithelial cells (EECs). Increasing concentrations of TGF-β were added to primary cultures of normal EECs under serum-free conditions and the cells harvested after 48 h. The immunoblot shows that TGF-β increased p27^{kip1} while simultaneously decreasing Cks1 indicating an inverse relationship between Cks1 and p27^{kip1} in response to TGF-β. β-actin was used as a loading control ($n = 3$).

regulated by proteasomal degradation *(247)*. In addition, it was shown that cell cycle arrest by TGF-β requires the proteasome pathway *(248)*. Nonetheless, as overexpression of p27 in human breast epithelial cells and cancer cell lines lengthened G1 phase and suppressed tumorigenicity in vivo *(249)*, a means to increase intranuclear levels of p27 with specific inhibitors of its degradation appears to be a feasible therapeutic approach. However, a superior method may be to design inhibitors specific for the substrate-specific E3 ligases, such as Skp2, which would disable p27 proteasomal degradation.

Whereas a high rate of ubiquitin-mediated degradation destroys p27 so that ECA cells can proliferate, our data demonstrates that, in addition, because these cells lack the TGF-β signaling machinery, p27 cannot be induced by TGF-β. However, induction of TGF-β appears to be a normal physiological event in SE, as illustrated by the high levels of p27 in the nucleus of the glands (Fig. 3, Panel B). Our studies further show that the mechanism by which TGF-β increases p27 levels is by preventing ubiquitin-mediated degradation of p27 by decreasing the level of Cks1, the critical rate limiting component of the SCF^{skp2} complex that targets p27 to the proteasome for degradation (Fig. 4). Importantly, the physiological significance of this inverse relationship is recapitulated in patients in vivo as the glands of both proliferative phase endometrium and the carcinomas show increased nuclear Cks1 and negligible levels of p27, whereas secretory phase glands express high levels of p27 but, little to no Cks1. As blocking TβRI kinase activity (Smad activation) with a small molecule inhibitor (SD208, Scios, Inc.) obviated TGF-β-mediated accumulation of p27 and growth inhibition, taken together, these results suggest that TGF-β signaling, at least in part through Smads, prevents p27 degradation by decreasing Skp2 and Cks1, as a mechanism involved in the inhibition of normal endometrial cell growth.

Inasmuch as primary ECA cells lack TGF-β signaling *(217)*, a well-differentiated ECA cell line, HEC-1A, possesses both TGF-β receptors (by RT PCR) and is growth inhibited by TGF-β (33% of untreated control). TGF-β (100 pM) reduced the expression of Cks1and Skp2 protein in HEC-1A cells on slides by immunocytochemistry. Similar to the results obtained with normal EECs, TGF-β (100 pM) induced a dose-dependent increase in p27 protein, which was blocked by the TβRI kinase inhibitor and further, enforced expression of Smad2 also decreased Cks1 protein confirming that the TGF-β-mediated response was via classic Smad signaling. Finally, knock-down of p27 by transient transfection of p27 siRNA nucleotides, blocked TGF-β-mediated growth inhibition and instead, the cells were growth stimulated (Fig. 5). As depicted by the schematic model in Figure 6, together, these data

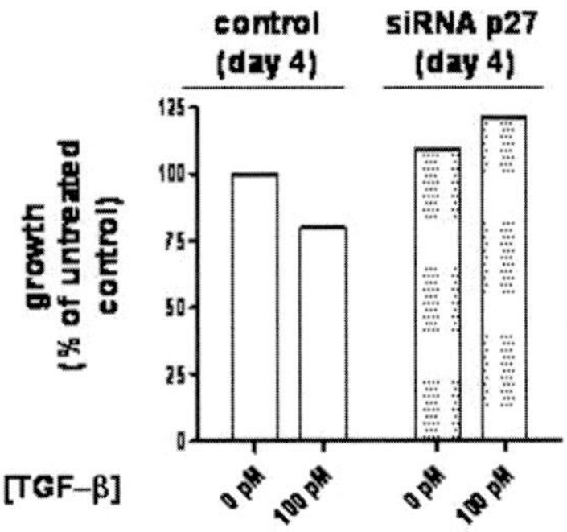

Fig. 5. Knocking-down p27^{kip1} with siRNA abolishes TGF-β growth inhibition in HEC-1A cells. The endometrial carcinoma cell line, HEC-1A (well-differentiated endometriod type I endometrial carcinoma) possesses TGF-β receptors and are growth inhibited by TGF-β at 25–33% of untreated controls (open boxes). Following transfection of siRNA p27 with HiPerFect (Qiagen) transfection reagent once a day for 3 d, causing a 30% decrease in p27^{kip1} by Western blot, the cells were treated with 100 pM TGF-β for 72 h and growth inhibition assessed by the MTS assay. The cells treated with siRNA p27^{kip1} were growth stimulated by TGF-β (dotted bars). These data suggest that p27^{kip1} mediates TGF-β induced growth inhibition (n = 2).

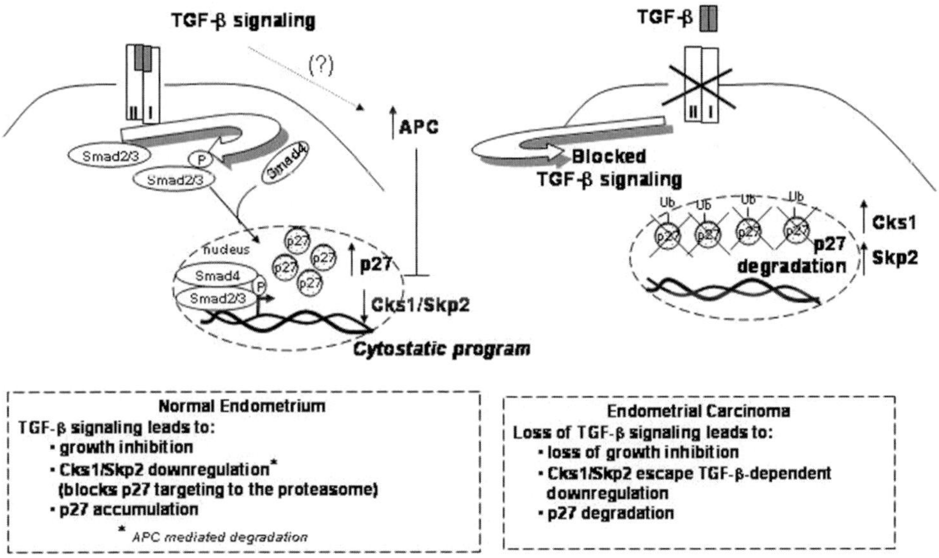

Fig. 6. Schematic model of TGF-β signaling effects on p27^{kip1} in normal endometrium and loss thereof in endometrial carcinoma.

provide proof of principle that TGF-β-induced accumulation of p27 functionally confers TGF-β-mediated growth inhibition and that lack of TGF-β signaling in ECA obviates this mechanism, permitting proliferation.

Because we also found that TGF-β affected Cks1 and Skp2 only at the protein level, we entertained the idea that TGF-β might be an effector of the degradation of these proteins. As the APC^{cdh1} complex has been shown to have E3 ligase activity for Skp2 and Cks1, causing their proteasomal degradation *(167,168)*, this complex is a potential target for upregulation by TGF-β. TGF-β has been shown to downregulate both Skp2 and Cks1 in other studies. The downregulation of Cks1 (shown at the mRNA level) by TGF-β in a number of cell lines simultaneously led to Skp2 autoubiquitylation saving p27 from Skp2-mediated ubiquitylation during G1 to S transition so that it could bind to cyclin E–cdk2 to inhibit kinase activity *(250)*. Thus, the induction of APC may be involved in Skp2 degradation in these studies and others, where Skp2 is shown to be degraded by the proteasomes. As we have shown in our studies herein, 13-*cis*-retinoic acid (RA) appears to act in a similar way to TGF-β in that it prevented p27 from being degraded owing to proteasomal degradation of Skp2 and Cks1 in a B-cell line that was growth inhibited by RA *(157)*. The implication of RA treatment for B-cell lymphomas in the maintenance of high intranuclear levels of p27 may have therapeutic potential *(157)*. Similarly, G1 arrest by all-trans retinoic acid (ATRA) in breast cancer cell lines was achieved by increasing the levels of p27 for inhibition of cyclinE–cdk2 activity through ubiquitin-mediated degradation of Skp2 *(251)*. In addition, both the mRNA and protein levels of Skp2 and Cks1 were decreased by Vitamin D treatment of a human head and neck squamous carcinoma cell line leading to decreased p27 turnover *(252)*. In another study, TGF-β increased p27 (but not p21 or p16) levels to effectuate growth inhibition in a human biliary cell line, H69, by a mechanism proposed to involve oxidative inactivation of the proteasome subunits *(253)*.

Similar to our studies, Skp2 and Cks1 are overexpressed in a number of human cancers with a concomitant decrease in p27, including oral cancer, lymphomas, colorectal, breast, prostate, and gastric cancers *(7,9,172–177,180,254,255)*. This inverse relationship between Skp2/Cks1and p27 is associated with rapid growth, higher tumor grade and/or poor prognosis. Interestingly, the activation of Notch1 stem cell differentiation factor receptor signaling was shown to induce Skp2 mRNA synthesis causing proteasome-mediated degradation of p27 and p21 and S phase progression as a mechanism of action *(256)*. The significance of the regulation of Skp2 and its influence on levels of p27 in growth control, therefore, also extends into pathways that determine cell fate.

3.2.3. p27 IS A MAJOR TARGET FOR GROWTH REGULATION IN THE ENDOMETRIUM: OVARIAN HORMONES AND STROMAL CELL INTERACTIONS

The significance of p27 as an important CDKI for growth regulation in the endometrium was further shown by the effect of ovarian hormones on intracellular p27 levels. Primary cultures of EECs, treated with 100 nM 17-β-estradiol, serum-free, for 4 days, resulted in a 60% diminution of p27 compared to untreated controls. Both the Mek1 inhibitor, PD98059, and the ubiquitin-proteasome inhibitor, lactacystin, restored p27 to control levels and furthermore, Erk1 was specifically activated by estrogen. Therefore, because estrogens have been shown to activate the MAPK pathway *(257,258)*, we ascertain that estrogen induces Erk1-mediated phosphorylation of p27, most likely at T187, to signal recognition of p27 by Skp2 for proteasomal degradation in the nucleus or on S10, for nuclear exclusion and subsequent (non-Skp2-related) degradation in the cytoplasm *(8,12,134)*. The significance of these findings is that p27 may be a major molecular target for estrogen-induced type I endometriod ECA. Future studies should determine the site(s) on p27 that are phosphorylated, the subcellular distribution, and whether Ras activation, upstream from MAPK, is involved in estrogen-induced ubiquitin-mediated degradation of p27 in primary endometrial cells as was shown to be the case in the MCF-7 breast cancer cell line *(259)*. Conversely, other studies have shown that Ras-MAPK activation initiates the binding of p27 to cyclinD–cdk4 reducing its availability for inhibition of cdk2 and not by proteasomal degradation *(260)*. Nonetheless, the regulation of cell cycle proteins, including cyclinD, Cdc25A phosphatase, and the CDKIs, by estrogens is dependent on the timing within the cell cycle *(261–264)* and likely, estrogens may regulate p27 levels by more than one mechanism. Importantly, the estrogen receptor antagonists Tamoxifen and ICI 182,780 (pure estrogen receptor blocker) have been shown to raise the levels of p27 and p21 and increase their binding to cyclin E–cdk complexes to cause growth arrest *(128,265,266)*. The increased levels of p27 are considered to be responsible for the high success rate of Tamoxifen treatment for recurrent breast cancer. Alternatively, constitutive activation of MAPK mediated proteasomal degradation of p27 is associated with antiestrogen resistance *(128)*. These results point to p27 as an important protein that determines success or failure of antiestrogen therapy.

Further establishing the importance of p27 in growth regulation of the endometrium, we show that 100 nM medroxyprogesterone treatment increases p27 levels by 6.7- and 8.5-fold, in primary cultures of normal and malignant endometrial cells, respectively. The mechanism of the progesterone-induced increase in p27 is not known at this time. However, unlike TGF-β, progesterone has been shown to regulate p27 at the transcriptional level in one study *(267)*. In contrast, p27 mRNA was not increased in response to progesterone in an ECA cell line or primary EECs *(268)*. Instead, p27 breakdown was decreased with increased binding to cdk2 for growth inhibition. Similarly, progesterone increased the binding of p27 to cyclin E–cdk2 reducing both cyclin D–cdk4 and E–cdk2 activity to inhibit growth of a breast cancer cell line *(269)*. Thus, one can consider that the highly successful therapeutic action of progesterone in endometrial hyperplasia and carcinoma is indeed owing to directly increasing the (nuclear) levels of p27 for cell growth arrest. Taken together, TGF-β and ovarian hormones appear to regulate normal endometrial growth by converging on the regulation of intracellular levels and subcellular distribution of p27 *(292)*. Because we have shown earlier that ECA cells have disrupted TGF-β signaling, our data suggests that at least one lesion in the pathogenesis of endometrial cancer is the inability of TGF-β to exert cell cycle control by preventing the degradation of p27 in the presence of uncontrolled estrogen-driven proteasomal degradation.

It has been long recognized that the role of the stromal cells that surround the glands in hormone-regulated organs, such as the endometrium, breast, and prostate, play a critical role in the function of glandular epithelial cells particularly with respect to growth regulation *(270–274)*. We envision that genetically unstable epithelial/carcinomas hijack and modify stromal cells as accessory cells to promote and perpetuate the carcinogenic process, early in progression, before they are eventually overgrown by the carcinoma. In preliminary experiments, we have shown that stromal cells derived from malignant prostate, that lose their normal smooth muscle cell phenotype and become fibroblastic, downregulate TGF-β receptors with concomitant acquisition of tumorigenicity in tissue recombinants with a nontumorigenic prostate epithelial cell line, BPH-1 (unpublished data). Similarly, we show that stromal cells derived from malignant endometrial tumors, but not from normal tissue, can stimulate the growth of normal EECs when grown in coculture and that soluble mediators from cell cultures of endometrial stromal, derived from malignant tumors, but not normal tissue, downregulate p27 with a concomitant increase in Cks1 in HEC-1A cells *(275)*. Therefore, it appears that we have uncovered a common mechanism for dysregulated growth in the endometrium whereby p27 levels are lowered through the ubiquitin–proteasome pathway. Support for p27 as the central regulator of normal endometrial growth and a focal point for malignant conversion is supported by studies using p27 null mice that are predisposed to endometrial hyperplasia and carcinoma (and of other organs) following gamma irradiation and urethane exposure *(12,70)*.

3.3. Parallel Roles of TGF-β and p27 in Tumor Suppression and Progression

An interesting parallel is emerging among a number of proteins that have dual functional roles related to tumor suppressor activity and antithetically, prooncogenicity. Since TGF-β was first discovered as the most potent endogenous inhibitor of growth, it was surprising to find that its role had reversed into mediating metastasis *(1–4,277,278)*. In addition to mutations in components of the TGF-β-signaling machinery, many human cancers are resistant to TGF-β-mediated growth inhibition and it appears that certain cellular switches harness the TGF-β pathway to function in EMT leading to metastasis *(1,2)*. TGF-β signaling in the MCF10A cells with p21 knocked down, but with intact p15, was shown to induce a number of characteristics associated with EMT *(276)*. For this reason, p21 was proposed to protect TGF-β tumor suppressor function and may be one mechanism involved in the switch of TGF-β function from tumor suppressor to a promoter of progression *(74)*. Interestingly, other means of subverting TGF-β growth arrest, such as upregulation of c-*myc*, does not lead to EMT (2,3 of Seoane 2004). The classic Smad signaling pathways support both growth suppression and EMT/metastasis *(3,79,277,278)*, as well as context and cell-specific responses in which MAPK and other pathways that can cross talk with Smads are involved *(1)*. In addition, because Ras has been shown to mislocalize p27, thereby blocking TGF-β-mediated growth inhibition, it has been proposed that activation of prooncogenic Ras may be one mechanism to explain the switch of TGF-β from a growth suppressor to becoming metastatic *(139)*.

With other proteins, such as p27, p21, and the phosphatase tensin homologue (PTEN) (inhibits the Akt pathway), it is becoming apparent that their subcellular localization dictates different functions, such that nuclear localization is tumor suppressive and cytoplasmic localization permits interaction with the cytoskeleton for migration or metastasis *(84,279–282)*. As described (Part 3.1.1.), phosphorylation of p27 at T157 and T198 by PKB/Akt and S10 by MAPK drives p27 into the cytoplasm providing the critical switch from its role in the nucleus as a tumor suppressor (inhibiting the cell cycle) to a role in migration/metastasis. Similarly, phosphorylation of p21 by PKB/Akt on T145 also drives this CDKI into the cytoplasm, where it has been shown to inhibit apoptosis thereby contributing to tumorigenesis *(136)*. Further, complex interactions between other cell cycle proteins and the cytoskeleton, which are not only related to the regulation of mitosis, have been revealed and thus, a stronger link between the cell cycle and cell motility is evolving *(280)*.

Albeit less frequently than the decrease of p27 found in human cancers, elevated p27 is found associated with poor prognosis and increased metastasis in subsets of carcinomas such as breast, cervix, and the uterus *(188,189,196,283)*. The lack of mutations associated with p27 may underscore its function as a tumor suppressor in the nucleus while being oncogenic in the cytoplasm. Considering that TGF-β regulates p27 with respect to growth inhibition with cdk2 activity as the meter of control, we speculate, in light of the effects of TGF-β and p27 on the cytoskeleton *(280,281,284,285)*, that TGF-β may potentially regulate or interact with p27 in the cytoplasm to promote metastasis. Interestingly, both p27 and TGF-β impact RhoGTPase/ROCK: this pathway is used for TGF-β/Smad signaling to downregulate c-*myc* for growth inhibition *(58)* and, p27 inhibits Rho in the cytoplasm for regulation of the cytoskeleton and migration *(281)*. As tumor metastasis is the key problem for patient survival, increasing our understanding of how an increasing number of tumor suppressors switch to mediating metastasis will provide important molecular targets for new therapeutic approaches. While others more recently have taken a proteomic approach to search for proteins/pathways in which TGF-β signaling impacts certain cancers *(286)*, TGF-βRI kinase inhibitors are actively being used to clinically combat high grade metastatic cancers *(287–291)*. Equally important is the prevention of neoplastic development. In this context, the reconstitution of TGF-β signaling to restore growth inhibition, which for certain cancers (low grade and stage), translates, as we have shown here, to preventing p27 degradation to recapture growth control could be considered. Furthermore, directly downstream from loss of growth inhibition by TGF-β, our studies suggest that this lesion in ECA obviates the ability of TGF-β to increase p27 levels which is achieved by downregulating Cks1 and Skp2, critical components of the SCF^{Skp2} complex that target p27 for ubiquitin-proteasome mediated degradation. Therefore, our results implicate a novel means by which TGF-β mediates growth inhibition by maintaining abundant levels of nuclear p27, to block cell cycle progression.

REFERENCES

1. Zavadil J, Bottinger EP. TGF-beta and epithelial-to-mesenchymal transitions. Oncogene 2005;24: 5764–5774.
2. Bachman KE, Park BH. Duel nature of TGF-beta signaling: tumor suppressor vs. tumor promoter. Curr Opin Oncol 2005;17:49–54.
3. Valcourt U, Kowanetz M, Niimi H, Heldin C-H, Moustakas A. TGF-beta and the Smad signaling pathway support transcriptomic reprogramming during epithelial-mesenchymal cell transition. Mol Biol Cell 2005;16:1987–2002.
4. Roberts AB, Wakefield LM. The two faces of transforming growth factor-beta in carcinogenesis. Proc Natl Acad Sci USA 2003;100:8621–8623.
5. Massagué J. G1 cell-cycle control and cancer. Nature 2004;432:298–306.
6. Kurland JF, Tansey WP. Crashing waves of destruction: the cell cycle and APC(Cdh1) regulation of SCF(Skp2). Cancer Cell 2004;5:305–306.
7. Latres E, Chiarle R, Schulman BA, et al. Role of the F-box protein Skp2 in lymphomagenesis. Proc Natl Acad Sci USA 2001;98:2515–2520.
8. Bloom J, Pagano M. Deregulated degradation of the cdk inhibitor p27 and malignant transformation. Semin Cancer Biol 2003;13:41–47.
9. Ben-Izhak O, Lahav-Baratz S, Meretyk S, et al. Inverse relationship between Skp2 ubiquitin ligase and the cyclin dependent kinase inhibitor p27Kip1 in prostate cancer. J Urol 2003;170:241–245.
10. Carrano AC, Eytan E, Hershko A, Pagano M. SKP2 is required for ubiquitin-mediated degradation of the CDK inhibitor p27. Nat Cell Biol 1999;1:193–199.
11. Sutterluty H, Chatelain E, Marti A, et al. p45SKP2 promotes p27Kip1 degradation and induces S phase in quiescent cells. Nat Cell Biol 1999;1:207–214.
12. Besson A, Gurian-West M, Chen X, Kelly-Spratt KS, Kemp CJ, Roberts JM. A pathway in quiescent cells that controls p27Kip1 stability, subcellular localization, and tumor suppression. Genes Dev 2006;20:47–64.
13. Siegel PM, Massagué J. Cytostatic and apoptotic actions of TGF-beta in homeostasis and cancer. Nat Rev Cancer 2003;3:807–821.
14. Dumont N, Arteaga CL. Targeting the TGF beta signaling network in human neoplasia. Cancer Cell 2003;3:531–536.
15. Philipp-Staheli J, Payne SR, Kemp CJ. p27(Kip1): regulation and function of a haploinsufficient tumor suppressor and its misregulation in cancer. Exp Cell Res 2001;264:148–168.
16. Hata A, Shi Y, Massagué J. TGF-beta signaling and cancer: structural and functional consequences of mutations in Smads. Mol Med Today 1998;4:257–262.
17. Massagué J, Blain SW, Lo RS. TGF-beta signaling in growth control, cancer, and heritable disorders. Cell 2000;103:295–309.
18. Gold LI. The role for transforming growth factor-beta (TGF-beta) in human cancer. Crit Rev Oncog 1999;10:303–360.
19. Munger JS, Harpel JG, Gleizes PE, Mazzieri R, Nunes I, Rifkin DB. Latent transforming growth factor-beta: structural features and mechanisms of activation. Kidney Int 1997;51:1376–1382.
20. Flaumenhaft R, Kojima S, Abe M, Rifkin DB. Activation of latent transforming growth factor beta. Adv Pharmacol 1993;24:51–76.
21. Shi Y, Massagué J. Mechanisms of TGF-beta signaling from cell membrane to the nucleus. Cell 2003;113:685–700.
22. Feng XH, Derynck R. Specificity and versatility in TGF-beta signaling through Smads. Annu Rev Cell Dev Biol 2005;21:659–693.
23. Javelaud D, Mauviel A. Crosstalk mechanisms between the mitogen-activated protein kinase pathways and Smad signaling downstream of TGF-beta: implications for carcinogenesis. Oncogene 2005;24: 5742–5750.
24. Pierreux CE, Nicolas FJ, Hill CS. Transforming growth factor-beta-independent shuttling of Smad4 between the cytoplasm and nucleus. Mol Cell Biol 2000;20:9041–9054.
25. Xu L, Chen YG, Massagué J. The nuclear import function of Smad2 is masked by SARA and unmasked by TGF-beta-dependent phosphorylation. Nat Cell Biol 2000;2:559–562.
26. Xu L, Massagué J. Nucleocytoplasmic shuttling of signal transducers. Nat Rev Mol Cell Biol 2004;5: 209–219.
27. Xu L, Alarcon C, Col S, Massagué J. Distinct domain utilization by Smad3 and Smad4 for nucleoporin interaction and nuclear import. J Biol Chem 2003;278:42,569–42,577.

28. Xu L, Kang Y, Col S, Massagué J. Smad2 nucleocytoplasmic shuttling by nucleoporins CAN/Nup214 and Nup153 feeds TGF beta signaling complexes in the cytoplasm and nucleus. Mol Cell 2002;10: 271–282.
29. Di Guglielmo GM, Le Roy C, Goodfellow AF, Wrana JL. Distinct endocytic pathways regulate TGF-beta receptor signalling and turnover. Nat Cell Biol 2003;5:410–421.
30. Abdel-Wahab N, Wicks SJ, Mason RM, Chantry A. Decorin suppresses transforming growth factor-beta-induced expression of plasminogen activator inhibitor-1 in human mesangial cells through a mechanism that involves Ca^{2+}-dependent phosphorylation of Smad2 at serine-240. Biochem J 2002; 362:643–649.
31. Yakymovych I, ten Dijke P, Heldin C-H, Souchelnytskyi S. Regulation of Smad signaling by protein kinase C. FASEB J 2001;15:553–555.
32. Ulloa L, Doody J, Massagué J. Inhibition of transforming growth factor-beta/SMAD signalling by the interferon-gamma/STAT pathway. Nature 1999;397:710–713.
33. Hayashi H, Abdollah S, Qiu Y, et al. The MAD-related protein Smad7 associates with the TGF beta receptor and functions as an antagonist of TGF beta signaling. Cell 1997;89:1165–1173.
34. Shin I, Bakin AV, Rodeck U, Brunet A, Arteaga CL. Transforming growth factor beta enhances epithelial cell survival via Akt-dependent regulation of FKHRL1. Mol Biol Cell 2001;12:3328–3339.
35. Akhurst RJ, Derynck R, Balmain A. TGF-beta signaling in cancer—a double-edged sword. Trends Cell Biol 2001;11:S44–S51.
36. Bierie B, Moses HL. TGF-beta and cancer. Cytokine Growth Factor Rev 2006;17:29–40.
37. Zhang HS, Postigo AA, Dean DC. Active transcriptional repression by the Rb-E2F complex mediates G1 arrest triggered by p16INK4a, TGF-beta, and contact inhibition. Cell 1999;97:53–61.
38. Balciunaite E, Spektor A, Lents NH, et al. Pocket protein complexes are recruited to distinct targets in quiescent and proliferating cells. Mol Cell Biol 2005;25:8166–8178.
39. Cloud JE, Rogers C, Reza TL, et al. Mutant mouse models reveal the relative roles of E2F1 and E2F3 in vivo. Mol Cell Biol 2002;22:2663–2672.
40. Takahashi Y, Rayman JB, Dynlacht BD. Analysis of promoter binding by the E2F and pRB families in vivo: distinct E2F proteins mediate activation and repression. Genes Dev 2000;14:804–816.
41. Gaubatz S, Lindeman GJ, Ishida S, et al. E2F4 and E2F5 play an essential role in pocket protein-mediated G1 control. Mol Cell 2000;6:729–735.
42. Sherr CJ, Roberts JM. Living with or without cyclins and cyclin-dependent kinases. Genes Dev 2004; 18:2699–2711.
43. Ortega S, Prieto I, Odajima J, et al. Cyclin-dependent kinase 2 is essential for meiosis but not for mitotic cell division in mice. Nat Genet 2003;35:25–31.
44. Sherr CJ. Cancer cell cycles. Science 1996;274:1672–1677.
45. Guadagno TM, Assoian RK. G1/S control of anchorage-independent growth in the fibroblast cell cycle. J Cell Biol 1991;115:1419–1425.
46. Han EK, Guadagno TM, Dalton SL, Assoian RK. A cell cycle and mutational analysis of anchorage-independent growth: cell adhesion and TGF-beta1 control G1/S transit specifically. J Cell Biol 1993; 122:461–471.
47. Assoian RK, Zhu X. Cell anchorage and the cytoskeleton as partners in growth factor dependent cell cycle progression. Curr Opin Cell Biol 1997;9:93–98.
48. Laiho M, DeCaprio JA, Ludlow JW, Livingston DM, Massagué J. Growth inhibition by TGF-beta linked to suppression of retinoblastoma protein phosphorylation. Cell 1990;62:175–185.
49. Satterwhite DJ, Moses HL. Mechanisms of transforming growth factor-beta1-induced cell cycle arrest. Invasion Metastasis 1994;14:309–318.
50. Matsuura I, Denissova NG, Wang G, He D, Long J, Liu F. Cyclin-dependent kinases regulate the antiproliferative function of Smads. Nature 2004;430:226–231.
51. Hall M, Peters G. Genetic alterations of cyclins, cyclin-dependent kinases, and Cdk inhibitors in human cancer. Adv Cancer Res 1996;68:67–108.
52. Ewen ME. The cell cycle and the retinoblastoma protein family. Cancer Metastasis Rev 1994;13:45–66.
53. Ewen ME. Regulation of the cell cycle by the Rb tumor suppressor family. Results Probl Cell Differ 1998;22:149–179.
54. Iavarone A, Massagué J. Repression of the CDK activator Cdc25A and cell-cycle arrest by cytokine TGF-beta in cells lacking the CDK inhibitor p15. Nature 1997;387:417–422.
55. Kimura K, Tsuji T, Takada Y, Miki T, Narumiya S. Accumulation of GTP-bound RhoA during cytokinesis and a critical role of ECT2 in this accumulation. J Biol Chem 2000;275:17,233–17,236.

56. Bhowmick NA, Ghiassi M, Aakre M, Brown K, Singh V, Moses HL. TGF-beta-induced RhoA and p160ROCK activation is involved in the inhibition of Cdc25A with resultant cell-cycle arrest. Proc Natl Acad Sci USA 2003;100:15,548–15,553.
57. Brown K, Bhowmick NA. Linking TGF-beta-mediated Cdc25A inhibition and cytoskeletal regulation through RhoA/p160(ROCK) signaling. Cell Cycle 2004;3:408–410.
58. Kamaraju AK, Roberts AB. Role of Rho/ROCK and p38 MAP kinase pathways in transforming growth factor-beta-mediated Smad-dependent growth inhibition of human breast carcinoma cells in vivo. J Biol Chem 2005;280:1024–1036.
59. Bhowmick NA, Ghiassi M, Bakin A, et al. Transforming growth factor-beta1 mediates epithelial to mesenchymal transdifferentiation through a RhoA-dependent mechanism. Mol Biol Cell 2001; 12:27–36.
60. Nagahara H, Ezhevsky SA, Vocero-Akbani AM, Kaldis P, Solomon MJ, Dowdy SF. Transforming growth factor beta targeted inactivation of cyclin E:cyclin-dependent kinase 2 (Cdk2) complexes by inhibition of Cdk2 activating kinase activity. Proc Natl Acad Sci USA 1999;96:14,961–14,966.
61. Ortega S, Malumbres M, Barbacid M. Cyclin D-dependent kinases, INK4 inhibitors and cancer. Biochim Biophys Acta 2002;1602:73–87.
62. Hunter T, Pines J. Cyclins and cancer. II: Cyclin D and CDK inhibitors come of age [see comments]. Cell 1994;79:573–582.
63. Motokura T, Bloom T, Kim HG, et al. A novel cyclin encoded by a bcl1-linked candidate oncogene [see comments]. Nature 1991;350:512–515.
64. Okamoto A, Jiang W, Kim SJ, et al. Overexpression of human cyclin D1 reduces the transforming growth factor-beta (TGF-beta) type II receptor and growth inhibition by TGF-beta1 in an immortalized human esophageal epithelial cell line. Proc Natl Acad Sci USA 1994;91:11,576–11,580.
65. Sandhu C, Garbe J, Bhattacharya N, et al. Transforming growth factor beta stabilizes p15INK4B protein, increases p15INK4B-cdk4 complexes, and inhibits cyclin D1-cdk4 association in human mammary epithelial cells. Mol Cell Biol 1997;17:2458–2467.
66. Hannon GJ, Beach D. p15INK4B is a potential effector of TGF-beta-induced cell cycle arrest. Nature 1994;371:257–261.
67. Reynisdottir I, Polyak K, Iavarone A, Massagué J. Kip/Cip and Ink4 Cdk inhibitors cooperate to induce cell cycle arrest in response to TGF-beta. Genes Dev 1995;9:1831–1845.
68. Feng XH, Lin X, Derynck R. Smad2, Smad3 and Smad4 cooperate with Sp1 to induce p15(Ink4B) transcription in response to TGF-beta. EMBO J 2000;19:5178–5193.
69. Sherr CJ. Principles of tumor suppression. Cell 2004;116:235–246.
70. Payne SR, Kemp CJ. p27(Kip1) (Cdkn1b)-deficient mice are susceptible to chemical carcinogenesis and may be a useful model for carcinogen screening. Toxicol Pathol 2003;31:355–363.
71. Ekholm SV, Reed SI. Regulation of G(1) cyclin-dependent kinases in the mammalian cell cycle. Curr Opin Cell Biol 2000;12:676–684.
72. Ahmed MM, Alcock RA, Chendil D, et al. Restoration of transforming growth factor-beta signaling enhances radiosensitivity by altering the Bcl-2/Bax ratio in the p53 mutant pancreatic cancer cell line MIA PaCa-2. J Biol Chem 2002;277:2234–2246.
73. Michael D, Oren M. The p53 and Mdm2 families in cancer. Curr Opin Genet Dev 2002;12:53–59.
74. Seoane J. p21(WAF1/CIP1) at the switch between the anti-oncogenic and oncogenic faces of TGFbeta. Cancer Biol Ther 2004;3:226–227.
75. Seoane J, Le HV, Shen L, Anderson SA, Massagué J. Integration of Smad and forkhead pathways in the control of neuroepithelial and glioblastoma cell proliferation. Cell 2004;117:211–223.
76. Zhu Y, Parada LF. The molecular and genetic basis of neurological tumours. Nat Rev Cancer 2002;2: 616–626.
77. Holland EC. Gliomagenesis: genetic alterations and mouse models. Nat Rev Genet 2001;2:120–129.
78. Ijichi H, Otsuka M, Tateishi K, et al. Smad4-independent regulation of p21/WAF1 by transforming growth factor-beta. Oncogene 2004;23:1043–1051.
79. Levy L, Hill CS. Smad4 dependency defines two classes of transforming growth factor-β (TGF-β) target genes and distinguishes TGF-β-induced epithelial-mesenchymal transition from its antiproliferative and migratory responses. Mol Cell Biol 2005;25:8108–8125.
80. Pardali K, Kowanetz M, Heldin C-H, Moustakas A. Smad pathway-specific transcriptional regulation of the cell cycle inhibitor p21(WAF1/Cip1). J Cell Physiol 2005;204:260–272.
81. Ammanamanchi S, Tillekeratne MP, Ko TC, Brattain MG. Endogenous control of cell cycle progression by autocrine transforming growth factor beta in breast cancer cells. Cancer Res 2004;64:2509–2515.

82. Jen J, Harper JW, Bigner SH, et al. Deletion of p16 and p15 genes in brain tumors. Cancer Res 1994; 54:6353–6358.
83. Herman JG, Jen J, Merlo A, Baylin SB. Hypermethylation-associated inactivation indicates a tumor suppressor role for p15INK4B. Cancer Res 1996;56:722–727.
84. Denicourt C, Dowdy SF. Cip/Kip proteins: more than just CDKs inhibitors. Genes Dev 2004;18:851–855.
85. Russo AA, Jeffrey PD, Patten AK, et al. Crystal structure of the p27Kip1 cyclin-dependent-kinase inhibitor bound to the cyclin A-Cdk2 complex. Nature 1996;382:325–331.
86. Scandura JM, Boccuni P, Massagué J, Nimer SD. Transforming growth factorbeta-induced cell cycle arrest of human hematopoietic cells requires p57KIP2 up-regulation. Proc Natl Acad Sci USA 2004; 101:15,231–15,236.
87. Shen L, Toyota M, Kondo Y, et al. Aberrant DNA methylation of p57KIP2 identifies a cell-cycle regulatory pathway with prognostic impact in adult acute lymphocytic leukemia. Blood 2003;101:4131–4136.
88. Kikuchi T, Toyota M, Itoh F, et al. Inactivation of p57KIP2 by regional promoter hypermethylation and histone deacetylation in human tumors. Oncogene 2002;21:2741–2749.
89. Li Y, Nagai H, Ohno T, et al. Aberrant DNA methylation of p57(KIP2) gene in the promoter region in lymphoid malignancies of B-cell phenotype. Blood 2002;100:2572–2577.
90. Nishimori S, Tanaka Y, Chiba T, et al. Smad-mediated transcription is required for transforming growth factor-beta1-induced p57(Kip2) proteolysis in osteoblastic cells. J Biol Chem 2001;276:10,700–10,705.
91. Urano T, Yashiroda H, Muraoka M, et al. p57(Kip2) is degraded through the proteasome in osteoblasts stimulated to proliferation by transforming growth factor beta1. J Biol Chem 1999;274:12,197–12,200.
92. Claassen GF, Hann SR. A role for transcriptional repression of p21CIP1 by c-Myc in overcoming transforming growth factor beta-induced cell-cycle arrest. Proc Natl Acad Sci USA 2000;97:9498–9503.
93. Seoane J, Pouponnot C, Staller P, Schader M, Eilers M, Massagué J. TGFbeta influences Myc, Miz-1 and Smad to control the CDK inhibitor p15INK4b. Nat Cell Biol 2001;3:400–408.
94. Staller P, Peukert K, Kiermaier A, et al. Repression of p15INK4b expression by Myc through association with Miz-1. Nat Cell Biol 2001;3:392–399.
95. Warner BJ, Blain SW, Seoane J, Massagué J. Myc downregulation by transforming growth factor beta required for activation of the p15(Ink4b) G(1) arrest pathway. Mol Cell Biol 1999;19:5913–5922.
96. Alexandrow MG, Moses HL. Kips off to Myc: implications for TGF beta signaling. Journal of Cellular Biochemistry 1997;66:427–432.
97. Sun P, Dong P, Dai K, Hannon GJ, Beach D. p53-independent role of MDM2 in TGF-beta1 resistance. Science 1998;282:2270–2272.
98. Yagi K, Furuhashi M, Aoki H, et al. c-myc is a downstream target of the Smad pathway. J Biol Chem 2002;277:854–861.
99. Pardali K, Kurisaki A, Moren A, ten Dijke P, Kardassis D, Moustakas A. Role of Smad proteins and transcription factor Sp1 in p21 (Waf1/Cip1) regulation by transforming growth factor-beta. J Biol Chem 2000;275:29,244–29,256.
100. Frederick JP, Liberati NT, Waddell DS, Shi Y, Wang XF. Transforming growth factor beta-mediated transcriptional repression of c-myc is dependent on direct binding of Smad3 to a novel repressive Smad binding element. Mol Cell Biol 2004;24:2546–2559.
101. Chen CR, Kang Y, Siegel PM, Massagué J. E2F4/5 and p107 as Smad cofactors linking the TGFbeta receptor to c-myc repression. Cell 2002;110:19–32.
102. Feng XH, Liang YY, Liang M, Zhai W, Lin X. Direct interaction of c-Myc with Smad2 and Smad3 to inhibit TGF-beta-mediated induction of the CDK inhibitor p15(Ink4B). Mol Cell 2002;9:133–143.
103. Grandori C, Cowley SM, James LP, Eisenman RN. The Myc/Max/Mad network and the transcriptional control of cell behavior. Annu Rev Cell Dev Biol 2000;16:653–699.
104. Chen CR, Kang Y, Massagué J. Defective repression of c-myc in breast cancer cells: A loss at the core of the transforming growth factor beta growth arrest program. Proc Natl Acad Sci USA 2001;98:992–999.
105. Nesbit CE, Tersak JM, Prochownik EV. MYC oncogenes and human neoplastic disease. Oncogene 1999;18:3004–3016.

106. Erisman MD, Litwin S, Keidan RD, Comis RL, Astrin SM. Noncorrelation of the expression of the c-myc oncogene in colorectal carcinoma with recurrence of disease or patient survival. Cancer Res 1988;48:1350–1355.
107. Erisman MD, Rothberg PG, Diehl RE, Morse CC, Spandorfer JM, Astrin SM. Deregulation of c-myc gene expression in human colon carcinoma is not accompanied by amplification or rearrangement of the gene. Mol Cell Biol 1985;5:1969–1976.
108. Lutz W, Leon J, Eilers M. Contributions of Myc to tumorigenesis. Biochim Biophys Acta 2002; 1602:61–71.
109. Henriksson M, Luscher B. Proteins of the Myc network: essential regulators of cell growth and differentiation. Adv Cancer Res 1996;68:109–182.
110. Smith DR, Myint T, Goh HS. Over-expression of the c-myc proto-oncogene in colorectal carcinoma. Br J Cancer 1993;68:407–413.
111. Bonilla M, Ramirez M, Lopez-Cueto J, Gariglio P. In vivo amplification and rearrangement of c-myc oncogene in human breast tumors. J Natl Cancer Inst 1988;80:665–671.
112. Stein CA. Antitumor effects of antisense phosphorothioate c-myc oligodeoxynucleotides: a question of mechanism. J Natl Cancer Inst 1996;88:391–393.
113. Akie K, Dosaka-Akita H, Murakami A, Kawakami Y. A combination treatment of c-myc antisense DNA with all-trans-retinoic acid inhibits cell proliferation by downregulating c-myc expression in small cell lung cancer. Antisense Nucleic Acid Drug Dev 2000;10:243–249.
114. Chen JP, Lin C, Xu CP, et al. Molecular therapy with recombinant antisense c-myc adenovirus for human gastric carcinoma cells in vitro and in vivo. J Gastroenterol Hepatol 2001;16:22–28.
115. Siegel PM, Shu W, Massagué J. Mad upregulation and Id2 repression accompany transforming growth factor (TGF)-beta-mediated epithelial cell growth suppression. J Biol Chem 2003;278: 35,444–35,450.
116. Kang Y, Chen CR, Massagué J. A self-enabling TGFbeta response coupled to stress signaling: Smad engages stress response factor ATF3 for Id1 repression in epithelial cells. Mol Cell 2003;11:915–926.
117. Alexandrow MG, Moses HL. Transforming growth factor beta and cell cycle regulation. Cancer Res 1995;55:1452–1457.
118. Miyazono K, Maeda S, Imamura T. Coordinate regulation of cell growth and differentiation by TGF-beta superfamily and Runx proteins. Oncogene 2004;23:4232–4237.
119. LaBaer J, Garrett MD, Stevenson LF, et al. New functional activities for the p21 family of CDK inhibitors. Genes Dev 1997;11:847–862.
120. Cheng M, Olivier P, Diehl JA, et al. The p21(Cip1) and p27(Kip1) CDK 'inhibitors' are essential activators of cyclin D-dependent kinases in murine fibroblasts. EMBO J 1999;18:1571–1583.
121. Koff A, Ohtsuki M, Polyak K, Roberts JM, Massagué J. Negative regulation of G1 in mammalian cells: inhibition of cyclin E-dependent kinase by TGF-beta. Science 1993;260:536–539.
122. Polyak K, Kato JY, Solomon MJ, et al. p27Kip1, a cyclin-Cdk inhibitor, links transforming growth factor-beta and contact inhibition to cell cycle arrest. Genes Dev 1994;8:9–22.
123. Slingerland JM, Hengst L, Pan CH, Alexander D, Stampfer MR, Reed SI. A novel inhibitor of cyclin-Cdk activity detected in transforming growth factor beta-arrested epithelial cells. Mol Cell Biol 1994;14:3683–3694.
124. Nourse J, Firpo E, Flanagan WM, et al. Interleukin-2-mediated elimination of the p27Kip1 cyclin-dependent kinase inhibitor prevented by rapamycin. Nature 1994;372:570–573.
125. Coats S, Flanagan WM, Nourse J, Roberts JM. Requirement of p27Kip1 for restriction point control of the fibroblast cell cycle. Science 1996;272:877–880.
126. Fang F, Orend G, Watanabe N, Hunter T, Ruoslahti E. Dependence of cyclin E-CDK2 kinase activity on cell anchorage. Science 1996;271:499–502.
127. Massagué J, Chen YG. Controlling TGF-beta signaling. Genes Dev 2000;14:627–644.
128. Donovan JC, Milic A, Slingerland JM. Constitutive MEK/MAPK activation leads to p27(Kip1) deregulation and antiestrogen resistance in human breast cancer cells. J Biol Chem 2001;276: 40,888–40,895.
129. Florenes VA, Bhattacharya N, Bani MR, Ben-David Y, Kerbel RS, Slingerland JM. TGF-beta mediated G1 arrest in a human melanoma cell line lacking p15INK4B: evidence for cooperation between p21Cip1/WAF1 and p27Kip1. Oncogene 1996;13:2447–2457.
130. Donovan JC, Rothenstein JM, Slingerland JM. Non-malignant and tumor-derived cells differ in their requirement for p27Kip1 in transforming growth factor-beta-mediated G1 arrest. J Biol Chem 2002;277:41,686–41,692.

131. Ciarallo S, Subramaniam V, Hung W, et al. Altered p27(Kip1) phosphorylation, localization, and function in human epithelial cells resistant to transforming growth factor beta-mediated G(1) arrest. Mol Cell Biol 2002;22:2993–3002.
132. Hara T, Kamura T, Kotoshiba S, et al. Role of the UBL-UBA protein KPC2 in degradation of p27 at G1 phase of the cell cycle. Mol Cell Biol 2005;25:9292–9303.
133. Hara T, Kamura T, Nakayama K, Oshikawa K, Hatakeyama S, Nakayama K. Degradation of p27(Kip1) at the G(0)-G(1) transition mediated by a Skp2-independent ubiquitination pathway. J Biol Chem 2001;276:48,937–48,943.
134. Ishida N, Hara T, Kamura T, Yoshida M, Nakayama K, Nakayama KI. Phosphorylation of p27Kip1 on serine 10 is required for its binding to CRM1 and nuclear export. J Biol Chem 2002;277:14,355–14,358.
135. Kotoshiba S, Kamura T, Hara T, Ishida N, Nakayama KI. Molecular dissection of the interaction between p27 and Kip1 ubiquitylation-promoting complex, the ubiquitin ligase that regulates proteolysis of p27 in G1 phase. J Biol Chem 2005;280:17,694–17,700.
136. Fujita N, Sato S, Katayama K, Tsuruo T. Akt-dependent phosphorylation of p27Kip1 promotes binding to 14-3-3 and cytoplasmic localization. J Biol Chem 2002;277:28,706–28,713.
137. Shin I, Rotty J, Wu FY, Arteaga CL. Phosphorylation of p27Kip1 at Thr-157 interferes with its association with importin alpha during G1 and prevents nuclear re-entry. J Biol Chem 2005;280:6055–6063.
138. Shin I, Yakes FM, Rojo F, et al. PKB/Akt mediates cell-cycle progression by phosphorylation of p27(Kip1) at threonine 157 and modulation of its cellular localization. Nat Med 2002;8:1145–1152.
139. Kfir S, Ehrlich M, Goldshmid A, Liu X, Kloog Y, Henis YI. Pathway- and expression level-dependent effects of oncogenic N-Ras: p27(Kip1) mislocalization by the Ral-GEF pathway and Erk-mediated interference with Smad signaling. Mol Cell Biol 2005;25:8239–8250.
140. Hengst L, Dulic V, Slingerland JM, Lees E, Reed SI. A cell cycle-regulated inhibitor of cyclin-dependent kinases. Proc Natl Acad Sci USA 1994;91:5291–5295.
141. Sheaff RJ, Groudine M, Gordon M, Roberts JM, Clurman BE. Cyclin E-CDK2 is a regulator of p27Kip1. Genes Dev 1997;11:1464–1478.
142. Vlach J, Hennecke S, Amati B. Phosphorylation-dependent degradation of the cyclin-dependent kinase inhibitor p27. EMBO J 1997;16:5334–5344.
143. Montagnoli A, Fiore F, Eytan E, et al. Ubiquitination of p27 is regulated by Cdk-dependent phosphorylation and trimeric complex formation. Genes Dev 1999;13:1181–1189.
144. Malek NP, Sundberg H, McGrew S, et al. A mouse knock-in model exposes sequential proteolytic pathways that regulate p27Kip1 in G1 and S phase. Nature 2001;413:323–327.
145. Nguyen H, Gitig DM, Koff A. Cell-free degradation of p27(kip1), a G1 cyclin-dependent kinase inhibitor, is dependent on CDK2 activity and the proteasome. Mol Cell Biol 1999;19:1190–1201.
146. Zhu XH, Nguyen H, Halicka HD, Traganos F, Koff A. Noncatalytic requirement for cyclin A-cdk2 in p27 turnover. Mol Cell Biol 2004;24:6058–6066.
147. Hershko A, Ciechanover A. The ubiquitin system. Annu Rev Biochem 1998;67:425–479.
148. Shirane M, Harumiya Y, Ishida N, et al. Down-regulation of p27(Kip1) by two mechanisms, ubiquitin-mediated degradation and proteolytic processing. J Biol Chem 1999;274:13,886–13,893.
149. Krek W. Proteolysis and the G1-S transition: the SCF connection. Curr Opin Genet Dev 1998;8:36–42.
150. Patton EE, Willems AR, Tyers M. Combinatorial control in ubiquitin-dependent proteolysis: don't Skp the F-box hypothesis. Trends Genet 1998;14:236–243.
151. Ganoth D, Bornstein G, Ko TK, et al. The cell-cycle regulatory protein Cks1 is required for SCF(Skp2)-mediated ubiquitinylation of p27. Nat Cell Biol 2001;3:321–324.
152. Spruck C, Strohmaier H, Watson M, et al. A CDK-independent function of mammalian Cks1: targeting of SCF(Skp2) to the CDK inhibitor p27Kip1. Mol Cell 2001;7:639–650.
153. DeSalle LM, Pagano M. Regulation of the G1 to S transition by the ubiquitin pathway. FEBS Lett 2001;490:179–189.
154. Sitry D, Seeliger MA, Ko TK, et al. Three different binding sites of Cks1 are required for p27-ubiquitin ligation. J Biol Chem 2002;277:42,233–42,240.
155. Seeliger MA, Breward SE, Friedler A, Schon O, Itzhaki LS. Cooperative organization in a macromolecular complex. Nat Struct Biol 2003;10:718–724.
156. Nakayama K, Nagahama H, Minamishima YA, et al. Skp2-mediated degradation of p27 regulates progression into mitosis. Dev Cell 2004;6:661–672.

157. Zancai P, Dal Col J, Piccinin S, et al. Retinoic acid stabilizes p27(Kip1) in EBV-immortalized lymphoblastoid B cell lines through enhanced proteasome-dependent degradation of the p45(Skp2) and Cks1 proteins. Oncogene 2005;24:2483–2494.
158. Minamishima YA, Nakayama K. Recovery of liver mass without proliferation of hepatocytes after partial hepatectomy in Skp2-deficient mice. Cancer Res 2002;62:995–999.
159. Nakayama K, Nagahama H, Minamishima YA, et al. Targeted disruption of Skp2 results in accumulation of cyclin E and p27(Kip1), polyploidy and centrosome overduplication. EMBO J 2000;19:2069–2081.
160. Nakayama KI, Hatakeyama S, Nakayama K. Regulation of the cell cycle at the G1-S transition by proteolysis of cyclin E and p27Kip1. Biochem Biophys Res Commun 2001;282:853–860.
161. Lenferink AE, Busse D, Flanagan WM, Yakes FM, Arteaga CL. ErbB2/neu kinase modulates cellular p27(Kip1) and cyclin D1 through multiple signaling pathways. Cancer Res 2001;61:6583–6591.
162. Lenferink AE, Simpson JF, Shawver LK, Coffey RJ, Forbes JT, Arteaga CL. Blockade of the epidermal growth factor receptor tyrosine kinase suppresses tumorigenesis in MMTV/Neu + MMTV/TGF-alpha bigenic mice. Proc Natl Acad Sci USA 2000;97:9609–9614.
163. Aktas H, Cai H, Cooper GM. Ras links growth factor signaling to the cell cycle machinery via regulation of cyclin D1 and the Cdk inhibitor p27KIP1. Mol Cell Biol 1997;17:3850–3857.
164. Hu W, Bellone CJ, Baldassare JJ. RhoA stimulates p27(Kip) degradation through its regulation of cyclin E/CDK2 activity. J Biol Chem 1999;274:3396–3401.
165. Kawada M, Yamagoe S, Murakami Y, Suzuki K, Mizuno S, Uehara Y. Induction of p27Kip1 degradation and anchorage independence by Ras through the MAP kinase signaling pathway. Oncogene 1997; 15:629–637.
166. Wang Y, Hengst L, Waddell MB, et al. Cdk-inhibitory activity and stability of $p27^{Kip1}$ are directly regulated by oncogenic tyrosine kinases. Cell 2007;128:269–280.
167. Bashir T, Pagano M. Don't skip the G1 phase: how APC/CCdh1 keeps SCFSKP2 in check. Cell Cycle 2004;3:850–852.
168. Bashir T, Dorrello NV, Amador V, Guardavaccaro D, Pagano M. Control of the SCF(Skp2-Cks1) ubiquitin ligase by the APC/C(Cdh1) ubiquitin ligase. Nature 2004;428:190–193.
169. Tedesco D, Lukas J, Reed SI. The pRb related protein p130 is regulated by phosphorylation-dependent proteolysis via the protein-ubiquitin ligase SCF(Skp2). Genes Dev 2002;16:2946–2957.
170. Ganiatsas S, Dow R, Thompson A, Schulman B, Germain D. A splice variant of Skp2 is retained in the cytoplasm and fails to direct cyclin D1 ubiquitination in the uterine cancer cell line SK-UT. Oncogene 2001;20:3641–3650.
171. Dehan E, Pagano M. Skp2, the FoxO1 hunter. Cancer Cell 2005;7:209–210.
172. Kitajima S, Kudo Y, Ogawa I, et al. Role of Cks1 overexpression in oral squamous cell carcinomas: cooperation with Skp2 in promoting p27 degradation. Am J Pathol 2004;165:2147–2155.
173. Kudo Y, Kitajima S, Sato S, Miyauchi M, Ogawa I, Takata T. High expression of S-phase kinase-interacting protein 2, human F-box protein, correlates with poor prognosis in oral squamous cell carcinomas. Cancer Res 2001;61:7044–7047.
174. Gstaiger M, Jordan R, Lim M, et al. Skp2 is oncogenic and overexpressed in human cancers. Proc Natl Acad Sci USA 2001;98:5043–5048.
175. Hershko D, Bornstein G, Ben-Izhak O, et al. Inverse relation between levels of p27(Kip1) and of its ubiquitin ligase subunit Skp2 in colorectal carcinomas. Cancer 2001;91:1745–1751.
176. Chiarle R, Fan Y, Piva R, et al. S-phase kinase-associated protein 2 expression in non-Hodgkin's lymphoma inversely correlates with p27 expression and defines cells in S phase. Am J Pathol 2002;160:1457–1466.
177. Signoretti S, Di Marcotullio L, Richardson A, et al. Oncogenic role of the ubiquitin ligase subunit Skp2 in human breast cancer. J Clin Invest 2002;110:633–641.
178. Yokoi S, Yasui K, Mori M, Iizasa T, Fujisawa T, Inazawa J. Amplification and overexpression of SKP2 are associated with metastasis of non-small-cell lung cancers to lymph nodes. Am J Pathol 2004;165:175–180.
179. Yokoi S, Yasui K, Saito-Ohara F, et al. A novel target gene, SKP2, within the 5p13 amplicon that is frequently detected in small cell lung cancers. Am J Pathol 2002;161:207–216.
180. Masuda TA, Inoue H, Nishida K, et al. Cyclin-dependent kinase 1 gene expression is associated with poor prognosis in gastric carcinoma. Clin Cancer Res 2003;9:5693–5698.
181. Boehm M, Yoshimoto T, Crook MF, et al. A growth factor-dependent nuclear kinase phosphorylates p27(Kip1) and regulates cell cycle progression. EMBO J 2002;21:3390–3401.

182. Ishida N, Kitagawa M, Hatakeyama S, Nakayama K. Phosphorylation at serine 10, a major phosphorylation site of p27(Kip1), increases its protein stability. J Biol Chem 2000;275:25,146–25,154.
183. Rodier G, Montagnoli A, Di Marcotullio L, et al. p27 cytoplasmic localization is regulated by phosphorylation on Ser10 and is not a prerequisite for its proteolysis. EMBO J 2001;20:6672–6682.
184. McAllister SS, Becker-Hapak M, Pintucci G, Pagano M, Dowdy SF. Novel p27(kip1) C-terminal scatter domain mediates Rac-dependent cell migration independent of cell cycle arrest functions. Mol Cell Biol 2003;23:216–228.
185. Connor MK, Kotchetkov R, Cariou S, et al. CRM1/Ran-mediated nuclear export of p27(Kip1) involves a nuclear export signal and links p27 export and proteolysis. Mol Biol Cell 2003;14:201–213.
186. Liu X, Sun Y, Ehrlich M, et al. Disruption of TGF-beta growth inhibition by oncogenic ras is linked to p27Kip1 mislocalization. Oncogene 2000;19:5926–5935.
187. Borriello A, Cucciolla V, Criscuolo M, et al. Retinoic acid induces p27Kip1 nuclear accumulation by modulating its phosphorylation. Cancer Res 2006;66:4240–4248.
188. Viglietto G, Motti ML, Bruni P, et al. Cytoplasmic relocalization and inhibition of the cyclin-dependent kinase inhibitor p27(Kip1) by PKB/Akt-mediated phosphorylation in breast cancer. Nat Med 2002;8:1136–1144.
189. Liang J, Zubovitz J, Petrocelli T, et al. PKB/Akt phosphorylates p27, impairs nuclear import of p27 and opposes p27-mediated G1 arrest. Nat Med 2002;8:1153–1160.
190. Motti ML, Califano D, Troncone G, et al. Complex regulation of the cyclin-dependent kinase inhibitor p27kip1 in thyroid cancer cells by the PI3K/AKT pathway: regulation of p27kip1 expression and localization. Am J Pathol 2005;166:737–749.
191. Singh SP, Lipman J, Goldman H, et al. Loss or altered subcellular localization of p27 in Barrett's associated adenocarcinoma. Cancer Res 1998;58:1730–1735.
192. Ciaparrone M, Yamamoto H, Yao Y, et al. Localization and expression of p27KIP1 in multistage colorectal carcinogenesis. Cancer Res 1998;58:114–122.
193. Masciullo V, Sgambato A, Pacilio C, et al. Frequent loss of expression of the cyclin-dependent kinase inhibitor p27 in epithelial ovarian cancer. Cancer Res 1999;59:3790–3794.
194. Rosen DG, Yang G, Cai KQ, et al. Subcellular localization of p27kip1 expression predicts poor prognosis in human ovarian cancer. Clin Cancer Res 2005;11:632–637.
195. Slingerland J, Pagano M. Regulation of the cdk inhibitor p27 and its deregulation in cancer. J Cell Physiol 2000;183:10–17.
196. Viglietto G, Motti ML, Fusco A. Understanding p27(kip1) deregulation in cancer: down-regulation or mislocalization. Cell Cycle 2002;1:394–400.
197. Sekimoto T, Fukumoto M, Yoneda Y. 14-3-3 suppresses the nuclear localization of threonine 157-phosphorylated p27(Kip1). EMBO J 2004;23:1934–1942.
198. Leone G, DeGregori J, Sears R, Jakoi L, Nevins JR. Myc and Ras collaborate in inducing accumulation of active cyclin E/Cdk2 and E2F. Nature 1997;387:422–426.
199. West MJ, Stoneley M, Willis AE. Translational induction of the c-myc oncogene via activation of the FRAP/TOR signalling pathway. Oncogene 1998;17:769–780.
200. Tomoda K, Kubota Y, Kato J. Degradation of the cyclin-dependent-kinase inhibitor p27Kip1 is instigated by Jab1. Nature 1999;398:160–165.
201. Tomoda K, Kubota Y, Arata Y, et al. The cytoplasmic shuttling and subsequent degradation of p27Kip1 mediated by Jab1/CSN5 and the COP9 signalosome complex. J Biol Chem 2002;277:2302–2310.
202. Jeon JH, Lee KN, Hwang CY, et al. Tumor suppressor VDUP1 increases p27(Kip1) stability by inhibiting JAB1 Cancer Res 2005;65:4485–4489.
203. Esteva FJ, Sahin AA, Rassidakis GZ, et al. Jun activation domain binding protein 1 expression is associated with low p27(Kip1)levels in node-negative breast cancer. Clin Cancer Res 2003;9:5652–5659.
204. Muller D, Thieke K, Burgin A, Dickmanns A, Eilers M. Cyclin E-mediated elimination of p27 requires its interaction with the nuclear pore-associated protein mNPAP60. EMBO J 2000;19:2168–2180.
205. Enenkel C, Lehmann A, Kloetzel PM. Subcellular distribution of proteasomes implicates a major location of protein degradation in the nuclear envelope-ER network in yeast. EMBO J 1998;17:6144–6154.
206. Creasman WT. Endometrial cancer: incidence, prognostic factors, diagnosis, and treatment. Semin Oncol 1997;24:S1-140–S1-150.

207. Bandera CA, Boyd J. The molecular genetics of endometrial cancer. In: Etiology of Breast and Gynecological Cancers, Wiley-Liss, Inc., New York, NY 1997.
208. Gold LI, Parekh TV. Loss of growth regulation by transforming growth factor-beta (TGF-beta) in human cancers: studies on endometrial carcinoma. Semin Reprod Endocrinol 1999;17:73–92.
209. Cohen CJ. Tamoxifen and endometrial cancer: tamoxifen effects on the human female genital tract. Semin Oncol 1997;24:S1-55–S1-64.
210. Cohen I, Rosen DJ, Shapira J, et al. Endometrial changes with tamoxifen: comparison between tamoxifen- treated and nontreated asymptomatic, postmenopausal breast cancer patients [see comments]. Gynecol Oncol 1994;52:185–190.
211. Cheng WF, Lin HH, Torng PL. Huang SC. Comparison of endometrial changes among symptomatic tamoxifen-treated and nontreated premenopausal and postmenopausal breast cancer patients. Gynecol Oncol 1997;66:233–237.
212. Gurpide E. Endometrial cancer: biochemical and clinical correlates. J Natl Cancer Inst 1991;83: 405–416.
213. Morrow CP, Curtin JP, Townsend DE. Tumors of the endometirum. Fourth edit. Synopsis of Gynecologic Oncology, Churchill Livingstone, New York 1993.
214. Kistner RW, Griffiths CT, Craig JM. Use of progestational agents in the management of endometrial cancer. Cancer 1965;18:1563–1579.
215. Bandera CA, Boyd J. The molecular genetics of endometrial carcinoma. Prog Clin Biol Res 1997; 396:185–203.
216. Sulak PJ. Endometrial cancer and hormone replacement therapy. Appropriate use of progestins to oppose endogenous and exogenous estrogen. Endocrinol Metabol Clin N Am 1997;26:399–412.
217. Parekh TV, Gama P, Wen X, et al. Transforming growth factor beta signaling is disabled early in human endometrial carcinogenesis concomitant with loss of growth inhibition. Cancer Res 2002;62: 2778–2790.
218. Farley J, Gray K, Nycum L, Prentice M, Birrer MJ, Jakowlew SB. Endocervical cancer is associated with an increase in the ligands and receptors for transforming growth factor-beta and a contrasting decrease in p27(Kip1). Gynecol Oncol 2000;78:113–122.
219. Macri E, Loda M. Role of p27 in prostate carcinogenesis. Cancer Metastasis Rev 1998;17:337–344.
220. Tsuchiya A, Zhang GJ, Kanno M. Prognostic impact of cyclin-dependent kinase inhibitor p27kip1 in node-positive breast cancer. J Surg Oncol 1999;70:230–234.
221. Wu J, Shen ZZ, Lu JS, et al. Prognostic role of p27Kip1 and apoptosis in human breast cancer. Br J Cancer 1999;79:1572–1578.
222. Catzavelos C, Tsao MS, DeBoer G, Bhattacharya N, Shepherd FA, Slingerland JM. Reduced expression of the cell cycle inhibitor p27Kip1 in non-small cell lung carcinoma: a prognostic factor independent of Ras. Cancer Res 1999;59,684–688.
223. Nycum LR, Smith LM, Farley JH, Kost ER, Method MW, Birrer MJ. The role of p27 in endometrial carcinoma. Gynecol Oncol 2001;81:242–246.
224. Oshita T, Shigemasa K, Nagai N, Ohama K. p27, cyclin E, and CDK2 expression in normal and cancerous endometrium. Int J Oncol 2002;21:737–743.
225. Ozkara SK, Corakci A. Significantly decreased P27 expression in endometrial carcinoma compared to complex hyperplasia with atypia (correlation with p53 expression). Pathol Oncol Res 2004;10:89–97.
226. Masciullo V, Susini T, Zamparelli A, et al. Frequent loss of expression of the cyclin-dependent kinase inhibitor p27(Kip1) in estrogen-related Endometrial adenocarcinomas. Clin Cancer Res 2003;9: 5332–5338.
227. Piva R, Cancelli I, Cavalla P, et al. Proteasome-dependent degradation of p27/kip1 in gliomas. J Neuropathol Exp Neurol 1999;58:691–696.
228. Cariou S, Catzavelos C, Slingerland JM. Prognostic implications of expression of the cell cycle inhibitor protein p27Kip1. Breast Cancer Res Treat 1998;52:29–41.
229. Lee MH, Yang HY. Regulators of G1 cyclin-dependent kinases and cancers. Cancer Metastasis Rev 2003;22:435–449.
230. Pietenpol JA, Bohlander SK, Sato Y, et al. Assignment of the human p27Kip1 gene to 12p13 and its analysis in leukemias. Cancer Res 1995;55:1206–1210.
231. Kawamata N, Morosetti R, Miller CW, et al. Molecular analysis of the cyclin-dependent kinase inhibitor gene p27/Kip1 in human malignancies. Cancer Res 1995;55:2266–2269.
232. Ponce-Castaneda MV, Lee MH, Latres E, et al. p27Kip1: chromosomal mapping to 12p12-12p13.1 and absence of mutations in human tumors. Cancer Res 1995;55:1211–1214.

233. Spirin KS, Simpson JF, Takeuchi S, Kawamata N, Miller CW, Koeffler HP. p27/Kip1 mutation found in breast cancer. Cancer Res 1996;56:2400–2404.
234. Kobayashi H, Montgomery KT, Bohlander SK, et al. Fluorescence in situ hybridization mapping of translocations and deletions involving the short arm of human chromosome 12 in malignant hematologic diseases. Blood 1994;84:3473–3482.
235. Fero ML, Randel E, Gurley KE, Roberts JM, Kemp CJ. The murine gene p27Kip1 is haplo-insufficient for tumour suppression. Nature 1998;396:177–180.
236. Muraoka RS, Lenferink AE, Law B, et al. ErbB2/Neu-induced, cyclin D1-dependent transformation is accelerated in p27-haploinsufficient mammary epithelial cells but impaired in p27-null cells. Mol Cell Biol 2002;22:2204–2219.
237. Fero ML, Rivkin M, Tasch M, et al. A syndrome of multiorgan hyperplasia with features of gigantism, tumorigenesis, and female sterility in p27(Kip1)-deficient mice. Cell 1996;85:733–744.
238. Nakayama K, Ishida N, Shirane M, et al. Mice lacking p27(Kip1) display increased body size, multiple organ hyperplasia, retinal dysplasia, and pituitary tumors. Cell 1996;85:707–720.
239. Hengst L, Reed SI. Translational control of p27Kip1 accumulation during the cell cycle. Science 1996;271:1861–1864.
240. Millard SS, Yan JS, Nguyen H, Pagano M, Kiyokawa H, Koff A. Enhanced ribosomal association of p27(Kip1) mRNA is a mechanism contributing to accumulation during growth arrest. J Biol Chem 1997;272:7093–7098.
241. Pagano M, Tam SW, Theodoras AM, et al. Role of the ubiquitin-proteasome pathway in regulating abundance of the cyclin-dependent kinase inhibitor p27. Science 1995;269:682–685.
242. Lloyd RV, Erickson LA, Jin L, et al. p27kip1: a multifunctional cyclin-dependent kinase inhibitor with prognostic significance in human cancers. Am J Pathol 1999;154:313–323.
243. Adams J, Palombella VJ, Sausville EA, et al. Proteasome inhibitors: a novel class of potent and effective antitumor agents. Cancer Res, 1999;59:2615–2622.
244. Kane RC, Farrell AT, Sridhara R, Pazdur R. United States food and drug administration approval summary: bortezomib for the treatment of progressive multiple myeloma after one prior therapy. Clin Cancer Res 2006;12:2955–2960.
245. O'Connor OA, Smith EA, Toner LE, et al. The combination of the proteasome inhibitor bortezomib and the bcl-2 antisense molecule oblimersen sensitizes human B-cell lymphomas to cyclophosphamide. Clin Cancer Res 2006;12:2902–2911.
246. Kaufman JL, Lonial S. Proteasome inhibition: novel therapy for multiple myeloma. Onkologie 2006; 29:162–168.
247. Ciechanover A. The ubiquitin-proteasome pathway: on protein death and cell life. EMBO J 1998;17: 7151–7160.
248. Zhang F, Monkkonen M, Roth S, Laiho M. TGF-beta induced G(1) cell cycle arrest requires the activity of the proteasome pathway. Exp Cell Res 2002;281:190–196.
249. Sgambato A, Zhang YJ, Ciaparrone M, et al. Overexpression of p27Kip1 inhibits the growth of both normal and transformed human mammary epithelial cells. Cancer Res 1998;58:3448–3454.
250. Wang W, Ungermannova D, Jin J, Harper JW, Liu X. Negative regulation of SCFSkp2 ubiquitin ligase by TGF-beta signaling. Oncogene 2004;23:1064–1075.
251. Dow R, Hendley J, Pirkmaier A, Musgrove EA, Germain D. Retinoic acid-mediated growth arrest requires ubiquitylation and degradation of the F-box protein Skp2. J Biol Chem 2001;276:45,945–45,951.
252. Lin R, Wang TT, Miller WH, Jr, White JH. Inhibition of F-Box protein p45(SKP2) expression and stabilization of cyclin-dependent kinase inhibitor p27(KIP1) in vitamin D analog-treated cancer cells. Endocrinology 2003;144:749–753.
253. Tadlock L, Yamagiwa Y, Hawke J, Marienfeld C, Patel T. Transforming growth factor-beta inhibition of proteasomal activity: a potential mechanism of growth arrest. Am J Physiol Cell Physiol 2003; 285:C277–C285.
254. Radke S, Pirkmaier A, Germain D. Differential expression of the F-box proteins Skp2 and Skp2B in breast cancer. Oncogene 2005;24:3448–3458.
255. Drobnjak M, Melamed J, Taneja S, et al. Altered expression of p27 and Skp2 proteins in prostate cancer of African-American patients. Clin Cancer Res 2003;9:2613–2619.
256. Sarmento LM, Huang H, Limon A, et al. Notch1 modulates timing of G1-S progression by inducing SKP2 transcription and p27 Kip1 degradation. J Exp Med 2005;202:157–168.
257. Migliaccio A, Di Domenico M, Castoria G, et al. Tyrosine kinase/p21ras/MAP-kinase pathway activation by estradiol-receptor complex in MCF-7 cells. EMBO J 1996;15:1292–1300.
258. Collins P, Webb C. Estrogen hits the surface. Nat Med 1999;5:1130–1131.

259. Foster JS, Fernando RI, Ishida N, Nakayama KI, Wimalasena J. Estrogens down-regulate p27Kip1 in breast cancer cells through Skp2 and through nuclear export mediated by the ERK pathway. J Biol Chem 2003;278:41,355–41,366.
260. Cheng M, Sexl V, Sherr CJ, Roussel MF. Assembly of cyclin D-dependent kinase and titration of p27Kip1 regulated by mitogen-activated protein kinase kinase (MEK1). Proc Natl Acad Sci USA 1998;95:1091–1096.
261. Foster JS, Henley DC, Ahamed S, Wimalasena J. Estrogens and cell-cycle regulation in breast cancer. Trends Endocrinol Metab 2001;12:320–327.
262. Foster JS, Wimalasena J. Estrogen regulates activity of cyclin-dependent kinases and retinoblastoma protein phosphorylation in breast cancer cells. Mol Endocrinol 1996;10:488–498.
263. Foster JS, Henley DC, Bukovsky A, Seth P, Wimalasena J. Multifaceted regulation of cell cycle progression by estrogen: regulation of Cdk inhibitors and Cdc25A independent of cyclin D1-Cdk4 function. Mol Cell Biol 2001;21:794–810.
264. Lamb J, Ladha MH, McMahon C, Sutherland RL, Ewen ME. Regulation of the functional interaction between cyclin D1 and the estrogen receptor. Mol Cell Biol 2000;20:8667–8675.
265. Cariou S, Donovan JC, Flanagan WM, Milic A, Bhattacharya N, Slingerland JM. Down-regulation of p21WAF1/CIP1 or p27Kip1 abrogates antiestrogen-mediated cell cycle arrest in human breast cancer cells. Proc Natl Acad Sci USA 2000;97:9042–9046.
266. Sutherland RL, Prall OW, Watts CK, Musgrove EA. Estrogen and progestin regulation of cell cycle progression. J Mammary Gland Biol Neoplasia 1998;3:63–72.
267. Gizard F, Robillard R, Gervois P, et al. Progesterone inhibits human breast cancer cell growth through transcriptional upregulation of the cyclin-dependent kinase inhibitor p27Kip1 gene. FEBS Lett 2005; 579:5535–5541.
268. Shiozawa T, Horiuchi A, Kato K, et al. Up-regulation of p27Kip1 by progestins is involved in the growth suppression of the normal and malignant human endometrial glandular cells. Endocrinology 2001;142:4182–4188.
269. Musgrove EA, Hunter LJ, Lee CS, Swarbrick A, Hui R, Sutherland RL. Cyclin D1 overexpression induces progestin resistance in T-47D breast cancer cells despite p27(Kip1) association with cyclin E-Cdk2. J Biol Chem 2001;276:47,675–47,683.
270. Cunha GR, Cooke PS, Kurita T. Role of stromal-epithelial interactions in hormonal responses. Arch Histol Cytol 2004;67:417–434.
271. Cooke PS, Buchanan DL, Young P, et al. Stromal estrogen receptors mediate mitogenic effects of estradiol on uterine epithelium. Proc Natl Acad Sci USA 1997;94:6535–6540.
272. Hayward SW, Cunha GR, Dahiya R. Normal development and carcinogenesis of the prostate—A unifying hypothesis. In Basis For Cancer Management,1996;Vol. 784:pp. 50–62.
273. Hom YK, Young P, Wiesen JF, et al. Uterine and vaginal organ growth requires epidermal growth factor receptor signaling from stroma. Endocrinology 1998;139:913–921.
274. Kurita T, Young P, Brody JR, Lydon JP, O'Malley BW, Cunha GR. Stromal progesterone receptors mediate the inhibitory effects of progesterone on estrogen-induced uterine epithelial cell deoxyribonucleic acid synthesis. Endocrinology 1998;139:4708–4713.
275. Lecanda J, Parekh TV, et al. TGF-β and ovarian hormones control endometrial growth by mechanisms that converge on the cyclin-dependent kinase inhibitor, $p27^{Kip1}$. Proc Am Assoc Cancer Res 2006;47: 1322 Abstract#5627.
276. Bachman KE, Blair BG, Brenner K, et al. p21(WAF1/CIP1) mediates the growth response to TGF-beta in human epithelial cells. Cancer Biol Ther 2004;3:221–225.
277. Deckers M, van Dinther M, Buijs J, et al. The tumor suppressor Smad4 is required for transforming growth factor beta-induced epithelial to mesenchymal transition and bone metastasis of breast cancer cells. Cancer Res 2006;66:2202–2209.
278. Tian F, DaCosta Byfield S, Parks WT, et al. Reduction in Smad2/3 signaling enhances tumorigenesis but suppresses metastasis of breast cancer cell lines. Cancer Res 2003;63:8284–8292.
279. Nagahara H, Vocero-Akbani AM, Snyder EL, et al. Transduction of full-length TAT fusion proteins into mammalian cells: TAT-p27Kip1 induces cell migration. Nat Med 1998;4:1449–1452.
280. Besson A, Assoian RK, Roberts JM. Regulation of the cytoskeleton: an oncogenic function for CDK inhibitors? Nat Rev Cancer 2004;4:948–955.
281. Besson A, Gurian-West M, Schmidt A, Hall A, Roberts JM. p27Kip1 modulates cell migration through the regulation of RhoA activation. Genes Dev 2004;18:862–876.
282. Hlobilkova A, Knillova J, Bartek J, Lukas J, Kolar Z. The mechanism of action of the tumour suppressor gene PTEN. Biomed Pap Med Fac Univ Palacky Olomouc Czech Repub 2003;147:19–25.

283. Watanabe J, Sato H, Kanai T, et al. Paradoxical expression of cell cycle inhibitor p27 in endometrioid adenocarcinoma of the uterine corpus—correlation with proliferation and clinicopathological parameters. Br J Cancer 2002;87:81–85.
284. Bakin AV, Rinehart C, Tomlinson AK, Arteaga CL. p38 mitogen-activated protein kinase is required for TGF-beta-mediated fibroblastic transdifferentiation and cell migration. J Cell Sci 2002;115:3193–3206.
285. Bakin AV, Safina A, Rinehart C, Daroqui C, Darbary H, Helfman DM. A critical role of tropomyosins in TGF-beta regulation of the actin cytoskeleton and cell motility in epithelial cells. Mol Biol Cell 2004;15:4682–4694.
286. Souchelnytskyi S. Proteomics of TGF-beta signaling and its impact on breast cancer. Expert Rev Proteomics 2005;2:925–935.
287. Laping NJ. Therapeutic uses of smad protein inhibitors: Selective inhibition of specific TGF-beta activities. IDrugs 1999;2:907–914.
288. Peng SB, Yan L, Xia X, et al. Kinetic characterization of novel pyrazole TGF-beta receptor I kinase inhibitors and their blockade of the epithelial-mesenchymal transition. Biochemistry 2005;44:2293–2304.
289. Singh J, Ling LE, Sawyer JS, Lee WC, Zhang F, Yingling JM. Transforming the TGFbeta pathway: convergence of distinct lead generation strategies on a novel kinase pharmacophore for TbetaRI (ALK5). Curr Opin Drug Discov Dev 2004;7:437–445.
290. Yingling JM, Blanchard KL, Sawyer JS. Development of TGF-beta signalling inhibitors for cancer therapy. Nat Rev Drug Discov 2004;3:1011–1022.
291. Pinkas J, Teicher BA. TGF-beta in cancer and as a therapeutic target. Biochem Pharmacol 2006;72: 523–529.
292. Lecanda J, Parekh TV, et al. Transforming growth factor-β, estrogen, and progesterone coverage on the regulation of $p27^{Kip1}$ in the normal and malignant endometrium. Cancer Res 2007;67: 1007–1018.

15 Transforming Growth Factor-β in Cancer Therapy

Proteasome in the Control of TGF-β Signaling During Development and Cancer

Arja Band and Marikki Laiho

CONTENTS

Introduction
Targeted Proteasomal Destruction of the TGF-β Pathway Signaling Components
Regulation of TGF-β Pathway Associated Proteasomally Targeted Proteins During Development
Dysregulation of TGF-β Signaling in Cancer by the Proteasome Pathway
Concluding Remarks
Acknowledgments
References

Abstract

Several transforming growth factor-β (TGF-β) signaling pathway proteins undergo proteasomal degradation. Controlled proteasomal degradation provides a means for the cell to time and fine-tune the duration of both positive and negative physiological signals. Importantly, this physiological control appears to be aberrant in for example tumor cells carrying mutations of the Smad signaling molecules. We summarize here the recent striking observations on the action of the ubiquitin-proteasome pathway in controlling TGF-β signal transduction in both physiological and pathological conditions, such as cancer.

Key Words: TGF-β; Smad; signal transduction; SnoN; Ski; proteasome; ubiquitin; SUMO, degradation.

1. INTRODUCTION

The proteasome pathway has evolved for the rapid destruction of misfolded and damaged proteins. It degrades abnormal proteins, which result from errors in transcription, translation, and folding or physical protein damage *(1)*. The control of protein quality is an essential function for the cell to preserve itself from the toxicity generated by the protein waste. During evolution, the proteasome system has been fine-tuned to serve another purpose – as a rapid regulatory mechanism to control the half-lives and levels of cellular proteins. Thus,

From: *Cancer Drug Discovery and Development: Transforming Growth Factor-β in Cancer Therapy, Vol I: Basic and Clinical Biology*
Edited by: Sonia B. Jakowlew © Humana Press Inc., Totowa, NJ

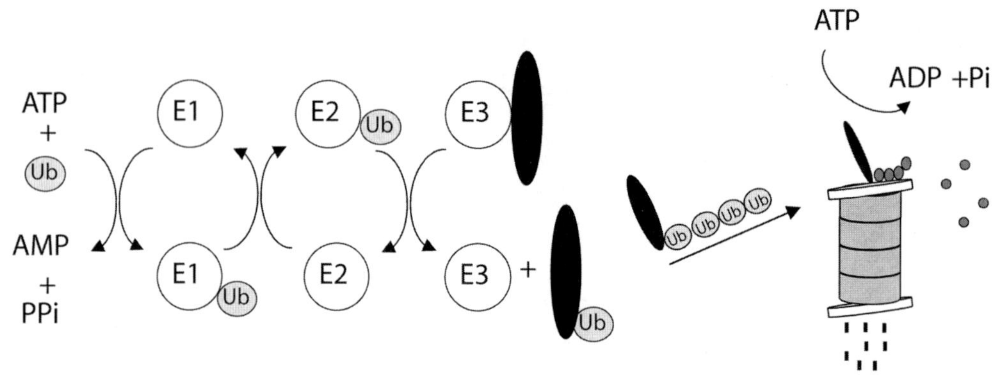

Fig. 1. Ubiquitin-proteasome machinery. Proteins targeted for degradation are marked by a polyubiquitin chain of four or more ubiquitin proteins. This occurs by repetition of the following three steps: Ubiquitin is conjugated to an ubiquitin-activating enzyme (E1) in an ATP-dependent reaction. The ubiquitin is further transferred to an ubiquitin-conjugating enzyme (E2) which transfers it to the target protein with the aid of an ubiquitin protein ligase enzyme (E3). The polyubiquitin chain serves as a recognition signal for the 26S proteasome. The target protein is unfolded and fed to the proteolytic barrel in an ATP-dependent manner. Ubiquitin side chains are removed by a deubiquitylating enzyme and reused.

the protein life spans can be kept as short as 2–20 min, and allowed to peak for a very narrow window of time for example during cell cycle progression. The proteasome machinery consists of a cylindrical 50-subunit barrel with protease activity (Fig. 1). Proteins targeted for destruction are recognized by their modification through attachment of multiple ubiquitin molecules, i.e., multiubiquitylation. Multiubiquitylation, the formation of a chain of four or more ubiquitin molecules linked through lysines 11, 29, 48, and 63 on ubiquitin, serves as a signal for protein destruction, the predominant chain-formation occuring through lysine 48. Monoubiquitylation, on the other hand, refers to a single attached ubiquitin moiety on either single or several lysines of the substrate protein. Monoubiquitylation is not sufficient for protein destruction and instead may generate signals for altered protein localization or function. Finally yet importantly, proteins may be modified by ubiquitin on their aminotermini independent of lysine residues *(2–4)*.

The covalent attachment of ubiquitin is catalyzed by an ATP-dependent sequential process involving ubiquitin-activating enzyme (E1), ubiquitin-conjugating enzyme (E2), and ubiquitin protein ligase (E3) (Fig. 1) *(5,6)*. Whereas there are around ten mammalian E1s, and one hundred E2s, there may be over one thousand E3s. The specificity of the reaction arises largely from the recognition of the substrate by the E3 ligases, and additionally by pairing of an E3 with an E2. There are three major types of E3 ligases: homology to E6-AP carboxyl-terminal (HECT) domain E3s, a really interesting new gene (RING)-motif containing E3s, and E4/U-box containing proteins (Fig. 2).

HECT-type E3 ligases form a family of proteins which include E6-AP, neural precursor cell expressed, developmentally down-regulated 4 (NEDD4), Smad ubiquitylation-related factor (Smurf) 1, Smurf2, WW-domain containing protein (WWP1), and atropin-interacting protein 4 (AIP4) (Fig. 2). They contain a C-terminal HECT-domain, which harbors the catalytically active cysteine residue forming a thiol–ester intermediate with ubiquitin. Several of the HECT E3 ligases contain a WW-domain, made of conserved tryptophan residues forming a hydrophobic pocket. These domains can bind to proline-rich sequences known as PY motifs *(2)*. Of these, Smurf ligases ubiquitylate essentially all members of the Smad family proteins: receptor/inhibitor/co-Smads, and the TGF-β type 1 receptor (*see* Section 2.1.). RING-finger E3s are single subunit ligases in which the E3 both recognizes the

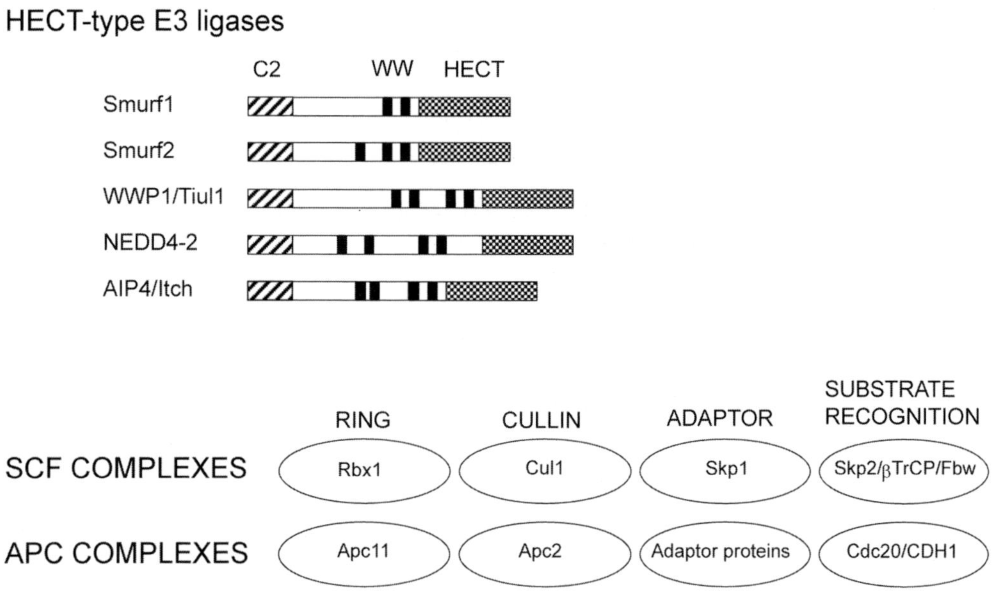

Fig. 2. A schematic picture of HECT and multisubunit E3 ligases involved in the proteasomal degradation of TGF-β signaling molecules. HECT-type E3 ligases contain a N-terminal C2 domain (hatched) which targets a ligase to the cell membrane. Two to four WW-domains (black) bind to the substrate PY motif and mediate the interaction with R-Smads. A C-terminal HECT domain (double-hatched) contains the catalytic activity. A multisubunit E3 complex, SCF, contains a ring finger protein (Rbx1), a cullin subunit (Cul1), an adaptor protein (Skp1), and variable F-box proteins which recognize the target substrates. APC complex contains a ring finger protein, a cullin subunit, several adaptor proteins and activators, Cdc20 and Cdh1.

substrate and has catalytic activity, or complex multisubunit E3s, which are composed of proteins acting in substrate recognition and a set of proteins responsible for the catalytic activity. Two such multisubunit RING E3s are involved in the control of the TGF-β signaling proteins, the Skp, Cullin, and F-box (SCF), and APC complexes (Fig. 2). The SCF complex is composed of a F-box protein, which recognizes the target substrate, RING-finger protein (Rbx1/Roc1), cullin subunit, and an adaptor protein (Skp1). Skp2 is a F-box protein, which recognizes targets of the SCF complex, like p27Kip1, Rb-family protein p130, E2F1, and Myc *(7)*. Its levels are frequently increased in human cancer, for example in breast and prostate tumors *(7,8)*. The APC complex is composed of at least 12 subunits, including the RING protein (Apc11), Cullin (Apc2), multiple adaptor proteins, and substrate recognition proteins Cdc20 or CDH1. Substrate recognition can additionally be enforced by protein modifications such as phosphorylation and acetylation. For example, the SCF F-box protein βTrCP, which targets the destruction of Cdc25A, Wee1, β-catenin, IκBα, and Snail, recognizes preferentially phosphorylated targets containing a DSGφXS motif *(8,9)*.

Defects in the activity of the proteasome, or an unfavorable cellular environment, can cause excessive protein denaturation and lead to the accumulation of abnormal amounts of protein and protein aggregates. In some cases, this results directly from mutations, like the tripeptide accumulation diseases causing neurodegeneration, mutations in the substrate proteins leading to their escape from proteolytic degradation, or from mutations or hyperactivation of the E3 machineries themselves *(10)*.

Several TGF-β signaling pathway proteins undergo proteasomal degradation. Controlled proteasomal degradation provides a means for the cell to time and fine-tune the duration of both positive and negative physiological signals. Importantly, this physiological control appears to be aberrant in for example tumor cells carrying mutations of the Smad signaling molecules. We will summarize here the recent striking observations on the action of the ubiquitin-proteasome pathway in controlling TGF-β signal transduction in both physiological and pathological conditions, such as cancer.

2. TARGETED PROTEASOMAL DESTRUCTION OF THE TGF-β PATHWAY SIGNALING COMPONENTS

Activation of the TGF-β signaling pathway is instigated by ligand binding which provokes TGF-β type II receptor association with and phosphorylation of the type I receptor and promotes the phosphorylation of receptor (R)-Smads, Smad2, and Smad3 *(11–13)*. The R-Smads shuttle rapidly between the cytoplasm and nucleus and this allows their complex formation with the co-Smad, Smad4. The complex so formed binds its transcriptional target sequence and activates the TGF-β specific target genes *(12)*. This process is effectively competed by the inhibitory (I)-Smads, Smad6, and Smad7. Smad7 binds directly to the activated receptor complex, thus preventing the access of R-Smads for receptor-mediated phosphorylation *(14,15)*.

2.1. TGF-β and BMP R-Smad Levels are Controlled both by Ligand-Dependent and Independent Manner

The duration of TGF-β signaling is controlled by the availability, phosphorylation and dephosphorylation, and subcellular redistribution of the R-Smads. Degradation via the ubiquitin-proteasome system leads to irreversible termination of TGF-β signaling *(16,17)*. Lo and Massagué *(18)* were the first to report that in human immortal keratinocytes the levels of Smad were controlled in a ligand and proteasome activity dependent manner. They further demonstrated that Smad2 was ubiquitylated and accumulated in the nucleus *(18)*. However, also basal Smad2 levels are controlled by the proteasome. The ubiquitylation of Smad2 was found to be carried out by UbcH5b/c E2 ubiquitin conjugating enzymes and Smurf2 E3 ubiquitin ligase *(18–20)* which decreases the Smad2-mediated transcription *(19)* (Fig. 3).

Similarly to Smad2, Smad3 is rapidly degraded following TGF-β treatment, an event which is effectively blocked by treatment of the cells with a proteasomal inhibitor *(21)*. SCF complex protein Fbw1a, interacting with ROC1 RING finger protein through the Cullin-subunit, recognizes Smad3 *(21)* (Fig. 3). The interaction of Smad3 with ROC1-SCF^{Fbw1a} and its consequent ubiquitylation is dependent on its ligand-induced phosphorylation causing Smad3 cytoplasmic translocation *(21)*. In contrast to Smad3, the proteasomal degradation of Smad2 is likely to occur in the nucleus as suggested by the preferential interaction of Smurf2 with phosphorylated Smad2 *(19)*, and by experiments showing that leptomycin B, a nuclear export inhibitor, does not prevent the degradation of phosphorylated Smad2 *(18)*.

Smad1 and Smad5 are R-Smads that mediate bone morphogenetic protein (BMP) signaling *(22)*. They are degraded through Smurf1 and Smurf2-mediated ubiquitylation in both a ligand-dependent and independent manner *(19,20,23,24)* (Table 1). An U-box dependent E3 ligase carboxyl terminus of Hsc70-interacting protein (CHIP) has recently been found to target Smad1 and unstimulated Smad3 for destruction *(25,26)*. Thus, the proteasome activity causes the destruction of both TGF-β and BMP R-Smads, effectively limiting their signal transduction pathways. Because Smad4 acts as a common Smad for both TGF-β and BMP-induced Smads, a competition between the respective pathways may exist depending on the levels of the individual Smads and Smurf1. For example, increased expression of Smurf1 or Smurf2 in *Xenopus* embryos represses BMP responses and enhances activin responses *(20,23)*.

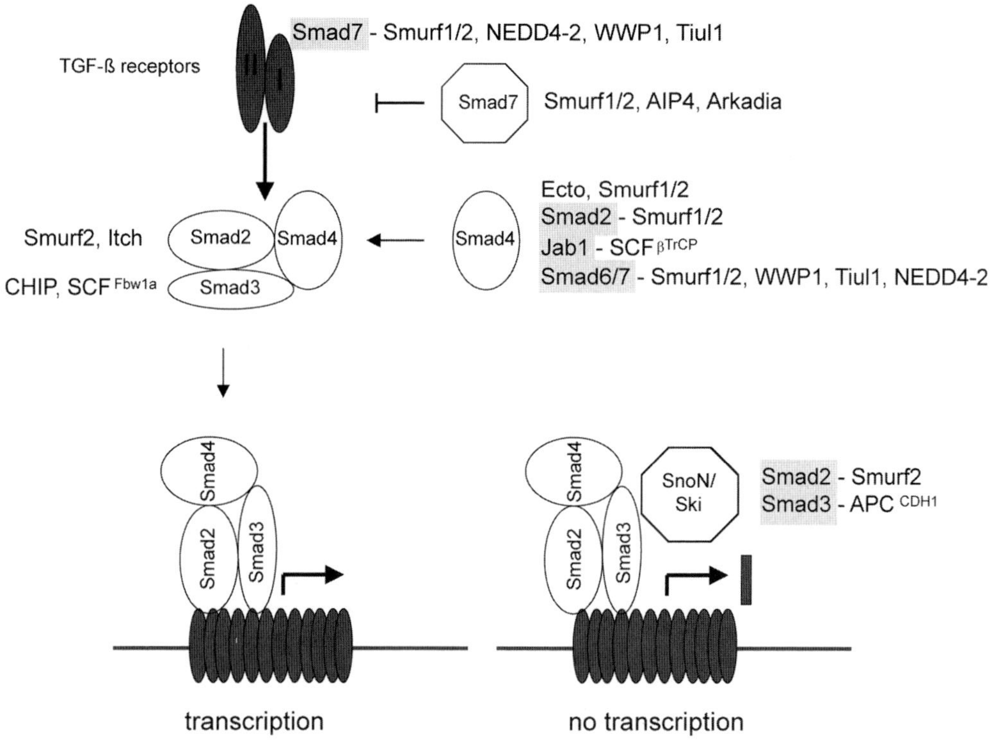

Fig. 3. Proteasome mediated degradation of the TGF-β signal transduction components. The relevant E3 ligases, F-box proteins, or possible adaptor proteins (gray boxes) involved in the ubiquitylation and recognition of the substrates are indicated.

Finally, the interactions of E3 ligases with R-Smads may be affected by other binding proteins. p300, a transcriptional coactivator and histone acetylase, is recruited to Smad3 and enhances the interaction of Smad3 with the ROC1-SCF^{Fbw1a} complex *(21)*. Subsequent ubiquitylation and degradation of Smad3 switches off the Smad3 signal *(21)*. As the ROC1-SCF^{Fbw1a} is also an E3-ligase for IκB and β-catenin, regulation of its levels and activity presents major intercepting points between TGF-β, NF-κB, and Wingless-signaling pathways.

2.2. Levels of Common Smad, Smad4, are Regulated by Multiple Proteasome Dependent Events

Smad4 is a key partner of R-Smad proteins and acts by recruiting transcription coactivators to R-Smads and by stabilizing the interaction between R-Smads and the coactivators *(11–13)*. The turnover of Smad4 in epithelial cells is very slow and independent of antiproliferative (TGF-β) or, conversely, of growth stimuli (serum, EGF, and PDGF). However, oncogenic Ras induces the ubiquitylation and proteasomal degradation of Smad4 *(27)*. In addition, treatment of cells with inhibitors of p38 or Jun N-terminal kinase (JNK) leads to an increase in the level of Smad4 implying that excessive kinase activity through these pathways may decrease TGF-β signaling *(28)*.

Wild type Smad4 undergoes mono/oligoubiquitylation of the MH2 domain Lys507, which participates in the recognition of phosphorylated R-Smad *(29)*. The monoubiquitylated Smad4 has increased affinity toward R-Smads and therefore enhances Smad transcriptional activity *(29)*.

Table 1
Ubiquitin-Proteasome System Mediated Degradation of TGF-β Family Signaling Components

Signaling component	*E3 complex*	*E3 ligase/ F-box protein*	*Modi- fication*	*Ligand- depen- dency*	*Adaptor*	*Effect of modi- fication on signaling*	*Ref.*
Receptor complex	HECT	Smurf2	Ub	–/+	Smad7	Inhibition	*(49,51)*
Type I receptor	HECT	Smurf1	Ub	+	Smad7	Inhibition	*(50)*
	HECT	NEDD4-2	Ub	+	Smad7	Inhibition	*(54)*
	HECT	WW-P1, Tiul1	Ub		Smad7	Inhibition	*(52,53)*
Smad2	HECT	Smurf2	Ub	+		Inhibition	*(18–20)*
	HECT	Itch	MonoUb			Stimulation	*(100)*
Smad3	SCF	ROCI, Fbw1a	Ub	+		Inhibition	*(21)*
		CHIP	Ub	–		Inhibition	*(26)*
		PIASy	Sumo		Smad7	Inhibition	*(41,101)*
Smad1, Smad5	HECT	Smurf1,2	Ub	–/+		Inhibition (BMP) Stimulation (activin)	*(19,20, 23,24)*
	U-box	CHIP	Ub			Inhibition	*(25)*
Smad4		PIASy	Sumo	–		Stimulation	*(37,38)*
		PIAS1	Sumo	+		Stimulation	*(40)*
		PIASxβ	Sumo			No Effect	*(39)*
		Unknown	MonoUb			Stimulation	*(29)*
	SCF	βTrCP	Ub	–	CSN5/ Jab1	Inhibition	*(30,33,98)*
	HECT	Smurf1,2	Ub	+	Smad2		*(31)*
	HECT	Smurf1,2	Ub	–	Smad6/7	Inhibition	*(31)*
	HECT	WW-P1, Tiul1	Ub	–	Smad6/7	Inhibition	*(31)*
	HECT	NEDD4-2	Ub	–	Smad6/7	Inhibition	*(31)*
		Ecto	Ub			Inhibition	*(78)*
Mutant Smad4	SCF	βTrCP	Ub	–		Inhibition	*(30,98)*
	SCF	Skp2	Ub				*(28)*
Smad7	HECT	Smurf1	Ub	+		Stimulation	*(50)*
	HECT	Smurf2		–/+		Stimulation	*(49)*
	HECT	AIP4	Ub			Inhibition	*(57)*
		Arkadia	Ub			Stimulation	*(56)*
SnoN	APC	CDH1	Ub	+	Smad3	Stimulation	*(68,69)*
	HECT	Smurf2	Ub	+	Smad2	Stimulation	*(70)*

In addition to monoubiquitylation, Smad4 is polyubiquitylated by several E3 ligases, including SCF-complex F-box proteins Skp2 and βTrCP *(28,30)* and HECT-domain E3s Smurf1/2, WWP1/TGIF interacting ligase (Tiul1), and NEDD4-2 *(31)* (Table 1). Some F-box proteins, like Skp2, interact with Smad4 constitutively, but the interaction does not promote Smad4 destruction *(28)*. βTrCP, another F-box protein, promotes Smad4 ubiquitylation and degradation, and it decreases TGF-β signaling *(30)*. Similarly, overexpressed CSN5/Jab1, a signalosome component, and coactivator of c-*Jun* *(32)*, interacts with Smad4 and induces its ubiquitylation and degradation through the proteasome *(33)*. Jab1, in addition to limiting the levels of Smad4, has powerful oncogenic potential by acting as an adaptor that promotes the proteasomal degradation of several tumor suppressor pathways including TGF-β, p27Kip1 *(34)*, and p53 *(35)*. Collectively, the destruction of Smad4 leads to a decrease in TGF-β induced transcriptional responses (Fig. 3) (Table 1).

HECT-domain E3s, which recognize PY-domains, are unable to directly bind Smad4, lacking this domain, and instead utilize Smad6/7 as adaptor proteins *(31)*. Efficient degradation by the HECT E3s therefore requires the formation of trimeric complexes between the I-Smads, HECT E3s and Smad4. However, the complex formation is independent of ligand binding. Alternatively, Smad2 can be used as an additional adaptor protein, but this interaction and the consequent Smad4 degradation is significantly enhanced by a constitutively active type I receptor *(31)* (Fig. 3).

The extreme complexity of Smad4 regulation (mono, polyubiquitylation), a large number of possible E3s, and the ligand dependency/independency, suggest that Smad4 levels and its function are carefully controlled in the cells. Some of the complexity may be a reflection of physiological settings, i.e., other signals affecting the TGF-β pathway proteins or E3s, and cell type specificities in the expression levels of the E3 or adaptor proteins, points which need clarification in future studies.

2.3. Sumoylation of R-Smads and Smad4

Sumoylation is a process which covalently attaches a small ubiquitin-like modifier (SUMO), a ubiquitin-related polypeptide, to lysine residues of the target proteins. Sumoylation is mediated through a process involving an E2 ubiquitin carrier protein (Ubc9) and ubiquitin ligases, and it may affect target protein interactions, localization, and function *(4)*. Smad4 is sumoylated on its linker region lysines in a process involving Ubc9 and several related E3 ligases, protein inhibitor of activated STAT (PIAS), PIASy *(36–38)*, PIAS1 *(39,40)*, and PIASx beta *(39)* (Table 1). The sumoylation of Smad4 alters its subnuclear localization, enhances its stability, and increases Smad4-mediated transcriptional activation *(36–38)*. Furthermore, the interaction between PIAS1 and Smad4 is positively regulated by TGF-β, thus creating a positive feedback loop *(40)*. Smad3 and Smad7 associate with PIASy leading to increased sumoylation of Smad3 *(41)*. Sumoylation of Smad4 by PIAS family members, PIAS1 and PIASx-beta, is enhanced by the p38 MAP kinase pathway *(39)*. The sumoylation of Smad4 may also lead to its novel protein interactions. Daxx has been described to recognize sumoylated Smad4 and, through this interaction, to suppress Smad4 mediated transcription *(42)*.

2.4. R-Smads as Adaptor Proteins for E3 Ligases

Smad3 frequently acts as an adaptor protein for several E3 ligases targeting the destruction of other proteins such as human enhancer of filamentation 1 (HEF1) and the protein phosphatase Cdc25A *(43–46)* (Table 2). HEF1 belongs to the family of Cas proteins which are implicated in a variety of cellular processes such as cellular attachment, motility, growth factor responses, apoptosis, and oncogenic transformation. HEF1 interaction with Smad3 promotes the rapid destruction of HEF1 in response to TGF-β stimulation *(43)*. This event is mediated by AIP4, which is the human homolog of mouse Itch, an E3 ligase for Notch and JunB *(47,48)*. AIP4 targets the destruction of atrophin-1 and HEF-1 through its interaction with Smad3 *(45)*. In addition, HEF1 harbors a D-box signature recognized by CDH1 on APC substrates. HEF1 constitutively interacts with CDH1, while Smad3 interacts with APC10

Table 2
Smads as Adaptor Proteins for Proteasomal Degradation

Substrate	*E3 class*	*E3 ligase/F-box protein*	*Modification*	*Adaptor*	*Ref.*
Cdc25A	SCF	βTrCP	Ub	Smad3	*(46)*
HEF1	APC	CDH1	Ub	Smad3	*(44)*
	HECT	AIP4 (hItch)	Ub	Smad3	*(45)*

subunit which depends on activation of the TGF-β receptor *(44)*. Thus, the availability of Smad3 coordinates the levels of HEF1 through two E3 ligases and impinges on other cellular signaling pathways.

2.5. Smad7 Acts as an Adaptor to Promote the Destruction of TGF-β Receptors and R-Smads

Smad7 has frequently been observed to act as an adaptor protein promoting the degradation of the type I-II receptor complex and R-Smads. Kavsak and coauthors reported that the receptor complex turnover was dependent on proteasomal activity and found that Smad7 mediated the interaction and degradation of the receptor complex by the E3 ligase Smurf2 *(49)*. Subsequently, also Smurf1 has been shown to be an E3 ligase for the type I receptor *(50)*. Smad7 stimulates Smurf activity by recruiting UbcH7, an E2, to Smurf2 HECT domain *(51)*.

Several other E3 ligases act through Smad7 to target the receptor complex (Table 1). These include WWP1, Tiul1, and NEDD4-2 *(52–54)* (Fig. 3). WWP1 binds and enhances Smad7 nuclear export and receptor degradation, but does not promote the degradation of associated R-Smads or SnoN *(52)*. Tiul1 targets the destruction of TGF-β receptors and Smad2 in a ligand-dependent manner through constitutive association with Smad7 and it suppresses TGF-β signaling *(53)*. NEDD4-2 also causes the degradation of the type I receptor through Smad7 *(54)*. NEDD4-2 appears to be a promiscuous protein interacting with the receptor-Smads in a ligand-dependent manner to form multiprotein complexes containing NEDD4-2, Smad2, and SnoN. Of these, NEDD4-2 has the capacity to cause the degradation of only Smad2, and to prevent TGF-β signaling *(54)*.

Because Smad7 is a direct inhibitor of TGF-β signaling by binding R-Smads and blocking R-Smad access to the receptor, it is vital to control its levels. The expression of Smad7 is stimulated by TGF-β and other cytokines, like interleukin and interferon-gamma *(13)* and these therefore provide potent ways to limit TGF-β signaling. TGF-β also stimulates Smurf2 expression, while PI3K/Akt kinase activated pathway suppresses it *(55)*. It is more than likely that these are only a few examples of the complex regulatory network determining the expression and levels of TGF-β signaling molecules and their inhibitors. In addition, when acting as an adaptor, Smad7 itself is also degraded via the proteasome *(49,50)*. Smad7 is polyubiquitinated by Arkadia, which selectively degrades Smad7, but not the receptors *(56)*. AIP4 is another E3 ligase specifically causing the proteolytic degradation of Smad7 (Fig. 3). Surprisingly, despite Smad7 degradation, the levels of type I receptor remain unchanged and AIP4 inhibits TGF-β signaling *(57)*.

Smad7 specific deubiquitinating (DUB) enzyme, ubiquitin C-terminal hydrolase UCH37, reverses Smurf-mediating ubiquitylation of type I receptor and increases TGF-β mediated transcription *(58)*. Interestingly, the stability of Smad7 protein is balanced by competition between acetylation, deacetylation, and ubiquitylation *(59,60)*. Acetyl transferase p300 acetylates and specific histone deacetylases (HDAC) deacetylate Smad7 on the same lysines ubiquitinated by Smurf thus protecting Smad7 protein against proteasomal degradation *(59,60)*.

Smad7 levels are also decreased through ubiquitin-mediated degradation in experimental models of induced renal fibrosis *(61)*. Conversely, in scleroderma fibroblasts Smad7 levels are increased and it forms stable complexes with type I receptor potently repressing TGF-β signaling *(62)*. Surprisingly, the type I receptor was stabilized, indicating that Smad7 also potently suppresses the receptor activity independent of receptor degradation. Nevertheless, dysfunction of proteasomal turnover may underlie the pathology of diseases associated with TGF-β pathway, like fibrotic diseases and cancer.

2.6. TGF-β Signaling Negative Regulators SnoN and Ski are Destroyed by the Proteasome

Ski-related novel gene N (SnoN) and Ski belong to the Ski family of protooncogenes *(63,64)*. They are structurally and functionally very similar, form hetero- and homodimers, and function as negative regulators of TGF-β signaling *(63–65)*. They can repress transcription of TGF-β responsive genes by directly associating with the Smad2/3-Smad4 complex at the promoter site and by recruiting a nuclear corepressor N-CoR to the complex. This leads to the association of histone deacetylase (HDAC) to the Smad complex and deacetylation of histones resulting in transcriptional repression *(63,64)*. Similarly to TGF-β signaling, Ski represses BMP signaling by interacting with Smad1, Smad4, and Smad5 *(66)*.

TGF-β stimulation induces rapid degradation of SnoN therefore allowing unhindered signaling. However, following prolonged exposure to TGF-β, SnoN mRNA and protein levels are increased, and TGF-β dependent transcription is turned off. Therefore, SnoN forms a negative feedback loop for TGF-β signaling. Additionally, SnoN/Ski represses TGF-β signaling by sequestering Smad3 and Smad4, and thus preventing them from forming a transcriptionally active complex *(64,67)*.

SnoN and Ski levels are tightly regulated by the ubiquitin mediated proteasomal pathway. The half-life of SnoN in mink lung epithelial cells is 60 min and upon TGF-β treatment is reduced to 10 min *(67)*. Smad3 acts as an adaptor by facilitating the degradation of SnoN by the APC complex *(68,69)*. Although, SnoN degradation is primed by ligand binding to the receptors, the APC-CDH1 complex can polyubiquitylate SnoN to some extent even in the absence of TGF-β signaling *(68,69)*. This low level of ubiquitin-mediated degradation may maintain a low steady-state level of SnoN in mammalian cells *(68)*. TGF-β can also induce the association of Smurf2 with Smad2 to mediate the polyubiquitylation of SnoN and its subsequent degradation by the proteasome *(70)*. Thus, both the APC and the Smurf2 pathways are involved in the degradation of SnoN upon TGF-β stimulation although the relative importance of each of them is yet unknown (Table 1). They may have a role in regulating SnoN levels either synergistically or differentially depending on tissue type, developmental stage or levels of the adaptor proteins *(68)*. The degradation of Ski has been studied to a lesser extent. The half-life of Ski is 100 min and upon TGF-β treatment it is reduced to 30 min in mink lung epithelial cells, indicating that its levels are also controlled through proteasome mediated events *(67)*.

The level of SnoN and Ski in mammalian cells is directly linked to their ability to repress TGF-β signaling. Overexpression of Ski /SnoN or stabilization of SnoN, by mutation of its three lysine residues that are ubiquitylated, leads to resistance of TGF-β-induced growth arrest *(67,68)*. The inability of TGF-β to cause SnoN degradation results in resistance to TGF-β-induced growth arrest in esophageal cancer cells *(71)*. High levels of SnoN/Ski have been detected in several lymphomas and carcinoma cells of stomach, thyroid, lung, breast, and colorectal cancer *(72–76)*. Furthermore, a high steady-state protein level of SnoN or Ski correlates with tumor progression in mammalian cells. In melanomas a high level of cytosolic Ski is associated with primary invasive and metastatic melanomas *(77)*. Conversely, in breast cancer low levels of SnoN are associated with good prognosis *(75)*. These studies suggest that the control of SnoN and Ski levels is important for the regulation of TGF-β signaling.

3. REGULATION OF TGF-β PATHWAY ASSOCIATED PROTEASOMALLY TARGETED PROTEINS DURING DEVELOPMENT

A prominent feature of TGF-β pathway proteins is their participation in vertebrate development. This finely tuned process involves organization of the ectoderm, mesoderm, and endoderm depending on controlled external cues. Nodal, a TGF-β superfamily ligand in *Xenopus*, is a maternally secreted factor which determines mesoderm formation. Recently, a novel E3 ligase, Ectodermin (Ecto) was found, which binds and targets Smad4 for destruction and consequently blocks TGF-β signaling *(78)*. Thus, Ecto blocks the formation of mesoderm and promotes instead ectoderm formation. Additionally, Ecto is expressed in adult somatic cells, and is a potent inhibitor of TGF-β cytostatic process e.g., in the colonic epithelial cells *(78)*. Expression of Ecto in the colonic crypts is limited to the basal, proliferative zone of colonic epithelial cells, suggesting that it can act as a master controller attenuating TGF-β signaling *(78)*. This fine balance may be disturbed in cancer through overexpression of Ecto.

Smurf1 overexpression inhibits BMP signaling and osteoblast differentiation *(79)*. To address the relevance of Smurf1 in mouse development, Zhao *et al. (79)* generated Smurf1 transgenic mice and found that overexpression of Smurf1 significantly reduced bone formation *(79)*. In another study where Smurf1 and Smad6 transgenic mice were generated, it was found that the Smad6 transgenic mice developed osteopenia and endochondral ossification leading to dwarfism. Whereas the phenotype of Smurf1 transgenic animals was normal, the Smad6/Smurf1 double transgenic animals had a more severe phenotype *(80)*. Conversely, the targeted disruption of Smurf1 does not disturb mouse embryogenesis, but leads to an increase of bone mass owing to, surprisingly, constitutive activation of osteoblasts through MEKK2-JNK kinase pathway *(81)*. However, the levels of TGF-β receptors and Smads were unaffected, apparently through compensatory upregulation of Smurf-2. Smurf-1 was found to act as an E3 ligase toward phosphorylated, activated MEKK2, and therefore to control AP1 activity and osteoblast functions *(81)*. Furthermore, *Xenopus* and *Drosophila* Smurfs (Dsmurf, a homolog of vertebrate Smurf1 and Smurf2) are also able to bind to phosphorylated Smad1 and mothers against dpp (MAD), respectively, and promote their proteasomal degradation, hence modulating the TGF-β family signaling during *Xenopus* and *Drosophila* embryogenesis *(23,82,83)*.

4. DYSREGULATION OF TGF-β SIGNALING IN CANCER BY THE PROTEASOME PATHWAY

Smad-pathway is activated during both tumor suppression and promotion *(84–87)*. The conversion of TGF-β from being a cytostatic growth factor to a factor promoting metastasis is not fully understood, but is at least partly owing to altered gene regulation and increased selection pressure toward the metastatic phenotype *(88)*. These events are often associated with the promotion of epithelial-mesenchymal transition by TGF-β *(84,86)*. Mutations in TGF-β signaling pathway molecules (type I, II receptors, Smad2, Smad4) are common in human carcinomas and gliomas *(84)*. While decreased expression of type II receptor may correlate with higher tumor grade *(84)*, its inactivating mutations may also be a favorable prognostic indicator in metastatic colon cancer *(89)*. Interestingly, mutations in Smad2 and Smad4 increase their turnover rate through the proteasome pathway in colorectal cancer and in hepatoma *(90,91)*. Furthermore, high levels of Smurf2 correlate with poor histopathological features and with outcome of patients who have esophageal carcinoma, and inversely correlate with the levels of Smad2 *(92)*.

Smad4 is frequently mutated in colorectal and pancreatic tumors leading to its missense or nonsense mutations, and homozygous deletion *(93–95)*. Interestingly, the Smad4 missense mutants have reduced stability and undergo polyubiquitylation and proteasomal degradation *(29,90,96,97)*. Several F-box proteins, like SCF Skp2 or βTrCP1, are involved in recognition of the mutant Smad4 *(28,98)*. Instead, Smurf, efficiently causing wild-type Smad4 polyubiquitylation, does not promote ubiquitylation of the Smad4 mutants *(31)*. Skp2 and βTrCP interact constitutively with Smad4 N-terminal MH1 domain, but they have higher binding affinity toward mutants residing in this area *(28,98)*. While some mutations in the MH2 domain lead to decreased protein half-life, some like (V370H, I383K) do not affect protein stability *(98)*. Downregulation of βTrCP by siRNA leads to an increase in the levels of several Smad4 mutants as well as an increase in TGF-β stimulated reporter activity *(98)*. The interaction between Skp2 and Smad4 mutants is further increased by phosphorylation of the Smad4 R100T mutant by JNK and p38 kinases, indicating that the mutant Smad4 stability is further decreased in the presence of high kinase activity *(28)*. Furthermore, Smad4 mutations prohibit its sumoylation, suggesting that sumoylation and ubiquitylation are inversely correlated *(36)*.

Smad4 mutations either directly inactivate the transcriptional capacity of Smad4 or cause its conformational changes. The increased propensity of cancer cells to eliminate Smad4, including its mutants with residual transcriptional activity, indicates a strong pressure for its elimination during tumorigenesis. Indeed, pancreas adenocarcinoma, a fatal tumor type which has the highest frequency of Smad4 mutations (50%), tends to have a worse prognosis when the Smad4 levels are low *(99)*. Furthermore, Smad4 and βTrCP levels are inversely correlated in a panel of 35 pancreas adenocarcinoma *(98)*. The observation that βTrCP promotes Smad4 degradation strongly suggests that βTrCP, combined with Smad4 mutations, acts to abrogate TGF-β signaling in pancreas adenocarcinomas *(98)*. This pathway may be further amplified by Smad4 phosphorylation, as βTrCP has an increased affinity toward phosphorylated substrates.

5. CONCLUDING REMARKS

Relatively little is known of the regulation of TGF-β signaling pathway components during organismal development, normal cellular physiology or during tumorigenesis. While several examples of the involvement of the proteasome pathway exist, like those emanating from other model organisms, it can be predicted that this powerful machinery must be involved in the regulation of signaling intensities of several TGF-β family protein cascades. The proteasome provides immensely complex opportunities to both fine-tune and time the duration of the signals with great speed and accuracy. It is not surprising that during tumorigenesis these pathways are altered and lead to dysfunction of TGF-β signaling. It is quite likely that our understanding of the multiple small molecule modifications, ubiquitylation, sumoylation, neddylation, perhaps even ISGylation, and how they are coupled with other modifications like phosphorylation and acetylation, will significantly increase, and provide new tools and possibilities to intercept dysfunctional TGF-β signaling in cancer.

ACKNOWLEDGMENTS

We thank the members of the Laiho lab for fruitful discussions.

REFERENCES

1. Goldberg AL. Protein degradation and protection against misfolded or damaged proteins. Nature 2003;426:895–899.
2. Weissman AM. Themes and variations on ubiquitylation. Nat Rev Mol Cell Biol 2001;2:169–178.
3. Hicke L. Protein regulation by monoubiquitin. Nat Rev Mol Cell Biol 2001;2:195–201.

4. Welchman RL, Gordon C, Mayer RJ. Ubiquitin and ubiquitin-like proteins as multifunctional signals. Nat Rev Mol Cell Biol 2005;6:599–609.
5. Ciechanover A. The ubiquitin-proteasome pathway: on protein death and cell life. EMBO J 1998;17: 7151–7160.
6. Hershko A. The ubiquitin system for protein degradation and some of its roles in the control of the cell division cycle. Cell Death Differ 2005;12:1191–1197.
7. Pagano M. Control of DNA synthesis and mitosis by the Skp2-p27-Cdk1/2 axis. Mol Cell 2004; 14:414–416.
8. Nakayama KI, Nakayama K. Regulation of the cell cycle by SCF-type ubiquitin ligases. Semin Cell Dev Biol 2005;16:323–333.
9. Fuchs SY, Spiegelman VS, Kumar KG. The many faces of beta-TrCP E3 ubiquitin ligases: reflections in the magic mirror of cancer. Oncogene 2004;23:2028–2036.
10. Roos-Mattjus P, Sistonen L. The ubiquitin-proteasome pathway. Ann Med 2004;36:285–295.
11. Attisano L, Wrana JL. Signal transduction by the TGF-beta superfamily. Science 2002;296:1646–1647.
12. ten Dijke P, Hill CS. New insights into TGF-beta-Smad signaling. Trends Biochem Sci 2004;29: 265–273.
13. Feng XH, Derynck R. Specificity and versatility in TGF-β signaling through Smads. Annu Rev Cell Dev Biol 2005;21:659–693.
14. Hayashi H, Abdollah S, Qiu Y, et al. The MAD-related protein Smad7 associates with the TGFbeta receptor and functions as an antagonist of TGFβ signaling. Cell 1997;89:1165–1173.
15. Nakao A, Afrakhte M, Moren A, et al. Identification of Smad7, a TGFβ-inducible antagonist of TGF-beta signaling. Nature 1997;389:631–635.
16. Zhang F, Laiho M. On and off: proteasome and TGF-beta signaling. Exp Cell Res 2003;291:275–281.
17. Izzi L, Attisano L. Regulation of the TGFbeta signaling pathway by ubiquitin-mediated degradation. Oncogene 2004;23:2071–2078.
18. Lo RS, Massagué J. Ubiquitin-dependent degradation of TGF-beta-activated smad2. Nat Cell Biol 1999;1:472–478.
19. Lin X, Liang M, Feng XH. Smurf2 is a ubiquitin E3 ligase mediating proteasome-dependent degradation of Smad2 in transforming growth factor-beta signaling. J Biol Chem 2000;275: 36,818–36,822.
20. Zhang Y, Chang C, Gehling DJ, Hemmati-Brivanlou A, Derynck R. Regulation of Smad degradation and activity by Smurf2, an E3 ubiquitin ligase. Proc Natl Acad Sci USA 2001;98:974–979.
21. Fukuchi M, Imamura T, Chiba T, et al. Ligand-dependent degradation of Smad3 by a ubiquitin ligase complex of ROC1 and associated proteins. Mol Biol Cell 2001;12:1431–1443.
22. Waite KA, Eng C. From developmental disorder to heritable cancer: it's all in the BMP/TGF-beta family. Nat Rev Genet 2003;4:763–773.
23. Zhu H, Kavsak P, Abdollah S, Wrana JL, Thomsen GH. A SMAD ubiquitin ligase targets the BMP pathway and affects embryonic pattern formation. Nature 1999;400:687–693.
24. Gruendler C, Lin Y, Farley J, Wang T. Proteasomal degradation of Smad1 induced by bone morphogenetic proteins. J Biol Chem 2001;276:46,533–46,543.
25. Li L, Xin H, Xu X, et al. CHIP mediates degradation of Smad proteins and potentially regulates Smad-induced transcription. Mol Cell Biol 2004;24:856–864.
26. Xin H, Xu X, Li L, Ning H, et al. CHIP controls the sensitivity of transforming growth factor-beta signaling by modulating the basal level of Smad3 through ubiquitin-mediated degradation. J Biol Chem 2005;280:20,842–20,850.
27. Saha D, Datta PK, Beauchamp RD. Oncogenic ras represses transforming growth factor-beta /Smad signaling by degrading tumor suppressor Smad4. J Biol Chem 2001;276:29,531–29,537.
28. Liang M, Liang YY, Wrighton K, et al. Ubiquitylation and proteolysis of cancer-derived Smad4 mutants by SCFSkp2. Mol Cell Biol 2004;24:7524–7537.
29. Moren A, Hellman U, Inada Y, Imamura T, Heldin CH, Moustakas A. Differential ubiquitylation defines the functional status of the tumor suppressor Smad4. J Biol Chem 2003;278:33,571–33,582.
30. Wan M, Tang Y, Tytler EM, et al. Smad4 protein stability is regulated by ubiquitin ligase SCF β-TrCP1. J Biol Chem 2004;279:14,484–14,487.
31. Moren A, Imamura T, Miyazono K, Heldin C-H, Moustakas A. Degradation of the tumor suppressor Smad4 by WW and HECT domain ubiquitin ligases. J Biol Chem 2005;280:22,115–22,123.
32. Claret FX, Hibi M, Dhut S, Toda T, Karin M. A new group of conserved coactivators that increase the specificity of AP-1 transcription factors. Nature 1996;383:453–457.

33. Wan M, Cao X, Wu Y, et al. Jab1 antagonizes TGF-beta signaling by inducing Smad4 degradation. EMBO Rep 2002;3:171–176.
34. Tomoda K, Kubota Y, Kato J. Degradation of the cyclin-dependent-kinase inhibitor p27Kip1 is instigated by Jab1. Nature 1999;398:160–165.
35. Bech-Otschir D, Kraft R, Huang X, et al. COP9 signalosome-specific phosphorylation targets p53 to degradation by the ubiquitin system. EMBO J 2001;20:1630–1639.
36. Lin X, Liang M, Liang YY, Brunicardi FC, Feng XH. SUMO-1/Ubc9 promotes nuclear accumulation and metabolic stability of tumor suppressor Smad4. J Biol Chem 2003;278:31,043–31,048.
37. Lin X, Liang M, Liang YY, Brunicardi FC, Melchior F, Feng XH. Activation of transforming growth factor-beta signaling by SUMO-1 modification of tumor suppressor Smad4/DPC4. J Biol Chem 2003;278:18,714–18,719.
38. Lee PS, Chang C, Liu D, Derynck R. Sumoylation of Smad4, the common Smad mediator of transforming growth factor-beta family signaling. J Biol Chem 2003;278:27,853–27,863.
39. Ohshima T, Shimotohno K. Transforming growth factor-beta-mediated signaling via the p38 MAP kinase pathway activates Smad-dependent transcription through SUMO-1 modification of Smad4. J Biol Chem 2003;278:50,833–50,842.
40. Liang M, Melchior F, Feng XH, Lin X. Regulation of Smad4 sumoylation and transforming growth factor-beta signaling by protein inhibitor of activated STAT1. J Biol Chem 2004;279: 22,857–22,865.
41. Imoto S, Sugiyama K, Muromoto R, Sato N, Yamamoto T, Matsuda T. Regulation of transforming growth factor-beta signaling by protein inhibitor of activated STAT, PIASy through Smad3. J Biol Chem 2003;278:34,253–34,258.
42. Chang CC, Lin DY, Fang HI, Chen RH, Shih HM. Daxx mediates the small ubiquitin-like modifier-dependent transcriptional repression of Smad4. J Biol Chem 2005;280:10,164–10,173.
43. Liu X, Elia AE, Law SF, Golemis EA, Farley J, Wang T. A novel ability of Smad3 to regulate proteasomal degradation of a Cas family member HEF1. EMBO J 2000;19:6759–6769.
44. Nourry C, Maksumova L, Pang M, Liu X, Wang T. Direct interaction between Smad3, APC10, CDH1 and HEF1 in proteasomal degradation of HEF1. BMC Cell Biol 2004;5:20.
45. Feng L, Guedes S, Wang T. Atrophin-1-interacting protein 4/human Itch is a ubiquitin E3 ligase for human enhancer of filamentation 1 in transforming growth factor-beta signaling pathways. J Biol Chem 2004;279:29,681–29,690.
46. Ray D, Terao Y, Nimbalkar D, et al. Transforming growth factor beta facilitates β-TrCP-mediated degradation of Cdc25A in a Smad3-dependent manner. Mol Cell Biol 2005;25:3338–3347.
47. Qiu L, Joazeiro C, Fang N, et al. Recognition and ubiquitination of Notch by Itch, a hect-type E3 ubiquitin ligase. J Biol Chem 2000;275:35,734–35,737.
48. Fang D, Elly C, Gao B, et al. Dysregulation of T lymphocyte function in itchy mice: a role for Itch in TH2 differentiation. Nat Immunol 2002;3:281–287.
49. Kavsak P, Rasmussen RK, Causing CG, et al. Smad7 binds to Smurf2 to form an E3 ubiquitin ligase that targets the TGF β receptor for degradation. Mol Cell 2000;6:1365–1375.
50. Ebisawa T, Fukuchi M, Murakami G, et al. Smurf1 interacts with transforming growth factor-beta type I receptor through Smad7 and induces receptor degradation. J Biol Chem 2001;276:12,477–12,480.
51. Ogunjimi AA, Briant DJ, Pece-Barbara N, et al. Regulation of Smurf2 ubiquitin ligase activity by anchoring the E2 to the HECT domain. Mol Cell 2005;19:297–308.
52. Komuro A, Imamura T, Saitoh M, et al. Negative regulation of transforming growth factor-beta (TGF-beta) signaling by WW domain-containing protein 1 (WWP1). Oncogene 2004;23:6914–6923.
53. Seo SR, Lallemand F, Ferrand N, et al. The novel E3 ubiquitin ligase Tiul1 associates with TGIF to target Smad2 for degradation. EMBO J 2004;23:3780–3792.
54. Kuratomi G, Komuro A, Goto K, et al. NEDD4-2 (neural precursor cell expressed, developmentally down-regulated 4-2) negatively regulates TGF-beta (transforming growth factor-beta) signaling by inducing ubiquitin-mediated degradation of Smad2 and TGF-beta type I receptor. Biochem J 2005;386: 461–470.
55. Ohashi N, Yamamoto T, Uchida C, et al. Transcriptional induction of Smurf2 ubiquitin ligase by TGF-beta. FEBS Lett 2005;579:2557–2563.
56. Koinuma D, Shinozaki M, Komuro A, et al. Arkadia amplifies TGF-beta superfamily signaling through degradation of Smad7. EMBO J 2003;22:6458–6470.
57. Lallemand F, Seo SR, Ferrand N, et al. AIP4 restricts transforming growth factor-beta signaling through a ubiquitylation-independent mechanism. J Biol Chem 2005;280:27,645–27,653.

58. Wicks SJ, Haros K, Maillard M, et al. The deubiquitinating enzyme UCH37 interacts with Smads and regulates TGF-beta signaling. Oncogene.
59. Gronroos E, Hellman U, Heldin C-H, Ericsson J. Control of Smad7 stability by competition between acetylation and ubiquitylation. Mol Cell 2002;10:483–493.
60. Simonsson M, Heldin C-H, Ericsson J, Gronroos E. The balance between acetylation and deacetylation controls Smad7 stability. J Biol Chem 2005;280:21,797–21,803.
61. Fukasawa H, Yamamoto T, Togawa A, et al. Down-regulation of Smad7 expression by ubiquitin-dependent degradation contributes to renal fibrosis in obstructive nephropathy in mice. Proc Natl Acad Sci USA 2004;101:8687–8692.
62. Asano Y, Ihn H, Yamane K, Kubo M, Tamaki K. Impaired Smad7-Smurf-mediated negative regulation of TGF-beta signaling in scleroderma fibroblasts. J Clin Invest 2004;113:253–264.
63. Liu X, Sun Y, Weinberg RA, Lodish HF. Ski/Sno and TGF-beta signaling. Cytokine Growth Factor Rev 2001;12:1–8.
64. Luo K. Ski and SnoN: negative regulators of TGF-beta signaling. Curr Opin Genet Dev 2004;14:65–70.
65. Cohen SB, Zheng G, Heyman HC, Stavnezer E. Heterodimers of the SnoN and Ski oncoproteins form preferentially over homodimers and are more potent transforming agents. Nucleic Acids Res 1999;27:1006–1014.
66. Wang W, Mariani FV, Harland RM, Luo K. Ski represses bone morphogenic protein signaling in Xenopus and mammalian cells. Proc Natl Acad Sci USA 2000;97:14,394–14,399.
67. Sun Y, Liu X, Ng-Eaton E, Lodish HF, Weinberg RA. SnoN and Ski protooncoproteins are rapidly degraded in response to transforming growth factor beta signaling. Proc Natl Acad Sci USA 1999;96: 12,442–12,447.
68. Stroschein SL, Bonni S, Wrana JL, Luo K. Smad3 recruits the anaphase-promoting complex for ubiquitylation and degradation of SnoN. Genes Dev 2001;15:2822–2836.
69. Wan Y, Liu X, Kirschner MW. The anaphase-promoting complex mediates TGF-beta signaling by targeting SnoN for destruction. Mol Cell 2001;8:1027–1039.
70. Bonni S, Wang HR, Causing CG, et al. TGF-beta induces assembly of a Smad2-Smurf2 ubiquitin ligase complex that targets SnoN for degradation. Nat Cell Biol 2001;3:587–595.
71. Edmiston JS, Yeudall WA, Chung TD, Lebman DA. Inability of transforming growth factor-beta to cause SnoN degradation leads to resistance to transforming growth factor-beta-induced growth arrest in esophageal cancer cells. Cancer Res 2005;65:4782–4788.
72. Nomura N, Sasamoto S, Ishii S, Date T, Matsui M, Ishizaki R. Isolation of human cDNA clones of ski and the ski-related gene, sno. Nucleic Acids Res 1989;17:5489–5500.
73. Colmenares C, Stavnezer E. Structure and activities of the ski oncogene. Semin Cancer Biol 1990;1: 383–387.
74. Fumagalli S, Doneda L, Nomura N, Larizza L. Expression of the c-ski proto-oncogene in human melanoma cell lines. Melanoma Res 1993;3:23–27.
75. Zhang F, Lundin M, Ristimaki A, et al. Ski-related novel protein N (SnoN), a negative controller of transforming growth factor-beta signaling, is a prognostic marker in estrogen receptor-positive breast carcinomas. Cancer Res 2003;63:5005–5010.
76. Buess M, Terracciano L, Reuter J, et al. Amplification of SKI is a prognostic marker in early colorectal cancer. Neoplasia 2004;6:207–212.
77. Reed JA, Bales E, Xu W, Okan NA, Bandyopadhyay D, Medrano EE. Cytoplasmic localization of the oncogenic protein Ski in human cutaneous melanomas in vivo: functional implications for transforming growth factor β signaling. Cancer Res 2001;61:8074–8078.
78. Dupont S, Zacchigna L, Cordenonsi M, et al. Germ-layer specification and control of cell growth by Ectodermin, a Smad4 ubiquitin ligase. Cell 2005;121:87–99.
79. Zhao M, Qiao M, Harris SE, Oyajobi BO, Mundy GR, Chen D. Smurf1 inhibits osteoblast differentiation and bone formation in vitro and in vivo. J Biol Chem 2004;279:12,854–12,859.
80. Horiki M, Imamura T, Okamoto M, et al. Smad6/Smurf1 overexpression in cartilage delays chondrocyte hypertrophy and causes dwarfism with osteopenia. J Cell Biol 2004;165:433–445.
81. Yamashita M, Ying SX, Zhang GM, et al. Ubiquitin ligase Smurf1 controls osteoblast activity and bone homeostasis by targeting MEKK2 for degradation. Cell 2005;121:101–113.
82. Podos SD, Hanson KK, Wang YC, Ferguson EL. The DSmurf ubiquitin-protein ligase restricts BMP signaling spatially and temporally during Drosophila embryogenesis. Dev Cell 2001;1:567–578.
83. Liang YY, Lin X, Liang M, et al. dSmurf selectively degrades decapentaplegic-activated MAD, and its overexpression disrupts imaginal disc development. J Biol Chem 2003;278:26,307–26,310.

84. Derynck R, Akhurst RJ, Balmain A. TGF-beta signaling in tumor suppression and cancer progression. Nat Genet 2001;29:117–129.
85. Siegel PM, Massagué J. Cytostatic and apoptotic actions of TGF-beta in homeostasis and cancer. Nat Rev Cancer 2003;3:807–821.
86. Kang Y, Massagué J. Epithelial-mesenchymal transitions: twist in development and metastasis. Cell 2004;118:277–279.
87. Muraoka-Cook RS, Dumont N, Arteaga CL. Dual role of transforming growth factor β in mammary tumorigenesis and metastatic progression. Clin Cancer Res 2005;11:937–943.
88. Kang Y, He W, Tulley S, et al. Breast cancer bone metastasis mediated by the Smad tumor suppressor pathway. Proc Natl Acad Sci USA 2005;102:13,909–13,914.
89. Watanabe T, Wu TT, Catalano PJ, et al. Molecular predictors of survival after adjuvant chemotherapy for colon cancer. N Engl J Med 2001;344:1196–1206.
90. Xu J, Attisano L. Mutations in the tumor suppressors Smad2 and Smad4 inactivate transforming growth factor beta signaling by targeting Smads to the ubiquitin-proteasome pathway. Proc Natl Acad Sci USA 2000;97:4820–4825.
91. Dumont E, Lallemand F, Prunier C, et al. Evidence for a role of Smad3 and Smad2 in stabilization of the tumor-derived mutant Smad2.Q407R. J Biol Chem 2003;278:24,881–24,887.
92. Fukuchi M, Fukai Y, Masuda N, et al. High-level expression of the Smad ubiquitin ligase Smurf2 correlates with poor prognosis in patients with esophageal squamous cell carcinoma. Cancer Res 2002;62:7162–7165.
93. Eppert K, Scherer SW, Ozcelik H, et al. MADR2 maps to 18q21 and encodes a TGFbeta-regulated MAD-related protein that is functionally mutated in colorectal carcinoma. Cell 1996;86:543–552.
94. Schutte M, Hruban RH, Hedrick L, et al. DPC4 gene in various tumor types. Cancer Res 1996;56: 2527–2530.
95. Hahn SA, Schutte M, Hoque AT, et al. DPC4, a candidate tumor suppressor gene at human chromosome 18q21.1. Science 1996;271:350–353.
96. Maurice D, Pierreux CE, Howell M, Wilentz RE, Owen MJ, Hill CS. Loss of Smad4 function in pancreatic tumors: C-terminal truncation leads to decreased stability. J Biol Chem 2001;276: 43,175–43,181.
97. Moren A, Itoh S, Moustakas A, ten Dijke P, Heldin C-H. Functional consequences of tumorigenic missense mutations in the amino-terminal domain of Smad4. Oncogene 2000;19:4396–4404.
98. Wan M, Huang J, Jhala NC, et al. SCF(beta-TrCP1) controls Smad4 protein stability in pancreatic cancer cells. Am J Pathol 2005;166:1379–1392.
99. Tascilar M, Skinner HG, Rosty C, et al. The SMAD4 protein and prognosis of pancreatic ductal adenocarcinoma. Clin Cancer Res 2001;7:4115–4121.
100. Bai Y, Yang C, Hu K, Elly C, Liu YC. Itch E3 ligase-mediated regulation of TGF-beta signaling by modulating smad2 phosphorylation. Mol Cell 2004;15:825–831.
101. Long J, Matsuura I, He D, Wang G, Shuai K, Liu F. Repression of Smad transcriptional activity by PIASy, an inhibitor of activated STAT. Proc Natl Acad Sci USA 2003;100:9791–9796.

16 TGF-β and Smad Signaling in Transcriptome Reprogramming During EMT

Sylvie Thuault, Ulrich Valcourt, Marcin Kowanetz, and Aristidis Moustakas

CONTENTS

INTRODUCTION
SIGNALING INPUTS FROM TGF-β SUPERFAMILY MEMBERS CONTROL EMT
SIGNALING MECHANISMS THAT SUPPORT EMT DOWNSTREAM OF TGF-β
TGF-β ENFORCES TRANSCRIPTOMIC REPROGRAMMING DURING EMT
CONCLUDING REMARKS
ACKNOWLEDGMENTS
REFERENCES

Abstract

Epithelial–mesenchymal transition (EMT) is a differentiation switch between two major cell types, polarized immobile epithelial and contractile and motile mesenchymal/fibroblastic cells. EMT is critical for proper embryonic development and is also relevant to vascular remodeling processes during which endothelial cells generate myoepithelial cells. Similar to all other developmentally relevant differentiation processes, EMT is governed by the concerted action of extracellular polypeptide/morphogenetic factors, transforming growth factor-β (TGF-β) representing one of them. Because of the relevance the EMT process has for tumor cell invasiveness, metastasis, and tissue fibrosis, considerable research effort has recently focused on the mechanism by which TGF-β elicits EMT of normal epithelial and carcinoma cells. Here, we summarize the state of the art with respect to signaling mechanisms and critical effectors of EMT downstream of TGF-β. We also discuss our recent work on this topic and present a wider perspective for the future of this thriving field.

Key Words: Adherens junction; EMT; metastasis; polarity; tight junction.

1. INTRODUCTION

During epithelial–mesenchymal transition (EMT), epithelial cells change their differentiation in a manner that destroys fully polarized epithelia and creates mesenchymal, migratory cells *(1)*. This differentiation switch is intimately linked to the disorganization of cell–cell

From: *Cancer Drug Discovery and Development: Transforming Growth Factor-β in Cancer Therapy, Vol I: Basic and Clinical Biology*
Edited by: Sonia B. Jakowlew © Humana Press Inc., Totowa, NJ

adhesion and cell–matrix adhesion and degradation of the extracellular matrix associated with the epithelial layer, events that allow invasiveness and motility of the emerging mesenchymal cells (Fig. 1). EMT therefore represents a basic morphogenetic process utilized in many instances not only during normal development, but also during disease pathogenesis, including carcinoma invasiveness and tissue fibrosis *(1–3)*. Whereas the role of EMT in the pathogenesis of fibrotic disorders, such as kidney fibrosis, is widely accepted by basic researchers and pathologists *(3)*, the importance of EMT during carcinoma progression and metastasis is currently viewed with both enthusiasm and skepticism *(1)*, the latter mainly reflecting the views of tumor pathologists *(4)*. At least during normal development and fibrotic injury of tissues, EMT is often coupled to the inverse phenomenon of mesenchymal–epithelial transition (MET), based on which transitory mesenchymal cells generate polarized epithelial layers after migration and homing into new territories of tissue organization *(3)*.

The majority of signal transduction pathways that are commonly used during embryonic development *(5)* have been implicated in the control of EMT in vitro and in vivo. These include receptor tyrosine kinase pathways such as hepatocyte growth factor/scatter factor or fibroblast growth factor signaling, Wnt, Notch, and receptor serine/threonine kinase (TGF-β) signaling (Fig. 1) *(6)*. We will discuss exclusively the TGF-β superfamily and the involvement of its various members in the control of EMT.

2. SIGNALING INPUTS FROM TGF-β SUPERFAMILY MEMBERS CONTROL EMT

2.1. EMT During Embryonic Development

The transient and reversible nature of EMT that is followed by MET makes it often difficult to study in vivo. However, recent advances in imaging technology and transgenic mouse models offer promising solutions to this problem *(7)*. Development of various organs in the embryo provides interesting examples of EMT in response to TGF-β in vivo. For example, TGF-β3 causes mouse palatal EMT and thus forms the connective tissue across the palate *(8)*. Accordingly, when TGF-β3 is knocked out, mice exhibit a cleft palate phenotype. Another prominent example of EMT of endothelial cells that respond sequentially to TGF-β2 and TGF-β3 leads to the formation of the heart valve and septa in chicken and mice *(9)*. These two TGF-β ligands signal by binding to the accessory TGF-β type III receptor/betaglycan, the TGF-β type II receptor, and the activin receptor-like 2 type I receptor, which is known to activate intracellular bone morphogenetic protein (BMP)-specific Smads *(9,10)*. However, the exact role of Smads of the BMP pathways during in vivo EMT in the heart awaits further investigation.

2.2. EMT During Cancer Progression

In addition to normal organogenesis, TGF-β induces EMT in vivo during progression of various tumor models. TGF-β1 accelerates spindle carcinoma progression in transgenic mice by creating more invasive carcinoma cells via EMT *(11)*. The enhanced aggressiveness of the skin tumors of this model was attributed to the ectopic expression of TGF-β1 in the keratinocytes, which secondarily led to TGF-β3 secretion and persistence of the spindle carcinoma phenotype *(11,12)*. It is possible that a similar scenario might occur in human malignancies of the skin because human skin carcinomas oversecrete TGF-β1 and deregulate their TGF-β type II receptor *(13)*. Spindle cell carcinomas developed by exposure of mice to chemical carcinogens, mouse mammary carcinomas created by Ras or Raf oncogenic transformation, and both colon carcinomas and hepatocarcinomas led to the understanding that oncogenes such as *Ras* or *Raf* induce EMT and tumor cell invasiveness in vitro and in vivo

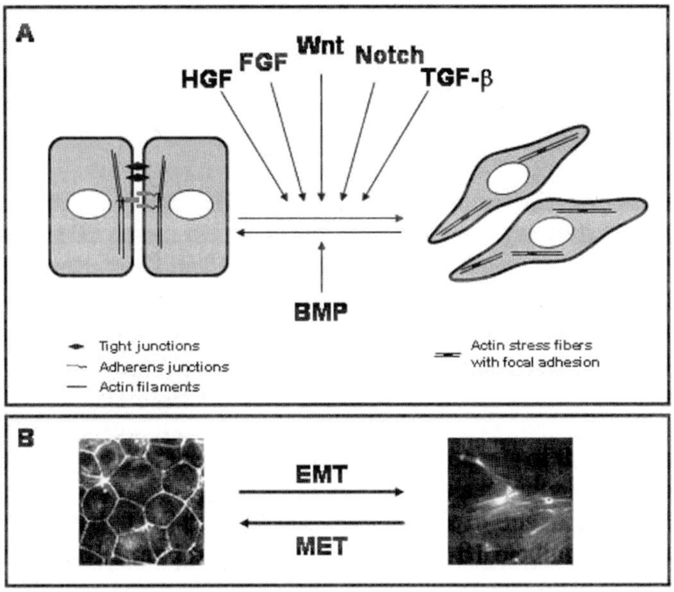

Fig. 1. EMT is induced by many signaling pathways. **(A)** A pair of polarized epithelial cells is drawn, with tight (black diamonds) and adherens junctions (transmembrane gray cadherins tethered to actin microfilaments in the cytoplasm). Hepatocyte growth factor (HGF), fibroblast growth factor (FGF), Wnt, Notch, and TGF-β signaling generate mesenchymal cells (actin stress fibers with focal adhesions are shown). BMP signaling promotes the reverse process toward epithelial differentiation. **(B)** Direct fluorescence microscopy of the actin cytoskeleton of polarized epithelial NMuMG cells undergoing EMT in response to TGF-β1 toward scattered fibroblasts, or reverting back to epithelial cells (MET) in response to BMP-7.

in a TGF-β-dependent manner *(12,14–17)*. Indeed, when the TGF-β pathway is inactivated, Ras alone cannot promote tumor cell invasiveness and metastasis. Conversely, on activation of the TGF-β pathway, carcinoma cells that exhibit more overt and irreversible EMT lead to more aggressive and metastatic tumors *(17)*. Overall, it is accepted that mouse and human carcinomas oversecrete TGF-β and are sensitized to TGF-β signaling, which leads to loss of growth inhibitory or apoptotic responses in response to TGF-β, yet allows the development of EMT (Fig. 1B). Thus, carcinoma cell EMT differs from the response of normal epithelial cells, which are growth arrested in response to TGF-β concomitantly with the onset of EMT *(18)*. It is therefore possible that the difference between reversible EMT of normal epithelial cells and irreversible EMT of oncogenically transformed cells may be linked to mechanisms controlling cell cycle progression. This is relevant to the dual role TGF-β seems to play in cancer progression *(19)*. Accordingly, TGF-β inhibits normal epithelial cell growth, which makes TGF-β a tumor suppressor at early, relatively benign adenoma stages. TGF-β acting on carcinoma cells promotes irreversible EMT and contributes to tumor aggressiveness and metastasis owing to its cooperation with oncogenes such as *Ras*.

2.3. EMT During Fibrotic Disorders

Not only cancer, but also tissue fibrosis, such as kidney fibrosis during chronic renal failure, involves in vivo processes of EMT that are controlled by TGF-β (reviewed in Ref. *[3]*). For example, bioactive TGF-β levels increase dramatically upon experimental ureteral obstruction or diabetic nephropathy, leading to EMT of renal tubular epithelia and enhanced extracellular matrix deposition, which cumulatively lead to the development of tubulointerstitial fibrosis *(20,21)*. In contrast to cancer, analysis of EMT during fibrotic disorders has already provided efficacious pharmacological treatment of these disorders at least in mouse models (see Concluding Remarks).

2.4. Commonly Used In Vitro Models of EMT and Ligand Specificity

Despite the well-established in vivo models of EMT in response to TGF-β outlined earlier, most signal transduction studies are performed in cell cultures in vitro or in 3D cultures within an extracellular matrix, where cells differentiate into tubular or spherical epithelia, undergo branching morphogenesis, or even invade the matrix. In vitro culture systems gave the first evidence that TGF-β could induce EMT, and today, many studies have described this widespread mammalian epithelial response in cells of mammary, lung, kidney, lens, skin, and liver origin *(18,20–28)*. As epithelial cells undergo reversible EMT in response to TGF-β1, they downregulate the expression of epithelial markers such as E-cadherin, specific keratins, and ZO-1 and upregulate the expression of mesenchymal markers such as fibronectin, Fsp1, α-smooth muscle actin, and vimentin (Fig. 1A). EMT is also correlated with a reorganization of the actin cytoskeleton from a cortical skeleton, which supports the polarized plasma membrane, to a skeleton rich in stress fibers connected to focal adhesions (Fig. 1B). But whether the reorganization of actin contributes to the development of EMT is disputed *(1)*. The above in vitro culture examples include primary and immortalized normal epithelial cells of human or mouse and rat origin. In addition, in vitro cultures of tumor and fibrotic epithelial cells derived from colon, kidney, liver, lung, mammary gland, or skin undergo irreversible EMT *(12–14,17,29)*, suggesting that transformed cells cannot undergo MET.

Most of the EMT examples summarized so far are based on studies of TGF-β1, the prototype member of the superfamily. In addition, TGF-β2 and TGF-β3 can induce EMT in vitro *(18,30)*. These are the major ligands known so far to be secreted by carcinoma or fibrotic cells and they are also linked to EMT during development. Our studies on TGF-β superfamily ligand specificity during EMT showed that those signaling pathways that activate Smad2 and Smad3, i.e., TGF-β/activin/nodal/myostatin and other ligands, should be able to elicit EMT because their corresponding type I receptors can induce EMT in vitro after ectopic expression (Fig. 2) *(18)*. This awaits validation by direct analysis of multiple ligands of the superfamily. On the other hand, ligands that activate Smad1, Smad5, and Smad8 signaling pathways, i.e., BMPs/GDFs/MIS, fail to elicit robust EMT (Fig. 2) *(18)*. We explain below the mechanism for this specificity. In addition, BMPs can antagonize the TGF-βs and induce MET, as reported for normal mammary and lens and fibrotic kidney epithelial cells in vitro *(23,31)* and during kidney fibrosis in vivo (Fig. 1) *(3)*. For example, BMP-7 induces MET of adult renal fibroblasts and acts therapeutically because BMP-7 promotes regeneration and healing of the injured kidney *(3)*. Identification of other BMP/growth differentiation factor (GDF) members capable of inducing MET in other organs, and in particular during tumor progression, will be of clinical importance.

3. SIGNALING MECHANISMS THAT SUPPORT EMT DOWNSTREAM OF TGF-β

The number of intracellular mechanisms induced by TGF-β and contributing to the establishment of EMT is ever expanding and includes Smad proteins primarily regulating gene expression and various alternative signaling proteins that regulate not only gene expression but also nongenomic targets such as the actin cytoskeleton (Fig. 3).

	Ligand	Type II receptor	Type I receptor	R-Smad	EMT (NMuMG)	EMT/MET
TGF-β branch	TGF-β1,2,3	TβR-II	ALK-5	Smad2 Smad3	+	• Palate/heart development • Mammary, lung, kidney, lens, skin EMT (in vitro)
	Activins	Actr-IIB	ALK-4			• EMT ?
	Nodal	Actr-IIB	ALK-7			• Gastrulation: EMT ?
BMP branch	BMP2/4/7	BMPR-II	ALK-3,6	Smad1 Smad5 Smad8	-	• Kidney MET
	GDFs/BMPs	ActR-II/B	ALK-3,6			• MET/EMT?
	BMP7	ActR-II/B	ALK-2			• Kidney MET
	AMH	AMHR-II	ALK-2,3,6			• MET/EMT?
	TGF-β1	TβR-II	ALK-1			• Endothelial EMT?

Fig. 2. TGF-β superfamily specificity during EMT. Table of various TGF-β superfamily ligands, receptors, and Smads, divided into the two major branches of the superfamily. The ability of these pathways to elicit EMT in the in vitro NMuMG cell model (based on the data of Ref. *[18]*) is tabulated together with a summary of known examples of EMT or MET to which the various pathways are implicated. Question marks indicate ligand–receptor pathways not yet analyzed in vivo or in vitro beyond the NMuMG cell system.

3.1. EMT Is Mediated by Smads

A series of in vitro and in vivo studies in mouse models have strongly established the role of Smad signaling in TGF-β-induced EMT *(18,20,21,23,25,30,32–36)*. A TGF-β type I receptor bearing point mutations at its L45 loop, the Smad-binding site of the receptor, cannot recruit and activate Smads and cannot induce EMT, supporting a general role of Smad signaling during EMT *(32)*. The same L45 mutant receptor enhances the growth of relatively benign mammary tumor cells and inhibits the metastasis of more aggressive mammary tumor cells to the lung of a mouse xenograft model *(36)*. The L45 mutant receptor confirms that Smad signaling mediates both tumor suppressor functions (former case) and prometastatic functions (latter case) of TGF-β. In addition, the multifunctional transcription factor YY1 and the nuclear proto-oncogene *c-Ski*, which bind directly to Smad complexes and repress either Smad DNA binding (YY1) or Smad activity (c-Ski), both efficiently block TGF-β-driven EMT at least in mouse mammary NMuMG cells *(37,38)*.

We originally showed that Smad3 enhances EMT of the in vitro mouse mammary cell model NMuMG *(30)*. Recently, we analyzed all Smads and their contribution to EMT of the same in vitro cell model *(18)*. Only Smads of the TGF-β branch, i.e., Smad2 and Smad3 together with the Co-Smad, Smad4, could elicit EMT. Additional studies have firmly placed Smad3 as a critical mediator of EMT in vivo during mammary tumor cell invasiveness, lens epithelial fibrosis caused by injury in postcataract surgery, and renal tubulointestinal fibrosis *(21,25,35)*. In a model of renal fibrosis, Smads were reported to induce expression of integrin-linked kinase (ILK) (Fig. 3), which supports both establishment of EMT and overproduction of extracellular matrix and metalloproteases by kidney cells that have undergone EMT *(20)*. Using the squamous carcinoma mouse model in which Ras and TGF-β signaling cooperate, Smad2 was established as a critical mediator of tumor cell invasiveness and metastasis

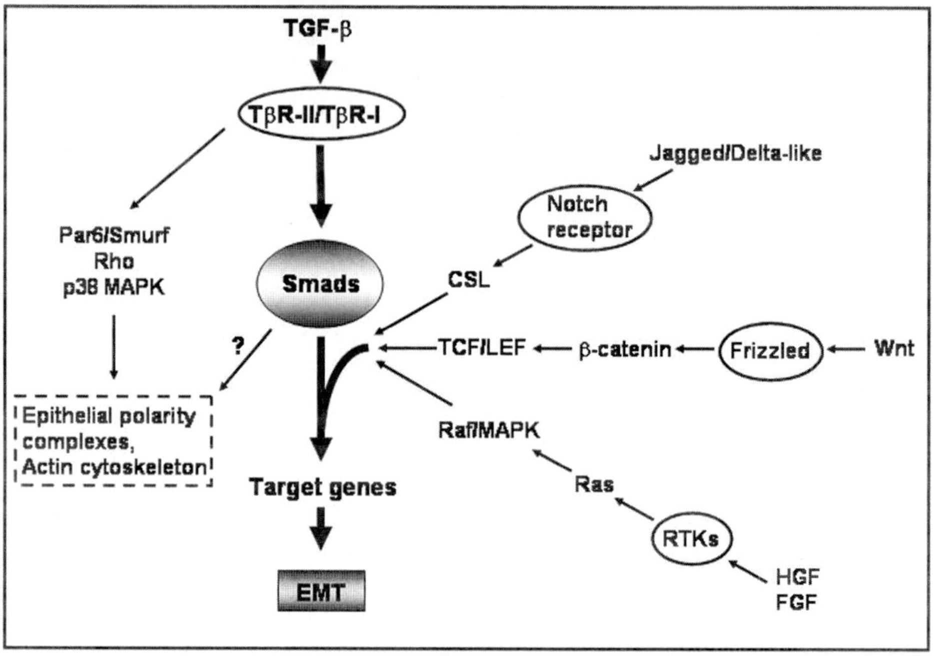

Fig. 3. Signaling pathways acting downstream and in crosstalk with TGF-β during EMT. The canonical TGF-β receptor/Smad pathway is shown with thick arrows. Non-Smad effectors and signaling pathways that crosstalk with Smad signaling are shown with thin arrows. The various plasma membrane receptors utilized by each pathway are shown in ovals. Abbreviations not defined in the main text: TGF-β type II receptor (TβR-II), TGF-β type I receptor (TβR-I), Rho GTPase (Rho), receptor tyrosine kinases (RTKs), transcriptional repressor CSL acting downstream of Notch signaling.

downstream of TGF-β *(34)*. When Smad4 was deleted in a mammary gland-specific knockout, mice developed squamous carcinomas, but the explanted carcinoma cells could not undergo EMT in response to TGF-β *(33)*. Surprisingly, partial depletion of Smad4 using RNA interference (RNAi) in human HaCaT keratinocytes and colon carcinoma cells could not block TGF-β-induced EMT despite very low Smad4 levels *(39)*. This suggests that low levels of endogenous Smad signaling are sufficient for EMT to occur. If proven true, this hypothesis may assist in the development of pharmacologic inhibitors of aggressive or metastatic tumors.

3.2. TGF-β Employs Additional Signaling Proteins During EMT

In addition to activation of the Smad pathway, TGF-β affects the activity of various non-Smad signaling effectors *(40,41)*. Accordingly, such non-Smad signal transducers seem to have an impact on EMT. In carcinoma cells expressing an activated *Ras* oncogene, mitogen-activated protein kinases (MAPK), e.g., extracellular-regulated kinase (Erk), and phospho-inositol 3′ kinase (PI3K) activities play critical roles in the establishment of EMT *(14,15)*. Erk signaling is also important in immortalized human keratinocytes HaCaT *(28)*. Alternatively, in Ras-transformed breast cancer models, EMT in vitro and metastasis in vivo depend on endogenous NF-κB signaling *(42)*. In this model, TGF-β activates both Smads and the inhibitor of NF-κB (IκB) kinase 2 (IKK-2), which phosphorylates and induces degradation of IκBα, thus releasing active NF-κB. The most plausible mechanism of activation of IKK-2 by the TGF-β receptor involves the TGF-β-activated kinase 1 (TAK1), which directly phosphorylates IKK-2 *(43)*. In mouse mammary epithelial NMuMG cells,

MAPKs such as Erk, p38, the phospholipid kinase PI3K, the small GTPase RhoA, and its downstream kinase ROCK1 have been shown to participate in eliciting EMT downstream of TGF-β *(44–46)*. Unfortunately, none of these signal transducers have yet been linked directly to the TGF-β receptors during EMT. Experiments with the L45 mutant type I receptor, which cannot activate the Smads but can activate endogenous p38 or Jun N-terminal kinase (JNK) pathways, supported the role of p38 in EMT *(47)*. In addition, ILK was thought to represent a non-Smad effector during TGF-β-induced EMT *(48)*, but we now appreciate that ILK expression is induced at the gene level by the Smad pathway first, prior to its co-operation with the PI3K-Akt pathway to promote renal cell survival and EMT *(20)*. ILK activity is also required downstream of BMP-7, which stimulates ureteric bud formation, a morphogenetic process of renal epithelial cells *(49)*. Finally, integrin signaling participates in EMT caused by TGF-β *(50,51)*. Integrin β1 function is required for p38 MAPK activation in NMuMG cells, whereas inhibition of integrin β1 effectively blocks EMT *(51)*. Similarly, integrin β6 expression is induced during progression of human colon carcinoma cells, which undergo EMT as they become more invasive and migratory, leading to the activation of autocrine TGF-β, which enhances further EMT and extracellular matrix secretion *(50)*. Of course, in this case, integrin β6 functions upstream of TGF-β mediating its activation, instead of transmitting TGF-β signals toward EMT.

Recently, a new mechanism that directly links TGF-β receptors and a polarity complex, which regulates the Rho GTPase pathway, provides compelling evidence for non-Smad effectors downstream of TGF-β during EMT of NMuMG cells *(52)*. In polarized cells, the TGF-β type I receptor localizes to tight junctions by interacting with the integral membrane protein occludin, where it also interacts with the polarity protein Par6 *(52)*. The type II receptor may localize in adherens junctions by forming complexes with E-cadherin and β-catenin, as determined in proximal tubular epithelial HK-2 cells that also undergo EMT in response to TGF-β *(53)*. Upon TGF-β signaling, the type II receptor is recruited to tight junctions and phosphorylates the type I receptor together with the type I receptor-tethered Par6. Phosphorylated Par6 can subsequently recruit the ubiquitin ligase Smurf1, which ubiquitylates and degrades RhoA *(52)*. This results in disassembly of the actin cytoskeleton locally and dissolution of tight junction, one of the hallmarks of EMT. Direct phosphorylation of Par6 by the type II receptor kinase is the first example of a non-TGF-β receptor protein substrate for this receptor kinase. This opens the exciting possibility that the type II receptor serine/threonine kinase may phosphorylate other signaling proteins, thus offering mechanistic links between the TGF-β receptors and various non-Smad effectors.

In conclusion, we currently consider a complex signaling network downstream of TGF-β as the mediator of EMT responses. The direct dissolution of epithelial cell adhesion seems to depend on both non-Smad and Smad signals, whereas changes in gene expression must be critical for the differentiation change and the generation of the mesenchymal cell. This emphasizes the role of gene targets of the TGF-β pathway in EMT.

4. TGF-β ENFORCES TRANSCRIPTOMIC REPROGRAMMING DURING EMT

In addition to the elucidation of signaling pathways that mediate EMT by TGF-β, recent work has focused on target genes of these pathways that may act as effectors of EMT or that may be required for the maintenance and full deployment of EMT. Such work could provide the long-sought markers of the EMT process, which could serve as screening tools in the laboratory and the clinic in order to unequivocally identify EMT in various pathophysiological processes *(1)*. Large-scale transcriptomic analyses of EMT under in vitro and in vivo conditions were reported over the past 4 yr *(18,28,54,55)*. Microarray screens for genes contributing to carcinoma invasiveness and metastasis are also informative for EMT, and it

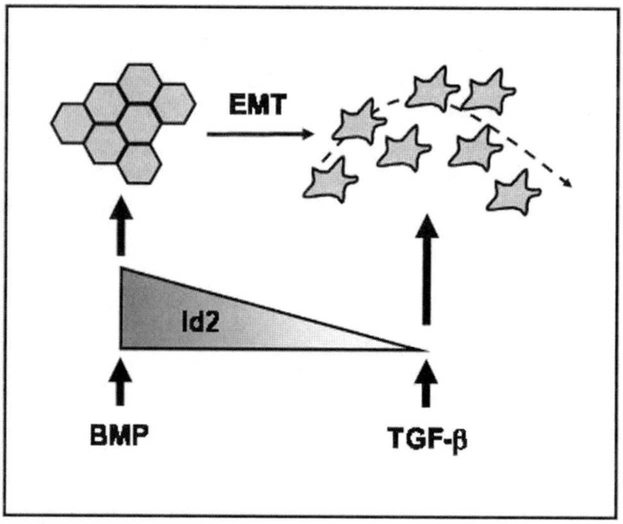

Fig. 4. Central role of Id2 in regulation of EMT by TGF-β superfamily members. BMP signaling induces expression of Id2 whereas TGF-β signaling represses expression of Id2. High Id2 levels preserve epithelial differentiation and inhibit EMT. Low Id2 levels are required for EMT to occur.

has been possible to directly implicate TGF-β signaling as a major regulator of tumor cell invasiveness *(56,57)*. Transcriptomic screens provide interesting gene lists that correlate with the processes of EMT, tumor cell invasiveness, or metastasis, whereas sometimes, short gene lists, commonly called gene signatures, can be derived. Despite the availability of large-scale gene expression data, the number of genes that are functionally linked to EMT remains short. We will discuss specific TGF-β gene targets that provide a deeper understanding of the mechanism of EMT.

4.1. Ids at the Crossroad Between EMT and Epithelial Growth Suppression by TGF-β

Others and we have recently reported that transcriptional repression of *Id* (*Id2*, *Id3*) gene expression by TGF-β is linked to the process of EMT *(23,58)*. TGF-β downregulates the expression of all three *Id* genes (*Id1–3*) in epithelial cells, yet only *Id2* and *Id3* have been rigorously analyzed. TGF-β-specific Smads transcriptionally repress *Id* genes, whereas BMP-specific Smads induce robust levels of Ids in epithelial cells (Fig. 4). Sustained repression of Id2 and Id3 is critical for TGF-β to downregulate E-cadherin and ZO-1 and eventually establish EMT *(23)*. A decrease in Id2 levels is critical, as it releases the E12/E47 basic helix-loop-helix (bHLH) factors so they can bind to the *E-cadherin* promoter and repress the gene *(58)*. In contrast, the high levels of all Ids achieved after BMP signaling enforce global inhibition of various bHLH proteins and failure to induce EMT. When endogenous Id2 and Id3 are knocked down by RNAi, BMPs can induce EMT in mammary and lens epithelial cells *(23)*. We have therefore proposed that regulation of *Id* gene expression explains the aforementioned antagonism between TGF-β and BMP, whereby BMP dominantly antagonizes TGF-β-induced EMT and promotes MET (Fig. 4) *(3,23,31)*. This concept has been recently verified in vivo on delivery of BMP-7, Id2, and Id3 via adenoviral vectors to injured lens epithelia *(59)*. All three factors blocked EMT of the fibrotic lens and when provided in combination, the therapeutic effect on the epithelium was increased dramatically. Extension of this approach to carcinomas of various tissue origins will most certainly result in equally positive therapeutic effects, an experiment much awaited in this field.

On the other hand, Ids also define the differential growth inhibitory response of epithelial cells to TGF-βs vs BMPs *(23)*. Transcriptional regulation of *Id* genes provides a molecular mechanism by which EMT and growth arrest become linked at least in normal epithelial cells. Transcriptional repression of *Ids* and of the proto-oncogene *myc* and induction of the cell cycle inhibitors *p15* and *p21* are hallmark events during epithelial cytostasis mediated by TGF-β *(60)*. Not only TGF-βs, but also BMPs repress *c-myc* and induce robust *p21* expression, whereas BMPs induce high Id levels. We therefore demonstrated that Ids play dominant roles over *c-myc* and *p21*, and in the presence of high Id levels, BMPs cannot establish efficient cytostatic responses in epithelial cells *(61)*. This suggests the involvement of additional, currently unknown genes that govern the cytostatic and EMT responses to TGF-β superfamily members. We propose that expression of such hypothetical genes must be regulated by Id proteins, and they may include novel regulators of EMT and cytostasis.

The Id model suggests that at early stages of carcinoma invasiveness when EMT occurs, Id levels must be reduced. However, upon metastasis and homing into a new site of tumor growth, Id levels must gradually increase as the transitory mesenchymal cells undergo MET and establish a new (semi-) differentiated tumor. In agreement with this model, Id2 levels are abnormally high in carcinomas that have metastasized to bone or lung *(62)*. However, measuring the transitory reduction of Id expression in carcinomas that prepare for metastasis in vivo is an open challenge for the future.

4.2. Regulation of E-cadherin *Gene Expression During EMT*

The integral protein E-cadherin represents a central component of adherens junctions, whose dissolution during EMT leads to cell detachment and is a prerequisite of cell motility *(63)*. Accordingly, EMT leads to a reduction of *E-cadherin* mRNA and protein levels and to a delocalization of the protein in cytoplasmic vesicles. E-cadherin is also involved (at least indirectly) in signal transduction events of the Wnt pathway, as it directly interacts with the intracellular protein β-catenin, which is a central signal transducer downstream of Wnt. Upon Wnt signaling the E-cadherin/β-catenin protein complex is disrupted and β-catenin is released to enter the nucleus and activate gene expression by modulating the activity of the high mobility group transcription factors, lymphoid enhancer factor/T-cell factor (LEF/TCF) (Fig. 2) *(8)*. The mechanisms of *E-cadherin* gene repression by several signaling pathways that mediate EMT, including TGF-β, have been investigated intensely *(63)*. This led to the discovery of several transcriptional repressors of the *E-cadherin* gene, such as members of the Snail family of zinc finger proteins (Snail, Slug), two-handed zinc finger/homeodomain proteins (ZEB1, ZEB2), bHLH proteins (E12/E47, Twist), and high mobility group proteins (LEF-1) (Fig. 5) *(63)*. These repressors recognize the DNA sequence motif known as E-box, located near the transcriptional initiation site of the *E-cadherin* gene, and recruit transcriptional co-repressors and histone deacetylases to the promoter of the gene. Transcriptomic studies have identified some of the transcriptional repressors of *E-cadherin* (e.g., Snail, Slug, LEF-1) as novel TGF-β-responsive genes in epithelial cells *(28,54)*.

Multiple mechanisms of TGF-β-mediated repression of *E-cadherin* gene expression during EMT have been reported (Fig. 5). When Id levels are reduced in response to TGF-β (see the discussion above), the E12/E47 repressors are relieved from inhibition and effectively repress *E-cadherin* expression *(58)*. In addition, Smads directly interact with ZEB1 and ZEB2 in response to TGF-β, thus forming repressor complexes on the same E-box sequences to which Snail, Slug, and E47 bind on the *E-cadherin* gene *(63,64)*. On the other hand, TGF-β can affect the expression of *E-cadherin* repressors, thus leading indirectly to the regulation of this gene. In the chicken, TGF-β2 causes EMT during normal heart vulval development via Slug *(65,66)*. TGF-β employs Smad3 during *Snail* induction because TGF-β cannot

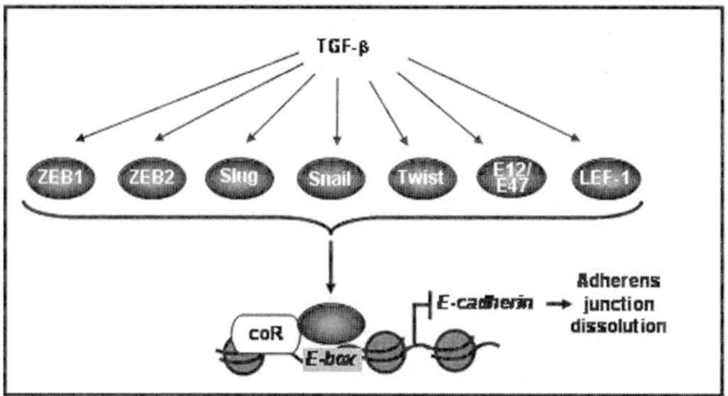

Upregulated genes			Downregulated genes
Snail	Tropomyosins	Net-1 ?	Id2, Id3
Slug	α-smooth muscle actin	Cutl1 ?	Id1 ?
ZEB2	N-cadherin	Zyxin ?	E-cadherin
Hmga2	Vimentin	Moesin ?	ZO-1
Lef-1	Fibronectin	Fibulin-2 ?	Desmoplakin
Hey-1	PAI-1	RhoB ?	β-catenin
Jagged-1	MMP-2		Mucin-1 ?
Disabled-2	MMP-9		Keratin1-18
ILK	Collagens type I & III		Keratin1-19 ?

Fig. 5. The EMT transcriptome downstream of TGF-β. TGF-β/Smad signaling induces expression or directly activates several transcriptional repressors. All these repressors bind to the E-boxes of the *E-cadherin* promoter, thus recruiting corepressors (coR) and inhibiting expression of this gene. Reduction of E-cadherin protein levels leads to dissolution of adherens junctions. Table of genes up- or downregulated (gray letters) by TGF-β during EMT. The functional role of these genes in EMT has been demonstrated experimentally. Gene names followed by question mark indicate established targets of the TGF-β pathway whose role on EMT is only suggested because on the functional properties of these genes. Those genes of the table that are not discussed in the text are based on our recent transcriptomic analysis of EMT in NMuMG cells *(18)*. Abbreviations not used in the main text: plasminogen activator inhibitor 1 (PAI-1), matrix metalloprotease 2 or 9 (MMP-2, MMP-9).

induce *Snail* gene expression in renal epithelial cells derived from Smad3 knockout mice *(21)*. Snail expression can also be induced after activation of the Erk and PI3K pathways by the synergistic signals of TGF-β and oncogenic Ras *(67)*. During normal palate development, TGF-β3/Smad signaling induces expression of the high mobility group factor LEF-1 *(8)*, which interacts with Smads (instead of its canonical transactivator β-catenin), and together repress *E-cadherin* expression and regulate several other genes involved in EMT *(54)*. In agreement with these models, when TGF-β3 is knocked out in mice, *Snail* expression is misregulated in the palate leading to the cleft phenotype *(65)*. A similar mechanism operates in transformed mammary epithelial cells by a synthetic *Fos-estrogen receptor* oncogene, which activates autocrine TGF-β/Smad signaling and β-catenin/LEF-1 to repress *E-cadherin* during EMT *(68)*. Finally, we have recently found that the bHLH factor Twist, a novel and critical repressor of *E-cadherin* expression that is involved in tumor cell invasiveness and metastasis *(69)*, is also transcriptionally induced by TGF-β during mammary epithelial EMT (Fig. 5).

Whether all the above transcriptional mechanisms of *E-cadherin* repression act in concert downstream of TGF-β or whether they represent tissue-specific scenarios remains currently unknown. It is also possible that TGF-β might also induce local destruction of E-cadherin protein at adherens junctions, possibly via proteasome-dependent degradation mechanisms. Finally, despite the abundance of transcriptomic data and the plethora of studies around the mechanism of transcriptional repression of *E-cadherin* during EMT, we currently know very little about the identity or the transcriptional regulation of those genes that are required for the establishment of the mesenchymal phenotype.

4.3. Other Gene Targets of TGF-β Signaling with Relevance to EMT

Despite the central role of *E-cadherin* and its regulators during EMT, the latter is a polygenic differentiation response to TGF-β involving regulation of many gene targets of the pathway (Fig. 5). Transcriptomic analysis of the TGF-β response of immortalized HaCaT keratinocytes uncovered that TGF-β, via the Smad pathway, induces expression of the ligand of Notch signaling, Jagged1 (Fig. 3) *(28)*. In this model, activation of Notch signaling is primed by TGF-β and subsequently, TGF-β cooperates with Notch signaling in regulating the expression of the *Hey1* transcriptional repressor *(70)*. The coordinate regulation of *Hey1* also occurs in mammary and kidney epithelial cells providing a general mechanism of crosstalk between TGF-β/Smad3 and Jagged1/Notch signaling during the establishment of EMT. Other microarray screens identified regulators of actin dynamics. TGF-β, via Smad3, induces expression of the guanine exchange factor NET1, which leads to sustained activation of Rho GTPases, thus supporting actin reorganization *(71)*. Whether NET1 directly contributes to TGF-β-induced EMT remains unknown. In addition, TGF-β, via the coordinate action of Smads and the p38 MAPK, induces the expression of several *tropomyosin* genes *(72)*. High tropomyosin levels contribute to cytoskeletal contractility and metastatic carcinoma cell motility. However, it remains unclear whether tropomyosins play direct effector roles on EMT or whether their role follows the differentiation switch toward the mesenchymal fate. TGF-β also induces the expression of the homeobox transcription factor CUTL1 that functions in a similar manner, as CUTL1 activates the expression of many other genes that regulate cell motility, tumor cell invasiveness, and extracellular matrix deposition *(73)*. In addition, CUTL1 serves as a poor prognosis marker for metastatic breast carcinoma, whereas it is unclear whether CUTL1 contributes to mammary epithelial EMT. Finally, TGF-β signaling induces the expression of the adaptor protein disabled-2 (Dab2), which participates in clathrin-mediated endocytosis *(74)*. In NMuMG cells undergoing TGF-β-induced EMT, Dab2 seems to protect the cells from apoptosis whereas permitting a differentiation switch to the mesenchymal phenotype. The above approaches promise that several new gene targets of the TGF-β/Smad pathway will soon be uncovered. The current expectation is to identify the most immediate and direct effectors of the EMT process downstream of TGF-β.

5. CONCLUDING REMARKS

Although the process of EMT downstream of TGF-β is not fully understood, here we attempted to present relatively complete models of signal transduction and gene regulation that describe the onset and establishment of EMT. A challenge for the future is to understand the complex signaling networks operating during EMT in vivo. The complexity of Smad and alternative pathways employed during the EMT response suggest the existence of additional important players. A focus on EMT promises the generation of new drugs not only against cancer cell invasiveness and metastasis, but also against tissue fibrosis. This has already been discussed heavily in the field *(75)*, and new discoveries are anxiously awaited in order to launch new therapeutic strategies focused on EMT. The central role Smad3 plays in mediating EMT led to the proposal of development of drugs that would target the function of this protein *(21)*. However, a Smad3-targeted therapy approach might affect many processes downstream of TGF-β signaling, as Smad3 seems to be involved in all aspects of TGF-β physiology at least in the adult cells of all tissue types examined so far. We therefore believe that the design of EMT-specific drugs awaits the discovery of novel mediators of the process that are amenable to pharmaceutical intervention and play critical roles during the disease.

The emerging technologies of single cell microdissection coupled to bioluminescence analysis of tagged cell types and large-scale gene expression analysis are a major promise in this field and have already provided exciting new discoveries in the realm of metastatic tumors *(7)*.

The role of non-Smad signaling pathways in EMT has provided interesting examples that have already or could potentially be translated to therapeutic application. Most examples so far focus on EMT during tissue fibrosis, but similar approaches are worth considering as cancer therapy. For example, based on the understanding that TGF-β utilizes the c-Abl kinase to signal in certain fibroblasts of injured kidneys, imatinib, the inhibitor of a small group of tyrosine kinases that include c-Abl, successfully blocked the progression of lung and kidney fibrosis in experimental mouse and rat models *(76)*. ILK is another attractive target for intervention because at least dominant-negative mutants of ILK are capable of blocking kidney fibrosis in mouse models *(20)*. In the case of breast cancer EMT, invasiveness, and metastasis induced by Ras-TGF-β cooperativity, the IKK-2 kinase of the NF-κB pathway is another attractive site for intervention *(42)*. Inhibition of NF-κB activity is sufficient to dramatically abrogate the metastatic potential of breast cancer cells to lung in mice *(42)*. These examples suggest that combinatorial pharmacologic regimes that target c-Abl, ILK, and IKKs might present a preferred method of treatment of tissue fibrosis or cancer.

We will close by emphasizing the current need for the definition of new molecular markers that define EMT and that can offer either prognostic or diagnostic tools to the clinic *(6)*. At least two of the proteins discussed above are now considered as important indicators of cancer progression: Integrin β6, whose levels correlate with the aggressiveness of human colon carcinomas and which can be used as a prognostic marker for patient life expectancy *(50)*, and Snail, which can also serve as an indicator of breast cancer recurrence after primary tumor surgery, as high Snail levels correlate with significantly lower relapse-free survival *(77)*. Because the gene targets of the TGF-β pathway that act as direct EMT effectors remain poorly understood, transcriptomic approaches are promising and should be followed by systematic functional analysis of the regulated genes, with the final aim of identifying new and more reliable markers of the mesenchymal cell phenotype and of the EMT process overall.

ACKNOWLEDGMENTS

Due to space limitations, we cite only selected literature in this chapter. Funding of the authors' work was provided by the Ludwig Institute for Cancer Research, the Swedish Cancer Society (project number: 4855-B03-01XAC), the Swedish Research Council (project number K2004-32XD-14936-01A), and the Marie Curie Research Training Network (RTN) "EpiPlastCarcinoma" under the European Union FP6 program. S.T. was supported by the Marie Curie RTN "EpiPlastCarcinoma". U.V. was supported by postdoctoral fellowships from the French Association pour la Recherche sur le Cancer and the Swedish Cancer Society, project number 4812-B03-01VAA. We thank our colleagues A. Gaal, A. Kurisaki, H. Niimi, and K. Pardali for their contributions to the scientific work emanating from our laboratory and Carl-Henrik Heldin (Ludwig Institute for Cancer Research, Uppsala) for continuous encouragement and intellectual support.

REFERENCES

1. Hay ED. The mesenchymal cell, its role in the embryo, and the remarkable signaling mechanisms that create it. Dev Dyn 2005;233:706–720.
2. Thiery JP. Epithelial–mesenchymal transitions in development and pathologies. Curr Opin Cell Biol 2003;15:740–746.

3. Zeisberg M, Kalluri R. The role of epithelial-to-mesenchymal transition in renal fibrosis. J Mol Med 2004;82:175–181.
4. Tarin D, Thompson EW, Newgreen DF. The fallacy of epithelial mesenchymal transition in neoplasia. Cancer Res 2005;65:5996–6000; Discussion 6000-1.
5. Barolo S, Posakony JW. Three habits of highly effective signaling pathways: principles of transcriptional control by developmental cell signaling. Genes Dev 2002;16:1167–1181.
6. Huber MA, Kraut N, Beug H. Molecular requirements for epithelial–mesenchymal transition during tumor progression. Curr Opin Cell Biol 2005;17:548–558.
7. Wang W, Goswami S, Sahai E, Wyckoff JB, Segall JE, Condeelis JS. Tumor cells caught in the act of invading: their strategy for enhanced cell motility. Trends Cell Biol 2005;15:138–145.
8. Nawshad A, LaGamba D, Hay ED. Transforming growth factor β (TGFβ) signalling in palatal growth, apoptosis and epithelial mesenchymal transformation (EMT). Arch Oral Biol 2004;49:675–689.
9. Person AD, Klewer SE, Runyan RB. Cell biology of cardiac cushion development. Int Rev Cytol 2005;243:287–335.
10. Desgrosellier JS, Mundell NA, McDonnell MA, Moses HL, Barnett JV. Activin receptor-like kinase 2 and Smad6 regulate epithelial–mesenchymal transformation during cardiac valve formation. Dev Biol 2005;280:201–210.
11. Cui W, Fowlis DJ, Bryson S, et al. TGFβ1 inhibits the formation of benign skin tumors, but enhances progression to invasive spindle carcinomas in transgenic mice. Cell 1996;86:531–542.
12. Portella G, Cumming SA, Liddell J, et al. Transforming growth factor β is essential for spindle cell conversion of mouse skin carcinoma *in vivo*: implications for tumor invasion. Cell Growth Differ 1998;9:393–404.
13. Han G, Lu SL, Li AG, et al. Distinct mechanisms of TGF-β1-mediated epithelial-to-mesenchymal transition and metastasis during skin carcinogenesis. J Clin Invest 2005;115:1714–1723.
14. Gotzmann J, Huber H, Thallinger C, et al. Hepatocytes convert to a fibroblastoid phenotype through the cooperation of TGF-β1 and Ha-Ras: steps towards invasiveness. J Cell Sci 2002;115: 1189–1202.
15. Janda E, Lehmann K, Killisch I, et al. Ras and TGFβ cooperatively regulate epithelial cell plasticity and metastasis: dissection of Ras signaling pathways. J Cell Biol 2002;156:299–313.
16. Lehmann K, Janda E, Pierreux CE, et al. Raf induces TGFβ production while blocking its apoptotic but not invasive responses: a mechanism leading to increased malignancy in epithelial cells. Genes Dev 2000;14:2610–2622.
17. Oft M, Heider KH, Beug H. TGFβ signaling is necessary for carcinoma cell invasiveness and metastasis. Curr Biol 1998;8:1243–1252.
18. Valcourt U, Kowanetz M, Niimi H, Heldin C-H, Moustakas A. TGF-β and the Smad signaling pathway support transcriptomic reprogramming during epithelial–mesenchymal cell transition. Mol Biol Cell 2005;16:1987–2002.
19. Akhurst RJ, Derynck R. TGF-β signaling in cancer—a double-edged sword. Trends Cell Biol 2001; 11:S44–S51.
20. Li Y, Yang J, Dai C, Wu C, Liu Y. Role for integrin-linked kinase in mediating tubular epithelial to mesenchymal transition and renal interstitial fibrogenesis. J Clin Invest 2003;112:503–516.
21. Sato M, Muragaki Y, Saika S, Roberts AB, Ooshima A. Targeted disruption of TGF-β1/Smad3 signaling protects against renal tubulointerstitial fibrosis induced by unilateral ureteral obstruction. J Clin Invest 2003;112:1486–1494.
22. Brown KA, Aakre ME, Gorska AE, et al. Induction by transforming growth factor-β1 of epithelial to mesenchymal transition is a rare event in vitro. Breast Cancer Res 2004;6:R215–R231.
23. Kowanetz M, Valcourt U, Bergström R, Heldin C-H, Moustakas A. Id2 and Id3 define the potency of cell proliferation and differentiation responses to tranforming growth factor β and bone morphogenetic protein. Mol Cell Biol 2004;24:4241–4254.
24. Miettinen PJ, Ebner R, Lopez AR, Derynck R. TGF-β induced transdifferentiation of mammary epithelial cells to mesenchymal cells: involvement of type I receptors. J Cell Biol 1994;127: 2021–2036.
25. Saika S, Kono-Saika S, Ohnishi Y, et al. Smad3 signaling is required for epithelial-mesenchymal transition of lens epithelium after injury. Am J Pathol 2004;164:651–663.
26. Valdes F, Alvarez AM, Locascio A, et al. The epithelial mesenchymal transition confers resistance to the apoptotic effects of transforming growth factor β in fetal rat hepatocytes. Mol Cancer Res 2002;1:68–78.

27. Willis BC, Liebler JM, Luby-Phelps K, et al. Induction of epithelial–mesenchymal transition in alveolar epithelial cells by transforming growth factor-β1: potential role in idiopathic pulmonary fibrosis. Am J Pathol 2005;166:1321–1332.
28. Zavadil J, Bitzer M, Liang D, et al. Genetic programs of epithelial cell plasticity directed by transforming growth factor-β. Proc Natl Acad Sci USA 2001;98:6686–6691.
29. Bates RC, Mercurio AM. The epithelial-mesenchymal transition (EMT) and colorectal cancer progression. Cancer Biol Ther 2005;4:365–370.
30. Piek E, Moustakas A, Kurisaki A, Heldin C-H, ten Dijke P. TGF-β type I receptor/ALK-5 and Smad proteins mediate epithelial to mesenchymal transdifferentiation in NMuMG breast epithelial cells. J Cell Sci 1999;112:4557–4568.
31. Zeisberg M, Shah AA, Kalluri R. Bone morphogenic protein-7 induces mesenchymal to epithelial transition in adult renal fibroblasts and facilitates regeneration of injured kidney. J Biol Chem 2005; 280:8094–8100.
32. Itoh S, Thorikay M, Kowanetz M, et al. Elucidation of Smad requirement in transforming growth factor-β type I receptor-induced responses. J Biol Chem 2003;278:3751–3761.
33. Li W, Qiao W, Chen L, et al. Squamous cell carcinoma and mammary abscess formation through squamous metaplasia in Smad4/Dpc4 conditional knockout mice. Development 2003;130:6143–6153.
34. Oft M, Akhurst RJ, Balmain A. Metastasis is driven by sequential elevation of H-ras and Smad2 levels. Nat Cell Biol 2002;4:487–494.
35. Tian F, DaCosta Byfield S, Parks WT, et al. Reduction in Smad2/3 signaling enhances tumorigenesis but suppresses metastasis of breast cancer cell lines. Cancer Res 2003;63:8284–8292.
36. Tian F, Byfield SD, Parks WT, et al. Smad-binding defective mutant of transforming growth factor β type I receptor enhances tumorigenesis but suppresses metastasis of breast cancer cell lines. Cancer Res 2004;64:4523–4530.
37. Kurisaki K, Kurisaki A, Valcourt U, et al. Nuclear factor YY1 inhibits transforming growth factorβ- and bone morphogenetic protein-induced cell differentiation. Mol Cell Biol 2003;23:4494–4510.
38. Takeda M, Mizuide M, Oka M, et al. Interaction with Smad4 is indispensable for suppression of BMP signaling by c-Ski. Mol Biol Cell 2004;15:963–972.
39. Levy L, Hill CS. Smad4 dependency defines two classes of transforming growth factor β (TGF-β) target genes and distinguishes TGF-β-induced epithelial–mesenchymal transition from its antiproliferative and migratory responses. Mol Cell Biol 2005;25:8108–8125.
40. Massagué J, Seoane J, Wotton D. Smad transcription factors. Genes Dev 2005;19:2783-2810.
41. Moustakas A, Heldin C-H. Non-Smad TGF-β signals. J Cell Sci 2005;118:3573–3584.
42. Huber MA, Azoitei N, Baumann B, et al. NF-kB is essential for epithelial-mesenchymal transition and metastasis in a model of breast cancer progression. J Clin Invest 2004;114:569–581.
43. Shim JH, Xiao C, Paschal AE, et al. TAK1, but not TAB1 or TAB2, plays an essential role in multiple signaling pathways in vivo. Genes Dev 2005;19:2668–2681.
44. Bakin AV, Rinehart C, Tomlinson AK, Arteaga CL. p38 mitogen-activated protein kinase is required for TGFβ-mediated fibroblastic transdifferentiation and cell migration. J Cell Sci 2002;115:3193–3206.
45. Bhowmick NA, Ghiassi M, Bakin A, et al. Transforming growth factor-β1 mediates epithelial to mesenchymal transdifferentiation through a RhoA-dependent mechanism. Mol Biol Cell 2001;12:27–36.
46. Xie L, Law BK, Chytil AM, Brown KA, Aakre ME, Moses HL. Activation of the Erk pathway is required for TGF-β1-induced EMT in vitro. Neoplasia 2004;6:603–610.
47. Yu L, Hebert MC, Zhang YE. TGF-β receptor-activated p38 MAP kinase mediates Smad-independent TGF-β responses. EMBO J 2002;21:3749–3759.
48. Lee YI, Kwon YJ, Joo CK. Integrin-linked kinase function is required for transforming growth factor β-mediated epithelial to mesenchymal transition. Biochem Biophys Res Commun 2004;316:997–1001.
49. Leung-Hagesteijn C, Hu MC, Mahendra AS, et al. Integrin-linked kinase mediates bone morphogenetic protein 7-dependent renal epithelial cell morphogenesis. Mol Cell Biol 2005;25:3648–3657.
50. Bates RC, Bellovin DI, Brown C, et al. Transcriptional activation of integrin β6 during the epithelial–mesenchymal transition defines a novel prognostic indicator of aggressive colon carcinoma. J Clin Invest 2005;115:339–347.
51. Bhowmick NA, Zent R, Ghiassi M, McDonnell M, Moses HL. Integrin β1 signaling is necessary for transforming growth factor-β activation of p38 MAPK and epithelial plasticity. J Biol Chem 2001;276:46,707–46,713.
52. Ozdamar B, Bose R, Barrios-Rodiles M, Wang HR, Zhang Y, Wrana JL. Regulation of the polarity protein Par6 by TGFβ receptors controls epithelial cell plasticity. Science 2005;307:1603–1609.

53. Tian YC, Phillips AO. Interaction between the transforming growth factor-β type II receptor/Smad pathway and β-catenin during transforming growth factor-β1-mediated adherens junction disassembly. Am J Pathol 2002;160:1619–1628.
54. LaGamba D, Nawshad A, Hay ED. Microarray analysis of gene expression during epithelial–mesenchymal transformation. Dev Dyn 2005;234:132–142.
55. Xie L, Law BK, Aakre ME, et al. Transforming growth factor β-regulated gene expression in a mouse mammary gland epithelial cell line. Breast Cancer Res 2003;5:R187–R198.
56. Jechlinger M, Grunert S, Tamir IH, et al. Expression profiling of epithelial plasticity in tumor progression. Oncogene 2003;22:7155–7169.
57. Kang Y, Siegel PM, Shu W, et al. A multigenic program mediating breast cancer metastasis to bone. Cancer Cell 2003;3:537–549.
58. Kondo M, Cubillo E, Tobiume K, et al. A role for Id in the regulation of TGF-β-induced epithelial-mesenchymal transdifferentiation. Cell Death Differ 2004;11:1092–1101.
59. Saika S, Ikeda K, Yamanaka O, et al. Adenoviral gene transfer of BMP-7, Id2, or Id3 suppresses injury-induced epithelial-to-mesenchymal transition of lens epithelium in mice. Am J Physiol Cell Physiol 2006;290:C282–C289.
60. Siegel PM, Massagué J. Cytostatic and apoptotic actions of TGF-β in homeostasis and cancer. Nat Rev Cancer 2003;3:807–820.
61. Pardali K, Kowanetz M, Heldin C-H, Moustakas A. Smad pathway-specific transcriptional regulation of the cell cycle inhibitor $p21^{WAF1/Cip1}$. J Cell Physiol 2005;204:260–272.
62. Perk J, Iavarone A, Benezra R. Id family of helix-loop-helix proteins in cancer. Nat Rev Cancer 2005;5:603–614.
63. Peinado H, Portillo F, Cano A. Transcriptional regulation of cadherins during development and carcinogenesis. Int J Dev Biol 2004;48:365–375.
64. Comijn J, Berx G, Vermassen P, et al. The two-handed E box binding zinc finger protein SIP1 downregulates E-cadherin and induces invasion. Mol Cell 2001;7:1267–1278.
65. Martinez-Alvarez C, Blanco MJ, Perez R, et al. Snail family members and cell survival in physiological and pathological cleft palates. Dev Biol 2004;265:207–218.
66. Romano LA, Runyan RB. Slug is an essential target of TGFβ2 signaling in the developing chicken heart. Dev Biol 2000;223:91–102.
67. Peinado H, Quintanilla M, Cano A. Transforming growth factor β-1 induces snail transcription factor in epithelial cell lines: mechanisms for epithelial mesenchymal transitions. J Biol Chem 2003;278: 21,113–21,123.
68. Eger A, Stockinger A, Park J, et al. β-Catenin and TGFβ signalling cooperate to maintain a mesenchymal phenotype after FosER-induced epithelial to mesenchymal transition. Oncogene 2004;23:2672–2680.
69. Yang J, Mani SA, Donaher JL, et al. Twist, a master regulator of morphogenesis, plays an essential role in tumor metastasis. Cell 2004;117:927–939.
70. Zavadil J, Cermak L, Soto-Nieves N, Böttinger EP. Integration of TGF-β/Smad and Jagged1/Notch signalling in epithelial-to-mesenchymal transition. EMBO J 2004;23:1155–1165.
71. Shen X, Li J, Hu PP, Waddell D, Zhang J, Wang X-F. The activity of guanine exchange factor NET1 is essential for transforming growth factor-β-mediated stress fiber formation. J Biol Chem 2001;276: 15,362–15,368.
72. Bakin AV, Safina A, Rinehart C, Daroqui C, Darbary H, Helfman DM. A critical role of tropomyosins in TGF-β regulation of the actin cytoskeleton and cell motility in epithelial cells. Mol Biol Cell 2004;15:4682–4694.
73. Michl P, Ramjaun AR, Pardo OE, et al. CUTL1 is a target of TGFβ signaling that enhances cancer cell motility and invasiveness. Cancer Cell 2005;7:521–532.
74. Prunier C, Howe PH. Disabled-2 (Dab2) is required for transforming growth factor β-induced epithelial to mesenchymal transition (EMT). J Biol Chem 2005;280:17,540–17,548.
75. Yingling JM, Blanchard KL, Sawyer JS. Development of TGF-β signalling inhibitors for cancer therapy. Nat Rev Drug Discov 2004;3:1011–1022.
76. Wang S, Wilkes MC, Leof EB, Hirschberg R. Imatinib mesylate blocks a non-Smad TGF-β pathway and reduces renal fibrogenesis in vivo. FASEB J 2005;19:1–11.
77. Moody SE, Perez D, Pan TC, et al. The transcriptional repressor Snail promotes mammary tumor recurrence. Cancer Cell 2005;8:197–209.

17 Functional Interactions Between MMPs and TGF-β in Normal and Tumor Tissue

Christina H. Stuelten, Anita B. Roberts, and William G. Stetler-Stevenson

CONTENTS

INTRODUCTION
MATRIX METALLOPROTEINASES
TGF-β ACTIVATION AND SIGNALING
SIMILARITIES AND UNKNOWNS OF TGF-β AND MMPS
REGULATION OF TGF-β ACTIVITY BY MMPS
REGULATION OF MMP ACTIVITY BY TGF-β
INTERACTIONS OF TGF-β AND MMPS IN TUMOR TISSUE
ACKNOWLEDGMENTS
REFERENCES

Abstract

Matrix metalloproteinases (MMPs) and TGF-β play a crucial role not only in regulating growth and development, tissue maintenance, and wound healing, but also in pathological processes like fibrosis and tumor development, progression, and metastases. Expression and activation of MMPs and TGF-β are tightly regulated. Interestingly, MMPs are involved in proteolytic activation of TGF-β, and TGF-β regulates expression levels in MMPs. Here, we discuss the functional interactions of MMPs and TGF-β in normal and tumor tissues.

Key Words: TGF-β; MMP; tumor; tumor–stroma interaction; cell signaling.

1. INTRODUCTION

In complex organisms a variety of different cell types are embedded in the extracellular matrix (ECM), which organizes cellular compartments during the formation of various tissues and organs. Localization and proliferation of cells as well as the composition of the surrounding matrix is tightly regulated by intercellular signaling via bioactive molecules. Whereas maintaining functionality in healthy organisms, dysregulation of cell signaling and matrix composition causes pathological features like fibrosis because of erroneous matrix production or invasive tumors, which exhibit increased cell proliferation, migration, and cell growth at distant sites.

From: *Cancer Drug Discovery and Development: Transforming Growth Factor-β in Cancer Therapy, Vol I: Basic and Clinical Biology*
Edited by: Sonia B. Jakowlew © Humana Press Inc., Totowa, NJ

For many years cancer researchers have focused on the progressive genetic and metabolic alterations within cancer cells during tumor initiation and progression. However, in the past decade investigators have come to appreciate that the tumor microenvironment, which includes tumor cells as well as host stromal cells and surrounding ECM, has a pivotal role in regulating tumor behavior *(1–3)*. A tumor is a complex mix of cancer cells as well as a variety of nontumorigenic cells, such as lymphocytes, mast cells, macrophages, endothelial cells, vascular smooth muscle cells, and fibroblasts. All cell types found in tumor tissue are embedded in a dynamic, continuous remodeling of ECM, which contains not only structural and scaffolding proteins such as collagens and laminins, but also soluble and diffusible elements such as enzymes, growth factors, and cytokines.

Recent studies show evidence that tumor progression is mediated by a tissue- and cell-type-specific stromal-carcinoma cell crosstalk. This crosstalk is bidirectional. It has long been recognized that tumor cells produce various growth factors, cytokines, and proteases, which induce significant remodeling of the ECM. These tumor cell products also influence the behavior and recruitment of host cells that produce growth factors and cytokines that can recruit new blood vessels and inflammatory cells.

Matrix metalloproteinases (MMPs) are proteases secreted into the extracellular environment by both tumor cells and tumor-associated fibroblasts. Collectively, these proteases can degrade most, if not all, structural components of the ECM. However, the role of MMPs in tumor progression is more complex than simply removing barriers to cellular invasion. It has been demonstrated that MMP activity can activate and/or inactivate growth factors, growth factor receptors, as well as cytokines and their receptors *(4)*. Thus the role of MMPs in cancer progression is extremely complex and depends on the specific MMP of interest, as well as the cellular interactions that regulate their activity. The cytokine TGF-β is well known for regulating the expression of MMPs, their endogenous inhibitors, tissue inhibitors of metalloproteinases (TIMP), as well as scaffolding matrix proteins like collagens *(4–8)*. Interestingly, MMPs and TGF-β have a considerable overlap in their mode of action. They both are involved in embryogenesis, development, inflammation, maintenance of tissues and matrix turnover, wound healing, and tumorigenesis *(9–14)*. For both MMPs and TGF-β, expression by stromal and tumor cells has been shown to act as both tumor promoters and tumor suppressors *(15–17)*.

More recent is the demonstration that stromal fibroblasts influence epithelial transformation and/or growth of carcinoma cells via production of paracrine factors and proteases *(2,3)*. Understanding the signals exchanged between cellular elements in the tumor microenvironment, as well as identification of effectors responsible for mediating the associated changes in cell–matrix interactions, should provide new strategies for cancer therapy. Here we will discuss how MMPs and TGF-β interact in normal and tumor tissues and how this interaction may influence tumor growth and progression.

2. MATRIX METALLOPROTEINASES

The MMPs are members of the metzincin superfamily of metalloenzymes, which also include the ADAM (a metalloproteinase with a disintegrin and metalloproteinase domain), ADAM-TS (ADAM with a thrombospondin-like domain), and astacin proteases. The mammalian MMP family, also known as the Matrixins, consists of 26 members, which include both soluble secreted proteases and cell membrane associated forms, referred to as membrane type- or MT-MMPs *(18,19)*. These two general classes are further divided into several subclasses based on distinct structural features. This general classification and functional subdomains of MMPs are summarized in Fig. 1.

All members of the MMP family share a conserved zinc-atom binding site within the catalytic domain and are initially produced in zymogen form requiring proteolytic removal of an amino-terminal prodomain to achieve enzymatic activity *(18,20)*. Most MMPs are activated extracellularly by serine- or metalloproteinase-mediated cleavage of their prodomains and may require further autoproteolytic processing of the new amino terminus. Proteolytic activation of MMP family members disrupts the critical interaction of the sulfhydryl side chain donated by a highly conserved cysteine in the "activation locus" with the zinc atom bound at the catalytic active site. Activation of soluble MMPs is associated with a decrease in molecular mass corresponding to the loss of the profragment (~10–12 kDa). Many "activated" MMP species are thus identified by this shift in molecular weight when analyzed by Western blot or zymogram analysis. However, MT-MMPs, as well as MMP-11 and MMP-28, are activated prior to their appearance on the cell surface by a mechanism involving furin-like proteases that cleave the MMP profragment resulting in a decrease in their molecular mass *(18,20)*.

Under physiologic conditions endogenous protease inhibitors tightly regulate the proteolytic action of MMP family members. These inhibitors include α2-macroglobulin, a ubiquitous plasma protein that may also be an important component of the provisional matrix associated with wound healing and early events in angiogenesis. MMPs have been shown to cleave the bait region in α2-macroglobulin, which results in sequestration of active MMPs into the "cage" region of this inhibitor *(18)*. The most extensively studied MMP inhibitors are the endogenous inhibitors known as the tissue inhibitors of metalloproteinases or TIMPs. The TIMP family consists of four mammalian genes, *TIMP-1, -2, -3,* and *-4 (5,21)*. The TIMP family members are distinguished by differential expression in various tissues. TIMP-2 is unique in that expression of this inhibitor is constitutive in most tissues, whereas the expression of TIMP-1, TIMP-3, and TIMP-4 protein is inducible *(5,21)*. Furthermore, recent works suggest that the protease inhibitory activity of TIMP-1 is poor against members of the MT-MMP class, and that TIMP-1 and TIMP-3 inhibit MMPs, as well as members of the related ADAM protease family *(5,21)*. In addition to α2-macroglobulin and TIMPs, a novel membrane-bound MMP inhibitor known as RECK (reversion-inducing cysteine-rich protein with kazal motifs) has been identified *(22,23)*. RECK deficiency in mice is embryonically lethal suggesting an important role for this inhibitor in the regulation of ECM turnover *(23)*.

The details of at least one cellular mechanism for MMP activation have been worked out. Pro-MMP-2 is activated at the cell surface by a unique pathway that involves MT1-MMP and TIMP-2 *(18)*. This mechanism involves the binding of TIMP-2 to both MT1-MMP and pro-MMP-2 to form a cell surface ternary complex, in which pro-MMP-2 is activated by a second, adjacent molecule of MT1-MMP that is not inhibited by TIMP-2. Following the cleavage of the pro-MMP-2 profragment the active MMP-2 is released from the cell surface. This model implies that the inhibitor-free MT1-MMP binds TIMP-2, which then acts as a cell surface receptor for TIMP-2-free pro-MMP-2. However, recent studies have not confirmed that tumor cells or microvascular endothelial cells bind TIMP-2 exclusively via MT1-MMP and in many tumor cell cultures pro-MMP-2 is found exclusively in complex with TIMP-2 *(24,25)*.

Recent data also challenge the concept that MMP activation is irreversibly associated with the loss of the profragment. Fedarko et al. *(26)* demonstrated that small integrin-binding ligand N-linked glycoproteins, also known as the SIBLING family, can bind and activate members of the MMP family, and that this activation can occur reversibly and without proteolytic processing of the MMP profragment. These authors demonstrate specificity in the interaction of members of the SIBLING family with MMPs. For example, bone sialoprotein copurified with MMP-2, osteopontin with MMP-3, and DMP1 with MMP-9. Interestingly, SIBLING interaction with MMPs could also reverse the inhibition of these proteases by the endogenous TIMPs. These findings suggest that members of the SIBLING family can control MMP activity by activation of latent pro-MMPs without prodomain cleavage, as well as disruption of MMP–TIMP complexes resulting in the recovery of proteolytic activity *(26)*. This is an important finding because it was previously thought that MMP–TIMP inhibitory complexes were tight binding and essentially irreversible.

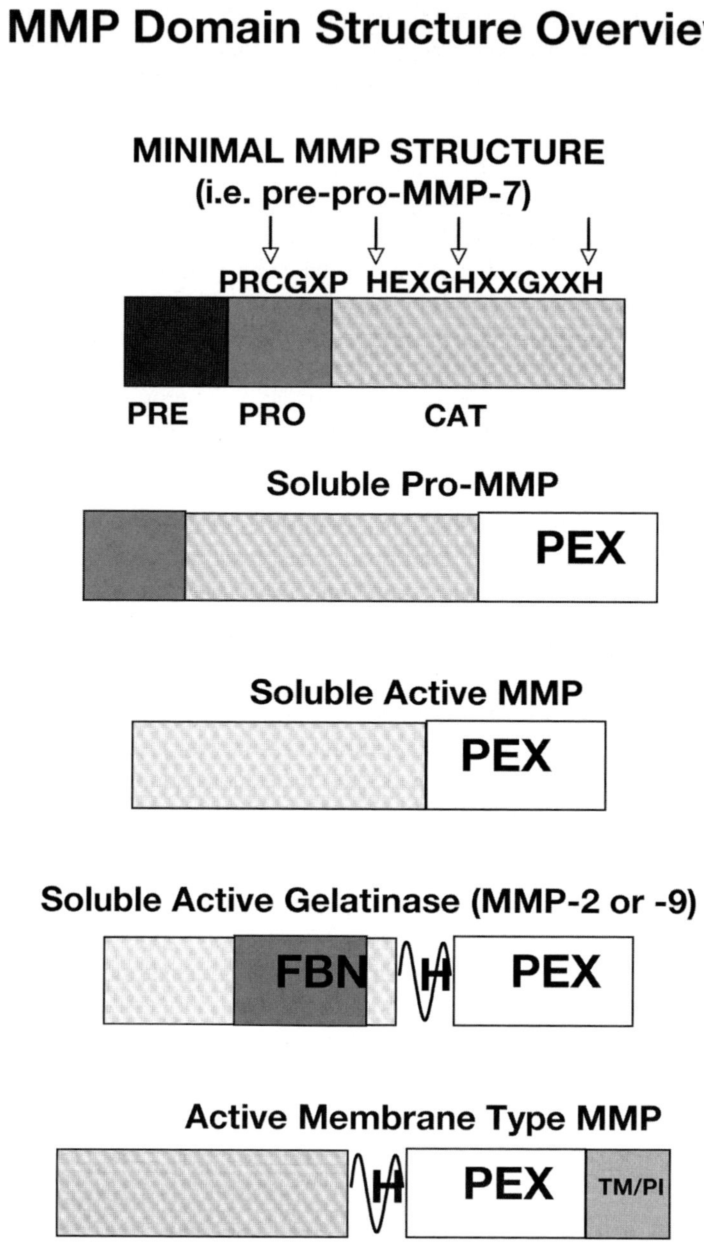

Fig. 1. Simplified overview of MMP domain structure. The minimal structure of an MMP is represented in the top schematic by pre-pro-MMP-7. During processing prior to secretion the pre-enzyme consisting of the signal peptide sequence directing molecular secretion is removed. Thus the secreted mature form of this minimal MMP molecule consists of a latent proenzyme. The position of the residues responsible for coordination of the catalytic zinc atom in this latent form of the enzyme are indicated by arrows. These include three histidine residues located in the catalytic domain, as well as a single cystine residue located in the enzyme profragment. Enzyme activation results in removal or displacement of the cystine residue from coordination with the catalytic zinc atom. However, most secreted MMPs also contain a C-terminal domain referred to as the PEX domain, which is important in determining substrate specificity. Again these enzymes are secreted in a latent form, and removal of the profragment is required to attain proteolytic activity. The gelatinases (MMP-2 and MMP-9) contain an additional domain (FBN), which is located immediately adjacent to the N-terminal region

The ECM sequesters a variety of growth factors and cytokines, some in latent forms. A second mechanism in which MMPs may contribute to tumor growth is through the release and/or activation of these matrix-associated factors, *see* Fig. 2A. MMP-9 has been implicated in the release of vascular endothelial growth factor (VEGF), which results in enhanced tumor angiogenesis and growth. This was first demonstrated in the RIP-Tag2 model of pancreatic islet cell carcinoma in which MMP-9 expression correlated with initiation of tumor angiogenesis, and either chemical or genetic ablation of MMP-9 expression resulted in tumors that failed to undergo vascularization *(27)*. The authors demonstrate that MMP-9, but not MMP-2, specifically enhanced the release of VEGF from the ECM. Similarly, MMP-1 and MMP-3, as well as plasmin and heparanase, degrade endothelial-derived perlecan causing release of bound bFGF *(28)*. These findings demonstrate MMP-mediated release of pro-angiogenic growth factors that mediate tumor-induced angiogenesis and enhance tumor growth (Fig. 2A).

MMPs may also regulate growth factor availability in other ways. For example, transgenic animal studies have demonstrated that alteration of the MMP/TIMP balance in vivo in favor of TIMP-1 activity can block neoplastic proliferation in the SV40 T-antigen-induced model of murine hepatocellular carcinoma *(29)*. The mechanism of this TIMP-1 effect was mediated by direct inhibition of MMP processing of insulin-like growth factor binding protein-3 (IGFBP-3) preventing release of insulin-like growth factor II (IGFII) mitogenic activity. Subsequent studies showed that IGFBPs are substrates for both MMP-7 and MMP-9 *(29,30)*. Thus, cleavage of IGFBP, resulting in the release of IGFs, may contribute to enhanced tumor growth. In addition to these two examples, MMPs have also been shown to mediate the release and/or activation of TGF-β as we will discuss later.

3. TGF-β ACTIVATION AND SIGNALING

The three known isoforms of TGF-β, TGF-β1, TGF-β2, and TGF-β3, are highly conserved, multipotent cytokines that are involved in a panoply of normal and pathological processes such as cell growth, development, matrix synthesis, wound healing, fibrosis, inflammation, and tumorigenesis *(31–33)*. TGF-β inhibits proliferation of normal epithelial cells and has also been found to inhibit growth of tumor cells and tumors. However, it has been shown to increase metastases by several mechanisms such as induction of epithelial–mesenchymal transition, alteration of cell–cell contacts, and modulation of ECM composition (Fig. 2B), thus displaying characteristics of both tumor suppressor and tumor promoter *(16,33,34)*.

Concordant with its complex biological functions, expression, activation, and signal transduction of TGF-β are tightly regulated. TGF-β is synthesized as a preproprotein and secreted as an inactive complex. This inactive TGF-β complex consists of the "active" TGF-β dimer, its propeptide latency-associated peptide (LAP), and latent TGF-β-binding protein (LTBP), a product of a second gene. In this complex, also called large latent complex (LLC), active TGF-β is noncovalently bound to LAP forming the small latent complex (SLC), and LAP is covalently bound to LTBP, which can also be linked covalently to components of the ECM *(31,35)*. Typically, the majority of extracellular TGF-β is inactive, and only a small fraction of this pool is locally activated.

of the catalytic domain. This region consists of type II Fibronectin repeats and is responsible for the powerful gelatinolytic activity of these proteases. In addition, the gelatinases (MMP-2 and MMP-9) incorporate a hing region (H) between the catalytic and PEX domain, possibly for further refining the substrate binding activity of these proteases by allowing substrate coordination between the FBN and PEX domains. The bottom figure represents the membrane type- or MT-MMPs. These enzymes may contain true transmembrane domains or are tethered to the cell surface via a phosphotidyl inositol type linkage.

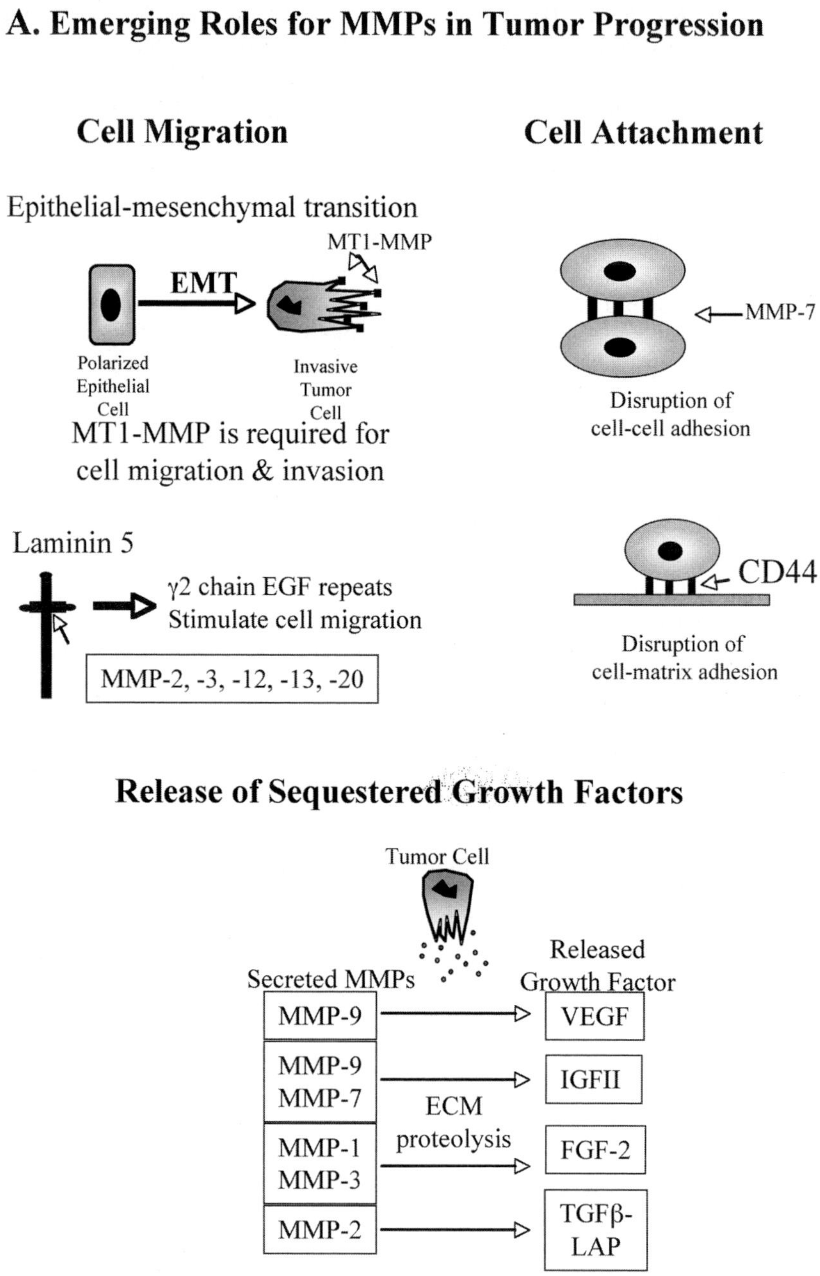

Fig. 2. (*Continued*)

On activation of the latent complex, the mature, dimeric TGF-β peptide is released. Several mechanisms of TGF-β activation have been described, all involving proteolytic degradation or conformational changes of LAP. Plasmin or MMPs proteolytically degrade LAP and thereby cause release of mature TGF-β. Alternatively, binding of LAP to integrins or thrombospondin or a locally acidic milieu, as it is physiologically found close to osteoclasts, causes conformational changes of LAP and release of active TGF-β *(31,36,37)*.

Active TGF-β binds to its specific receptor, TGF-β-receptor II (TβRII), which recruits and activates TGF-β-receptor I (TβRI) by transphosphorylation. Activated TβRI then phosphorylates

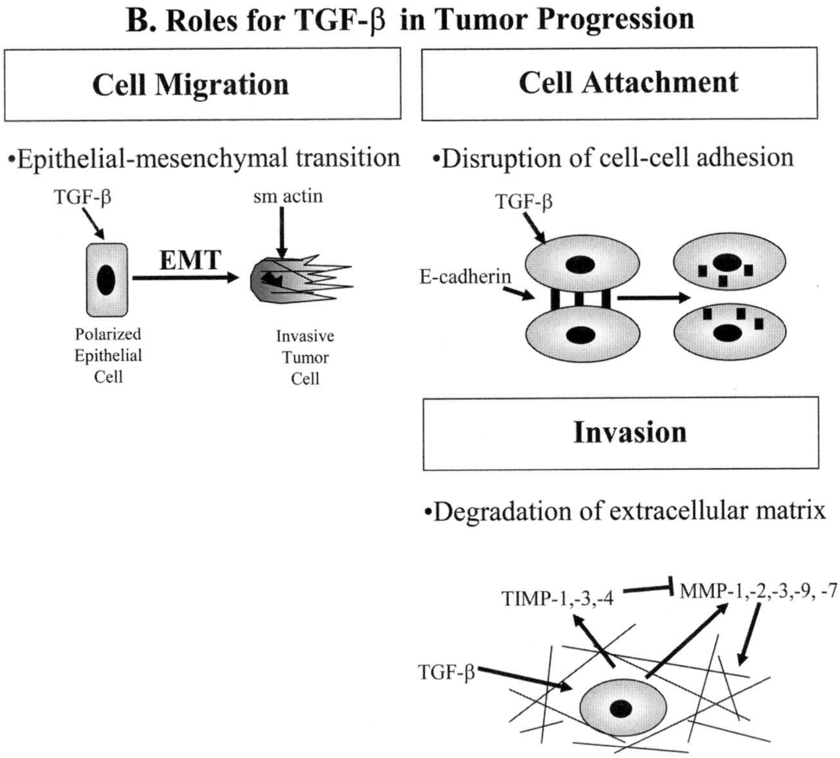

Fig. 2. Emerging roles for MMPs (**A**) and TGF-β (**B**) in tumor progression. MMPs and TGF-β both influence tumor progression and metastasis by a variety of mechanisms including modulation of cell migration and invasion, matrix degradation, release of growth factors, and cell attachment. However, MMPs typically modulate the microenvironment by enzymatic cleavage of their substrates, whereas TGF-β typically modulates the microenvironment by up- and downregulation of protein expression levels and relocalization of proteins like E-cadherin.

a highly conserved C-terminal SSXS motif of Smad2 and Smad3, which then form a complex with Smad4, translocate into the nucleus, and act as transcription factors for TGF-β-dependent genes *(38–40)*. Canonical Smad signaling leads to inhibition of cell proliferation, apoptosis, increased TGF-β expression, and altered expression of proteins involved in ECM composition. Additionally, TGF-β can activate noncanonical MAPK- and rho signaling, and MAPK signaling vice versa can alter canonical TGF-β signal transduction *(34,41,42)*. This cross-signaling of TGF-β with other signaling cascades results in context-dependent effects of TGF-β and contributes to such characteristic effects as cell migration and epithelial–mesenchymal transition *(33,34,38)*.

Active TGF-β that is not bound to its specific receptor is rapidly cleared from the intracellular space by binding to extracellular proteins like decorin or α2-macroglobulin, internalization, and subsequent degradation.

4. SIMILARITIES AND UNKNOWNS OF TGF-β AND MMPs

Although belonging to different protein families, TGF-β and MMPs exhibit many similarities. Both are secreted as proproteins that are activated in the extracellular space. This activation is tightly regulated, and typically more than one mode of activation is known. Additionally, both proteins can shift between an activated and inactivated status, and the presence of the protein as determined by immunostaining does not necessarily correlate with activity. Although under thorough investigation, determining specific activity of MMPs or TGF-β in vivo and in vitro is still difficult, and typically indirect methods are used. ELISA and Western blot determine protein levels, but not the activity. MMPs can be investigated by zymography, but the process of electrophoresis activates pro-MMPs irreversibly and disrupts MMP–TIMP complexes. Although in situ zymography correctly indicates locations of high MMP activity, the substrate specificity is not high enough to correlate this activity specifically with one enzyme, and extensive experiments including MMP-knockout cells or animals are required to overcome this problem. An ELISA-coupled enzyme assay specifically determines MMP activity, but is unable to localize the activity of the enzyme in the tissue specimen. Similarly, TGF-β levels can be determined by ELISA, but levels of active TGF-β are still most precisely investigated in luciferase reporter assays, with the drawback that the cells necessary for this system might activate TGF-β in the conditioned medium themselves. Very few antibodies are specific for active TGF-β, but even if immunostaining correctly identifies active TGF-β, it cannot specify its bioavailability. With these limitations in mind, we will now try to summarize and discuss the functional interaction of TGF-β and MMPs.

5. Regulation of TGF-β Activity by MMPs

5.1. MMPs Proteolytically Activate TGF-β

Activation of TGF-β occurs via an activation cascade including release of LLC from the ECM, release of SLC, and release of active TGF-β and/or LAP from this complex. MMPs have been shown to be potentially involved in all these changes (Fig. 3), although different MMPs might be involved in the activation of different TGF-β isoforms *(10)*. Stromelysin-1 (MMP-3) has been shown to release a truncated LLC from the chondrocyte ECM by cleavage of LTBP1 *(43)*. It is not yet clear how this complex is activated, but it was speculated that similar to activation of LLC by plasmin, the truncated LLC might be activated by binding to mannose-6-phosphate receptors on the cell surface *(43)*. Furthermore, MMP-2, but not the closely related MMP-9, can specifically cleave the ECM-bound form of LTBP, suggesting that this protease may also function in the release of latent TGF-β complexes (TGF-β-LAP), which are then activated by MMP cleavage of LAP *(44)*.

Release of TGF-β from the SLC mainly seems to take place after binding of the complex to a cell surface. Several members of the MMP family have been shown to directly cleave LAP resulting in release of active TGF-β *(45–47)*. LAP-β1, a component of SLC, has been shown to bind to integrin αv by an RGD site, and αvβ6 and αvβ8 have been shown to bind and activate latent TGF-β *(31,45,48,49)*. Interestingly, activation of TGF-β1 by αvβ8 requires recruitment of MT1-MMP into this complex whereas activation by αvβ6 does not *(45)*. Cleavage of LAP by this mechanism inactivates LAP, i.e., it can no longer bind mature TGF-β1, and releases mature TGF-β1.

Similarly, TGF-β2 can be activated by cell-surface-bound MMP-9 and possibly MMP-2. MMP-9 is recruited to the cell surface by the hyaluron receptor CD44, and LAP-β2 is proteolyzed by this complex *(46)*. Although it was shown that MMP-2 activates TGF-β in vitro and that a hybrid of CD44/MMP-2 also can activate TGF-β2, it is not known whether CD44 binds MMP-2 in vivo. However, the integrin αvβ3 was shown to localize MMP-2 to the cell surface in vivo *(46,50)* and might function in MMP-2-mediated TGF-β activation analogous to CD44 in MMP-9-mediated TGF-β activation.

5.2. MMPs also Influence Activity of TGF-β by Influencing Extracellular Levels of TGF-β Binding Proteins

Active TGF-β that is not bound to a specific receptor is quickly removed by binding to matrix proteins. MMPs exert indirect effects on the levels of active TGF-β by influencing levels of matrix proteins that can bind TGF-β, e.g., decorin and α2-macroglobulin. The

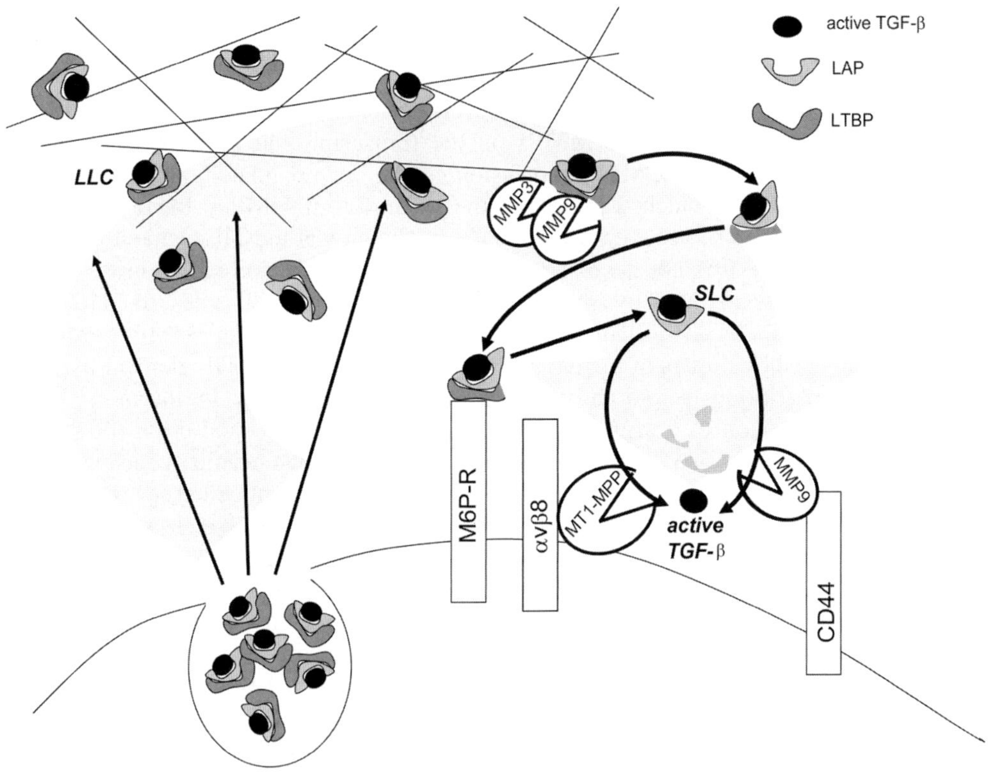

Fig. 3. Activation of TGF-β by MMPs. The latent large complex (LLC) consists of active TGF-β, LAP, and LTBP and is typically bound to extracellular matrix proteins. The small latent complex, which consists of LAP and active TGF-β, can be released on cleavage of LTBP by MMP-3 or MMP-9. Subsequently LAP is cleaved by cell-bound MMPs such as MT1-MMP or MMP-9, and active TGF-β is released.

small leucine-rich proteoglycan decorin was shown to be a substrate of MMP-2, -3, and -7. Degradation of decorin by MMPs leads to an increased bioavailability of active TGF-β *(4,51)*. α2-macroglobulin is a substrate of MMP-1, -3, -8, -9, -11, -12, -13, and -14. In contrast, although as a short-term effect, α2-macroglobulin prevents binding of TGF-β to its specific receptors, it releases active TGF-β from the latent complex as a long-term effect *(52)*. Degradation of α2-macroglobulin by MMPs might therefore result in temporarily increasing levels of bioactive TGF-β by lowering clearance of TGF-β, but in the longer term might result in reduced levels of bioactive TGF-β by lowering the activation of TGF-β. The example of α2-macroglobulin emphasizes that the bioavailability of active TGF-β is regulated not only by the mere binding to matrix proteins, but also by long-term effects of altered protein levels on TGF-β activation.

In summary, MMPs influence activation of TGF-β and clearance of TGF-β on multiple levels and in a complex fashion.

6. REGULATION OF MMP ACTIVITY BY TGF-β

6.1. TGF-β Regulates Expression of MMPs

TGF-β is a key regulator of fibrosis and influences MMP expression on the transcriptional and the translational level. In vitro, TGF-β is well known for altering MMP expression in a variety of cell lines including tumor and stromal cell lines *(53–60)*. Although as a rule of thumb TGF-β increases expression of collagens, decreases expression of MMPs, and increases expression of TIMPs in fibrosis, its role in tumor biology is less clear-cut, and TGF-β effects on MMPs depend on the cell type and the specific MMP expressed. TGF-β might be able to exert these effects on MMP expression by several mechanisms.

TGF-β regulates expression of MMPs on the transcriptional level through several promoter elements. In general, MMP expression can be downregulated through a TGF-β-inhibitory element (TIE), which is present in the promoters of MMP-1, MMP-3, MMP-9, and MMP-13 *(35,61–63)*. However, it has also been shown that the TIE element present in the promoter of MMP-9 is not necessary for TGF-β-induced downregulation of MMP-9 but rather this effect is mediated by a NF-κB binding site *(61)*. TGF-β is described to upregulate MMP expression by AP-1 elements, which are present in all but the MMP-2 promoter, and can be activated by dimers of the jun and fos families *(126)*. The TIE element itself can bind several transcription factors and interacts with AP-1 elements. Furthermore TGF-β itself induces expression of transcription factors like c-fos *(64,65)* and c-jun *(66)*. These direct and indirect effects of TGF-β on MMP promoters determine whether gene expression is finally inhibited or induced and might also explain why different cell types respond differently to stimulation with TGF-β. We will illustrate the effects of TGF-β on MMP expression in more detail for MMP-1 and MMP-9.

MMP-1 or collagenase-1 is one of the few enzymes that can degrade the ubiquitous and relatively biochemically inert native fibrillary collagen. TGF-β mainly suppresses MMP-1 expression, but TGF-β-induced MMP-1 expression has also been described *(67)*. Effects of TGF-β on MMP-1 expression are Smad3 and Smad4 dependent but do not require Smad1 or Smad2 *(68)*. Smad3 and Smad4 are shown to bind to the AP-1 site in the MMP-1 promoter that is required for TGF-β-induced MMP-1 repression *(69)*. Additionally, the binding of c-Fos, JunD, JunB, and c-jun to the AP-1 site is essential for the mediation of TGF-β effects on MMP-1 expression *(63,69)*. The TIE element of the MMP-1 promoter mediates TGF-β-induced MMP-1 suppression by c-fos binding; however, it can also bind c-fos in the absence of TGF-β and then basally repress MMP-1 transcription. TGF-β has been shown to change the protein binding pattern of the TIE element and alter protein/DNA interaction, and by this mechanism represses induction of MMP-1 by agents such as phorbol esters *(62,70)*. Such an alteration of protein/DNA interactions can integrate signaling cascades on the promoter level and explain the context-dependent effects of TGF-β on protein expression.

MMP-9 expression levels can be decreased or increased by TGF-β. Typically, TGF-β inhibits LPS- or cytokine-induced MMP-9 expression especially in hematopoietic cells, but itself increases MMP-9 expression in epithelial cells or fibroblasts. In human monocytic leukemia cells and macrophages, TGF-β suppresses LPS-induced MMP-9 expression by a NF-κB site in the promoter rather than by the TIE site *(61)*, and TGF-β attenuates kit-ligand-induced MMP-9 expression in human mast cells *(71)*. On the other hand TGF-β can increase MMP-9 expression in colon cancer cells, fibroblasts, and keratinocytes *(72–74)*, alone or together with other mediators like TNF-α *(75)* or tenascin *(76,77)*. Additionally, although TGF-β itself stimulates MMP-9 expression in human monocytes, a neutralizing antibody to TGF-β has been shown to increase MMP-9 expression in the very same cell line, indicating that endogenous TGF-β might increase, whereas exogenous TGF-β might decrease MMP-9 expression levels in THP-1 cells *(78)*.

Regulation of MMP-9 expression by TGF-β requires both canonical and noncanonical TGF-β signaling. Impairment of canonical TGF-β signaling such as Smad3 elimination in knockout fibroblasts results in lower MMP-9 expression levels as compared to wild-type cells *(73)*, but TGF-β-induced MMP-9 expression was shown to also depend on unimpaired ras-, MEK-, and p38 signaling *(73,74,79)*. Interestingly, TGF-β not only regulates MMP expression at the promoter level but also has been shown to cause increased stability of MMP-9 mRNA and to increase the stability of secreted pro-MMP-2 *(59,80)*.

In summary, TGF-β signaling via canonical and noncanonical signaling cascades influences MMP expression in a context-dependent manner at the transcriptional level through at least two promoter elements, AP-1 and TIE, on the translational level by modulation of mRNA stability and at the protein level by altering the stability of the proenzyme.

6.2. TGF-β Influences Enzymatic Activity of MMPs by Regulating Expression Levels of Members of the TIMP- and SIBLING Families

MMP activity is tightly regulated in vivo by proteolytic activation or inactivation as well as by binding of MMPs to specific inhibitors, TIMPs, or to proteins of the SIBLING family that can activate pro-MMPs without proteolytic cleavage or disrupt MMP–TIMP complexes *(26)*. TIMP-1, TIMP-3, and TIMP-4 have been shown to be induced by TGF-β *(81–85)*. Two members of the SIBLING family, bone sialoprotein and osteopontin, are upregulated by TGF-β, whereas a third, dentin matrix protein, is downregulated by TGF-β *(86–91)*. In contrast to the direct effect of TGF-β on expression levels of MMPs, TIMPS, and proteins of the SIBLING family, it also seems to be involved in the regulation of local enzymatic activity of MMPs and therefore contributes significantly to the resulting net MMP activity observed in vivo.

7. INTERACTIONS OF TGF-β AND MMPs IN TUMOR TISSUE

7.1. TGF-β and MMP Expression Is Altered in Tumors

Dysregulation of TGF-β and MMP expression and activation are frequently observed in tumor tissues and have been shown to contribute to tumor progression. TGF-β is a potent tumor suppressor at early stages of tumor development, whereas in later stages it promotes metastases. MMPs are involved in the removal of the basement membrane that is a hallmark of progression of benign to malignant tumors. At later tumor stages, increased secretion and activation of MMPs by tumor cells are commonly believed to promote metastases by enabling cells to migrate through ECM. Clinical data show that increased tissue and serum levels of MMPs or TGF-β frequently correlate with a poor clinical prognosis *(92–101)*. However, it is not clear whether these increased MMPs and TGF-β levels are the cause of tumor progression or rather the consequence of a larger tumor burden or alternatively representative of a more aggressive tumor phenotype.

7.2. Tumor Cells Modulate MMP Expression of Stromal Cells

In vivo, MMP expression has been described for tumor cells as well as a variety of stromal cell types depending on the tumor type *(102)*. MMP-9 mRNA and immunoreactivity have been detected in tumor, stroma, and vascular cells within tumor tissues *(103,104)* as well as in cocultures *(105,106)*. Increased MMP-2 levels have been described for both tumor-derived or normal fibroblasts and breast cancer cells in various coculture systems *(107–109)*. However, none of these parameters reflects the activity of MMPs in tissues, and in situ zymography of cocultures and tumor tissues showed that gelatinolytic activity is localized around stromal cells whereas the gelatinolytic activity of the tumor cells is negligible as compared to the activity localized around fibroblasts, possibly because of coexpression of TIMPs *(73,110)*.

Stromal MMPs can be induced by tumor cells via several mechanisms. Tumor cells secrete extracellular matrix metalloproteinase inducer (EMMPRIN or CD147), which induces MMP expression in tumor and stromal cells *(111,112)*. Recently we have shown that tumor cell derived TGF-β and TNF-α induces MMP-9 secretion in stromal fibroblasts and that MMP-9 induction was dependent on unimpaired Smad-, ras-, and PI3K signaling. Interestingly, TGF-β-induced MMP-9 expression in fibroblasts required intact EGF- and HGF signaling transduction, although these individual cytokines had only a minimal effect on MMP-9 levels *(73)*.

These data imply that HGF- and EGF signaling might be secondary contributors to TGF-β-mediated MMP-9 induction and emphasize the tight regulation of MMP expression.

In summary, tumor cells modulate their environment not only directly by secretion of ECM components, enzymes, and inhibitors, but also by stimulating and inhibiting protein synthesis and secretion by stromal cells.

7.3. Interaction of TGF-β and MMPs Can Modulate Biological Behavior of Cells

Increased expression of TGF-β or MMPs has been shown to promote tumor cell migration in vitro and invasiveness and metastases in vivo *(56,77,79,113,114)*. Multiple mechanisms are involved. Altered MMP expression can change migratory behavior of cells by degradation of the ECM that then enables tumor cells to move through this cleared space. However, interaction of membrane-bound MMPs and matrix proteins as in the binding of MT1-MMP and MMP-2 to laminin-5 can also promote cell migration *(115–117)*, enabling cells to move through the matrix on a protein track. As a third mechanism, altered levels of MMPs alter levels of bioactive factors (growth factors, cytokines) in the extracellular space. As an example, proteolysis of latent TGF-β by MMPs results in multiple degradation products. The biological effects of active TGF-β are well described but LAP and LTBP also have biological functions. LAP was recently shown to induce cell migration by binding to integrins *(118,119)*. LTBPs have been shown to increase adhesion and migration of cells *(120,121)*. MMPs also can activate or inactivate cytokines like interleukins and release bioactive molecules from the cell surface, an effect known as "shedding" *(122,123)*, or degrade cell surface protein like E-cadherin *(124,125)* resulting in the release of β-catenin from the complex and subsequently increased Wnt signaling.

Additionally, degradation of matrix proteins by MMPs leads to increased levels of bioactive matrix fragments (matrikines) like angiostatin or tumstatin *(123)*, which have been shown to inhibit angiogenesis and regulation of tumor growth. Interestingly, both angiostatin and tumstatin inhibit tumor growth rather than accelerating it. TGF-β-modified MMP expression might thus not only influence matrix composition but also regulate exposure of cryptic sites and levels of bioactive substances (Fig. 2A).

In summary, we have shown that TGF-β signaling and MMP expression and activation form an intricate network, which is regulated on many levels. Many details of this network remain to be elucidated. However, it is clear that perturbations in the network can contribute to the pathogenesis of malignant tumors, and more complex understanding of the network may offer therapeutic opportunities in cancer treatment.

ACKNOWLEDGMENTS

The authors truly appreciate the helpful discussions with Lalage Wakefield and the comments on the manuscript from Stacey Byfield DaCosta, Astrid Baege, and Glenn Merlino.

REFERENCES

1. Bhowmick NA, Chytil A, Plieth D, et al. TGF-beta signaling in fibroblasts modulates the oncogenic potential of adjacent epithelia. Science 2004;303:848–851.
2. Bhowmick NA, Neilson EG, Moses HL. Stromal fibroblasts in cancer initiation and progression. Nature 2004;432:332–337.
3. Littlepage LE, Egeblad M, Werb Z. Coevolution of cancer and stromal cellular responses. Cancer Cell 2005;7:499–500.
4. Sternlicht MD, Werb Z. How matrix metalloproteinases regulate cell behavior. Annu Rev Cell Dev Biol 2001;17:463–516.

5. Lambert E, Dasse E, Haye B, Petitfrere E. TIMPs as multifacial proteins. Crit Rev Oncol Hematol 2004;49:187–198.
6. Matrisian LM Metalloproteinases and their inhibitors in matrix remodeling. Trends Genet 1990;6: 121–125.
7. Roberts AB, Sporn MB, Assoian RK, et al. Transforming growth factor type beta: rapid induction of fibrosis and angiogenesis in vivo and stimulation of collagen formation in vitro. Proc Natl Acad Sci USA 1986;83:4167–4171.
8. Roberts AB, Tian F, Byfield SD, et al. Smad3 is key to TGF-beta-mediated epithelial-to-mesenchymal transition, fibrosis, tumor suppression and metastasis. Cytokine Growth Factor Rev 2006; 17:19–27.
9. Goetzl EJ, Banda MJ, Leppert D. Matrix metalloproteinases in immunity. J Immunol 1996;156:1–4.
10. Mott JD, Werb Z. Regulation of matrix biology by matrix metalloproteinases. Curr Opin Cell Biol 2004;16:558–564.
11. Mandal M, Mandal A, Das S, Chakraborti T, Sajal C. Clinical implications of matrix metalloproteinases. Mol Cell Biochem 2003;252:305–329.
12. Letterio JJ. TGF-beta signaling in T cells: roles in lymphoid and epithelial neoplasia. Oncogene 2005;24:5701–5712.
13. Ashcroft GS, Roberts AB. Loss of Smad3 modulates wound healing. Cytokine Growth Factor Rev 2000;11:125-131.
14. Koch RM, Roche NS, Parks WT, Ashcroft GS, Letterio JJ, Roberts AB. Incisional wound healing in transforming growth factor-beta1 null mice. Wound Repair Regen 2000;8:179–191.
15. Blavier L, Lazaryev A, Dorey F, Shackleford GM, DeClerck YA. Matrix metalloproteinases play an active role in Wnt1-induced mammary tumorigenesis. Cancer Res 2006;66:2691–2699.
16. Wakefield LM, Letterio JJ, Geiser AG, et al. Transforming growth factor-beta s in mammary tumorigenesis: promoters or antipromoters? Prog Clin Biol Res 1995;391:133-148.
17. Egeblad M, Werb Z. New functions for the matrix metalloproteinases in cancer progression. Nat Rev Cancer 2002;2:161–174.
18. Visse R, Nagase H. Matrix metalloproteinases and tissue inhibitors of metalloproteinases: structure, function, and biochemistry. Circ Res 2003;92:827–839.
19. Overall CM. Molecular determinants of metalloproteinase substrate specificity: matrix metalloproteinase substrate binding domains, modules, and exosites. Mol Biotechnol 2002;22:51–86.
20. Bode W, Maskos K. Structural basis of the matrix metalloproteinases and their physiological inhibitors, the tissue inhibitors of metalloproteinases. Biol Chem 2003;384:863–872.
21. Brew K, Dinakarpandian D, Nagase H. Tissue inhibitors of metalloproteinases: evolution, structure and function. Biochim Biophys Acta 2000;1477:267–283.
22. Takahashi C, Sheng Z, Horan TP, et al. Regulation of matrix metalloproteinase-9 and inhibition of tumor invasion by the membrane-anchored glycoprotein RECK. Proc Natl Acad Sci USA 1998;95: 13,221–13,226.
23. Oh J, Takahashi R, Kondo S, et al. The membrane-anchored MMP inhibitor RECK is a key regulator of extracellular matrix integrity and angiogenesis. Cell 2001;107:789–800.
24. Seo DW, Li H, Guedez L, et al. TIMP-2 mediated inhibition of angiogenesis: an MMP-independent mechanism. Cell 2003;114:171–180.
25. Hoegy SE, Oh HR, Corcoran ML, Stetler-Stevenson WG. Tissue inhibitor of metalloproteinases-2 (TIMP-2) suppresses TKR-growth factor signaling independent of metalloproteinase inhibition. J Biol Chem 2001;276:3203–3214.
26. Fedarko NS, Jain A, Karadag A, Fisher LW. Three small integrin binding ligand N-linked glycoproteins (SIBLINGs) bind and activate specific matrix metalloproteinases. FASEB J 2004;18:734–736.
27. Bergers G, Brekken R, McMahon G, et al. Matrix metalloproteinase-9 triggers the angiogenic switch during carcinogenesis. Nat Cell Biol 2000;2:737–744.
28. Whitelock JM, Murdoch AD, Iozzo RV, Underwood PA. The degradation of human endothelial cell-derived perlecan and release of bound basic fibroblast growth factor by stromelysin, collagenase, plasmin, and heparanases. J Biol Chem 1996;271:10079–10086.
29. Manes S, Llorente M, Lacalle RA, et al. The matrix metalloproteinase-9 regulates the insulin-like growth factor-triggered autocrine response in DU-145 carcinoma cells. J Biol Chem 1999;274: 6935–6945.
30. Manes S, Mira E, Barbacid MM, et al. Identification of insulin-like growth factor-binding protein-1 as a potential physiological substrate for human stromelysin-3. J Biol Chem 1997;272:25706–25712.
31. Annes JP, Munger JS, Rifkin DB. Making sense of latent TGFbeta activation. J Cell Sci 2003;116: 217–224.

32. Roberts AB. Molecular and cell biology of TGF-beta. Miner Electrolyte Metab 1998;24:111–119.
33. Roberts AB, Wakefield LM. The two faces of transforming growth factor beta in carcinogenesis. Proc Natl Acad Sci U S A 2003;100:8621–8623.
34. Wakefield LM, Roberts AB. TGF-beta signaling: positive and negative effects on tumorigenesis. Curr Opin Genet Dev 2002;12:22–29.
35. Hyytiäinen M, Penttinen C, Keski-Oja J. Latent TGF-beta binding proteins: extracellular matrix association and roles in TGF-beta activation. Crit Rev Clin Lab Sci 2004;41:233–264.
36. Koli K, Saharinen J, Hyytiäinen M, Penttinen C, Keski-Oja J Latency, activation, and binding proteins of TGF-beta. Microsc Res Tech 2001;52:354–362.
37. Munger JS, Harpel JG, Gleizes PE, Mazzieri R, Nunes I, Rifkin DB. Latent transforming growth factor-beta: structural features and mechanisms of activation. Kidney Int 1997;51: 1376–1382.
38. Massagué J. How cells read TGF-beta signals. Nat Rev Mol Cell Biol 2000;1:169–178.
39. Wrana JL, Attisano L, Wieser R, Ventura F, Massagué J Mechanism of activation of the TGF-beta receptor. Nature 1994;370:341–347.
40. Heldin C-H, Miyazono K, ten Dijke P. TGF-beta signalling from cell membrane to nucleus through SMAD proteins. Nature 1997;390:465–471.
41. Mori S, Matsuzaki K, Yoshida K, et al. TGF-beta and HGF transmit the signals through JNK-dependent Smad2/3 phosphorylation at the linker regions. Oncogene 2004;23:7416–7429.
42. Kamaraju AK, Roberts AB. Role of Rho/ROCK and p38 MAP kinase pathways in transforming growth factor-beta-mediated Smad-dependent growth inhibition of human breast carcinoma cells in vivo. J Biol Chem 2005;280:1024–1036.
43. Maeda S, Dean DD, Gomez R, Schwartz Z, Boyan BD. The first stage of transforming growth factor beta1 activation is release of the large latent complex from the extracellular matrix of growth plate chondrocytes by matrix vesicle stromelysin-1 (MMP-3). Calcif Tissue Int 2002;70:54–65.
44. Dallas SL, Rosser JL, Mundy GR, Bonewald LF. Proteolysis of latent transforming growth factor-beta (TGF-beta)-binding protein-1 by osteoclasts. A cellular mechanism for release of TGF-beta from bone matrix. J Biol Chem 2002;277:21,352–21,360.
45. Mu D, Cambier S, Fjellbirkeland L, et al. The integrin alpha(v)beta8 mediates epithelial homeostasis through MT1-MMP-dependent activation of TGF-beta1. J Cell Biol 2002;157:493–507.
46. Yu Q, Stamenkovic I. Cell surface-localized matrix metalloproteinase-9 proteolytically activates TGF-beta and promotes tumor invasion and angiogenesis. Genes Dev 2000;14:163–176.
47. Yu Q, Stamenkovic I. Localization of matrix metalloproteinase 9 to the cell surface provides a mechanism for CD44-mediated tumor invasion. Genes Dev 1999;13:35–48.
48. Annes JP, Chen Y, Munger JS, Rifkin DB. Integrin alphaVbeta6-mediated activation of latent TGF-beta requires the latent TGF-beta binding protein-1. J Cell Biol 2004;165:723-734.
49. Annes JP, Rifkin DB, Munger JS. The integrin alphaVbeta6 binds and activates latent TGFbeta3. FEBS Lett 2002;511:65–68.
50. Brooks PC, Stromblad S, Sanders LC, et al. Localization of matrix metalloproteinase MMP-2 to the surface of invasive cells by interaction with integrin alpha v beta 3. Cell 1996;85:683–693.
51. Imai K, Hiramatsu A, Fukushima D, Pierschbacher MD, Okada Y. Degradation of decorin by matrix metalloproteinases: identification of the cleavage sites, kinetic analyses and transforming growth factor-beta1 release. Biochem J 1997;322 (Pt 3):809–814.
52. Feige JJ, Negoescu A, Keramidas M, Souchelnitskiy S, Chambaz EM. Alpha 2-macroglobulin: a binding protein for transforming growth factor-beta and various cytokines. Horm Res 1996;45: 227–232.
53. Duivenvoorden WC, Hirte HW, Singh G. Transforming growth factor beta1 acts as an inducer of matrix metalloproteinase expression and activity in human bone-metastasizing cancer cells. Clin Exp Metastasis 1999;17:27–34.
54. Rooprai HK, Rucklidge GJ, Panou C, Pilkington GJ. The effects of exogenous growth factors on matrix metalloproteinase secretion by human brain tumour cells. Br J Cancer 2000;82:52–55.
55. Javelaud D, Delmas V, Moller M, et al. Stable overexpression of Smad7 in human melanoma cells inhibits their tumorigenicity in vitro and in vivo. Oncogene 2005;24:7624–7629.
56. Lin SW, Lee MT, Ke FC, et al. TGFbeta1 stimulates the secretion of matrix metalloproteinase 2 (MMP2) and the invasive behavior in human ovarian cancer cells, which is suppressed by MMP inhibitor BB3103. Clin Exp Metastasis 2000;18:493–499.
57. Ellenrieder V, Hendler SF, Ruhland C, Boeck W, Adler G, Gress TM. TGF-beta-induced invasiveness of pancreatic cancer cells is mediated by matrix metalloproteinase-2 and the urokinase plasminogen activator system. Int J Cancer 2001;93:204–211.

58. Kim ES, Kim MS, Moon A. TGF-beta-induced upregulation of MMP-2 and MMP-9 depends on p38 MAPK, but not ERK signaling in MCF10A human breast epithelial cells. Int J Oncol 2004;25: 1375–1382.
59. Sehgal I, Thompson TC. Novel regulation of type IV collagenase (matrix metalloproteinase-9 and -2) activities by transforming growth factor-beta1 in human prostate cancer cell lines. Mol Biol Cell 1999;10:407–416.
60. Wick W, Platten M, Weller M. Glioma cell invasion: regulation of metalloproteinase activity by TGF-beta. J Neurooncol 2001;53:177–185.
61. Ogawa K, Chen F, Kuang C, Chen Y. Suppression of matrix metalloproteinase-9 transcription by transforming growth factor-beta is mediated by a nuclear factor-kappaB site. Biochem J 2004;381: 413–422.
62. White LA, Mitchell TI, Brinckerhoff CE. Transforming growth factor beta inhibitory element in the rabbit matrix metalloproteinase-1 (collagenase-1) gene functions as a repressor of constitutive transcription. Biochim Biophys Acta 2000;1490:259–268.
63. Benbow U, Brinckerhoff CE. The AP-1 site and MMP gene regulation: what is all the fuss about? Matrix Biol 1997;15:519–526.
64. Jin G, Howe PH. Transforming growth factor beta regulates clusterin gene expression via modulation of transcription factor c-Fos. Eur J Biochem 1999;263:534–542.
65. Bergman CA, Talavera F, Christman GM, Baker VV, Roberts JA, Menon KM. Transforming growth factor-beta negatively modulates proliferation and c-fos expression of the human endometrial adenocarcinoma cell line HEC-1-A. Gynecol Oncol 1997;65:63–68.
66. Wong C, Rougier-Chapman EM, Frederick JP, et al. Smad3-Smad4 and AP-1 complexes synergize in transcriptional activation of the c-Jun promoter by transforming growth factor beta. Mol Cell Biol 1999;19:1821–1830.
67. Duivenvoorden WC, Hirte HW, Singh G. Transforming growth factor beta1 acts as an inducer of matrix metalloproteinase expression and activity in human bone-metastasizing cancer cells. Clin Exp Metastasis 1999;17:27–34.
68. Yuan W, Varga J. Transforming growth factor-beta repression of matrix metalloproteinase-1 in dermal fibroblasts involves Smad3. J Biol Chem 2001;276:38,502–38,510.
69. Hall MC, Young DA, Waters JG, et al. The comparative role of activator protein 1 and Smad factors in the regulation of Timp-1 and MMP-1 gene expression by transforming growth factor-beta 1. J Biol Chem 2003;278:10,304–10,313.
70. Kerr LD, Miller DB, Matrisian LM. TGF-beta 1 inhibition of transin/stromelysin gene expression is mediated through a Fos binding sequence. Cell 1990;61:267–278.
71. Fang KC, Wolters PJ, Steinhoff M, Bidgol A, Blount JL, Caughey GH. Mast cell expression of gelatinases A and B is regulated by kit ligand and TGF-beta. J Immunol 1999;162:5528–5535.
72. Shimizu S, Nishikawa Y, Kuroda K, et al. Involvement of transforming growth factor beta1 in autocrine enhancement of gelatinase B secretion by murine metastatic colon carcinoma cells. Cancer Res 1996;56:3366–3370.
73. Stuelten CH, Byfield SD, Arany PR, Karpova TS, Stetler-Stevenson WG, Roberts AB. Breast cancer cells induce stromal fibroblasts to express MMP-9 via secretion of TNF-α and TGF-β. J Cell Sci 2005;118 (Pt 10):2143–2153.
74. Santibanez JF, Guerrero J, Quintanilla M, Fabra A, Martinez J. Transforming growth factor-beta1 modulates matrix metalloproteinase-9 production through the Ras/MAPK signaling pathway in transformed keratinocytes. Biochem Biophys Res Commun 2002;296:267–273.
75. Han YP, Tuan TL, Hughes M, Wu H, Garner WL. Transforming growth factor-beta - and tumor necrosis factor-alpha-mediated induction and proteolytic activation of MMP-9 in human skin. J Biol Chem 2001;276:22,341–22,350.
76. Kalembeyi I, Inada H, Nishiura R, Imanaka-Yoshida K, Sakakura T, Yoshida T. Tenascin-C upregulates matrix metalloproteinase-9 in breast cancer cells: direct and synergistic effects with transforming growth factor beta1. Int J Cancer 2003;105:53–60.
77. Ilunga K, Nishiura R, Inada H, El Karef A, Imanaka-Yoshida K, Sakakura T, Yoshida T. Co-stimulation of human breast cancer cells with transforming growth factor-beta and tenascin-C enhances matrix metalloproteinase-9 expression and cancer cell invasion. Int J Exp Pathol 2004;85:373–379.
78. Tajirian T, Dennis JW, Swallow CJ. Regulation of human monocyte proMMP-9 production by fetuin, an endogenous TGF-beta antagonist. J Cell Physiol 2000;185:174–183.

79. Suarez-Cuervo C, Merrell MA, Watson L, et al. Breast cancer cells with inhibition of p38alpha have decreased MMP-9 activity and exhibit decreased bone metastasis in mice. Clin Exp Metastasis 2004;21:525–533.
80. Overall CM, Wrana JL, Sodek J. Transcriptional and post-transcriptional regulation of 72-kDa gelatinase/type IV collagenase by transforming growth factor-beta 1 in human fibroblasts. Comparisons with collagenase and tissue inhibitor of matrix metalloproteinase gene expression. J Biol Chem 1991;266:14,064–14,071.
81. Su S, DiBattista JA, Sun Y, Li WQ, Zafarullah M. Up-regulation of tissue inhibitor of metalloproteinases-3 gene expression by TGF-beta in articular chondrocytes is mediated by serine/threonine and tyrosine kinases. J Cell Biochem 1998;70:517–527.
82. Edwards DR, Leco KJ, Beaudry PP, Atadja PW, Veillette C, Riabowol KT. Differential effects of transforming growth factor-beta 1 on the expression of matrix metalloproteinases and tissue inhibitors of metalloproteinases in young and old human fibroblasts. Exp Gerontol 1996;31:207–223.
83. Huang HY, Wen Y, Irwin JC, Kruessel JS, Soong YK, Polan ML. Cytokine-mediated regulation of 92-kilodalton type IV collagenase, tissue inhibitor or metalloproteinase-1 (TIMP-1), and TIMP-3 messenger ribonucleic acid expression in human endometrial stromal cells. J Clin Endocrinol Metab 1998;83:1721–1729.
84. Kwak HJ, Park MJ, Cho H, et al. Transforming growth factor-beta1 induces tissue inhibitor of metalloproteinase-1 expression via activation of extracellular signal-regulated kinase and Sp1 in human fibrosarcoma cells. Mol Cancer Res 2006;4:209–220.
85. Ma C, Chegini N. Regulation of matrix metalloproteinases (MMPs) and their tissue inhibitors in human myometrial smooth muscle cells by TGF-beta1. Mol Hum Reprod 1999;5:950–954.
86. Unterbrink A, O'Sullivan M, Chen S, MacDougall M. TGF beta-1 downregulates DMP-1 and DSPP in odontoblasts. Connect Tissue Res 2002;43:354–358.
87. Shi X, Bai S, Li L, Cao X. Hoxa-9 represses transforming growth factor-beta-induced osteopontin gene transcription. J Biol Chem 2001;276:850–855.
88. Saygin NE, Tokiyasu Y, Giannobile WV, Somerman MJ. Growth factors regulate expression of mineral associated genes in cementoblasts. J Periodontol 2000;71:1591–1600.
89. Hullinger TG, Panek RL, Xu X, Karathanasis SK. p21-activated kinase-1 (PAK1) inhibition of the human scavenger receptor class B, type I promoter in macrophages is independent of PAK1 kinase activity, but requires the GTPase-binding domain. J Biol Chem 2001;276:46,807–46,814.
90. Ogata Y, Niisato N, Furuyama S, et al. Transforming growth factor-beta 1 regulation of bone sialoprotein gene transcription: identification of a TGF-beta activation element in the rat BSP gene promoter. J Cell Biochem 1997;65:501–512.
91. Nam JS, Suchar AM, Kang MJ, et al. Bone sialoprotein mediates the tumor cell-targeted prometastatic activity of transforming growth factor β in a mouse model of breast cancer. Cancer Res 2006;66:6327–6335.
92. Turpeenniemi-Hujanen T. Gelatinases (MMP-2 and -9) and their natural inhibitors as prognostic indicators in solid cancers. Biochimie 2005;87:287–297.
93. Desruisseau S, Palmari J, Giusti C, Romain S, Martin PM, Berthois Y. Determination of TGFbeta1 protein level in human primary breast cancers and its relationship with survival. Br J Cancer 2006;94: 239–246.
94. Okumoto K, Hattori E, Tamura K, et al. Possible contribution of circulating transforming growth factor-beta1 to immunity and prognosis in unresectable hepatocellular carcinoma. Liver Int 2004;24: 21–28.
95. Ivanovic V, Demajo M, Krtolica K, et al. Elevated plasma TGF-beta(1) levels correlate with decreased survival of metastatic breast cancer patients. Clin Chim Acta 2006;371(1-2):191–193.
96. Kubben FJ, Sier CF, van Duijn W, et al. Matrix metalloproteinase-2 is a consistent prognostic factor in gastric cancer. Br J Cancer 2006;94:1035–1040.
97. Ruokolainen H, Paakko P, Turpeenniemi-Hujanen T. Serum matrix metalloproteinase-9 in head and neck squamous cell carcinoma is a prognostic marker. Int J Cancer 2005;116:422–427.
98. Ruokolainen H, Paakko P, Turpeenniemi-Hujanen T. Tissue and circulating immunoreactive protein for MMP-2 and TIMP-2 in head and neck squamous cell carcinoma—tissue immunoreactivity predicts aggressive clinical course. Mod Pathol 2006;19:208–217.
99. Nikkola J, Vihinen P, Vuoristo MS, Kellokumpu-Lehtinen P, Kahari VM, Pyrhonen S. High serum levels of matrix metalloproteinase-9 and matrix metalloproteinase-1 are associated with rapid progression in patients with metastatic melanoma. Clin Cancer Res 2005;11:5158–5166.

100. Gentilini F, Calzolari C, Turba ME, Agnoli C, Fava D, Forni M, Bergamini PF. Prognostic value of serum vascular endothelial growth factor (VEGF) and plasma activity of matrix metalloproteinase (MMP) 2 and 9 in lymphoma-affected dogs. Leuk Res 2005;29:1263–1269.
101. Fukuchi M, Miyazaki T, Fukai Y, et al. Plasma level of transforming growth factor beta1 measured from the azygos vein predicts prognosis in patients with esophageal cancer. Clin Cancer Res 2004; 10:2738–2741.
102. Jodele S, Blavier L, Yoon JM, DeClerck YA. Modifying the soil to affect the seed: role of stromal-derived matrix metalloproteinases in cancer progression. Cancer Metastasis Rev 2006;25:35–43.
103. Scorilas A, Karameris A, Arnogiannaki N, et al. Overexpression of matrix-metalloproteinase-9 in human breast cancer: a potential favourable indicator in node-negative patients. Br J Cancer 2001;84: 1488–1496.
104. Giannelli G, Fransvea E, Marinosci F, et al. Gelatinase levels in male and female breast cancer. Biochem Biophys Res Commun 2002;292:161–166.
105. Wang TN, Albo D, Tuszynski GP. Fibroblasts promote breast cancer cell invasion by upregulating tumor matrix metalloproteinase-9 production. Surgery 2002;132:220–225.
106. Singer CF, Kronsteiner N, Marton E, et al. MMP-2 and MMP-9 expression in breast cancer-derived human fibroblasts is differentially regulated by stromal-epithelial interactions. Breast Cancer Res Treat 2002;72:69–77.
107. Petersen OW, Nielsen HL, Gudjonsson T, et al. Epithelial to mesenchymal transition in human breast cancer can provide a nonmalignant stroma. Am J Pathol 2003;162:391–402.
108. Saad S, Bendall LJ, James A, Gottlieb DJ, Bradstock KF. Induction of matrix metalloproteinases MMP-1 and MMP-2 by co-culture of breast cancer cells and bone marrow fibroblasts. Breast Cancer Res Treat 2000;63:105–115.
109. Saad S, Gottlieb DJ, Bradstock KF, Overall CM, Bendall LJ. Cancer cell-associated fibronectin induces release of matrix metalloproteinase-2 from normal fibroblasts. Cancer Res 2002;62: 283–289.
110. Mook OR, Van Overbeek C, Ackema EG, Van Maldegem F, Frederiks WM. In situ localization of gelatinolytic activity in the extracellular matrix of metastases of colon cancer in rat liver using quenched fluorogenic DQ-gelatin. J Histochem Cytochem 2003;51:821–829.
111. Biswas C. Collagenase stimulation in cocultures of human fibroblasts and human tumor cells. Cancer Lett 1984;24:201–207.
112. Kataoka H, DeCastro R, Zucker S, Biswas C. Tumor cell-derived collagenase-stimulatory factor increases expression of interstitial collagenase, stromelysin, and 72-kDa gelatinase. Cancer Res 1993;53:3154–3158.
113. Johansson N, Ala-aho R, Uitto V, Grenman R, Fusenig NE, Lopez-Otin C, Kahari VM. Expression of collagenase-3 (MMP-13) and collagenase-1 (MMP-1) by transformed keratinocytes is dependent on the activity of p38 mitogen-activated protein kinase. J Cell Sci 2000;113 Pt 2:227–235.
114. Leivonen SK, Ala-aho R, Koli K, Grenman R, Peltonen J, Kahari VM. Activation of Smad signaling enhances collagenase-3 (MMP-13) expression and invasion of head and neck squamous carcinoma cells. Oncogene 2006;25:2588–2600.
115. Koshikawa N, Giannelli G, Cirulli V, Miyazaki K, Quaranta V. Role of cell surface metalloprotease MT1-MMP in epithelial cell migration over laminin-5. J Cell Biol 2000;148:615–624.
116. Giannelli G, Brassard J, Foti C, et al. Altered expression of basement membrane proteins and their integrin receptors in lichen planus: possible pathogenetic role of gelatinases A and B. Lab Invest 1996;74:1091–1104.
117. Udayakumar TS, Chen ML, Bair EL, et al. Membrane type-1-matrix metalloproteinase expressed by prostate carcinoma cells cleaves human laminin-5 beta3 chain and induces cell migration. Cancer Res 2003;63:2292–2299.
118. Thomas GJ, Hart IR, Speight PM, Marshall JF. Binding of TGF-beta1 latency-associated peptide (LAP) to alpha(v)beta6 integrin modulates behaviour of squamous carcinoma cells. Br J Cancer 2002;87:859–867.
119. Munger JS, Harpel JG, Giancotti FG, Rifkin DB. Interactions between growth factors and integrins: latent forms of transforming growth factor-beta are ligands for the integrin alphavbeta1. Mol Biol Cell 1998;9:2627–2638.
120. Kanzaki T, Otabe M. Latent transforming growth factor-beta binding protein-1, a component of latent transforming growth factor-beta complex, accelerates the migration of aortic smooth muscle cells in diabetic rats through integrin-beta3. Diabetes 2003;52:824–828.

121. Vehvilainen P, Hyytiainen M, Keski-Oja J. Latent transforming growth factor-beta-binding protein 2 is an adhesion protein for melanoma cells. J Biol Chem 2003;278:24,705–24,713.
122. Lynch CC, Matrisian LM. Matrix metalloproteinases in tumor-host cell communication. Differentiation 2002;70:561–573.
123. Hornebeck W, Maquart FX. Proteolyzed matrix as a template for the regulation of tumor progression. Biomed Pharmacother 2003;57:223–230.
124. Noe V, Fingleton B, Jacobs K, et al. Release of an invasion promoter E-cadherin fragment by matrilysin and stromelysin-1. J Cell Sci 2001;114:111–118.
125. Davies G, Jiang WG, Mason MD. Matrilysin mediates extracellular cleavage of E-cadherin from prostate cancer cells: a key mechanism in hepatocyte growth factor/scatter factor-induced cell-cell dissociation and in vitro invasion. Clin Cancer Res 2001;7:3289–3297.
126. Vincenti MP, Brinckerhoff CE. Transcriptional regulation of collagenase (MMP-1, MMP-13) genes in arthritis: integration of complex signaling pathways for the recruitment of gene-specific transcription factors. Arthritis Res 2002;4(3):157–164.

18 Cooperation Between TGF-β and Wnt Pathway Components in Regulating Mesenchymal Stem Cell Function

Irwin M. Liu, Stephen H. Schilling, and Xiao-Fan Wang

CONTENTS

INTRODUCTION
OVERVIEW OF THE TGF-β SIGNALING PATHWAY
OVERVIEW OF THE WNT SIGNALING PATHWAY
REGULATION OF MESENCHYMAL STEM CELLS BY THE TGF-β AND WNT SIGNALING PATHWAYS
CROSSTALK BETWEEN THE TGF-β AND WNT SIGNALING PATHWAYS
TGF-β STIMULATES NUCLEAR ACCUMULATION OF β-CATENIN IN MESENCHYMAL STEM CELLS
SMAD3 IS REQUIRED FOR TGF-β-STIMULATED TRANSLOCATION OF β-CATENIN TO THE NUCLEUS
NUCLEAR TRANSLOCATION OF β-CATENIN IS REQUIRED FOR EFFECTS OF TGF-β ON MESENCHYMAL STEM CELLS
β-CATENIN AND SMAD3 COOPERATIVELY REGULATE SELECT TARGET GENES IN MESENCHYMAL STEM CELLS
CONCLUDING REMARKS
ACKNOWLEDGMENTS
REFERENCES

Abstract

Stem cells have become a subject of intensive research based on their potential for varied clinical applications. One such stem cell, the mesenchymal stem cell, has attracted attention owing to its ability to differentiate into several mesenchymal lineages such as bone, cartilage, and fat. The molecular pathways which regulate the functions of mesenchymal stem cells are not well-characterized, but the TGF-β and Wnt signaling pathways are thought to play critical roles. Here, we describe the importance of TGF-β and Wnt signaling pathway crosstalk in governing mesenchymal stem cell function.

Key Words: Mesenchymal stem cells; Smad3; β-catenin; crosstalk; proliferation; differentiation.

From: *Cancer Drug Discovery and Development: Transforming Growth Factor-β in Cancer Therapy, Vol I: Basic and Clinical Biology*
Edited by: Sonia B. Jakowlew © Humana Press Inc., Totowa, NJ

1. INTRODUCTION

Stem cells are a unique type of cell with the capacity not only to self-renew, but also to differentiate into cell types of multiple different lineages. Although embryonic stem cells have the greatest plasticity, many adult tissues also contain stem cells that have the ability to differentiate into more defined subsets of cell types in situations such as trauma or aging. One of these types of adult stem cells is the mesenchymal stem cell (MSC). MSCs are multipotent cells present in the adult bone marrow with the potential to differentiate into mesenchymal lineages including bone, cartilage, and fat *(1)*. Owing to the potential plasticity and self-renewal capacity of MSCs in tissue culture, many clinical applications are now emerging for MSCs, including many that would aid in the treatment of cancer. Obviously, the ability of MSCs to differentiate into a number of mesenchymal lineages makes tissue engineering applications possible, such as the reconstruction of bone following resection of malignant bone tumors *(2,3)*. However, the clinical potential of these cells is much more wide-ranging than just tissue engineering. For example, there is evidence that MSCs can be used to improve the rate and quality of hematopoietic engraftment in cancer patients following stromal-damaging treatments *(4)*. Furthermore, the predisposition for MSCs to migrate to microenvironments that are injured or stressed may be able to be exploited to deliver therapies to tumors *(5,6)*. Despite this increasing interest in MSCs, much remains unknown about what regulates the self-renewal and cell fate determination of these cells. To truly harness the clinical potential of MSCs, a more detailed understanding of the mechanisms driving the proliferation and differentiation of these cells is needed. Here, we review how a unique form of crosstalk between the TGF-β and Wnt signaling pathways serves to regulate some of these MSC functions.

2. OVERVIEW OF THE TGF-β SIGNALING PATHWAY

The TGF-β signaling pathway is initiated upon binding of the TGF-β ligand to its receptors at the cell surface *(7)* (Fig. 1). This binding induces phosphorylation of the type I receptor by the type II receptor kinase, thereby activating the type I receptor. The activated type I receptor then phosphorylates the receptor-regulated Smads, Smad2 and Smad3, increasing their affinity for Smad4. The resulting Smad complexes then translocate to the nucleus, where they regulate a number of target genes and induce a biological effect.

3. OVERVIEW OF THE WNT SIGNALING PATHWAY

In the absence of a Wnt signal, β-catenin is phosphorylated and targeted for degradation by a complex containing adenomatous polyposis coli (APC), axin, and glycogen synthase kinase 3β (GSK3β), and casein kinase I epsilon (CKIε) *(8)* (Fig. 2 left). Upon Wnt binding to the frizzled receptor (Fz) and low-density lipoprotein receptor-related protein (LRP) coreceptors, activation of downstream components such as the axin-binding molecule Dishevelled (DVL) and the LRP coreceptors leads to disruption of APC/Axin/GSK3β complex-mediated phosphorylation of β-catenin (Fig. 2 right). Hypophosphorylated β-catenin then accumulates and translocates to the nucleus, where it interacts with T-cell factor/lymphoid enhancer binding factor (TCF/LEF) transcription factors to regulate target genes.

4. REGULATION OF MESENCHYMAL STEM CELLS BY THE TGF-β AND WNT SIGNALING PATHWAYS

Among the many active signaling pathways in the bone marrow, the TGF-β signaling pathway is one of the most important. The first hint that TGF-β might be an important factor in this context came with the discovery that, although it is found in many tissue types,

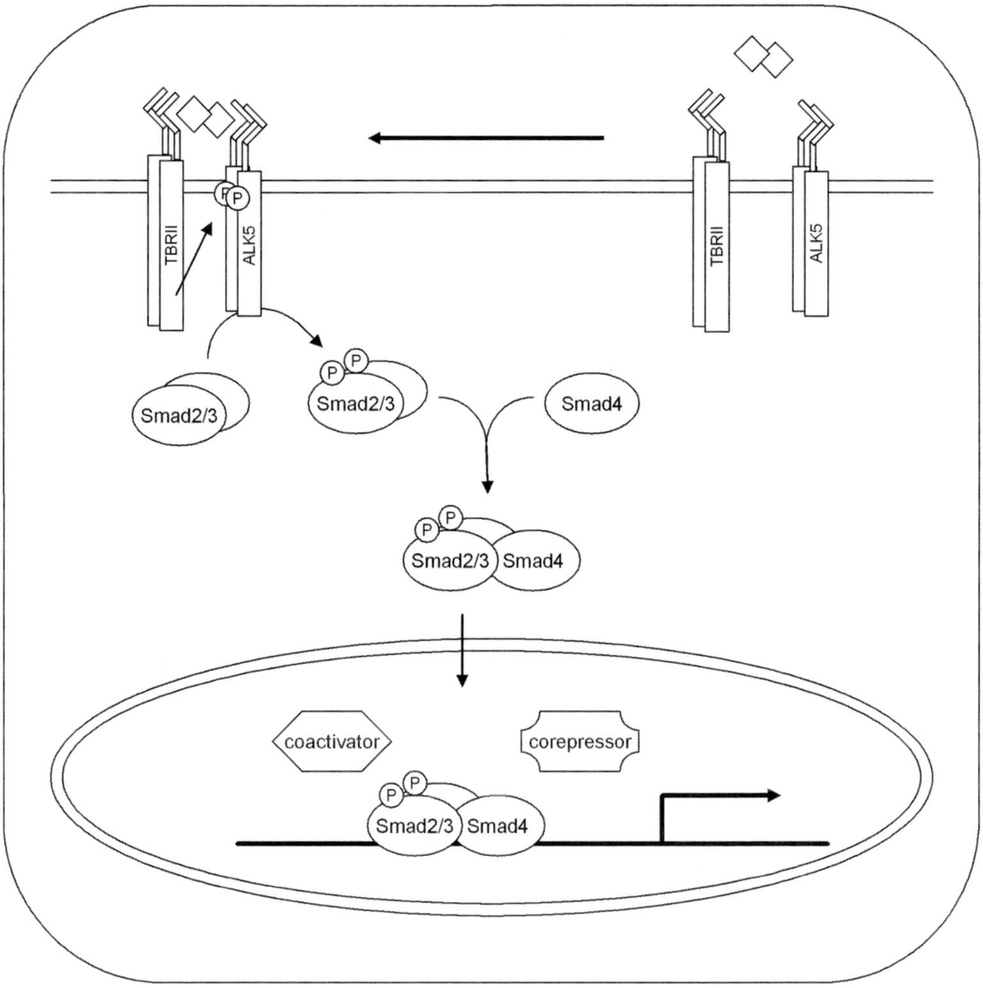

Fig. 1. The canonical TGF-β signaling pathway.

TGF-β is found at the highest level in the bone *(9)*. Not surprisingly, TGF-β was subsequently shown to regulate the function of several types of cells in this environment, including mesenchymal progenitor cells. Not only does TGF-β induce proliferation of these cells *(8)*, but it also regulates their differentiation, inhibiting osteoblast and adipocyte differentiation and influencing the differentiation of other mesenchymal lineages as well *(11–14)*.

The Wnt signaling pathway also contributes to the regulation of mesenchymal progenitor cells in the bone marrow. Surprisingly, despite having opposite effects in many other cell types *(15)*, the downstream effects of the Wnt and TGF-β signaling pathways largely mirror each other in MSCs. Wnt, like TGF-β, has antiosteogenic and antiadipogenic effects, as well as proliferative effects on MSCs *(16–18)*.

In accordance with these previous studies using various primary mesenchymal progenitor cells and cell lines, we found that the TGF-β and Wnt signaling pathways exert similar effects on primary human mesenchymal stem cells. To study activation of the Wnt pathway, a constitutively active mutant of β-catenin was used. This mutant contains alanine substitutions at the four serine phosphorylation sites, thereby preventing phosphorylation and the subsequent ubiquitin-mediated degradation of β-catenin *(19)*. When ectopically expressed in MSCs,

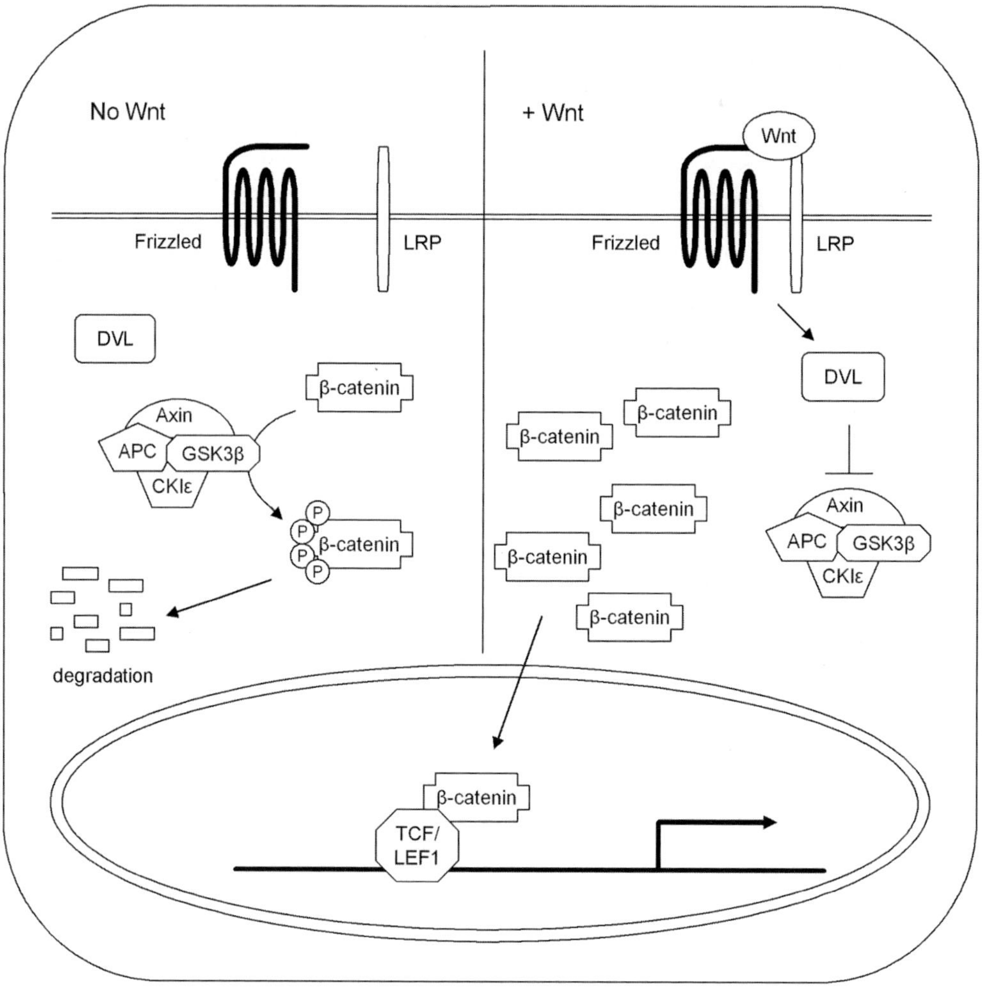

Fig. 2. The canonical Wnt signaling pathway.

this constitutively active β-catenin mutant is predominantly localized in the nucleus *(20)*. As expected, based on the proliferative effects of Wnt in most other contexts *(8)*, ectopic expression of this mutant induces proliferation in MSCs *(20)*. Overexpression of this β-catenin mutant also blocks the ability of MSCs to differentiate into osteoblasts in the presence of osteogenic supplement media *(20)*. Similar effects were observed in the presence of TGF-β. TGF-β treatment inhibits osteogenic differentiation and induces proliferation in MSCs *(20)*. Taken together, these data suggest that the TGF-β and Wnt pathways cooperate to regulate the function of MSCs.

5. CROSSTALK BETWEEN THE TGF-β AND WNT SIGNALING PATHWAYS

It has been known for many years that TGF-β induces proliferation of mesenchymal progenitor cells *(10)*. However, in contrast to the extensive data on the mechanism by which TGF-β potently inhibits the proliferation of most other cell types *(7)*, the downstream events mediating the induction of proliferation by TGF-β are largely unknown. Because the TGF-β and Wnt signaling pathways have very similar effects in the context of MSCs, we speculated that there might be crosstalk between these two pathways in this context.

Several different levels of crosstalk between these two pathways have been described in other cell types and systems. In the nucleus, TGF-β-activated Smads have been shown to cooperate with β-catenin or its nuclear binding partners TCF/LEF1 to induce the expression of downstream target genes such as gastrin *(21)* or *Xtwn* during mouse development *(22,23)*. However, crosstalk between these two pathways is not limited to the nucleus and has been shown further upstream as well. During chondrocyte differentiation, TGF-β has been shown to increase stability and nuclear accumulation of β-catenin *(24,25)*. Additionally, Axin, a member of the β-catenin destruction complex, has been shown to interact with Smad3, thereby modulating TGF-β signaling intensity *(26)* and Smad3 turnover (Guo, X. and Wang, X. F., unpublished data). Taken together, these previous studies indicate that the TGF-β and Wnt signaling pathways can intersect at many critical junctions to alter the outputs of their respective signals.

6. TGF-β STIMULATES NUCLEAR ACCUMULATION OF β-CATENIN IN MESENCHYMAL STEM CELLS

Given the enrichment of TGF-β and Wnt in the bone microenvironment *(9)*, the results of the experiments described above, and the precedent for crosstalk between the TGF-β and Wnt signaling pathways, it seemed likely that crosstalk between these pathways could be an important determinant of the ability of TGF-β to induce proliferation and inhibit differentiation of MSCs. To first confirm the existence of canonical Wnt signaling in MSCs, we monitored β-catenin localization in Wnt-treated MSCs. Using a nuclear/cytoplasmic fractionation assay, we observed the expected accumulation of β-catenin in the nucleus following a 6 h treatment with Wnt3A. We also found that β-catenin accumulated in the nucleus within 2 h of TGF-β treatment *(20)*. Using immunofluorescence imaging, we confirmed the nuclear localization of β-catenin following TGF-β treatment in greater than 90% of MSCs. We were particularly intrigued by this rapid TGF-β-induced effect because it differed from the slower TGF-β-induced nuclear accumulation of β-catenin described previously *(24,25)*.

We next examined whether TGF-β induces this effect through mechanisms parallel to Wnt-induced regulation of β-catenin. We found that treatment of MSCs with Wnt3A, but not TGF-β, elevates steady-state levels of β-catenin *(20)* (Fig. 3A). This suggests that, unlike Wnt, TGF-β does not induce nuclear localization of β-catenin by stabilizing it. Because TGF-β is known to induce the expression of various Wnt isoforms in bone marrow progenitor cells *(24,25)*, one possible explanation is that the TGF-β-stimulated nuclear accumulation of β-catenin we observed is an indirect autocrine effect mediated predominantly by Wnt ligand induction and signaling (Fig. 3B). However, we determined that this scenario is unlikely. By pretreating MSCs with the protein translation inhibitor cycloheximide (CHX) prior to addition of TGF-β, we demonstrated that the accumulation of β-catenin does not require *de novo* protein synthesis *(20)*. Additionally, we tested the effect of two inhibitors of Wnt signaling: Fz8CRD, a soluble dominant negative inhibitor of the Wnt receptor Frizzled *(27)*, and DVL-ΔPDZ, a dominant negative form of Dishevelled *(28)*. Addition of Fz8CRD to the culture medium of MSCs or ectopic expression of DVL-ΔPDZ in MSCs inhibited nuclear accumulation of β-catenin in response to Wnt3A, but not TGF-β *(20)*. Taken together, these data indicate that the ability of TGF-β treatment to induce β-catenin nuclear accumulation in MSCs is independent of Wnt ligand production and upstream activators of canonical Wnt signaling. Interestingly, we did not observe the rapid TGF-β-stimulated nuclear accumulation of β-catenin in other cell types, such as HaCaT

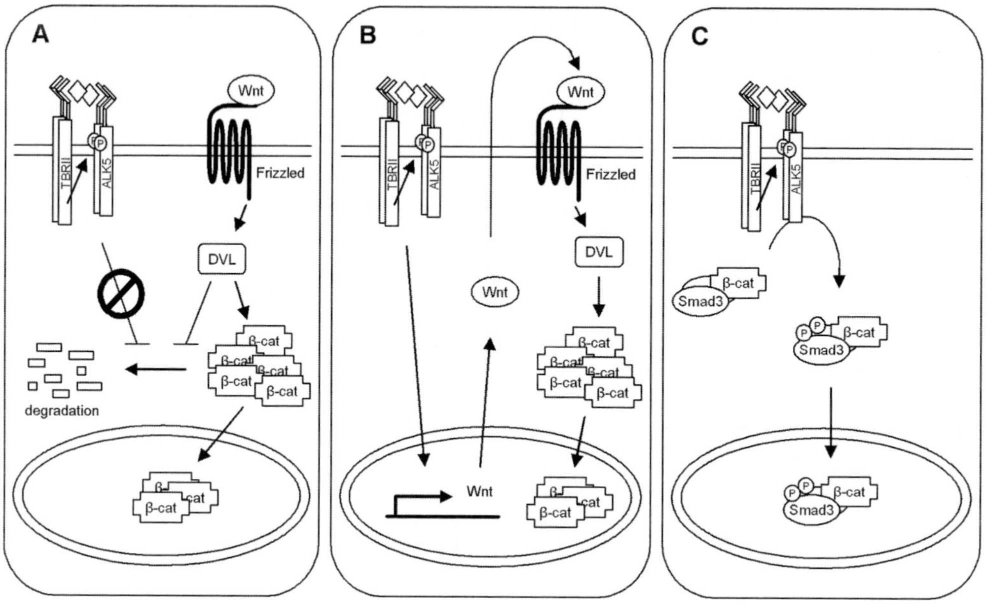

Fig. 3. Possible mechanisms of TGF-β stimulated nuclear accumulation of β-catenin.

human keratinocytes or BJ human fibroblasts, suggesting that this effect may be found only in permissive cellular contexts such as MSCs.

7. SMAD3 IS REQUIRED FOR TGF-β-STIMULATED TRANSLOCATION OF β-CATENIN TO THE NUCLEUS

To further investigate the mechanism by which TGF-β induces nuclear accumulation of β-catenin, we addressed the possibility that β-catenin nuclear accumulation may be a direct consequence of type I TGF-β receptor (TβRI) activation by TGF-β (Fig. 3C). To do this, we treated MSCs with SD208, a small molecule TβRI kinase inhibitor *(29)*, prior to the addition of TGF-β. SD208 effectively inhibited canonical TGF-β signaling, as revealed by the absence Smad2 phosphorylation, and also dose-dependently inhibited accumulation of β-catenin in the nucleus *(20)*, indicating that TβRI activity is required for this effect.

Because Smad3 is a primary effector of the TGF-β signaling pathway, we tested whether the nuclear accumulation of β-catenin in response to TGF-β is Smad3-dependent through the use of shRNA. Infection of MSCs with a pSUPER-retro shRNA construct targeting Smad3 resulted in greater than 90% knockdown of Smad3 protein levels *(20)*. Interestingly, Smad3 knockdown inhibits the accumulation of β-catenin in the nucleus in response to TGF-β but not Wnt3A *(20)*. This result indicates a requirement for Smad3 and further reinforces the notion that the TGF-β-stimulated effect mechanistically distinct from previously described mechanisms of β-catenin nuclear accumulation.

Previous reports have shown that Smads interact with β-catenin and its binding partners *(22,23,26,30–32)*. We also found that Smad3 interacts with β-catenin in MSCs and that TGF-β treatment does not appear to significantly disrupt this association. This observation, taken together with the well-characterized translocation of Smad3 to the nucleus following TGF-β treatment, suggests that activated Smad3 may cooperatively shuttle with β-catenin into the nucleus. This seems likely given that other studies have implicated a similar function for Smad3 when preassembled in complexes containing ATF3 or E2F4/5/DP/p107 *(33,34)*.

8. NUCLEAR TRANSLOCATION OF β-CATENIN IS REQUIRED FOR EFFECTS OF TGF-β ON MESENCHYMAL STEM CELLS

To more directly address the possibility that β-catenin is responsible for the TGF-β-stimulated enhancement of proliferation and inhibition of differentiation in MSCs, we attempted to knock down β-catenin levels next. Several stable shRNA expression constructs were generated using targeting sequences designed by our lab and others *(35,36)*. However, this approach did not produce a significant level of knockdown, perhaps owing to the relatively greater stability of β-catenin in MSCs as compared to other cell types *(20)*.

In the absence of a reliable gene silencing approach, we decided to study the function of β-catenin in TGF-β-stimulated MSC responses by excluding β-catenin from the nucleus. To accomplish this, we used LEF1ΔC, a C-terminal truncation mutant of the LEF1 protein (Fig. 4, top). LEF1 is a transcription factor known to complex with Smad3 and β-catenin and coordinately regulates the transcription of various target genes *(37)*. Whereas the LEF1/Smad3 interaction is mediated by the HMG box of LEF1 *(23)*, the LEF1/β-catenin interaction is mediated by the N-terminal region of LEF1 *(38)*. Notably, the LEF1ΔC mutant lacks the HMG box and the nuclear localization signal of LEF1 *(39)*, but retains the β-catenin interaction site in the N-terminus. Thus, when overexpressed, the LEF1ΔC mutant is predicted to sequester β-catenin within the cytoplasm without altering the localization of Smad3, thereby functioning as a dominant-negative inhibitor of β-catenin function in the nucleus (Fig. 4, bottom). In agreement with this, we found that LEF1ΔC localizes to the cytoplasm and complexes with β-catenin, but not Smad3 *(20)*. LEF1ΔC also inhibits the nuclear translocation of β-catenin in response to TGF-β *(20)*. Importantly, our data show that overexpression of LEF1ΔC eliminates the ability of TGF-β to stimulate proliferation and inhibit osteogenic differentiation of MSCs *(20)*, thereby suggesting a requirement for β-catenin in these responses. Overall, these data support the idea that the rapid nuclear translocation of β-catenin plays a key role in mediating the TGF-β induced cellular responses in MSCs.

9. β-CATENIN AND SMAD3 COOPERATIVELY REGULATE SELECT TARGET GENES IN MESENCHYMAL STEM CELLS

To identify target genes that lie downstream of this Smad3-dependent β-catenin-mediated pathway, we performed microarray analysis on RNA samples isolated from LEF1ΔC-expressing or control MSCs that were treated with or without TGF-β for 2 h. The expression of several TGF-β-regulated genes is affected by LEF1ΔC expression. Notably, we found that B Lymphoid Tyrosine Kinase (BLK) mRNA expression is induced more than 40-fold by TGF-β in control MSCs, but is induced less than 1.5-fold in LEF1ΔC-expressing MSCs *(20)*, suggesting the potential requirement for β-catenin in the induction of BLK by TGF-β. This result was verified by quantitative real-time RT-PCR analysis. On the other hand, we also found that many TGF-β target genes, including well-characterized TGF-β target genes such as JunB and PAI-I, are not affected by the expression of LEF1ΔC. Therefore, the expression of LEF1ΔC in MSCs alters a small subset of TGF-β target genes, indicating that only some TGF-β target genes in MSCs are β-catenin-dependent. Interestingly, we discovered that well-known β-catenin-inducible growth promoters such as cyclin D1 and c-*myc* are not upregulated by TGF-β in MSCs. It is possible, then, that β-catenin-dependent genes that were identified in this array such as BLK may functionally mediate TGF-β-induced proliferation through novel mechanisms in the unique context of MSCs.

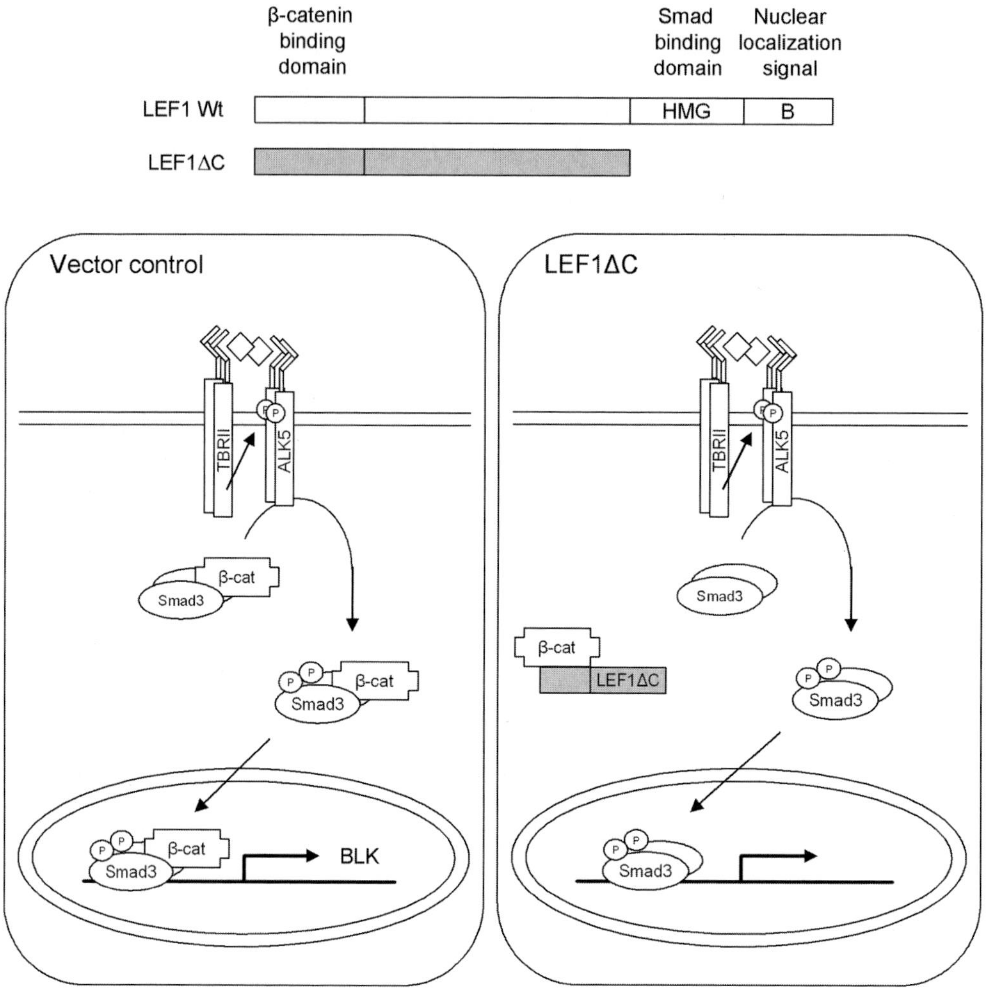

Fig. 4. Effect of LEF1ΔC on TGF-β-stimulated target gene expression in MSCs.

10. CONCLUDING REMARKS

In summary, we have found that TGF-β induces rapid, Smad3-dependent nuclear translocation of β-catenin in MSCs. This event is required for the stimulation of MSC proliferation and the inhibition of MSC osteogenic differentiation by TGF-β (Fig. 5). Intriguingly, similar mechanisms seem to be at play in more differentiated mesenchymal lineages as well. Whereas the absence of β-catenin has no effect on the ability of TGF-β to inhibit the proliferation of keratinocytes, loss of β-catenin abolishes the ability of TGF-β to induce the proliferation of fibroblasts *(40)*. Similarly, the presence of β-catenin is essential for the ability of TGF-β to regulate proliferation in primary murine chondrocytes, another mesenchymal lineage *(41)*. In these cells, TGF-β treatment increases β-catenin protein levels in a Smad3-dependent manner, and β-catenin colocalizes with Smad3 in the nucleus following TGF-β treatment. Once in the nucleus, Smad3 and β-catenin cooperatively induce the expression of cyclin D1, and the presence of both is required for this induction. Although there are some differences between this Smad3/β-catenin pathway in MSCs and more differentiated lineages, the fact that very similar downstream mechanisms exist in differentiated mesenchymal cells strengthens our findings in MSCs. Despite this, several questions remain unanswered. This pathway seems to be restricted to mesenchymal

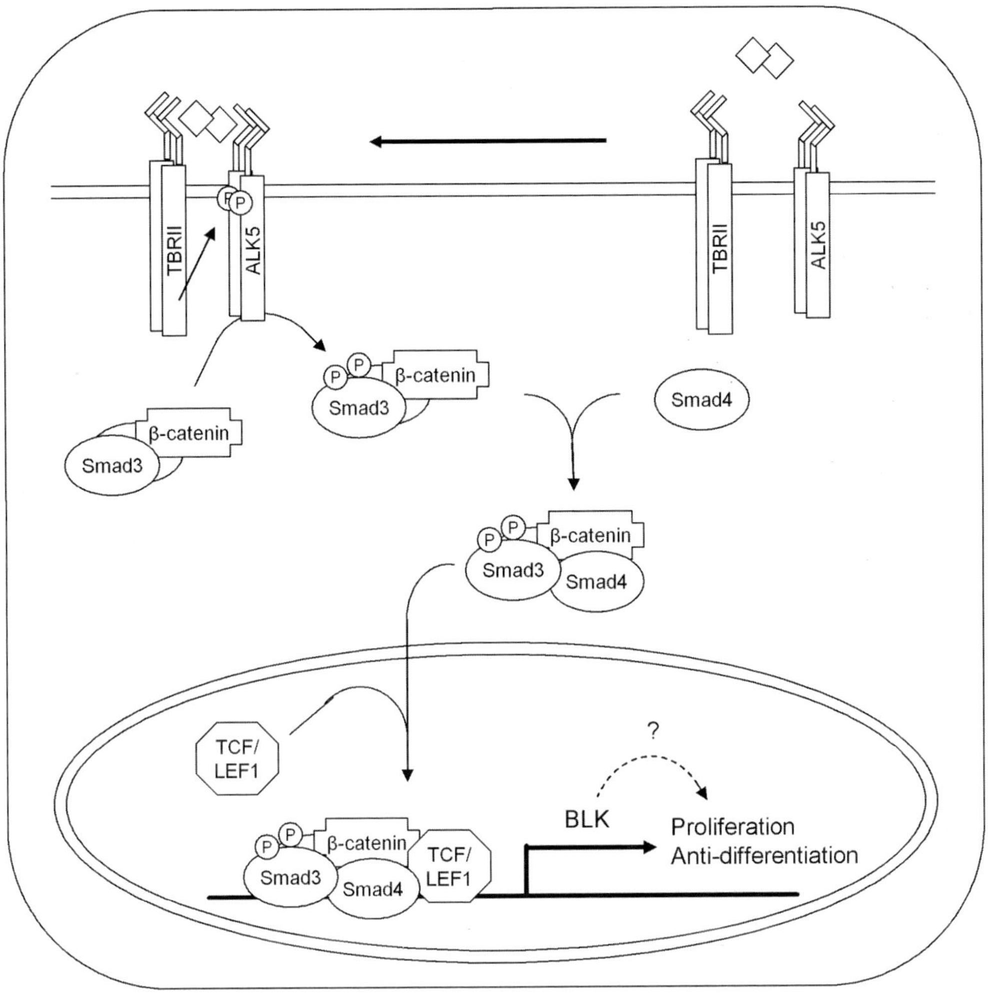

Fig. 5. Hypothesized pathway and effects of TGF-β stimulated nuclear translocation of β-catenin.

stem cells and mesenchymal lineages, but it is still not clear what about these cells is unique, permitting this pathway to exist. It is tempting to speculate that factors regulating the stability of β-catenin may be important because we found that β-catenin levels were relatively insensitive to proteasome inhibitor treatment in MSCs but increased with the same treatment in other cell types *(20)*. Furthermore, one has to wonder how β-catenin nuclear translocation is induced. It is likely that Smad3 and Smad4 are chaperones for β-catenin in this context, but this still has not been proven. Finally, the question of what downstream gene responses are mediating the induction of proliferation by TGF-β in MSCs remains to be seen. Although we have shown that BLK is regulated by TGF-β in a β-catenin-dependent manner, it is unclear whether this gene is required for the induction of proliferation by this signaling pathway. Obtaining answers to such questions will help us gain an understanding of this unique cellular context and perhaps aid in our attempts to exploit these cells for clinical use in the future.

ACKNOWLEDGMENTS

I.M.L. is supported by a DOD BCRP predoctoral traineeship award. S.H.S. is supported by an NSF graduate research fellowship. X.F.W. is supported by an NIH grant.

REFERENCES

1. Pittenger MF, Mackay AM, Beck SC, et al. Multilineage potential of adult human mesenchymal stem cells. Science 1999;284(5411):143–147.
2. Tuan RS, Boland G, Tuli R. Adult mesenchymal stem cells and cell-based tissue engineering. Arthritis Res Ther 2003;5(1):32–45.
3. Lee OK, Coathup MJ, Goodship AE, Blunn GW. Use of mesenchymal stem cells to facilitate bone regeneration in normal and chemotherapy-treated rats. Tissue Eng 2005;11(11–12):1727–1735.
4. Koc ON, Gerson SL, Cooper BW, et al. Rapid hematopoietic recovery after coinfusion of autologous-blood stem cells and culture-expanded marrow mesenchymal stem cells in advanced breast cancer patients receiving high-dose chemotherapy. J Clin Oncol 2000;18(2):307–316.
5. Nakamizo A, Marini F, Amano T, et al. Human bone marrow-derived mesenchymal stem cells in the treatment of gliomas. Cancer Res 2005;65(8):3307–3318.
6. Studeny M, Marini FC, Dembinski JL, et al. Mesenchymal stem cells: potential precursors for tumor stroma and targeted-delivery vehicles for anticancer agents. J Natl Cancer Inst 2004;96(21):1593–1603.
7. Siegel PM, Massagué J. Cytostatic and apoptotic actions of TGF-beta in homeostasis and cancer. Nat Rev Cancer 2003;3(11):807–821.
8. Reya T, Clevers H. Wnt signalling in stem cells and cancer. Nature 2005;434(7035):843–850.
9. Bonewald LF, Mundy GR. Role of transforming growth factor-beta in bone remodeling. Clin Orthop Relat Res 1990;(250):261–276.
10. Cassiede P, Dennis JE, Ma F, Caplan AI. Osteochondrogenic potential of marrow mesenchymal progenitor cells exposed to TGF-beta1 or PDGF-BB as assayed in vivo and in vitro. J Bone Miner Res 1996;11(9):1264–1273.
11. Borton AJ, Frederick JP, Datto MB, Wang X-F, Weinstein RS. The loss of Smad3 results in a lower rate of bone formation and osteopenia through dysregulation of osteoblast differentiation and apoptosis. J Bone Miner Res 2001;16(10):1754–1764.
12. Alliston T, Choy L, Ducy P, Karsenty G, Derynck R. TGF-beta-induced repression of CBFA1 by Smad3 decreases cbfa1 and osteocalcin expression and inhibits osteoblast differentiation. EMBO J 2001;20(9): 2254–2272.
13. Choy L, Skillington J, Derynck R. Roles of autocrine TGF-beta receptor and Smad signaling in adipocyte differentiation. J Cell Biol 2000;149(3):667–682.
14. Liu D, Black BL, Derynck R. TGF-beta inhibits muscle differentiation through functional repression of myogenic transcription factors by Smad3. Genes Dev 2001;15(22):2950–2966.
15. Mishra L, Shetty K, Tang Y, Stuart A, Byers SW. The role of TGF-beta and Wnt signaling in gastrointestinal stem cells and cancer. Oncogene 2005;24(37):5775–5789.
16. de Boer J, Siddappa R, Gaspar C, van Apeldoorn A, Fodde R, van Blitterswijk C. Wnt signaling inhibits osteogenic differentiation of human mesenchymal stem cells. Bone 2004;34(5):818–826.
17. de Boer J, Wang HJ, Van Blitterswijk C. Effects of Wnt signaling on proliferation and differentiation of human mesenchymal stem cells. Tissue Eng 2004;10(3–4):393–401.
18. Boland GM, Perkins G, Hall DJ, Tuan RS. Wnt 3a promotes proliferation and suppresses osteogenic differentiation of adult human mesenchymal stem cells. J Cell Biochem 2004;93(6):1210–1230.
19. Barth AI, Stewart DB, Nelson WJ. T cell factor-activated transcription is not sufficient to induce anchorage-independent growth of epithelial cells expressing mutant beta-catenin. Proc Natl Acad Sci USA 1999;96(9):4947–4952.
20. Jian H, Shen X, Liu I, Semenov M, He X, Wang X-F. Smad3-dependent nuclear translocation of beta-catenin is required for TGF-β1-induced proliferation of bone marrow-derived adult human mesenchymal stem cells. Genes Dev 2006;20(6):666–674.
21. Lei S, Dubeykovskiy A, Chakladar A, Wojtukiewicz L, Wang TC. The murine gastrin promoter is synergistically activated by transforming growth factor-beta/Smad and Wnt signaling pathways. J Biol Chem 2004;279(41):42,492–42,502.
22. Nishita M, Hashimoto MK, Ogata S, et al. Interaction between Wnt and TGF-beta signalling pathways during formation of Spemann's organizer. Nature 2000;403(6771):781–785.
23. Labbe E, Letamendia A, Attisano L. Association of Smads with lymphoid enhancer binding factor 1/T cell-specific factor mediates cooperative signaling by the transforming growth factor-beta and Wnt pathways. Proc Natl Acad Sci USA 2000;97(15):8358–8363.
24. Zhou S, Eid K, Glowacki J. Cooperation between TGF-beta and Wnt pathways during chondrocyte and adipocyte differentiation of human marrow stromal cells. J Bone Miner Res 2004;19(3):463–470.

25. Tuli R, Tuli S, Nandi S, et al. Transforming growth factor-beta-mediated chondrogenesis of human mesenchymal progenitor cells involves N-cadherin and mitogen-activated protein kinase and Wnt signaling cross-talk. J Biol Chem 2003;278(42):41,227–41,236.
26. Furuhashi M, Yagi K, Yamamoto H, et al. Axin facilitates Smad3 activation in the transforming growth factor beta signaling pathway. Mol Cell Biol 2001;21(15):5132–5141.
27. Hsieh JC, Rattner A, Smallwood PM, Nathans J. Biochemical characterization of Wnt-frizzled interactions using a soluble, biologically active vertebrate Wnt protein. Proc Natl Acad Sci USA 1999;96(7):3546–3551.
28. Hino S, Michiue T, Asashima M, Kikuchi A. Casein kinase I epsilon enhances the binding of Dvl-1 to Frat-1 and is essential for Wnt-3a-induced accumulation of beta-catenin. J Biol Chem 2003;278(16): 14,066–14,073.
29. Uhl M, Aulwurm S, Wischhusen J, et al. SD-208, a novel transforming growth factor beta receptor I kinase inhibitor, inhibits growth and invasiveness and enhances immunogenicity of murine and human glioma cells in vitro and in vivo. Cancer Res 2004;64(21):7954–7961.
30. Waddell DS, Liberati NT, Guo X, Frederick JP, Wang X-F. Casein kinase Iepsilon plays a functional role in the transforming growth factor-beta signaling pathway. J Biol Chem 2004;279(28): 29,236–29,246.
31. Hu MC, Rosenblum ND. Smad1, beta-catenin and Tcf4 associate in a molecular complex with the Myc promoter in dysplastic renal tissue and cooperate to control Myc transcription. Development 2005;132(1):215–225.
32. Edlund S, Lee SY, Grimsby S, et al. Interaction between Smad7 and beta-catenin: importance for transforming growth factor β-induced apoptosis. Mol Cell Biol 2005;25(4):1475–1488.
33. Kang Y, Chen CR, Massagué J. A self-enabling TGFbeta response coupled to stress signaling: Smad engages stress response factor ATF3 for Id1 repression in epithelial cells. Mol Cell 2003;11(4):915–926.
34. Chen CR, Kang Y, Siegel PM, Massagué J. E2F4/5 and p107 as Smad cofactors linking the TGFbeta receptor to c-myc repression. Cell 2002;110(1):19–32.
35. Thompson B, Townsley F, Rosin-Arbesfeld R, Musisi H, Bienz M. A new nuclear component of the Wnt signalling pathway. Nat Cell Biol 2002;4(5):367–373.
36. van de Wetering M, Oving I, Muncan V, et al. Specific inhibition of gene expression using a stably integrated, inducible small-interfering-RNA vector. EMBO Rep 2003;4(6):609–615.
37. Logan CY, Nusse R. The Wnt signaling pathway in development and disease. Annu Rev Cell Dev Biol 2004;20:781–810.
38. Hsu SC, Galceran J, Grosschedl R. Modulation of transcriptional regulation by LEF-1 in response to Wnt-1 signaling and association with beta-catenin. Mol Cell Biol 1998;18(8):4807–4818.
39. Prieve MG, Guttridge KL, Munguia JE, Waterman ML. The nuclear localization signal of lymphoid enhancer factor-1 is recognized by two differentially expressed Srp1-nuclear localization sequence receptor proteins. J Biol Chem 1996;271(13):7654–7658.
40. Cheon SS, Wei Q, Gurung A, et al. beta-catenin regulates wound size and mediates the effect of TGF-β in cutaneous healing. FASEB J 2006;20(6):692–701.
41. Li TF, Chen D, Wu Q, et al. TGF-beta stimulates cyclin D1 expression through activation of beta-catenin signaling in chondrocytes. J Biol Chem 2006;281(30):21,296–21,304.

19 Interaction of Oncogenic Ras and TGFβ1 Signaling in Cancer: Lessons From the Multistage Skin Carcinogenesis Model

Adam Glick, Dong-Soon Bae, Xin Liu, and Kinnimulki Vijayachandra

CONTENTS

OVERVIEW OF TGFβ1 SIGNALING AND CANCER DEVELOPMENT
RAS ACTIVATION AND MULTISTAGE SKIN CARCINOGENESIS
CHANGES IN EXPRESSION AND RESPONSE TO TGFβ1 DURING SKIN CARCINOGENESIS
TGFβ1 ENHANCES MALIGNANT PHENOTYPE
INTERACTIONS BETWEEN RAS AND TGFβ1 SIGNALING PATHWAYS
POTENTIAL CANCER THERAPY WITH SMALL MOLECULE INHIBITORS OF TGFβ1 SIGNALING
REFERENCES

Abstract

The multistep nature of cancer derives from the stepwise activation and loss of signaling pathways, which govern the behavior of the tumor cell, its interaction with neighboring cells, local microenvironment, and organism. Rather than acting independently these pathways are likely to synergize or antagonize each other in ways that produce novel phenotypes for the cancer cell. Two pathways that are important in many human cancers and that interact to produce novel phenotypes are those regulated by transforming growth factor β1 (TGFβ1), which controls a major growth inhibitory pathway for epithelial cells, and c-Ha-ras, a prototypical transforming oncogene, which is central to pathways that control cell proliferation and survival. The mouse epidermis has been a useful model both for understanding basic mechanisms of multistage carcinogenesis and for analysis of the interaction of these two signaling pathways. This review discusses our current understanding of the role of TGFβ1 in the multistage skin carcinogenesis model and how interactions with oncogenic ras signaling are likely to modulate cancer development and tumor phenotype.

Key Words: TGFβ1; ras oncogene; skin; keratinocytes; multistage carcinogenesis; skin cancer.

From: *Cancer Drug Discovery and Development: Transforming Growth Factor-β in Cancer Therapy, Vol I: Basic and Clinical Biology*
Edited by: Sonia B. Jakowlew © Humana Press Inc., Totowa, NJ

1. OVERVIEW OF TGFβ1 SIGNALING AND CANCER DEVELOPMENT

Transforming Growth Factor β1 (TGFβ1) is a member of a large family of structurally related cytokines that play an important role in normal tissue homeostasis and tumor development. TGFβ1 binds to a cell surface serine–threonine receptor complex, which is present on most cell types and signals primarily through intracellular Smads to regulate gene expression *(1,2)*. For most epithelial cells, TGFβ1 negatively regulates the cell cycle, controls extracellular matrix gene expression, and regulates aspects of cell phenotype such as differentiation state, cell shape, attachment, and migration *(3)*. This cytokine plays a dual role in cancer development acting as both tumor suppressor and prometastatic factor, but the molecular basis of this switch remains unresolved. Inactivating mutations have been identified in the TGFβ type II receptor, Smad4, or Smad2 indicating that these genes and the TGFβ1 signaling pathway function as a tumor suppressor pathway *(4)*, but at the same time many human solid tumors including breast *(5)*, lung *(6)*, and colon cancers *(7)* overexpress TGFβ1 and this is associated with poor clinical prognosis with significantly shorter time to relapse, disease progression, and metastases. In vitro studies with human cancer cell lines engineered to overexpress TGFβ1 or to have a specific blockade in TGFβ1 signaling support the idea that overexpression of TGFβ1 in tumor cells enhances the malignant phenotype *(8–10)*. Proposed mechanisms to account for this switch in function include alterations in the balance between extracellular matrix production and degradation, induction of angiogenesis, immune suppression, and acquisition of a mesenchymal phenotype *(11)*. Whereas the TGFβ1 pathway is an attractive target for therapeutics, a better understanding of the molecular mechanism of these stage-specific effects of TGFβ1 will greatly enhance rational use of this signaling pathway as a therapeutic modality. One possible mechanism that could account for changes in TGFβ1 signaling is through alterations in signaling pathways that interact and modify the readout of TGFβ1 signaling. Whereas there are a number of potential signaling pathways and transcription factors that can interact with TGFβ1 pathway, there is substantial evidence that the interaction of TGFβ1 and ras oncogene pathway is important in carcinogenesis.

2. RAS ACTIVATION AND MULTISTAGE SKIN CARCINOGENESIS

Approximately 30% of all human tumors contain a mutation in one of the four ras family members (Ha-ras, N-ras, Ki-ras A, B) with Ki-ras mutations present in 95% of pancreatic cancers *(12)*. Mutational activation of the GTPase activity of ras constitutively activates several effector pathways that are important in proliferation, survival, and motility, including the raf-erk-mapk pathway, PI3kinase pathway, Rho-GTPase *(13,14)*, and transcription factors such as AP-1 and ETS *(15–17)*. Numerous studies have implicated the activation of one or more of these pathways by ras in the altered cell adhesion, motility, protease production, and induction of angiogenesis that enhance tumor cell invasion and metastasis *(18)*.

The mouse epidermis has been an important model for analysis of the role of ras in multistage carcinogenesis and to examine how interaction between ras and TGFβ1 shapes cancer development. Squamous cancers developing on mouse skin in a two-stage initiation promotion protocol evolve through a series of predictable stages: initiation, papilloma, dysplastic papilloma, and carcinoma. Treatment of the mouse dorsal skin with the polycyclic aromatic hydrocarbon carcinogen DMBA (7,12-dimethylbenz[*a*]anthracene) or urethane causes AT transition mutations at many sites but invariably the benign squamous papillomas, which form following repeated application of a tumor promoter such as TPA, contain activating mutations of the c-Ha-ras gene *(19)*. Other carcinogens such as *N*-methyl-*N*′-nitro-*N*-nitrosoguanidine (MNNG), methylnitrosourea (MNU), 3-methylcholanthrene (MCA) induce mutations in the c-Ha-ras gene, although the mutation spectrum is distinct *(20)*. As early as 1 wk after carcinogen exposure, small population of epidermal cells that contain the signature ras mutation can be detected *(21)*, and these persist and expand in frequency in the hyperplasia induced by tumor promoters *(22)*. Activation of ras is a sufficient genetic change for benign tumor formation as overexpression of activated ras in suprabasal keratinocytes of transgenic mice *(23)* or transduction and grafting of primary keratinocytes with replication-defective v-ras^{Ha} retrovirus onto athymic mice reproduces the papilloma phenotype. These tumors exhibit elevated proliferation along the basal layer, increased number of cell layers, and delayed differentiation, but retain the stratified architecture of the epidermis. With time some papillomas progress to locally invasive squamous cell carcinomas and then more aggressive highly undifferentiated spindle cell carcinomas.

Premalignant progression and malignant conversion are associated with decreased differentiation, elevated proliferation, and aneuploidy *(24)*. Trisomy of chromosome 7 is observed relatively early during progression, and squamous cell carcinomas frequently have loss of heterozygosity at the c-Ha-ras locus and amplification of the mutant ras allele *(25–27)*. Thus increased levels of the ras oncoprotein and activation of its signaling pathways may drive progression. Interestingly, expression of mutant ras in the basal layer and stem cell compartment of transgenic epidermis causes rapid development of squamous cell carcinomas, suggesting that the developmental state/fate of the keratinocyte receiving the genotoxic insult may play an important role in the ultimate rate of progression *(28)*.

3. CHANGES IN EXPRESSION AND RESPONSE TO TGFβ1 DURING SKIN CARCINOGENESIS

TGFβ1 causes cell cycle arrest of primary mouse and human keratinocytes *(29,30)*. Although some immortalized cell lines expressing a ras oncogene are resistant to TGFβ1-mediated growth inhibition, expression of oncogenic v-ras^{Ha} and associated hyperactivation of the Erk1/2 MAPK pathway do not alter sensitivity to TGFβ1 in primary mouse keratinocytes *(30)*. Similarly, examination of a series of cell lines derived from chemically induced squamous tumors suggests that loss of responsiveness to the growth factor is a late event in squamous cell tumor development *(31)*. However, it must be noted that even slight changes in responsiveness could alter proliferative rates enough to enhance tumor growth.

Localization of TGFβ1 expression in the epidermis is dependent on the antibody used. Because mRNA and protein expression are frequently discordant, interpretation of in situ hybridization results is difficult. Using antibodies directed against the latency-associated peptide and the extracellular/secreted form of TGFβ1, we showed that TGFβ1 is expressed predominantly in the basal layer of the epidermis *(32)*. TGFβ2 expression is confined to the suprabasal differentiating cell layers, and this correlates with differentiation-associated induction of TGFβ2 in vitro *(32,33)*. Other groups using an antibody detecting the intracellular form have documented suprabasal expression of TGFβ1. Which localization pattern is the most biologically relevant is uncertain, although there is a strong correlation in benign tumors between lack of localized basal layer staining for TGFβ1 and focal increase in BrdU labeling *(32)*.

Tumor promoters induce a complex response in the epidermis, which includes hyperproliferation, hyperplasia, differentiation, and an inflammatory cell infiltrate. When the mouse epidermis is treated with the classical PKC-activating tumor promoter TPA (12-*O*-tetradecanoylphorbol-13-acetate), TGFβ1 mRNA is increased in the suprabasal differentiating cell layers *(35)* Other tumor promoters such as anthralin, benzoyl peroxide, cumene hydroperoxide, and mezerein also cause a transient induction of TGFβ1 mRNA in the epidermis, although the extent of induction does not correlate with hyperplastic responses to the specific promoting agents. Much of the secreted TGFβ1 is found in the dermis *(36)*. However, full thickness grafts of TGFβ1–/– skin onto athymic mice have a similar hyperplastic response to acute TPA treatment as TGFβ1+/+ grafts *(37)*. This suggests that the rapid upregulation of TGFβ1 in keratinocytes is not essential for hyperplasia, but does not clarify what the effect of the TGFβ1 null genotype would be on chronic TPA treatment and tumor promotion. More recent studies using Smad3 null mice suggest that some level of TGFβ1 signaling is required for the tumor-promoting effects of TPA. The observation that TPA promotion alone produces papillomas in transgenic mice expressing a DNTβRII in the epidermis *(38)* suggests that tumor promotion is dependent on Smad3 signaling in cell types other than keratinocytes.

3.1. Genetic Manipulation of TGFβ1 Signaling in Multistage Carcinogenesis

Transgenic mouse models have been very important for analysis of the role of TGFβ1 in skin carcinogenesis and indirectly on keratinocytes harboring activating ras mutations. Overexpression of the constitutively active ser$^{223\ 225}$TGFβ1 in the epidermis of transgenic mice substantially blocked the formation of benign papillomas in a two-stage DMBA–TPA carcinogenesis protocol *(39)*. Similarly, overexpression of BMP4 blocked TPA-induced hyperplasia and prevented the formation of MNNG-induced benign or malignant tumors *(40)*. It is difficult to determine from this type of study whether the major effect of TGFβ1 overexpression at this stage is to block TPA-induced hyperplasia or more directly on the clonal outgrowth of ras-initiated keratinocytes or a combination of the two. Because primary keratinocytes transduced with the v-ras^{Ha} retrovirus remain equally sensitive to the growth inhibitory effects of TGFβ1 as control cells, it is likely that overexpression of TGFβ1 in these transgenic models suppress outgrowth of ras-initiated keratinocytes. To more directly address questions of how TGFβ1 and ras interact at the stage of initiation we have compared gene expression profiles of primary keratinocytes and v-ras^{Ha} keratinocytes treated with TGFβ1. We find, as expected, that most cell cycle regulatory genes are suppressed by TGFβ1 regardless of the presence of ras. However, we observed a reproducible block in the induction of extracellular matrix genes by TGFβ1 in cells expressing v-ras^{Ha}, suggesting that ras-initiated keratinocytes have altered matrix responses to TGFβ1 *(34)*.

More direct studies demonstrating that TGFβ1 acts to negatively regulate clonal expansion and papilloma formation by carcinogen-initiated keratinocytes were conducted on transgenic mice with an epidermally targeted dominant negative TGFβ type II receptor. In studies with these mice produced by two different groups, inhibition of TGFβ1 signaling in keratinocytes decreased the latency and increased the yield of benign papillomas, as well as squamous cell carcinomas *(38,41)*. Most of the papillomas that formed in the DNTβR11 mice became rapidly TPA independent, i.e., they did not regress when TPA was stopped whereas many in the wild-type mice did *(38)*. In addition, the papillomas that formed in the DNTβRII transgenic mice had increased cell proliferation and reduced expression of key cell cycle regulatory molecules, p15ink4b, p21waf1, and p27, but no detectable change in chromosome number *(42)*. Interestingly, tumors expressing the DNTβRII transgene had increased levels of TGFβ1 expression and this was associated with elevated angiogenesis and expression of the angiogenic factor VEGFA and decreased expression of thrombospondin-1 *(38)*.

Initial studies of TGFβ1 expression during tumor progression revealed that benign tumor with high frequency of malignant conversion had a paradoxical increase in message levels but reduced or absent levels of cytokine in tissue sections *(32,43)*. Similarly, benign tumors induced p53 null mice by a two-stage protocol have a high risk for rapid progression to squamous cell carcinoma, and these also do not have detectable TGFβ1 expression using immunohistochemistry *(43)*. This suggests that reduced levels of TGFβ1 in the extracellular environment are associated with rapid progression in this model. In keeping with this, the expression of Smads 1–5 decreases in carcinomas *(44)*, suggesting a general reduction in activity of the TGFβ1 signaling pathway both through reduced ligand secretion and signaling intermediates. Both transgenic and skin graft models demonstrate that TGFβ1 signaling acts to suppress premalignant progression of keratinocytes expressing a ras oncogene. When primary mouse keratinocytes from the TGFβ1–/– line are transduced with the replication defective v-*ras*^{Ha} retrovirus and grafted onto the dorsal epidermis of athymic mice, they initially form papillomas, but these rapidly convert to squamous cell carcinomas, whereas TGFβ1+/+ keratinocytes remain as benign tumors *(45)*. Similarly, Smad3–/– v-*ras*^{Ha} keratinocytes *(46)* and keratinocytes expressing Smad7 and v-*ras*^{Ha} *(47)* rapidly progress to squamous cell carcinoma in the graft model indicating that the accelerated malignancy is related to inhibition of TGFβ1 signaling within the keratinocytes. Similarly, the frequency of SCC is increased in transgenic mice expressing a DNTβRII *(41,48)*. Taken together these data support the concept that TGFβ1 signaling acts as a tumor suppressor pathway in skin carcinogenesis.

3.2. Mechanisms of Suppression of Ras Carcinogenesis by TGFβ1

Although cell cycle inhibition is an obvious mechanism through which TGFβ1 could inhibit ras-initiated carcinogenesis and tumor outgrowth, other pathways may be important in the suppression of tumor progression. When oncogenic ras is introduced into primary mouse keratinocytes via a replication-defective retrovirus, the cells become hyperproliferative relative to the mock or uninfected cells *(49)*. These v-ras-transduced keratinocytes persist in this proliferative state longer than the primary cells, but ultimately the growth of these cells arrest and they express markers of senescence *(49)*. In contrast primary dermal fibroblasts undergo a premature growth arrest when transduced with this retrovirus *(32)*, but these results suggest that the in vitro arrest of the primary keratinocytes cannot be overcome by the ras oncogene. As the v-*ras*^{Ha}-transduced keratinocytes senesce there is an increase in the secretion of TGFβ1 and arrest with a 2N or 4N chromosome compliment *(49,50)*. Inhibition of TGFβ1 signaling either with neutralizing TGFβ1 antibodies, a dominant negative TβRII adenovirus, Smad7 overexpression, or a TGFβ1 and Smad3 null genotype blocks senescence and causes rapid generation of aneuploid cells *(46,47,49,50)*. Conversely, TGFβ1 induces senescence in v-*ras*^{Ha}-transduced TGFβ1 KO keratinocytes, or keratinocytes from inbred strains with delays in vitro senescence, and suppresses the generation of aneuploidy. Whereas other pathways including p53, Arf, and p16 are clearly involved in senescence *(51,52)*, these results demonstrate that ras-associated senescence in primary mouse keratinocytes is dependent on autocrine TGFβ1. The recent demonstration of senescent cells in vivo within benign melanocytic nevi containing a BRAF(V600E) mutation suggests that induction of senescence is not simply an in vitro artifact and may reflect a true in vivo mechanism of tumor suppression *(53)*. Although not yet demonstrated in squamous cells it is possible that TGFβ1 production by ras-initiated benign papilloma cells may suppress tumor progression through the induction of senescence.

4. TGFβ1 ENHANCES MALIGNANT PHENOTYPE

As with other mouse models and human cancers there is a paradoxical switch in TGFβ1 function during tumor progression. The first indication of this came from Cui et al. *(39)*, who showed that whereas overexpression of TGFβ1 largely blocked benign tumor formation, ultimately spindle cell carcinomas emerged, suggesting that TGFβ1 was promoting the development of the most malignant undifferentiated tumors. Using an inducible model that allowed expression of TGFβ1 only after benign tumors had formed, Weeks et al. *(54)* showed that overexpression of TGFβ1 in benign squamous tumors for 15 wk enhanced a metastatic phenotype in a high percentage of the persistent papillomas. Thus on the surface, these and other transgenic models *(55,56)* would support in vitro studies showing synergistic cooperation between ras or other oncogenes and TGFβ signaling, which causes a poorly differentiated metastatic phenotype. However, the interpretation of these studies is clouded by development of the transgenic tumors in an environment of high TGFβ1 secretion or constitutive TGFβ1 signaling. It is not clear whether the effects on tumor differentiation as well as frequency of metastases are directly because of TGFβ1 overexpression or indirectly from selection for more aggressive variants within the primary tumor with altered or reduced responses to TGFβ1. Thus papillomas expressing TGFβ1 for prolonged periods had reduced expression of Smads *(54)*; metastases from MMTV-neu/MMTV-TβRI bigenic mice were no longer sensitive to TGFβ1-induced growth control compared to the primary tumors *(56)*; and bigenic MMTV-neu/MMTV-TGFβ1 mammary tumors had constitutively phosphorylated Smad2, AKT, and Erk1/2 whereas bigenic mammary glands did not *(55)*. Furthermore, if metastasis and EMT phenotype are manifest only in the genetic background of a tumor cell, it is possible that other undefined genetic changes or the malignant microenvironment are essential for this phenotype. These are critical distinctions because clonal selection and synergy between signaling pathways represent distinct mechanisms with distinct possibilities for therapeutic targeting.

5. INTERACTIONS BETWEEN RAS AND TGFβ1 SIGNALING PATHWAYS

Antagonistic and synergistic interactions exist between TGFβ1 and *ras* signaling pathways, which may define the phenotype of the cancer cell. In normal cells ras family members or downstream effectors may be important for some components of TGFβ1 signaling. Thus, blocking ras activity or MAPK activity in certain epithelial cells inhibits TGFβ1 signaling *(57,58)*. On the other hand overexpression of an oncogenic *ras* in cell lines *(59)* or activation of the Erk1/2 pathway by mitogens blocks TGFβ1 or TGFβ superfamily signaling, and this occurs through phosphorylation of Smads in the linker region by Erk1/2 *(60,61)*. Other reports link oncogenic *ras* to the degradation of Smad4 *(60,62)* and an activated *ras*-raf-erk pathway to the stabilization of specific corepressors of the TGFβ1 signaling pathway such as SnoN or TG interacting factor (TGIF) *(63,64)*. These reports suggest that ras can overcome the tumor suppressor effects of TGFβ1 by blocking its growth inhibitory signaling pathway. However, these in vitro studies must be weighed against numerous transgenic mouse models where, as described previously, ras is the likely transforming oncogene and TGFβ1 over-expression blocks tumor formation or loss of TGFβ1 signaling accelerates premalignant progression *(39,40,45)*. One interpretation of these results is that the level of ras oncogene expression may alter the TGFβ1 response, or the stage of progression (immortalized normal or malignant cell lines vs normal epithelia) or the in vivo environment favors tumor suppressor functions of TGFβ1 over the oncogenic activity of ras.

Epithelial to mesenchymal transition (EMT) is a frequently used in vitro surrogate for the acquisition of invasive metastatic phenotype. Although in a few cell types TGFβ1 can induce EMT by itself *(65)*, *ras* and TGFβ1 cooperate to enhance EMT and invasive behavior of immortalized and neoplastic human breast and mouse squamous cell carcinoma lines *(66, 67)*. Both the EpH4 normal mouse mammary epithelial cell line and a ras-transformed clone Ep/Ras form organoid structures when grown in a collagen gel. When treated with TGFβ1 the EpH4 line responds with growth arrest and apoptosis, but the presence of ras blocks this and causes an EMT response to TGFβ1 in which E-cadherin expression is lost, vimentin is induced, and the cells become invasive in collagen gels *(68,69)*. The sequential elevation of ras levels either by LOH or amplification may alter the response to TGFβ1 in malignant progression of squamous cancers. Mouse spindle cell carcinoma lines, in contrast to squamous cell or benign tumor lines, secrete high levels of TGFβ1 and have constitutively activated TGFβ1 signaling and nuclear localization of Smad2. By itself, expression of a constitutively active Smad2 mutant in squamous cell carcinoma cells causes a migratory phenotype, but does not cause EMT, whereas a dominant negative

Smad2 induces a squamous cuboidal morphology in spindle cell lines *(66)*. Interestingly, overexpression of mutant Ha-ras in squamous cell lines causes nuclear translocation of Smad2 and activation of TGFβ1-mediated gene expression. Overexpression of mutant Ha-ras alone does not cause an EMT, but coexpression of mutant H-ras and activated Smad2 caused a spindle-like morphology, expression of vimentin, and loss of keratin filaments. An activated Smad3 mutant also cooperated with mutant H-ras to cause EMT but this did not occur if activated H-ras was coexpressed with wild-type Smad2, Smad3, or Smad4. Thus, increased levels of H-ras and TGFβ1 signaling are sufficient to induce EMT in squamous tumor cells that do not undergo EMT with activation of either signaling pathway alone. More importantly, the spindle cells expressing the activated Smad2 mutant became highly invasive and formed lung metastases *(66)*. Given the lack of DNA-binding capacity of Smad2 and the few genes that have been identified as Smad2 dependent in microarray studies with Smad3 and Smad2 null cells *(70,71)* these data suggest that Smad2 or Smad3 may target different gene subsets in the presence of activated Ha-ras. These results show that cooperativity between TGFβ1 and oncogenic pathways are critical in malignant progression and metastasis of squamous tumors.

6. POTENTIAL CANCER THERAPY WITH SMALL MOLECULE INHIBITORS OF TGFβ1 SIGNALING

Recently, several small molecule, orally bioavailable inhibitors have been developed with specificity for the kinase of ALK receptors of the TGFβ1 superfamily *(72–75)*. SB431542 (4-[4-(3,4 methylenedioxyphenyl)-5-(2-pyridyl)-1H-imidazol-2-yl]benzamide) is structurally related to MAPK p38 inhibitors and shows selectivity for the kinase activity of ALK5 (the TGFβ1 type I receptor), ALK4, and the nodal receptor ALK7 *(72)*, but no effect on the activity of BMP receptors. *(72,73,76)*. Other structurally related derivatives such as SB525334 appear to have more selectivity for the ALK5 receptor *(77)*. Recent publications show that these inhibitors can block induction of extracellular matrix genes in vitro and inhibit experimentally induced pulmonary, renal, and lung fibrosis in vivo *(74,75,77)*. The ALK5 inhibitor SD-208 has shown promise in its ability to reduce growth of a xenotransplanted human glioblastoma cell line *(78)*. In this study systemic SD-208 treatment prolonged survival of tumor-bearing mice, but did not alter angiogenesis or tumor cell proliferation or apoptosis. Instead, there was increased CD8 T-cell, NK cell, and macrophage infiltration into the tumor, suggesting that reversal of paracrine immunosuppressive effects of TGFβ1 secreted by the tumor cells was important for the efficacy of this inhibitor *(78)*. Both the pharmacological and DNA-based approaches show promise as therapies for TGFβ1-induced pathologies. However, current models suffer from the complexity of the initiating insult (i.e., bleomycin-induced fibrosis) or as poor models of multistage cancer development (i.e., xenotransplantation of human tumor lines). Given the bifunctional nature of TGFβ1 in cancer development the impact of long-term treatment with these inhibitors on occult preneoplastic is unknown. Identification of inhibitors that could selectively block some of the biological readouts from the TGFβ1 pathway would be of great therapeutic utility. For instance, it would be of therapeutic value in the treatment of fibrotic disease with TGFβ1 pathway inhibitors to block induction of extracellular matrix gene expression in fibroblasts without inhibiting cell cycle regulation of normal or preneoplastic epithelial cells. Conversely inhibition of tumor cell produced TGFβ1 effects on normal fibroblast, endothelial, and lymphocyte populations within the tumor microenvironment is likely to be an important use of these inhibitors in the treatment of malignancy. However, given the bifunctional effects of TGFβ1 in cancer pathogenesis, it will be important to test what effects chronic treatment with these inhibitors would have on the outgrowth of preneoplastic cells.

REFERENCES

1. Massagué J. TGF-beta signal transduction. Annu Rev Biochem 1998;67:753–791.
2. Derynck R, Zhang YE. Smad-dependent and Smad-independent pathways in TGF-beta family signalling. Nature 2003;425(6958):577–584.
3. de Caestecker MP, Piek E, Roberts AB. Role of transforming growth factor-beta signaling in cancer. J Natl Cancer Inst 2000;92(17):1388–1402.
4. Markowitz S, Wang J, Myeroff L, et al. Inactivation of the type II TGF-β receptor in colon cancer cells with microsatellite instability. Science 1995;268:1336–1338.
5. Gorsch SM, Memoli VA, Stukel TA, Gold LI, Arrick BA. Immunohistochemical staining for transforming growth factor β1 associates with disease progression in human breast cancer. Cancer Res 1992;52:6949–6952.
6. Takanami I, Tanaka F, Hashizume T, Kodaira S. Roles of the transforming growth factor beta1 and its type I and II receptors in the development of a pulmonary adenocarcinoma: results of an immunohistochemical study. J Surg Oncol 1997;64(4):262–267.
7. Friedman E, Gold LI, Klimstra D, Zeng ZS, Winawer S, Cohen A. High levels of transforming growth factor β1 correlate with disease progression in human colon cancer. Cancer Epidemiol Biomarkers Prev 1995;4:549–554.
8. Welch D, Fabra A, Nakajima M. Transforming growth factor β stimulates mammary adenocarcinoma cell invasion and metastatic potential. Proc Natl Acad Sci USA 1990;87:7678–7682.
9. Ueki N, Nakazato M, Ohkawa T, et al. Excessive production of transforming growth-factor beta1 can play an important role in the development of tumorigenesis by its action for angiogenesis: validity of neutralizing antibodies to block tumor growth. Biochim Biophys Acta 1992;1137(2):189–196.
10. Arteaga CL, Carty-Dugger T, Moses HL, Hurd SD, Pietenpol JA. Transforming growth factor beta1 can induce estrogen- independent tumorigenicity of human breast cancer cells in athymic mice. Cell Growth Differ 1993;4:193–201.
11. Glick AB. TGFbeta1, back to the future: revisiting its role as a transforming growth factor. Cancer Biol Ther 2004;3(3):276–283.
12. Hamad NM, Elconin JH, Karnoub AE, et al. Distinct requirements for Ras oncogenesis in human versus mouse cells. Genes Dev 2002;16(16):2045–2057.
13. Vries-Smits AM, Burgering BM, Leevers SJ, Marshall CJ, Bos JL. Involvement of p21ras in activation of extracellular signal-regulated kinase 22. Nature 1992;357(6379):602–604.
14. Frost JA, Geppert TD, Cobb MH, Feramisco JR. A requirement for extracellular signal-regulated kinase (ERK) function in the activation of AP-1 by Ha-Ras, phorbol 12-myristate 13-acetate, and serum. Proc Natl Acad Sci USA 1994;91(9):3844–3848.
15. Angel P, Karin M. The role of Jun, Fos and the AP-1 complex in cell-proliferation and transformation. Biochim Biophys Acta 1991;1072(2–3):129–157.
16. Behrens A, Jochum W, Sibilia M, Wagner EF. Oncogenic transformation by ras and fos is mediated by c-Jun N-terminal phosphorylation. Oncogene 2000;19(22):2657–2663.
17. Mechta F, Lallemand D, Pfarr CM, Yaniv M. Transformation by ras modifies AP1 composition and activity. Oncogene 1997;14(7):837–847.
18. Campbell PM, Der CJ. Oncogenic Ras and its role in tumor cell invasion and metastasis. Semin Cancer Biol 2004;14(2):105–114.
19. Balmain A, Brown K. Oncogene activation in chemical carcinogenesis. Adv Cancer Res 1988;51:147–182.
20. Brown K, Buchmann A, Balmain A. Carcinogen-induced mutations in the mouse c-Ha-*ras* gene provide evidence of multiple pathways for tumor progression. Proc Natl Acad Sci USA 1990;87:538–542.
21. Nelson MA, Futscher BW, Kinsella T, Wymer J, Bowden GT. Detection of mutant Ha-ras genes in chemically initiated mouse skin epidermis before the development of benign tumors. Proc Natl Acad Sci USA 1992;89:6398–6402.
22. Finch JS, Albino HE, Bowden GT. Quantitation of early clonal expansion of two mutant 61st codon c-Ha-*ras* alleles in DMBA/TPA treated mouse skin by nested PCR/RFLP. Carcinogenesis 1996; 17(12):2551–2557.
23. Bailleul B, Surani MA, White S, et al. Skin hyperkeratosis and papilloma formation in transgenic mice expressing a *ras* oncogene from a suprabasal keratin promoter. Cell 1990;62:697–708.
24. Aldaz CM, Conti CJ, Klein-Szanto AJP, Slaga TJ. Progressive dysplasia and aneuploidy are hallmarks of mouse skin papillomas: relevance to malignancy. Proc Natl Acad Sci USA 1987;84:2029–2032.
25. Aldaz CM, Trono D, Larcher F, Slaga TJ, Conti CJ. Sequential trisomization of chromosomes 6 and 7 in mouse skin premalignant lesions. Mol Carcinog 1989;2:22–26.

26. Bremner R, Balmain A. Genetic changes in skin tumor progression: correlation between presence of a mutant ras gene and loss of heterozygosity on mouse chromosome 7. Cell 1990;61:407–417.
27. Buchmann A, Ruggeri B, Klein-Szanto AJP, Balmain A. Progression of squamous carcinoma cells to spindle carcinomas of mouse skin is associated with an imbalance of H-ras alleles on chromosome 7. Cancer Res 1991;51:4097–4101.
28. Brown K, Strathdee D, Bryson S, Lambie W, Balmain A. The malignant capacity of skin tumours induced by expression of a mutant H-ras transgene depends on the cell type targeted. Curr Biol 1998; 8(9):516–524.
29. Shipley GD, Pittelkow MR, Wille JJ, Jr., Scott RE, Moses HL. Reversible inhibition of normal human prokeratinocyte proliferation by type beta transforming growth factor-growth inhibitor in serum-free medium. Cancer Res 1986;46:2068–2071.
30. Glick AB, Sporn MB, Yuspa SH. Altered regulation of TGF–β1 and TGF-α in primary keratinocytes and papillomas expressing v-Ha-*ras*. Mol Carcinog 1991;4:210–219.
31. Haddow S, Fowlis DJ, Parkinson K, Akhurst RJ, Balmain A. Loss of growth control by TGF-beta occurs at a late stage of mouse skin carcinogenesis and is independent of ras gene activation [published erratum appears in Oncogene 1991 Dec;6(12): 2377-8]. Oncogene 1991;6:1465–1470.
32. Glick AB, Kulkarni AB, Tennenbaum T, et al. Loss of expression of transforming growth factor β in skin and skin tumors is associated with hyperproliferation and a high risk for malignant conversion. Proc Natl Acad Sci USA 1993;90:6076–6080.
33. Glick AB, Danielpour D, Morgan D, Sporn MB, Yuspa SH. Induction and autocrine receptor binding of transforming growth factor-β2 during terminal differentiation of primary mouse keratinocytes. Mol Endocrinol 1990;4:46–52.
34. Bae D, Blazanin N, Glick A. Angiogenesis and disrupted epidermal structure in mice expressing TGFβ1 and oncogenic ras. J Invest Derm 2006;126(54):21.
35. Akhurst RJ, Fee F, Balmain A. Localized production of TGF-beta mRNA in tumour promoter-stimulated mouse epidermis. Nature 1988;331:363–365.
36. Patamalai B, Burow DL, Gimenez-Conti I, et al. Altered expression of transforming growth factor-beta1 mRNA and protein in mouse skin carcinogenesis. Mol Carcinog 1994;9:220–229.
37. Li J, Foitzik K, Calautti E, Baden H, Doetschman T, Dotto GP. TGF-beta3, but not TGF-beta1, protects keratinocytes against 12-O-tetradecanoylphorbol-13-acetate-induced cell death in vitro and in vivo. J Biol Chem 1999;274(7):4213–4219.
38. Go C, Li P, Wang XJ. Blocking transforming growth factor β signaling in transgenic epidermis accelerates chemical carcinogenesis: a mechanism associated with increased angiogenesis. Cancer Res 1999;59(12):2861–2868.
39. Cui W, Fowlis DJ, Bryson S, et al. TGFβ1 inhibits the formation of benign skin tumors, but enhances progression to invasive spindle carcinomas in transgenic mice. Cell 1996;86:531–542.
40. Blessing M, Nanney LB, King LE, Hogan BL. Chemical skin carcinogenesis is prevented in mice by the induced expression of a TGF-β related transgene. Teratog Carcinog Mutagen 1995; 15:11–21.
41. Amendt C, Schirmacher P, Weber H, Blessing M. Expression of a dominant negative type II TGF-β receptor in mouse skin results in an increase in carcinoma incidence and an acceleration of carcinoma development. Oncogene 1998;17:25–34.
42. Li AG, Wang D, Feng XH, Wang XJ. Latent TGFβ1 overexpression in keratinocytes results in a severe psoriasis-like skin disorder. EMBO J 2004;23(8):1770–1781.
43. Cui W, Kemp CJ, Duffie E, Balmain A, Akhurst RJ. Lack of transforming growth factor-β1 expression in benign skin tumors of p53null mice is prognostic for a high risk of malignant conversion. Cancer Res 1994;54:5831–5836.
44. He W, Cao T, Smith DA, Myers TE, Wang XJ. Smads mediate signaling of the TGFbeta superfamily in normal keratinocytes but are lost during skin chemical carcinogenesis. Oncogene 2001;20(4): 471–483.
45. Glick AB, Lee MM, Darwiche N, Kulkarni AB, Karlsson S, Yuspa SH. Targeted deletion of the TGF-β1 gene causes rapid progression to squamous cell carcinoma. Genes Dev 1994;8:2429–2440.
46. Vijayachandra K, Lee J, Glick AB. Smad3 regulates senescence and malignant conversion in a mouse multistage skin carcinogenesis model. Cancer Res 2003;63(13):3447–3452.
47. Liu X, Lee J, Cooley M, Bhogte E, Hartley S, Glick A. Smad7 but not Smad6 cooperates with oncogenic ras to cause malignant conversion in a mouse model for squamous cell carcinoma. Cancer Res 2003;63(22):7760–7768.
48. Go C, He W, Zhong L, et al. Aberrant cell cycle progression contributes to the early-stage accelerated carcinogenesis in transgenic epidermis expressing the dominant negative TGFbetaRII. Oncogene 2000;19(32):3623–3631.

49. Tremain R, Marko M, Kinnimulki V, Ueno H, Bottinger E, Glick A. Defects in TGFβ signaling overcome senescence of mouse keratinocytes expressing v-ras. Oncogene 2000;19(13):1698–1709.
50. Glick A, Popescu N, Alexander V, Ueno H, Bottinger E, Yuspa SH. Defects in transforming growth factor-β signaling cooperate with a ras oncogene to cause rapid aneuploidy and malignant transformation of mouse keratinocytes. Proc Natl Acad Sci USA 1999;96(26):14,949–14,954.
51. Ferbeyre G, de Stanchina E, Lin AW, et al. Oncogenic ras and p53 cooperate to induce cellular senescence. Mol Cell Biol 2002;22(10):3497–3508.
52. Serrano M, Lin AW, McCurrach ME, Beach D, Lowe SW. Oncogenic ras provokes premature cell senescence associated with accumulation of p53 and $p16^{INK4a}$. Cell 1997;88:593–602.
53. Michaloglou C, Vredeveld LC, Soengas MS, et al. BRAFE600-associated senescence-like cell cycle arrest of human naevi. Nature 2005;436(7051):720–724.
54. Weeks BH, He W, Olson KL, Wang XJ. Inducible expression of transforming growth factor beta1 in papillomas causes rapid metastasis. Cancer Res 2001;61(20):7435–7443.
55. Muraoka RS, Koh Y, Roebuck LR, et al. Increased malignancy of Neu-induced mammary tumors overexpressing active transforming growth factor beta1. Mol Cell Biol 2003;23(23):8691–8703.
56. Siegel PM, Shu W, Cardiff RD, Muller WJ, Massagué J. Transforming growth factor beta signaling impairs Neu-induced mammary tumorigenesis while promoting pulmonary metastasis. Proc Natl Acad Sci U S A 2003;100(14):8430–8435.
57. Hartsough MT, Frey RS, Zipfel PA, et al. Altered transforming growth factor beta signaling in epithelial cells when ras activation is blocked. J Biol Chem 1996;271:22,368–22,375.
58. Mulder KM, Morris SL. Activation of p21ras by transforming growth factor beta in epithelial cells. J Biol Chem 1992;267(8):5029–5031.
59. Kretzschmar M, Doody J, Timokhina I, Massagué J. A mechanism of repression of TGFbeta/Smad signaling by oncogenic Ras. Genes Dev 1999;13(7):804–816.
60. Kretzschmar M, Doody J, Massagué J. Opposing BMP and EGF signalling pathways converge on the TGF-β family mediator Smad1. Nature 1997;389:618–622.
61. Lo RS, Wotton D, Massagué J. Epidermal growth factor signaling via Ras controls the Smad transcriptional co-repressor TGIF. EMBO J 2001;20(1–2):128–136.
62. Saha D, Datta PK, Beauchamp RD. Oncogenic ras represses transforming growth factor-beta/Smad signaling by degrading tumor suppressor Smad4. J Biol Chem 2001;276(31):29,531–29,537.
63. Liu X, Sun Y, Weinberg RA, Lodish HF. Ski/Sno and TGF-beta signaling. Cytokine Growth Factor Rev 2001;12(1):1–8.
64. Dai C, Liu Y. Hepatocyte growth factor antagonizes the profibrotic action of TGF-beta1 in mesangial cells by stabilizing Smad transcriptional corepressor TGIF. J Am Soc Nephrol 2004;15(6): 1402–1412.
65. Brown KA, Aakre ME, Gorska AE et al. Induction by transforming growth factor-beta1 of epithelial to mesenchymal transition is a rare event in vitro. Breast Cancer Res 2004;6(3):R215–R231.
66. Oft M, Akhurst RJ, Balmain A. Metastasis is driven by sequential elevation of H-ras and Smad2 levels. Nat Cell Biol 2002;4(7):487–494.
67. Shi Y, Massagué J. Mechanisms of TGF-beta signaling from cell membrane to the nucleus. Cell 2003; 113(6):685–700.
68. Oft M, Peli J, Rudaz C, Schwarz H, Beug H, Reichmann E. TGF-beta1 and Ha-Ras collaborate in modulating the phenotypic plasticity and invasiveness of epithelial tumor cells. Genes Dev 1996; 10(19):2462–2477.
69. Janda E, Lehmann K, Killisch I, et al. Ras and TGFβ cooperatively regulate epithelial cell plasticity and metastasis: dissection of Ras signaling pathways. J Cell Biol 2002;156(2):299–313.
70. Piek E, Ju WJ, Heyer J et al. Functional characterization of transforming growth factor beta signaling in Smad2- and Smad3-deficient fibroblasts. J Biol Chem 2001;276(23):19,945–19,953.
71. Yang YC, Piek E, Zavadil J, et al. Hierarchical model of gene regulation by transforming growth factor beta1. Proc Natl Acad Sci U S A 2003;100(18):10,269–10,274.
72. Callahan JF, Burgess JL, Fornwald JA, et al. Identification of novel inhibitors of the transforming growth factor beta1 (TGF-beta1) type 1 receptor (ALK5). J Med Chem 2002;45(5):999–1001.
73. Inman GJ, Nicolas FJ, Callahan JF, et al. SB-431542 is a potent and specific inhibitor of transforming growth factor-beta superfamily type I activin receptor-like kinase (ALK) receptors ALK4, ALK5, and ALK7. Mol Pharmacol 2002;62(1):65–74.
74. Bonniaud P, Margetts PJ, Kolb M, et al. Progressive TGF-β1-induced lung fibrosis is blocked by an orally active ALK5 kinase inhibitor. Am J Respir Crit Care Med 2005;171:889–898.

75. de Gouville AC, Boullay V, Krysa G, et al. Inhibition of TGF-beta signaling by an ALK5 inhibitor protects rats from dimethylnitrosamine-induced liver fibrosis. Br J Pharmacol 2005;145:166–177.
76. Laping NJ, Grygielko E, Mathur A, et al. Inhibition of transforming growth factor (TGF)-beta1-induced extracellular matrix with a novel inhibitor of the TGF-beta type I receptor kinase activity: SB-431542. Mol Pharmacol 2002;62(1):58–64.
77. Grygielko ET, Martin WM, Tweed CW, et al. Inhibition of gene markers of fibrosis with a novel inhibitor of TGFβ-type I receptor kinase in puromycin-induced nephritis. J Pharmacol Exp Ther 2005; 313:943–951.
78. Uhl M, Aulwurm S, Wischhusen J, et al. SD-208, a novel transforming growth factor beta receptor I kinase inhibitor, inhibits growth and invasiveness and enhances immunogenicity of murine and human glioma cells in vitro and in vivo. Cancer Res 2004;64(21):7954–7961.

II TGF-β Superfamily Members in Normal and Tumor Biology

20 Proteomics Analysis of TGFβ Superfamily Members

Serhiy Souchelnytskyi

CONTENTS

STRUCTURAL FEATURES OF LIGANDS, RECEPTORS, AND SMAD PROTEINS DEFINE SPECIFICITY IN TGFβ FAMILY SIGNALING
PROTEIN–PROTEIN INTERACTION NETWORKS IN TGFβ FAMILY SIGNALING
POST-TRANSLATIONAL MODIFICATIONS OF RECEPTORS AND SMAD PROTEINS
FUNCTIONAL PROTEOMICS OF TGFβ FAMILY SIGNALING
SYSTEMIC ANALYSIS OF TGFβ FAMILY SIGNALING
FUTURE PROSPECTIVE
ACKNOWLEDGMENTS
REFERENCES

Abstract

Signaling by transforming growth factor-β (TGFβ) family members is a chain of protein–protein interactions and various post-translational modifications (PTMs), leading to protein translocations and regulation of cellular activities. Protein–protein interactions are defined by the structural features of ligands, receptors, and Smad proteins. TGFβ family ligands also regulate a number of proteins that are not directly associated with TGFβ signaling, e.g., various transcription factors, enzymes, components of cytoskeleton, and scaffold proteins. To explore TGFβ-initiated signaling in an unbiased and comprehensive way, large-scale technologies have to be employed. Proteomics, or a comprehensive study of proteomes, is one of such technologies. Proteomics approaches analysis of 3D structure of proteins and study of PTMs, e.g., phosphorylation, glycosylation, acetylation, ubiquitylation. Dissecting of protein–protein interaction networks in combination with functional proteomics provides an overview of proteome changes initiated by TGFβ family members. In this chapter, recent progress in proteome studies of TGFβ family signaling is discussed. Studies of protein interaction networks and multiple targets of TGFβ, which have provided insights into a number of novel functions and regulatory mechanisms, are emphasized.

Key Word: Proteomics; signaling; transforming growth factor-β; bone morphogenetic proteins; Smad.

1. STRUCTURAL FEATURES OF LIGANDS, RECEPTORS, AND SMAD PROTEINS DEFINE SPECIFICITY IN TGFβ FAMILY SIGNALING

Protein–protein interactions of ligands, receptors, and Smad proteins are the backbone of TGFβ/BMP/Smad-dependent signaling. Specificity and dynamics of the ligand–receptor,

From: *Cancer Drug Discovery and Development: Transforming Growth Factor-β in Cancer Therapy, Vol I: Basic and Clinical Biology*
Edited by: Sonia B. Jakowlew © Humana Press Inc., Totowa, NJ

receptor–receptor, receptor–Smad, and Smad–Smad interactions are defined by the structural features of interactors. Together with a number of post-translational modifications (PTMs) and interactions with other proteins and nucleic acids, it creates a variety of regulatory processes known as TGFβ family signaling *(1,2)*.

TGFβ family members are secreted in latent forms. The most studied are complexes of TGFβ with the latency-associated protein (LAP). LAP–TGFβ complex often contains another protein e.g., the latent TGFβ binding protein (LTBP). Generation of an active TGFβ requires changes in the structure of latent complexes. This conformational change can be induced by a proteolytic cleavage, low pH, or interaction with other proteins *(3)*. However, structural changes on activation of TGFβ have not been described. Structural studies have shown that ligands of TGFβ family have conserved cysteine knot *(4)*. Specificity of ligands is provided by the secondary structures of ligand dimers. It is also suggested that more than one surface of the ligand contributes to the interaction with specific receptors. As an example, specificity of TGFβ can be provided by the N-terminal segment, the central α3-helix, and the external sequences of each monomer *(5)*.

Contact surfaces between ligands and receptors have been described for TGFβ3 and TβR-II, BMP2 and BMPR-IA, BMP7 and ActRII, and activin A and ActRIIB *(6–10)*. The main conclusion from these works is that surfaces of both monomers contribute to interaction with the receptors. However, these surfaces differ. For BMP2, the contact is made by the finger-helix grooves and "wrist" epitope of BMP2 and the knob-and-pocket motif in BMPR-IA *(7)*. For TGFβ3, the fingertip epitope is involved in the interaction with acidic residues of TβR-II *(6)*, and the "knuckle" epitopes of BMP7 and activin A are involved in the interaction with ActRII and ActRIIB, respectively *(8–10)*.

Specific signaling receptors of TGFβ family form heterotetrameric complexes upon binding of ligand and subsequent activation *(1,11,12)*. Receptors may also form preassembled complexes, which on interaction with the ligand form heterotetrameric complex *(11,12)*. In this complex, two type II and two type I receptors have been described. Receptor surfaces which mediate receptor–receptor complex formation have not been identified. However, it has been shown that the ligand-dependent activation, and probably complex formation, may change intramolecular structural features of the receptors. This has been shown for the long form of type II BMP receptor (BMPR-II(L)). Notably, BMPR-II(L) activation includes changes of conformation and/or position of the long cytoplasmic tail with respect to the kinase domain of the receptor. This conformational change may regulate repertoire of interacting proteins and affect signaling *(13,14)*.

Phosphorylation of the Gly-Ser-rich region in type I TGFβ receptor (TβR-I) has been proposed to create a site for interaction with TβR-I substrate, Smad2 and Smad3 *(15)*. Multiple phosphorylation of this Gly-Ser-rich region at serine and threonine residues generates a negatively charged surface, which interacts with a positively charged groove next to the L3 loop of Smad2 *(15,16)*. Smad–type I receptor interaction has been found to be mediated by L45 loop in type I receptors and L3 loop in the MAD homology domain-2 (MH2) of receptor-regulated Smads *(16)*. αH1-helix and the C-terminal tail of Smads also define the specificity of interaction *(17–19)*. Notably, Arg462 and Cys463 provide specificity of substrate recognition by the TβR-I *(18)*.

Domain structures in Smad proteins have been described for the N-terminal MH1 and the C-terminal MH2 domains, whereas no defined structural features were observed for the central linker region *(20)*. Phosphorylation of receptor-activated Smad proteins at the C-terminal serine residues was found to trigger intracellular signaling *(21,22)*. This phosphorylation introduces certain rigidness to the conformation of the C-terminal tail of the Smads *(19,23)*. There is also a suggestion that the C-terminal phosphorylation defavors intramolecular interaction between the MH1 and the MH2 domains, leading to "opening" of the Smad protein *(20–22)*. However, structural changes in Smads corresponding to the phosphorylation-dependent "opening" have not been described. Neither effects of ubiquitylation and acetylation on the conformation of Smads have been explored in structural studies.

Structural studies showed that receptor-regulated Smads form homo- or heterotrimeric complexes, with inclusion of one Smad4 and two Smad2 or Smad3 in heterotrimers *(19,23, 24)*. Identification of more than 50 Smad-interacting proteins is the strong indication that practically all surfaces of Smads may be involved in interactions with other proteins *(25–27)*. A number of motifs have been described; examples are Smad activation domain (SAD), nuclear localization (NLS) and nuclear export (NES) signals, DNA-binding region in the MH1 domain, and PY region for binding of E3-ligases *(25–28)*. Many interacting proteins use multiple structures in Smads to enhance the affinity and specificity of complex formation. For example, Smad anchor for receptor activation (SARA) interacts with three different regions in the MH2 domain of Smad2, e.g., αH2-, αH5-helixes, and β8 sheet *(27,28)*.

The reports of phosphoryl group mediating an interaction of Smad2 and Smad4 *(22,23)* and structural similarity of the MH2 domain with the forkhead domain *(29)* indicated that the MH2 domain may have properties of a phosphoserine/phosphothreonine binding domain. To confirm this suggestion, structural studies revealed that the C-terminal tail of activated receptor-regulated Smads may interact with phosphoserine binding pocket formed by the L3 loop and β8 sheet of the MH2 domain of Smad4 *(23,28,30)*. Thus, the MH2 domain may represent a novel type of phosphoamino acid binding domains.

This brief overview of interactions between components of TGFβ family signaling illustrates multiplicity of modes of interactions by exposure and access of structural features, as well as induction of conformational changes on interactions and PTMs. It has to be noted that some conclusions are based on approximation of results obtained with few selected proteins to other components of TGFβ family signaling. Though it allows predictions of overall structure and functional activities, it certainly will not detect all specific features. This urges for high-throughput structure determination. Today proteomics and structural biology fields provide tools that make such studies possible (Fig. 1) *(31)*.

Inclusion of components of TGFβ signaling in this pipeline raises hope that in the nearest years structures of single proteins involved in this signaling, as well as protein complexes will be solved.

2. PROTEIN–PROTEIN INTERACTION NETWORKS IN TGFβ FAMILY SIGNALING

Treatment of cells with TGFβ initiates a number of interactions between cellular components of TGFβ signaling, e.g., receptors and Smads and proteins that are not exclusively specific to TGFβ signaling. Notably, Smads form complexes with various transcription factors, kinases, and proteins regulating transport and degradation *(25–28)*. Studies of selected TGFβ receptor- and Smad-interacting proteins have shown that these proteins can significantly modulate TGFβ signaling. As an example, an interaction of TGFβ receptors with E3 ubiquitin ligases, e.g., Smurf2 or Smurf1, may inhibit signaling *(32,33)*, whereas an interaction of receptors with Smad-presenting protein SARA may promote activation of Smad2 *(34)*. Smad-interacting proteins may have even more profound effects, showing a capability to change the activity of a Smad protein from a stimulator to an inhibitor. For example, Smad3 interaction with transcriptional cofactors CBP/p300 or transcriptional repressors c-Ski and SnoN may result in stimulation or inhibition of transcriptional activity, respectively *(35–39)*. Identification of receptor-and Smad-interacting proteins showed firmly that knowledge of formed complexes and understanding of their dynamics are essential for delineating biological effects of TGFβ family members. However, many of the interacting proteins have been identified

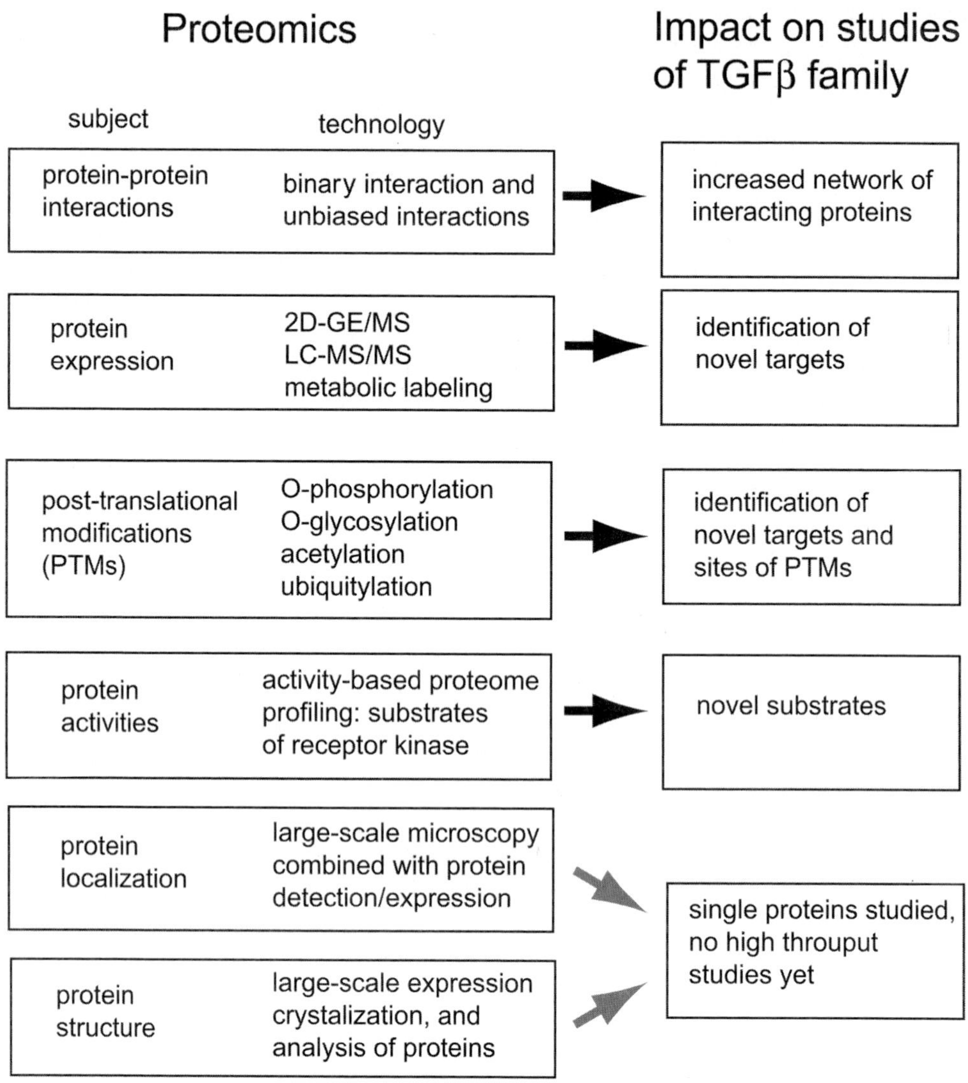

Fig. 1. Application of proteomics to studies of TGFβ family signaling. Impact of proteomics on delineating protein–protein interactions and studies of protein expression, activities, and PTMs are indicated in connection with the employed tools. Proteomics is represented by subjects of studies and most often used technologies. Black arrows indicate fields of research where progress in generation of results has been achieved. Gray arrows indicate fields of research where TGFβ family signaling has not been studied extensively.

in studies of a limited number of interactors, often only two selected proteins *(25–28)*. Such small scale does not allow a comprehensive overview of the interaction network and does not provide insights into the dynamics of various interactions.

To obtain a comprehensive overview of protein–protein interactions, studies of interaction networks at large scale and in unbiased way have to be performed. Possibilities of such studies have been provided by proteomics (Fig. 1). Two types of techniques have been applied to unveil protein–protein interactions in TGFβ signaling. Techniques of the first type detect binary interactions by coprecipitation of two preselected proteins and by yeast two-hybrid screens.

Techniques of the second type identify protein complexes by an unbiased coprecipitation of mammalian proteins with bait constructs representing components of TGFβ signaling *(26)*.

Binary interactions using yeast two-hybrid screens have delivered a set of data suitable for a large-scale analysis. Colland et al. *(40)* studied interactions with 11 Smad and proteins involved in TGFβ signaling, using the human cDNA library. They established a network of 755 interactions, which involves 591 proteins. As a proof of efficiency, 18 proteins known to be involved in TGFβ signaling were detected. One of the detected novel interactors with Smad8 was LMO4 protein, which has LIM domains and serves as an adaptor for protein–protein interactions. Increased LMO4 expression induced mammary hyperplasia, promoted cell invasion, and predicted a poor outcome for breast cancer *(41)*. Another interacting protein with described involvement in human cancers is LAPTm5. Its inactivation was observed in multiple myeloma *(42)*, and it is upregulated in B-lymphomas *(43)*. Downregulation of LAPTm5 level led to enhanced sensitivity of cells to TGFβ. A number of novel or hypothetical proteins were also identified. Thus, KIAA1196, which contains seven zinc fingers and a leucine zipper, may participate in transcriptional regulation. Downregulation of this protein resulted in the inhibition of TGFβ-dependent transcription. Identification by Colland et al. *(40)* of 179 proteins with poor annotation of functions may lead to the identification of novel signaling processes initiated by TGFβ and Smad proteins. This illustrates the efficiency of yeast two-hybrid screens in the discovery of interaction networks.

Yeast two-hybrid screen was also used to explore the interaction network of *C. elegans* DAF7/TGFβ signaling. Seventy-one interactions of 59 proteins have been described using *C. elegans* cDNA library *(44)*. Genetic perturbation analysis of identified interactions confirmed the modulatory character of a number of identified interactors. Validation study with the DAF-3/Smad4 interacting protein W01G7.1/DAF-5 showed its functional homology to Ski/ Sno. As was expected for a Ski/Sno homolog, W01G7.1/DAF-5 inhibited TGFβ-dependent transcriptional activity *(44)*. Though screens with *C. elegans* cDNA library are important for understanding TGFβ signaling in worm development, links of the identified interactors to mammalian systems have to be explored.

To explore binary interactions inside of mammalian cells, Barrios-Rodiles et al. *(45)* developed a technique that is based on the detection of luciferase activity linked to a bait, whereas tagged prey is enriched by precipitation with antitag antibodies. Screening of 518 cDNA prey clones with core members of the TGFβ pathway revealed 947 interactions. Links were detected between TGFβ signaling and the p21-activated kinase network, to the polarity complex and to occludin *(45)*. TGFβ type I receptor interaction with occludin may affect dissolution of tight junctions during invasiveness of tumor cells, suggesting the biological relevance of observed interactions. However, binary interaction screens are not able to evaluate cooperativity of interactions between multiple proteins in a single complex.

Proteins function in complexes and identification of such complexes may be more informative for biological relevance, as compared to studies of binary interactions. In the TGFβ signaling field, two studies have described Smad3 and BMPR-II complexes with mammalian proteins *(13,46)*. Purified complexes were resolved by two-dimensional gel electrophoresis (2D-GE) and proteins were identified by mass spectrometry (MS). Pull-down with GST fusion constructs of Smad3 and BMPR-II revealed 26 proteins in complex with Smad3 *(46)* and 33 proteins in complexes with BMPR-II (13) constructs. Smad3 interaction with SREBP-2 was found to inhibit Smad3-dependent transcriptional activation. SREBP-2 is a potent stimulator of sterol synthesis, and its interaction with Smad3 indicates that enhanced sterol production may be coordinated with downregulation of TGFβ signaling. Some of the identified Smad3-interacting proteins are also known as regulators of cellular functions. As an example, chain A of ornithine decarboxylase interacted with Smad3 suggesting a novel physical connection of TGFβ signaling to the regulation of the cell migration by affecting metabolic processes *(46)*.

Twenty-four of the identified Smad3-interacting proteins have not been previously reported as involved in TGFβ signaling. Seven of these proteins have ascribed functions in transcriptional regulation, RNA processing, and regulation of protein degradation. Thus, an exploration of proteins that form complexes with a nonactivated Smad3 has already significantly enlarged the Smad3 interaction network *(46)*. Studies of changes in composition of intracellular complexes on Smad activation are required for understanding their dynamics.

The proteomics-based studies discovered a novel paradigm in the crosstalk between serine/threonine and tyrosine kinase receptors. The identification of tyrosine kinase receptors in a complex with serine/threonine kinase BMPR-II receptor showed that these two types of signaling systems already interact physically on the receptor level *(13,14,47)*. Two tyrosine kinase receptors were found to interact with BMPR-II (c-kit and an orphan receptor; 13,14) and one with BMPR-Ib (Ror2; 47). An in-depth study of BMPR-II interaction with the c-kit receptor showed that this interaction is important for the cooperation between BMP and the c-kit ligand stem cell factor in regulation of the transcription and differentiation of cells *(14)*. Other identified proteins provided physical links to the regulation of apoptosis, ion transport, trafficking, transcription, and cytoskeleton rearrangements. An involvement of Ras and mitogen-activated protein kinase (MAPK) in protumorigenic signaling is well established *(1,2)*. Detection of MOS kinase, MAPKKK8, and protein kinase Cβ (PKCβ) in a complex with BMPR-II indicated physical links to BMP-dependent regulation of MAPK activity *(13)*. Four BMPR-II interacting proteins have been described as regulators of transcription, e.g., CtBP, forkhead L1 transcription factor, LIM hd1, and p050b. These proteins indicate that BMPR-II may have direct effect on transcriptional regulators beside Smad proteins. Identification of BMPR-II interacting proteins showed that the long C-terminal tail of the receptor provides binding sites for a number of proteins. Similar complexity of signaling is also expected for other receptors of TGFβ family, and further large-scale protein–protein interaction studies are awaited.

Thus, binary and protein complex interaction screens have uncovered a number of novel activities associated with TGFβ family signaling. However, to delineate protein–protein interaction networks in their dynamics, protein complexes have to be studied during a defined time with identification of complexes at a number of functionally relevant time points. Such studies are much more laborious and have not yet been performed with components of TGFβ signaling.

3. POST-TRANSLATIONAL MODIFICATIONS OF RECEPTORS AND SMAD PROTEINS

The following PTMs have been described for the components of TGFβ family signaling: regulated proteolytic cleavage, N-glycosylation, phosphorylation, acetylation, ubiquitylation, and sumoylation *(1–3,25,27,30)*.

Phosphorylation of receptors and Smads changes the structural features of these proteins in such a way that it initiates receptor–Smad and Smad–Smad interactions *(1,30)*. The receptors and Smad proteins are phosphorylated at multiple sites. Though most of the phosphorylation is at the serine and threonine residues, tyrosine phosphorylation has also been described for receptors *(48)*. Another PTM described for both receptors and Smad proteins is ubiquitylation, which has been proposed as the main mechanism of degradation for these proteins *(49)*. Sumoylation in contrary may stabilize Smad proteins *(50)*. Acetylation has been described for Smads, as a modification regulating proteins' stability; for Smad7, acetylation and ubiquitylation were shown to have a reciprocal role in the regulation of Smad7 stability *(51)*. These PTMs have been studied in selected proteins, with approaching one PTM or even one site of a modification at the time.

Proteomics has developed techniques, which allow such a comprehensive overview of PTMs (Fig. 1) *(52)*. For example, phosphorylation can be studied by enrichment of phosphoproteins followed by detection of loss of phosphoryl group from the peptide or by chemical modifications of phosphorylated residues *(53)*. Acetylation is another PTM that is stable during separation and identification of proteins. It is manifested in a characteristic shift of modified peptides, as compared to nonmodified peptides. Glycosylation can be detected using deglycosylation enzymes, antibodies, chemical modification of glycosylated residues, and by mass shifts of glycosylated vs nonglycosylated peptides, which can be detected by mass spectrometry *(52)*. A number of observations suggest that Smad proteins may have more than five types of PTMs in one polypeptide chain, and the extent and sites of these PTMs change on treatment of the cells with TGFβ and Smad activation (Bhaskaran, Iwahana, and Souchelnytskyi, unpublished observation). Further studies of the dynamics of PTMs are expected to provide insights into receptor and Smad activation, localization, and degradation.

4. FUNCTIONAL PROTEOMICS OF TGFβ FAMILY SIGNALING

Proteins crucial for TGFβ signaling may change their expression and/or activities without formation of physical complexes with TGFβ receptors or Smads. Proteome profiling of TGFβ signaling in various cell lines have described a number of such proteins affected by TGFβ *(26)*. The striking outcome of these studies has been the identification of a significant number of novel targets for TGFβ. The identified proteins have enlarged the TGFβ signaling network and unveiled activities that were not previously associated with TGFβ.

The growth inhibitory effect of TGFβ has been associated with its regulation of cyclin-dependent kinases (CDK), CDK inhibitors, cyclins, and cdc25A phosphatase *(1,25,27)*. In addition, proteomics studies identified the far upstream element-binding protein (FUSEBP) as a TGFβ target *(54)*. FUSEBP is an important activator of c-myc transcription, and c-myc downregulation is one of the key activities of TGFβ required for inhibition of cell proliferation. It has been observed that downregulation of FUSEBP is a mechanism that may contribute to shut off c-myc expression. Additional pathways that affect cell proliferation can be attributed to novel TGFβ-dependent regulators of transcription. As an example, TGFβ-induced phosphorylation of general transcription factor II-I (TFII-I) was observed in phosphoproteomics studies *(55)*. This TFII-I phosphorylation modulated TGFβ-dependent regulation of expression of cyclin D2, cyclin D3, and E2F2. Proteomics provides also strong indications that cell proliferation can be affected by TGFβ-dependent phosphorylation of components of translational machinery, e.g., elongation factors and ribosomal proteins *(55)*. Proteome profiling of Smad4-dependent changes in pancreatic carcinoma cells identified 14-3-3ζ, CSN8, and stathmin, which have been described as cell cycle regulators *(56)*. Thus, proteomics has shown a multiplicity of TGFβ effects on cell proliferation.

TGFβ is a potent regulator of cellular morphology. Therefore identification of a significant number of cytoskeleton components as TGFβ targets has been expected. Twenty-eight out of the 81 protein spots affected by TGFβ in protein expression maps of human trabecular meshwork cells are components of the cytoskeleton *(57)*. Trabecular meshwork cells regulate intraocular pressure. An increase of intraocular pressure may be associated with higher levels of TGFβ, thus implicating TGFβ effects on these cells in the development of glaucoma.

TGFβ potently stimulates chondrogenic differentiation of bone marrow mesenchymal stem cells. Proteomics analysis showed that TGFβ regulates a number of proteins involved in cytoskeleton rearrangements, matrix deposition, and metabolism in these cells *(58)*. Functional studies confirmed that the TGFβ-dependent increase of α-actin and the decrease of gelsolin are important for TGFβ-induced chondrogenic differentiation of cells. This insight into mechanisms of differentiation may open up clinical applications of bone marrow stem cells by providing means to direct their differentiation.

TGFβ-dependent changes in extracellular matrix deposition are thought to be crucial for the development of fibrosis. Asthma is one such fibrotic conditions, which involves remodeling of connective tissue. Proteome profiling of TGFβ action on human lung cells revealed expression changes for four protein spots in 2D gels. This correlated with changes observed on transformation of normal fibroblasts to myofibroblasts. An increased expression of actin and tropomyosin on TGFβ treatment was observed, suggesting a mechanism of matrix deposition in asthma *(59)*.

Cytoskeleton changes were also observed in proteome profiling of pancreatic cancer cells upon silencing of Smad4 expression *(56)*. TGFβ inhibition of vimentin expression and effects on HSP27, profiling-II, and ARP3 are proposed to be of importance for cytoskeleton rearrangements in these cells. However, confirmation of this suggestion requires further functional studies.

Rearrangement of the cytoskeleton is one of the hallmarks of tumorigenesis-related remodeling of epithelial cells and affects invasiveness of cancer cells. Among the identified TGFβ-regulated proteins in epithelial cells are various keratins (K2, K8, K10, K18), regulatory light chain of myosin, GFAP, lamin A, and α-chain of tropomyocin *(54,55)*. TGFβ-dependent inhibition of protease MMP13 expression may contribute to the formation of an extracellular matrix *(60)*, which is of importance for the survival and formation of tumors by carcinogenic cells. Thus, changes in the expression and phosphorylation of proteins involved in cytoskeleton remodeling have been observed in various cell lines, e.g., of epithelial and mesenchymal origin, as well as in the functional context of proliferation, migration, and differentiation of cells.

TGFβ is also a potent regulator of apoptosis, and in many studied cells TGFβ promoted apoptosis. Regulation of apoptosis is especially crucial for morphogentic activities of bone morphogenetic proteins *(61)*. In tumorigenesis, escape from proapoptotic signals is an essential feature of malignant cells. Observed upregulation of caspase 10 and downregulation of caspase 3 by TGFβ may contribute to the execution phase of apoptosis *(54)*. TGFβ-dependent decrease of glycosylation of CIDE-A protein is of importance for stimulation of apoptosis. CIDE-A role in cell death is thought to initiate degradation of genomic DNA *(62)*, and glycosylation of CIDE-A was found to regulate nuclear localization of CIDE-A, thus modulating TGFβ-dependent apoptosis *(63)*.

The strong immunosuppressive effect of TGFβ has been described in a number of studies *(64)*. Proteomics data add to the list of TGFβ targets in immunomodulation. Proteomics has shown that the downregulation of caspase 1 *(54)* could contribute to the effect of TGFβ on inflammatory responses. Caspase 1 regulates maturation of proinflammatory cytokines *(65)*, and its inhibition may affect levels of cytokines and subsequently inhibit immune responses.

Proteomics studies have also unveiled the direct effect of TGFβ on DNA damage repair, which is a novel pathway for TGFβ. Downregulation of Rad51 and Rag1 and interaction of Smad3 with BRCA1 have shown that TGFβ can inhibit repair of DNA double-strand breaks *(54,66)*. Upon low-intensity DNA damage it may promote chromosomal instability, whereas upon substantial DNA damage, e.g., on treatment with anti-cancer drugs, TGFβ can enhance DNA damage-induced signaling leading to cell death. Proteomics-based findings have shown that TGFβ can directly influence the efficiency of DNA damage repair mechanisms and therefore regulate chromosomal instability *(54,66)*.

TGFβ affects the expression of a number of metabolic enzymes. Proteomics studies identified upregulation of fructose biphosphate aldolase and glyceraldehyde 3-phosphate dehydrogenase, which are involved in glycolysis in breast ductal carcinomas and in cultured mammary epithelial cells *(67,68)*. TGFβ-dependent regulation of phosphorylation of glucose-6-phosphate dehydrogenase and enolase-1, as well as glycosylation of alpha 1,2-fucosyl transferase, has been observed in human mammary epithelial cells *(55,63)*. TGFβ also increased the expression of phosphoglycerate kinase-1 and downregulated the expression of inosine-5-monophosphate dehydrogenase in epithelial cells *(54)*. TGFβ-dependent changes in expression and PTMs of metabolic enzymes, together with the identification of enzymes as BMPR-II- and Smad3-interacting proteins, are therefore the novel way to influence cell functions.

Changes in expression of metabolic enzymes were observed in response to deletion of myostatin. Myostatin is a member of TGFβ family and belongs to the bone morphogenetic protein group. Deletion of myostation may cause increased muscle mass *(69)*. Using 2D-GE and mass spectrometry, Picard and colleagues identified 13 proteins with altered expression in muscle in response to myostatin deletion *(70)*. Observed increase in expression of phosphoglycomutase and decrease of fatty acid-binding protein indicated changes of energy production in muscle. In addition, deletion of myostatin resulted in significant changes in contractile apparatus of muscle *(70)*. Eight of the 13 affected proteins are involved in muscle contraction. The pattern of changes in protein spots migration indicated that myostatin deletion affected not only the expression of proteins, but also their PTM *(70)*.

Another cell type that showed potential changes in metabolism regulation on TGFβ treatment is human trabecular meshwork cells. In these cells, TGFβ downregulated glucose-6-phosphate 1-dehydrogenase and changed the expression of pyruvate kinase isoforms *(57)*. However, no functional studies of the possible TGFβ-induced metabolism changes have been performed with trabecular meshwork cells.

Genetic ablation of TGFβ family members often resulted in defects of angiogenesis in development, indicating an important role of TGFβ in angiogenesis *(71)*. Analysis of TGFβ1-regulated proteins in human microvascular endothelial cells showed that TGFβ1 affected cell morphology and proliferation *(72)*. Regulatory processes affected by TGFβ include transcription, RNA processing, translation, trafficking, stress response, and metabolism regulation *(72)*. All these activities are common targets for TGFβ1 in endothelial and epithelial cells *(54,71)*. TGFβ is known to modulate the environment in tumors, including inhibiting the recognition of tumor cells by immune surveillance system. An inhibition of expression of MHC class II antigens, CSF-1, and protein similar to Chic1 and changes in expression of T-cell receptor chains may contribute to the TGFβ1 anti-inflammatory activity. Proteomics of endothelial cells has also shown that E2F6 downregulation may counteract induction of CDK inhibitors *(72)*. This is an example of initiation of both growth-promoting and growth-inhibiting pathways by TGFβ, which suggests that the balance between these two pathways defines the level of inhibition of cell proliferation.

Proteomics data allow analysis of TGFβ effects on selected cellular activities, e.g., proliferation, apoptosis, cytoskeleton rearrangement, or differentiation. Such clustering of TGFβ targets shows multiplicity of mechanisms that TGFβ employs to regulate a single selected activity. The overlap of activities regulated by TGFβ in various cells was also observed. This is expected, as the main signal transducers of TGFβ in these cells are the same, e.g., receptors and Smad proteins.

Another way of analysis of TGFβ signaling is by using specific cellular models. For example, tumorigenesis studies are often concentrated on TGFβ targets in epithelial cells, whereas BMPs have been studied using osteoblasts *(1,2,11–14)*. Model-oriented studies have shown that TGFβ regulates multiple activities in one given cell line. As an example, in human epithelial cells TGFβ regulates proteins involved in proliferation, migration, cytoskeleton rearrangements, apoptosis, DNA damage repair, and immunomodulation. Thus, functional proteomics has shown that TGFβ regulates multiple signaling pathways, with a significant similarity in the sets of its targeted functions in various cell lines.

5. SYSTEMIC ANALYSIS OF TGFβ FAMILY SIGNALING

Constantly growing number of proteins regulated by TGFβ family members urges for tools for a comprehensive analysis. Times when one protein or gene was selected for an in-depth study and was claimed as a main target in a certain activity are to fade away, as the concept of network signaling is entering the stage of biology *(73)*. For the TGFβ family signaling, discovery of as many targets as possible is the first step. The next step will be to define relations between identified valuables/proteins and genes. Attempts to extract such dependencies from microarray data have been described, but not for TGFβ signaling *(74)*. However, correlations of mRNA and corresponding protein expression have been reported to be below 50%. This suggests that microarray data have to be validated on the protein level or in functional studies. Proteome profiling directly describes changes in protein status, e.g., expression, interactions, modifications, and activity. No large-scale modeling has been performed with proteomics data so far, though unpublished studies indicate that proteomics provides sufficiently rich datasets to build an informative model (Souchelnytskyi, Woksepp, and Bhaskaran, unpublished data).

6. FUTURE PROSPECTIVE

Despite the fact that proteomics studies of TGFβ signaling are in their infancy, they have already provided significant insights into the complexity of signaling. Proteomics has shown that TGFβ family members regulate cellular functions via multiple pathways and targets. Both positive and negative signals can be extracted from the analysis of proteomics data, indicating existence of feed-forward and feedback loops. Richness of proteomics data and development of tools for large-scale modeling provide a ground for successful modeling of TGFβ signaling. Such modeling will facilitate search for sensitive targets, which can be used for drug development and therapeutic interventions. This is a thrilling possibility which can be realized in the nearest years.

ACKNOWLEDGMENTS

The work in the author's group is supported in part by grants from the Swedish Cancer Society, the Swedish Research Council, the EU program RTN "EpiPlastCarcinoma", INTAS, UICC, Hiroshima University, and Merck KGaA.

REFERENCES

1. Derynck R, Zhang YE. Smad-dependent and Smad-independent pathways in TGF-beta family signalling. Nature 2003;425:577–584.
2. Souchelnytskyi S. TGF-beta signalling and its role in cancer. Exp Oncology 2002;24:3–12.
3. Koli K, Saharinen J, Hyytiäinen M, Penttinen C, Keski-Oja J. Latency, activation, and binding proteins of TGF-β. Microsc Res Tech 2001;52:354–362.
4. Vitt UA, Hsu SY, Hsueh AJW. Evolution and classification of cyteine knot-containing hormones and related extracellular signaling molecules. Mol Endocrinol 2001;15:681–694.
5. Daopin S, Piez KA, Ogawa Y, Davies DR. Crystal structure of transforming growth factor-beta2: an unusual fold for the superfamily. Science 1992;257:369–373.
6. Hart PJ, Deep S, Taylor AB, Shu Z, Hinck CS, Hinck AP. Crystal structure of the human TβR2 ectodomain—TGF-beta3 complex. Nat Struct Biol 2002;9:203–208.
7. Kirsch T, Sebald W, Dreyer MK. Crystal structure of the BMP-2-BRIA ectodomain complex. Nat Struct Biol 2000;7:492–496.
8. Greenwald J, Groppe J, Gray P, et al. The BMP7/ActRII extracellular domain complex provides new insights into cooperative nature of receptor assembly. Mol Cell 2003;11:605–617.
9. Thompson TB, Woodruff TK, Jardetzky TS. Structures of an ActRIIB:activin A complex reveal a novel binding mode for TGF-beta ligand:receptor interactions. EMBO J 2003;22:1555–1566.
10. Sebald W, Mueller TD. The interaction of BMP-7 and ActRII implicates a new mode of receptor assembly. Trends Biochem Sci 2003;28:518–521.
11. Massagué J. TGF-beta signal transduction. Annu Rev Biochem 1998;67:753–791.
12. Nohe A, Keating E, Knaus P, Petersen NO. Signal transduction of bone morphogenetic protein receptors. Cell Signal 2004;16:291–299.

13. Hassel S, Eichner A, Yakymovych M, Hellman U, Knaus P, Souchelnytskyi S. Proteins associated with type II bone morphogenetic protein receptor (BMPR-II) and identified by two-dimensional gel electrophoresis and mass spectrometry. Proteomics 2004;4:1346–1358.
14. Hassel S, Yakymovych M, Hellman U, Rönnstrand L, Knaus P, Souchelnytskyi S. Interaction and functional cooperation between the serine/threonine kinase bone morphogenetic protein type II receptor with the tyrosine kinase stem cell factor receptor. J Cell Physiol 2006;206:457–467.
15. Huse M, Muir TW, Xu L, Chen YG, Kuriyan J, Massagué J. The TGF-beta receptor activation process: an inhibitor- to substrate-binding switch. Mol Cell 2001;8:671–682.
16. Lo RS, Chen YG, Shi Y, Pavletich NP, Massagué J. The L3 loop: a structural motif determining specific interactions between SMAD proteins and TGF-beta receptors. EMBO J 1998;17:996–1005.
17. Chen YG, Massagué J. Smad1 recognition and activation by the ALK1 group of transforming growth factor-beta family receptors. J Biol Chem 1999;274:3672–3677.
18. Yakymovych I, Heldin C-H, Souchelnytskyi S. Smad2 phosphorylation by type I receptor: contribution of arginine 462 and cysteine 463 in the C terminus of Smad2 for specificity. J Biol Chem 2004;279: 35,781–35,787.
19. Chacko BM, Qin B, Correia JJ, Lam SS, de Caestecker MP, Lin K. The L3 loop and C-terminal phosphorylation jointly define Smad protein trimerization. Nat Struct Biol 2001;8:248–253.
20. Shi Y. Structural insights on Smad function in TGF-beta signaling. BioEssays 2001;23:223–232.
21. Macias-Silva M, Abdollah S, Hoodless PA, Pirone R, Attisano L, Wrana JL. MADR2 is a substrate of the TGF-beta receptor and its phosphorylation is required for nuclear accumulation and signaling. Cell 1996;87:1215–1224.
22. Souchelnytskyi S, Tamaki K, Engström U, Wernstedt C, ten Dijke P, Heldin C-H. Phosphorylation of Ser465 and Ser467 in the C terminus of Smad2 mediates interaction with Smad4 and is required for transforming growth factor-β signaling. J Biol Chem1997;272:28,107–28,115.
23. Wu JW, Hu M, Chai J, et al. Crystal structure of a phosphorylated Smad2. Recognition of phosphoserine by the MH2 domain and insights on Smad function in TGF-beta signaling. Mol Cell 2001;8:1277–1289.
24. Qin BY, Chacko BM, Lam SS, de Caestecker MP, Correia JJ, Lin K. Structural basis of Smad1 activation by receptor kinase phosphorylation. Mol Cell 2001;8:1303–1312.
25. Itoh S, Itoh F, Goumans MJ, ten Dijke P. Signaling of transforming growth factor-beta family members through Smad proteins. Eur J Biochem 2000;267:6954–6967.
26. Souchelnytskyi S. Proteomics of TGF-beta signalling and its impact on breast cancer. Expert Rev Proteomics 2005;6:925–935.
27. ten Dijke P, Hill C. New insights into TGF-beta-Smad signalling. Trends Biochem Sci 2004;29:265–273.
28. Souchelnytskyi S, Moustakas A, Heldin C-H. TGFbeta signalling from a 3-dimensional perspective: insight into selection of partners. Trends Cell Biol 2002;12:304–307.
29. Durocher D, Taylor IA, Sarbassova D, et al. The molecular basis of FHA domain:phosphopeptide binding specificity and implications for phospho-dependent signaling mechanisms. Mol Cell 2000;6: 1169–1182.
30. Yakymovych I, Souchelnytskyi S. Regulation of Smads functions by phosphorylation. In: Smad Signal Transduction-Smads in Proliferation, Differentiation and Disease. Heldin C-H, ten Dijke P. (Springer Verlag, Dordrecht, Germany) 2006;pp.235–252.
31. Liu HL, Hsu JP. Recent developments in structural proteomics for protein structure determination. Proteomics 2005;5:2056–2068.
32. Kavsak P, Rasmussen RK, Causing CG, et al. Smad7 binds to Smurf2 to form an E3 ubiquitin ligase that targets the TGF-beta receptor for degradation. Mol Cell 2000;6:1365–1375.
33. Ebisawa T, Fukuchi M, Murakami G, et al. Smurf1 interacts with transforming growth factor-beta type I receptor through Smad7 and induces receptor degradation. J Biol Chem 2001;276:12,477–12,480.
34. Tsukazaki T, Chiang TA, Davison AF, Attisano L, Wrana JL, SARA, a FYVE domain protein that recruits Smad2 to the TGFbeta receptor. Cell 1998;95:779–791.
35. Janknecht R, Wells NJ, Hunter T. TGF-beta-stimulated cooperation of smad proteins with the coactivators CBP/p300. Genes Dev 1998;12:2114–2119.
36. Feng XH, Zhang Y, Wu RY, Derynck R. The tumor suppressor Smad4/DPC4 and transcriptional adaptor CBP/p300 are coactivators for Smad3 in TGF-beta-induced transcriptional activation. Genes Dev 1998;12:2153–2163.
37. Nishihara A, Hanai JI, Okamoto N, et al. Role of p300, a transcriptional coactivator, in signalling of TGF-beta. Genes Cells 1998;3:613–623.

38. Liberati NT, Moniwa M, Borton AJ, Davie JR, Wang X-F. An essential role for Mad homology domain 1 in the association of Smad3 with histone deacetylase activity. J Biol Chem 2001;276:22,595–22,603.
39. Mizuide M, Hara T, Furuya T, Takeda M, et al. Two short segments of Smad3 are important for specific interaction of Smad3 with c-Ski and SnoN. J Biol Chem 2003;278:531–536.
40. Colland F, Jacq X, Trouplin V, et al. Functional proteomics mapping of a human signaling pathway. Genome Res 2004;14:1324–1332.
41. Sum EY, Peng B, Yu X, et al. The LIM domain protein LMO4 interacts with the cofactor CtIP and the tumor suppressor BRCA1 and inhibits BRCA1 activity. J Biol Chem 2002;277:7849–7856.
42. Hayami Y, Iida S, Nakazawa N, et al. Inactivation of the E3/LAPTm5 gene by chromosomal rearrangement and DNA methylation in human multiple myeloma. Leukemia 2003;17:1650–1657.
43. Seimiya M, O-Wang J, Bahar R, et al. Stage-specific expression of Clast6/E3/LAPTM5 during B cell differentiation: elevated expression in human B lymphomas. Int J Oncol 2003;22:301–304.
44. Tewari M, Hu PJ, Ahn JS, et al. Systemic interaction mapping and genetic perturbation analysis of a C.elegans TGF-beta signaling network. Mol Cell 2004;13:469–482.
45. Barrios-Rodiles M, Brown KR, Ozdamar B, et al. High-throughput mapping of a dynamic signaling network in mammalian cells. Science 2005;307:1621–1625.
46. Grimsby S, Jaensson H, Dubrovska A, Lomnytska M, Hellman U, Souchelnytskyi S. Proteomics-based identification of proteins interacting with Smad3: SREBP-2 forms a complex with Smad3 and inhibits its transcriptional activity. FEBS Letters 2004;577:93–100.
47. Sammar M, Stricker S, Schwabe GC, et al. Modulation of GDF5/BRI-b signalling through interaction with the tyrosine kinase receptor Ror2. Genes Cells 2004;9:1227–1238.
48. Lawler S, Feng XH, Chen RH, et al. The type II transforming growth factor-beta receptor autophosphorylates not only on serine and threonine but also on tyrosine residues. J Biol Chem 1997; 272:14,850–14,859.
49. Izzi L, Attisano L. Regulation of the TGFbeta signalling pathway by ubiquitin-mediated degradation. Oncogene 2004;23:2071–2078.
50. Lee PS, Chang C, Liu D, Derynck R. Sumoylation of Smad4, the common Smad mediator of transforming growth factor-beta family signaling. J Biol Chem 2003;278:27,853–27,863.
51. Gronroos E, Hellman U, Heldin C-H, Ericsson J. Control of Smad7 stability by competition between acetylation and ubiquitination. Mol Cell 2002;10:483–493.
52. de Hoog CL, Mann M. Proteomics. Annu Rev Genomics Hum Genet 2004;5:267–293.
53. Reinders J, Sickmann A. State-of-the-art in phosphoproteomics. Proteomics 2005;5:4052–4061.
54. Kanamoto T, Hellman U, Heldin C-H, Souchelnytskyi S. Functional proteomics of transforming growth factor-β1-stimulated Mv1Lu epithelial cells: Rad51 as a target of TGFβ1-dependent regulation of DNA repair. EMBO J 2002;21:1219–1230.
55. Stasyk T, Dubrovska A, Lomnytska M, et al. Phosphoproteome profiling of transforming growth factor-β signalling: abrogation of TGFβ1-dependent phosphorylation of TFII-I enhances cooperation of TFII-I and Smad3 in transcription. Mol Biol Cell 2005;16:4765–4780.
56. Imamura T, Kanai F, Kawakami T, et al. Proteomic analysis of the TGF-β signaling pathway in pancreatic carcinoma cells using stable RNA interference to silence Smad4 expression. Biochem Biophys Res Com 2004;318:289–296.
57. Zhao X, Ramsey KE, Stephan DA, Russell P. Gene and protein expression changes in human trabecular meshwork cells treated with transforming growth factor-beta. Invest Ophthalmol Vis Sci 2004;45:4023–4034.
58. Wang D, Park JS, Chu JS, et al. Proteomic profiling of bone marrow mesenchymal stem cells upon transforming growth factor β1 stimulation. J Biol Chem 2004;279:43,725–43,734.
59. Westergren-Thorsson G, Malmstrom J, Marko-Varga G. Proteomics–the protein expression technology to study connective tissue biology. J Pharm Biomed Anal 2001;24:815–824.
60. Pendas AM, Uria JA, Limenez MG, Balbin M, Freije JP, Lopez-Otin C. An overview of collagenase-3 expression in malignant tumors and analysis of its potential value as a target in antitumor therapies. Clin Chim Acta 2000;291:137–155.
61. Schuster N, Krieglstein K. Mechanisms of TGF-beta-mediated apoptosis. Cell Tissue Res 2002;307:1–14.
62. Inohara N, Koseki T, Chen S, Wu X, Nunez G. CIDE, a novel family of cell death activators with homology to the 45 kDa subunit of the DNA fragmentation factor. EMBO J 1998;17:2526–2533.
63. Iwahana H, Yakymovych I, Dubrovska A, Hellman U, Souchelnytskyi S. Glycoproteome profiling of transforming growth factor-β (TGFβ) signaling: non-glycosylated cell death-inducing DFF-like effector A. inhibits TGFβ1-dependent apoptosis. Proteomics 2006;6:6168–6180.

64. Wahl SM, Swisher J, McCartney-Francis N, Chen W. TGF-beta: the perpetrator of immune suppression by regulatory T cells and suicidal T cells. J Leukoc Biol 2004;76:15–24.
65. Nicholson DW, Thornberry NA. Caspases: killer proteases. Trends Biochem Sci 1997;22:299–306.
66. Dubrovska A, Kanamoto T, Lomnytska M, Heldin C-H, Volodko N, Souchelnytskyi S. TGFβ/Smad3 counteracts BRCA1-dependent repair of DNA damage. Oncogene 2005;21:2289–2297.
67. Pucci-Minafra I, Fontana S, Cancemi P, Alaimo G, Minafra S. Proteomic patterns of cultured breast cancer cells and epithelial mammary cells. Ann NY Acad Sci 2002;963:122–139.
68. Bini L, Magi B, Marzocchi B, et al. Protein expression profiles in human breast ductal carcinoma and histologically normal tissue. Electrophoresis 1997;18:2832–2841.
69. Tobin JF, Celeste AJ. Myostatin, a negative regulator of muscle mass: implications for muscle degenerative diseases. Curr Opin Pharmacol 2005;5:328–332.
70. Bouley J, Meunier B, Chambon C, De Smet S, Hocquette JF, Picard B. Proteomic analysis of bovine skeletal muscle hypertrophy. Proteomics 2005;5:490–500.
71. Goumans MJ, Mummery C. Functional analysis of the TGFbeta receptor/Smad pathway through gene ablation in mice. Int J Dev Biol 2000;44:253–265.
72. Lomnytska M, Lukiyanchuk V, Hellman U, Souchelnytskyi S. Transforming growth factor-β1-regulated proteins in human endothelial cells identified by two-dimensional gel electrophoresis and mass spectrometry. Proteomics 2004;4:995–1006.
73. Souchelnytskyi S. Bridging proteomics and systems biology: what are the roads to be traveled? Proteomics 2005;5:4123–4137.
74. Wong DJ, Chang HY. Learning more from microarrays: insights from modules and networks. J Investig Dermatol 2005;125:175–182.

21 Mullerian Inhibiting Substance: A Potential Nontoxic, Naturally Occurring Therapeutic Agent for Certain Human Cancers

David T. MacLaughlin and Patricia K. Donahoe

CONTENTS

INTRODUCTION
MIS PROTEIN
MIS RECEPTORS
SERUM MIS: CLINICAL UTILITY
MIS AND CANCER
CONCLUDING REMARKS
ACKNOWLEDGMENTS
REFERENCES

Abstract

A considerable body of in vitro and animal in vivo data support our hypothesis that Mullerian Inhibiting Substance (MIS), a biological response modifier of the transforming growth factor-β (TGF-β) superfamily of genes, could be an effective, potentially nontoxic treatment for tumors of Mullerian origin and others expressing the MIS type II receptor (MISRII). Epithelial ovarian cancer will be the first targeted disease because this cancer affects approx 25,000 North American women each year and the high five-year mortality of over 70% results in over 16,000 deaths every year. Preclinical growth inhibition data suggest other much more frequent cancers such as cervical, endometrial, breast, and prostate cancers may be MIS targets as well.

The molecular tools to identify MIS receptor expressing tumors and, therefore, the ability to select patients for treatment are in place. An MIS enzyme-linked immunosorbent assay (ELISA) exists to follow administered doses of MIS, as well. Clinical trials await the production of sufficient supplies of recombinant human MIS for this purpose.

Key Words: Mullerian Inhibiting Substance; cancer treatment.

1. INTRODUCTION

Much of the research effort of the Pediatric Surgical Research Laboratories at the Massachusetts General Hospital has focused on uncovering the molecular mechanism of action of Mullerian Inhibiting Substance, (MIS), also known as anti-Mullerian hormone (AMH) (*for reviews see [1–4]*). Our laboratory examines the process of sexually dimorphic

From: *Cancer Drug Discovery and Development: Transforming Growth Factor-β in Cancer Therapy, Vol I: Basic and Clinical Biology*
Edited by: Sonia B. Jakowlew © Humana Press Inc., Totowa, NJ

development of the normal mammalian reproductive tract and then we seek to apply this new information to reveal some of the molecular pathophysiology of human reproductive tract anomalies. A major goal of this work is to translate our findings into new therapies or diagnostic tools for the management of the diseases dealt with so often by our pediatric clinical colleagues requiring medical intervention and/or surgical reconstruction. In fact, as a result of our findings and those of several other MIS and AMH laboratories around the world, a diagnostic tool was developed that has proven to be very useful clinically, namely a specific immunoassay to measure the MIS protein in serum in a variety of circumstances impacting the practice of pediatric endocrinology, pediatric surgery and gynecologic oncology *(5–7)*. In addition, our group has added a newly identified MIS type II receptor (MISRII) mutant *(8)* to the list of the known MIS and MIS receptor mutants that contribute to the persistent Mullerian duct syndrome (PMDS) phenotype *(9–11)*. In many respects these clinical applications of MIS research were logical extensions of MIS physiology and they will be outlined only briefly below.

The main topic of this chapter, however, relates to an unanticipated consequence of the knowledge gained form basic research into MIS mechanism of action in mammalian embryos, namely, the use of human MIS as a novel, naturally occurring, potentially nontoxic, cancer chemotherapeutic administered exogenously to patients with MIS receptor(s) expressing malignancies. As it became clear that a primary function of MIS is to inhibit cell division and/or induce apoptosis in a highly specific manner, it was suggested by a colleague of ours, Professor Robert E. Scully of the Massachusetts General Hospital Pathology Department and an expert in ovarian neoplasias, that MIS might be an inhibitor of cancers of Mullerian duct origin in adults. His observation that certain epithelial ovarian cancers recapitulate embryonal Mullerian histology *(12)* lead to the logical question of whether these lesions retained or reacquired MIS responsiveness as is observed in embryos. If so, administration of exogenous MIS could be an effective, highly specific reagent to include in existing cancer treatments for this devastating disease. After years of studying this question we conclude that MIS may indeed be a useful biological agent to treat Mullerian and other MISRII expressing cancers and we propose that phase I trials be initiated as soon as sufficient clinical grade material is available.

Before a review of the preclinical MIS cancer studies is presented, a summary of relevant MIS biochemistry will be reviewed as useful background information to better understand MIS as a biological response modifier with therapeutic potential.

2. MIS PROTEIN

Mammalian embryos begin development with the capacity to produce both female and male complete reproductive tracts. The Mullerian ducts, named for Professor Johannes Müller a German anatomist (1801–1858) who first described the tissue, will become the upper third of the vagina, the cervix, uterus, Fallopian tubes and the lining of the ovaries. The Wolffian ducts, identified nearly a century earlier by Dr. Caspar F. Wolff, will become the seminal vesicles, vas deferens and epididymides. Early in embryogenesis both ductal systems grow independently of one another. However, once the genetic sex of the embryo is declared, based upon its chromosomal makeup, a genetic switch is activated, sex-determining region Y gene (SRY) in the case of genetic males *(13)*, and the undifferentiated embryos become either testes or ovaries. As gonads begin to differentiate, about 10 wk of gestation in the human, one of the reproductive tract primordia must be destroyed and the other must be allowed to proliferate and differentiate. How this selection is accomplished at a molecular level has been the subject of research for decades but one of the most important finding was that of Professor Alfred Jost nearly 60 yr ago in Paris *(14,15)*. In the 1940s, it was widely held that the key to normal sexual development before birth was whether testes were present in the embryo. The testes would make testosterone which would stimulate the Wolffian ducts and also destroy the Mullerian ducts. To test this hypothesis, Professor Jost performed an endocrine replacement experiment in which the undifferentiated gonads of rabbit embryos were removed and replaced with crystals of testosterone and the pregnancy allowed to go to completion. As expected, testosterone induced Wolffian duct development but Jost was surprised to observe that the Mullerian ducts persisted and grew rather than regressing *(15)*. In fact, Mullerian ducts persisted even in the absence of an ovary, thus there seemed to be no ovarian contribution to Mullerian duct development *in utero*. Dr. Jost then speculated that some other testicular factor, a MIS also called AMH was responsible for the ablation of the Mullerian structures, later shown to be due mainly to apoptosis, autophagocytosis, disruption of basement membranes, and epithelial mesenchymal transformation of cells followed by migration toward the mesonephros *(16–21)*.

Over two decades after Jost's discovery a critical milestone in MIS history was reached when several groups developed a rat based in vitro organ culture assay to substitute for the in vivo rabbit bioassay Jost employed *(22,23)*. With a reliable and specific bioassay in place, it became possible to identify the fetal source of MIS as Sertoli cells *(24–25)* in rat *(26)*, bovine *(27,28)* and human testis *(29,30)*. The fact that MIS recovered from a variety of animals was active in the rat embryonic tissue suggested the MIS structure is not species specific. Furthermore, MIS expression is not limited to the fetus or to males. MIS is, in fact, a sexually dimorphically expressed gene. MIS gene transcription appears to be regulated by a number of factors including SF1, SOX 9 and WT-1 *in utero* but after birth androgens have been implicated in suppression of MIS expression *(31–37)* while there are data showing MIS can inhibit androgen synthesis as well *(38–40)*.

As stated above, MIS is only expressed in significant amounts by males during fetal life but postnatally both granulosa cells in the female and Sertoli cells in the male produce the protein. In male serum MIS levels remain very high until puberty, dropping to basal levels, thereafter. Although, MIS mRNA is detectable in ovaries shortly after birth *(41,42)* serum MIS is undetectable until puberty where its levels reach those of adult males (Fig. 1). The detection of MIS in males and females serum, even after the regression and differentiation of the Mullerian duct in the embryo and after birth *(5–7)* suggests multifunctional roles for MIS.

While the function of MIS in the fetus is well established, its role after birth is the topic of several ongoing studies. MIS blocks meiosis II in the ovary *(43)*, inhibits ovarian granulosa cell division and progesterone production *(44)*, and modulates follicular development *(45)*. Furthermore, MIS levels in follicular fluid are inversely correlated with granulosa cell proliferative index *(46,47)*. Similarly in males, MIS affects Leydig cell development and blocks transcriptional activation of the steroidogenic enzyme, CYP17 *(38,39,48,49)*. Its stage-specific expression in Sertoli cells in seminiferous tubules indicates a role in control of spermatogenesis *(50)*. Most recent findings showing MIS and its receptor expression in pituitary and certain motor neurons suggest even more wide spread non-Mullerian roles for the substance *(51,52)*. In addition, the prostate gland *(53,54)* and ductal epithelium of the mammary gland *(55,56)* and some of their cancers *(56)* are possible targets for MIS. These findings have led to the possibility of treating prostate and breast cancers, as well as those of Mullerian origin, and encourage the future study of MIS in normal breast and prostate development.

The Donahoe and Josso groups were able to purify the MIS protein, initially from bovine testes *(57–61)* and later from recombinant human sources *(62,63)* using the bioassay to monitor progress. Initially, rodent fetal and neonatal testes were selected as a source of MIS for purification but because of availability and larger size, calf testes were later chosen for this important task.

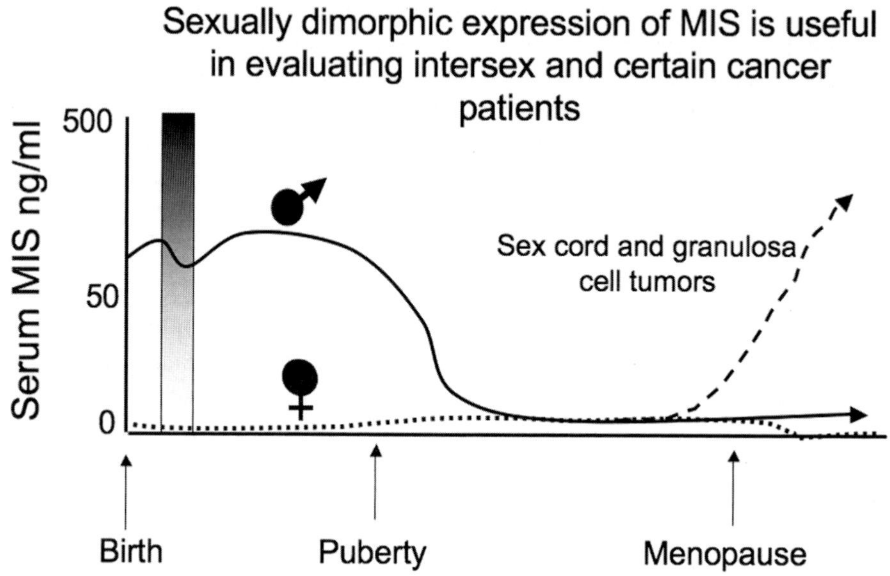

Fig. 1. The pattern of MIS in human serum varies with sex and age. MIS in males is high at birth and, after a transient drop at approx 2 y of age, levels are maintained until puberty. Serum MIS in females becomes measurable in the peripubertal period and is sustained until the menopause when it falls to undetectable levels. The shaded bar demonstrates why serum MIS is useful in the perinatal period to evaluate cases of ambiguous genitalia. MIS is absent in genetic females and positive in cases with testicular tissue. The higher the values are, the more normal the tissue. Interestingly, serum testosterone is undetectable in normal males at this time. Rising serum MIS over normal limits in adult females (dashed line) is consistent with granulosa or sex cord tumors. Response to therapy leads to serum MIS falling toward normal limits.

Conventional biochemical approaches were taken by these laboratories to purify MIS/AMH from testicular secretions in vitro *(64)* or protein extracts *(57–64)* including dye and carbohydrate affinity, anion and cation exchange chromatography and, once specific antibodies were produced, immunoaffinity chromatography was employed *(65–66)*.

Compositional analyses of the purified MIS revealed it to be a 140 kDa glycoprotein of approx 15% carbohydrate by weight. Western analyses of the protein polyacrylamide gels in the presence of sodium dodecylsulfate and disulfide bond reduction suggested that MIS was actually a disulfide linked dimer that was partially cleaved into smaller species perhaps during the biosynthetic process. The significance of these findings became more apparent when the MIS gene was cloned *(62)*.

Based on the partial amino acid sequence data from the purified bovine molecule a series of degenerate oligonucleotide primers were designed to enable the cloning of the bovine complementary DNA and then the human genomic sequence from DNA libraries of bovine testes and human placentae, respectively *(62,63)*. Sequence analysis of the genes revealed them to be related, albeit weakly, to the transforming beta family of biological modifiers with the most striking homology (28%) residing in the carboxyterminal domain. The 2.8 kb human gene contains five exons and four introns and is located on the short arm of chromosome 19 *(67)*. The deduced protein sequence of this gene was shown to contain a 25 amino acid secretion specific signal peptide and a monomeric protein of 535 amino acids that, when glycosylated at two putative *N*-linked glycosylation sites, has a molecular weight of 70,000 daltons (Fig. 2). Examination of the primary human MIS sequence revealed the presence of a cleavage motif at residue 427 that would, if acted upon, explain

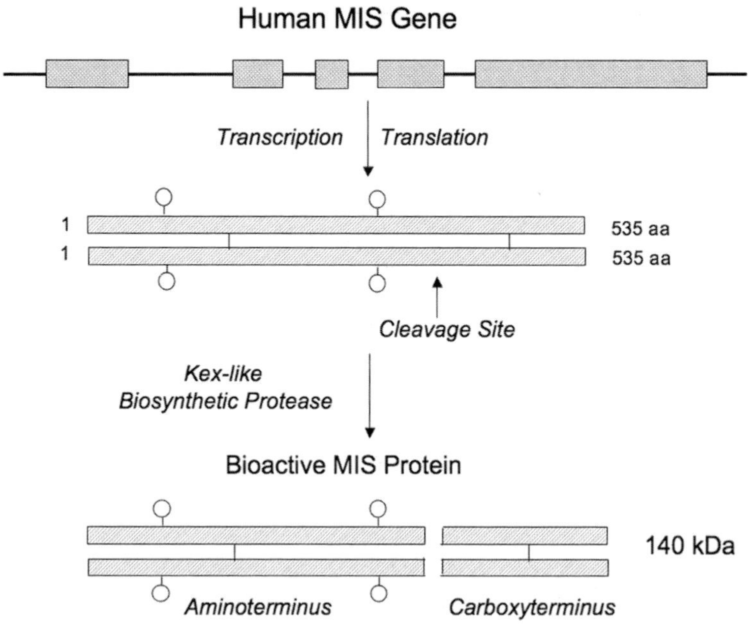

Fig. 2. The human MIS gene has five exons and four introns and encodes a monomer of 70 kDa with two presumed *N*-linked glycosylation sites (circles). The monomer forms a homodimer via disulfide bond formation and is activated by biosynthetic kex-like proteases to produce the 25-kDa carboxy-terminal dimer, which is the domain of the molecule responsible for its bioactivity. The 110-kDa amino-terminus stays associated with the carboxyterminus via noncovalent forces.

the origins of the major cleavage products of the MIS fragments (12.5 and 55 kDa) appearing on reduced polyacrylamide gels of the purified protein. A weaker motif at residue 229 is also present but its role in MIS action is unclear. Several studies have clarified the role of MIS carboxyterminal cleavage in the scheme of MIS action. The MIS carboxyterminus results from processing by a biosynthetic protease, most likely a kex-like enzyme *(68,69)*, into an amino-terminal dimer (110 kDa) and a carbohydrate-free carboxyterminal dimer (25 kDa) held together in noncovalent association (Fig. 2). It is the carboxyterminal domain which possesses the biological activity *(68,70,71)*. The amino-terminal domain may contribute to proper protein folding and assembly during synthesis and/or it may increase the serum half-life of the carboxyterminus and enhance bioactivity *(71)*. The carboxy-terminal sequence of MIS is extremely highly conserved, a fact which explains why MIS from many different species are all active in the rat in vitro bioassay used for MIS purification. A recombinant preparation of MIS carboxyterminus is available commercially for study but this material is not bioactive in regression assays and it is not suited for in vivo work because of its short half-life in serum.

The human gene was transfected into Chinese Hamster ovary cells and its conditioned medium is used as a source of recombinant human MIS secreted from these cells *(72,73)* for purification of the MIS needed for all of the cancer studies summarized in the sections that follow. The MIS as purified from serum free conditioned media is proteolytically processed, free of potential contaminants from bovine serum, hamster cells, mouse monoclonal antibodies *(73)*, and significant amounts of endotoxins *(74)*.

At present, no commercial source of bioactive holo-MIS/AMH exists; therefore, the protein must be acquired from academic laboratories. The increased production for large scale preclinical, pharmacology, and clinical trials will require commercial partnerships.

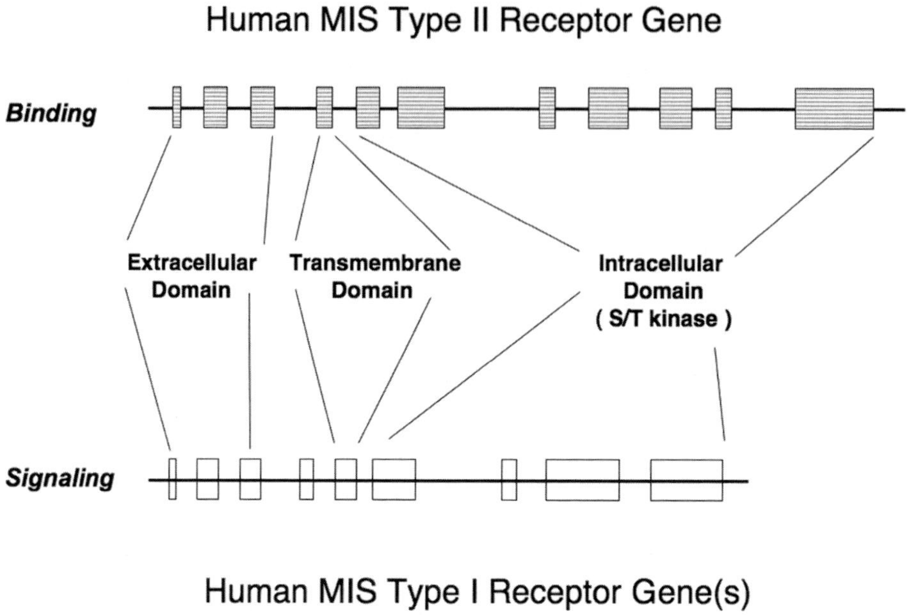

Fig. 3. The human MISRII gene, with its 11 exons is larger than the nine exon type I gene. They share several similar but not identical features including extracellular ligand binding domains, transmembrane spanning regions and intracellular serine-threonine kinase domains. Both types are required for MIS signaling and mutations in the type II receptor are associated with phenotypic changes in humans.

3. MIS RECEPTORS

MIS biological activity requires interaction with two very similar receptors termed type I and II. Each is a single membrane spanning, serine-threonine kinase that, after cross phosphorylation events, initiates a series of intracellular cascades leading to the control of cell cycle regulating proteins and the altered transcription of a number of genes, depending upon the target tissues. The type II receptor is believed to be responsible for ligand binding and the type I heteromer is the signaling receptor (Fig. 3). The MISRII was cloned in the laboratories of Baarends and Themmen, Josso and Cate, and Behringer, as well as our own laboratory *(75–78)*. The human gene, located on chromosome 12q13 *(9)* has 11 exons and 10 introns.

The MISRII is expressed in Mullerian duct mesenchymal cells surrounding the adjacent ductal epithelium, as one might expect because the fetal mesenchyme directs Mullerian duct regression *(79)*, and in fetal and adult gonadal Leydig and granulosa cells *(75–77)* where MIS regulates testosterone *(38,40,48,49)*, estradiol and progesterone synthesis, respectively *(44)*. Functional MISRII has also been found in adult rodent uterus *(77,80)*, the human endometrium *(80)*, breast and prostate tissues *(81,82)* and, unexpectedly, in motor neurons in the mouse brain *(52)*. It is not yet understood what receptor mediated actions MIS has on these tissues in normal adults, but MIS inhibits the proliferation of tumor cell lines derived from them (*see* Section 5) supporting our hypothesis that MIS may be a useful adjuvant agent in the treatment of these diseases.

Josso and colleagues have studied a large number of patients with persistent Mullerian duct syndrome (PMDS) resulting from loss of function mutations in the MISRII (Fig. 4) or the MIS molecule itself (for a review *see [84]*). In so doing, they have defined a number of hot spots in the gene as well as polymorphisms present in the normal sequence. Recently, our laboratory identified a novel intron mutation associated with PMDS as well *(8)*.

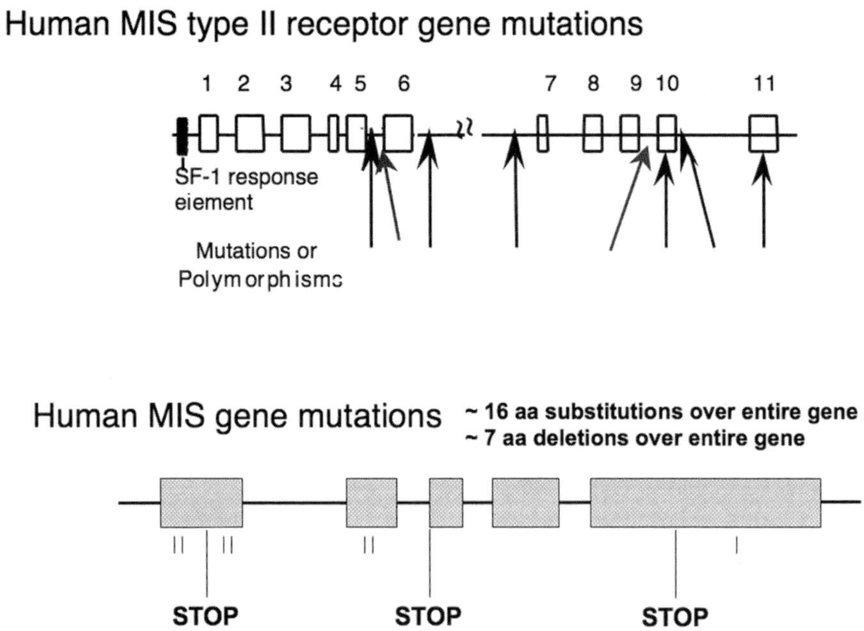

Fig. 4. Mutations in either the MISRII (top panel) or the MIS gene (bottom panel) are associated with persistent Mullerian duct syndrome (PMDS). The arrows and ticks indicate where some of the common mutations have been localized. Please *see* refs. *(8)* and *(84)* for more detailed descriptions.

Mutations in the MIS gene (Fig. 4) also contribute to the cause of the syndrome *(84)*. A common defect in the MIS gene is a truncation mutation which produces a protein lacking the carboxyterminal domain and is therefore, unable to cause Mullerian duct regression *in utero*. Heterozygotes of either mutation type, i.e., receptor or ligand have normal phenotypes; affected individuals are either homozygous for a given mutation or compound heterozygotes.

Several different type I receptors have been identified as interacting with the type II receptor (Alk3, Alk2, and Alk6 *[85–90]*). How each type I receptor interacts with the type II receptor in human tissues is not yet know although all are expressed in the human cancer cell lines studied thus far.

4. SERUM MIS: CLINICAL UTILITY

The discovery that MIS was expressed and secreted postnatally in both males and females prompted numerous studies which demonstrated the utility of measuring MIS in serum in a variety of clinical scenarios including ambiguous genitalia, undescended testes, delayed puberty, sex cord, and granulosa cell tumors. Fortunately, three very useful human MIS specific and highly sensitive immunoassays were developed concurrently by the Donahoe, Josso and Hutson laboratories *(5–7)* for this purpose (*see* Fig. 1).

Serum MIS, for example, is an excellent predictor of the presence of testicular tissue in cases of nonpalpable testes *(91)* as well as a useful marker of testicular function in these cases and in cases of sexual ambiguity. Undetectable MIS suggests absent testes, normal MIS levels for age predict normal testes and the cause of ambiguity may be androgen insensitivity whereas, lower than normal MIS is most likely a form of gonadal dysgenesis. In many instances, serum MIS data are more valuable than human chorionic gonadotropin (hCG) stimulation tests to evaluate gonadal status.

In adult females, on the other hand, serum MIS is a useful predictor of ovarian reserve as well as a marker of response to IVF protocols *(92–98)*. In the case of MIS secreting tumors including sex cord *(99)* and granulosa cell tumors *(100)* serum MIS predicts recurrences and response to therapy. These tumors may express the MIS receptor *(101)* but the fact that these tumors are unresponsive to the MIS they produce suggests a loss of function downstream from the receptor itself; therefore, these tumors are not considered targets for MIS therapy. Serum concentrations of MIS above the upper limit of normal for age is consistent with the presence of either one of these tumors; rising serum MIS reflects tumor growth and decreases show response to therapy *(99)*. In numerous cases serum MIS rises to thousands fold the normal limit for age without any significant adverse reactions that could be attributed to the endogenous MIS molecule. It is for this reason that we predict that exogenous MIS administered to cancer patients as a therapy at doses designed to produce much lower serum concentrations may have a very favorable toxicity profile. Whether recombinant human MIS is toxic will be the topic of detailed toxicology studies planned for the future when clinical grade MIS is available.

Several MIS assays similar to the ELISAs developed by research laboratories have been commercialized, thus serum MIS data in the management of pediatric endocrine, reproductive medicine and gynecologic oncology are becoming more widely used. Having widespread access to reliable and sensitive MIS/AMH immunoassays is critical for the analysis of human clinical trials where serum levels of MIS must be measured after administration to cancer patients.

5. MIS AND CANCER

The concept that a naturally occurring growth inhibitor such as MIS could be an effective adjuvant treatment for cancer is attractive for a number of reasons. Because MISRII expression is restricted to so few tissues, for example, MIS effects could be extremely specific thus minimizing adverse side effects. For the same reason, MIS may be used as a delivery system for more toxic drugs, again limiting exposure to nontarget tissues. Our choice to focus initially on certain of the ovarian cancers arises from the fact that they have, when compared with other potential MIS targets, the worst prognosis and might benefit significantly from this novel biological reagent, thus offering a new approach to augment the efforts of others to find useful new strategies to employ.

5.1. Ovary

MIS as an anticancer agent for certain ovarian cancers has been the focus of our laboratory because we hypothesized that any cancer of Mullerian origin could be a target for MIS treatment *(102,103)*. The idea that MIS could be used to treat ovarian cancer is predicted by the fact that the histology of the embryonic Mullerian ducts is recapitulated in the common ovarian adenocarcinomas that arise from the outer ovarian coelomic epithelium, which in the embryo, invaginates to form the Mullerian duct *(104–107)*. Results of more recent experiments, however, identify cervical and endometrial cancers as other potential targets, as well as several non-Mullerian cancers, including breast and prostrate (*see* Sections 5.4. and 5.5.). Thus, the original hypothesis can be extended to include any tumor that expresses a functional MISRII.

Epithelial ovarian cancer affects nearly 25,000 North American women each year and is the fifth most common malignancy in women with a five-year mortality of over 70% with over 16,000 deaths per year *(108)*. Although the mortality rate is significantly lower in women with stage Ia or Ib disease, early peritoneal seeding and metastatic spread accounts for the fact that less than 25% of women are diagnosed at stage I. Surgery and cytotoxic drug therapy results in favorable clinical response in 50–80% of patients but the majority will relapse *(109)*. Thus, ovarian cancer becomes a chronic disease and it is timely to search for novel therapies.

We tested our hypothesis by beginning with in vitro studies using human ovarian cancer cell lines followed by a series of ex-vivo experiments on these lines and thereafter, the work progressed to the examination of human ovarian cancer cells in ascites collected from patients with recurrent disease *(103)*.

These initial studies began with partially purified bovine MIS, which suppressed growth of a human ovarian carcinoma cell lines in vitro, compared to vehicle controls *(102,110)* and in nude mice *(111)*. In addition, a large number of primary tumor cells taken from patients with human ovarian, and other reproductive cancers *(112)* were growth inhibited in vitro. Importantly, over 50% of the Stage III ovarian patients from whom abdominal ascites were collected had cells that bound recombinant human MIS, expressed MISRII, and were growth inhibited by MIS when treated ex vivo *(103)*. Therefore, the incidence of receptor positive and MIS responsive cases should be sufficiently high to warrant further investigation. Highly purified human MIS also inhibited other human carcinoma cell lines of Mullerian origin including OVCAR 3, 5 and 8, IGROV-1, and HOC-1 in vitro *(113)* and/or in vivo *(114)*. The observations on fresh surgical specimens and ascitic fluid, however, are perhaps the most significant to date as they reflect what is true for patients and not cell lines. Receptor expression is frequent in a random sample of ovarian cancer cases and the receptor appears to be functional as MIS inhibits cancer cell proliferation in vitro *(103)*.

After phase I trials are completed MIS will most likely be administered to patients in combination with other drugs. It is important, therefore, to study downstream signaling mechanisms by which MIS inhibits proliferation as a prelude to using MIS in combination with commonly used cytotoxic drugs. It is important to test for synergy or additivity between MIS and any drug, to make certain that they do not counteract one another. Studies to date show that MIS downstream signal transduction pathways include type I receptor(s), Smads, cyclin dependent kinase inhibitors, and cytokine inducible pathways. MIS also increased expression of p16, p107, and p130 protein *(115)*. An inhibitor of cell cycle progression, p16 functions by binding cyclin/CDK complexes, preventing nuclear translocation; p16 is mutated in a number of patients with ovarian cancer and in fact it is mutated in the SKOV 3 cell line that fails to respond to MIS *(103)*. p107 and p130 are pocket proteins like Rb, which is mutated in OVCAR 8 cells. Expression of the downstream protein, E2F1, which is associated with apoptosis, was also enhanced by MIS treatment of OVCAR 8 cells *(80)*.

These results suggest mechanisms different from those of most cytotoxic drugs. If MIS and drugs can function in combination it may be able to decrease the dose needed for either agent alone, resulting in decreased toxicity. Preliminary studies to examine these questions are now underway in our laboratory and this possibility will be examined in vivo in greater detail when clinical grade MIS is produced.

The ultimate translation of MIS as a therapeutic, from studies on cancers in vitro, to preclinical trials in animals, and eventually to phase I clinical trials, demands sufficient quantities of a uniform preparation of recombinant human MIS. Accordingly, our laboratory has spent considerable effort optimizing purification protocols from mammalian cells and is exploring a number of alternative sources for enhanced production of MIS. In addition, we have developed MIS-specific assays to detect the protein and measure its bioactivity with a high degree of precision.

The preclinical in vivo trials for MIS in ovarian cancer were advanced greatly by the development of a new mouse model for the disease by Dr. Connolly and colleagues at the Fox Chase Cancer Center in Philadelphia, PA. *(116)*. She used the MISRII promoter to

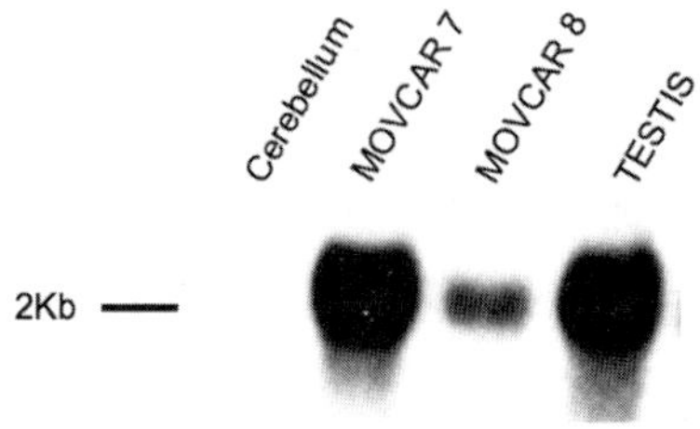

From Pieretti-Vanmarcke et al 2006

Fig. 5. Northern analysis of MISRII mRNA expression in MOVCAR cells. The migration of the λ/HinD III size marker is indicated on the left. As expected, the 2 kb MISRII mRNA is expressed in two MOVCAR cell lines (called 7 and 8) and testis but not in the cerebellum. MOVCAR 7 had more receptor than MOVCAR 8 was preferentially for the in vitro and in vivo MIS growth inhibition studies.

drive the expression of oncogenic T antigen of SV 40 that inactivates p53 and the pocket proteins, Rb, p107, and p130 in transgenic mice. These animals formed ovarian carcinomas first seen as papillary outgrowths from the surface epithelium of the ovary where they express both T antigen and the MISRII as determined by PCR *(116)*. The tumors then faithfully recapitulate the phenotype of the most common human ovarian cancer, serous cystadenocarcinoma. This animal model is extremely important because it is one of the first animal models of human ovarian cancer with a relatively large percentage (50%) of mice developing cancers in a short period of time, (6–13 wk). This is also one of the first animal models in immunocompetent mice so advantage can be taken to evaluate the role of the immune response in tumor regression, which may be important because MIS activates the NFκB pathway in breast and prostate tissues and cancers (*see* Sections 5.4. and 5.5.).

We began testing MIS against cell lines, called MOVCAR cells that were harvested from the ascites of these mice. They express the MISRII mRNA (Fig. 5), they grew rapidly in culture and responded to MIS dramatically in a dose dependent manner with up to 95% growth inhibition (Fig. 6). It should be noted that this is the most vigorous degree of inhibition produced in vitro and in vivo by MIS in any cell lines studied *(117)*.

To study the effects of MIS in vivo, growth curves of MOVCAR cells were established because these cells responded so well to MIS in vitro. T cell deficient female nude mice (nu/nu) outbred on a Swiss background were injected with tumor cells at 6 wk of age into the subcutaneous tissue just over the femoral muscles of the right hindlimb or into the dorsal fat pad. Tumors appeared in a dose dependent manner in both implantation sites. For example, when injected with approx 10,000 cells, tumors became visible and palpable on average in either the dorsal fat pad or left leg in 10 wk, 100,000 cells produced tumors in 8 wk, tumors were visible in 9 wk after injection of 3×0^6 cells, and 12×10^6 cells produced tumors in approx 4 wk. Large tumors that appeared on the back or in the leg were dissected and shown to express both the type II MIS receptor (Fig. 7) and TAg as detected by immunohistochemical assays using a commercial monoclonal antibody to TAg and the MIS receptor polyclonal antibody developed in our laboratory for this purpose. Results from a representative experiment show the time to tumor appearance was significantly greater and tumor volume significantly smaller in animals treated with MIS as compared to vehicle control (Fig. 8). Intraperitoneal treatment with MIS (10–20 μg/d, 5 or 7 d/wk) inhibited fat pad or leg MOVCAR tumor equally well. Tumors became apparent at 35 d with a statistically significant difference in volume between the PBS and the MIS groups ($p < 0.05$) at that time and when measured thereafter for up to 82 d of treatment (Fig. 8, asterisks). No

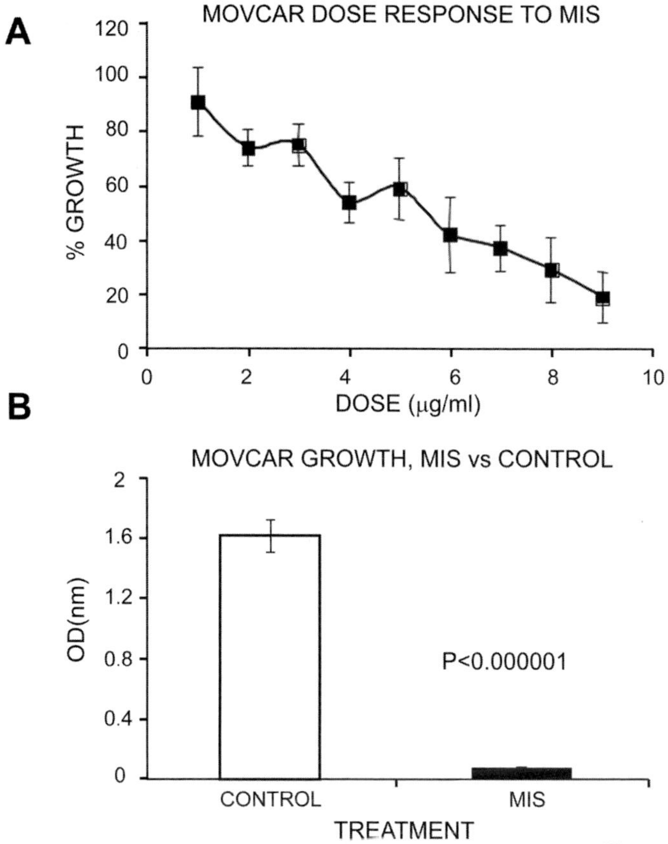

From Pieretti-Vanmarcke et al 2006

Fig. 6. Purified recombinant human MIS inhibits MOVCAR cell lines in vitro. Panel (**A**) shows the MIS dose dependent inhibition of MOVCAR cells in a monolayer cell culture assay. The IC50 was approx 5 μg/ml (35 nM) and approx 80% inhibition was observed. The graph shows changes in percentage of growth +/– SD compared to PBS treated controls. Panel (**B**) shows the mean level of inhibition of MOVCAR cells by 70 nM MIS in six different experiments; approx 95% compared to controls in a colorimetric cell count assay.

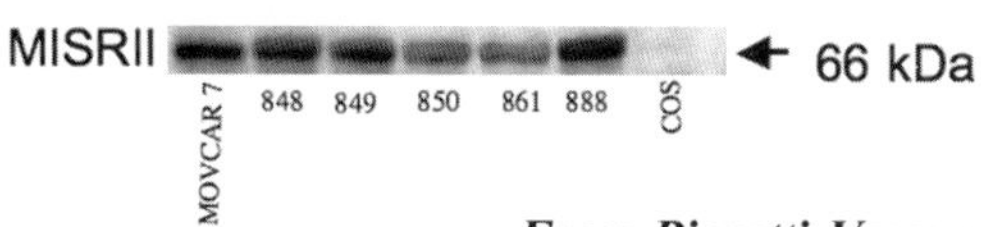

From Pieretti-Vanmarcke et al 2006

Fig. 7. MOVCAR cells express the MISRII protein as seen in this Western analysis of extracts of MOVCAR cells and the tumors they cause in nude mice (848, 849, 850, 861, and 888) but not the monkey kidney derived COS negative controls cells.

apparent toxicity was noted in these experiments and considering the fact that this duration of MIS exposure in the mouse is roughly equivalent to about 7 yr of continuous cancer treatment in humans, we are encouraged MIS may be relatively safe.

In one preliminary experiment, a MOVCAR tumor was allowed to grow for 151 d in the fat-pad before MIS treatment which completed regressed this tumor after 38 d of treatment (Fig. 9), as documented by histology. A vehicle treated animal continued to show growth of the tumor, supporting the conclusion that the shrinking tumor was the result of MIS treatments.

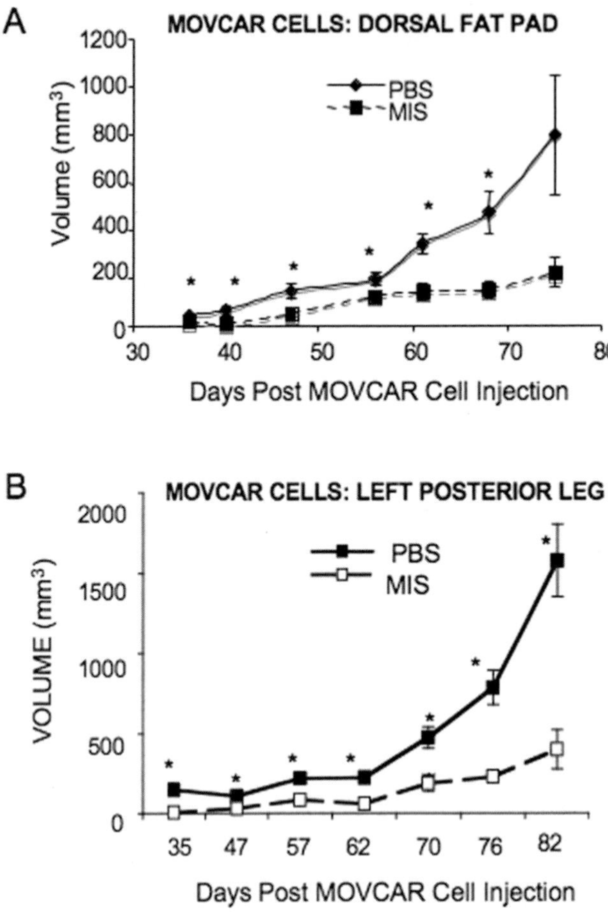

From Pieretti-Vanmarcke et al 2006

Fig. 8. Recombinant human MIS decreases tumor growth of MOVCAR allografts in nude mice in vivo. MOVCAR cells were injected in the dorsal fat pad (Panel **A**) and animals were treated with PBS or MIS ($n = 5$ per group) and tumor volumes were measured. The graph shows tumor volume +/– SEM between 35 and 75 d after injection of 6×10^6 cells. *Significant from MIS. Panel (**B**) shows the results of another type of in vivo experiment in which MOVCAR cells were injected in the left posterior leg of nude mice. Animals were treated with PBS or MIS ($n = 5$ per group) and tumor volumes measured. The graph demonstrates changes in volume +/– SEM between 35 and 82 d after injection of 6×10^6 cells. *Significant from MIS.

5.2. Cervix and Cervical Cancer

The cervix is of Mullerian duct origin, therefore, it is logical to study whether normal cervix and cervical tumors express the MISRII and if human cervical cancers respond to MIS. Three human cervical carcinoma cell lines, CaSki, SiHa and C33A, kindly provided by Dr. Karl Munger of the Brigham and Women's Hospital in Boston, MA., express the MISRII protein by western analysis and they were all growth inhibited in vitro by MIS added to the growth media or produced by the cells themselves after transient transfection with a transcript encoding bioactive MIS (K2) as compared with growth of cells transfected with an empty vector or an inactive mutant of MIS (K1) *(115)*. Representative results for these cell lines are given in Figure 10. The receptor is seen in the normal rat cervix by western analysis but, as expected, it was not detected in the

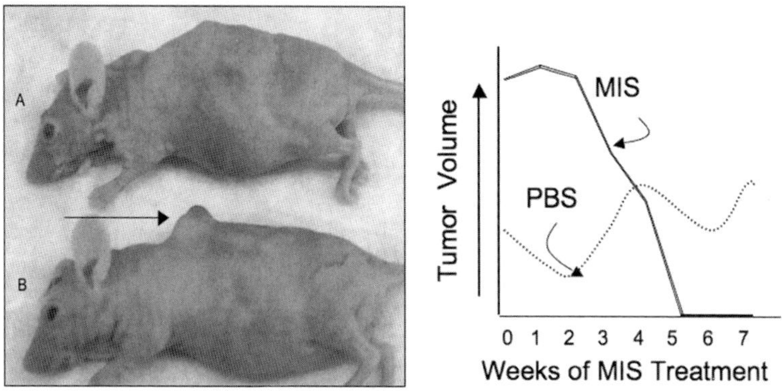

Fig. 9. MIS given intraperitoneally to a mouse after a dorsal fat pad tumor had formed ablated the mass after 38 d of treatment. The left panel shows the treated animal (top) and the control animal (bottom) with its tumor (arrow). Serial tumor volume measurements for the two animals in this experiment are given in the right panel.

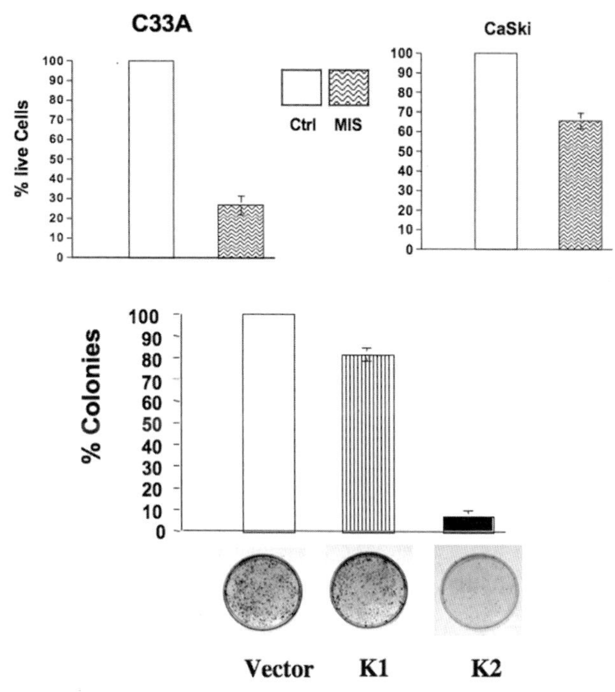

Fig. 10. Top panel. Inhibition of cervical cancer cell growth by MIS for C33A cells, 70–80%, and CaSki cells, 30–40%, after 4 d of MIS treatment ($p < 0.001$). Bottom panel. C33A cells were stably transfected with empty vector, a leaderless inactive form of MIS (K1), and bioactive MIS (K2). Minimal reduction in colony number is seen after transfection with inactive MIS, compared to 90–95% inhibition of colony growth in cells transfected with active MIS.

non-Mullerian lower vagina or in small intestine (*115*). As was seen with the ovarian cancer cells MIS specifically upregulated p16, p130, p107, as well as E2F1, and E2F4 in the C33A cell line without affecting E2F2, 3, or 5. Because p53 is mutated in C33A, and Rb is absent, the MIS affect appears to be independent of both. When p16, p130, p107, or E2F1

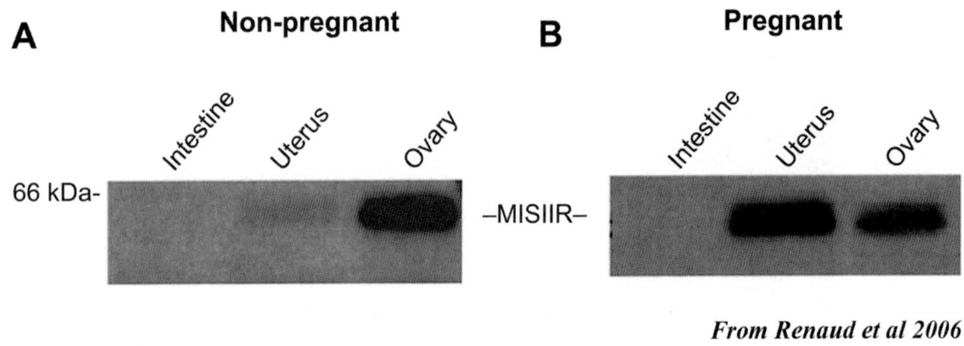

Fig. 11. Expression of the MISRII protein in the rat uterus: Samples of intestine, uterus, and ovary were harvested from nonpregnant (Panel **A**) and pregnant (Panel **B**) female rats. Western blots with the 153P antibody specific for the MISRII detected the 63 kDa protein in nonpregnant and pregnant uterus; expression of the MISIIR protein appeared to increase in the uterus during pregnancy. All lanes were loaded with 75 μg protein lysate; rat ovary served as a positive control, intestine as a negative control.

were overexpressed in these cells growth was dramatically inhibited compared to vectors alone *(115)*. These findings in cervical cancer further broaden the targets for MIS.

5.3. Endometrium and Endometrial Cancer

In addition to the cervix, Fallopian tubes, and the ovarian coelomic epithelial lining, the uterus also arises from the Mullerian ducts and associated coelomic epithelium. The MISRII mRNA is expressed in the Mullerian ducts of the female rat embryo between days 13 and 19 of development and in the uterine mesenchyme during the first postnatal days. The MISRII protein was also detected in a human endometrial cancer cell line AN3CA *(80)* and normal and pregnant adult rat uterus by western analysis using the MIS receptor antibody developed in our laboratory *(80)* but not in small intestine (Fig. 11). In the human, MISIIR mRNA was detected in myometrium and endometrium of a postmenopausal uterus (*not shown*) and the protein observed by western analysis in endometrial tumors.

Two cell lines which express the receptor mRNA and protein were inhibited by MIS as well (Fig. 12). AN3CA, a human cell line derived from an endometrial carcinoma metastasis, showed significant growth inhibition (57% to 67% inhibition relative to control, 0%). KLE, a human cell line derived from a poorly differentiated endometrial cancer, also demonstrated significant growth inhibition (44.9%) after treatment with MIS. As was true of the cervical cancer cell line, transfection of an MIS construct also produced growth inhibition of an endometrial cancer cell line (Fig. 12). The cervical and endometrial transfection data support the conclusion that the growth inhibition effects are owing to the MIS molecule itself rather than a contaminant in the purified protein preparation. Studies of the molecular mechanism by which MIS inhibits growth indicate that these cells are arrested in G1 and may initiate apoptosis, as indicated by an increase in Caspase 3 cleavage products. Unlike the ovarian and cervical cancer cell lines AN3CA cells do not express p16 or p21. MIS treatment does, however, increase p107 and p130 mRNA at 48 h while E2F1 initially increases, and then decreases at 72 h *(80)*.

5.4. Breast and Breast Cancer

Of interest is the fact that the MISRII is expressed in locations other than Mullerian duct tissues and gonads. MISRII is expressed in the normal rat breast; the level of expression is

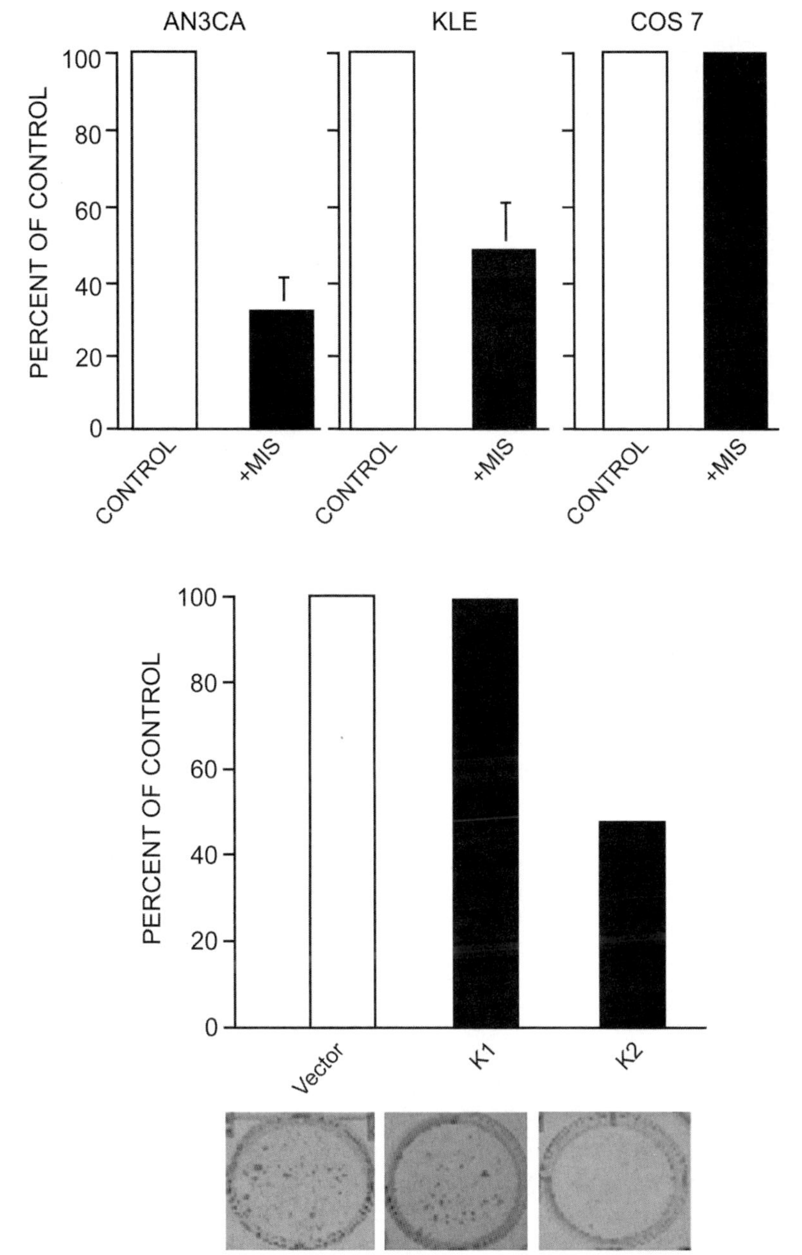

From Renaud et al 2006

Fig. 12. Top panel: MIS inhibits the growth of human endometrial cancer cell lines AN3CA and KLE. Treatment with 70 nM recombinant human MIS inhibited the growth of AN3CA cells in monolayer culture by 67% compared with cells treated with buffer control ($p < 0.0001$, $n = 4$). KLE was likewise inhibited by MIS treatment (52.5%, $p = 0.0041$, $n = 3$). MIS did not inhibit the growth of COS 7 cells, which do not express the MISRII ($n = 3$). Bottom panel: Stable transfection of AN3CA cells with cDNA for bioactive MIS (K2) or inactive MIS (K1) confirms the MIS growth inhibitory effect on these endometrial cells. MIS secreted from the transfected cells significantly inhibits AN3CA colony formation (K2; *see* photograph of colonies below) to a similar degree as the monolayer cell inhibition (Top panel). Nonsecreted MIS (K1; *see* photograph of colony cultures below) had no effect compared to the empty vector control (vector).

inversely proportional to the state of proliferation of the breast. Receptor expression was found in the virgin breast but was undetectable in the lactating proliferative breast and expression returned to basal levels after the pups were weaned *(118)*. In a number of estrogen receptor positive and negative human breast cancer cell lines MIS enhanced IκB dependent DNA binding of NFκB that resulted in induction of IEX-1 mRNA *(118)*, an immediate early gene induced by radiation, interferon-γ, or tumor necrosis factor alpha (TNF-α) but not by TGF-β, via an NFκB pathway *(56,119,120)*. When five-week old female mice were injected with MIS twice daily for 7 d, apoptosis was enhanced in the mammary epithelium compared to control animals *(118)*. Thus, it is likely that breast cancer could be a target for MIS therapy.

5.5. Prostate and Prostate Cancer

MIS treatment of cancer may also include the prostate as discussed earlier *(53,54)*. Furthermore, because MIS suppresses testosterone production *(38,39,40,48)* it might indeed exert a double effect on prostatic cancer, a direct growth inhibition and an indirect effect by lowering testosterone. The MISRII and two candidate MIS type I receptors *(85–89)* are expressed in a prostatic cancer cell line as well as in multiple human prostate tumors *(53)*. MIS inhibits the growth of human prostate cancer cell lines, furthermore, MIS also induced IRF1, and interferon γ was found to enhance dramatically the inhibiting effects of MIS in vitro and in vivo. Dr. Shyamala Maheswaran of the Massachusetts General Hospital Cancer Center continues to pursue this important work because these findings in breast and prostate significantly broaden the target population and the indications for the clinical uses of MIS.

6. CONCLUDING REMARKS

The present body of work suggests that MIS, a biological response modifier with potentially very little toxicity can be harnessed to treat tumors expressing the MIS receptor. There are in vitro and in vivo data on MIS inhibition of growth in human and animal cell lines mainly for ovarian cancers but recently similar data are being reported for cervical, endometrial, breast and prostate cancers. MIS activity is restricted to cells that express the MISRII. Potentially, there are a large number of people in the United States alone who may benefit from MIS therapy. Approximately 25,000 new ovarian cancer patients are diagnosed each year, breast cancer strikes nearly 300,000 women each year, and prostate cancer another 200,000 cases. This larger target population, without even including endometrial and cervical cancers, makes development of MIS as an anticancer therapeutic for more than ovarian cancers more compelling.

The molecular tools to identify MIS receptor expressing tumors and, therefore, the ability to select patients for treatment are in place. An MIS ELISA exists to follow administered doses of MIS, as well. Clinical trials await the production of sufficient supplies of recombinant human MIS for this purpose.

ACKNOWLEDGMENTS

The authors wish to thank all of the fellows and collaborators who have worked with us on this project over the years, as well as the technical staff who made this work possible. We have not referred directly to the entire primary source papers in this field but we have tried to provide a numerous review articles that cite these important reports wherever possible. The bulk of the research was funded by the National Institutes of Health (CA 17393-30) with additional support from the American Cancer Society, and the generous philanthropic support of the McBride family and The Commons Development Group Inc. Finally, the authors wish to thank Dr. Rafael Piertetti-Vanmarcke for his assistance in preparing this manuscript.

REFERENCES

1. Donahoe PK, Clarke T, Teixeira J, Maheswaran S, MacLaughlin DT. Enhanced purification and production of Mullerian inhibiting substance for diagnostic therapeutic applications. Mol Cell Endo 2003;211:37–41.
2. Teixeira J, Maheswaran S, Donahoe PK. Mullerian inhibiting substance: An Instructive Developmental Hormone with Diagnostic and Possible Therapeutic Application. Endocr Rev 2001;22:657–674.
3. DiClemente N, Josso N, Gouedard L, Belville C. Components of the anti-Mullerian hormone signaling pathway in gonads. Mol Cell Endocrinol 2003;211:9–14. Review.
4. Josso N, diClemente N. Transduction pathway of anti-Mullerian hormone, a sex-specific member of the TGF-beta family. Trends Endocrinol Metab 2003;14:91–97. Review.
5. Hudson PL, Dougas I, Donahoe PK, et al. An immunoassay to detect human Müllerian inhibiting substance in males and females during normal development. J Clin Endo Metab 1990;70:16–22.
6. Josso N, Legeai L, Forest MG, Chaussain JL, Brauner R. An enzyme linked immunoassay for anti-Mullerian hormone: a new tool for the evaluation of testicular function in infants and children. J Clin Endo Metab 1990;70:23–27.
7. Baker ML, Metcalfe SA, Hutson JM. Serum levels of Mullerian inhibiting substance in boys from birth to 18, as determined by enzyme immunoassay. J Clin Endo Metab 1990;70:11–15.
8. Hoshiya M, Christian B, Cromie W, Zhan Y, MacLaughlin DT, Donahoe PK. Persistent Mullerian duct syndrome with a deletion and a novel splicing Mutation in the MIS type II receptor. Birth Def Res 2003;67:868–874.
9. Imbeaud S, Faure E, Lamarre I, et al. Insensitivity to anti-mullerian hormone due to a mutation in the human anti-mullerian hormone receptor. Nat Genet 1995;11:382–388.
10. Belville C, Van Vlijmen H, Ehrenfels C, et al. Mutations of the anti-Müllerian hormone gene in patients with persistent Müllerian duct syndrome: biosynthesis, secretion and processing of the abnormal proteins and analysis using a three-dimensional model. Mol Endocrinol 2004;18:708–721.
11. Josso N, Belville C, di Clemente N, Picard JY. AMH and AMH receptor defects in persistent Mullerian duct syndrome. Hum Reprod Update 2005;11:351–356. Review.
12. Scully RE. Recent progress in ovarian cancer. 1970 Hum Pathol 1970:1:73 98.
13. Goodfellow PN, Lovell-Badge R. SRY and sex determination in mammals. Annu Rev Genet 1993: 27:71–92.
14. Jost A. Recherches sue la differentiation sexulle de l'embryonde lapen. Arch Anat Microsc Morphol Exp 1947;8:379–418.
15. Jost A. Problems of Fetal Endocrinology: the gonadal and hypophyseal hormones. Rec Prog Horm Res 1953;8:379–418.
16. Price JM, Donahoe PK, Ito Y, Hendren WH, III. Programmed cell death in the Müllerian duct induced by Müllerian inhibiting substance. Am J Anat 1977;149:353–375.
17. Price JM, Donahoe PK, Ito Y. Involution of the female Müllerian duct of the fetal rat in the organ-culture assay for the detection of Müllerian inhibiting substance. Am J Anat 1979;156:265–284.
18. Trelstad RL, Hayashi A, Hayashi K, Donahoe PK. The epithelial-mesenchymal interface of the male Müllerian duct: loss of basement membrane integrity and ductal regression. Dev Biol 1982;92:27–40.
19. Hayashi A, Donahoe PK, Budzik GP, Trelstad RL. Periductal and matrix glycosaminoglycans in rat Müllerian duct development and regression. Dev Biol 1982;92:16–26.
20. Ikawa H, Trelstad RL, Hutson JM, Manganaro TF, Donahoe PK. Changing patterns of fibronectin, laminin, type IV collagen, and a basement membrane proteoglycan during rat Müllerian duct regression. Dev Biol 1984;102:260–263.
21. Dyche WJ. A comparative study of the differentiation and involution of the Müllerian duct and Wolffian duct in the male and female fetal mouse. J Morphol 1979;162:175–209.
22. Picon R. Action du testicule foetal sur le développement in vitro des canaux de Müller chez le rat. Arch Anat Microsc Morph Exp 1969;58:1–19.
23. Donahoe PK, Ito Y, Marfatia S, Hendren WH, III. The production of Mullerian inhibiting substance by the fetal, neonatal and adult rat. Biol Reprod 1976;15:329–334.
24. Josso N. In vitro synthesis of Mullerian inhibiting hormone by seminiferous tubules isolated from the calf fetal testis. Endocrinology 1973;93:829–834.

25. Blanchard MG, Josso N. Source of the anti-Müllerian hormone synthesized by the fetal testis: Müllerian-inhibiting activity of fetal bovine Sertoli cells in tissue culture. Pediatr Res 1974;8:968–971.
26. Picon R. Action of the fetal testis on the development in vitro of the Mullerian ducts in the rat. Arch Anat Microsc Morphol 1969;Exp58:1–19.
27. Josso N, Forest MG, Picard JY. Müllerian inhibiting activity of calf fetal testis : relationship to testosterone and protein synthesis. Biol Reprod 1975;13:163–167.
28. Donahoe PK, Ito Y, Price JM, Hendren WH. Müllerian inhibiting substance activity in bovine fetal, newborn, and prepubertal testes. Biol Reprod 1977;16:238–243.
29. Josso N. Permeability of membranes to the Müllerian inhibiting substance synthesized by the human fetal testis. J Clin Endocrinol Metab 1972;34:265–270.
30. Donahoe PK, Ito Y, Morikawa Y, Hendren WH, III. Müllerian inhibiting substance in human testes after birth. J Pediatr Surg 1977;12:323–330.
31. Shen WH, Moore CCD, Ikeda Y, Parker KL, Ingraham HA. Nuclear receptor steroidogenic factor 1 regulates the mullerian inhibiting substance gene: A link to the sex determination cascade. Cell 1994; 77:651–661.
32. Giuili G, Shen WH, Ingraham HA. The nuclear receptor SF-1 mediates sexually dimorphic expression of Mullerian inhibiting substance, in vivo. Development 1997;124:1799–1807.
33. Arango NA, Lovell-Badge R, Behringer RR. Targeted mutagenesis of the endogenous mouse Mis gene promoter: In vivo definition of genetic pathways of vertebrate sexual development. Cell 1999; 99:409–419.
34. Hossain A, Saunders GF. Role of Wilms tumor 1 (WT1) in the transcriptional regulation of the mullerian-inhibiting substance promoter. Biol Reprod 2003;69:1808–1814.
35. Nachtigal MW, Hirokawa Y, EnyeartVan Houten DL, Flanagan JN, Hammer GD, Ingraham HA. Wilms' tumor 1 and DAX-1 modulate the orphan nuclear receptor SF-1 in sex-specific gene expression. Cell 1998;93:445–454.
36. Tremblay JJ, Viger RS. Nuclear receptor Dax-1 represses the transcriptional cooperation between GATA-4 and SF-1 in Sertoli cells. Biol Reprod 2001;64:1191–1199.
37. Lasala C, Carre-Eusebe, Picard JY Rey R. Subcellular and molecular mechanisms regulating anti-Mullerian hormone gene expression in mammalian and nonmammalian species. DNA Cell Biol 2004;23:572–585. Review.
38. Teixeira J, Fynn-Thompson E, Payne A, Donahoe PK. Müllerian-inhibiting substance regulates androgen synthesis at the transcriptional level. Endocrinology 1999;140:4732–4738.
39. Laurich VM, Trbovich AM, O'Neill FH, et al. Müllerian inhibiting substance blocks the protein kinase A-induced expression of cytochrome P450 17alpha-hydroxylase/C17-20 lyase mRNA in a mouse Leydig cell line independent of cAMP responsive element binding protein phosphorylation. Endocrinology 2002;143:3351–3360.
40. Sriraman V, Niu E, Matias JR, et al. Müllerian inhibiting substance inhibits testosterone synthesis in adult rats. J Androl 2001;22:750–758.
41. Bezard J, Vigier B, Tran D, Mauleon P, Josso N. Immunocytochemical study of anti-Mullerian hormone in sheep ovarian follicles during fetal and post-natal development. J Reprod Fertil 1987;80:509–516.
42. Munsterberg A, Lovell-Badge R. Expression of the mouse anti-Müllerian hormone gene suggests a role in both male and female sexual differentiation. Development 1991;113:613–624.
43. Ueno S, Manganaro TF, Donahoe PK. Human recombinant Müllerian inhibiting substance inhibition of rat oocyte meiosis is reversed by epidermal growth factor in vitro. Endocrinology 1988;123: 1652–1659.
44. Kim JH, Seibel MM, MacLaughlin DT, et al. The inhibitory effects of Müllerian inhibiting substance on epidermal growth factor induced proliferation and progesterone production of human granulosa-luteal cells. J Clin Endo Metab 1992;75:911–917.
45. Durlinger AL, Kramer P, Karel B, et al. Control of primordial follicle recruitment by anti-Mullerian hormone in the mouse ovary. Endocrinology 1999;140:5789–5796.
46. Seifer DB, MacLaughlin DT, Penzias AS, et al. Gonadotropin-releasing hormone agonist-induced differences in granulosa cell cycle kinetics are associated with alterations in follicular fluid Mullerian inhibiting substance and androgen content. J Clin Endo Metab 1993;76:711–714.
47. Seifer DB, MacLaughlin DT, Christian BP, Feng B, Shelden RM. Early follicular serum Mullerian inhibiting substance levels are associated with ovarian response during assisted reproductive technology cycles. Fertil Steril 2002;77:468–471.
48. Trbovich AM, Sluss PM, Laurich VM, et al. Mullerian inhibiting substance lowers testosterone in luteinizing hormone-stimulated rodents. Proc Natl Acad Sci USA 2001;98:3393–3397.

49. Lee MM, Seah CC, Masiakos PT, et al. Mullerian inhibiting substance type II receptor expression and function in purified rat Leydig cells. Endocrinology 1999;140:2819–2827.
50. Baarends WM, Hoogerbrugge JW, Post M, et al. Anti-Mullerian hormone and anti-Mullerian hormone type II receptor messenger ribonucleic acid expression during postnatal testis development and in the adult testis of the rat. Endocrinology 1995;136:5614–5622.
51. Bedecarrats GY, O'Neill FH, Norwitz ER, Kaiser UB, Teixeira J. Regulation of gonadotropin gene expression by Mullerian inhibiting substance. Proc Natl Acad Sci USA 2003;100:9348–9353.
52. Wang PY, Koishi K, McGeachie AB, et al. Mullerian inhibiting substance acts as a motor neuron survival factor in vitro. Proc Natl Acad Sci USA 2005;102:16,421–16,425.
53. Segev DL, Hoshiya Y, Hoshiya M, et al. Mullerian inhibiting substance regulates NF-κB signaling in the prostate in vitro and in vivo. Proc Natl Acad Sci USA 2002;99:239–244.
54. Hoshiya Y, Gupta V, Segev DL, et al. Mullerian substance induces NFκB signaling in breast and prostate cancer cells. Mol Cell Endocrinol 2003;211:43–49.
55. Segev DL, Hoshiya Y, Stephen AE, Hoshiya M, Tran TT, MacLaughlin DT, Donahoe PK, Maheswaran S. Müllerian inhibiting substance regulates NFκB signaling and growth of mammary epithelial cells in vivo. J Biol Chem 2001;276:26,799–26,806.
56. Hoshiya Y, Gupta V, Kawakubo H, et al. MIS Promotes INF-γ-induced gene expression and apoptosis in breast cancer cells. J Biol Chem 2003;278:51,703–51,712.
57. Picard JY, Josso N. Anti-Mullerian hormone: estimation of molecular weight by gel filtration. Biomedicine 1976;25:147–150.
58. Picard JY, Josso N. Purification of testicular anti-Mullerian hormone allowing direct visualization of the pure glycoprotein and determination of yield and purification factor. Mol Cell Endocrinol 1984; 34:23–119.
59. Swann DA, Donahoe PK, Ito Y, Morikawa Y, Hendren WH. Extraction of Müllerian inhibiting substance from newborn calf testis. Dev Biol 1979;69:73–84.
60. Budzik GP, Swann DA, Hayashi A, Donahoe PK. Enhanced purification of Müllerian inhibiting substance by lectin affinity chromatography. Cell 1980;21:909–915.
61. Budzik GP, Powell SM, Kamagata S, Donahoe PK. Mullerian inhibiting substance fractionation by dye affinity chromatography. Cell 1983;34:307–314.
62. Cate RL, Mattaliano RJ, Hession C, et al. Isolation of the bovine and human genes for Müllerian inhibiting substance and expression of the human gene in animal cells. Cell 1986;45:685–698.
63. Picard JY, Benarous R, Guerrier D, Josso N, Kahn A. Cloning and expression of cDNA for anti-Mullerian hormone. Proc Natl Acad Sci USA 1986;83:5464–5468.
64. Josso N, Forest MG, Picard JY. Müllerian inhibiting activity of calf fetal testis : relationship to testosterone and protein synthesis. Biol Reprod 1975;13:163–167.
65. Vigier B, Picard JY, Josso N. A monoclonal antibody against bovine anti-Müllerian hormone. Endocrinology 1982;110:131–137.
66. Shima H, Donahoe PK, Budzik GP, Kamagata S, Hudson P, Mudgett-Hunter M. Production of monoclonal antibodies for affinity purification of bovine Müllerian inhibiting substance activity. Hybridoma 1984;3:201–214.
67. Cohen-Haguenauer O, Pcard JY, Mattei MG, et al. Mapping of the gene for antimullerian hormone to the short arm of human chromosome 19. Cytogenet Cell Genet 1987;44:2–6.
68. Pepinsky RB, Sinclair LK, Chow EP, et al. Proteolytic processing of Müllerian inhibiting substance produces a transforming growth factor-á-like fragment. J Biol Chem 1988;263:18,961–18,965.
69. Nachtigal MW, Ingraham HA. Bioactivation of Mullerian inhibiting substance during gonadal development by a kex2/subtilisin-like endoprotease. Proc Natl Acad Sci USA 1996;93:7711–7716.
70. MacLaughlin DT, Hudson PL, Graciano AL, et al. Müllerian duct regression and antiproliferative bioactivities of Müllerian inhibiting substance reside in its carboxy-terminal domain. Endocrinology 1992;131:291–296.
71. Wilson CA, di Clemente N, Ehrenfels C, et al. Müllerian inhibiting substance requires its N-terminal domain for maintenance of biological activity, a novel finding within the transforming growth factor-ß superfamily. Mol Endocrinol 1993;7:247–257.
72. Ragin RC, Donahoe PK, Kenneally MK, Ahmad M, MacLaughlin DT. Human Müllerian inhibiting substance: Enhanced purification imparts biochemical stability and restores antiproliferative effects. Protein Expres Purification 1992;3:236–245.
73. Lorenzo HK, Teixeira J, Pahlavan N, Laurich VM, Donahoe PK, MacLaughlin DT. New approaches for high-yield purification of Müllerian inhibiting substance improve its bioactivity. J Chromato B 2002;766:89–98.

74. Hoshiya Y, Gupta V, Kawakubo H, et al. MIS Promotes INF-γ–induced gene expression and apoptosis in breast cancer cells. J Biol Chem 2003;278:51,703–51,712.
75. Baarends WM, van Helmond JM, Post M, et al. A Novel member of the transmembrane serine/threonine kinase receptor family is specifically expressed in the gonads and in mesenchymal cells adjacent to the Mullerian duct. Development 1994;120:189–197.
76. DiClemente N, Wilson C, Faure E, et al. Cloning, expression, and alternative splicing of the receptor for anit-Mullerian hormone. Mol Endocrinol 1994;8:1006–1020.
77. Teixeira J, He WW, Shah PC, et al. Developmental expression of a candidate Mullerian inhibiting substance type II receptor. Endocrinology 1996;137:160–165.
78. Mishina Y, Tizard R, Deng JM, et al. Sequence, genomic organization and chromosomal location of the mouse Mullerian inhibiting substance type II receptor gene. Biochem Res Commun 1997;237:741–746.
79. Tsuji M, Shima H, Yonemura CY, Brody J, Donahoe PK, Cunha GR. Effect of human recombinant Mullerian inhibiting substance on isolated epithelial and mesenchymal cells during mullerian duct regression in the rat. Endocrinology 1992;131:1481–1488.
80. Renaud EJ, MacLaughlin DT, Oliva E, Rueda BR, Donahoe, PK. Endometrial cancer is a receptor mediated target for Mullerian inhibiting substance. Proc Natl Acad Sci USA 2006;102:111–116. Erratum in: Proc Natl Acad Sci USA 2005;102:6513.
81. Hoshiya Y, Gupta V, Segev DL, et al. Mullerian inhibiting substance induces NF kappa B signaling in breast and prostate cancer cells. Mol Cell Endocrinol 2003;211:43–49.
82. Gupta V, Carey JL, Kawakubo H, Muzikansky A, et al. Mullerian inhibiting substance suppresses tumor growth in the C3(1)T antigen transgenic mouse mammary carcinoma model. Proc Natl Acad Sci USA 2005;102:3219–3224.
83. Barbie TU, Barbie DA, MacLaughlin DT, Maheswaran S, Donahoe PK. Mullerian inhibiting substance inhibits cervical cancer cell growth via a pathway involving p130 and p107. Proc Natl Acad Sci USA 2003;100:15,601–15,906.
84. Josso N, Belville C, di Clemente N, Picard YP. AMH and AMH receptor defects in persistent Mullerian duct syndrome. Hum Reprod Update. 2005;11:351–356. Review.
85. Gouedard L, Chen YG, Thevenet L, et al. Engagement of bone morphogenetic protein type IB receptor and Smad1 signaling by anti-Mullerian hormone and its type II receptor. J Biol Chem 2000;275: 27,973–27,978.
86. Visser JA, Olaso R, Verhoef-Post M, Kramer P, Themmen AP, Inraham HA. The serine/threonine transmembrane receptor ALK2 mediates Mullerian inhibiting substance signaling. Mol Endocrinol 2001;15:936–945.
87. Jamin S, Arango NA, Mishina Y, Hanks MC, Behringer RR. Requirement of BMPR1a for Mullerian duct Clarke TR, Hoshiya Y, Yi SE, Liu X, Lyons KM, Donahoe PK. Müllerian inhibiting substance signaling uses a BMP-like pathway mediated by ALK2 and induces Smad6 expression. Mol Endocrinol 2001;15:946–995.
88. He WW, Gustafson ML, Hirobe S, Donahoe PK. The developmental expression of four novel serine/threonine kinase receptors homologous to the activin/TGF-b II receptor family. Dev Dynamics 1993; 196:133–142.
89. Clarke TR, Hoshiya Y, Yi SE, Liu X, Lyons KM, Donahoe PK. Müllerian inhibiting substance Signaling uses a BMP-like pathway mediated by ALK2 and induces Smad6 expression. Mol Endocrinol 2001;15:946–959.
90. Gu Z, Reynolds EM, Song J, Lei H, Donahoe PK, Li E. The type I serine-threonine kinase receptor ActRIA functions in extraembryonic cells to regulate mouse gastrulation. Development 1999;126: 2551–2561.
91. Lee MM, Donahoe PK, Silverman BL, et al. Mullerian inhibiting substance in the evaluation of children with nonpalpable gonads. New Engl J Med 1997;336:1480–1486.
92. Seifer DB, MacLaughlin DT, Christian BP, Feng B, Shelden RM. Early follicular serum mullerian-inhibiting substance levels are associated with ovarian response during assisted reproductive technology cycles. Fertil Steril 2002;77:468–471.
93. van Rooij IA, Broekmans FJ, te Velde ER, et al. Serum anti-Mullerian hormone levels: a novel measure of ovarian reserve. Hum Reprod 2002;17:3065–3071.
94. Muttukrishna S, Suharjono H, McGarrigle H, Sathanandan M. Inhibin B and anti-Mullerian hormone: markers of ovarian response in IVF/ICSI patients? Brit J Obstet Gynaecol 2004;111:1248–1253.
95. Hazout A, Bouchard P, Seifer DB, Aussage P, Junca AM, Cohen-Barcie P. Serum antimullerian hormone/mullerian-inhibiting substance appears to be a more discriminatory marker of assisted reproductive technology outcome than follicle-stimulating hormone, inhibin B, or estradiol. Fertil Steril 2004;82:1323–1329.

96. Penarrubia J, Fabregues F, Manau D, et al. Basal and stimulation day 5 anti-Mullerian hormone serum concentrations as predictors of ovarian response and pregnancy in assisted reproductive technology cycles stimulated with gonadotropin-releasing hormone agonist—gonadotropin treatment. Hum Reprod 2005;20:915–922.
97. Eldar-Geva T, Ben-Chetrit A, Spitz IM, et al. Dynamic assays of inhibin B, anti-Mullerian hormone and estradiol following FSH stimulation and ovarian ultrasonography as predictors of IVF outcome. Hum Reprod 2005;20:3178–3183.
98. Silberstein T, MacLaughlin DT, Shai I, et al. Mullerian inhibiting substance levels at the time of HCG administration in IVF cycles predict both ovarian reserve and embryo morphology. Hum Reprod 2006;21:159–163.
99. Gustafson ML, Lee MM, Scully RE, et al. Mullerian inhibiting substance as a marker for ovarian sex-cord tumor. New Engl J Med 1992;326:466–471.
100. Gustafson ML, Lee MM, Asmundson L, MacLaughlin DT, Donahoe PK. Mullerian inhibiting substance in the diagnosis and management of intersex and gonadal abnormalities. J Pedi Surg 1993; 28:439–444.
101. Dutertre M, Gouedard L, Xavier F, et al . Ovarian granulosa cell tumors express a functional membrane receptor for anti-Mullerian hormone in transgenic mice. Endocrinology 2001;142:4040–4046.
102. Donahoe PK, Swann DA, Hayashi A, Sullivan MD. Müllerian duct regression in the embryo is correlated with cytotoxic activity against a human ovarian cancer. Science 1979;205:913–915.
103. Masiakos PT, MacLaughlin DT, Maheswaran S, et al. Human ovarian cancer, cell lines and primary ascites cells express the human MIS type II receptor, bind, and are responsive to MIS. Clin Cancer Res 1999;5:488–499.
104. Price JM, Donahoe PK, Ito Y. Involution of the female Müllerian duct of the fetal rat in the organ culture assay for the detection of Müllerian inhibiting substance. Am J Anat 1979;156:265–284.
105. Price JM, Donahoe PK, Ito Y, Hendren WH. Programmed cell death in the Müllerian duct induced by Müllerian inhibiting substance. Am J Anat 1977;149:353–375.
106. Trelstad RL, Hayashi A, Hayashi K, Donahoe PK. The epithelial-mesenchymal interface of the male rat Mullerian duct: loss of basement membrane integrity and ductal regression. Dev Biol 1982;92:27–40.
107. Hayashi A, Donahoe PK, Budzik GP, Trelstad RL. Periductal and matrix glycosaminoglycans in rat Mullerian duct development and regression. Develop Biol 1982;92:16–26.
108. Berkenblit A, Cannistra SA. Advances in the management of epithelial ovarian cancer. J Reprod Med 2005;50:426–438.
109. Jemal A, Murray T, Samuels A, Ghafoor A, Ward E, Thun M. Cancer Statistics. J Clin Cancer 2003; 53:5–26.
110. Fuller AF Jr, Guy SR, Budzik GP, Donahoe PK. Müllerian inhibiting substance inhibits colony growth of a human ovarian cancer cell line. J Clin Endocrinol Metab 1982;54:1051–1055.
111. Donahoe PK, Fuller AF Jr, Scully RE, Guy SR, Budzik GP. Müllerian inhibiting substance inhibits growth of a human ovarian cancer in nude mice. Ann Surg 1981;194:472–480.
112. Fuller AF, Jr, Krane IM, Budzik GP, Donahoe PK. Müllerian inhibiting substance reduction of colony growth of human gynecologic cancers in a stem cell assay. Gynecol Oncol 1985;22:135–148.
113. Chin T, Parry RL, Donahoe PK. Human Müllerian inhibiting substance inhibits tumor growth in vitro and in vivo. Cancer Res 1991;51:2101–2106.
114. Stephen AE, Pearsall LA, Christian BP, Donahoe PK, Vacanti JP, MacLaughlin DT. Highly purified Mullerian inhibiting substance inhibits ovarian cancer in vivo. Clin Cancer Res 2002;8:2640–2646.
115. Barbie TU, Barbie DA, MacLaughlin DT, Maheswaran S, Donahoe PK. Mullerian inhibiting substance inhibits cervical cancer cell growth via a pathway involving p130 and p107. Proc Natl Acad Sci USA 2003;100:15,601–15,606.
116. Connolly DC, Bao R, Nikitin AY, et al. Female mice chimeric for expression of the Simian Virus 40 TAg under control of the MISIIR promoter develop epithelial ovarian cancer. Cancer Res 2003; 63:1389–1397.
117. Pieretti-Vanmarcke R, Donahoe PK, Szotek P, et al. Recombinant human Mullerian inhibiting substance inhibits long term growth of MISRII-directed transgenic mouse ovarian cancers in vivo. Clin Cancer Res 2006;12:1593–1598.
118. Segev DL, Hoshiya Y, Stephen AE, et al. Müllerian inhibiting substance regulates NFκB signaling and growth of mammary epithelial cells in vivo. J Biol Chem 2001;276:26,799–26,806.
119. Sovak MA, Bellas RE, Kim DW, et al. Aberrant nuclear factor-κB/Rel Expression and the pathogenesis of breast cancer. J Clin Invest 1997;100:2952–2960.
120. Nakshatri H, Bhat-Nakshatri P, Martin DA, Goulet RJ, Sledge GW. Constitutive activation of NF-kB during progression of breast cancer to hormone-independent growth. Mol Cell Biol 1997;17:3629–3639.

22 Activins, Inhibins, and Bone Morphogenetic Proteins as Modulators and Biomarkers of Prostate Cancer Progression

Tetsu Hayashida, Vandana Gupta, Sam Thiagalingam, and Shyamala Maheswaran

CONTENTS

Development of the Prostate Gland, Benign Prostate Hyperplasia, and Prostate Cancer
Components of the TGF-β Family Signaling Cascade
Expression and Function of Activins and Inhibins in the Normal Prostate and Prostate Cancer
Expression and Function of BMPs in the Normal Prostate and Prostate Cancer
Conclusions
References

Abstract

The transforming growth factor-β (TGF-β) superfamily consists of over 35 structurally related proteins, which regulate a myriad of cellular processes. Signaling by these growth regulatory molecules is initiated by ligand-induced hetero-oligomerization of distinct type II and type I serine/threonine kinase receptors that activate receptor-activated Smad proteins as well as Smad-independent pathways. TGF-β-related ligands and receptors are expressed in the prostate gland and aberrant expression and function of this class of signaling molecules are likely to regulate development and tumorigenesis of the prostate. The functional significance of TGF-β in prostate cancer has been reviewed in several articles. The objective of this review is to evaluate the expression and functional significance of activins, inhibins, and BMP signaling in prostate development and tumorigenesis, and their role as modulators and biomarkers of prostate cancer progression.

Key Words: Prostate cancer; activin; bone morphogenetic proteins; inhibins; ALK2; ALK3; ALK6; activin type II receptor; BMP type II receptor.

1. DEVELOPMENT OF THE PROSTATE GLAND, BENIGN PROSTATE HYPERPLASIA, AND PROSTATE CANCER

The prostate, a male accessory sex gland, arises as an epithelial out growth from the urethra and its formation involves epithelial mesenchymal interactions and androgenic

From: *Cancer Drug Discovery and Development: Transforming Growth Factor-β in Cancer Therapy, Vol I: Basic and Clinical Biology*
Edited by: Sonia B. Jakowlew © Humana Press Inc., Totowa, NJ

stimulation. The urogenital mesenchymal cells act on epithelial outgrowths to induce duct formation, and regulate the expression of epithelial androgen receptors, epithelial cell proliferation, and the synthesis of prostate specific secretory proteins. The epithelium in turn promotes the surrounding mesenchyme to differentiate into smooth muscle cells *(1–5)*. In the adult, the prostate is divided into four biologically and anatomically distinct zones, the peripheral, central, transitional and periurethral zones. Most hyperplasias arise in the transitional and peripheral zones, whereas carcinomas mostly originate in the peripheral zone *(6)*. Benign prostate hyperplasia (BPH) is characterized by hyperplasia of prostatic stromal and epithelial cells. Accumulation of dihydrotestosterone (DHT) within the prostate may serve as the hormonal mediator for hyperplasia *(7)*. Age-related changes in prostatic stromal-epithelial interactions and/or an increase in the number of prostatic stem cells have also been implicated in the etiology of BPH *(8,9)*.

Prostate carcinoma is the most common form of cancer in men. Although little is known about the etiology of prostate cancer; several factors including age, race, family history, hormone levels and environment may be involved in this process. Prostate cancer is stratified into five grades according to the Gleason system, which is based on the glandular patterns and degree of differentiation *(10)*. There is generally a good correlation between prognosis and the Gleason grade and it serves as an important parameter in the selection of therapy. Androgen, a key regulator of prostate growth, is also required for the maintenance of prostate cancer cells. Thus, androgen ablation therapy is used to treat patients with advanced prostate cancer and metastatic tumors. Although most men with prostate cancer initially respond to androgen-withdrawal, they subsequently relapse and the recurring tumors become androgen-independent *(11–15)*. This process may involve androgen-independent survival of some carcinoma cells, which may eventually acquire the ability to proliferate depending on the balance between positive and negative growth regulatory polypeptides in the prostate, especially members of the fibroblast growth factor, epidermal growth factor, Insulin-like growth factor and the transforming growth factor β (TGF-β) family. These growth factors, which play a critical role in the maintenance of normal prostatic growth and differentiation, and in tumorigenesis, may enable the tumors to continue to proliferate in an androgen-independent manner *(16–18)*.

Advanced prostate cancer avidly metastasizes to the bone. Prostate cancer skeletal metastasis radiographically characterized is often osteoblastic rather than osteolytic. However, histologically they are a heterogeneous mixture of osteolytic and osteoblastic lesions with the latter being predominant *(19–22)*. The mechanism by which prostate cancer cells promotes bone mineralization and/or resorption in unclear, however, prostate cancer cells produce several peptides with osteogenic properties including members of the TGF-β superfamily *(23,24)*.

2. COMPONENTS OF THE TGF-β FAMILY SIGNALING CASCADE

The TGF-β superfamily consists of over 35 polypeptides including TGF-β, activins, inhibins, bone morphogenetic proteins (BMPs), growth and differentiation factors (GDFs), Mullerian inhibiting substance (MIS), Nodal and others that regulate a multitude of cellular processes. TGF-βs, activins and MIS transduce their signals by binding to a type II receptor, a transmembrane serine-threonine kinase, that in turn complexes with another distinct transmembrane serine-threonine kinase known as the type I receptor. In contrast, some BMPs appear to bind directly to both type I and type II BMP receptors. Thus far in vertebrates five type II receptors and seven type I receptors have been identified indicating receptor promiscuity in order to accommodate all the ligands. Biological activity and signaling by ligands is also influenced by membrane-associated proteins such as endoglin, betaglycan and Cripto-1 and by extracellular proteins decorin, follistatin, noggin, chordin, Gremlin, differential screening selected gene aberrant in neuroblastoma (DAN), and cerberus *(25–31)*.

Ligand-induced heteromeric complex formation of type I and type II receptors leads to phosphorylation of the type I receptor and induction of its latent kinase activity that subsequently propagates a signaling cascade mediated by phosphorylation of receptor activated Smad (R-Smad) proteins or through Smad-independent pathways. The R-Smad family is divided into ligand or receptor specific pathways; Smads-1, -5, and -8 transmit BMP and MIS-stimulated signals and Smad-2 and -3 mediate TGF-β and activin-induced signals. The phosphorylated R-Smads heteromerize with the common Smad4, translocate to the nucleus and associate with other transcriptional regulators to positively or negatively modulate transcription. Another class of Smads, Smad-6 and -7 inhibit signaling by the TGF-β family by either binding to R-Smad or by blocking their access to the type I receptors *(29,31,32)*. Although the major signaling route for the TGF-β family appears to be through receptor-activated Smad proteins, Smad-independent pathways are also activated following ligand binding *(31–34)*.

Because many facets of the role of TGF-β in prostate cancer have been reviewed in several articles *(35–39)*, the main objective of this review is to evaluate the expression and functional significance of activins, inhibins and BMP signaling in prostate development and tumorigenesis. We will review their role as modulators and biomarkers of prostate cancer progression.

3. EXPRESSION AND FUNCTION OF ACTIVINS AND INHIBINS IN THE NORMAL PROSTATE AND PROSTATE CANCER

3.1. Expression and Function of Activins and Inhibins in the Normal Prostate

Activins and inhibins are dimeric glycoproteins that consist of two common β subunits, βA and βB. Activins are dimers of β subunits: activin A (βAβA), activin B (βBβB) and activin AB (βAβB) *(40)*. Inhibins are heterodimers consisting of an α subunit and either a βA subunit (inhibin A) or a βB subunit (inhibin B). Activin βC, βD, and βE are newly identified members of this family *(41–44)*, for which very little functional data are available. Activin signaling is dependent on the presence of ActRIIA and ActRIIB, of the type II receptor family, and ALK2 (Activin-like kinase2; ActRIA) and ALK4 (ActRIB) of the type I receptor subclass *(29,31,32)*. An important modulator of activin function is follistatin (FS), an extracellular matrix protein that binds to activin with high affinity and irreversibly blocks its activation *(45,46)*. Target cells of activin produce FS and its synthesis is regulated by activins. All naturally occurring FS isoforms bind activins and inhibit their biological activity *(28,30,47)*. Although, the most recognized role for activins and inhibins is in the regulation of gonadal function *(48)*, they have also been implicated in control of cell growth and tumorigenesis *(49–51)*.

The activin ligands and receptors are expressed in epithelial cells of the rat prostate *(52–54)*. Activin βA is expressed throughout rat prostate development during branching morphogenesis, differentiation, and maturation of secretary function. During development, activin βA localizes to the cap of the undifferentiated mesenchyme surrounding the epithelial bud tips, and during puberty and adulthood expression is maintained in the epithelial compartment. Activin βB is present only in some cells of the mesenchyme and stroma but is strongly expressed in the columnar epithelial cells of the mature prostate *(54)*. In rats, activin A expression in the prostate was not altered 3 d after castration *(53)* whereas in dogs, activin was readily detectable in the prostate glandular epithelial cells and completely disappeared 1–2 wk after castration *(55)*. Thus, it is unlikely that androgens are direct regulators of activin expression in the prostate. FS mRNA is expressed in the prostatic epithelium during development and differentiation and its expression is reduced in the maturing epithelium *(54)* suggesting that locally produced activin might be biologically active in the mature prostate. The prostate also expresses inhibin protein and changes in androgen levels do not influence inhibin expression *(53)*.

Of the activin receptors, ALK 2 localizes to the prostatic epithelium and stroma and ActRIIA and ActRIIB are expressed throughout the epithelium *(54)*. The dorsolateral prostate also expresses a novel 55 kDa serine threonine kinase receptor, which shares 80% homology with ALK4 and ALK5 whose ligands are activin and TGF-β, respectively *(56)*. However, their functional significance remains unknown. The dynamic temporal and spatial patterns of activin, activin receptor, and FS expression in the prostate during development suggest that activins may play an important paracrine/autocrine role in regulating the growth and differentiation processes in the prostate.

Development of glandular organs such as the kidney, lung, and prostate involves the process of epithelial branching morphogenesis and activins may play a key role in this process *(57)*. Activin inhibits prostatic branching and growth in organ cultures of rat prostates without increasing apoptosis. FS, the activin antagonist, enhanced branching in the absence but not presence of testosterone *(54)*. Sonic hedgehog (Shh), the vertebrate homologue of the *Drosophila* segment-polarity gene hedgehog, inhibits cell proliferation and promotes differentiation of luminal epithelial cells in organotypic cultures of developing rat prostates. This process was associated with induction of activin A and TGF-β *(58)*. Although, BMP4 is also implicated in regulating epithelial branching in the prostate *(59)*, its expression remained essentially unchanged following Shh treatment *(58)*.

3.2. Expression and Functional Significance of Activin in Prostate Cancer

Activin, activin receptors and FS are also expressed in BPH, prostate cancer cell lines and tissues *(60–64)*, and activin suppresses androgen-induced growth and androgen-independent survival of prostate cancer cells *(60,61,65–67)*. It can also block IL-10, FS, and DHT stimulated growth and PSA production in high-grade prostatic intraepithelial neoplasia (PIN) cultures immortalized by HPV-18 infection *(68)*.

Prostate explants and primary prostate epithelial cells established from tissues resected from prostate carcinoma patients secrete both activin A and FS, the concentrations of which ranged from 2.5 to 21 ng/ml and 2.4 to 48.3 ng/ml, respectively. Because only free FS was measured in these assays, and the FS-activin complex is essentially nondissociable, the measurement of free FS in the culture medium indicates that FS was secreted in excess of activin by all of the prostate cancer explants. Hence it is likely that the overproduction of FS by prostate carcinoma renders the locally produced activin biologically inactive. In fact exposure of prostate epithelial cells established from these prostate cancer explants, respond to the growth inhibitory effects of exogenous activin in a dose dependent manner, an effect neutralized by increasing concentrations of recombinant FS *(69)*.

Despite overexpressing endogenous FS, the prostate epithelial cells established from prostate cancer explants respond to exogenous FS, which enhances thymidine incorporation in these cells suggesting that FS might be exerting activin-independent effects on the growth of prostate cancer cells *(69)*. In fact exogenous FS promotes prostatic cancer cell line (LNCaP) cell growth *(60)* and enhances the proliferation and colony forming ability of PIN cultures immortalized by HPV-18 *(68)*. PC3, an androgen-independent prostate cancer cell line is resistant to the growth inhibitory effects of activin owing to overexpression of FS; neutralization of FS with an antibody restored activin sensitivity *(62)*. Whether FS induced proliferation of prostate cancer cells occurs owing to inhibition of endogenous activin and/or activin-independent effects remains to be determined.

The effects of activin and androgen on prostate cancer cell growth are mutually antagonistic. Activin induces apoptosis in the androgen-dependent LNCaP grown in androgen-depleted medium and androgen suppresses this effect. Moreover, activin does not alter the cell cycle distribution of LNCaP cells but suppresses androgen-induced cell cycle progression and androgen-stimulated growth. The expression cell cycle regulatory proteins such as Rb, E2F-1, and p27 demonstrate a strong correlation with the mutually antagonistic growth effects of activin and androgen. Interestingly, activin mediated inhibition of prostate cancer cell growth was independent of Smad3 phosphorylation *(65)*.

Consistent with its growth inhibitory function in LNCaP cells, activin induced the expression of the tumor suppressor p53 and apoptosin, a gene that triggers DNA fragmentation in cells. Exposure of LNCaP cells to antisense apoptosin prevented activin induced apoptosis *(70,71)*. The cyclin dependent kinase inhibitor, p16, and Siva, a proapoptotic protein, are also stimulated by activin *(70)*. Microarray analysis of RNA isolated from laser-capture microdissected activin-positive and activin-negative PIN cells, led to the identification of several differentially expressed genes implicated in growth regulation and carcinogenesis. Compared with expression in activin-negative PIN cells, human prostate-specific Ets (hPSE), suppression of tumorigenicity 14 (St14), and Patched (ptc)-like receptor, were upregulated in activin-positive PIN cells, whereas the cell division cycle 6 (CDC6), ROK1, and Heat-shock protein70 (hsp70), were suppressed *(72)*. hPSE is a transcriptional activator belonging to the Ets family, which regulates the proliferation, differentiation, and development of prostate epithelial cells and the hPSE mRNA is translated in normal glands but not in prostate cancer cell lines *(73)*, approx 30% of prostate cancers do not immunostain for hPSE, whereas normal glands, hyperplastic glands, and PIN lesions stain with similar intensities suggesting that loss of hPSE protein may suggest malignancy in the prostate glands *(74)*. It is also a potentially novel marker for detection of metastatic breast cancer in axillary lymph nodes *(75)*. St14 is aberrantly methylated in pancreatic carcinoma *(76)* and its expression in the prostate remains to be investigated.

Of the genes suppressed in activin-positive PIN lesions, CDC6 is a regulator of the onset of DNA replication in eukaryotic cells *(77)*. Hsp70 expression is elevated in prostate cancer cells *(78)* and depletion of Hsp70 expression in cells led to apoptosis *(79)*. Moreover, morphologically advanced cancers demonstrated diminished expression of Hsp70 compared with nonneoplastic prostatic epithelium *(80)*. Although the functional properties of the repertoire of genes either up- or downregulated in activin positive PIN lesions are consistent with the hypothesis that activin may be an autocrine growth inhibitory factor in the prostate, whether these genes are directly regulated in response to activin exposure remains to be determined.

In addition, activin also induces the expression of the prostatic markers, prostate-specific antigen (PSA) and prostatic acid phosphatase, in LNCaP cells *(65,67,81,82)*. Upregulation of PSA by activin was androgen receptor-independent but dependent on Smad3 phosphorylation *(65)*. Although activins and androgen have opposing effects on prostate cancer cell growth, both additively upregulate PSA expression in LNCaP cells *(65,81,82)*. Interestingly, PSA can cleave parathyroid-hormone-related protein (PTHRP), an endocrine hormone that stimulates bone resorption *(83,84)*. Whether activin, by inducing PSA in prostate cancer cells, regulates the bone microenvironment remains to be determined. The expression of PSA is primarily regulated by androgens and it serves as a very important tumor marker for prostate cancer, and modulation of PSA expression by activin, an inhibitor of prostate cancer cell growth, could in part contribute to the lack of specificity of PSA in prostate cancer screening.

Because locally produced activins may potentially be negative regulators of proliferation in the prostate, several groups have determined whether activin signaling would be aberrantly regulated in prostate cancer. Analysis of nine human prostate tumors xenografted into nude mice demonstrated that activin type II and type I receptors were present in all the tumors while inhibin α was expressed in none. Activin βA was undetectable in androgen-dependent xenografts, whereas activin βB was abundantly expressed in androgen-dependent xenografts and significantly lower in androgen-independent tumors. FS expression ranged from undetectable to relatively high levels in androgen-independent prostate tumor xenografts *(64)*. Analysis of high-grade prostate cancer demonstrated that the nonmalignant epithelium expresses activin βA and βB proteins whereas basal epithelial cells and stroma express FS mRNA. Both activin βA and βB and FS mRNAs were coexpressed in poorly differentiated tumor cells suggesting that FS may render the tumor cells resistant to the growth inhibitory effects of activin *(63)* suggesting that prostate tumors may exhibit decreased growth sensitivity to activin signaling.

A tumor suppressor role for activin is evident in many human tumor types. The ActRIIA gene, which has two 8-base pair polyadenine tracts experiences mutation in 58% of colorectal microsatellite instability neoplasms *(85)*. Human pituitary tumors specifically express alternatively spliced ActRIB mRNAs and overexpression of wild type ActR1B in a human pituitary tumor cell line restored growth inhibitory response to activin *(86)*. About 40% of rat and mouse liver tumors display increased levels of FS as a strategy to decrease local bioavailability of activin *(87)*. Decrease in activin, receptors and signaling intermediates such as Smads has also been associated with tumor progression in the breast *(88)*. Although, existing data strongly support that activin and FS may be important regulators of prostate development and prostate cancer growth (Table 1), the precise role they play in these processes is yet to be established.

3.3. Inhibins in Prostate Cancer

The incidence of gonadal and adrenal tumors in inhibin α null mice suggests a tumor suppressor role for inhibin *(89)*. The inhibin α gene localizes to chromosome 2q (2q33-q36) and LOH at 2q32-36 has been observed in 42% of prostate carcinomas *(90,91)*. Inhibin α is low or undetectable in several prostate cancer cell lines *(64, 92)* owing to hypermethylation of the promoter *(93)* and inhibin α gene promoter hypermethylation has also been demonstrated in prostate tumors of Gleason grade 3, 4, and 5 *(90)*. These observations suggest that inhibin α may indeed be a tumor suppressor in the prostate.

However, analysis of a large cohort of patients with multiple antibodies against inhibin α led to the observation that inhibin α expression is upregulated in 23–66% of patients with prostate cancer *(94)*. Consistent with this observation, the mean serum inhibin concentration in patients with nonmalignant and malignant prostate cancer was higher than that observed in normal men. Serum inhibin concentration has been reported to be higher in women with breast or lung cancer compared with women with benign disease of breast or lung and normal women *(95)*. Measurement of serum inhibin levels help in diagnosis and follow up of particular granulosa cell tumors, and mucinous epithelial adenocarcinomas *(96)*. Further investigation of the conflicting data on inhibin α expression in prostate cancer are needed to determine whether inhibin α plays an oncogenic and/or tumor suppressor role during prostate cancer progression (Table 1).

4. EXPRESSION AND FUNCTION OF BMPS IN THE NORMAL PROSTATE AND PROSTATE CANCER

4.1. Expression and Function of BMPs in the Normal Prostate

BMPs are functionally pleiotropic proteins with important regulatory roles in embryonal development and tissue morphogenesis and play a pivotal role in skeletal morphogenesis. The BMP family includes BMPs, osteogenic proteins, GDFs and the cartilage-derived morphogenetic proteins *(97)*. BMP type I (BMPRI) and type II (BMPRII) receptors bind BMPs with high affinity when expressed together, and with lower affinity when expressed separately. In addition to BMPRII, ActRIIA, and ActRIIB also bind BMPs cooperatively when ALK3 or ALK6 is coexpressed in the cell *(29,31,32)*.

Table 1
Activin, Inhibin – Expression and Putative Function in Normal Prostate and Prostate Cancer

Ligand	*Site of expression in the prostate*	*Putative function in normal prostate*
Activin βA	Undifferentiated mesenchyme, and epithelium of the mature prostate *(54,55)*	Activin A inhibits ductal branching of prostate explants in vitro *(54)*, may also mediate Shh-induced inhibition of ductal branching *(59)*
Activin βB	Some cells of the mensenchyme and stroma and columnar epithelium of the mature prostate *(54)*	Unknown
Follistatin	Epithelium *(54)*	Promotes branching morphogenesis of prostate explants in vitro *(54)*
	Expression in prostate cancer	***Putative function in prostate cancer***
Activin βA	Secreted by prostate tumor explants *(69)* Undetectable in androgen-independent prostate tumor xenografts *(64)*	Activin A inhibits androgen-independent independent survival and androgen-induced growth of prostate cancer cells in vitro *(60,61,65–67,69)* and may be a negative regulator of proliferation in the prostate
Activin βB	Expression is significantly lower in androgen-independent prostate tumor xenografts compared to androgen-dependent tumors *(64)*	Unknown
Follistatin	Undetectable to relatively high levels in androgen-independent prostate tumor xenografts *(64)*. Human prostate tumor explants secrete high amounts of FS in excess of activin *(69)*	FS promotes prostate cancer cell proliferation *(60,62,68,69)*, which may occur through activin dependent and/or independent mechanisms
Inhibin	Undetectable in prostate cancer cells *(64,92)*; promoter is hypermethylated in prostate tumors *(92)*. There is also evidence for overexpression in tumors *(94)*, and serum inhibin levels are higher in patients with prostate cancer *(95)*	Unknown

The normal rat ventral prostate expresses BMP 2, 3, 4, and BMP6 mRNAs and the normal human prostate predominantly expresses BMP4 *(98–101)*. Lamm et al. demonstrated that BMP4 is a mesenchymal factor that regulates ductal morphogenesis of the prostate. BMP4 expression was high in the mouse urogenital sinus at embryonic d14 through birth, during which time the main prostatic ducts are formed and ductal branching begins. BMP4 was expressed throughout the prostatic anlage of the urogenital sinus and eventually became restricted to the mesenchyme surrounding the nascent prostatic ducts and branches. Exogenous BMP4 inhibited epithelial cell proliferation and blocked ductal budding in urogenital sinus tissues cultured in vitro. Moreover, BMP4 haploinsufficient mice had an increased number of ductal tips in the ventral prostate *(59)*. Quantification of prostate branching in BMP4 haplo-insufficient mice, which develop enlarged prostate glands in adulthood, demonstrated that the prostate was significantly larger by d3, well before the emergence of the phenotype in older animals. The enlargement of the ventral prostate was owing to increased number of main epithelial ducts whereas the enlargement of the anterior prostate in mutant animals occurred owing to increased branching *(102)*. Thus BMP4 is a mesenchymal factor that inhibits prostate ductal budding and branching morphogenesis.

BMP6 is present in the normal prostate and in neoplastic human prostate cancer *(99–101, 103–106)*. It localizes to the cytoplasm and is expressed in the basal cells and in areas of basal cell hyperplasia in samples of benign prostatic hyperplasia *(99)*. In rats, BMP6 expression in the ventral prostate was not affected by castration suggesting that its regulation is not influenced by androgens *(100)*. The normal prostate glandular tissue also expressed high levels of BMP7 *(107)*, and its expression in the dorsal and ventral prostates declines sharply upon castration induced androgen-withdrawal; treatment of castrated mice with (DHT) restored BMP7 mRNA expression *(108)*. Similarly, DHT induced BMP7 expression in human prostate epithelial cells *(107)* suggesting that BMP-7 expression is regulated by androgens.

The placental bone morphogenetic protein, also known as GDF15, placental TGF-β, macrophage inhibitory cytokine-1, prostate-derived factor or nonsteroidal antiinflammatory drug-activated protein-1, is expressed in many tissues including the prostate *(109,110)*; GDF-15 localizes predominantly to the epithelial cells *(111,112)* and androgens positively regulate its expression in the ventral prostates of rats *(112)*.

The presence of multiple members of the BMP family in the normal prostate and the currently available functional data suggest that BMPs influence development, differentiation and function of the normal prostate (Table 2). However, how they influence the prostate microenvironment and their precise role in prostate development remains unknown.

4.2. Expression and Function of BMPs in Prostate Cancer

4.2.1. BMP2

BMP2 inhibits LNCaP cell proliferation by preventing cell cycle progression *(113,114)*. Inhibition of DHT induced LNCaP growth by BMP2 correlated with decreased Rb phosphorylation, reduction in cyclin A, E2F1 and CDK-2 expression and the induction of p21 *(114)*. Dai et al. demonstrated that BMP2 does not alter the growth of LuCaP and C4-2B cells in vitro but increases their in vitro invasiveness *(104)*.

Clinically, organ-confined prostate cancers strongly express BMP-2 whereas expression is low in infiltrating and metastatic sites in patients with advanced prostate cancer. Horvath et al. demonstrated that BMP2 expression was high in normal and hyperplastic prostates but low in prostate cancer. Loss of BMP2 expression correlated with increase in Gleason Score with 75% of prostate tumors with Gleason scores ≥8 being BMP2 negative. Absence of BMP2 in prostate tumors was associated with a significant decrease in relapse-free patient survival compared to those with BMP2 expression. Other clinical markers of prognosis such as pathological stage and preoperative PSA did not correlate with BMP2 expression *(115)*.

BMP2 induces RUNX2, a transcription factor essential for osteoblast differentiation *(116)*, and osterix, another transcription factor which controls osteogenesis *(117)*. In PC3 cells, BMP2 stimulates the expression of osteoprotegerin, a molecule, which inhibits osteoclastogenesis *(113)*. Bone remodeling process relies on the coupling between bone resorption and formation that involves osteoclasts, osteoblasts, and osteocytes. Because prostate cancers are mostly osteoblastic, and osteoblasts have been implicated in the progression of prostate cancer to the bone *(118,119)*, it is unclear why the expression of BMP-2, an inducer of RUNX2, osterix and osteoprotegerin, decreases during prostate cancer progression. Whether suppression of BMP2 expression in metastatic prostate cancer indicates that its predominant role is to inhibit cell proliferation needs further investigation.

4.2.2. BMP6

In the human prostate, higher levels of BMP6 protein is detectable in tumors compared to adjacent uninvolved prostate tissue *(99–101,103–106)*. Levels increased with Gleason

Table 2
BMPs – Expression and Putative Function in Normal Prostate and Prostate Cancer

Ligand	*Site of expression in the prostate*	*Putative function in normal prostate*
BMP2	Ventral prostate *(98)*	Unknown
BMP3	Ventral prostate *(98)*	Unknown
BMP4	Mesenchyme *(59)*	Inhibits epithelial proliferation and ductal budding *(59,102)*
BMP6	Ventral prostate *(98,100)* basal epithelial cells *(99)*	Unknown
BMP7	Glandular tissue *(107)*	Unknown
GDF-15	Epithelium *(111,112)*	Unknown
	Expression in prostate cancer	***Putative function in prostate cancer***
BMP2	Expression is high in normal and hyperplastic prostate but low in advanced prostate tumors. Loss of BMP2 in prostate tumors correlates with decreased relapse-free survival *(115)*	BMP2 inhibits prostate cancer cell proliferation *in vitro (113)*
BMP6	Expression increases with disease progression and correlates with increased recurrence, skeletal metastasis and decreased survival *(99–101,103–106)*	Increases invasiveness and promotes intraosseous growth of prostate tumor cells by inducing osteoblastic bone remodeling *(104)*. Hence may facilitate invasion and skeletal metastatis of prostate cancer
BMP7	Expression declines during early stages of the disease but is high in skeletal metastasis *(107,126)*	Inhibits prostate cancer cell growth *(127)* and induces epithelial-mesenchymal transition *(128)*
GDF-15	Expression is lower in BPH and primary tumors *(111,113)*; reexpression during disease progression correlates positively with Gleason score. GDF-15 levels in serum of prostate cancer patients is elevated *(129,134,135)*	Induces apoptosis *(130)* and can stimulate cartilage and bone formation *(112)*

score where BMP6 staining in prostate tumors with a score ≥6 was higher than tumors of score ≤ 4 *(100)*. Hamdy et al. demonstrated that BMP6 was expressed in malignant epithelial cells in 95% of patients with metastases and in 18% of patients with localized cancer; benign samples did not express BMP6 *(99)*. In another study, Dai et al. demonstrated that BMP6 was detectable in nonneoplastic prostate, primary tumors and in metastasis to the liver, bone and adrenal with the levels of expression increasing with disease progression *(104)*. Comparison of BMP6 expression in bone metastasis from prostate carcinoma and skeletal deposits from nonprostatic malignancies demonstrated that BMP6 was strongly expressed both in the primary tumor and skeletal metastasis of prostate adenocarcinomas and less frequently in skeletal metastasis from other human carcinomas *(103)*. Recently, expression of BMP6 in radical prostectomy specimens was shown to correlate with increased recurrence and decreased survival *(106)*.

Analysis of BMP6 and bone sialoprotein, a bone related protein, and the angiogenic factor thymidine phosphorylase expression in patients with localized prostate cancer demonstrated that expression of BMP6 and bone sialoprotein was significantly associated, and that their expression correlated with bone metastasis. Thymidine phosphorylase expression was not related to bone sialoprotein and/or BMP6 positivity but was related to local recurrence. This study indicated that determining bone sialoprotein, BMP6 and thymidine phosphorylase expression at an early stage of the disease could serve as prognostic markers to identify the subgroups of patients that are at different risks of bone metastasis or recurrence *(120)*.

Interestingly, in rats, BMP6 expression was comparable in the normal and malignant prostate and expression did not vary depending on metastatic potential *(100)*. This may be attributed to the preferred tropism of prostate cancer cells to different metastatic sites in the two species. The invasive and distally metastatic prostate adenocarcinomas induced in rats generally metastasize to the abdominal cavity, liver, lung and/or lymph nodes and rarely to the bone, a site commonly favored in humans *(121)*. The increase in BMP6 during prostate cancer progression in human and its correlation with skeletal metastasis *(99–101,103–106)* suggests that BMP6 may be responsible, in part, for the osteoblastic lesions secondary to human prostate cancer.

The mechanism by which BMP6 is upregulated during prostate cancer progression is unclear. Tel-2, a member of the Ets transcription factor family, suppresses BMP6 *(122)* and is expressed in the prostate at readily detectable levels. Analysis of Tel-2 expression in prostate cancer may help determine whether increased expression of BMP6 during prostate cancer progression results from declining Tel-2 levels. In addition, promoter methylation also regulates BMP6 expression in prostate tissues; methylation of the CpG loci around the Sp1 site of the BMP6 promoter represents an important epigenetic mechanism regulating BMP6 expression in prostate cancer cells *(105,123)*. The BMP6 promoter was frequently methylated in organ-confined prostate cancer and benign prostate hyperplasia whereas at metastatic sites the CpG loci in 5′ flanking region of the BMP6 promoter demonstrated global demethylation *(105)*.

Consistent with the clinical studies reporting increased BMP6 expression during disease progression, experiments using prostate cancer cell lines suggest that BMP6 may indeed be a key player in promoting invasiveness and osseous metastasis. LNCaP and its bone metastatic variant C4-2B cells produce BMP-2, -4, -6, and -7, and C4-2B conditioned medium induced mineralization of the osteoblast precursor cells, MC3T3. Utilizing antibodies against individual BMPs, the mineralizing potency was determined to be BMP-6 > BMP-7 > BMP-4 *(104)*. BMP6 did not alter the growth of the prostate cancer cells lines LuCaP and C4-2B in vitro but increased their in vitro invasiveness. Moreover, inhibiting BMP6 with an anti-BMP6 antibody did not alter the subcutaneous growth of LuCaP xenografts in mice. However, BMP6 promoted intraosseous growth of these cells in vivo *(104)*.

Prostate cancer metastasis to the bone is often osteoblastic than osteolytic *(124,125)*. In an experiment to characterize the contribution of BMP6 to osteoblastic lesions, LuCaP 23.1 cells were inoculated into fetal human bone hemisections implanted subcutaneoulsy in mice. LuCaP 23.1 induced osteosclerotic bone lesions, which was partially inhibited by an anti-BMP6 antibody. Further characterization of this model demonstrated that LuCaP cells increased the rate of bone deposition by increasing the number of osteoblasts through a BMP6 mediated process. Analysis of systemic markers for bone remodeling demonstrated that LuCaP 23.1 cells increased osteocalcin and bone alkaline phosphatase levels. On the other hand, LuCaP-induced increase in urinary Ntx and serum TRACP 5b, markers of bone resorption, was not affected by the BMP6 antibody suggesting that LuCAP cells induce both osteoblastic and osteoclastic bone remodeling and that BMP6 is not required for the latter process *(104)*.

There is mounting evidence to suggest that BMP6 plays a role in promoting osseous metastasis and that it may serve as a prognostic marker for skeletal metastasis of prostate cancer. Characterization of BMP6 expression in prostate cancer samples, and analysis of BMP6 promoter methylation in prostate cancer may present rapid and sensitive screening techniques for early detection and prediction of skeletal metastasis. Moreover, targeting

BMP6 may also be of therapeutic utility to prevent and treat osteoblastic component of prostate cancer metastases.

4.2.3. BMP7

Characterization of BMP7 expression in clinical samples demonstrated that BMP7 was lower in prostate tumors compared to normal prostate; expression was detectable in 91% of normal prostates, in 64% of newly identified prostate cancers, and in 30% of recurrent prostate cancers *(107)*. In metastatic bone lesions of prostate cancer, BMP7 is detectable, and unlike BMP6, the expression of which is comparable in skeletal metastasis and normal bone, BMP7 is expressed at higher levels in osseous metastasis of prostate cancer compared to that in normal bone tissue *(126)* suggesting that BMP7 may be reexpressed during prostate cancer progression. It is likely that the decline in BMP7 levels during early stages of the disease *(107)* renders metastatic prostate cancer cells refractory to its growth inhibitory effects *(127)*, whereas increased BMP7 expression at later stages may promote the ability of prostate cancer cells to modulate the bone microenvironment.

BMP-7 prevents cell cycle progression of BPH-1, a cell line representing benign prostatic epithelial hyperplasia, and induces epithelial-mesenchymal transdifferentiation (EMT) in PC-3 cells with associated changes in morphology, and expression of molecular markers of motility and invasiveness. However, not all cells responded similarly; treatment of BPH-1, DU145, LAPC-4, LNCaP, and C4-2B with BMP7 did not induce EMT *(128)*. Supporting these observations, Dai et al. demonstrated that BMP7 does not increase the invasiveness of LuCaP and C4-2B cells in vitro *(104)*.

Miyazaki et al. on the other hand demonstrated that high concentrations of BMP7 (500 ng/ml) inhibits PC-3 and DU-145 proliferation with an associated increase in p21, and decreased the activity of CDK2, leading to hypophosphorylation of Rb. Expression of a constitutively active BMP type I receptor, ALK-6, in PC3 cells inhibited proliferation in vitro and led to a decrease in tumor size suggesting that BMP signaling may suppress the growth of androgen-insensitive prostate tumors *(127)*.

4.2.4. GDF-15/ Placental BMP

GDF-15 is overexpressed in androgen-independent and metastatic variants of LNCaP cells compared with androgen-dependent LNCaP cells and was absent or low in androgen-independent DU145 and PC3 cells *(110,111,129)*. Neither GDF-15 treatment nor blocking autocrine GDF-15 had any effect on LNCaP cell growth whereas GDF-15 induced apoptosis in DU145 cells; induction of apoptosis was associated with stimulation of the antiapoptotic gene metallothionein 1E *(130)*.

In benign prostatic tissues, GDF-15 was expressed in the epithelium with higher levels in luminal cells compared with the basal cells *(110)*. Genome wide scans for prostate cancer susceptibility genes in familial prostate cancer identified a linkage on chromosome 19p13, the locus to which GDF-15 is localized *(131,132)* and DNA microarray analysis demonstrated downregulation of GDF-15 in BPH compared to normal prostate *(111,133)*. However, quantitative analysis of GDF-15 expression in 66 prostate cancer patients demonstrated GDF-15 was overexpressed in tumors in comparison with noncancerous tissues in 88% of patients whereas only 8% demonstrated a decline in the tumor compared to uninvolved tissue. GDF-15 overexpression did not correlate with tumor stage or tumor grade but demonstrated a positive correlation with Gleason Score *(134)*. Thus as with BMP7, GDF-15 may be reexpressed during prostate cancer progression *(129,134,135)*.

Consistent with these observations, GDF-15 mRNA was lower in primary prostate tumors compared with matched normal prostatic epithelium. In metastatic prostate tumors, GDF-15 levels varied depending on the host tissue to which metastasis occurred; GDF-15 was reexpressed in >20% of tumors cells in osseous metastases, whereas no expression was detected in nonosseous metastasis *(110)*. GDF-15, like other members of the BMP family can induce cartilage and bone formation *(112)* and its reexpression following metastasis to the bone suggests that GDF-15 may be an important factor in this process.

In agreement with GDF-15 overexpression in high-grade prostate tumors *(110,134)*, GDF-15 concentration was elevated in the serum of patients with metastatic prostate cancer and these differences correlated with increased GDF-15 expression in the tumors. Although serum GDF-15 levels in patients with metastatic breast and colorectal cancers were also significantly increased compared with levels in the normal serum, the levels in metastatic prostate cancer patients were significantly higher than those with breast and colorectal cancer *(136)* suggesting that measuring GDF-15 levels in the serum may be of potential clinical utility as tissue and serum biomarkers.

4.2.5. BMP Receptors

In the benign prostate, BMPR1A, BMPR1B and BMPRII proteins localize predominantly to the epithelium *(137)* and several prostate cancer cell lines express BMPRII, BMPR1A and BMPR1B *(104)*. The expression of BMPR1B, which is highest in the prostate, is regulated by androgen. In LNCaP cells, androgen induces BMPR1B while the levels of BMPR11 and BMPR1A remain unchanged *(138)*.

In prostate tumors, the loss of BMP receptor expression correlated with increasing tumor grade. In well-differentiated cancers of Gleason score 2–4, >70% of the tumors expressed BMPRII, BMPR1A, and BMPRIB. In 20 moderately differentiated cancers of Gleason score 5–7, BMPRII, BMPR1A and BMPR1B expression was absent in 11, 5, and 12 samples, respectively. In poorly differentiated cancers of Gleason score 8–10 only one out of ten tumors expressed each of the three receptors *(137)* suggesting that loss of BMP receptors may confer a growth advantage to these tumors.

In summary, the expression of BMP6 increases and that of BMP2 decreases during disease advancement, whereas BMP7 and GDF-15 demonstrate a biphasic regulation during prostate cancer progression. BMP receptor expression inversely correlates with tumor grade (Table 2) but paradoxically, loss of BMP-RII expression correlated significantly with 5-yr survival and biochemical recurrence-free rate following radical prostatectomy *(137)*. Because BMPs inhibit prostate cancer growth and/or promote skeletal metastasis, a fine regulation of BMP ligands and receptor expression during prostate cancer progression may enable prostate tumor cells to escape the growth inhibitory effects of the BMPs during early stages of the disease and facilitate the bone forming skeletal metastases during advanced prostate cancer.

5. CONCLUSIONS

Review of the data on the role of activins, inhibins and BMPs in the prostate demonstrate that studies are focused predominantly on expression analysis. The expression patterns of activin, inhibins and BMPs in the normal prostate suggest that these polypeptides may be key players in maintaining a suitable microenvironment for proper development and function; clinical studies demonstrate that they may be intricately involved in cancer progression and osseous metastasis (Table 2). Future studies using appropriate cell culture model systems, and animal models in which these targets are genetically manipulated specifically in the prostate are required to elucidate the mechanistic role of these ligands in the normal prostate and prostate cancer progression. Such studies together with microarray and proteomic technologies may lead to the development of diagnostic and prognostic assays as well designing therapeutics against prostate cancer.

REFERENCES

1. Hayward SW, Cunha GR. The prostate: development and physiology. Radiol Clin North Am 2000; 38:1–14.
2. Aumuller G. Morphologic and endocrine aspects of prostatic function. Prostate 1983;4:195–214.
3. Aumuller G. Morphologic and regulatory aspects of prostatic function. Anat Embryol (Berl) 1989; 179:519–531.
4. Cunha GR, Donjacour AA, Sugimura Y. Stromal-epithelial interactions and heterogeneity of proliferative activity within the prostate. Biochem Cell Biol 1986;64:608–614.
5. Cunha GR. Role of mesenchymal-epithelial interactions in normal and abnormal development of the mammary gland and prostate. Cancer 1994;74:1030–1044.
6. Kumar VL, Majumder PK. Prostate gland: structure, functions and regulation. Int Urol Nephrol 1995;27:231–243.
7. Wilson JD. The pathogenesis of benign prostatic hyperplasia. Am J Med 1980;68:745–756.
8. Griffiths K, Eaton CL, Harper ME, Peeling B, Davies B. Steroid hormones and the pathogenesis of benign prostatic hyperplasia. Eur Urol 20 Suppl 1991;1:68–77.
9. Isaacs JT, Coffey DS. Etiology and disease process of benign prostatic hyperplasia. Prostate Suppl 1989;2:33–50.
10. Cotran R, Kumar V, Collins T. Pathologic Basis of Disease, 6th Edition ed., Saunders WB Company 1974.
11. Taneja SS. The role of androgen receptor coactivators in prostate cancer growth. Prostate Cancer Prostatic Dis 2000;3:S38.
12. Huang H, Tindall DJ. The role of the androgen receptor in prostate cancer. Crit Rev Eukaryot Gene Expr 2002;12:193–207.
13. Debes JD, Tindall DJ. The role of androgens and the androgen receptor in prostate cancer, Cancer Lett 2002;187:1–7.
14. Suzuki H, Ito H. Role of androgen receptor in prostate cancer. Asian J Androl 1999;1:81–85.
15. Jenster G. The role of the androgen receptor in the development and progression of prostate cancer, Semin Oncol 1999;26:407–421.
16. Kim HG, Kassis J, Souto JC, Turner T, Wells A. EGF receptor signaling in prostate morphogenesis and tumorigenesis, Histol Histopathol 1999;14:1175–1182.
17. Thompson TC. Growth factors and oncogenes in prostate cancer. Cancer Cells 1990;2:345–354.
18. Wong YC, Wang YZ. Growth factors and epithelial-stromal interactions in prostate cancer development. Int Rev Cytol 2000;199:65–116.
19. Urwin GH, Percival RC, Harris S, Beneton MN, Williams JL, Kanis JA. Generalised increase in bone resorption in carcinoma of the prostate. Br J Urol 1985;57:721–723.
20. Percival RC, Urwin GH, Harris S, et al. Biochemical and histological evidence that carcinoma of the prostate is associated with increased bone resorption. Eur J Surg Oncol 1987;13:41–49.
21. Vinholes J, Coleman R, Eastell R. Effects of bone metastases on bone metabolism: implications for diagnosis, imaging and assessment of response to cancer treatment. Cancer Treat Rev 1996;22: 289–331.
22. Berruti A, Piovesan A, Torta M, et al. Biochemical evaluation of bone turnover in cancer patients with bone metastases: relationship with radiograph appearances and disease extension. Br J Cancer 1996;73:1581–1587.
23. Logothetis CJ, Lin SH. Osteoblasts in prostate cancer metastasis to bone. Nat Rev Cancer 2005; 5:21–28.
24. Keller ET, Brown J. Prostate cancer bone metastases promote both osteolytic and osteoblastic activity. J Cell Biochem 2004;91:718–729.
25. Sakuma R, Ohnishi Yi Y, Meno C, et al. Inhibition of Nodal signalling by Lefty mediated through interaction with common receptors and efficient diffusion. Genes Cells 2002;7:401–412.
26. Ying SY. Inhibins, activins follistatins J. Steroid Biochem 1989;33:705–713.
27. Yamaguchi Y, Mann DM, Ruoslahti E, Negative regulation of transforming growth factor-beta by the proteoglycan decorin. Nature 1990;346:281–284.
28. Schneyer A, Tortoriello D, Sidis Y, Keutmann H, Matsuzaki T, Holmes W. Follistatin-related protein (FSRP): a new member of the follistatin gene family, Mol Cell Endocrinol 2001;180:33–38.
29. Massagué J. TGF-beta signal transduction. Annu Rev Biochem 1998;67:753–791.
30. Balemans W, Van Hul W. Extracellular regulation of BMP signaling in vertebrates: a cocktail of modulators. Dev Biol 2002;250:231–250.

31. Chang H, Brown CW, Matzuk MM. Genetic analysis of the mammalian transforming growth factor-beta superfamily. Endocr Rev 2002;23:787–823.
32. Attisano L, Wrana JL. Signal transduction by the TGF-beta superfamily. Science 2002;296:1646–1647.
33. Segev DL, Ha TU, Tran TT, et al. Mullerian inhibiting substance inhibits breast cancer cell growth through an NFkappa B-mediated pathway. J Biol Chem 2000;275:28,371–28,379.
34. Sovak MA, Arsura M, Zanieski G, Kavanagh KT, Sonenshein GE. The inhibitory effects of transforming growth factor beta1 on breast cancer cell proliferation are mediated through regulation of aberrant nuclear factor-kappaB/Rel expression. Cell Growth Differ 1999;10:537–544.
35. Bello-DeOcampo D, Tindall DJ. TGF-betal/Smad signaling in prostate cancer. Curr Drug Targets 2003;4:197–207.
36. Danielpour D. Functions and regulation of transforming growth factor-beta (TGF-beta) in the prostate. Eur J Cancer 2005;41:846–857.
37. Wikstrom P, Bergh A, Damber JE. Transforming growth factor-beta1 and prostate cancer. Scand J Urol Nephrol 2000;34:85–94.
38. Wikstrom P, Damber J, Bergh A. Role of transforming growth factor-beta1 in prostate cancer. Microsc Res Tech 2001;52:411–419.
39. Lee C, Sintich SM, Mathews EP, et al. Transforming growth factor-beta in benign and malignant prostate. Prostate 1999;39:285–290.
40. Ying SY. Inhibins, activins, and follistatins: gonadal proteins modulating the secretion of follicle-stimulating hormone. Endocr Rev 1988;9:267–293.
41. Fang J, Yin W, Smiley E, Wang SQ, Bonadio J. Molecular cloning of the mouse activin beta E subunit gene. Biochem Biophys Res Commun 1996;228:669–674.
42. Lau AL, Kumar TR, Nishimori K, Bonadio J, Matzuk MM. Activin betaC and betaE genes are not essential for mouse liver growth, differentiation, and regeneration. Mol Cell Biol 2000;20:6127–6137.
43. Lau AL, Nishimori K, Matzuk MM. Structural analysis of the mouse activin beta C gene. Biochim Biophys Acta 1996;1307:145–148.
44. Oda S, Nishimatsu S, Murakami K, Ueno N. Molecular cloning and functional analysis of a new activin beta subunit: a dorsal mesoderm-inducing activity in Xenopus. Biochem Biophys Res Commun 1995; 210:581–588.
45. de JP Winter, ten Dijke P, de CJ Vries, et al. Follistatins neutralize activin bioactivity by inhibition of activin binding to its type II receptors. Mol Cell Endocrinol 1996;116:105–114.
46. Schneyer AL, Rzucidlo DA, Sluss PM, Crowley WF, Jr. Characterization of unique binding kinetics of follistatin and activin or inhibin in serum. Endocrinology 1994;135:667–674.
47. Welt C, YSidis, Keutmann H, Schneyer A. Activins, inhibins, and follistatins: from endocrinology to signaling. A paradigm for the new millennium, Exp Biol Med (Maywood) 2002;227:724–752.
48. de DM Kretser, Meinhardt A, Meehan T, Phillips DJ, O'Bryan MK, Loveland KA. The roles of inhibin and related peptides in gonadal function. Mol Cell Endocrinol 2000;161:43–46.
49. Chen YG, Lui HM, Lin SL, Lee JM, Ying SY. Regulation of cell proliferation, apoptosis, and carcinogenesis by activin. Exp Biol Med (Maywood) 2002;227:75–87.
50. Risbridger GP, Schmitt JF, Robertson DM. Activins and inhibins in endocrine and other tumors. Endocr Rev 2001;22:836–858.
51. Ying SY, Zhang Z, Furst B, Batres Y, Huang G, Li G. Activins and activin receptors in cell growth. Proc Soc Exp Biol Med 1997;214:114–122.
52. Feng ZM, Madigan MB, Chen CL. Expression of type II activin receptor genes in the male and female reproductive tissues of the rat. Endocrinology 1993;132:2593–2600.
53. Risbridger GP, Thomas, Gurusinghe CJ, McFarlane JR. Inhibin-related proteins in rat prostate. J Endocrinol 1996;149:93–99.
54. Cancilla B, Jarred RA, Wang H, Mellor SL, Cunha GR, Risbridger GP. Regulation of prostate branching morphogenesis by activin A and follistatin. Dev Biol 2001;237:145–158.
55. Al-Omari A, Shidaifat F, Dardaka M. Castration induced changes in dog prostate gland associated with diminished activin and activin receptor expression. Life Sci 2005;77:2752–2759.
56. Kang Y, Reddi AH. Identification and cloning of a novel type I serine/threonine kinase receptor of the TGF-beta/BMP superfamily in rat prostate. Biochem Mol Biol Int 1996;40:993–1001.
57. Ball EM, Risbridger GP. Activins as regulators of branching morphogenesis. Dev Biol 2001;238: 1–12.
58. Wang BE, Shou J, Ross S, Koeppen H, De Sauvage FJ, Gao WQ. Inhibition of epithelial ductal branching in the prostate by sonic hedgehog is indirectly mediated by stromal cells. J Biol Chem 2003;278:18,506–18,513.

59. Lamm ML, Podlasek CA, Barnett DH, et al. Mesenchymal factor bone morphogenetic protein 4 restricts ductal budding and branching morphogenesis in the developing prostate. Dev Biol 2001; 232:301–314.
60. Dalkin AC, Gilrain JT, Bradshaw D, Myers CE. Activin inhibition of prostate cancer cell growth: selective actions on androgen-responsive LNCaP cells. Endocrinology 1996;137:5230–5235.
61. McPherson SJ, Thomas TZ, Wang H, Gurusinghe CJ, Risbridger GP. Growth inhibitory response to activin A and B by human prostate tumour cell lines, LNCaP and DU145. J Endocrinol 1997;154: 535–545.
62. McPherson SJ, Mellor SL, Wang H, Evans LW, Groome NP, Risbridger GP. Expression of activin A and follistatin core proteins by human prostate tumor cell lines. Endocrinology 1999;140:5303–5309.
63. Thomas TZ, Wang H, Niclasen P, et al. Expression and localization of activin subunits and follistatins in tissues from men with high grade prostate cancer. J Clin Endocrinol Metab 1997;82:3851–3858.
64. van Schaik RH, Wierikx CD, Timmerman MA, et al.Variations in activin receptor, inhibin/activin subunit and follistatin mRNAs in human prostate tumour tissues. Br J Cancer 2000;82:112–117.
65. Carey JL, Sasur LM, Kawakubo H, et al. Mutually antagonistic effects of androgen and activin in the regulation of prostate cancer cell growth. Mol Endocrinol 2004;18:696–707.
66. Wang QF, Tilly KI, Tilly JL, et al. Activin inhibits basal and androgen-stimulated proliferation and induces apoptosis in the human prostatic cancer cell line. LNCaP, Endocrinology 1996;137:5476–5483.
67. Zhang Z, Zhao Y, Batres Y, Lin MF, Ying SY. Regulation of growth and prostatic marker expression by activin A in an androgen-sensitive prostate cancer cell line LNCAP. Biochem Biophys Res Commun 1997;234:362–365.
68. Wang M, Liu A, Garcia FU, Rhim JS, Stearns ME. Growth of HPV-18 immortalized human prostatic intraepithelial neoplasia cell lines. Influence of IL-10, follistatin, activin-A, and DHT. Int J Oncol 1999;14:1185–1195.
69. Wang Q, Tabatabaei S, Planz B, Lin CW, Sluss PM. Identification of an activin-follistatin growth modulatory system in the human prostate: secretion and biological activity in primary cultures of prostatic epithelial cells. J Urol 1999;161:1378–1384.
70. Lin S, Ying SY. Differentially expressed genes in activin-induced apoptotic LNCaP cells. Biochem Biophys Res Commun 1999;257:187–192.
71. Ying SY, Chuong CM, Lin S. Suppression of activin-induced apoptosis by novel antisense strategy in human prostate cancer cells. Biochem Biophys Res Commun 1999;265:669–673.
72. Ying SY, Lin SL. Gene expression in precursor cells of prostate cancer associated with activin by combination of subtractive hybridization and microarray technologies. Biochem Biophys Res Commun 2004;313:104–109.
73. Nozawa M, Yomogida K, Kanno N, et al. Prostate-specific transcription factor hPSE is translated only in normal prostate epithelial cells. Cancer Res 2000;60:1348–1352.
74. Tsujimoto Y, Nonomura N, Takayama H, et al. Utility of immunohistochemical detection of prostate-specific Ets for the diagnosis of benign and malignant prostatic epithelial lesions. Int J Urol 2002; 9:167–172.
75. Mitas M, Mikhitarian K, Hoover L, et al. Prostate-Specific Ets (PSE) factor: a novel marker for detection of metastatic breast cancer in axillary lymph nodes. Br J Cancer 2002;86:899–904.
76. Sato N, Fukushima N, Maitra A, et al. Discovery of novel targets for aberrant methylation in pancreatic carcinoma using high-throughput microarrays. Cancer Res 2003;63:3735–3742.
77. Dutta A, Bell SP. Initiation of DNA replication in eukaryotic cells. Annu Rev Cell Dev Biol 1997;13: 293–332.
78. Tang D, Khaleque MA, Jones EL, et al. Expression of heat shock proteins and heat shock protein messenger ribonucleic acid in human prostate carcinoma in vitro and in tumors in vivo. Cell Stress Chaperones 2005;10:46–58.
79. Jones EL, Zhao MJ, Stevenson MA, Calderwood SK. The 70 kilodalton heat shock protein is an inhibitor of apoptosis in prostate cancer. Int J Hyperthermia 2004;20:835–849.
80. Cornford PA, Dodson AR, Parsons KF, et al. Heat shock protein expression independently predicts clinical outcome in prostate cancer. Cancer Res 2000;60:7099–7105.
81. Fujii Y, Kawakami S, Okada Y, Kageyama Y, Kihara K. Regulation of prostate-specific antigen by activin A in prostate cancer LNCaP cells. Am J Physiol Endocrinol Metab 2004;286:E927–E931.
82. Chen G, Nomura M, Morinaga H, et al. Modulation of androgen receptor transactivation by FoxH1. A newly identified androgen receptor corepressor. J Biol Chem 2005;280:36,355–36,363.

83. Iwamura M, Hellman J, Cockett AT, Lilja H, Gershagen S. Alteration of the hormonal bioactivity of parathyroid hormone-related protein (PTHrP) as a result of limited proteolysis by prostate-specific antigen. Urology 1996;48:317–325.
84. Cramer SD, Chen Z, Peehl DM. Prostate specific antigen cleaves parathyroid hormone-related protein in the PTH-like domain: inactivation of PTHrP-stimulated cAMP accumulation in mouse osteoblasts, J Urol 1996;156:526–531.
85. Hempen PM, Zhang L, Bansal RK, et al. Evidence of selection for clones having genetic inactivation of the activin A type II receptor (ACVR2) gene in gastrointestinal cancers. Cancer Res 2003;63:994–999.
86. Danila DC, Inder WJ, Zhang X, et al. Activin effects on neoplastic proliferation of human pituitary tumors. J Clin Endocrinol Metab 2000;85:1009–1015.
87. Rossmanith W, Chabicovsky M, Grasl-Kraupp B, Peter B, Schausberger E, Schulte-Hermann R. Follistatin overexpression in rodent liver tumors: a possible mechanism to overcome activin growth control. Mol Carcinog 2002;35:1–5.
88. Jeruss JS, Sturgis CD, Rademaker AW, Woodruff TK. Down-regulation of activin, activin receptors, and Smads in high-grade breast cancer. Cancer Res 2003;63:3783–3790.
89. Matzuk MM, Finegold MJ, Su JG, Hsueh AJ, Bradley A. Alpha-inhibin is a tumour-suppressor gene with gonadal specificity in mice. Nature 1992;360:313–319.
90. Schmitt JF, Millar DS, Pedersen JS, et al. Hypermethylation of the inhibin alpha-subunit gene in prostate carcinoma. Mol Endocrinol 2002;16:213–220.
91. Risbridger GP, Mellor SL, McPherson SJ, Schmitt JF. The contribution of inhibins and activins to malignant prostate disease. Mol Cell Endocrinol 2001;180:149–153.
92. Ying SY, Zhang Z, Huang G. Expression and localization of inhibin/activin subunits and activin receptors in the normal rat prostate. Life Sci 1997;60:397–401.
93. Balanathan P, Ball EM, Wang H, Harris SE, Shelling AN, Risbridger GP. Epigenetic regulation of inhibin alpha-subunit gene in prostate cancer cell lines. J Mol Endocrinol 2004;32:55–67.
94. Risbridger GP, Ball EM, Wang H, Mellor SL, Peehl DM. Re-evaluation of inhibin alpha subunit as a tumour suppressor in prostate cancer. Mol Cell Endocrinol 2004;225:73–76.
95. Sheth NA, Hurkadli KS, Sathe VS, Sheth AR. Circulating levels of inhibin in cancer. Neoplasma 1984;31:315–321.
96. Lappohn RE, Burger HG, Bouma J, Bangah M, Krans M, de HW Bruijn. Inhibin as a marker for granulosa-cell tumors. N Engl J Med 1989;321:790–793.
97. Chen D, Zhao M, Mundy GR. Bone morphogenetic proteins. Growth Factors 2004;22:233–241.
98. Harris SE, Harris MA, Mahy P, Wozney J, Feng JQ, Mundy GR. Prostate 1994;24:204–211.
99. Hamdy FC, Autzen P, Robinson MC, Horne CH, Neal DE, Robson CN. Immunolocalization and messenger RNA expression of bone morphogenetic protein-6 in human benign and malignant prostatic tissue. Cancer Res 1997;57:4427–4431.
100. Barnes J, Anthony CT, Wall N, Steiner MS. Bone morphogenetic protein-6 expression in normal and malignant prostate. World J Urol 1995;13:337–343.
101. Bentley H, Hamdy FC, Hart KA, et al. Expression of bone morphogenetic proteins in human prostatic adenocarcinoma and benign prostatic hyperplasia. Br J Cancer 1992;66:1159–1163.
102. Almahbobi G, Hedwards S, Fricout G, Jeulin D, Bertram JF, Risbridger GP. Computer-based detection of neonatal changes to branching morphogenesis reveals different mechanisms of and predicts prostate enlargement in mice haplo-insufficient for bone morphogenetic protein 4. J Pathol 2005;206:52–61.
103. Autzen P, Robson CN, Bjartell A, et al. Bone morphogenetic protein 6 in skeletal metastases from prostate cancer and other common human malignancies, Br J Cancer 1998;78:1219–1223.
104. Dai J, Keller J, Zhang J, Lu Y, Yao Z, Keller ET. Bone morphogenetic protein-6 promotes osteoblastic prostate cancer bone metastases through a dual mechanism. Cancer Res 2005;65:8274–8285.
105. Tamada H, Kitazawa R, Gohji K, Kitazawa S. Epigenetic regulation of human bone morphogenetic protein 6 gene expression in prostate cancer. J Bone Miner Res 2001;16:487–496.
106. Thomas BG, Hamdy FC. Bone morphogenetic protein-6: potential mediator of osteoblastic metastases in prostate cancer. Prostate Cancer Prostatic Dis 2000;3:283–285.
107. Masuda H, Fukabori Y, Nakano K, Shimizu N, Yamanaka H. Expression of bone morphogenetic protein-7 (BMP-7) in human prostate. Prostate 2004;59:101–106.
108. Thomas R, Anderson WA, Raman V, Reddi AH. Androgen-dependent gene expression of bone morphogenetic protein 7 in mouse prostate. Prostate 1998;37:236–245.

109. Bottner M, Suter-Crazzolara C, Schober A, Unsicker K. Expression of a novel member of the TGF-beta superfamily, growth/differentiation factor-15/macrophage-inhibiting cytokine-1 (GDF-15/MIC-1) in adult rat tissues. Cell Tissue Res 1999;297:103–110.
110. Thomas R, True LD, Lange PH, Vessella RL. Placental bone morphogenetic protein (PLAB) gene expression in normal, pre-malignant and malignant human prostate: relation to tumor development and progression. Int J Cancer 2001;93:47–52.
111. Kakehi Y, Segawa T, Wu XX, Kulkarni P, Dhir R, Getzenberg RH. Down-regulation of macrophage inhibitory cytokine-1/prostate derived factor in benign prostatic hyperplasia. Prostate 2004;59:351–356.
112. Paralkar VM, Vail AL, Grasser WA, et al. Cloning and characterization of a novel member of the transforming growth factor-beta/bone morphogenetic protein family. J Biol Chem 1998;273: 13,760–13,767.
113. Brubaker KD, Corey E, Brown LG, Vessella RL. Bone morphogenetic protein signaling in prostate cancer cell lines. J Cell Biochem 2004;91:151–160.
114. Tomari K, Kumagai T, Shimizu T, Takeda K. Bone morphogenetic protein-2 induces hypophosphorylation of Rb protein and repression of E2F in androgen-treated LNCaP human prostate cancer cells. Int J Mol Med 2005;15:253–258.
115. Horvath LG, Henshall SM, Kench JG, et al. Loss of BMP2, Smad8, and Smad4 expression in prostate cancer progression. Prostate 2004;59:234–242.
116. Lee KS, Kim HJ, Li QL, et al. Runx2 is a common target of transforming growth factor beta1 and bone morphogenetic protein 2, and cooperation between Runx2 and Smad5 induces osteoblast-specific gene expression in the pluripotent mesenchymal precursor cell line C2C12. Mol Cell Biol 2000;20:8783–8792.
117. Nakashima K, Zhou X, Kunkel G, et al. The novel zinc finger-containing transcription factor osterix is required for osteoblast differentiation and bone formation. Cell 2002;108:17–29.
118. Gleave M, Hsieh JT, Gao CA, von AC Eschenbach, Chung LW. Acceleration of human prostate cancer growth in vivo by factors produced by prostate and bone fibroblasts. Cancer Res 1991;51: 3753–3761.
119. Yang J, Fizazi K, Peleg S, et al. Prostate cancer cells induce osteoblast differentiation through a Cbfa1-dependent pathway. Cancer Res 2001;61:5652–5659.
120. De Pinieux G, Flam T, Zerbib M, et al. Bone sialoprotein, bone morphogenetic protein 6 and thymidine phosphorylase expression in localized human prostatic adenocarcinoma as predictors of clinical outcome: a clinicopathological and immunohistochemical study of 43 cases. J Urol 2001;166:1924–1930.
121. Shirari T, Tamano S, Takahashi S, Ito N. Animal prostate carcinoma models: limited potential for vertebral metastasis. Adv Exp Med Biol 1992;324:151–158.
122. Gu X, Shin BH, Akbarali Y, Weiss A, Boltax J, Oettgen P, Libermann TA. Tel-2 is a novel transcriptional repressor related to the Ets factor Tel/ETV-6. J Biol Chem 2001;276:9421–9436.
123. Tamada H, Kitazawa R, Gohji K, Kamidono S, Maeda S, Kitazawa S. Molecular cloning and analysis of the 5′-flanking region of the human bone morphogenetic protein-6 (BMP-6). Biochim Biophys Acta 1998;1395:247–251.
124. Rana A, Chisholm GD, Khan M, Sekharjit SS, Merrick MV, Elton RA. Patterns of bone metastasis and their prognostic significance in patients with carcinoma of the prostate. Br J Urol 1993;72:933–936.
125. Bubendorf L, Schopfer A, Wagner U, et al. Metastatic patterns of prostate cancer: an autopsy study of 1,589 patients. Hum Pathol 2000;31:578–583.
126. Masuda H, Fukabori Y, Nakano K, Takezawa Y, T CS, Yamanaka H. Increased expression of bone morphogenetic protein-7 in bone metastatic prostate cancer. Prostate 2003;54:268–274.
127. Miyazaki H, Watabe T, Kitamura T, Miyazono K. BMP signals inhibit proliferation and in vivo tumor growth of androgen-insensitive prostate carcinoma cells. Oncogene 2004;23:9326–9335.
128. Yang S, Zhong C, Frenkel B, Reddi AH, Roy-Burman P. Diverse biological effect and Smad signaling of bone morphogenetic protein 7 in prostate tumor cells. Cancer Res 2005;65:5769–5777.
129. Karan D, Chen SJ, Johansson SL, et al. Dysregulated expression of MIC-1/PDF in human prostate tumor cells. Biochem Biophys Res Commun 2003;305:598–604.
130. Liu T, Bauskin AR, Zaunders J, et al. Macrophage inhibitory cytokine 1 reduces cell adhesion and induces apoptosis in prostate cancer cells. Cancer Res 2003;63:5034–5040.
131. Wiklund F, Gillanders EM, Albertus JA, et al. Genome-wide scan of Swedish families with hereditary prostate cancer: suggestive evidence of linkage at 5q11.2 and 19p13.3. Prostate 2003;57: 290–297.

132. Hsieh CL, Oakley -Girvan I, Balise RR, et al. A genome screen of families with multiple cases of prostate cancer: evidence of genetic heterogeneity. Am J Hum Genet 2001;69:148–158.
133. Taoka R, Tsukuda F, Ishikawa M, Haba R, Kakehi Y. Association of prostatic inflammation with down-regulation of macrophage inhibitory cytokine-1 gene in symptomatic benign prostatic hyperplasia. J Urol 2004;171:2330–2335.
134. Nakamura T, Scorilas A, Stephan C, et al. Quantitative analysis of macrophage inhibitory cytokine-1 (MIC-1) gene expression in human prostatic tissues. Br J Cancer 2003;88:1101–1104.
135. Welsh JB, Sapinoso LM, Su AI, et al. Analysis of gene expression identifies candidate markers and pharmacological targets in prostate cancer. Cancer Res 2001;61:5974–5978.
136. Welsh JB, Sapinoso LM, Kern SG, et al. Large-scale delineation of secreted protein biomarkers overexpressed in cancer tissue and serum. Proc Natl Acad Sci USA 2003;100:3410–3415.
137. Kim IY, Lee DH, Ahn HJ, et al. Expression of bone morphogenetic protein receptors type-IA, -IB and -II correlates with tumor grade in human prostate cancer tissues. Cancer Res 2000;60:2840–2844.
138. Ide H, Yoshida T, Matsumoto N, et al. Growth regulation of human prostate cancer cells by bone morphogenetic protein-2. Cancer Res 1997;57:5022–5027.

23 Bone Morphogenic Proteins and Angiogenesis in Developing Tumors

George Berberian, Deepak Malhotra, and John E. Langenfeld

CONTENTS

INTRODUCTION
BMP SIGNALING
BMP-2/4 REGULATION OF VASCULOGENESIS
BMP-2 IN CANCER
BMP-2 STIMULATES ANGIOGENESIS IN TUMORS
BMPS DIRECTLY ACTIVATE ENDOTHELIAL CELLS
BMPS INDUCE VEGF
OTHER MECHANISMS BMPS MAY INDUCE ANGIOGENESIS
REFERENCES

Abstract

Bone Morphogenetic Proteins (BMPs) are multipurpose cytokines that act in both a paracrine and self-autonomous manner. BMP-2 was originally identified for its ability to induce bone formation. BMP-2 and BMP-4 have also been shown to have a critical role in the regulation of stem cells promoting their self-renewal and differentiation. More recently, BMP-2 was demonstrated to be aberrantly expressed in lung and other carcinomas. Studies have suggested that BMP-2 has an important role in cancer biology promoting tumor invasion and metastasis. One essential mechanism may be its ability to induce a neovasculature in tumors. This work reviews the literature that highlights the role of BMPs inducing angiogenesis in tumors.

Key Words: Bone Morphogenetic Protein (BMP-2); angiogenesis; carcinomas; lung; breast.

1. INTRODUCTION

Bone Morphogenetic Protein (BMP) is a member of the transforming growth factor-β (TGF-β) superfamily. BMPs have been shown to be essential for the development of the lung *(1)*, heart *(2)*, bone *(3)*, digits *(4)*, teeth *(5)*, feather *(6)*, and central nervous system *(7)*. Functional knockout of BMP-4 *(8)* and mice and BMP-2 in zebrafish *(9)* is lethal with animals containing no primordial germ cells. As might be expected from these complex in vivo functions, BMPs also play key roles in regulating fate choices during stem cell differentiation. For example, BMP-2 directs the pluripotent C3H10T/1/2 stem cells to differentiate into osteoblastic and adipocytic lineages *(10)*. BMPs have also been shown to initiate neural

From: *Cancer Drug Discovery and Development: Transforming Growth Factor-β in Cancer Therapy, Vol I: Basic and Clinical Biology*
Edited by: Sonia B. Jakowlew © Humana Press Inc., Totowa, NJ

crest differentiation from embryonic rat central nervous stem cells *(11)*. BMP-2 and BMP-4 in collaboration with STAT3 can block the differentiation of embryonic stem (ES) cells thus sustaining self-renewal *(12)*. In postnatal tissue, BMP-2/4 has been shown to promote bone formation, induce cell migration and differentiation of several cell types *(13–15)*. BMP-2 has also been shown to stimulate ovarian granulosa cell differentiation and induce estradiol production *(16)*. More recently, the BMPs have been shown to regulate angiogenesis *(17)* and promote tumorigenesis *(18)*.

2. BMP SIGNALING

The BMPs are synthesized as inactive precursor proteins which are proteolytically cleaved producing a mature protein that is the active form *(19)*. The mature protein is secreted from the cell, acting in a paracrine and/or autocrine manner. BMP signals are mediated by type I and type II serine/threonine kinase receptors. Three forms of the type I BMP receptors have been identified, the type IA and IB receptors (BMPR-IA and BMPR IB) and the type IA activin receptor (ActR-IA) *(20,21)*. Three type II BMP receptors have also been identified, type II BMP receptor (BMPRII) and type II and IIB activin receptors (ActR-II and ActR-IIB). Upon BMP binding, the type I and II receptors form a hetero-multimer complex *(22)*. Type II receptors phosphorylate a stretch of amino acids on type I receptor. The BMP receptor complex phosphorylates and thus activates Smads 1/5 and/or 8 transcription factors. Smad proteins play a central role in BMP signaling. Smads 1, 5, and 8 directly interact with activated type I BMP receptors, which phosphorylate the C-terminal SSXS motif of Smad in a ligand-dependent manner *(23)*. The activated Smad 1/5/8 translocates into the nucleus and activates downstream target genes.

A key regulator of the BMP/Smad signaling pathway is the inhibitor of differentiation (Id-1). Id-1 has been shown to be a direct downstream target of BMP signaling in early development and in postnatal cells *(24)*. Id-1 protein belongs to the Id family of helix–loop–helix proteins. It lacks the basic domain for DNA binding and functions mainly as a dominant inhibitor of the bHLH transcription factor through heterodimerization *(25)*. During development, Id-1 has an essential role in mediating BMP signaling. BMP-2/4 regulation of stem cell self-renewal occurs through a Smad 1/5 mediated expression of Id-1 *(12)*. BMP-2 has been shown to induce the expression of Id-1 in osteoblasts *(26)*, endothelial cells *(15,17)*, and cancer cells *(27,18)*. Id-1 has been shown to have an essential role in regulating vasculogenesis and angiogenesis. Described below is emerging data suggesting that the BMPs have a critical role in regulating vasculogenesis and angiogenesis. Of particular importance is recent data showing that BMP-2 regulates angiogenesis in developing tumors.

3. BMP-2/4 REGULATION OF VASCULOGENESIS

Vasculogenesis, the formation of blood vessels de novo, has recently been shown to have an important role in establishing the neovasculature of tumors. Although the BMPs have not been characterized as being critical regulators of vasculogenesis several studies would implicate they have an essential role. Mice with a function knockout of Smad 1 or Smad 5 die at approx 9.5 to 10.5 d post conception (p.c.). and have defects in angiogenesis *(28,29)*. Smad 5 mutant embryos had enlarged blood vessels, a decrease in smooth muscle cells and contained mesenchymal cells, which were unable to direct angiogenesis *(29)*. Id mutant mice also have defect in angiogenesis *(30)*. Id knockout mice fail to sustain growth of tumors because of a block in angiogenesis *(30)*. Interestingly, BMP-4 knockout mice *(8)* and BMP-2 knockout in zebrafish *(9)* are both lethal prior to the development of mesenchymal precursors. During embryonic development, hematopoietic and endothelial precursors are

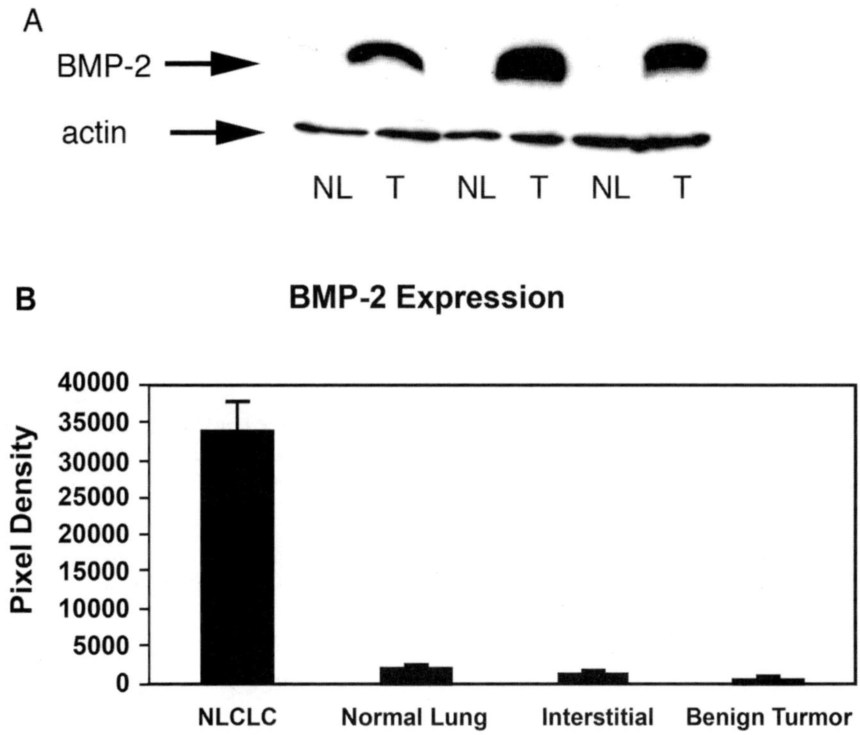

Fig. 1. BMP-2 expression in nonsmall cell lung carcinomas (NSCLC). (**A**) BMP-2 expression was examined in 42 NSCLC by Westernblot analysis. A representative immunoblot of normal lung tissue (NL) and NSCLC (T) is shown. (**B**) The immunoblots were analyzed by NIH imaging and level of BMP-2 expression determined in NSCLC, normal lung, benign lung tumors and benign interstitial lung diseases.

derived from a common precursor called the hemangioblast, which is derived from mesenchymal precursors *(31,32)*. Thus, BMP-2/4 mediates signaling prior to the formation of the hemangioblast and Flk-1 expression. This suggests that BMP-2/4 may regulate endothelial progenitors at an early stage during development.

The BMP-binding endothelial cell precursor-derived regulator (BMPER) is a secreted protein found in mouse embryos that directly interacts with BMP-2, BMP-4, and BMP-6 and antagonizes BMP dependent Smad 5 activation *(33)*. In an embryoid body differentiation assay using mouse ES cells BMP-4 induced the expression of the endothelial cell marker, VE-cadherin, which was antagonized by BMPER *(33)*. These data further highlight that the BMPs may have a role in vasculogenesis and perhaps angiogenesis.

4. BMP-2 IN CANCER

Recent studies suggest that BMPs promote the growth of tumors. BMP-2 is highly expressed in approx 98% of lung carcinomas with little to no expression in benign tumors and normal lung tissue *(34)* (Fig. 1). BMP-2 and BMP-4 have been shown to be overexpressed in other solid tumors including breast *(27)*, melanoma *(35)*, pancreatic *(36)* gastric *(37)*, ovarian *(38)*, and oral epithelium *(39)*. BMP-2/4 are also expressed in sarcomas *(40)*. Forced expression of BMP-2 in the A549 lung cancer cells greatly enhanced tumor growth in the lung following iv injection into nude mice *(18)* (Fig. 2). BMP-2 is reported to stimulate invasion and migration of lung *(41)*, melanoma *(35)*, and breast cancer cell lines *(42)*.

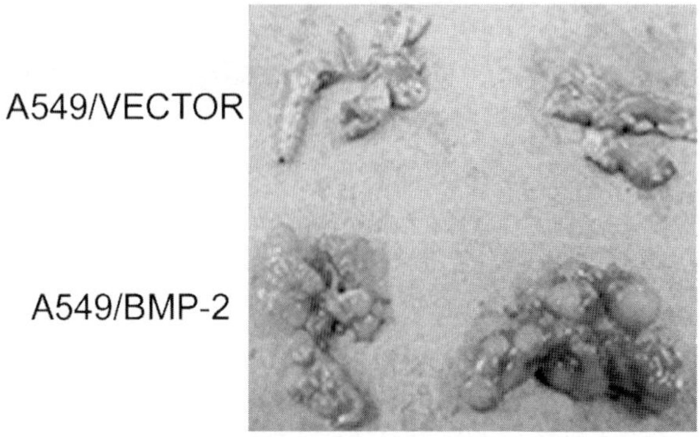

Fig. 2. BMP-2 enhances the development of lung metastasis. A549 cells transfected with empty vector (A549/Vector) or BMP-2 (A549/BMP-2) was injected into the tail veins of nude mice. A representative photograph showing large bulky tumor in the lungs of mice injected with A549/BMP-2 cells. The A549/Vector cells produced only small subcentemeter tumor growth in the lungs.

High expression of BMP-2 in stage I lung cancer is associated with worse survival *(43)*. Studies suggest that one mechanism by which BMPs may enhance tumor growth is by stimulating the production of a neovasculature.

5. BMP-2 STIMULATES ANGIOGENESIS IN TUMORS

The ability of BMPs to induce an angiogenic response in postnatal tissue has only recently been described. Two studies thus far have linked the expression of BMP-2 to the formation of a blood supply in developing tumors. Co-injecting Affi-Blue agarose beads coated with recombinant BMP-2 with A549 cells into nude mice significantly enhanced the development of neovasculature *(17)* (Fig. 3). Recombinant BMP-2 was also shown to promote the growth of A549 tumors in nude mice while the BMP-2 antogonist, noggin, caused a significant reduction in tumor growth *(41)*. Adding recombinant BMP-2 to Matrigel plugs containing A549 also demonstrated an increase in the number of blood vessels *(17)* (Fig. 4). The effects of BMP-2 on angiogenesis were abrogated by the addition of the BMP-2 noggin (Fig. 4). Decreasing the basal level expression of BMP-2 in A549 cells using antisense transfection also led to a decrease in the number of blood vessels in the Matrigel plugs *(17)* (Fig. 4).

Corroborating these findings is a report using the MCF-7 breast cancer cell line *(44)*. The authors showed that daily injections of recombinant BMP-2 into sponges that had been placed subcutaneously in mice significantly enhanced angiogenesis *(44)*. The BMP-2 induced angiogenic response was equivalent to that produced by daily injections of recombinant vascular endothelial growth factor (VEGF). The addition of both BMP-2 and VEGF had an additive effect on angiogenesis. BMP-2 transfected MCF-7 cells formed tumors in mice with a well-developed vasculature *(44)*. MCF-7 cells transfected with empty vector did not form tumors. This study also suggests that BMP-2 induced angiogenesis may be essential in tumors.

6. BMPs DIRECTLY ACTIVATE ENDOTHELIAL CELLS

Further support that the BMPs stimulate angiogenesis postnatally is demonstrated by their ability to activate endothelial cells. The BMP IA, IB, and type II receptors are expressed on endothelial and smooth muscle cells *(15)*. The BMP-2 and BMP-4 ligands are also expressed in endothelial cells. Recombinant BMP-2 has been shown to

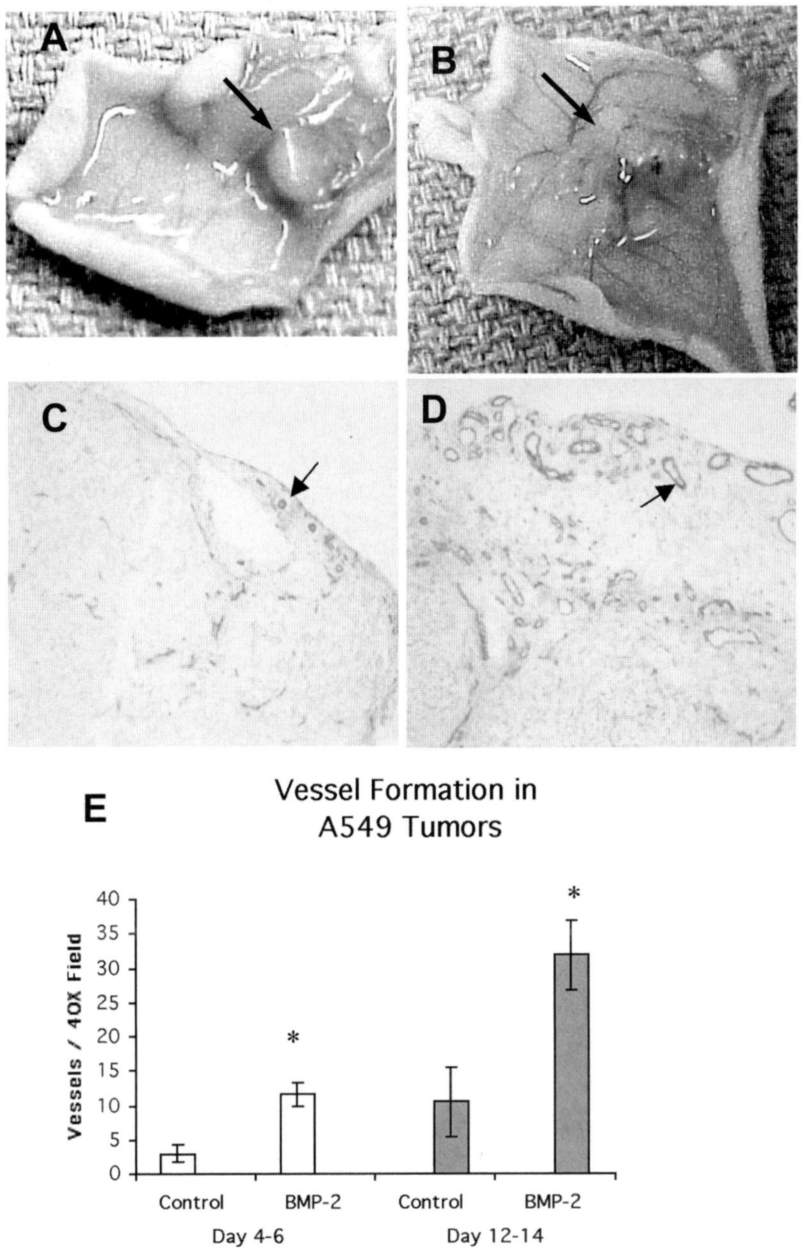

Fig. 3. BMP-2 stimulates tumor neovasculature. A549 cells were coinjected subcutaneously into nude mice with Affi-Blue agarose beads coated with (**A**) BSA or (**B**) recombinant human BMP-2. Bold arrow highlights the tumor. Endothelial cells within the tumors were measured by immunohistochemistry. Photograph of representative immunohistochemical study showing CD31 (+) blood vessels in tumors treated with (**C**) BSA or (**D**) recombinant BMP-2. Arrow demonstrating blood vessels that are staining brown. (**E**) Data showing the number of blood vessels per 40X field in tumors treated for 4–6 or 12–14 d. * $p < 0.05$ compared to controls.

phosphorylate its transcription factor, Smad 1/5 and its downstream target, Id-1, in human umbilical vein, human aortic, and human dermal microvascular endothelial cells *(17,44)*. BMP-6 has also been shown to activate Smad 1/5 and induce Id-1 expression in bovine

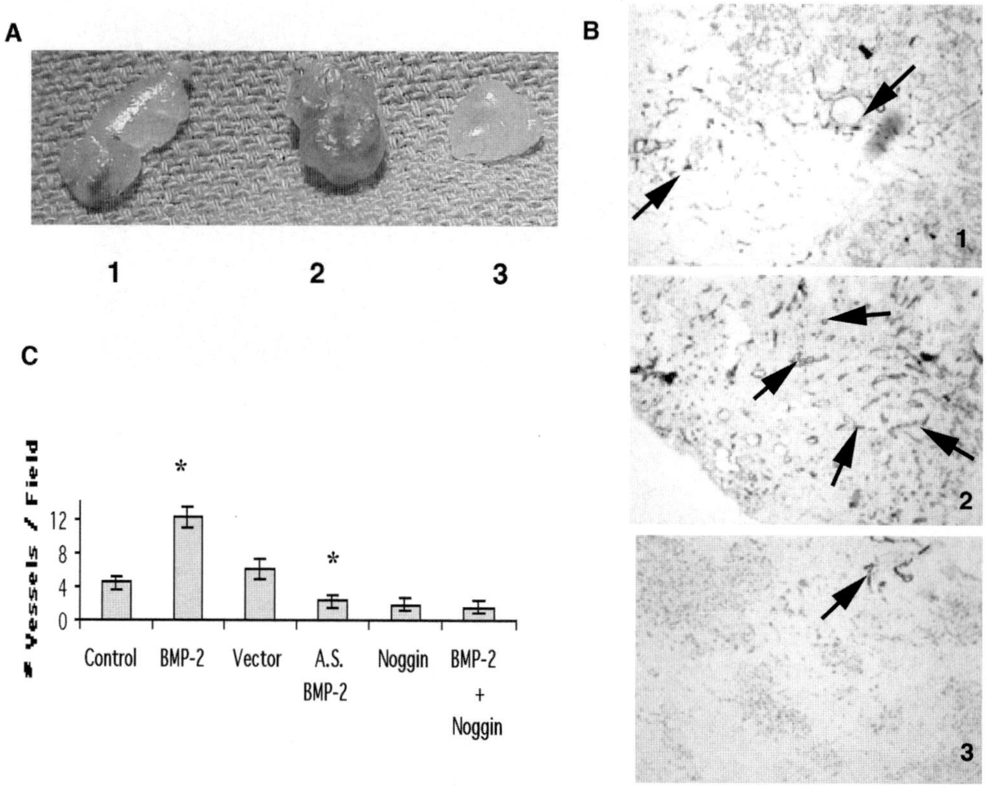

Fig. 4. Matrigel angiogenesis assay. (**A**) Photograph of Matrigel plugs containing A549 cells supplemented with (A 1) BSA or (A 2) recombinant BMP-2 or (A 3) or A549 cells transfected with antisense BMP-2 (A549/AS). (**B**) Representative photograph of an immunohistochemical study demonstrating endothelial cells within the Matrigel plugs. Blood vessels are staining brown. Matrigel plugs containing A549 cells supplemented with (B 1) BSA (B 2) recombinant BMP-2 or (3) A549-AS alone. (**C**) Graphic representation of the number of blood vessels per high-powered field. Also shown is the number of blood vessels in Matrigel plugs containing BMP-2 preincubated with recombinant noggin and A549 cells transfected with pcDNA3 (vector). * $p < 0.05$ compared to control.

endothelial cells *(15)*. In all the above endothelial cells, BMP-2 or BMP-6 stimulated tube formation on Matrigel *(13–15)*. In some endothelial cells the BMPs also stimulated migration and proliferation *(15)*.

7. BMPs INDUCE VEGF

The BMPs likely induce an angiogenic response by more than one mechanism. BMPs were originally described for their ability to stimulate the entire cascade of endochondral bone formation at ectopic sites in animals *(45)*. The formation of new bone involves the development of neovasculature. The angiogenic response promoted by BMP-2 during bone formation likely involves the induction of VEGF. BMP-2/4 stimulates the expression of VEGF-A in the murine preosteoblast-like cell line KS483 *(46)*. BMP-2 and BMP-4 stimulated an angiogenic response in fetal bone extracts, which was arrested by noggin and soluble BMP IA receptor *(46)*. The presence of a blocking VEGF-A antibody abrogated BMP induced angiogenesis *(46)*. This data supported that VEGF is an essential downstream target of BMP. BMP-7 has also been shown to increase the steady-state level of VEGF expression in fetal rat calvaria cells *(47)*.

VEGF may be a direct target of the BMP signaling cascade. The zebrafish VEGF promoter contains activated Smad binding elements (SBE) *(48)*. The SBE bound Smad 1 and 5 protein. The Smad 1, but not the Smad 5, SBE were required for BMP-4 to activate the VEGF promoter in zebrafish embryos *(48)*. The ectopic expression of BMP-4 expanded posterior inner cell mass (ICM) compartment with an increase in VEGF and flk-1 expression.

8. OTHER MECHANISMS BMPS MAY INDUCE ANGIOGENESIS

The activation of Id-1 may represent another mechanism by which BMPs induce angiogenesis. BMP-2 and BMP-4 have been shown to be strong inducers of Id-1 (Id/DNA synthesis) in both solid tumors and endothelial cells *(15,17,27,44)*. Studies suggest that Id-1 functions as an oncogene *(49)*. Forced expression of Id-1 in prostate cancer cells increase VEGF expression *(50)*. Id-1 expressing prostate cancer cells induced an angiogenic response that was dependent on VEGF expression *(50)*.

BMP-2 is a chemotactic factor for smooth muscle cells *(51)*, as well as monocytes *(52)*. The incorporation of circulating moncytes in tumors can also significantly enhance blood vessel formation through the liberation of cytokines *(53)*. Postnatal stem cells are also thought to have a role in the development of a tumor neovasculature. Postnatal stem cells isolated from the peripheral circulation or bone marrow retain the capacity to differentiate into endothelial cells that can incorporate into functional blood vessels *(30)*. Ninety percent of the tumor's endothelial cells were derived from donor postnatal stem/progenitor cells *(30)*. Studies have suggested that the BMPs may have a role during development in promoting hematopoietic/endothelial differentiation. Whether BMPs produced from tumors induces postnatal stem cells to differentiate into endothelial cells remains to be determined.

REFERENCES

1. Weaver M, Yingling JM, Dunn NR, Bellusci S, Hogan BL. Bmp signaling regulates proximal-distal differentiation of endoderm in mouse lung development. Development 1999;126(18):4005–4015.
2. Jiao K, Kulessa H, Tompkins K, et al. An essential role of Bmp4 in the atrioventricular septation of the mouse heart. Genes Dev 2003;17(19):2362–2367.
3. Wozney JM. Bone morphogenetic proteins. Prog Growth Factor Res 1989;1(4):267–280.
4. Dahn RD, Fallon JF. Interdigital regulation of digit identity and homeotic transformation by modulated BMP signaling. Science 2000;289(5478):438–441.
5. Tucker AS, Matthews KL, Sharpe PT. Transformation of tooth type induced by inhibition of BMP signaling. Science 1998;282(5391):1136–1138.
6. Yu M, Wu P, Widelitz RB, Chuong CM. The morphogenesis of feathers. Nature 2002;420(6913): 308–312.
7. Yung SY, Gokhan S, Jurcsak J, Molero AE, Abrajano JJ, Mehler MF. Differential modulation of BMP signaling promotes the elaboration of cerebral cortical GABAergic neurons or oligodendrocytes from a common sonic hedgehog-responsive ventral forebrain progenitor species. Proc Natl Acad Sci USA 2002;99(25):16,273–16,278.
8. Winnier G, Blessing M, Labosky PA, Hogan BL. Bone morphogenetic protein-4 is required for mesoderm formation and patterning in the mouse. Genes Dev 1995;9(17):2105–2116.
9. Kishimoto Y, Lee KH, Zon L, Hammerschmidt M, Schulte-Merker S. The molecular nature of zebrafish swirl: BMP2 function is essential during early dorsoventral patterning. Development 1997; 124(22):4457–4466.
10. Spinella-Jaegle S, Rawadi G, Kawai S, et al. Sonic hedgehog increases the commitment of pluripotent mesenchymal cells into the osteoblastic lineage and abolishes adipocytic differentiation. J Cell Sci 2001;114(Pt 11):2085–2094.
11. Gajavelli S, Wood PM, Pennica D, Whittemore SR, Tsoulfas P. BMP signaling initiates a neural crest differentiation program in embryonic rat CNS stem cells. Exp Neurol 2004;188(2):205–223.
12. Ying QL, Nichols J, Chambers I, Smith A. BMP induction of Id proteins suppresses differentiation and sustains embryonic stem cell self-renewal in collaboration with STAT3. Cell 2003;115(3): 281–292.

13. Kim WJ, Mohan RR, Wilson SE. Effect of PDGF, IL-1alpha, and BMP2/4 on corneal fibroblast chemotaxis: expression of the platelet-derived growth factor system in the cornea. Invest Ophthalmol Vis Sci 1999;40(7):1364–1372.
14. Abe E, Yamamoto M, Taguchi Y, et al. Essential requirement of BMPs-2/4 for both osteoblast and osteoclast formation in murine bone marrow cultures from adult mice: antagonism by noggin. J Bone Miner Res 2000;15(4):663–673.
15. Valdimarsdottir G, Goumans MJ, Rosendahl A, et al. Stimulation of Id1 expression by bone morphogenetic protein is sufficient and necessary for bone morphogenetic protein-induced activation of endothelial cells. Circulation 2002;106(17):2263–2270.
16. Souza CJ, Campbell BK, McNeilly AS, Baird DT. Effect of bone morphogenetic protein 2 (BMP2) on oestradiol and inhibin A production by sheep granulosa cells, and localization of BMP receptors in the ovary by immunohistochemistry. Reproduction 2002;123(3):363–369.
17. Langenfeld EM, Langenfeld J. Bone morphogenetic protein-2 stimulates angiogenesis in developing tumors. Mol Cancer Res 2004;2(3):141–149.
18. Langenfeld EM, Kong Y, Langenfed J. Bone morphogenetic protein 2 stimulation of tumor growth involves the activation of Smad-1/5. Oncogene 2006;25(5):685–692.
19. Cui Y, Jean F, Thomas G, Christian JL. BMP-4 is proteolytically activated by furin and/or PC6 during vertebrate embryonic development. EMBO J 1998;17(16):4735–4743.
20. ten Dijke P, Yamashita H, Sampath TK, et al. Identification of type I receptors for osteogenic protein-1 and bone morphogenetic protein-4. J Biol Chem 1994;269(25):16,985–16,988.
21. Ikeda T, Takahashi H, Suzuki A, et al. Cloning of rat type I receptor cDNA for bone morphogenetic protein-2 and bone morphogenetic protein-4, and the localization compared with that of the ligands. Dev Dyn 1996;206(3):318–329.
22. Liu F, Ventura F, Doody J, Massagué J. Human type II receptor for bone morphogenic proteins (BMPs): extension of the two-kinase receptor model to the BMPs. Mol Cell Biol 1995;15(7): 3479–3486.
23. Nishimura R, Kato Y, Chen D, Harris SE, Mundy GR, Yoneda T. Smad5 and DPC4 are key molecules in mediating BMP-2-induced osteoblastic differentiation of the pluripotent mesenchymal precursor cell line C2C12. J Biol Chem 1998;273(4):1872–1879.
24. Hollnagel A, Oehlmann V, Heymer J, Ruther U, Nordheim A. Id genes are direct targets of bone morphogenetic protein induction in embryonic stem cells. J Biol Chem 1999;274(28):19,838–19,845.
25. Alani RM, Hasskarl J, Grace M, Hernandez MC, Israel MA, Munger K. Immortalization of primary human keratinocytes by the helix-loop-helix protein, Id-1. Proc Natl Acad Sci USA 1999;96(17): 9637–9641.
26. Ogata T, Wozney JM, Benezra R, Noda M. Bone morphogenetic protein 2 transiently enhances expression of a gene, Id (inhibitor of differentiation), encoding a helix-loop-helix molecule in osteoblast-like cells. Proc Natl Acad Sci USA 1993;90(19):9219–9222.
27. Clement JH, Marr N, Meissner A, et al. Bone morphogenetic protein 2 (BMP-2) induces sequential changes of Id gene expression in the breast cancer cell line MCF-7. J Cancer Res Clin Oncol 2000; 126(5):271–279.
28. Lechleider RJ, Ryan JL, Garrett L, et al. Targeted mutagenesis of Smad1 reveals an essential role in chorioallantoic fusion. Dev Biol 2001;240(1):157–167.
29. Yang X, Castilla LH, Xu X, et al. Angiogenesis defects and mesenchymal apoptosis in mice lacking SMAD5. Development 1999;126(8):1571–1580.
30. Lyden D, Young AZ, Zagzag D, et al. Id1 and Id3 are required for neurogenesis, angiogenesis and vascularization of tumour xenografts. Nature 1999;401(6754):670–677.
31. Choi K. The hemangioblast: a common progenitor of hematopoietic and endothelial cells. J Hematother Stem Cell Res 2002;11(1):91–101.
32. Eichmann A, Corbel C, Nataf V, Vaigot P, Breant C, Le Douarin NM. Ligand-dependent development of the endothelial and hemopoietic lineages from embryonic mesodermal cells expressing vascular endothelial growth factor receptor 2. Proc Natl Acad Sci USA 1997;94(10):5141–5146.
33. Moser M, Binder O, Wu Y, et al. BMPER, a novel endothelial cell precursor-derived protein, antagonizes bone morphogenetic protein signaling and endothelial cell differentiation. Mol Cell Biol 2003; 23(16):5664–5679.
34. Langenfeld EM, Bojnowski J, Perone J, Langenfeld J. Expression of bone morphogenetic proteins in human lung carcinomas. Ann Thorac Surg 2005;80(3):1028–1032.
35. Rothhammer T, Poser I, Soncin F, Bataille F, Moser M, Bosserhoff AK. Bone morphogenic proteins are overexpressed in malignant melanoma and promote cell invasion and migration. Cancer Res 2005; 65(2):448–456.

36. Kleeff J, Maruyama H, Ishiwata T, et al. Bone morphogenetic protein 2 exerts diverse effects on cell growth in vitro and is expressed in human pancreatic cancer in vivo. Gastroenterology 1999;116(5): 1202–1216.
37. Katoh M, Terada M. Overexpression of bone morphogenic protein (BMP)-4 mRNA in gastric cancer cell lines of poorly differentiated type. J Gastroenterol 1996;31(1):137–139.
38. Kiyozuka Y, Nakagawa H, Senzaki H, et al. Bone morphogenetic protein-2 and type IV collagen expression in psammoma body forming ovarian cancer. Anticancer Res 2001;21(3B):1723–1730.
39. Jin Y, Tipoe GL, Liong EC, Lau TY, Fung PC, Leung KM. Overexpression of BMP-2/4, -5 and BMPR-IA associated with malignancy of oral epithelium. Oral Oncol 2001;37(3):225–233.
40. Yoshikawa H, Rettig WJ, Lane JM, et al. Immunohistochemical detection of bone morphogenetic proteins in bone and soft-tissue sarcomas. Cancer 1994;74(3):842–847.
41. Langenfeld EM, Calvano SE, Abou-Nukta F, Lowry SF, Amenta P, Langenfeld J. The mature bone morphogenetic protein-2 is aberrantly expressed in non-small cell lung carcinomas and stimulates tumor growth of A549 cells. Carcinogenesis 2003;24(9):1445–1454.
42. Clement JH, Raida M, Sanger J, et al. Bone morphogenetic protein 2 (BMP-2) induces in vitro invasion and in vivo hormone independent growth of breast carcinoma cells. Int J Oncol 2005;27(2): 401–407.
43. Beer DG, Kardia SL, Huang CC, et al. Gene-expression profiles predict survival of patients with lung adenocarcinoma. Nat Med 2002;8(8):816–824.
44. Raida M, Clement JH, Leek RD, et al. Bone morphogenetic protein 2 (BMP-2) and induction of tumor angiogenesis. J Cancer Res Clin Oncol 2005;131(11):741–750.
45. Wozney JM, Rosen V, Celeste AJ, et al. Novel regulators of bone formation: molecular clones and activities. Science 1988;242(4885):1528–1534.
46. Deckers MM, van Bezooijen RL, van der Horst G, et al. Bone morphogenetic proteins stimulate angiogenesis through osteoblast-derived vascular endothelial growth factor A. Endocrinology 2002;143(4): 1545–1553.
47. Yeh LC, Lee JC. Osteogenic protein-1 increases gene expression of vascular endothelial growth factor in primary cultures of fetal rat calvaria cells. Mol Cell Endocrinol 1999;153(1–2):113–124.
48. He C, Chen X. Transcription regulation of the vegf gene by the BMP/Smad pathway in the angioblast of zebrafish embryos. Biochem Biophys Res Commun 2005;329(1):324–330.
49. Fong S, Itahana Y, Sumida T, et al. Id-1 as a molecular target in therapy for breast cancer cell invasion and metastasis. Proc Natl Acad Sci USA 2003;100(23):13,543–13,548.
50. Ling MT, Lau TC, Zhou C, et al. Overexpression of Id-1 in prostate cancer cells promotes angiogenesis through the activation of vascular endothelial growth factor (VEGF). Carcinogenesis 2005;26(10): 1668–1676.
51. Willette RN, Gu JL, Lysko PG, Anderson KM, Minehart H, Yue T. BMP-2 gene expression and effects on human vascular smooth muscle cells. J Vasc Res 1999;36(2):120–125.
52. Cunningham NS, Paralkar V, Reddi AH. Osteogenin and recombinant bone morphogenetic protein 2B are chemotactic for human monocytes and stimulate transforming growth factor beta 1 mRNA expression. Proc Natl Acad Sci USA 1992;89(24):11,740–11,744.
53. Leek RD, Lewis CE, Whitehouse R, Greenall M, Clarke J, Harris AL. Association of macrophage infiltration with angiogenesis and prognosis in invasive breast carcinoma. Cancer Res 1996;56(20): 4625–4629.

24 Histone Methyltransferase and Smad Interactions in BMP Signaling

Jennifer E. Leader, M. Pilar Frontelo, and Robert J. Lechleider

CONTENTS

INTRODUCTION
BMP R-SMADS INTERACT WITH SUV39H HISTONE METHYLTRANSFERASES
ENHANCEMENT OF BMP-INDUCED TRANSCRIPTIONAL REPRESSION BY A HISTONE METHYLTRANSFERASE
BMP-INDUCE GLOBAL CHANGES IN HISTONE METHYLATION PATTERNS
CONCLUSIONS
ACKNOWLEDGMENTS
REFERENCES

Abstract

Although the role of transforming growth factor-β (TGF-β) signaling in the development and progression of cancer is now beginning to be understood, the importance of signaling by bone morphogenetic protein (BMP) ligands is just becoming clear, and much less is known about the mechanisms by which BMP signaling may mediate carcinogenesis. We have shown that the Smad proteins which mediate signals from the BMP cell surface receptors can interact with histone methyltransferases of the Suv39h family, and that this interaction can contribute to transcriptional repression mediated by BMP signaling. This provides the first evidence that BMP signaling can modulate chromatin structure through methylation, and may provide a mechanism for maintenance of suppression originally initiated by Smad signals. Aberrant regulation of this phenomenon is likely to be important both during tumorigenesis and subsequent malignant progression.

Key Words: Histone methyltransferase; Smad; BMP; cell surface receptors; Suv39h family; chromatin.

1. INTRODUCTION

The transforming growth factor-β (TGF-β) superfamily has long been known to play important roles in carcinogenesis. This has best been studied in the prototypical member of the family, TGF-β *(1)*. Although the role bone morphogenetic protein (BMP) ligands play in the development of cancer is not as well known as that of TGF-β, there is clear evidence that BMPs play critical roles in this disease as well. Studies from human hereditary cancer

From: *Cancer Drug Discovery and Development: Transforming Growth Factor-β in Cancer Therapy, Vol I: Basic and Clinical Biology*
Edited by: Sonia B. Jakowlew © Humana Press Inc., Totowa, NJ

syndromes have been illustrative in this regard, in that mutations in the BMP type I receptor, BMPR-IA (Alk3) have been found in patients with hereditary juvenile polyposis *(2)*. BMPs have also been shown to play a role in the regulation of breast *(3,4)* and prostate *(5,6)* cancer cell growth in vitro. Future work will undoubtedly provide more clues regarding roles of BMPs in the development and progression of cancer.

The most studied pathway that mediates the effects of the TGF-β superfamily is the Smad signaling pathway. As for BMP itself, BMPs bind to the Type II receptor serine threonine kinase, which recruits and phosphorylates the Type I receptor. The Type I receptor can then recruit one of the BMP R-Smads, Smad1, 5 or 8. These R-Smads bind to the common Smad, Smad4, and together this complex translocates into the nucleus where Smads bind with other protein coregulators of transcription and modulate gene expression. Many Smad-interacting transcription factors are already known, and it is the combinations of these proteins that likely account for the regulation of transcription. Among others, CBP/p300 and P/CAF proteins, which contain histone acetyltransferase activity, have already been shown to bind to Smads and together these complexes work to activate target gene expression *(7,8)*.

Our data *(9)* demonstrate that Suv39h proteins are involved in a Smad-mediated transcriptional repression response. Suv39h proteins are histone methyltransferases that specifically trimethylate histone H3 on lysine 9. They are part of a diverse group of proteins that regulate gene expression through modification of histone tails. Numerous recent studies have shown that the order and combination of histone tail modifications, including methylation, phosphorylation, acetylation, and ubiquitination, play important roles in regulating gene expression *(10)*. Under this model, known as the "histone code hypothesis *(11)*," histone modifications are read by histone binding proteins which translate the 'code' into either an active or repressed chromatin state. Depending on the modifications of histones at specific target genes, these genes may be either permissive or repressive for gene transcription. This epigenetic regulation of gene expression likely plays important roles in development and cancer. Several drugs that modify the state of histone modification, and subsequently alter gene expression profiles, are currently in clinical trials for treatment of various human cancers *(12)*.

One of the most important regulatory histone modifications is methylation. Methylation has been identified on specific lysine or arginine residues of both histone H3 and H4. Arginine methylation on the histone tail typically leads to transcriptional activation, while lysine methylation typically leads to repression. To date, all histone methyltransferases identified have a so-called SET domain, which is the site of histone methyltransferase activity *(13)*. Numerous proteins contain SET domains, and not all have been shown to contain protein methyltransferase activity. Whether all proteins containing a SET domain exhibit methyltransferase activity is yet to be determined. Histone methyltransferases are also specific for the degree of methylation that they produce; there are specific mono-, di-, and trimethyltransferases. These levels of histone lysine methylation dictate relative levels of repression. Current thinking suggests that trimethylation, which is mediated most prominently by Suv39h proteins, leads to a relatively permanent modification that can be inherited in an epigenetic manner, whereas di- or monomethylation is more easily reversible and contributes to short-term repression *(14,15)*.

There are two known human Suv39h histone methyltransferases, Suv39h1 and Suv39h2. These genes were first identified in the fruit fly, where mutations in the *Drosophila* orthologs lead to suppression of the phenomenon of variegation, where the same gene inserted into different chromosomal locations can display different levels of expression *(16)*. Suv39h proteins were characterized as histone methyltransferases that can modify heterochromatic regions to maintain epigenetic silencing. They do this in part by subsequent recruitment of proteins such as heterochromatic protein 1 (HP1) that stabilize the repressed chromatic state

(17). Suv39h proteins have also been shown to modify euchromatin, and so to participate in regulation of gene expression on a more dynamic scale. This was first identified by the interaction of human Suv39h1 with the tumor suppressor product of the RB gene *(18,19)*. We have subsequently demonstrated that BMP signaling recruits Suv39h proteins to genes repressed by SMAD signaling, providing a mechanism for BMP mediated transcriptional repression and silencing.

2. BMP R-SMADS INTERACT WITH SUV39H HISTONE METHYLTRANSFERASES

We sought to understand the mechanisms of BMP signaling by investigating interactions of R-Smads regulated predominantly by BMP activation using a yeast two-hybrid assay to identify novel Smad5 binding partners. We probed a murine E9.5 cDNA library using Smad5 as bait. We chose this library as prey because mice homozygous null at the Smad5 locus die at about E9.5, suggesting an important role for Smad5 signaling during this stage of murine development *(20)*. Using this system, we identified Suv39h2 as a Smad5-interacting protein. Once the interaction between Smad5 and Suv39h2 was identified in the yeast two-hybrid system, it became important to determine whether this was a specific association between these two proteins, or whether there was a more generalized interaction between Smad proteins and the Suv39h histone methyltransferases. We first sought to determine if the other known member of the Suv39h family, Suv39h1, could bind to Smad5. Suv39h1 and Suv39h2 share significant sequence identity, but their patterns of expression in the developing mouse differ markedly *(21)*. Interestingly, loss of neither Suv39H1 nor Suv39H2 leads to embryonic death as determined in mouse knockout experiments *(22)*, suggesting significant functional overlap between these two proteins. This is in sharp contrast to Smad1 and Smad5, where loss of either gene during development leads to embryonic death in mid-gestation. In order to determine if Smad5 could bind to either Suv39h family member, we performed co-immunoprecipitation (IP) experiments in transiently transfected 293 cells to confirm the yeast two-hybrid results. This system has been used by numerous laboratories to characterize the interactions of Smad proteins during signaling in mammalian cells. In these experiments we used a constitutively active Type I receptor construct to induce phosphorylation of the Smads, as 293 cells do not express a full complement of BMP receptors. Additionally, we used epitope tagged constructs of the Smads and Suv39h proteins in order to identify the transiently transfected constructs and determine the mechanisms of interaction. We transiently transfected 293 cells with construct for Smad5 and either Suv39h1 or Suv39h2, either with or without the activated BMP type I receptor BMPR1A. As previously published *(9)*, following IP with an antibody to the *myc* epitope tag on the Suv39h constructs and Western blotting for flag-tagged Smad5, we demonstrated that Smad5 bound to both Suv39h1 and h2. In fact, these and subsequent experiments suggested that binding to Suv39h1 was more robust than was binding to Suv39h2. In subsequent experiments we therefore used Suv39h1 as the interaction partner with the Smads.

We next examined whether the Suv39h histone methyltransferases would bind to both TGF-β R-Smads and BMP R-Smads. BMPs and TGF-β can elicit different responses in the same cells in culture, and although the mechanism for this is poorly understood, there are likely to be differences in the binding affinities of associated cofactors for the various Smad proteins. There has been significant understanding of how this occurs between the two TGF-β R-Smads, Smad2 and 3, but little is known about how the BMP R-Smads differ in their ability to bind specific factors compared to the highly similar Smad3. We felt it would be important to determine if Smad1 and Smad5, the two best characterized BMP R-Smads, could bind Suv39h proteins with different affinity from the TGF-β R-Smads. In order to do this, we performed a similar co-IP experiment using transiently transfected 293 cells. In this experiment, we used Suv39h1 as the partner for Smads1, 2, 3 and 5. Activated type I receptors, Alk5 for Smads 2 and 3 or Alk3 for Smads1 and 5, were used to stimulate Smad activity. IP of the Smads and subsequent immunoblotting for associated Suv39h1 demonstrated that neither Smad2 nor Smad3 bound to Suv39h1 with any significant degree or with specificity for activation. In contrast, both Smad1 and Smad5 strongly bound Suv39h1 in a ligand dependent manner. While these data do not preclude a role for Suv39h association with TGF-β regulated Smads, they strongly suggest that the primary binding partners for Suv39h proteins in the Smad family are the BMP-regulated Smads.

One criticism of such overexpression and IP experiments is that because all of the potential partners are overexpressed, the associations are possibly nonphysiological, and will not be witnessed in intact cells. In order to address this point, we went on to analyze the association of BMP R-Smads and Suv39h1 in cells expressing endogenous levels of both proteins. We did this in two different ways. First, we confirmed the association seen by IP-Western experiments using antibodies directed against the endogenous proteins. In these experiments, cells were treated with BMP for various times or left untreated, and cell lysates were immunoprecipitated with antibodies that recognized both Smad1 and Smad5, and then blotted for Suv39h1. These experiments demonstrated that endogenous Smads and Suv39h1 associated in a ligand dependent manner. To confirm this result, we also examined the intracellular localization of Smads and Suv39h1 proteins before and after stimulation with BMP. Suv39h proteins are found constitutively in the nucleus, while Smads accumulate in the nucleus only as a result of activation following ligand binding to the cell surface receptors or expression of a constitutively active receptor construct. To determine if Smads and Suv39h proteins could colocalize in the nucleus after ligand stimulation, we used indirect immunofluorescence and confocal microscopy to visualize and localize both proteins. Following ligand stimulation, we saw a significant increase in the amount of Smad1/5 found in the nucleus, as was expected. To determine if these Smad proteins were associated with Suv39h1, we determined whether Smad1/5, Suv39h1 or both were contained within pixels in the nucleus. This analysis showed that there was a significant increase in colocalization of Smad1/5 and Suv39h1 following BMP treatment of the cells. These data lend further support to the idea that Smads1 and 5 associate with Suv39h1 in a regulatable and physiologic manner.

In order to further investigate the binding properties of these proteins we went on to perform mapping experiments. The Suv39h proteins are made up of three main domains. The N-terminal region known as the chromo domain is typically a site for histone-tail binding and can also be a site for general protein binding. The poorly defined middle region of the protein is known as the linker region. Finally, the SET domain, so named after the first three proteins identified to contain this domain (**S**uv39h, **E**nhancer of Zeste and **T**rithorax) is the site of histone methyltransferase activity. We used various mutants of Suv39h1 in co-IP/Western blotting experiments to identify the region of Suv39h1 that bound to Smad1. Using these constructs, we found that Smad1 bound specifically to the poorly characterized linker region of Suv39h1. Interestingly, a construct that had a four amino acid deletion that abolished SET domain methyltransferase activity (ΔNHSC) consistently bound with higher affinity to Smad1 then the wild type or other deletion constructs. The role methyltransferase activity plays in the affinity of Suv39h1 for Smad1 remains undetermined, although functional data indicate it is essential for Smad-Suv mediated transcriptional repression (*see* Section 3).

Suv39h proteins interact with HP-1 to mediate transcriptional silencing *(17)*. Suv39 proteins are known to interact with HP-1 directly, and HP-1 associates with methylated histone H3 to maintain the repressive chromatin state. We felt it would be important to know if Smads, HP-1 and Suv39h proteins all interacted in a tripartite complex. This would lend

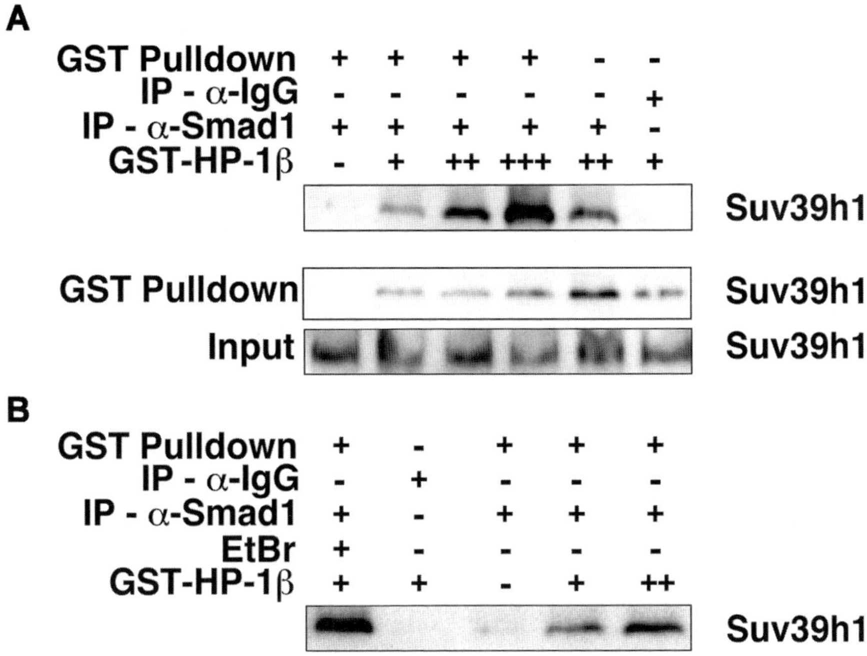

Fig. 1. Association of Smad1, Suv39H1 and HP-1 in a tripartite complex. (**A**) All three components exist together in a complex. (**B**) DNA binding is not necessary for association of the three proteins.

support to the idea that the association of Smads and Suv39h was important in mediating transcriptional silencing. In order to determine if this was the case, we used transient transfection and sequential affinity purification to determine whether all three proteins exist together in a complex. To do these experiments, we used glutathione S-transferase tagged HP-1 and the previously described epitope tagged Suv39h and Smad constructs. All three constructs were cotransfected in 293 cells in various combinations, and the complexes initially purified using glutathione Sepharose beads. We then eluted the complexes using reduced glutathione, and precipitated the eluates using antibodies to the Flag-tagged Smad construct. The immunoprecipitates were then blotted for the presence of Suv39h using the *myc* tag. Using this system, we were able to demonstrate that all three proteins exist together in a complex (Fig. 1A). Because Suv39h and HP-1 have been previously identified as interacting proteins, it strongly suggests that binding of Smad and HP-1 to Suv39h is not mutually exclusive, and that Suv39h may help to recruit HP-1 to regions of chromatin containing genes repressed by Smad signaling. Using ethidium bromide in the cell lysates to disrupt DNA-protein interactions, we further showed that the Smad-Suv39h interaction was DNA-independent, and that HP1 stabilizes this interaction in the absence of DNA (Fig. 1B). This suggests that the effect is owing to conformational changes in the complex leading to altered affinity, and not to an indirect association between Smads, Suv39h and HP1 on chromatin complexes. Whether the complex exists as a direct tripartite molecular entity, or requires other partners for efficient formation is yet to be determined and will be a focus of future work.

Once the association between Smads, Suv39h and HP1 was known, we sought to determine the role of HP1 in formation of the Suv39h-Smad complex. We hypothesized that addition of HP1 would stabilize the complex. This might lead to a more robust methylation at target genes sites. It might also provide a mechanism of regulation. If HP1 levels are low, the Smad-Suv39h complex might be less stable, while if HP1 is high, this might lead

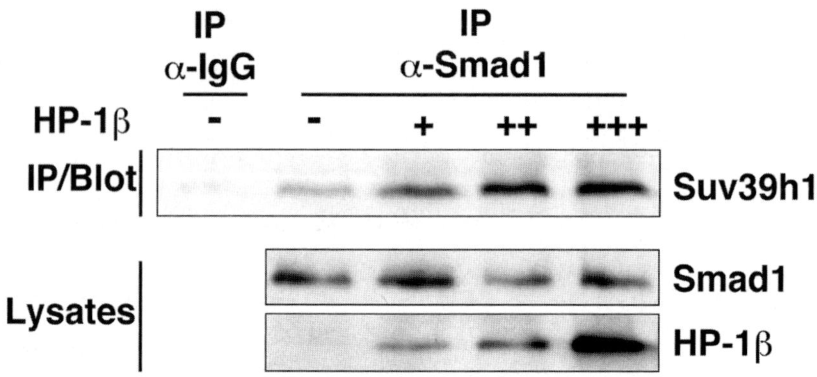

Fig. 2. HP-1 stabilizes the Smad-Suv39h complex. The increasing amounts of HP-1 cause a greater association of Smad1 with Suv39h1.

to stabilization of the Smad-Suv39h complex and more robust methylation of histones at target genes. To test this hypothesis, we transiently transfected 293 cells with Smad and Suv39h constructs along with increasing amounts of HP1. We then immunoprecipitated for Smad and blotted for Suv39h using epitope tags. As seen in Figure 2, addition of exogenous HP1 to the transfectants led to an increase in the amount of Suv39h co-IP with Smad1, while Smad and Suv39h levels within the cells remained the same. We interpret this data to mean that HP1 can stabilize the Smad-Suv39h interaction in the absence of DNA binding.

3. ENHANCEMENT OF BMP-INDUCED TRANSCRIPTIONAL REPRESSION BY A HISTONE METHYLTRANSFERASE

Both Suv39h histone methyltransferases and BMP R-Smads have been shown to act as transcriptional repressors. Suv39h proteins have long been known as being associated with transcriptionally silent, heterochromatic regions of DNA. Much recent work has determined that trimethylation of histone H3 on lysine 9 by Suv39h, and subsequent recruitment of HP1 leads to creation of such transcriptionally silent regions *(17)*. HP1 and Suv39h1 work together to form a structurally compact region of chromatin, which does not allow for access of transcriptional activators to the DNA and therefore allows for maintenance of the silent state. Other recent data has shown, however, that Suv39h proteins can also be recruited to specific target genes within euchromatin, and repress expression of transcriptionally active genes. This was first demonstrated when it was shown that Suv39h can bind to the retinoblastoma protein (pRB) and inhibit expression of target genes regulated by pRB *(18,19)*. This raised the possibility that Suv39h proteins might play a broader role in transcriptional regulation by influencing expression of genes found in both heterochromatic and euchromatic regions.

Smad proteins serve as scaffolds to assemble transcriptional activating or repressing complexes on genes regulated by TGF-β family signaling. Multiple studies have shown that the relative levels of various coactivators or corepressors can regulate Smad target gene expression (reviewed in ref. *[23]*). We sought to determine if the Suv39h-Smad association directly regulated target genes, and whether this was a generalized or specific phenomenon. We initially sought to determine whether overexpression of Suv39h proteins could repress genes known to be activated by BMP signaling. We used transient transfection and promoter luciferase assays to determine the level of expression of various target genes. Surprisingly, of several genes that we examined, none were repressed by overexpression of Suv39h1. This suggested to us that the mechanism of Suv39h1 repression was unique, and did not simply depend on altering the balance of transcriptional activators and repressors present at the Smad-DNA complex. We then sought a system where BMP signaling served to downregulate gene expression, arguing that there was a specific set of proteins assembled at a repressor complex that might then recruit Suv39h for an additional level of transcriptional control. Previous work has demonstrated that the muscle creatine kinase (MCK) gene is downregulated by BMP signaling during inhibition of myogenic differentiation. The regulation of the MCK gene by BMP was first examined in C2C12 myoblasts *(24)*. In this system, C2C12 mouse fibroblasts are induced to form myotubes by culture in low serum medium. During myotube differentiation, MCK is upregulated. Treatment of the cells with either TGF-β or BMP inhibits the formation of myotubes, and MCK upregulation is repressed. Although for TGF-β signaling it is known that Smad3 can sequester the muscle specific transcription factor MyoD and prevent MCK expression *(25)*, it is not known at the molecular level how BMP signaling can inhibit MCK expression.

For our studies we took advantage of this BMP-induced repression of the MCK gene at the promoter level to examine transcriptional repression by Suv39h and BMP signaling. We used two cell systems in our studies. The first was the murine C2C12 cell line described above. To examine if Suv39h1 plays a role in the BMP-mediated transcriptional repression of MCK promoter activity, we cotransfected the Suv39h1 expression construct, along with the activated BMP Type I receptor and a MCK promoter luciferase construct, into C2C12 cells. Cells were then induced to undergo myogenic differentiation by growth in low serum medium. We analyzed luciferase activity as a marker for promoter activation using standard assays. We showed in these assays that when both Suv39h1 and the activated receptor are cotransfected, we see an additive, dose-dependent repression. Transfection of Suv39h alone leads to a moderate inhibition of MCK promoter activity. Whether this is a nonspecific effect owing to Suv39h expression leading to a block in myogenic differentiation or a direct effect on MCK promoter expression through interaction with endogenous Smads found in the nucleus is uncertain. This suggests that both Smads and Suv39h proteins interact at the promoter to inhibit expression of the MCK gene.

In the second system we used murine C3H10T1/2 fibroblasts, where expression of the transcription factor MyoD is enough to induce myotube formation. This morphological change, along with expression of muscle specific genes, can be blocked by BMP signaling. We used this system to confirm the results we had obtained in the C2C12 myoblasts. As we have shown *(9)*, Suv39h1 and BMP signaling together repress MCK promoter activity, and in this case the repression appears to be synergistic. Taken together, our experiments demonstrating cooperative repression of specific targets of BMP signaling and a lack of generalized repression demonstrate that the Smad-Suv39h interaction is very specific, and only directly regulates genes repressed by BMP signaling. This is in contrast to most other corepressors previously shown to interact with Smad signaling. In those other cases, repression can be instituted at any target if the level of corepressor is sufficient. Our data suggest that the Smad-Suv39h interaction requires other cofactors to specify repression, at which point Suv39h activity can serve to initiate histone methylation and epigenetic silencing.

Although our data demonstrate that Suv39h and Smad proteins can contribute to repression of target genes where expression is downregulated by BMP signaling, it was not clear that either histone methyltransferase or Smad activation were required for this effect. We first sought to determine if the methyltransferase activity of Suv39h1 was necessary for the observed repression. In order to do this, we used a Suv39h1 construct with a deletion in the SET domain, the site of histone methyltransferase activity. We again employed the C3H10T1/2 cell system and the MCK promoter activated by MyoD expression. When expressed alone, this functional mutant had only a modest effect on repression of MCK activity and expression of a constitutively activated type I receptor did not change basal activity levels.

In contrast, in the same experiment, overexpression of wild-type Suv39h1 caused a more pronounced basal repression, and coexpression of activated type I receptor led to even further repression of MCK promoter activity. These data demonstrate that the methyltransferase activity of Suv39h1 is important in the repression of MCK promoter activity brought about by BMP signaling. Our next question was whether Smad signaling was also important. To address this question, we performed a similar experiment using a mutant Smad1 construct that could not be phosphorylated and activated by BMP signaling. As we have previously published *(9)*, expression of the Smad1 3S > A construct led to a reversal of MCK repression in a dose dependent manner. When repression is augmented by overexpression of Suv39h1, we can reverse this phenomenon by overexpression of mutant Smad1 which blocks Smad signaling from the receptor. Together these data indicate that the histone methyltransferase activity of Suv39h1 is necessary for the full effect of Smad mediated repression. There may be a small amount of repression mediated independently of methyltransferase activity, as there is a modest repression when the ΔSET construct is expressed alone that is not augmented by Smad signaling. Methyltransferase independent repression has previously been reported for Suv39h proteins in other contexts *(26)*. It is also clear that Smad signaling is necessary for the observed repression.

4. BMP-INDUCE GLOBAL CHANGES IN HISTONE METHYLATION PATTERNS

The Suv39h family is just one family of histone methyltransferases that may be important for regulating gene expression. As discussed above, histone H3K9 can be mono-, di- or trimethylated. Mono- and dimethylation are presumed more easily reversible, and specific demethylating enzymes for these modifications have been identified for Histone H3K4 *(14,15)*. To extend our studies of the role of BMP signaling in regulating histone methylation, we looked at the patterns of histone methylation in cells following treatment with BMP. We used three different cell lines that express BMP receptors, and have previously been shown to respond to these signals. Either NT2D1 teratocarcinoma, MDA-MB-231 human breast cancer or PC-3 prostate cancer cells were examined for histone methylation patterns before and after BMP treatment. Cells were treated with BMP or left untreated, and then processed for indirect immunofluorescence using antibodies specific to di- or trimethyl histone H3K9. We then used confocal microscopy to determine the amount and quality of histone methylation following treatment. As seen in Figure 3A,B, treatment with BMP had a profound effect on the pattern of histone methylation following BMP treatment. Treatment of cells with BMP led to a dramatic increase in dimethyl histone H3K9 shortly after addition of BMP. In contrast, the overall amount of trimethyl histone H3 did not change, but there was a redistribution of expression to a more generalized pattern. These data were confirmed for dimethylation by examining histone methylation in NT2D1 cells using Western blotting. Cells were treated with BMP and histones extracted by acid solubilization. We then blotted for dimethyl histone H3K9 and total histone content. As seen in Figure 3C, total amounts of histone did not change with BMP treatment, but dimethyl histone H3K9 dramatically increased. These data demonstrate that BMPs play a more general role in regulating gene expression through histone modifications. How BMPs generate an increase in dimethyl histone H3 is not currently known, but it may be through an increase in expression of histone methyltransferases such as G9a, an increase in activity, a change in targeting or a combination of these mechanisms. The elucidation of this mechanism is the focus of current studies.

5. CONCLUSIONS

Members of the BMP family have been well characterized for their roles in regulating important processes in development in organisms from the flatworm *Caenorhabditis elegans*

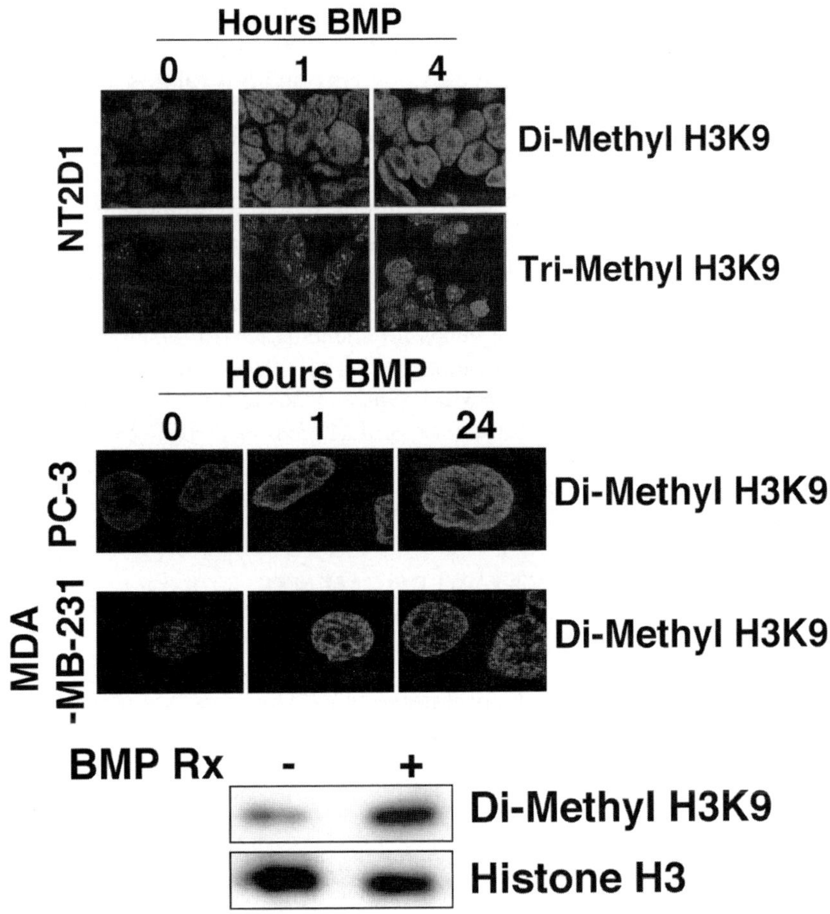

Fig. 3. BMP signaling leads to changes in histone methylation. (**A**) NT-2D1 human teratocarcinoma cells demonstrate increased dimethylation and a redistribution of trimethylation with BMP treatment. (**B**) Both MDA-MB-231 human breast carcinoma and PC3 human prostate cancer cells show increases in dimethyl histone H3K9 staining with BMP addition. (**C**) Histone H3 extracted from MDA-MB-231 cells shows an increase in dimethyl-H3K9 following BMP treatment.

to mammals. Less is known about the role of BMP signaling during the process of tumor formation, growth and metastasis. Already there are important hints that BMPs will play important roles in these processes, as demonstrated by in vitro studies of BMP effects on cell growth, and by the demonstration that mutations in components of BMP signaling pathways contribute to the generation of tumors in humans *(2)*. In addition, it is quite clear from many systems that genes and signaling pathways involved in developmental processes are often perturbed during carcinogenesis and cancer progression. Thus, study of the BMP signaling pathway is likely to play an increasingly prominent role in the understanding of how TGF-β superfamily members contribute to human disease.

We have demonstrated a mechanism of transcriptional repression induced by BMP signaling that is likely to play a prominent role in carcinogenesis. In unpublished microarray studies of TGF-β and BMP signaling, we have seen that half as many genes are downregulated as are upregulated. Developmentally, repression of specific targets is critical for induction of differentiated phenotypes. Thus, understanding the mechanisms of transcriptional repression, and in particular how repression is maintained in the absence of persistent signals, will no doubt lead to important insights into understanding the role of BMP signaling in normal development and carcinogenesis. Such understanding will lead to new therapies aimed at reversing aberrant epigenetic repression. Our discovery that the Smad signaling molecules interact with the histone methyltransferases Suv39h1 and Suv39h2 is an important step in understanding this process. This interaction may lead to an enduring repression of genes targeted by BMP signaling.

Suv39h histone methyltransferases have also demonstrated important roles in cancer development. Inactivation of Suv39h1 and h2 in mice by targeted mutagenesis leads to the spontaneous development of tumors resembling diffuse large B cell lymphomas *(22)*. A recent study demonstrated that part of this mechanism may be the maintenance of cellular senescence by Suv39h1 activity which may be coupled with the ability of Suv39h1 to interact with pRB and target growth promoting genes for silencing *(27)*. It is tempting to postulate that a similar mechanism may be operating in the association of Suv39h proteins with Smads. In this model, interaction with BMP R-Smads leads to transcriptional silencing and maintenance of a differentiated phenotype. Loss of Suv39h or BMP R-Smad activity leads to derepression and dedifferentiation leading to cell growth or disease progression. Whether this or other mechanisms of BMP regulation of histone methylation in human cancers are active is yet to be determined.

ACKNOWLEDGMENTS

We thank T Jenuwein, T Kouzarides, J Wrana, L Attisano, D Trouche, E Olsen, A Lassar, R Derynck, and P ten Dijke for constructs used in various experiments. We would also like to thank AB Roberts and CS Hill for helpful suggestions. This work was supported by grants from both the NIH and American Heart Association.

REFERENCES

1. Akhurst RJ, Derynck R. TGF-β signaling in cancer- a double-edged sword. Trends Cell Biol 2001;11: S44–S51.
2. Howe JR, Bair JL, Sayed MG, et al. Germline mutations of the gene encoding bone morphogenetic protein receptor 1A in juvenile polyposis. Nat Genet 2001;28:184–187.
3. Pouliot F, Labrie C. Role of Smad1 and Smad4 proteins in the induction of p21WAF1, Cip1 during bone morphogenetic protein-induced growth arrest in human breast cancer cells. J Endocrinol 2002; 172:187–198.
4. Ghosh-Choudhury N, Woodruff K, Qi W, Celeste A, Abboud SL, Ghosh Choudhury G. Bone morphogenetic protein-2 blocks MDA MB 231 human breast cancer cell proliferation by inhibiting cyclin dependent kinase- mediated retinoblastoma protein phosphorylation. Biochem Biophy Res Commun 2000;272:705–711.
5. Miyazaki H, Watabe T, Kitamura T, Miyazono K. BMP signals inhibit proliferation and in vivo tumor growth of androgen-insensitive prostate carcinoma cells. Oncogene 2004;23:9326–9335.
6. Haudenschild DR, Palmer SM, Moseley TA, You Z, Reddi AH. Bone morphogenetic protein (BMP)-6 signaling and BMP antagonist noggin in prostate cancer. Cancer Res 2004;64:8276–8284.
7. Feng XH, Zhang Y, Wu RY, Derynck R. The tumor suppressor Smad4/DPC4 and transcriptional adaptor CBP/p300 are coactivators for Smad3 in TGF-beta –induced transcriptional activation. Genes Dev 1998;12:2153–2163.
8. Janknecht R, Wells NJ, Hunter T. TGF-beta stimulated cooperation of Smad proteins with the coactivators CBP/p300. Genes Dev 1998;12:2114–2119.
9. Frontelo P, Leader JE, Yoo N, et al. Suv39h histone methyltransferases interact with Smads and cooperate in BMP-induced repression. Oncogene 2004; 23:5242–5251.
10. Jenuwein T, Allis CD. Translating the histone code. Science 2001;293:1074–1080.
11. Strahl BD, Allis CD. The language of covalent histone modifications. Nature 2000;403:41–45.
12. Piekarz R, Bates S. A review of depsipeptide and other histone deacetylase inhibitors in clinical trials. Curr Pharm Res 2004;10:2289–2298.
13. Jenuwein T, Laible G, Dorn R, Reuter G. SET domain proteins modulate chromatin domains in eu- and heterochromatin. Nature 1998;406:593–599.

14. Forneris F, Binda C, Vanoni MA, Mattevi A, Battagliolo E. Histone demethylation catalysed by LSD1 is a flavin-dependent oxidative process. FEBS Lett 2005;579:2203–2207.
15. Shi T, Lan F, Matson C, et al. Histone demethylation mediated by the nuclear amine oxidase homolog LSD1. Cell 2004;119:941–953.
16. Reuter G, Spierer P. Position effect variegation and chromatin proteins. Bioessays 1992;14:605.
17. Lachner M, O'Carroll D, Rea S, Mechtler K, Jenuwein T. Methylation of Histone H3 lysine 9 creates a binding site for HP1 proteins. Nature 2001;406:593–599.
18. Nielsen SJ, Schneider R, Bauer UM, et al. Rb targets histone H3 methylation and HP1 to promoters. Nature 2001;412:561–565.
19. Vandel L, Nicolas E, Vaute O, Ferreira R, Ait-Si-Ali S, Trouche D. Transcriptional repression by the retinoblastoma protein through the recruitment of a histone methyltransferase. Mol Cell Biol 2001; 21:6484–6494.
20. Chang H, Huylebroeck D, Verschueren K, Guo Q, Matzuk MM, Zwijsen A. Smad5 knockout mice die at mid-gestation due to multiple embryonic and extraembryonic defects. Development 1999;126: 1631–1642.
21. O'Carroll D, Scherthan H, Peters AH, et al. Isolation and characterization of Suv39h2, a second histone H3 methyltransferase gene that displays testis-specific expression. Cell Biol 2000;20:9423–9433.
22. Peters AH, O'Carroll D, Scherthan H, et al. Loss of the Suv39h histone methyltransferases impairs mammalian heterochromatin and genome stability. Cell 2001;107:323–327.
23. Massagué J, Seoane J, Wotton D. Smad transcription factors. Genes Dev 2005;9:2783–2810.
24. Akiyama S, Katagiri T, Namiki M, et al. Constitutively active BMP type I receptors transduce BMP-2 signals without the ligand in C2C12 myoblasts. Exp Cell Res 1997;235:362–369.
25. Liu D, Black L, Derynck R. TGF-beta inhibits muscle differentiation through functional repression of myogenic transcription factors by Smad3. Genes Dev 2001;15:2950–2966.
26. Chakraborty S, Sinha KK, Senyuk V, Nucifora G. SUV39H1 interacts with AML1 and abrogates AML1 transactivity. Oncogene 2003;22:5229–5237.
27. Braig M, Lee S, Loddenkemper C, Rudolph C, et al. Oncogene-induced senescence as an initial barrier in lymphoma development. Nature 2005;436:660–665.

25 Endoglin (CD105): A Strong Candidate for Immunologic Targeting of Tumor Neovasculature in Human Malignancies

Ester Fonsatti and Michele Maio

CONTENTS

INTRODUCTION
STRUCTURAL FEATURES AND INTERACTIONS WITH TGF-β
EXPRESSION, TISSUE DISTRIBUTION, AND IMPLICATIONS IN ANGIOGENESIS AND CARCINOGENESIS
MODULATION OF CD105 EXPRESSION
ROLE IN VASCULAR DEVELOPMENT AND PHYSIOLOGY
PROGNOSTIC POTENTIAL
TARGETING FOR TUMOR IMAGING AND IMMUNOTHERAPY
CD105-BASED VACCINES
CONCLUDING REMARKS
ACKNOWLEDGMENT
REFERENCES

Abstract

Tumor-associated angiogenesis is a well-acknowledged therapeutic target for human malignancies, and different markers of tumor neovasculature are actively investigated as potential candidates for antiangiogenetic therapy in cancer. Among these is endoglin (CD105), a homodimeric transmembrane glycoprotein overexpressed on proliferating endothelial cells, which has been identified as a functional component of the transforming growth factor-β (TGF-β) receptor system about a decade ago. The large body of experimental data accumulated so far indicates that CD105 plays a crucial role in angiogenesis and vascular integrity; furthermore, intratumoral microvessel density as evaluated by staining for CD105 represents a strong prognostic factor in solid malignancies of different histology. In this work, we review the biologic features of CD105 and the complex of evidences highlighting its great potentialities as novel target for diagnostic and therapeutic approaches in human malignancies.

Key Words: Endoglin; CD105; angiogenesis; carcinogenesis; endothelial cells; vascular targeting.

1. INTRODUCTION

Angiogenesis is a tightly regulated process that supports local tumor growth and disease progression; furthermore, increasing evidences demonstrate a strict correlation between

From: *Cancer Drug Discovery and Development: Transforming Growth Factor-β in Cancer Therapy, Vol I: Basic and Clinical Biology*
Edited by: Sonia B. Jakowlew © Humana Press Inc., Totowa, NJ

tumor microvascular density, distant metastasis, and prognosis of human malignancies of different histotype. These findings suggest that therapeutic strategies interfering with tumor blood supply represent promising modalities for cancer treatment *(1,2)*. Among the different therapeutic approaches aiming to the inhibition of tumor angiogenesis, great interest is focused on the molecular targeting of tumor vasculature, a concept proposed almost thirty years ago *(3,4)* and subsequently validated in animal models *(5)*. Within tumor blood vessels, endothelial cells are intriguing therapeutic targets because: (i) represent a genetically stable cell population; (ii) are easily accessible to drugs through the bloodstream; (iii) destruction of a limited number of endothelial cells can affect a great number of neoplastic cells that depend on blood supply; (iv) are suitable targets shared by malignancies of different histotype; (v) neoplastic cells can enter into the bloodstream and reach distant organs through endothelial cells. A specific marker for endothelial cells associated to tumor vasculature remains to be found. However, angiogenic and quiescent endothelia differ in their antigenic profile. In particular, endothelial cells lining tumor vasculature express high levels of activation and proliferation markers, which are currently under active investigation as potential candidates for the targeting of tumor-associated vasculature *(2)*.

Endoglin was first identified on a pre-B leukemic cell line *(6,7)* and subsequently described as a proliferation-associated cell surface antigen of endothelial cells *(8)*. In the course of the Fifth International Workshop on Human Leukocytes Differentiation Antigens, endoglin was assigned the Cluster of Differentiation (CD) number CD105 *(9)*. CD105 is a transmembrane glycoprotein that modulates transforming growth factor β (TGF-β) signaling by interacting with TGF-β receptors I and II. As demonstrated by different studies, CD105 has functional implications in angiogenesis, vasculature integrity, and tumor progression. In the present chapter, we will review and discuss the complex of experimental evidences that identify CD105 as a marker of angiogenesis and support its clinical potential in human malignancies.

2. STRUCTURAL FEATURES AND INTERACTIONS WITH TGF-β

CD105 is a homodimeric transmembrane glycoprotein made up of disulphide-linked subunits of 95 kDa *(7,10)*. Two isoforms of CD105, arising by alternative splicing and with differences in the amino acid sequence of their cytoplasmic tail, have been described in humans *(11)*. Consistently, two isoforms of CD105, with a different pattern of tissue distribution that suggests different functions, have also been discovered in mice *(12)*. Interestingly, the two murine isoforms, L- and S-CD105, interact to form heterodimers through the common extracellular domain *(12)*. L-CD105 was found to be predominantly expressed on endothelial cells *(12–14)* with a higher extent of phosphorylation compared to S-CD105 *(15)*, and it has been characterized as a protein of 633 amino acids, with a transmembrane region of 25 amino acids and a cytoplasmic tail of 47 amino acids. Potential N-linked glycosylation sites, a region rich in serine and threonine, and an arginine-glycine-aspartic acid tripeptide motif were identified *(7,16)*. High amino acid sequence homology was observed among human, porcine, and murine CD105 protein *(17,18)* and the major differences were found in the extracellular domain *(18)*.

Human *CD105* gene is approx 40 kb long and it is arranged into 14 exons *(19,20)*. It has been mapped to chromosome 9q34→ter, in a telomeric position respect to the Philadelphia breakpoint *(19)*.

Different mutations in *CD105* gene have been described and found to be associated with hereditary hemorrhagic telangiectasia type-1 (HHT-1), a dominant-inherited disease characterized by arterio-venous malformations and frequent bleeding, and CD105 haplo-insufficiency is the model for this vascular disorder *(21–23)*. Heterogeneous symptoms among different families, as well as among members of a single family have been reported; however, no correlations have been found between disease severity and type or position of CD105 mutations *(23–25)*. Instead, evidences in mice suggest that besides CD105 haploinsufficiency, HHT-1 severity is dependent on the genetic background and/or epigenetic factors *(26,27)*. Interestingly, circulating endothelial cells from HHT-1 patients show decreased levels of CD105, impaired TGF-β signaling, and failure to form cord-like structures *(28)*. Genetic analysis of patients with HHT-1 indicates that CD105 mutations are mostly truncating and preferentially located in the external domain *(24,25)*. Recently, the study of a family with clinical features of both HHT-1 and hereditary nonpolyposis colorectal cancer, that segregate independently within the kindred, suggested that CD105 haploinsufficiency could have a protective role on the development of solid tumors *(29)*.

CD105 has been functionally identified as a component of the TGF-β receptor system *(30,31)* that binds different components of the TGF-β superfamily including activin-A, bone morphogenetic protein-7 (BMP-7) and BMP-2 *(31)*. In particular, CD105 binds TGF-β1 and TGF-β3 with high affinity but it does not bind TGF-β2 *(30–32)*. Nevertheless, only a very small fraction of CD105 expressed on endothelial cells binds TGF-β (30). Furthermore, it has been demonstrated that CD105 requires the association with TGF-β signaling receptors type-II (TβR-II) to bind ligands *(31–33)*, and that it can interact with TβR-I or TβR-II in the absence of ligand *(33)*. CD105 functions in TGF-β signaling are still poorly understood; it has been proposed that through its interactions with the TβR-I and TβR-II, CD105 regulates their phosphorylation status and subsequently their signaling ability *(33)*.

TGF-β is a pleiotropic cytokine that regulates different cellular functions involved in vascular integrity, vasculogenesis and angiogenesis, such as proliferation, migration, adhesion, synthesis of extracellular matrix, and apoptosis of different cell types *(34)*.

In endothelial cells, two TβR-I pathways with opposite effects have been identified: the activin receptor-like kinases-5 (ALK-5)-inducing Smad2/3 phosphorylation, and the ALK-1-inducing Smad1/5 phosphorylation *(34–36)*. The latter pathway is known to promote cell proliferation and migration, whilst the first one inhibits these cellular responses to TGF-β *(36)*. Downregulation of CD105 expression through short interfering RNA, has recently allowed to demonstrate that CD105 promotes TGF-β-induced Smad 1/5 phosphorylation, and proliferation and migration of murine endothelial cells via ALK-1 receptor. Along this line, it has been shown that loss of CD105 abrogates ALK-1 signaling and endothelial cells proliferation, while CD105 haploinsufficiency leads to a downregulation of surface ALK-5 expression, probably as an adaptation mechanism *(37)*. These results suggest that CD105 is required for efficient TGF-β/ALK-1 signaling and that it acts as a modulator factor of the balance between TGF-β/ALK-1 and TGF-β/ALK-5 signaling pathways *(37)*. Consistently, it has been demonstrated that CD105 promotes TGF-β/ALK-1 signaling through its extracellular domain *(38)*, and that CD105 overexpression inhibits Smad3 transcriptional activity *(39)*.

In stark contrast with these findings, an enhanced proliferation and a strong activation of ALK-1-signaling pathway in CD105-null murine endothelial cells, together with increased levels of surface ALK-5, in response to TGF-β has been recently demonstrated *(40)*. These results suggested that CD105 is not required for TGF-β-dependent activation of ALK-1 pathway *(40)*.

The different findings obtained in these studies might be owing to the different technologies utilized to delete CD105 *(40)*; however, the emerging results point to a key role of CD105 in modulating the levels of surface TβR-I. Furthermore, the discovery that mutations in the gene coding for ALK-1 are associated with HHT-2, a disease with clinical manifestations comparable to those caused by HHT-1 *(23)*, further supports the notion that CD105 and ALK-1 are involved in the same signaling pathway.

Despite limited information available on the mechanism(s) of action of CD105, several evidences support the notion that different levels of CD105 expression differentially modulate cellular responses to TGF-β. In this regard, inhibition of cell proliferation, downregulation of c-*myc* mRNA, fibronectin synthesis, cellular adhesion, and phosphorylation of PECAM-1/CD31 were observed in human U937 monocytic leukemia cells transfected with L- or S-CD105 cDNA following TGF-β1 treatment *(41)*. Consistently, CD105 transfection of rat myoblasts reduced their growth inhibition and synthesis of plasminogen activator inhibitor-1 (PAI-1) in response to TGF-β1 *(32)*. Instead, deletion of CD105 by an antisense approach enhanced the ability of TGF-β1 to suppress growth, migration or microvessels formation in human umbilical vein endothelial cells (HUVEC) and smooth muscle cells *(13,42)*. Furthermore, the modulatory effect of CD105 to TGF-β responses is enforced by the observation that anti-CD105 monoclonal antibodies (mAb) and TGF-β have a synergic effect in suppressing the growth of human endothelial cells *(43)*.

Additionally, it has been reported that CD105 overexpression can affect cellular functions *per se*, in the absence of TGF-β. Along this line, reduced synthesis of fibronectin, PAI-1 and collagene, decreased migration in chemotactic and wound healing assays, as well as changes in the cellular morphology, were described in murine fibroblasts transfected with CD105 cDNA *(32,44,45)*.

Furthermore, myoblasts stably-transfected with CD105 showed a decreased expression, both at RNA and protein levels, of lumican, an extracellular matrix proteoglycan involved in the maintenance of tissue structure. This inverse correlation between CD105 and lumican expression was validated by immunohistochemical staining of blood vessels in human tissues. These results indicate that selected CD105 functions are independent from TGF-β signaling, and are suggestive for a crucial role of CD105 in cellular transmigration *(46)*.

3. EXPRESSION, TISSUE DISTRIBUTION, AND IMPLICATIONS IN ANGIOGENESIS AND CARCINOGENESIS

CD105 is differentially expressed on normal and neoplastic cells of different histotype however, consistent with its described functional role in angiogenesis, CD105 is highly expressed in endothelial cells *(for review see 47)*. In particular, elevated levels of CD105 expression are detectable on human microvascular endothelial cells that play a major role in angiogenesis *(48)*.

The human promoter of CD105 has been cloned, and isolated genomic fragments showed a strong and selective promoter activity in endothelial cells *(49)*. Supporting this finding, a vector containing the promoter of CD105 was efficiently utilized to deliver gene expression specifically to the endothelial cells of mouse blood vessels *(50)*. Additionally, human CD105 promoter fragments were successfully utilized in pigs to drive the expression of complement-regulatory molecules in the endothelium of small vessels of heart, kidney, and lung, but not in the large vessels of these organs *(51)*.

During human embryo development, CD105 is exclusively expressed on vascular endothelium of embryos at 4–8 wk of gestation, and transiently upregulated during heart septation and heart valve formation *(52)*. In addition, CD105 expression is associated with vasculogenesis and angiogenesis in mouse embryogenesis *(53)*.

The first evidence pointing to CD105 as a good marker of neovascularization, and thus as a potential candidate for vascular targeting approaches, derived from the observation that CD105 is overexpressed in proliferating HUVEC *(8,54)* and in the vascular endothelium of tissues undergoing angiogenesis, including tumor tissues of different histotype *(8,55,56)*. Furthermore, it was demonstrated that opposite to those of CD105, levels of expression of CD31, the "golden standard" for the assessment of angiogenetic activity, inversely correlated with HUVEC proliferation and cell density in vitro *(54,57)*. Consistently, when compared to a panel of endothelial cell markers, a stronger staining for CD105 was observed in blood vessels around and within tumor tissues *(58,59)*. As shown by different studies, the association of CD105 expression with tumor angiogenetic vessels is also highlighted by the finding that CD105 is largely expressed in small and likely immature blood vessels *(60)*, and that a correlation between intensity of CD105 staining and density of vascularization in the tumor lesions was found *(55,61)*. However, discrepancies concerning CD105 distribution in human normal and tumor tissues arised when different anti-CD105 mAb were utilized *(14,62,63)*, suggesting for their differential recognition of distinct epitopes of CD105 *(14,62,63)*.

Interestingly, a different pattern of vascular distribution was described for the pan endothelial markers CD31, CD34 and CD105 in oral squamous carcinoma tissues. Concerning CD105 staining, it was detected in neovessels with strong remodeling activity and in immature neovessels, while it was not detected on endothelial cells with weak remodeling activity *(64)*.

In solid neoplasia, CD105 staining was observed in peri- and intratumoral blood vessels and on tumor stromal components *(8,57–59,65)* while a weak expression was occasionally observed in the cytoplasm of neoplastic cells *(65)*. Furthermore, a comprehensive study of CD105 distribution on melanocytic cells revealed a variable expression of the molecule both in benign and malignant lesions *(66)*. Additionally, a weak and heterogeneous cell surface expression of CD105 was observed in a large panel of melanoma cell lines *(66)*. The analysis of different human neoplastic cell lines demonstrated that CD105 is poorly expressed in the majority of human carcinoma cell lines, highly expressed in human sarcoma cell lines and differently expressed by miscellaneous tumor cells *(67)*. Interestingly, tumor cells with higher levels of CD105 expression showed a higher rate of proliferation *(67)*.

In prostate cancer cells, staining for CD105 showed a focal localization to the areas of cell contact, and its transcript levels were found to be significantly lower in transformed compared to normal prostate cells. In addition, both levels of CD105 mRNA and of CD105 protein were slightly lower in the metastatic variant of prostate carcinoma cells PC3-M as compared to the parental PC3 cell line *(68)*. These findings, together with the observation that unique among 4000 genes evaluated *(69)* CD105 was downregulated during detachment of metastatic prostate cancer cells, suggested that loss of CD105 expression could affect prostate cancer progression through the modulation of cell adhesion and motility *(68)*.

CD105 is constitutively expressed in mouse and human epidermis and in skin appendages *(70)*. Supporting the hypothesis that CD105 is involved in benign and malignant neoplasms, a dual role of CD105 haploinsufficiency was demonstrated in multistage mouse skin carcinogenesis. In fact, heterozygous *CD105+/–* mice, subjected to long-term topical application of two carcinogenic drugs (7,12-dimethylbenz[α]anthracene and 12-*o*-tetradecanoylphorbol-13-acetate), showed a reduced incidence of tumors compared to control homozygous *CD105+/+* mice. However, more than half of tumors developed by *CD105+/–* mice were carcinomas, while all tumors investigated in *CD105+/+* mice were benign papillomas *(70)*. Noteworthy, similar results were obtained in transgenic mice overexpressing TGF-β1 in the epidermis *(71)*.

Studies with a transgenic mouse model suggested that opposite to the L-CD105 isoform that exerts a proangiogenic activity, the S isoform could act as an antiangiogenetic molecule *(12)*. In particular, it has been reported that S-CD105-positive transgenic mice bearing Lewis lung carcinoma, a model for tumor angiogenesis and metastatic infiltration, showed a reduction in tumor growth and vascularization *(12)*. Consistent with this observation, when subjected to a standard two-step chemical carcinogenesis protocol, a reduction in the number of benign papillomas *per* mouse was also found in S-CD105-positive mice compared to control mice *(12)*. Interestingly, the transgenic expression of S-CD105 did not rescue the lethal phenotype of homozygous CD105 knockout mice indicating that the S-CD105 isoform cannot substitute the L-CD105 functions in endothelial cells *(12)*.

Further evidence that CD105 is involved in tumor progression derives by the observation that treatment of human choriocarcinoma cells with methotrexate, an antagonist of folic acid, induced their morphological differentiation into nonproliferating syncytiotrophoblast-like cells, in association with increased levels of CD105 expression *(72)*.

In hematopoietic tumors, CD105 was detected in stem-cell-derived disorders such as refractory anemia with excess of blasts and blast crisis of chronic myelogeneous leukemia *(6,73)*. In different subtypes of acute myeloid leukemia, CD105 was invariably detected in most immature cellular subtypes but not in more differentiated ones *(57)*. Additionally, CD105 was detected on pro-B and pre-B leukemia cell lines *(7,10,74,75)*. Noteworthy, the expression profile of CD105 was found consistent among fresh and cultured leukemia or lymphoma cells of the corresponding phenotypes *(6)*.

4. MODULATION OF CD105 EXPRESSION

Compelling evidences indicate that CD105 expression is modulated by different angiogenetic stimuli and/or factors involved in vascular remodeling.

Levels of CD105 protein, mRNA, and promoter activity have been found to be upregulated by hypoxic stimuli *(76–79)*. In this regard, different mechanisms of CD105 induction have been proposed, including binding of hypoxia-inducible factor-1α to the hypoxia response elements in the CD105 promoter, or regulation via mitogen-activated protein kinases pathways *(76–79)*. Based on these evidences, it appears interesting that in angiopoietin-2 positive non-small cell lung cancer samples, elevated intratumoral microvascular density (IMVD) was detected by immunohistochemical staining for CD105 in the presence of elevated levels of VEGF, a proangiogenic factor for which transcription is also regulated by hypoxia *(80)*.

Interestingly, TGF-β1 stimulates the activity of CD105 promoter, reduces the levels of CD105 phosphorylation, increases the levels of CD105 expression *(41,81,82)*, and synergizes with hypoxia to induce transcriptional activation of CD105 in endothelial cells *(76–78)*. Similar results were observed in U937 monocytic leukemia cells and fibroblasts *(41,76)*. Furthermore, an upregulation of CD105 expression was observed on HUVEC infected with a recombinant adenovirus carrying a constitutively active form of ALK-1. This finding further supports the idea that CD105 and ALK-1 cooperate in regulating TGF-β signaling *(83)*. Instead, a downmodulation of CD105 at protein level but not at transcriptional level, was found to be mediated by tumor necrosis factor-α *(84)*.

Recently, it has been shown that thrombin-protease-activated receptor-1 (PAR-1) pathway downregulates TGF-β signaling through the PKC-ζ-dependent internalization of both TβRII and CD105, and the sequestering of Smad2/3 in primary human endothelial cells *(85)*. This finding appears of interest because thrombin-PAR-1 signaling pathway is involved in angiogenesis and vascular development, and thrombin inhibits capillary tube formation in endothelial cells.

5. ROLE IN VASCULAR DEVELOPMENT AND PHYSIOLOGY

Besides the evidences reported above, the assumption that CD105 plays a crucial role in vascular development and physiology is strongly enforced by the discovery that *CD105* knockout mice show defective vascular development that leads to dead during early gestation *(86)*, as previously reported for *TGF-β1*- and *ALK-1*-null mice *(87–89)*. In particular, the observation that important structural defects are involved in the primitive vascular plexus of the yolk sac suggests that CD105 has a role in the formation of normal mature blood vessels *(86)*.

Table 1
Intratumor Microvessel Density as Determined by Immunohistochemical Staining for CD105 Correlates with Prognosis in Breast Cancer

Prognostic factors	*References*
Overall survival	*(118–122)*
Disease free-survival	*(118)*
Negative lymph-node in patients with poorer survival	*(119)*
Higher risk of metastasis in lymph-node negative patients	*(119)*
Higher risk of metastasis	*(121)*
High local recurrence risk	*(120)*

Recently, a role for CD105 in regulating nitric oxide (NO)-dependent vasodilatation has been suggested by the finding that CD105-deficient vessels show reduced levels of endothelial NO-synthase (eNOS) *(90)*. Successively, a more in-depth study identified CD105 as an essential component of eNOS activation pathway that facilitates the association between eNOS and Hsp90 and regulates the vascular tone *(91)*.

Inhibition of the apoptosis of endothelial cells is crucial during angiogenesis and vasculogenesis, and it is well-known that proangiogenetic factors also promote survival of endothelial cells. Recently, it has been demonstrated that CD105 protects hypoxic endothelial cells from apoptosis either in the presence or absence of TGF-β1 *(78)*. Supporting this finding, an inverse correlation between the incidence of apoptosis and intratumoral vascular density as assessed by anti-CD105 antibodies but not by anti-CD34 antibodies was found in nonsmall cell lung cancer patients *(92)*.

6. PROGNOSTIC POTENTIAL

A number of studies have reported that the assessment of neovascularization by CD105 staining is a predictor of poor prognosis in different solid and hematopoietic malignancies, providing support to the usefulness of CD105 targeting in antiangiogenetic therapy of cancer.

The count of CD105-positive blood vessels correlates with survival in prostate cancer patients with Gleason score 5–7 *(60)*, with overall survival of node-negative breast carcinoma patients (Table 1), and with lymph node metastasis in cervical cancer *(93)*. Additionally, CD105 staining shows a positive correlation with angiolymphatic invasion, lymph node metastasis and tumor stage in endometrial carcinoma *(94)*. A lower intensity of CD105 staining was detected in T1 as compared to T2–T4 squamous cell carcinomas of the oral cavity *(56)*. In head and neck squamous cell carcinoma patients, a high CD105-positive IMVD was reported as an independent marker of tumor recurrence or death that might provide useful information for the clinical management of lymph-node negative patients *(95)*. In hepatocellular carcinoma, an elevated IMVD-CD105 staining associated with larger and more aggressive tumors *(96)*, and detection of diffuse CD105-staining within the microvessels in the adjacent tissues to the tumor was suggested to be predictive of early disease recurrence *(96)*.

Altogether, different studies reported that CD105 expression is a better marker to evaluate IMVD and to predict prognosis of cancer patients when compared to the expression of different pan endothelial markers (Table 2). However, at variance with studies at protein level, a preliminary investigation of CD105 expression at mRNA level did not find correlation with prognostic factors in breast cancer *(97)*.

Consistent with the evidence that anti-CD105 antibodies strongly react with angiogenic vessels, staining of CD105 by immunochemistry showed an increase in microvessel density

Table 2
Intratumor Microvessel Density in Solid Neoplasia Assessed by Immunohistochemical Staining for CD105 as Compared to Different Endothelial Cell Markers

Tumor histotype	*Endothelial cell markers*	*References*
Astrocytic tumors	CD31	*(123)*
Breast cancer	CD31, VEGF-R1/R2, Tie2/Tek	*(118–121)*
Colorectal cancer	CD31, CD34	*(102,124)*
Endometrial carcinoma	CD31	*(94)*
Head and neck squamous cell carcinoma	CD34	*(95)*
Hepatocellular carcinoma	CD34	*(96)*
Nonseminomatous testicular germ cell tumors	CD34	*(125)*
Nonsmall cell lung cancer	CD34, vWF	*(92,126)*
Prostate cancer	vWF	*(60,127)*
Gastrointestinal tract neoplasia	CD31, VIII-RA	*(128)*

during the progressive stages of colorectal carcinogenesis from low- to high-grade dysplasia, and from high-grade dysplasia to carcinoma *(98)*.

The limited data available on hematopoietic malignancies indicate that CD105 is a good surrogate marker to quantify microvessel density in the bone marrow of patients with multiple myeloma or hairy cell leukemia *(99,100)*.

A soluble form of CD105 (sCD105) was found in sera of healthy subjects and in conditioned medium of cultured endothelial and CD105-positive neoplastic cells *(57)*, suggesting that CD105 release is a physiological mechanism. Compared to healthy donors, higher levels of a circulating form of CD105 have been reported to be detectable in serum and plasma of cancer patients. In breast and colorectal cancer, the highest amounts of circulating sCD105 have been found in patients with metastatic disease, and increased levels of sCD105 were detected in sera of patients who did not receive chemotherapy *(101,102)*. Preliminary data in hematopoietic malignancies demonstrated elevated levels of sCD105 in serum or plasma of patients affected by acute myeloid leukemia and chronic myeloproliferative disorders. Low levels of CD105 have been detected in serum of patients with essential thrombocythemia associated with a history of thrombotic complications. Noteworthy, these patients showed an inverse correlation between levels of sCD105 and levels of soluble TGF-β1 *(103)*. Altogether, these evidences point to sCD105 as a surrogate marker of angiogenetic activity, and indicate that quantification of sCD105 levels may be useful in monitoring disease progression in solid *(101,102)* and hematopoietic malignancies *(103)*.

7. TARGETING FOR TUMOR IMAGING AND IMMUNOTHERAPY

7.1. In Vitro Evidences

Supporting the large body of data pointing to CD105 as a proliferation marker of endothelial cells, selected mAb were found to inhibit the proliferation of microvascular and macrovascular endothelial cells *(8,43,104)* and showed synergistic activity with TGF-β1 to inhibit the proliferation of HUVEC *(43)*.

In light of the potential clinical application of anti-CD105 antibodies, different efforts are currently addressed to generate engineered mAb that specifically target and/or destroy tumor-associated vasculature. Along this line, a bispecific single-chain diabody directed to the adenovirus fiber knob domain and to CD105, efficiently enhanced adenovirus transduction in HUVEC, suggesting that CD105 protein is a promising target for therapeutic gene transfer in endothelial cells *(105)*. Furthermore, a bispecific antibody directed to CD105 and CD3, mediated killing of CD105-positive endothelial cells by cytotoxic T lymphocytes *(106)*. Additionally, single-chain Fv fragments directed to CD105 were used to generate immunoliposome of encapsulated therapeutic drugs to target endothelial cells *(107)*. These complexes, which exhibited a specific binding to endothelial cells, were efficiently internalized and were able to improve cytotoxicity *(107)*.

Another strategy of CD105 targeting for anti-angiogenetic treatment of cancer might derive by the use of conditionally replicating adenovirus (CRADS), generated utilizing Flk-1 and CD105 regulatory elements. These CRADs are transcriptionally targeted to dividing endothelial cells and efficiently kill HUVEC *(108)*.

Microbubbles are suitable contrast agents for nonspecific and passive enhancement of ultrasound images. Recently, it has been demonstrated that avidin can be incorporated into the shell of microbubbles, acting as a direct ligand for biotinylated mAb. Using this technique, it was demonstrated that microbubbles targeted to CD105 specifically bind to polyoma middle T-transformed mouse brain capillary endothelial cells expressing high levels of CD105, but not to murine fibroblasts expressing low levels of CD105. Instead, untargeted or control IgG-targeted microbubbles failed to bind both cell types. Because ultrasound is emerging as an attractive tool to monitor entity of angiogenesis, these findings suggest that microbubbles directed to CD105, combined with ultrasound, might be used for noninvasive characterization of the tumor vasculature in vivo *(109)*.

7.2. Ex Vivo Evidences

A ^{99}Tcm-labeled anti-CD105 mAb was utilized for the perfusion of renal artery in freshly excised kidneys from seven patients with renal carcinoma. The scintigraphy invariantly identified hot spots of radioactivity, which matched with the position of the neoplastic lesions. The specificity of the localization of labeled anti-CD105 mAb into tumor mass was confirmed by the observation that a prior perfusion of unlabeled mAb completely blocked the localization of ^{99}Tcm-conjugated mAb. Interestingly, in one patient, two tumor masses were detected, but only one was identified by the presurgery magnetic resonance imaging scan, suggesting important implications in improving cancer diagnosis *(110)*.

7.3. In Vivo Evidences

Studies performed in different animal models demonstrated that CD105 targeting by radiolabeled mAb is a suitable and safe procedure for tumor imaging, regardless of tumor histologic type and independently from CD105 expression on neoplastic cells. In this regard, iv injection of a ^{125}I-labeled anti-CD105 mAb efficiently imaged spontaneous mammary carcinomas in a canine model. The uptake of the radiolabeled mAb into the tumor areas was rapid and intense. No systemic side effects were observed during a 3 mo follow-up after imaging procedure *(54)*. In line with these findings, the scintigraphy performed after the administration of an ^{111}In-labeled anti-CD105 mAb demonstrated accumulation of radioactivity in xenografts of human melanoma in mice. As demonstrated by autoradiography and immunohistology, the radiolabeled mAb was concentrated in the periphery of the tumor mass, with an heterogeneous distribution in the centre. Furthermore, the antibody was quickly cleared from the circulation and its blood half-life was less than 1 min. *(61)*.

The therapeutic potential of CD105 targeting has been extensively investigated in mice. All these studies clearly demonstrated a long-lasting suppression of tumor growth and metastasis by naked, radiolabeled or immunotoxin-conjugated anti-CD105 mAb, likely mediated by the inhibition of tumor-associated angiogenesis, and/or by the destruction of tumor-associated vasculature *(111–114)*. Interestingly, a naked anti-CD105 mAb completely suppressed human blood vessels but poorly repressed murine vessels in large established tumors of a human skin/SCID mouse chimera model *(114)*. In addition, the results of this study indicated that the combination of anti-CD105 mAb and cyclophosphamide had synergistic antitumor efficacy and completely suppressed established tumors in some chimeras *(114)*.

In the perspective of assessing the antitumor efficacy and safety of anti-CD105 mAb in cancer patients, a human/mouse chimeric antibody of IgG1 isotype was generated, and its pharmacokinetic and immunogenicity were investigated in monkeys. This mAb showed pharmacokinetic parameters comparable with those reported in humans for different mAb currently utilized for cancer therapy *(115)*.

8. CD105-BASED VACCINES

Vaccine strategies were also designed to investigate the therapeutic feasibility of the induction of an autoimmune response against tumor angiogenesis by immunization with a xenogeneic CD105 protein. Active immunotherapy with porcine CD105 protein as vaccine was shown to break immunotolerance against self CD105 and to mediate both protective and therapeutic anti-tumor immunity in mice *(116)*. In particular, decreased angiogenesis and a marked increase of apoptosis within tumor tissues were described. Furthermore, the antitumor activity and the production of autoantibodies against mouse CD105 were abrogated by depletion of CD4-positive T lymphocytes *(116)*. Additionally, low-dose cisplatin, one of the most extensively used cytotoxic drug, was effective in combination with recombinant xenogenic CD105 vaccine in inducing antitumor activity without increasing host toxicity *(117)*.

9. CONCLUDING REMARKS

Different evidences point to CD105 as an attractive target to design innovative diagnostic and therapeutic strategies in cancer. In particular, it has been well established that CD105 modulates several TGF-β-mediated functions related to angiogenesis and vascular development, and that CD105 is strongly overexpressed in proliferating endothelial cells. Consistently, selected anti-CD105 mAb strongly and specifically localize to tumor mass in the areas of major vascularization and induce tumor-regression by suppression of tumor-associated angiogenesis in animal models.

Much remains to be understood on the functional role of CD105 in angiogenesis and on the mechanisms regulating its interactions with TGF-β. However, the available information undoubtedly suggest that CD105 is a strong candidate for vascular targeting, and warrants additional efforts to translate the encouraging preclinical results so far obtained into the most appropriate clinical settings.

ACKNOWLEDGMENT

This work was supported by the Associazione Italiana per la Ricerca sul Cancro.

REFERENCES

1. Carmeliet P. Angiogenesis in life, disease and medicine. Nature 2005;438:932–936.
2. Neri D, Bicknell R. Tumour vascular targeting. Nat Rev Cancer 2005;5:436–446.
3. Folkman J. Tumor angiogenesis: therapeutic implications. N Engl J Med 1971;285:1182–1186.
4. Denekamp J. Vascular endothelium as the vulnerable element in tumours. Acta Radiol Oncol 1984;23:217–225.
5. Burrows FJ, Thorpe PE. Eradication of large solid tumors in mice with an immunotoxin directed against tumor vasculature. Proc Natl Acad Sci USA 1993;90:8996–9000.

6. Haruta Y, Seon BK. Distinct human leukemia-associated cell surface glycoprotein GP160 defined by monoclonal antibody SN6. Proc Natl Acad Sci USA 1986;83:7898–7902.
7. Gougos A, Letarte M. Identification of a human endothelial cell antigen with monoclonal antibody 44G4 produced against a pre-B leukemic cell line. J Immunol 1988;141:1925–1933.
8. Burrows FJ, Derbyshire EJ, Tazzari PL, et al. Up-regulation of endoglin on vascular endothelial cells in human solid tumors: implications for diagnosis and therapy. Clin Cancer Res 1995;1:1623–1634.
9. Letarte M, Greaves A, Vera S. CD105 (endoglin) cluster report. In: White Cell Differentiation Antigens. Schlossman SF, et al. Oxford, New York, Tokyo: Oxford University Press 1995:1756–1759.
10. Gougos A, Letarte M. Biochemical characterization of the 44G4 antigen from the HOON pre-B leukemic cell line. J Immunol 1988;141:1934–1940.
11. Bellon T, Corbi A, Lastres P, Cales C, et al. Identification and expression of two forms of the human transforming growth factor-beta-binding protein endoglin with distinct cytoplasmic regions. Eur J Immunol 1993;23:2340–2345.
12. Perez-Gomez E, Eleno N, Lopez-Novoa JM, et al. Characterization of murine S-endoglin isoform and its effects on tumor development. Oncogene 2005;24:4450–4461.
13. Li C, Hampson IN, Hampson L, Kumar P, Bernabeu C, Kumar S. CD105 antagonizes the inhibitory signaling of transforming growth factor beta1 on human vascular endothelial cells. FASEB J 2000; 14:55–64.
14. Duff SE, Li C, Garland JM, Kumar S. CD105 is important for angiogenesis: evidence and potential applications. FASEB J 2003;17:984–992.
15. Lastres P, Martin-Perez J, Langa C, Bernabeu C. Phosphorylation of the human-transforming-growth-factor-beta-binding protein endoglin. Biochem J 1994;301:765–768.
16. Gougos A, Letarte M. Primary structure of endoglin, an RGD-containing glycoprotein of human endothelial cells. J Biol Chem 1990;265:8361–8364.
17. Ge AZ, Butcher EC. Cloning and expression of a cDNA encoding mouse endoglin, an endothelial cell TGF-beta ligand. Gene 1994;138:201–206.
18. Yamashita H, Ichijo H, Grimsby S, Moren A, ten Dijke P, Miyazono K. Endoglin forms a heteromeric complex with the signaling receptors for transforming growth factor-beta. J Biol Chem 1994; 269:1995–2001.
19. Fernandez-Ruiz E, St-Jacques S, Bellon T, Letarte M, Bernabeu C. Assignment of the human endoglin gene (END) to 9q34→qter. Cytogenet Cell Genet 1993;64:204–207.
20. McAllister KA, Baldwin MA, Thukkani AK, et al. Six novel mutations in the endoglin gene in hereditary hemorrhagic telangiectasia type 1 suggest a dominant-negative effect of receptor function. Hum Mol Genet 1995;4:1983–1985.
21. McAllister KA, Grogg KM, Johnson DW, et al. Endoglin, a TGF-beta binding protein of endothelial cells, is the gene for hereditary haemorrhagic telangiectasia type 1. Nat Genet 1994;8:345–351.
22. Pece N, Vera S, Cymerman U, White RI Jr, Wrana JL, Letarte M. Mutant endoglin in hereditary hemorrhagic telangiectasia type 1 is transiently expressed intracellularly and is not a dominant negative. J Clin Invest 1997;100:2568–2579.
23. van den Driesche S, Mummery CL, Westermann CJ. Hereditary hemorrhagic telangiectasia: an update on transforming growth factor beta signaling in vasculogenesis and angiogenesis. Cardiovasc Res 2003;58:20–31.
24. Pece-Barbara N, Cymerman U, Vera S, Marchuk DA, Letarte M. Expression analysis of four endoglin missense mutations suggests that haploinsufficiency is the predominant mechanism for hereditary hemorrhagic telangiectasia type 1. Hum Mol Genet 1999;8:2171–2181.
25. Paquet ME, Pece-Barbara N, Vera S, et al. Analysis of several endoglin mutants reveals no endogenous mature or secreted protein capable of interfering with normal endoglin function. Hum Mol Genet 2001;10:1347–1357.
26. Bourdeau A, Dumont DJ, Letarte M. A murine model of hereditary hemorrhagic telangiectasia. J Clin Invest 1999;104:1343–1351.
27. Bourdeau A, Faughnan ME, McDonald ML, Paterson AD, Wanless IR, Letarte M. Potential role of modifier genes influencing transforming growth factor-beta1 levels in the development of vascular defects in endoglin heterozygous mice with hereditary hemorrhagic telangiectasia. Am J Pathol 2001; 158:2011–2020.
28. Fernandez-L A, Sanz-Rodriguez F, Zarrabeitia R, et al. Blood outgrowth endothelial cells from Hereditary Haemorrhagic Telangiectasia patients reveal abnormalities compatible with vascular lesions. Cardiovasc Res 2005;68:235–248.

29. Goldschmidt N, Metzger S, Wexler ID, Goldshmidt O, Hershcovici T, Chajek-Shaul T. Association of hereditary hemorrhagic telangiectasia and hereditary nonpolyposis colorectal cancer in the same kindred. Int J Cancer 2005;116:808–812.
30. Cheifetz S, Bellon T, Cales C, et al. Endoglin is a component of the transforming growth factor-beta receptor system in human endothelial cells. J Biol Chem 1992;267:19,027–19,030.
31. Barbara NP, Wrana JL, Letarte M. Endoglin is an accessory protein that interacts with the signaling receptor complex of multiple members of the transforming growth factor-beta superfamily. J Biol Chem 1999;274:584–594.
32. Letamendia A, Lastres P, Botella LM, et al. Role of endoglin in cellular responses to transforming growth factor-beta. A comparative study with betaglycan. J Biol Chem 1998;273:33,011–33,019.
33. Guerrero-Esteo M, Sanchez-Elsner T, Letamendia A, Bernabeu C. Extracellular and cytoplasmic domains of endoglin interact with the transforming growth factor-beta receptors I and II. J Biol Chem 2002;277:29,197–29,209.
34. Govinden R, Bhoola KD. Genealogy, expression, and cellular function of transforming growth factor-beta. Pharmacol Ther 2003;98:257–265.
35. Lebrin F, Deckers M, Bertolino P, ten Dijke P. TGF-beta receptor function in the endothelium. Cardiovasc Res 2005;65:599–608.
36. Goumans MJ, Valdimarsdottir G, Itoh S, Rosendahl A, Sideras P, ten Dijke P. Balancing the activation state of the endothelium via two distinct TGF-beta type I receptors. EMBO J 2002;21:1743–1753.
37. Lebrin F, Goumans MJ, Jonker L, et al. Endoglin promotes endothelial cell proliferation and TGF-beta/ALK1 signal transduction. EMBO J 2004;23:4018–4028.
38. Blanco FJ, Santibanez JF, Guerrero-Esteo M, Langa C, Vary CP, Bernabeu C. Interaction and functional interplay between endoglin and ALK-1, two components of the endothelial transforming growth factor-beta receptor complex. J Cell Physiol 2005;204:574–584.
39. Guo B, Slevin M, Li C, et al. CD105 inhibits transforming growth factor-beta-Smad3 signalling. Anticancer Res 2004;24:1337–1345.
40. Pece-Barbara N, Vera S, Kathirkamathamby K, et al. Endoglin null endothelial cells proliferate faster and are more responsive to transforming growth factor beta1 with higher affinity receptors and an activated Alk1 pathway. J Biol Chem 2005;280:27,800–27,808.
41. Lastres P, Letamendia A, Zhang H, et al. Endoglin modulates cellular responses to TGF-beta 1. J Cell Biol 1996;133:1109–1121.
42. Ma X, Labinaz M, Goldstein J, et al. Endoglin is overexpressed after arterial injury and is required for transforming growth factor-beta-induced inhibition of smooth muscle cell migration. Arterioscler Thromb Vasc Biol 2000;20:2546–2552.
43. She X, Matsuno F, Harada N, Tsai H, Seon BK. Synergy between anti-endoglin (CD105) monoclonal antibodies and TGF-beta in suppression of growth of human endothelial cells. Int J Cancer 2004; 108:251–257.
44. Guerrero-Esteo M, Lastres P, Letamendia A, et al. Endoglin overexpression modulates cellular morphology, migration, and adhesion of mouse fibroblasts. Eur J Cell Biol 1999;78:614–623.
45. Diez-Marques L, Ortega-Velazquez R, Langa C, et al. Expression of endoglin in human mesangial cells: modulation of extracellular matrix synthesis. Biochim Biophys Acta 2002;1587:36–44.
46. Botella LM, Sanz-Rodriguez F, Sanchez-Elsner T, et al. Lumican is down-regulated in cells expressing endoglin. Evidence for an inverse correlationship between Endoglin and Lumican expression. Matrix Biol 2004;22:561–572.
47. Fonsatti E, Maio M. Highlights on endoglin (CD105): from basic findings towards clinical applications in human cancer. J Transl Med 2004;2:18.
48. Wong SH, Hamel L, Chevalier S, Philip A. Endoglin expression on human microvascular endothelial cells association with betaglycan and formation of higher order complexes with TGF-beta signalling receptors. Eur J Biochem 2000;267:5550–5560.
49. Graulich W, Nettelbeck DM, Fischer D, Kissel T, Muller R. Cell type specificity of the human endoglin promoter. Gene 1999;227:55–62.
50. Velasco B, Ramirez JR, Relloso M, et al. Vascular gene transfer driven by endoglin and ICAM-2 endothelial-specific promoters. Gene Ther 2001;8:897–904.
51. Cowan PJ, Shinkel TA, Fisicaro N, et al. Targeting gene expression to endothelium in transgenic animals: a comparison of the human ICAM-2, PECAM-1 and endoglin promoters. Xenotransplantation 2003;10:223–231.
52. Qu R, Silver MM, Letarte M. Distribution of endoglin in early human development reveals high levels on endocardial cushion tissue mesenchyme during valve formation. Cell Tissue Res 1998;292: 333–343.

53. Jonker L, Arthur HM. Endoglin expression in early development is associated with vasculogenesis and angiogenesis. Mech Dev 2002;110:193–196.
54. Fonsatti E, Jekunen AP, Kairemo KJ, et al. Endoglin is a suitable target for efficient imaging of solid tumors: in vivo evidence in a canine mammary carcinoma model. Clin Cancer Res 2000;6: 2037–2043.
55. Miller DW, Graulich W, Karges B, et al. Elevated expression of endoglin, a component of the TGF-beta-receptor complex, correlates with proliferation of tumor endothelial cells. Int J Cancer 1999;81: 568–572.
56. Schimming R, Marme D. Endoglin (CD105) expression in squamous cell carcinoma of the oral cavity. Head Neck 2002;24:151–156.
57. Fonsatti E, Del Vecchio L, Altomonte M, et al. Endoglin: An accessory component of the TGF-beta-binding receptor-complex with diagnostic, prognostic, and bioimmunotherapeutic potential in human malignancies. J Cell Physiol 2001;188:1–7.
58. Wang JM, Kumar S, Pye D, van Agthoven AJ, Krupinski J, Hunter RD. A monoclonal antibody detects heterogeneity in vascular endothelium of tumours and normal tissues. Int J Cancer 1993;54: 363–370.
59. Wang JM, Kumar S, Pye D, Haboubi N, al-Nakib L. Breast carcinoma: comparative study of tumor vasculature using two endothelial cell markers. J Natl Cancer Inst 1994;86:386–388.
60. Wikstrom P, Lissbrant IF, Stattin P, Egevad L, Bergh A. Endoglin (CD105) is expressed on immature blood vessels and is a marker for survival in prostate cancer. Prostate 2002;51:268–275.
61. Bredow S, Lewin M, Hofmann B, Marecos E, Weissleder R. Imaging of tumour neovasculature by targeting the TGF-beta binding receptor endoglin. Eur J Cancer 2000;36:675–681.
62. Balza E, Castellani P, Zijlstra A, Neri D, Zardi L, Siri A. Lack of specificity of endoglin expression for tumor blood vessels. Int J Cancer 2001;94:579–585.
63. Seon BK. Expression of endoglin (CD105) in tumor blood vessels. Int J Cancer 2002;99:310–311.
64. Nagatsuka H, Hibi K, Gunduz M, et al. Various immunostaining patterns of CD31, CD34 and endoglin and their relationship with lymph node metastasis in oral squamous cell carcinomas. J Oral Pathol Med 2005;34:70–76.
65. Fonsatti E, Altomonte M, Nicotra MR, Natali PG, Maio M. Endoglin (CD105): a powerful therapeutic target on tumor-associated angiogenetic blood vessels. Oncogene 2003;22:6557–6563.
66. Altomonte M, Montagner R, Fonsatti E, et al. Expression and structural features of endoglin (CD105), a transforming growth factor beta1 and beta3 binding protein, in human melanoma. Br J Cancer 1996;74:1586–1591.
67. Postiglione L, Di Domenico G, Caraglia M, et al. Differential expression and cytoplasm/membrane distribution of endoglin (CD105) in human tumour cell lines: Implications in the modulation of cell proliferation. Int J Oncol 2005;26:1193–1201.
68. Liu Y, Jovanovic B, Pins M, Lee C, Bergan RC. Over expression of endoglin in human prostate cancer suppresses cell detachment, migration and invasion. Oncogene 2002;21:8272–8281.
69. Jovanovic BD, Huang S, Liu Y, Naguib KN, Bergan RC. A simple analysis of gene expression and variability in gene arrays based on repeated observations. Am J Pharmacogenomics 2001;1: 145–152.
70. Quintanilla M, Ramirez JR, Perez-Gomez E, et al. Expression of the TGF-beta coreceptor endoglin in epidermal keratinocytes and its dual role in multistage mouse skin carcinogenesis. Oncogene 2003;22:5976–5985.
71. Cui W, Fowlis DJ, Bryson S, et al. TGFbeta1 inhibits the formation of benign skin tumors, but enhances progression to invasive spindle carcinomas in transgenic mice. Cell 1996;86:531–542.
72. Letamendia A, Lastres P, Almendro N, et al. Endoglin, a component of the TGF-beta receptor system, is a differentiation marker of human choriocarcinoma cells. Int J Cancer 1998;76:541–546.
73. Lo Pardo C, Altomonte M, Boccuni P, et al. CD105 Workshop: Expression of CD105 on acute leucemia blast cells: application for minimal residual disease detection. In: Kishimoto T, Kikutani H, Kr von dem Borne AEG, Goyert SM, Mason DY, Miyasaka M, Moretta L, Okumura K, Shaw S Springer TA, Sugamura K, Zola H. eds, Leucocyte typing VI White Cell Differentiation Antigens. New York, London: Garland Publishing, Inc 1997:711–712.
74. O'Connell PJ, McKenzie A, Fisicaro N, Rockman SP, Pearse MJ, d'Apice AJ. Endoglin: a 180-kD endothelial cell and macrophage restricted differentiation molecule. Clin Exp Immunol 1992; 90:154–159.
75. Rokhlin OW, Cohen MB, Kubagawa H, Letarte M, Cooper MD. Differential expression of endoglin on fetal and adult hematopoietic cells in human bone marrow. J Immunol 1995;154:4456–4465.

76. Sanchez-Elsner T, Botella LM, Velasco B, Langa C, Bernabeu C. Endoglin expression is regulated by transcriptional cooperation between the hypoxia and transforming growth factor-beta pathways. J Biol Chem 2002;277:43,799–43,808.
77. Zhu Y, Sun Y, Xie L, Jin K, Sheibani N, Greenberg DA. Hypoxic induction of endoglin via mitogen-activated protein kinases in mouse brain microvascular endothelial cells. Stroke 2003;34: 2483–2488.
78. Li C, Issa R, Kumar P, Hampson IN, Lopez-Novoa JM, Bernabeu C, Kumar S. CD105 prevents apoptosis in hypoxic endothelial cells. J Cell Sci 2003;116:2677–2685.
79. Guo B, Kumar S, Li C, Slevin M, Kumar P. CD105 (endoglin), apoptosis, and stroke. Stroke 2004;35:94–95.
80. Tanaka F, Ishikawa S, Yanagihara K, et al. Expression of angiopoietins and its clinical significance in non-small cell lung cancer. Cancer Res 2002;62:7124–7129.
81. Rius C, Smith JD, Almendro N, et al. Cloning of the promoter region of human endoglin, the target gene for hereditary hemorrhagic telangiectasia type 1. Blood 1998;92:4677–4690.
82. Botella LM, Sanchez-Elsner T, Rius C, Corbi A, Bernabeu C. Identification of a critical Sp1 site within the endoglin promoter and its involvement in the transforming growth factor-beta stimulation. J Biol Chem 2001;276:34,486–34,494.
83. Ota T, Fujii M, Sugizaki T, et al. Targets of transcriptional regulation by two distinct type I receptors for transforming growth factor-beta in human umbilical vein endothelial cells. J Cell Physiol 2002;193:299–318.
84. Li C, Guo B, Ding S, Rius C, Langa C, Kumar P, Bernabeu C, Kumar S. TNF alpha down-regulates CD105 expression in vascular endothelial cells: a comparative study with TGF beta 1. Anticancer Res 2003;23:1189–1196.
85. Tang H, Low B, Rutherford SA, Hao Q. Thrombin induces endocytosis of endoglin and type-II TGF-beta receptor and down-regulation of TGF-beta signaling in endothelial cells. Blood 2005; 105:1977–1985.
86. Li DY, Sorensen LK, Brooke BS, et al. Defective angiogenesis in mice lacking endoglin. Science 1999;284:1534–1537.
87. Dickson MC, Martin JS, Cousins FM, Kulkarni AB, Karlsson S, Akhurst RJ. Defective haematopoiesis and vasculogenesis in transforming growth factor-beta 1 knock out mice. Development 1995; 121:1845–1854.
88. Goumans MJ, Zwijsen A, van Rooijen MA, Huylebroeck D, Roelen BA, Mummery CL. Transforming growth factor-beta signalling in extraembryonic mesoderm is required for yolk sac vasculogenesis in mice. Development 1999;126:3473–3483.
89. Urness LD, Sorensen LK, Li DY. Arteriovenous malformations in mice lacking activin receptor-like kinase-1. Nat Genet 2000;26:328–331.
90. Jerkic M, Rivas-Elena JV, Prieto M, et al. Endoglin regulates nitric oxide-dependent vasodilatation. FASEB J 2004;18:609–611.
91. Toporsian M, Gros R, Kabir MG, et al. A role for endoglin in coupling eNOS activity and regulating vascular tone revealed in hereditary hemorrhagic telangiectasia. Circ Res 2005;96:684–692.
92. Tanaka F, Otake Y, Yanagihara K, et al. Correlation between apoptotic index and angiogenesis in non-small cell lung cancer: comparison between CD105 and CD34 as a marker of angiogenesis. Lung Cancer 2003;39:289–296.
93. Brewer CA, Setterdahl JJ, Li MJ, Johnston JM, Mann JL, McAsey ME. Endoglin expression as a measure of microvessel density in cervical cancer. Obstet Gynecol 2000;96:224–228.
94. Saad RS, Jasnosz KM, Tung MY, Silverman JF. Endoglin (CD105) expression in endometrial carcinoma. Int J Gynecol Pathol 2003;22:248–253.
95. Martone T, Rosso P, Albera R, et al. Prognostic relevance of CD105+ microvessel density in HNSCC patient outcome. Oral Oncol 2005;41:147–155.
96. Ho JW, Poon RT, Sun CK, Xue WC, Fan ST. Clinicopathological and prognostic implications of endoglin (CD105) expression in hepatocellular carcinoma and its adjacent non-tumorous liver. World J Gastroenterol 2005;11:176–181.
97. Gomez-Esquer F, Agudo D, Martinez-Arribas F, Nunez-Villar MJ, Schneider J. mRNA expression of the angiogenesis markers VEGF and CD105 (endoglin) in human breast cancer. Anticancer Res 2004;24:1581–1585.
98. Akagi K, Ikeda Y, Sumiyoshi Y, et al. Estimation of angiogenesis with anti-CD105 immunostaining in the process of colorectal cancer development. Surgery 2002;31:S109–S113.

99. Pruneri G, Ponzoni M, Ferreri AJ, et al. Microvessel density, a surrogate marker of angiogenesis, is significantly related to survival in multiple myeloma patients. Br J Haematol 2002;118:817–820.
100. Pruneri G, Bertolini F, Baldini L, et al. Angiogenesis occurs in hairy cell leukaemia (HCL) and in NOD/SCID mice transplanted with the HCL line Bonna-12. Br J Haematol 2003;120: 695–698.
101. Takahashi N, Kawanishi-Tabata R, Haba A, et al. Association of serum endoglin with metastasis in patients with colorectal, breast, and other solid tumors, and suppressive effect of chemotherapy on the serum endoglin. Clin Cancer Res 2001;7:524–532.
102. Li C, Gardy R, Seon BK, et al. Both high intratumoral microvessel density determined using CD105 antibody and elevated plasma levels of CD105 in colorectal cancer patients correlate with poor prognosis. Br J Cancer 2003;88:1424–1431.
103. Calabrò L, Fonsatti E, Bellomo G, et al. Differential levels of soluble endoglin (CD105) in myeloid malignancies. J Cell Physiol 2003;194:171–175.
104. Maier JA, Delia D, Thorpe PE, Gasparini G. In vitro inhibition of endothelial cell growth by the antiangiogenic drug AGM-1470 (TNP-470) and the anti-endoglin antibody TEC-11. Anticancer Drugs 1997;8:238–244.
105. Nettelbeck DM, Miller DW, Jerome V, et al. Targeting of adenovirus to endothelial cells by a bispecific single-chain diabody directed against the adenovirus fiber knob domain and human endoglin (CD105). Mol Ther 2001;3:882–891.
106. Korn T, Muller R, Kontermann RE. Bispecific single-chain diabody-mediated killing of endoglin-positive endothelial cells by cytotoxic T lymphocytes. J Immunother 2004;27:99–106.
107. Volkel T, Holig P, Merdan T, Muller R, Kontermann RE. Targeting of immunoliposomes to endothelial cells using a single-chain Fv fragment directed against human endoglin (CD105). Biochim Biophys Acta 2004;1663:158–166.
108. Savontaus MJ, Sauter BV, Huang TG, Woo SL. Transcriptional targeting of conditionally replicating adenovirus to dividing endothelial cells. Gene Ther 2002;9:972–979.
109. Korpanty G, Grayburn PA, Shohet RV, Brekken RA. Targeting vascular endothelium with avidin microbubbles. Ultrasound Med Biol 2005;31:1279–1283.
110. Costello B, Li C, Duff S, et al. Perfusion of 99Tcm-labeled CD105 Mab into kidneys from patients with renal carcinoma suggests that CD105 is a promising vascular target. Int J Cancer 2004;109: 436–441.
111. Seon BK, Matsuno F, Haruta Y, Kondo M, Barcos M. Long-lasting complete inhibition of human solid tumors in SCID mice by targeting endothelial cells of tumor vasculature with antihuman endoglin immunotoxin. Clin Cancer Res 1997;3:1031–1044.
112. Matsuno F, Haruta Y, Kondo M, Tsai H, Barcos M, Seon BK. Induction of lasting complete regression of preformed distinct solid tumors by targeting the tumor vasculature using two new anti-endoglin monoclonal antibodies. Clin Cancer Res 1999;5:371–382.
113. Tabata M, Kondo M, Haruta Y, Seon BK. Antiangiogenic radioimmunotherapy of human solid tumors in SCID mice using (125)I-labeled anti-endoglin monoclonal antibodies. Int J Cancer 1999;82:737–742.
114. Takahashi N, Haba A, Matsuno F, Seon BK. Antiangiogenic therapy of established tumors in human skin/severe combined immunodeficiency mouse chimeras by anti-endoglin (CD105) monoclonal antibodies, and synergy between anti-endoglin antibody and cyclophosphamide. Cancer Res 2001;61: 7846–7854.
115. Shiozaki K, Harada N, Greco WR, et al. Antiangiogenic chimeric anti-endoglin (CD105) antibody: pharmacokinetics and immunogenicity in nonhuman primates and effects of doxorubicin. Cancer Immunol Immunother 2006;55:140–50.
116. Tan GH, Wei YQ, Tian L, et al. Active immunotherapy of tumors with a recombinant xenogeneic endoglin as a model antigen. Eur J Immunol 2004;34:2012–2021.
117. Tan GH, Tian L, Wei YQ, et al. Combination of low-dose cisplatin and recombinant xenogeneic endoglin as a vaccine induces synergistic antitumor activities. Int J Cancer 2004;112: 701–706.
118. Kumar S, Ghellal A, Li C, et al. Breast carcinoma: vascular density determined using CD105 antibody correlates with tumor prognosis. Cancer Res 1999;59:856–861.
119. Dales JP, Garcia S, Carpentier S, et al. Long-term prognostic significance of neoangiogenesis in breast carcinomas: comparison of Tie-2/Tek, CD105, and CD31 immunocytochemical expression. Hum Pathol 2004;35:176–183.

120. Dales JP, Garcia S, Carpentier S, Andrac L, et al. Prediction of metastasis risk (11 year follow-up) using VEGF-R1, VEGF-R2, Tie-2/Tek and CD105 expression in breast cancer (n=905). Br J Cancer 2004;90:1216–1221.
121. Dales JP, Garcia S, Bonnier P, et al. CD105 expression is a marker of high metastatic risk and poor outcome in breast carcinomas. Correlations between immunohistochemical analysis and long-term follow-up in a series of 929 patients. Am J Clin Pathol 2003;119:374–380.
122. Dales JP, Garcia S, Andrac L, et al. Prognostic significance of angiogenesis evaluated by CD105 expression compared to CD31 in 905 breast carcinomas: correlation with long-term patient outcome. Int J Oncol 2004;24:1197–204.
123. Yao Y, Kubota T, Takeuchi H, Sato K. Prognostic significance of microvessel density determined by an anti-CD105/endoglin monoclonal antibody in astrocytic tumors: comparison with an anti-CD31 monoclonal antibody. Neuropathology 2005;25:201–206.
124. Saad RS, Liu YL, Nathan G, Celebrezze J, Medich D, Silverman JF. Endoglin (CD105) and vascular endothelial growth factor as prognostic markers in colorectal cancer. Mod Pathol 2004;17:197–203.
125. Adam M, Schmidt D, Wardelmann E, Wernert N, Albers P. Angiogenetic protooncogene ets-1 induced neovascularization is involved in the metastatic process of testicular germ cell tumors. Eur Urol 2003;44:329–336.
126. Tanaka F, Otake Y, Yanagihara K, et al. Evaluation of angiogenesis in non-small cell lung cancer: comparison between anti-CD34 antibody and anti-CD105 antibody. Clin Cancer Res 2001;7: 3410–3415.
127. Josefsson A, Wikstrom P, Granfors T, et al. Tumor size, vascular density and proliferation as prognostic markers in GS 6 and GS 7 prostate tumors in patients with long follow-up and non-curative treatment. Eur Urol 2005;48:577–583.
128. Yu JX, Zhang XT, Liao YQ, et al. Relationship between expression of CD105 and growth factors in malignant tumors of gastrointestinal tract and its significance. World J Gastroenterol 2003;9: 2866–2869.

26 Activins and Inhibins in Cancer Progression

Gail P. Risbridger and Christopher Butler

CONTENTS

INTRODUCTION
ACTIVIN SIGNALING
ROLE OF ACTIVINS IN CANCER
CONCLUSIONS
REFERENCES

Abstract

Activins, and their antagonists inhibins, are members of the transforming growth factor (TGF-β) growth factor superfamily with essential regulatory effects in diverse physiological systems. Like TGF-β, activins can affect the initiation and progression of cancers by influencing angiogenesis, metastasis and immune suppression. Although activins signal via the Smad signaling pathway used by TGF-β, activin ligand activities are regulated by a number of extracellular binding proteins such as follistatin, follistatin-like protein and BAMBI (for bone morphogenetic protein and activin membrane-bound inhibitor). The importance of the regulatory proteins is far better validated than for the activin ligands *per se*. For example, the inhibin α-subunit is a proven tumor suppressor and widely used to monitor ovarian cancer.

Frequently, inhibin and activin have opposing actions and because the inhibin α-subunit is a tumor suppressor it would be plausible to predict that activin has the opposite action of promoting tumor progression. However, the evidence to support this concept is far from convincing and often contradictory. In this situation, where the evidence that activins promote tumor formation is equivocal, this laboratory reviewed the role of its antagonist, inhibin, and reexamined the evidence that inhibin is a tumor suppressor. Inhibin α-subunit is a proven tumor suppressor in mice, but is used as a marker for advanced disease and recurrence in human ovarian cancers. In prostate cancer inhibin α-subunit is downregulated or silenced in moderate cancers but upregulated in tumors from patients with increased risk of recurrence. These data, and their similarity to the dual nature of TGF-β in breast cancer, led us to postulate that inhibin α-subunit has dual roles and is both tumor suppressives and prometastatic in a single cell lineage. Further investigation is needed to establish whether the dual functions in carcinogenesis, postulated for both inhibin α-subunit and TGF-β, can also be demonstrated for activin ligands and/or other members of the TGF-β growth factor family.

Key Words: Activin; inhibin; tumorigenesis; carcinogenesis; α-subunit; βA subunit; metastasis; angiogenesis; immune suppression.

From: *Cancer Drug Discovery and Development: Transforming Growth Factor-β in Cancer Therapy, Vol I: Basic and Clinical Biology*
Edited by: Sonia B. Jakowlew © Humana Press Inc., Totowa, NJ

1. INTRODUCTION

Activins and inhibins are members of the transforming growth factor (TGF-β) superfamily of signaling molecules *(1)*. First identified as regulators of follicle-stimulating hormone (FSH) secretion, activins were shown to be essential regulators of diverse systems in physiology, with effects on development, somatic growth, cell proliferation and apoptosis, branching morphogenesis, inflammation and reproduction *(2–17)*. The related proteins, inhibins, were also identified by their effect on FSH secretion, which opposed that of activin action acting as competitive ligands for activin receptors *(3,18,19)*. Inhibins do not appear to play a major role in the developing mammalian embryo because mice homozygous for the null allele of the inhibin α-subunit develop normally. However, the inhibin-null adults are infertile and rapidly develop gonadal tumors of high penetrance identifying the inhibin α-subunit as a tumor suppressor *(20)*.

Similar to other members of the TGF-β superfamily, activins and inhibins are dimeric glycoproteins composed of either two β-subunits (Activins) or an α-subunit with either a βA or βB-subunit (Inhibin A or B). There is one α-subunit (although a variant form has recently been identified in monkey testis) and five activin β-subunits; βA, βB, βC, βD, βE *(21)*. The activin βA and βB-subunits and their homo-/heterodimers Activin A (βAβA), Activin B (βBβB-) and Activin AB (βAβB-) are well characterized. To date, the βD-subunit has only been found in *Xenopus laevis* where it acts as a mesoderm-inducing factor *(22)*. The other two related activin subunits, located in mammalian cells, are activin βC and βE (*see [23]* for review). The nomenclature of activins and inhibins is often confusing, with ambiguity arising from reference to both the subunits and the functional dimeric proteins with the single terms "activins" or "inhibins". For the purposes of this review we will use the terms activin β-subunit or inhibin α-subunit to refer to the subunit components of the protein and Activin A, AB or B, and Inhibin A or B to refer to the dimeric proteins.

Like TGF-β, activins and inhibins are secreted dimeric proteins that signal through cell surface receptors (*see* Section 2). Both TGF-β and activin/inhibin subunits are initially produced as glycosylated preprohormones. The subunits consist of a signal sequence, a prodomain of varying size and a mature C-terminal segment; this precursor then dimerizes with another subunit before export from the cell. Posttranslational modification of activin/inhibin dimers includes cleaving of these prohormone dimers intracellularly before secretion from the cell. Unlike TGF-β, which is exported from the cell bound to latency-associated peptide as an inactive protein, the activin and inhibin ligands are exported as mature bioactive dimeric proteins *(24)*.

2. ACTIVIN SIGNALING

The mode of action of the activin subfamily bears a number of similarities to that of TGF-β, but also displays some critical differences that are noteworthy, particularly those that relate to the regulation of ligand bioactivity by extracellular binding proteins such as follistatin and betaglycan.

Like TGF-β, activins signal through a serine/threonine kinase pathway utilizing two receptors: activin type I and activin type II similar to TGF-βRI and TGF-βRII *(24,25)*. In signaling, extracellular activin ligand binds to an activin type II receptor (ActRII A or B). Activin receptor type I (ActRI A or B) is then recruited to form a heterodimer and phosphorylated by ActRII. This triggers the same Smad protein signaling cascade utilized in TGF-β signaling. Smad2 or 3, called R-Smads, (receptor-regulated Smads) are phosphorylated and bind the Co-Smad, Smad4 (common signaling Smad). This complex translocates to the nucleus and binds with other factors such as Fox HI (Fast 1), which then activate the activin response element of the gene to initiate transcription. Activation of Smad6 or 7 (iSmads or inhibitory Smads) inhibits the signal cascade either by binding to ActRI or competitively binding Smad4 *(26,27)*.

Inhibin A or B also binds ActRII but this complex fails to recruit or phosphorylate Act RI and therefore inhibin acts as a competitive antagonist of activin signal transduction *(18,19)*. Betaglycan is a coreceptor molecule for inhibin as well as a type III TGF-β receptor; it enhances the binding affinity of inhibin for ActRII and confers sensitivity to inhibin on cells previously unable to respond *(24,28)*.

Different to TGF-β, there is a suite of extracellular molecules that function as regulators of activin signaling and hence bioactivity of the activin ligands *(29,30)*. Follistatin, an activin binding protein, is one of the most prominent and important regulators of activin ligand bioactivity. It occurs in two molecular weight isoforms: FS288 and FS315, which are the cell membrane bound and free-circulating forms respectively. These molecules bind activin dimers with high affinity and also bind inhibin, BMPs 2, 4, 6, 7, 11, and 15, and myostatin but with lower affinity. Binding of follistatin to activin prevents association with the type II receptor and also masks the type I receptor site *(31,32)*. Another member of the follistatin family of proteins characterized more recently, is follistatin related gene (FLRG), also known as follistatin-related protein or follistatin-like 3. FLRG, like FS315, acts as a circulating activin-binding protein *(33)*. The transmembrane protein pseudoreceptor BAMBI (for bone morphogenetic protein and activin membrane-bound inhibitor) is similar in structure to type I receptors but lacks an intracellular kinase domain. BAMBI attenuates signaling by TGF-β family ligands including TGF-β, BMPs and activin by stably associating with both type I and type II receptors *(34)*.

As well as the external controls of binding proteins and pseudoreceptors there is increasing evidence that the expression levels of other activin β-subunits within the cell represent an internal restraint on activin action. Both βC and βE-subunits are known to colocalize and can form heterodimers with βA and βB-subunits effectively lowering levels of Activin A in the cell *(35,36)*. Sequestration of βA-subunits by dimerization with βC-subunits to form Activin AC, represents an intracellular regulator of Activin A bioactivity (*[23]*, Fig. 1).

3. ROLE OF ACTIVINS IN CANCER

Differences and similarities between activin and TGF-β extend to their roles in the initiation and progression of cancers. A relatively unexplored field, the role of activins (and inhibins) is important to the TGF-β field because of the shared Smad signaling system and consequent opportunity for activins and inhibins to modify TGF-β signaling *(37–39)*. Any therapies based on disrupting TGF-β signaling must address the possible need to concomitantly abrogate and inactivate the activin ligands, in order to prevent their alternate action. In many instances activin ligands play similar roles to TGF-β and regulate cell proliferation and differentiation, angiogenesis and immunosuppression; actions that are crucial to initiation and progression of cancers *(40)*.

3.1. Inhibin is a Tumor Suppressor Gene in Mice

In 1992 Matzuk and colleagues published a seminal paper describing the inhibin-α deficient mouse model and showed the α-subunit of inhibin was a tumor suppressor gene *(20)*. This was the first secreted protein shown to be tumor suppressive. Mice lacking the inhibin-α-subunit develop normally but have reduced fertility and develop gonadal stromal tumors as early as 4 wk with virtually 100% penetrance *(20)*. Tumor progression is accompanied by a wasting syndrome similar to cachexia in humans with cancer; most mice die within 12 wk of birth. Inhibin deficient mice have increased serum activins, estradiol and FSH levels compared to controls *(41)*.

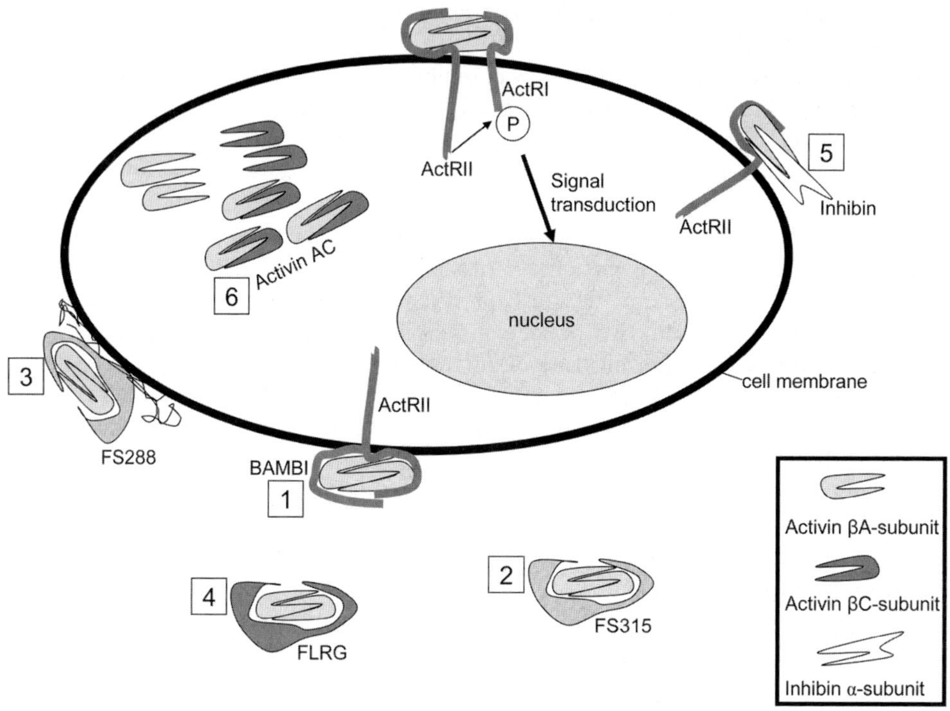

Fig. 1. Activation and abrogation of Activin signalling. The Activin A dimer can bind to ActRII, which recruits and phosphorylates ActRI and induces the signaling cascade. There are several inhibiting alternate ligands for activin binding. Extracellularly, if Activin A binds to BAMBI (1) this pseudo-receptor stably associates with ActRII but lacks the intracellular domain necessary for signal induction. Circulating (FS315) (2) or membrane bound (FS288) (3) follistatin bind activin A and prevent activin reaching the receptor. FS288 is also thought to present activin for lysosomal degradation. Follistatin related gene (FLRG) (4) also acts to bind dimeric activin in circulation. Alternately, dimeric inhibin (5) competitively binds to ActRII but fails to recruit ActRI and excludes activin A from the receptor. Intracellularly, the presence of alternative β-subunits such as activin βC (6) gives rise to heterodimers (Activin AC) and these sequester activin βA-subunits preventing formation of activin A. Inhibin-α-subunits sequester βA-subunits in a similar way to form inhibin A also reducing activin A levels (not shown).

The molecular pathways involved in this model of gonadal tumor formation have been dissected in an elegant series of genetic intercrosses, breeding inhibin-α-mutant mice with mice deficient in a range of endocrine factors. Thus, mice with double-null mutations of both inhibin α and activin receptor type II demonstrated that the elevated levels of activin are responsible for the cachexia syndrome observed in inhibin deficient mice. Since these mice still develop gonadal tumors, this experiment also showed that activin is not responsible for tumor formation *(42)*. Conversely, mice overexpressing follistatin and deficient in inhibin-α develop tumors have reduced activin levels and cachexia is less severe. *(43)*. The lethal tumor phenotype is completely rescued in bigenic mice overexpressing inhibin A on an inhibin-α null background *(44)*.

As well as the interplay between activin and inhibin, other factors such as GnRH and gonadotrophins are modifiers of inhibin gene tumor suppression. Mice lacking GnRH and inhibin-α do not develop tumors demonstrating that gonadotropins directly influence tumor development *(45)*. Mice expressing the SV40 T-antigen driven by the inhibin promoter also develop gonadal tumors of high penetrance *(46)*. When these mice are crossed with *hpg* mice which lack GnRH they fail to develop tumors supporting the concept that gonadotropins are required for gonadal tumor formation *(47)*.

Interestingly, mice lacking only one of the gonadotropins, FSH, still develop tumors on an inhibin-α deficient background although these are of late onset and much less aggressive *(48)*. At present, there has been no corresponding experiment to explore the effect of luteinizing hormone (LH) on the inhibin-α mutant. LH stimulates Leydig cells of the testis to produce testosterone and the effect of testosterone on tumor development was investigated by crossing mice bearing the *Tfm* mutation, which lack functional androgen receptors, with inhibin-α deficient mice. Again, there was modification of tumor formation and the mice survive longer because the tumors progress less rapidly and are less hemorrhagic *(49)*.

Sertoli cells produce anti-Müllerian hormone (AMH). Crossing mice deleted for AMH with inhibin-α null mouse gave mice, which developed more aggressive tumors and Leydig cell hyperplasia. Given that AR deficient mice show less aggressive tumorigenesis, the increased Leydig cell number in these mice suggest that tumor progression is driven by androgens and inhibited by AMH *(50)*. Although there is significant information on the various endocrine pathways that modify the development of gonadal tumors in inhibin-α deficient mouse, the direct links between inhibin and carcinogenesis and the mechanism by which inhibin causes tumor formation, remain to be elucidated.

3.2. Activins in Tumor Progression

In many tissues and cells, inhibin and activin have dual opposing actions and because inhibin is a tumor suppressor it would be plausible to predict that activin has the opposite action of promoting tumor progression *(51)*. However the evidence to support this concept is far from convincing and often contradictory. Thus, although activin A promotes proliferation of a subset of ovarian cancer cell lines without affecting normal ovarian surface epithelial cells, activin is implicated as a potential tumor suppressor in a variety of other cancers including neuroblastoma, liver, pituitary, endometrium, prostate, breast, and colon *(38,52–69)*.

Generally activins are growth inhibiting and Activin A inhibits proliferation of prostate cancer, breast cancer and B-cell leukemia cells, as well as vascular endothelium, vascular smooth muscle, fibroblasts, hepatocytes, liver cells and fetal adrenal cells *(19,58,67,70–79)*. Commonly, growth inhibition by activin involves cell cycle arrest in the G1 phase or induction of apoptosis by caspase activation (*[67,80,81], see [82]* for review). Activin induces apoptosis through Smad-dependent expression of the inositol phosphatase SHIP and can induce cell cycle arrest by activation of the p38 MAPK pathway and by Smad activation of p15 and repression of cyclin A and Rb phosphorylation *(37,67,83,84)*. Breast cancer cell proliferation is inhibited by activin A arresting cells in the G_0-G_1 cell cycle phase, repressing cyclin A and Rb phosphorylation, but, in high-grade breast cancer, activin signal transduction components are downregulated *(67)*.

By analogy to TGF-β, a key event in tumor progression may relate to the acquisition of resistance to activin by the tumor cell. There is limited evidence that some specimens from men with prostate cancer had inactivating mutations of the ActRII receptor *(85)*. The expression of follistatin provides another means of acquiring resistance to activin (because follistatin blocks the binding of activin ligands to the activin receptor); upregulation of follistatin occurs in prostate and breast cancer specimens *(66,86,87)*.

As well as regulating cell proliferation, activins are known to promote angiogenesis and metastases as well as suppressing the immune system and stimulating inflammation. These actions implicate that activins in the promotion of tumorigenesis and suggest a similarity to the role of TGF-β in other cancers, such as colon cancer but once again, the evidence is equivocal.

3.3. Effects on Angiogenesis

Blood vessel formation or angiogenesis is a crucial process in cancer. Activin A stimulates inflammatory corneal angiogenesis by increasing VEGF levels and induces vascular endothelial growth factor via Sp1-dependent stimulation of VEGF promoter activity in hepatocellular carcinoma *(61,88)*. However, activin A inhibits vascular endothelial cell growth and angiogenesis in vivo and endothelial cell proliferation in xenograft tumors *(60,62,63,89)*. Follistatin, an activin binding protein, induces proliferation of endothelial cells *(90)*. These results suggest activin suppresses tumor development by restricting blood vessel growth *(82)*.

3.4. Effects on Metastasis

The effects of activins on metastasis are varied. Consistent with activins promoting tumor progression, Activin A is overexpressed in esophageal carcinoma, and mRNA expression of the activin-βA-subunit is positively associated with tumor aggressiveness and lymph node metastasis; patients with high levels of activin-βA-subunit mRNA show poor prognosis *(91,92)*. In other tumors this correlation is not clear cut and although activins are elevated in breast cancer, this does not correlate to metastatic disease in the patients. Serum activin A levels are higher in breast cancer patients and tissue levels are higher in breast cancer tissue, and invasive breast tumors in bone (but not liver) metastases; but levels of activin A fail to correlate with lymph node metastasis, grade or tumor size *(93,94)*. Activin βA-subunit expression is associated with primary breast tumors with subsequent recurrence, but both the activin βA-subunit and inhibin α-subunit are downregulated in invasive breast cancer and associated metastases *(95)*. Activin increases NCAM expression which itself is negatively correlated with tumor progression and activin βB-subunit is expressed in poorly-metastatic melanoma cells but is absent from highly-metastatic cells *(96,97)*.

3.5. Immunosuppressive Effects

Members of the TGF-β growth factor family, including TGF-β and activins, regulate the immune system separately and interactively. Activin A inhibits proliferation and induces apoptosis in immune cells including B and T cells (*see* also discussion on angiogenesis in Section 3) *(98–101)*.

Activin A is also a significant component of the innate immune system and activin A concentrations in the bloodstream are acutely responsive to inflammatory challenge in postnatal life *(102)*. Elevated in a number of inflammatory diseases including inflammatory bowel disease, rheumatoid arthritis and gout, activin is associated with inflammatory changes in the brain and spinal fluid in a rabbit model of meningitis *(103–105)*. In a sheep model of systemic inflammation Activin A levels are elevated within 30 min of lipopolysaccharide (LPS) treatment, preceding the release of IL-6 *(9)*. Both pro and antiinflammatory effects of Activin A have been observed in placental tissues with low concentrations of Activin A stimulating IL-6 production and high concentrations suppressing it *(106)*. Further clarification of the specific role that activin plays in inflammation is needed, particularly in relation to diagnosis and treatment of diseases as well as cancer *(8)*. The emerging association between inflammation and cancer, together with the ability of activin to contribute to the inflammatory cascade, makes it imperative that we understand the implications of elevated activin in cancer.

In reviewing the evidence to support the case for activin being tumor promoting, a confusing picture emerges. For every instance in which one might argue that activin promotes tumorigenesis and affects progression by regulating angiogenesis and metastases, as well as the immune response, there are alternative data to refute its role. In this situation, where the evidence that activins promote tumor formation is equivocal, this laboratory reviewed the role of its antagonist, inhibin, and reexamined the evidence that inhibin is a tumor suppressor.

3.6. Evidence that the Inhibin α-subunit is more than a Tumor Suppressor Gene

For many years, it has been known that the measurement of inhibin in serum reliably and reproducibly predicts the recurrence of ovarian cancer *(107,108)*. This observation is well validated and used for the detection and monitoring for recurrence of granulosa cell tumors and mucinous carcinoma of the ovary *(109–111)*. Yet these observations are not consistent with the role of inhibin as a tumor suppressor as demonstrated in mice null for the inhibin α-subunit *(20)*. Initially explained by the fact that mice and women are different, research in other cancers has made this concept untenable.

In a subset of ovarian cancer patients peritoneal inhibin A levels decreased in late stage disease and decreased inhibin expression in ovarian tumors was associated with poor patient outcome *(112,113)*. In prostate carcinoma, consistent with a tumor suppressive role, inhibin-α-subunit expression is downregulated in human prostate cancer tissues and cell lines owing to hypermethylation of the promoter region of the gene or loss of heterozygosity *(114,115)*.

The field was further complicated by a large retrospective study of inhibin expression in radical prostatectomy specimens from prostate cancer patients carried out recently which gave surprising results. Although loss of inhibin expression by methylation or LOH had been shown previously, this study demonstrated that in advanced cancer increased expression of inhibin was associated with high-grade cancer and higher risk of recurrence *(116)*. These results were later confirmed in studies at four independent centers in Australia which showed upregulation of inhibin-α-subunit expression in 23–66% of high-grade prostate cancer biopsies *(117)*. These results also clarified the findings of a subset of patients in the molecular analysis of methylation of the inhibin-α-promoter region in prostate cancer specimens which did not display inactivation of the promoter. These patients were found to have cribriform carcinoma, a pathology commonly associated with poor patient outcome *(115)*.These contradictory and paradoxical data sets prompted our laboratory to reevaluate the role of inhibin in cancer.

A plausible explanation for the differences between mice, men, and women, based on the emerging evidence of a dual role of TGF-β in breast cancer, is that the inhibin-α-subunit has dual roles and is both tumor suppressive and prometastatic in a single cell lineage *(117,118)*. Cancer is a multistep process involving a transition from nonmalignant to malignant status via premalignant lesions and localised cancer. This is followed by metastasis and alterations, which result in poorly differentiated tumors and advanced cancers. In the nonmalignant state the activities of tumor suppressors dominate and loss of inhibin-α predicts a loss of tumor suppressor activity in low to moderate grade cancer. In support of this tumors used in our initial studies were commonly from men with localized prostate cancer and moderate Gleason scores and showed downregulation of inhibin-α-subunit expression *(119)*. We postulate that, as disease progresses and metastasis occurs, there is a switch in function and expression of the inhibin-α-subunit. Upregulation of inhibin α-subunit in late stage cancers becomes indicative of advanced disease and metastasis and a prometastatic function of the inhibin α-subunit.

A switch in gene function as proposed here may account for the contradictory findings in ovarian cancer. In mice, where the inhibin α-subunit is deleted, loss of inhibin-α-subunit expression is associated with initiation of carcinogenesis and development of ovarian tumors *(20)*. In women, ovarian cancer is most frequently diagnosed at a late stage of the disease

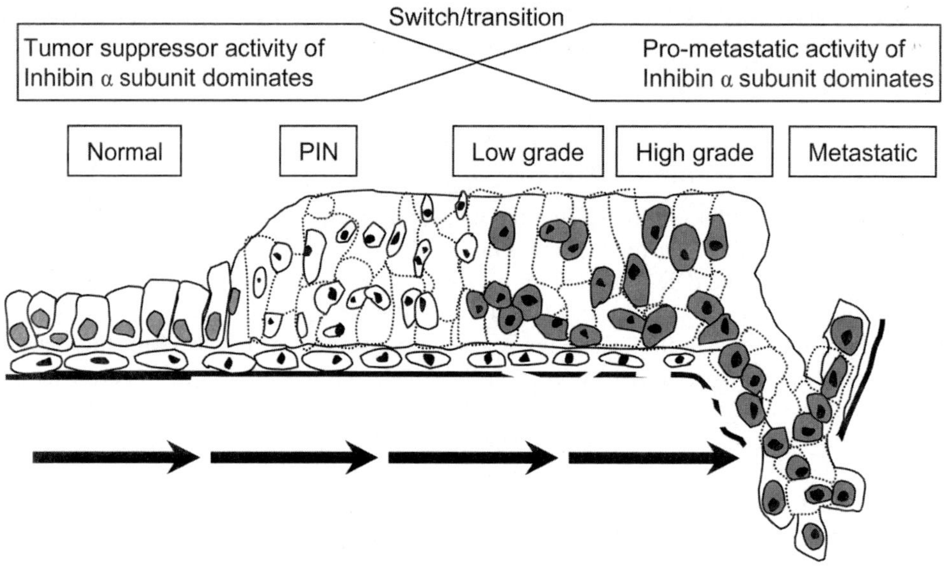

Fig. 2. Schematic diagram of the proposed switch hypothesis for inhibin-α in prostate cancer progression. In nonmalignant tissues tumor suppressor activities predominate but decline as premalignant and malignant progression proceeds. There is a switch in inhibin-α-subunit activity from initially tumor suppressive to oncogenic and prometastatic in advanced stage disease. This switch, which is also known to occur with TGF-β, may also be common to other members of the TGF-β superfamily of growth factors. PIN, prostatic epithelial hyperplasia.

and inhibin-α-subunit expression is associated with advanced disease and recurrence *(40)*. If these two examples represent two ends of a continuum rather than discrepant results, the apparent paradox is resolved (Fig. 2). Whether this hypothesis also holds for other types of cancer warrants further investigation.

4. CONCLUSIONS

Activins and their regulatory proteins are powerful effectors of many biological processes and disease including cancer. Investigation of the activities of modifiers of activin action, such as inhibin, has yielded provocative hypotheses. The inhibin α- subunit is implicated as a tumor suppressor but used as a marker of advanced carcinogenesis. A postulated switch in inhibin-α-subunit action during cancer progression provides an explanation for the apparently paradoxical nature of inhibin-α-subunit activity in endocrine cancers. In early stages of malignancy inhibin α-subunit is a tumor suppressor and is downregulated. As the disease progresses a prometastatic function of inhibin α-subunit emerges and expression levels increase. The resemblance of this hypothesis to recent findings that TGF-β has dual functions in carcinoma suggests that the close structural similarity of the TGF-β family members is also reflected in their biological activity in disease processes. The main biological function of inhibin is to oppose the action of activin. Given the tumor suppressive action of the inhibin α-subunit, this might suggest that activin has a tumorogenic role. However, evidence to support this is not forthcoming. For every example of an oncogenic role of activin there are counter examples demonstrating a tumor suppressor role. It remains to be seen if the dual functions in carcinoma outlined for both inhibin α-subunit and TGF-β can also be demonstrated for activin ligands or other members of the TGF-β growth factor family.

REFERENCES

1. Chang H, Brown CW, Matzuk MM. Genetic analysis of the mammalian transforming growth factor-beta superfamily. Endocr Rev 2002;23(6):787–823.
2. Robertson DM, Foulds LM, Leversha L, et al. Isolation of inhibin from bovine follicular fluid. Biochem Biophys Res Commun 1985;126(1):220–226.
3. Ying SY. Inhibins and activins: chemical properties and biological activity. Proc Soc Exp Biol Med 1987;186(3):253–264.
4. Ling N, Ying SY, Ueno N, et al. Pituitary FSH is released by a heterodimer of the beta-subunits from the two forms of inhibin. Nature 1986;321(6072):779–782.
5. Brown CW, Li L, Houston-Hawkins DE, Matzuk MM. Activins are critical modulators of growth and survival. Mol Endocrinol 2003;17(12):2404–2417.
6. Hully JR, Chang L, Schwall RH, Widmer HR, Terrell TG, Gillett NA. Induction of apoptosis in the murine liver with recombinant human activin A. Hepatology 1994;20(4 Pt 1):854–862.
7. Ball EM, Risbridger GP. Activins as regulators of branching morphogenesis. Dev Biol 2001; 238(1):1–12.
8. Jones KL, de Kretser DM, Patella S, Phillips DJ. Activin A and follistatin in systemic inflammation. Mol Cell Endocrinol 2004;225(1–2):119–125.
9. Phillips DJ, Jones KL, Scheerlinck JY, Hedger MP, de Kretser DM. Evidence for activin A and follistatin involvement in the systemic inflammatory response. Mol Cell Endocrinol 2001;180(1–2): 155–162.
10. Risbridger GP, Cancilla B. Role of activins in the male reproductive tract. Rev Reprod 2000; 5(2):99–104.
11. de Kretser DM, Hedger MP, Loveland KL, Phillips DJ. Inhibins, activins and follistatin in reproduction. Hum Reprod Update 2002;8(6):529–541.
12. Piepenburg O, Grimmer D, Williams PH, Smith JC. Activin redux: specification of mesodermal pattern in Xenopus by graded concentrations of endogenous activin B. Development 2004;131(20): 4977–4986.
13. McDowell N, Gurdon JB. Activin as a morphogen in Xenopus mesoderm induction. Semin Cell Dev Biol 1999;10(3):311–317.
14. Albertson RC, Payne-Ferreira TL, Postlethwait J, Yelick PC. Zebrafish acvr2a and acvr2b exhibit distinct roles in craniofacial development. Dev Dyn 2005;233(4):1405–1418.
15. Lambert-Messerlian GM, Pinar H, Laprade E, Tantravahi U, Schneyer A, Canick JA. Inhibins and activins in human fetal abnormalities. Mol Cell Endocrinol 2004;225(1–2):101–108.
16. Ferguson CA, Tucker AS, Christensen L, Lau AL, Matzuk MM, Sharpe PT. Activin is an essential early mesenchymal signal in tooth development that is required for patterning of the murine dentition. Genes Dev 1998;12(16):2636–2649.
17. Ferguson CA, Tucker AS, Heikinheimo K, et al. The role of effectors of the activin signalling pathway, activin receptors IIA and IIB, and Smad2, in patterning of tooth development. Development 2001;128(22):4605–4613.
18. Lebrun JJ, Vale WW. Activin and inhibin have antagonistic effects on ligand-dependent heteromerization of the type I and type II activin receptors and human erythroid differentiation. Mol Cell Biol 1997;17(3):1682–1691.
19. Xu J, McKeehan K, Matsuzaki K, McKeehan WL. Inhibin antagonizes inhibition of liver cell growth by activin by a dominant-negative mechanism. J Biol Chem 1995;270(11):6308–6313.
20. Matzuk MM, Finegold MJ, Su JG, Hsueh AJ, Bradley A. Alpha-inhibin is a tumour-suppressor gene with gonadal specificity in mice. Nature 1992;360(6402):313–319.
21. Bernard DJ, Woodruff TK, Plant TM. Cloning of a novel inhibin alpha cDNA from rhesus monkey testis. Reprod Biol Endocrinol 2004;2:71.
22. Oda S, Nishimatsu S, Murakami K, Ueno N. Molecular cloning and functional analysis of a new activin beta subunit: a dorsal mesoderm-inducing activity in Xenopus. Biochem Biophys Res Commun 1995;210(2):581–588.
23. Butler CM, Gold EJ, Risbridger GP. Should activin betaC be more than a fading snapshot in the activin/TGFbeta family album? Cytokine Growth Factor Rev 2005;16(4–5):377–385.
24. Massagué J. TGF-beta signal transduction. Annu Rev Biochem 1998;67:753–789.
25. Phillips DJ, Woodruff TK. Inhibin: actions and signalling. Growth Factors 2004;22(1):13–18.
26. Hata A, Lagna G, Massagué J, Hemmati-Brivanlou A. Smad6 inhibits BMP/Smad1 signaling by specifically competing with the Smad4 tumor suppressor. Genes Dev 1998;12(2):186–197.

27. Pangas SA, Woodruff TK. Activin Signal Transduction Pathways. Trends Endocrinol Metab 2000; 11(8):309–314.
28. Lewis KA, Gray PC, Blount AL, et al. Betaglycan binds inhibin and can mediate functional antagonism of activin signalling. Nature 2000;404(6776):411–414.
29. Phillips DJ. Regulation of activin's access to the cell: why is mother nature such a control freak? Bioessays 2000;22(8):689–696.
30. Tsuchida K. Activins, myostatin and related TGF-beta family members as novel therapeutic targets for endocrine, metabolic and immune disorders. Curr Drug Targets Immune Endocr Metabol Disord 2004;4(2):157–166.
31. de Winter JP, ten Dijke P, de Vries CJ, et al. Follistatins neutralize activin bioactivity by inhibition of activin binding to its type II receptors. Mol Cell Endocrinol 1996;116(1):105–114.
32. Thompson TB, Lerch TF, Cook RW, Woodruff TK, Jardetzky TS. The structure of the follistatin: activin complex reveals antagonism of both type I and type II receptor binding. Dev Cell 2005; 9(4):535–543.
33. Harrison CA, Gray PC, Vale WW, Robertson DM. Antagonists of activin signaling: mechanisms and potential biological applications. Trends Endocrinol Metab 2005;16(2):73–78.
34. Onichtchouk D, Chen YG, Dosch R, et al. Silencing of TGF-beta signalling by the pseudoreceptor BAMBI. Nature 1999;401(6752):480–485.
35. Mellor SL, Ball EM, O'Connor AE, et al. Activin betaC-subunit heterodimers provide a new mechanism of regulating activin levels in the prostate. Endocrinology 2003;144(10):4410–4419.
36. Mellor SL, Cranfield M, Ries R, et al. Localization of activin betaA, betaA and betaC subunits in human prostate and evidence for formation of new activin heterodimers of betaC subunit. J Clin Endocrinol Metab 2000;85(12):4851–4858.
37. Valderrama-Carvajal H, Cocolakis E, Lacerte A, et al. Activin/TGF-beta induce apoptosis through Smad-dependent expression of the lipid phosphatase SHIP. Nat Cell Biol 2002;4(12):963–969.
38. Deacu E, Mori Y, Sato F, et al. Activin type II receptor restoration in ACVR2-deficient colon cancer cells induces transforming growth factor-beta response pathway genes. Cancer Res 2004;64(21): 7690–7696.
39. Ying SY, Becker A. Inhibin and Activin Modulate Transforming growth factor-beta-induced immunosuppression. J Biomed Sci 1995;2(3):237–241.
40. Robertson DM, Burger HG, Fuller PJ. Inhibin/activin and ovarian cancer. Endocr Relat Cancer 2004;11(1):35–49.
41. Matzuk MM, Finegold MJ, Mather JP, Krummen L, Lu H, Bradley A. Development of cancer cachexia-like syndrome and adrenal tumors in inhibin-deficient mice. Proc Natl Acad Sci USA 1994;91(19):8817–8821.
42. Coerver KA, Woodruff TK, Finegold MJ, Mather J, Bradley A, Matzuk MM. Activin signaling through activin receptor type II causes the cachexia- like symptoms in inhibin-deficient mice. Mol Endocrinol 1996;10(5):534–543.
43. Cipriano SC, Chen L, Kumar TR, Matzuk MM. Follistatin is a modulator of gonadal tumor progression and the activin-induced wasting syndrome in inhibin-deficient mice. Endocrinology 2000; 141(7):2319–2327.
44. Pierson TM, Wang Y, DeMayo FJ, Matzuk MM, Tsai SY, Omalley BW. Regulable expression of inhibin A in wild-type and inhibin alpha null mice. Mol Endocrinol 2000;14(7):1075–1085.
45. Kumar TR, Wang Y, Matzuk MM. Gonadotropins are essential modifier factors for gonadal tumor development in inhibin-deficient mice. Endocrinology 1996;137(10):4210–4216.
46. Kananen K, Markkula M, Rainio E, Su JG, Hsueh AJ, Huhtaniemi IT. Gonadal tumorigenesis in transgenic mice bearing the mouse inhibin alpha-subunit promoter/simian virus T-antigen fusion gene: characterization of ovarian tumors and establishment of gonadotropin- responsive granulosa cell lines. Mol Endocrinol 1995;9(5):616–627.
47. Kananen K, Rilianawati, Paukku T, Markkula M, Rainio EM, Huhtanemi I. Suppression of gonadotropins inhibits gonadal tumorigenesis in mice transgenic for the mouse inhibin alpha-subunit promoter/simian virus 40 T-antigen fusion gene. Endocrinology 1997;138(8):3521–3531.
48. Kumar TR, Palapattu G, Wang P, et al. Transgenic models to study gonadotropin function: the role of follicle- stimulating hormone in gonadal growth and tumorigenesis. Mol Endocrinol 1999;13(6): 851–865.
49. Shou W, Woodruff TK, Matzuk MM. Role of androgens in testicular tumor development in inhibin-deficient mice. Endocrinology 1997;138(11):5000–5005.

50. Matzuk MM, Finegold MJ, Mishina Y, Bradley A, Behringer RR. Synergistic effects of inhibins and mullerian-inhibiting substance on testicular tumorigenesis. Mol Endocrinol 1995;9(10):1337–1345.
51. Hsueh AJ, Dahl KD, Vaughan J, et al. Heterodimers and homodimers of inhibin subunits have different paracrine action in the modulation of luteinizing hormone-stimulated androgen biosynthesis. Proc Natl Acad Sci USA 1987;84(14):5082–5086.
52. Steller MD, Shaw TJ, Vanderhyden BC, Ethier JF. Inhibin resistance is associated with aggressive tumorigenicity of ovarian cancer cells. Mol Cancer Res 2005;3(1):50–61.
53. Cho SH, Yao Z, Wang SW, et al. Regulation of activin A expression in mast cells and asthma: its effect on the proliferation of human airway smooth muscle cells. J Immunol 2003;170(8):4045–4052.
54. Pawlowski JE, Taylor DS, Valentine M, et al. Stimulation of activin A expression in rat aortic smooth muscle cells by thrombin and angiotensin II correlates with neointimal formation in vivo. J Clin Invest 1997;100(3):639–648.
55. Antin PB, Yatskievych T, Dominguez JL, Chieffi P. Regulation of avian precardiac mesoderm development by insulin and insulin-like growth factors. J Cell Physiol 1996;168(1):42–50.
56. Jung B, Doctolero RT, Tajima A, et al. Loss of activin receptor type 2 protein expression in microsatellite unstable colon cancers. Gastroenterology 2004;126(3):654–659.
57. Jeruss JS, Sturgis CD, Rademaker AW, Woodruff TK. Down-regulation of activin, activin receptors, and Smads in high-grade breast cancer. Cancer Res 2003;63(13):3783–3790.
58. McPherson SJ, Thomas TZ, Wang H, Gurusinghe CJ, Risbridger GP. Growth inhibitory response to activin A and B by human prostate tumour cell lines, LNCaP and DU145. J Endocrinol 1997; 154(3):535–545.
59. Danila DC, Inder WJ, Zhang X, et al. Activin effects on neoplastic proliferation of human pituitary tumors. J Clin Endocrinol Metab 2000;85(3):1009–1015.
60. Panopoulou E, Murphy C, Rasmussen H, Bagli E, Rofstad EK, Fotsis T. Activin A suppresses neuroblastoma xenograft tumor growth via antimitotic and antiangiogenic mechanisms. Cancer Res 2005;65(5):1877–1886.
61. Wagner K, Peters M, Scholz A, et al. Activin A stimulates vascular endothelial growth factor gene transcription in human hepatocellular carcinoma cells. Gastroenterology 2004;126(7):1828–1843.
62. Breit S, Rossler J, Fotsis T, Schweigerer L. N-myc down-regulates activin A. Biochem Biophys Res Commun 2000;274(2):405–409.
63. Schramm A, von Schuetz V, Christiansen H, et al. High activin A-expression in human neuroblastoma: suppression of malignant potential and correlation with favourable clinical outcome. Oncogene 2005;24(4):680–687.
64. Rossmanith W, Chabicovsky M, Grasl-Kraupp B, Peter B, Schausberger E, Schulte-Hermann R. Follistatin overexpression in rodent liver tumors: a possible mechanism to overcome activin growth control. Mol Carcinog 2002;35(1):1–5.
65. Alexander JM, Bikkal HA, Zervas NT, Laws ER, Jr., Klibanski A. Tumor-specific expression and alternate splicing of messenger ribonucleic acid encoding activin/transforming growth factor-beta receptors in human pituitary adenomas. J Clin Endocrinol Metab 1996;81(2):783–790.
66. Thomas TZ, Wang H, Niclasen P, et al. Expression and localization of activin subunits and follistatins in tissues from men with high grade prostate cancer. J Clin Endocrinol Metab 1997;82(11): 3851–3858.
67. Burdette JE, Jeruss JS, Kurley SJ, Lee EJ, Woodruff TK. Activin A mediates growth inhibition and cell cycle arrest through Smads in human breast cancer cells. Cancer Res 2005;65(17):7968–7975.
68. Danila DC, Zhang X, Zhou Y, Haidar JN, Klibanski A. Overexpression of wild-type activin receptor alk4-1 restores activin antiproliferative effects in human pituitary tumor cells. J Clin Endocrinol Metab 2002;87(10):4741–4746.
69. Di Simone N, Schneyer AL, Caliandro D, Castellani R, Caruso A. Regulation of endometrial adenocarcinoma cell proliferation by Activin-A and its modulation by 17beta-estradiol. Mol Cell Endocrinol 2002;192(1–2):187–195.
70. Nishihara T, Okahashi N, Ueda N. Activin A induces apoptotic cell death. Biochem Biophys Res Commun 1993;197(2):985–991.
71. Wang QF, Tilly KI, Tilly JL, et al. Activin inhibits basal and androgen-stimulated proliferation and induces apoptosis in the human prostatic cancer cell line, LNCaP. Endocrinology 1996;137(12): 5476–5483.
72. Schwall RH, Robbins K, Jardieu P, Chang L, Lai C, Terrell TG. Activin induces cell death in hepatocytes in vivo and in vitro. Hepatology 1993;18(2):347–356.

73. Dalkin AC, Gilrain JT, Bradshaw D, Myers CE. Activin inhibition of prostate cancer cell growth: selective actions on androgen-responsive LNCaP cells. Endocrinology 1996;137(12):5230–5235.
74. Russell CE, Hedger MP, Brauman JN, de Kretser DM, Phillips DJ. Activin A regulates growth and acute phase proteins in the human liver cell line, HepG2. Mol Cell Endocrinol 1999;148(1–2): 129–136.
75. Fujii Y, Kawakami S, Okada Y, Kageyama Y, Kihara K. Regulation of prostate-specific antigen by activin A in prostate cancer LNCaP cells. Am J Physiol Endocrinol Metab 2004;286(6):E927–E931.
76. Planz B, Oltean H, Deix T, et al. Effect of keratinocyte growth factor and activin on cell growth in the human prostatic cancer cell line LNCaP. World J Urol 2004;22(2):140–144.
77. Kojima I, Mogami H, Kawamura N, Yasuda H, Shibata H. Modulation of growth of vascular smooth muscle cells by activin A. Exp Cell Res 1993;206(1):152–156.
78. Spencer SJ, Rabinovici J, Jaffe RB. Human recombinant activin-A inhibits proliferation of human fetal adrenal cells in vitro. J Clin Endocrinol Metab 1990;71(6):1678–1680.
79. Kojima I, Mogami H, Kawamura N, Shibata H. Activin A inhibits cell-cycle progression induced by insulin-like growth factor-I in Balb/c 3T3 cells. J Endocrinol 1993;137(1):99–105.
80. Sternberg D, Honigwachs-sha'anani J, Brosh N, Malik Z, Burstein Y, Zipori D. Restrictin-P/stromal activin A, kills its target cells via an apoptotic mechanism. Growth Factors 1995;12(4):277–287.
81. Sehy DW, Shao LE, Yu AL, Tsai WM, Yu J. Activin A-induced differentiation in K562 cells is associated with a transient hypophosphorylation of RB protein and the concomitant block of cell cycle at G1 phase. J Cell Biochem 1992;50(3):255–265.
82. Chen YG, Lui HM, Lin SL, Lee JM, Ying SY. Regulation of cell proliferation, apoptosis, and carcinogenesis by activin. Exp Biol Med (Maywood) 2002;227(2):75–87.
83. Cocolakis E, Lemay S, Ali S, Lebrun JJ. The p38 MAPK pathway is required for cell growth inhibition of human breast cancer cells in response to activin. J Biol Chem 2001;276(21):18,430–18,436.
84. Zauberman A, Oren M, Zipori D. Involvement of p21(WAF1/Cip1), CDK4 and Rb in activin A mediated signaling leading to hepatoma cell growth inhibition. Oncogene 1997;15(14):1705–1711.
85. van Schaik RH, Wierikx CD, Timmerman MA, et al. Variations in activin receptor, inhibin/activin subunit and follistatin mRNAs in human prostate tumour tissues. Br J Cancer 2000;82(1):112–117.
86. McPherson SJ, Mellor SL, Wang H, Evans LW, Groome NP, Risbridger GP. Expression of activin A and follistatin core proteins by human prostate tumor cell lines. Endocrinology 1999;140(11):5303–5309.
87. Chen CR, Kang Y, Massagué J. Defective repression of c-myc in breast cancer cells: A loss at the core of the transforming growth factor beta growth arrest program. Proc Natl Acad Sci USA 2001; 98(3):992–999.
88. Poulaki V, Mitsiades N, Kruse FE, et al. Activin a in the regulation of corneal neovascularization and vascular endothelial growth factor expression. Am J Pathol 2004;164(4):1293–1302.
89. McCarthy SA, Bicknell R. Inhibition of vascular endothelial cell growth by activin-A. J Biol Chem 1993;268(31):23,066–23,071.
90. Kozian DH, Ziche M, Augustin HG. The activin-binding protein follistatin regulates autocrine endothelial cell activity and induces angiogenesis. Lab Invest 1997;76(2):267–276.
91. Yoshinaga K, Mimori K, Yamashita K, Utsunomiya T, Inoue H, Mori M. Clinical significance of the expression of activin A in esophageal carcinoma. Int J Oncol 2003;22(1):75–80.
92. Yoshinaga K, Inoue H, Utsunomiya T, et al. N-cadherin is regulated by activin A and associated with tumor aggressiveness in esophageal carcinoma. Clin Cancer Res 2004;10(17):5702–5707.
93. Reis FM, Cobellis L, Tameirao LC, et al. Serum and tissue expression of activin a in postmenopausal women with breast cancer. J Clin Endocrinol Metab 2002;87(5):2277–2282.
94. Reinholz MM, Iturria SJ, Ingle JN, Roche PC. Differential gene expression of TGF-beta family members and osteopontin in breast tumor tissue: analysis by real-time quantitative PCR. Breast Cancer Res Treat 2002;74(3):255–269.
95. Mylonas I, Jeschke U, Shabani N, Kuhn C, Friese K, Gerber B. Inhibin/activin subunits (inhibin-alpha, -betaA and -betaB) are differentially expressed in human breast cancer and their metastasis. Oncol Rep 2005;13(1):81–88.
96. Jiang TX, Yi JR, Ying SY, Chuong CM. Activin enhances chondrogenesis of limb bud cells: stimulation of precartilaginous mesenchymal condensations and expression of NCAM. Dev Biol 1993; 155(2):545–557.
97. Hashimoto Y, Shindo-Okada N, Tani M, Takeuchi K, Toma H, Yokota J. Identification of genes differentially expressed in association with metastatic potential of K-1735 murine melanoma by messenger RNA differential display. Cancer Res 1996;56(22):5266–5271.

98. Hedger MP, Clarke L. Isolation of rat blood lymphocytes using a two-step Percoll density gradient. Effect of activin (erythroid differentiation factor) on peripheral T lymphocyte proliferation in vitro. J Immunol Methods 1993;163(1):133–136.
99. Hedger MP, Drummond AE, Robertson DM, Risbridger GP, de Kretser DM. Inhibin and activin regulate [3H]thymidine uptake by rat thymocytes and 3T3 cells in vitro. Mol Cell Endocrinol 1989;61(1):133–138.
100. Brosh N, Sternberg D, Honigwachs-Sha'anani J, et al. The plasmacytoma growth inhibitor restrictin-P is an antagonist of interleukin 6 and interleukin 11. Identification as a stroma-derived activin A. J Biol Chem 1995;270(49):29,594–29,600.
101. Nishihara T, Ohsaki Y, Ueda N, Koseki T, Eto Y. Induction of apoptosis in B lineage cells by activin A derived from macrophages. J Interferon Cytokine Res 1995;15(6):509–516.
102. McClure L, O'Connor AE, Hayward S, Jenkin G, Walker DW, Phillips DJ. Effects of age and pregnancy on the circulatory activin response of sheep to acute inflammatory challenge by lipopolysaccharide. J Endocrinol 2005;185(1):139–149.
103. Hubner G, Brauchle M, Gregor M, Werner S. Activin A: a novel player and inflammatory marker in inflammatory bowel disease? Lab Invest 1997;77(4):311–318.
104. Yu EW, Dolter KE, Shao LE, Yu J. Suppression of IL-6 biological activities by activin A and implications for inflammatory arthropathies. Clin Exp Immunol 1998;112(1):126–132.
105. Michel U, Gerber J, E O'Connor A, Bunkowski S, Bruck W, Nau R, Phillips D. Increased activin levels in cerebrospinal fluid of rabbits with bacterial meningitis are associated with activation of microglia. J Neurochem 2003;86(1):238–245.
106. Keelan JA, Zhou RL, Mitchell MD. Activin A exerts both pro- and anti-inflammatory effects on human term gestational tissues. Placenta 2000;21(1):38–43.
107. Healy DL, Mamers P, Bangah M, Burger HG. Clinical and pathophysiological aspects of inhibin. Hum Reprod 1993;8 Suppl 2:138–140.
108. Healy DL, Burger HG, Mamers P, et al. Elevated serum inhibin concentrations in postmenopausal women with ovarian tumors. N Engl J Med 1993;329(21):1539–1542.
109. Robertson DM, Stephenson T, Pruysers E, et al. Characterization of inhibin forms and their measurement by an inhibin alpha-subunit ELISA in serum from postmenopausal women with ovarian cancer. J Clin Endocrinol Metab 2002;87(2):816–824.
110. Robertson DM, Stephenson T, Pruysers E, et al. Inhibins/activins as diagnostic markers for ovarian cancer. Mol Cell Endocrinol 2002;191(1):97–103.
111. Cooke I, O'Brien M, Charnock FM, Groome N, Ganesan TS. Inhibin as a marker for ovarian cancer. Br J Cancer 1995;71(5):1046–1050.
112. Ala-Fossi SL, Aine R, Punnonen R, Maenpaa J. Is potential to produce inhibins related to prognosis in ovarian granulosa cell tumors? Eur J Gynaecol Oncol 2000;21(2):187–189.
113. Ala-Fossi SL, Maenpaa J, Blauer M, Tuohimaa P, Punnonen R. Inhibin A, B and pro-alphaC in serum and peritoneal fluid in postmenopausal patients with ovarian tumors. Eur J Endocrinol 2000;142(4): 334–339.
114. Balanathan P, Ball EMA, Wang H, Harris SE, Shelling AN, Risbridger GP. Epigenetic regulation of inhibin a-subunit gene in prostate cancer cell lines. J Mol Endocrinol 2004;32(1): 55–67.
115. Schmitt JF, Millar DS, Pedersen JS, et al. Hypermethylation of the inhibin alpha-subunit gene in prostate carcinoma. Mol Endocrinol 2002;16(2):213–220.
116. Risbridger GP, Shibata A, Ferguson KL, Stamey TA, McNeal JE, Peehl DM. Elevated expression of inhibin alpha in prostate cancer. J Urol 2004;171(1):192–196.
117. Risbridger GP, Ball EM, Wang H, Mellor SL, Peehl DM. Re-evaluation of inhibin alpha subunit as a tumour suppressor in prostate cancer. Mol Cell Endocrinol 2004;225:73–76.
118. Wakefield LM, Roberts AB. TGF-beta signaling: positive and negative effects on tumorigenesis. Curr Opin Genet Dev 2002;12(1):22–29.
119. Mellor SL, Richards MG, Pedersen JS, Robertson DM, Risbridger GP. Loss of the expression and localization of inhibin alpha-subunit in high grade prostate cancer. J Clin Endocrinol Metab 1998;83(3):969–975.

27 Bone Morphogenetic Proteins/Growth Differentiation Factors and Smad Activation in Ovarian Granulosa Cells and Carcinoma

Sabine Mazerbourg

CONTENTS

LIGANDS BMPS AND GDFS IN THE OVARY: EXPRESSION PATTERN AND FUNCTION
BMP/GDF SIGNALING PATHWAYS IN GRANULOSA CELLS
BMP/GDF IN OVARIAN CARCINOMA
CONCLUSIVE REMARKS
ACKNOWLEDGMENTS
REFERENCES

Abstract

The ligands of the transforming growth factor β (TGF-β) superfamily are important paracrine/autocrine regulators of the ovarian follicular development. This group includes TGF-β, activins, bone morphogenetic proteins (BMPs), growth differentiation factors (GDFs), and others. Nine BMPs and GDFs are expressed in the ovary, with a well-conserved pattern of expression among mammalian species. Oocytes express BMP6, BMP15, and GDF9; theca cells express BMP3, BMP3b, BMP4, and BMP7; and granulosa cells express BMP2 and BMP5. In vivo studies have revealed the importance of GDF9 and BMP15 in the control of early follicular growth and ovulation. In vitro, most of the BMPs and GDFs induce granulosa cell proliferation but reduce follicle stimulating hormone (FSH)-induced progesterone production, thus delaying the differentiation process.

The general mechanism of the signaling pathway for TGF-β family members has been well described. The ligands interact with type II and type I serine/threonine kinase receptors leading to the activation of two sets of downstream Smad-transcriptional factors. Owing to their common evolutionary origin, more than thirty related members of the TGF-β superfamily likely interact with seven type I (ALK) and five type II receptors, thereby activating Smad proteins. Instead of purifying the GDF9 receptors from granulosa cells for their identification, we hypothesized that GDF9 activates the known serine/threonine kinase receptors. We demonstrated the ability of the BMP receptor type II (BMPRII) ectodomain to block GDF9 stimulation of rat granulosa cell proliferation. In a GDF9-nonresponsive cell line, overexpression of the TGF-β type I receptor, ALK5, but not any other six type I receptors, conferred GDF9 responsiveness. The predicted role of BMPRII and ALK5 as type II and type I receptors, respectively, for GDF9 was validated in granulosa cells following the suppression of endogenous BMPRII and ALK5 expression using gene "knockdown" approaches. In parallel, Moore et al. have shown that BMP15, the GDF9 paralog, is likely to signal in granulosa

From: *Cancer Drug Discovery and Development: Transforming Growth Factor-β in Cancer Therapy, Vol I: Basic and Clinical Biology*
Edited by: Sonia B. Jakowlew © Humana Press Inc., Totowa, NJ

cells by binding to BMPRII and ALK6. Less is known about the receptors involved in the signaling of the other ovarian BMPs and GDFs in granulosa cells. Ligands of the TGF-β superfamily occupy a central position in the signaling networks that control the growth and differentiation of cells. However, in the ovary, data are limited on the involvement and the disruption of BMPs and GDFs pathways in granulosa and epithelium cell tumor formation.

Identification of GDF9 receptors in granulosa cells based on the coevolution of ligands and receptors has proved that evolutionary tracing of polypeptide ligands, receptors, and downstream signaling molecules in their respective subgenomes verifies a new paradigm for hormone research. These studies suggested that more than thirty ligands of the TGF-β family interact with a limited number of receptors in a combinatorial manner to activate two downstream Smad pathways. In granulosa cells and ovarian cancer cells, identification of the components of the Smad-dependent signaling pathway for the nine ovarian BMPs and GDFs and its multilevel regulations should be pursued. These studies will be essential for identification of potential alterations in the BMPs and GDFs signaling leading to carcinogenesis.

Key Words: TGF-β; BMP; GDF; Smad; ovary; granulosa cell; TGF-β receptors; signaling pathway; ovarian carcinoma.

Ovarian folliculogenesis involves the development of follicles through primordial, primary, and preantral stages before acquiring an antral cavity. Once selected, antral follicles continue to grow and acquire specific functional characteristics that permit differentiation to the preovulatory stage. Increased serum estradiol triggers the surge of luteinizing hormone (LH) that, in turn, stimulates preovulatory follicle(s) to ovulate and luteinize thus forming the corpus luteum. The process of follicle development is characterized by the proliferation and differentiation of the theca and granulosa cells. It is controlled by pituitary gonadotrophins, follicle stimulating hormone (FSH) and LH, and intraovarian factors *(1,2)*. It is now evident that TGF-β-related ligands, bone morphogenetic proteins (BMPs), and growth differentiation factors (GDFs), regulate the granulosa cell functions by paracrine and/or autocrine mechanisms.

After a presentation on the role of the BMPs and GDFs in the ovary, this review will describe their signaling pathways in granulosa cells. The identification of the receptors and Smads for GDF9 will be emphasized. Because disruption in TGF-β signaling has been implicated in cancer development, our final section will report on the limited data for the potential role of the BMPs and GDFs and their signaling pathways in ovarian cell carcinogenesis.

1. LIGANDS BMPs AND GDFs IN THE OVARY: EXPRESSION PATTERN AND FUNCTION

1.1. General Presentation of the TGF-β Superfamily

BMPs and GDFs are members of the transforming growth factor β (TGF-β) superfamily with thirty three ligands including TGF-β proteins, activins/inhibins, antimullerian hormone (AMH), and others (Fig. 1) *(3,4)*. The nomenclature for BMP and GDF is not referring to two different subgroups of ligands. Because some ligands are named as both BMP and GDF (BMP15 = GDF9b; BMP3b = GDF10), one can refer to this group as BMP/GDF. The ligands are synthesized as large precursor molecules that are cleaved proteolytically, by members of the subtilin-like proprotein convertase family and the BMP1/Tolloid-like proteinase, to release a C-terminal peptide of 110–140 amino acids *(5–7)*. Despite the low degree of sequence similarity (~35%), the C-terminal peptide shares a conserved structure of six cysteines arrangement known as cystine knot conformation *(8)*, thereby allowing the formation of a common structural scaffold *(9,10)*. In most cases, an extra cysteine is engaged in an intermolecular disulfide bond that is necessary for the assembly of biologically active dimers. GDF9, BMP15, GDF3, Lefty1 and Lefty2 are missing the extracysteine and are likely noncovalently

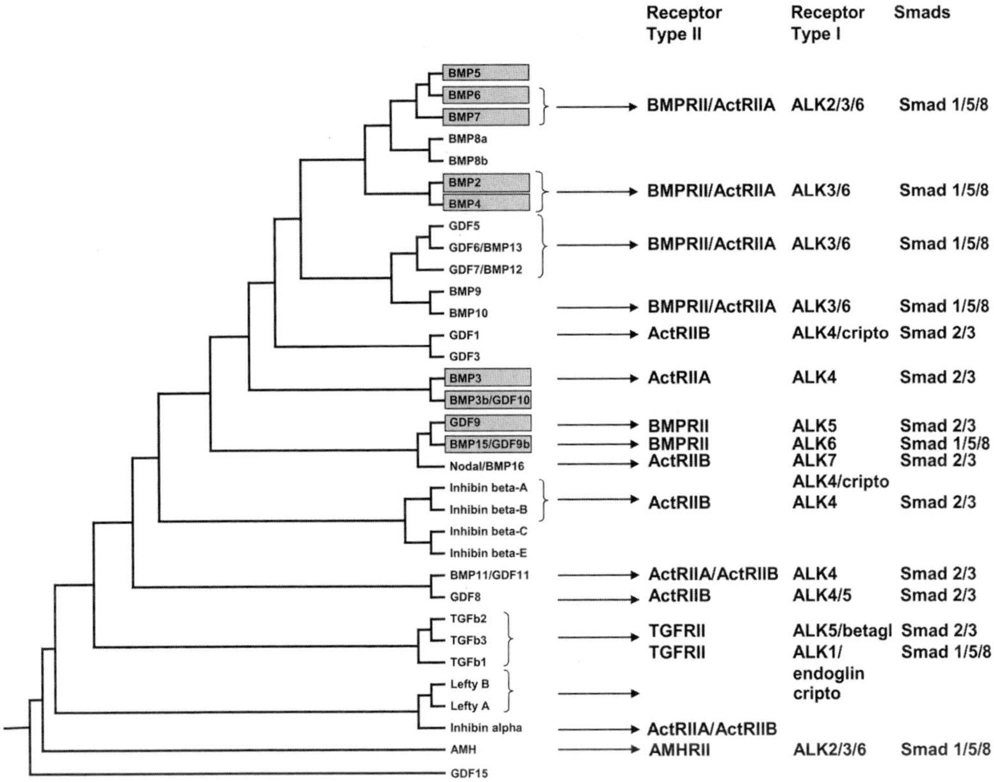

Fig. 1. Phylogenetic relationship of paralogous TGF-β/BMP/GDF ligands, as well as characterized receptors and signaling pathways for individual ligands. The alignment of thirty three TGF-β-related ligands was performed using the C-terminal region containing the cystine-knot structure, starting from the first invariant cysteine residue. Taken from published literature (for references, see Mazerbourg et al., 2005 *(4)*) is a list of the type II and type I receptors as well as the intracellular signaling Smad proteins for individual ligands. The ligands expressed in the ovary are highlighted. Betag: betaglycan. Reproduced with permission from Mazerbourg et al., J Biol Chem 2005;280: 32,122–32,132 *(4)*.

associated. Although the structural scaffold of different monomers is conserved, comparative analyses of the crystal structure of different dimers showed that the conformational arrangement differs between TGF-βs, activin, and BMPs *(9,10)*. TGF-β2, TGF-β3, BMP2, and BMP7 dimers present an extended symmetric arrangement described as an "open form", whereas activin dimers are described as a "closed form" with a compact folded-back conformation *(10)*. Specific conformation of the dimers that are associated with a distinct pattern of surface charge and hydrophobicity suggests a particular binding interaction of the ligands with the receptors.

1.2. BMP/GDF Expression Pattern in the Ovary

Among the thirty three ligands of the TGF-β superfamily, nine BMP/GDF factors are expressed in the mammalian ovary in temporally and spatially restricted patterns (Table 1A). They show a well-conserved pattern of expression among the different species studied (rat, mouse, sheep, cattle, and human). Oocytes from early follicular stages express BMP6, BMP15, and GDF9. Theca cells are the major site of expression for BMP4, BMP7, and BMP3b whereas granulosa cells mainly express BMP2 and BMP5. Only BMP3 appears to

Table 1
Cellular Localization of the BMP/GDF Ligands (A), Type II Receptors (B), Type I Receptors (C), and Smads (D) in the Ovary

	Species	*Cell type*	*Follicular stage*	*Reference*
A-Ligand				
BMP2	Rat	GC	Primary	*(76)*
		T	Secondary	
BMP3	Rat	T	Preovulatory	*(76)*
	Human	GC	Preovulatory	*(136)*
BMP3b	Rat	T	Preovulatory	*(76)*
		OSE		
BMP4	Rat	T	Primary	*(76)*
		OSE		
	Cattle	T		*(31)*
BMP5	Rat	GC, O	Early antral	*(25)*
BMP6	Rat	O, GC	Secondary	*(76)*
		OSE		
	Mouse	O, GC	Primary	*(137,138)*
	Sheep	O	Primordial	*(139)*
	Cattle	O, GC		*(31)*
BMP7	Rat	T	Secondary	*(76)*
	Cattle	T		*(31)*
BMP15	Rat	O	Primary	
	Mouse	O	Primary	
	Human	O	Primary	
	Sheep	O	Primary	*(16,76,140–144)*
	Goat	O	Primary	
	Cattle	O	Primary	
	Human	O	Primary	
GDF9	Rat	O	Primary	*(142,143,145–150)*
	Mouse	O	Primary	
	Human	O, Cm	Primary	
	Sheep	O	Primordial	
	Goat	O	Primordial	
	Cattle	O	Primordial	
	Pig	O, GC		
B-Type II Receptors				
BMPRII	Rat	O, GC	Primary	*(76)*
	Sheep	O, GC	Primary	*(32,89)*
		TC	Antral	
		OSE		
	Cattle	O, GC, TC		*(31)*
	Human	GCL		*(151)*
ActRIIA	Mouse	O		*(150,152)*
	Rat	O	Primordial	*(153–155)*
		GC,TC	Primary	
		OSE		
	Sheep	GC	Antral	*(139)*
	Cattle	O, GC, TC		*(31)*
	Human	GC	Antral	*(156)*
		O, Cm		*(150)*
		GCL		*(157)*

(*Continued*)

Table 1 (*Continued*)

	Species	*Cell type*	*Follicular stage*	*Reference*
ActRIIB	Mouse	O		*(150)*
	Rat	O	Primordial to primary	*(155)*
		GC	Primary	
	Sheep	O	Primary to antral	*(139)*
		GC		
		T	Small preantral	
	Cattle	O, GC, T		*(31)*
	Human	GC	Antral	*(156)*
		O, Cm		*(150)*
		GCL		*(157)*
TGFRII	Human	O	Primordial to primary	*(77)*
		O	Post-hCG	*(158)*
		O		*(75)*
		Pregranulosa		
C-Type I Receptors				
ALK2	Mouse	O		*(150)*
	Rat	O	Primordial	*(155)*
		GC	Secondary	*(159)*
	Human	O, Cm		*(150)*
		GCL		*(157)*
	Cattle	O		*(31)*
		GC		
		TC		
ALK3	Rat	O, GC	Primordial	*(76)*
		TC	Preantral	
	Sheep	O, GC	Primary	*(32)*
		TC	Preantral	
		OSE		
	Human	GCL		*(151)*
ALK4	Mouse	O		*(150)*
	Rat	O	Primordial	*(155)*
		GC	Primordial	
	Human	O, Cm		*(150)*
		GCL		*(157)*
ALK5	Human	O		*(75,158)*
		Pregranulosa		
		O, GC	Primordial	*(77)*
ALK6	Mouse	O, GC	Antral	*(93)*
	Rat	O, GC	Primordial	*(76)*
		TC	Preantral	
	Sheep	O, GC	Primary	*(32,89)*
		TC	Antral	
		OSE		
	Cattle	O, GC, T		*(31)*
D-Smads				
Smad1	Cattle	GC		*(31)*
	Human	GCL		*(151)*

(*Continued*)

Table 1 (*Continued*)

	Species	Cell type	Follicular stage	Reference
Smad2	Rat	O	Primordial	*(155)*
		GC	Primary	
		T	Secondary	
		O	Primordial	*(78)*
		GC, T	Preantral	
	Cattle	GC		*(31)*
	Human	O		*(151,158)*
		GCL		
		GC, T	Preantral	*(156)*
Smad3	Mouse	GC		*(160)*
	Rat	O	Primordial	*(78)*
		GC	Preantral	
	Human	O		*(151,158)*
		GCL		
Smad4	Rat	O	Primordial	*(155)*
		GC	Primary	
		T	Secondary	
	Human	GC, T	Preantral	*(151,156)*
		GCL		
Smad5	Rat	O	Primordial	*(155)*
	Human	GCL		*(151)*
Smad8	Rat	O	Primordial	*(155)*

The tables summarize the mRNA or protein expression sites in the ovary of different species. When expression had been followed over different follicular stages (primordial to preovulatory), only the first follicular stage at which the expression was detected is mentioned.

Abbreviation: O, oocyte; GC, granulosa cell; GCL, luteinized granulosa cell; T, theca cell; Cm, cumulus; OSE, ovarian superficial epithelium.

present a divergent cellular localization among species. BMP4, BMP3, and BMP6 expression also is detected in rat ovarian surface epithelium (OSE). Overall, most of the BMP/GDF production is confined to oocytes and theca cells. These factors are likely to influence the granulosa cell function in a paracrine manner.

1.3. BMP/GDF Function in the Ovary

1.3.1. In vivo Studies

Homozygous GDF9 knockout mice and homozygous GDF9 mutant sheep (allele *Fec*G) are infertile, revealing the important role of this oocyte-derived factor in the stimulation of the primary-preantral follicle transition *(11–13)*. Furthermore, Vitt et al. have shown that recombinant GDF9 is able to stimulate initial follicle recruitment in vivo in rat *(14)*. Thus, GDF9 treatment specifically increases the number of primary and small preantral follicles *(14)* in contrast to FSH that mainly stimulates preantral follicular growth *(15)*. Similarly to mutation in GDF9 gene, female sheep carrying homozygous mutations in the oocyte-derived BMP15 gene (allele *Fec*X) are infertile with a blockage of folliculogenesis at the primary stage *(12,16)*. In contrast, fertility of heterozygous mutant BMP15 or heterozygous mutant GDF9 sheep increases with a higher ovulation rate *(12,13,16,17)*. Thus, the absence or very low concentration of GDF9 or BMP15 leads to an inhibition of ovarian follicular development, whereas a partial reduction in concentration increases ovulation rate and litter size. Indeed, different degrees of immunization of sheep against BMP15 and GDF9 to monitor their levels can increase or decrease fertility *(18)*. The underlying cause of the higher ovulation rate appears to be the precocious maturation of small follicles becoming respondent to an LH surge *(17)*. The importance of BMP15 in ovarian function has been confirmed in humans. Females with a heterozygous mutation of the BMP15 gene are infertile owing to a hypergonadotropic ovarian failure *(19)*. In contrast, mice lacking a functional BMP15 gene exhibit only subfertility with impaired differentiation of the cumulus cells *(20,21)*. Nevertheless, as in sheep where both GDF9 and BMP15 dosage is essential, in vivo data support the importance of both factors and their respective dosages in mice. BMP15-deficient mice with one copy of GDF9 show a greater fertility defect than the BMP15-deficient mice with both GDF9 copies. This suggests a synergistic role for GDF9 and BMP15 in the development and function of the mouse cumulus-oocyte complex *(20)*. The discrepancies observed between the reproductive phenotypes described support the notion that the relative importance of the two factors is likely to be different between monoovulatory and polyovulatory species. Furthermore, these phenotypic data raise the question of the mechanism of action of GDF9 and BMP15 as monomers, dimers, or both.

BMP7 is another factor able to promote early folliculogenesis. In rat, in vivo injection of BMP7 increases the numbers of primordial, preantral, and antral follicles and decreases the number of primordial follicles *(22)*. In contrast, BMP7 inhibits progesterone production and decreased the ovulation rate *(22)*. These data suggest that BMP7 promotes follicular growth while preventing luteinization. The importance of BMP7 in early follicular development has not been confirmed in vivo as BMP7 null mice are not viable *(23,24)*. Despite the expression of BMP5 and BMP6 in the mouse ovary, *(25,26)*, mice exhibiting a natural or experimental loss of function mutation in BMP5 and BMP6 genes, respectively, are viable and have no apparent ovarian abnormalities *(26,27)*. This suggests that, in contrast to GDF9, these BMPs are dispensable for normal ovarian function, their absence being likely compensated by other members of the BMP/GDF subfamily. Alternatively, one could hypothesize that invalidation of one of these genes leads to a minor alteration of ovarian function, without clear consequences on fertility.

1.3.2. In vitro Studies

1.3.2.1. Granulosa Cell Proliferation and Differentiation. The effects of BMP/GDF ligands on granulosa cell proliferation have been tested in vitro (Table 2). Using rat primary granulosa cell cultures, recombinant BMP5, BMP7, BMP15, and GDF9 have been shown to stimulate cell proliferation *(22,25,28,29)*. In contrast, BMP2, BMP4, and BMP6 were ineffective in promoting the division of granulosa cells from sheep, cattle, and rat ovaries respectively *(30–32)*.

Several studies report the action of BMP/GDFs on the regulation of granulosa cell steroidogenesis (Table 2). In general, BMP/GDFs have limited effect on basal secretion of progesterone and estradiol by granulosa cells. However, recombinant BMP4, BMP5, BMP6, BMP7, BMP15, and GDF9 strongly inhibit FSH-induced progesterone secretion by rat or sheep granulosa cells in culture *(22,28–30,32–34)*. In contrast, BMP2, BMP4, and BMP7 enhanced FSH-dependent estradiol production whereas BMP6 and BMP15 were without effect *(28–30,32,33,35)*. GDF9 was the only factor to downregulate the FSH-induced estradiol production *(29)*. BMP/GDFs also have been shown to modulate IGF-I-induced steroidogenesis *(25,31)*.

At the molecular level, the BMP/GDFs selectively downregulate expression of the FSH receptor, the LH receptor, and inhibin/activin subunits by granulosa cells (Table 2). Furthermore, they modulate FSH-induced steroidogenic enzyme gene expression (Table 2). The downregulation of the genes encoding the key enzymes steroidogenic acute regulatory (StAR) and/or p450scc (p450 side-chain cleavage) is likely to induce a decrease in

Table 2
In Vitro Effects of BMP/GDFs on Proliferation and Differentiation of Granulosa Cells in Rodent, Ruminant, and Human

			Secretion								
		Prolif	*P4*			*E2*			*Inhibin/Activin*		
Ligand	*GC species*	*Basal*	*Basal*	*FSH*	*IGF-I*	*Basal*	*FSH*	*IGF-I*	*Basal*	*FSH-induced gene mRNA expression*	*Ref.*
BMP2	Sheep	⇔					↗			Inhibin βA ↗	*(32)*
	Human									Inhibin βB ↗ Inhibin α,βA ⇔	*(151)*
BMP4	Rat			↘			↗				*(33)*
	Sheep		↘	↘						FSHR ⇔ StAR ↘ P450scc ↘ 3βHSD ⇔	*(88)*
	Cattle	⇔	↘		↘	↗		↗	Inhibin A↗ Activin A↗		*(31)*
BMP5	Rat	↗	↘	↘	↘		⇔			StAR ↘ P450scc ⇔	*(25)*
BMP6	Rat	⇔		↘			⇔			FSHR ⇔ LHR ↘ Inhibin α,βA,βB ↘ StAR ↘ P450scc ↘ P450 arom ⇔	*(30)*
	Cattle	↗	↘		↘	↗		↗	Inhibin A↗ Activin A↗		*(31)*

BMP7	Rat	↗		↘			↗			StAR ↘ P450 arom ↗ P450scc ⇔ 3βHSD ⇔	*(22,33)*
	Cattle	↗	↘		↘	↗		↗	Inhibin A↗ Activin A↗		*(31)*
BMP15	Rat	↗		↘			⇔			FSHR ↘ LHR ↘ Inhibin α,βA,βB ↘ StAR ↘ P450scc ↘ 3βHSD ↘ Kit ligand ↗	*(28,35)*
GDF9	Rat	↗	↗	↘		↗	↘			LHR ↘ Inhibin α,βA,βB ↗	*(29,73, 146)*
	Mouse		↗							LHR ↘ StAR ↗ Kit ligand ↘	*(36,38)*
	Human		⇔	↘					Inhibin B↗	Inhibin βB ↗ Inhibin α,βA ⇔ StAR ↘ P450scc ↘ P450 arom ↘	*(74,161)*

The secretion of progesterone and estradiol by granulosa cells was observed after treatment either with the BMP/GDFs alone (basal conditions) or in combination with FSH or IGF-I. Only the effects of BMP/GDFs on FSH-induced gene expression by granulosa cells are reported here.

Abbreviation: GC, granulosa cells; Prolif, proliferation; P4, progesterone; E2 estradiol; FSHR, FSH receptor; LHR, LH receptor; StAR, steroidogenic acute regulatory protein; p450 scc, p450 side-chain cleavage; 3βHSD, 3β hydroxysteroid dehydrogenase; p450 arom, P450 aromatase; ⇔, no change ; ↗, increase; ↘, decrease.

FSH-induced progesterone production by BMP4, BMP5, BMP6, BMP7, and BMP15 *(22, 25,30, 34,35)*. Similarly, BMP7 stimulation of FSH-induced estradiol production is associated with an increase in p450 aromatase gene expression *(22)*. Although the regulation of FSH signaling by BMP4, BMP6, BMP15, and GDF9 is associated with a decrease in cAMP production *(29,30,34,35)*, the underlying mechanism is likely to be specific for each BMP/GDF factor. BMP15 suppresses FSH receptor expression *(35)* whereas BMP6 most likely downregulates the adenylate cyclase activity *(30)*. In sheep granulosa cells, the BMP4 inhibitory effect on cyclic adenosine monophosphate (cAMP) production is exerted both upstream and downstream of cAMP signaling *(34)*. In particular, BMP4 induces the inhibition of steroidogenic factor-1 (SF-1) activity, a cAMP-dependent transcription factor and a key regulator of granulosa cell differentiation *(34)*. Overall, these results suggest that BMP/GDF treatment of cultured granulosa cells from both small antral and preovulatory follicles induces cell proliferation but reduces FSH-induced differentiation in delaying the luteinization process.

1.3.2.2. Cumulus Cell Differentiation. Although BMP15 null mice show an impaired differentiation of the cumulus cells *(20)*, GDF9 appears to be the key modulator of the periovulatory response in granulosa cells. Unlike the other oocyte-expressed factors BMP15 and BMP6, recombinant GDF9 can mimic the effect of oocytes in stimulating cumulus cell differentiation and expansion in vitro *(36)*. The importance of GDF9 in cumulus expansion had been confirmed recently by using RNA interference *(37)*. The GDF9 knockdown oocytes are unable to stimulate FSH-induced cumulus cell expansion, unlike control oocytes. In mouse granulosa cell culture, recombinant GDF9 induces the expression of several genes detected specifically in cumulus cells in vivo, such as hyaluronan synthase 2, cyclooxygenase 2 *(36)*, StAR, prostaglandin endoperoxide synthase 2 (Ptgs2), prostaglandin E2 receptor *(38)*, pentraxin 3, tumor necrosis factor-induced protein 6 *(39)*, and gremlin *(40)*. Combined in vivo and in vitro studies demonstrated the importance of both GDF9 and BMP15 in the differentiation of cumulus cells during the periovulatory period in rodents.

2. BMP/GDF SIGNALING PATHWAYS IN GRANULOSA CELLS

2.1. Characteristics of the TGF-β Signaling Pathways

Before focusing on the BMP/GDF signaling pathways in granulosa cells, the characteristics of the signaling pathway of the TGF-β superfamily of ligands will be summarized.

2.1.1. TGF-β Family of Ligands Interact with a Limited Number of Serine/Threonine Kinase Receptors

The molecular signaling pathways for several ligands of the TGF-β superfamily (TGF-β, activins, BMP2, and BMP7) have been investigated intensively. Members of the TGF-β superfamily have been shown to initiate signaling by assembling serine/threonine kinase receptor complexes that activate downstream Smad transcription factors *(41)* (Fig. 2). In the human genome, there are five type II serine-threonine kinase receptors: the BMP receptor type II (BMPRII), the anti-Mullerian hormone receptor type II (AMHRII), the TGF-β receptor type II (TGFRII), and the activin receptors type II (ActRIIA and ActRIIB) (Fig. 3). In addition, there are seven-type I receptors designated as activin receptor-like kinases (ALK1 to ALK7) *(42)* (Fig. 3). Both types of receptors consist of about 500 amino acids organized into an amino-terminal extracellular ligand-binding domain with 10 or more cysteines, a transmembrane region and a carboxyl-terminal serine/threonine kinase domain. A group of membrane-anchored proteins functions as enhancers of ligand binding to the receptors. The proteoglycan betaglycan enhances TGF-β binding to the receptor complex

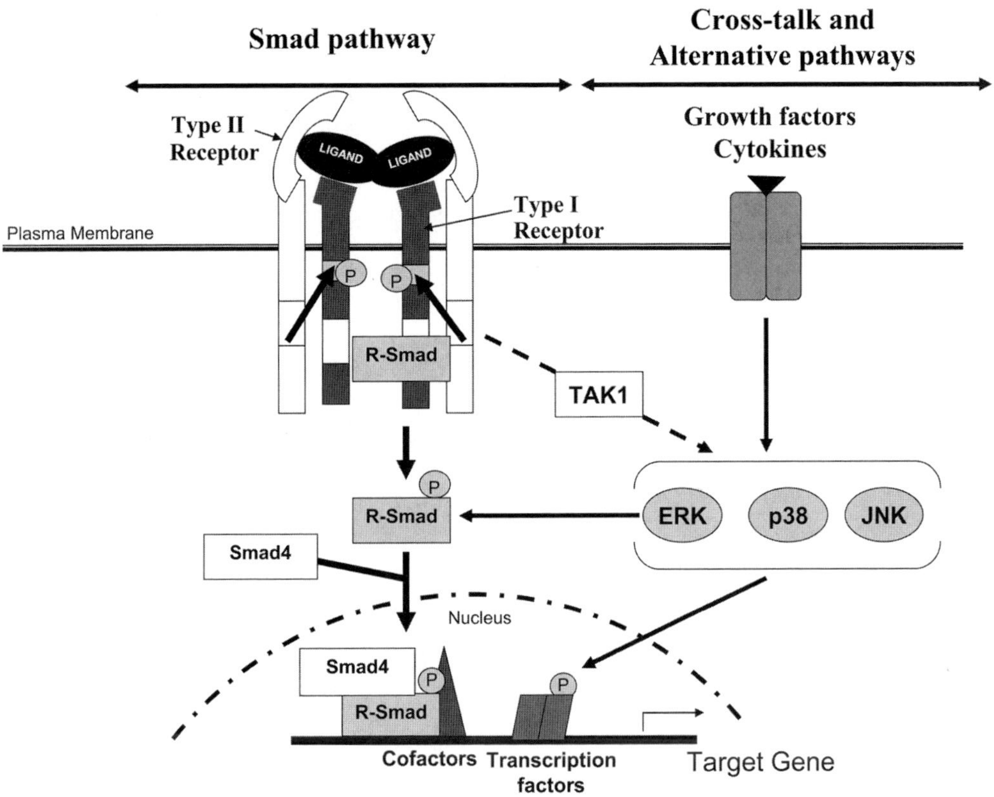

Fig. 2. Schematic representation of the signaling pathways for TGF-β family ligands. Dimeric ligands bind to two types of serine-threonine receptors named type I and type II receptors. Formation of the tetramer receptor (2-type I and 2-type II) allows phosphorylation of the type I receptor by the type II receptor on the glycine/serine rich domain resulting in activation of the type I receptor kinase. Type I receptors specifically recognize and phosphorylate R-Smad. Phosphorylated R-Smads, in turn, associate with a common Smad (Co-Smad), Smad4. The complexes of R-Smads/Smad4 translocate to the nucleus and interact with specific DNA motifs. However, effective binding to particular gene regulatory sites is enabled and modulated by diverse DNA-binding cofactors and transcription factors. TGF-β responses are cell-type specific and may depend on interactions of Smad signaling with the three principal MAPK pathways (ERK, p38, JNK). ERK can phosphorylate R-Smads and block its accumulation in the nucleus. Growth factors and cytokines may activate the MAPK pathways and enhance the activity of transcription factors that cooperate with Smads to dictate the precise response to the ligand. TGF-β and BMP/GDFs themselves also could activate the MAPK pathways, likely through the receptor- associated protein TAK1.
Abbreviation: R-Smad, receptor-Smads; MAPK, mitogen-activated protein kinase; ERK, extracellular signal-regulated kinase; JNK, Jun amino-terminal kinase; TAK1, TGFβ-activated kinase.

(43), and enables the activin antagonist, inhibin, to bind to activin receptors *(44)*. Endoglin and cripto are accessory receptors for TGF-β and nodal, respectively *(45,46)*.

The type II receptors have autophosphorylation activity *(47–49)*. After forming a complex with the ligand, the type II receptor phosphorylates the type I receptor at a glycine- and serine-rich motif (the GS domain) nearby upstream of the C-terminal kinase domain. The phosphorylation of the GS domain activates the type I receptor kinase, thus leading to the phosphorylation of downstream receptor Smad (R-Smad) proteins *(48,50)* (Fig. 2). Structural

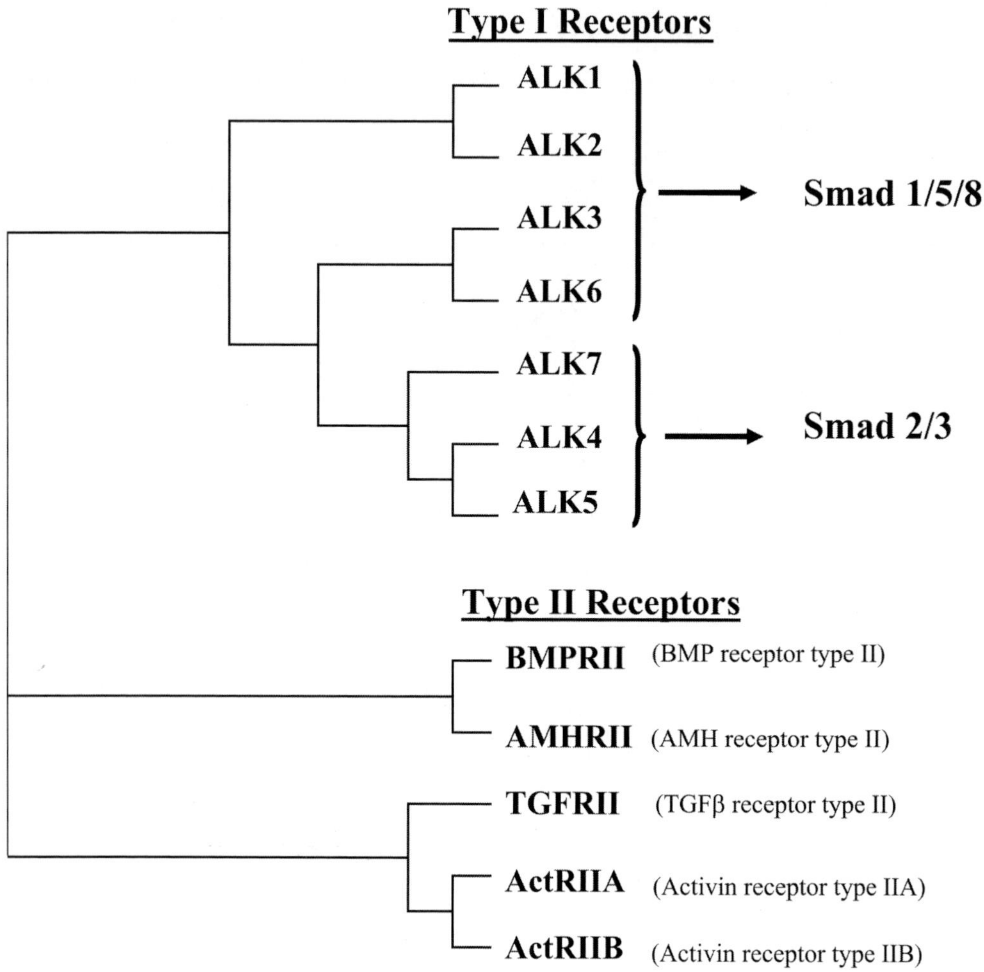

Fig. 3. Phylogenetic relationship of the type II and type I serine/threonine kinase receptors. Phylogenetic analyses of all known human serine/threonine kinase receptors were performed based on multiple full-length sequence alignment using the ClusterW algorithm (http://www.ch.embnet.org/software/ClustalW.html) and the TreeView drawing software (http://taxonomy.zoology.com). Smad proteins involved in the signaling pathway associated with individual type I receptors (ALK) are indicated. The GenBank accession numbers for individual human receptor protein sequence are: ALK1: NP_000011; ALK2: NP_001096; ALK3: NP_004320; ALK6: NP_001194; ALK4: NP_004293; ALK5: NP_004603; ALK7: AAM93495; BMPRII: NP_001195; AMHRII: Q16671; TGFRII: NP_001020018; ActRIIB: NP_001097; ActRIIA: NP_001607.

data on the TGF-β ligands and their receptor complexes support a model of oligomeric receptor assembly that does not involve a direct receptor–receptor interaction *(51,52)*.

2.1.2. Two Downstream Smad Signaling Pathways

The intracellular TGF-β signaling mediators are a group of phylogenetically related proteins called the Smads *(53)*. Smad proteins are divided into three structural domains *(53)*. The N-terminal MH1 domain exhibits sequence-specific DNA binding activity, except in the major spliced form of Smad2. The C-terminal MH2 domain is involved in the interaction with the type I receptor and the formation of the Smad complexes. The intermediate domain is divergent among Smads. It contains multiple sites of phosphorylation and allows specific cross-talk with other signaling pathways.

Functional studies have demonstrated that Smads can be grouped into three subfamilies, the receptor regulated Smads (R-Smads), the common Smad (Co-Smad) and the inhibitory Smads (I-Smads). The R-Smads are phosphorylated by the type I receptor kinases on a conserved carboxyl-terminal SSXS motif. ALK1, ALK2, ALK3, and ALK6 phosphorylate Smad1, Smad5, and Smad8 whereas ALK4, ALK5, and ALK7 phosphorylate Smad2 and Smad3 (Fig. 3) *(54–56)*. R-Smads form heteromeric complexes with the Co-Smad, Smad4. Smad4 is a shared partner of the R-Smads and is not phosphorylated in response to ligands. The activated Smad complexes are translocated into the nucleus, bind DNA on promoter sequences defined as Smad-binding element (Fig. 2). Because Smads bind DNA with low affinity and low specificity, they require cooperation with other sequence-specific binding factors to interact efficiently with promoters of target genes. Both the MH1 and MH2 domains interact with a large number of proteins (transcription factors, coactivators, and corepressors) in the nucleus *(57,58)*. In addition to R-Smads and Co-Smads, I-Smads (Smad7 and Smad6) form a distinct subclass of Smads that antagonize TGF-β signaling. Although I-Smads contain a C-terminal MH2 domain, their N-terminal region has a low similarity to the canonical MH1 domain. Smad7 stably interacts with all activated type I receptors to prevent R-Smad activation and downstream transcriptional modulation *(59,60)*. In contrast, Smad6 specifically competes with R-Smad1 for complex formation with Smad4, thus preferentially inhibiting the BMP pathway *(61,62)*.

Based on the genomic analysis of the entire repertoire of thirty-three TGF-β/BMP/GDF ligands *(3,4)* (Fig. 1), the TGF-β family members interact with a limited number of receptors and activate only two major intracellular signaling pathways characterized by the activation of the two different groups of intracellular Smad proteins, Smad1/5/8 and Smad2/3 (Fig. 1).

2.1.3. Cross-Talks and Alternative Pathways

TGF-β responses are not only the result of the activation of the Smad cascade, but depend upon interactions of Smad signaling with a variety of other intracellular pathways that may or may not be initiated by the ligand (Fig. 2). TGF-β and BMP/GDFs can activate mitogen-activated protein kinases (MAPK) (ERK, p38, JNK) likely through receptor interacting proteins (TAK1, Rho) *(63,64)*. MAPK can regulate the Smad activity. Indeed, the Ras-ERK pathway can phosphorylate R-Smads in the linker region and block the translocation of the R-Smad/Smad4 complex in the nucleus *(65)*. Moreover, activated R-Smads can interact with coactivators, corepressors, and transcriptional cofactors, which themselves may be regulated by other signaling cascades *(58,65)*. The ligand-induced response may be amplified or antagonized. The physiological importance of the MAPK pathways in TGF-β signaling remains uncertain.

2.2. BMP/GDF Signaling Pathways in Granulosa Cells

In situ hybridization and immunohistochemistry studies report the expression of all the components of the BMP signaling pathway in the ovary (Table 1B–D). Type I and type II receptors, as well as the Smads, are expressed in the granulosa cells and oocytes of most species studied. Some are detected in theca cell layers. When their expression was followed over the different follicular stages (primordial to preovulatory), the first follicular stage of expression is reported in Table 1. Indeed, most of the receptors and Smads start to be expressed in the granulosa cells of primordial or primary follicles. Altogether, these observations confirmed the presence of a complete BMP/GDF signaling pathway in the ovary, enabling autocrine, and paracrine regulation. However, data on the signaling pathways in the granulosa cells of most of the ovarian BMP/GDFs are still fragmentary.

2.2.1. GDF9 Signaling Pathway

GDF9 expression and function are restricted to the ovary and no data are available on the identity of its receptors in common cell lines. Instead of purifying the receptor proteins for identification, we hypothesized that GDF9, like other ligands in the same family, activates type II and type I serine/threonine kinase receptors. Because searches of human genes with sequence homology to already known serine/threonine kinase receptors failed to reveal uncharacterized receptor genes, GDF9 likely interacts with the known type II and type I receptors in granulosa cells. Phylogenetically, it is placed between the BMP2/4/6/7 and the activin/TGF-β subgroups of ligands, but with a closer relationship to the BMPs *(66)* (Fig. 1). In order to identify the type II receptor for GDF9, the soluble forms (ectodomain of the receptor fused to the Fc-binding region of human IgG) of the type II receptors, BMPRII and ActRIIA, were used to study potential interactions with GDF9 *(66)*. Of interest, the stimulatory effects of GDF9 on granulosa cell proliferation were completely blocked following coincubation with the soluble form of BMPRII. Similarly, the BMPRII ectodomain was capable of blocking the inhibitory effect of GDF9 on FSH-induced progesterone production. In addition, direct interactions between GDF9 and BMPRII were demonstrated by coimmunoprecipitation of GDF9 with the ectodomain of BMPRII *(66)* whereas the ActRIIA was only minimally efficient in binding GDF9. Furthermore, suppression of endogenous BMPRII biosynthesis using an antisense RNA approach completely blocked the stimulatory effects of GDF9 on the proliferation of rat granulosa cells in vitro *(66)*. These results showed that BMPRII is a receptor that is essential for GDF9 signaling in granulosa cells.

To identify the type I receptor for GDF9 in rat granulosa cells, we took advantage of the availability of different promoter-luciferase constructs for analyses of downstream pathways of the TGF-β family ligands. The CAGA promoter is known to be activated by the TGF-β/activin pathway mediated by Smad3 *(67)*, whereas the activation of BMP Response Element (BRE) and GCCG promoters are mediated by Smad1 and 5 *(68–70)*. We transfected individual promoter-reporter constructs into cultured rat granulosa cells and found that GDF9 treatment induced the activation of the CAGA promoter, but not the BRE or GCCG promoters *(71)*. Similar results also were found for human granulosa cells *(72)*. Because the transcriptional activities of all R-Smad proteins (Smad1, 2, 3, 5, and 8) are blocked by the inhibitory Smad7 *(59,60)*, whereas those of Smad1, 5, and 8 are blocked by the inhibitory Smad6 *(61, 62)*, we tested the inhibitory activities of the two inhibitory Smads on GDF9 stimulation of the CAGA promoter. Cotransfection with Smad7, but not Smad6, led to the suppression of the GDF9 stimulation of the CAGA promoter confirming that GDF9 signaling does not involve the BMP-responsive pathway mediated by Smad1, 5, and 8. We further demonstrated that treatment with GDF9, like activin, increased the level of phospho-Smad3 and phospho-Smad2 in rat and human granulosa cells *(73,74)*. Following identification of the downstream pathway for GDF9 in granulosa cells, we selected a cell line with minimal responsiveness to GDF9 but containing BMPRII in order to search for the type I receptor for GDF9. We overexpressed each of the seven-type I receptors in the minimally responsive Cos7 cells and found that the expression of ALK5, but not any other type I receptors, conferred GDF9 activation of the CAGA promoter (Fig. 4). We further performed interference RNA experiments to conclusively demonstrate the important role of ALK5 as the type I GDF9 receptor in granulosa cells *(71)*. Our data suggested cross-talk between the classic BMPRII and ALK5 together with downstream Smad3 and Smad2 proteins (Fig. 1). Earlier reports showed the expression of ALK5, BMPRII, and Smad2 in granulosa cells of follicles from the primordial/primary stage *(75–77)* as well as the expression of Smad3 in the granulosa cells of preantral and antral follicles and in the oocytes of primordial and primary follicles *(78)*. Although the fertility status of ALK5 null mice cannot be studied owing to embryonic lethality at midgestation *(79)*, the Smad3 null mice are viable but have reduced fertility compared to wild-type

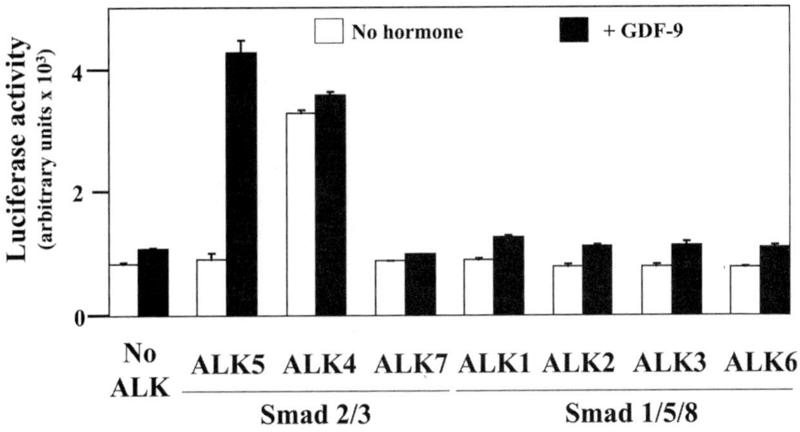

Fig. 4. Overexpression of ALK5 confers GDF9 responsiveness in Cos7 cells. Cells were transfected with 500 ng of the CAGA reporter and 30 ng of the plasmids encoding individual ALK proteins. Cells were incubated for 24 h without or with GDF9. The relative luciferase activity was normalized based on β-galactosidase activity. ALK5, 4, and 7 are type I receptors inducing Smad2/3 phosphorylation; ALK1, 2, 3, and 6 are type I receptors inducing Smad1/5/8 phosphorylation. Modified with permission from Mazerbourg et al., Mol Endocrinol 2004;18:653–665 *(71)*. Copyright 2004, The Endocrine Society.

mice *(80,81)*. Consistent with the important role of the GDF9 pathway in initial follicle recruitment, decreases in Smad3 expression in Smad3 null mice did not affect the size of the primordial follicle pool at birth but did reduce the growth of primordial follicles to the antral stage.

2.2.2. BMP15

Following identification of GDF9 receptors in granulosa cells, the BMP15 signaling pathway has been investigated in rat granulosa cells and the human granulosa cell line COV434 *(82)*. Treatment of the cells with recombinant BMP15 induces the phosphorylation of Smad1 and the activation of the XVent-2 luciferase reporter, which specifically responds to BMP stimulation. Thus, BMP15 is likely to interact with the known type I and type II receptors implicated in the Smad1/5/8 pathway. The soluble form (ectodomain of the receptor fused to the Fc-binding region of human IgG) of the type II receptors, BMPRII and ActRIIA, and the type I receptors, ALK2, ALK3, and ALK6 were tested for their potential interactions with BMP15 *(82)*. BMPRII, ActRIIA, ALK2, and ALK6 coimmunoprecipitate with BMP15, ALK6 being the most efficiently coimmunoprecipitated *(82)*. In contrast to the weak binding activity of the BMPRII soluble form, it is the most efficient in inhibiting the action of BMP15 on both FSH-induced progesterone production and ^{3}H-thymidine incorporation. These results suggest that BMP15 preferentially binds BMPRII and ALK6 to activate the Smad1/5/8 pathway in granulosa cells *(82)*.

Although GDF9 and BMP15 are the closest paralogs, they appear to activate two different pathways in vitro. This result is supported by the ability of TGF-β to stimulate two pathways by binding to ALK1 or ALK5, depending on the cellular context *(83)*. The use of distinct signaling pathways by GDF9 and BMP15 homodimers could explain their unique roles in follicular development. However, GDF9 and BMP15 also could work as heterodimers. In vitro, they can form homo and/or heterodimers when produced in the same cell, likely through noncovalent interactions *(84)*. Formation of the GDF9/BMP15 heterodimers could modify their affinity for a given receptor complex and induce distinct physiological responses *(85,86)*. The ability of the heterodimer GDF9/BMP15 to bind a receptor complex and activate the Smad pathway remains to be determined.

2.2.3. BMP2, BMP3, BMP3b, BMP4, BMP5, and BMP7 Signaling Pathways

In various cell types, BMP2 has been shown to bind to the type II receptors, BMPRII and ActRIIA, and the type I receptors, ALK3 and ALK6, leading to the activation of Smad1, 5, and 8. In addition to interacting with these BMP receptors, BMP6 and BMP7 also can signal through ALK2 *(41)*. BMP5 receptors are not defined; however, BMP5 belongs to the same group as BMP6 and BMP7 and is likely to share the same receptors. This subfamily of BMP/GDF ligands induces the phosphorylation of the intracellular factors Smad1, 5, and 8 (Fig. 1).

In agreement with previous in vitro data, treatment of human and rat granulosa cells with BMP2 induces the phosphorylation of Smad1, but not Smad2 *(73,87)*. Furthermore, Bondestam et al. suggest that ALK3 is one of the type I receptors for BMP2 in human luteinized granulosa cells *(87)*. Overexpression of the constitutive active form of ALK3 induces the secretion of inhibin B whereas the dominant negative form of the receptor blocks BMP2 stimulation of inhibin production *(87)*. The Smad1-dependant pathway also is activated by recombinant BMP6 and BMP7 in bovine and rat granulosa cell cultures *(31,82)*. Similarly, BMP4 treatment induces the phosphorylation of Smad1 protein in bovine and ovine granulosa cells *(31,34)*. Moreover, Pierre et al. showed that BMP4 treatment stimulates the activity of a Smad1-dependent luciferase construct *(34)*. Although the identity of the receptors involved in BMP2, 4, 6, and 7 signaling in granulosa cells is uncertain, they are likely to be the receptors activating the Smad1/5/8 pathways: ALK2, ALK3, and/or ALK6 (Fig. 3).

The ovarian function of BMP3 and BMP3b is unknown and no data are available on their signaling pathway in granulosa cells.

2.2.4. ALK6

The importance of the BMP/GDF system in ovarian function is confirmed by the ovarian phenotype in the Booroola Merino ewes. A point mutation in the ALK6 gene (allele *Fec*B) is associated with increased ovulation rate and litter size *(88–90)*. Booroola follicles mature and ovulate at a smaller size than those of the wild-type phenotype *(91)*. Fabre et al. suggest that the Q249R substitution in the serine/threonine kinase domain of the receptor is associated with a loss of responsiveness of the ALK6 receptor to BMP/GDFs. In Booroola sheep, the substitution would impair the antiluteinizing function of the ALK6 pathway, leading to advanced differentiation of granulosa cells and advanced maturation of follicles *(92)*. The ligand involved in the Booroola phenotype remains to be determined. However, this phenotype is reminiscent of the increase in ovulation rate observed in the heterozygous mutant BMP15 sheep *(16)*. Furthermore, the granulosa cells of Booroola sheep produce higher amounts of FSH-induced progesterone, but not estradiol. This response is the opposite of BMP15 and BMP6, which specifically inhibit FSH-induced progesterone production, but not estradiol *(28,30,35)*. These observations suggest that in the granulosa cells from Booroola sheep, the BMP15 (or BMP6)-ALK6-Smad1/5/8 cascade may be altered leading to early maturation.

In contrast to the hyperprolificity of Booroola sheep, ALK6 knockout mice are infertile owing to alterations in cumulus cell expansion and embryo implantation *(93)*. Similar to BMP15, the discrepancies in the phenotype between sheep and mouse could be owing to species differences (monoovulatory vs polyovulatory) and to the nature of the alteration (point mutation vs deletion).

2.2.5. Alternative Signaling Pathways

In granulosa cells, only few data on the activation of kinase pathways by BMP/GDFs are available. Administration of inhibitors of ERK phosphorylation inhibits mouse cumulus cell expansion induced by GDF9 *(94)* and rat granulosa cell proliferation induced by BMP15 *(82)*. These data on BMP15 point out that the Smad-dependent signaling cascade involved in the proliferation process is differently modulated than the one involved in differentiation.

3. BMP/GDF IN OVARIAN CARCINOMA

TGF-β superfamily members occupy a central position in the signaling networks that control cell growth and differentiation. Disruption in their signaling pathways is a common feature of many cancers *(95,96)*. BMP signaling is known to affect cell adhesion *(97)*. It can be prometastatic or growth inhibitory depending on the cancer cell type and experimental systems *(98–100)*. In the ovary, data on the role of the BMP/GDFs and their signaling pathways in tumor formation are limited and their involvement remains to be shown.

3.1. Granulosa Cell Tumor (GCT)

Among ovarian cancers, GCT account for 5% of primary ovarian tumors *(101)*. The molecular mechanisms of GCT tumor formation are poorly understood *(102)*. Few strains of transgenic mice have been shown to develop GCT *(103–106)*. In null mice for the inhibin α-subunit gene, GCT develops in a gonadotropin-dependent manner *(105,107)*. It has been postulated that inhibin acts in normal ovaries as a defense mechanism against the proliferative effects of elevated gonadotropins *(105,107)*. The relevance of this model to the study of GCT is unclear as inhibin production is a characteristic of human granulosa cell tumors *(108,109)*. Transgenic mice that overexpress the gene that encodes LH or the Simian virus (SV) 40 T-antigens (Tag) also develop GCT *(103,104,110)*. Recently, mice expressing a dominant stable β-catenin mutant in granulosa cells developed GCT with high penetrance *(106)*, suggesting that inappropriate activation of the Wnt/β-catenin pathway plays a role in the etiology of GCT.

Alteration of the BMP/GDF pathways in GCT has been underinvestigated. Granulosa-like cell tumors have been observed in ovaries of homozygous BMP15 mutant sheep. Those tumors share some characteristics with human GCT. They strongly express the receptor for FSH and produce inhibin *(111)*. An important question is how the absence of the granulosa cell proliferation factor, BMP15, leads to tumor-like structure. In vitro, BMP15 can inhibit FSH-induced granulosa cells differentiation by suppressing FSH receptor expression *(28,35)*. It has been proposed that in the absence of BMP15, granulosa cells of immature small follicles become strongly reactive to FSH, leading to an abnormal differentiation process. However, because of the increase in inhibin production in the BMP15–/– sheep, the action of FSH may be limited and other FSH-independent mechanisms may be involved.

Human GCT expresses inhibin α, βA, βB-subunits genes, as well as the activin receptor type IIA, IIB, ALK2, ALK4, and betaglycan *(109,112)*. No data are available on the expression of the other components of the BMP/GDF pathways. In humans, inhibin is presented as a granulosa cell tumor marker *(108)*. The role of inhibin in the development of the tumors is unknown. Inhibin functions as a competitive antagonist of not only activin but also of the BMPs in the presence of betaglycan. It has been reported that betaglycan potentiates inhibin binding to ActRIIA and is necessary for inhibin binding to BMPRII *(44,113)*. Inhibin could interfere with the proliferative and antiluteinizing BMP/GDF signaling pathways during follicular development and participates in granulosa cell transformation. Moreover, change in inhibin production could be the consequence of the disruption of the BMP2, BMP6,

BMP15, or GDF9 signaling involved in the regulation of the expression of inhibin α, βA, βB subunits in the ovary (Table 2). The identification of aberrant BMP/GDF signaling pathways in granulosa cells could provide new information on the molecular mechanisms leading to carcinogenesis.

3.2. Ovarian Epithelial Tumor

The majority of ovarian cancers originate in the epithelial cells on the surface of the ovary. The OSE is a single layer of epithelial cells that are separated from underlying ovarian stromal tissue by an extracellular matrix. During normal ovarian cycles, the OSE undergoes periodic changes and has the capacity to remodel the ovarian cortex. OSE cells often form inclusion bodies within the ovarian cortex, and these are the predominant sites of epithelial dysplasia and cancer formation *(114)*. Ovarian tumor development likely involves alteration in the signaling pathways regulating OSE proliferation and transformation. OSE expresses the receptors for FSH and LH as well as many of the receptors for growth factors produced by the different ovarian cell types *(114)*.

TGF-β isoforms, ALK5, TGF-β type II receptors, and Smad2, 3, and 4 are expressed by normal OSE, ovarian cancer cell lines, and primary ovarian cancer cells *(115–119)*. Normal OSE cell growth is inhibited by TGF-β. Some studies report that deletion or inactivating mutations of some TGF-β pathway components occur in human ovarian cancers *(120–122)*. Normal and transformed OSE also expresses the activin/inhibin α, βA, βB-subunits, as well as the different receptors for activin (ActRIIA, ActRIIB, ALK2, and ALK4) *(119,123–125)*. Activin A stimulates the proliferation of various ovarian cancer cell lines but not the growth of normal OSE cells *(119,124,125)*. BMP signaling pathway components also are present in normal and transformed human OSE. They express ALK3, ALK6, BMPRII, and Smad1, 4, 5, and 8 *(126)* and produce BMP4, but not BMP7. In normal rat OSE, Shimasaki et al. also showed the expression of BMP4, BMP3b, BMP6, but not BMP7 *(76)*. The functional significance of these expressions in the ovary remains unknown. In vitro, treatment of primary ovarian cancer cells with BMP4 induces a minor inhibitory effect on their proliferation but enhances their adhesion *(126)*. Furthermore, addition of BMP4 leads to the phosphorylation of Smad1 and to the increase of expression of the BMP/GDF target genes, Id1 (Inhibitor of DNA binding) and Id3 *(126)*. Interestingly, Shepherd et al. report that the BMP4-induction of Id genes expression is higher in ovarian cancer cell lines than in normal OSE cells *(126)*. Moreover, Id expression in human ovarian cancer is associated with more aggressive tumor cells in vivo and poor patient outcome *(127)*. Id proteins interact with a variety of transcription factors involved in growth regulation. Then, dysregulation of Id expression could contribute to abnormal OSE cell growth control *(128)*.

Ovarian epithelial tumor development likely involves abnormal expression of growth factors or signaling pathway components that affect normal OSE proliferation. The specific growth factors involved remain unknown; however, the BMP/GDFs are potential candidates.

4. CONCLUSIVE REMARKS

The identification of the receptors and signaling pathway for GDF9 *(71)* was based on an original genomic approach. Owing to the sequencing of diverse genomes, we can trace the evolution of complete families of genes of polypeptide ligands and receptors. Because ligand and receptor families have coevolved, analyses of the subgenomes of extracellular protein ligands and their transmembrane receptors provide a new paradigm to match orphan ligands with their cognate receptors *(129)*. This approach allowed us to identify receptors for the orphan ligands GDF6, GDF7, and BMP10 *(4)* and can be transposed to the study of other receptor/ligand subfamilies.

Ovarian BMP/GDFs are important autocrine and paracrine regulators of granulosa cell function during multiple stages of follicular development. However, they interact with a limited number of receptors and activate only two major intracellular signaling pathways. In fact, their physiological effects result from the modulation of their action at various points. Rat granulosa cells express the BMP antagonists follistatin, gremlin, and PRDC (protein related to DAN and cerberus) *(40,76,130,131)*. In the extracellular compartment, antagonists regulate initiation of the signaling cascade and may participate in modulating the stimulation. Different amplitudes or durations of signal may lead to the activation of different sets of genes *(132,133)*. At the receptor level, BAMBI (BMP and activin membrane-bound inhibitor), expressed by rat granulosa cells, can function as a dominant negative receptor of the BMP/GDFs and prevent the formation of functional receptor complexes *(134,135)*. In the nucleus, the activation of the target genes depends as much on ligand activity as the activity of cofactors and other transcription factors. In vivo, granulosa cell response to BMP/GDFs will be the result of the integration of several hormonal stimulations, pointing out the potential limitation of a cell culture approach.

Considering the multiple roles BMP/GDFs play in different biological process and the limited number of signaling components, the complexity of regulation and interaction increases the diversity of cellular response. In granulosa cells, those interactions need to be examined. A better understanding of the BMP/GDF role in the development of granulosa and epithelial cell-derived tumors is necessary. The identification of BMP/GDF alternative signaling pathways, cross-talks, and altered gene expression is critical to the development of better diagnostics and more efficacious therapeutics.

ACKNOWLEDGMENTS

I wish to thank Dr. Aaron JW Hsueh for his confidence and support in the writing of this review. I am grateful to Drs. Leon J Spicer and Stéphane Flament for critical comments on the manuscript and Caren Spencer for editorial assistance. The work on the identification of GDF9 receptors was performed in Dr. Aaron JW Hsueh's laboratory (Stanford University, USA). It was supported by the National Institute of Child Health and Human Development, and the National Institutes of Health, through Cooperative Agreement U54 HD31398 as part of the Specialized Cooperative Centers Program in Reproduction Research.

REFERENCES

1. McGee EA, Hsueh AJ. Initial and cyclic recruitment of ovarian follicles. Endocr Rev 2000;21:200–214.
2. Richards JS. Hormonal control of gene expression in the ovary. Endocr Rev 1994;15:725–751.
3. Chang H, Brown CW, Matzuk MM. Genetic analysis of the mammalian transforming growth factor-beta superfamily. Endocr Rev 2002;23:787–823.
4. Mazerbourg S, Sangkuhl K, Luo CW, Sudo S, Klein C, Hsueh AJ. Identification of receptors and signaling pathways for orphan bone morphogenetic protein/growth differentiation factor ligands based on genomic analyses. J Biol Chem 2005;280:32,122–32,132.
5. Constam DB, Robertson EJ. Regulation of bone morphogenetic protein activity by pro domains and proprotein convertases. J Cell Biol 1999;144:139–149.
6. Wolfman NM, McPherron AC, Pappano WN, et al. Activation of latent myostatin by the BMP-1/tolloid family of metalloproteinases. Proc Natl Acad Sci USA 2003;100:15,842–15,846.
7. Ge G, Hopkins DR, Ho WB, Greenspan DS. GDF11 forms a bone morphogenetic protein 1-activated latent complex that can modulate nerve growth factor-induced differentiation of PC12 cells. Mol Cell Biol 2005;25:5846–5858.
8. Vitt UA, Hsu SY, Hsueh AJ. Evolution and classification of cystine knot-containing hormones and related extracellular signaling molecules. Mol Endocrinol 2001;15:681–694.
9. Scheufler C, Sebald W, Hulsmeyer M. Crystal structure of human bone morphogenetic protein-2 at 2.7 A resolution. J Mol Biol 1999;287:103–115.

10. Thompson TB, Woodruff TK, Jardetzky TS. Structures of an ActRIIB:activin A complex reveal a novel binding mode for TGF-beta ligand:receptor interactions. EMBO J 2003;22:1555–1566.
11. Dong J, Albertini DF, Nishimori K, Kumar TR, Lu N, Matzuk MM. Growth differentiation factor-9 is required during early ovarian folliculogenesis. Nature 1996;383:531–535.
12. McNatty KP, Moore LG, Hudson NL, et al. The oocyte and its role in regulating ovulation rate: a new paradigm in reproductive biology. Reproduction 2004;128:379–386.
13. Hanrahan JP, Gregan SM, Mulsant P, et al. Mutations in the genes for oocyte-derived growth factors GDF9 and BMP15 are associated with both increased ovulation rate and sterility in Cambridge and Belclare sheep (Ovis aries). Biol Reprod 2004;70:900–909.
14. Vitt UA, McGee EA, Hayashi M, Hsueh AJ. In vivo treatment with GDF-9 stimulates primordial and primary follicle progression and theca cell marker CYP17 in ovaries of immature rats. Endocrinology 2000;141:3814–3820.
15. McGee EA, Perlas E, LaPolt PS, Tsafriri A, Hsueh AJ. Follicle-stimulating hormone enhances the development of preantral follicles in juvenile rats. Biol Rcprod 1997;57:990 998.
16. Galloway SM, McNatty KP, Cambridge LM, et al. Mutations in an oocyte-derived growth factor gene (BMP15) cause increased ovulation rate and infertility in a dosage-sensitive manner. Nat Genet 2000;25:279–283.
17. Moore RK, Erickson GF, Shimasaki S. Are BMP-15 and GDF-9 primary determinants of ovulation quota in mammals? Trends Endocrinol Metab 2004;15:356–361.
18. Juengel JL, Hudson NL, Whiting L, McNatty KP. Effects of immunization against bone morphogenetic protein 15 and growth differentiation factor 9 on ovulation rate, fertilization, and pregnancy in ewes. Biol Reprod 2004;70:557–561.
19. Di Pasquale E, Beck-Peccoz P, Persani L. Hypergonadotropic ovarian failure associated with an inherited mutation of human bone morphogenetic protein-15 (BMP15) gene. Am J Hum Genet 2004;75:106–111.
20. Yan C, Wang P, DeMayo J, et al. Synergistic roles of bone morphogenetic protein 15 and growth differentiation factor 9 in ovarian function. Mol Endocrinol 2001;15:854–866.
21. Su YQ, Wu X, O'Brien MJ, et al. Synergistic roles of BMP15 and GDF9 in the development and function of the oocyte-cumulus cell complex in mice: genetic evidence for an oocyte-granulosa cell regulatory loop. Dev Biol 2004;276:64–73.
22. Lee WS, Otsuka F, Moore RK, Shimasaki S. Effect of bone morphogenetic protein-7 on folliculogenesis and ovulation in the rat. Biol Reprod 2001;65:994–999.
23. Luo G, Hofmann C, Bronckers AL, Sohocki M, Bradley A, Karsenty G. BMP-7 is an inducer of nephrogenesis, and is also required for eye development and skeletal patterning. Genes Dev 1995;9: 2808–2820.
24. Dudley AT, Lyons KM, Robertson EJ. A requirement for bone morphogenetic protein-7 during development of the mammalian kidney and eye. Genes Dev 1995;9:2795–2807.
25. Pierre A, Pisselet C, Dupont J, Bontoux M, Monget P. Bone morphogenetic protein 5 expression in the rat ovary: biological effects on granulosa cell proliferation and steroidogenesis. Biol Reprod 2005;73:1102–1108.
26. Solloway MJ, Dudley AT, Bikoff EK, Lyons KM, Hogan BL, Robertson EJ. Mice lacking Bmp6 function. Dev Genet 1998;22:321–339.
27. Kingsley DM, Bland AE, Grubber JM, et al. The mouse short ear skeletal morphogenesis locus is associated with defects in a bone morphogenetic member of the TGF beta superfamily. Cell 1992;71: 399–410.
28. Otsuka F, Yao Z, Lee T, Yamamoto S, Erickson GF, Shimasaki S. Bone morphogenetic protein-15. Identification of target cells and biological functions. J Biol Chem 2000;275:39,523–39,528.
29. Vitt UA, Hayashi M, Klein C, Hsueh AJ. Growth differentiation factor-9 stimulates proliferation but suppresses the follicle-stimulating hormone-induced differentiation of cultured granulosa cells from small antral and preovulatory rat follicles. Biol Reprod 2000;62:370–377.
30. Otsuka F, Moore RK, Shimasaki S. Biological function and cellular mechanism of bone morphogenetic protein-6 in the ovary. J Biol Chem 2001;276:32,889–32,895.
31. Glister C, Kemp CF, Knight PG. Bone morphogenetic protein (BMP) ligands and receptors in bovine ovarian follicle cells: actions of BMP-4, -6 and -7 on granulosa cells and differential modulation of Smad-1 phosphorylation by follistatin. Reproduction 2004;127:239–254.
32. Souza CJ, Campbell BK, McNeilly AS, Baird DT. Effect of bone morphogenetic protein 2 (BMP2) on oestradiol and inhibin A production by sheep granulosa cells, and localization of BMP receptors in the ovary by immunohistochemistry. Reproduction 2002;123:363–369.

33. Shimasaki S, Zachow RJ, Li D, et al. A functional bone morphogenetic protein system in the ovary. Proc Natl Acad Sci USA 1999;96:7282–7287.
34. Pierre A, Pisselet C, Dupont J, et al. Molecular basis of bone morphogenetic protein-4 inhibitory action on progesterone secretion by ovine granulosa cells. J Mol Endocrinol 2004;33:805–817.
35. Otsuka F, Yamamoto S, Erickson GF, Shimasaki S. Bone morphogenetic protein-15 inhibits follicle-stimulating hormone (FSH) action by suppressing FSH receptor expression. J Biol Chem 2001;276: 11,387–11,392.
36. Elvin JA, Clark AT, Wang P, Wolfman NM, Matzuk MM. Paracrine actions of growth differentiation factor-9 in the mammalian ovary. Mol Endocrinol 1999;13:1035–1048.
37. Gui LM, Joyce IM. RNA interference evidence that growth differentiation factor-9 mediates oocyte regulation of cumulus expansion in mice. Biol Reprod 2005;72:195–199.
38. Elvin JA, Yan C, Matzuk MM. Growth differentiation factor-9 stimulates progesterone synthesis in granulosa cells via a prostaglandin E2/EP2 receptor pathway. Proc Natl Acad Sci USA 2000;97: 10,288–10,293.
39. Varani S, Elvin JA, Yan C, et al. Knockout of pentraxin 3, a downstream target of growth differentiation factor-9, causes female subfertility. Mol Endocrinol 2002;16:1154–1167.
40. Pangas SA, Jorgez CJ, Matzuk MM. Growth differentiation factor 9 regulates expression of the bone morphogenetic protein antagonist gremlin. J Biol Chem 2004;279:32,281–32,286.
41. Massagué J. TGF-beta signal transduction. Annu Rev Biochem 1998;67:753–791.
42. ten Dijke P, Ichijo H, Franzen P, et al. Activin receptor-like kinases: a novel subclass of cell-surface receptors with predicted serine/threonine kinase activity. Oncogene 1993;8:2879–2887.
43. Lopez-Casillas F, Wrana JL, Massagué J. Betaglycan presents ligand to the TGF beta signaling receptor. Cell 1993;73:1435–1444.
44. Lewis KA, Gray PC, Blount AL, et al. Betaglycan binds inhibin and can mediate functional antagonism of activin signalling. Nature 2000;404:411–414.
45. Lebrin F, Goumans MJ, Jonker L, et al. Endoglin promotes endothelial cell proliferation and TGF-beta/ ALK1 signal transduction. EMBO J 2004;23:4018–4028.
46. Schier AF, Shen MM. Nodal signalling in vertebrate development. Nature 2000;403:385–389.
47. Attisano L, Wrana JL, Montalvo E, Massagué J. Activation of signalling by the activin receptor complex. Mol Cell Biol 1996;16:1066–1073.
48. Wrana JL, Attisano L, Wieser R, Ventura F, Massagué J. Mechanism of activation of the TGF-beta receptor. Nature 1994;370:341–347.
49. Mathews LS, Vale WW. Characterization of type II activin receptors. Binding, processing, and phosphorylation. J Biol Chem 1993;268:19,013–19,018.
50. Wieser R, Wrana JL, Massagué J. GS domain mutations that constitutively activate T beta R-I, the downstream signaling component in the TGF-beta receptor complex. EMBO J 1995;14: 2199–2208.
51. Greenwald J, Groppe J, Gray P, et al. The BMP7/ActRII extracellular domain complex provides new insights into the cooperative nature of receptor assembly. Mol Cell 2003;11:605–617.
52. Kirsch T, Sebald W, Dreyer MK. Crystal structure of the BMP-2-BRIA ectodomain complex. Nat Struct Biol 2000;7:492–496.
53. Attisano L, Tuen Lee-Hoeflich S. The Smads. Genome Biol 2001;2:REVIEWS3010.1–3010.8.
54. Kretzschmar M, Massagué J. SMADs: mediators and regulators of TGF-beta signaling. Curr Opin Genet Dev 1998;8:103–111.
55. Chen YG, Massagué J. Smad1 recognition and activation by the ALK1 group of transforming growth factor-beta family receptors. J Biol Chem 1999;274:3672–3677.
56. Watanabe R, Yamada Y, Ihara Y, et al. The MH1 domains of smad2 and smad3 are involved in the regulation of the ALK7 signals. Biochem Biophys Res Commun 1999;254:707–712.
57. Massagué J, Wotton D. Transcriptional control by the TGF-beta/Smad signaling system. EMBO J 2000;19:1745–1754.
58. Wrana JL. Crossing Smads. Sci STKE 2000;2000:RE1–RE9.
59. Nakao A, Afrakhte M, Moren A, et al. Identification of Smad7, a TGFbeta-inducible antagonist of TGF-beta signalling. Nature 1997;389:631–635.
60. Hayashi H, Abdollah S, Qiu Y, et al. The MAD-related protein Smad7 associates with the TGFbeta receptor and functions as an antagonist of TGFbeta signaling. Cell 1997;89:1165–1173.
61. Hata A, Lagna G, Massagué J, Hemmati-Brivanlou A. Smad6 inhibits BMP/Smad1 signaling by specifically competing with the Smad4 tumor suppressor. Genes Dev 1998;12:186–197.

62. Ishisaki A, Yamato K, Hashimoto S, et al. Differential inhibition of Smad6 and Smad7 on bone morphogenetic protein- and activin-mediated growth arrest and apoptosis in B cells. J Biol Chem 1999;274: 13,637–13,642.
63. Engel ME, McDonnell MA, Law BK, Moses HL. Interdependent SMAD and JNK signaling in transforming growth factor-beta-mediated transcription. J Biol Chem 1999;274:37,413–37,420.
64. Lutz M, Knaus P. Integration of the TGF-beta pathway into the cellular signalling network. Cell Signal 2002;14:977–988.
65. Derynck R, Zhang YE. Smad-dependent and Smad-independent pathways in TGF-beta family signalling. Nature 2003;425:577–584.
66. Vitt UA, Mazerbourg S, Klein C, Hsueh AJ. Bone morphogenetic protein receptor type II is a receptor for growth differentiation factor-9. Biol Reprod 2002;67:473–480.
67. Dennler S, Itoh S, Vivien D, ten Dijke P, Huet S, Gauthier JM. Direct binding of Smad3 and Smad4 to critical TGF beta-inducible elements in the promoter of human plasminogen activator inhibitor-type 1 gene. EMBO J 1998;17:3091–3100.
68. Korchynskyi O, ten Dijke P. Identification and functional characterization of distinct critically important bone morphogenetic protein-specific response elements in the Id1 promoter. J Biol Chem 2002; 277:4883–4891.
69. Monteiro RM, de Sousa Lopes SM, Korchynskyi O, ten Dijke P, Mummery CL. Spatio-temporal activation of Smad1 and Smad5 in vivo: monitoring transcriptional activity of Smad proteins. J Cell Sci 2004;117:4653–4663.
70. Kusanagi K, Inoue H, Ishidou Y, Mishima HK, Kawabata M, Miyazono K. Characterization of a bone morphogenetic protein-responsive Smad-binding element. Mol Biol Cell 2000;11:555–565.
71. Mazerbourg S, Klein C, Roh J, et al. Growth differentiation factor-9 signaling is mediated by the type I receptor, activin receptor-like kinase 5. Mol Endocrinol 2004;18:653–665.
72. Kaivo-Oja N, Mottershead DG, Mazerbourg S, et al. Adenoviral gene transfer allows Smad-responsive gene promoter analyses and delineation of type I receptor usage of transforming growth factor-beta family ligands in cultured human granulosa luteal cells. J Clin Endocrinol Metab 2005;90:271–278.
73. Roh JS, Bondestam J, Mazerbourg S, et al. Growth differentiation factor-9 stimulates inhibin production and activates Smad2 in cultured rat granulosa cells. Endocrinology 2003;144:172–178.
74. Kaivo-Oja N, Bondestam J, Kamarainen M, et al. Growth differentiation factor-9 induces Smad2 activation and inhibin B production in cultured human granulosa-luteal cells. J Clin Endocrinol Metab 2003;88:755–762.
75. Schilling B, Yeh J. Expression of transforming growth factor (TGF)-beta1, TGF-beta2, and TGF-beta3 and of type I and II TGF-beta receptors during the development of the human fetal ovary. Fertil Steril 1999;72:147–153.
76. Erickson GF, Shimasaki S. The spatiotemporal expression pattern of the bone morphogenetic protein family in rat ovary cell types during the estrous cycle. Reprod Biol Endocrinol 2003;1:9.
77. Qu J, Godin PA, Nisolle M, Donnez J. Expression of receptors for insulin-like growth factor-I and transforming growth factor-beta in human follicles. Mol Hum Reprod 2000;6:137–145.
78. Xu J, Oakley J, McGee EA. Stage-specific expression of Smad2 and Smad3 during folliculogenesis. Biol Reprod 2002;66:1571–1578.
79. Larsson J, Goumans MJ, Sjostrand LJ, et al. Abnormal angiogenesis but intact hematopoietic potential in TGF-beta type I receptor-deficient mice. EMBO J 2001;20:1663–1673.
80. Tomic D, Brodie SG, Deng C, et al. Smad 3 may regulate follicular growth in the mouse ovary. Biol Reprod 2002;66:917–923.
81. Tomic D, Miller KP, Kenny HA, Woodruff TK, Hoyer P, Flaws JA. Ovarian follicle development requires Smad3. Mol Endocrinol 2004;18:2224–2240.
82. Moore RK, Otsuka F, Shimasaki S. Molecular basis of bone morphogenetic protein-15 signaling in granulosa cells. J Biol Chem 2003;278:304–310.
83. Goumans MJ, Valdimarsdottir G, Itoh S, Rosendahl A, Sideras P, ten Dijke P. Balancing the activation state of the endothelium via two distinct TGF-beta type I receptors. EMBO J 2002;21:1743–1753.
84. Liao WX, Moore RK, Otsuka F, Shimasaki S. Effect of intracellular interactions on the processing and secretion of bone morphogenetic protein-15 (BMP-15) and growth and differentiation factor-9. Implication of the aberrant ovarian phenotype of BMP-15 mutant sheep. J Biol Chem 2003;278: 3713–3719.
85. Nishimatsu S, Thomsen GH. Ventral mesoderm induction and patterning by bone morphogenetic protein heterodimers in Xenopus embryos. Mech Dev 1998;74:75–88.

86. Butler SJ, Dodd J. A role for BMP heterodimers in roof plate-mediated repulsion of commissural axons. Neuron 2003;38:389–401.
87. Bondestam J, Kaivo-oja N, Kallio J, et al. Engagement of activin and bone morphogenetic protein signaling pathway Smad proteins in the induction of inhibin B production in ovarian granulosa cells. Mol Cell Endocrinol 2002;195:79–88.
88. Mulsant P, Lecerf F, Fabre S, et al. Mutation in bone morphogenetic protein receptor-IB is associated with increased ovulation rate in Booroola Merino ewes. Proc Natl Acad Sci USA 2001;98:5104–5109.
89. Wilson T, Wu XY, Juengel JL, et al. Highly prolific Booroola sheep have a mutation in the intracellular kinase domain of bone morphogenetic protein IB receptor (ALK-6) that is expressed in both oocytes and granulosa cells. Biol Reprod 2001;64:1225–1235.
90. Souza CJ, MacDougall C, Campbell BK, McNeilly AS, Baird DT. The Booroola (FecB) phenotype is associated with a mutation in the bone morphogenetic receptor type 1 B (BMPR1B) gene. J Endocrinol 2001;169:R1–R6.
91. McNatty KP, Kieboom LE, McDiarmid J, Heath DA, Lun S. Adenosine cyclic 3′,5′-monophosphate and steroid production by small ovarian follicles from Booroola ewes with and without a fecundity gene. J Reprod Fertil 1986;76:471–480.
92. Fabre S, Pierre A, Pisselet C, et al. The Booroola mutation in sheep is associated with an alteration of the bone morphogenetic protein receptor-IB functionality. J Endocrinol 2003;177:435–444.
93. Yi SE, LaPolt PS, Yoon BS, Chen JY, Lu JK, Lyons KM. The type I BMP receptor BmprIB is essential for female reproductive function. Proc Natl Acad Sci USA 2001;98:7994–7999.
94. Su YQ, Wigglesworth K, Pendola FL, O'Brien MJ, Eppig JJ. Mitogen-activated protein kinase activity in cumulus cells is essential for gonadotropin-induced oocyte meiotic resumption and cumulus expansion in the mouse. Endocrinology 2002;143:2221–2232.
95. Eng C. To be or not to BMP. Nat Genet 2001;28:105–107.
96. Massagué J, Blain SW, Lo RS. TGFbeta signaling in growth control, cancer, and heritable disorders. Cell 2000;103:295–309.
97. Nissinen L, Pirila L, Heino J. Bone morphogenetic protein-2 is a regulator of cell adhesion. Exp Cell Res 1997;230:377–385.
98. Ide H, Yoshida T, Matsumoto N, et al. Growth regulation of human prostate cancer cells by bone morphogenetic protein-2. Cancer Res 1997;57:5022–5027.
99. Hjertner O, Hjorth-Hansen H, Borset M, Seidel C, Waage A, Sundan A. Bone morphogenetic protein-4 inhibits proliferation and induces apoptosis of multiple myeloma cells. Blood 2001;97:516–522.
100. Kim IY, Lee DH, Ahn HJ, et al. Expression of bone morphogenetic protein receptors type-IA, -IB and -II correlates with tumor grade in human prostate cancer tissues. Cancer Res 2000;60: 2840–2844.
101. Schumer ST, Cannistra SA. Granulosa cell tumor of the ovary. J Clin Oncol 2003;21:1180–1189.
102. Fuller PJ, Chu S. Signalling pathways in the molecular pathogenesis of ovarian granulosa cell tumours. Trends Endocrinol Metab 2004;15:122–128.
103. Nilson JH, Abbud RA, Keri RA, Quirk CC. Chronic hypersecretion of luteinizing hormone in transgenic mice disrupts both ovarian and pituitary function, with some effects modified by the genetic background. Recent Prog Horm Res 2000;55:69–89; discussion 89–91.
104. Kananen K, Markkula M, Rainio E, Su JG, Hsueh AJ, Huhtaniemi IT. Gonadal tumorigenesis in transgenic mice bearing the mouse inhibin alpha-subunit promoter/simian virus T-antigen fusion gene: characterization of ovarian tumors and establishment of gonadotropin-responsive granulosa cell lines. Mol Endocrinol 1995;9:616–627.
105. Matzuk MM, Finegold MJ, Su JG, Hsueh AJ, Bradley A. Alpha-inhibin is a tumour-suppressor gene with gonadal specificity in mice. Nature 1992;360:313–319.
106. Boerboom D, Paquet M, Hsieh M, et al. Misregulated Wnt/beta-catenin signaling leads to ovarian granulosa cell tumor development. Cancer Res 2005;65:9206–9215.
107. Matzuk MM, Kumar TR, Shou W, et al. Transgenic models to study the roles of inhibins and activins in reproduction, oncogenesis, and development. Recent Prog Horm Res 1996;51:123–154; discussion 155–157.
108. Lappohn RE, Burger HG, Bouma J, Bangah M, Krans M, de Bruijn HW. Inhibin as a marker for granulosa-cell tumors. N Engl J Med 1989;321:790–793.
109. Fuller PJ, Chu S, Jobling T, Mamers P, Healy DL, Burger HG. Inhibin subunit gene expression in ovarian cancer. Gynecol Oncol 1999;73:273–279.
110. Kananen K, Markkula M, Mikola M, Rainio EM, McNeilly A, Huhtaniemi I. Gonadectomy permits adrenocortical tumorigenesis in mice transgenic for the mouse inhibin alpha-subunit promoter/

simian virus 40 T-antigen fusion gene: evidence for negative autoregulation of the inhibin alpha-subunit gene. Mol Endocrinol 1996;10:1667–1677.

111. Juengel JL, Quirke LD, Tisdall DJ, Smith P, Hudson NL, McNatty KP. Gene expression in abnormal ovarian structures of ewes homozygous for the inverdale prolificacy gene. Biol Reprod 2000;62: 1467–1478.
112. Fuller PJ, Zumpe ET, Chu S, Mamers P, Burger HG. Inhibin-activin receptor subunit gene expression in ovarian tumors. J Clin Endocrinol Metab 2002;87:1395–1401.
113. Wiater E, Vale W. Inhibin is an antagonist of bone morphogenetic protein signaling. J Biol Chem 2003;278:7934–7941.
114. Auersperg N, Wong AS, Choi KC, Kang SK, Leung PC. Ovarian surface epithelium: biology, endocrinology, and pathology. Endocr Rev 2001;22:255–288.
115. Henriksen R, Gobl A, Wilander E, Oberg K, Miyazono K, Funa K. Expression and prognostic significance of TGF-beta isotypes, latent TGF-beta 1 binding protein, TGF-beta type I and type II receptors, and endoglin in normal ovary and ovarian neoplasms. Lab Invest 1995;73:213–220.
116. Yamada SD, Baldwin RL, Karlan BY. Ovarian carcinoma cell cultures are resistant to TGF-beta1-mediated growth inhibition despite expression of functional receptors. Gynecol Oncol 1999;75:72–77.
117. Ito I, Minegishi T, Fukuda J, Shinozaki H, Auersperg N, Leung PC. Presence of activin signal transduction in normal ovarian cells and epithelial ovarian carcinoma. Br J Cancer 2000;82:1415–1420.
118. Dunfield LD, Dwyer EJ, Nachtigal MW. TGF beta-induced Smad signaling remains intact in primary human ovarian cancer cells. Endocrinology 2002;143:1174–1181.
119. Steller MD, Shaw TJ, Vanderhyden BC, Ethier JF. Inhibin resistance is associated with aggressive tumorigenicity of ovarian cancer cells. Mol Cancer Res 2005;3:50–61.
120. Lynch MA, Nakashima R, Song H, et al. Mutational analysis of the transforming growth factor beta receptor type II gene in human ovarian carcinoma. Cancer Res 1998;58:4227–4232.
121. Wang D, Kanuma T, Mizunuma H, et al. Analysis of specific gene mutations in the transforming growth factor-beta signal transduction pathway in human ovarian cancer. Cancer Res 2000;60:4507–4512.
122. Chen T, Triplett J, Dehner B, et al. Transforming growth factor-beta receptor type I gene is frequently mutated in ovarian carcinomas. Cancer Res 2001;61:4679–4682.
123. Welt CK, Lambert-Messerlian G, Zheng W, Crowley WF, Jr., Schneyer AL. Presence of activin, inhibin, and follistatin in epithelial ovarian carcinoma. J Clin Endocrinol Metab 1997;82:3720–3727.
124. Choi KC, Kang SK, Nathwani PS, Cheng KW, Auersperg N, Leung PC. Differential expression of activin/inhibin subunit and activin receptor mRNAs in normal and neoplastic ovarian surface epithelium (OSE). Mol Cell Endocrinol 2001;174:99–110.
125. Di Simone N, Crowley WF, Jr., Wang QF, Sluss PM, Schneyer AL. Characterization of inhibin/activin subunit, follistatin, and activin type II receptors in human ovarian cancer cell lines: a potential role in autocrine growth regulation. Endocrinology 1996;137:486–494.
126. Shepherd TG, Nachtigal MW. Identification of a putative autocrine bone morphogenetic protein-signaling pathway in human ovarian surface epithelium and ovarian cancer cells. Endocrinology 2003;144:3306–3314.
127. Schindl M, Schoppmann SF, Strobel T, et al. Level of Id-1 protein expression correlates with poor differentiation, enhanced malignant potential, and more aggressive clinical behavior of epithelial ovarian tumors. Clin Cancer Res 2003;9:779–785.
128. Norton JD. ID helix-loop-helix proteins in cell growth, differentiation and tumorigenesis. J Cell Sci 2000;113(Pt 22):3897–3905.
129. Leo CP, Hsu SY, Hsueh AJ. Hormonal genomics. Endocr Rev 2002;23:369–381.
130. Balemans W, Van Hul W. Extracellular regulation of BMP signaling in vertebrates: a cocktail of modulators. Dev Biol 2002;250:231–250.
131. Sudo S, Avsian-Kretchmer O, Wang LS, Hsueh AJ. Protein related to DAN and cerberus is a bone morphogenetic protein antagonist that participates in ovarian paracrine regulation. J Biol Chem 2004; 279:23,134–23,141.
132. Green JB, New HV, Smith JC. Responses of embryonic Xenopus cells to activin and FGF are separated by multiple dose thresholds and correspond to distinct axes of the mesoderm. Cell 1992;71:731–739.
133. Dyson S, Gurdon JB. The interpretation of position in a morphogen gradient as revealed by occupancy of activin receptors. Cell 1998;93:557–568.
134. Onichtchouk D, Chen YG, Dosch R, et al. Silencing of TGF-beta signalling by the pseudoreceptor BAMBI. Nature 1999;401:480–485.
135. Loveland KL, Bakker M, Meehan T, et al. Expression of Bambi is widespread in juvenile and adult rat tissues and is regulated in male germ cells. Endocrinology 2003;144:4180–4186.

136. Jaatinen R, Rosen V, Tuuri T, Ritvos O. Identification of ovarian granulosa cells as a novel site of expression for bone morphogenetic protein-3 (BMP-3/osteogenin) and regulation of BMP-3 messenger ribonucleic acids by chorionic gonadotropin in cultured human granulosa-luteal cells. J Clin Endocrinol Metab 1996;81:3877–3882.
137. Elvin JA, Yan C, Matzuk MM. Oocyte-expressed TGF-beta superfamily members in female fertility. Mol Cell Endocrinol 2000;159:1–5.
138. Lyons KM, Pelton RW, Hogan BL. Patterns of expression of murine Vgr-1 and BMP-2a RNA suggest that transforming growth factor-beta-like genes coordinately regulate aspects of embryonic development. Genes Dev 1989;3:1657–1668.
139. Juengel JL, McNatty KP. The role of proteins of the transforming growth factor-beta superfamily in the intraovarian regulation of follicular development. Hum Reprod Update 2005;11:143–160.
140. Dube JL, Wang P, Elvin J, Lyons KM, Celeste AJ, Matzuk MM. The bone morphogenetic protein 15 gene is X-linked and expressed in oocytes. Mol Endocrinol 1998;12:1809–1817.
141. Laitinen M, Vuojolainen K, Jaatinen R, et al. A novel growth differentiation factor-9 (GDF-9) related factor is co-expressed with GDF-9 in mouse oocytes during folliculogenesis. Mech Dev 1998;78: 135–140.
142. Aaltonen J, Laitinen MP, Vuojolainen K, et al . Human growth differentiation factor 9 (GDF-9) and its novel homolog GDF-9B are expressed in oocytes during early folliculogenesis. J Clin Endocrinol Metab 1999;84:2744–2750.
143. Silva JR, van den Hurk R, van Tol HT, Roelen BA, Figueiredo JR. Expression of growth differentiation factor 9 (GDF9), bone morphogenetic protein 15 (BMP15), and BMP receptors in the ovaries of goats. Mol Reprod Dev 2005;70:11–19.
144. Pennetier S, Uzbekova S, Perreau C, Papillier P, Mermillod P, Dalbies-Tran R. Spatio-temporal expression of the germ cell marker genes MATER, ZAR1, GDF9, BMP15, andVASA in adult bovine tissues, oocytes, and preimplantation embryos. Biol Reprod 2004;71:1359–1366.
145. McGrath SA, Esquela AF, Lee SJ. Oocyte-specific expression of growth/differentiation factor-9. Mol Endocrinol 1995;9:131–136.
146. Hayashi M, McGee EA, Min G, et al. Recombinant growth differentiation factor-9 (GDF-9) enhances growth and differentiation of cultured early ovarian follicles. Endocrinology 1999;140: 1236–1244.
147. Jaatinen R, Laitinen MP, Vuojolainen K, et al. Localization of growth differentiation factor-9 (GDF-9) mRNA and protein in rat ovaries and cDNA cloning of rat GDF-9 and its novel homolog GDF-9B. Mol Cell Endocrinol 1999;156:189–193.
148. Bodensteiner KJ, Clay CM, Moeller CL, Sawyer HR. Molecular cloning of the ovine growth/ differentiation factor-9 gene and expression of growth/differentiation factor-9 in ovine and bovine ovaries. Biol Reprod 1999;60:381–386.
149. Prochazka R, Nemcova L, Nagyova E, Kanka J. Expression of growth differentiation factor 9 messenger RNA in porcine growing and preovulatory ovarian follicles. Biol Reprod 2004;71:1290–1295.
150. Sidis Y, Fujiwara T, Leykin L, Isaacson K, Toth T, Schneyer AL. Characterization of inhibin/activin subunit, activin receptor, and follistatin messenger ribonucleic acid in human and mouse oocytes: evidence for activin's paracrine signaling from granulosa cells to oocytes. Biol Reprod 1998;59: 807–812.
151. Jaatinen R, Bondestam J, Raivio T, et al. Activation of the bone morphogenetic protein signaling pathway induces inhibin beta(B)-subunit mRNA and secreted inhibin B levels in cultured human granulosa-luteal cells. J Clin Endocrinol Metab 2002;87:1254–1261.
152. Wu TC, Jih MH, Wang L, Wan YJ. Expression of activin receptor II and IIB mRNA isoforms in mouse reproductive organs and oocytes. Mol Reprod Dev 1994;38:9–15.
153. Cameron VA, Nishimura E, Mathews LS, Lewis KA, Sawchenko PE, Vale WW. Hybridization histochemical localization of activin receptor subtypes in rat brain, pituitary, ovary, and testis. Endocrinology 1994;134:799–808.
154. Zhao J, Taverne MA, van der Weijden GC, Bevers MM, van den Hurk R. Effect of activin A on in vitro development of rat preantral follicles and localization of activin A and activin receptor II. Biol Reprod 2001;65:967–977.
155. Drummond AE, Le MT, Ethier JF, Dyson M, Findlay JK. Expression and localization of activin receptors, Smads, and beta glycan to the postnatal rat ovary. Endocrinology 2002;143:1423–1433.
156. Pangas SA, Rademaker AW, Fishman DA, Woodruff TK. Localization of the activin signal transduction components in normal human ovarian follicles: implications for autocrine and paracrine signaling in the ovary. J Clin Endocrinol Metab 2002;87:2644–2657.

157. Eramaa M, Hilden K, Tuuri T, Ritvos O. Regulation of inhibin/activin subunit messenger ribonucleic acids (mRNAs) by activin A and expression of activin receptor mRNAs in cultured human granulosa-luteal cells. Endocrinology 1995;136:4382–4389.
158. Osterlund C, Fried G. TGFbeta receptor types I and II and the substrate proteins Smad 2 and 3 are present in human oocytes. Mol Hum Reprod 2000;6:498–503.
159. He WW, Gustafson ML, Hirobe S, Donahoe PK. Developmental expression of four novel serine/threonine kinase receptors homologous to the activin/transforming growth factor-beta type II receptor family. Dev Dyn 1993;196:133–142.
160. Kano K, Notani A, Nam SY, Fujisawa M, Kurohmaru M, Hayashi Y. Cloning and studies of the mouse cDNA encoding Smad3. J Vet Med Sci 1999;61:213–219.
161. Yamamoto N, Christenson LK, McAllister JM, Strauss JF, 3rd. Growth differentiation factor-9 inhibits 3′5′-adenosine monophosphate-stimulated steroidogenesis in human granulosa and theca cells. J Clin Endocrinol Metab 2002;87:2849–2856.

28 Regulation of the Transforming Growth Factor-β Superfamily by Betaglycan

Fernando López-Casillas, M. Magdalena Vilchis-Landeros, José Esparza-López, M. Verónica Ponce-Castañeda, N. Gabriela Velasco-Loyden, Patricia Juárez, Jose Luis Montiel, and Valentín Mendoza

CONTENTS

TRANSFORMING GROWTH FACTOR-β SUPERFAMILY, RECEPTORS AND SIGNALING
BETAGLYCAN AND ENDOGLIN AS TRANSFORMING GROWTH FACTOR-β CORECEPTORS
BETAGLYCAN GENE AND ITS EXPRESSION
BETAGLYCAN STRUCTURAL AND FUNCTIONAL FEATURES
BETAGLYCAN LIGAND BINDING DOMAINS
BETAGLYCAN INTERACTIONS WITH TYPE II RECEPTORS
BETAGLYCAN TRANSMEMBRANE AND CYTOPLASMIC REGIONS
BETAGLYCAN AS A HEPARAN AND CHONDROITIN SULFATE PROTEOGLYCAN
BETAGLYCAN NULL MICE
SOLUBLE BETAGLYCAN AND ITS POTENT TRANSFORMING GROWTH FACTOR-β NEUTRALIZING ACTIVITY
THE SHEDDING OF BETAGLYCAN ECTODOMAIN AS A POTENTIAL ANTI-TGF-β TARGET
CONCLUDING REMARKS
ACKNOWLEDGMENTS
REFERENCES

Abstract

The type III transforming growth factor-β (TGF-β) receptor (TβRIII), also known as betaglycan (BG), is a membrane proteoglycan coreceptor that modulates the action of diverse members of the TGF-β superfamily of growth factors. Membrane bound BG is the precursor of a soluble receptor

From: *Cancer Drug Discovery and Development: Transforming Growth Factor-β in Cancer Therapy, Vol I: Basic and Clinical Biology*
Edited by: Sonia B. Jakowlew © Humana Press Inc., Totowa, NJ

form generated by the proteolytic cleavage (shedding) of its extracellular region. While membrane BG enhances TGF-β binding to the type II TGF-β receptor, thereby increasing the activity of the factor *(1)*, the soluble form sequesters the ligand, acting as a neutralizing agent of TGF-β *(2)*. Recombinant soluble BG (rSBG) has been used successfully for treatment of diverse experimental pathologies in which TGF-β plays a physiopathological role. Given the recent discovery that BG shedding can be regulated, TGF-β modulatory properties of BG make it an attractive target for pharmacological regulation of TGF-β action. Despite its reputation as an "accessory" TGF-β receptor, the mice lacking BG (BG-null or TβRIII$^{-/-}$) exhibited an embryonic lethal phenotype *(3)*, indicating that this coreceptor has essential, not yet identified, cellular functions. Here, we review the current knowledge of BG and discuss future scenarios for BG research and applications.

Key Words: Betaglycan; betaglycan-null mice; ectodomain shedding; heart development; inhibin; nephropathies; type III TGF-β receptor; TGF-β; TGF-β inhibitors; tumor xenografts; tuberculosis.

1. TRANSFORMING GROWTH FACTOR-β SUPERFAMILY, RECEPTORS AND SIGNALING

Transforming growth factor-β (TGF-β) is a multifunctional cytokine that regulates cell proliferation, differentiation, development, embryogenesis, tissue repair, and the immune response *(4–7)*. Given the many physiological processes in which TGF-β plays a central role, it is not surprising that disturbances in its regulation or signaling mechanisms lead to diseases encompassing almost every medical field *(8,9)*. TGF-β, and in general all other members of the superfamily, signals through type I and type II receptors. These are single spanning membrane glycoproteins with extracellular regions that bind ligand and cytoplasmic regions containing serine/threonine protein kinases. Ligand binding promotes the association of the type I and the type II receptors. In this complex, type I receptor kinase is phosphorylated and activated by the constitutively active type II receptor kinase. Active type I kinase phosphorylates R-Smads, receptor-regulated members of a family of transcriptional factors, the Smads. Complexes of phospho-R-Smads and common Smads (Co-Smads) migrate to the cell nucleus to regulate the expression of TGF-β target genes. Because comprehensive reviews discussing the TGF-β signaling pathways have been published recently (*see [10–12]*), they will not be discussed any further.

2. BETAGLYCAN AND ENDOGLIN AS TRANSFORMING GROWTH FACTOR-β CORECEPTORS

Growth factor coreceptors are molecules capable of binding the factor but lacking any catalytic signaling activity on their own. Instead, they assist the authentic signaling receptor to initiate the signaling cascade. Well-known examples of receptors of this type are the cell surface heparan sulfate proteoglycans, which are necessary for the functional interaction of fibroblast growth factor (FGF) with its signaling tyrosine kinase receptor *(13,14)*. TGF-β has two coreceptors, betaglycan (BG) and endoglin, which occasionally are named together as the "type III TGF-β receptors", however, this designation is more commonly associated with BG. Endoglin and BG are transmembrane glycoproteins with large extracellular regions that bind TGF-β, and small cytoplasmatic regions with no clearly identifiable signaling motif *(15)*. Tissue culture cells lacking BG or endoglin are responsive to TGF-β, thereby their fame as "accessory" and nonessential TGF-β coreceptors. When BG and endoglin are present, they modulate the interaction of TGF-β with the type II and I receptors thereby regulating the outcome of the TGF-β stimulation *(16,17)*. The in vivo effects of the absence or mutation of BG and endoglin revealed the fundamental biological relevance of these TGF-β coreceptors. Endoglin is one of the 2 target genes for the dominantly inherited vascular disorder Hereditary Hemorrhagic Telangiectasia type 1 *(18)*. Accordingly, the endoglin-null mice exhibit a lethal phenotype, dying at 10.5 d of gestational age owing to defects in blood vessel and heart formation *(19)*. The knockout of BG gene in mice also results in embryonic lethality, which is owing to malformations in heart, liver and erythropoiesis *(3)*.

3. BETAGLYCAN GENE AND ITS EXPRESSION

Human betaglycan is encoded by a single gene located in chromosome 1 (p32–33) (Ref: *[20]* and NCBI genomic database). The human BG gene extends over a region 272,000 bp and it is expressed as a 7000 nucleotides long mRNA *(21)*. The primary structures of human, rat, mouse, pig and chicken betaglycans indicate that it is a much conserved protein *(22)*. Searches at NCBI genomic databases have not revealed any BG ortholog in nonvertebrate species. Specifically, the *Drosophila melanogaster* and *Caenorhabditis elegans* genomes do not contain any gene that could be considered a BG ortholog. This absence is unfortunate because prevents the use of these powerful model organisms to explore the genetics of BG. On the other hand, it suggests a crucial role for BG in vertebrate biology. The mouse BG gene promoter region has been identified and used to show that BG gene is transcriptionally upregulated during the C2C12 myoblast differentiation *(23)*. The BG gene promoter activity is stimulated by MyoD and retinoic acid and repressed by TGF-β *(23)*, all of them factors involved in the regulation of myogenesis. The rat BG gene promoter region, which is very similar to its murine counterpart, has been used to demonstrate regulation of BG expression in osteoblasts *(24)*.

4. BETAGLYCAN STRUCTURAL AND FUNCTIONAL FEATURES

Betaglycan is a membrane proteoglycan, containing heparan and chondroitin sulfate chains. It is present in epithelial cell types, but absent in endothelial, myoblasts and hematopoietic lineages *(21,25)*. Betaglycan is a versatile coreceptor that binds TGF-β, Inhibin A and bFGF, however, experimental evidence of its function as a coreceptor have only been demonstrated for members of the TGF-β superfamily *(16,26–28)*. Betaglycan is a highly conserved membrane protein with an evident modular design, made up of a large ectodomain (785 residues in the rat), a single spanning transmembrane region and a short (43 residues) cytoplasmic region (Fig. 1). Binding of its various ligands is a function of the betaglycan ectodomain, a region that exhibits segments of well-defined similarity with other coreceptors and extracellular proteins *(15,29)*. TGF-β and inhibin A bind betaglycan ectodomain via its core protein, while the binding of bFGF is through the heparan sulfate chains. Two conserved Ser-Gly sequences (Serines 535 and 546 in the rat, and 533 and 544 in the mouse) have been identified as the glycosaminoglycan (GAG) attachment sites. Mutation of these serines for alanines results in a receptor that despite the lack of these complex carbohydrates, still binds TGF-β and inhibin *(16,22,30)*.

5. BETAGLYCAN LIGAND BINDING DOMAINS

Betaglycan binds all three TGF-β isoforms with high affinity, however, the affinity for TGF-β2 is the highest, being one order of magnitude higher than for TGF-β1 *(31–33)*. When betaglycan ectodomain is expressed as a recombinant soluble receptor, it exhibits the same TGF-β isoform binding selectivity *(2)*. The unusually large extracellular region of betaglycan contains two independent ligand binding domains that split the region into approximately equal halves *(16,30,34)*. The E domain (for Endoglin-similarity) corresponds to residues 45–409 and the U domain (for Uromodulin-similarity) to residues 410–781 of the rat wild-type receptor *(16)*. In spite of their primary sequence dissimilarity, when these

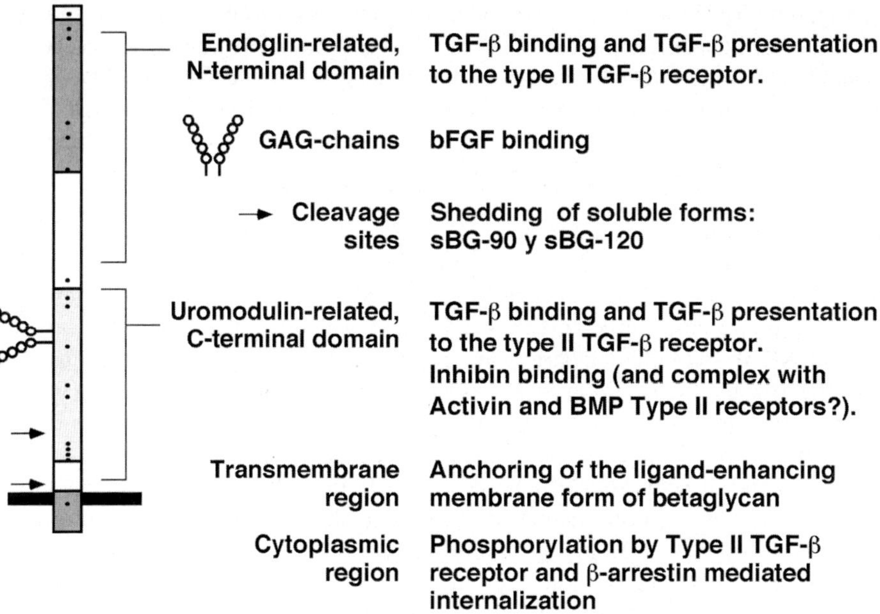

Fig. 1. Betaglycan structural and functional modules. Regions of betaglycan and their assigned functions are shown. Uromodulin related domain has inhibin binding activity, however, it has not been shown to mediate on its own inhibin action. The precise location of the cleavage sites resulting in the shedding of sBG-90 and sBG-120 has not been determined. Glycosaminoglycan (GAG) chains are not shown to scale. The mass of the core protein (including its several N-linked glycosilations) is 120–130 kDa, however, when GAGs are attached it has a heterogeneous mass greater than 300 kDa *(26)*.

domains are expressed as truncated membrane receptors, or as soluble recombinant proteins, they bind TGF-β similarly to each other and to the wild-type receptor *(16)*. Betaglycan core protein also binds inhibin *(28)* and this ability resides in the U domain *(16)*. The existence of two *bona fide*, independently folded, TGF-β-binding domains in betaglycan ectodomain is further supported by partial proteolysis experiments. Mild plasmin digestion of both the membrane bound and the recombinant soluble betaglycan yields two major fragments that nicely match with the E and U domains identified by mutagenesis of the cDNA (Montiel et al., in preparation).

6. BETAGLYCAN INTERACTIONS WITH TYPE II RECEPTORS

One of the most interesting properties of BG is its ability to form ligand-induced complexes with Type II receptors for several members of the TGF-β superfamily, namely, the TGF-β, Activin and bone morphogenetic protein (BMP) Type II receptors. The complex formed by BG + TGF-β + Type II TGF-β receptor results in the enhancement of the TGF-β signaling, an effect that has been named the "TGF-β presentation" function of BG *(1)*. This function, which is especially noticeable for the TGF-β2 isoform, is exhibited by both betaglycan TGF-β binding domains *(16)*. Likewise, BG also associates with the Activin and BMP Type II receptors. This association depends on the presence of inhibin and is strong enough to permit the immunoprecipitation of complexes formed by BG + Inhibin + Activin Type II receptor and by BG + Inhibin + BMP Type II receptor *(28,35)*. The presence of the Activin and BMP Type II receptors in these heterogeneous complexes prevents their participation in their regular signaling complexes with Activin or BMP and their respective Type I receptors. This "sequestration" results in the inhibin-mediated abolishment of the BMP and Activin actions. So far, this is the best explanation for the inhibin antagonism of Activin action, an important regulatory component in the hypothalamic-pituitary-gonad axis *(28,36)*. Because of these versatile interactions, it is likely that the presence of BG could affect the outcome of physiological or developmental processes in which diverse members of the TGF-β superfamily act in concert.

7. BETAGLYCAN TRANSMEMBRANE AND CYTOPLASMIC REGIONS

Betaglycan transmembrane and intracellular regions are its best conserved portions and, surprisingly, they have a high identity (63%) to the same domains of endoglin *(21,22,25)*. These features suggested that these regions in BG and endoglin should be endowed with important functions. It has been reported that GIPC, a PDZ-domain containing protein interacted with BG cytoplasmatic region; however, the functional relevance of this interaction has not been established *(37)*. The cytoplasmatic regions of all sequenced betaglycans have a high content (42%) of serines and threonines that are amenable of phosphorylation. During the formation of the TGF-β presentation complex the Type II TGF-β receptor phosphorylates Thr841, a modification that allows the binding of β-arrestin *(38)*. The functional consequence of this series of events is the endocytosis of BG and the Type II receptor complex, thereby downregulating the TGF-β signal. This work proved that upon phosphorylation, the BG tail could perform as a docking site for cytosolic proteins, opening the possibility that other receptors that associate with BG in a ligand dependent manner (e.g., Activin or BMP Type II receptors) could phosphorylate and activate a similar docking function in the BG tail.

8. BETAGLYCAN AS A HEPARAN AND CHONDROITIN SULFATE PROTEOGLYCAN

The functions of BG glycosaminoglycan (GAG) chains are not well understood. The interactions of BG with the members of the TGF-β superfamily discussed above do not require the presence of GAG chains. As a matter of fact, BG is considered a "part-time" proteoglycan because a given fraction of cell surface BG is not modified with these complex carbohydrates *(26)*. Nonetheless, it has been shown that in renal epithelial LLC-PK1 cells the BG GAG chains have a negative effect in the TGF-β signaling *(39)*. Apparently, in this particular cell line the large amounts of GAG chains that are attached to BG play a sterical hindrance role that prevents the formation of a signaling complex between the type I and II TGF-β receptors. This finding is relevant because, as it has been shown for other heparans sulfate proteoglycans *(40)*, it demonstrates that the quantity and/or quality of the GAG chains in BG could be a variable of the cell lineage in which it is expressed. Finally, despite the fact that BG binds basic FGF (bFGF) through its heparan sulfate chains there is no cellular function attributed to this union *(27)*. However, given the fact that the two types of GAGs in BG have been involved in regulation of many developmental factors and processes *(41,42)*, it is imperative to study whether or not the BG GAG chains have such functional role. If they do, they would help to account for the lethal phenotype of the BG-null mouse.

9. BETAGLYCAN NULL MICE

The embryonic lethal phenotype exhibited by the mice lacking BG (BG-null or $T\beta RIII^{-/-}$) was very surprising, because it was not expected for an apparently dispensable receptor *(3)*. $T\beta RIII^{-/-}$ mice die at d 13.5 of embryonic development owing to proliferative defects in heart and liver apoptosis. A secondary effect of hepatic apoptosis was the disruption of erythropoiesis. In the BG-null embryo the myocardium fails to proliferate resulting in thin myocardial walls and defective septum. As expected, murine embryonic fibroblasts derived form the TβRIII$^{-/-}$ embryos had a decreased sensitivity to TGF-β2 stimulation, which was postulated as a major mechanism accounting for the phenotype *(3)*. However, this explanation is not totally satisfactory. If the only in vivo function of BG was its role in the TGF-β2 signaling, then the BG-null and TGF-β2-null mice should had overlapping phenotypes, which is not the case. The TGF-β2-null mice exhibit perinatal mortality and only one third of the TGF-β2$^{-/-}$ is born alive; however, they die shortly after because of collapsed airway passages *(43)*. Most of the developmental defects in the TGF-β2$^{-/-}$ mice are evident late in embryogenesis. At d 18.5E these defects are mainly craniofacial, skeletal, in the eye, ear, and urogenital tract. The heart defects observed in these animals are not fully penetrant and affect mainly the valves and the origin of major vessels like aorta and pulmonary arteries, and importantly, no liver defects were reported *(43)*. The discrepancies between the BG-null and TGF-β2-null mice demonstrate that in vivo, BG function is not limited to TGF-β2 signaling. Furthermore, they strongly suggest that BG is required for other physiological and/or developmental processes that determine the development of heart and liver.

10. SOLUBLE BETAGLYCAN AND ITS POTENT TGF-β NEUTRALIZING ACTIVITY

A naturally occurring soluble form of betaglycan is found in serum and extracelular matrices *(44)*. As is the case for many other soluble receptor and growth factors, membrane betaglycan is the precursor of the soluble form, which is generated by a proteolytic cleavage of its extracellular region, a process commonly referred as "ectodomain shedding" *(45,46)*. The function of the soluble form of BG has been explored using a recombinant version, which is produced, in insect cells via baculoviral infection *(30)*. In contrast to its membrane-bound precursor, the soluble betaglycan is a potent TGF-β antagonist *(2)*. Human and rat rSBG are expressed in insect cells as core proteins totally devoid of GAG chains, despite the fact that their encoding cDNAs are wild type at the GAG attachment sites *(2,47)*. rSBG binds TGF-β with affinities and isoform selectivity that are similar to its membrane-bound counterpart. Using saturation binding kinetics we have demonstrated that rSBG binds TGF-β1 with a Kd of 3.5 nM, which is in close agreement with that of the soluble BG found in natural sources *(2,44)*. Heterologous TGF-β binding competition assays have shown rSBG binds the other isoforms with the following relative binding affinities: TGF-β2 > TGF-β3 > TGF-β1. Interestingly, rSBG does not bind inhibin on its own, suggesting that for inhibin binding, BG requires the presence of the Activin Type II receptor. The TGF-β neutralizing potency of rSBG corresponds to its TGF-β binding affinities, being 10 times more potent against TGF-β2. This neutralizing activity is better than, or at least comparable with, those reported for other proteins with anti-TGF-β activity *(2)*. This property of rSBG has been used in several animal models of disease in which the overproduction of TGF-β plays a pathogenic role. The group of Lu Zhe Sun in San Antonio has successfully used rSBG to achieve a decrease in the growth and metastasic potential of breast and prostate cancer xenografts *(47,48)*. These rSBG effects depend on its ability to block angiogenesis and MMP9 activity in the vicinity and within the tumor, prooncogenic activities of TGF-β *(49)*. In addition to its great potential in cancer therapeutics, rSBG has proven extremely useful for disease prevention in other animal models. The kidney damage observed in the genetically obese and diabetic db/db mouse has been regarded as a good model of the human diabetic nephropathy *(50)*. As in practically all forms of nephropathy, in the one developed by the db/db mouse, TGF-β has a substantial physiopathological role *(51–53)*. We have administered rSBG to db/db mice and significantly prevented pathological and functional kidney damage *(54)*. Finally, other model in which rSBG has shown a tremendous therapeutic potential is pulmonary tuberculosis. The control of this infectious disease is heavily dependent on the cell-mediated immune response, which in turn is regulated by a local microenvironment of cytokines in which TGF-β has a prominent role *(55,56)*. In a murine model of tuberculosis that resembles the human disease, it has been shown that an increase in the levels of lung TGF-β correlates with a decrease of the Th1 and an increase of the Th2 pattern of lung cytokines as well as with the progression of the disease *(57)*. Administration of rSBG during this phase of the disease prevents the shift of the Th1/Th2 cytokines, restores the cell-mediated immune response and decreases the pulmonary bacilli load *(58)*. This later example demonstrates the wide spectrum of TGF-β physiopathological engagements, a fact that widens the range of diseases and applications for the many anti-TGF-β agents that have designed for cancer treatment *(59)*.

11. THE SHEDDING OF BETAGLYCAN ECTODOMAIN AS A POTENTIAL ANTI-TGF-β TARGET

Compared to the many other anti-TGF-β agents currently available, soluble betaglycan scores very high because of its unique biochemical and cellular properties. The first is its high affinities for TGF-β, discussed in Sections 5 and 10. Secondly, being a molecule that is normally found in serum and extracellular matrices, it shall not elicit an immune response, a shortcoming of neutralizing antibodies. However, the crucial property of soluble betaglycan that makes it stand out above other TGF-β inhibitors is the fact that its in vivo generation is amenable to regulation. Ever because the discovery that soluble betaglycan was, in opposition to its membrane form, an inhibitor of TGF-β *(30)*, it was proposed that this receptor could play a relevant role in the physiological regulation of TGF-β, working like a "switch" that could turn-on its functions as a membrane bound receptor, or could turn them off as a soluble one *(60)*. With the discovery that the shedding of membrane betaglycan is a regulated process *(46)*, and given BG widespread expression in almost every cell lineage, it is not difficult to foresee new pharmacological strategies for an anti-TGF-β therapy based on the regulation of BG shedding. If betaglycan shedding could be subjected to pharmacological regulation then, it would be possible to control the relative ratio of the membrane and soluble forms of the receptor, providing a way to switch on or off TGF-β action. Because of this possibility, we have been seeking for the proteases that could perform the regulated shedding of BG (Fig. 2). So far, we have identified membrane type-1 matrix metalloprotease (MT1-MMP) as a BG sheddase regulated by pervanadate, a general tyrosine phosphatase inhibitor *(46)*. MT1-MMP is the sheddase that generates sBG-90, a short form of soluble betaglycan with a mass of approx 90 kDa. When compared to the recombinant soluble BG, sBG-90 has a limited TGF-β neutralizing activity, probably because of the loss of portions of the U domain (Velasco-Loyden et al., unpublished observations). Other still unknown pervanadate-sensitive sheddase(s) generate sBG-120, a longer form of soluble betaglycan, which is predicted to have both ligand binding domains intact, and therefore, a full anti-TGF-β neutralizing activity. The identification of the sBG-120 sheddase, as well as the mechanisms and molecules that permit its regulation, shall provide the tools for the design of novel anti-TGF-β agents based betaglycan shedding regulation.

12. CONCLUDING REMARKS

At the time of its cloning *(21,25)*, nobody could have foreseen the many pathways that betaglycan would travel in these 15 yr. Its identification as a functional inhibin receptor unveiled betaglycan's versatility. The in vivo anti-TGF-β activity of rSBG in several animal models of disease points to its great therapeutical potential. However, the discovery that betaglycan gene absence results in a lethal phenotype indicates that the most exciting part of betaglycan research is yet to come.

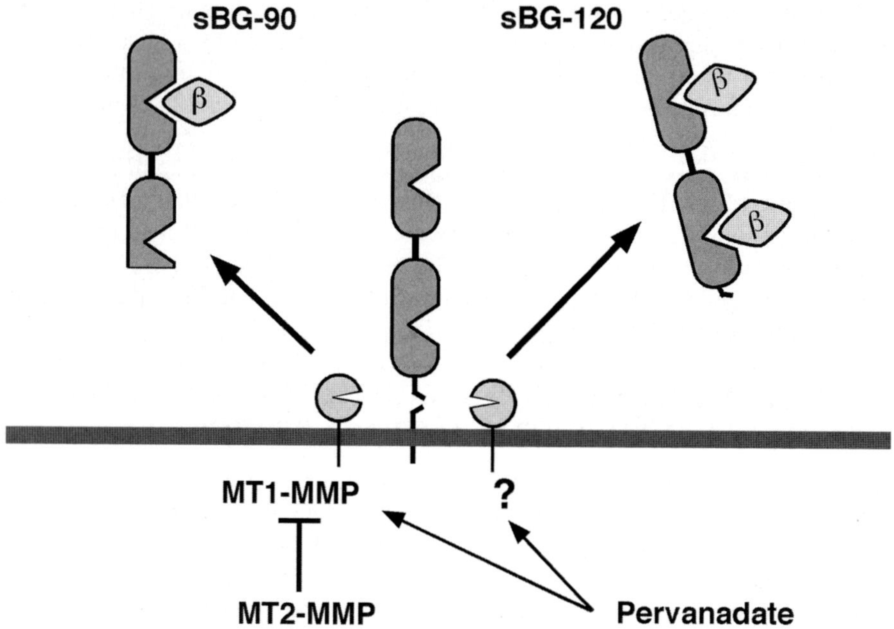

Fig. 2. The betaglycan sheddases. Two pervanadate-sensitive sheddases have been found to cleave BG ectodomain, resulting in a soluble BG of 90 kDa (sBG-90) and another of 120 kDa (sBG-120). Membrane type 1 Matrix metalloprotease (MT1-MMP) has been shown to produce sBG-90. MT1-MMP is downregulated by MT2-MMP. The metalloprotease generating sBG-120 has not been identified. Modified from Velasco-Loyden et al. *(46)*.

ACKNOWLEDGMENTS

Work in the laboratory has been supported by grants from Howard Hughes Medical Institute, International Centre for Genetic Engineering and Biotechnology, Universidad Nacional Autonóma de México and Consejo Nacional de Ciencia y Tecnología.

REFERENCES

1. López-Casillas F, Wrana JL, Massagué J. Betaglycan presents ligand to the TGFβ signaling receptor. Cell 1993;73:1435–1444.
2. Vilchis-Landeros MM, Montiel JL, Mendoza V, Mendoza-Hernández G, López-Casillas F. Recombinant soluble betaglycan is a potent and isoform-selective transforming growth factor-β neutralizing agent. Biochem J 2001;355:215–222.
3. Stenvers KL, Tursky ML, Harder KW, et al. Heart and liver defects and reduced transforming growth factor β2 sensitivity in transforming growth factor β type III receptor-deficient embryos. Molecular and Cellular Biology 2003;23:4371–4385.
4. Massagué J. The transforming growth factor-β family. Annu Rev Cell Biol 1990;6:597–641.
5. Roberts AB, Sporn MB. Physiological actions and clinical applications of transforming growth factor-β (TGF-β). Growth Factors 1993;8:1–9.
6. Kingsley DM. The TGF-β superfamily:new members, new receptors, and new genetic tests of function in different organisms. Genes and Development 1994;8:133–146.
7. Letterio JJ, Roberts AB. Regulation of immune responses by TGF-β. Annu Rev Immunol 1998;16: 137–154.
8. Blobe GC, Schiemann WP, Lodish HF. Role of transforming growth factor β in human disease. N Engl J Med 2000;342:1350–1358.

9. Massagué J, Blain SW, Lo RS. TGFβ signaling in growth control, cancer, and heritable disorders. Cell 2000;103:295–309.
10. Derynck, R, Zhang YE. Smad-dependent and Smad-independent pathways in TGF-β family signalling. Nature 2003;425:577–584.
11. Massagué J, Chen Y-G. Controlling TGF-β signaling. Genes and Development 2000;14:627–644.
12. Massagué J. G1 cell-cycle control and cancer. Nature 2004;432:298–306.
13. Park PW, Reizes O, Bernfield M. Cell surface heparan sulfate proteoglycans: selective regulators of ligand-receptor encounters. J Biol Chem 2000;275:29,923–29,926.
14. Ornitz DM. FGFs, heparan sulfate and FGFRs: complex interactions essential for development. BioEssays 2000;22:108–112.
15. Cheifetz S, Bellón T, Calés C, et al. Endoglin is a component of the transforming growth factor-β receptor system in human endothelial cells. J Biol Chem 1992;267:19,027–19,030.
16. Esparza-López J, Montiel JL, Vilchis-Landeros MM, Okadome T, Miyazono K, López-Casillas F. Ligand binding and functional properties of betaglycan, a co-receptor of the transforming growth factor-β superfamily. J Biol Chem 2001;276:14,588–14,596.
17. Pece-Barbara N, Vera S, Kathirkamathamby K, et al. Endoglin null endothelial cells proliferate faster and are more responsive to TGF-β1 with higher affinity receptors and an activated Alk1 pathway. J Biol Chem 2005;280:27,800–27,808.
18. McAllister KA, Grogg KM, Johnson DW, et al. Endoglin, a TGF-β binding protein of endothelial cells, is the gene for hereditary haemorrhagic telangiectasia type 1. Nat Genet 1994;8:345–351.
19. Bourdeau A, Dumont DJ, Letarte MA. Murine model of hereditary hemorrhagic telangiectasia. J Clin Invest 1999;104:1343–1351.
20. Jonhson DW, Qumsiyeh M, Benkhalifa. M, Marchuck DA. Assignment of human transforming growth factor-β type I and type III receptor genes (TGFR1 and TGFRIII) to 9q33-q34 and 1p32-p33, respectively. Genomics 1995;28:356–357.
21. Wang X-F, Lin HY, Ng-Eaton E, Downward J, Lodish HF, Weinberg RA. Expression cloning and characterization of the TGF-β type III receptor. Cell 1991;67:797–805.
22. Ponce-Castañeda MV, Esparza-López J, Vilchis-Landeros MM, Mendoza RV, López-Casillas F. Murine betaglycan primary structure, expression and glycosaminoglycan attachment sites. Biochim Biophys Acta 1998;1384:189–196.
23. López-Casillas F, Riquelme C, Pérez-Kato Y, et al. Betaglycan expression is transcriptionally up-regulated during skeletal muscle differentiation. J Biol Chem 2003;278:382–390.
24. Ji C, Chen Y, McCarthy TL, Centrella M. Cloning the promoter for transformimg growth factor-β type III receptor. Basal and conditional expression in fetal rat osteoblasts. J Biol Chem 1999;274: 30,487–30,494.
25. López-Casillas F, Cheifetz S, Doody J, Andres JL, Lane WS, Massagué J. Structure and expression of the membrane proteglycan betaglycan, a component of the TGF-β receptor system. Cell 1991;67: 785–795.
26. Cheifetz S, Andres JL, Massagué J. The transforming growth factor-β receptor type III is a membrane proteoglycan. Domain structure of the receptor. J Biol Chem 1988;263:16,984–16,991.
27. Andres JL, DeFalcis D, Noda M, Massagué Y. Binding of two growth factor families to separated domains of the proteoglycan betaglycan. J Biol Chem 1992;267:5727–5930.
28. Lewis KA, Gray PC, Blount AL, et al. Betaglycan binds inhibin and can mediate functional antagonism of activin signalling. Nature 2000;404:411–414.
29. Bork P, Sander CA. large domain common to sperm receptors (ZP2 and Zp3) and TGF-β type III receptor. FEBS Lett 1992;300:237–240.
30. López-Casillas F, Payne HM, Andres JL, Massagué J. Betaglycan can act as a dual modulator of TGF-β access to signaling receptors:mapping of ligand binding and GAG attachment sites. J Cell Biol 1994;124:557–568.
31. Segarini PR, Roberts AB, Rosen DM, Seyedin SM. Membrane binding characteristics of two forms of transforming growth factor-β. J Biol Chem 1987;262:14,655–14,662.
32. Mitchell EJ, Fitz-Gibbon L, O'Connor-McCourt MD. Subtypes of betaglycan and of type I and type II transforming growth factor-β (TGF-β) receptors with different affinities for TGF-β1 and TGF-β2 are exhibited by human placental trophoblast cells. J Cell Physiol 1992; 150: 334–343.
33. Andres JL, Rönnstrand L, Cheifetz S. Massagué J. Purification of the transforming growth factor-β (TGF-β) binding proteoglycan betaglycan. J Biol Chem 1991;266:23,282–23,287.

34. Pepin MC, Beauchemin M, Plamondon J, O'Connor-McCourt MD. Mapping of the ligand binding domain of the transforming growth factor β receptor type III by deletion mutagenesis. Proc Natl Acad Sci USA 1994;91:6997–7001.
35. Wiater E, Vale W. Inhibin is an antagonist of bone morphogenetic protein signaling. J Biol Chem 2003;278:7934–7941.
36. Gray PC, Bilezikjian LM, Vale W. Antagonism of activin by inhibin and inhibin receptors: a functional role for betaglycan. Mol Cell Endocrinol 2002;188:254–260.
37. Blobe GC, Liu X, Fang SJ, How T, Lodish HF. A novel mechanism for regulating transforming growth factor β (TGF-β) signaling. J Biol Chem 2001;276:39,608–39,617.
38. Chen W, Kirkbride KC, How T, et al. β-arrestin 2 mediates endocytosis of type III TGF-β receptor and down-regulation of its signaling. Science 2003;301:1394–1397.
39. Eickelberg O, Centrella M, Reiss M, Kashgarian M, Wells RG. Betaglycan inhibits TGF-β signaling by preventing type I-type II receptor complex formation. J Biol Chem 2002;277:823–829.
40. Allen BL, Rapraeger AC. Spatial and temporal expression of heparan sulfate in mouse development regulates FGF and FGF receptor assembly. J Cell Biol 2003;163:637–648.
41. Hacker U, Nybakken K, Perrimon N. Heparan sulphate proteoglycans: the sweet side of development. Nature Reviews: Mol Cell Endocrinol 2005;6:530–541.
42. Klüppel M, Wigth TN, Chan C, Hinek A, Wrana JL. Maintenance of chondroitin sulfation balance by chondroitin-4-sulfotransferase 1 is required for chondrocyte development and growth factor signaling during cartilage morphogenesis. Development 2005;132:3989–4003.
43. Sanford LP, Ormsby I, Gittenberger-de Groot AC, et al. TGFβ2 knockout mice have multiple developmental defects that are non-overlapping with other TGFβ knockout phenotypes. Development 1997; 124:2659–2670.
44. Andres JL, Stanley K, Cheifetz S, Massagué J. Membrane-anchored and soluble forms of betaglycan, a polymorphic proteoglycan that binds transforming growth factor-β. J Cell Biol 1989; 109:3137–3145.
45. Arribas J, López.-Casillas F, Massagué J. Role of the juxtamembrane domains of the transforming growth factor-α precursor and the β-amyloid precursor protein in regulated ectodomain shedding. J Biol Chem 1997;272:17,160–17,165.
46. Velasco-Loyden G, Arribas J, López-Casillas F. The shedding of betaglycan is regulated by pervanadate and mediated by membrane type matrix metalloprotease-1. J Biol Chem 2004;279:7721–7733.
47. Bandyopadhyay A, López-Casillas F, Malik SN, et al. Antitumor activity of a recombinant soluble betaglycan in human breast cancer xenograft. Cancer Research 2002;62:4690–4695.
48. Bandyopadhyay A, Wang L, López-Casillas F, Mendoza V, Yeh IT, Sun LZ. Systemic administration of a soluble betaglycan suppresses tumor growth, angiogenesis, and matrix metalloproteinase-9 expression in a human xenograft model of prostate cancer. The Prostate 2005;63:81–90.
49. Roberts AB,Wakefield LM. The two faces of transforming growth factor β in carcinogenesis. Proc Natl Acad Sci USA 2003;100:8621–8623.
50. Sharma K, McCue P, Dunn SR. Diabetic kidney disease in the *db/db* mouse. Am J Physiol Renal Physiol 2003;284:F1138–F1144.0
51. Sharma K, Ziyadeth FN. The emerging role of transforming growth factor-β in kidney diseases. Am J Physiol 1994;266:F829–F842.
52. Ziyadeh FN, Hoffman BB, Han DC, Iglesias de la Cruz MC. Long-term prevetion of renal insufficiency, excess matrix gene expresion, and glomerular mesangial matrix expansion by treatment with monoclonal antitransforming growth factor-β antibody in db/db diabetic mice. Proc Natl Acad Sci USA 2000;97:8015–8020.
53. Ziyadeh FN. Mediators of diabetic renal disease: The case for TGF-β as the major mediator. J Am Soc Nephrol 2004;15:S55–S57.
54. Juaréz P, Vilchis-Landeros MM, Ponce-Corcia J, et al. Soluble betaglycon reduces renal damage progression in db/db mice. AM J Physiol Renal Physiol 2007;292:F321–F329.
55. Rook GA, Hernández-Pando R. The pathogenesis of tuberculosis. Annu Rev Microbiol 1996;50: 259–284.
56. Toossi Z, Ellner JJ. The role of TGF-β in the pathogenesis of human tuberculosis. Clin Immunol Immunopathol 1998;87:107–114.
57. Hernandez-Pando R, Orozco EH, Arriaga K, Sampieri A, Larriva-sahd J, Madrid-Marina V. Analysis of the local kinetics and localization of interleukin-1α, tumour necrosis factor-α, tumour necrosis factor-α and transforming growth factor-β during the course of experimental pulmonary tuberculosis. Immunology 1997;90:607–617.

58. Hernández-Pando R, Orozco-Esteves H, Maldonado HA, et al. A combination of a transforming growth factor -beta antagonist and an inhibitor of cyclooxygenase is an effective treatment for murine pulmonary tuberculosis. Clin Exp Immunol 2006;144(2):264–272.
59. Dumont N, Arteaga CL. Targeting the TGFβ signaling network in human neoplasia. Cancer Cell 2003;3:531–535.
60. Attisano L, Wrana JL, López-Casillas F, Massagué J. TGF-β receptors and actions. Biochimica et Biophysica Acta 1994;1222:71–80.

29 Uterine Sensitization Associated Gene-1: A Bone Morphogenetic Protein Antagonist with a Role in Kidney

Motoko Yanagita

CONTENTS

INTRODUCTION: BMP-7 AND KIDNEY DISEASES
REGULATION OF LOCAL BMP ACTIVITIES
UTERINE SENSITIZATION-ASSOCIATED GENE-1: NEGATIVE MODULATOR OF BMP-7 ACTIVITY IN ADULT KIDNEY
SUMMARY AND FUTURE DIRECIONS
REFERENCES

Abstract

Tubular damage and interstitial fibrosis is a final common pathway leading to ESRD, and once tubular damage is established, it cannot be reversed by currently available treatment. The administration of bone morphogenetic protein-7 (BMP-7) in pharmacological doses repairs established tubular damages and improves renal function in several kidney disease models, however, pathophysiological role of endogenous BMP-7 and regulatory mechanism of its activities remain elusive.

Here, we show that uterine sensitization-associated gene-1 (USAG-1), novel BMP antagonist abundantly expressed in the kidney, is the central negative regulator of BMP-7 in the kidney, and that mice lacking USAG-1 (*USAG-1*$^{-/-}$ mice) are resistant to kidney injuries. *USAG-1*$^{-/-}$ mice exhibited markedly prolonged survival and preserved renal function in acute and chronic renal injuries. Renal BMP signaling, assessed by phosphorylation of Smad proteins, is significantly enhanced in *USAG-1*$^{-/-}$ mice during renal injury, indicating that the preservation of renal function is attributed to enhancement of endogenous BMP-7 signaling. Although many other BMP antagonists are expressed in injured kidney, solitary inhibition of USAG-1 markedly slows the disease progression, suggesting that USAG-1 plays a critical role in the modulation of renoprotective action of BMP-7, and that inhibition of USAG-1 will be promising means of development of novel treatment for kidney diseases.

Key Words: BMP-7; ectodin; Wise; sclerostin; Wnt; kidney diseases.

1. INTRODUCTION: BMP-7 AND KIDNEY DISEASES

Despite a significant increase in understanding of the pathophysiology of renal diseases, the incidence of end-stage renal disease (ESRD) is still increasing. Tubular damage and interstitial fibrosis is the final common pathway leading to ESRD *(1)* irrespective of the nature of the initial renal injury and the degree of tubular damage parallels the impairment of renal function *(2)*. Once tubular damage is established, it cannot be reversed or repaired

From: *Cancer Drug Discovery and Development: Transforming Growth Factor-β in Cancer Therapy, Vol I: Basic and Clinical Biology*
Edited by: Sonia B. Jakowlew © Humana Press Inc., Totowa, NJ

by currently available treatment, and renal function deteriorates to renal failure, which is often life threatening *(3)*. If we can come up with an agent that can reverse established tubular damage, it would significantly reduce the need for dialysis, and bone morphogenetic protein (BMP) is one candidate for such an agent.

BMPs are phylogenetically conserved signaling molecules that belong to the transforming growth factor β (TGF-β) superfamily *(4–7)*. Although these proteins were first identified by their capacity to promote endochondral bone formation *(8–10)*, they are involved in the cascades of body patterning and morphogenesis *(11)*. Furthermore, BMPs play important roles after birth in pathophysiology of several diseases including osteoporosis *(12)*, arthritis *(8)*, pulmonary hypertension *(13,14)*, cerebrovascular diseases *(15)*, cancer *(16)*, and kidney diseases *(17–19)*.

BMP-7, also known as osteogenic protein-1 (OP-1), is a 35-kDa homodimeric protein, and kidney is the major site of BMP-7 synthesis *(20–23)* during embryogenesis as well as postnatal development. Its genetic deletion in mice leads to severe impairment of kidney development resulting in perinatal death *(24,25)*. Expression of BMP-7 in adult kidney is confined to distal collecting tubules and podocytes of glomeruli *(26)*, and the expression decreases in several kidney disease models *(27–31)*. Recently, several reports indicate that the administration of pharmacological doses of BMP-7 inhibits and repairs chronic kidney injury in animal models *(28–30,32–36)*. The administration of BMP-7 reverses TGF-β1-induced EMT and induces mesenchymal-to-epithelial transition in vitro, inhibits the induction of inflammatory cytokine expression in the kidney *(26)*, attenuates inflammatory cell infiltration *(33)*, and reduces apoptosis of tubular epithelial cells in renal disease models *(37)* (Fig. 1). Collectively, BMP-7 plays critical roles in repairing processes of the renal tubular damage in kidney diseases. However, the physiological role and precise regulatory mechanism of endogenous BMP-7 activity remain elusive.

2. REGULATION OF LOCAL BMP ACTIVITIES

The local activity of endogenous BMP is controlled by at least two different ways. First, the expression pattern of BMP and its cell surface receptors controls local activity of BMP. Oxburgh et al. recently reported that BMP-4 knock-in in the locus of BMP-7 in *BMP-7* deficient mice rescues the phenotype of BMP-7 deficiency, and that kidney develops normally in these mice. This result demonstrates that structurally divergent BMP family members, sharing only minimal sequence similarity can function interchangeably to activate all the essential signaling pathways for growth and morphogenesis of the kidney. Therefore it is likely that expression pattern of each BMP, but not its biochemical property defines its biological functions.

Second, BMP signaling is precisely regulated by certain classes of molecules termed as BMP antagonists *(4–6,38)*. BMP antagonists function through direct association with BMPs, thus prohibiting BMPs from binding their cognate receptors. The interplay between BMP and their antagonists fine-tunes the level of available BMPs, and governs developmental and cellular processes as diverse as establishment of the embryonic dorsal–ventral axis *(39)*, induction of neural tissue *(40)*, formation of joints in the skeletal system *(8)* and neurogenesis in the adult brain *(41)*. The indispensable roles of BMP-7 in the kidney led us to postulate the existence of some BMP antagonist that modulates the activities of BMP-7 in the kidney.

3. UTERINE SENSITIZATION-ASSOCIATED GENE-1: NEGATIVE MODULATOR OF BMP-7 ACTIVITY IN ADULT KIDNEY

3.1. Discovery and Characterization of Uterine Sensitization-Associated Gene-1

Through a genome-wide search for kidney-specific transcripts, our group found a novel gene, which encodes a secretory protein with a signal peptide and cysteine-rich domain *(42)*.

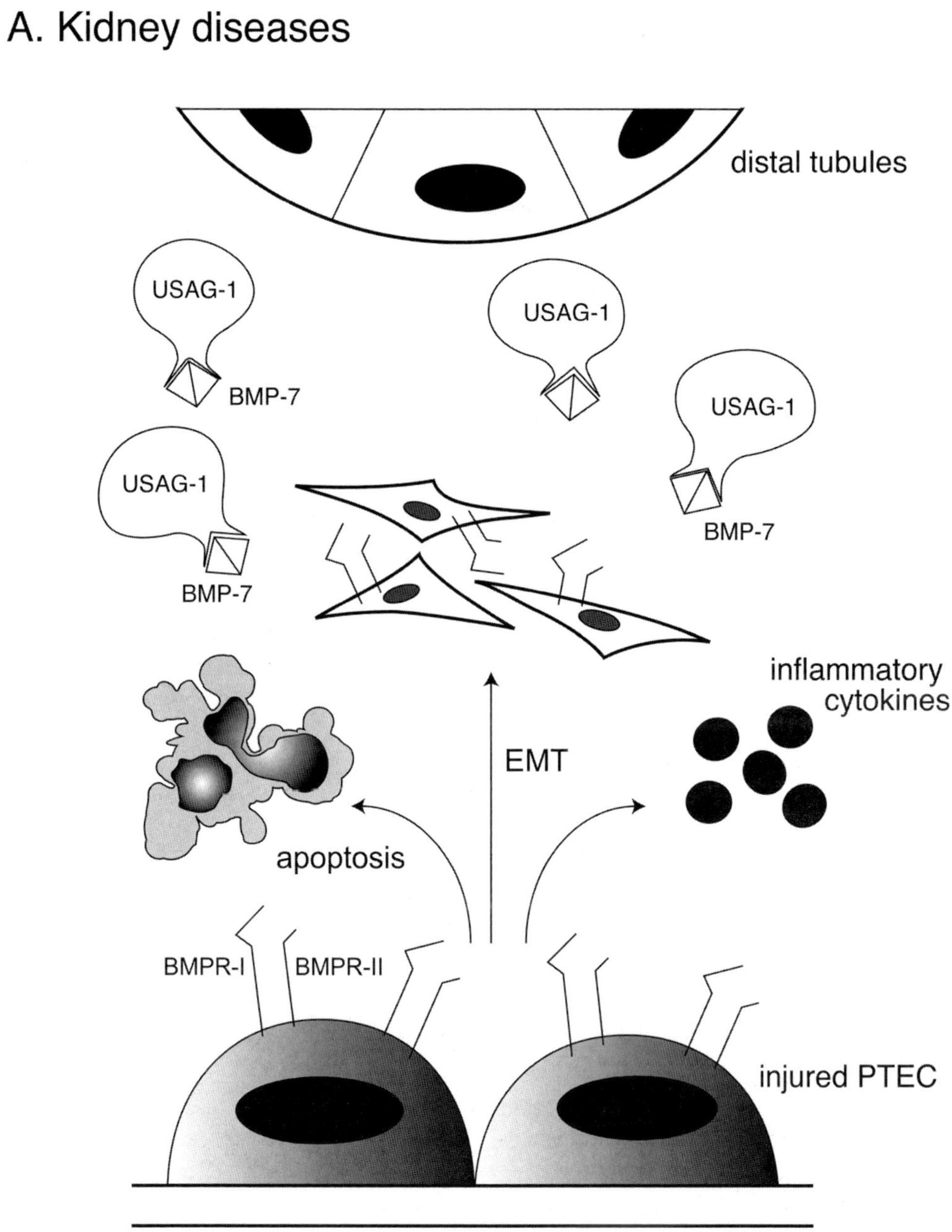

Fig. 1. (*Continued*)

The rat orthologue of the gene was previously reported as a gene of unknown function that was preferentially expressed in sensitized endometrium of rat uterus, termed uterine sensitization-associated gene-1 (USAG-1) *(43)*. Amino acid sequences encoded in rat and mouse cDNAs are 97 and 98% identical to the human sequence, respectively, indicating high degrees of sequence conservation.

Domain search predicted this protein to be a member of the cystine-knot superfamily, which comprises of growth factors, BMPs, and BMP antagonists. Homology search revealed that USAG-1 has significant amino acid identities (38%) to sclerostin, the product

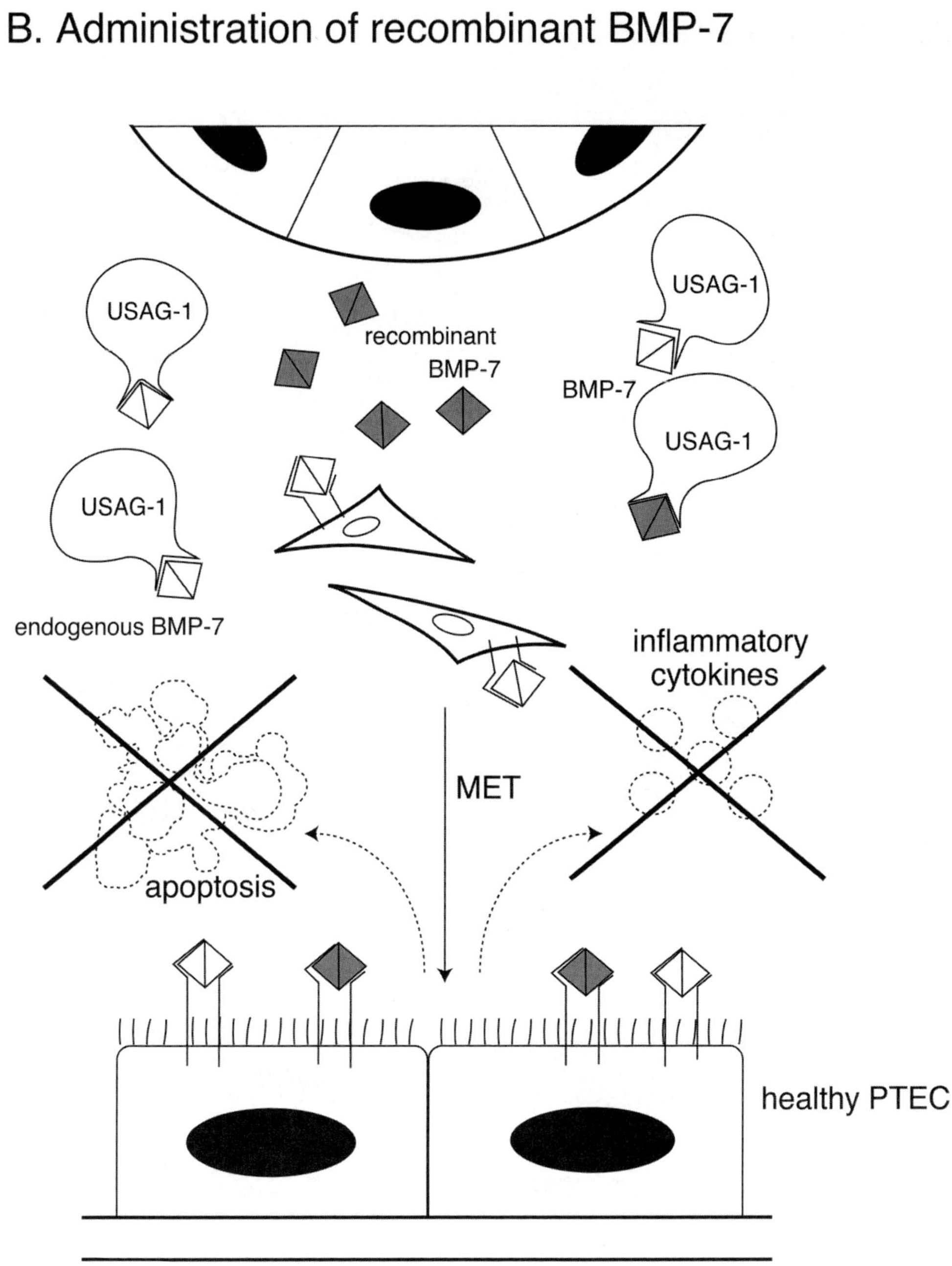

Fig. 1. (*Continued*)

of the SOST gene (*see* Section 3.6.). Mutations of SOST are found in patients with sclerosteosis, a syndrome of sclerosing skeletal dysplasia *(44)*. Because sclerostin was subsequently shown to be a new member of BMP antagonist expressed in bones and cartilages *(45–47)*, USAG-1 is postulated to be a BMP antagonist expressed in the kidney *(42)*.

USAG-1 protein is a 28–30 kDa secretory protein and behaves as a monomer *(42,48)*, in spite that a number of BMP antagonists form disulfide-bridged dimers. This is consistent with the fact that USAG-1 protein does not have the extra cysteine residues present in

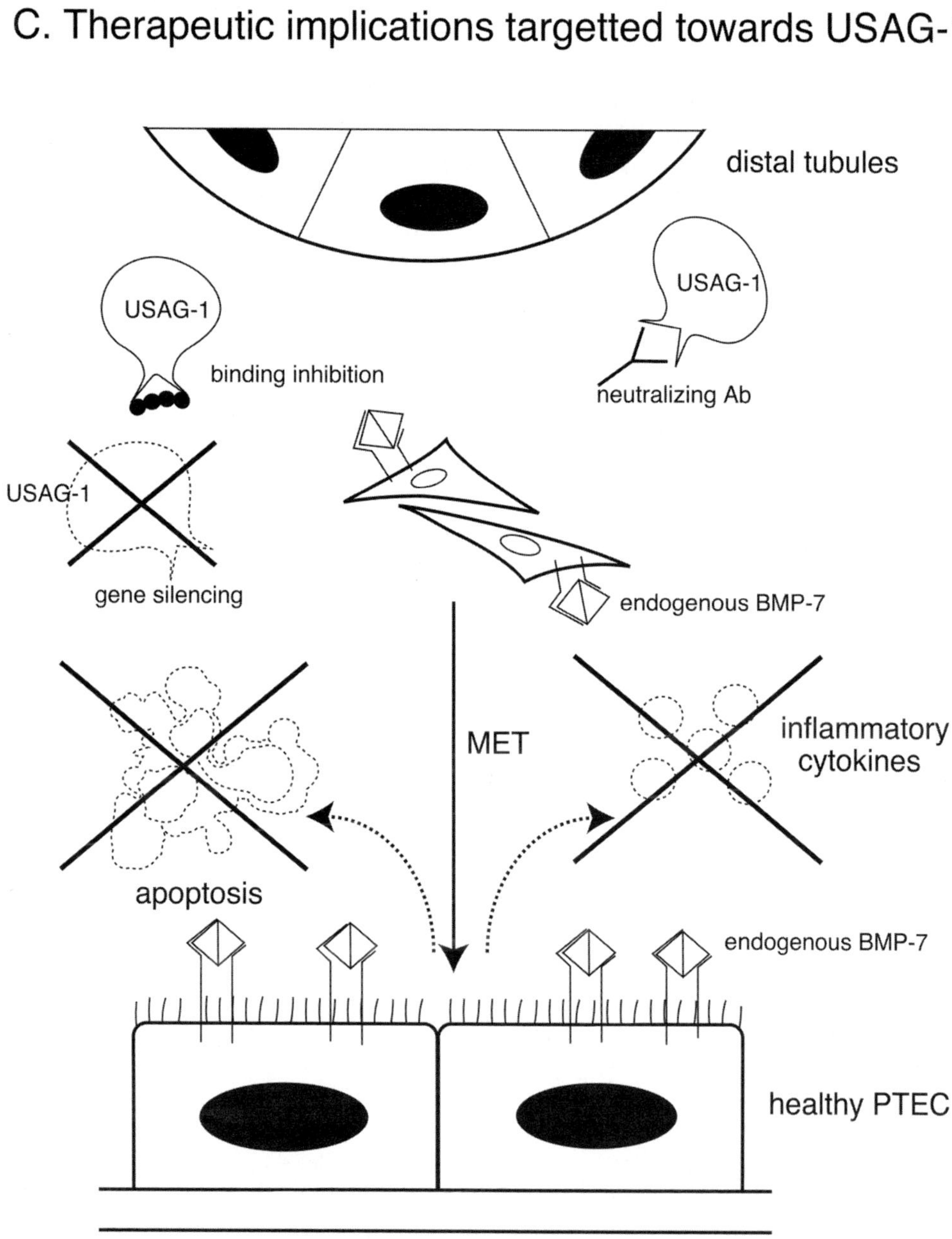

Fig. 1. In kidney diseases, injured proximal tubule epithelial cell undertake apoptosis and epithelial-to-mesenchymal transition (EMT), and produce inflammatory cytokines (**A**). Administration of pharmacological doses of BMP-7 inhibits apoptosis, EMT, and production of cytokines of PTEC (**B**). USAG-1 is secreted from distal tubules, binds to BMP-7, and inhibits the binding of BMP-7 to its receptors (**C**). Therefore, drugs or neutralizing antibodies that inhibits binding between USAG-1 and BMP, or gene-silencing therapy for USAG-1 would increase available endogenous BMP, and might be a promising way to develop novel therapeutic methods for severe renal diseases.

noggin and DAN, which are necessary to make intermolecular disulfide bridges *(38)*. Recombinant USAG-1 protein binds to BMP-2, -4, -6, and -7, leading to the inhibition of alkaline phosphatase activities induced by each BMP in C2C12 cells *(42)* and

MC3T3-E1 cells *(48)* dose-dependently, while sclerostin only inhibits BMP6 and BMP7 activities *(46)*.

Furthermore, the activity of USAG-1 as a BMP antagonist was confirmed in vivo using *Xenopus* embryos *(42)*. Injection of synthetic RNA encoding BMP antagonists to the ventral portion of *Xenopus* embryos inhibits the ventralizing signal of endogenous BMP, and induces secondary axis formation and hyperdorsalization of the embryos *(49)*. The injection 100 pg USAG-1 mRNA was sufficient to cause secondary axis formation, and injection of increasing doses of mRNA up to 1000 pg led to a corresponding increase in the frequency of dorsalization phenotypes, while embryos developed normally when irrelevant mRNA was injected. Furthermore, injection of USAG-1 mRNA into the animal pole of *Xenopus* embryos induced expression of NCAM in animal caps dose-dependently, indicating that USAG-1 directly antagonizes BMP signaling in *Xenopus* embryos.

3.2. Expression of USAG-1 in the Kidney

In mouse embryogenesis, expression of USAG-1 mRNA was first detected on embryonic Day 11.5 (E11.5) and increased toward E17.5 *(42)*. *In situ* hybridization of mouse embryos on E11.5 revealed the moderate expression of USAG-1 mRNA in branchial arches and pharynx. On E17.5, strong expression of USAG-1 mRNA was confined to kidney tubules and ameloblasts of teeth. In addition, moderate expression was observed in hair follicles, choroids plexus and ependymal cells in the ventricles of the brain.

In adult tissues, the expression was by far most abundant in the kidney and is restricted to the epithelial cells of distal collecting tubules. No expression was observed in proximal tubules, glomeruli or blood vessels in the kidney. Thus, the cellular distribution of USAG-1 is overlapping with that of BMP-7 in the kidney *(26)*. Taken together with the fact that proximal tubule epithelial cells (PTECs) are the site of injuries in many types of kidney diseases, and they express the receptors for BMP-7*(26)*, we hypothesized the working model about the regulation of renoprotective action of BMP-7 (Fig. 1): in renal injuries, PTECs are mainly damaged and undertake apoptosis or EMT to fibroblast-like mesenchymal cells. BMP-7 secreted from distal tubules binds to the receptors in the PTECs, and inhibits apoptosis and EMT *(36)*. USAG-1 is also secreted from distal tubules, binds to BMP-7, and inhibits the renoprotective actions of BMP-7 by reducing the amount of available BMP-7.

To evaluate this working model, our group generated *USAG-1* knockout (*USAG-1*$^{-/-}$) mice, and induced acute and chronic renal disease models in which the renal tubules were mainly damaged.

3.3. USAG-1$^{-/-}$ *Mice are Resistant to Kidney Tubular Injury*

USAG-1$^{-/-}$ mice were born at the ratio expected by Mendel's law of heredity, and were viable, fertile and appeared healthy except that *USAG-1*$^{-/-}$ mice exhibit supernumerary teeth, both in the incisors and molars, and fused teeth in the molar teeth region *(49a)*. Although there was variation in the sites of extra teeth and fused teeth, these teeth phenotype was fully penetrant (*see* Section 3.5.).

Because the renal function and histology of *USAG-1*$^{-/-}$ mice appears normal, our group challenged the mice with two different kidney disease models and found that *USAG-1*$^{-/-}$ mice are resistant to renal injury.

As a model for acute renal failure, we utilized cisplatin nephrotoxicity model. Administration of cisplatin to wild-type littermates causes acute tubular injuries that result in severe renal failure. Within the first three days, 54% of wild-type mice died, while 92% of *USAG-1*$^{-/-}$ mice survived the period. Renal function and histology of *USAG-1*$^{-/-}$ mice at d 3 was significantly preserved when compared to wild-type littermate *(49a)*. Tubular apoptosis, a characteristic feature of tubular injuries in cisplatin nephrotoxicity, was also significantly reduced in *USAG-1*$^{-/-}$ mice.

As a model of chronic renal injury, our group performed unilateral ureteral obstruction *(17,50)* in both *USAG-1*$^{-/-}$ mice and wild-type mice, and the kidneys were harvested 14 d after the operation. In wild-type mice, the obstructed kidney showed degeneration of renal tubules and interstitial fibrosis, whereas the normal architecture was preserved in *USAG-1*$^{-/-}$ mice except for mild dilatation of tubules. Expression of E-cadherin, a marker for tubular epithelial integrity *(51)*, was severely reduced in the kidney of wild-type mice, while its expression was preserved in *USAG-1*$^{-/-}$ mice.

Renal BMP signaling, assessed by phosphorylation of Smad proteins, is significantly enhanced in *USAG-1*$^{-/-}$ mice during renal injury, indicating that the preservation of renal function is attributed to enhancement of endogenous BMP signaling.

Furthermore, the administration of neutralizing antibody against BMP-7 abolished renoprotection in *USAG-1*$^{-/-}$ mice. These results strongly support the working model, and BMP-7 is the potent candidate for the counterpart of USAG-1.

Interestingly, the expression of USAG-1 decreases during the course of disease models. We assume that the reduction of USAG-1 in kidney diseases is a kind of self-defense mechanism to minimize the inhibitory effect on BMP signaling. Because the reduction of USAG-1 expression in WT mice is not enough to overcome the reduction of BMP-7 expression, further reduction or abolishment of the action of USAG-1 is desirable for the preservation of renal function, and the results in the present study justify the therapy targeted toward USAG-1. For example, drugs or neutralizing antibodies that inhibits binding between USAG-1 and BMP, or gene-silencing therapy for USAG-1 would enhance the activities of endogenous BMP, and might be a promising way to develop novel therapeutic methods for severe renal diseases (Fig. 1C). Because the expression of USAG-1 is confined to the kidney in adult mice and humans *(42)*, it would be a better target for kidney-specific therapeutic trials. On the contrary, the administration of recombinant BMP-7 protein, whose target cells are widely distributed throughout the body, might produces some additional extrarenal actions, which includes beneficial effects, such as actions on renal osteodystrophy *(18,52–55)* and vascular calcification *(56,57)*. Furthermore, these therapy targeted toward USAG-1 might protect the kidney during administration of nephrotoxic agents such as cisplatin.

3.4. USAG-1 is the Most Abundant BMP Antagonist in Adult Kidney

Our group demonstrated that USAG-1 is by far the most abundant BMP antagonist in the kidney (Yanagita et al. JCI in press). The expression of USAG-1 and other BMP antagonists in adult kidneys were analyzed by modified real-time PCR with the standard curve using various concentration of plasmid encoding each BMP antagonist, and the copy number of each genes contained in kidney cDNA were determined.

Among known BMP antagonists, USAG-1 was by far the most abundant in the kidneys. Because other BMP antagonists also antagonize BMP-7 activities, it is concluded that USAG-1 plays important role in the modulation of BMP activities in the kidney not because of its ligand specificity, but because of its high expression among other BMP antagonists. In addition, localization of USAG-1 is quite similar to that of BMP-7 *(26)*, so that USAG-1 can effectively access to and inactivate BMP-7 at the site of production. Although USAG-1/BMP-7 binding is illustrated in the outside of the PTECs (Fig. 1), it might be possible that the binding occurs intracellulary within the secretory pathway in PTECs, and that USAG-1 and BMP-7 are secreted in the complex form. Further investigations are necessary to clarify the point.

3.5. USAG-1 in Teeth Morphogenesis

USAG-1 is also expressed in developing teeth, and a USAG-1-positive area surrounds the enamel knot signaling centers where BMPs are expressed *(48)*. As mentioned earlier,

USAG-1$^{-/-}$ mice exhibit supernumerary teeth, both in the incisors and molars, and fused teeth in the molar teeth region. Because BMP-4 is known to be involved in the induction of the enamel knot signaling centers, loss of the inhibitory effect of USAG-1 might induce extra signaling centers, resulting in supernumerary teeth. Kassai et al. independently reported that USAG-1/ectodin (they renamed USAG-1 as ectodin) deficient mice have enlarged enamel knots, altered cusp patterns, and extra teeth *(58)*. They also reported that excess BMP accelerates patterning in *USAG-1* deficient teeth, and proposed that USAG-1 is critical for robust spatial delineation of enamel knots and cusps.

Furthermore, USAG-1 is intensely expressed in the ameloblasts of teeth *(42)*. Ameloblasts secrete enamel matrix extracts, of which the mineralization remains incomplete until apathetic crystals accumulate. It is recently reported that BMPs in enamel extracts induce mineralization of teeth and exogenous administration of recombinant noggin inhibits the mineralizing activity of BMP in tissue cultures. Thus far, however, no endogenous BMP antagonist has been described in developing tooth tissues. USAG-1 secreted from ameloblasts may modulate the activities of BMPs and regulate mineralization of enamel matrix extracts.

3.6. USAG-1 and Sclerostin Form a New Family of BMP Antagonists

Based on the size of the cystine knot and overall amino acid sequence similarity, USAG-1 and sclerostin (*see* section 3.1) seem to form a new family of BMP antagonists *(59)*. In addition, there are several evidences to support the idea that USAG-1 and sclerostin form a distinct family of BMP antagonists. One reason is the evolutionary origin of the genes; genes for USAG-1 and sclerostin are missing in fly and nematode, however, a single ortholog for both USAG-1 and sclerostin was found in *Fugu rubripes* and *Ciona intestinalis (38)*. Second, exon-intron arrangements of all orthologous genes for USAG-1 and sclerostin are highly conserved *(38)*. In addition, USAG-1 and sclerostin are secreted as monomers *(42,46)*, while many other BMP antagonists form dimers *(60)*. This observation is consistent with the findings that many BMP antagonists possess extracysteine residues that could allow the formation of homodimers, which are missing in USAG-1 and sclerostin.

3.7. USAG-1 as a Modulator of Wnt Signaling

Recently, Itasaki et al. reported another function of USAG-1 as a context-dependent activator and inhibitor of Wnt signaling *(61)*. They demonstrated that Wise/USAG-1 activated and inhibited Wnt pathways in *Xenopus* embryogenesis in a context dependent way, as well as the physical interaction between Wise/USAG-1 and Wnt coreceptor, lipoprotein receptor-related protein 6 *(62)*.

Close relationships between the Wnt and BMP pathways have been recently reported: for instance, the canonical Wnt antagonist DKK1 and noggin cooperate in head induction *(63)*, while the expression of DKK1 is regulated by BMP-4 in limb development *(64)*. Furthermore, a BMP antagonist called Cerberus has a binding site for Wnt proteins that is distinct from the BMP binding site, and antagonizes Wnt activities by directly binding to Wnt *(65)*. USAG-1 might also have dual activities, and play as a molecular link between Wnt and BMP signaling pathway.

4. SUMMARY AND FUTURE DIRECTIONS

In conclusion, USAG-1 plays important roles in the progression of renal diseases, and might be a potent negative regulator of the renoprotective action of endogenous BMP signaling. Recently, Lin et al. identified a positive regulator of BMP-7 named kielin/chordin-like protein (KCP), and demonstrated that *KCP*$^{-/-}$ mice are susceptible to tubular injury and interstitial fibrosis *(66)*. Because these negative and positive modulators of BMP signaling regulate and edge the boundaries of BMP activity, further understanding of these modulators would give valuable information about their pathophysiological functions and provide a rationale for a therapeutic approach against these proteins.

ACKNOWLEDGEMENT

I would like to give my sincere thanks to Sonia for giving me the opportunity to contribute to this book. This study was supported by Grants-in Aid from the Ministry of Education, Culture, Science, Sports, and Technology of Japan (177090551), Center of Excellence grant from the Ministry of Education, Culture, Science, Sports, and Technology of Japan, a research grant for health sciences from the Japanese Ministry of Health, Labor and Welfare, and partially by a grant from Astellas foundation for Research on Metabolic Disorders.

REFERENCES

1. Eddy AA. Molecular insights into renal interstitial fibrosis. J Am Soc Nephrol 1996;7(12):2495–2508.
2. van Kooten C, Daha MR, van Es LA. Tubular epithelial cells: A critical cell type in the regulation of renal inflammatory processes. Exp Nephrol 1999;7(5–6):429–437.
3. Neilson EG. Setting a trap for tissue fibrosis. Nat Med 2005;11(4):373–374.
4. Massagué J, Chen YG. Controlling TGF-beta signaling. Genes Dev 2000;14(6):627–644.
5. Canalis E, Economides AN, Gazzerro E. Bone morphogenetic proteins, their antagonists, and the skeleton. Endocr Rev 2003;24(2):218–235.
6. Reddi AH. Interplay between bone morphogenetic proteins and cognate binding proteins in bone and cartilage development: noggin, chordin and DAN. Arthritis Res 2001;3(1):1–5.
7. Attisano L, Wrana JL. Signal transduction by members of the transforming growth factor-beta superfamily. Cytokine Growth Factor Rev 1996;7(4):327–339.
8. Reddi AH. Bone morphogenetic proteins and skeletal development: the kidney-bone connection. Pediatr Nephrol 2000;14(7):598–601.
9. Urist MR. Bone: formation by autoinduction. Science 1965;150(698):893–899.
10. Wozney JM, Rosen V, Celeste AJ, et al. Novel regulators of bone formation: molecular clones and activities. Science 1988;242(4885):1528–1534.
11. Hogan BL. Bone morphogenetic proteins in development. Curr Opin Genet Dev 1996;6(4):432–438.
12. Wang EA. Bone morphogenetic proteins (BMPs): therapeutic potential in healing bony defects. Trends Biotechnol 1993;11(9):379–383.
13. Miyazono K, Kusanagi K, Inoue H. Divergence and convergence of TGF-beta/BMP signaling. J Cell Physiol 2001;187(3):265–276.
14. Morse JH, Deng Z, Knowles JA. Genetic aspects of pulmonary arterial hypertension. Ann Med 2001; 33(9):596–603.
15. Chang CF, Lin SZ, Chiang YH, et al. Intravenous administration of bone morphogenetic protein-7 after ischemia improves motor function in stroke rats. Stroke 2003;34(2):558–564.
16. Howe JR, Bair JL, Sayed MG, et al. Germline mutations of the gene encoding bone morphogenetic protein receptor 1A in juvenile polyposis. Nat Genet 2001;28(2):184–187.
17. Klahr S. The bone morphogenetic proteins (BMPs). Their role in renal fibrosis and renal function. J Nephrol 2003;16(2):179–185.
18. Hruska KA, Saab G, Chaudhary LR, Quinn CO, Lund RJ, Surendran K. Kidney-bone, bone-kidney, and cell-cell communications in renal osteodystrophy. Semin Nephrol 2004;24(1):25–38.
19. Zeisberg M, Muller GA, Kalluri R. Are there endogenous molecules that protect kidneys from injury? The case for bone morphogenic protein-7 (BMP-7). Nephrol Dial Transplant 2004; 19(4):759–761.
20. Helder MN, Ozkaynak E, Sampath KT, et al. Expression pattern of osteogenic protein-1 (bone morphogenetic protein-7) in human and mouse development. J Histochem Cytochem 1995;43(10):1035–1044.
21. Vukicevic S, Stavljenic A, Pecina M. Discovery and clinical applications of bone morphogenetic proteins. Eur J Clin Chem Clin Biochem 1995;33(10):661–671.
22. Vukicevic S, Kopp JB, Luyten FP, Sampath TK. Induction of nephrogenic mesenchyme by osteogenic protein 1 (bone morphogenetic protein 7). Proc Natl Acad Sci USA 1996;93(17):9021–9026.

23. Ozkaynak E, Schnegelsberg PN, Oppermann H. Murine osteogenic protein (OP-1): high levels of mRNA in kidney. Biochem Biophys Res Commun 1991;179(1):116–123.
24. Luo G, Hofmann C, Bronckers AL, Sohocki M, Bradley A, Karsenty G. BMP-7 is an inducer of nephrogenesis, and is also required for eye development and skeletal patterning. Genes Dev 1995; 9(22):2808–2820.
25. Dudley AT, Lyons KM, Robertson EJ. A requirement for bone morphogenetic protein-7 during development of the mammalian kidney and eye. Genes Dev 1995;9(22):2795–2807.
26. Gould SE, Day M, Jones SS, Dorai H. BMP-7 regulates chemokine, cytokine, and hemodynamic gene expression in proximal tubule cells. Kidney Int 2002;61(1):51–60.
27. Wang SN, Lapage J, Hirschberg R. Loss of tubular bone morphogenetic protein-7 in diabetic nephropathy. J Am Soc Nephrol 2001;12(11):2392–2399.
28. Hruska KA. Treatment of chronic tubulointerstitial disease: a new concept. Kidney Int 2002;61(5): 1911–1922.
29. Morrissey J, Hruska K, Guo G, Wang S, Chen Q, Klahr S. Bone morphogenetic protein-7 improves renal fibrosis and accelerates the return of renal function. J Am Soc Nephrol 2002;13 Suppl 1:S14–S21.
30. Dube PH, Almanzar MM, Frazier KS, Jones WK, Charette MF, Paredes A. Osteogenic Protein-1: gene expression and treatment in rat remnant kidney model. Toxicol Pathol 2004;32(4):384–392.
31. Almanzar MM, Frazier KS, Dube PH, et al. Osteogenic protein-1 mRNA expression is selectively modulated after acute ischemic renal injury. J Am Soc Nephrol 1998;9(8):1456–1463.
32. Vukicevic S, Basic V, Rogic D, et al. Osteogenic protein-1 (bone morphogenetic protein-7) reduces severity of injury after ischemic acute renal failure in rat. J Clin Invest 1998;102(1):202–214.
33. Hruska KA, Guo G, Wozniak M, et al. Osteogenic protein-1 prevents renal fibrogenesis associated with ureteral obstruction. Am J Physiol Renal Physiol 2000;279(1):F130–F143.
34. Wang S, Chen Q, Simon TC, et al. Bone morphogenic protein-7 (BMP-7), a novel therapy for diabetic nephropathy. Kidney Int 2003;63(6):2037–2049.
35. Zeisberg M, Bottiglio C, Kumar N, et al. Bone morphogenic protein-7 inhibits progression of chronic renal fibrosis associated with two genetic mouse models. Am J Physiol Renal Physiol 2003;285(6): F1060–F1067.
36. Zeisberg M, Hanai J, Sugimoto H, et al. BMP-7 counteracts TGF-beta1-induced epithelial-to-mesenchymal transition and reverses chronic renal injury. Nat Med 2003;9(7):948–964.
37. Li T, Surendran K, Zawaideh MA, Mathew S, Hruska KA. Bone morphogenetic protein 7: a novel treatment for chronic renal and bone disease. Curr Opin Nephrol Hypertens 2004;13(4):417–422.
38. Avsian-Kretchmer O, Hsueh AJ. Comparative genomic analysis of the eight-membered ring cystine knot-containing bone morphogenetic protein antagonists. Mol Endocrinol 2004;18(1):1–12.
39. Wagner DS, Mullins MC. Modulation of BMP activity in dorsal-ventral pattern formation by the chordin and ogon antagonists. Dev Biol 2002;245(1):109–123.
40. Wessely O, Agius E, Oelgeschlager M, Pera EM, De Robertis EM. Neural induction in the absence of mesoderm: beta-catenin-dependent expression of secreted BMP antagonists at the blastula stage in Xenopus. Dev Biol 2001;234(1):161–173.
41. Lim DA, Tramontin AD, Trevejo JM, Herrera DG, Garcia-Verdugo JM, Alvarez-Buylla A. Noggin antagonizes BMP signaling to create a niche for adult neurogenesis. Neuron 2000;28(3):713–726.
42. Yanagita M, Oka M, Watabe T, et al. USAG-1: a bone morphogenetic protein antagonist abundantly expressed in the kidney. Biochem Biophys Res Commun 2004;316(2):490–500.
43. Simmons DG, Kennedy TG. Uterine sensitization-associated gene-1: a novel gene induced within the rat endometrium at the time of uterine receptivity/sensitization for the decidual cell reaction. Biol Reprod 2002;67(5):1638–1645.
44. Brunkow ME, Gardner JC, Van Ness J, et al. Bone dysplasia sclerosteosis results from loss of the SOST gene product, a novel cystine knot-containing protein. Am J Hum Genet 2001;68(3):577–589.
45. Winkler DG, Sutherland MK, Geoghegan JC, et al. Osteocyte control of bone formation via sclerostin, a novel BMP antagonist. Embo J 2003;22(23):6267–6276.
46. Kusu N, Laurikkala J, Imanishi M, et al. Sclerostin is a novel secreted osteoclast-derived bone morphogenetic protein antagonist with unique ligand specificity. J Biol Chem 2003;278(26): 24,113–24,117.
47. van Bezooijen RL, Roelen BA, Visser A, et al. Sclerostin is an osteocyte-expressed negative regulator of bone formation, but not a classical BMP antagonist. J Exp Med 2004;199(6):805–814.
48. Laurikkala J, Kassai Y, Pakkasjarvi L, Thesleff I, Itoh N. Identification of a secreted BMP antagonist, ectodin, integrating BMP, FGF, and SHH signals from the tooth enamel knot. Dev Biol 2003;264(1): 91–105.

49. Harland R, Gerhart J. Formation and function of Spemann's organizer. Annu Rev Cell Dev Biol 1997;13:611–667.

49a. Yanagita M, Okuda T, Endo S, et al. Uterine sensitization-associated gene-1 (USAG-1), a novel BMP antagonist expressed in the kidney, accelerates tubular injury. J Clin Invest 2006;116(1):70–79.

50. Chevalier RL. Molecular and cellular pathophysiology of obstructive nephropathy. Pediatr Nephrol 1999;13(7):612–619.

51. Yang J, Liu Y. Dissection of key events in tubular epithelial to myofibroblast transition and its implications in renal interstitial fibrosis. Am J Pathol 2001;159(4):1465–1475.

52. Davies MR, Lund RJ, Mathew S, Hruska KA. Low turnover osteodystrophy and vascular calcification are amenable to skeletal anabolism in an animal model of chronic kidney disease and the metabolic syndrome. J Am Soc Nephrol 2005;16(4):917–928.

53. Gonzalez EA, Lund RJ, Martin KJ, et al. Treatment of a murine model of high-turnover renal osteodystrophy by exogenous BMP-7. Kidney Int 2002;61(4):1322–1331.

54. Lund RJ, Davies MR, Brown AJ, Hruska KA. Successful treatment of an adynamic bone disorder with bone morphogenetic protein-7 in a renal ablation model. J Am Soc Nephrol 2004;15(2):359–369.

55. Simic P, Vukicevic S. Bone morphogenetic proteins in development and homeostasis of kidney. Cytokine Growth Factor Rev 2005;16(3):299–308.

56. Hruska KA, Mathew S, Saab G. Bone morphogenetic proteins in vascular calcification. Circ Res 2005;97(2):105–114.

57. Davies MR, Lund RJ, Hruska KA. BMP-7 is an efficacious treatment of vascular calcification in a murine model of atherosclerosis and chronic renal failure. J Am Soc Nephrol 2003;14(6):1559–1567.

58. Kassai Y, Munne P, Hotta Y, et al. Regulation of mammalian tooth cusp patterning by ectodin. Science 2005;309(5743):2067–2070.

59. Yanagita M. BMP antagonists: Their roles in development and involvement in pathophysiology. Cytokine Growth Factor Rev 2005;16(3):309–317.

60. Groppe J, Greenwald J, Wiater E, et al. Structural basis of BMP signalling inhibition by the cystine knot protein Noggin. Nature 2002;420(6916):636–642.

61. Itasaki N, Jones CM, Mercurio S, et al. Wise, a context-dependent activator and inhibitor of Wnt signalling. Development 2003;130(18):4295–4305.

62. Tamai K, Semenov M, Kato Y, et al. LDL-receptor-related proteins in Wnt signal transduction. Nature 2000;407(6803):530–535.

63. del Barco Barrantes I, Davidson G, Grone HJ, Westphal H, Niehrs C. Dkk1 and noggin cooperate in mammalian head induction. Genes Dev 2003;17(18):2239–2244.

64. Grotewold L, Ruther U. The Wnt antagonist Dickkopf-1 is regulated by Bmp signaling and c-Jun and modulates programmed cell death. Embo J 2002;21(5):966–975.

65. Piccolo S, Agius E, Leyns L, et al. The head inducer Cerberus is a multifunctional antagonist of Nodal, BMP and Wnt signals. Nature 1999;397(6721):707–710.

66. Lin J, Patel SR, Cheng X, et al. Kielin/chordin-like protein, a novel enhancer of BMP signaling, attenuates renal fibrotic disease. Nat Med 2005;11(4):387–393.

30 Role of Betaig-h3 Gene in Carcinogenesis

Yongliang Zhao and Tom K. Hei

CONTENTS

INTRODUCTION
BETAIG-H3 IN CELLULAR ADHESION AND APOPTOSIS
FREQUENT LOSS OF BETAIG-H3 EXPRESSION IN HUMAN CANCER CELLS
SUPPRESSION OF TUMORIGENICITY AFTER REEXPRESSION OF BETAIG-H3 GENE
HYPERMETHYLATION OF BETAIG-H3 PROMOTER REGION AND GENE SILENCING
CONCLUDING REMARKS
ACKNOWLEDGMENTS
REFERENCES

Abstract

Betaig-h3, as a secreted protein induced by transforming growth factor-β (TGF-β), has been shown to modulate cell adhesion, wound healing, apoptosis and tumorigenicity. Mutations of this gene result in single amino-acid changes in the Betaig-h3 protein and are related to the development of human corneal dystrophies. In addition, Betaig-h3 promoter is frequently silenced because of promoter methylation in human cancer cells. The data suggest that *Betaig-h3* gene might play an important role in mediating the pivotal cellular functions of TGF-β signalings such as proliferation, apoptosis and anti-tumor activities.

Key Words: Betaig-h3; carcinogenesis; cornea dystrophies; methylation; proliferation; apoptosis; antitumor.

1. INTRODUCTION

Transforming growth factor-β (TGF-β) is a member of a large family of disulfide-bonded cytokines with both tumor suppressor and tumor promoting activities *(1–3)*. Although, TGF-β inhibits the proliferation of normal epithelial cells and functions as a tumor suppressor in early tumorigenesis, it acts as a tumor promoter in later stages of tumor progression *(4)*. Numerous studies have shown that high level of TGF-β in late stage of tumor development promotes tumor progression by increasing activities for angiogenesis, immunosuppression and synthesis of extracellular matrix (ECM) which provide an appropriate microenvironment for rapid tumor growth and metastasis *(4)*.

From: *Cancer Drug Discovery and Development: Transforming Growth Factor-β in Cancer Therapy, Vol I: Basic and Clinical Biology*
Edited by: Sonia B. Jakowlew © Humana Press Inc., Totowa, NJ

Betaig-h3, as a secreted protein, can be induced by TGF-β in many human cell types *(5)*. Mutations of this gene result in single amino-acid changes in the Betaig-h3 protein and are related to the development of human corneal dystrophies *(6)*. Betaig-h3 protein contains four homologous internal domains and, last of which contains an integrin recognition sequence at its carboxyl end. Sequence analysis of Betaig-h3 reveals significant evolutional conservation between murine and human forms of the protein. Betaig-h3 protein is a fascilin-like protein and has been shown to be a component of ECM in a variety of tissues including lung, bladder, kidney, skin, cornea, and bone *(7,8)*. In addition, Betaig-h3 protein is involved in wound healing, apoptosis and tumorigenesis *(9–12)*. The hypothesis is that *Betaig-h3* gene might serve as one of the important downstream factors of TGF-β to regulate pivotal cellular processes such as proliferation, differentiation and apoptosis. This chapter will discuss the mechanisms of dysregulation of *Betaig-h3* gene in tumor progression.

2. BETAIG-H3 IN CELLULAR ADHESION AND APOPTOSIS

Tumor growth and metastasis is a multistep process involving cell adhesion, proteolytic enzyme degradation of the ECM and motility factors that influence cell migration *(13)*. The interaction of cells with the ECM is mediated by contacts between cellular surface proteins called integrins and components of the ECM including laminin, fibronectin, and collagen *(14)*. Integrins are cell surface adhesive receptors made up of α- and β-chain heterocomplexes. Both subunits transverse the membrane and mediate the physical and functional interactions between cell and its surrounding ECM, thus serving as bidirectional transducers of extra- and intracellular signals which ultimately lead to regulation of adhesion, proliferation, differentiation, antiapoptosis, and tumor progression *(15)*.

Betaig-h3 gene encodes a highly conserved 683 amino-acid protein that contains a secretary signal sequence and four internal homologous domains of 140 amino-acids, the last of which contains an Arg-Gly-Asp (RGD) sequence which can serve as a ligand recognition site for several integrins *(5)*. In vitro experimental studies showed that it binds to a number of other matrix components including fibronectin, laminin, and several collagen types *(16,17)*. Betaig-h3 protein contains multiple cell adhesion motifs within the fascilin-like domains that can mediate cell adhesion and migration via integrin receptors α3β1, αvβ3, and αvβ5 *(18–20)*. It is expressed principally in collagen-rich tissue where it may interact with cells and ECM molecules *(21)*. Therefore, Betaig-h3 has been proposed to act as a bifunctional linker protein interconnecting collagens with other ECM molecules and the cells *(22)*. ECM plays an important role not only as a barrier to prevent the spread of tumor cells, but a reservoir of cell binding protein and growth factors that affect tumor cell behavior *(23)*. Therefore, dysregulation of Betaig-h3 expression may results in alteration of cell-ECM interaction that in turn leads to regulation of adhesion, proliferation, antiapoptosis and tumor progression.

Betaig-h3 protein induces apoptosis through integrin-related pathway. Overexpression of *Betaig-h3* gene induced a strong apoptotic response in human corneal epithelial and Hela cells *(24)*. However, transfecting Betaig-h3 construct that lack the integrin recognition sites, an RGD and Asp-Ile (DI) at C-terminal, had no effect on apoptosis in these cells *(24)*. There is evidence that RGD peptides released from Betaig-h3 may facilitate TGF-β-induced apoptosis, which could be abolished by either deletion of the RGD sequence or mutation of RGD *(25)*. This is further confirmed by the results that synthetic peptides of E**RGD**EL and E**RGD**SP derived from Betaig-h3 and fibronectin, respectively, induced apoptosis in Chinese hamster ovary (CHO) and H1299 cells *(25)*.

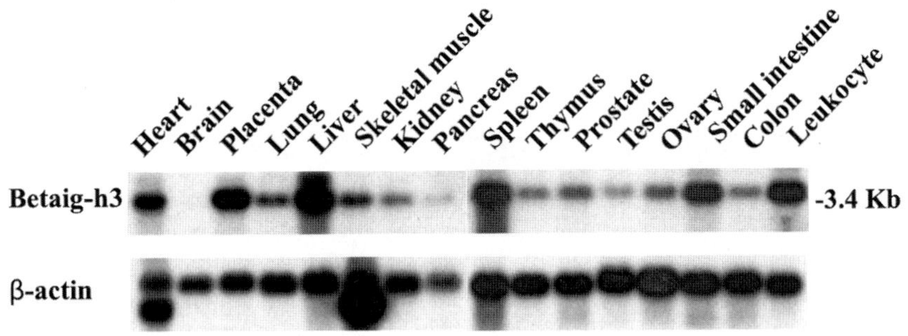

Fig. 1. mRNA levels of *Betaig-h3* gene in a variety of human normal tissues. High-expression levels of Betaig-h3 were found in heart, placenta, liver, spleen, small intestine, and leukocyte.

3. FREQUENT LOSS OF BETAIG-H3 EXPRESSION IN HUMAN CANCER CELLS

Betaig-h3 gene has been found to be ubiquitously expressed in human normal tissues, with high-expression levels in heart, placenta, liver, spleen, small intestine, and peripheral blood leukocytes. Other tissues such as pancreas, kidney, thymus, testis, and ovary expressed comparably lower levels of the gene, with the exception of brain that showed no Betaig-h3 expression (Fig. 1). The results from immunohistochemical staining have shown that intense matrix staining of Betaig-h3 protein within wall of a bronchiole and alveoli, and more localized staining of cell nuclei in human lung tissue *(24)*. In addition, consistent nuclear staining was also demonstrated in human bladder smooth muscle and fibroblast cells as well as in lamina propria fibroblasts, urothelial and smooth muscle cells of human bladder tissue *(26,27)*. Western analysis demonstrated the presence of Betaig-h3 protein in purified nuclei, conditioned medium, and solubilized matrix of bladder cells *(27)*. A similar staining pattern has recently been demonstrated by our laboratory *(38)*. We found intense staining of Betaig-h3 protein both in the stroma and in the cytoplasm and nuclei of bronchial epithelial cells. Because Betaig-h3 is a secreted protein that is localized to ECM and associated with collagen fibers *(21)*, it is expected that high-staining level of this protein exists in the stroma of lung tissue but not in the nuclei of cells. However, existence of Betaig-h3 protein in nuclei suggests that it may have other intracellular functions in addition to its role as an ECM component.

Interstitial deletion or loss of chromosome 5q is one of the most frequent occurrences in human cancers *(28–30)*. This locus contains the adenomatous polyposis coli (APC) tumor suppressor gene, originally identified as the target of germline mutations that cause familial adenomatous polyposis *(31)*. However, infrequent APC mutations in human cancers suggest the involvement of other possible tumor suppressor genes *(31)*. Chromosome 5q31, where *Betaig-h3* gene has been regionally mapped to, is often deleted in leukemias, myelodysplastic syndromes and many human cancers such as renal cell, esophageal and lung carcinomas *(32–34)*. In addition, downregulation of this gene was found in a variety of human tumor cell lines including lung, breast, prostate, embryonic rhabdomyosarcoma, and mesenchymal tumor cells *(12,35)* (Fig. 2). Our laboratory has screened the level of Betaig-h3 protein in primary lung carcinomas by using lung tumor tissues array and immunohistochemical staining approach. We found that Betaig-h3 protein was absent or reduced by more than twofold in 45 of 130 primary lung carcinomas relative to normal lung tissues examined *(38)*. These data provide strong evidence that loss of Betaig-h3 expression is a frequent event during the progression of human cancers.

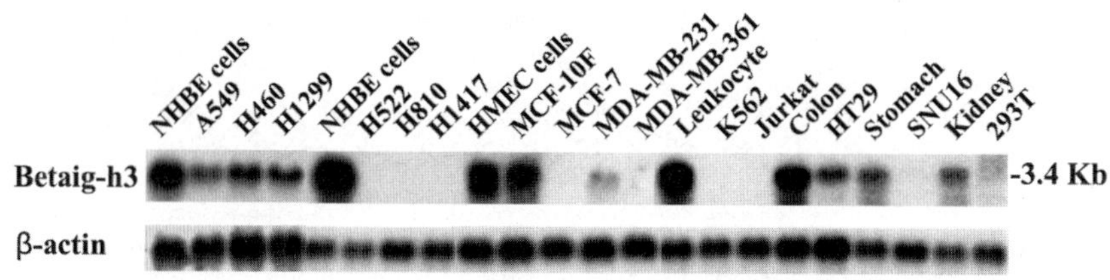

Fig. 2. Total mRNA levels of *Betaig-h3* gene in various human tumor cell lines. 2.5 μg mRNAs were isolated from lung cancer cell lines (A549, H460, H1419, H522, H810, H1417), leukemia cell lines (K562, Juekat), colon cancer (HT29), stomach cancer (SNU16), breast cancer (MCF-7, MDA-MB-231, MDA-MB-361) and kidney cancer (293T). Normal human bronchial epithelial (NHBE) cells, human mammary epithelial cell (HMEC), peripheral blood leukocyte, colon, stomach and kidney tissues were used as various controls. The blots were hybridized to ^{32}P-labeled cDNA probes. After stripping, the membranes were rehybridized to human beta-actin, which is used as control.

4. SUPPRESSION OF TUMORIGENICITY AFTER REEXPRESSION OF *BETAIG-H3* GENE

It has been shown that transfection of *Betaig-h3* gene into CHO fibroblast cells resulted in alteration of morphology, decrease in growth rate, and suppression of tumorigenicity in nude mice when compared with the parental cells *(36)*. By using in vitro malignantly transformed cell lines, *Betaig-h3* gene was demonstrated to be significantly downregulated during the process of malignant transformation in human papillamavirus-immortalized human bronchial epithelial (BEP2D) cells treated with either radiation or asbestos fiber. Recovered expression of this gene in the tumorigenic cells resulted in a marked decrease of tumorigenicity in nude mice *(11,12,37)*. Furthermore, the tumorigenic potential of human lung adenocarcinoma-derived H522 cells that lack the endogenous Betaig-h3 expression could be abrogated by ectopic expression of wild-type *Betaig-h3* gene (Fig. 3) *(38)*. The data provide strong evidence that loss of *Betaig-h3* gene correlates with the tumorigenic phenotype in human cancer cells.

The mechanism(s) of antitumor function of Betaig-h3 is not well understood. It has been postulated that *Betaig-h3* gene may be involved in the tumor progression by regulating cell-ECM interaction. This is supported by the report that acquisition of tumorigenic phenotype of BEP2D cells is accompanied by an increased expression α5β1 integrin receptor at both the mRNA and protein levels *(37)*. Overexpression of *Betaig-h3* gene in tumorigenic cells led to downregulation of integrin and suppression of tumorigenicity. The data suggest that *Betaig-h3* gene is involved in tumorigenic process by regulating α5β1 expression. The observation is consistent with other reports that α5β1, while undetectable in normal lung epithelial, is significantly elevated in SV40 large T-transformed human bronchial epithelial cells *(14,39)*. In nonsmall lung carcinoma cells, higher level of α5β1 expression represents a negative prognostic factor *(40)*. Similar results have also been shown with other human tissues that high level of α5β1 integrin is associated with more malignant phenotype in melanoma, transitional and colon cell carcinomas *(41,42)*. The α5β1 integrin favors cell survival and protects cells from apoptosis in vitro via upregulation of antiapoptotic Bcl-2, whereas resistance to apoptosis is a feature of many malignant cells *(43)*. These data suggest that *Betaig-h3* gene is involved in tumor progression of human bronchial epithelial cell model by regulating integrin receptor α5β1. Previously reports have shown that integrins are expressed in a cell-type and stage-specific manner *(44)*. One group of integrins that is associated with migration and proliferation in various types of cells is called "emergency integrins" including α5β1, αvβ3, and αvβ6 *(45)*. These integrins are particularly important in cancer.

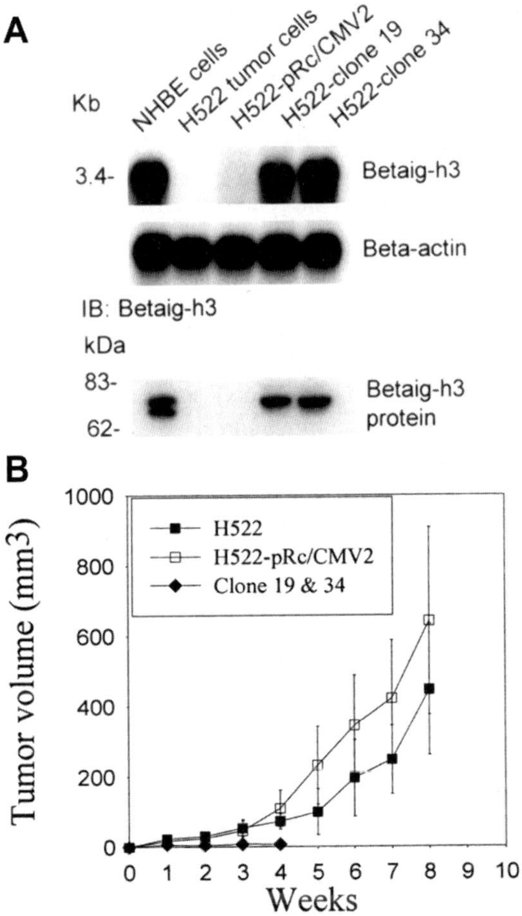

Fig. 3. Expression of *Betaig-h3* gene determined by Northern and Western blotting. (**A**) mRNA and protein levels in NHBE, parental H522, vector and pRc/CMV2-Betaigh3 transfected H522 clone 19 and 34 cells. (**B**) In vivo tumor growth of Betaig-h3 transfected H522 lung tumor cells. Tumor volume was calculated using the formula (longest diameter × shortest diameter^2) original magnification ×0.5. Results are expressed as the mean ± s.d. from three independent experiments. At each timepoint, the tumor volumes of Betaig-h3 transfected cells were significantly smaller than parental H522 lung tumor cells ($p < 0.01$).

Therefore, in addition to α5β1, other integrins may also important in mediating the antitumor function of *Betaig-h3* gene.

5. HYPERMETHYLATION OF BETAIG-H3 PROMOTER REGION AND GENE SILENCING

Betaig-h3 gene is one of the downstream effectors of TGF-β signaling pathway, which involves lots of signaling molecules including TGF-β receptor type I, II (TGFβR I, II), Smad2, 3, and 4 *(46)*. Any blockade or inactivation of this pathway caused by loss of the TGFβR and mutations of Smads would lead to lack of response to TGF-β, which in turn would affect expression of *Betaig-h3* gene. Downregulation of type II TGF-β receptor has been reported in many cancers including small cell lung carcinomas, retinoblastoma, gastric cancer and hereditary forms of nonpolyposis colorectal cancer *(47)*. Similarly, a large percentage of pancreatic cancers (50%) and colorectal cancers (30%) have mutations in Smad4 *(48,49)*. However, except the colon and pancreatic tumors, the majority of tumors with loss of growth inhibition do not acquire the above genetic defects. The data suggest that in addition to the dysregulation of TGF receptor and smads molecules, other yet unidentified mechanism(s) might responsible for the downregulation of Betaig-h3 expression.

DNA methylation, one of the most commonly occurring epigenetic events in the mammalian genome, is an important regulator of gene transcription, and plays an important role in cancer biology, genetic imprinting, developmental abnormalities and X-chromosome inactivation *(50,51)*. CpG islands located at the promoter area of certain tumor suppressor genes undergo aberrant hypermethylation in cancer cells which is recognized as a crucial factor responsible for the gene silencing *(50)*. It has been well-documented that some known tumor suppressor genes are modified and transcriptionally silenced by promoter hypermethylation including cell cycle inhibitor (p16INK4a) and DNA repair genes (hMLH1 and BRCA1) *(52)*. Because *Betaig-h3* gene functions as a tumor suppressor, it is expected that silence of *Betaig-h3* gene may result from promoter hypermethylation. Our laboratory has screened the methylation profile of *Betaig-h3* promoter in 17 human tumor cell lines by using bisulfite treatment and sequencing approach *(52a)*. Among all of the cell lines tested, tumor cells with negative or low level of *Betaig-h3* gene showed a densely methylation pattern. However, normal, immortalized and some tumor cell lines that expressed high levels of *Betaig-h3* were not methylated in the promoter region. Furthermore, treatment with the demethylation agent, 5-Aza-CdR, resulted in the reexpression or upregulation of the gene in *Betaig-h3*-negative tumor cell lines *(52a)*. The data give us indication that methylation of CpG island on the promoter region is correlated with Betaig-h3 silencing in human tumor cells, which may constitute an important mechanism for silencing of the gene in tumor progression.

6. CONCLUDING REMARKS

Although the Betaig-h3 protein has been identified for more than ten years, the detailed physiological functions of this gene are largely unknown. Based on the limited data published, Betaig-h3 protein seems to have important intracellular functions in addition to its regulation of cellular adhesion because it is present in both stroma and nuclei of epithelial cells. In addition, downregulation of this gene has been shown to be correlated with the tumorigenic phenotype by using in vitro cell culturing systems. These data suggest that Betaig-h3 might act as one of the important downstream effectors of TGF-β in mediating its pivotal cellular functions including proliferation, differentiation, apoptosis, and deposition of extracellular matrix, immunosuppression and antitumor activities *(53)*. However, further studies by using Betaig-h3 knock-out mice model will be needed to define the physiological functions of this gene, especially the role in tumor progression.

ACKNOWLEDGMENTS

The authors would like to thank the members of our laboratory for the helpful comments.

REFERENCES

1. Wakefield LM, Roberts AB. TGF-beta signaling: positive and negative effects on tumorigenesis. Curr Opin Genet Dev 2002;2(1):22–29.
2. Akhurst RJ, Derynck R. TGF-beta signaling in cancer-a double-edged sword. Trends Cell Biol 2001;11(11):S44–S51.
3. de Caestecker MP, Piek E, Roberts AB. Role of transforming growth factor-beta signaling in cancer. J Natl Cancer Inst 2000;92(17):1388–1402.

4. Siegel PM, Massagué J. Cytostatic and apoptotic actions of TGF-beta in homeostasis and cancer. Nat Rev Cancer 2003;3(11):807–821.
5. Skonier J, Neubauer M, Madisen L, Bennett K, Plowman GD, Purchio AF. cDNA cloning and sequence analysis of beta ig-h3, a novel gene induced in a human adenocarcinoma cell line after treatment with transforming growth factor-beta. DNA Cell Biol 1992;11(7):511–522.
6. Munier FL, Korvatska E, Djemai A, et al. Kerato-epithelin mutations in four 5q31-linked corneal dystrophies. Nat Genet 1997;15(3):247–251.
7. LeBaron RG, Bezverkov KI, Zimber MP, Pavelec R, Skonier J, Purchio AF. Beta IG-H3, a novel secretory protein inducible by transforming growth factor-beta, is present in normal skin and promotes the adhesion and spreading of dermal fibroblasts in vitro. J Invest Dermatol 1995;104(5):844–849.
8. Kim JE, Kim SJ, Lee BH, Park RW, Kim KS, Kim IS. Identification of motifs for cell adhesion within the repeated domains of transforming growth factor-beta-induced gene, betaig-h3. J Biol Chem 2000; 275(40):30,907–30,915.
9. Kim JE, Kim SJ, Jeong HW, et al. RGD peptides released from beta ig-h3, a TGF-beta-induced cell-adhesive molecule, mediate apoptosis. Oncogene 2003;22:2045–2053.
10. Kim JE, Kim EH, Han EH, et al. A TGF-beta-inducible cell adhesion molecule, big-h3, is downregulated in melorheostosis and involved in osteogenesis. J Cell Biochem 2000;77:169–178.
11. Zhao Y, Shao G, Piao CQ, Berenguer J, Hei TK. Down-regulation of Betaig-h3 gene is involved in the tumorigenesis in human bronchial epithelial cells induced by heavy-ion radiation. Radiat Res 2004; 162:655–659.
12. Zhao YL, Piao CQ, Hei TK. Downregulation of Betaig-h3 gene is causally linked to tumorigenic phenotype in asbestos treated immortalized human bronchial epithelial cells. Oncogene 2002;21: 7471–7477.
13. Tlsty TD. Cell-adhesion-dependent influences on genomic instability and carcinogenesis. Curr Opin Cell Biol 1998;10(5):647–653.
14. Albelda SM, Buck CA. Integrin and other cell adhesion molecules. FASEB J 1990;4(11):2868–2880.
15. Giancotti FG, Ruoslahti E. Integrin signaling. Science 1999;285(5430):1028–1032.
16. Billings PC, Whitbeck JC, Adams CS, et al. The transforming growth factor-beta-inducible matrix protein (beta)ig-h3 interacts with fibronectin. J Biol Chem 2002;277(31):28,003–28,009.
17. Kim JE, Park RW, Choi JY, et al. Molecular properties of wild-type and mutant betaIG-H3 proteins. Invest Ophthalmol Vis Sci 2002;43(3):656–661.
18. Jeong HW, Kim IS. TGF-beta1 enhances betaig-h3-mediated keratinocyte cell migration through the alpha3beta1 integrin and PI3K. J Cell Biochem 2004;92(4):770–780.
19. Nam JO, Kim JE, Jeong HW, et al. Identification of the alphavbeta3 integrin-interacting motif of betaig-h3 and its anti-angiogenic effect. J Biol Chem 2003;278(28):25,902–25,909.
20. Kim JE, Jeong HW, Nam JO, et al. Identification of motifs in the fasciclin domains of the transforming growth factor-beta-induced matrix protein betaig-h3 that interact with the alphavbeta5 integrin. J Biol Chem 2002;277(48):46,159–46,165.
21. Ferguson JW, Mikesh MF, Wheeler EF, LeBaron RG. Developmental expression patterns of Beta-ig (betaIG-H3) and its function as a cell adhesion protein. Mech Dev 2003;120(8):851–864.
22. Billings PC, Herrick DJ, Kucich U, et al. Extracellular matrix and nuclear localization of beta ig-h3 in human bladder smooth muscle and fibroblast cells. J Cell Biochem 2000;79(2):261–273.
23. Pupa SM, Menard S, Forti S, Tagliabue E. New insights into the role of extracellular matrix during tumor onset and progression. J Cell Physiol 2002;192(3):259–267.
24. Morand S, Buchillier V, Maurer F, et al. Induction of apoptosis in human corneal and HeLa cells by mutated BIGH3. Invest Ophthalmol Vis Sci 2003;44(7):2973–2979.
25. Kim JE, Kim SJ, Jeong HW, et al. RGD peptides released from beta ig-h3, a TGF-beta-induced cell-adhesive molecule, mediate apoptosis. Oncogene 2003;22(13):2045–2053.
26. Billings PC, Herrick DJ, Kucich U, et al. Extracellular matrix and nuclear localization of betaig-h3 in human bladder smooth muscle and fibroblast cells. J Cell Biochem 2000;79(2):261–273.
27. Billings PC, Herrick DJ, Howard PS, Kucich U, Engelsberg BN, Rosenbloom J. Expression of betaig-h3 by human bronchial smooth muscle cells: localization To the extracellular matrix and nucleus. Am J Respir Cell Mol Biol 2000;22(3):352–359.
28. Petersen I, Langreck H, Wolf G, et al. Small-cell lung cancer is characterized by a high incidence of deletions on chromosomes 3p, 4q, 5q, 10q, 13q and 17p. Br J Cancer 1997;75(1):79–86.
29. Brezinova J, Zemanova Z, Cermak J, Michalova K. Fluorescence in situ hybridization confirmation of 5q deletions in patients with hematological malignancies. Cancer Genet Cytogenet 2000;117(1):45–49.

30. Peralta RC, Casson AG, Wang RN, Keshavjee S, Redston M, Bapat B. Distinct regions of frequent loss of heterozygosity of chromosome 5p and 5q in human esophageal cancer. Int J Cancer 1998;78(5): 600–605.
31. Ohgaki H, Kros JM, Okamoto Y, Gaspert A, Huang H, Kurrer MO. APC mutations are infrequent but present in human lung cancer. Cancer Lett 2004;207(2):197–203.
32. Brezinova J, Zemanova Z, Cermak J, Michalova K. Fluorescence in situ hybridization confirmation of 5q deletions in patients with hematological malignancies. Cancer Genet Cytogenet 2000;117(1): 45–49.
33. Peralta RC, Casson AG, Wang RN, Keshavjee S, Redston M, Bapat B. Distinct regions of frequent loss of heterozygosity of chromosome 5p and 5q in human esophageal cancer. Int J Cancer 1998;78(5): 600–605.
34. Wu X, Zhao Y, Kemp BL, Amos CI, Siciliano MJ, Spitz MR. Chromosome 5 aberrations and genetic predisposition to lung cancer. Int J Cancer 1998;79(5):490–493.
35. Genini M, Schwalbe P, Scholl FA, Schafer BW. Isolation of genes differentially expressed in human primary myoblasts and embryonal rhabdomyosarcoma. Int J Cancer 1996;66(4):571–577.
36. Skonier J, Bennett K, Rothwell V, et al. Beta ig-h3: a transforming growth factor-beta-responsive gene encoding a secreted protein that inhibits cell attachment in vitro and suppresses the growth of CHO cells in nude mice. DNA Cell Biol 1994;13(6):571–584.
37. Zhao YL, Piao CQ, Hei TK. Overexpression of Betaig-h3 gene downregulates integrin alpha5beta1 and suppresses tumorigenicity in radiation-induced tumorigenic human bronchial epithelial cells. Br J Cancer 2002;86(12):1923–1928.
38. Zhao YL, El-Gabry M, Hei TK. Loss of Betaig-h3 protein is frequent in primary lung carcinoma and related to tumorigenic phenotype in lung cancer cells. Mol Carcinogen 2006;45(2):84–92.
39. Schiller JH, Bittner G. Loss of the tumorigenic phenotype with in vitro, but not in vivo, passaging of a novel series of human bronchial epithelial cell lines: possible role of an alpha 5/beta 1-integrin-fibronectin interaction. Cancer Res 1995;55(24):6215–6221.
40. Adachi M, Taki T, Higashiyama M, Kohno N, Inufusa H, Miyake M. Significance of integrin alpha5 gene expression as a prognostic factor in node-negative non-small cell lung cancer. Clin Cancer Res 2000;6(1):96–101.
41. Saito T, Kimura M, Kawasaki T, Sato S, Tomita Y. Correlation between integrin alpha 5 expression and the malignant phenotype of transitional cell carcinoma. Br J Cancer 1996;73(3):327–331.
42. Beliveau A, Berube M, Rousseau A, Pelletier G, Guerin SL. Expression of integrin alpha5beta1 and MMPs associated with epithelioid morphology and malignancy of uveal melanoma. Invest Ophthalmol Vis Sci 2000;41(8):2363–2372.
43. Zhang Z, Vuori K, Reed JC, Ruoslahti E. The alpha 5 beta 1 integrin supports survival of cells on fibronectin and up-regulates Bcl-2 expression. Proc Natl Acad Sci USA 1995;92(13):6161–6165.
44. Ruoslahti E. Fibronectin and its integrin receptors in cancer. Adv Cancer Res 1999;76:1–20.
45. Sheppard D. Epithelial integrins. Bioessays 1996;18(8):655–660.
46. Shi Y, Massagué J. Mechanisms of TGF-beta signaling from cell membrane to the nucleus. Cell 2003; 113:685–700.
47. Kelly DL, Rizzino A. Growth regulatory factors and carcinogenesis: the roles played by transforming growth factor β, its receptors and signaling pathways. Anticancer Res 1999;19:4791–4808.
48. Hahn SA, Schutte M, Hoque AT, et al. DPC4, a candidate tumor suppressor gene at human chromosome 18q21.1. Science 1996;271(5247):350–353.
49. Miyaki M, Iijima T, Konishi M, et al. Higher frequency of Smad4 gene mutation in human colorectal cancer with distant metastasis. Oncogene 1999;18(20):3098–3103.
50. Das PM, Singal R. DNA methylation and cancer. J Clin Oncol 2004;22:4632–4642.
51. Laird PW. The power and the promise of DNA methylation markers. Nat Rev Cancer 2003;3:253–266.
52. Esteller M. CpG island hypermethylation and tumor suppressor genes: a booming present, a brighter future. Oncogene 2002;21:5427–5440.
52a. Shao G, Berenguer J, Borczuk AC, Powell CA, Hei TK, Zho YL. Epigenetic inactivation of TGFBI gene in human cancer cells. Cancer Res 2006;66(9):4566–4573.
53. Derynck R, Akhurst RJ, Balmain A. TGF-beta signaling in tumor suppression and cancer progression. Nat Genet 2001;29(2):117–129.

31 Activins and the Liver

Michael Grusch, Chantal Rodgarkia-Dara, Wilfried Bursch, and Rolf Schulte-Hermann

CONTENTS

INTRODUCTION
STRUCTURE AND PROCESSING
ACTIVIN ANTAGONISTS
ACTIVIN FUNCTIONS IN THE LIVER
ACTIVIN IN LIVER TUMORS
ACTIVINS AS POTENTIAL THERAPEUTIC TARGETS
ACKNOWLEDGMENTS
REFERENCES

Abstract

Activins are members of the transforming growth factor-beta (TGF-β) family. They are dimers composed of β subunits four of which are known in mammals: βA, βB, βC, and βE; a fifth subunit, βD, was found in *Xenopus*. As other TGF-β family members, activins are synthesised as proforms and are intracellularly processed to the mature peptides. They contain a characteristic knot of nine conserved cysteines. Activin A has been quite-well characterized. It is expressed in many organs and cell types, its receptors and binding proteins (follistatins) as well as the binding regions in the mature peptide are known. The liver activin A, like TGF-β1, is a negative regulator of growth and contributes to the adjustment of liver size according to physiological requirements. In addition, activin A appears to exert proinflammatory and fibrogenic activities. In rodent liver tumors levels of activin A are decreased and those of follistatins are increased suggesting that deregulation of this growth inhibitory system contributes to carcinogenesis. Activins C and E are predominantly expressed in the liver. Little definitive evidence is available on their function. Activin C has been described as proapoptotic and growth inhibitory in liver and hepatoma cells, while other studies have reported growth stimulatory effects. Activin E was found to inhibit DNA synthesis and stimulate apoptosis in normal liver and hepatoma cells.

Key Words: Activins, inhibin; TGF β; follistatins; hepatic apoptosis; hepatic DNA synthesis; inflammatory effects; liver fibrosis; chemical mitogens; liver tumors.

1. INTRODUCTION

Activins and inhibins are members of the transforming growth factor-β (TGF-β) family, a family involved in regulation of embryonic development, reproductive processes, inflammation, cell proliferation, and apoptosis. The existence of inhibin has been postulated as biological basis for gonadal regulation of pituitary function more than 80 years ago *(1)*.

From: *Cancer Drug Discovery and Development: Transforming Growth Factor-β in Cancer Therapy, Vol I: Basic and Clinical Biology*
Edited by: Sonia B. Jakowlew © Humana Press Inc., Totowa, NJ

A few years later a substance was isolated from bovine testes that inhibited hypertrophy of pituitary cells, and was named inhibin *(2)*. Inhibin is a component of a feedback loop mechanism to concert gonadal and pituitary growth and function, mainly by inhibiting the release of follicle stimulating hormone (FSH). About 20 yr ago, purification and characterization of inhibin from various species began *(reviewed in [3])*, and cloning of the genes started. Inhibin turned out to be a dimeric factor, consisting of an α- and a β-subunit linked by a single disulfide bridge, as other TGF-β family members. During these studies a further factor consisting of two β-chains was isolated. It was found to stimulate FSH release, thereby antagonizing the action of inhibin and therefore was named activin *(4)*. Until now, one α-subunit and four β-subunits have been described in mammals. The expression of α-, βA and βB-subunits is almost ubiquitous, whereas expression of βC- and βE-subunits is more restricted with the highest expression in the liver. Additionally, in *Xenopus* a unique subunit only distantly related to the other β-subunits, has been identified and was named βD *(5)*. The current state of knowledge about structure and processing of activins as well as their role in liver physiology and liver tumors is reviewed in the present chapter.

2. STRUCTURE AND PROCESSING

2.1. Protein and Gene Identification

Inhibin was found to be a gonadal hormone, which exerts a negative feedback action on the pituitary secretion of FSH in male and female mammals. In the late 70s, a few groups started to purify inhibin from ovarian *(6)* and testicular fluids *(7,8)*. At this time two fractions with inhibin-like activity were isolated, a larger protein with apparent molecular mass between 40 and 70 kDa found in gonadal extracts and fluids, and a smaller fraction, with molecular mass between 5 and 20 kDa, found in seminal plasma *(reviewed in [3])*. Inhibin purified from porcine follicular fluid was characterized as a 32 kDa heterodimeric protein. The subunits were named α and β chain *(9)*. Soon after that, two different β-subunits were identified, βA and βB and evidence for β-chain homodimers arose, counteracting inhibin activity *(4,10,11)*.

Thus far, the subunits βA and βB mRNA were found to be easily detectable in almost all tissues analyzed *(12,13)*. An additional β subunit designated activin βC was first identified from a liver cDNA library in human *(14)* and subsequently in mouse *(15)* and rat *(13)*. Soon after the discovery of activin βC, the subunit βE was cloned from liver transcripts in mouse, rat, and human *(16–18)*. Interestingly, activin subunits βC and βE are predominantly expressed in the liver, much lower levels of both mRNAs were detected in several other organs *(13,18–24)*.

Depending on the cellular expression levels of the respective subunits a variety of activin/inhibin combinations is possible. Inhibins A (αβA) and B (αβB) have been purified from natural sources *(11)*, whereas the existence of inhibin C (αβC) was recently described after ectopic coexpression of both subunits in Chinese hamster ovary (CHO) cells *(25)*. Besides activins composed of identical β-subunits like activin A (βAβA), also heterodimers between the different β-subunits occur, e.g., activins AB (βAβB) and AC (βAβC) *(13,26,27)*.

2.2. Gene Structure

The gene of the α-subunit is named *Inha* and the genes of the β-subunits are termed *InhbA, InhbB, InhbC and InhbE*, respectively (Table 1). The genomic organization of inhibin subunits is simpler and more consistent, as of other TGF-β-family members, which vary in number of exons and splice variants. Inhibin coding regions are distributed on 2 exons separated by an intron, ranging from 0.2 to 14 kb (Table 1). A third exon in βA was identified upstream of the designated exon 1 encoding for a 5′ untranslated sequence in human *(28)* and rat *(29)*. Interestingly, the genes of the two liver-expressed subunits βC and βE are closely linked, as for example in mouse both genes are separated by 5.5 kb only *(30)*. Also in other species these two TGF-β-family members are located in proximity on the same chromosome,

Table 1
Human Activin/Inhibin Subunits. Gene Identification (GI) Numbers of Full Length cDNAs and Chromosomal Location Were Derived From NCBI Databases (http://www.ncbi.nlm.nih.gov). Exon–Intron Information was Derived From Original Publications and the NCBI Databases.

Gene	*GI*	*Chromosome*	*Exon-Intron*	*Lit.*
βA (InhbA)	62953137	7p15-p13	3 exons, 2 and 9.5 kb intron	Tanimoto 1996
βB (InhbB)	9257224	2cen-q13	2 exons, 2.4 kb intron	Mason 1989
βC (InhbC)	15718678	12q13.1	2 exons, 14 kb intron	Schmitt 1996
βE (InhbE)	37622353	12q13.3	2 exons, 0.2 kb intron	—
α (InhA)	9257223	2q33-q36	2 exons, 2 kb intron	Stewart 1986

e.g., in human and chimpanzee (GI 59709462) on chromosome 12, in rat (GI 59709462) on chromosome 7, in cattle (GI 76618875) on chromosome 5, and in mouse (GI 20070688) and dog (GI 73968465) on chromosome 10 (http://www.ncbi.nlm.nih.gov/mapview, and entrez gene). All URLS were accessed in May 2007.

2.3. Protein Structure

Inhibin/activin subunits are synthesized as precursors with 350–426 amino acids and a molecular weight between 38 and 50 kDa. Their respective prodomain is removed to release mature peptides ranging from 112 to 134 amino acids.

The amino acid sequences of the mature peptides are approximately 50% conserved among the four human β-subunits, whereas the sequence homology in the prodomain is only about 20%. Phylogenetic analysis of the mature human peptides predict a close relationship between βA/βB and βC/βE. Within the TGF-β family, TGF-β1 is more closely related to the β-subunits than to the α-subunit. Also within the prodomains relationship is closest between βA and βB, followed by βC and βE (Fig. 1). An extensive phylogenetic comparison of the mouse sequences of the TGF-β superfamily is reported in *(31)*.

2.4. Prodomains

The prodomains of TGF-β, inhibin and activin subunits determine the folding, dimerization and export of the respective mature cytokine *(32)*. TGF-β1 is secreted as a latent complex consisting of the TGF-β homodimer, its prodomain, also termed latency-associated propeptide (LAP), and the latent TGF-β binding protein (LTBP) *(reviewed in [33])*. The first cysteine residue in the TGF-β prodomain is required for covalent binding to LTBPs *(34)*. Therefore loss of this cysteine bridge leads to enhanced secretion of bioactive TGF-β1 dimer. Mutation of one cysteine in the CxC-motif at the C-terminal region of LAP has no effect, whereas mutation of both prevents LAP dimerization, but has no effect on TGF-β1 bioactivity *(35)*.

In contrast, in the prodomain of βA mutation of all four cysteine residues does neither alter intracellular processing, dimer assembly nor secretion *(36)*. No similar studies have been performed for βC and βE subunits, and a conclusion by analogy would be inappropriate because of the different cysteine distribution in their prodomains (Fig. 2).

By means of computational analysis and database comparisons a few unexpected motives were detected in the individual human prodomains and will be annotated here in brief. A classical nuclear localization signal consisting of four lysine (K) or arginine (R) residues, identical to potential convertase cleavage sites, are found in the βA and βE prodomain by the PSORT II Prediction (http://psort.nibb.ac.jp/form2.html). An unusual proline-rich region is found in the N-terminus of the human βB-prodomain *(37)* and is conserved in simian, bovine, mouse and rat βB. Proline-rich domains are often dimerization structures, e.g., in SAP49 for

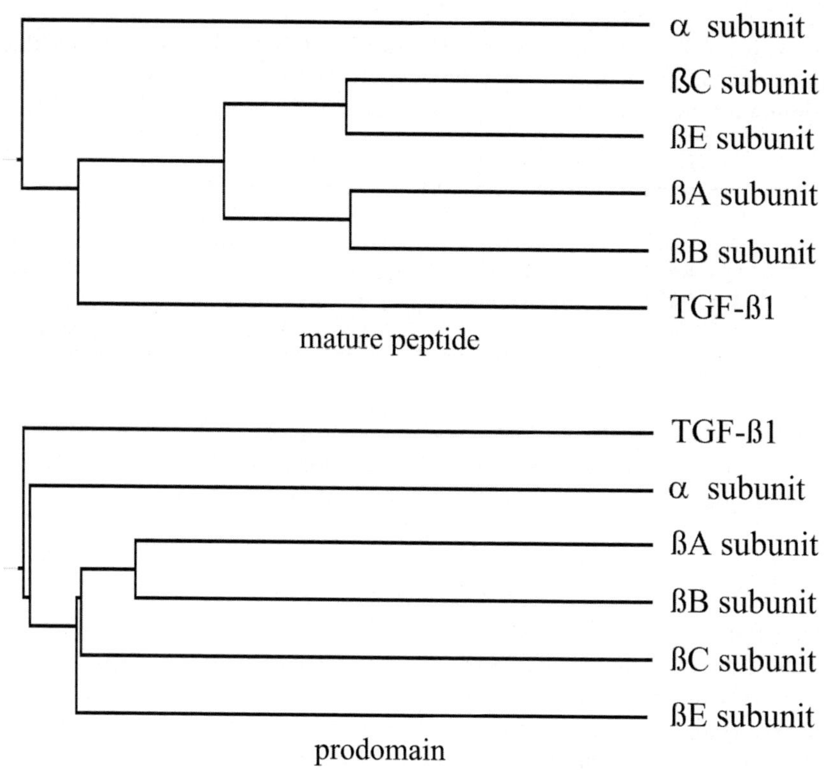

Fig. 1. Phylogenetic trees of amino acid sequences of prodomains and mature human activin/inhibin subunits and TGF-β1 were calculated by multiway protein alignment with BLOSUM62 as scoring matrix.

its binding to BMPR-IA *(38)*. Moreover, this domain is interpreted as Src-homology3 (SH3) ligand region by functional-site-analysis with ELM (http://elm.eu.org/). A classical leucine zipper, a further dimerization motif, was detected in the prodomain of human βC by PSORT II prediction. Interestingly, this domain is also present in simian and bovine, but not in mouse and rat βC. Last but not least, a WHxL motif is detected by ELM in the prodomain of both human liver-expressed β-subunits C and E, whereas in mouse and rat this motif is found in βE only. In synaptic transmembrane proteins a WHxL motif is essential for the docking of synaptic vesicles to plasma membrane *(39)*.

The relevance of these computationally predicted, unexpected structures for processing and function of the activin subunits remains to be elucidated.

2.5. Glycosylation

LAP, the TGF-β prodomain, contains three potential asparagine-linked (N-linked) glycosylation sites. The first and second N-linked carbohydrates contain mannose-6-phosphate and stabilize the mature TGF-β1 dimer in the latent complex. N-linked glycosylation is necessary for translocation to the Golgi apparatus. Mutation of one or two N-glycosylation sites in the TGF-β1 prodomain still leads to secretion of LAP, but not to bioactive TGF-β1. Mutation of all three N-glycosylation sites impairs also secretion of LAP *(40)*. Treatment with tunicamycin, which prevents the formation of the dolichol intermediate necessary for oligosaccharide addition, caused an intracellular accumulation of proTGF-β1 *(41)* and inhibits activin βA secretion *(42)*. TGF-β1 is a substrate for the mannose-6-phosphate (M6P) receptor

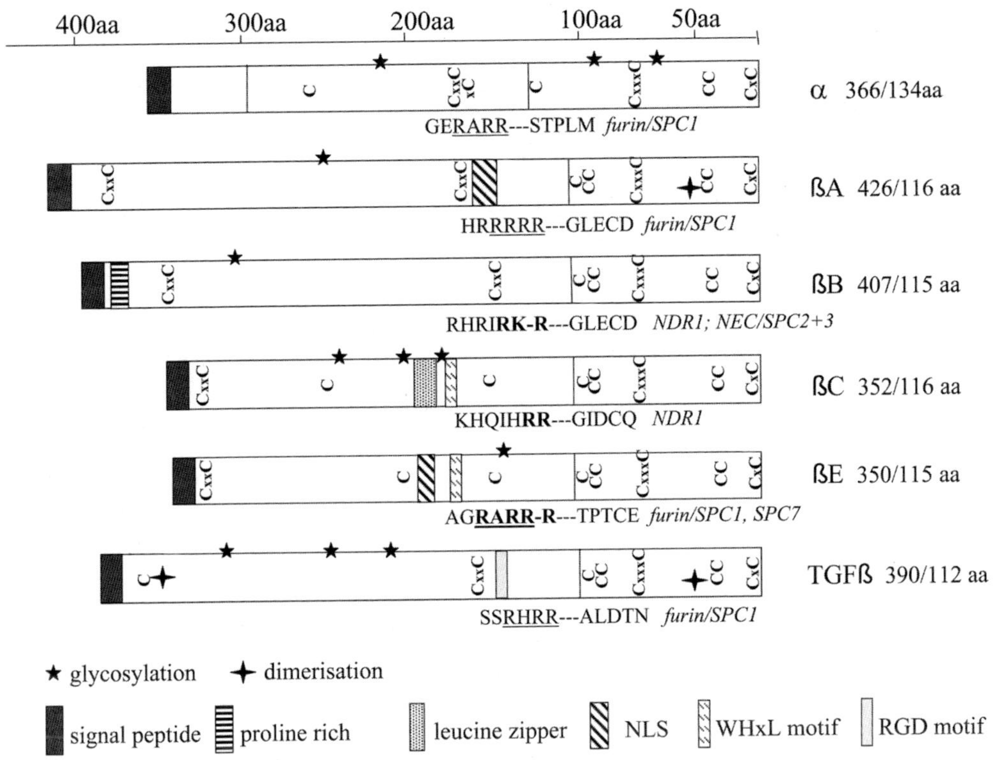

Fig. 2. Protein domain structure of activin/inhibin subunits and TGF-β1. The number of AAs of full length and the mature protein is annotated for the respective protein. The positions of cysteines are marked with C. Prodomains and the mature peptides are separated by a bar and the predicted convertases, cleavage motifs and the first five AAs of the mature peptide are annotated beneath the respective protein. The furin/SPC1cleavage motif is underlined; Cleavage motifs for NDR1, NEC/SPC2+3 and SPC7 are marked with bold characters.

during processing in the Golgi apparatus. Treatment with 1-deoxymanojirimycin, an inhibitor of α-mannosidase-I, abolished TGF-β1 secretion. In contrast to TGF-β1, activin βA processing, secretion, or bioactivity is not effected by deoxymanojirimycin *(42)*.

The inhibin α-subunit is the only one with two N-linked glycosylation sites in the mature peptide. Mutational analysis of these two sites revealed no changes in expression, processing, dimerization, secretion, and biological activity *(43)*. Vice versa the insertion of such an N-linked glycosylation site into the mature βA-subunit does not interfere with receptor binding *(43)*.

The βA, βB, and βE-prodomains contain one N-linked glycosylation site, like the α-subunit. The βC-prodomain exhibits three potential N-linked glycosylation sites similar to LAP. The function of these glycosylation sites has not been examined more closely.

2.6. Mature Peptides: Cysteine Knot

A characteristic feature of TGF-β1 and the activin β subunits are nine conserved cysteines in the mature peptides, whereas the mature inhibin α-subunit exhibits only seven cysteines (Fig. 2). These cysteines determine the 3-dimensional structure by disulfide bridges of the mature monomers and dimers *(reviewed in [44,45])*. Most structural data are obtained from extensive mutational studies on activin βA *(36)* and TGF-β1 *(40)*. In both cytokines the sixth cysteine is responsible for dimerization, whereas the others are responsible for

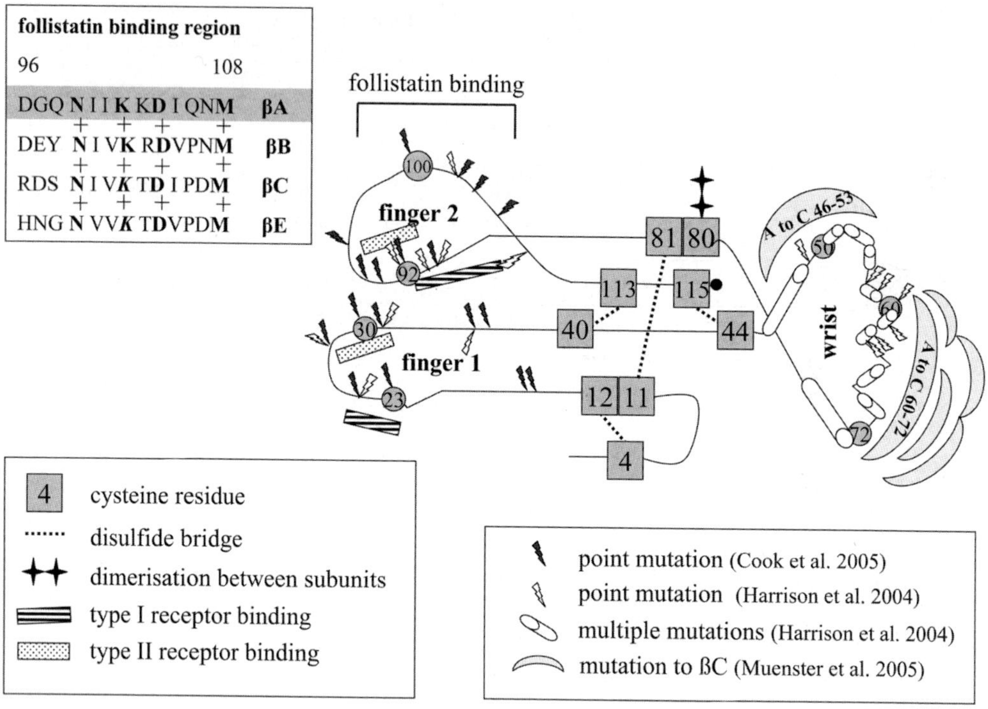

Fig. 3. Mature peptide structure of activin βA and experimentally introduced mutations. The complex cysteine knot, the two fingers, the wrist region, follistatin- and receptor-binding regions are depicted. Amino acid numbers in circles are annotated to assist orientation. Mutations are indicated as arrows, barrels and arcs. The amino acid sequences of the follistatin binding region of βA (gray) and a comparison with corrsponding sequences of βB, βC and βE is shown in the inlet. Identical AAs are shown in bold and marked with +.

proper protein folding. This results in two antiparallel β-sheet structures named activin fingers on the one side, and an α-helical structure named wrist region on the other side of the cyteine knot (Fig. 3).

Mutation of the first and the third cysteine leads to the secretion of dimeric proteins with reduced bioactivity, whereas all other cysteine mutations lead to secretion of the unprocessed proprotein or no secretion at all *(36)*. This results from incorrect protein folding and degradation of the newly synthesized, nascent proteins in the endoplasmic reticulum (ER). Owing to the conserved cysteine distribution pattern (Fig. 2) in the mature subunits a conclusion by analogy for the protein structure of βB, βC, and βE is presumed in the literature.

2.7. Mature Peptides: Receptor and Follistatin Binding

Activin A, like other members of the TGF-β superfamily, initiates signaling by assembling a complex of two types of transmembrane serine/threonine receptor kinases classified as Type II and Type I *(reviewed in [46])*. First, the activin dimer binds to Type II activin receptors, which are single-transmembrane domain serine/threonine kinase receptors. Then Type I receptors are recruited and activated by phosphorylation. The Type II receptor binds to the concave site whereas Type I receptor binding occurs on the convex site of the activin fingers *(47,48)* as depicted in Fig. 3. The α-helix/wrist region seems to be relevant for the proper bending of the activin dimer and therefore for recruitment of the receptor complex *(49)*. Chimeric βA/C, where the original βA sequence of the α-helix/wrist region is exchanged against the respective βC sequence, retains Type II binding but inhibits Type I binding and subsequently has no transcriptional activity and can even act as a competitive inhibitor of activin A. Also the exchange from the βA- against the βC sequence of the prehelix loop (aa 46–53) and parts of the α-helix/wrist region reduced transcriptional activity in 239T cells *(45)*.

Extensive mutational studies exchanging βA amino acid residues by the respective α-subunit amino acids *(50)* and mutation of highly conserved residues in BMP-2 and βA *(48)* provide an insight into receptor binding ability and biological activity of the mutant activin βA proteins. Inhibin binds Type II receptor through its single β-subunit, whereas the α-subunit is unable to bind to the receptor. Single amino acid exchanges in positions 28 to 35 and 90 to 93 by the respective α-subunit residue repress all kinds of receptor binding and biological activity, measured in the gonadotrope cell line LβT2 by a luciferase construct driven by the FSHβ promoter *(50)*. The disruption of the α-helical structure in the wrist region by inserting proline decreased biological activity, measured in HEK293T cells by A3 lux reporter gene assay *(48)*.

All amino acid residues critical for receptor binding of activin A are identical in βB and βC, and similar in βE, where only conservative changes from isoleucine to leucine or valine are present. Surprisingly however, activin C is not able to replace receptor-bound activin A from its receptor tested in the hepatocyte cell line AML12 *(51)*. A comparison of activin A, B, and C transcriptional activity performed in CHO cells and in the pituitary cell line LβT2 confirm the incapability of activin C to induce gene expression driven by different activin A-responsive elements, whereas activin B exhibited similar or even higher activity than activin A *(27)*. No experimental data for activin E transcriptional activity are available.

Binding of follistatin, a binding partner (discussed in Section 3) and thereby regulator of activin A activity, takes place at amino acid positions 100 to 109 in βA *(52)*, which exhibit only weak similarity between the human β-subunits (Fig. 3, inlet). βE was reported to bind follistatin and is competed by an excess of βA *(18)*, whereas βC was not able to replace βA-binding from follistatin *(51)*. This seems to be a surprising result, because the putative βB-, βC- and βE-follistatin-binding sequences have similarity to each other rather than to βA sequence. With respect to different follistatin-binding assays used in the above cited publications, a reevaluation of follistatin-binding capacity may be necessary to shed light on this topic.

2.8. Processing

Inhibins/activins are translated as inactive proforms and need to be processed to gain activity. Like other TGF-β ligands, they are biologically active as dimers only, but in detail there are also many differences compared to TGF-β in processing, storage, and activation.

Most secreted proteins contain a signal peptide for translocation to the ER. In eukaryotic cells, proper protein folding and dimerization take place in the lumen of the rough ER. Thus disulfide bonds are found only in secreted proteins and in the exoplasmic domains of membrane proteins *(53)*. The formation and rearrangement of disulfide bonds is accelerated by the members of the protein disulfide isomerase (PDI) family *(reviewed in [54,55])*. PDIs catalyze the formation of native disulfide bonds, isomerization in compact folding intermediates and formation of homo- or heterodimers of the respective subunits. Whereas proper folding of the C-terminal peptide is a prerequisite for further translocation to the Golgi apparatus and for secretion, cysteines in the N-terminal prodomain and the dimerization-specific cysteine are dispensable for further processing *(36)*.

Generally, the prodomains of TGF-β1 and βA are necessary for their secretion. Fusion proteins of the preprodomain of parathyroid hormone and the mature TGF-β1 or βA domain were not secreted. Coexpression of a prodomain construct partially rescued secretion of bioactive TGF-β1 and activin A. This indicates that the separated prodomain/mature domain can be processed in the ER, translocated to Golgi and to the secretory pathway *(32)*.

Cleavage of the prodomain from the mature peptide may occur in the ER and in the early Golgi, and is performed by the family of subtilase-like proprotein convertases (SPC) *(56)*. SPC1 and 7 are ubiquitously expressed, with highest expression in liver and epidermis, whereas SPC2 and 3 are expressed in neuroendocrine tissues (*reviewed in [57]*). Both, TGF-β1 and βA subunit are substrates for SPC1/furin *(58,59)* with the recognition motif RxRR (annotated in Fig. 3, underlined). α- and βE-subunit exhibit the same recognition motif, whereas in βB and βC cleavage motifs are optimal for nardilysin (NDR1), a metalloendopeptidase, or SPC2/3 (annotated in Fig. 3, bold letters). An additional cleavage motif for SPC7 has been detected in βE by ELM (http://elm.eu.org/). Interestingly, in contrast to the human and chimpanzee βC cleavage motif, rodent, bovine, and canine βC subunits possess an ideal SPC1/furin cleavage motif (RvRR). However, no experimental studies are available on processing of βC and βE subunits from human and rodent origin.

2.9. Secretion and Activation

TGF-β1 is secreted as a latent complex consisting of the TGF-β homodimer, its prodomain, LAP, and LTBP1 and stored in the extracellular matrix coupled to integrins as reviewed in *(60)*. The bioactive homodimer is actively released from this complex for example by urokinase-type (uPA) plasminogen activators in osteoclasts *(61)*, matrix metalloproteinases (MMPs) in cartilage *(62)* and furin-like convertase in platelets *(63)*.

In the case of inhibin/activin subunits, the mode of secretion is not well characterized. After infection of mammalian cells with vaccinia virus expressing βA the recombinant protein is secreted within five hours. The percentage of processed mature peptide varied from 4 to 98 percent dependent on the cell line *(42)*. When these subunits are overexpressed, the proform and the prodomain are secreted *(13,36,37,43)*. Under physiological conditions proinhibin α is secreted into the culture medium *(24,64)* and this proform isolated from human plasma was reported as biologically active *(65)*.

Activin A can be stored in the extracellular matrix via follistatin *(66)*. Circulating activin A in the blood plasma is reported to be associated with follistatin, but also with α2-macroglobulin *(67)*, similar to TGF-β *(68)*, basic fibroblast growth factor *(69)* and platelet derived growth factor *(68)*. Intravenous application of lipopolysaccharide *(70)* or heparin *(71)* in sheep led to a dramatic increase in plasma activin A levels within 30 minutes, leading the authors to speculate that activin A is released from the extracellular matrix of vascular endothelium. In-vitro assays suggest the release of activin A from human umbilical endothelial cells and peripheral blood monocytes after LPS, TNFα and IL1-treatment *(72)*. Vascular smooth muscle cells are reported as a further source of circulating activin A *(73)*. The presence of the βC subunit in plasma was reported recently *(74)*.

3. ACTIVIN ANTAGONISTS

As discussed in Section 4 activin A has important functions in the liver and in many other organs. Consequently, its activity has to be tightly controlled. Recent years have revealed an intricate network of factors that control activin function at different extra- and intracellular sites (Fig. 4).

The production of mature dimeric activin depends on the relative expression levels of several β-subunits and of the α-subunit. In the pituitary inhibin A and activin A have clearly opposing roles with respect to FSH secretion *(11)*. In HepG2 hepatoma cells inhibin A had no activity of its own but antagonized the inhibitory effect of activin A on DNA synthesis

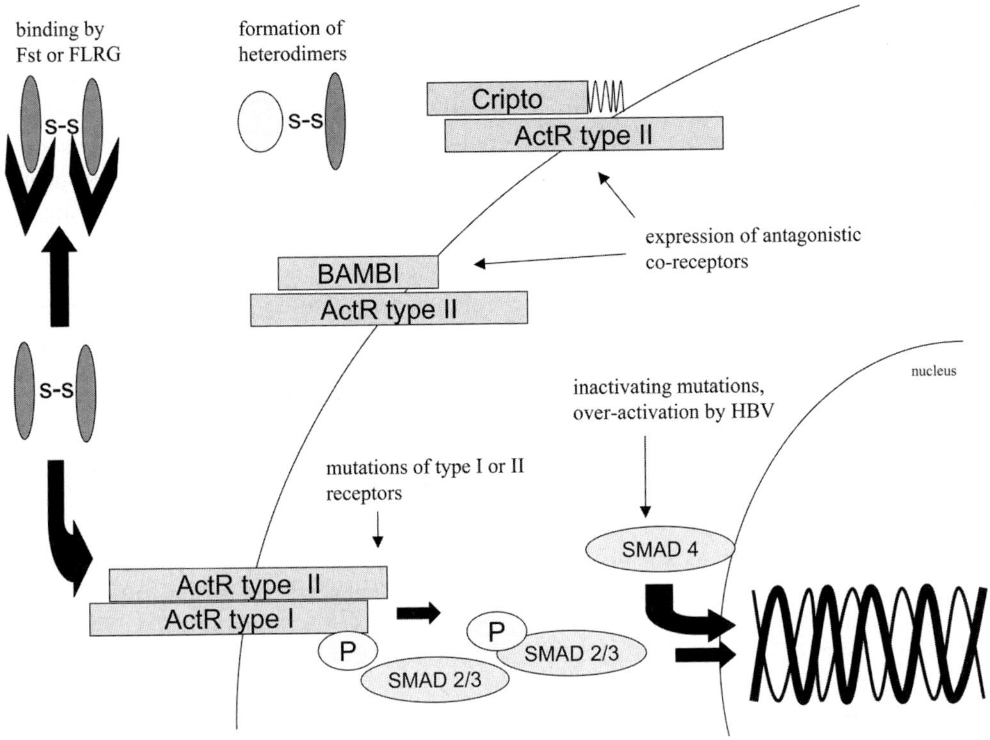

Fig. 4. Graphic model of activin A signaling. Activin A transduces signals to the cell nucleus via type II and type I activin receptors and Smad proteins. Under physiological conditions activin A signaling is strictly controlled by several mechanisms: (i) altered expression of subunits leads to formation of activins or inhibins with different capacities for receptor binding and activation; (ii) extracellular binding proteins for activin A like follistatin or FLRG can block interaction of activin A with activin receptors; (iii) expression of inhibitory coreceptors like cripto or BAMBI can block receptor activation; (iv) intracellular proteins interacting with the Smad pathway can modulate activin A signals. All these levels of control can become deregulated during carcinogenesis.

by blocking recruitment of the type I receptor *(75)*. Increased expression of the α-subunit may lead to an increased formation of inhibin A and, correspondingly, a decrease of activin A. Because, the four mammalian β-subunits are capable of forming hetero- as well as homodimeric activins *(13,24,37)*, there is also ample opportunity for regulation at the level of dimerization by varying the expression of each β-subunit. Functional differences between distinct activin homo- and heterodimers, however, are only incompletely understood. βC, for instance, has been demonstrated to form AC heterodimers which are incapable of activating transcription from activin A-responsive transcriptional elements. Overexpression of the βC subunit in the PC3 human prostate tumor cell line increased activin AC levels, decreased activin A levels, and reduced activin A signaling *(27)*. Thus, increased expression of the βC-subunit could be a means of decreasing activin A activity.

Because, dimerization of activin subunits is a strictly intracellular process, regulation by dimerization depends on coexpression of both subunits in the same cell. Several other factors, in contrast, can control activin A action after its secretion into the extracellular space. The first of these to be discovered was follistatin *(76,77)*. In contrast to activins, follistatin is a monomeric protein which is structurally unrelated to the TGF-β family. The functions described so far for follistatin all depend on its interaction with activin A and other members of the TGF-β family. Follistatin is expressed in most organs *(12,78)* and binds mature dimeric activin A with high affinity (K_d 50–680 pM), and a 2:1 stoichiometry *(79–81)*. Incubation of activin A with follistatin completely abolished the binding of activin A to its type II receptors, and thus blocked activin signaling *(82)*. Schneyer et al. *(80)* showed that follistatin (i) is a binding protein for activin A in human serum, (ii) has an affinity for activin A slightly greater than that of the activin receptor, and (iii) binds activin A with nearly irreversible kinetics under physiological conditions. These properties of follistatin suggest that its function is to limit the systemic and para/autocrine availability of activin A.

Three forms of secreted follistatin have been described resulting from alternative splicing/ processing of a single gene. These are termed follistatin 288, 303, and 315, respectively according to the number of amino acids in the processed protein *(79)*. All forms of follistatin contain a unique N-terminal domain and three follistatin domains *(83)*, and have a similar affinity for activin A. The first two follistatin domains as well as amino acids in the N-terminal domain have been implicated in activin binding *(84)*. The crystal structure of a complex of activin A bound to a fragment of follistatin has recently been resolved and it has been suggested that the insertion of arginine 192 of follistatin between the fingers of activin A is crucial for the interaction *(81)* (compare structure of activins described in Section 2). Follistatin 288 binds to heparan sulfates, whereas this binding is blocked by the acidic tail of follistatin 315 *(79)*. Follistatin 288 can thus attach to cell surface heparan sulfate proteoglycans and has been suggested to act as local regulator of activin A function *(85)*. Moreover, cell surface-bound follistatin 288 has been shown to lead to endocytosis and degradation of activin A in rat pituitary cells *(66)*.

The affinity of follistatin for activin B was shown to be about 10-fold lower than for activin A *(86)*. Recently, it has been suggested that activin E may bind follistatin, because, it could be coprecipitated with follistatin an in vitro assay *(18)*. A similar assay performed with activin C in our laboratory suggested that activin C does not bind follistatin (unpublished observation). This is in agreement with data from Wada et al. showing that activin C was unable to replace activin A from follistatin *(51)*. Follistatin binds several other members of the TGF-β family with lower affinity than activin A, for instance bone morphogenetic proteins (BMPs) 2, 4, 6, and 7, as well as myostatin *(87–89)*.

Analysis of the follistatin sequence revealed three "follistatin domains", each of which contained 10 conserved cysteine residues. Follistatin domains are present in varying numbers in several other proteins, for instance in the extracellular matrix-associated proteins SPARC (secreted protein, acidic, rich in cysteines) and agrin *(83,90)*. However, only one of these proteins known as follistatin-related gene (FLRG), follistatin-related protein (FSRP), or follistatin-like 3 (FSTL-3) has a high overall similarity with follistatin (but has only two instead of three follistatin domains) and shares its ability to bind activin A (and BMP-2) with comparable affinity (estimated K_d = 850 pM) *(91)*. The FLRG gene has originally been identified as a target of chromosomal rearrangement in a case of B-cell chronic lymphocytic leukemia *(92)*. Despite the high similarity and the capacity for activin binding, differences between follistatin and FLRG have been described with respect to tissue expression, intracellular trafficking, and function *(93–95)*. While follistatin expression was highest in ovary, testis, and pituitary, FLRG expression was highest in placenta *(91,96)*. A considerable fraction of newly synthesized FLRG was not secreted but transported to the cell nucleus *(93,96)*. A specific nuclear function of FLRG, however, has not yet been described.

Furthermore, regulation of activin activity can also occur at the receptor level. Apart from the expression regulation of either Type I or II activin receptors, additional factors appear to modulate activin activity at this level. Cripto/TDGF1 is a cell surface-associated protein of the EGF-CFC family of growth factor-like molecules. It is secreted and attached to the outer cell membrane through a glycosylphosphatidylinositol (GPI) anchor. During early embryogenesis it acts as a crucial coreceptor for nodal signaling, and it is overexpressed in 75–80% of human breast, colon, and lung cancers, and 50–60% of testicular, stomach, pancreatic, and ovarian cancers *(97)*. Recently, it has been reported by two different groups to inhibit activin/ activin receptor interactions *(98,99)*. BAMBI (bone morphogenetic protein and activin membrane bound inhibitor) is a pseudoreceptor related to TGF-β-family Type I receptors but lacking an intracellular kinase domain. It associates with TGF-β-family receptors and inhibits BMP, activin A, and TGF-β signaling *(100)*.

4. ACTIVIN FUNCTIONS IN THE LIVER

The major intracellular signaling pathways of activin A and TGF-β1 are partially identical *(101)*. Downstream from their respective receptors both cytokines activate Smads 2 and 3 and regulate transcription of an overlapping set of target genes. Therefore, it is not surprising that activin A and TGF-β1 have many biological effects in common.

Both factors are expressed in many organs and are involved in the regulation of physiological processes as diverse as mesoderm formation *(102,103)*, somatic growth *(104)*, cell proliferation and apoptosis (see this section), branching morphogenesis *(105)*, inflammation *(106,107)* and reproduction *(108,109)*. There are, however, also important differences between activin A and TGF-β1 with respect to regulation and cell type of expression, protein processing, secretion, and storage or trapping by extracellular binding proteins. These differences probably account for some of the differences in biological actions of the two cytokines.

In the liver, TGF-β1 is produced mainly by nonparenchymal cells, whereas activin A is expressed predominantly in hepatocytes *(13)*. However, also nonparenchymal cells were reported to express activin βA mRNA *(110,111)*, and TGF-β1 expression was found in hepatocytes of hyperplastic livers induced by mitogen treatment, in culture, or after malignant transformation *(112–116)* (Fig. 5). For activin βC and βE mRNA, hepatocytes constitute the major source in the body *(13,15)* (Fig. 5). Weak βB immunoreactivity was reported in normal and CCl_4-treated rat livers *(111)*.

TGF-β1 and activin A are critically involved in several important aspects of liver biology including hepatic inflammation, fibrosis, and, most importantly, regulation of liver growth. Homeostasis of hepatic cell number is maintained by a concerted action of positive and negative growth factors. TGF-β1 and activin A appear to constitute a major paracrine/ autocrine growth regulatory system by tonic inhibition of hepatocellular proliferation. Their importance is suggested by the following lines of evidence:

A. *TGF-β1:*

1. TGF-β1 was found to inhibit DNA synthesis in hepatocytes in vitro and in vivo as well as in regenerating rat liver *(117–123)*.
2. TGF-β1 was found to induce apoptosis of hepatocytes in vitro and in vivo *(113–115,121,124–126)*. Notably, mouse hepatocytes were much less sensitive to the proapoptotic action of TGF-β1 than rat hepatocytes *(123,127,128)*.
3. Liver-specific conditional knockout mice for TGF-β receptor II showed accelerated liver regeneration after partial hepatectomy (PHX). In hepatocytes isolated from these knockout mice the growth inhibitory effect of TGF-β1 was attenuated *(129)*.
4. TGF-β1 induced apoptosis in vivo in a model of liver regression following mitogen-induced hyperplasia *(114,115,125)*. In this model, chemical hepatomitogens such as cyproterone acetate, peroxisome proliferators, or follistatin are used to induce hyperplastic growth in normal rodent liver *(114,115,130)*. Organ enlargement and hyperplasia are maintained as long as the hepatomitogen is administered. When the treatment is discontinued liver mass begins to regress by apoptotic elimination of cells *(114,115)*. This involution was found to involve TGF-β1, from paracrine sources as

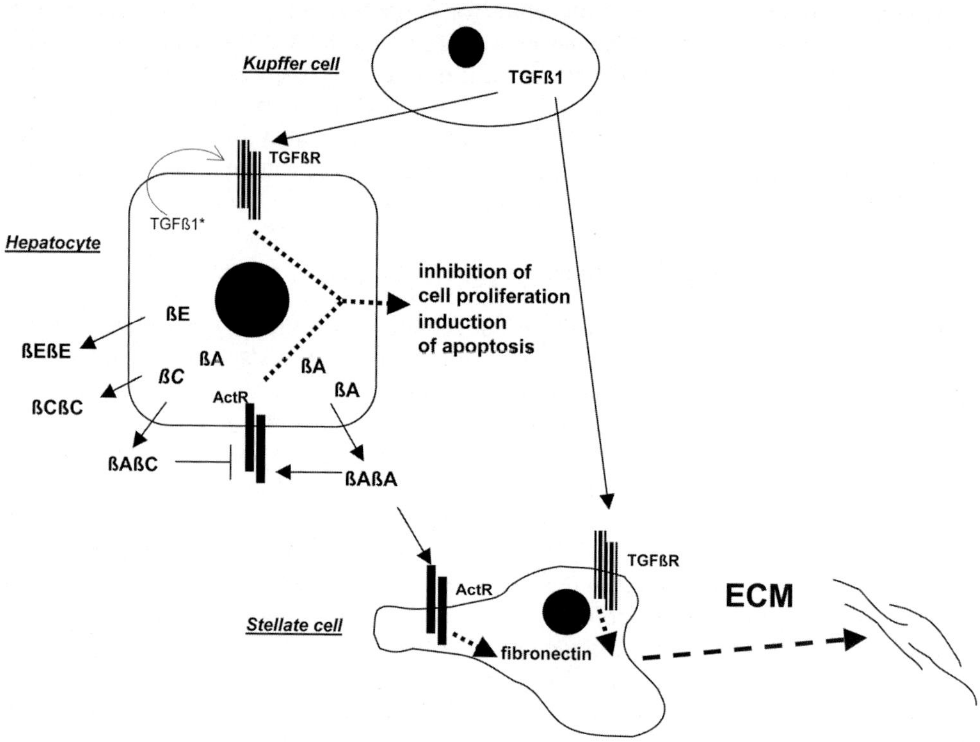

Fig. 5. Expression and effects of TGF-β1 and activin A in the liver. The figure is explained in the text. TGFβR: TGFβ receptor; ActR: activin A receptor; *hepatocelluar TGF-β1 expression in the hyperplastic state of the liver; → stimulation; —| inhibition; dashed lines indicate multistep processes not described in detail. ECM: extracellular matrix.

well as by specific expression in hepatocytes destined to die (Fig. 5) *(113,114,130)*. The anticipated proapoptotic function of endogenous TGF-β1 was corroborated by injection of TGF-β1 into rats, which induced apoptosis much more effectively in the regressing liver than in the resting liver *(125)*.

B. Activin A:

1. Activin A was found to inhibit DNA synthesis and to induce apoptosis in vitro and in vivo either by injection of the protein or ectopic expression *(122,131–134)*. Inhibin α-deficient mice exhibited 10-fold elevated serum levels of activin A, leading to cell death of hepatocytes around the central veins *(135)*. The proapoptotic function of activin A was supported by recent findings showing that in rats apoptotic hepatocytes were immunoreactive for activin βA *(74)*.
2. In a serum-starved human hepatoma cell line HLF treatment with activin A-antisense oligonucleotides stimulated cell proliferation, supporting a growth inhibitory function of endogenous activin A *(136)*.
3. Administration of follistatin in vivo caused liver growth in normal rats. Adenovirus-mediated overexpression of follistatin resulted in an increase of liver weight by about 50% after 12 days, whereas injection of the protein was less effective with only 10% increase. This effect was transient with a peak after three days and subsequently surplus cells were removed by apoptosis *(130,137)*. Because the antagonism to activin A is the main function of follistatin (see section 3), these increases are obviously owing to abolition of the anti-proliferative action of activin A.

4. Adenovirus-mediated overexpression of dominant negative type II receptors for activin A or TGF-β resulted in stimulation of DNA synthesis in normal rat liver *(138)*.
5. In a number of studies, the expression of activin A was followed during rodent liver regeneration after PHX. Upon PHX, almost all of the normally quiescent hepatocytes and nonparenchymal cells participate in restoring liver size, architecture and function within a few days *(119,120)*. Activin βA mRNA was reported to drop to less than 50% of sham operated control values at 12 h and remained at this level until at least 96 h. At 168 h post-PHX it was increased to three times above controls *(74)*. Conversely, an increase in the expression of follistatin with a peak at 24–48 h post-PHX was observed, coincident with the increase in hepatocyte replication. The expression of ActRIIB, which has a high affinity for activin A, paralleled that of βA-activin mRNA; no changes were observed in ActRIIA mRNA *(74)*. Takamura et al. observed that increased activin receptor expression peaked between 48 h and 72 h after PHX, i.e., after the DNA synthesis peak *(139)*. In line with these findings, injection of follistatin into rats was found to accelerate liver regeneration upon PHX *(140,141)*. However, a transient increase between 12 h and 24 h of the βA subunit mRNA upon PHX has been reported as well *(142,143)*, coincidentally with a decrease in its receptor mRNA expression *(142)*.

During liver regeneration, besides the coordinated compensatory growth of liver cells, also liver architecture needs to be restored. Activin A was found to stimulate fibronectin production *(142)* and, like TGF-β1, collagen synthesis in hepatic stellate cells *(144)*. Likewise, activin A cooperated with vascular endothelial growth factor (VEGF) for sinusoid restoration *(145)*. Endo et al reported that administration of follistatin after PHX accelerated liver regeneration but led to impaired restoration of normal tissue architecture *(146)*. These findings suggest that activin A supports the formation of extracellular matrix required for complete restitution of the liver architecture.

In summary, TGF-β1 and activin A constitute an essential part of a growth regulatory network to adjust liver size according to the physiological requirements of the organism. A hypothetical scheme of their interaction with other endogenous and with exogenous growth regulatory factors is depicted in Figure 6. When comparing these data with the biological features of the CD95 (APO-1/Fas)-network *(147–149)* it is tempting to speculate that the CD95-network predominantly serves to control survival/death of damaged, or virus infected cells, whereas the TGF-β1/activin A system predominates in the control of homeostasis of cell number.

Much less is known about the functions of the activin subunits βB, βC, and βE in liver biology. Elucidation of the roles of the two liver-enriched subunits βC and βE is impeded by the lack of appropriate molecular tools, i.e., availability of sufficient amounts of purified activins, specific antibodies, and reliable bioassays. Activin B (βBβB), unlike activin A, showed no influence on DNA synthesis in primary rat hepatocytes *(131,150)*. In contrast to activin βA and βB, the βC and βE subunits seem to be dispensable for mouse embryonic development. Both of them are expressed rather late in embryonic development. βC-expression occurred between day 11.5 to 14.5 depending on the detection method and reached a maximum a few weeks after birth *(14,151,152)*. βE expression started at day 17.5 and reached a maximum at birth *(152)*. Mice in which either one or both subunits had been knocked-out were viable without obvious abnormalities in liver size and function *(152)*.

Published data on the role of activin βC in liver growth regulation are conflicting *(51,134,153,154)*. Activin C has been postulated to act as a liver chalone, because following PHX transient downregulation of activin βC mRNA was observed *(143,151,152)*. A decreased expression of activin C would reduce its assumed inhibitory effect on cell proliferation to allow liver regeneration, similar to the role of activin A (cf. Fig. 6). In addition, transfection of the βC subunit gene in vivo seemed to reduce DNA synthesis in mouse liver

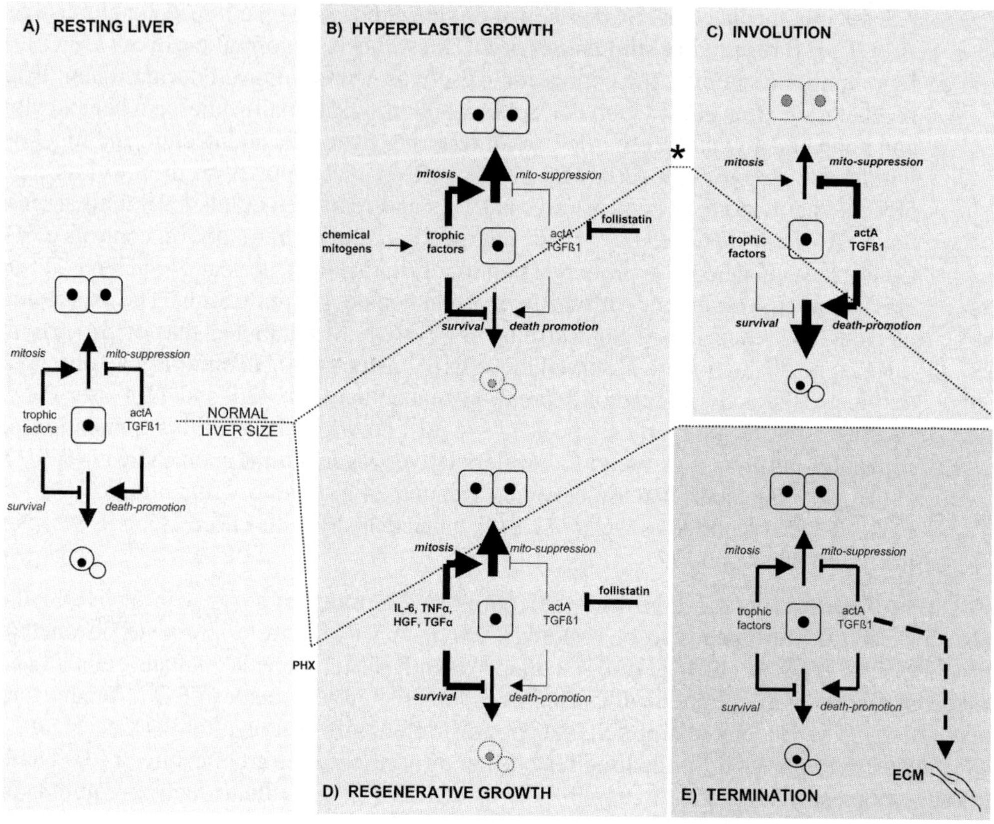

Fig. 6. Hypothetic model of the role of TGF-β1 and activin A in the regulation of liver growth. Five different states are considered. (**A**) *Resting liver,* balance between positive and negative growth factors. (**B**) *Hyperplastic growth,* administration of mitogens/activation of trophic factors or block of activin A by follistatin outweighs negative growth regulation, resulting in stimulation of cell multiplication and suppression of apoptosis. *During repeated mitogen treatment cell replication is enhanced only during the first few days; subsequently liver cells become resistant to mitogenic stimulation. This *mitosuppressed* state may reflect the operation of a feedback system that monitors the DNA content or cell number of the liver *(114,115)*. (**C**) *Involution,* withdrawal of mitogens and enhanced expression of proapoptotic factors results in downregulation of size and cell number of the liver to basal levels. Restriction of TGF-β1 and activin A expression to dying cells may explain why enhanced production of extracellular matrix (ECM) has not been observed during involution. (**D**) *Regenerative growth,* upon partial hepatectomy (PHX) renewal of hepatocytes and of nonparenchymal cells is initiated by growth factors including HGF, TGF-α, IL-6, and TNF-α, by blockade of antiproliferative factors and their downstream signalling. (**E**) *Termination,* regain of balance between positive and negative growth factors at completion of regenerative growth. TGF- β1 and activin A also interact to support ECM restitution of extracellular matrix (cf. Fig. 5). → stimulation; —| inhibition, thickness indicates level of expression; thin dotted lines indicate changes of liver size.

(154). In support of these observations, ectopic expression of rat activin C in HepG2, Hep3B, and H4IIe hepatoma cell lines (all with activins downregulated, see section 5) reduced proliferation, enhanced apoptosis rates and elevated levels of active caspases. Likewise, coculture with a CHO cell clone secreting human activin C also inhibited HepG2 cell proliferation. These results suggested that activin C has the potential to induce apoptosis in hepatoma-derived cell lines *(134)*.

In contrast to these studies, recombinant human activin C as well as conditioned medium of CHO cells expressing FLAG-tagged mouse activin βC, or adenovirus-mediated over-expression of that construct was found to promote cell proliferation and survival of AML12 hepatocytes, a nontumorigenic immortalized cell line derived from TGF-α transgenic mice *(51,155)*. A slight increase in DNA synthesis in response to the activin C containing conditioned medium was also found in primary rat hepatocyte *(51)*. In vivo, liver regeneration upon PHX was found to be accelerated in rats overexpressing βC from the adenovirus vector described above *(153)*. Gold et al. *(74)* reported that immunoreactivity for activin βC is frequently found in mitotic hepatocytes after PHX, suggesting that activin βC expression may favor cell replication in this model.

Less contradictory, but even more sparse information is available on the function of activin βE. As already mentioned above βE knock-out mice had no obvious phenotypic abnormalities. However in the same study, it was also found that activin βE mRNA was increased up to 10-fold in wt and βC knock-out mice within 6 h to 12 h after PHX implying some regulatory role for activin βE *(152)*. Transient transfection of hepatoma cells with βE-cDNA (as with βC, see above) reduced their growth and enhanced apoptosis *(134)*. Likewise, adenovirus mediated overexpression of the βE subunit reduced DNA synthesis and induced apoptosis in AML12 hepatocytes *(155)*. In mouse liver transient overexpression of βE was reported to inhibit regenerative DNA synthesis *(154)*. Recently, transgenic mice expressing human βE were found to develop defects in pancreatic exocrine cells leading to reduced size of the pancreas and deposition of adipose tissue, but no abnormalities in the liver were observed *(156)*.

In addition to their growth regulatory function activins have an important role in inflammation and liver fibrosis *(106,107,157)*. Thus, activin A may act proinflammatory by activating macrophages *(158)*. In the liver, activin A expression was found to be increased in CCl_4-induced hepatic fibrosis *(142,159–161)*. Likewise, bacterial lipopolysaccharide induced activin βE-mRNA expression in rats *(17)*. Gold et al. *(161)* found an increase in activin βC- and βA-mRNA as well as in βC protein in association with an increase in apoptosis during the development of rat liver cirrhosis.

Activin A exerted an anti-inflammatory effect in the hepatoma cell line HepG2 by inhibiting the induction of acute-phase-protein (APP) synthesis by IL-6 *(107)*. Interestingly, a similar action has been found with activin C. Proteome analysis of primary rat hepatocytes treated with inducers of APP followed by human recombinant activin C revealed alterations in synthesis of numerous proteins. These included decreases in secreted APP T-kininogen, fibrinogen β- and γ-chain and fetuin A while synthesis of transthyretin, albumin and apolipoprotein A-1 was increased *(162)*. These finding may provide a new access to understanding the biology of activin C.

In human, activin βA mRNA or activin A protein is elevated in inflammatory bowel disease and inflammatory arthropathies, and circulating concentrations of follistatin are elevated in patients with sepsis *(107,163)*. Elevated levels of circulating activin A and follistatin have also been found in chronic viral hepatitis, alcoholic cirrhosis, and acute liver failure *(164–166)*. Lin et al. *(167)* have recently suggested that a decreased follistatin/activin A ratio in the serum may be an indicator for the severity of liver injury in hepatitis related acute liver disease. The functions of activin A in tissue repair, fibrosis, and inflammatory processes have been recently reviewed by Werner and Alzheimer *(107)*.

Finally, it may be worth to note that the expression and function of activins occurs within the network of TGF-β superfamily members and therefore, it is difficult to sort out the contribution of any one factor. Contradictory findings reflect that the functions of activins are yet insufficiently explored and may be due at least in part to differences in experimental models and conditions.

5. ACTIVIN IN LIVER TUMORS

Deregulation of pathways controlling cell growth and apoptosis is considered one of the hallmarks of tumor development *(168)*. Therefore, from the important effects that activins appear to have on cell replication and cell death, one would easily be able to predict a role of these factors in tumor development. Indeed, in reproductive and endocrine organ tumors the importance of activins for tumorigenesis has long been recognized. These studies have been extensively reviewed by Risbridger et al *(169)*. In recent years, new tumor entities have been added to the list of cancers, where an impact of activin signaling pathways has been demonstrated or suggested. In breast cancer, activin and activin receptors are downregulated in the tumor tissue *(170)*, whereas cripto, which has been suggested to block activin signaling is upregulated in mammary tumors *(171)*. In stage IV colon cancer, in contrast, increased expression of activin A has been reported *(172)*. Inactivating mutations of activin receptors were found in microsatellite instable colon cancers and pancreatic cancer *(173–175)*. Secretion of high amounts of the activin antagonist follistatin was found in melanoma cell lines *(176)*.

Primary tumors of the liver are additional candidates for an impact of activin signals on carcinogenesis. On the one hand, the role of activin A as negative regulator of hepatocyte growth is well established, as discussed above. Therefore one would hypothesize that a developing tumor in the liver should either (i) minimize its exposure to activin A, (ii) become resistant to the growth inhibitory effects of activin, or (iii) turn around activin signals for its own benefit. On the other hand, human liver tumors mainly (but not exclusively) develop in a setting of chronic inflammation, and activin A was shown to have both pro- and anti-inflammatory effects *(163,177)*.

Several observations suggest a role of deregulated activin/follistatin signals in human and rodent liver carcinogenesis. In patients with cirrhosis and with hepatocellular carcinoma (HCC) increased serum levels of both activin A and follistatin have been found *(166,178)*. In experimental tumor models in mice and rats follistatin was upregulated in the tumor tissue *(179)*. Human hepatoma cell lines express follistatin but have reduced expression levels of activin β subunits compared to normal hepatocytes *(134,179,180)*. Recent studies in our group suggested that overexpression of follistatin also occurs in human liver tumors. FLRG (FSTL-3) was enhanced in chemically induced rat liver tumors but not in human tumors. The activin subunits βA, and βE, in contrast, were downregulated in human tumors and in the rat models. Addition of follistatin to normal and preneoplastic hepatocytes in primary culture enhanced, whereas addition of activin A eliminated the growth advantage of the preneoplastic hepatocytes. H4IIe hepatoma cells showed strong overexpression of FLRG, and consequently were insensitive to addition of follistatin *(181)*. These data suggest that deregulation of the balance between expression of activins and activin antagonists frequently occurs during, and may contribute to, carcinogenesis of the liver. In agreement with the data on H4IIe cells, Fuwii et al. have found that treatment with follistatin did not enhance growth of AH130 liver tumor cells in vitro or after intraportal injection in rats *(182)*. Autocrine production of activin antagonists may offer a possible explanation for the unresponsiveness of liver tumor cells to follistatin. Expression changes or mutations of activin receptors in liver tumors have yet to be investigated.

Downstream from receptor activation activin A signals and TGF-β signals are transduced to the cell nucleus through Smad proteins. Inactivating mutations in Smad 2 or 4 genes, which are both considered tumor suppressors, have been found in diverse human cancers, for instance, pancreatic and colon carcinoma *(183,184)*. Somatic mutations in either Smad 2 or Smad 4 have also been detected in about 10% of human HCC samples *(185)*. On the other hand, Lee et al. *(186)* have suggested that overactivation of the Smad pathway by interaction of the hepatitis B virus oncoprotein pX with Smad 4 contributes to hepatic fibrosis.

Thus, there is increasing evidence that deregulation of activin signals may be a key event in liver carcinogenesis via a combination of two mechanisms (i) the escape of hepatocytes from activin-mediated growth control, and (ii) contribution of deregulated activin signals to hepatic inflammation and fibrosis. These data suggest that activins and activin antagonists might be feasible targets for intervention in a tumor-preventive or therapeutic setting.

6. ACTIVINS AS POTENTIAL THERAPEUTIC TARGETS

Activin A shares the Smad signaling pathway with TGF-β1 *(187)* and both factors have a considerable functional overlap. TGF-β1 has a well-recognized dual role in carcinogenesis *(188)*. In early stages of hepatocarcinogenesis, it acts as a tumor suppressor by inducing apoptosis and eliminating precursor lesions *(121,127)*. In later stages, however, liver tumor cells become resistant to the growth inhibitory potential of TGF-β1, and often produce large amounts of this cytokine themselves *(189)*. TGF-β1 has been shown to enhance migration and invasiveness of transformed hepatocytes *(190)*.

The dual role of TGF-β poses a considerable challenge with regard to its use as a therapeutic target. In advanced tumors targeting TGF-β or TGF-β receptors with neutralizing antibodies, antisense oligonucleotides, or small molecule inhibitors may be of considerable therapeutic benefit. Consequently, several agents inhibiting TGF-β or its receptors have been developed for clinical use and are currently in various stages of preclinical and clinical evaluation for therapy of fibrotic and malignant disease *(191,192)*. On the other hand, there is a concern that inhibition of TGF-β signaling will also abrogate its tumor suppressive function on early tumors and premalignant lesions. Similar considerations also apply to activins and related factors. Several components of the activin signaling pathway may hold great promise for consideration and further investigation as rational targets for therapeutic applications and will be discussed below. Careful consideration of possible adverse effects will, however, be necessary, as activin signals may have different functions depending on cell type and tumor stage.

6.1. Activin A

The proapoptotic function of activin A and the apparent loss of β A expression in tumor cells and tumor tissue *(134,170,180)* suggest that restoration of activin A expression could be of potential therapeutic benefit. This approach, however, may be hampered by the pleiotrophic effects of activin A especially by its proinflammatory *(193,194)* and profibrotic activities *(163,193,194)*. Moreover, a recent investigation has described induction of VEGF expression by activin A in human hepatoma cell lines *(195)*, suggesting a potential proangiogenic role of activin A in HCC.

6.2. Activin Receptors

Small molecule inhibitors (SB-431542, SB-505124) that were developed as competitive inhibitors of the ATP-binding site of the TGF-β type I receptor (ALK5) also inhibit the activin type I receptor (ALK4) *(196,197)*. In addition, Harrison et al. have developed a mutated form of activin A (M108A) which binds activin receptor type II but does not initiate signal transduction, thus blocking the type II receptor *(48)*. However, it is presently not clear, whether activin A plays a role in tumor progression, similar to that described for TGF-β, and consequently whether blocking activin signals could be beneficial at late stages of tumor development. With regard to a possible stimulation of early tumors and tumor pre-stages in the liver, concerns would be similar as for TGF-β receptor inhibitors.

6.3. Activin C

The biological function of the activin βC subunit is still quite incompletely understood, as discussed above. With regard to its potential as therapy target, it may be interesting, that overexpression of the βC subunit induced apoptosis in hepatoma cells *(134)* but stimulated proliferation of normal hepatocytes after PHX *(153)*. A possible explanation may be that βC overexpression in normal hepatocytes, which also express βA, may reduce the growth inhibitory effect of activin A by preferential formation of inactive AC heterodimers *(27)*. Hepatoma cells expressed much less activin βA, thus the main activin formed would be homodimeric activin C. Along these lines, gene therapy approaches based on hepatic activin βC expression might offer a chance for selective elimination of tumor cells. However, a much more complete picture of the functions of the βC subunit will be required before any clinical studies can be performed. Especially, potential effects of activin βC on hepatic fibrosis, inflammation and angiogenesis have yet to be investigated.

6.4. Follistatin

Overexpression of follistatin and FLRG has been found in experimentally induced liver tumors in animal models *(179,181)*. Because, follistatin 288 binds to cell surface heparan sulfate proteoglycans *(79)* and may thus shield tumor cells from growth inhibition by activin A, antibodies against follistatin might restore sensitivity of tumor cells to activin signals or mark follistatin overexpressing tumor cells for T-cell mediated destruction.

On the other hand, infusion of follistatin could limit the profibrotic activity of activin A and accelerate liver regeneration after surgical liver resection or toxic cell damage. Indeed, attenuation of CCl_4 induced liver fibrosis by follistatin infusion has been demonstrated in a rat model *(110)*. Endo et al. however, have recently added a cautionary note to using follistatin for speeding-up liver regeneration. They show that while follistatin infusion did accelerate liver regeneration after 90% hepatectomy in rats, it also lead to incomplete restoration of tissue architecture and reduced liver function *(146)*.

6.5. Cripto/TDGF1

Cripto may be an especially attractive target. It is overexpressed in many tumors, is not significantly expressed in most adult tissues, and has a clear protumorigenic activity *(97,198)*. Part of that activity has been demonstrated to result from its inhibition of activin signaling *(98,99)*, and blocking antibodies to cripto were shown to reduce tumor growth in a xenograft mouse model *(98)*.

In conclusion, several factors involved in activin signaling show promise for therapeutic application. These factors should be further investigated as candidates for novel approaches in the fight against liver cancer.

ACKNOWLEDGMENTS

The authors wish to apologize to those investigators whose experimental work has only been cited indirectly.

REFERENCES

1. Mottram C, Cramer W. On the general aspects of exposure to radium on metabolism and tumour growth in the rat and the special effects on the testis and pituitary. Q J Exp Physiol 1923;13:209–229.
2. McCullagh D. Dual endocrine activity of testes. Science 1932;76:19–22.
3. De Jong FH, Robertson DM. Inhibin: 1985 update on action and purification. Mol Cell Endocrinol 1985;42:95–103.
4. Ying SY, Becker A, Baird A, et al. Type beta transforming growth factor (TGF-beta) is a potent stimulator of the basal secretion of follicle stimulating hormone (FSH) in a pituitary monolayer system. Biochem Biophys Res Commun 1986;135:950–956.
5. Oda S, Nishimatsu S, Murakami K, Ueno N. Molecular cloning and functional analysis of a new activin beta subunit: a dorsal mesoderm-inducing activity in Xenopus. Biochem Biophys Res Commun 1995;210:581–588.

6. Chari S, Hopkinson CR, Daume E, Sturm G. Purification of "inhibin" from human ovarian follicular fluid. Acta Endocrinol (Copenh) 1979;90:157–166.
7. Murthy HM, Ramasharma K, Moudgal NR. Studies on purification and characterization of sheep testicular inhibin. J Reprod Fertil Suppl 1979;26:61–70.
8. Thakur AN, Vaze AY, Dattatreyamurty B, et al. Isolation and characterization of inhibin from human seminal plasma. Indian J Exp Biol 1978;16:854–856.
9. Rivier J, Spiess J, McClintock R, et al. Purification and partial characterization of inhibin from porcine follicular fluid. Biochem Biophys Res Commun 1985;133:120–127.
10. Ling N, Ying SY, Ueno N, et al. A homodimer of the beta-subunits of inhibin A stimulates the secretion of pituitary follicle stimulating hormone. Biochem Biophys Res Commun 1986;138:1129–1137.
11. Vale W, Rivier J, Vaughan J, et al. Purification and characterization of an FSH releasing protein from porcine ovarian follicular fluid. Nature 1986;321:776–779.
12. Tuuri T, Eramaa M, Hilden K, Ritvos O. The tissue distribution of activin beta A- and beta B-subunit and follistatin messenger ribonucleic acids suggests multiple sites of action for the activin-follistatin system during human development. J Clin Endocrinol Metab 1994;78:1521–1524.
13. Vejda S, Cranfield M, Peter B, et al. Expression and dimerization of the rat activin subunits betaC and betaE: evidence for the formation of novel activin dimers. J Mol Endocrinol 2002;28:137–148.
14. Hoetten G, Neidhardt H, Schneider C, Pohl J. Cloning of a new member of the TGF-beta family: a putative new activin beta C chain. Biochem Biophys Res Commun 1995;206:608–613.
15. Lau AL, Nishimori K, Matzuk MM. Structural analysis of the mouse activin beta C gene. Biochim Biophys Acta 1996;1307:145–148.
16. Fang J, Yin W, Smiley E, et al. Molecular cloning of the mouse activin beta E subunit gene. Biochem Biophys Res Commun 1996;228:669–674.
17. O'Bryan MK, Sebire KL, Gerdprasert O, et al. Cloning and regulation of the rat activin betaE subunit. J Mol Endocrinol 2000;24:409–418.
18. Hashimoto O, Tsuchida K, Ushiro Y, et al. cDNA cloning and expression of human activin betaE subunit. Mol Cell Endocrinol 2002;194:117–122.
19. Rosendahl A, Checchin D, Fehniger TE, et al. Activation of the TGF-beta/activin-Smad2 pathway during allergic airway inflammation. Am J Respir Cell Mol Biol 2001;25:60–68.
20. Loveland KL, McFarlane JR, de Kretser DM. Expression of activin beta C subunit mRNA in reproductive tissues. J Mol Endocrinol 1996;17:61–65.
21. Gold EJ, O'Bryan MK, Mellor SL, et al. Cell-specific expression of betaC-activin in the rat reproductive tract, adrenal and liver. Mol Cell Endocrinol 2004;222:61–69.
22. Casagrandi D, Bearfield C, Geary J, et al. Inhibin, activin, follistatin, activin receptors and betaglycan gene expression in the placental tissue of patients with pre-eclampsia. Mol Hum Reprod 2003;9:199–203.
23. Thomas TZ, Chapman SM, Hong W, et al. Inhibins, activins, and follistatins: expression of mRNAs and cellular localization in tissues from men with benign prostatic hyperplasia. Prostate 1998;34:34–43.
24. Mellor SL, Cranfield M, Ries R, et al. Localization of activin beta(A)-, beta(B)-, and beta(C)-subunits in humanprostate and evidence for formation of new activin heterodimers of beta(C)-subunit. J Clin Endocrinol Metab 2000;85:4851–4858.
25. Ushiro Y, Hashimoto O, Seki M, et al. Analysis of the function of Activin beta(C) subunit using recombinant protein. J Reprod Dev 2006;52:487–495.
26. Ling N, Ying SY, Ueno N, et al. Pituitary FSH is released by a heterodimer of the beta-subunits from the two forms of inhibin. Nature 1986;321:779–782.
27. Mellor SL, Ball EM, O'Connor AE, et al. Activin betaC-subunit heterodimers provide a new mechanism of regulating activin levels in the prostate. Endocrinology 2003;144:4410–4419.
28. Tanimoto K, Yoshida E, Mita S, et al. Human activin betaA gene. Identification of novel 5′ exon, functional promoter, and enhancers. J Biol Chem 1996;271:32,760–32,769.
29. Ardekani AM, Romanelli JC, Mayo KE. Structure of the rat inhibin and activin betaA-subunit gene and regulation in an ovarian granulosa cell line. Endocrinology 1998;139:3271–3279.
30. Fang J, Wang SQ, Smiley E, Bonadio J. Genes coding for mouse activin beta C and beta E are closely linked and exhibit a liver-specific expression pattern in adult tissues. Biochem Biophys Res Commun 1997;231:655–661.
31. Chang H, Brown CW, Matzuk MM. Genetic analysis of the mammalian transforming growth factor-beta superfamily. Endocr Rev 2002;23:787–823.

32. Gray AM, Mason AJ. Requirement for activin A and transforming growth factor-beta 1 pro-regions in homodimer assembly. Science 1990;247:1328–1330.
33. Todorovic V, Jurukovski V, Chen Y, et al. Latent TGF-beta binding proteins. Int J Biochem Cell Biol 2005;37:38–41.
34. Saharinen J, Taipale J, Keski-Oja J. Association of the small latent transforming growth factor-beta with an eight cysteine repeat of its binding protein LTBP-1. EMBO J 1996;15:245–253.
35. Brunner AM, Marquardt H, Malacko AR, et al. Site-directed mutagenesis of cysteine residues in the pro region of the transforming growth factor beta 1 precursor. Expression and characterization of mutant proteins. J Biol Chem 1989;264:13,660–13,664.
36. Mason AJ. Functional analysis of the cysteine residues of activin A. Mol Endocrinol 1994;8:325–332.
37. Mason AJ, Berkemeier LM, Schmelzer CH, Schwall RH. Activin B: precursor sequences, genomic structure and in vitro activities. Mol Endocrinol 1989;3:1352–1358.
38. Nishanian TG, Waldman T. Interaction of the BMPR-IA tumor suppressor with a developmentally relevant splicing factor. Biochem Biophys Res Commun 2004;323:91–97.
39. Fukuda M, Moreira JE, Liu V, et al. Role of the conserved WHXL motif in the C terminus of synaptotagmin in synaptic vesicle docking. Proc Natl Acad Sci USA 2000;97:14,715–14,719.
40. Brunner AM, Lioubin MN, Marquardt H, et al. Site-directed mutagenesis of glycosylation sites in the transforming growth factor-beta 1 (TGF beta 1) and TGF beta 2 (414) precursors and of cysteine residues within mature TGF beta 1: effects on secretion and bioactivity. Mol Endocrinol 1992;6:1691–1700.
41. Sha X, Brunner AM, Purchio AF, Gentry LE. Transforming growth factor beta 1: importance of glycosylation and acidic proteases for processing and secretion. Mol Endocrinol 1989;3:1090–1098.
42. Huylebroeck D, Van Nimmen K, Waheed A, et al. Expression and processing of the activin-A/erythroid differentiation factor precursor: a member of the transforming growth factor-beta superfamily. Mol Endocrinol 1990;4:1153–1165.
43. Mason AJ, Farnworth PG, Sullivan J. Characterization and determination of the biological activities of noncleavable high molecular weight forms of inhibin A and activin A. Mol Endocrinol 1996;10:1055–1065.
44. McDonald NQ, Hendrickson WA. A structural superfamily of growth factors containing a cystine knot motif. Cell 1993;73:421–424.
45. Muenster U, Harrison CA, Donaldson C, et al. An activin-A/C chimera exhibits activin and myostatin antagonistic properties. J Biol Chem 2005;280:36,626–36,632.
46. Keah HH, Hearn MT. A molecular recognition paradigm: promiscuity associated with the ligand-receptor interactions of the activin members of the TGF-beta superfamily. J Mol Recognit 2005;18:385–403.
47. Thompson TB, Woodruff TK, Jardetzky TS. Structures of an ActRIIB:activin A complex reveal a novel binding mode for TGF-beta ligand:receptor interactions. EMBO J 2003;22:1555–1566.
48. Harrison CA, Gray PC, Fischer WH, et al. An activin mutant with disrupted ALK4 binding blocks signaling via type II receptors. J Biol Chem 2004;279:28,036–28,044.
49. Greenwald J, Vega ME, Allendorph GP, et al. A flexible activin explains the membrane-dependent cooperative assembly of TGF-beta family receptors. Mol Cell 2004;15:485–489.
50. Cook RW, Thompson TB, Kurup SP, et al. Structural basis for a functional antagonist in the transforming growth factor beta superfamily. J Biol Chem 2005;280:40,177–40,186.
51. Wada W, Maeshima A, Zhang YQ, et al. Assessment of the function of the betaC-subunit of activin in cultured hepatocytes. Am J Physiol Endocrinol Metab 2004;287:E247–E254.
52. Fischer WH, Park M, Donaldson C, et al. Residues in the C-terminal region of activin A determine specificity for follistatin and type II receptor binding. J Endocrinol 2003;176:61–68.
53. Lodish H, Arnold Berk A, Zipursky LS, et al. Molecular Cell Biology. 4th Edition, New York: Freeman, W.H. & Co., 2000; p. 1184.
54. Freedman RB, Hirst TR, Tuite MF. Protein disulphide isomerase: building bridges in protein folding. Trends Biochem Sci 1994;19:331–336.
55. Ellgaard L, Ruddock LW. The human protein disulphide isomerase family: substrate interactions and functional properties. EMBO Rep 2005;6:28–32.
56. Salvas A, Benjannet S, Reudelhuber TL, et al. Evidence for proprotein convertase activity in the endoplasmic reticulum/early Golgi. FEBS Lett 2005;579:5621–5625.
57. Rouille Y, Duguay SJ, Lund K, et al. Proteolytic processing mechanisms in the biosynthesis of neuroendocrine peptides: the subtilisin-like proprotein convertases. Front Neuroendocrinol 1995;16:322–361.

58. Dubois CM, Laprise MH, Blanchette F, et al. Processing of transforming growth factor beta 1 precursor by human furin convertase. J Biol Chem 1995;270:10,618–10,624.
59. Constam DB, Robertson EJ. Regulation of bone morphogenetic protein activity by pro domains and proprotein convertases. J Cell Biol 1999;144:139–149.
60. Keski-Oja J, Koli K, von Melchner H. TGF-beta activation by traction? Trends Cell Biol 2004;14: 657–659.
61. Yee JA, Yan L, Dominguez JC, et al. Plasminogen-dependent activation of latent transforming growth factor beta (TGF beta) by growing cultures of osteoblast-like cells. J Cell Physiol 1993;157:528–534.
62. D'Angelo M, Billings PC, Pacifici M, et al. Authentic matrix vesicles contain active metalloproteases (MMP). a role for matrix vesicle-associated MMP-13 in activation of transforming growth factor-beta. J Biol Chem 2001;276:11,347–11,353.
63. Blakytny R, Ludlow A, Martin GE, et al. Latent TGF-beta1 activation by platelets. J Cell Physiol 2004;199:67–76.
64. Bilezikjian LM, Vaughan JM, Vale WW. Characterization and the regulation of inhibin/activin subunit proteins of cultured rat anterior pituitary cells. Endocrinology 1993;133:2545–2553.
65. Robertson DM, Sullivan J, Watson M, Cahir N. Inhibin forms in human plasma. J Endocrinol 1995;144:261–269.
66. Hashimoto O, Nakamura T, Shoji H, et al. A novel role of follistatin, an activin-binding protein, in the inhibition of activin action in rat pituitary cells. Endocytotic degradation of activin and its acceleration by follistatin associated with cell-surface heparan sulfate. J Biol Chem 1997;272: 13,835–13,842.
67. Vaughan JM, Vale WW. Alpha 2-macroglobulin is a binding protein of inhibin and activin. Endocrinology 1993;132:2038–2050.
68. Huang SS, O'Grady P, Huang JS. Human transforming growth factor beta. Alpha 2-macroglobulin complex is a latent form of transforming growth factor beta. J Biol Chem 1988;263:1535–1541.
69. Dennis PA, Saksela O, Harpel P, Rifkin DB. Alpha 2-macroglobulin is a binding protein for basic fibroblast growth factor. J Biol Chem 1989;264:7210–7216.
70. Jones KL, de Kretser DM, Clarke IJ, et al. Characterisation of the rapid release of activin A following acute lipopolysaccharide challenge in the ewe. J Endocrinol 2004;182:69–80.
71. Jones KL, De Kretser DM, Phillips DJ. Effect of heparin administration to sheep on the release profiles of circulating activin A and follistatin. J Endocrinol 2004;181:307–314.
72. Tannetta DS, Muttukrishna S, Groome NP, et al. Endothelial cells and peripheral blood mononuclear cells are a potential source of extraplacental activin a in preeclampsia. J Clin Endocrinol Metab 2003;88:5995–6001.
73. Kanzaki M, Nobusawa R, Mogami H, et al. Production of activin A and follistatin in cultured rat vascular smooth muscle cells. Mol Cell Endocrinol 1995;108:11–16.
74. Gold EJ, Zhang X, Wheatley AM, et al. betaA- and betaC-activin, follistatin, activin receptor mRNA and betaC-activin peptide expression during rat liver regeneration. J Mol Endocrinol 2005;34:505–515.
75. Xu J, McKeehan K, Matsuzaki K, McKeehan WL. Inhibin antagonizes inhibition of liver cell growth by activin by a dominant-negative mechanism. J Biol Chem 1995;270:6308–6313.
76. Nakamura T, Takio K, Eto Y, et al. Activin-binding protein from rat ovary is follistatin. Science 1990;247:836–838.
77. Ueno N, Ling N, Ying SY, et al. Isolation and partial characterization of follistatin: a single-chain Mr 35,000 monomeric protein that inhibits the release of follicle-stimulating hormone. Proc Natl Acad Sci USA 1987;84:8282–8286.
78. Michel U, Rao A, Findlay JK. Rat follistatin: ontogeny of steady-state mRNA levels in different tissues predicts organ-specific functions. Biochem Biophys Res Commun 1991;180:223–230.
79. Sugino K, Kurosawa N, Nakamura T, et al. Molecular heterogeneity of follistatin, an activin-binding protein. Higher affinity of the carboxyl-terminal truncated forms for heparan sulfate proteoglycans on the ovarian granulosa cell. J Biol Chem 1993;268:15,579–15,587.
80. Schneyer AL, Rzucidlo DA, Sluss PM, Crowley WF, Jr. Characterization of unique binding kinetics of follistatin and activin or inhibin in serum. Endocrinology 1994;135:667–674.
81. Harrington AE, Morris-Triggs SA, Ruotolo BT, et al. Structural basis for the inhibition of activin signalling by follistatin. Embo J 2006;25:1035–1045.
82. de Winter JP, ten Dijke P, de Vries CJ, et al. Follistatins neutralize activin bioactivity by inhibition of activin binding to its type II receptors. Mol Cell Endocrinol 1996;116:105–114.

83. Ullman CG, Perkins SJ. The Factor I and follistatin domain families: the return of a prodigal son. Biochem J 1997;326(Pt 3):939–941.
84. Keutmann HT, Schneyer AL, Sidis Y. The role of follistatin domains in follistatin biological action. Mol Endocrinol 2004;18:228–240.
85. Nakamura T, Sugino K, Titani K, Sugino H. Follistatin, an activin-binding protein, associates with heparan sulfate chains of proteoglycans on follicular granulosa cells. J Biol Chem 1991;266: 19,432–19,437.
86. Schneyer A, Schoen A, Quigg A, Sidis Y. Differential binding and neutralization of activins A and B by follistatin and follistatin like-3 (FSTL-3/FSRP/FLRG). Endocrinology 2003;144:1671–1674.
87. Iemura S, Yamamoto TS, Takagi C, et al. Direct binding of follistatin to a complex of bone-morphogenetic protein and its receptor inhibits ventral and epidermal cell fates in early Xenopus embryo. Proc Natl Acad Sci USA 1998;95:9337–9342.
88. Amthor H, Nicholas G, McKinnell I, et al. Follistatin complexes Myostatin and antagonises Myostatin-mediated inhibition of myogenesis. Dev Biol 2004;270:19–30.
89. Glister C, Kemp CF, Knight PG. Bone morphogenetic protein (BMP) ligands and receptors in bovine ovarian follicle cells: actions of BMP-4, -6 and -7 on granulosa cells and differential modulation of Smad-1 phosphorylation by follistatin. Reproduction 2004;127:239–254.
90. Shimasaki S, Koga M, Esch F, et al. Primary structure of the human follistatin precursor and its genomic organization. Proc Natl Acad Sci USA 1988;85:4218–4222.
91. Tsuchida K, Arai KY, Kuramoto Y, et al. Identification and characterization of a novel follistatin-like protein as a binding protein for the TGF-beta family. J Biol Chem 2000;275:40,788–40,796.
92. Hayette S, Gadoux M, Martel S, et al. FLRG (follistatin-related gene), a new target of chromosomal rearrangement in malignant blood disorders. Oncogene 1998;16:2949–2954.
93. Saito S, Sidis Y, Mukherjee A, et al. Differential biosynthesis and intracellular transport of follistatin isoforms and follistatin-like-3. Endocrinology 2005;146:5052–5062.
94. Schneyer A, Sidis Y, Xia Y, et al. Differential actions of follistatin and follistatin-like 3. Mol Cell Endocrinol 2004;225:25–28.
95. Schneyer A, Tortoriello D, Sidis Y, et al. Follistatin-related protein (FSRP): a new member of the follistatin gene family. Mol Cell Endocrinol 2001;180:33–38.
96. Tortoriello DV, Sidis Y, Holtzman DA, et al. Human follistatin-related protein: a structural homologue of follistatin with nuclear localization. Endocrinology 2001;142:3426–3434.
97. Adamson ED, Minchiotti G, Salomon DS. Cripto: a tumor growth factor and more. J Cell Physiol 2002;190:267–278.
98. Adkins HB, Bianco C, Schiffer SG, et al. Antibody blockade of the Cripto CFC domain suppresses tumor cell growth in vivo. J Clin Invest 2003;112:575–587.
99. Gray PC, Harrison CA, Vale W. Cripto forms a complex with activin and type II activin receptors and can block activin signaling. Proc Natl Acad Sci USA 2003;100:5193–5198.
100. Onichtchouk D, Chen YG, Dosch R, et al. Silencing of TGF-beta signalling by the pseudoreceptor BAMBI. Nature 1999;401:480–485.
101. Feng XH, Derynck R. Specificity and versatility in tgf-beta signaling through Smads. Annu Rev Cell Dev Biol 2005;21:659–693.
102. McDowell N, Gurdon JB. Activin as a morphogen in Xenopus mesoderm induction. Semin Cell Dev Biol 1999;10:311–317.
103. Piepenburg O, Grimmer D, Williams PH, Smith JC. Activin redux: specification of mesodermal pattern in Xenopus by graded concentrations of endogenous activin B. Development 2004;131: 4977–4986.
104. Brown CW, Li L, Houston-Hawkins DE, Matzuk MM. Activins are critical modulators of growth and survival. Mol Endocrinol 2003;17:2404–2417.
105. Ball EM, Risbridger GP. Activins as regulators of branching morphogenesis. Dev Biol 2001;238:1–12.
106. Phillips DJ, Jones KL, Scheerlinck JY, et al. Evidence for activin A and follistatin involvement in the systemic inflammatory response. Mol Cell Endocrinol 2001;180:155–162.
107. Werner S, Alzheimer C. Roles of activin in tissue repair, fibrosis, and inflammatory disease. Cytokine Growth Factor Rev 2006;17:157–171.
108. Risbridger GP, Cancilla B. Role of activins in the male reproductive tract. Rev Reprod 2000;5: 99–104.
109. Phillips DJ, Jones KL, Clarke IJ, et al. Activin A: from sometime reproductive factor to genuine cytokine. Vet Immunol Immunopathol 2005;108:23–27.

110. Patella S, Phillips DJ, Tchongue J, et al. Follistatin attenuates early liver fibrosis: effects on hepatic stellate cell activation and hepatocyte apoptosis. Am J Physiol Gastrointest Liver Physiol 2006;290:G137–G144.
111. De Bleser PJ, Niki T, Xu G, et al. Localization and cellular sources of activins in normal and fibrotic rat liver. Hepatology 1997;26:905–912.
112. Oberhammer F, Nagy P, Tiefenbacher R, et al. The antiandrogen cyproterone acetate induces synthesis of transforming growth factor beta 1 in the parenchymal cells of the liver accompanied by an enhanced sensitivity to undergo apoptosis and necrosis without inflammation. Hepatology 1996;23:329–337.
113. Bursch W, Oberhammer F, Jirtle RL, et al. Transforming growth factor-beta 1 as a signal for induction of cell death by apoptosis. Br J Cancer 1993;67:531–536.
114. Schulte-Hermann R, Bursch W, Grasl-Kraupp B, et al. Apoptotic cytokines as regulators of anti-oncogenic responses and hepatocellular carcinoma growth. In: Moreno-Otreo R, Albillos A, Garcia-Mozon C, eds. Immunology and the Liver: Cytokines. Madrid: Association Liver Higado de Investigacion Cientifica, 2002; 281–290.
115. Schulte-Hermann R, Bursch W, Grasl-Kraupp B. Active cell death (apoptosis) in liver biology and disease. Prog Liver Dis 1995;13:1–35.
116. Gao C, Gressner G, Zoremba M, Gressner AM. Transforming growth factor beta (TGF-beta) expression in isolated and cultured rat hepatocytes. J Cell Physiol 1996;167:394–405.
117. Russell WE, Coffey RJ, Jr., Ouellette AJ, Moses HL. Type beta transforming growth factor reversibly inhibits the early proliferative response to partial hepatectomy in the rat. Proc Natl Acad Sci USA 1988;85:5126–5130.
118. Russell WE. Transforming growth factor beta (TGF-beta) inhibits hepatocyte DNA synthesis independently of EGF binding and EGF receptor autophosphorylation. J Cell Physiol 1988;135:253–261.
119. Taub R. Liver regeneration: from myth to mechanism. Nat Rev Mol Cell Biol 2004;5:836-847.
120. Romero-Gallo J, Sozmen EG, Chytil A, et al. Inactivation of TGF-beta signaling in hepatocytes results in an increased proliferative response after partial hepatectomy. Oncogene 2005;24:3028–3041.
121. Mullauer L, Grasl-Kraupp B, Bursch W, Schulte-Hermann R. Transforming growth factor beta 1-induced cell death in preneoplastic foci of rat liver and sensitization by the antiestrogen tamoxifen. Hepatology 1996;23:840–847.
122. Bissell DM, Roulot D, George J. Transforming growth factor beta and the liver. Hepatology 2001;34: 859–867.
123. Parzefall W, Kainzbauer E, Qin HM, et al. Response of isolated hepatocytes from carcinogen sensitive (C3H) and insensitive (C57BL) mice to signals inducing replication or apoptosis. Arch Toxicol 2002;76:699–706.
124. Lin JK, Chou CK. In vitro apoptosis in the human hepatoma cell line induced by transforming growth factor beta 1. Cancer Res 1992;52:385–388.
125. Oberhammer FA, Pavelka M, Sharma S, et al. Induction of apoptosis in cultured hepatocytes and in regressing liver by transforming growth factor beta 1. Proc Natl Acad Sci USA 1992;89: 5408–5412.
126. Oberhammer F, Bursch W, Parzefall W, et al. Effect of transforming growth factor beta on cell death of cultured rat hepatocytes. Cancer Res 1991;51:2478–2485.
127. Chabicovsky M, Wastl U, Taper H, et al. Induction of apoptosis in mouse liver adenoma and carcinoma in vivo by transforming growth factor-beta1. J Cancer Res Clin Oncol 2003;129:536–542.
128. Schulte-Hermann R, Bursch W, Gotzmann J, et al. TGF-beta-induced liver cell apoptosis. In: Gressner AM, Heinrich PC, Matern S, editors. Cytokines in liver injury and repair. Hannover: Kluwer Academic Publishers; 2002; pp. 31–43.
129. Oe S, Lemmer ER, Conner EA, et al. Intact signaling by transforming growth factor beta is not required for termination of liver regeneration in mice. Hepatology 2004;40:1098–1105.
130. Kogure K, Zhang YQ, Maeshima A, et al. The role of activin and transforming growth factor-beta in the regulation of organ mass in the rat liver. Hepatology 2000;31:916–921.
131. Yasuda H, Mine T, Shibata H, et al. Activin A: an autocrine inhibitor of initiation of DNA synthesis in rat hepatocytes. J Clin Invest 1993;92:1491–1496.
132. Hully JR, Chang L, Schwall RH, et al. Induction of apoptosis in the murine liver with recombinant human activin A. Hepatology 1994;20:854–862.
133. Schwall RH, Robbins K, Jardieu P, et al. Activin induces cell death in hepatocytes in vivo and in vitro. Hepatology 1993;18:347–356.

134. Vejda S, Erlach N, Peter B, et al. Expression of activins C and E induces apoptosis in human and rat hepatoma cells. Carcinogenesis 2003;24:1801–1809.
135. Matzuk MM, Finegold MJ, Mather JP, et al. Development of cancer cachexia-like syndrome and adrenal tumors in inhibin-deficient mice. Proc Natl Acad Sci USA 1994;91:8817–8821.
136. Takabe K, Lebrun JJ, Nagashima Y, et al. Interruption of activin A autocrine regulation by antisense oligodeoxynucleotides accelerates liver tumor cell proliferation. Endocrinology 1999;140:3125–3132.
137. Takabe K, Wang L, Leal AM, et al. Adenovirus-mediated overexpression of follistatin enlarges intact liver of adult rats. Hepatology 2003;38:1107–1115.
138. Ichikawa T, Zhang YQ, Kogure K, et al. Transforming growth factor beta and activin tonically inhibit DNA synthesis in the rat liver. Hepatology 2001;34:918–925.
139. Takamura K, Tsuchida K, Miyake H, et al. Activin and activin receptor expression changes in liver regeneration in rat. J Surg Res 2005;126:3–11.
140. Kogure K, Omata W, Kanzaki M, et al. A single intraportal administration of follistatin accelerates liver regeneration in partially hepatectomized rats. Gastroenterology 1995;108:1136–1142.
141. Kogure K, Zhang YQ, Shibata H, Kojima I. Immediate onset of DNA synthesis in remnant rat liver after 90% hepatectomy by an administration of follistatin. J Hepatol 1998;29:977–984.
142. Date M, Matsuzaki K, Matsushita M, et al. Differential regulation of activin A for hepatocyte growth and fibronectin synthesis in rat liver injury. J Hepatol 2000;32:251–260.
143. Zhang YQ, Shibata H, Schrewe H, Kojima I. Reciprocal expression of mRNA for inhibin betaC and betaA subunits in hepatocytes. Endocr J 1997;44:759–764.
144. Wada W, Kuwano H, Hasegawa Y, Kojima I. The dependence of transforming growth factor-beta-induced collagen production on autocrine factor activin A in hepatic stellate cells. Endocrinology 2004;145:2753–2759.
145. Endo D, Kogure K, Hasegawa Y, et al. Activin A augments vascular endothelial growth factor activity in promoting branching tubulogenesis in hepatic sinusoidal endothelial cells. J Hepatol 2004;40:399–404.
146. Endo D, Maku-Uchi M, Kojima I. Activin or follistatin: Which is more beneficial to support liver regeneration after massive hepatectomy? Endocr J 2006;53:73–78.
147. Galle PR, Krammer PH. CD95-induced apoptosis in human liver disease. Semin Liver Dis 1998;18:141–151.
148. Kanzler S, Galle PR. Apoptosis and the liver. Semin Cancer Biol 2000;10:173–184.
149. Strand S, Vollmer P, van den Abeelen L, et al. Cleavage of CD95 by matrix metalloproteinase-7 induces apoptosis resistance in tumour cells. Oncogene 2004;23:3732–3736.
150. Niimi S, Horikawa M, Seki T, et al. Effect of activins AB and B on DNA synthesis stimulated by epidermal growth factor in primary cultured rat hepatocytes. Biol Pharm Bull 2002;25:437–440.
151. Esquela AF, Zimmers TA, Koniaris LG, et al. Transient down-regulation of inhibin-betaC expression following partial hepatectomy. Biochem Biophys Res Commun 1997;235:553–556.
152. Lau AL, Kumar TR, Nishimori K, et al. Activin betaC and betaE genes are not essential for mouse liver growth, differentiation, and regeneration. Mol Cell Biol 2000;20:6127–6137.
153. Wada W, Medina J, Hasegawa Y, et al. Adenovirus-mediated overexpression of the activin betaC subunit accelerates liver regeneration in partially hepatectomized rats. J Hepatol 2005;43:823–828.
154. Chabicovsky M, Herkner K, Rossmanith W. Overexpression of activin beta(C) or activin beta(E) in the mouse liver inhibits regenerative deoxyribonucleic acid synthesis of hepatic cells. Endocrinology 2003;144:3497–3504.
155. Wada W, Medina JJ, Kuwano H, Kojima I. Comparison of the function of the beta(C) and beta(E) subunits of activin in AML12 hepatocytes. Endocr J 2005;52:169–175.
156. Hashimoto O, Ushiro Y, Sekiyama K, et al. Impaired growth of pancreatic exocrine cells in transgenic mice expressing human activin betaE subunit. Biochem Biophys Res Commun 2006;341:416–424.
157. Schmidt-Weber CB, Blaser K. Regulation and role of transforming growth factor-beta in immune tolerance induction and inflammation. Curr Opin Immunol 2004;16:709–716.
158. Zhang XJ, Li Y, Tai GX, et al. Effects of activin A on the activities of the mouse peritoneal macrophages. Cell Mol Immunol 2005;2:63–67.
159. Huang X, Li DG, Wang ZR, et al. Expression changes of activin A in the development of hepatic fibrosis. World J Gastroenterol 2001;7:37–41.
160. Kobayashi T, Niimi S, Hashimoto O, Hayakawa T. Expression of inhibin betaA, betaB and follistatin mRNAs in the carbon tetrachloride induced rat liver regeneration model. Biol Pharm Bull 2000;23:755–757.

161. Gold EJ, Francis RJ, Zimmermann A, et al. Changes in activin and activin receptor subunit expression in rat liver during the development of CCl4-induced cirrhosis. Mol Cell Endocrinol 2003; 201:143–153.
162. Erlach N, Drucker C, Teufelhofer O, et al. Activin C has anti-inflammatory properties in rat hepatocytes. In preparation 2006.
163. Jones KL, de Kretser DM, Patella S, Phillips DJ. Activin A and follistatin in systemic inflammation. Mol Cell Endocrinol 2004;225:119–125.
164. Patella S, Phillips DJ, de Kretser DM, et al. Characterization of serum activin-A and follistatin and their relation to virological and histological determinants in chronic viral hepatitis. J Hepatol 2001;34:576–583.
165. Hughes RD, Evans LW. Activin A and follistatin in acute liver failure. Eur J Gastroenterol Hepatol 2003;15:127–131.
166. Yuen MF, Norris S, Evans LW, et al. Transforming growth factor-beta 1, activin and follistatin in patients with hepatocellular carcinoma and patients with alcoholic cirrhosis. Scand J Gastroenterol 2002;37:233–238.
167. Lin SD, Kawakami T, Ushio A, et al. Ratio of circulating follistatin and activin A reflects the severity of acute liver injury and prognosis in patients with acute liver failure. J Gastroenterol Hepatol 2006;21:374–380.
168. Hanahan D, Weinberg RA. The hallmarks of cancer. Cell 2000;100:57–70.
169. Risbridger GP, Schmitt JF, Robertson DM. Activins and inhibins in endocrine and other tumors. Endocr Rev 2001;22:836–858.
170. Jeruss JS, Sturgis CD, Rademaker AW, Woodruff TK. Down-regulation of activin, activin receptors, and Smads in high-grade breast cancer. Cancer Res 2003;63:3783–3790.
171. Panico L, D'Antonio A, Salvatore G, et al. Differential immunohistochemical detection of transforming growth factor alpha, amphiregulin and CRIPTO in human normal and malignant breast tissues. Int J Cancer 1996;65:51–56.
172. Wildi S, Kleeff J, Maruyama H, et al. Overexpression of activin A in stage IV colorectal cancer. Gut 2001;49:409–417.
173. Hempen PM, Zhang L, Bansal RK, et al. Evidence of selection for clones having genetic inactivation of the activin A type II receptor (ACVR2) gene in gastrointestinal cancers. Cancer Res 2003;63:994–999.
174. Su GH, Bansal R, Murphy KM, et al. ACVR1B (ALK4, activin receptor type 1B) gene mutations in pancreatic carcinoma. Proc Natl Acad Sci USA 2001;98:3254–3257.
175. Jung B, Doctolero RT, Tajima A, et al. Loss of activin receptor type 2 protein expression in microsatellite unstable colon cancers. Gastroenterology 2004;126:654–659.
176. Stove C, Vanrobaeys F, Devreese B, et al. Melanoma cells secrete follistatin, an antagonist of activin-mediated growth inhibition. Oncogene 2004;23:5330–5339.
177. Keelan JA, Zhou RL, Mitchell MD. Activin A exerts both pro- and anti-inflammatory effects on human term gestational tissues. Placenta 2000;21:38–43.
178. Pirisi M, Fabris C, Luisi S, et al. Evaluation of circulating activin-A as a serum marker of hepatocellular carcinoma. Cancer Detect Prev 2000;24:150–155.
179. Rossmanith W, Chabicovsky M, Grasl-Kraupp B, et al. Follistatin overexpression in rodent liver tumors: a possible mechanism to overcome activin growth control. Mol Carcinog 2002;35:1–5.
180. Mashima H, Kanzaki M, Nobusawa R, et al. Derangements in the activin-follistatin system in hepatoma cells. Gastroenterology 1995;108:834–840.
181. Grusch M, Drucker C, Peter-Vörösmary B, et al. Deregulation of the activin-follistatin system in hepatocarcinognesis. Heparol 2006;45:673–680.
182. Fuwii M, Ishikawa M, Iuchi M, Tashiro S. Effect of follistatin on rat liver regeneration and tumor growth after portal occlusion. Hepatogastroenterology 2005;52:833–838.
183. Eppert K, Scherer SW, Ozcelik H, et al. MADR2 maps to 18q21 and encodes a TGFbeta-regulated MAD-related protein that is functionally mutated in colorectal carcinoma. Cell 1996;86:543–552.
184. Hahn SA, Schutte M, Hoque AT, et al. DPC4, a candidate tumor suppressor gene at human chromosome 18q21.1. Science 1996;271:350–353.
185. Yakicier MC, Irmak MB, Romano A, et al. Smad2 and Smad4 gene mutations in hepatocellular carcinoma. Oncogene 1999;18:4879–4883.
186. Lee DK, Park SH, Yi Y, et al. The hepatitis B virus encoded oncoprotein pX amplifies TGF-beta family signaling through direct interaction with Smad4: potential mechanism of hepatitis B virus-induced liver fibrosis. Genes Dev 2001;15:455–466.

187. Kanamaru C, Yasuda H, Fujita T. Involvement of Smad proteins in TGF-beta and activin A-induced apoptosis and growth inhibition of liver cells. Hepatol Res 2002;23:211–219.
188. Piek E, Roberts AB. Suppressor and oncogenic roles of transforming growth factor-beta and its signaling pathways in tumorigenesis. Adv Cancer Res 2001;83:1–54.
189. Abou-Shady M, Baer HU, Friess H, et al. Transforming growth factor betas and their signaling receptors in human hepatocellular carcinoma. Am J Surg 1999;177:209–215.
190. Gotzmann J, Huber H, Thallinger C, et al. Hepatocytes convert to a fibroblastoid phenotype through the cooperation of TGF-beta1 and Ha-Ras: steps towards invasiveness. J Cell Sci 2002; 115:1189–1202.
191. Lahn M, Kloeker S, Berry BS. TGF-beta inhibitors for the treatment of cancer. Expert Opin Invest Drugs 2005;14:629–643.
192. Yingling JM, Blanchard KL, Sawyer JS. Development of TGF-beta signalling inhibitors for cancer therapy. Nat Rev Drug Discov 2004;3:1011–1022.
193. Munz B, Smola H, Engelhardt F, et al. Overexpression of activin A in the skin of transgenic mice reveals new activities of activin in epidermal morphogenesis, dermal fibrosis and wound repair. EMBO J 1999;18:5205–5215.
194. Sugiyama M, Ichida T, Sato T, et al. Expression of activin A is increased in cirrhotic and fibrotic rat livers. Gastroenterology 1998;114:550–558.
195. Wagner K, Peters M, Scholz A, et al. Activin A stimulates vascular endothelial growth factor gene transcription in human hepatocellular carcinoma cells. Gastroenterology 2004;126:1828–1843.
196. DaCosta Byfield S, Major C, Laping NJ, Roberts AB. SB-505124 is a selective inhibitor of transforming growth factor-beta type I receptors ALK4, ALK5, and ALK7. Mol Pharmacol 2004;65: 744–752.
197. Inman GJ, Nicolas FJ, Callahan JF, et al. SB-431542 is a potent and specific inhibitor of transforming growth factor-beta superfamily type I activin receptor-like kinase (ALK) receptors ALK4, ALK5, and ALK7. Mol Pharmacol 2002;62:65–74.
198. Strizzi L, Bianco C, Normanno N, Salomon D. Cripto-1: a multifunctional modulator during embryogenesis and oncogenesis. Oncogene 2005;24:5731–5741.

32 Myostatin Signaling in Normal and Tumor Cells

Xiangyang Zhu, Li-fang Liang, and Ronald L. Stotish

CONTENTS

INTRODUCTION
PROTEOLYTIC PROCESSING OF MYOSTATIN
THE RECEPTORS FOR MYOSTATIN SIGNALING
SMAD2–4 PARTICIPATE IN MYOSTATIN SIGNALING
SMAD7 INHIBITS MYOSTATIN SIGNALING THROUGH A NEGATIVE FEEDBACK MECHANISM
NON-SMAD MYOSTATIN SIGNALS
SKI AND SNON
MYOSTATIN INHIBITS MUSCLE CELL PROLIFERATION AND DIFFERENTIATION
ROLE OF MYOSTATIN IN ADIPOSE TISSUE AND GLUCOSE METABOLISM
MYOSTATIN AND RHABDOMYOSARCOMA TUMORS
CONCLUDING REMARKS
ACKNOWLEDGMENTS
REFERENCES

Abstract

As a member of the TGF-β superfamily, myostatin is a specific negative regulator of skeletal muscle development. Loss of myostatin function in knock-out mice results in "double-muscle" phenotype observed in cattle breeds such as the Belgian Blue and the Piedmontese. In this chapter, we have included most recent observations on myostatin research, including the proteolytic processing of myostatin, its receptors, and the components in myostatin signaling pathway. In addition, we also discuss the role of myostatin in adipose tissue and glucose metabolism, and its relationship with rhabdomyosarcoma tumors.

Key Word: Myostatin; transforming growth factor-β; skeletal muscle; signal pathway; Smad; adipose tissue; rhabdomyosarcoma tumors.

1. INTRODUCTION

Myostatin, a growth and differentiation factor belonging to the TGF-β superfamily, acts as a negative regulator of skeletal muscle mass *(1–3)*. Deletion of myostatin in mice causes

From: *Cancer Drug Discovery and Development: Transforming Growth Factor-β in Cancer Therapy, Vol I: Basic and Clinical Biology*
Edited by: Sonia B. Jakowlew © Humana Press Inc., Totowa, NJ

a dramatic increase in skeletal muscle mass as a result of hyperplasia and hypertrophy *(1–5)*. The *double-muscled* phenotype identified in certain breeds of cattle, such as Belgian Blue and Piedmontese has been shown to be owing to a loss-of-function of the myostatin protein *(6–8)*. Recently, a human baby with a *double-muscled* phenotype has been shown to be homozygous negative for myostatin protein *(9)*, owing to a point mutation in the myostatin gene results in a misspliced myostatin mRNA transcript. Since its initial discovery in 1997, myostatin has emerged as a key regulator of skeletal muscle development and a major target for the treatment of muscle wasting diseases. Many aspects of myostatin have been well studied, and compared with TGF-β. This work will discuss myostatin signaling in both normal and tumor cells.

2. PROTEOLYTIC PROCESSING OF MYOSTATIN

Myostatin, like other members of the TGF-β superfamily, is synthesized as a precursor protein containing an N-terminal propeptide domain, and a conserved pattern of cysteine residues in the C-terminal mature region *(10,11)*. Proteolytic processing between the propeptide domain and the C-terminal domain releases the propeptide, and mature myostatin *(12–14)*. The proteolytic processing of myostatin is carried out by the proprotein convertase furin, a calcium-dependent serine protease *(13)*. Like TGF-β, the mature myostatin dimer and two molecules of propeptide remain noncovalently associated after cleavage, producing a latent complex. In the latent complex, myostatin is prevented from interacting with its receptor.

So far seven members in the proprotein convertase family have been identified *(15)*. McFarlane et al. *(16)* show that the mature myostatin autoregulates its processing during myogensis by controlling the furin protease promoter activity. In cultured C_2C_{12} cells, no appreciable changes in myostatin mRNA levels are detected between proliferating myoblasts and differentiated myotubes. However, Western blot analysis showed that no precursor myostatin was observed in myoblasts culture in proliferation media. In the myoblasts undergoing differentiation, a relatively high level of precursor myostatin is observed. These results suggest that there is a change in the level of proteolytic processing of myostatin during myogenic differentiation. In vivo studies demonstrate that low levels of myostatin processing occur during fetal muscle development in comparison with postnatal or adult muscle; whereas high levels of circulating mature myostatin are observed in postnatal serum, fetal circulatory myostatin level is undetectable. In the same studies, furin promoter activity is negatively regulated by myostatin. Taken together, the amount of circulating mature myostatin that can initiate myostatin signaling is controlled at the level of proteolytic processing. Myostatin-mediated signaling negatively regulates furin expression, which then reduces the proteolytic processing of myostatin from its propeptide. This autoregulatory loop appears to be an important mechanism for the regulation of myostatin processing during development.

The cleaved propeptide molecules remain noncovalently bound to the mature dimer, forming a latent complex and inhibiting its biological activity. In addition to interacting with the propeptide, myostatin can also interact with a number of other proteins. Using affinity purification and mass spectrometry techniques, follistatin-related gene (FLRG), and GDF-associated serum protein-1 (GASP-1) proteins have been identified to interact with myostatin in human and mouse serum *(17,18)*. Interestingly, GASP-1 can also interact with the propeptide molecule and thus represents a novel class of inhibitory TGF-β binding proteins *(18)*. Titin cap and human small glutamine-rich tetratricopeptide repeat-containing protein (hSGT) have also been shown to interact with myostatin in yeast two-hybrid screens *(19,20)*; however, it is not clear whether these proteins play a physiological role in vivo. Wolfman et al. *(21)* presented evidence that members of the bone morphogenetic protein-1/

tolloid (BMP-1/TLD) family of metalloproteinases may be involved in the activation of the latent protein complex. In vitro assays showed that BMP-1, mTLD, mTLL-1, and mTLL-2 cleaved between Arg-75 and Asp-76 of the propeptide and thereby activating the latent complex.

3. THE RECEPTORS FOR MYOSTATIN SIGNALING

It is widely accepted that TGF-β and related proteins initiate cellular responses by binding to two different types of serine/threonine kinase receptors (type I and type II) *(22–25)*. The signaling pathway is initiated by the ligand binding directly to the Type II receptor, which leads to the recruitment of the Type I receptor. The Type I receptor is activated by phosphorylation and initiates a specific intracellular signal cascade mediated by the Smad proteins *(26,27)*. Smad proteins are a group of molecules that function as intracellular signal transducers downstream of the receptors of the TGF-β superfamily. Eight different Smad proteins are divided into three subfamilies based upon their function: receptor-regulated Smads (R-Smads), common-partner Smads (Co-Smads), and inhibitory Smads (I-Smads) *(28–31)*. R-Smads are activated by phosphorylation by the type I receptor serine kinase. This family consists of Smad 1, 2, 3, 5, and 8. Smad1, Smad5, and Smad8 mediate signaling for bone morphogenetic proteins (BMPs) and anti-Müllerian hormone pathways *(31–33)*, while Smad2 and Smad3 act in the TGF-β and activin pathways *(28–30)*. Smad4 is the Co-Smad, which positively regulates all the above pathways *(30,31)*. In contrast to R-Smads and Co-Smad, I-Smads, including Smad6, and Smad7, bind to the intracellular domain of type I receptors. They compete with R-Smads for activation by the type I receptors, resulting in inhibition of TGF-β superfamily signaling *(34–36)*. Smad6 inhibits BMP signaling *(37,38)*, while Smad7 inhibits TGF-β and activin signaling *(39–41)*.

Previous studies have reported that ACVR2B (ActRIIB) is the primary type II receptor for myostatin *(13,42)*. Cross-linking experiments using ^{125}I-myostatin demonstrated that the activated form of myostatin binds with the ACVR2B receptor, and less strong to ACVR2 (ActRIIA). No binding is observed to either TGF-βRII or BMPRII *(13)*. The binding between myostatin and the ActR2B receptor is specific. Transgenic mice overexpressing a dominant negative form of ACVR2B show increased skeletal muscle mass *(13)*. Currently, both the ACVR1B (ALK4) and TGF-β type I receptor (ALK5) are reported as possible candidates for the type I receptor for myostatin signaling *(13,42)*. Crossing-link experiments show both ALK4 and ALK5 can bind with the myostatin-ACRV2B complex *(13,42,43)*. When a *dead mutation* of type I receptor, ALK5K232R, is expressed in C_2C_{12} cells, it inhibits myostatin signaling *(44)*. Data from our in vitro experiments suggest that both ALK4 and ALK5 play roles in myostatin signaling. In experiments utilizing a p(CAGA)$_{12}$-MLP-Luc reporter construct transiently transfected into a CCL-64-derived mutant cell line, R1B (lacking the TGF-β type I receptor), weak luciferase activity is induced by myostatin, but no induction is observed with TGF-β *(45)*. When the TGF-β type I (ALK5) gene was cotransfected with the p(CAGA)$_{12}$-MLP-Luc reporter construct into R1B cells, the transfected cells responded very well to TGF-β and moderately to myostatin. On the other hand, when the ALK4 gene was cotransfected with the p(CAGA)$_{12}$-MLP-Luc reporter construct, R1B cells respond to myostatin extremely well, with no response to TGF-β. These results indicate that both the ALK4 and ALK5 receptors can play a role in myostatin signaling.

The interaction between myostatin and its receptors is a focus for the development of target molecules to block myostatin signaling. Molecules such as the myostatin propeptide, follistatin, and FLRG that can interfere with the binding between myostatin and ACVR2B receptors have been proposed as potential inhibitors for use against muscular dystrophy (*mdx*) *(17,18)*. Follistatin, FLRG, and growth and differentiation factor associated serum protein-1 (GASP-1) have all been shown to bind to the extracellular form of myostatin and interfere with the interaction between myostatin and ACVR2B. Muscle specific overexpression of follistatin, the soluble extracellular domain of ACVR2B, or the myostatin propeptide in transgenic mice result in a 200–300% increase in skeletal muscle *(13)*. Recently, Muenster et al. *(43)* have shown that an activin-A/C chimera can also function as a myostatin antagonist by interfering with the interaction between myostatin and ACVR2B.

4. SMAD2–4 PARTICIPATE IN MYOSTATIN SIGNALING

Several recent studies have indicated that Smad 2/3 are the R-Smad proteins for myostatin signaling *(42,43)* and that Smad4 is required for the signal pathway *(46)*. We have examined the R-Smad components involved in the myostatin signal transduction by investigating the roles of Smad 1, 2, 3, and 5 in our assay system. The p(CAGA)$_{12}$-MLP-Luc plasmid was transiently cotransfected into A204 cells with R-Smads (Smad1, 2, 3, or 5) or/and the common Smad4. As shown in Figure 1. Smad 2, 3, and 4 enhance the myostatin-induced luciferase activity of the reporter construct, while both Smad1 and 5 show no effect *(46)*. These results are consistent with previous reports describing TGF-β and activin signaling *(20,22–23)*. Cotransfection of Smad2/Smad4 or Smad3/Smad4 results in an increased luciferase induction compared to each respective Smad alone. This synergistic effect has been previously reported in TGF-β signaling *(22)*. However, when Smad1 or 5 is cotransfected with Smad4, no such synergy is observed.

The participation of Smad2 and Smad3 in the myostatin-mediated signaling has been confirmed by immunoblot analysis. When cell lysates from HepG2 cells (Human heptoma cells responsive to TGF-β through a Smad signaling system) are immunoblotted using an antibody against phosphorylated-Smad2/3, the level of phosphorylated Smad2 and Smad3 are dramatically increased upon induction with myostatin *(46)*.

The role of Smad4 in myostatin-mediated signaling is further shown by studies in MDA-MB-468 cells, a human epithelial cell line derived from a breast cancer and shown to be deficient for endogenous Smad4 *(47)*. No myostatin-mediated luciferase activity is detected when p(CAGA)$_{12}$-MLP-Luc is transiently transfected into MDA-MB-468 cells. However, cotransfection of Smad4 with p(CAGA)$_{12}$-MLP-Luc in those cells restored the activity of myostatin on p(CAGA)$_{12}$ promoter activation *(46)*, clearly demonstrating Smad4 requirement for myostatin signaling.

5. SMAD7 INHIBITS MYOSTATIN SIGNALING THROUGH A NEGATIVE FEEDBACK MECHANISM

As described in Section 3, Smad6 and Smad7 are the inhibitory Smad proteins for signal transduction of the TGF-β family. When plasmids encoding either Smad6 or Smad7 are co-transfected with p(CAGA)$_{12}$-MLP-Luc into A204 cells, Smad7 inhibited the myostatin-induced luciferase activity of the p(CAGA)$_{12}$ promoter dramatically, while Smad6 gives a slight inhibition *(46)*. However, a Smad7 mutation (Smad7-408stop, Smad7M, a deletion of 19 amino acids in the C-terminal end) abolishes the inhibition of the myostatin-induced activity. This Smad7 inhibitory activity is enhanced in the presence of Smad Ubiquitin Regulatory Factor 1 (Smurf1) *(48)*. It has been demonstrated that Smurf1 binds with Smad7 and induces Smad7 ubiqutination and translocation into the cytoplasm, binding to the type I receptor via Smad7 *(48,49)*. When the Smurf1 expression vector was included into the co-transfection of Smad7 and p(CAGA)$_{12}$-MLP-Luc, the combined action of both Smad7 and Smurf1 completely inhibited the myostatin-induced p(CAGA)$_{12}$ luciferase activity *(46)*.

Smad7 expression is initiated by TGF-β and activin stimulation, providing a regulatory feedback mechanism to terminate signaling through the activated receptors *(39,40)*. The

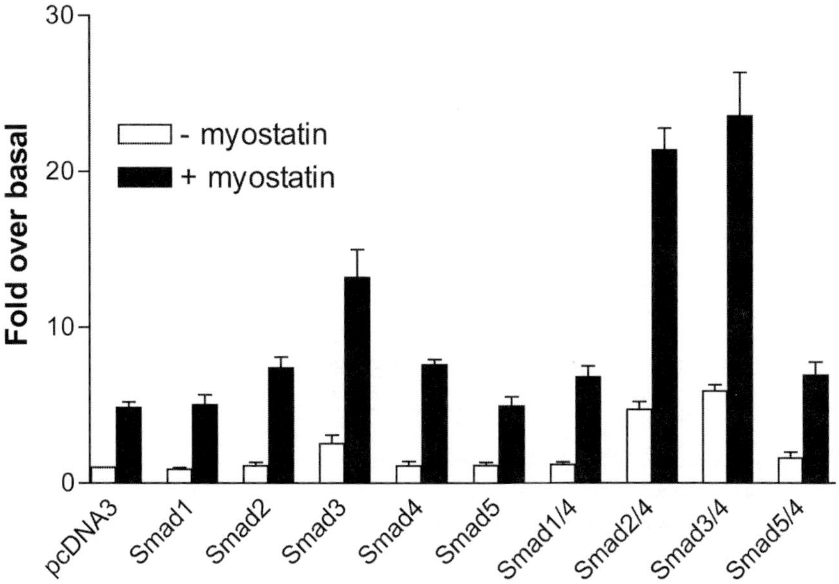

Fig. 1. Smad 2/3/4 are involved in the myostatin-induced transcription. (**A**) A204 cells were cotransfected with the $p(CAGA)_{12}$-MLP-Luc construct (1 μg/well) and expression vectors encoding different Smad proteins. Each well had an equal amount of DNA, adjusted by vector control pcDNA3. In each case (except for pcDNA3 control), an identical amount of Smad expression vector was transfected into the cells. In single transfections, 1.5 μg of the indicated Smad was used; in double transfections 0.75 μg of each Smad was used, in triple transfections 0.5 μg of each Smad was used. Luciferase activities are shown by fold over noninduced pcDNA3 control cells. The results are from at least three independent experiments, each carried out in triplicate.

transcription profile of endogenous Smad7 in myostatin-treated HepG2 cells was examined using Northern blot analysis. A basal level of Smad7 transcripts was detected in untreated HepG2 cells (Fig. 2). Myostatin induction (20 ng/ml) increased the level of Smad7 transcripts, peaking around 1.5 h postinduction, which then decreased to basal levels by 8 h postinduction. We have also demonstrated that phosphorylated Smad2/3 complex binds to the Smad7 promoter to initiate transcription upon stimulation by myostatin *(46)*. Those results show that Smad7 is one of the earliest genes upregulated by myostatin. The newly produced Smad7 protein can form a complex with Smurf1 to travel from the nucleus to the cytoplasm and interact with the type I receptor, inhibiting myostatin signaling.

6. NON-SMAD MYOSTATIN SIGNALS

Although the Smad pathway is the main mediator of TGF-β signaling, more and more cellular and genetic evidence suggest that non-Smad TGF-β signal pathways play important roles in TGF-β initiated cellular activities *(50–54)*. Moustakas and Heldin *(50)* classifed the non-Smad signaling into three distinct signaling mechanisms: (1) non-Smad signaling pathways that modify Smad function; (2) non-Smad effectors whose function is directly modulated by Smads and which transmit signals to other pathways; and (3) non-Smad proteins that directly interact with or become phosporylated by TGF-β receptors and do not interact with Smad proteins. Recent studies have shown that members of the TGF-β family can activate the mitogen-activated protein kinase (MAPK) pathways. The three MAPK have been identified as MAPK/ERK (extracellular signal-related kinase) *(54)*, SAPK (Stress-

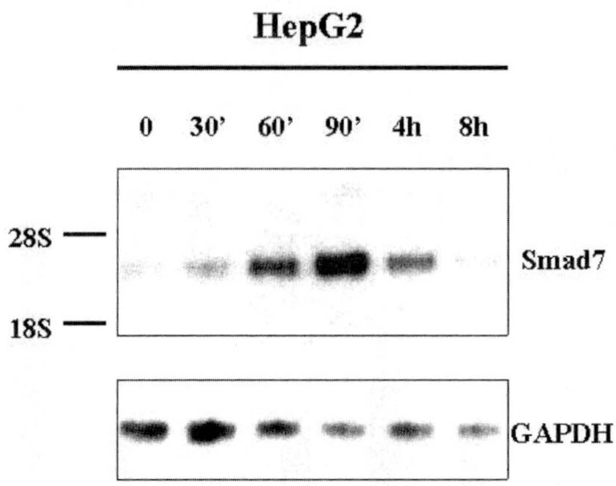

Fig. 2. The endogenous Smad7 gene expression in HepG2 cells is regulated by myostatin. Total RNA (10 μg) isolated from HepG2 cells stimulated with myostatin (20 ng/ml) for various times as indicated was probed with HRP-labeled Smad7 and GAPDH.

activated Protein Kinases)/JNK (c-*Jun* N-terminal kinase) (JNK) pathway *(55)*, and the p38 MAPK (stress response pathway) *(52–56)*. Philip et al. *(44)* discovered that myostatin can activate the p38 MAPK pathway through the TGF-β-activated kinase 1 (TAK1). The p38 MAPK pathway is activated by a number of environmental stress factors and inflammatory cytokines, and is independent of the Smad family of proteins. They also demonstrated that the p38 MAPK pathway plays an important role in the myostatin-induced inhibition of proliferation and upregulation of the cyclin kinase inhibitor p21. When a *dead* mutation of the type I receptor, ALK5K232R, was expressed in transfected cells, it inhibited both the Smad and p38 signal pathways. This suggests both the Smad and p38 pathways utilize the same type I receptor. In the same report, they also demonstrated that a common transcription factor, ATF-2, is involved in both the Smad and p38 signal pathways for transcription initiation.

7. SKI AND SNON

Ski and SnoN are members of Ski family of nuclear protooncogenes. Both Ski and SnoN have been suggested to play important roles in the control of cell growth and skeletal muscle differentiation *(57–59)*. They are incorporated into the histone deacetylase-1 complex through the binding to the nuclear hormone receptor corepressor N-CoR and mSin3A, and mediate transcription of many genes *(60,61)*. In the nucleus, both Ski and SnoN can directly interact with the complexes of Smad2/3/4 and down regulate TGF-β initiated gene transcription *(62–64)*. Homozygous Ski mutants exhibited a dramatic decrease in skeletal muscle mass *(65)*. Overexpression of Ski increased skeletal muscle development in transgenice mice, resulting in hypertrophy of Type II fast fibers *(66)*. Overexpression of Ski and/or SnoN in TGF-β-responsive cells blocks TGF-β induced growth arrest, causes oncogenic transformation, anchorage independence, enhanced cell proliferation and viability, and skeletal muscle differentiation *(67,68)*.

To address the role of Ski and SnoN in myostatin signal transduction, the Ski/SnoN expression vectors were cotransfected with a myostatin inducible promoter-fused luciferase reporter construct. Both Ski and SnoN were able to fully inhibit the myostatin-mediated luciferase activity (Fig. 3). The cotransfection of Ski and/or SnoN with various Smad expression vectors demonstrated that both Ski and SnoN could inhibit the Smad2/Smad4, and Smad3/

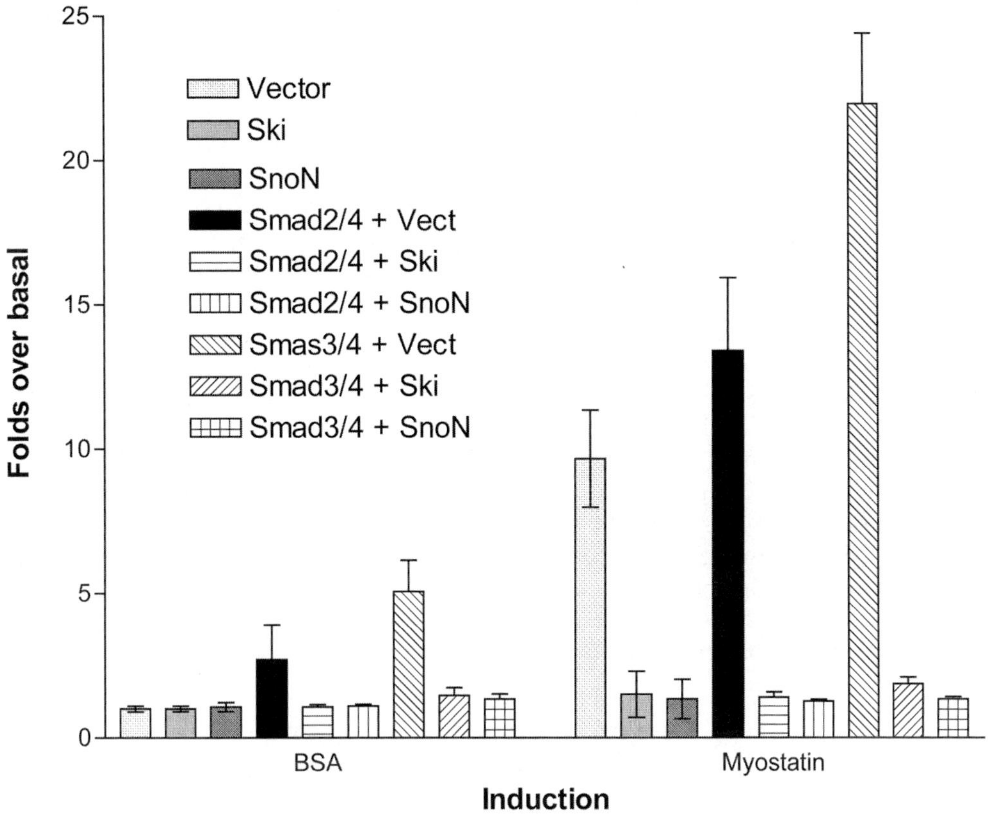

Fig. 3. Ski and SnoN inhibit myostatin-mediated luciferase activity. (**A**) A204 cells were cotransfected with the p(CAGA)$_{12}$-MLP-Luc construct (1 µg/well) and expression vectors encoding different Ski or SnoN and Smad proteins. Each well had an equal amount of DNA, adjusted by vector control pcDNA3. In each case, identical total amounts of DNA were transfected into the cells as indicated in Figure 1. Transfected cells were induced by myostatin or BSA (negative control). Luciferase activities are shown by fold over noninduced pcDNA3 control cells. The results are from at least three independent experiments, each carried out in triplicate.

Smad4 activation of myostatin signaling. Those results indicate that Ski/SnoN are negative effectors for myostatin signaling, and their downregulation activity is through Smad 2/3/4 complexes as shown for TGF-β signaling. It remains to be shown whether Ski and/or SnoN are involved in myostatin action on muscle cell proliferation and differentiation.

8. MYOSTATIN INHIBITS MUSCLE CELL PROLIFERATION AND DIFFERENTIATION

During the last ten years, many studies have addressed the mechanism precisely how myostatin regulates the muscle mass. When the myostatin gene is knocked-out in mice, those mice show a two- to threefold increase in skeletal muscle mass *(3)*. Muscle specific over-expression of myostatin propeptide, follistatin, or soluble ACVRIIB results in dramatically increased muscle mass *(3–13)*. The same phenomenon has been observed in Belgian Blue and Piedmontese cattle breeds and a human carrying a mutation in the myostatin gene *(9)*. A mouse model of *mdx* has been used to demonstrate that inhibiting myostatin attenuates the severity of *mdx* and enhances muscle gorwth *(69–70)*. On the other hand, Zimmer et al. *(5)*

show that systemic overexpression of myostatin in adult mice can induce profound muscle and fat loss analogous to that seen in cachexia. During skeletal muscle fiber growth, quiescent satellite cells enter again the cell cycle, proliferate, and fuse to existing muscle fibers. The observation that loss of myostatin function results in increased muscle fiber size suggests that myostatin targets satellite cells *(71–73)*. Using cultured C_2C_{12} myoblast cells and fluorescence-activated cell sorting analysis, Thomas et al. *(72)* showed that myostatin regulates cell cycle progression of myoblasts by controlling the G1 to S-phase and G2- to M-phase transition. Western blot analysis indicated that addition of myostatin to C_2C_{12} cells decreased the expression level of cyclin-dependent kinase 2 (Cdk2), and increased the expression of p21, a cyclin-dependent kinase inhibitor *(71–73)*. Whether the expression levels of both p21 and Cdk2 are regulated through the Smad signaling pathway or the p-38 MAPK pathway remains to be investigated. It is clear that the hyperplasia observed with the loss of myostatin is owing to increased cell proliferation, resulting in more cells. However, the hypertrophy observed with the loss of myostatin is caused by the stimulation of myoblast differentiation. Myostatin inhibits myoblast differentiation through Smad3 *(71)*. On one hand, myostatin signaling through the Smad pathway downregulates the expression of the family of basic helix-loop-helix factors, MRFs, which include MyoD, Myf5, myogenin and MRF4. MRFs are a group of factors, which control the process of myogenic specification and differentiation. On the other hand, myostatin signaling specifically induces Smad3 phosphorylation and increases Smad3:MyoD association, further inhibiting the skeletal muscle gene expression program.

9. ROLE OF MYOSTATIN IN ADIPOSE TISSUE AND GLUCOSE METABOLISM

Although myostatin is mainly expressed in skeletal muscle cells, it is also expressed in adipose tissue, although to a far lesser degree than skeletal muscle *(74)*. The major effect of the loss of myostatin function in myostatin null mice is an increase in skeletal muscle mass; however, those mice also accumulate much less adipose tissue compared to wild-type controls *(1–3)*. Leptin, an adipocyte-secreted hormone that regulates feed intake and energy homeostasis, is downregulated in the serum and adipose tissues of myostatin null mice *(4)*. Generation of transgenice mice overexpressing the myostatin propeptide effectively depressed myostatin function, resulting in significant *double muscled* phenotypes with 20% faster growth rate and up to 44% more muscle mass than wild-type mice *(72–76)*. Interestingly, when these transgenic mice were fed a high-fat diet (45% Kcal fat), they maintained normal blood glucose, insulin sensitivity, and fat mass. Control wild-type mice on the same high-fat diet produced 170–214% more fat mass and developed impaired glucose tolerance and insulin resistance. After 8 wk of dietary treatment, fasting blood glucose levels were similar in transgenic mice fed either a high-fat diet or a normal diet, and were comparable to those of wild-type mice fed a normal diet. These three groups of mice all showed a normal glucose tolerance test, their glucose concentration returning to initial levels 2 h after glucose injection. However, wild-type mice fed a high-fat diet showed significantly higher fasting blood glucose levels and grossly abnormal responses to glucose injection. Therefore, overexpression of myostatin propeptide in transgenic mice enhances insulin sensitivity in animals fed a high-fat diet, suggesting loss of myostatin function can affect the incidence of diet-induced obesity. Studies in genetically obese mice such as A^y and $lep^{ob/ob}$ give a similar result *(3)*. A^y mice display abnormal expression of the agouti protein, leading to increased food intake and obesity. $lep^{ob/ob}$ mice lack leptin expression and are severely obese. Abnormalities in glucose metabolism are observed in both of these mouse obesity models. When both A^y and $lep^{ob/ob}$ mice were crossed with myostatin null mice, the resulting mice show less severe abnormalities in glucose metabolism. This suggests that reduced myostatin expression may have beneficial effects with respect to the development of hyperglycemia and type II diabetes in these obesity models *(3)*. Rebbapragada et al. *(42)* show that myostatin serves as an antagonist to block other signal pathways, such as BMP7. Bone Morphogentic proteins, a group of proteins belong to the TGF-β superfamily, play an important role in determination and differentiation of mesenchymal progenitors along a variety of lineages, including the adipogenic pathways *(77,78)*. Both BMP2 and BMP7 induce adipogenic conversion of the pluripotent mesenchymal precursor cells, C3H 10t1/2 *(42)*. Using these cells, they demonstrated that myostatin specifically antagonizes BMP7 but not BMP2. Myostatin and BMP7 share the same ACVR2B receptor.

10. MYOSTATIN AND RHABDOMYOSARCOMA TUMORS

Rhabdomyosarcoma (RMS) is the most common childhood soft tissue sarcoma and the third most common extracranial childhood solid tumor *(79,80)*. RMSs are malignant tumors of mesenchymal origin thought to arise from cells that are committed to skeletal muscle lineage, but fail to complete the differentiation process *(81,82)*. MyoD and myogenin are two of the factors controlling the process of myogenic specification and differentiation in RMS. Although both MyoD and myogenin are expressed and can bind with DNA, rhabdomyoblasts fail to complete the differentiation program, suggesting a deficiency of factor(s) required for MyoD or myogenin activation *(82)*. Riacud et al. *(83)* demonstrated that myostatin is expressed at a higher level in RMS lines such as embryonal RD line, and the alveolar Rh30 line when compared to a normal human myoblast line. Like normal myoblasts, RD cell proliferation is inhibited by myostatin, owing to the downregulation of the cell cycle regulatory complex cyclin-E and Cdk2 *(82)*. Inhibition of myostatin expression through an antisense expression vector as well as inactivation of myostatin activity through follistatin and antimyostatin antibody resulted in the restoration of terminal myogenic differentiation and normal cycle withdrawal in RMS lines *(83)*. This inhibition is myostatin dose-dependent, reversible and can control RMSs progression. In the presence of myostatin, the transcriptional activity of MyoD is decreased dramatically. Conversely, inhibition of myostatin by follistatin (a myostatin antagonist) restores the transcriptional activity of MyoD. These results suggest that myostatin acts as an autocrine in the formation of RMSs, and inhibition of myostatin activity or myostatin expression in these cells may provide an opportunity to disrupt RMS tumor development.

11. CONCLUDING REMARKS

Myostatin binds to the type II receptor ACVR2B first, and then recruits the ALK4/ALK5 receptors to form a complex to initiate signaling. As shown in Figure 4, this complex can phosphorylate Smad2/3 in the cytoplasm, which binds with Smad4 to form another complex. The phosphorylated Smad2/3 and Smad4 complex then translocates into the nucleus to bind with myostatin-regulated promoters with the help of other transcription factors. Smad7 is one of the earliest genes activated by myostatin, and newly synthesized Smad7 protein interacts with Smurf1 to form an inhibitory complex, which interacts with the type I receptor to control myostatin signaling in a negative feedback loop mechanism. The alternative signal pathway for myostatin, the p38 pathway can coexist with the Smad pathway or operate in special conditions, such as cell stress. Both of the Smad and p38 signal pathways require the same type I receptor. Myostatin also negatively regulates furin promoter activity to control the levels of furin in the serum, and autoregulates its own proteolytic processing. Myostatin is involved in at least three different physiological functions: controlling skeletal muscle mass, regulating adipose tissue development, and regulating glucose uptake. Loss of myostatin

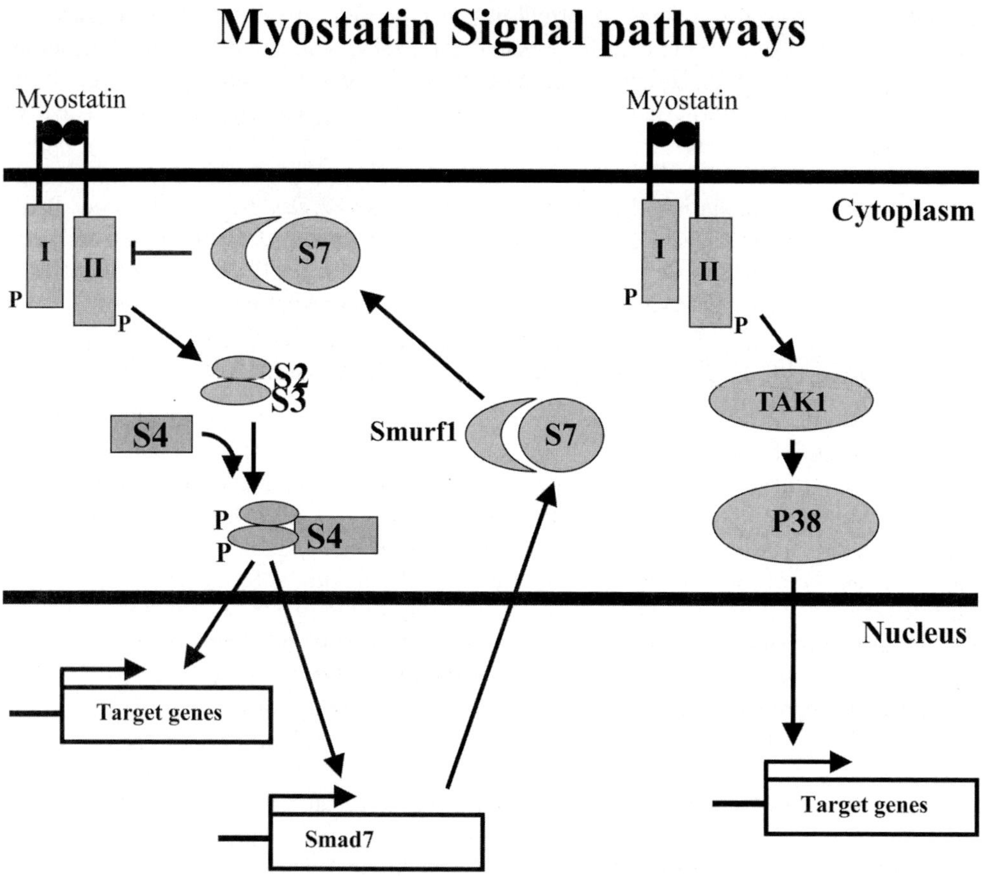

Fig. 4. Smad and non-Smad Myostatin Signal Pathway. Myostatin bound to both type II and type I receptors initiates signaling throught the phosphorylation of Smad2/Smad3 and formation of the Smad2/3/4 complex. This complex can translocate into the nucleus and cooperate with other transcription factors to regulate gene transcription. Smad7 is one of the earliest genes upregulated by the Smad signaling. The newly synthesized Smad7 interacts with Smurf1 to form an inhibitory complex, which exits the nucleus in response to the incoming myostatin signal and binds the receptor complex, leading to inhibition of the Smad-mediated signaling. Myostatin can also activate the p38 MARK pathway through the TAK1-MKK6 pathway, independent of Smad activation to initiate gene transcription in the nucleus.

signaling results in skeletal mass increase, adipose tissue decrease, and enhanced glucose uptake and utilization. Myostatin regulates the proliferation and differentiation of myoblasts. When the expression of myostatin or any other components of the signal transduction pathway is impaired, it can cause abnormal development of myoblasts (e.g., RMS). New therapeutic approaches may be explored for the treatment of RMS based on modulation the myostatin levels.

ACKNOWLEDGMENTS

We thank Dr. Xuedong Liu for providing us with Ski and SnoN expression vectors and Drs. Takeshi Imamura and Kohei Miyazono for providing us with Smurf1 expression vector. We also thank Dr. Man-Shiow Jiang for helpful discussions.

REFERENCES

1. McPherron AC, Lawler AM, Lee SJ. Regulation of skeletal muscle mass in mice by a new TGF-β superfamily member. Nature 1997;387:83–90.
2. Arnold H, Della-Fera MA, Baile CA. Review of myostatin history, physiology and applications. LifeXY 2001;1:1001–1013.
3. McPherron AC, Lee SJ. Suppression of body fat accumulation in myostatin-deficient mice. J Clin Invest 2002;109:595–601.
4. Lin J, Arnold HB, Della-Fera MA, Azain MJ, Hartzell DL, Baile CA. Myostatin knockout in mice increases myogenesis, and decreases adipogenesis. Biochem Biophys Res Commun 2002;291: 701–706.
5. Zimmers TA, Davies MV, Koniaris LG, et al. Induction of cachexia in mice by systemically administered myostatin. Science 2002;296:1486–1488.
6. McPherron AC, Lee SJ. Double muscling in cattle due to mutations in the myostatin gene. Proc Natl Acad Sci USA 1997;94:12,457–12,461.
7. Kambadur R, Sharma M, Smith TP, Bass JJ. Mutations in myostatin (GDF8) in double-muscled Belgian Blue and Piedmontese cattle. Genome Res 1997;7:910–916.
8. Grobet L, Martin LJ, Poncelet D, et al. A deletion in the bovine myostatin gene causes the double-muscled phenotype in cattle. Nat Genet. 1997;17:71–74.
9. Schuelke M, Wagner KR, Stolz LE, et al. Myostatin mutation associated with gross muscle hypertrophy in a child. N Engl J Med. 2004;350:2682–2688.
10. Derynck R, Jarrett JA, Chen EY, et al. Human transforming growth factor-β complementary DNA sequence and expression in normal and transformed cells. Nature 1985;316:701–705.
11. Wozney JM, Rosen V, Celeste AJ, et al. Novel regulators of bone formation: molecular clones and activities. Science 1988;242:1528–1534.
12. Thies RS, Chen T, Davies MV, et al. GDF-8 propeptide binds to GDF-8 and antagonizes biological activity by inhibiting GDF-8 receptor binding. Growth Factors 2001;18:251–259.
13. Lee SJ, McPherron AC. Regulation of myostatin activity and muscle growth. Proc Natl Acad Sci USA 2001;98:9306–9311.
14. Jiang MS, Liang LF, Wang S, et al. Characterization and identification of the inhibitory domain of GDF-8 propeptide. Biochem Biophys Res Commun 2004;315:525–531.
15. Thomad G. Furin at the cutting edge: from protein traffic to embryogenesis and disease. Nat Rev Mol Cell Biol. 2002;3:753–766.
16. McFarlane C, Langley B, Thomas A, et al. Proteolytic processing of myostatin is auto-regulated during myogenesis. Dev Biol 2005;283:58–69.
17. Hill JJ, Davies MV, Pearson AA, et al. The myostatin propeptide and the follistatin-related gene are inhibitory binding proteins of myostatin in normal serum. J Biol Chem 2002;277:40,735–40,741.
18. Hill JJ, Qiu Y, Hewick RM, Wolfman NM. Regulation of myostatin in vivo by growth and differentiation factor-associated serum protein-1: A novel protein with protease inhibitor and follistatin domains. Mol Endocrinol 2003;17:1144–1154.
19. Nicholas G, Thomas M, Lasngley B, et al. Titin-cap associates with, and regulates secretion of, Myostatin. J Cell Physiol 2002;193:120–131.
20. Wang H, Zhang Q, Zhu D. hSGT interacts with the N-terminal region of myostatin. Biochem Biophys Res Commun 2003; 311:877–883.
21. Wolfman NM, McPherron AC, Pappano WN, et al. Activation of latent myostatin by the BMP-1/tolloid family of metalloproteinases. Proc Natl Acad Sci USA 2003; 100:15,842–15,846.
22. Zhang Y, Derynck R. Regulation of Smad signalling by protein associations and signalling crosstalk. Trends Cell Biol 1999;9:274–279.
23. Massagué J, Wotton D. Transcriptional control by the TGF-beta/Smad signaling system. EMBO J 2000;19:1745–1754.
24. Roberts AB. The ever-increasing complexity of TGF-beta signaling. Cytokine Growth Factor Rev 2002;13:3–5.
25. Mehra A, Wrana JL. TGF-beta and the Smad signal transduction pathway. Biochem Cell Bio 2002; 180:605–622.
26. Franzen P, ten Dijke P, Ichijo H, et al. Cloning of a TGF beta type I receptor that forms a heteromeric complex with the TGF beta type II receptor. Cell 1993;75:681–692.
27. Nakao A, Imamura T, Souchelnytskyi S, et al. TGF-beta receptor-mediated signalling through Smad2, Smad3 and Smad4. EMBO J 1997;16:5353–5362.

28. Roberts AB. TGF-beta signaling from receptors to the nucleus. Microbes Infect 1999;1:1265–1273.
29. Dennler S, Itoh S, Vivien D, ten Dijke P, Huet S, Gauthier JM. Direct binding of Smad3 and Smad4 to critical TGF beta-inducible elements in the promoter of human plasminogen activator inhibitor-type 1 gene. EMBO J 1998;17:3091–3100.
30. Feng XH, Lin X, Derymck R. Smad2, Smad3 and Smad4 cooperate with Sp1 to induce p15(Ink4B) transcription in response to TGF-beta. EMBO J 2000;19:5178–5193.
31. Attisano L, Wrana JL. Smads as transcriptional co-modulators. Curr Opin Cell Biol 2000;12:235–243.
32. Hoodless PA, Haerry T, Abdollah S, et al. MADR1, a MAD-related protein that functions in BMP2 signaling pathways. Cell 1996;85:489–500.
33. Liu F, Hata A, Baker JC, et al. A human Mad protein acting as a BMP-regulated transcriptional activator. Nature 1996;381:620–623.
34. Imamura T, Takase M, Nishihara A, et al. Smad6 inhibits signalling by the TGF-beta superfamily. Nature 1997;389:622–626.
35. Zhao H, Shi W, Chen H, Warburton D. Smad7 and Smad6 differentially modulate transforming growth factor beta -induced inhibition of embryonic lung morphogenesis. J Biol Chem. 2000; 275: 23,992–23,997.
36. Chen Y, Lebrun JJ, Vale W. Regulation of transforming growth factor beta- and activin-induced transcription by mammalian Mad proteins. Proc Natl Acad Sci USA 1996;93:12,992–12,997.
37. Hata A, Lagna G, Massagué J, Hemmati P, Hemmati-Brivanlou A. Smad6 inhibits BMP/Smad1 signaling by specifically competing with the Smad4 tumor suppressor. Genes Dev 1998;12:186–197.
38. Ishisaki A, Yamato K, Hashimoto S, et al. Differential inhibition of Smad6 and Smad7 on bone morphogenetic protein- and activin-mediated growth arrest and apoptosis in B cells. J Biol Chem 1999;274:13637–13642.
39. Nakao A, Afrakhte M, Moren A, et al. Identification of Smad7, a TGFbeta-inducible antagonist of TGF-beta signalling. Nature 1997;389:631–635.
40. Hayashi H, Abdollah S, Qiu Y, et al. The MAD-related protein Smad7 associates with the TGFbeta receptor and functions as an antagonist of TGFbeta signaling. Cell 1997;89:1165–1173.
41. Ishisaki A, Yamato K, Nakao A, et al. Smad7 is an activin-inducible inhibitor of activin-induced growth arrest and apoptosis in mouse B cells. J Biol Chem 1998;273:24,293–24,296.
42. Rebbapragada A, Benchabane H, Wrana JL, Celeste AJ, Attisano L. Myostatin signals through a transforming growth factor beta-like signaling pathway to block adipogenesis. Mol Cell Biol 2003; 20:7230–7242.
43. Muenster U, Harrison CA, Donaldson C, Vale W, Fisher WH. An activin-A/C chimera exhibits activin and myostatin antagonistic properties. J Biol Chem 2005;280:36,623–36,632.
44. Philip B, Lu Z, Gao Y. Regulation of GDF-8 signaling by the p38 MAPK. Cell Signal 2005;17:365–375.
45. Zhu X, Liang LF, Stotish RL. Unpublished data.
46. Zhu X, Topouzis S, Liang LF, Stotish RL. Myostatin signaling through Smad2, Smad3 and Smad4 is regulated by the inhibitory Smad7 by a negative feedback mechanism. Cytokine 2004;26:262–272.
47. von Gersdorff G, Susztak K, Rezvani F, Bitzer M, Liang D, Bottinger EP. Smad3 and Smad4 mediate transcriptional activation of the human Smad7 promoter by transforming growth factor beta. J Biol Chem 2000;275:11,320–11,326.
48. Suzuki C, Murakami G, Fukuchi M, et al. Smurf1 regulates the inhibitory activity of Smad7 by targeting Smad7 to the plasma membrane. J Biol Chem 2002;277:39,919–39,925.
49. Tajima Y, Goto K, Yoshida M, et al. Chromosomal region maintenance 1 (CRM1)-dependent nuclear export of Smad ubiquitin regulatory factor 1 (Smurf1) is essential for negative regulation of transforming growth factor-beta signaling by Smad7. J Biol Chem 2003;278:10,716–10,721.
50. Moustakas A, Heldin C-H. Non-Smad TGF-beta signals. J Cell Sci 2005;118:3573–3584.
51. Edlund S, Landstrom M, Heldin C-H, Aspenstrom P. Transforming growth factor-beta-induced mobilization of actin cytoskeleton requires signaling by small GTPases Cdc42 and RhoA. Mol Biol Cell 2002;13:902–914.
52. Edlund S, Bu S, Schuster N, et al. Transforming growth factor-beta1 (TGF-beta)-induced apoptosis of prostate cancer cells involves Smad7-dependent activation of p38 by TGF-beta-activated kinase 1 and mitogen-activated protein kinase kinase. Mol Biol Cell 2002;14:529–544.
53. Engel ME, MaDonnell MA, Law RK, Moses HL. Interdependent SMAD and JNK signaling in transforming growth factor-beta-mediated transcription. J Biol Chem 1999;274:37,413–37,420.
54. Yue J, Mulder KM. Activation of the mitogen-activated protein kinase pathway by transforming growth factor-beta. Methods Mol Biol. 2000;142:125–131.

55. Guicheux J, Lemonnier J, Ghayor C, et al. Activation of p38 mitogen-activated protein kinase and c-Jun-NH2-terminal kinase by BMP-2 and their implication in the stimulation of osteoblastic cell differentiation. J Bone Miner Res 2003;11:2060–2068.
56. Hanafusa H, Ninomiya-Tsuji J, Masuyama N, et al. Involvement of the p38 Mitogen-activated protein kinase pathway in transformation growth factor-β induced gene expression. J Biol Chem 1999;274: 27,161–27,167.
57. Ichikawa K, Nagase T, Ishii S, Asano A, Mimura N. Trans-regulation of myogenin promoter/enhancer activity by c-ski during skeletal-muscle differentiation: the C-terminus of the c-Ski protein is essential for transcriptional regulatory activity in myotubes. Biochem J 1997;328:607–613.
58. Heyman HC, Stavnezer E. A carboxyl-terminal region of the ski oncoprotein mediates homodimerization as well as heterodimerization with the related protein SnoN. J Biol Chem 1994;269: 26,996–27,003.
59. Nomura N, Sasamoto S, Ishii S, Date T, Matsui M, Ishizki R. Isolation of human cDNA clones of ski and the ski-related gene, sno. Nucleic Acids Res 1989;17:5489–5500.
60. Liberati NT, Noniwa M, Borton AJ, Davie JR, Wang XF. An essential role for Mad homology domain 1 in the association of Smad3 with histone deacetylase activity. J Biol Chem 2001;276: 22,595–22,603.
61. Cohen SB, Nicol R, Stavnezer E. A domain necessary for the transforming activity of SnoN is required for specific DNA binding, transcriptional repression and interaction with TAF(II)110. Oncogene 1998;17:2505–2513.
62. He J, Tegen SB, Krawitz AR, Matin GS, Luo K. The transforming activity of Ski and SnoN is dependent on their ability to repress the activity of Smad proteins. J Biol Chem 2003;278:30,540–30,547.
63. Macias-Silva M, Li W, Liu JI, Crissey MA, Tauh R. Up-regulated transcriptional repressors SnoN and Ski bind Smad proteins to antagonize transforming growth factor-beta signals during liver regeneration. J Biol Chem 2002;277:28,483–28,490.
64. Liu X, Sun Y, Weinberg RA, Lodish HF. Ski/Sno and TGF-beta signaling. Cytokine Growth Factor Rev 2001;12:1–8.
65. Berk M, Desal SY, Heyman HC, Colmanares C. Mice lacking the ski proto-oncogene have defects in neurulation, craniofacial patterning, and skeletal muscle development. Genes Dev 1997;16: 2029–2039.
66. Sutrave P, Kelly AM, Hughes, SH. Ski can cause selective growth of skeletal muscle in transgenic mice. Genes Dev 1990;4:1462–1472.
67. Reed JA, Lin Q, Chen D, Main IS, Medrano EE. SKI pathways inducing progression of human melanoma. Cancer Metastasis Rev 2005;242:265–272.
68. Luo K. Ski and SnoN: negative regulators of TGF-beta signaling. Curr Opin Genet Dev. 2004;14:65–70.
69. Wagner KR, McPherron AC, Winik N, Lee SJ. Loss of myostatin attenuates severity of muscular dystrophy in mdx mice. Ann Neurol 2002;52:832–836.
70. Wagner KR, Liu X, Chang X, Aleen RE. Muscle regeneration in the prolonged absence of myostatin. Proc Natl Acad Sci USA 2005;102:2519–2524.
71. Langley B, Thomas M, Bishop A, Shama M, Gilmour S, Karnbadur R. Myostatin inhibits myoblast differentiation by down-regulating MyoD expression. J Biol Chem 2002;277:49,831–49,840.
72. Thomas M, Langley B, Berry C, Shama M, Karnbadur R. Myostatin, a negative regulator of muscle growth, functions by inhibiting myoblast proliferation. J Biol Chem 2000;275:40,235–40,243.
73. Sharma M, Langley B, Bass J, Karnbadur R. Myostatin in muscle growth and repair. Exerc Sport Sci Rev 2001;29:155–158.
74. Gonzalez-Cadavid NF, Bhasin S. Role of myostatin in metabolism. Curr Opin Clin Nut Metab Care 2004;7:451–457.
75. Zhao B, Wall RJ, Yang J. Transgenic expression of myostatin propeptide prevents diet-induced obesity and insulin resistance. Biochem Biophys Res Commun 2005;337:248–255.
76. Yang J, Ratovitski T, Brady JP, Solomon MB, Wells KD, Wall TJ. Expression of myostatin pro domain results in muscular transgenic mice. Mol Reprod Dev 2001;60:351–361.
77. Hogan BL. Bone morphogenetic proteins in development. Curr Opin Genet Dev 1996;4:432–438.
78. Hoffman A, Gross G. BMP signaling pathways in cartilage and bone formation. Crit Rev Eukaryot Gene Expr 2001;11:23–45. Review.
79. Merlino G, Helmon LJ. Rhabdomyosarcoma—working out the pathways. Oncogene 1999;18: 5340–5348.
80. Dagher R, Helmon LJ. Rhabdomyosarcoma: an overview. Oncologist 1999;4:34–44.

81. Bouche M, Senni MI, Grossi AM, et al. TPA-induced differentiation of human rhabdomyosarcoma cells: expression of the myogenic regulatory factors. Exp Cell Res. 1993;208(1):209–217.
82. Langley B, Thomas M, McFarlane C, Gilmour S, Shama M, Karnbadur R. Myostatin inhibits rhabdomyosarcoma cell proliferation through an Rb-independent pathway. Oncogene 2004;23: 524–534.
83. Ricaud S, Vernus B, Duclos M, et al. Inhibition of autocrine secretion of myostatin enhances terminal differentiation in human rhabdomyosarcoma cells. Oncogene 2003;22:8221–8232.

III TGF-β in Inflammation and Fibrosis

33 Triterpenoids, Vitamin D, and TGF-β/Smad Signaling in Inflammation and Cancer

Nanjoo Suh

CONTENTS

TGF-β/SMAD SIGNALING, INFLAMMATION, AND CANCER
INTERACTIONS BETWEEN THE TGF-β/SMAD SYSTEM AND THE NUCLEAR RECEPTOR SUPERFAMILY
REGULATION OF THE TGF-β/SMAD SYSTEM BY VITAMIN D, DELTANOIDS, AND THE VDR
REGULATION OF THE TGF-β/SMAD SYSTEM BY TRITERPENOIDS
OTHER NUCLEAR RECEPTOR LIGANDS CONTRIBUTING TO THE MODULATION OF TGF-β/SMAD SIGNALING
IMPLICATIONS OF NUCLEAR RECEPTOR LIGANDS FOR MODULATING TGF-β/SMAD SIGNALING IN THE TREATMENT AND PREVENTION OF INFLAMMATION AND CANCER
CONCLUDING REMARKS
ACKNOWLEDGMENT
REFERENCES

Abstract

The transforming growth factor-β (TGF-β) superfamily interacts with the nuclear receptor superfamily in a variety of cell types. The vitamin D receptor (VDR) and the peroxisome proliferator activated receptor-γ (PPAR-γ) are among the members of the nuclear receptor superfamily that after activation by their ligands interact with the TGF-β/Smad system to modulate cellular responses. This chapter discusses the importance of interactions between nuclear receptors and the TGF-β/Smad system in inflammation as well as in cancer. Crosstalk between the TGF-β/Smad system and nuclear receptors occurs at multiple levels and will be discussed later in this chapter when describing the role of nuclear receptors and their ligands such as vitamin D derivatives and triterpenoids on (a) modulation of TGF-β/Smad signaling, (b) interactions of nuclear receptors with Smads, and (c) the consequences of their interactions in diseases such as inflammation and cancer. Understanding crosstalk between the TGF-β/Smad system and the nuclear receptor superfamily may give new insights for the possible use of nuclear receptor ligands for the treatment and prevention of these and other diseases.

Key Words: TGF-β; Smad; cancer; inflammation; nuclear receptors.

From: *Cancer Drug Discovery and Development: Transforming Growth Factor-β in Cancer Therapy, Vol I: Basic and Clinical Biology*
Edited by: Sonia B. Jakowlew © Humana Press Inc., Totowa, NJ

1. TGF-β/SMAD SIGNALING, INFLAMMATION, AND CANCER

The transforming growth factor-β (TGF-β) superfamily, including TGF-βs, activins, and bone morphogenetic proteins (BMPs), are multifunctional cytokines that affect inflammation, the immune response, cell growth, differentiation, apoptosis, and development as well as carcinogenesis *(1,2)*. Upon stimulation by the ligands of the TGF-βs, activins, and BMPs, cells induce the formation of heteromeric complexes of transmembrane serine/threonine kinase type II and type I receptors, which then initiate signaling involving (a) receptor-regulated Smads (Smad2/3 for TGF-β and Smad1/5/8 for BMPs), (b) a co-Smad (Smad4), and (c) inhibitory Smads (Smad6 and Smad7). Smad complexes then translocate to the nucleus, where together with a coactivator or corepressor, regulate transcription of target genes *(2–4)*. Although the Smad intracellular signaling molecules confer TGF-β signaling, biological effects of TGF-β in a variety of cells occur by either a Smad-dependent or Smad-independent pathway *(2)*. Among many potential actions of TGF-β in different disease processes, this chapter will describe two disease processes that are regulated by both the nuclear receptor superfamily and the TGF-β/Smad system, namely carcinogenesis and inflammation.

1.1. TGF-β/Smad Signaling and Cancer

During carcinogenesis, the role of the TGF-β/Smad superfamily is complex. TGF-β is a potent inducer of apoptosis and growth inhibition in many epithelial cell types *(2,5,6)*. Conversely, TGF-β stimulates the growth of certain stroma-fibroblast cell types *(5,7)*. During mammary gland development, TGF-β regulates branching morphogenesis and differentiation by acting on both epithelial and stromal cells. At the end of pregnancy, it is known that TGF-β is a key molecule that induces apoptosis in mammary epithelial cells and stimulates matrix remodeling during involution *(5)*. Maintaining TGF-β function as an inducer of apoptosis in the mammary gland is critical. Interestingly, during the early stages of breast cancer development, the transformed epithelial cells are sensitive to growth arrest by TGF-β, and TGF-β can act as a tumor suppressor *(8,9)*. However, the growth inhibition and induction of apoptosis in epithelial cell types are lost with mutations and loss of gene expression for the formation of molecules of the TGF-β/Smad signaling pathway, resulting in uncontrolled cell proliferation with tumor formation *(5,8,9)*. With the loss of growth inhibition that results from somatic mutations in the TGF-β/Smad pathway, unregulated proliferation of cells and surrounding stromal cells then increase their production of TGF-β with a resultant increase in angiogenesis, immunosuppression, invasion, and metastasis *(7,10)*. Over 50% of pancreatic cancers and 30% of colorectal cancers are associated with mutations in Smad4, and many other cancers have mutations of other Smads in the pathway *(11–15)*. Mutations in the TGF-β type II receptor have also been shown to occur in several types of cancer, including colorectal cancer, hereditary nonpolyposis colon cancer, gastric cancer, and endometrial cancer *(16–18)*.

In addition to the mutations or deletions of Smads, loss of TGF-β type II receptor is also known to contribute to reduced TGF-β-induced cellular differentiation and growth inhibition *(19)*. The loss of expression of type II receptor has also been associated with progressive malignant phenotype in gastric cancer and in T-cell lymphomas *(19,20)*. It is clear that aberrant regulation or mutations in TGF-β/Smad signaling, either in Smads, TGF-βs, or receptors, is critical in carcinogenesis as well as in other disease processes *(9,21–24)*. Even in the absence of mutations or deletions of players in the TGF-β/Smad system, the receptor-regulated Smad of TGF-β (Smad2/3) plays a role as a tumor suppressor at an early stage of breast carcinogenesis, but enhances metastasis at later stages in breast carcinogenesis *(8,9,25)*. Although we understand that Smads, such as Smad2/3, may play different roles during the different stages of carcinogenesis, it is not yet known which regulatory factors or signals trigger the switch that converts Smads from tumor suppressors to metastasis promoters. Advanced and late-stage breast cancers are mostly refractory to TGF-β-mediated growth inhibition and produce large amounts of TGF-β, which may enhance tumor cell invasion and metastasis *(6)*. Therefore, the tumor promoting effects of TGF-β may provide a therapeutic target for late stage breast cancer through TGF-β antagonists, as shown with soluble receptor antagonists or antibodies to neutralize TGF-βs *(26–28)*. In contrast to the effects of TGF-β antagonists on late stage carcinogenesis, any agent that can restore the responsiveness to TGF-β for growth inhibition or maintain the role of Smads as tumor suppressors, either by induction of the TGF-β type II receptor or regulating Smad signaling, may hold promise for the prevention of early carcinogenesis. Future studies are needed to uncover the precise role of TGF-β action in cancer.

1.2. TGF-β/Smad Signaling and Inflammation

Not only is TGF-β a potent growth inhibitor and inducer of apoptosis in cancer, but it is also a very potent inhibitor of the inflammatory process. Inflammation progresses by the action of proinflammatory cytokines, including interleukin (IL)-1, IL-6, tumor necrosis factor (TNF)-α, and interferon (IFN)-γ, and is resolved by antiinflammatory cytokines, such as IL-4, IL-10, IL-13, IFN-α, and TGF-β. Although, the role of TGF-βs is complex during carcinogenesis because TGF-βs have both tumor suppressor and oncogenic characteristics depending on the cellular context and the progression stage *(9,29)*, the suppressive effect of TGF-β during the inflammation process is clear *(30–34)*. TGF-β is also a potent immunosuppressor, and perturbation of TGF-β signaling is linked to autoimmunity as well as inflammation and cancer *(30–33)*. TGF-β has been shown to inhibit the formation of inflammatory cytokines and regulate the inducible nitric oxide synthase (iNOS) system, which can contribute to the antiinflammatory action of TGF-β *(21,23,24,33–35)*. The antiinflammatory role of TGF-β/ Smad molecules was demonstrated using IL-10 gene-deficient mice after colonization with *Enterococcus faecalis (36)*. IL-10 gene-deficient mice, but not wild-type littermates, showed defects in TGF-β/Smad signaling (lack of phosphorylation of Smad2) and failed to inhibit proinflammatory gene expression in intestinal epithelial cells *(36)*, suggesting a critical role for antiinflammatory cytokine TGF-β in maintaining normal epithelial cell homeostasis. Several studies have shown that a TGF-β1 null mutation or Smad3 mutation in mice causes an excessive inflammatory response and multifocal inflammatory disease *(31,33,37,38)*, suggesting that TGF-β and Smad3 have a protective role during inflammatory processes. Furthermore, in inflammatory bowel disease (IBD) with upregulated Smad7 (an inhibitory Smad in TGF-β/Smad signaling), excessive inflammation occurs, which is owing to suppression of the TGF-β response by the inhibitory Smad7 *(24,39)*. These results indicate that it is important to maintain TGF-β function to suppress inflammation and that TGF-β is an antiinflammatory cytokine.

Notably, over the last decade, there has been major interest in studies on the mechanisms of inflammation that relate to carcinogenesis *(40)*. The concept that inflammation and carcinogenesis are related phenomena has been proposed and reviewed for many decades *(30,41–45)*. Several studies on the correlation of inflammation and carcinogenesis have focused on the role of inflammatory enzymes, such as iNOS and inducible cyclooxygenase (COX-2) *(46–48)*, as well as inflammatory cytokines, such as TNFα, ILs, and matrix metalloproteinases (MMPs), which are also known to be regulated by TGF-β/Smad signaling pathways *(43–45,49)*. While the physiological activity of inflammatory enzymes and cytokines (such as iNOS, COX-2, TNFα, MMPs, and ILs) have importance and may provide a definite benefit at normal physiological levels, aberrant overexpression or dysregulation of these inflammatory enzymes and cytokines has been implicated in the pathogenesis of cancer *(46–48,50)*. In addition, there has been a particularly strong interest in the possibility that overexpression of COX-2 or iNOS or its upstream transcription factors, such as NF-κB or AP-1, might be directly involved in the causation of cancers including those in the lung, breast, and prostate *(51,52)*. Therefore, understanding the role of TGF-β in regulating inflammatory enzymes and cytokines to suppress inflammation is critical not only in inflammation but also in cancer. Because chronic inflammation is now well linked to carcinogenesis by many different mechanisms of action, such as controlling inflammatory enzymes, cytokines, growth factors, NF-κB, and infections *(43,53,54)*, novel agents that regulate TGF-β/Smad signaling and restore TGF-β responsiveness in tumor tissues may be of great importance to the fields of inflammation as well as carcinogenesis *(55,56)*. Nuclear receptor ligands have been known to modulate the TGF-β superfamily, and it will be important to understand whether nuclear receptor ligands modulate the TGF-β/Smad system by interacting with their receptors, regulating intracellular Smad signaling, or interacting with coregulators of the transcription machinery complex in the nucleus to modulate transcriptional activation of TGF-β target genes. An understanding of the detailed mechanisms of interactions between the TGF-β/ Smad system and the nuclear receptor superfamily will be important for the practical use of nuclear receptor ligands for preventing or treating cancer and other inflammation related diseases.

2. INTERACTIONS BETWEEN THE TGF-β/SMAD SYSTEM AND THE NUCLEAR RECEPTOR SUPERFAMILY

The nuclear receptor superfamily consists of transcription factors that bind to responsive sequences in DNA and regulate the transcription of target genes *(57,58)*. There are at least 45 nuclear receptors identified so far, and these nuclear receptors can be categorized into three groups depending on their ligands. They are (a) classical endocrine receptors (their ligands are steroid hormones), (b) orphan receptors (nuclear receptors without known ligands) or (c) adopted orphan receptors (their ligands are now identified, mostly xenobiotics and lipids) *(57,58)*. The nuclear receptors contain several functional domains, such as activation function domains (for transactivation and protein–protein interaction), DNA binding domain (for DNA binding and dimerization), and ligand binding domain (for ligand binding, corepressor–coactivator interaction, and nuclear localization). The activation of gene transcription requires multiple steps, including ligand binding to the receptor, conformation changes of the receptor, receptor binding to a specific sequence on DNA, release of corepressors, and recruitment of coactivators to the transcriptional machinery. In turn, the activation of nuclear receptors leads to regulation of gene expression *(57–60)*.

Interactions between the nuclear receptor family and the TGF-β/Smad superfamily have been investigated for many years *(61–64)*, and so far multiple levels of interactions have been shown between steroid receptors and the TGF-β/Smad superfamily (Table 1). Steroids, such as 1α,25-dihydroxyvitamin D_3, retinoids, or estrogen receptor (ER) ligands, have been shown to enhance the TGF-β system by inducing the synthesis of TGF-β ligands or its receptors in leukemia and breast cancer cells *(62–68)*. Furthermore, there is now an abundant literature on the interaction of steroids and their receptors with Smads and other intracellular signaling molecules *(69–75)*. A simple generalized diagram implicating the possible interaction and modulation of the TGF-β system by nuclear receptors and their ligands is shown in Figure 1. In studies on the interaction between nuclear receptors and Smads, it was reported that 1α,25$(OH)_2D_3$, in combination with TGF-β, activated Smad signaling pathways by an interaction with intracellular effectors of TGF-β, the Smad3/4 complex, and the vitamin D receptor (VDR) in the nucleus, and potentiated VDR-dependent transcription *(69,76)*. This indicates that Smad3 is an

Table 1
Regulation of the TGF-β/Smad System by Nuclear Receptors and Their Ligands

Nuclear receptor	*Nuclear receptor ligands*	*Effects on TGF-β ligands or receptor synthesis*	*Smad interaction (interacting domain)*
Vitamin D receptor (VDR)	1α,25-$(OH)_2D_3$ EB1089 Calcipotriol	Increased TGF-β1, 2 or 3 ligands *(63,67,99,100,101)* Increased TGF-β receptor I, II *(64,68,102)*	Synergistic (MH1 domain of Smad3) *(69,77,95)*
Retinoic acid receptor (RAR)	All-trans-retinoic acid	Increased TGF-β or BMP ligands *(100,101,163)* Increased TGF-β receptor II *(62)*	RARγ acts as a co-activator for Smad3/4 (MH2 domain of Smad3) *(145)*
Glucocorticoid receptor (GR)	Dexamethasone, Budesonide, Hydrocortisone	Decreased TGF-β1 ligands *(146)*	Smad3/4 Antagonistic (MH2 domain of Smad3/4) *(70)*
Peroxisome proliferator activated receptor (PPAR)-γ	Triterpenoids, Rosiglitazone, Pioglitazone, GW7845, 15-Deoxyprosta-glandin J2	Increased TGF-β1 ligands *(81)* Increased TGF-β receptor II *(78)*	Smad3 Antagonistic *(73,79)*
Estrogen receptor (ER)	Estradiol, Tamoxifen, Raloxifene, Arzoxifene	Increased TGF-β1, 3 ligands *(66,142,164)*	Smad3 Antagonistic (MH2 domain of Smad3) *(71)* Smad4/ERα interaction (MH1 domain of Smad4) *(139,165,166)*
Androgen receptor (AR)	Testosterone, Dihydrotesto-sterone	Decreased TGF-β1, 2 or 3 ligands *(167)* Increased TGF-β1 ligands *(168)* Decreased TGF-β receptor I, II *(169)*	Antagonistic (MH2 domain of Smad3) *(72,74,170)*

important mediator of the crosstalk between vitamin D and TGF-β signaling pathways. Interestingly, among many nuclear receptors, the VDR has been shown to synergize with Smad3 by interaction with the MH-1 domain of Smad3 and to further activate vitamin D-induced transcription *(69,76,77)*, while most other nuclear receptors investigated have been shown to interact with the MH-2 domain of Smad3 and result in an antagonistic action *(70–74)*.

Additional crosstalk between the nuclear receptor superfamily and the TGF-β/Smad system has been shown with novel triterpenoids, which are ligands for the nuclear receptor peroxisome proliferator-activated receptor-γ (PPAR-γ). Triterpenoids are known to induce TGF-β receptor II and to further enhance the TGF-β response to exhibit antiinflammatory activities in macrophages *(78–80)*. It was also demonstrated that the induction of the ligand TGF-β1 by a triterpenoid, 2-cyano-3,12-dioxooleana-1,9-dien-28-oic acid (CDDO), in the gut was responsible for the protection by CDDO in the *Toxoplasma gondii*-induced model of inflammation in mice *(81)*. Not only have VDR ligands and triterpenoids been shown to regulate the TGF-β/Smad system, but other classes of nuclear receptor ligands, such as

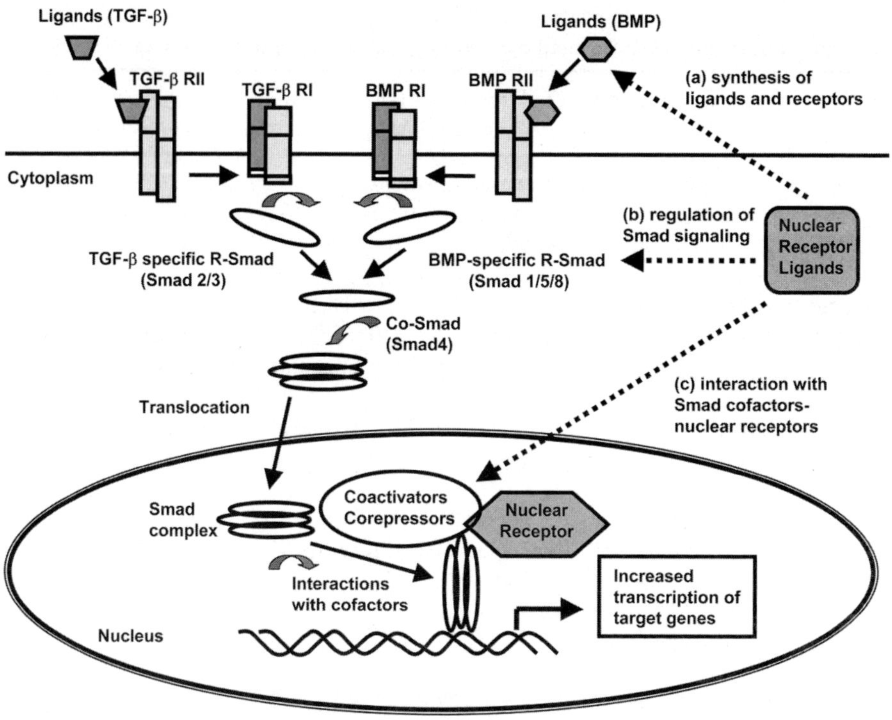

Fig. 1. Crosstalk between the TGF-β/Smad signaling and nuclear receptors. Ligands for nuclear receptors possibly modulate the TGF-β/Smad system at multiple sites, as shown in the figure, (**A**) synthesis of ligands and receptors (**B**) regulation of Smad signaling, and (**C**) interaction with Smad cofactors-nuclear receptors.

selective estrogen receptor modulators (SERMs) and RXR ligands (rexinoids), also enhance the TGF-β/Smad system. It was shown that nuclear receptor ligands of two classes, a SERM (arzoxifene) and a rexinoid (LG100268), prevent experimental breast cancer in animals *(65)*. The ability of the two nuclear receptor ligands to induce apoptosis and growth suppression in breast cancer in vivo and in vitro is the combined result of induction of TGF-β by arzoxifene, together with the inhibition of the prosurvival nuclear factor-κB (NF-κB) and phosphatidylinositol 3′ kinase (PI3K) signaling pathways by LG100268 *(65,66)*. In the next section, the discussion will focus on the interaction between the TGF-β/Smad system and two nuclear receptor superfamily members, VDR and PPAR-γ, and the role of their ligands in diseases such as cancer and inflammation.

3. REGULATION OF THE TGF-β/SMAD SYSTEM BY VITAMIN D, DELTANOIDS, AND THE VDR

Nuclear receptors control the expression of a large number of genes relevant to the process of carcinogenesis *(58)*, and members of the nuclear receptor superfamily have been extensively studied as excellent examples of molecular targets in cancer. This nuclear receptor family includes the ERs, the retinoic acid receptors (RARs), the retinoid X receptors (RXRs), the VDR, and PPAR-γ, and ligands for these receptors have been used to prevent or treat experimental and clinical cancer *(51,65,66,81,83–90)*. Among many ligands for nuclear receptors, 1α,25$(OH)_2D_3$ and its synthetic analogs (also known as deltanoids) have been extensively studied and developed for the treatment and prevention of cancer, inflammation, and proliferative skin diseases *(91,92)*. The hormonally active form of vitamin D_3, 1α, $25(OH)_2D_3$, functions in the maintenance of calcium homeostasis through regulation of genes in the intestine, kidney and bone, and it also controls immune cells and hormone secretion as well as inflammatory responses *(93,94)*. 1α,$25(OH)_2D_3$ and deltanoids are potent regulators of growth and differentiation of various cell types *(91,92)*. Most of the effects of 1α,$25(OH)_2D_3$ and synthetic deltanoids are mediated through the VDR-regulated transcription of genes that are involved in cell proliferation, apoptosis, differentiation, angiogenesis, and metastasis *(91,92)*. Further, 1α,$25(OH)_2D_3$ and synthetic deltanoids have been shown to induce apoptosis and cell cycle arrest in the G1 phase by increasing the level of cyclin-dependent kinase inhibitors, such as p21 and p27, in breast cancer cells *(92,94)*. Structures of the active metabolite of vitamin D_3, 1α,$25(OH)_2D_3$ and some synthetic deltanoids are shown in Figure 2A.

It has been demonstrated in recent studies that 1α,$25(OH)_2D_3$ and many synthetic deltanoids enhance the secretion of TGF-β ligands and interact with the TGF-β/Smad system in various target cells *(64,69,95,96)*. Moreover, identification of the vitamin D response element (VDRE) in the TGF-β2 gene promoter has been reported *(67)*. 1α,$25(OH)_2D_3$ and deltanoids are potent regulators of the TGF-β and plasminogen activator systems in cells of epithelial and mesenchymal origin, and this is known to be owing to the increased formation of active TGF-β *(97)*. In another study, 1α,$25(OH)_2D_3$ caused a dose-dependent induction of latent and active TGF-β1 proteins in keratinocytes. TGF-β has been suggested to be an effective inhibitor for abnormal keratinocyte growth in psoriasis *(98)*, indicating that antiproliferative and antiinflammatory effects of 1α,$25(OH)_2D_3$ on psoriatic lesional skin may be mediated by a complex TGF-β regulation in local dermal fibroblasts. It is evident that 1α,$25(OH)_2D_3$ and other related analogs induce the synthesis of TGF-β ligands and TGF-β type I or II receptors at both the RNA and protein levels in leukemia cells, prostate cancer cells as well as in breast cancer cells *(63,64, 68, 99–102)*.

Importantly, two-way crosstalk between nuclear receptors and the TGF-β/Smad system has been postulated *(103)*. Not only does 1α,$25(OH)_2D_3$ and synthetic deltanoids regulate the synthesis of TGF-β ligands and receptors, but TGF-β1 is also known to induce VDR expression and inhibit cell proliferation synergistically when combined with a vitamin D analog, EB1089, by increasing the CDK inhibitor, p27 protein *(68)*. Furthermore, many other factors, such as parathyroid hormone, epidermal growth factor, TGF-β, estrogen, and glucocorticoids, have been shown to induce VDR in many different cell types *(104)*. Although VDR, IGF-1, ER-α, and TGF-β are among the known bone genetic factors *(105)*, TGF-β is one of the most abundant growth factors secreted by bone cells, and the regulation of TGF-β expression or TGF-β/Smad signaling is crucial for bone development and growth *(67,106)*. Therefore, it is natural to postulate that there must be close interaction between TGF-β and vitamin D (its role for calcium/phosphate homeostasis and bone metabolism is well-known), in particular in bone cells. In a study by Borton et al., it is demonstrated that Smad3 null mice have decreased bone density with a lower rate of bone formation, a reduced rate of mineral apposition, and decreased osteoid width, indicating aberrant osteoblast-mediated bone formation *(106)*. Because mice with targeted deletion of Smad3 were shown to be osteopenic with less cortical and cancellous bone compared with wild type littermates, it is evident that the TGF-β/Smad3 pathway is critical for maintaining bone density as well as bone formation *(106)*.

Both TGF-β and 1α,$25(OH)_2D_3$ are effective in regulating proliferation, differentiation, bone matrix maturation and cell-specific gene expression in bone cells. Synergy between vitamin D ligands and TGF-β1 was shown on the osteocalcin promoter and it was reported

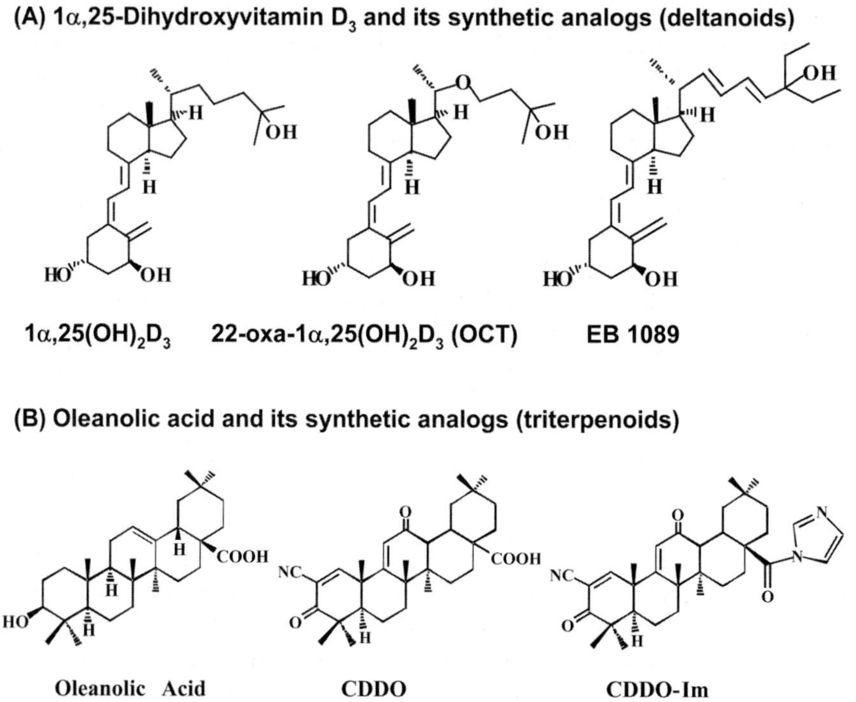

Fig. 2. (**A**) Structures of the naturally active metabolite of vitamin D, 1α,25-dihydroxyvitamin D_3 (1α,25$(OH)_2D_3$) and its synthetic analogs, 22-oxa-1α,25$(OH)_2D_3$ (OCT, 22-oxacholecalciferol) and EB1089 (1α,25-dihydroxy-22,24-diene-24,26,27-trishomovitamin D_3). (**B**) Structures of oleanolic acid (OA), and its synthetic analogs, CDDO (2-cyano-3,12-dioxooleana-1,9,dien-28-oic acid) and CDDO-Imidazolide (CDDO-Im).

that the synergy required both the VDRE and the Smad binding element (SBE) to be located in close proximity to the target promoter *(77)*. It was also shown that TGF-β induced VDR mRNA and protein in bone cells, and that TGF-β inhibited the response of 1α,25$(OH)_2D_3$ on osteopontin and osteocalcin production, suggesting that TGF-β plays an important role in bone metabolism *(107)*. Moreover, TGF-β stimulates osteoblast proliferation and growth, but the expression of the markers characteristic of the osteoblast phenotype, such as osteocalcin, is inhibited *(108)*. In contrast to the effects of TGF-β, 1α,25$(OH)_2D_3$ causes inhibition of the proliferation of osteoblasts, arrests their growth, and stimulates expression of specific markers. Complex interactions have been demonstrated between TGF-β and 1α,25$(OH)_2D_3$ that modulate the expression of the receptors, and effects on osteoblast-specific gene expression *(108)*. Therefore, the cooperative actions of TGF-β and 1α,25$(OH)_2D_3$ can be either synergistic or antagonistic in bone cells. TGF-β has been shown to induce CYP24 (a major deactivating enzyme of 1α,25$(OH)_2D_3$) in MG63 human osteoblast-like cells such that TGF-β can induce metabolism and downregulate vitamin D_3 action, suggesting a regulatory feedback for the action of vitamin D_3 in target tissues *(104,107)*.

It was recently established that Smad proteins that transduce signals downstream from TGF-β stimulation may mediate crosstalk between TGF-β and 1α,25$(OH)_2D_3$ signaling *(108)*. Yanagisawa et al. demonstrated that the VDR synergizes with Smad3 through interaction with the MH-1 domain of Smad3 and further activates vitamin D-induced transcription *(69,76)*. 1α,25$(OH)_2D_3$, in combination with TGF-β, activated Smad signaling pathways by an interaction with the Smad3/4 complex and the VDR in the nucleus. This interaction resulted in increased VDR-dependent transcription. Among intracellular effector Smads, Smad3 was identified to mediate crosstalk between the vitamin D and TGF-β signaling pathways *(69,76,77,109)*. However, most other nuclear receptors investigated, such as androgen receptor, PPAR-γ, ER-α, and glucocorticoid receptor (GR), have been shown to interact with the MH-2 domain of Smad3 and to result in an antagonistic action *(70–74)* (Table 1). In addition to the cooperative interaction between Smad3 and the VDR in the nucleus, a recent study of crosstalk between the TGF-β/Smad system and the nuclear receptor superfamily revealed important interactions with transcriptional corepressors in the nucleus *(110)*. Ski, one of the well-known TGF-β coregulators, has been proposed to recruit the corepressor N-CoR to either the TGF-β-regulated Smad transcription factors or nuclear receptors. It was shown that Ski negatively regulates vitamin D-mediated transcription by directly interacting with VDR, and the Ski/N-CoR interaction is essential for repression of vitamin D signaling, but not TGF-β signaling *(110)*. Further studies on the interaction of TGF-β specific coactivators or corepressors with the VDR should be investigated to determine the role of coregulators (both corepressors and coactivators of TGF-β) in VDR-mediated action in cancer and inflammation.

Importantly, VDR ligands, 1α,25$(OH)_2D_3$ and 1α-(OH)-D_5, have been shown to prevent breast cancer, and it may be owing to the induction of TGF-β1 in mammary tissues *(111)*. In addition, Yang et al. showed that VDR ligands (EB1089) required TGF-β sensitivity in breast cancer to inhibit cell proliferation, as shown in the primary breast cell line 184A1 as well as in MCF-7 early- and late-passage cells *(96)*. In this report, MCF-7 late passage cells lacking TGF-β response lost the sensitivity to VDR ligands, and further, TGF-β type II receptor restored responsiveness to VDR ligand, EB1089. Another study showed that 1α,25$(OH)_2D_3$ together with TGF-β synergistically increased AP-1 activity, and this was a VDR-dependent genomic action, and supports the novel positive regulation by vitamin D action on TGF-β-induced AP-1 activity *(112)*. In summary, the regulation of TGF-β/Smad signaling by nuclear receptors (and their ligands) are critical for the use of these ligands in the treatment and prevention of cancer. It will be important to have a better mechanistic understanding of these effects.

The antiinflammatory role of vitamin D and deltanoids has been investigated extensively. Vitamin D deficiency has been shown to be associated with increased autoimmune diseases including IBD and experimental autoimmune encephylomyelitis *(93,113)*. 1α,25$(OH)_2D_3$ suppressed the IBD via a VDR-dependent mechanism in appropriate animal models, suggesting the antiinflammatory role of VDR in IBD. More severe colitis was observed in VDR knockout (KO) mice than in wild-type mice when infected with *Schistosoma mansoni* to induce colitis *(93)*. In the same study, another model of IBD used was the spontaneous colitis that develops in IL-10 KO mice. It was shown that VDR/IL-10 double KO mice developed even more accelerated IBD than either IL-10 or VDR single KO mice. The severe IBD in VDR/IL-10 double KO mice resulted from a defect in the immune system, not a result of altered calcium homeostasis or gastrointestinal tract function, suggesting an essential role for VDR signaling in the regulation of immune function and inflammation *(93)*. So far, it is not known whether the antiinflammatory action of the VDR (and its ligands) in this disease is through the antiinflammatory action of TGF-β. Because it has been shown that IL-10 deficient mice lack TGF-β/Smad signaling *(36)*, it is possible that VDR signaling requires active TGF-β/Smad system to suppress inflammatory process. Interestingly, constitutively upregulated Smad7 (an inhibitory Smad in TGF-β/Smad signaling) has been shown in human IBD *(39)*. Because, it is important to maintain TGF-β function to suppress inflammation in IBD *(24,39)*, we can predict that the action of VDR and its ligands in inflammatory diseases may act through a TGF-β dependent mechanism. The possible overlap between the VDR and the TGF-β/Smad system in cancer and inflammation needs to be further investigated to determine whether the antiinflammatory and anti-cancer actions of VDR ligands are TGF-β/ Smad-dependent.

4. REGULATION OF THE TGF-β/SMAD SYSTEM BY TRITERPENOIDS

Triterpenoids are a large family of substances synthesized in plants by the cyclization of squalene, a precursor of steroids, but triterpenoids are different from steroids because triterpenoids retain all 30-carbon atoms. Structures of a naturally occurring triterpenoid (oleanolic acid) and its synthetic analogs CDDO and CDDO-imidazolide (CDDO-Im) are shown in Figure 2B. Some natural triterpenoids, such as oleanolic acid, ursolic acid, asiatic acid, madecassic acid, and boswellic acid, have been shown to have wound healing, antiinflammatory, or anticarcinogenic activities *(114–120)*. Naturally occurring triterpenoids are relatively weak agents so that it would be desirable to develop more potent agents for the practical treatment of inflammation and cancer. To improve their antiinflammatory and anticancer activities, many derivatives of triterpenoids have been synthesized and assayed for their ability to suppress the expression of inflammatory genes such as COX-2 and iNOS *(80,121,122)*. Many synthetic analogs of triterpenoids were shown to be significantly more potent than the parent compounds, either ursolic or oleanolic acid *(80,121,123)*. CDDO and its analogs, CDDO-imidazolide or CDDO-methyl ester, have shown exceptional activity in suppressing the ability of several inflammatory cytokines from inducing the *de novo* formation of iNOS and COX-2 as well as in blocking the growth of many human cancer cells *(78,80,121–123)*. These antiinflammatory effects of triterpenoids have been seen in primary macrophages, in a macrophage-like tumor cell-line, and in nonneoplastic colon fibroblasts *(80,121)*. A synthetic oleanane triterpenoid, CDDO, has also been shown to induce adipogenic differentiation in 3T3 L1 preadipocytes, and this is associated with its binding to PPAR-γ, a nuclear receptor, establishing the triterpenoid CDDO as a member of a new class of PPAR-γ ligands *(124)*.

Interestingly, a classical overlap between triterpenoids and TGF-β has been shown with their suppression of the inflammatory mediator, iNOS *(35,80,121,125,126)*. It has been reported that TGF-β (at picomolar concentrations) is a potent inhibitor of iNOS action at multiple levels, such as by decreasing mRNA stability, decreasing protein translation, and inhibiting enzyme activity, and that TGF-β plays a role as an antiinflammatory cytokine *(35)*. The triterpenoids, CDDO and CDDO-Im, have been shown to inhibit the *de novo* synthesis of iNOS protein and iNOS mRNA, and this inhibition was shown at nanomolar concentrations of either CDDO or CDDO-Im *(80,121,127)*. Moreover, CDDO induces cell differentiation in human promonocytic leukemia cell lines, as measured by increased staining for α-naphthyl acetate esterase, a marker for monocyte/macrophage differentiation. CDDO was shown to synergize with TGF-β1 in this assay, suggesting important interactions between triterpenoids and the TGF-β system *(80,121)*. TGF-β, a multifunctional cytokine, regulates cellular proliferation and differentiation in many cell types, but genetic mutations of the TGF-β type II receptor are rarely found in leukemia cells. However, low levels of expression of this receptor are a frequent finding in leukemia and other cancers, and it is suggested that this is owing to transcriptional repression of receptor expression *(19,64)*. Because, TGF-β responsiveness in cancer cells and leukemia cells may be restored by the induction of the TGF-β type II receptor *(56,128)*, many nuclear receptor ligands have been tested to determine whether they induce TGF-β type II receptor and to restore the differentiating effects of TGF-β. It was shown that CDDO induces the TGF-β type II receptor in THP-1 and U937 promonocytes, which may explain why CDDO synergizes with TGF-β to induce monocytic differentiation *(80)*.

The interaction between triterpenoids and the TGF-β/Smad system has been shown at multiple levels, including the effects on Smad2/3 signaling as well as on the inhibitory Smad7 *(78,79)*. In a study of triterpenoids on Smad signaling in U937 leukemia cells *(78)*, it was shown that treatment with TGF-β activated its signaling molecule, Smad2, within 20 min (phosphorylated Smad2 Ser465/Ser467), but phosphorylation of Smad2 was diminished by 2 h because TGF-β induced its own inhibitor, Smad7, in a negative feedback loop. CDDO alone did not affect the phosphorylation of Smad2 but CDDO sustained the activation of Smad2 that was induced by TGF-β at 2 h, suggesting that CDDO may prevent deactivation of phosphorylated Smad2 and maintain active Smad signaling. Functional effects of triterpenoids on transactivation of Smad-dependent transcription, using a SBE linked to a luciferase (SBE-Luc, [CAGA]12-Luc) specific for response to TGF-β and activins, were also studied *(78)*. The triterpenoid CDDO-Im had no significant effect on transcription in the absence of Smads. Only when cells in which SBE-Luc activity had been stimulated, either by adding exogenous TGF-β1 or by cotransfection with ALK5 (constitutively activated TGF-β type I receptor), was a remarkable stimulatory effect of CDDO-Im shown. When an inhibitory Smad, Smad7, was overexpressed to inhibit the TGF-β response, CDDO-Im reversed the inhibitory effects of Smad7 and restored the TGF-β response *(78)*. In summary, these findings suggest that CDDO not only upregulates a receptor for TGF-β, but also causes sustained activation of TGF-β/Smad signaling possibly by modulating the activity of the Smad7-Smurf-receptor complex.

The ubiquitin-mediated proteasomal degradation pathway tightly regulates TGF-β superfamily signaling, and the Smad/receptor complex is degraded by E3 ubiquitin ligases such as the Smad ubiquitination-related factor 1/2 (Smurf1/2) *(129,130)*. Because, Smurf1/2 interacts with the TGF-β type I receptor through Smad7 and induces receptor degradation *(131)*, it will be interesting to determine the effects of CDDO or CDDO-Im on ubiquitination pathways. Further, Smurf1/2 is involved in the degradation of Smad/SnoN complex when stimulated with TGF-β *(129–134)*. The Smurfs and related ubiquitin ligases are of particular importance to TGF-β signaling, as Smads also function as adapters to recruit Smurfs to the TGF-β receptor complex and to the transcriptional repressor, SnoN, and thereby regulate the degradation of these Smad-associating proteins *(129,135)*. Thus, by controlling the level of Smads as well as positive and negative regulators of the pathway, Smurfs and other related ubiquitin ligases may control TGF-β signaling, and triterpenoids may interact on this side of the TGF-β pathway.

TGF-β is known to enhance the degradation of its repressor protein SnoN and to increase TGF-β dependent transcription. Interestingly, the triterpenoid, CDDO, also enhances degradation of SnoN and further synergizes with TGF-β to induce the degradation of SnoN (N. Suh, unpublished). The enhanced degradation of SnoN by TGF-β and triterpenoids can be blocked by the proteasome inhibitor, MG132, indicating that the effects of triterpenoids to enhance TGF-β/Smad signaling may result from enhanced degradation of key repressor molecules in TGF-β/Smad signaling (N. Suh, unpublished). The activation of TGF-β/Smad signaling, such as phosphorylation of Smads, nuclear localization of Smads, and increased transcription in the nucleus, are critical processes for regulating the signaling pathway, but ubiquitination and proteasomal degradation of the Smad signaling molecules are also important steps for the regulation of the TGF-β/Smad system as well. Ubiquitination and proteasomal degradation are implicated in the turnover of tumor-derived Smad mutants and may thus play a role in disease progression, especially in cancer as well as in inflammation *(130)*. Therefore, the detailed mechanism of action of triterpenoids in the regulation of TGF-β/Smad signaling, particularly for the degradation of SnoN, needs to be further investigated. Other non-Smad signaling proteins, such as FLIP or cyclin D1, are also regulated by triterpenoids, and these are proteins that are regulated by ubiquitination and degradation *(136,137)*. Therefore, it is possible that the effects of triterpenoids on ubiquitination and degradation might not be specific to TGF-β/Smad ubiquitin ligases, but rather represent more universal effects on ubiquitin ligases and proteasomes for a wide variety of proteins.

Significant impact of crosstalk between triterpenoids and the TGF-β/Smad system has been shown as Smad interaction with inflammatory cytokines in disease models, such as arthritis and IBDs *(79,81,138)*. Minns et al. showed that triterpenoids prevented ileitis development through the global downregulation of inflammatory cytokines and chemokines *(81)*. More importantly, total TGF-β1 production and Smad2 expression were increased by triterpenoids in intraepithelial lymphocytes. Blocking TGF-β reversed triterpenoid-induced protection and prevented the up-regulation of Smad2 in the small intestine, suggesting that triterpenoids are a novel antiinflammatory class capable of preventing ileitis by activating the TGF-β signaling pathway in a pathogen-driven ileitis model *(81)*. Further, it has been demonstrated that a triterpenoid CDDO requires Smads for the repression of MMP-1. Specifically, MMP-1 is inhibited neither by CDDO in the absence of TGF-β receptor-activated Smad3 nor when a negative regulator, Smad7, attenuates TGF-β/Smad signaling *(79)*, suggesting that antiinflammatory action of triterpenoids may be through a TGF-β dependent mechanism.

Angiogenesis is another pathological process where triterpenoids may act through a TGF-β dependent pathway. Angiogenesis is a strictly controlled process under normal conditions. It is highly regulated by a variety of endogenous angiogenic and angiostatic factors. The process can be turned on during wound healing and pathological processes, such as cancer, chronic inflammation, or atherosclerosis. Angiogenesis inhibitors are able to interfere with various steps of angiogenesis, as determined by basement destruction of blood vessels, proliferation and migration of endothelial cells, or lumen formation. Triterpenoids, such as asiatic acid and madecassic acid, are among the known angiogenesis inhibitor compounds derived from natural sources *(117,118)*. Because, angiogenesis and wound healing processes are closely related to the effects of TGF-β in fibroblasts and endothelial cells, it may be postulated that triterpenoids induce wound healing through TGF-β dependent mechanisms. There are only limited studies on whether growth inhibition and apoptosis effects of triterpenoids may act via TGF-β dependent pathways. The detailed molecular mechanisms of action of triterpenoids in cancer and inflammation should be further investigated.

5. OTHER NUCLEAR RECEPTOR LIGANDS CONTRIBUTING TO THE MODULATION OF TGF-β/SMAD SIGNALING

5.1. Estrogen Receptors

Several other nuclear receptors, such as ERs, the GR, PPAR-γ, and RXRs, also have intimate interaction with the TGF-β/Smad system in many biological systems. For example, the influence of ER-α/Smad4 signaling crosstalk on pituitary tumor prolactinoma pathogenesis is a well-known interaction between estrogens and the TGF-β pathway *(139)*. Physical and functional interactions between Smad4 and ERs take place in prolactinoma cells, providing a molecular explanation to link the tumorigenic action of these two important players for prolactinoma pathogenesis *(140)*. TGF-β is known to be induced by antiestrogens in MCF-7 cells, and a TGF-β and antiestrogen-sensitive reporter gene assay has been used to show that p38 activation precedes TGF-β activation. Antiestrogen-induced transcription of TGF-β2 and TGF-β type II receptor has also been shown *(140)*. It was reported that antiestrogens induce phosphorylation of Smad2 through p38, suggesting an antiestrogen signal transduction pathway involving sequential activation of p38 and TGF-β pathways to mediate growth inhibition *(141)*. In human breast, ER-α cells rarely colocalize with markers of proliferation, but their increased frequency correlates with increased breast cancer risk. Nuclear Smad colocalized with nuclear ER-α, and colocalization of ER-α with markers of proliferation, such as Ki-67 or bromodeoxyuridine labeling index, at estrus has been shown to be significantly increased in the mammary glands of Tgf β1 C57/bl/129SV genetically engineered heterozygous mice *(142)*. It has been shown that the activation of TGF-β1 can functionally restrain ER-α-positive cells from proliferating in the adult mammary gland *(142)*, suggesting that TGF-β1 dysregulation may promote proliferation of ER-α-positive cells associated with breast cancer risk in humans.

In addition to the TGF-β/Smad system, BMPs, another member of the TGF-β superfamily, are known to play a central role in differentiation, development, and physiological tissue remodeling. Estrogens have been shown to suppress BMP function by repressing BMP gene expression. BMP-2-induced activation of Smad activity and BMP-2-mediated gene expression were suppressed by 17β-estradiol in breast cancer cells, and 17β-estradiol-mediated inhibition of Smad activation was reversed by tamoxifen, an ER antagonist *(143)*. Interestingly, in the same study, it was shown that the inhibitory action of ER-α on Smad activity was owing to direct physical interactions between the Smads and ER-α, which represents a novel mechanism for the crosstalk between BMP and ER-α signaling pathways *(143)*. A direct physical interaction between ER-α and Smad1/4 has also been shown in association with a specific BMP-4-induced crosstalk mechanism that stimulates cell proliferation *(139)*. While TGF-β has been shown to induce a physical interaction between ER-α and Smad3 *(71)*, BMP-4 has been shown to induce an interaction between ER-α and Smad1/4 *(139)*. The rapid induction of c-*myc* by BMP-4 and estrogens suggest that the c-*myc* promoter might be a direct target of these ER–Smad complexes *(139)*, suggesting that an interaction between ER and the TGF-β/Smad system may contribute to the significant effects in several tumor types and in bone physiology.

5.2. Retinoic Acid Receptors and Retinoid X Receptors

Two other classes of nuclear receptors, the RARs and the RXRs (the heterodimer partner of RARs and many other receptors), have been shown to be involved in mediating transcriptional responses to their ligands, such as retinoic acids or rexinoids. Retinoic acids are important regulators of cell growth and differentiation in many different cell types acting through RARs and RXRs. RAR ligands, such as all-trans-retinoic acid, have been known to induce TGF-β ligands and its receptors in many cell types *(62,100–102,144)*. Furthermore, it was reported that overexpression of RARγ enhanced Smad3/4 transactivation, as shown by a direct interaction between RARγ and the MH2 domain of Smad3 *(145)*. However, this RARγ-enhanced Smad3/Smad4-dependent transcription was inhibited by natural and synthetic RAR agonists, but potentiated by synthetic RAR antagonists *(145)*, suggesting that RARs may function as coactivators of the Smad pathway in the absence of RAR agonists or in the presence of their antagonists. More detailed studies are needed to understand the complex role of RARs and their ligands on TGF-β/Smad signaling.

5.3. Glucocorticoid Receptor

Crosstalk between the TGF-β/Smad system and GR has also been reported *(70,146)*. Glucocorticoids can regulate diverse biological processes through modulation of the expression of target genes of the TGF-β/Smad system. GR is known to repress TGF-β transcriptional activation of the type-1 plasminogen activator inhibitor gene in a ligand-dependent manner *(70)*. Similarly, GR represses TGF-β activation of the TGF-β responsive sequence containing Smad3/4-binding sites, and GR inhibits transcriptional activation by both Smad3 and Smad4 C-terminal activation domains. In addition, it has been shown that GR interacts with Smad3 both in vitro and in vivo *(70)*, suggesting a molecular basis for the crossregulation between glucocorticoid and TGF-β signaling pathways. Interestingly, TGF-β1 and glucocorticoids are known to play an important role in deactivating monocytes/

macrophages and both work as potent antiinflammatory agents. The effects of TGF-β1 may be partially achieved through modulation of the sensitivity of these cells to glucocorticoids, and TGF-β1 increases glucocorticoid binding and signaling in inflammatory cells through a Smad 2/3- and AP-1-mediated process *(147)*, suggesting a strong interaction between GR and the TGF-β/Smad system. More detailed studies are needed to determine either synergistic or antagonistic effects of GR (and its ligands) on TGF-β/Smad signaling in cancer and inflammation.

5.4. Peroxisome Proliferator-Activated Receptor-g

PPAR-γ was originally discovered as an orphan receptor, as was the case for RXRs and liver X receptors. The known ligands for PPAR-γ are 15-deoxyprostaglandin J_2, certain thiazolidinediones (TZDs) (rosiglitazone and pioglitazone), a tyrosine analog GW7845, and triterpenoids, such as CDDO *(82,124,148)*. Many widely used antidiabetic TZD drugs are ligands for PPAR-γ, although the chemical structure of the endogenous ligand for this receptor remains uncertain *(148)*. PPAR-γ ligands, such as TZDs, have been shown to exhibit antiatherosclerotic effects in type 2 diabetes, owing to an induction of vascular smooth muscle cell apoptosis by PPAR-γ activation, rapid induction of TGF-β1, and selective phospho-Smad2 nuclear recruitment *(149)*. Understanding the role of Smad signaling on vascular smooth muscle cell might provide future clinical benefits by PPAR-γ ligands in vascular diseases *(149,150)*. In another study, it was shown that the a PPAR-γ natural ligand (15-deoxyprostaglandin J_2) and a synthetic ligand (GW7845) significantly inhibit TGF-β-induced connective tissue growth factor (CTGF), a key factor regulating extracellular matrix, production in human aortic smooth muscle cells *(73)*. It was also shown that PPAR-γ activation inhibited TGF-β-induced CTGF promoter activity, and suppression of CTGF promoter activity by PPAR-γ activation is completely rescued by overexpression of Smad3, but not by Smad4. This study clearly demonstrates that PPAR-γ physically interacts with Smad3 but not Smad4 in vitro in human aortic smooth muscle cells, and these data imply that abrogation of TGF-β-induced CTGF production by PPAR-γ activation may be one of the mechanisms through which PPAR-γ agonists inhibit neointimal formation after vascular injury *(73)*. Further interaction between PPAR-γ and the TGF-β/Smad system has been postulated in regulating genes for proteins (such as resistin, CD36, SR-A, liver X receptors-α, MMP 9, iNOS, and the tumor suppressor, PTEN) that are directly relevant to cancer, diabetes, and atherosclerosis as well as inflammatory diseases *(148)*. In conclusion, although only a handful of nuclear receptors have been discussed in this section, an ever-growing list of nuclear receptors and their crosstalk with the TGF-β/ Smad system should be expanded because this information will be critical for a thorough understanding of the nuclear receptor system as a whole.

6. IMPLICATIONS OF NUCLEAR RECEPTOR LIGANDS FOR MODULATING TGF-B/SMAD SIGNALING IN THE TREATMENT AND PREVENTION OF INFLAMMATION AND CANCER

The TGF-β/Smad system is dysregulated in carcinogenesis and inflammation, and restoring normal functions of the TGF-β/Smad system by using novel agents may be an approach for the prevention or treatment of the disease process. We have indicated that the TGF-β/Smad system can be regulated by nuclear receptors and their ligands at multiple levels, including the activation of intracellular signaling molecules, induction of nuclear translocation of Smads, and further activation of TGF-β dependent transcriptional activity. As summarized in Table 1, there are many other nuclear receptor ligands involved in the regulation of TGF-β/Smad signaling at multiple levels in various cell types. These ligands can regulate the TGF-β/Smad system by modulating TGF-β ligands and receptor synthesis, as well as modulating interacting with Smads. Also, it is known that some of these ligands will interact with inhibitors or activators of TGF-β/Smad signaling, at the cell surface as well as in the nucleus, to change signaling pathways of the TGF-β/Smad system. The TGF-β/Smad system can also be controlled by several mechanisms such as ubiquitination and degradation pathways that involve SnoN, Ski and other negative regulators. Nuclear receptor ligands may modify the ubiquitination and degradation of Smad, Smurf, TGF-β receptor, and/or SnoN to regulate TGF-β/Smad signaling. Furthermore, it will be important to determine whether antiinflammatory or anticancer activity by nuclear receptor ligands are closely linked to TGF-β/Smad signaling and act through a Smad-dependent mechanism (such as by regulating the expression of TGF-β, receptors, and Smads, or by the interaction with Smad7, Smurfs, SnoN, or other Smad signaling molecules). Some potential sites of action of nuclear receptor ligands are shown in Figure 1.

Because aberrant function or signaling of the TGF-β/Smad system is closely linked to the pathogenesis of cancer and inflammation, it is important to investigate nuclear receptor ligands for their ability to modulate TGF-β/Smad signaling and to further determine their use in cancer and inflammation. For example, inflammatory cytokines can upregulate inhibitory Smad (Smad7) to suppress TGF-β/Smad signaling. Therefore, nuclear receptor ligands that modulate TGF-β/Smad signaling, in particular by inhibiting Smad7, may be interesting agents for the treatment of cancer and inflammation. Nuclear receptor ligands with anti-inflammatory activity may be tested for targeting this regulation of Smad7 in IBD, and the results from both in vitro and in vivo animal studies in this regard should be of great benefit for translational clinical studies in the future. Understanding the mechanisms of action of how nuclear receptor ligands regulate the TGF-β/Smad system, in particular for combinations of several different nuclear receptor ligands (possible synergistic effects), should contribute to the better use of these drugs for the prevention of cancer as well as for inflammatory diseases *(151,152)*. Although, we have emphasized potentially beneficial effects of the activation of TGF-β/Smad signaling by vitamin D_3 analogs or triterpenoids, it is important to acknowledge that there may also be undesirable effects that result from the ability of this pathway to enhance fibrosis, angiogenesis, invasion, and metastasis *(9,22,153)*. However, there are many reports showing the effects of nuclear receptor ligands, such as vitamin D_3 analogs, as antiangiogenic and antimetastatic agents *(154–162)*, it is likely that beneficial effects will be greater than adverse effects, but further studies are needed to determine relative benefits and risks of modulating TGF-β signaling under varying conditions.

7. CONCLUDING REMARKS

It is now well documented that the TGF-β/Smad system can be regulated by nuclear receptors and their ligands at multiple levels. The interaction between the nuclear receptor superfamily and the TGF-β/Smad system also occurs as two-way crosstalk, and this can be cell type specific and depends on the progression of the disease as well. Careful investigation of the interaction between the nuclear receptor superfamily and the TGF-β/Smad system in animal models will determine the potential benefit of using nuclear receptor ligands in relevant diseases. The importance of interactions between nuclear receptors and the TGF-β/ Smad system for cancer as well as for inflammatory diseases has been discussed in this chapter. In conclusion, a better understanding of crosstalk between the TGF-β/Smad system and the nuclear receptor superfamily will be helpful for investigating mechanisms of inflammation and carcinogenesis.

ACKNOWLEDGMENT

The author is supported by NIH K22 CA 99990, NIH R03 CA112642, and a CINJ New Investigator Award. The author is grateful to Drs. Allan H. Conney and Fang Liu for helpful comments on this chapter, and to the administrative staff at the Department of Chemical Biology for their assistance in preparing this manuscript.

REFERENCES

1. ten Dijke P, Goumans MJ, Itoh F, Itoh S. Regulation of cell proliferation by Smad proteins. J Cell Physiol 2002;191:1–16.
2. Derynck R, Zhang YE. Smad-dependent and Smad-independent pathways in TGF-β family signalling. Nature 2003; 425:577–584.
3. Massagué J, Wotton D. Transcriptional control by the TGF-β/Smad signaling system. EMBO J 2000;19:1745–1754.
4. Miyazono K, Kusanagi K, Inoue H. Divergence and convergence of TGF-β/BMP signaling. J Cell Physiol 2001;187:265–276.
5. Wakefield LM, Piek E, Bottinger EP. TGF-β signaling in mammary gland development and tumorigenesis. J Mammary Gland Biol Neoplasia 2001;6:67–82.
6. Siegel PM, Massagué J. Cytostatic and apoptotic actions of TGF-β in homeostasis and cancer. Nat Rev Cancer 2003;3:807–821.
7. Elliott RL, Blobe GC. Role of transforming growth factor β in human cancer. J Clin Oncol 2005;23:2078–2093.
8. Tang B, Vu M, Booker T, et al. TGF-β switches from tumor suppressor to prometastatic factor in a model of breast cancer progression. J Clin Invest 2003;112:1116–1124.
9. Wakefield LM, Roberts AB. TGF-β signaling: positive and negative effects on tumorigenesis. Curr Opin Genet Dev 2002;12:22–29.
10. Blobe GC, Schiemann WP, Lodish HF. Role of transforming growth factor β in human disease. N Engl J Med 2000;342:1350–1358.
11. Hahn SA, Schmiegel WH. Recent discoveries in cancer genetics of exocrine pancreatic neoplasia. Digestion 1998;59:493–501.
12. Maurice D, Pierreux C E, Howell M, Wilentz RE, Owen MJ, Hill CS. Loss of Smad4 function in pancreatic tumors: C-terminal truncation leads to decreased stability. J Biol Chem 2001;276:43,175–43,181.
13. Baldus SE, Schwarz E, Lohrey C, et al. Smad4 deficiency in cervical carcinoma cells. Oncogene 2005;24:810–819.
14. Riggins GJ, Kinzler KW, Vogelstein B, Thiagalingam S. Frequency of Smad gene mutations in human cancers. Cancer Res 1997;57:2578–2580.
15. Grady WM, Markowitz SD. Genetic and epigenetic alterations in colon cancer. Annu Rev Genomics Hum Genet 2002;3:101–128.
16. Markowitz S, Wang J, Myeroff L, et al. Inactivation of the type II TGF-β receptor in colon cancer cells with microsatellite instability. Science 1995;268:1336–1338.
17. Myeroff LL, Markowitz S. Detection of TGF-β type II receptor hot-spot mutations: the BAT-RII assay. Methods Mol Biol 2000;142:133–137.
18. Myeroff LL, Parsons R, Kim SJ, et al. A transforming growth factor β receptor type II gene mutation common in colon and gastric but rare in endometrial cancers with microsatellite instability. Cancer Res 1995;55:5545–5547.
19. Kim S, Im Y, Markowitz SD, Bang Y. Molecular mechanisms of inactivation of TGF-β receptors during carcinogenesis. Cytokine Growth Factor Rev 2000;11:159–168.
20. Knaus PI, Lindemann D, DeCoteau JF, et al. A dominant inhibitory mutant of the type II transforming growth factor β receptor in the malignant progression of a cutaneous T-cell lymphoma. Mol Cell Biol 1996;16:3480–3489.
21. Fiocchi C. TGF-β/Smad signaling defects in inflammatory bowel disease: mechanisms and possible novel therapies for chronic inflammation. J Clin Invest 2001;108:523–526.
22. Derynck R, Akhurst RJ, Balmain A. TGF-β signaling in tumor suppression and cancer progression. Nat Genet 2001;29:117–129.

23. Nakao A, Okumura K, Ogawa H. Smad7: a new key player in TGF-β-associated disease. Trends Mol Med 2002;8:361–363.
24. Monteleone G, Pallone F, MacDonald TT. Smad7 in TGF-β-mediated negative regulation of gut inflammation. Trends Immunol 2004;25:513–517.
25. Tian F, DaCosta Byfield S, Parks WT, et al. Reduction in Smad2/3 signaling enhances tumorigenesis but suppresses metastasis of breast cancer cell lines. Cancer Res 2003;63:8284–8292.
26. Tsang ML, Zhou L, Zheng BL, et al. Characterization of recombinant soluble human transforming growth factor-β receptor type II (rhTGF-β sRII). Cytokine 1995;7:389–397.
27. Komesli S, Vivien D, Dutartre P. Chimeric extracellular domain type II transforming growth factor (TGF)-β receptor fused to the Fc region of human immunoglobulin as a TGF-β antagonist. Eur J Biochem 1998;254:505–513.
28. Yang YA, Dukhanina O, Tang B, et al. Lifetime exposure to a soluble TGF-β antagonist protects mice against metastasis without adverse side effects. J Clin Invest 2002;109:1607–1615.
29. Siegel PM, Shu W, Cardiff RD, Muller WJ, Massagué J. Transforming growth factor β signaling impairs Neu-induced mammary tumorigenesis while promoting pulmonary metastasis. Proc Natl Acad Sci USA 2003;100:8430–8435.
30. Sporn MB, Roberts AB. Peptide growth factors and inflammation, tissue repair, and cancer. J Clin Invest 1986;78:329–332.
31. Kulkarni AB, Huh CG, Becker D, et al. Transforming growth factor β1 null mutation in mice causes excessive inflammatory response and early death. Proc Natl Acad Sci USA 1993;90:770–774.
32. Ashcroft GS, Yang X, Glick AB, et al. Mice lacking Smad3 show accelerated wound healing and an impaired local inflammatory response. Nat Cell Biol 1999;1:260–266.
33. Kulkarni AB, Thyagarajan T, Letterio JJ. Function of cytokines within the TGF-β superfamily as determined from transgenic and gene knockout studies in mice. Curr Mol Med 2002;2:303–327.
34. Moustakas A, Pardali K, Gaal A, Heldin C-H. Mechanisms of TGF-β signaling in regulation of cell growth and differentiation. Immunol Lett 2002;82:85–91.
35. Vodovotz Y. Control of nitric oxide production by transforming growth factor-β1: mechanistic insights and potential relevance to human disease. Nitric Oxide 1997;1:3–17.
36. Ruiz PA, Shkoda A, Kim SC, Sartor RB, Haller D. IL-10 gene-deficient mice lack TGF-β/Smad signaling and fail to inhibit proinflammatory gene expression in intestinal epithelial cells after the colonization with colitogenic Enterococcus faecalis. J Immunol 2005;174:2990–2999.
37. Shull MM, Ormsby I, Kier AB, et al. Targeted disruption of the mouse transforming growth factor-β1 gene results in multifocal inflammatory disease. Nature 1992;359:693–699.
38. Kulkarni AB, Ward JM, Yaswen L, et al. Transforming growth factor-β1 null mice. An animal model for inflammatory disorders. Am J Pathol 1995;146:264–275.
39. Monteleone G, Kumberova A, Croft NM, McKenzie C, Steer HW, MacDonald TT. Blocking Smad7 restores TGF-β1 signaling in chronic inflammatory bowel disease. J Clin Invest 2001;108: 601–609.
40. Nathan C. Points of control in inflammation. Nature 2002;420:846–852.
41. Ohshima H, Bartsch H. Chronic infections and inflammatory processes as cancer risk factors: possible role of nitric oxide in carcinogenesis. Mutat Res 1994;305:253–264.
42. Coussens LM, Werb Z. Inflammation and cancer. Nature 2002;420:860–867.
43. Pikarsky E, Porat RM, Stein I, et al. NF-kB functions as a tumour promoter in inflammation-associated cancer. Nature 2004;431:461–466.
44. Balkwill F, Coussens LM. Cancer: an inflammatory link. Nature 2004;431:405–406.
45. Karin M, Greten FR. NF-kB: linking inflammation and immunity to cancer development and progression. Nat Rev Immunol 2005;5:749–759.
46. Tamir S, Tannenbaum SR. The role of nitric oxide (NO) in the carcinogenic process. Biochim Biophys Acta 1996;1288:F31–F36.
47. Oshima M, Dinchuk JE, Kargman SL, et al. Suppression of intestinal polyposis in Apc delta716 knockout mice by inhibition of cyclooxygenase 2 (COX-2). Cell 1996;87:803–809.
48. Ambs S, Merriam WG, Bennett WP, et al. Frequent nitric oxide synthase-2 expression in human colon adenomas: implication for tumor angiogenesis and colon cancer progression. Cancer Res 1998;58:334–341.
49. Letterio JJ. TGF-β signaling in T cells: roles in lymphoid and epithelial neoplasia. Oncogene 2005;24: 5701–5712.

50. Takahashi M, Fukuda K, Ohata T, Sugimura T, Wakabayashi K. Increased expression of inducible and endothelial constitutive nitric oxide synthases in rat colon tumors induced by azoxymethane. Cancer Res 1997;57:1233–1237.
51. Shen Q, Brown PH. Novel agents for the prevention of breast cancer: targeting transcription factors and signal transduction pathways. J Mammary Gland Biol Neoplasia 2003;8:45–73.
52. Lechner M, Lirk P, Rieder J. Inducible nitric oxide synthase (iNOS) in tumor biology: the two sides of the same coin. Semin Cancer Biol 2005;15:277–289.
53. Clevers H. At the crossroads of inflammation and cancer. Cell 2004;118:671–674.
54. Viatour P, Merville MP, Bours V, Chariot A. Phosphorylation of NF-kB and IkB proteins: implications in cancer and inflammation. Trends Biochem Sci 2005;30:43–52.
55. Lee YC, Lane KB. The many faces of transforming growth factor-β in pleural diseases. Curr Opin Pulm Med 2001;7:173–179.
56. Adnane J, Bizouarn FA, Chen Z, et al. Inhibition of farnesyltransferase increases TGFβ type II receptor expression and enhances the responsiveness of human cancer cells to TGFβ. Oncogene 2000;19:5525–5533.
57. Chawla A, Repa JJ, Evans RM, Mangelsdorf DJ. Nuclear receptors and lipid physiology: opening the X-files. Science, 2001;294:1866–1870.
58. Mangelsdorf DJ, Thummel C, Beato M, et al. The nuclear receptor superfamily: the second decade. Cell 1995;83:835–839.
59. Xu L, Glass CK, Rosenfeld MG. Coactivator and corepressor complexes in nuclear receptor function. Curr Opin Genet Dev 1999;9:140–147.
60. MacDonald PN, Baudino TA, Tokumaru H, Dowd DR, Zhang C. Vitamin D receptor and nuclear receptor coactivators: crucial interactions in vitamin D-mediated transcription. Steroids 2001;66: 171–176.
61. Wakefield L, Kim SJ, Glick A, Winokur T, Colletta A, Sporn M. Regulation of transforming growth factor-β subtypes by members of the steroid hormone superfamily. J Cell Sci Suppl 1990;13:139–148.
62. Falk LA, De Benedetti F, Lohrey N, et al. Induction of transforming growth factor-β1 (TGF-β1), receptor expression and TGF-β1 protein production in retinoic acid-treated HL-60 cells: possible TGF-β1-mediated autocrine inhibition. Blood 1991;77:1248–1255.
63. Koli K, Keski-Oja J. 1, 25-Dihydroxyvitamin D3 enhances the expression of transforming growth factor β1 and its latent form binding protein in cultured breast carcinoma cells. Cancer Res 1995;55: 1540–1546.
64. Wu G, Fan RS, Li W, Srinivas V, Brattain MG. Regulation of transforming growth factor-β type II receptor expression in human breast cancer MCF-7 cells by vitamin D3 and its analogues. J Biol Chem 1998;273:7749–7756.
65. Suh N, Lamph WW, Glasebrook AL, Grese TA, et al. Prevention and treatment of experimental breast cancer with the combination of a new selective estrogen receptor modulator, arzoxifene, and a new rexinoid, LG 100268. Clin Cancer Res 2002;8:3270–3275.
66. Rendi MH, Suh N, Lamph WW, et al. The selective estrogen receptor modulator arzoxifene and the rexinoid LG100268 cooperate to promote transforming growth factor β-dependent apoptosis in breast cancer. Cancer Res 2004;64:3566–3571.
67. Wu Y, Craig TA, Lutz WH, Kumar R. Identification of 1α,25-dihydroxyvitamin D3 response elements in the human transforming growth factor β2 gene. Biochemistry 1999;38:2654–2660.
68. Jung CW, Kim ES, Seol JG, et al. Antiproliferative effect of a vitamin D3 analog, EB1089, on HL-60 cells by the induction of TGF-β receptor. Leuk Res 1999;23:1105–1112.
69. Yanagisawa J, Yanagi Y, Masuhiro Y, et al. Convergence of transforming growth factor-β and vitamin D signaling pathways on SMAD transcriptional coactivators. Science 1999;283:1317–1321.
70. Song CZ, Tian X, Gelehrter TD. Glucocorticoid receptor inhibits transforming growth factor-β signaling by directly targeting the transcriptional activation function of Smad3. Proc Natl Acad Sci USA 1999;96:11,776–11,781.
71. Matsuda T, Yamamoto T, Muraguchi A, Saatcioglu F. Cross-talk between transforming growth factor-β and estrogen receptor signaling through Smad3. J Biol Chem 2001;276:42,908–42,914.
72. Kang HY, Lin HK, Hu YC, Yeh S, Huang KE, Chang C. From transforming growth factor-β signaling to androgen action: Identification of Smad3 as an androgen receptor coregulator in prostate cancer cells. Proc Natl Acad Sci USA 2001;98:3018–3023.

73. Fu M, Zhang J, Zhu X, et al. Peroxisome proliferator-activated receptor γ inhibits transforming growth factor β-induced connective tissue growth factor expression in human aortic smooth muscle cells by interfering with Smad3. J Biol Chem 2001;276:45,888–45,894.
74. Chipuk JE, Cornelius SC, Pultz NJ, et al. The androgen receptor represses transforming growth factor-β signaling through interaction with Smad3. J Biol Chem 2002;277:1240–1248.
75. Cao Z, Flanders KC, Bertolette D, et al. Levels of phospho-Smad2/3 are sensors of the interplay between effects of TGF-β and retinoic acid on monocytic and granulocytic differentiation of HL-60 cells. Blood 2003;101:498–507.
76. Yanagi Y, Suzawa M, Kawabata M, Miyazono K, Yanagisawa, J, Kato S. Positive and negative modulation of vitamin D receptor function by transforming growth factor-β signaling through smad proteins. J Biol Chem 1999;274:12,971–12,974.
77. Subramaniam N, Leong GM, Cock TA, et al. Cross-talk between 1,25-dihydroxyvitamin D3 and transforming growth factor-β signaling requires binding of VDR and Smad3 proteins to their cognate DNA recognition elements. J Biol Chem 2001;276:15,741–15,746.
78. Suh N, Roberts AB, Birkey Reffey S, et al. Synthetic triterpenoids enhance transforming growth factor β/Smad signaling. Cancer Res 2003;63:1371–1376.
79. Mix KS, Coon CI, Rosen ED, Suh N, Sporn MB, Brinckerhoff, CE. Peroxisome proliferator-activated receptor-γ-independent repression of collagenase gene expression by 2-cyano-3,12-dioxooleana-1,9-dien-28-oic acid and prostaglandin 15-deoxy-delta(12,14) J2: a role for Smad signaling. Mol Pharmacol 2004;65:309–318.
80. Suh N, Wang Y, Honda T, et al. A novel synthetic oleanane triterpenoid, 2-cyano-3,12-dioxoolean-1, 9-dien-28-oic acid, with potent differentiating, antiproliferative, and anti-inflammatory activity. Cancer Res 1999;59:336–341.
81. Minns LA, Buzoni-Gatel D, Ely KH, Rachinel N, Luangsay S, Kasper LH. A novel triterpenoid induces transforming growth factor β production by intraepithelial lymphocytes to prevent ileitis. Gastroenterology 2004;127:119–126.
82. Suh N, Wang Y, Williams CR, et al. A new ligand for the peroxisome proliferator-activated receptor-γ (PPAR-γ), GW7845, inhibits rat mammary carcinogenesis. Cancer Res 1999;59:5671–5673.
83. Anzano MA, Byers SW, Smith JM, et al. Prevention of breast cancer in the rat with 9-cis-retinoic acid as a single agent and in combination with tamoifen. Cancer Res 1994;54:4614–4617.
84. Anzano MA, Smith JM, Uskokovic MR, et al. 1α,25-Dihydroxy-16-ene-23-yne-26,27-hexafluoro-cholecalciferol (Ro24-5531), a new deltanoid (vitamin D analogue) for prevention of breast cancer in the rat. Cancer Res 1994;54:1653–1656.
85. Bischoff ED, Gottardis MM, Moon TE, Heyman RA, Lamph WW. Beyond tamoxifen: the retinoid X receptor-selective ligand LGD1069 (TARGRETIN) causes complete regression of mammary carcinoma. Cancer Res 1998;58:479–484.
86. Fisher B, Costantino JP, Wickerham DL, et al. Tamoxifen for prevention of breast cancer: report of the National Surgical Adjuvant Breast and Bowel Project P-1 Study. J Natl Cancer Inst 1998;90: 1371–1388.
87. Veronesi U, De Palo G, Marubini E, et al. Randomized trial of fenretinide to prevent second breast malignancy in women with early breast cancer. J Natl Cancer Inst 1999;91:1847–1856.
88. Cummings SR, Eckert S, Krueger KA, et al. The effect of raloxifene on risk of breast cancer in postmenopausal women: results from the MORE randomized trial. Multiple Outcomes of Raloxifene Evaluation. JAMA 1999;281:2189–2197.
89. Suh N, Glasebrook AL, Palkowitz AD, et al. Arzoxifene, a new selective estrogen receptor modulator for chemoprevention of experimental breast cancer. Cancer Res 2001;61:8412–8415.
90. Cauley JA, Norton L, Lippman ME, et al. Continued breast cancer risk reduction in postmenopausal women treated with raloxifene: 4-year results from the MORE trial. Multiple outcomes of raloxifene evaluation. Breast Cancer Res Treat 2001;65:125–134.
91. Guyton KZ, Kensler TW, Posner GH. Vitamin D and vitamin D analogs as cancer chemopreventive agents. Nutr Rev 2003;61:227–238.
92. Lamprecht SA, Lipkin M. Chemoprevention of colon cancer by calcium, vitamin D and folate: molecular mechanisms. Nat Rev Cancer 2003;3:601–614.
93. Froicu M, Weaver V, Wynn TA, McDowell, MA, Welsh, JE, and Cantorna, MT. A crucial role for the vitamin D receptor in experimental inflammatory bowel diseases. Mol Endocrinol 2003;17:2386–2392.

94. Sutton AL, MacDonald PN. Vitamin D: more than a "bone-a-fide" hormone. Mol Endocrinol 2003; 17:777–791.
95. Yanagi Y, Suzawa M, Kawabata M, Miyazono K, Yanagisawa, J, Kato S. Positive and negative modulation of vitamin D receptor function by transforming growth factor-β signaling through Smad proteins. J Biol Chem 1999;274:12,971–12,974.
96. Yang L, Yang J, Venkateswarlu S, Ko T, Brattain MG. Autocrine TGFβ signaling mediates vitamin D3 analog-induced growth inhibition in breast cells. J Cell Physiol 2001;188:383–393.
97. Koli K, Keski-Oja J. Vitamin D3 regulation of transforming growth factor-β system in epithelial and fibroblastic cells—relationships to plasminogen activation. J Invest Dermatol Symp Proc 1996;1:33–38.
98. Oyama N, Iwatsuki K, Satoh M, Akiba H, Kaneko F. Dermal fibroblasts are one of the therapeutic targets for topical application of 1α,25-dihydroxyvitamin D3: the possible involvement of transforming growth factor-β induction. Br J Dermatol 2000;143:1140–1148.
99. Koli K, Keski-Oja J. Vitamin D3 and calcipotriol enhance the secretion of transforming growth factor-β1 and -β2 in cultured murine keratinocytes. Growth Factors 1993;8:153–163.
100. Danielpour D. Induction of transforming growth factor-β autocrine activity by all-trans-retinoic acid and 1α,25-dihydroxyvitamin D3 in NRP-152 rat prostatic epithelial cells. J Cell Physiol 1996;166: 231–239.
101. Defacque H, Piquemal D, Basset A, Marti J, Commes T. Transforming growth factor-β1 is an autocrine mediator of U937 cell growth arrest and differentiation induced by vitamin D3 and retinoids. J Cell Physiol 1999;178:109–119.
102. Wu Y, Haugen JD, Zinsmeister AR, Kumar R. 1α,25-Dihydroxyvitamin D3 increases transforming growth factor and transforming growth factor receptor type I and II synthesis in human bone cells. Biochem Biophys Res Commun 1997;239:734–739.
103. Roberts AB, Sporn MB. Mechanistic interrelationships between two superfamilies: the steroid/ retinoid receptors and transforming growth factor-β. Cancer Surv 1992;14:205–220.
104. Staal A, van den Bemd GJ, Birkenhager JC, Pols HA, van Leeuwen JP. Consequences of vitamin D receptor regulation for the 1,25-dihydroxyvitamin D3-induced 24-hydroxylase activity in osteoblast-like cells: initiation of the C24-oxidation pathway. Bone 1997;20:237–243.
105. Spector TD, MacGregor AJ. Risk factors for osteoarthritis: genetics. Osteoarthritis Cartilage 2004; 12:S39–S44.
106. Borton AJ, Frederick JP, Datto MB, Wang XF, Weinstein RS. The loss of Smad3 results in a lower rate of bone formation and osteopenia through dysregulation of osteoblast differentiation and apoptosis. J Bone Miner Res 2001;16:1754–1764.
107. Staal A, Birkenhager JC, Pols HA, et al. Transforming growth factor β-induced dissociation between vitamin D receptor level and 1,25-dihydroxyvitamin D3 action in osteoblast-like cells. Bone Miner 1994;26:27–42.
108. Gurlek A, Kumar R. Regulation of osteoblast growth by interactions between transforming growth factor-β and 1α,25-dihydroxyvitamin D3. Crit Rev Eukaryot Gene Expr 2001;11:299–317.
109. MacDonald PN, Baudino TA, Tokumaru H, Dowd DR, Zhang C. Vitamin D receptor and nuclear receptor coactivators: crucial interactions in vitamin D-mediated transcription. Steroids 2001;66: 171–176.
110. Ueki N, Hayman MJ. Signal-dependent N-CoR requirement for repression by the Ski oncoprotein. J Biol Chem 2003;278:24,858–24,864.
111. Mehta RG, Moriarty RM, Mehta RR, et al. Prevention of preneoplastic mammary lesion development by a novel vitamin D analogue, 1α-hydroxyvitamin D5. J Natl Cancer Inst 1997;89:212–218.
112. Takeshita A, Imai K, Kato S, Kitano S, Hanazawa S. 1α,25-dehydroxyvitamin D3 synergism toward transforming growth factor-β1-induced AP-1 transcriptional activity in mouse osteoblastic cells via its nuclear receptor. J Biol Chem 1998;273:14,738–14,744.
113. Meehan TF, DeLuca HF. The vitamin D receptor is necessary for 1α,25-dihydroxyvitamin D3 to suppress experimental autoimmune encephalomyelitis in mice. Arch Biochem Biophys 2002;408: 200–204.
114. Nishino H, Nishino A, Takayasu J, et al. Inhibition of the tumor-promoting action of 12-O-tetradecanoylphorbol-13-acetate by some oleanane-type triterpenoid compounds. Cancer Res 1988;48: 5210–5215.
115. Huang MT, Ho CT, Wang ZY, et al. Inhibition of skin tumorigenesis by rosemary and its constituents carnosol and ursolic acid. Cancer Res 1994;54:701–708.

116. Huang MT, Badmaev V, Ding Y, Liu Y, Xie JG, Ho CT. Anti-tumor and anti-carcinogenic activities of triterpenoid, β-boswellic acid. Biofactors 2000;13:225–230.
117. Paper DH. Natural products as angiogenesis inhibitors. Planta Med 1998;64:686–695.
118. Coldren CD, Hashim P, Ali JM, Oh SK, Sinskey AJ, Rha C. Gene expression changes in the human fibroblast induced by Centella asiatica triterpenoids. Planta Med 2003;69:725–732.
119. Manez S, Recio MC, Giner RM, Rios JL. Effect of selected triterpenoids on chronic dermal inflammation. Eur J Pharmacol, 1997;334:103–105.
120. Giner-Larza EM, Manez S, Recio MC, et al. Oleanonic acid, a 3-oxotriterpene from Pistacia, inhibits leukotriene synthesis and has anti-inflammatory activity. Eur J Pharmacol 2001;428:137–143.
121. Suh N, Honda T, Finlay HJ, et al. Novel triterpenoids suppress inducible nitric oxide synthase (iNOS) and inducible cyclooxygenase (COX-2) in mouse macrophages. Cancer Res 1998;58:717–723.
122. Dinkova-Kostova AT, Liby KT, Stephenson KK, et al. Extremely potent triterpenoid inducers of the phase 2 response: correlations of protection against oxidant and inflammatory stress. Proc Natl Acad Sci USA 2005;102:4584–4589.
123. Honda T, Rounds BV, Bore L, et al. Synthetic oleanane and ursane triterpenoids with modified rings A and C: a series of highly active inhibitors of nitric oxide production in mouse macrophages. J Med Chem 2001;43:4233–4246.
124. Wang Y, Porter WW, Suh N, et al. A synthetic triterpenoid, 2-cyano-3,12-dioxooleana-1,9-dien-28-oic acid (CDDO), is a ligand for the peroxisome proliferator-activated receptor γ. Mol Endocrinol 2000;14:1550–1556.
125. Sporn MB, Suh N. Chemoprevention: an essential approach to controlling cancer. Nat Rev Cancer 2002;2:537–543.
126. Sporn MB, Liby KT. Cancer chemoprevention: scientific promise, clinical uncertainty. Nat Clin Pract Oncol 2005;2:518–525.
127. Place AE, Suh N, Williams CR, et al. The novel synthetic triterpenoid, CDDO-imidazolide, inhibits inflammatory response and tumor growth in vivo. Clin Cancer Res 2003;9:2798–2806.
128. Miyajima A, Asano T, Seta K, Asano T, Kakoi N, Hayakawa M. Loss of expression of transforming growth factor-b receptor as a prognostic factor in patients with renal cell carcinoma. Urology 2003; 61:1072–1077.
129. Bonni S, Wang HR, Causing CG, et al. TGF-β induces assembly of a Smad2-Smurf2 ubiquitin ligase complex that targets SnoN for degradation. Nat Cell Biol 20013:587–595.
130. Izzi L, Attisano L. Regulation of the TGFβ signalling pathway by ubiquitin-mediated degradation. Oncogene 2004;23:2071–2078.
131. Ebisawa T, Fukuchi M, Murakami G, et al. Smurf1 interacts with transforming growth factor-β type I receptor through Smad7 and induces receptor degradation. J Biol Chem 2001;276: 12,477–12,480.
132. Zhu H, Kavsak P, Abdollah S, Wrana JL, Thomsen GH. A SMAD ubiquitin ligase targets the BMP pathway and affects embryonic pattern formation. Nature 1999;400:687–693.
133. Kavsak P, Rasmussen RK, Causing CG, et al. Smad7 binds to Smurf2 to form an E3 ubiquitin ligase that targets the TGFβ receptor for degradation. Mol Cell 2000;6:1365–1375.
134. Zhang F, Lundin M, Ristimaki A, et al. Ski-related novel protein N (SnoN), a negative controller of transforming growth factor-β signaling, is a prognostic marker in estrogen receptor-positive breast carcinomas. Cancer Res 2003;63:5005–5010.
135. Kuratomi G, Komuro A, Goto K, et al. NEDD4-2 (neural precursor cell expressed, developmentally down-regulated 4-2) negatively regulates TGF-β (transforming growth factor-β) signalling by inducing ubiquitin-mediated degradation of Smad2 and TGF-β type I receptor. Biochem J 2005;386: 461–470.
136. Kim Y, Suh N, Sporn M, Reed JC. An inducible pathway for degradation of FLIP protein sensitizes tumor cells to TRAIL-induced apoptosis. J Biol Chem 2002;277:22,320–22,329.
137. Dragnev KH, Pitha-Rowe I, Ma Y, et al. Specific chemopreventive agents trigger proteasomal degradation of G1 cyclins: implications for combination therapy. Clin Cancer Res 2004;10:2570–2577.
138. Mix KS, Sporn MB, Brinckerhoff CE, Eyre D, Schurman DJ. Novel inhibitors of matrix metalloproteinase gene expression as potential therapies for arthritis. Clin Orthop Relat Res 2004;427S:S129–S137.
139. Paez-Pereda M, Giacomini D, Refojo D, et al. Involvement of bone morphogenetic protein 4 (BMP-4) in pituitary prolactinoma pathogenesis through a Smad/estrogen receptor crosstalk. Proc Natl Acad Sci USA 2003;100:1034–1039.

140. Refojo D, Liberman AC, Giacomini D, et al. Integrating systemic information at the molecular level: cross-talk between steroid receptors and cytokine signaling on different target cells. Ann N Y Acad Sci 2003;992:196–204.
141. Buck MB, Pfizenmaier K, Knabbe C. Antiestrogens induce growth inhibition by sequential activation of p38 mitogen-activated protein kinase and transforming growth factor-β pathways in human breast cancer cells. Mol Endocrinol 2004;18:1643–1657.
142. Ewan KB, Oketch-Rabah HA, Ravani SA, Shyamala G, Moses HL, Barcellos-Hoff MH. Proliferation of estrogen receptor-α-positive mammary epithelial cells is restrained by transforming growth factor-β1 in adult mice. Am J Pathol 2005;167:409–417.
143. Yamamoto T, Saatcioglu F, Matsuda T. Cross-talk between bone morphogenic proteins and estrogen receptor signaling. Endocrinology 2002;143:2635–2642.
144. Testa U, Grignani F, Barberi T, et al. PML/RARα+ U937 mutant and NB4 cell lines: retinoic acid restores the monocytic differentiation response to vitamin D3. Cancer Res 1994;54:4508–4515.
145. Pendaries V, Verrecchia F, Michel S, Mauviel A. Retinoic acid receptors interfere with the TGF-β/Smad signaling pathway in a ligand-specific manner. Oncogene 2003;22:8212–8220.
146. Wen FQ, Kohyama T, Skold CM, et al. Glucocorticoids modulate TGF-β production. Inflammation 2002;26:279–290.
147. Peltier J, Perez J, Bellocq A, Escoubet B, Fouqueray B, Baud L. Transforming growth factor-β1 increases glucocorticoid binding and signaling in macrophages through a Smad- and activated protein-1-mediated process. Kidney Int 2003;63:2028–2036.
148. Sporn MB, Suh N, Mangelsdorf DJ. Prospects for prevention and treatment of cancer with selective PPARg modulators (SPARMs). Trends Mol Med 2001;7:395–400.
149. Redondo S, Ruiz E, Santos-Gallego CG, Padilla E, Tejerina T. Pioglitazone induces vascular smooth muscle cell apoptosis through a peroxisome proliferator-activated receptor-γ, transforming growth factor-β1, and a Smad2-dependent mechanism. Diabetes 2005;54:811–817.
150. Han C, Demetris AJ, Liu Y, Shelhamer JH, Wu T. Transforming growth factor-β (TGF-β) activates cytosolic phospholipase A2α (cPLA2α)-mediated prostaglandin E2 (PGE)2/EP1 and peroxisome proliferator-activated receptor-γ (PPAR-γ)/Smad signaling pathways in human liver cancer cells. A novel mechanism for subversion of TGF-β-induced mitoinhibition. J Biol Chem 2004;279: 44,344–44,354.
151. Nagpal S, Na S, Rathnachalam R. Noncalcemic actions of vitamin D receptor ligands. Endocr Rev 2005;26:662–687.
152. Yee YK, Chintalacharuvu SR, Lu J, Nagpal S. Vitamin D receptor modulators for inflammation and cancer. Mini Rev Med Chem 2005;5:761–778.
153. Massagué J, Blain SW, Lo RS. TGFβ signaling in growth control, cancer, and heritable disorders. Cell 2000;103:295–309.
154. Getzenberg RH, Light BW, Lapco PE, et al. Vitamin D inhibition of prostate adenocarcinoma growth and metastasis in the Dunning rat prostate model system. Urology 1997;50:999–1006.
155. Fujioka T, Hasegawa M, Ishikura K, Matsushita Y, Sato M, Tanji S. Inhibition of tumor growth and angiogenesis by vitamin D3 agents in murine renal cell carcinoma. J Urol 1998;160:247–251.
156. Iseki K, Tatsuta M, Uehara H, et al. Inhibition of angiogenesis as a mechanism for inhibition by 1α-hydroxyvitamin D3 and 1,25-dihydroxyvitamin D3 of colon carcinogenesis induced by azoxymethane in Wistar rats. Int J Cancer 1999;81:730–733.
157. Lokeshwar BL, Schwartz GG, Selzer MG, et al. Inhibition of prostate cancer metastasis in vivo: a comparison of 1,25-dihydroxyvitamin D (calcitriol) and EB1089. Cancer Epidemiol Biomarkers Prev 1999;8:241–248.
158. Yudoh K, Matsuno H, Kimura T. 1α,25-dihydroxyvitamin D3 inhibits in vitro invasiveness through the extracellular matrix and in vivo pulmonary metastasis of B16 mouse melanoma. J Lab Clin Med 1999;133:120–128.
159. El Abdaimi K, Dion N, Papavasiliou V, et al. The vitamin D analogue EB 1089 prevents skeletal metastasis and prolongs survival time in nude mice transplanted with human breast cancer cells. Cancer Res 2000;60:4412–4418.
160. Evans SR, Shchepotin EI, Young H, Rochon J, Uskokovic M, Shchepotin IB. 1,25-dihydroxyvitamin D3 synthetic analogs inhibit spontaneous metastases in a 1,2-dimethylhydrazine-induced colon carcinogenesis model. Int J Oncol 2000;16:1249–1254.
161. Mantell DJ, Owens PE, Bundred NJ, Mawer EB, Canfield AE. 1α,25-dihydroxyvitamin D3 inhibits angiogenesis in vitro and in vivo. Circ Res 2000;87:214–220.

162. Nakagawa K, Kawaura A, Kato S, Takeda E, Okano T. 1α,25-Dihydroxyvitamin D3 is a preventive factor in the metastasis of lung cancer. Carcinogenesis 2005;26:429–440.
163. Li X, Schwarz EM, Zuscik MJ, et al. Retinoic acid stimulates chondrocyte differentiation and enhances bone morphogenetic protein effects through induction of Smad1 and Smad5. Endocrinology 2003;144:2514–2523.
164. Yang NN, Bryant HU, Hardikar S, et al. Estrogen and raloxifene stimulate transforming growth factor-β3 gene expression in rat bone: a potential mechanism for estrogen- or raloxifene-mediated bone maintenance. Endocrinology 1996;137:2075–2084.
165. Refojo D, Liberman AC, Giacomini D, et al. Integrating systemic information at the molecular level: cross-talk between steroid receptors and cytokine signaling on different target cells. Ann N Y Acad Sci 2003;992:196–204.
166. Wu L, Wu Y, Gathings B, et al. Smad4 as a transcription corepressor for estrogen receptor α. J Biol Chem 2003;278:15,192–15,200.
167. Lucia MS, Sporn MB, Roberts AB, Stewart LV, Danielpour D. The role of transforming growth factor-β1, -β2, and -β3 in androgen-responsive growth of NRP-152 rat prostatic epithelial cells. J Cell Physiol 1998;175:184–192.
168. Olsen NJ, Zhou P, Ong H, Kovacs WJ. Testosterone induces expression of transforming growth factor-β1 in the murine thymus. J Steroid Biochem Mol Biol 1993;45:327–332.
169. Kim IY, Ahn HJ, Zelner DJ, Park L, Sensibar JA, Lee C. Expression and localization of transforming growth factor-β receptors type I and type II in the rat ventral prostate during regression. Mol Endocrinol 1996;10:107–115.
170. Hayes SA, Zarnegar M, Sharma M, et al. SMAD3 represses androgen receptor-mediated transcription. Cancer Res 2001;61:2112–2118.

34 Thrombospondin-Dependent Activation of Latent TGF-β in Fibrosis and Disease

Joanne E. Murphy-Ullrich

CONTENTS

TGF-β: Biologic Properties
Mechanisms of TGF-β Activation
Thrombospondin: An Important Modulator of Matrix Structure and Cellular Function
TSP1 Binds and Activates Latent TGF-β
TSP1 Dependent TGF-β Activation: Physiology and Disease
Conclusions
Acknowledgments
References

Abstract

TGF-β is a pleotrophic cytokine with effects on cell growth and survival, immune cell regulation and inflammation, and extracellular matrix remodeling. Its activity must be carefully regulated for homeostasis of the organism. Control of latent TGF-β activation is a major means of regulating appropriate levels of TGF-β activity. The matricellular protein, thrombospondin-1 (TSP1), is an important physiologic regulator of latent TGF-β activation. TSP1 binding to the latent complex through the lysine-arginine-phenylalanine-lysine (KRFK) sequence of TSP1 competes for binding of the leucine-serine-lysine-leucine (LSKL) sequence in the latency associated peptide (LAP) of TGF-β to the mature domain of TGF-β, altering folding of the latent complex which renders TGF-β active. In this review, we discuss this and other mechanisms by which TSP1 induces activation of latent TGF-β. We also review the evidence for participation of TSP1 in the regulation of TGF-β activation in renal, pulmonary, hepatic, cardiovascular, and dermal fibrotic diseases. The use of peptide antagonists of TSP1-dependent TGF-β activation as a therapeutic strategy to treat fibrotic disease will be considered. The role of TSP1 in regulating TGF-β activity in cancer and wound healing and in host responses to infectious agents will also be discussed.

Key Words: Thrombospondin; latent TGF-β; activation; fibrosis; cancer; wound healing.

1. TGF-β: BIOLOGIC PROPERTIES

Transforming growth factors-β are a family of cytokines that influence cell growth, differentiation, apoptosis, inflammatory processes, immune regulation, and gene expression *(1–3)*. Appropriate levels of TGF-β activity are critical as too much TGF-β results in fibrosis and immune suppression, whereas deficiencies result in inflammation, epithelial hyperplasia, and defective wound healing *(3)*. TGF-β activity can be controlled at the level of protein

From: *Cancer Drug Discovery and Development: Transforming Growth Factor-β in Cancer Therapy, Vol I: Basic and Clinical Biology*
Edited by: Sonia B. Jakowlew © Humana Press Inc., Totowa, NJ

synthesis and secretion, regulation of bioavailability and matrix deposition by latent TGF-β binding proteins (LTBP), control of signaling and accessory receptor expression, by endogenous inhibitors such as decorin and α_2-macroglobulin, and by complex regulation of downstream signaling through Smad, ERK-dependent pathways, and c-*Abl* *(3–8)*. A major point of control lies in regulation of the conversion from its inactive proform to the biologically active form: this process is referred to as latent TGF-β activation *(4,9–13)*.

The three mammalian isoforms of TGF-β *(1,3)* are each secreted by cells as a disulfide linked homodimer consisting of a 278 aa propeptide known as the latency associated peptide (LAP) that is noncovalently associated with a 112 aa bioactive mature domain. The LAP domain is cleaved intracellularly from mature TGF-β by an endoprotease furin: however, the LAP and mature domains remain associated and are secreted as a noncovalent complex, termed the small latent complex. LAP binding to the mature TGF-β peptide is critical for latency, preventing TGF-β receptor binding and signaling. The latent complex must be modified to be biologically active—a process that involves altering the interactions between LAP and mature TGF-β, either by proteolysis of the LAP or by changing LAP folding and/or binding to the mature domain *(4,13,14)*.

The small latent complex can be disulfide-bonded to a fibrillin-like protein, the LTBP *(15,16)*. Latent TGF-β bound to LTBP is known as the large latent complex. There are four LTBP proteins (LTBP 1–4), each with multiple splice variants: knockout animals of each LTBP show distinct phenotypes suggesting differential expression and function. LTBPs are important for secretion of the small latent complex and they target the latent complex to the extracellular matrix and regulate bioavailability *(4,5)*. Mutations in the LTBP hinge region can hinder TGF-β activation and mutations in the extracellular matrix (ECM)-binding sites of LTBP-1 increase TGF-β bioavailability and activity *(17,18)*. LTBP-2 does not bind the small latent complex.

2. MECHANISMS OF TGF-β ACTIVATION

Expression of latent TGF-β protein is insufficient to drive TGF-β's biological effects, such as fibrosis. In elegant studies by Sime et al., intratracheal delivery of adenovirus expressing latent TGF-β protein failed to induce pulmonary fibrosis *(19)*. In contrast, delivery of adenovirus expressing a constitutively active form of TGF-β induced severe pulmonary fibrosis. These studies showed for the first time that TGF-β must be in its biologically active form to induce fibrosis. Thus understanding how TGF-β activation is regulated is critical for regulating TGF-β action in health and in disease.

The physiologic mechanisms that regulate latent TGF-β activation are not well understood, although multiple mechanisms have been identified. There appears to be some cell and tissue type specificity as to which mechanism is engaged. Moreover, the various activators of latent TGF-β are themselves differentially regulated, suggesting that there is also stimulus and disease-related specificity governing which activation mechanism is operative.

Proteolysis of the LAP by plasmin was the first mechanism of activation identified *(20,21)*. In endothelial-smooth muscle coculture systems, plasmin mediates activation directly by cleaving the LAP from the mature domain *(20)* or indirectly by liberating matrix bound latent TGF-β *(22)*. Plasmin preferentially activates the large latent complex *(23)*. The presence of active TGF-β in plasminogen knockout animals suggests that other factors are also operative in vivo *(24)*. Other proteases such as matrix metalloproteinases (MMP-9, MMP-2), calpain, and cathepsins can also activate latent TGF-β *(25–29)*. Viral neuraminidase has also been shown to activate latent TGF-β and influence survival of virally-infected cells *(30)*.

Latent TGF-β1 and 3 can bind to $\alpha_v\beta_6$ integrins through the RGD sequence in the β1 and β3 LAPs of TGF-β *(31,32)*. Binding of this integrin to the large latent TGF-β complex results in activation of TGF-β *(31)*. Conformational alterations of integrin-bound latent complex owing to mechanical forces from both the cytosketelon and the extracellular matrix are thought to be involved. Binding of the latent complex to a fibronectin matrix though fibronectin-LTBP-1 interactions is critical for integrin dependent activation *(17,33)*. $\alpha_v\beta_6$ integrin expression is restricted to epithelium. It is present at low levels on normal epithelium and unregulated in inflammation and in response to bleomycin. Integrin-dependent TGF-β activation has been shown to be important in tubulointerstitial fibrosis in a rodent model of unilateral ureteral obstruction and in bleomycin-induced pulmonary fibrosis *(34–37)*. Activation of latent TGF-β by $\alpha_v\beta_6$ integrin on airway epithelial cells is critical for the development of bleomycin-induced pulmonary fibrosis, because Munger et al. showed that β_6 knockout mice do not develop fibrosis with bleomycin treatment *(38–40)*. The $\alpha_v\beta_8$ integrin on airway epithelium and on astrocytes also activates latent TGF-β, but requires MT1-MMP for activation *(41–43)*.

Modification of the LAP by reactive oxygen species can activate latent TGF-β *(44)*. Barcellos-Hoff showed an increase in immunostaining specific for active TGF-β in breast tissues following exposure to ionizing radiation *(45)*. Activation likely occurs through site-specific oxidation of certain amino acids in the LAP, eliciting a conformational change in the latent complex that releases free active TGF-β. Although, it has yet to be shown that physiologic changes in redox state are sufficient to mediate TGF-β activation, this mechanism could be consistent with TGF-β activity in chronic disease or exposure to environmental toxins like asbestos *(46)*.

3. THROMBOSPONDIN: AN IMPORTANT MODULATOR OF MATRIX STRUCTURE AND CELLULAR FUNCTION

Thrombospondins (TSP1) are a family of multifunctional, modular glycoproteins that exist as both a secreted protein and as an insoluble extracellular matrix molecule (reviewed in *[47–49]*). TSP1, a disulfide-linked trimer, is the best-characterized member of the family of 5 isoforms. It is a major component of platelet α-granules and it is widely expressed during tissue remodeling. The ability to activate latent TGF-β is restricted to the TSP1 isoform. TSP1 acts as an immediate early response gene and it is rapidly upregulated in response to serum, glucose, and growth factors such as TGF-β and PDGF *(50–54)*. Increased levels of TSP1 are expressed in response to injury and during development *(55–57)*. TSP1 expression is regulated by p53 and c-*myc* *(58,59)*. There is substantial evidence that TSP1 acts as a tumor suppressor through inhibition of angiogenesis, although there is also evidence that TSP or certain domains of TSP can be proangiogenic *(60–66)*. TSP1 plays important and diverse roles in modulating cell–ECM and cell–cell adhesion, cell migration, apoptosis and survival, and gene expression *(49,67–69)*. TSP1 is also important in wound healing responses: TSP1 null mice have delayed wound healing, persistence of granulation tissue, and impaired macrophage recruitment *(70,71)*. We and others show that TSP1 is a major physiologic regulator of TGF-β activation (discussed in Section 5).

4. TSP1 BINDS AND ACTIVATES LATENT TGF-β

The mechanism of TSP1 activation of latent TGF-β has been determined primarily in systems using purified soluble proteins. In studies with purified large latent complex or small latent complex, we showed that TSP1 was capable of binding and activating the complex *(72)*. Activation does not involve proteolysis or require the presence of cells *(73)*. In addition, it does not appear that dissociation of latent TGF-β from TSP1 is necessary for TGF-β to be biologically active because active TGF-β is present in preparations of TSP1 purified from human platelets and TGF-β biological activity migrates as a high molecular weight complex when the small latent complex is incubated with TSP1 *(72,74)*. In these studies, only a minor fraction of the biologically active TGF-β was present as the free dimer. However, it is unknown whether active TGF-β remains associated with TSP1 in vivo. Recent data from Breitkopf et al. showed that the LSKL blocking peptide (*see* Section 5.5.) reduced TGF-β signaling by active TGF-β, suggesting that endogenous TSP1 might facilitate signaling of active TGF-β, possibly through a cofactor or possibly by altering TGF-β receptor interactions *(75)*.

There are two sequences in the TSP1 molecule that are required for binding and activation of latent TGF-β. Binding of the KRFK sequence, located between the first and second type 1 repeats of TSP1, to the LAP of the latent complex is necessary for activation *(76)*. Of the five isoforms of TSP, only TSPs 1 and 2 have the type 1 repeats and only TSP1 has the KRFK sequence. TSP2 does not activate latent TGF-β because the RFK homologue in TSP2 is RIR, which cannot activate latent TGF-β *(76)*. Binding of KRFK or TSP1 to the latent complex induces a conformational change in the latent complex and in the LAP as determined by alterations in the circular dichroism spectra (Jablonsky, Jackson, Su, Muccio, and Murphy-Ullrich, unpublished data). Furthermore, binding of TSP to free LAP prevents LAP from conferring latency on the mature domain *(77)*. TSP1 binds to the LAP at a sequence which is conserved in the LAP regions of TGF-βs1-3 (L_{54}SKL) *(77)*. Our lab recently showed that the LSKL sequence is critical for maintaining latency through binding to the arginine-lysine-proline-lysine (RKPK) sequence in the mature domain *(78,79)*. Thus, KRFK-TSP1 appears to activate the latent complex by competitively disrupting the latency interaction between the LSKL sequence in the LAP and the RKPK sequence in the mature domain *(79)*. The KRFK sequence, when expressed as a peptide, binds the LAP and the small latent complex and this interaction is sufficient to induce activation *(76)*. The KRFK recognition sequence in the LAP, LSKL, prevents TSP1-mediated TGF-β activation by competing for TSP1 binding to the LAP/latent complex. This peptide has been used successfully to block TSP1-dependent TGF-β activation in vitro *(77)* and in four different animal models in vivo (*see* Section 5) *(80–82)*.

There is a second motif present in each of the three thrombospondin type 1 repeats (TSRs), Tryptophan-x-x-Tryptophan (WxxW), that binds to the active domain of TGF-β and facilitates the ability of KRFK to activate latent TGF-β, possibly acting as a docking site to correctly orient the KRFK sequence with its complementary site on the latent TGF-β molecule *(78)* (Fig. 1).

WxxW peptides alone do not activate latent TGF-β, but can be used to competitively block both TSP1 binding to the mature domain and activation of the latent TGF-β complex by TSP1. The WxxW motif binds to the sequence *(61)* valine-leucine-alanine-leucine (VLAL) present in both the mature domain and the LAP of TGF-β1. This sequence is conserved across multiple species and the sequences in the β2 (VLSL) and β3 (VLGL) isoforms represent conservative substitutions, suggesting that TSP1 can bind to all mammalian isoforms. This is consistent with evidence that TSP1 activates both TGF-β1 and β2 *(76)*. Deletion of the VLAL site in the mature domain abrogates the ability of TSP1 to activate the latent complex. Similarly, VLAL peptides prevent TSP1 binding to the mature domain and activation by TSP1, but not by plasmin or the KRFK peptide. Interestingly, the *(77)* VLAL sequence is also present in the LAP and it is important not only for TSP1 binding, but also for secretion of the mature domain and to some extent, maintenance of latency *(78)*. Structural studies of the LAP and the small latent complex will be needed to establish the precise role of these interactions in both latency and activation mechanisms.

In vitro studies from our lab and others have shown that TSP1 activates latent TGF-β secreted by multiple cell types including endothelial cells, mesangial cells, hepatic stellate cells and skin, lung, and cardiac fibroblasts *(13,53,73,75,83–85)*. Peptides such as LSKL or WxxW which block TSP1 binding to the latent complex have been used to establish the involvement of endogenous TSP1 in TGF-β activation.

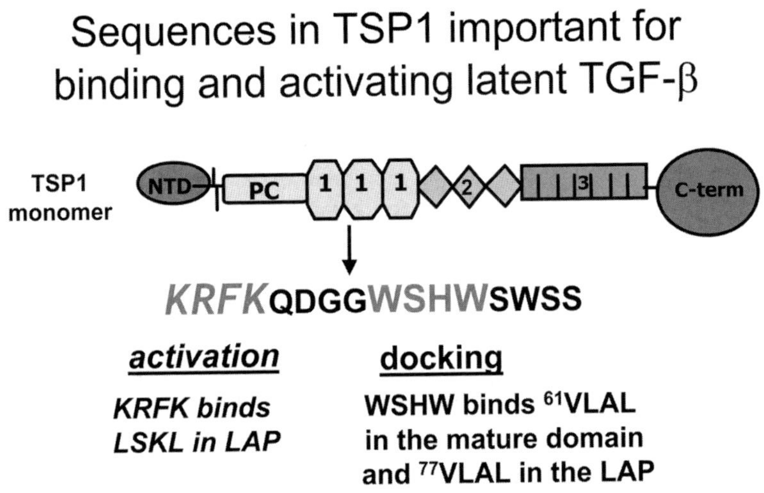

Fig. 1. Sequences of TSP1 involved in binding and activation of the latent TGF-β complex. The three type 1 repeats of TSP-1 are enlarged to show the KRFK and the three WxxW (*) motifs. The WxxW motifs bind to the VLAL sequence in mature TGF-β and possibly on LAP to facilitate KRFK binding to the LAP. The KRFK sequence binds LAP through the LSKL site. KRFK binding to LSKL in the LAP disrupts LAP interacts with the RKPK sequence in the mature domain. These interactions alter the folding of the LAP in relation to mature TGF-β, thus unmasking its receptor binding sites. Mature TGF-β which is active can remain in a heterotrimeric complex of TSP and LAP.

In addition to activation by direct binding of TSP1 to the soluble latent complex, there are conditions in which the conformation of either TSP1 or the latent complex might not be favorable for activation through these direct interactions. Assays in which either the latent complex, the LAP, or TSP1 are immobilized usually fail to show binding or activation, suggesting that conformational flexibility is required for optimal interactions between TSP1 and the latent complex (*[86]*, Poczatek and Murphy-Ullrich, unpublished results). For example, bleomycin-stimulated alveolar macrophages have increased levels of active TGF-β which is critical for the development of pulmonary fibrosis. Macrophage secreted latent TGF-β complex is bound to TSP1 which is localized to the macrophage surface through binding of TSP1 to its receptor, CD36. Binding of latent TGF-β to CD36-bound TSP1 is not sufficient to induce activation and activation requires plasmin. TSP binds the latent complex through its LAP and presents the latent complex to plasmin which cleaves and activates the latent molecule *(87)*. The CD36 binding site in TSP1 is also in the type 1 repeats *(88)*. Although TSP bound to CD36 still binds the latent complex, it is possible that the WxxW motifs are sterically prevented from binding to the latent complex in a manner which facilitates KRFK interactions with the LSKL sequence of the LAP or that CD36-bound TSP1 does not have sufficient conformational flexibility to effect the interactions necessary for activation.

It is not entirely clear whether TSP1 activates the latent complex only in the soluble phase or whether it can also activate the latent complex at the cell surface. Antibodies to cell surface receptors for TSP, $a_v\beta_3$ integrin and CD47, inhibit activation of TGF-β in tamoxifen-treated mammary carcinoma cells *(89)*, suggesting that in some conditions, latent TGF-β activation by TSP1 can occur at or near the cell surface. Consistent with this idea, is the recent observation that oxidized LDL binding to TSP1 reduces its capacity to activate latent TGF-β by binding to TSP1 and blocking TSP1 binding to the latent complex *(90)*. CD36 is a scavenger receptor for both oxidized low density lipoprotein (LDL) and TSP1 and oxidized LDL reduces TSP1 binding to mouse peritoneal macrophages and latent TGF-β activation by these cells *(90)*. However, oxidized LDL also reduced latent TGF-β activation by TSP1 in a cell-free system *(90)*, suggesting oxidized LDL acts potentially both by steric inhibition of soluble TSP1 binding to the latent complex and by blocking localization of the latent complex to the cell surface for activation. Further studies are needed to more clearly define the role of cell surface interactions in regulating TSP1-dependent TGF-β activation.

5. TSP1 DEPENDENT TGF-β ACTIVATION: PHYSIOLOGY AND DISEASE

We have used both in vitro and in vivo systems to understand the role of TSP1-dependent TGF-β activation in both physiologic and pathologic conditions. Here we will review the evidence for the involvement of TSP1 in control of latent TGF-β activation in multiple disease processes.

5.1. Neonatal Development

We first identified that TSP1 is a physiological activator of activation in studies of TSP1 null mouse pups. The phenotypes of TGF-β1 null and TSP1 null mice reveal similar histopathology in multiple organ systems, particularly the lung and pancreas, and include vascular smooth muscle cell hyperplasia, alveolar hemorrhage, bronchial and pancreatic islet cell hyperplasia. This similarity in phenotype suggested that a major factor in the histologic abnormalities in TSP1 null mice is the lack of active TGF-β1 *(80)*. Many aspects of the TSP1 null phenotype were rescued by intraperitoneal injection of the KRFK activating peptide. Furthermore, treatment of wild-type animals with the LSKL peptide from the LAP, which inhibits TSP-mediated activation, induced histopathologic changes in the proximal airway epithelium, pancreas and other tissues resembling those of the TSP1 and TGF-β1 null mice. Staining for active TGF-β in these tissues was reduced as compared to wild-type animals or animals treated with control SLLK peptide. The extent of inflammation in the TSP1 null mice was not as serious as in the TGF-β1 null mice, suggesting that other mechanisms of activation can compensate the lack of TSP1 at some level. A recent study comparing the phenotypes of the TSP-1 null mice with that of the β_6 integrin null and a double knockout animal lacking both TSP1 and the β_6 integrin suggests that these two activation mechanisms have both overlapping and distinct roles on controlling TGF-β activation, characterized primarily by inflammation and epithelial hyperplasia in multiple organs *(91)*. Interestingly, hyperplastic epithelium from the airways of the double knockout animals showed evidence of phosphorylated Smad2/3 staining indicating the role of additional alternative activation mechanisms *(91)*.

5.2. TSP1 and Renal Fibrosis

There is evidence from multiple disease models that TSP1-dependent TGF-β activation plays a role in regulation of TGF-β activation in renal fibrosis. TSP-dependent latent TGF-β activation is important in an anti-Thy-1 induced model of glomerulosclerosis in rats *(92)*. The expression of TSP-1 in the anti-Thy-1 model correlates temporally and spatially with up-regulation of TGF-β *(93)*. Rats treated with TSP1 antagonist peptides that competitively block activation (delivered by minipump infusion), either GGWSHW from TSP or LSKL from the LAP, had reduced active TGF-β, glomerular matrix accumulation and proteinuria *(82)*. Glomerular influx of macrophages was not affected by TSP1- antagonist peptide treatment *(82)*. Antisense oligonucleotides to reduce TSP1 expression in the kidney also reduced renal fibrosis in the anti-Thy-1 model *(94)*. Glomerular mesangial cells stimulated by high levels of amino acids also activate latent TGF-β through a TSP1-dependent mechanism in vitro *(95)*. In addition to mesangial derived TSP1, there is also evidence that injured podocytes upregulate TSP1 expression as determined by in situ hybridization of specimens from patients with idiopathic focal segmental glomerulosclerosis *(96)*. Although the expression of TSP1 in the tubulointerstitium precedes and correlates with subsequent development of tubulointerstitial fibrosis *(82,93)*, in another model of tubulointerstitial fibrosis generated by unilateral uretral obstruction, TSP1 levels were not upregulated and the integrin $\alpha_v\beta_6$ was shown to be important for local TGF-β activation in this mouse model of renal fibrosis *(34)*.

5.3. TSP1 in Fibrotic Complications of Diabetes and Hypertension

TSP1 is known to be upregulated in both animal models and in tissues from patients with diabetes *(54,97–100)*. In diabetes models, we identified that TSP1 activates TGF-β under high glucose conditions in both renal mesangial cells and in rat cardiac fibroblasts in vitro *(53,101)*. TSP1 antagonist peptides reduce both active TGF-β matrix protein expression (fibronectin, osteopontin) to levels observed with basal (5 mM) glucose. Glucose increases TSP1 transcription through stimulation of the transcription factor USF2 via PKC, p38 MAPK, and ERK signaling *(102,103)* (Fig. 2). Angiotensin II, a factor in fibrosis owing to hypertension, increases TGF-β activity and has been shown to be a factor in the development of cardiac and renal fibrosis *(104,105,106)*. Angiotensin II similarly upregulates TSP1 expression through a p38 MAPK and JNK-dependent pathway in mesangial cells and cardiac fibroblasts and TSP1 antagonist peptides block angiotensin II-mediated increases in TGF-β activity, implicating TSP1 in hypertensive sclerosis *(83,107)*. The TSP1 antagonist peptides also attenuate myocardial and renal fibrosis in a rat model of diabetes with hypertension. Intraperitoneal administration of the LSKL antagonist peptide, but not the control peptide (LSAL), reduced both fibrosis as determined by picrosirius red staining and hydroxyproline content and TGF-β activity as determined by phospho-Smad levels and immunohistochemistry for active TGF-β *(108)*. Furthermore, LSKL peptide reduced myocyte hypertrophy and restored left ventricular function in this model of hypertensive diabetic cardiomyopathy. The importance of TSP1 for control of TGF-β activation in diabetes is supported by studies showing increased TSP1 expression in human kidneys with diabetic nephropathy *(98)*: GGWSHW peptide treated human mesangial cell cultures grown in high-glucose conditions showed reduced CTGF levels, a key mediator of TGF-β-dependent matrix synthesis *(98)*.

There is conflicting evidence as to whether TSP1 regulates TGF-β activation downstream of advanced glycated end (AGE) product stimulation. Wahab et al. showed that the GGWSHW peptide failed to reduce CTGF expression by human mesangial cells stimulated with glycated albumen *(98)*. In contrast, there is evidence that TSP1 is transcriptionally upregulated by AGE product stimulation of renal distal tubule cells Madin-Darby canine kidney epithelial cells (MDCK) and that anti-TSP1 neutralizing antibodies block increases in TGF-β1 bioactivity and cellular hypertrophy induced by AGEs *(109)*. A transcription factor decoy for the AP-1 site in the TSP1 promoter attenuated the effects of AGEs on TSP1 expression and MDCK hypertrophy *(109)*.

5.4. Lung Fibrosis

The role of TSP1 in lung fibrosis has not been as extensively studied as have models of renal fibrosis. However, there is evidence for the involvement of TSP1 in control of TGF-β activation by several cell types relevant to the pathogenesis of pulmonary fibrosis. As mentioned previously, TSP1 is a critical player in localizing the latent complex to cell surface CD36 on bleomycin-stimulate macrophages for subsequent activation by plasmin *(87,109)*. Intratracheal administration of peptides that block TSP binding to CD36 attenuate bleomycin-induced fibrosis in rats *(110)*. TSP antagonist peptides (LSKL) can also block

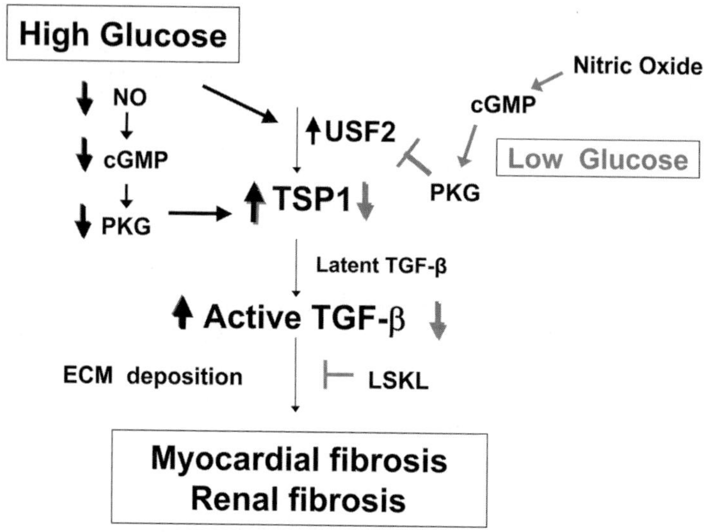

Fig. 2. TSP1 is a major regulator of TGF-β activation in diabetes. Under normal glucose conditions, TSP1 levels are repressed by nitric oxide-cGMP-PKG mediated down regulation of the transcription factor USF2. High glucose levels increase USF2 expression and decrease PKG activity because of decreased bioavailable nitric oxide. The increased TSP1 expression under high glucose conditions results in increased TGF-β activation and extracellular matrix accumulation which is key to both diabetic myocardial and renal fibrosis. In vitro and in vivo studies show that antagonism of TSP1-dependent TGF-β activation attenuates fibrotic progression in animal models of diabetes with hypertension.

TGF-β activation by rat lung fibroblasts stimulated in vitro with either bleomycin or IL-4 *(84)*. Consistent with the idea that TSP1 is one of the molecular regulators of TGF-β activation in pulmonary fibrosis, it has recently been shown that interferon-β, a drug in trials for the treatment of idiopathic pulmonary fibrosis, reduces both TSP1 and TGF-β expression *(111)*. These LSKL peptides also block TGF-β activation in vitro in airway epithelial cells in a model of airway remodeling in asthma *(112)*. More recently, in a mouse model of acute lung injury induced by a nickel aerosol, in which TGF-β is increased, a cDNA microarray showed that both TSP1 and the α_v integrin genes were upregulated as early as 24 h following nickel exposure *(113)*. The integrin α_v subunit was upregulated twofold over 24–72 h, whereas TSP1 was induced fourfold at 24 h and by eightfold at 72 h. This difference in kinetics indicates a role for TSP1 in TGF-β activation primarily by later effector cells such as the interstitial fibroblast rather than earlier effector cells such as the alveolar epithelium.

5.5. Liver Fibrosis

TSP1 has been implicated in regulating TGF-β activation in both human and animal models of hepatic fibrosis *(81,114)*. Congenital hepatic fibrosis is a rare disease associated with autosomal recessive polycystic kidney disease: livers from these patients have higher levels of both TSP1 and TGF-β1 than do normal livers. Hepatic stellate cells isolated from livers of patients with congenital hepatic fibrosis secreted both TSP1 and TGF-β1 in vitro, suggesting that TSP1 might represent an autocrine mechanism for regulating TGF-β activation and fibrosis in these patients *(114)*. Production of TSP1 by isolated hepatic stellate cells is highest immediately (1–3 d) following isolation of cells and was induced by PDGF-BB *(75)*. PDGF also stimulates increases in TGF-β activity by hepatic stellate cells in culture and these increases were blocked by the inhibitory LSKL peptide or monoclonal antibodies to TSP1 *(75)*. Interestingly, this group observed that the LSKL peptide also reduced TGF-β activity stimulated by the addition of active TGF-β to either hepatic stellate cell or dermal fibroblast cultures, suggesting an additional regulatory effect of endogenous TSP1 on TGF-β signaling, possibly through stabilization of PDGF. In support of a role for TSP1 in local control of TGF-β activation in fibrotic liver disease, Kondou et al., showed that intraperitoneal administration of the LSKL blocking peptide decreased liver atrophy, fibrosis scores, hydroxyproline content and levels of active TGF-β1 as measured by Smad 2 phosphorylation in a rat model of chemically induced (dimethylnitrosamine) hepatic fibrosis *(81)*.

5.6. Cardiovascular Disease

Increased TGF-β activity is associated with neointimal fibroproliferation, restenosis, as well as stabilization of the fibrous cap surrounding atherosclerotic lesions *(115–119)*. TSP1 is induced by PDGF and it has been localized to fibrous regions and microcalcification areas of atherosclerotic lesions *(120–122)*. TSP1 is also detected in porcine coronary arteries from 1 hr to 14 d following angioplasty which correlates with the increased levels of active TGF-β observed in the injured artery following this procedure *(123)*. Other activators such as mannose 6 phosphate/IGF-II receptor were also upregulated, indicating that multiple activation mechanisms might be involved. These results suggest that interventions that block TSP1-dependent TGF-β activation in the vasculature might be useful for retarding restenosis or progression of neointimal lesions. However, blockade of TGF-β activation could also destabilize the fibrous cap, leading to plaque rupture and embolism *(83)*. Furthermore, it remains to be determined whether TSP1 has a significant role in mediating local TGF-β activation in the arterial wall, because it was recently shown that oxidized LDL can inhibit TSP1 activation of latent TGF-β *(90)*.

TSP1 control of TGF-β activation appears to play a role in myocardial remodeling *(124)*. Frangogiannis et al. recently showed that TSP1 (–/–) mice had increased inflammation with infiltration of macrophages, inflammatory mediators, and myofibroblasts following myocardial infarction *(124)*. TSP1 is expressed at the border of the healing infarct and the absence of TSP1 was associated with an expansion of granulation tissue and fibrotic remodeling into the noninfarcted myocardium *(124)*. Although TSP1 has multiple actions, there was evidence for decreased TGF-β signaling (phosphorylated Smad2) in the infarct regions of TSP1 null mice, suggesting that local activation of TGF-β by TSP1 is important for limiting inflammation and paradoxically, fibrotic remodeling *(124)*. As mentioned previously, TSP1 control of latent TGF-β activation also is critical to myocardial fibrosis owing to hypertension with diabetes *(83,108)*.

5.7. Scleroderma

There is recent evidence that endogenous TSP1 secreted by scleroderma fibroblasts contributes an autocrine stimulation of TGF-β activity and extracellular matrix production *(85)*. TSP1 levels are increased in fibroblasts isolated from scleroderma patients as compared to isolates from normal donors and TSP1 was downregulated by TGF-β blocking antibody or antisense oligonucleotide only in the scleroderma fibroblasts. The LSKL peptide or antisense oligonucleotides to TSP1 reduced collagen expression and phosphorylated Smad3 levels in the scleroderma fibroblasts, indicating that this autocrine regulation of TSP1 by TGF-β contributes to fibrogenesis in scleroderma fibroblasts.

5.8. Wound Healing

The role of TSP1 in control of latent TGF-β activation during dermal wound is not clear. Higher levels of TSP1 and TGF-β are both present in keloid-derived fibroblasts as compared to normal fibroblasts *(125)*. The phenotype of the excisional wound healing in the TSP1 null mouse is consistent with a decrease in local TGF-β activation *(71)*. There is a delay in macrophage recruitment and capillary angiogenesis and a persistence of granulation tissue. Topical treatment of TSP1 null wounds with the KRFK activating peptide largely restored the TSP1 null wound phenotype to normal (Nor, Murphy-Ullrich, Polverini et al., unpublished data). Measurements of total TGF-β in these wounds showed that TGF-β levels were increased in wounds following KRFK treatment and that the effects of the KRFK peptide were blocked by a pan-specific anti-TGF-β antibody. While these data suggest that TSP1 plays a role in local activation of TGF-β during wounding, further analysis of active and total TGF-β levels in the wounds of TSP1 null mice showed that there was no difference in the fraction of active TGF-β in wounds between wild-type and TSP1 null mice *(71)*. The authors concluded that the decreased TGF-β in the wounds of TSP1 null mice is indirect and primarily owing to defects in macrophage recruitment to wounds (a major source of TGF-β in wounds) rather than to a defect in activation *(71)*. Despite this controversy, it is clear that TSP1 has the potential to modify the wound healing process: subcutaneous implantation of TSP1 soaked sponges had higher levels of active TGF-β, increased gel contraction and fibroblast migration (45). Others have used a derivative of the KRFK sequence, KFK coupled to a fatty acyl moiety *(126)*. In skin fibroblast cultures treated with this KFK peptide, there was increased active TGF-β and TIMP-1. The fatty acid linked-peptide inhibited MMP-induced elastin and collagen degradation when applied to ex vivo skin tissue sections *(126)*. It remains to be determined whether either positive or negative modulation of TSP1 action represents a viable therapeutic strategy to modify either defective or excessive wound repair processes in vivo.

5.9. Cancer

TGF-β plays a complex role in tumor initiation and metastatic progression (reviewed in *[127,128]*). TSP1 is generally regarded as a tumor suppressor because of its potent anti-angiogenic activity (reviewed in *[65]*). However, there is some evidence that TSP1 can influence tumor progression through control of TGF-β activation, independent of its activity as an angiogenesis inhibitor. There is evidence that TSP1 expression is increased in high-grade glioma tissue sections and cell lines as compared to low-grade gliomas *(129)*; increased TSP1 expression correlated with increased TGF-β1, 2, and 3 expression by these tumors. Further work from this group showed that two different antibodies to TSP1, including an antibody raised against a peptide containing the KRFK activation sequence of TSP1, blocked latent TGF-β activation in glioma cell conditioned media *(130)*. These studies suggest that endogenous TSP1 produced by malignant glioblastomas can induce autocrine TGF-β activation and promote tumor progression through immune suppression and/or enhanced invasion *(130)*. Systemic administration (intraperitoneal injection) of recombinant protein expressing the second TSP1 type 1 repeat with the RFK sequence to mice who had been subcutaneously inoculated with B16F10 melanoma cells induced increased TGF-β activation and increased TGF-β dependent tumor cell apoptosis *(131)*. Expression of a type 1 repeat protein which contains the KRFK sequence by a transfected human squamous cell carcinoma cell line A431 also caused TSP1-dependent TGF-β activation that similarly inhibited tumor growth in vivo, independent of the effects of the TSRs on inhibition of angiogenesis *(132)*. The importance of TSP1 control of TGF-β activation for regulation of tumor growth appears to be tumor cell specific. Growth inhibition of Lewis lung carcinomas by systemic administration of recombinant type 1 repeat proteins was independent of the presence or absence of the KRFK sequence *(131)* and injection of peptides containing the KRFK sequence did not affect the growth of MDA MB435 mammary carcinoma cells *(133)*. It should be noted that both of these cell types are not growth regulated by

TGF-β *(131)*. TSP1 can also indirectly control tumor progression by regulating stroma desmoplastic responses to tumors. Bleuel et al. showed that TSP1 overexpression halts tumor vascularization of human carcinoma cell lines by increasing matrix deposition at the tumor-stromal border *(134)*. TSP1 mRNA is also increased in the desmoplastic-rich stroma of human breast carcinomas *(135)*. These data suggest that local TGF-β activation by TSP1 in the stroma might influence tumor vasculogenesis and invasion, although this hypothesis needs to be tested directly. The level and duration of TSP1 expression by tumor and stroma is likely to influence tumor progression. Using a tet-repressive expression system for TSP1, Filleur et al. showed that continuous expression of TSP1 by tumor cells results in increased active TGF-β, suggesting that tumor angiogenesis driven by active TGF-β can potentially overcome the initial inhibitory effects of TSP's antiangiogeneic activity *(136)*. Finally, the increased TGF-β activity associated with tamoxifen treatment of estrogen-dependent mammary carcinoma cells (MCF-7 and TD-47) is activated by TSP1 in a manner that is blocked by both the WxxW peptide and antibodies to two different TSP receptors (CD47 and $\alpha_v\beta_3$ integrin), suggesting that activation might occur at the cell surface *(89)*. Thus, the role of TSP1 in control of TGF-β activation and its effect on tumor progression is likely to be complex and tumor and stage dependent.

5.10. Immune Regulation

TSP1 is a known chemoattractant for macrophages and has been shown to play a role in alveolar macrophage activation of latent TGF-β *(87,137)*. TSP1 also affects both T and NK cell activity through control of TGF-β activation. In the eye, TSP1 expressed by retinal pigmented epithelial cells is responsible for activation of latent TGF-β *(138)*. TGF-β activity in the eye is critical for inhibiting T-cell activation and maintaining the immune privilege status of the eye *(138)*. In TSP1 knockout mice, there was almost no active TGF-β in eye-cups isolated from these mice and there was a failure to suppress T-cell proliferation stimulated by anti-CD3, which resulted in unresolved retinal inflammation *(138)*. The early inhibition and late inhibition of human natural killer cell populations has also been shown to be regulated by TSP1 control of TGF-β activation *(139)*. TSP1 has complex roles in the regulation of both immune and inflammatory cells through both TGF-β dependent and independent mechanisms *(140–144)*. Like TGF-β, which has both anti- and proinflammatory roles, carefully designed experiments will be needed to definitively ascertain the role of TSP1 in TGF-β immune regulation.

5.11. Infection

TSP1 control of TGF-β activation might play a role in host response to infectious agents. Bovine monocytes infected with *Mycobacterium avium* subsp. *paratuberculosis*, a cause of a chronic granulomatous enteritis in wild and domestic ruminants, had significantly highly levels of both TSP1 and TGF-β gene expression than did monocytes infected with another stain of mycobacterium which is more easily killed by macrophages *(145)*. A significant aspect of the pathology of malaria is the mounting of an inflammatory response to induce macrophage ingestion of parasitized red blood cells but which also induces severe febrile episodes *(146)*. Depending on the lethality of the strain of *Plasmodium*, TGF-β activity can beneficially attenuate the inflammatory complications or exacerbate the parasite infection *(146)*. In intriguing studies, Omer et al. showed that plasmodium-infected erythrocytes activate latent TGF-β in vitro and that activation was partially blocked by antibodies to TSP1 and to the TRAP (thrombospondin-related anonymous protein) homologue, which contains sequences homologous to the WxxW motif in TSP1 and 2 but not the KRFK activating sequence *(146)*. Peptides of the WxxW motif failed to activate latent TGF-β, but blocked activation by lysates of infected red cells, consistent with our model of activation by the KRFK sequence. Thus, activation of latent TGF-β by plasmodium requires both latent complex binding to the TSP1 homologue TRAP and MMP activity *(146)*, similar to the two step mechanism observed with bleomycin-stimulated alveolar macrophages.

6. CONCLUSIONS

Since our initial findings that active TGF-β copurified with TSP1 isolated from human platelets, the idea that TSP1 is not only a TGF-β binding protein, but a major physiologic activator of latent TGF-β has been supported by evidence from diverse systems. Our understanding of how TSP1 binding to the latent complex can induce activation by disruption of LAP binding to the mature domain defines a mechanism for TSP1-induced activation. However, there is also other evidence that TSP1 can also participate in alternative activation mechanisms involving plasmin, MMP and possibly, integrins. It is still not clear whether activation by TSP1 occurs predominately in the soluble phase or at the cell surface. Evidence that TSP, the LAP, and the mature domain can exist as a biologically active ternary complex *(77)* suggests that there might be additional levels of regulation of TGF-β signaling by the localization to the extracellular matrix, half life, and processing of this large molecular weight complex in contrast to the free mature growth factor. The ability of TSP1 binding proteins such as oxidized LDL or CD36 to modulate the ability of TSP1 to either bind or activate the latent complex suggests that the ability of TSP1 to control activation is likely to be regulated by the molecular milieu of the microenvironment as well as the conformational state of TSP1 and the latent complex. Such an example occurs in platelets. Despite the high-molecular concentrations of both the latent complex and TSP1 in platelet alpha granules, furin and not TSP1 activate latent TGF-β released from stimulated platelets *(147)*. This finding is consistent with the low stoichiometry of active TGF-β associated with platelet TSP1 *(147)* and suggests that the conformation of TSP1 when present at high concentrations or the presence of other proteins which bind the TSRs prevents large scale activation of TGF-β during platelet release (which is a lucky thing). There is also intriguing evidence that c-mannosylated forms of TSP1 are present in the aortic wall of diabetic Zucker rats *(149)*: the WxxW motifs of TSP1 can be c-mannosylated *(150)*. The presence or absence of mannosylated WxxW motifs might affect the ability of TSP1 to properly bind the latent complex and induce activation, suggesting an additional level of regulation of TSP1 action. Clearly, there are situations in which TSP1 is not a significant regulator of TGF-β activation. Others have also shown that TSP1 does not affect TGF-β activation in smooth muscle explants or else that it activates only a small fraction of the total TGF-β complex *(96,151)*.

The involvement of TSP1 in control of TGF-β activation appears to occur primarily in disease conditions and possibly during development rather than during tissue homeostasis. This is consistent with the upregulation of TSP1 expression in development and disease as compared to its low-level expression in quiescent tissues. As reviewed in this chapter, it is clear that TSP1 plays a significant role in controlling TGF-β activation in various fibrotic conditions induced by both local and systemic metabolic factors. There is also evidence that TSP1 can regulate inflammation and immune responses and possibly, host responses to infectious agents, through control of latent TGF-β activation. TSP1 likely plays a role in the stromal desmoplastic response to tumors, regulation of tumor immune surveillance, and in tumor survival/apoptosis through its ability to regulate TGF-β activation. Ongoing studies in our lab strongly indicate that TSP1 can regulate mesenchymal stem cell differentiation in vitro through its regulation of TGF-β activation *(152)*.

The primary involvement of TSP1 in controlling TGF-β activation in disease processes rather than under normal homeostatic conditions suggests that targeting TSP1-dependent TGF-β activation is a more selective therapeutic strategy than agents which inhibit the

TGF-β molecule, its receptors, or its kinases. The use of stimulatory and inhibitory peptides to regulate TSP1 action on TGF-β activation represents an exciting therapeutic strategy to better control TGF-β activity with a decreased potential for undesirable side effects from abrogation of homeostatic levels of TGF-β activity.

The next several years should yield new data which further our understanding of the complex interactions, timing, and localization of TSP1- activated TGF-β and elucidation of the physiologic consequences in animal models.

ACKNOWLEDGMENTS

I would like to thank the many outstanding young scientists in my lab (Stacey Schultz-Cherry, Solange Ribeiro, Maria Poczatek, Shuxia Wang, Geoffrey Young, Yong Zhou, Kim Bailey Dubose, Melissa Talbert, and Yun Su) who have furthered this story and my many distinguished collaborators who have made critical and substantial contributions to the original studies described in this chapter. This work was supported by grants from the American Cancer Society, and NIH (HL50061, DK60658, DK54624) to JEMU and by a Juvenile Diabetes Foundation postdoctoral fellowship to Shuxia Wang. I would also like to thank Anita Roberts who served as a constant source of inspiration both scientifically and personally.

REFERENCES

1. Massagué J. The transforming growth factor-beta family. Annu Rev Cell Dev Biol 1990; 6:597–641.
2. Lawrence DA. Transforming growth factor-beta: a general review. Eur Cytokine Netw 1996;7:363–374.
3. Roberts AB. Molecular and cell biology of TGF-beta. Miner Electrolyte Metab 1998;24:111–119.
4. Annes JP, Munger JS, Rifkin DB. Making sense of latent TGFbeta activation. J Cell Sci 2003;116: 217–224.
5. Tapiale JMK, Heldin C-H, Keski-Oja J. Latent transforming growth factor-beta 1 associates to fibroblast extrcellular matrix via latent TGF-beta binding protein. J Cell Biol 1994;124:171–181.
6. LaMarre J, Hayes MA, Wollenberg GK, Hussaini I, Hall SW, Gonias SL. An alpha 2-macroglobulin receptor-dependent mechanism for the plasma clearance of transforming growth factor-beta 1 in mice. J Clin Invest 1991;87:39–44.
7. Thannickal VJ, Lee DY, White ES, et al. Myofibroblast differentiation by transforming growth factor-beta1 is dependent on cell adhesion and integrin signaling via focal adhesion kinase. J Biol Chem 2003;278:12,384–12,389.
8. Daniels CE, Wilkes MC, Edens M, et al. Imatinib mesylate inhibits the profibrogenic activity of TGF-beta and prevents bleomycin-mediated lung fibrosis. J Clin Invest 2004;114:1308–1316.
9. Barcellos-Hoff MH. Latency and activation in the control of TGF-beta. J Mammary Gland Biol Neoplasia 1996;1:353–363.
10. Khalil N. TGF-beta: from latent to active. Microbes Infect 1999;1:1255–1263.
11. Lawrence DA. Latent-TGF-beta: an overview. Mol Cell Biochem 2001;219:163–170.
12. Munger JS, Harpel JG, Gleizes PE, Mazzieri R, Nunes I, Rifkin DB. Latent transforming growth factor-beta: structural features and mechanisms of activation. Kidney Int 1997;51:1376–1382.
13. Murphy-Ullrich JE, Poczatek M. Activation of latent TGF-beta by thrombospondin-1: mechanisms and physiology. Cytokine Growth Factor Rev 2000;11:59–69.
14. Zhu HJ, Burgess AW. Regulation of transforming growth factor-beta signaling. Mol Cell Biol Res Commun 2001;4:321–330.
15. Koli K, Saharinen J, Hyytiainen M, Penttinen C, Keski-Oja J. Latency, activation, and binding proteins of TGF-beta. Microsc Res Tech 2001;52:354–362.
16. Nunes I, Gleizes PE, Metz CN, Rifkin DB. Latent transforming growth factor-beta binding protein domains involved in activation and transglutaminase-dependent cross-linking of latent transforming growth factor-beta. J Cell Biol 1997;136:1151–1163.
17. Annes JP, Chen Y, Munger JS, Rifkin DB. Integrin alphavbeta6-mediated activation of latent TGF-beta requires the latent TGF-beta binding protein-1. Cell Bio 2004;165:723–734.
18. Mazzieri R, Jurukovski V, Obata H, et al. Expression of truncated latent TGF-beta-binding protein modulates TGF-beta signaling. Cell Science 2005;118:2177–2187.

19. Sime PJ, Xing Z, Graham FL, Csaky KG, Gauldie J. Adenovector-mediated gene transfer of active transforming growth factor-beta1 induces prolonged severe fibrosis in rat lung. J Clin Invest 1997; 100:768–776.
20. Lyons RM, Gentry LE, Purchio AF, Moses HL. Mechanism of activation of latent recombinant transforming growth factor beta 1 by plasmin. J Cell Biol 1990;110:1361–1367.
21. Lyons RM, Keski-Oja J, Moses HL. Proteolytic activation of latent transforming growth factor-beta from fibroblast-conditioned medium. J Cell Biol 1988;106:1659–1665.
22. Tapiale J LJ, Saarinen J, Kovanen PT, Keski-Oja J. Human mast cell chymase andeukocyte elastase release latent transforming growth factor beta from the extracellular matrix of cultured human epithelial and endothelial cells. J Biol Chem 1995;270:4689–4696.
23. Kojima S, Nara K, Rifkin DB. Requirement for transglutaminase in the activation of latent transforming growth factor-beta in bovine endothelial cells. J Cell Biol 1993;121:439–448.
24. Brugge THKK, Flick MJ, Daugherty CC, Danton MJS, Degen JL. Loss of fibrinogen rescues mice from the pleiotropic effects of plasminogen deficiency. Cell 1996;87:709–719.
25. Guo M, Mathieu PA, Linebaugh B, Sloane BF, Reiners JJ, Jr. Phorbol ester activation of a proteolytic cascade capable of activating latent transforming growth factor-betaL a process initiated by the exocytosis of cathepsin B. J Biol Chem 2002;277:14,829–14,837.
26. Yu Q, Stamenkovic I. Cell surface-localized matrix metalloproteinase-9 proteolytically activates TGF-beta and promotes tumor invasion and angiogenesis. Genes Dev 2000;14:163–176.
27. Abe M, Oda N, Sato Y. Cell-associated activation of latent transforming growth factor-beta by calpain. J Cell Physiol 1998;174:186–193.
28. Gantt KR, Schultz-Cherry S, Rodriguez N, et al. Activation of TGF-beta by Leishmania chagasi: importance for parasite survival in macrophages. J Immunol 2003;170:2613–2620.
29. Somanna A, Mundodi V, Gedamu L. Functional analysis of cathepsin B-like cysteine proteases from Leishmania donovani complex. Evidence for the activation of latent transforming growth factor beta. J Biol Chem 2002;277:25,305–25,312.
30. Schultz-Cherry S, Hinshaw VS. Influenza virus neuraminidase activates latent transforming growth factor beta. J Virol 1996;70:8624–8629.
31. Munger JS, Huang X, Kawakatsu H, et al. The integrin alpha v beta 6 binds and activates latent TGF beta 1: a mechanism for regulating pulmonary inflammation and fibrosis. Cell 1999;96:319–328.
32. Annes JP, Rifkin DB, Munger JS. The integrin alphaVbeta6 binds and activates latent TGFbeta3. FEBS Lett 2002;511:65–68.
33. Rifkin DB. Latent TGF-beta binding proteins: Orchestrators of TGF-beta availability. J Biol Chem 2004;280:7409–7412.
34. Ma LJ, Yang H, Gaspert A, et al. Transforming growth factor-beta-dependent and -independent pathways of induction of tubulointerstitial fibrosis in beta6(-/-) mice. Am J Pathol 2003;163:1261–1273.
35. Munger JS, Huang X, Kawakatsu H, et al. The integrin alpha v beta 6 binds and activates latent TGF beta 1: a mechanism for regulating pulmonary inflammation and fibrosis. Cell 1999;96:319–328.
36. Sheppard D. Integrin-mediated activation of transforming growth factor-beta(1) in pulmonary fibrosis. Chest 2001;120:49S–53S.
37. Pittet JF, Griffiths MJ, Geiser T, et al. TGF-beta is a critical mediator of acute lung injury. J Clin Invest 2001;107:1537–1544.
38. Goicoechea S, Orr AW, Pallero MA, Eggleton P, Murphy-Ullrich JE. Thrombospondin mediates focal adhesion disassembly through interactions with cell surface calreticulin. J Biol Chem 2000; 275:36,358–36,368.
39. Goicoechea S, Pallero MA, Eggleton P, Michalak M, Murphy-Ullrich JE. The anti-adhesive activity of thrombospondin is mediated by the N-terminal domain of cell surface calreticulin. J Biol Chem 2002;277:37,219–37,228.
40. Orr AW, Pedraza CE, Pallero MA, et al. Low density lipoprotein receptor-related protein is a calreticulin coreceptor that signals focal adhesion disassembly. J Cell Biol 2003;161:1179–1189.
41. Mu D, Cambier S, Fjellbirkeland L, et al. The integrin alpha(v)beta8 mediates epithelial homeostasis through MT1-MMP-dependent activation of TGF-beta1. J Cell Biol 2002;157:493–507.
42. Cambier S, Gline S, Mu D, et al. Integrin alpha(v)beta8-mediated activation of transforming growth factor-beta by perivascular astrocytes: an angiogenic control switch. Am J Pathol 2005;166:1883–1894.
43. Fjellbirkeland L, Cambier S, Broaddus VC, et al. Integrin alphavbeta8-mediated activation of transforming growth factor-beta inhibits human airway epithelial proliferation in intact bronchial tissue. Am J Pathol 2003;163:533–542.

44. Orr AW, Elzie CA, Kucik DF, Murphy-Ullrich JE. Thrombospondin signaling through the calreticulin/LDL receptor-related protein co-complex stimulates random and directed cell migration. J Cell Sci 2003;116:2917–2927.
45. Sakai K, Sumi Y, Muramatsu H, Hata K, Muramatsu T, Ueda M. Thrombospondin-1 promotes fibroblast-mediated collagen gel contraction caused by activation of latent transforming growth factor beta-1. J Dermatol Sci 2003;31:99–109.
46. Pociask DA, Sime PJ, Brody AR. Asbestos-derived reactive oxygen species activate TGF-beta1. Lab Invest 2004;84:1013–1023.
47. Bornstein P. Thrombospondins as matricellular modulators of cell function. J Clin Invest 2001;107: 929–934.
48. Chen H, Herndon ME, Lawler J. The cell biology of thrombospondin-1. Matrix Biol 2000;19: 597–614.
49. Adams JC. Thrombospondins: multifunctional regulators of cell interactions. Annu Rev Cell Dev Biol 2001;17:25–51.
50. Janat MF, Liau G. Transforming growth factor beta 1 is a powerful modulator of platelet-derived growth factor action in vascular smooth muscle cells. J Cell Physiol 1992;150:232–242.
51. Majack RA, Mildbrandt J, Dixit VM. Induction of thrombospondin messenger RNA levels occurs as an immediate primary response to platelet-derived growth factor. J Biol Chem 1987;262:8821–8825.
52. Bornstein P. Thrombospondins: structure and regulation of expression. Faseb J 1992;6:3290–3299.
53. Poczatek MH, Hugo C, Darley-Usmar V, Murphy-Ullrich JE. Glucose stimulation of transforming growth factor-beta bioactivity in mesangial cells is mediated by thrombospondin-1. Am J Pathol 2000;157:1353–1363.
54. Tada H, Kuboki K, Nomura K, Inokuchi T. High glucose levels enhance TGF-beta1-thrombospondin-1 pathway in cultured human mesangial cells via mechanisms dependent on glucose-induced PKC activation. J Diabetes Complications 2001;15:193–197.
55. Raugi GJ, Olerud JE, Gown AM. Thrombospondin in early human wound tissue. J Invest Dermatol 1987;89:551–554.
56. O'Shea KS, Liu LH, Kinnunen LH, Dixit VM. Role of the extracellular matrix protein thrombospondin in the early development of the mouse embryo. J Cell Biol 1990;111:2713–2723.
57. Reed MJ, Iruela-Arispe L, O'Brien ER, et al. Expression of thrombospondins by endothelial cells. Injury is correlated with TSP-1. Am J Pathol 1995;147:1068–1080.
58. Dameron KM, Volpert OV, Tainsky MA, Bouck N. Control of angiogenesis in fibroblasts by p53 regulation of thrombospondin-1. Science 1994;265:1582–1584.
59. Janz A, Sevignani C, Kenyon K, Ngo CV, Thomas-Tikhonenko A. Activation of the myc oncoprotein leads to increased turnover of thrombospondin-1 mRNA. Nucleic Acids Res 2000;28:2268–2275.
60. Good DJ, Polverini PJ, Rastinejad F, et al. A tumor suppressor-dependent inhibitor of angiogenesis is immunologically and functionally indistinguishable from a fragment of thrombospondin. Proc Natl Acad Sci USA 1990;87:6624–6628.
61. DiPietro LA. Thrombospondin as a regulator of angiogenesis. Exs 1997;79:295–314.
62. Motegi K, Harada K, Pazouki S, Baillie R, Schor AM. Evidence of a bi-phasic effect of thrombospondin-1 on angiogenesis. Histochem J 2002;34:411–421.
63. de Fraipont F, Nicholson AC, Feige JJ, Van Meir EG. Thrombospondins and tumor angiogenesis. Trends Mol Med 2001;7:401–407.
64. Lawler J. Thrombospondin-1 as an endogenous inhibitor of angiogenesis and tumor growth. J Cell Mol Med 2002;6:1–12.
65. Lawler J, Detmar M. Tumor progression: the effects of thrombospondin-1 and -2. Int J Biochem Cell Biol 2004;36:1038–1045.
66. Ferrari do Outeiro-Bernstein MA, Nunes SS, Andrade AC, Alves TR, Legrand C, Morandi V. A recombinant NH(2)-terminal heparin-binding domain of the adhesive glycoprotein, thrombospondin-1, promotes endothelial tube formation and cell survival: a possible role for syndecan-4 proteoglycan. Matrix Biol 2002;21:311–324.
67. Elzie CA, Murphy-Ullrich JE. The N-terminus of thrombospondin: the domain stands apart. Int J Biochem Cell Biol 2004;36:1090–1101.
68. Bornstein P, Agah A, Kyriakides TR. The role of thrombospondins 1 and 2 in the regulation of cell-matrix interactions, collagen fibril formation, and the response to injury. Int J Biochem Cell Biol 2004;36:1115–1125.
69. Adams JC, Lawler J. The thrombospondins. Int J Biochem Cell Biol 2004;36:961–968.

70. DiPietro LA, Nissen NN, Gamelli RL, Koch AE, Pyle JM, Polverini PJ. Thrombospondin 1 synthesis and function in wound repair. Am J Pathol 1996;148:1851–1860.
71. Agah A, Kyriakides TR, Lawler J, Bornstein P. The lack of thrombospondin-1 (TSP1) dictates the course of wound healing in double-TSP1/TSP2-null mice. Am J Pathol 2002;161:831–839.
72. Schultz-Cherry S, Ribeiro S, Gentry L, Murphy-Ullrich JE. Thrombospondin binds and activates the small and large forms of latent transforming growth factor-beta in a chemically defined system. J Biol Chem 1994;269:26,775–26,782.
73. Schultz-Cherry S, Murphy-Ullrich JE. Thrombospondin causes activation of latent transforming growth factor-beta secreted by endothelial cells by a novel mechanism. J Cell Biol 1993;122:923–932.
74. Murphy-Ullrich JE, Schultz-Cherry S, Hook M. Transforming growth factor-beta complexes with thrombospondin. Mol Biol Cell 1992;3:181–188.
75. Breitkopf K, Sawitza I, Westhoff JH, Wickert L, Dooley S, Gressner AM. Thrombospondin 1 acts as a strong promoter of transforming growth factor beta effects via two distinct mechanisms in hepatic stellate cells. Gut 2005;54:673–681.
76. Schultz-Cherry S, Chen H, Mosher DF, et al. Regulation of transforming growth factor-beta activation by discrete sequences of thrombospondin 1. J Biol Chem 1995;270:7304–7310.
77. Ribeiro SM, Poczatek M, Schultz-Cherry S, Villain M, Murphy-Ullrich JE. The activation sequence of thrombospondin-1 interacts with the latency-associated peptide to regulate activation of latent transforming growth factor-beta. J Biol Chem 1999;274:13,586–13,593.
78. Young GD, Murphy-Ullrich JE. The Tryptophan-rich motifs of the thrombospondin type 1 repeats bind VLAL motifs in the latent transforming growth factor-beta complex. J Biol Chem 2004;279: 47,633–47,642.
79. Young GD, Murphy-Ullrich JE. Molecular interactions that confer latency to transforming growth factor-beta. J Biol Chem 2004;279:38,032–38,039.
80. Crawford SE, Stellmach V, Murphy-Ullrich JE, et al. Thrombospondin-1 is a major activator of TGF-beta1 in vivo. Cell 1998;93:1159–1170.
81. Kondou H, Mushiake S, Etani Y, Miyoshi Y, Michigami T, Ozono K. A blocking peptide for transforming growth factor-beta1 activation prevents hepatic fibrosis in vivo. J Hepatol 2003;39:742–748.
82. Daniel C, Wiede J, Krutzsch HC, et al. Thrombospondin-1 is a major activator of TGF-beta in fibrotic renal disease in the rat in vivo. Kidney Int 2004; 65:459–468.
83. Zhou YPM, Berecek KH, Murphy-Ullrich JE. Thrombospondin 1 mediates angiotensin II induction of TGF-beta activation by cardiac and renal cells under both high and low glucose conditions. Biochem Biophys Res Commun 2006;339:633–641.
84. Zhou Y, Hagood JS, Murphy-Ullrich JE. Thy-1 expression regulates the ability of rat lung fibroblasts to activate transforming growth factor-beta in response to fibrogenic stimuli. Am J Pathol 2004;165: 659–669.
85. Mimura Y, Ihn H, Jinnin M, Asano Y, Yamane K, Tamaki K. Constitutive thrombospondin-1 overexpression contributes to autocrine transforming growth factor-beta signaling in cultured scleroderma fibroblasts. Am J Pathol 2005;166:1451–1463.
86. Bailly S, Brand C, Chambaz EM, Feige JJ. Analysis of small latent transforming growth factor-beta complex formation and dissociation by surface plasmon resonance. Absence of direct interaction with thrombospondins. J Biol Chem 1997;272:16,329–16,334.
87. Yehualaeshet T, O'Connor R, Green-Johnson J, et al. Activation of rat alveolar macrophage-derived latent transforming growth factor beta-1 by plasmin requires interaction with thrombospondin-1 and its cell surface receptor, CD36. Am J Pathol 1999;155:841–851.
88. Asch AS, Silbiger S, Heimer E, Nachman RL. Thrombospondin sequence motif (CSVTCG) is responsible for CD36 binding. Biochem Biophys Res Commun 1992;182:1208–1217.
89. Harpel JG, Schultz-Cherry S, Murphy-Ullrich JE, Rifkin DB. Tamoxifen and estrogen effects on TGF-beta formation: role of thrombospondin-1, alphavbeta3, and integrin-associated protein. Biochem Biophys Res Commun 2001;284:11–14.
90. Sakamoto UMA, Tamagawa H, Wang G-P, Horiuchi S. Specific interaction of oxidized low-density lipoprotein with thrombospondin-1 inhibits transforming growth factor-beta from its activation. Atherosclerosis. 2005;18:85–93.
91. Ludlow A, Yee KO, Lipman R, et al. Characterization of integrin beta6 and thrombospondin-1 double-null mice. J Cell Mol Med 2005;9:421–437.
92. Hugo C. The thrombospondin 1-TGF-beta axis in fibrotic renal disease. Nephrol Dial Transplant 2003;18:1241–1245.

93. Hugo C, Kang DH, Johnson RJ. Sustained expression of thrombospondin-1 is associated with the development of glomerular and tubulointerstitial fibrosis in the remnant kidney model. Nephron 2002;90:460–470.
94. Daniel C, Takabatake Y, Mizui M, et al. Antisense oligonucleotides against thrombospondin-1 inhibit activation of TGF-beta in fibrotic renal disease in the rat in vivo. Am J Pathol 2003;163:1185–1192.
95. Meek RL, Cooney SK, Flynn SD, et al. Amino acids induce indicators of response to injury in glomerular mesangial cells. Am J Physiol Renal Physiol 2003;285:F79–F86.
96. Zhou L, Isenberg J, Cao Z, Roberts DD. Type I collagen is a molecular target for inhibiton of angiogenesis by endogenous thrombospondin-1. Oncogene 2005;1:1–10.
97. Bayraktar M, Dundar S, Kirazli S, Teletar F. Platelet factor 4, beta-thromboglobulin and thrombospondin levels in type I diabetes mellitus patients. J Int Med Res 1994;22:90–94.
98. Wahab NA, Schaefer L, Weston BS, et al. Glomerular expression of thrombospondin-1, transforming growth factor beta and connective tissue growth factor at different stages of diabetic nephropathy and their interdependent roles in mesangial response to diabetic stimuli. Diabetologia 2005;48: 2650–2660.
99. Zhang XMSF, Xv ZY, Yan ZY, Han S. Expression changes of thrombospondin-1 and neuropeptide Y in myocardium of STZ-induced rats. Int J Cardiol 2005;105:192–197.
100. Stenina OI, Krukovets I, Wang K, et al. Increased expression of thrombospondin-1 in vessel wall of diabetic Zucker rat. Circulation 2003;107:3209–3215.
101. Kim JHKB, Moon KC, Hong HK, Lee HS. Activation of the TGF-beta/Smad signaling pathway in focal segmental glomerulosclerosis. Kidney International 2003;64:1715–1721.
102. Wang S, Wu X, Lincoln TM, Murphy-Ullrich JE. Expression of constitutively active cGMP-dependent protein kinase prevents glucose stimulation of thrombospondin 1 expression and TGF-beta activity. Diabetes 2003;52:2144–2150.
103. Wang S, Shiva S, Poczatek MH, Darley-Usmar V, Murphy-Ullrich JE. Nitric oxide and cGMP-dependent protein kinase regulation of glucose-mediated thrombospondin 1-dependent transforming growth factor-beta activation in mesangial cells. J Biol Chem 2002;277:9880–9888.
104. Lim DS, Lutucuta S, Bachireddy P, et al. Angiotensin II blockade reverses myocardial fibrosis in a transgenic mouse model of human hypertrophic cardiomyopathy. Circulation 2001;103:789–791.
105. Noble NA, Border WA. Angiotensin II in renal fibrosis: should TGF-beta rather than blood pressure be the therapeutic target? Semin Nephrol 1997;17:455–466.
106. Gaedeke J, Peters H, Noble NA, Border WA. Angiotensin II, TGF-beta and renal fibrosis. Contrib Nephrol 2001:153–160.
107. Naito T, Masaki T, Nikolic-Paterson DJ, Tanji C, Yorioka N, Kohno N. Angiotensin II induces thrombospondin-1 production in human mesangial cells via p38 MAPK and JNK: a mechanism for activation of latent TGF-beta1. Am J Physiol Renal Physiol 2004;286:F278–F287.
108. Belmadani S, Bernol J, Wei C-C, et al. A thrombospondin-1 antagonist of TGF-β activation blocks cardiomyopathy in rats with diabetes and elevated angiotensin II, manuscript submitted.
109. Yang YL, Chuang LY, Guh JY, et al. Thrombospondin-1 mediates distal tubule hypertrophy induced by glycated albumin. Biochem J 2004;379:89–97.
110. Yehualaeshet T, O'Connor R, Begleiter A, Murphy-Ullrich JE, Silverstein R, Khalil N. A CD36 synthetic peptide inhibits bleomycin-induced pulmonary inflammation and connective tissue synthesis in the rat. Am J Respir Cell Mol Biol 2000;23:204–212.
111. Azuma A, Li YJ, Abe S, et al. Interferon-β inhibits bleomycin-induced lung fibrosis by decreasing TGF-β and thrombospondin. Am J Respir Cell Mol Biol 2004;32:93–98.
112. Morishima Y, Nomura A, Uchida Y, et al. Triggering the induction of myofibroblast and fibrogenesis by airway epithelial shedding. Am J Respir Cell Mol Biol 2001;24:1–11.
113. Wesselkamper SCCL, Henning LN, Borchers MT, et al. Gene expression changes during the development of acute lung injury: role of transforming growth factor beta. Am J Respir Crit Care Med 2005;172:1399–1411.
114. El-Youssef M, Mu Y, Huang L, Stellmach V, Crawford SE. Increased expression of transforming growth factor-beta1 and thrombospondin-1 in congenital hepatic fibrosis: possible role of the hepatic stellate cell. J Pediatr Gastroenterol Nutr 1999;28:386–392.
115. Wang XL, Liu SX, Wilcken DE. Circulating transforming growth factor beta 1 and coronary artery disease. Cardiovasc Res 1997;34:404–410.
116. Chamberlain J. Transforming growth factor-beta: a promising target for anti-stenosis therapy. Cardiovasc Drug Rev 2001;19:329–344.

117. Robertson AK, Rudling M, Zhou X, Gorelik L, Flavell RA, Hansson GK. Disruption of TGF-beta signaling in T cells accelerates atherosclerosis. J Clin Invest 2003;112:1342–1350.
118. Hosomi N, Noma T, Ohyama H, Takahashi T, Kohno M. Vascular proliferation and transforming growth factor-beta expression in pre- and early stage of diabetes mellitus in Otsuka Long-Evans Tokushima fatty rats. Atherosclerosis 2002;162:69–76.
119. Riessen R, Axel DI, Fenchel M, Herzog UU, Rossmann H, Karsch KR. Effect of HMG-CoA reductase inhibitors on extracellular matrix expression in human vascular smooth muscle cells. Basic Res Cardiol 1999;94:322–332.
120. Wight TN, Raugi GJ, Mumby SM, Bornstein P. Light microscopic immunolocation of thrombospondin in human tissues. J Histochem Cytochem 1985;33:295–302.
121. Riessen R, Kearney M, Lawler J, Isner JM. Immunolocalization of thrombospondin-1 in human atherosclerotic and restenotic arteries. Am Heart J 1998;135:357–364.
122. Canfield AE, Farrington C, Dziobon MD, et al. The involvement of matrix glycoproteins in vascular calcification and fibrosis: an immunohistochemical study. J Pathol 2002;196:228–234.
123. Chamberlain J, Gunn J, Francis SE, et al. TGFbeta is active, and correlates with activators of TGFbeta, following porcine coronary angioplasty. Cardiovasc Res 2001;50:125–136.
124. Frangogiannis NGRG, Dewald O, Zymek P, et al. Critical role of endogenous thrombospondin-1 in preventing expansion of healing myocardial infacts. Circulation 2005;111:2935–2942.
125. Chipev CC, Simman R, Hatch G, Katz AE, Siegel DM, Simon M. Myofibroblast phenotype and apoptosis in keloid and palmar fibroblasts in vitro. Cell Death Differ 2000;7:166–176.
126. Cauchard JH, Berton A, Godeau G, Hornebeck W, Bellon G. Activation of latent transforming growth factor beta 1 and inhibition of matrix metalloprotease activity by a thrombospondin-like tripeptide linked to elaidic acid. Biochem Pharmacol 2004;67:2013–2022.
127. Muraoka-Cook RS, Dumont N, Arteaga CL. Dual role of transforming growth factor beta in mammary tumorigenesis and metastatic progression. Clin Cancer Res 2005;11:937s–943s.
128. Wakefield LM, Roberts AB. TGF-beta signaling: positive and negative effects on tumorigenesis. Curr Opin Genet Dev 2002;12:22–29.
129. Kawataki T, Naganuma H, Sasaki A, Yoshikawa H, Tasaka K, Nukui H. Correlation of thrombospondin-1 and transforming growth factor-beta expression with malignancy of glioma. Neuropathology 2000;20:161–169.
130. Sasaki A, Naganuma H, Satoh E, Kawataki T, Amagasaki K, Nukui H. Participation of thrombospondin-1 in the activation of latent transforming growth factor-beta in malignant glioma cells. Neurol Med Chir (Tokyo) 2001;41:253–258; discussion 258–259.
131. Miao WM, Seng WL, Duquette M, Lawler P, Laus C, Lawler J. Thrombospondin-1 type 1 repeat recombinant proteins inhibit tumor growth through transforming growth factor-beta-dependent and -independent mechanisms. Cancer Res 2001;61:7830–7839.
132. Yee KO, Streit M, Hawighorst T, Detmar M, Lawler J. Expression of the type-1 repeats of thrombospondin-1 inhibits tumor growth through activation of transforming growth factor-beta. Am J Pathol 2004;165:541–552.
133. Iruela-Arispe ML, Lombardo M, Krutzsch HC, Lawler J, Roberts DD. Inhibition of angiogenesis by thrombospondin-1 is mediated by 2 independent regions within the type 1 repeats. Circulation 1999; 100:1423–1431.
134. Bleuel K, Popp S, Fusenig NE, Stanbridge EJ, Boukamp P. Tumor suppression in human skin carcinoma cells by chromosome 15 transfer or thrombospondin-1 overexpression through halted tumor vascularization. Proc Natl Acad Sci USA 1999;96:2065–2070.
135. Bertin N, Clezardin P, Kubiak R, Frappart L. Thrombospondin-1 and -2 messenger RNA expression in normal, benign, and neoplastic human breast tissues: correlation with prognostic factors, tumor angiogenesis, and fibroblastic desmoplasia. Cancer Res 1997;57:396–399.
136. Filleur S, Volpert OV, Degeorges A, et al. In vivo mechanisms by which tumors producing thrombospondin 1 bypass its inhibitory effects. Genes Dev 2001;15:1373–1382.
137. Mansfield PJ, Suchard SJ. Thrombospondin promotes chemotaxis and haptotaxis of human peripheral blood monocytes. J Immunol 1994;153:4219–4229.
138. Zamiri P, Masli S, Kitaichi N, Taylor AW, Streilein JW. Thrombospondin plays a vital role in the immune privilege of the eye. Invest Ophthalmol Vis Sci 2005;46:908–919.
139. Pierson BA, Gupta K, Hu WS, Miller JS. Human natural killer cell expansion is regulated by thrombospondin-mediated activation of transforming growth factor-beta 1 and independent accessory cell-derived contact and soluble factors. Blood 1996;87:180–189.

140. Doyen V, Rubio M, Braun D, et al. Thrombospondin 1 is an autocrine negative regulator of human dendritic cell activation. J Exp Med 2003;198:1277–1283.
141. Beppu R, Nakamura K, Miyajima-Uchida H, et al. Soluble thrombospondin-1 suppresses T cell proliferation and enhances IL-10 secretion by antigen presenting cells stimulated with phytohemagglutinin. Immunol Invest 2001;30:143–156.
142. Li SS, Forslow A, Sundqvist KG. Autocrine regulation of T cell motility by calreticulin-thrombospondin-1 interaction. J Immunol 2005;174:654–661.
143. Li Z, Calzada MJ, Sipes JM, et al. Interactions of thrombospondins with alpha4beta1 integrin and CD47 differentially modulate T cell behavior. J Cell Biol 2002;157:509–519.
144. Kuznetsova SA, Roberts DD. Functional regulation of T lymphocytes by modulatory extracellular matrix proteins. Int J Biochem Cell Biol 2004;36:1126–1134.
145. Weiss DJ, Evanson OA, Deng M, Abrahamsen MS. Sequential patterns of gene expression by bovine monocyte-derived macrophages associated with ingestion of mycobacterial organisms. Microb Pathog 2004;37:215–224.
146. Omer FM, de Souza JB, Corran PH, Sultan AA, Riley EM. Activation of transforming growth factor beta by malaria parasite-derived metalloproteinases and a thrombospondin-like molecule. J Exp Med 2003;198:1817–1827.
147. Blakytny R, Ludlow A, Martin GE, et al. Latent TGF-beta1 activation by platelets. J Cell Physiol 2004;199:67–76.
148. Tolsma SS, Volpert OV, Good DJ, Frazier WA, Polverini PJ, Bouck N. Peptides derived from two separate domains of the matrix protein thrombospondin-1 have anti-angiogenic activity. J Cell Biol 1993;122:497–511.
149. Ihara Y, Manabe S, Kanda M, et al. Increased expression of protein C-mannosylation in the aortic vessels of diabetic Zucker rats. Glycobiology 2005;15:383–392.
150. Hofsteenge J, Huwiler KG, Macek B, et al. C-mannosylation and O-fucosylation of the thrombospondin type 1 module. J Biol Chem 2001;276:6485–6498.
151. Grainger DJ, Frow EK. Thrombospondin 1 does not activate transforming growth factor beta1 in a chemically defined system or in smooth-muscle-cell cultures. Biochem J 2000;350(Pt 1):291–298.
152. DuBose KB, Talbert MA, Zayzafoon M, Murphy-Ullrich JE. Thrombospondin-1 regulates human mesenchymal stem cell differentiation through control of latent TGF-β activation, in revision.

35 TGF-β in Pulmonary Fibrosis

Masahito Ebina

CONTENTS

PATHOGENESIS OF PULMONARY FIBROSIS
TGF-β1 EXPRESSION IN THE LUNGS OF PATIENTS WITH IPF
TGF-β1 IN DISEASE MODEL OF LUNG INJURY AND FIBROSIS
ANTI-TGF-β TREATMENT AGAINST PULMONARY FIBROSIS
CONCLUSIONS
REFERENCES

Abstract

Transforming growth factor-β1 (TGF-β1) is a key regulator of pulmonary fibrosis as well as other fibrotic diseases of various organs. In the lung patients with idiopathic pulmonary disease, regarded as one of the most refractory inflammatory diseases of the lung, aggressive fibrosis with elevated expression of TGF-β1 is observed without remarkable preceeding inflammation. Because TGF-β1 is an obvious target for antifibrotic therapy, many experimental trials have been reported on successful reduction of fibrosis in disease models. However, in the event of inhibition of the potent antiinflammatory effects of TGF-β1 by anti-TGF-β1 therapy, active inflammation would likely be prolonged, which would delay tissue repair. In this chapter, the evidences reported on the essential roles of TGF-β1 in fibrogenesis in the patients with idiopathic pulmonary fibrosis (IPF) and in its disease models is reviewed. In addition, a pluripotent natural regulator is refered as a promising new therapy against pulmonary fibrosis by inhibition of the collagen synthesis of TGF-β1, inflammation, and apoptosis of alveolar epithelial cells simultaneously along with increases in the epithelial and endothelial cells available for tissue generation.

Key Words: TGF-β1; pulmonary fibrosis; lung injury; fibroblast; HGF.

1. PATHOGENESIS OF PULMONARY FIBROSIS

Idiopathic pulmonary fibrosis (IPF), a chronic but progressive disorder with poor prognosis, is recognized as one of the most refractory inflammatory diseases of the lung *(1)*. The generally accepted gold standard for the diagnosis of IPF is a light microscopy viewing of a surgical lung biopsy which reveals a pattern termed "usual interstitial pneumonia" (UIP), exhibiting temporal heterogeneity of fibrotic lesions *(2)*. Despite a long history of clinical and basic research in IPF, the pathogenesis has not been elucidated yet and no effective therapy has been established.

In general fibrosis is induced after severe inflammation and is characterized by the wound healing process of skin *(3)* or acute respiratory distress syndrome *(4)*. The inflammatory hypothesis in the pathogenesis of IPF *(5)* has fallen out of favor because scrutiny of the UIP pattern in patients fails to reveal evidence of significant interstitial or alveolar inflammation

From: *Cancer Drug Discovery and Development: Transforming Growth Factor-β in Cancer Therapy, Vol I: Basic and Clinical Biology*
Edited by: Sonia B. Jakowlew © Humana Press Inc., Totowa, NJ

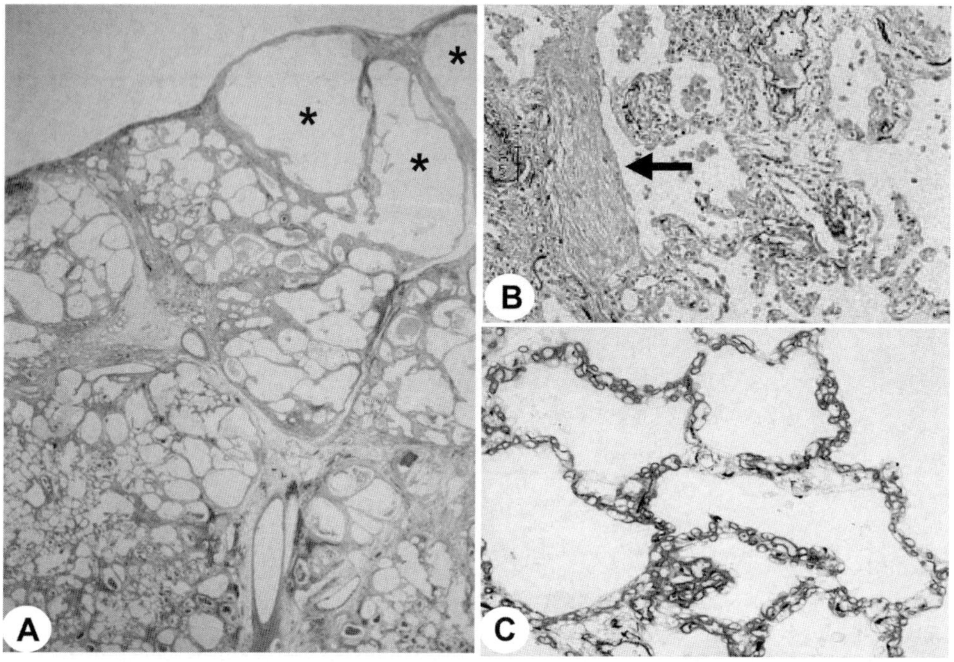

Fig. 1. Pathologic characteristics of the lungs of patients with idiopathic pulmonary fibrosis. (**A**) Increased fibrotic lesions are distributed chiefly in interlobular septa and subpleural regions where *honeycomb* lesions develop (*) (Elastica-Masson stain, ×12.5). (**B**) Fibroblastic foci (indicated by an arrow) contain no vessels inside (Elastica-Masson stain, ×400). (**C**) Alveolar capillaries are increased in nonfibrotic alveolar septa (Immunohistochemical stain for CD34, ×400).

(6). However, a recent hypothesis of inflammation, which is inducible without classic inflammatory cells, might be a key to the mystery of the fibrosing process in IPF. The alveolar macrophages activated in the lungs of patients with IPF secrete growth factors *(7,8)*, including transforming growth factor-β (TGF-β) *(9,10)*, in the process of fibrogenesis. In addition, alveolar epithelial cells *(11)* and endothelial cells *(12)* may also be an important source of these growth factors or TGF-β. Myofibroblasts, the leading player in fibrogenesis, may also be a key source of factors, which contribute to the fibrogenesis in IPF *(13)*.

The initial alveolar epithelial cell injury which induces apoptosis of these cells has been proposed to lead to pathogenic fibrosis *(14)*. It is unfortunate, however, that this hypothesis does not explain the pathogenesis of the heterogeneous distribution of fibrotic lesions, chiefly in the interlobular septa and in the subpleural regions in the lungs of patients with IPF (Fig. 1A). The demonstration of remodeling of alveolar capillaries, entailing a decrease in massive fibrotic lesions but increase in non-fibrotic alveolar septa (Fig. 1B&C), shed new light on the pathogenesis of this desperate disease *(15)*, considering the findings that TGF-β directly increases the permeability of endothelial monolayers and regulates angiogenesis *(16,17)*. Thus, there remains a great deal, which remains to be clarified regarding the pathogenesis of IPF. The current conceptual pathogenesis of IPF is schematically depicted in Figure 2.

2. TGF-B1 EXPRESSION IN THE LUNGS OF PATIENTS WITH IPF

The augmented expression of TGF-β in the lungs of patients with IPF was first described in 1988 *(9)* with evidence for the expression of extracellular matrix (ECM) genes *(10)* by lung fibroblasts *(18)*. According to subsequent additional information on the subtypes of TGF-β, TGF-β1 (but not TGF-β2 or TGF-β3) was exhibited to be differentially present in epithelial cells and advanced pulmonary fibrosis *(19)*.

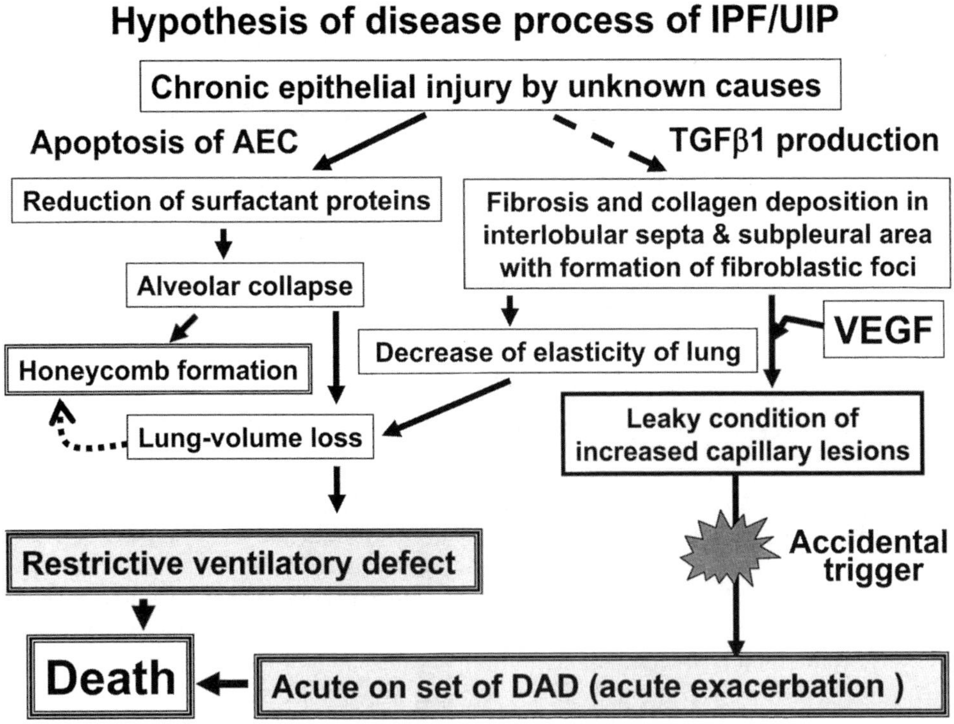

Fig. 2. Schema of the hypothesis of disease process of IPF/UIP. Chronic epithelial injury unknown etiology causes apoptotic condition of alveolar Type II epithelial cells, which reduces surfactant protein synthesis to proceed to alveolar collapse and increase TGF-β1 production to differentiate fibroblasts into myofibroblasts. The heterogeneity of fibrotic lesions distribute chiefly in interlobular septa and subpleural space, and the nonfibrotic alveolar septa increase the capillaries where remained leaky status to cause diffuse alveolar damage by an accidental trigger.

TGF-β1 is released in a biologically latent form (L-TGF-β1). In order to become biologically active, TGF-β must be converted to its active form and interact with both TGF-β receptors type I and II (TβR-I and TβR-II). The latent forms of TGF-β1 are related to the pathologic distribution of active TGF-β1 in the lungs of patients with IPF *(20)*. TGF-β latency binding protein-1, which facilitates the release and activation of L-TGF-β1, is also important in the biology of TGF-β1. L-TGF-β1 has been found to be present in all of the lung cells of patients with IPF, and TGF-β latency binding protein-1 is detected primarily in alveolar macrophages and epithelial cells lining honeycomb cysts in areas of advanced IPF *(20)*. Because TβR-I was markedly reduced in most cells in the areas of honeycomb cysts except for interstitial myofibroblasts in the lungs of patients with IPF, TGF-β1 is thought to inhibit epithelial proliferation, and a lack of TβR-I expression by the epithelial cells lining honeycomb cysts to facilitate epithelial cell proliferation *(20)*.

3. TGF-β1 IN DISEASE MODEL OF LUNG INJURY AND FIBROSIS

3.1. Cytokines and TGF-β1

The animal models of IPF have not been established yet, but the fibrotic changes in the lungs after acute lung injury caused by cytokines or chemical drugs, such as bleomycin, have been closely investigated. Following lung injury, TGF-α has been most thoroughly evaluated during the late phases of tissue repair, where it plays a critical role in the development of pulmonary fibrosis *(10)*. A central role as a regulator of fibrogenesis in bleomycin- treated lung is demonstrated by the effect of a blockade of TGF-β by its antibody *(21)*.

TGF-β1 knockout mice develop diffuse mononuclear cell infiltrates that prove lethal within a few weeks from birth *(22)*. In contrast, TGF-β2 and TGF-β3 knockout mice display only developmental defects. Major differences among TGF-β isoform functions in vivo are due at least in part to differences in the promoter regions of the various isoform genes *(23)*.

Although chemokines such as interleukin-8 (IL-8), RANTES, IP-10, MIG or lymphotactin do not induce fibrosis when overexpressed in rodent lung, there have been cytokines and growth factors found to indeed be profibrotic. IL-1β demonstrates marked inflammation, tissue damage and chronic fibrosis; TNFβ, which induces inflammation and mild fibrosis; and GM-CSF, which induces moderate inflammation and fibrosis *(24)*.

In addition to the proinflammatory activities of TNF-α in vivo *(25)*, its possible fibrogenic potential is supposed, given that it is both mitogenic and chemotactic for fibroblasts *(26)*, although its effects on collagen gene expression are inhibitory *(27)*. Upregulation of TNF-α in vivo does not always result in the same biological outcome. In particular, TNF-α is detected in many inflammatory and immune diseases that resolve without tissue fibrosis *(28)*, yet in certain pathologies TNF-α upregulation is reported to be associated with fibrotic sequelae *(29,30)*. Sime et al. demonstrated that local over expression of TNF-α induced by cDNA gene transfer by adenoviral vector (overexpression for 7–10 d) to the respiratory epithelium resulted in severe pulmonary inflammation with significant increases in neutrophils, macrophages, and lymphocytes and, to a lesser extent, eosinophils, with a peak (at) on day 7 *(31)*. By day 14, the inflammatory cell accumulation had declined, and fibrogenesis became evident, with fibroblast accumulation and deposition of ECM proteins. In bronchoalveolar fluids, TGF-β1 was transiently elevated from day 7 (peak at day 14) immediately preceding the onset of fibrogenesis. There was a striking accumulation of myofibroblasts from day 7 onwards, with the most extensive on day 14, which suggested that the fibrogenesis might be mediated by secondary upregulation of TGF-β1 and the induction of pulmonary myofibroblasts *(31)*.

Induction of IL-1β in the lung has also shown to cause fibrosis after severe inflammation in the lung *(32)*. IL-1β treatment of rats with a single dose of recombinant adenovirus encoding IL-1β, a central regulator of acute inflammation, leads to a progressive form of pulmonary fibrosis that continues over a period of at least 60 d. IL-1β expression increased only transiently, returning to near base-line values by 14 d after infection. In contrast, pulmonary fibrosis is not apparent until day 21 and dramatically increases thereafter. Thus, IL-1β can be added to a short list of agents, including the cytotoxic drug bleomycin, ligands of Fas death receptor, and TGF-β1 that can bypass the normal repair process and initiate a self-perpetuating cycle of pulmonary fibrosis *(32)*.

After the induction of injury, expression levels of several TGF-β-inducible genes dramatically increase as early as the second day, a time point that precedes the maximal increase in alveolar flooding *(21)*. Endothelial cells also produce TGF-β after bleomycin-induced lung injury *(7)*. By adenovector-mediated gene transfer, transient overexpression of active, but not latent, TGF-β1 resulted in prolonged and severe interstitial and pleural fibrosis characterized by extensive deposition of the ECM proteins collagen, fibronectin, and elastin, and by the emergence of cells with the myofibroblast phenotype *(33)*. With the activation of latent TGF-β1 by binding to αvβ6, sites in the β6 cytoplasmic domain become accessible for binding to the actin cytoskeleton *(34)*. This mechanism was also revealed in the finding that mice lacking αvβ6 were completely protected from pulmonary edema after bleomycin-induced acute lung injury *(35)*.

3.2. Collagen Synthesis by TGF-β1

TGF-β1 is the most potent direct stimulator of collagen production known. Connective tissue growth factor, discovered as a protein secreted by human endothelial cells *(36)*, is induced by TGF-β and is considered a downstream mediator of the effects of TGF-β on fibroblasts *(37)*. Similarly, TGF-β induces expression of an extracellular domain (ED) A splice variant of cellular fibronectin (ED-A FN), a variant of fibronectin that results from alternative splicing of the fibronectin transcript *(38)*. This induction of ED-A FN is required for TGF-β1-triggered enhancement of α-smooth muscle actin (α-SMA) and collagen type I expression *(39)*. In contrast, the proinflammatory cytokine TNF-α, which is expressed by macrophages during the wound healing response *(40)*, is antifibrotic in that this cytokine suppresses the expression of matrix genes *(41)*. The proinflammatory cytokine IFN-γ, released by T-cells immediately after injury, suppresses collagen synthesis *(42)*.

TNF-α and TGF-β induce prostaglandin production in fibroblasts, and COX-1 expression is induced by TGF-β, whereas COX-2 is induced by TNF-α *(43)*. Prostaglandin levels appear to be reduced in IPF, due in part to a reduced expression of COX2 by fibroblasts in this disorder *(44,45)*, suggesting the fibrosis may be exacerbated in IPF owing to a reduction in the ability of fibroblasts to synthesize this normally antifibrotic agent. This illustrates the capacity of exogenously added prostaglandins to suppress the fibrotic response may result from a rescue of this deficiency. Thus, it is tempting so speculate that fibrosis may develop in part owing to a failure of the normal precise control that suppresses and terminates the wound healing response to operate appropriately *(46)*.

3.3. Apoptosis and TGF-β

Apart from the direct influence on inflammation and the activation of collagen synthesis, TGF-β stimulates fibroblast differentiation to the myofibroblast phenotype and inhibits myofibroblast apoptosis *(47)*. TGF-β1 treatment also completely suppresses IL-1β-induced apoptosis in myofibroblasts. In contrast, TGF-β1 induces apoptosis of primary cultured bronchiolar epithelial cells via caspase-3 activation and downregulation of the cyclin-dependent kinase inhibitor p21 *(48)*. This might also occur in the alveolar epithelial cells in the lungs of patients of IPF, because TGF-β1 enhances Fas-mediated epithelial cell apoptosis and lung injury via caspase-3 activation in mice *(48)*. In gene transgenic experiments, TGF-β1 induces epithelial apoptosis, followed by mononuclear-rich inflammation, tissue fibrosis, myofibroblast and myocyte hyperplasia, and septal rupture with honeycombing, all of which is mediated by the early growth response gene (Egr)-1 *(49)*.

3.4. Mesenchymal Cell Transition by TGF-β1

Recent evidence suggests the possible transition of alveolar epithelial cells to mesenchymal cells by TGF-β1 in vitro *(50,51)*. It would be very difficult to show these phenomena in vivo and direct evidence has not been obtained in lungs of patients with IPF. Thus, the importance of the role of epithelial mesenchymal transmission (EMT) in the pathogenesis of IPF would be undetermined even if EMT were to be shown to be the mechanism of behind the fibroblastic foci frequently observed in the lungs of patients with IPF.

4. ANTI-TGF-β TREATMENT AGAINST PULMONARY FIBROSIS

Because fibrotic disease comprises one of the largest groups of disorders for which there are no effective therapy. And because TGF-β1 is known to be the most potent stimulator of collagen production in a variety of organs, including the lung, a number of experimental trials have been undertaken to block the function of TGF-β1 upstream or downstream of TGF-β1 or by blocking TGF-β1 itself.

In terms of the current clinical treatment of pulmonary fibrosis, combination therapy of immunosuppressive drugs and steroid is generally recommended *(52)*. Although both Cyclosporin A and IFN-γ have only rarely been used in IPF with limited data to support their effect, the molecular mechanisms of inhibitory effects of TGF-β1 have been demonstrated. Eickelberg, et al. found that Cyclosporin A and IFN-γ, but not glucocorticoids, cyclophosphamide, or azathioprine, inhibit TGF-β-induced signaling and collagen deposition in lung fibroblasts, mediated by JunD/AP-1 activation. The antagonism of TGF-β by Cyclosporin A is achieved via direct inhibition of JunD/AP-1 activation. In contrast, the effects of IFN-γ require STAT-1, which inhibits AP-1 transcriptional activity via competition for CBP/p300 *(53)*.

IL-7 has also both shown to inhibit both TGF-β production and signaling in fibroblasts and requires an intact JAK1/STAT1 signal transduction pathway *(54)*. IL-7-mediated inhibition of TGF-β signaling is associated with an increase in Smad7, a major inhibitory regulator in the SMAD family. In the presence of IL-7, Smad7 dominant negative fibroblasts restore TGF-β-induced collagen synthesis, indicating that an IL-7-mediated increase in Smad7 suppresses TGF-β signaling. Consistent with these findings in vitro, recombinant IL-7 decreases bleomycin-induced pulmonary fibrosis in vivo, independent of IFN-γ *(54)*.

Smad7 has been identified as an intracellular antagonist of TGF-β signaling; it inhibits TGF-β-induced transcriptional responses *(55)*. Smad7 associates with activated TGF-β receptors and interferes with the activation of Smad2 and Smad3 by preventing their receptor interaction and phosphorylation *(55)*. C57BL/6 mice with bleomycin-induced lungs receiving an intratracheal transfection of adenoviral vectors carrying Smad7 cDNA exhibited suppressed type I precollagen mRNA and hydroxyproline content without morphological fibrotic responses in the lungs *(56)*. Expression of the Smad7 transgene blocked Smad2 phosphorylation induced by bleomycin in mouse lungs *(56)*.

Decorin is a ubiquitous proteoglycan, one of the matrix components induced by TGF-β with a core protein of 45 kDa. Decorin contains two binding sites for TGF-β and neutralizes its biological activity. Therefore, decorin appears to be a natural regulator of TGF-β *(57)*. The therapeutic effect of decorin was first described in a kidney disease model *(57)*, and in a bleomycin-induced lung fibrosis *(58,59)*. We constructed a replication-defective recombinant adenovirus harboring the human decorin gene (AdCMV.DC) and administered 1×10^9 plaque- forming units of AdCMV.DC intratracheally or intravenously to C57BL/6 mice with a chronic intraperitoneal (i.p.) injection of bleomycin. This induced a subpleural fibroproliferation, mimicking UIP, by day 28 *(59)*. Intratracheal administration of AdCMV.DC alone increased decorin mRNA expression in the lung and decreased the hydroxyproline content augmented in bleomycin-induced pulmonary fibrosis (*[59]*, Fig. 3). Interestingly, transfection of the decorin gene to mice without bleomycin injury promoted inflammation induced by the adenoviral null vector alone (Fig. 4A,B). Moreover, the lungs injured by bleomycin and transfected with the decorin gene exhibited reduced fibrosis but prolonged inflammation (Fig. 4C). These phenomena are thought to be caused by the inhibitory effects of decorin administered against the various functions of TGF-β including its pronounced anti-inflammation, which should be taken into consideration prior to clinical applications.

Hepatocyte growth factor (HGF), a mesenchymal-derived factor which regulates the growth, motility, and morphogenesis of epithelial and endothelial cells, was found to be negatively regulated by TGF-β1 in human lung embryonic fibroblast cell line *(60)* at the posttranscriptional level *(61)*. Although HGF does not inhibit TGF-β1-induced ECM protein expression *(61)*, administration of recombinant HGF to the mice sensitized by ovalbumin challenge significantly suppressed TGF-β in BAL fluid and in lung tissues *(62)*.

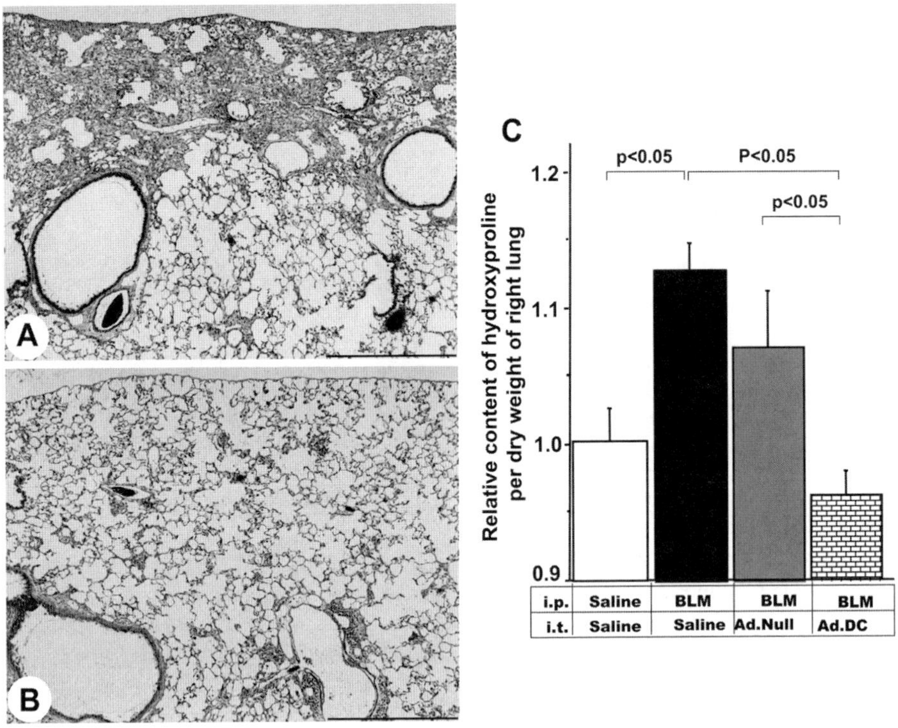

Fig. 3. Inhibitory effects of decorin gene transfection to the bleomycin-induced lung fibrosis of mice. (**A**) Subpleural distribution of fibrotic lesions after bleomycin-induced lung injury (bar = 500 μm). (**B**) Decorin gene transfer by intratracheal administration reduced the fibrotic changes (bar = 500 μm). (**C**) Intratracheal administration of AdCMV.DC (BLM/AdCMV.DC-i.t.) reduced the hydroxyproline content, in contrast to that of BLM/saline-i.t. and BLM/AdCMV.Null-i.t. with statistic difference.

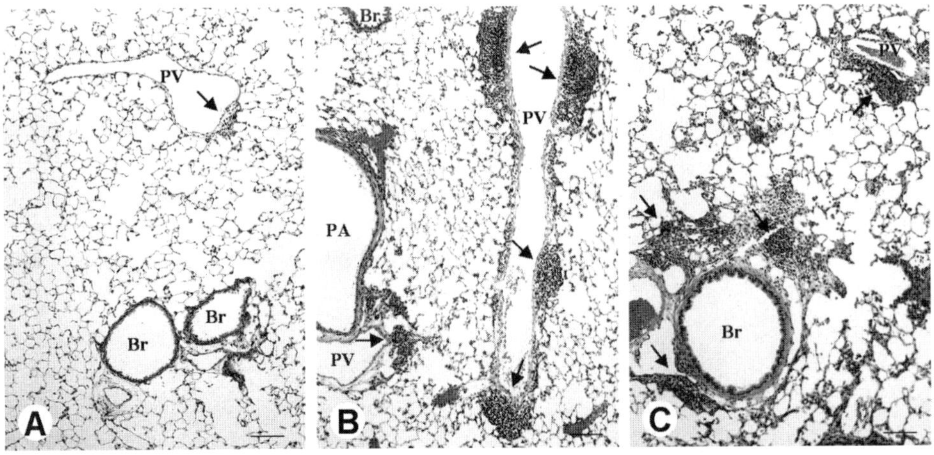

Fig. 4. Inflammation after intratracheal administration of AdCMV.DC. (H&E stain, bar = 100 μm). (**A**) Inflammatory cells (indicated by an arrow) are scarcely distributed in the lungs 28 days after intratracheal administration of AdCMV.Null. (**B**) Mononuclear cells are infiltrated around pulmonary veins (arrows) in the lungs 28 d after intratracheal administration of AdCMV.DC. (**C**) In the bleomycin-induced fibrotic lungs treated with intratracheal administration of AdCMV.DC., inflammatory cells were apparent, though fibrosis was remarkably decreased. (Br, broncho-bronchiole; PV, pulmonary vein; PA, pulmonary artery).

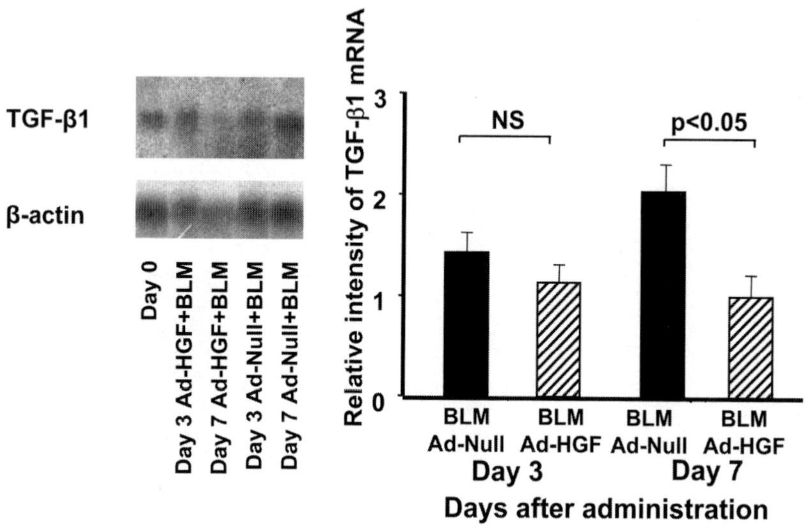

Fig. 5. Adenoviral transduction of HGF gene to bleomycin-induced lungs. Adenoviral transduction of HGF gene to bleomycin-induced lungs of mice reduced the expression of TGF-β1 mRNA in the lungs in day 7 after administration of HGF gene.

Figure 5 shows adenoviral gene transfer of HGF to mice after bleomycin-induced lung injury, and the mRNA expression of TGF-β in the lung tissues to be reduced by day 7. We also examined the therapeutic gene transfer of human hepatocyte growth factor (hHGF) to alveolar septa in mouse bleomycin-induced lung fibrosis using macroaggregated albumin-polyethylenimine complex (MAA-PEI) *(63)*. Intravenous administration of MAA-PEI along with 1 μg of pCAG.hHGF to C57BL/6 mice increased the uptake of plasmids into alveolar capillary endothelial cells and epithelial cells, prolonged hHGF expression in the lung, and induced a level of hHGF expression equal to that seen with 10 μg of hHGF-expression plasmids alone. The exogenously sourced hHGF gene expression increased the endogenous mouse HGF in the lungs and significantly decreased TNF-α, IL-6 and collagen synthesis after bleomycin injury. Because GFP-labeled bone marrow-derived stem cells after bleomycin injury were reduced in number by HGF, the primary mechanism of HGF is likely to be the prevention of apoptosis, as has been suggested by in vitro experiments *(63)*.

Other substances have been examined in the fibrotic diseases caused by TGF-β in other organs, such as soluble TβRII fragments *(64)*, tranilas *(65)*, neutralizing antibodies *(66)*, threonine kinase inhibitors *(67)*, and RNA expression inhibitors such as antisense expression vectors or blocking oligonucleotides *(68)*.

5. CONCLUSIONS

Because TGF-β1 has an established role in tissue fibrosis, effective inhibition of TGF-β1 is an obvious target for antifibrotic therapy. Such therapy is urgently required for the successful treatment of currently refractory diseases of various organs, including IPF. However, in the event of inhibition of the potent antiinflammatory effects of TGF-β by anti-TGF-β therapy, active inflammation would likely be prolonged, as shown in Figure 4, which would delay tissue repair. The opposite effects on myofibroblast apoptosis, which is stimulated by TGF-β but inhibited by IL-1β, may in part support this phenomenon. Therefore, a pluripotent natural regulator, such as HGF, which inhibits the collagen synthesis of TGF-β, inflammation, and apoptosis of alveolar epithelial cells simultaneously and more,

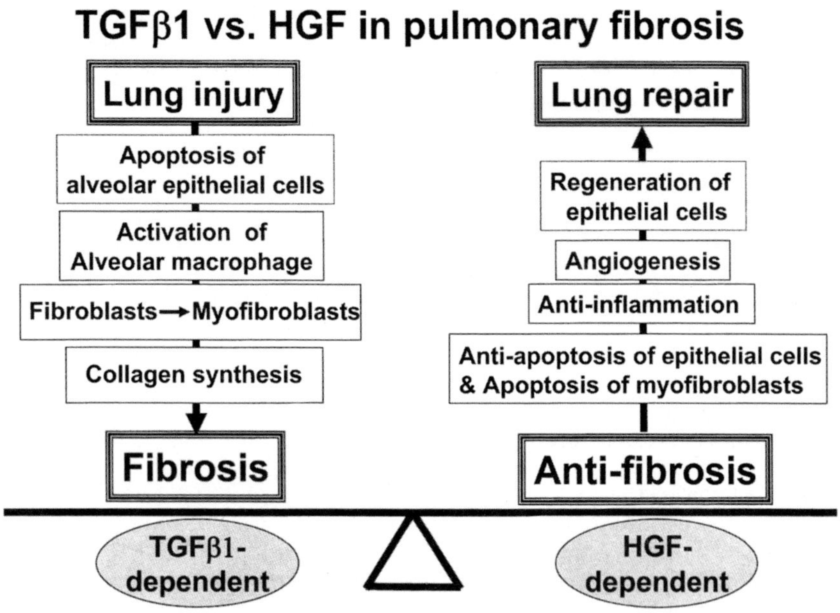

Fig. 6. Schema of pathway of TGF-β1 increased in injured lung to pulmonary fibrosis and antifibrotic pathway by HGF to lung repair.

increases the epithelial and endothelial cells available for tissue generation, is expected to be a more useful approach for the treatment of IPF (Fig. 6).

REFERENCES

1. Idiopathic pulmonary fibrosis: Diagnosis and treatment. International consensus statement. Am J Respir Crit Care Med 2000;161:646–664.
2. Katzenstein AL, Myers JL. Idiopathic pulmonary fibrosis: clinical relevance of pathologic classification. Am J Respir Crit Care Med 1998;157:1301–1315.
3. Singer AJ, Clark RA. Cutaneous wound healing. N Engl J Med 1999;341:738–746.
4. Ware LB, Matthay MA. The acute respiratory distress syndrome. N Engl J Med 2000;342:1334–1349.
5. Crystal RG, Fulmer JD, Roberts WC, Moss ML, Line BR, Reynolds HY. Idiopathic pulmonary fibrosis: clinical, histologic, radiographic, physiologic, scintigraphic, cytologic, and biochemical aspects. Ann Intern Med 1976;85:769–788.
6. Noble PW, Homer RJ. Historical perspective on the pathogenesis of idiopathic pulmonary fibrosis. Am J Respir Cell Mol Biol 2005;33:113–120.
7. Martinet Y, Bitterman PB, Mornex JF, Grotendorst GR, Martin GR, Crystal RG. Activated human monocytes express the c-sis proto-oncogene and release a mediator showing PDGF-like activity. Nature 1986;319:158–160.
8. Nagaoka I, Trapnell BC, Crystal RG. Upregulation of platelet-derived growth factor-A and -B gene expression in alveolar macrophages of individuals with idiopathic pulmonary fibrosis. J Clin Invest 1990;85:2023–2027.
9. Yamauchi K, Martinet Y, Basset P, Fells GA, Crystal RG. High levels of transforming growth factor-beta are present in the epithelial lining fluid of the normal human lower respiratory tract. Am Rev Respir Dis 1988;137:1360–1363.
10. Broekelmann TJ, Limper AH, Colby TV, McDonald JA. Transforming growth factor beta 1 is present at sites of extracellular matrix gene expression in human pulmonary fibrosis. Proc Natl Acad Sci USA 1991;88:6642–6646.
11. Antoniades HN, Bravo MA, Avila RE, et al. Platelet-derived growth factor in idiopathic pulmonary fibrosis. J Clin Invest 1990;86:1055–1064.

12. Phan SH, Gharaee-Kermani M, Wolber F, Ryan US. Stimulation of rat endothelial cell TGFβ production by bleomycin. J Clin Invest 1991;87:148–154.
13. Mizuno S, Matsumoto K, Li MY, Nakamura T. HGF reduces advancing lung fibrosis in mice: a potential role for MMP-dependent myofibroblast apoptosis. FASEB J 2005;19:580–582.
14. Kuwano K, Hagimoto N, Kawasaki M, et al. Essential roles of the Fas-Fas ligand pathway in the development of pulmonary fibrosis. J Clin Invest 1999;104:13–19.
15. Ebina M, Shimizukawa M, Shibata N, et al. Heterogeneous increase in CD34-positive alveolar capillaries in idiopathic pulmonary fibrosis. Am J Respir Crit Care Med 2004;169:1203–1208.
16. Hurst V IV, Goldberg PL, Minnear FL, Heimark RL, Vincent PA. Rearrangement of adherens junctions by TGFβ1: role of contraction. Am J Physiol 1999;276:L582–L595.
17. Larsson J, Goumans MJ, Sjostrand LJ, et al. Abnormal angiogenesis but intact hematopoietic potential TGF-β type I receptor-deficient mice. EMBO J 2001;20:1663–1673.
18. Raghu G, Masta S, Meyers D, Narayanan AS. Collagen synthesis by normal and fibrotic human lung fibroblasts and the effect of transforming growth factor-β. Am Rev Respir Dis 1989;140:95–100.
19. Khalil N, O'Connor RN, Flanders KC, Unruh H. TGFβ1, but not TGFβ2 or TGF-β3, is differentially present in epithelial cells or advanced pulmonary fibrosis: an immnohistochemical study. Am J Respir Cell Mol Biol 1996;14:131–138.
20. Khalil N, Parekh TV, O'Connor R, et al. Regulation of the effects of TGF-β1 by activation of latent TGF-β1 and differential expression of TGF-β receptors (TbR-I and TbR-II) in idiopathic pulmonary fibrosis. Thorax 2001;56:907–915.
21. Giri SN, Hyde DM, Hollinger MA. Effects of antibody to TGFβ on bleomycin induced accumulation of lung collagen in mice. Thorax 1993;48:959–966.
22. Shull MM, Ormsby I, Kier AB, et al. Targeted disruption of the mouse transforming growth factor-β1 gene results in multifocal inflammatory disease. Nature 1992;359:693–699.
23. Saharinen J, Taipale J, Monni O, Keski-Oja J. Identification and characterization of a new latent transforming growth factor-β-binding protein, LTBP-4. J Biol Chem. 1998;273:18,459–18,469.
24. Kelly M, Kolb M, Bonniaud P, Gauldie J. Re-evaluation of fibrogenic cytokines in lung fibrosis. Curr Pharm Des 2003;9:39–49.
25. Ulich TR. Tumore necrosis factor. In: Cytokines of the lung (Kelleys J, ed.), Marcel Dekker, New York. 1993;pp.:307–332.
26. Leibovich SJ, Polverini PJ, Shepard HM, Wiseman DM, Shively V, Nuseir N. Macrophage-induced angiogenesis is mediated by TNF-α. Nature 1987;329:630–632.
27. Mauviel A, Daireaux M, Redini F, Galera P, Loyau G, Pujol JP. TNF inhibits collagen and fibronectin synthesis in human dermal fibroblasts. FEBS Lett 1988;236:47–52.
28. Nelson S, Bagby GJ, Bainton BG, Wilson LA, Thompson JJ, Summer WR. Compartmentalization of intra-alveolar and systemic LPS-induced TNF and the pulmonary inflammatory response. J Infect Dis 1989;159:189–194.
29. Phan SH, Kunkel SL. Lung cytokine production in bleomycin-induced pulmonary fibrosis. Exp Lung Res 1992;18:29–43.
30. Piguet PF, Collart MA, Grau GE, Sappino AP, Vassalli P. Recruitment of TNF for development of silica-induced pulmonary fibrosis. Nature 1990;344:245–247.
31. Sime PJ, Marr RA, Gauldie D, et al. Transfer of Tumor Necrosis Factor-α to rat lung induces severe pulmonary inflammation and patchy interstitial fibrogenesis with induction of transforming growth factor-b1 and myofibroblasts. Am J Pathol 1998;153:825–832.
32. Kolb M, Margetts PJ, Anthony DC, Pitossi F, Gauldie J. Transient expression of IL-1b induces acute lung injury and chronic repair leading to pulmonary fibrosis. J Clin Invest 2001;107:1529–1536.
33. Sime PJ, Xing Z, Graham FL, Csaky KG, Gauldie J. Adenovector-mediated gene transfer of active TGFb1 induces prolonged severe fibrosis in rat lungs. J Clin Invest 1997;100:768–776.
34. Munger JS, Huang X, Kawakatsu H, et al. The integrin alpha v beta 6 binds and activates latent TGF beta 1: a mechanism for regulating pulmonary inflammation and fibrosis. Cell 1999;96:319–328.
35. Pittet JF, Griffiths MJ, Geiser T, et al. TGF-beta is a critical mediator of acute lung injury. J Clin Invest 2001;107:1537–1544.
36. Bradham DM, Igarashi A, Potter RL, Grotendorst GR. Connective tissue growth factor: a cysteine-rich mitogen secreted by human vascular endothelial cells is related to the SRC-induced immediate early gene product CEF-10. J Cell Biol 1991;114:1285–1294.
37. Leask A, Denton CP, Abraham DJ. Insights into the molecular mechanism of sustained fibrosis: The role of connective tissue growth factor in scleroderma. J Invest Dermatol 2004;122:1–6.

38. Grotendorst GR. Connective tissue growth factor: a mediator of TGF-beta action on fibroblasts. Cytokine Growth Factor Rev 1997;8:171–179.
39. Serini G, Bochaton-Piallat ML, Ropraz P, et al. The fibronectin domain ED-A is crucial for myofibroblastic phenotype induction by transforming growth factor-β. J. Cell Biol 1998;142:873–881.
40. Hill AG. Initiators and propagators of the metabolic response to injury. World J Surg 2000;24: 624–629.
41. Hatamochi A, Mori K, Ueki H. Role of cytokines in controlling connective tissue gene expression. Arch Dermatol Res 1994;287:115–121.
42. Ghosh A K, Yuan W, Mori Y, Chen SJ, Varga J. Antagonistic regulation of type I collagen gene expression by interferon-gamma and transforming growth factor-β. Integration at the level of p300/CBP transcriptional coactivators. J Biol Chem 2001;276:11,041–11,048.
43. Diaz A, Chepenik KP, Korn JH, Reginato AM, Jimenez SA. Differential regulation of cyclooxygenases 1 and 2 by interleukin-1 beta, tumor necrosis factor-alpha, and transforming growth factor-β1 in human lung fibroblasts. Exp Cell Res 1998;241:222–229.
44. Wilborn J, Crofford LJ, Burdick MD, Kunkel SL, Strieter RM, Peters-Golden M. Cultured lung fibroblasts isolated from patients with idiopathic pulmonary fibrosis have a diminished capacity to synthesize prostaglandin E2 and to express cyclooxygenase-2. J Clin Invest 1995;95:1861–1868.
45. Petkova DK, Clelland CA, Ronan JE, Lewis S, Knox AJ. Reduced expression of cyclooxygenase (COX) in idiopathic pulmonary fibrosis and sarcoidosis. Histopathology 2003;43:381–386.
46. Leask A, Abraham KDJ. TGF-β signaling and the fibrotic response. FASEB J 2004;18:816–827.
47. Zhang HY, Phan SH. Inhibition of myofibroblast apoptosis by transforming growth factor β1. Am J Respir Cell Mol Biol 1999;21:658–665.
48. Hagimoto N, Kuwano K, Inoshima I, et al. TGF- β1 as an enhancer of Fas-mediated apoptosis of lung epithelial cells. J Immunol 2002;168:6470–6478.
49. Lee CG, Cho SJ, Kang MJ, et al. Early growth response gene 1-mediated apoptosis is essential for transforming growth factor β1-induced pulmonary fibrosis. J Exp Med 2004;200:377–389.
50. Willis BC, Liebler JM, Luby-Phelps K, et al. Induction of epithelial-mesenchymal transition in alveolar epithelial cells by transforming growth factor-β1: potential role in idiopathic pulmonary fibrosis. Am J Pathol 2005;166:1321–1332.
51. Kasai H, Allen JT, Mason RM, Kamimura T, Zhang Z. TGF- β1 induces human alveolar epithelial to mesenchymal cell transition (EMT). Respir Res 2005;6:56.
52. American Thoracic Society. Idiopathic pulmonary fibrosis: diagnosis and treatment. International consensus statement. American Thoracic Society, and the European Respiratory Society. Am J Respir Crit Care Med 2000;161:646–664.
53. Eickelberg O, Pansky A, Koehler E, et al. Molecular mechanisms of TGF-(beta) antagonism by interferon (gamma) and cyclosporine A in lung fibroblasts. FASEB J 2001;15:797–806.
54. Huang M, Sharma S, Zhu LX, et al. IL-7 inhibits fibroblast TGF-beta production and signaling in pulmonary fibrosis. J Clin Invest 2002;109:931–937.
55. Nakao A, Afrakhte M, Moren A, et al. Identification of Smad7, a TGFbeta-inducible antagonist of TGF-beta signalling. Nature 1997;389:631–635.
56. Nakao A, Fujii M, Matsumura R, et al. Transient gene transfer and expression of Smad7 prevents bleomycin-induced lung fibrosis in mice. J Clin Invest 1999;104:5–11.
57. Border WA, Noble NA, Yamamoto T, et al. Natural inhibitor of transforming growth factor-β protects against scarring in experimental kidney disease. Nature 1992;360:361–364.
58. Kolb M, Margetts PJ, Galt T, et al. Transient transgene expression of decorin in the lung reduces the fibrotic response to Bleomycin. Am J Respir Crit Care Med 2001;163:770–777.
59. Shimizukawa M, Ebina M, Narumi K, Kikuchi T, Munakata H, Nukiwa T. Intratracheal gene transfer of decorin reduces subpleural fibroproliferation induced by bleomycin. Am J Physiol Lung Cell Mol Physiol 2003;284:L526–L532.
60. Matsumoto K, Tajima H, Okazaki H, Nakamura T. Negative regulation of hepatocyte growth factor gene expression in human lung fibroblasts and leukemic cells by transforming growth factor-beta 1 and glucocorticoids. J Biol Chem 1992;267:24,917–24,920.
61. Harrison P, Bradley L, Bomford A. Mechanism of regulation of HGF/SF gene expression in fibroblasts by TGF-beta1. Biochem Biophys Res Commun 2000;271:203–211.
62. Ito W, Kanehiro A, Matsumoto K, et al. Hepatocyte growth factor attenuates airway hyperresponsiveness, inflammation, and remodeling. Am J Respir Cell Mol Biol 2005;32:268–280.
63. Watanabe M, Ebina M, Orson FM, et al. Hepatocyte growth factor gene transfer to alveolar septa for effective suppression of lung fibrosis. Mol Ther 2005;12:58–67.

64. Ueno H, Sakamoto T, Nakamura T, et al. A soluble transforming growth factor beta receptor expressed in muscle prevents liver fibrogenesis and dysfunction in rats. Hum Gene Ther 2000;11: 33–42.
65. Kosuga K, Tamai H, Ueda K, et al. Effectiveness of tranilast on restenosis after directional coronary atherectomy. Am Heart J 1997;134:712–718.
66. Hill C, Flyvbjerg A, Rasch R, Bak M, Logan A. Transforming growth factor-β2 antibody attenuates fibrosis in the experimental diabetic rat kidney. J Endocrinol 2001;170:647–651.
67. Callahan JF, Burgess JL, Fornwald JA, et al. Identification of novel inhibitors of the transforming growth factor β1 (TGF-β1) Type 1 receptor (ALK5). J Med Chem 2002;45:999–1001.
68. Maggard M, Meng L, Ke B, Allen R, Devgan L, Imagawa DK. Antisense TGF-β2 immunotherapy for hepatocellular carcinoma: Treatment in a rat tumor model. Ann Surg Oncol 2001;8:32–37.

36 The Role of TGF-β in Bleomycin Induced Pulmonary Fibrosis

Nasreen Khalil and Robert O'Connor

CONTENTS

INTRODUCTION
MECHANISM OF BLEOMYCIN ACTION
TRANSFORMING GROWTH FACTOR-β
CLINICAL USE OF BLEOMYCIN
CONCLUSIONS
REFERENCES

Abstract

Bleomycin sulfate, a combination of two analogues of bleomycin, bleomycin A_2 and B_2 is an antineoplastic antibiotic used as a chemotherapeutic agent for a number of malignancies. Bleomycin has several functional domains but its most critical action, which is oxidative damage to DNA, is located in the N-terminal domain that binds to oxygen and then to DNA by intercalation. The clinical use of bleomycin is limited by the potential pulmonary toxicity and fibrosis associated with bleomycin. In animal models of bleomycin induced pulmonary fibrosis (BPF), TGF-β1, a potent fibrogenic cytokine is elevated in the alveolar macrophages, epithelial cells, endothelial cells, and interstitial fibroblasts after bleomycin administration. The release of transforming growth factor-β (TGF-β1) in an active form by alveolar macrophages and conversion of biologically latent-TGF-β1 to an active conformation by alveolar epithelial cells (AECs) are critical to the inflammatory and fibrotic sequelae of bleomycin lung toxicity (BLT). A number of agents can abrogate BLT in animal models by decreasing the release of active TGF-β1 or reducing the expression or effects of TGF-β1. In humans, the mainstay of management of BLT remains prevention by using a reduced dose of bleomycin and high dose corticosteroids in the event of BLT.

Key Words: Bleomycin; CD36; pneumonitis; pulmonary fibrosis; TGF-β1.

1. INTRODUCTION

The antineoplastic antibiotic, bleomycin, is a glycoprotein originally isolated by Umezawa et al. as a fermentation product from cultures of the actinomycete, *Streptomyces verticillus (1)*. The clinically used compound, bleomycin sulfate, is a combination of two analogues of bleomycin, whereby bleomycin A_2 is roughly 60% of the mixture while bleomycin B_2 is approx 30% of the mixture *(2)*. Bleomycin is used as a chemotherapeutic agent for a number of malignancies, namely lymphomas, germ cell tumors, and some squamous cell carcinomas *(3)*. The clinical use of bleomycin is limited by the potential pulmonary toxicity associated with bleomycin *(3,4)*. Although a number of agents can abrogate BLT in animal

From: *Cancer Drug Discovery and Development: Transforming Growth Factor-β in Cancer Therapy, Vol I: Basic and Clinical Biology*
Edited by: Sonia B. Jakowlew © Humana Press Inc., Totowa, NJ

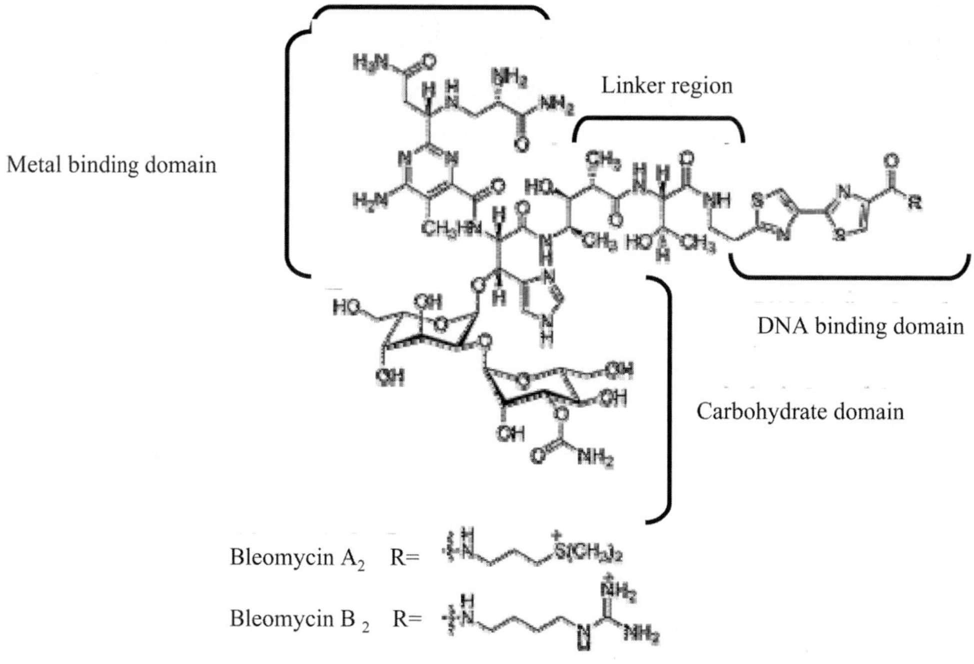

Fig. 1. Structure of bleomycin. Bleomycin has a number of domains, the metal binding domain, carbohydrate domain, DNA binding domain, and a linker region.

models, the mainstay of management of BLT in humans remains prevention by using a reduced dose of bleomycin and high dose corticosteroids in the event of BLT *(4)*.

2. MECHANISM OF BLEOMYCIN ACTION

2.1. Bleomycin Damage to DNA

Bleomycin A_2 and B_2 are members of a number of related compounds with several functional domains *(2)* (Fig. 1). These domains are the N-terminal domain, the metal binding domain, a linker region, the carbohydrate domain with disaccharides, a pyrimidine moiety and a bithiazole region at the C-terminus. Each region has an ascribed function *(2,5–7)*. The metal-binding domain binds with transition metals primarily Fe^{+2} derived from hemoglobin or Cu^{+2} from plasma, to become reduced and thereby activated *(3)*. Binding with a metal ion facilitates interaction between bleomycin and DNA and generation of reactive radicals, which augment cellular damage *(3)*. The N-terminal domain binds to oxygen and the bithiazole region binds to DNA by intercalation, a step critical to bleomycin induced DNA damage *(3)*. The linker region is required for cleaving the dsDNA. The carbohydrate region does not seem necessary for in vitro DNA cleavage and its purpose is unclear but could be important for facilitating entry of bleomycin into the cell *(3)*.

The proposed sequence of events is that bleomycin binds reduced iron (Fe^{+2}) that forms a complex with a molecule of oxygen *(1–6)*. The bleomycin-Fe^{+2}-O2 removes *(1–6)* a 4′- hydrogen *(2)* from C4′ of a (deoxyribose) pyrimidine 3′ and moves it to guanine *(2)*. This effect converts deoxyribose to an unstable sugar-carbon compound, which can result in lack of template formation by DNA polymerase *(3)*. Bleomycin mediated single strand breaks in the DNA at the 3′ occurs by conversion of deoxyribose to a 3′-phosphoglycolate, which interrupts DNA synthesis by DNA polymerase *(3)*. Bleomycin mediated oxidation of the apurinic/apyrimidine (AP) site results in loss of a base, which interferes with template formation by DNA polymerase of both ssDNA and dsDNA *(3)*. The reduced bleomycin complex, bleomycin-Fe^{+3}-OH also damages the DNA *(3,8)*. The cleavage of DNA results in breaks in the chromosome structure, deletions in the chromosomes, creation of chromosomes with two centemeres, or an aberrant ring formation *(2)* leading to cellular muation because of incorrect incorporation of neucleotides by DNA at the sites of DNA damage *(3)*. These effects of bleomycin may add to the antineoplastic effects of bleomycin.

2.2. Other Bleomycin-Induced Cellular Damage

Bleomycin cleaves mRNA, tRNA, rRNA by oxidation *(3)*. Because bleomycin RNA injury is protected by physiological concentrations of Mg^{+} *(3)*, it is unclear if the effects of bleomycin on RNA are therapeutically relevant *(3)*. The reactive oxygen products released by the effects of bleomycin have been shown to cause cell and nuclear membrane peroxidation, carbohydrate oxidation, and altered intracellular prostaglandin metabolism, which may also contribute to cell death and damage *(9)*. The aforementioned deleterious effects of oxygen radicals on bleomycin cellular toxicity can be inhibited by antioxidants *(9)*.

3. TRANSFORMING GROWTH FACTOR-β

3.1. General

The family of protein called transforming growth factor-β (TGF-β) are members of a super family with over 30 proteins *(15–18)* found in a number of species *(10–13)*. The TGF-βs are subclassified as inhibins, activins, bone morphogenic proteins (BMP), growth differentiation factors, and Mullerian-inhibiting substances *(10–13)*. In general, the TGF-β family of proteins is dimeric and well conserved proteins that participate in a spectrum of biological activities, many of which are physiological in nature, such as development at the embryonic level, wound healing, control of cellular proliferation, and homeostasis *(10–13)*. All isoforms of TGF-β are encoded as large precursor proteins that are 390–412 amino acids in size but the intracellular processing step of TGF-β mediated by the endopeptidase, furin, cleaves the TGF-β protein between amino acids 278 and 279 *(14)*. The proteolysis yields two products that assemble into dimers. The 65–75 kDa dimer protein from the N-terminal region is called the latency-associated peptide (LAP), while a second 25 kDa protein from the C-terminal portion of the precursor is called the mature TGF-β *(14)*. Despite the cleavage of the precursor, the N-terminal portion remains noncovalently associated with the rest of the protein *(15,16)*. The association of LAP with TGF-β makes the TGF-β biologically inactive and is referred to as latent-TGF-β (L-TGF-β) *(15,16)*. In vitro, the LAP from all isoforms of L-TGF-β can be removed by extremes of pH such as 2 or 8, heat, and a number of proteases *(15)*. The most commonly described protease to convert L-TGF-β1 to active TGF-β1 is the serine protease, plasmin *(15,16)*. Alternatively, L-TGF-β1 can be conformationally altered to expose the C-terminal portion by reactive oxygen intermediates, radiation *(15)*, the integrins $\alpha_v\beta_6$ *(17)*, $\alpha_v\beta 3$ *(18)*, or the glycoprotein, thrombospondin-1 (TSP-1) *(19)*. In some instances L-TGF-β is associated with a high molecular weight latent TGF-β binding protein (L-TBP) *(16)*. The binding of L-TGF-β to L-TBP sequesters L-TGF-β in the extracellularmatrix (ECM) by covalent cross-linking to the ECM *(16)*. In some instances L-TBP1 also participates in activation of L-TGF-β1 *(16)*.

3.2. TGF-β Receptors and Signaling

There are a number of proteins that bind TGF-β and of these there are three TGF-β receptors, TβR-I, TβR-II, and TβR-III *(12,13,17–21)*. TβR-I is a 55 kDa protein while TβR-II is 80 kDa and TβR-III is 280 kDa. TβR-I and TβR-II are glycoproteins with a cytoplasmic serine/threonine domain, a single hydrophobic domain, and an extracellular domain with cysteine residues *(20–21)*. TβR-III is a betaglycan which has no intracellular domain *(13)*. TβR-I does not bind TGF-β or signal on its own *(10–13)*. However, when TGF-β binds to TβR-II, which is constitutively autophosphorylated, it leads to TβR-II phosphorylating of the GS domain of the TβR-I *(12,13)*. Once phosphorylation occurs, TβR-I and TβR-II form a heteromeric complex, which is essential prior to any signaling *(12,13)*. TβR-III also binds TGF-β and facilitates the affinity of TGF-β for TβR-II but does not directly participate in TGF-β signaling *(13)*. Once the TβR-1 domain is activated there is subsequent phosphorylation of the intracellular proteins, the Smads *(10,12,13)*. There are eight mammalian Smads and are classified into three groups based on their structure and function *(10)* and they are: receptor-associated Smads (R-Smads) (Smads1,2,3,5, and 8) *(10)*, which interact with the TβR-1; a single common partner Smads (Co-Smads), Smad4, which associate with the R-Smads; and the inhibitor Smads (I-Smads), Smads6 and 7 *(10)*. Smad 8 is a substrate for the BMP receptor type 1 *(10)*. Smads2 and 3 are recruited to the activated TβR-I with help of Smad anchor for receptor activation *(20)* and are phosphorylated by the TβR-I kinase at a conserved SSXS motif located on the Smad2 and Smad3. The Smad2 and 3 heterocomplex associates with Smad4, which translocates to the nucleus *(10,20,21)*. Phosphorylated Smad2 from the R-Smads2/3-Smad4 complex binds to a DNA binding protein called the forkhead activin signal transducer 1(Fast-1). The Smad2/3-Smad4-Fast-1 complex *(10,12,13)* may activate or suppress gene expression *(11)*. Of the I-Smads, Smad6 inhibits the BMP pathway while Smad7 recruits Smad unbiquitination regulatory factors 1 and 2 (Smurf 1 and Smurf 2) to the proximity of TβR-I which degrade the receptor *(12,20,21)* resulting in prevention of phosphorylation of Smad2 and Smad3 and thus curtailing TGF-β mediated biological effects.

3.3. TGF-β1 and the ECM

TGF-β1 is the isoform most commonly associated with hyperproliferative and fibrotic diseases *(22)*. In fact, of the well-characterized growth factors and cytokines produced by pulmonary inflammatory and structural cells, TGF-β1 is one of the most potent regulators of ECM synthesis in vitro and in vivo *(11,23)*. During an injury response, TGF-β1 is expressed by monocytes and macrophages, is a potent chemoattractant for these cells and induces monocytes and macrophages to produce other fibrogenic growth factors TGF-β1 itself *(24)*. TGF-β1 is a chemoattractant *(25)* and a mitogen *(26,27)* for some types of fibroblasts, so that at sites of injury the presence of TGF-β1 could expand the population of fibroblasts, the main cellular source of connective tissue proteins *(28)*. TGF-β1 regulates the synthesis of collagens I, III, IV, fibronectin, proteogylcans, and other ECM proteins *(11,22,26,28,29)*. The induction of collagens by TGF-β is dependent on activation of Smads and occurs soon after interaction of TGF-β with mesenchymal cells *(12)*. In that regard, phosphorylated Smad2 activates type I and VII collagens, while Smad3 is required for induction of connective tissue growth factor, an intermediary of TGF-β1 mediated connective tissue synthesis *(32)*. Phosphorylated Smad-3 is also required for collagen I synthesis *(32)*. TGF-β1 inhibits the degradation of ECM by inhibiting production of matrix proteases while inducing protease inhibitors so that the presence of TGF-β1 stabilizes ECM proteins *(33,34)*.

3.4. Transforming Growth Factor-β in Bleomycin-induced Pulmonary Fibrosis

3.4.1. Animal Models of Bleomycin Lung Toxicity

Bleomycin invariably results in a dose dependent lung injury and fibrosis. For this reason bleomycin induced pulmonary fibrosis (BPF) is the single most utilized model for understanding the mechanisms of pulmonary fibrosis *(35)*. Although bleomycin lung toxicity (BLT) may occur with the administration of bleomycin by any route, the most common research model uses a singe dose of bleomycin administered by the intratracheal route into

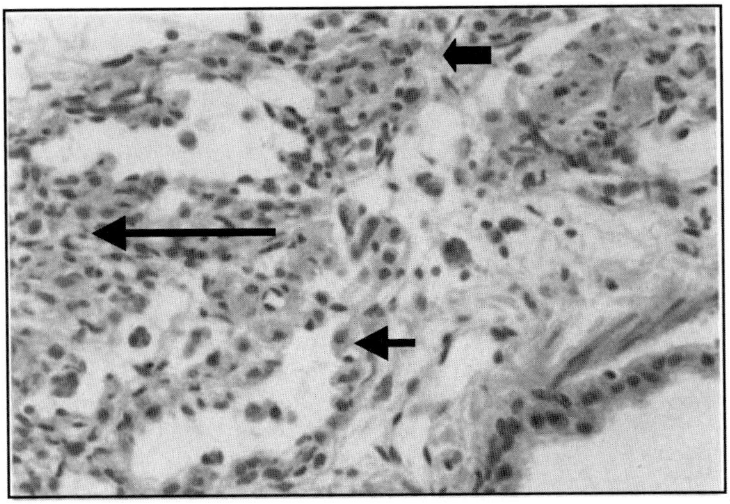

Fig. 2. Histological appearance of bleomycin lung injury. There is hyperplasia of type II pneumocytes (small arrow), interstitial inflammation (long arrow), and interstitial fibrosis (thick arrow). Hematoxylin and Eosin was used for staining and the magnification is ×40.

a rodent *(35–42)*. There follows an injury to the pulmonary alveolar epithelial cells (AECs) and endothelium, followed by influx of activated inflammatory cells *(35–42)* (Fig. 2). Thereafter the injured AECs are replaced by proliferating type 2 AECs, which are progenitors of the more differentiated type 1 AECs. In addition there is increased connective tissue synthesis and fibrosis in the interstitium *(35–42)*. All isoforms of TGF-β1 are expressed ubiquitously at the mRNA level in the lung *(43)*. After bleomycin administration there is an increase in the induction of TGF-β1 mRNA in the lung and the expression is elevated in pulmonary endothelial cells and infiltrating inflammatory cells *(44)*. Using immunohistochemistry the expression of TGF-β1 was demonstrated to be increased in alveolar macrophages in bleomycin treated rats but not normal saline treated controls *(36)* (Fig. 3). Additionally, immunofluorescence studies have demonstrated an increase in TGF-β1 in not only in the alveolar macrophages but also in AECs and interstitial fibroblasts *(45)*. However, the expression of TGF-β2 and TGF-β3 is ubiquitous (Fig. 4) and remains unchanged irrespective of the treatment of rats with bleomycin or normal saline *(39)*. It has been extensively demonstrated that bleomycin is a potent inhibitor of epithelial cells *(39,46)*. However at the time of maximal generation of active TGF-β1 from alveolar macrophages, which are in close contact with AECs, there is active proliferation of AECs *(39)*. The lack of response from AECs to the antiproliferative effects of TGF-β1 from the alveolar macrophages may be owing to the decreased expression of TβR-1 by the AECs at the time of active AEC proliferation *(42)* (Fig. 4). It is of interest that AECs isolated at regular times after bleomycin injury released small quantities of active TGF-β3 early in the injury response when there was proliferation of the AECs *(39)*. However, 28 d after bleomycin administration when there is differentiation of AECs there was a marked release of increased quantities of active TGF-β3 and return of the expression of TβR-I on AECs *(39)*. Because TGF-β3 is also an inhibitor of epithelial cells *(39)* and induces epithelial cell differentiation *(47)*, the concomitant release of TGF-β3 and the return of the TβR-I on AECs 28 d after bleomycin injury would result in inhibition of further proliferation of AECs and differentiation of type II AECs to type I AECs *(39)*. In bleomycin induced lung injury the expression of TβR-I and TβR-II was very prominent at all times on the interstitial fibroblasts *(42)* (Fig. 4) suggesting that after bleomycin lung injury the fibroblasts are responsive to TGF-βs released from any source. The emergence of fibroblasts

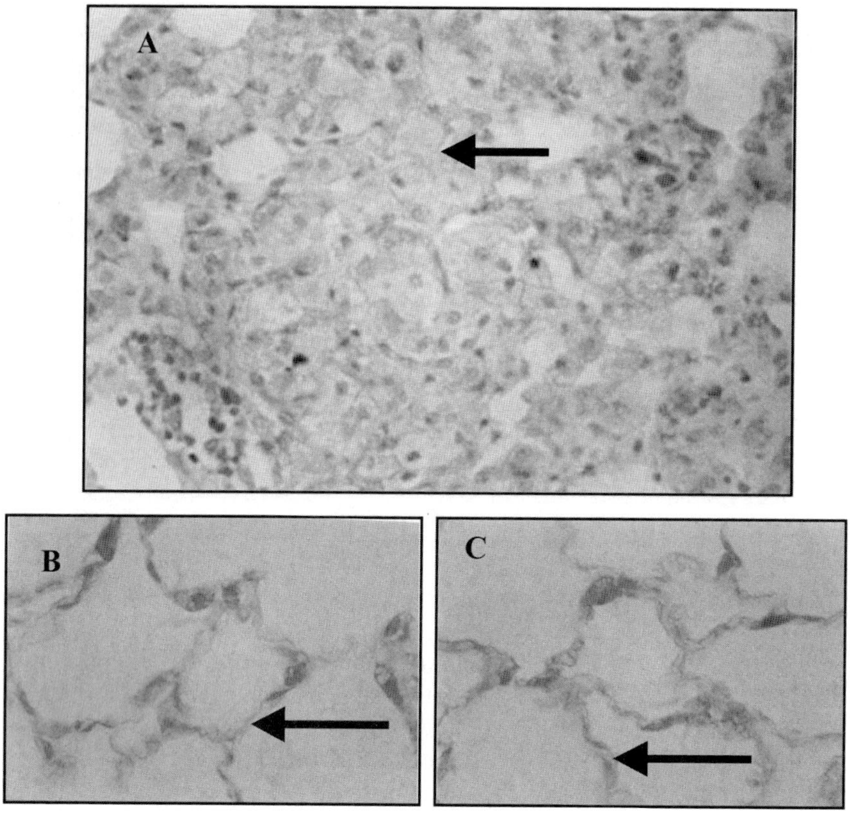

Fig. 3. Localization of TGF-β isoforms in rat lungs by immunohistochemistry. (**A**) Using immunohistochemistry and anti-LC-1-30-TGF-β1 antibody, TGF-β1 was prominently present in alveolar macrophages (arrow; detection was by avidin-biotin-alkaline-phosphatase and appears as red) seven days after bleomycin administration, but not seen in alveolar epithelial cells (AECs) of bleomycin injured rat lungs. Methyl green was used as a counterstain. (**B**) and (**C**) using anti-LC-50-75-TGF-β2 (**B**) antibody or anti-LC-50-60-TGF-β3 antibody (**C**) and avidin-biotin-peroxidase detection system TGF-β2 and TGF-β3 are present in AECs (arrow) of normal rat lungs.

in the interstitial pulmonary space is dependent on the release of TGF-β1. This was confirmed in another rat model where the presence of antibodies to TGF-β1 or fetuin, a competitive antagonist of the TβR-II *(26,27)* markedly reduced the emergence of fibroblasts, fibroblast proliferation, and TGF-β1 mediated connective tissue synthesis *(26,27)*.

It is important to understand that the expression of TGF-β1 at the mRNA or protein level is not critical to the action of TGF-β1. This is because the biological effects of TGF-β1 can only occur if the TGF-β1 is released in its biologically active form *(15,26,38)*. It has been demonstrated that the release of the biologically active form of TGF-β1 from alveolar macrophages is critical to the fibrotic response seen in BPF *(35,37,38)*. The activation process requires the complex of TSP-1-LAP-1-TGF-β1 to associate with the TSP-1 receptor, CD36 before a plasmin mediated proteolytic release of active TGF-β1 *(35,37,38)* (Fig. 5). The concomitant delivery of bleomycin with a CD36 synthetic peptide mimicking the ectodomain of the CD36 receptor between amino acids 93–110 (CD36 93–110) interfered with release of active TGF-β1 by alveolar macrophages and resulted in decreased bleomycin induced pulmonary inflammation and fibrosis *(41)*. In addition, a plasmin independent mechanism in a murine model of BPF is also important for converting the L-TGF-β1 into

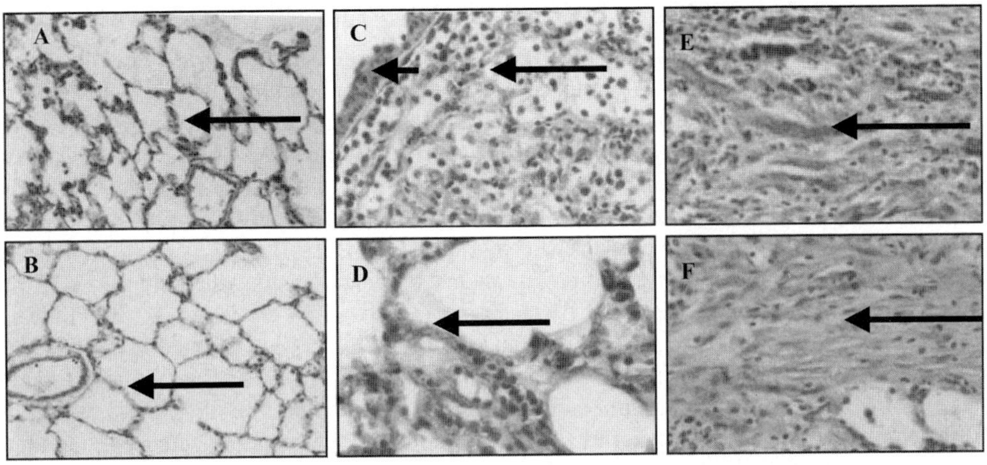

Fig. 4. Distribution of TGF-β receptors, TβR-I and TβR-II in rat lungs. (**A**), and (**D**) Using immunohistochemistry and antibodies to TβR-I, and TβR-II , both TβR-I and TβR-II are seen to be expressed by AECs (arrows). Four days after bleomycin administration, there is decrease in expression of TβR-I on AECs (arrow) (**B**) but TβR-II (**E**) staining remains unchanged (arrow). (**C**) and (**F**) Both TβR-I and TβR-II are expressed an interstitial fibroblasts (arrows).

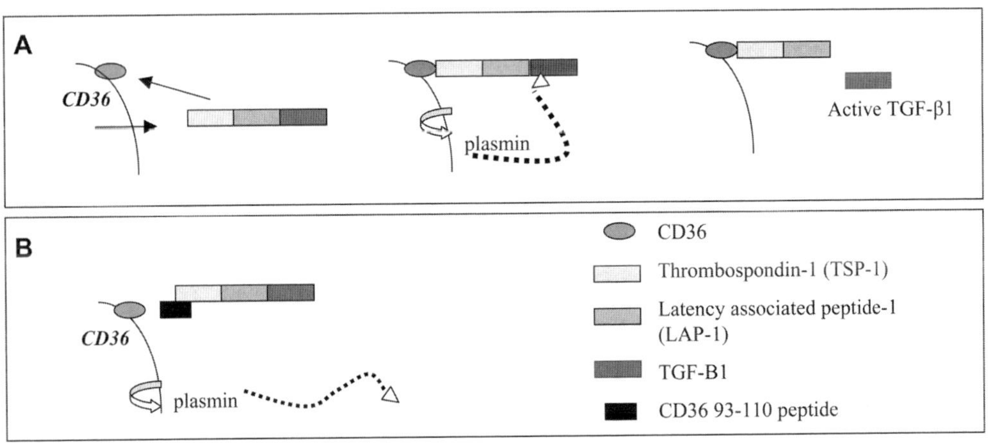

Fig. 5. Model for the conversion of L-TGF-β 1 to active TGF-β1. (**A**) Upon cell activation, a complex of LAP-1 associated with thrombospondin-1 (TSP-1), is released which localizes to the cell membrane by the interaction of the TSP-1 portion of the complex with the TSP-1 receptor, CD36. The association of the TSP-1/LAP-1/TGF-β1 with CD36 on the cell membrane brings the L-TGF-β1 in close proximity to plasminogen also present on the cell membrane. Plasminogen is the precursor of plasmin. The generation of plasmin in turn releases the TGF-β1 from it association with LAP-1. (**B**) CD36 93–110 synthetic peptide interferes with the attachment of TSP-1/LAP-1/TGF-β1 with the CD36 receptor. When CD36 93–110 synthetic peptide is present the L-TGF-β1 stays in solution and the plasmin does not release TGF-β1 from its association with the LAP-1 and no active TGF-β1 is detected.

its active conformation. After bleomycin administration there was an increased expression of the integrin, $\alpha_v\beta_6$ by AECs, association of the LAP-1 of L-TGF-β1 with $\alpha_v\beta_6$ that allowed the TGF-β1 to interact with TβR-II, resulting in the induction of a connective tissue response *(17,48)*.

3.5. Inhibition of TGF-β Mediated Fibrosis in BLT

In keeping with the observations that TGF-β1 is important in BPF, it has been demonstrated that anti-TGF-β1 antibodies administered with bleomycin diminish BPF *(49)*. In addition decorin, which binds the active form of TGF-β1 when administered intratracheally *(50)* or induced by gene mediated over expression of decorin in lungs, also diminished BPF *(51)*. Similarly, the administration of the soluble receptor to TβR-II, which may bind active TGF-β1 that is not associated with a receptor, also diminished BPF *(52,53)*. There are other substances have been described to prevent BPF by regulating TGF-β1 expression or the effects of TGF-β1. For example administration of antagonists of angiotensin, which induces TGF-β1 expression and conversion of L-TGF-β1 to active TGF-β1, inhibits BPF *(54–56)*. Taurine and niacin, which are antioxidants, reduced the total lung expression of mRNA for TGF-β1 and protein levels of TGF-β1 in the bronchoalveolar lavage fluid (BALF) of rodents treated with bleomycin *(57–59)*. Pirfenidone is a pyridone derivative that inhibits the expression of TGF-β, and when administered to rats receiving bleomycin there was a decrease in the total lung mRNA for TGF-β1, a decrease in TGF-β1 in BALF, and reduced inflammatory and fibrotic changes *(60,61)*. Interferon-β decreased bleomycin mediated induction of TGF-β1 protein by Western analysis and immunofluorescence and there was a concomitant reduction of connective tissue synthesis *(45)*. Interferon-γ also reduced total lung TGF-β1 mRNA and connective tissue synthesis induced by bleomycin *(62)*. The c-Abelson (c-*Abl*) protooncogene, which is induced by TGF-β1 and PDGF, mediates fibroblast proliferation, migration, and possibly connective tissue synthesis *(63)*. Imatinab mesylate, an inhibitor of c-*Abl* administered to mice receiving bleomycin, had decreased collagen synthesis owing to interruption of TGF-β1 and PDGF mediated signaling *(63)*. Pretreatment of rats with KGF, a heparin binding growth factor, markedly diminished TGF-β1 protein induction and inflammation and fibrosis from bleomycin *(64)*. Bleomycin administered to Smad-3 knockout (KO) mice had less fibrotic lesions with a decrease in collagen I and fibronectin synthesis compared to the effects of bleomycin in the wild-type control *(30–32)*. Smad7 interferes with activation of Smad2 and Smad3 *(31)*. When lungs of mice receiving bleomycin were infected with AdCMV-Smad7 there was decreased fibrosis *(31)*. The authors suggested that the decrease in collagen synthesis was owing to the exogenous Smad7 blockade of phosphorylation of Smad2 and subsequent TGF-β1 mediated pulmonary fibrosis *(31,65)*.

4. CLINICAL USE OF BLEOMYCIN

4.1. Malignancies Treated with Bleomycin

As an antineoplastic agent bleomycin is used to treat a number of malignancies (Table 1) such as Hodgkin's lymphoma *(66)*, non-Hodgkin's lymphomas *(67)*, germ cell tumors *(68)*, such as testicular carcinoma *(69)*, yolk sac tumors *(70)*, gestational trophoblastic tumors *(71)*, epidermoid anal cell carcinoma *(72)*, sex cord-stromal ovarian tumors *(73)*. Bleomycin is also used to treat a number of squamous cell tumors such as those of the head and neck *(74)*, penis *(75)*, cervix *(76)*, as well as Kaposi's sarcoma *(77)*. Other uses of bleomycin are by local instillation for treatment of cystic craniopharingiomas *(78)*, pleurodesis in malignant effusions *(79)*, and sclerosing vascular malformations *(80)*.

4.2. BLT in Humans

4.2.1. General

The responsiveness of certain malignancies to bleomycin and the lack of myelosuppression make bleomycin an excellent antineoplastic agent *(4,81–83)*. Unfortunately, even though bleomycin chemotherapy has clear advantages, the use of bleomycin is limited by the potential

Table 1
Malignancies Treated with Bleomycin

Cell type of malignancy	*Examples of malignancy*
Lymphoma	Hodgkin's lymphoma, non-Hodgkin's lymphoma
Germ cell tumors	Testicular carcinoma, yolk sac tumor, gestational trophoblastic tumor, sex cord-stromal ovarian tumor
Squamous cell carcinoma	Head and neck, penis, cervix, anus
Sarcoma	Kaposi's

Table 2
Dose of Bleomycin and Lung Toxicity

Accumulative dose	*Risk of bleomycin pulmonary toxicity (%)*
< 450 mgms	3–5
450–549 mgms	13
> 550 mgms	17

to cause BLT *(4,81–83)*. In animal models BLT is dose dependent, and in humans the relationship of dose with pulmonary toxicity is also seen (Table 2) *(82)*, but may occur with very minimal quantities of bleomycin or not at all even with the maximal allowable dose *(4,81–83)*. The incidence of bleomycin induced pulmonary toxicity ranges from 2% to 46% *(4,81–83)* and there is approx a 3% mortality of all patients treated with bleomycin *(4,81–83)*. The risk of developing BLT is increased with a high dose especially if it exceeds an accumulated dose of 450 mgms. Another risk factor is renal dysfunction because the kidneys illuminate bleomycin and renal impairment would impair the clearance of bleomycin *(4,81–83)*. Because BLT is based on generation of free radicals it has been suggested that other clinical conditions where there may be free radicals generated in the lung, such as the use of radiotherapy or high doses of oxygen BLT may be potentiated *(84)*. In animal models the concomitant administration of GM-CSF augments BLT *(85)* and in one study the administration of GM-CSF to patients having received bleomycin was associated with greater incidence of pulmonary dysfunction, recruitment of neutrophils to the lungs, worse pulmonary toxicity by histological criteria, and increased mortality *(85)*. Other risk factors for BLT is an age over 70 yr old, which may be associated with decreased renal clearance or comorbidity *(81)*, smoking, and route of administration whereby bolus intravenous infusion may be associated with a higher incidence of BLT than when bleomycin is given in a continuous infusion *(4,81,82)*. The role of TGF-β1 in the pathogenesis of BLT and BPF in humans is unknown. However, the acute lung injury of the acute respiratory distress syndrome (ARDS) can be associated with a wide spectrum of etiologies including bleomycin *(85)*. There is an overlap of the histological lesions seen in animal models of BLT, BPF *(36–42,)* and the human ARDS *(85)*. Because the BALF of patients with ARDS owing to a number of causes had increased quantities of TGF-β1 in the BALF *(86,87)*, it is possible that TGF-β1 is important in the pathogenesis of human BLT and BPF.

The clinical recognition of bleomycin toxicity may be difficult as the patients are generally ill, receiving other chemotherapeutic agents, and the symptoms, signs, and investigations reveal nonspecific information *(4,81–85)*. The symptoms are dyspnea, cough, and occasionally fever with inspiratory crackles upon auscultation *(4,81–85)*. The symptoms may suggest pneumonia, pulmonary edema, or pulmonary metastases *(4)*. A number of pulmonary reactions have been associated with BLT such as interstitial pneumonitis, the most common manifestation

(4), bronchiolitis obliterans organizing pneumonia (BOOP; *[4]*), and eosinophilic pneumonitis *(4)*. On plain film there may be patchy infiltrates or there may be consolidation *(4)*. High-resolution computed tomography of the chest demonstrates similar findings but better delineates the distribution and extent of involvement. A transbronchial biopsy is not helpful in making the diagnosis because none of these histological changes can be confidently diagnosed on a small piece of lung obtained by transbronchial biopsy. If a diagnostic lung biopsy is done characteristic features seen are type 2 pneumocyte hyperplasia, bronchiolar squamous metaplasia, inflammatory cell infiltration, edema, and connective tissue deposition (Fig. 1).

4.2.2. Treatment of BLT

Although a number of agents have been successfully demonstrated to inhibit bleomycin induced pulmonary inflammation and fibrosis in rodent models, the most common management for bleomycin induced pulmonary injury and fibrosis in humans is prevention by using less than an accumulative dose of 350–450 mgms and stopping bleomycin at the first sign of presumed toxicity *(4)*. In animal models *(89–91)* and humans use of high dose corticosteroids has been found to be helpful in treating bleomycin induced fibrosis *(4,89–91)*. In vitro it has been demonstrated that steroids decrease the TGF-β1 mediated collagen synthesis *(89)*. It was found in one study that patients improved faster when on steroids and had an improved survival *(88)*. It is suspected that BOOP or an eosinophilic reaction to bleomycin is likely to be the steroid responsive pulmonary response; while it is unclear if bleomycin induced pneumonitis is steroid responsive *(88)*.

5. CONCLUSIONS

The bleomycin mixture of bleomycin A and bleomycin B is an effective antineoplastic agent against a number of malignancies. However, because of the fear of BLT, the use of bleomycin is restricted to an accumulated doses of less than or equal to 350–450 mgms. Based on extensive research on BPF in animal models TGF-β1 is important in the pathogenesis of BPF. Several agents that regulate TGF-β1 mediated BPF in the animal models are potential therapeutic options for patients receiving bleomycin and may prevent BLT and make it possible to use higher doses of bleomycin to achieve an improved outcome.

REFERENCES

1. Umezawa H, Maeda K, Takeuchi T, Okami YJ. New antibiotics, bleomycin A and B. J Antibiot (Tokyo) 1966;19:200–209.
2. Chen J, Stubbe J. Bleomycins: towards better therapeutics. Nat Rev Cancer 2005;5:102–112.
3. Ramotar D, Wang H. Protective mechanisms against the antitumor agent bleomycin: lessons from Saccharomyces cerevisiae. Curr Genet 2003;43:213–224.
4. Sleijfer S. Bleomycin-induced pneumonitis. Chest 2001;120:617–624.
5. Xu ZD, Wang M, Xiao SL, Liu CL, Yang M. Synthesis, biological evaluation and DNA binding properties of novel bleomycin analogues. Bioorg Med Chem Lett 2003;13:2595–2599.
6. Deng JZ, Newman DJ, Hecht SM. Use of COMPARE analysis to discover functional analogues of bleomycin. J Nat Prod 2000;63:1269–1272.
7. Leitheiser CJ, Smith KL, Rishel MJ, et al. Solid-phase synthesis of bleomycin group antibiotics. Construction of a 108-member deglycobleomycin library. J Am Chem Soc 2003;125:8218–8227.
8. Ming LJ. Structure and function of "metalloantibiotics". Med Res Rev 2003;23:697–762.
9. Boyaci H, Maral H, Turan G, et al. Effects of erdosteine on bleomycin-induced lung fibrosis in rats. Mol Cell Biochem 2006;281:129–137.
10. Schiller M, Javelaud D, Mauviel A. TGF-beta-induced SMAD signaling and gene regulation: consequences for extracellular matrix remodeling and wound healing. J Dermatol Sci 2004;35:83–92.
11. Chin D, Boyle GM, Parsons PG, Coman WB. What is transforming growth factor-beta (TGF-beta)? Br J Plast Surg 2004;57:215–221.

12. Leask A, Abraham DJ. TGF-beta signaling and the fibrotic response. FASEB J 2004;18:816–827.
13. de Caestecker M. The transforming growth factor-beta superfamily of receptors. Cytokine Growth Factor Rev 2004;15:1–11.
14. Dubois CM, Laprise MH, Blanchette F, Gentry LE, Leduc R. Processing of transforming growth factor beta 1 precursor by human furin convertase. J Biol Chem 1995;270:10,618–10,624.
15. Khalil N. TGF-β: from latent to active. Microbes Infect 1999;1:1255–1265.
16. Khalil N. Post-translational Activation of L-TGF-β, Histology and Histopathology 2001;16:541–551.
17. Brunner G, Blakytny R. Extracellular regulation of TGF-beta activity in wound repair: growth factor latency as a sensor mechanism for injury. Thromb Haemost 2004;92:253–261.
18. Annes JP, Chen Y, Munger JS, Rifkin DB. Integrin alphaVbeta6-mediated activation of latent TGF-beta requires the latent TGF-beta binding protein-1. J Cell Biol 2004;165:723–734.
19. Ludbrook SB, Barry ST, Delves CJ, Horgan CM. The integrin alpha v beta3 is a receptor for the latency-associated peptides of transforming growth factors beta1 and beta3. Biochem J 2003;369 (Pt 2):311–318.
20. Murphy-Ullrich JE, Poczatek M. Activation of latent TGF-beta by thrombospondin-1: mechanisms and physiology. tokine Growth Factor Rev 2000;11:59–69.
21. Tsukazaki T, Chiang TA, Davison AF, Attisano L, Wrana JL. SARA, a FYVE domain protein that recruits Smad2 to the TGFbeta receptor. Cell 1998;95:779–791.
22. Massagué J, Wotton D. Transcriptional control by the TGF-beta/Smad signaling system. EMBO J 2000;19:1745–1754.
23. Border WA, Noble NA. Transforming growth factor beta in tissue fibrosis. N Engl J Med 1994;331: 1286–1292.
24. Mauviel A. Transforming growth factor-beta: a key mediator of fibrosis. Methods Mol Med 2005;117:69–80.
25. McCartney-Francis N, Mizel D, Wong H, Wahl L, Wahl S. TGF-beta regulates production of growth factors and TGF-beta by human peripheral blood monocytes. Growth Factors 1990;4:27–35.
26. Flanders KC, Major CD, Arabshahi A, et al. Interference with transforming growth factor-beta/ Smad3 signaling results in accelerated healing of wounds in previously irradiated skin. Am J Pathol 2003; 163:2247–2257.
27. Xu YD, Hua J, Mui A, O'Connor R, Grotendorst G, Khalil N. Release of biologically active TGF-beta1 by alveolar epithelial cells results in pulmonary fibrosis. Am J Physiol Lung Cell Mol Physiol 2003;285:L527–L539.
28. Khalil N, Xu YD, Duronio V, O'Connor R. Proliferation of interstitial pulmonary fibroblasts is mediated by TGF-β1 induced bFGF phosphorylation of p38 MAPK and JNK. J Biol Chem 2005;280: 43,000–43,009.
29. Carlson MA, Longaker MT. The fibroblast-populated collagen matrix as a model of wound healing: a review of the evidence. Wound Repair Regen 2004;12:134–147.
30. Bhattacharyya S, Ghosh AK, Pannu J, et al. Fibroblast expression of the coactivator p300 governs the intensity of profibrotic response to transforming growth factor beta. Arthritis Rheum 2005;52: 1248–1258.
31. Flanders KC. Smad3 as a mediator of the fibrotic response. Int J Exp Pathol 2004;85:47–64.
32. Cutroneo KR. Evidence for TGF-ss1 and bleomycin intracellular signaling through autocrine regulation of Smad 3 binding to the proximal promoter of the Smad 7 gene. J Cell Biochem 2006;97:933–939.
33. Zhao J, Shi W, Wang YL, et al. Smad3 deficiency attenuates bleomycin-induced pulmonary fibrosis in mice. Am J Physiol Lung Cell Mol Physiol 2002;282:L585–L593.
34. Siwik DA, Colucci WS. Regulation of matrix metalloproteinases by cytokines and reactive oxygen/nitrogen species in the myocardium Heart Fail Rev 2004;9:43–51.
35. Fernandes JC, Martel-Pelletier J, Pelletier JP. The role of cytokines in osteoarthritis pathophysiology. Biorheology 2002;39:237–246.
36. Gharaee-Kermani M, Ullenbruch M, Phan SH. Animal models of pulmonary fibrosis. Methods Mol Med 2005;117:251–259.
37. Khalil N, Bereznay O, Sporn M, Greenberg AH. Macrophage production of transforming growth factor-beta and fibroblast collagen synthesis in chronic pulmonary inflammation. J Exp Med 1989;170: 727–737.
38. Khalil N, Whitman C, Danielpour D, Greenberg AH. Regulation of alveolar macrophage TGF-beta secretion by corticosteroids in bleomycin induced pulmonary inflammation in the rat. J Clin Invest 1993;92:1812–1818.

39. Yehualaeshet T, O'Connor R, Mai S, et al. Activation of alveolar macrophage derived L-TGF-beta1 by plasmin requires interaction with TSP-1 and the cell surface receptor, CD36, Am J Pathol 1999;155:841–851.
40. Khalil N, O'Connor RN, Flanders KC, Shing W, Whitman CI. Regulation of type II alveolar epithelial-cell proliferation by TGF-beta during bleomycin- induced lung injury in rats. Am J Physiol –Lung Cel Mol Physiol 1994;11:L498–L507.
41. Khalil N, Corne S, Whitman C, Yacyshyn H. Plasmin regulates the activation of cell associated latent TGF-beta secreted by rat alveolar macrophages after in vivo bleomycin injury. Am J Resp Cell Mol Biol 1996;15:252–259.
42. Yehualaeshet T, O'Connor R, Begleiter A, Murphy-Ullrich JE, Silverstein R, Khalil N. A CD36 synthetic peptide inhibits bleomycin-induced pulmonary inflammation and connective tissue synthesis in the rat. Am J Resp Cell Mol Biol 2000;23:204–212.
43. Khalil N, Parekh TV, O'Connor RN, Gold LI. Differential expression of transforming growth factor-ß type I and II receptors by pulmonary cells in bleomycin-induced lung injury: correlation with repair and fibrosis. Exp Lung Res 2002;28:233–250.
44. Jakowlew SB, Mariano J. Detection of the transcripts and proteins for the transforming growth factor-beta isoforms and receptors in mouse lung tumorigenesis using in situ hybridization and immunohistochemistry in paraffin-embedded tissue sections. Methods Mol Med 2003;74:145–166.
45. Santana A, Saxena B, Noble NA, Gold LI, Marshall BC. Increased expression of transforming growth factor beta isoforms (beta 1, beta 2, beta 3) in bleomycin-induced pulmonary fibrosis. Am J Respir Cell Mol Biol 1995;13:34–44.
46. Azuma A, Li YJ, Abe S, et al. Interferon-β inhibits bleomycin-induced lung fibrosis by decreasing transforming growth factor-β and thrombospondin. Am J Respir Cell Mol Biol 2005;32:93–98.
47. Huang SS, Huang JS. TGF-beta control of cell proliferation. J Cell Biochem 2005;96:447–462.
48. Moustakas A, Pardali K, Gaal A, Heldin CH. Mechanisms of TGF-beta signaling in regulation of cell growth and differentiation. Immunol Lett 2002;82:85–91.
49. Munger JS, Huang X, Kawakatsu H, et al. The integrin alpha v beta 6 binds and activates latent TGF beta 1: a mechanism for regulating pulmonary inflammation and fibrosis. Cell 1999;96:319–328.
50. Giri SN, Hyde DM, Hollinger MA. Effect of antibody to transforming growth factor beta on bleomycin induced accumulation of lung collagen in mice. Thorax. 1993;48:959–966.
51. Giri SN, Hyde DM, Braun RK, Gaarde W, Harper JR, Pierschbacher MD. Antifibrotic effect of decorin in a bleomycin hamster model of lung fibrosis. Biochem Pharmacol 1997;54:1205–1216.
52. Shimizukawa M, Ebina M, Narumi K, Kikuchi T, Munakata H, Nukiwa T. Intratracheal gene transfer of decorin reduces subpleural fibroproliferation induced by bleomycin. Am J Physiol Lung Cell Mol Physiol 2003;284:L526–L532.
53. Wang Q, Hyde DM, Gotwals PJ, Giri SN. Effects of delayed treatment with transforming growth factor-beta soluble receptor in a three-dose bleomycin model of lung fibrosis in hamsters. Exp Lung Res 2002;28:405–417.
54. Wang Q, Wang Y, Hyde DM, et al. Reduction of bleomycin induced lung fibrosis by transforming growth factor beta soluble receptor in hamsters. Thorax 1999;54:801–805.
55. Mancini GB, Khalil N. Angiotensin type 1 receptor blocker inhibits pulmonary injury. Clin Invest Med 2005;28:118–126.
56. Otsuka M, Takahashi H, Shiratori M, Chiba H, Abe S. Reduction of bleomycin induced lung fibrosis by candesartan cilexetil, an angiotensin II type 1 receptor antagonist. Thorax 2004;59:31–38.
57. Marshall RP, Gohlke P, Chambers RC, et al. Angiotensin II and the fibroproliferative response to acute lung injury. Am J Physiol Lung Cell Mol Physiol 2004;286:L156–L164.
58. Gurujeyalakshmi G, Wang Y, Giri SN. Taurine and niacin block lung injury and fibrosis by down-regulating bleomycin-induced activation of transcription nuclear factor-kappaB in mice. Pharmacol Exp Ther 2000;293:82–90.
59. Giri SN, Gurujeyalakshmi G, Wang Y. Suppression of bleomycin-induced increased production of nitric oxide and NF-kB activation by treatment with taurine and niacin. Adv Exp Med Biol 2000;483:545–561.
60. Gurujeyalakshmi G, Hollinger MA, Giri SN. Regulation of transforming growth factor-beta1 mRNA expression by taurine and niacin in the bleomycin hamster model of lung fibrosis. Am J Respir Cell Mol Biol 1998;18:334–342.
61. Iyer SN, Gurujeyalakshmi G, Giri SN. Effects of pirfenidone on transforming growth factor-beta gene expression at the transcriptional level in bleomycin hamster model of lung fibrosis. J Pharmacol Exp Ther 1999;291:367–373.

62. Iyer SN, Hyde DM, Giri SN. Anti-inflammatory effect of pirfenidone in the bleomycin-hamster model of lung inflammation. mmation 2000;24:477–491.
63. Gurujeyalakshmi G, Giri SN. Molecular mechanisms of antifibrotic effect of interferon gamma in bleomycin-mouse model of lung fibrosis: downregulation of TGF-beta and procollagen I and III gene expression. Exp Lung Res 1995;21:791–808.
64. Daniels CE, Wilkes MC, Edens M, et al. Imatinib mesylate inhibits the profibrogenic activity of TGF-beta and prevents bleomycin-mediated lung fibrosis. J Clin Invest 2004;114:1308–1316.
65. Yi ES, Salgado M, Williams S, et al. Keratinocyte growth factor decreases pulmonary edema, transforming growth factor-beta and platelet-derived growth factor-BB expression, and alveolar type II cell loss in bleomycin-induced lung injury. Inflammation 1998;22(3):315–325.
66. Nakao A, Fujii M, Matsumura R, et al. Transient gene transfer and expression of Smad7 prevents bleomycin-induced lung fibrosis in mice. J Clin Invest 1999;104:5–11.
67. Connors JM. State-of-the-art therapeutics: Hodgkin's lymphoma. J Clin Oncol 2005;23(26):6400–6408.
68. Cheson BD. What is new in lymphoma? CA Cancer J Clin 2004;54(5):260–272.
69. Christian JA, Huddart RA, Norman A, et al. Intensive induction chemotherapy with CBOP/BEP in patients with poor prognosis germ cell tumors. J Clin Oncol 2003;21(5):871–877.
70. Jones RH, Vasey PA. Part II: testicular cancer—management of advanced disease. Lancet Oncol 2003;4(12):738–747.
71. Miller KD, Michael H, Jacobson L, Sutton GP. Primary yolk sac tumor of the rectum. Cancer Invest 2000;18(7):597–601.
72. Lurain JR. Pharmacotherapy of gestational trophoblastic disease. Expert Opin Pharmacother 2003;4: 2005–2017.
73. Nilsson PJ, Svensson C, Goldman S, Ljungqvist O, Glimelius B. Epidermoid anal cancer: a review of a population-based series of 308 consecutive patients treated according to prospective protocols. Int J Radiat Oncol Biol Phys 2005;61:92–102.
74. Brown J, Shvartsman HS, Deavers MT, et al. The activity of taxanes compared with bleomycin, etoposide, and cisplatin in the treatment of sex cord-stromal ovarian tumors. Gynecol Oncol 2005;97:489–496.
75. Mazeron JJ, Martin M, Brun B, et al. Induction chemotherapy in head and neck cancer: results of a phase III trial. Head Neck 1992;14:85–91.
76. Mobilio G, Ficarra V. Genital treatment of penile carcinoma. Curr Opin Urol 2001;11:299–304.
77. Petru E, Benedicic C, Seewann A, Pickel H. Palliative cytostatic treatment of cervical carcinoma. Eur J Gynaecol Oncol 2003;24:473–474.
78. Krown SE, Northfelt DW, Osoba D, Stewart JS. Use of liposomal anthracyclines in Kaposi's sarcoma. Semin Oncol 2004;31:36–52.
79. Hader WJ, Steinbok P, Hukin J, Fryer C. Intratumoral therapy with bleomycin for cystic craniopharyngiomas in children. Pediatr Neurosurg 2000;33:211–218.
80. Weatherhead M, Antunes G. Chemotherapeutic management of malignant pleural effusion. Expert Opin Pharmacother 2004;5:1233–1242.
81. Muir T, Kirsten M, Fourie P, Dippenaar N, Ionescu GO. Intralesional bleomycin injection (IBI) treatment for haemangiomas and congenital vascular malformations. Pediatr Surg Int 2004;19: 766–773.
82. Azambuja E, Fleck JF, Batista RG, Menna Barreto SS. Bleomycin lung toxicity: who are the patients with increased risk? Pulm Pharmacol Ther 2005;18:363–366.
83. Jules-Elysee K, White DA. Bleomycin-induced pulmonary toxicity. Clin Chest Med 1990;11:1–20.
84. Samuels ML, Johnson DE, Holoye PY, Lanzotti VJ. Large-dose bleomycin therapy and pulmonary toxicity. A possible role of prior radiotherapy. JAMA 1976;235:1117–1120.
85. Azoulay E, Herigault S, Levame M, et al. Effect of granulocyte colony-stimulating factor on bleomycin-induced acute lung injury and pulmonary fibrosis. Crit Care Med 2003;31:1442–1448.
86. Vourlekis JS, Brown KK, Schwarz MI. Acute interstitial pneumonitis: current understanding regarding diagnosis, pathogenesis, and natural history. Semin Respir Crit Care Med 2001;22:399–408.
87. Dhainaut JF, Charpentier J, Chiche JD. Transforming growth factor-beta: a mediator of cell regulation in acute respiratory distress syndrome. Crit Care Med 2003;31:S258–S264.
88. Danson S, Blackhall F, Hulse P, Ranson M. Interstitial lung disease in lung cancer: separating disease progression from treatment effects. Drug Saf 2005;28:103–113.
89. White DA, Stover DE. Severe bleomycin-induced pneumonitis. Clinical features and response to corticosteroids. Chest 1984;86:723–728.

90. Shukla A, Meisler N, Cutroneo KR. Perspective article: transforming growth factor-beta: crossroad of glucocorticoid and bleomycin regulation of collagen synthesis in lung fibroblasts. Wound Repair Regen 1999;7:133–140.
91. Dik WA, McAnulty RJ, Versnel MA, Naber BA, Zimmermann LJ, Laurent GJ, Mutsaers SE. Short course dexamethasone treatment following injury inhibits bleomycin-induced fibrosis in rats. Thorax 2003;58:765–771.

37 Smads in the Fibrotic Response: Findings in the Smad3 Knockout Mouse

Kathleen C. Flanders

CONTENTS

INTRODUCTION
TGF-β SIGNALING THROUGH SMADS
SMAD2 VS SMAD3
SMAD3 AND WOUND HEALING
SMAD3 AND CHRONIC FIBROTIC CONDITIONS
SMAD3 INHIBITORS AS POTENTIAL ANTIFIBROTIC AGENTS
INHIBITION OF SMAD3 THROUGH INDUCTION OF SMAD7
INFLAMMATION, TGF-β AND CANCER
REFERENCES

Abstract

Transforming growth factor-β (TGF-β) mediates both physiological and pathological fibrotic responses by inducing influx and activation of inflammatory cells, epithelial to mesenchymal transdifferentiation of cells, and influx of fibroblasts and secretion of extracellular matrix. Activated TGF-β receptors phosphorylate signaling intermediates called Smad proteins which translocate to the nucleus and alter gene transcription. Mice in which the gene for Smad3, one of the TGF-β signaling proteins, has been deleted show more rapid healing of incisional and excisional wounds with less inflammation and matrix deposition. Smad3 null inflammatory cells and fibroblasts do not respond to the chemotactic effects of TGF-β and do not autoinduce TGF-β or respond to TGF-β-mediated induction of extracellular matrix proteins. Smad3 also appears to modulate pathologic fibrosis in that Smad3 null mice are resistant to fibrosis in a number of animal models including radiation-induced cutaneous fibrosis, bleomycin-induced pulmonary fibrosis, glomerulosclerosis resulting from unilateral ureter obstruction, and proliferative vitreoretinopathy. Agents that inhibit phosphorylation of Smad3 by affecting the TGF-β Type I receptor kinase reduce collagen synthesis of cells in vitro. Inducing expression of Smad7, an inhibitory Smad, inhibits unwanted matrix deposition in animal models of cutaneous, pulmonary, ocular, and renal fibrosis, suggesting that Smad3 inhibitors may have clinical potential in the treatment of pathological fibrotic diseases.

Key Words: Extracellular matrix; fibrosis; inflammation; Smad3; TGF-β; wound healing.

1. INTRODUCTION

Transforming growth factor-β (TGF-β), a multifunctional cytokine which regulates cell growth and differentiation, apoptosis, cell migration, inflammation, and extracellular matrix

From: *Cancer Drug Discovery and Development: Transforming Growth Factor-β in Cancer Therapy, Vol I: Basic and Clinical Biology*
Edited by: Sonia B. Jakowlew © Humana Press Inc., Totowa, NJ

(ECM) production, is an important mediator of the physiological and pathological fibrotic response *(1–3)*. Of the three highly homologous mammalian TGF-β isoforms (TGF-βs 1, 2, and 3), elevated levels of TGF-β1 are most often associated with fibrotic conditions *(4)*. Additionally, TGF-β1 is abundant in platelets and so is released by degranulating platelets at the site of injury. This bolus of TGF-β1 recruits inflammatory cells and fibroblasts into the area and then stimulates these cells to produce cytokines and ECM, respectively. TGF-β also autoinduces its expression in these cells keeping TGF-β levels elevated in the wound area. Because TGF-β not only increases synthesis of matrix proteins, but also enhances secretion of protease inhibitors while reducing secretion of proteases, it is one of the most potent stimulators of matrix accumulation *(5)*. While the responses elicited by TGF-β play a role in the normal physiology of tissue repair following injury, too often this process does not properly resolve and chronic pathological conditions result. This review will focus on the central role of Smad3, a mediator of TGF-β signaling, in contributing to fibrosis in a number of organs followed by a brief discussion of possible therapeutic strategies for blocking unwanted fibrosis by inhibiting actions of Smad3.

2. TGF-β SIGNALING THROUGH SMADS

The major downstream mediators of TGF-β signaling are the Smad proteins. Smads are grouped into three subfamilies, the five receptor-activated Smads (R-Smads), the one common mediator Smad (Co-Smad), and the two inhibitory Smads (I-Smads) (reviewed in *[6–8]*). Of the R-Smads, Smads2 and 3 mediate signals for TGF-β and activin while Smads1, 5, and 8 transduce signals from bone morphogenetic proteins (BMPs) (Fig. 1). For TGF-β signaling, ligand binding to the constituitively active ser/thr kinase Type II receptor recruits the Type I receptor into the complex where it is phosphorylated by the Type II receptor resulting in its subsequent activation. Smads2 and 3 are recruited to the activated Type I receptor which directly phosphorylates the last two serines of a conserved SSXS motif located at the extreme carboxyl terminus of the R-Smads. The phosphorylated R-Smad is released from the receptor complex to form a heteromeric complex of two R-Smads and the co-Smad (Smad4) and this complex translocates to the nucleus where it can interact with various transcription factors and affect transcriptional responses. The I-Smads (Smad 6 for the BMP pathway and Smad7 for the TGF-β/activin pathway) function by binding to the Type I receptor and preventing recruitment and phosphorylation of R-Smads. It is now appreciated that the activated receptor complex can also signal through other pathways, such as those involving the mitogen-activated protein kinases (MAPKs) and phosphoinositiol-3 kinases (PI3K), although the molecular details are not completely understood *(9)*.

Smad activity can also be regulated by phosphorylation through nonreceptor kinases. The proline-rich linker regions of the R-Smads which connect the conserved N- and C-terminal domains contain a number of phosphorylation sites for proline-directed kinases which mediate the crosstalk between the Smad pathway and a variety of other signaling mechanisms *(8,10)*. For example, ERK MAP kinase which is activated by epidermal growth factor, phosphorylates 3 Ser/Thr residues in the linker region of Smad3 *(11)*, while cyclin-dependent kinases phosphorylate other Ser/Thr residues in the Smad3 linker region *(12)*. In both cases phosphorylation of residues in the linker region inhibits Smad3 transcriptional activity. Thus, other growth factors in the cellular milieu can influence TGF-β signaling and it is the sum of the positive and negative signaling inputs that will determine the net response, making it difficult to predict the affects of a single cytokine in the complex in vivo environments that exist in wound healing, chronic fibrosis, and carcinogenesis.

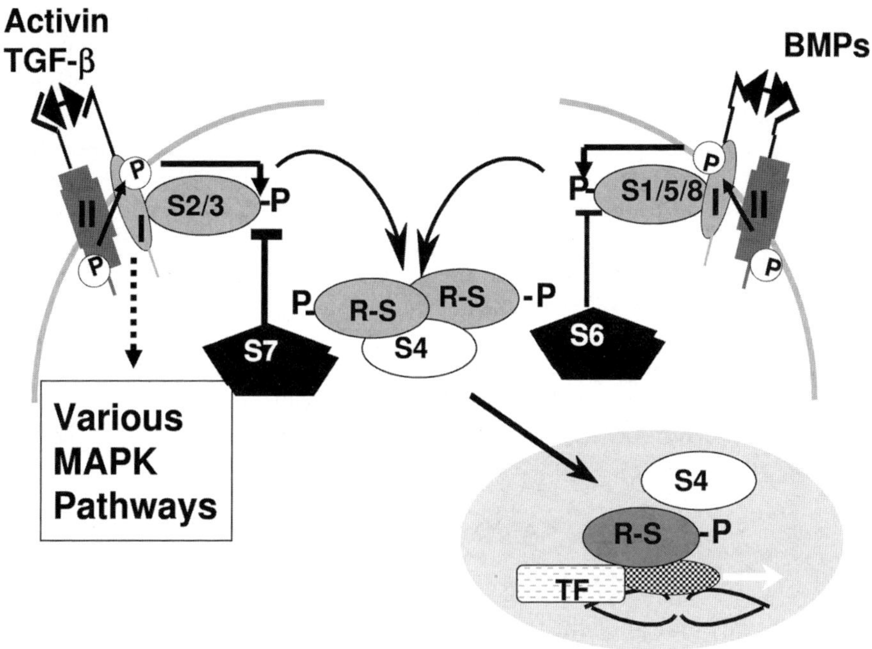

Fig. 1. Overview of the TGF-β signaling pathway. At the cell surface, binding of TGF-β to the constituitively active Type II receptor phosphorylates the Type I receptor. The activated Type I receptor then phosphorylates Smad2 or 3 (S2/3) at the C-terminal Ser residues. Activin also phosphorylates Smads2 and 3, while BMPs phsophorylate Smads1, 5, and 8 (S1/5/8). The receptor-activated Smads (R-S) then complex with the common mediator Smad4 (S4) and this complex translocates to the nucleus where it regulates transcription of target genes and binds to a variety of transcription factors (TF). Activation of R-Smads by Type I receptor kinases is inhibited by Smad6 (S6) for the BMP pathway and Smad7 (S7) for the TGF-β/activin pathway. The activated TGF-β receptor complex can also signal through various mitogen-activated kinase (MAPK) pathways.

3. SMAD2 VS SMAD3

Both Smads2 and 3 mediate signals from TGF-β but they have nonredundant functions in vivo. Smad2 knockout (KO) mice die between E7.5 and E8.5 with failure to gastrulate, and form mesoderm and an anterior-posterior axis *(13–15)*. On the other hand, Smad3 null mice are viable with diminished T-cell responsiveness to TGF-β and usually die at about 6 mo of age from defects in mucosal immunity with a progressive degenerative cartilage disease *(16–18)*. The embryonic lethality of Smad2 null mice suggests that Smad2 may be more involved in mediating developmental signals than Smad3. Furthermore, using Smad2- and Smad3-deficient mouse embryo fibroblasts (MEFs) Piek et al. *(19)* showed that activation of a smad binding element-Lux reporter required Smad3, but not Smad2, whereas activation of the activin response element-Lux reporter required Smad2, suggesting specific roles for Smads2 and 3 in signaling.

Structural differences in the amino-terminal regions of Smad2 and Smad3 also contribute to differences in activities mediated by these Smads. In the nucleus Smad3 binds DNA directly, while Smad2 does not bind to DNA and instead activates transcription indirectly by binding to transcription factors, suggesting these two Smads have distinct effects on regulation of target genes. This has been confirmed by showing that Smad2- and 3-deficient MEFs show different patterns of gene induction by TGF-β; matrix metalloprotease-2 (MMP-2) induction is Smad2 dependent, while c-*fos*, Smad7, and TGF-β1 induction is Smad3 dependent *(19)*. Microarray analyses using these cells and a 9K gene chip suggest that Smad3 is the essential mediator of TGF-β signaling and directly activates genes encoding transcriptional regulators and signal transducers through Smad3/4 DNA-binding motifs that are characteristic of immediate-early target genes of TGF-β. Smad2 instead predominantly transmodulates regulation of both immediate-early and intermediate genes by TGF-β/Smad3 *(20)*.

The ability of Smad3 to regulate transcription of genes involved in extracellular matrix (ECM) metabolism is a contributing factor to the central role of Smad3 in fibrosis. In human dermal fibroblasts a number of collagen gene promoters induced by TGF-β are Smad3 dependent *(21)*. TGF-β contributes to fibrosis both by enhancing ECM synthesis and inhibiting ECM degradation by downregulating expression of matrix-degrading enzymes and increasing expression of MMP inhibitors which also may be a Smad3-dependent process. In dermal fibroblasts negative regulation of the MMP-1 promoter is mediated by Smad3 (it does not occur in Smad3 null MEFs) *(22)*. The induction of the tissue inhibitor of metalloprotease-1 by TGF-β also appears to be Smad3 dependent *(21)*.

4. SMAD3 AND WOUND HEALING

4.1. Healing of Incisional Wounds

The seminal study indicating the importance of Smad3 in mediating the fibrotic response was a study of cutaneous incisional wound healing in Smad3 KO mice done by Ashcroft et al. *(23)*. Figure 2 shows the functions usually mediated by TGF-β in the wound healing process and how these are altered in Smad3 KO mice. Platelet degranulation at the site of injury normally recruits inflammatory cells and fibroblasts into the area and subsequently stimulates these cells to produce TGF-β. Fewer cells migrate into the Smad3 KO wound bed because Smad3-null neutrophils *(17)*, monocytes *(23)*, and fibroblasts *(24)* do not respond to the chemotactic effects of TGF-β. Smad3-null cells also have decreased ability to autoinduce TGF-β *(19,23)* and so levels of this fibrotic cytokine are lower in Smad3 KO wounds. Granulation tissue in Smad3 KO incisional wounds shows decreased accumulation of ECM owing to the reduced numbers of fibroblasts recruited into the area *(24)* and decreased synthesis of collagens, many of which are Smad3-dependent *(21)*. Additionally, incisions made in Smad3 KO skin re-epithelialize more quickly than in Smad3 wild-type (WT) littermates, because Smad3 null keratinocytes lose their response to the growth inhibitory effects of TGF-β and proliferate more quickly.

The cutaneous incisional wound healing model was used to investigate the crosstalk between Smad3 and hormone signaling reported in vitro *(25,26)*. Estrogens usually accelerate healing associated with a reduced inflammatory response, while androgens inhibit repair and enhance leukocyte recruitment *(27,28)*. Our studies suggest that Smad3 plays a role in androgen-mediated inhibition of wound healing, but not in the responses to estrogen modulation in vivo *(29)*. Both Smad3 WT and KO female mice exhibit delayed healing following ovariectomy, which was reversed by estrogen replacement. In contrast, castration accelerates healing in WT male mice and is reversible by exogenous androgen treatment, but modulation of androgen levels in Smad3 KO mice does not alter the healing response. Further study shows that Smad3 null macrophages have a muted response to stimulation by androgens, but not by estrogens. Additional work is needed to determine the molecular interactions underlying the crosstalk between Smad3 and androgens.

4.2. Healing of Excisional Wounds

Full thickness excisional skin wounds made with a 4-mm punch biopsy on the backs of Smad3 KO mice also show more rapid epithelial closure and reduced wound area compared

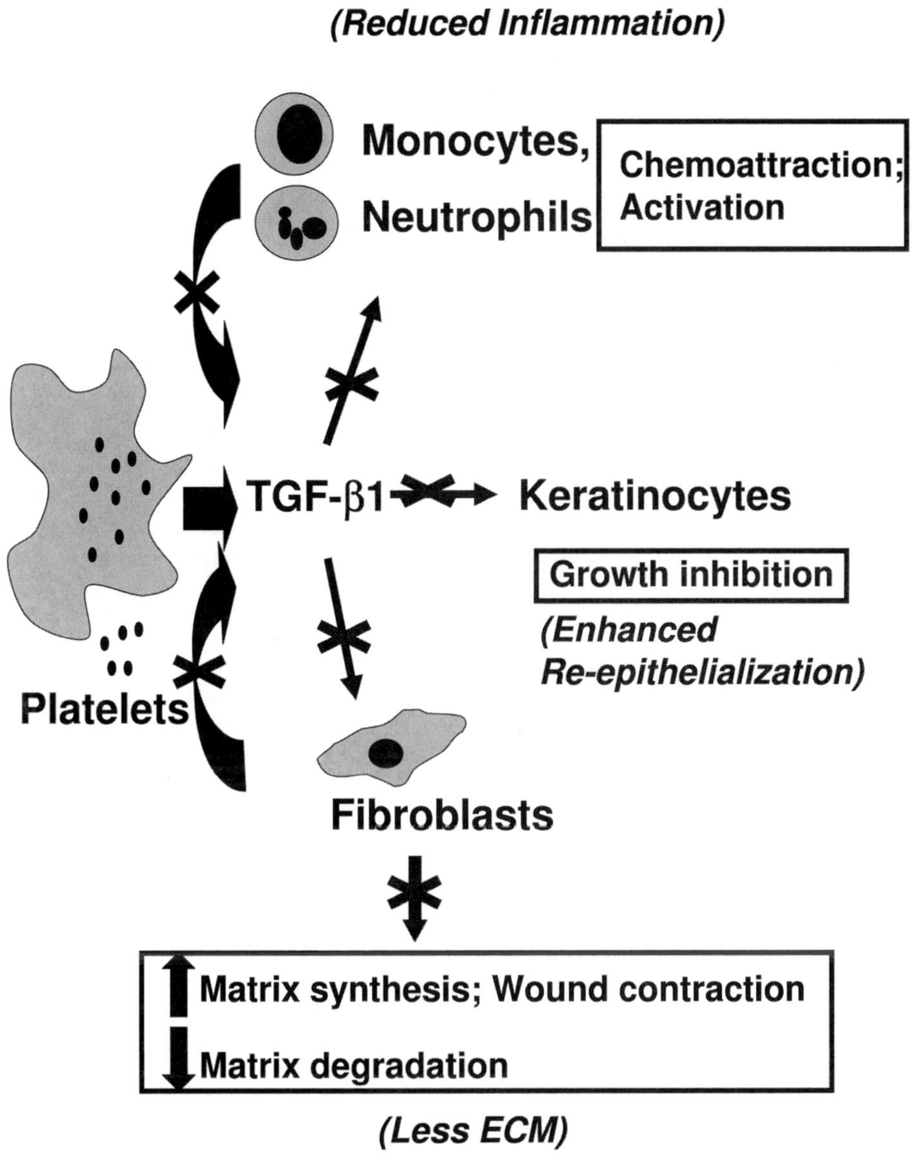

Fig. 2. Proposed role of TGF-β /Smad3 in wound healing. Degranulation of platelets releases TGF-β1 which is chemotactic for neutrophils, monocytes, and fibroblasts with subsequent autoinduction of TGF-β by these cells and induction of collagen and other matrix proteins by fibroblasts. Processes marked with an X are blocked by loss of Smad3 leading to wounds in Smad3 KO mice showing reduced inflammation and less matrix deposition. Additionally, Smad3 null keratinocytes don't respond to growth inhibitory effects of TGF-β so wounds close more quickly.

to WT littermates *(30)*. Additionally, in a model of delayed healing where 0.3×1.0 cm full thickness wounds are made down to the fascia on the dorsal aspect of the mouse tail and left to heal by secondary intention, wounds heal 30% faster in Smad3 KO mice *(31)*. Unexpectedly, excisional wounds made in the ear, which lacks an underlying matrix support, enlarge over time in Smad3 KO compared to WT mice (P. Arany, submitted). As opposed to incisional wounds where there is decreased inflammation in the KO mouse, there is a similar influx of inflammatory cells in ear wounds in the two genotypes. In Smad3 KO mice elastin and glycosaminoglycan levels are increased and collagen fibril diameter is decreased. These factors seem to contribute to altering the mechanotransduction properties of the wound with a possible increase in retractile forces in the KO mice. Additionally, Smad3 KO fibroblasts do not undergo contraction when treated with TGF-β in a floating collagen gel in vitro model of wound contraction *(32)*.

5. SMAD3 AND CHRONIC FIBROTIC CONDITIONS

Dysregulation of processes needed for a successful wound healing response, such as inflammation, angiogenesis, and fibrogenesis, occurs in pathological fibrotic conditions. Increased levels of TGF-β are often found in tissues undergoing an uncontrolled fibrotic response and administration of agents that block TGF-β activity, such as antibodies and antisense oligonucleotides, reduces fibrosis in a number of animal models *(33)*. Wound healing studies discussed in the last section have shown that Smad3 is essential for TGF-β-mediated migration of inflammatory cells and fibroblasts, TGF-β autoinduction, and ECM synthesis, suggesting that Smad3 KO mice might be resistant to pathological fibrosis. Indeed, there are now a number of studies in which Smad3 KO mice show a diminished response in different animal models of fibrosis. The results of these studies will be summarized in this section.

5.1. Skin

While radiation therapy is often successful in treatment of malignancies, many patients experience the side effects of a robust fibrotic response which compromises quality of life and complicates later surgery. We demonstrated that Smad3 is a central mediator of this fibrosis using the Smad3 KO mouse. Six weeks after exposure of flank skin to 30 Gy of localized ionizing irradiation, the skin from Smad3 KO mice shows: (1) less epidermal acanthosis; (2) fewer mast cells, neutrophils, and macrophages in the dermis; (3) decreased expression of TGF-β; and (4) less scarring as determined by picrosirius red staining compared to irradiated WT skin *(34)*. Additionally, even in previously irradiated skin, incisional wounds in the Smad3 KO mice are still able to heal more quickly than those in WT mice *(24)*. Because impaired healing of wounds often occurs in irradiated tissues *(35)* making surgical intervention in previously irradiated tissues difficult, these results suggest that Smad3 inhibitors might be useful therapeutic agents in this setting.

When Smad3 WT and KO dermal fibroblasts in vitro were treated with a combination of ionizing radiation and TGF-β (to mimic the in vivo tissue response where irradiation increases TGF-β levels), Smad3 WT, but not KO cells, show a synergistic induction of mRNA for both TGF-β1 and connective tissue growth factor (CTGF) *(24)*. Because the combination of TGF-β and CTGF may be necessary for optimal ECM production *(36)*, the decreased levels of these cytokines may partially explain the decreased scarring in irradiated Smad3 KO skin. Smad3 may also play a role in the fibrotic response in keloids. Using Smad2 and 3 KO and WT MEFs cocultured with keloid keratinocytes, Phan et al. *(37)* demonstrated that expression of Smad3, but not Smad2, may be crucial for fibroblast proliferation, collagen production and contraction in keloid fibroblasts when stimulated by profibrotic paracrine factors secreted by keloid keratinocytes.

5.2. Lung

Pulmonary fibrosis is characterized by mesenchymal cell proliferation and transdifferentiation of some of these cells to myofibroblasts with excessive collagen accumulation in alveolar and interstitial compartments in the lung. An inflammatory response resulting from injury sometimes initiates this fibroproliferative response. In four models of pulmonary fibrosis Smad3 KO mice exhibit decreased matrix accumulation. These models include:

(1) bleomycin-induced pulmonary fibrosis *(38)*; (2) a model of obliterative bronchiolitis accomplished by heterotopic tracheal transplantation *(39)*; inhalation of adenovirus expressing (3) active TGF-β1 *(40)* or (4) interleukin-1β (IL-1β) *(41)*. In each of these models Smad3 KO mice show markedly reduced levels of collagen and TGF-β compared to WT mice in spite of the fact that the initial inflammatory responses are similar in WT and KO mice. Whether inflammation is required for progression to a chronic fibrotic response in the lung is controversial and antiinflammatory drugs tend to have poor efficacy in the treatment of pulmonary fibrosis *(41)*. Following administration of adenoviral IL-1β, fibrosis develops 1 wk after the resolution of the initial acute inflammatory response. The fibrotic remodeling is associated with a persistent upregulation of endogenous TGF-β suggesting that progressive fibrosis is more related to impairment of the repair process and less to chronic inflammation *(41)*. The Smad3-mediated autoinduction of TGF-β in fibroblasts may be essential for the progression to chronic fibrosis in the lung.

Smad3 KO mice do develop enlargement of the peripheral airspace, a condition characteristic of emphysema *(40,42)*. In emphysema there is a progressive destruction of ECM thought to result from an imbalance in protease-antiprotease activity. Lungs from Smad3 KO mice exhibit higher levels of MMP-9 and MMP-12 than do lungs from WT mice. The loss of Smad3 signaling appears to abrogate the ability of TGF-β to negatively regulate the expression of these proteases, resulting in loss of ECM in the Smad3 KO lungs.

5.3. Liver

Injury to the liver can activate hepatic stellate cells (HSCs) which induces these cells to proliferate, undergo morphological changes to myofibroblast-like cells which express α-smooth muscle actin (α-SMA), and increase ECM production. When acute liver injury is induced by administration of CCl_4, Smad3 KO mice show one-half the induction of hepatic collagen type I mRNA as do Smad3 WT mice, even though there is similar inflammation and necrosis in each genotype *(43)*. This is another example of dissociation between the inflammatory and fibrotic responses where Smad3 seems more involved in the fibrotic process. It should be noted that this is a model of acute hepatic injury and that chronic hepatic fibrosis which is often associated with increased levels of TGF-β has not yet been evaluated in Smad3 KO mice.

5.4. Kidney

Progressive renal interstitial fibrosis in which glomerular tissue is replaced by ECM, can be induced by a number of conditions such as hypertension, diabetes, glomerulonephritis and urinary tract obstruction. Increased expression of TGF-β is associated with increased mesangial matrix in several glomerular diseases. Again Smad3 seems to play a central role in mediating fibrosis. When Smad3 KO mice are treated with streptozotocin to induce a type I-like diabetic glomerulopathy, there is attenuated thickening of the glomerular basement membrane, albuminuria and upregulation of fibronectin and collagen IV mRNA as compared to WT mice *(44)*. Unilateral ureteral obstruction (UUO) in Smad3 WT and KO mice results in decreased influx of inflammatory cells and collagen deposition in KO compared to WT mice *(45,46)*. Following UUO there is decreased apoptosis of renal tubular cells in obstructed kidneys of Smad3 KO mice *(46)*. Smad3 KO mice also exhibit reduced epithelial-mesenchymal transition (EMT) of renal tubules as evidenced by maintenance of E-cadherin expression and lack of induction of α-SMA and Snail *(45)*. Thus, TGF-β -induced EMT in which polarized epithelial cells acquire a more mesenchymal phenotype and are capable of synthesizing ECM, is also a Smad3-dependent process *(47)*. EMT is thought to be an important mechanism in ECM deposition in chronic fibrosis and is also associated with tumor invasiveness and intravasations and extravasations of metasatic cells.

5.5. Eye

Pathologic EMT is often involved in the healing of ocular wounds such as in the lens epithelium following cataract surgery and in retinal pigment epithelial (RPE) cells following retina reattachment surgery. The resulting fibrosis can lead to opacification of the implanted lens and proliferative vitreoretinopathy (PVR), respectively. In vitro TGF-β can induce EMT in both lens epithelial and RPE cells *(48)* and again Smad3 mediates this transdifferentiation. Following injury to the lens capsule *(49)* or retinal detachment *(50)* in Smad3 WT and KO mice, EMT is completely blocked in the KO mice as evidenced by lack of induction of snail, α-SMA, and collagen I. The resulting reduction in fibrosis in the Smad3 KO mice suggests that inhibition of the Smad-signaling pathway may prevent formation of secondary cataracts and PVR. In models of corneal repair however, loss of Smad3 seems to be only partially beneficial *(51)*. While Smad3 KO mice show a reduction of stromal cell fibrotic transformation during corneal healing, there was no reduction in fibronectin deposition which would still result in stromal haze after vision corrective surgery.

6. SMAD3 INHIBITORS AS POTENTIAL ANTIFIBROTIC AGENTS

The ability of TGF-β to mediate a number of biological responses through multiple signaling pathways, such as Smads, MAP kinases, and PI3K *(9)*, implies that therapies that block expression, activation, or receptor binding of TGF-β could potentially induce an array of unwanted side effects by blocking all TGF-β signaling pathways. Because Smad3 plays a critical role in mediating pathological fibrosis, inhibiting only Smad3 signaling could be a prime target for intervention in fibrotic diseases with minimal complications. Figure 3 shows that multiple steps in the fibrotic process could potentially be blocked by Smad3 inhibitors, making them especially effective therapeutic agents. While some of the inhibitors discussed in Section 7 would be expected to inhibit both Smads2 and 3, this may not present a clinical problem because it appears that Smad3 is the essential mediator of TGF-β signaling, while Smad2 instead predominantly transmodulates regulation of genes by TGF-β/Smad3 *(20)*.

7. INHIBITION OF SMAD3 THROUGH INDUCTION OF SMAD7

Smad7 inhibits TGF-β -induced phosphorylation of Smads2 and 3 (Fig. 1) and gene therapy approaches of introduction of vectors expressing Smad7 have successfully blocked fibrosis in several animal models. A single intratracheal injection of a Smad7 adenovirus in a murine model of bleomycin-induced pulmonary fibrosis suppresses both collagen expression and a morphological fibrotic response *(52)*. When fibrosis is induced in rat liver by ligation of the common bile duct, a 50% reduction in collagen and α-SMA expression is seen 3 wk after ligation when adenoviral Smad7 is injected into the portal vein during surgery and into the tail vein at later times *(53)*. In the UUO model of renal fibrosis, injection and electroporation of adenoviral CMV-Smad7 into the pelvic space *(54)* or doxycyline-induced Smad7 gene expression in the kidney *(55)* results in inhibition of tubulointerstitial fibrosis with decreased expression of collagen and α-SMA.

The antifibrotic properties of certain cytokines and small molecules appear to be mediated by induction of Smad7. In a mouse model of bleomycin-induced pulmonary fibrosis intraperitoneal injections of IL-7 decrease TGF-β and collagen expression in the lung and in cultured pulmonary fibroblasts IL-7 inhibits TGF-β-induced collagen production by induction of Smad7 *(56)*. Halofuginone, a low molecular weight plant alkaloid, inhibits collagen synthesis both in vitro and in several animal models of fibrotic disease *(57)*. In the tight skin mouse model of scleroderma halofuginone treatment inhibits spontaneous dermal fibrosis, as well as inhibiting TGF-β-induced upregulation of collagen and Smad3 phosphorylation in cultured fibroblasts *(58)*. Injection of halofuginone in mice also reduces ionizing

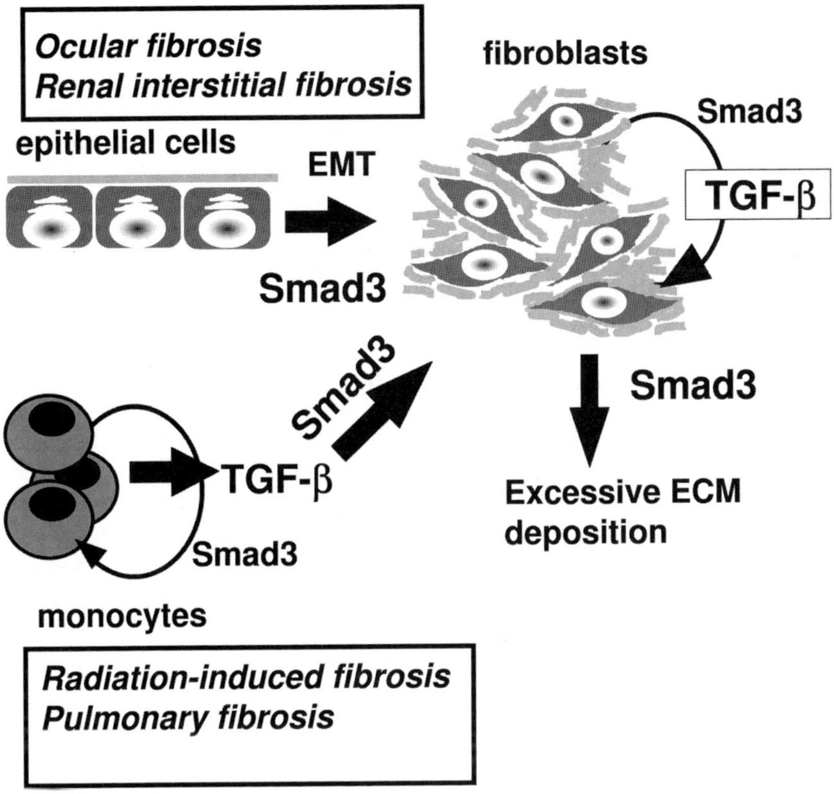

Fig. 3. Many processes in pathologic fibrosis are mediated by Smad3 and are potential targets for Smad3 inhibitors. These include: (1) EMT which is important in the induction of renal interstitial fibrosis and ocular fibrosis; (2) recruitment of inflammatory cells and fibroblasts and autoinduction of TGF-β in these cells which is involved in radiation-induced fibrosis and pulmonary fibrosis and (3) induction of collagen synthesis by TGF-β.

radiation-induced hind leg contracture resulting from fibrosis and improvement is maintained even after halofuginone treatment is discontinued *(59)*. In several cell lines halofuginone decreases TGF-β-induced phosphorylation of Smads2 and 3 and also rapidly induces expression of Smad7 mRNA. One important aspect of this study was its demonstration that the effectiveness of radiation treatment of subcutaneous tumors in mice was not affected by halofuginone, suggesting that halofuginone may have potential uses as a therapeutic agent to protect normal tissue from unwanted fibrosis in patients receiving radiation treatment of malignancies.

7.1. Direct Inhibition of Smad3

Several water-soluble small molecule inhibitors related to imidazole inhibitors of p38 inhibit the kinase activity of TGF-β Type I receptors, ALK4 and ALK5. The first inhibitor in this class, SB-431542, abrogates TGF-β-induced phosphorylation and nuclear import of Smads2 and 3 and prevents TGF-β-induced stimulation of collagen, fibronectin, CTGF, TGF-β autoinduction, and myofibroblast transdifferentiation in cultured cells without preventing TGF-β-induced activation of MAP kinases *(60,61)*. The inhibitor also suppresses TGF-β-induced contraction of collagen gels by both normal and keloid-derived fibroblasts *(62)*. A newly synthesized inhibitor, SB-505124, developed as a competitive inhibitor of the ATP binding site of ALK5, is 3–5 × more potent than SB-431542 in inhibiting Smad2/3 signaling and does not alter BMP-induced signaling through Smads1, 5, and 8 *(57,63)*. Another inhibitor, A-83-01 is also more potent than SB-431542 and prevents phosphorylation of Smads2 and 3 without altering other signaling pathways and inhibits TGF-β-induced EMT in NMuMG cells *(64)*.

Small interfering RNA (siRNA) targeted to Smad3 inhibits nuclear translocation of the Smad3/4 complex, as well as TGF-β-induced promoter activity and anti-proliferative effects in Mv1Lu cells *(65)*. These agents also inhibit collagen I and PAI-1 gene expression in rat HSCs *(66)*. Recently, several Smad-binding peptide aptamers have been developed and shown to selectively inhibit TGF-β -induced gene expression, suggesting that selective disruption of Smad transcriptional complexes may offer another mechanism to inhibit the fibrotic response *(67)*. Whether siRNA and peptide aptamer approaches to inhibit Smad3 will be successful in vivo and have potential uses as antifibrotic therapeutic agents remains to be seen.

8. INFLAMMATION, TGF-β AND CANCER

Because Dvorak *(68)* described the similarity between wound healing and generation of tumor stroma, a number of links between chronic inflammation and carcinogenesis have been observed *(69)*. Inflammation caused by agents such as cigarette smoking, gastroesophageal reflux, or chronic inflammatory bowel disease have been linked to induction of cancers of the lung, esophagus, and colon, respectively *(70)*. In addition *H. pylori*, hepatitis B and C virus, and human papilloma virus may contribute to the initiation of gastric, hepatocellular and cervical cancer, respectively. Activated inflammatory cells generate reactive oxygen and nitrogen species which damage DNA in proliferating epithelial cells resulting in permanent genomic alterations which may contribute to malignant transformation. The replacement of normal epithelial cells with variants suited to survive in the inflammatory milieu may produce cells with more malignant phenotypes. On the other hand, there are some types of inflammatory processes which do not lead to tumor formation *(71)*. Psoriasis, a chronic cutaneous inflammatory disease is seldom, if ever, accompanied by cancer. In Mdr2 KO mice in which hepatitis precedes hepatocellular carcinoma, bile duct tumors are rarely seen despite extensive inflammation and NF-κB activation in the bile ducts, the same processes which contribute to hepatocellular carcinoma *(72)*. Clearly, the properties of the responding epithelial cells and the composition of the inflammatory mediators may determine if inflammation will lead to carcinogenesis.

So what effects would long-term partial suppression of TGF-β signaling through inhibition of Smad3 have on carcinogenesis if a Smad3 inhibitor were used as an antifibrotic therapy? Mice expressing the chimeric Fc:TβRII TGF-β antagonistic protein are free of serious disease and have no changes in tumor incidence or progression *(73)*, suggesting that TGF-β inhibition in vivo may be tolerable. TGF-β has a dual role in carcinogenesis. In the early stages of carcinogenesis TGF-β appears to have tumor suppressor functions, while as the neoplastic process progresses, TGF-β has prooncogenic effects *(9)*. Likewise, interference with endogenous Smad2/3 signaling enhances the malignancy of xenografted tumors of premalignant and well differentiated tumor cells, while suppressing metastasis of more aggressive carcinoma cells *(74)*. Similarly a Smad3 null colonocyte cell line does not demonstrate growth inhibition or apoptosis in response to TGF-β *(75)* and one strain of Smad3 null mice develops metastatic colorectal cancer *(76)*, although other strains do not *(16,17)*. Thus loss of Smad3 might contribute to later stages of carcinogenesis in certain tissues. Clearly, the effects of Smad3 inhibitors on carcinogenesis will be context- and tissue-dependent and their use as antifibrotic agents must be carefully monitored. In certain fibroproliferative disease models in lung and liver *(38,41,43)*, Smad3 KO mice show reduced levels of fibrosis, even though the inflammatory response is similar to that of WT mice, suggesting that in some tissues Smad3 inhibitors may have greater antifibrotic than antiinflammatory effects. Fibrotic conditions where Smad3 inhibitors can be applied locally such as to minimize ocular fibrosis, improve cutaneous wound healing, and protect skin from radiation therapy, may be conditions to be treated initially.

REFERENCES

1. Roberts AB, Sporn MB. The transforming growth factors-β. In: Sporn MB, Roberts AB, eds. Handbook of Experimental Pharmacology. Peptide Growth Factors and Their Receptors. New York: Springer-Verlag 1990:419–472.
2. Massagué J. TGF-beta signal transduction. Annu Rev Biochem 1998;67:753–791.
3. Flanders KC. Smad3 as a mediator of the fibrotic response. Int J Exp Pathol 2004;85:47–64.
4. Border WA, Noble NA. Transforming growth factor beta in tissue fibrosis. N Engl J Med 1994;31: 1286–1292.
5. Roberts AB, Sporn MB. Transforming growth factor-β. In: Clark RAF, ed. The Molecular and Cellular Biology of Wound Repair. New York:Plenum Press 1996:275–308.
6. Moustakas A, Souchelnytskyi S, Heldin C-H. Smad regulation in TGF-beta signal transduction. J Cell Sci 2001;114:4359–4369.
7. Shi Y, Massagué J. Mechanisms of TGF-beta signaling from cell membrane to the nucleus. Cell 2003;113:685–700.
8. Derynck R, Zhang YE. Smad-dependent and Smad-independent pathways in TGF-beta family signalling. Nature 2003;425:577–584.
9. Wakefield LM, Roberts AB. TGF-beta signaling: positive and negative effects on tumorigenesis. Curr Opin Genet Dev 2002;12:22–29.
10. Javelaud D, Mauviel A. Crosstalk mechanisms between the mitogen-activated protein kinase pathways and Smad signaling downstream of TGF-beta: implications for carcinogenesis. Clin Exp Nephrol 2005;24:5742–5750.
11. Matsuura I, Wang G, He D, Liu F. Identification and characterization of ERK MAP kinase phosphorylation sites in Smad3. Biochemistry 2005;44:12,546–12,553.
12. Matsuura I, Denissova NG, Wang G, He D, Long J, Liu F. Cyclin-dependent kinases regulate the antiproliferative function of Smads. Nature 2004;430:226–231.
13. Nomura M, Li E. Smad2 role in mesoderm formation, left-right patterning and craniofacial development. Nature 1998;393:786–790.
14. Waldrip WR, Bikoff EK, Hoodless PA, Wrana JL, Robertson EJ. Smad2 signaling in extraembryonic tissues determines anterior-posterior polarity of the early mouse embryo. Cell 1998;92:797–808.
15. Weinstein M, Yang X, Li C, Xu X, Gotay J, Deng CX. Failure of egg cylinder elongation and mesoderm induction in mouse embryos lacking the tumor suppressor smad2. Proc Natl Acad Sci USA 1998;95:9378–9383.
16. Datto MB, Frederick JP, Pan L, Borton AJ, Zhuang Y, Wang XF. Targeted disruption of Smad3 reveals an essential role in transforming growth factor beta-mediated signal transduction. Mol Cell Biol 1999;19:2495–2504.
17. Yang X, Letterio JJ, Lechleider RJ, et al. Targeted disruption of SMAD3 results in impaired mucosal immunity and diminished T cell responsiveness to TGF-beta. EMBO J 1999;18:1280–1291.
18. Yang X, Chen L, Xu X, Li C, Huang C, Deng CX. TGF-beta/Smad3 signals repress chondrocyte hypertrophic differentiation and are required for maintaining articular cartilage. J Cell Biol 2001;153:35–46.
19. Piek E, Ju WJ, Heyer J, et al. Functional characterization of transforming growth factor beta signaling in Smad2- and Smad3-deficient fibroblasts. J Biol Chem 2001;276:19,945–19,953.
20. Yang YC, Piek E, Zavadil J, et al. Hierarchical model of gene regulation by transforming growth factor-beta. Proc Natl Acad Sci USA 2003;100:10,269–10,274.
21. Verrecchia F, Chu ML, Mauviel A. Identification of novel TGF-beta/Smad gene targets in dermal fibroblasts using a combined cDNA microarray/promoter transactivation approach. J Biol Chem 2001;276:17,058–17,062.
22. Yuan W, Varga J. Transforming growth factor-beta repression of matrix metalloproteinase-1 transcription in dermal fibroblasts involves Smad3. J Biol Chem 2001;276:38,502–38,510.
23. Ashcroft GS, Yang X, Glick AB, et al. Mice lacking Smad3 show accelerated wound healing and an impaired local inflammatory response. Nat Cell Biol 1999;1:260–266.
24. Flanders KC, Major CD, Arabshahi A, et al. Interference with transforming growth factor-beta/ Smad3 signaling results in accelerated healing of wounds in previously irradiated skin. Am J Pathol 2003; 163:2247–2257.

25. Matsuda T, Yamamoto T, Muraguchi A, Saatcioglu F. Cross-talk between transforming growth factor-beta and estrogen receptor signaling through Smad3. J Biol Chem 2001;276:42,908–42,914.
26. Chipuk JE, Cornelius SC, Pultz NJ, et al. The androgen receptor represses transforming growth factor-beta signaling through interaction with Smad3. J Biol Chem 2002;277:1240–1248.
27. Ashcroft GS, Dodsworth J, van Boxtel E, et al. Estrogen accelerates cutaneous wound healing associated with an increase in TGF-beta1 levels. Nat Med 1997;3:1209–1215.
28. Ashcroft GS, Mills SJ. Androgen receptor-mediated inhibition of cutaneous wound healing. J Clin Invest 2002;110:615–624.
29. Ashcroft GS, Mills SJ, Flanders KC, et al. Role of Smad3 in the hormonal modulation of in vivo wound healing responses. Wound Repair Regen 2003;11:468–473.
30. Arany PA, Flanders KC, Kobayashi T, et al. Smad3 deficiency alters key structural elements of the extracellular matrix and mechanotransduction of wound closure. Proc Natl Acad Sci USA 2006;103: 9250–9255.
31. Falanga V, Schrayer D, Cha J, et al. Full-thickness wounding of the mouse tail as a model for delayed wound healing: accelerated wound closure in Smad3 knock-out mice. Wound Repair Regen 2004;12: 320–326.
32. Liu X, Wen FQ, Kobayashi T, et al. Smad3 mediates the TGF-beta-induced contraction of type I collagen gels by mouse embryo fibroblasts. Cell Motil Cytoskeleton 2003;54:248–253.
33. Flanders KC, Burmester JK. Medical applications of transforming growth factor-beta. Clin Med Res 2003;1:13–20.
34. Flanders KC, Sullivan CD, Fujii M, et al. Mice lacking Smad3 are protected against cutaneous injury induced by ionizing radiation. Am J Pathol 2002;160:1057–1068.
35. Tibbs MK. Wound healing following radiation therapy: a review. Radiother Oncol 1997;42:99–106.
36. Mori T, Kawara S, Shinozaki M, et al. Role and interaction of connective tissue growth factor with transforming growth factor-beta in persistent fibrosis: A mouse fibrosis model. J Cell Physiol 1999;181:153–159.
37. Phan TT, Lim IJ, Aalami O, et al. Smad3 signalling plays an important role in keloid pathogenesis via epithelial-mesenchymal interactions. J Pathol 2005;207:232–242.
38. Zhao J, Shi W, Wang YL, et al. Smad3 deficiency attenuates bleomycin-induced pulmonary fibrosis in mice. Am J Physiol Lung Cell Mol Physiol 2002;282:L585–L593.
39. Ramirez AM, Takagawa S, Sekosan M, Jaffe HA, Varga J, Roman J. Smad3 deficiency ameliorates experimental obliterative bronchiolitis in a heterotopic tracheal transplantation model. Am J Pathol 2004;165:1223–1232.
40. Bonniaud P, Kolb M, Galt T, et al. Smad3 null mice develop airspace enlargement and are resistant to TGF-beta-mediated pulmonary fibrosis. J Immunol 2004;173:2099–2108.
41. Bonniaud P, Margetts PJ, Ask K, Flanders K, Gauldie J, Kolb M. TGF-beta and Smad3 signaling link inflammation to chronic fibrogenesis. J Immunol 2005;175:5390–5395.
42. Chen H, Sun J, Buckley S, et al. Abnormal mouse lung alveolarization caused by Smad3 deficiency is a developmental antecedent of centrilobular emphysema. Am J Physiol Lung Cell Mol Physiol 2005; 288:L683–L691.
43. Schnabl B, Kweon YO, Frederick JP, Wang XF, Rippe RA, Brenner DA. The role of Smad3 in mediating mouse hepatic stellate cell activation. Hepatology 2001;34:89–100.
44. Fujimoto M, Maezawa Y, Yokote K, et al. Mice lacking Smad3 are protected against streptozotocin-induced diabetic glomerulopathy. Biochem Biophys Res Commun 2003;305:1002–1007.
45. Sato M, Muragaki Y, Saika S, Roberts AB, Ooshima A. Targeted disruption of TGF-beta1/Smad3 signaling protects against renal tubulointerstitial fibrosis induced by unilateral ureteral obstruction. J Clin Invest 2003;112:1486–1494.
46. Inazaki K, Kanamaru Y, Kojima Y, et al. Smad3 deficiency attenuates renal fibrosis, inflammation, and apoptosis after unilateral ureteral obstruction. Kidney Int 2004;66:597–604.
47. Zavadil J, Bottinger EP. TGF-beta and epithelial-to-mesenchymal transitions. Clin Exp Nephrol 2005;24:5764–5774.
48. Saika S. TGF-beta pathobiology in the eye. Lab Investig 2006;86:106–115.
49. Saika S, Kono-Saika S, Ohnishi Y, et al. Smad3 signaling is required for epithelial-mesenchymal transition of lens epithelium after injury. Am J Pathol 2004;164:651–663.
50. Saika S, Kono-Saika S, Tanaka T, et al. Smad3 is required for dedifferentiation of retinal pigment epithelium following retinal detachment in mice. Lab Invest 2004;84:1245–1258.
51. Stramer BM, Austin JS, Roberts AB, Fini ME. Selective reduction of fibrotic markers in repairing corneas of mice deficient in Smad3. J Cell Physiol 2005;203:226–232.

52. Nakao A, Fujii M, Matsumura R, et al. Transient gene transfer and expression of Smad7 prevents bleomycin- induced lung fibrosis in mice. J Clin Invest 1999;104:5–11.
53. Dooley S, Hamzavi J, Breitkopf K, et al. Smad7 prevents activation of hepatic stellate cells and liver fibrosis in rats. Gastroenterology 2003;125:178–191.
54. Terada Y, Hanada S, Nakao A, Kuwahara M, Sasaki S, Marumo F. Gene transfer of Smad7 using electroporation of adenovirus prevents renal fibrosis in post-obstructed kidney. Kidney Int 2002;61:94–98.
55. Lan HY, Mu W, Tomita N, et al. Inhibition of renal fibrosis by gene transfer of inducible Smad7 using ultrasound-microbubble system in rat UUO model. J Am Soc Nephrol 2003;14:1535–1548.
56. Huang M, Sharma S, Zhu LX, et al. IL-7 inhibits fibroblast TGF-beta production and signaling in pulmonary fibrosis. J Clin Invest 2002;109:931–937.
57. Pines M, Nagler A. Halofuginone: a novel antifibrotic therapy. Gen Pharmacol 1998;30:445–450.
58. McGaha TL, Phelps RG, Spiera H, Bona C. Halofuginone, an inhibitor of type-I collagen synthesis and skin sclerosis, blocks transforming-growth-factor-beta-mediated Smad3 activation in fibroblasts. J Invest Dermatol 2002;118:461–470.
59. Xavier S, Pike E, Fujii M, et al. Amelioration of radiation-induced fibrosis: Inhibtion of transforming growth factor-β signaling by halofuginone. J Biol Chem 2004;279:15,167–15,176.
60. Laping NJ, Grygielko E, Mathur A, et al. Inhibition of transforming growth factor (TGF)-beta1-induced extracellular matrix with a novel inhibitor of the TGF-beta type I receptor kinase activity: SB-431542. Mol Pharmacol 2002;62:58–64.
61. Mori Y, Ishida W, Bhattacharyya S, Li Y, Platanias LC, Varga J. Selective inhibition of activin receptor-like kinase 5 signaling blocks profibrotic transforming growth factor beta responses in skin fibroblasts. Arthritis Rheum 2004;50:4008–4021.
62. Hasegawa T, Nakao A, Sumiyoshi K, Tsuchihashi H, Ogawa H. SB-431542 inhibits TGF-beta-induced contraction of collagen gel by normal and keloid fibroblasts. J Dermatol Sci 2005;39:33–38.
63. Byfield SD, Major C, Laping NJ, Roberts AB. SB-505124 is a selective inhibitor of transforming growth factor-beta type I receptors ALK4, ALK5, and ALK7. Mol Pharmacol 2004;65:744–752.
64. Tojo M, Hamashima Y, Hanyu A, et al. The ALK-5 inhibitor A-83-01 inhibits Smad signaling and epithelial-to-mesenchymal transition by transforming growth factor-beta. Cancer Sci 2005;96:791–800.
65. Yeom SY, Jeoung D, Ha KS, Kim PH. Small interfering RNA (siRNA) targetted to Smad3 inhibits transforming growth factor-beta signaling. Biotechnol Lett 2004;26:699 703.
66. Liu XJ, Ruan CM, Gong XF, et al. Antagonism of transforming growth factor-beta signaling inhibits fibrosis-related genes. Biotechnol Lett 2005;27:1609–1615.
67. Cui Q, Kyun LS, Zhao B, Michael HF. Selective inhibition of TGF-beta responsive genes by Smad-interacting peptide aptamers from FoxH1, Lef1 and CBP. Clin Exp Nephrol 2005;24:3864–3874.
68. Dvorak HF. Tumors: wounds that do not heal. Similarities between tumor stroma generation and wound healing. N Engl J Med 1986;315:1650–1659.
69. Coussens LM, Werb Z. Inflammation and cancer. Nature 2002;420:860–867.
70. Moss SF, Blaser MJ. Mechanisms of disease: Inflammation and the origins of cancer. Nat Clin Pract Oncol 2005;2:90–97.
71. Nickoloff BJ, Ben Neriah Y, Pikarsky E. Inflammation and cancer: is the link as simple as we think? J Invest Dermatol 2005;124:x–xiv.
72. Pikarsky E, Porat RM, Stein I, et al. NF-kappaB functions as a tumour promoter in inflammation-associated cancer. Nature 2004;431:461–466.
73. Yang YA, Dukhanina O, Tang B, et al. Lifetime exposure to a soluble TGF-beta antagonist protects mice against metastasis without adverse side effects. J Clin Invest 2002;109:1607–1615.
74. Tian F, Byfield SD, Parks WT, et al. Reduction in Smad2/3 signaling enhances tumorigenesis but suppresses metastasis of breast cancer cell lines. Cancer Res 2003;63:8284–8292.
75. Mithani SK, Balch GC, Shiou SR, Whitehead RH, Datta PK, Beauchamp RD. Smad3 has a critical role in TGF-beta-mediated growth inhibition and apoptosis in colonic epithelial cells. J Surg Res 2004;117:296–305.
76. Zhu Y, Richardson JA, Parada LF, Graff JM. Smad3 mutant mice develop metastatic colorectal cancer. Cell 1998;94:703–714.

38 TGF-β Signaling in Renal Fibrosis and Inflammation: Gene Therapy Using Ultrasound-Microbubble-Mediated Inducible Smad7

Hui Y. Lan

CONTENTS

TGF-β AND ITS SIGNALING PATHWAY IN RENAL FIBROSIS
TGF-β AND ITS SIGANLING PATHWAY IN RENAL INFLAMMATION
GENE THERAPY FOR RENAL FIBROSIS AND INFLAMMATION USING ULTRASOUND-MICROBUBBLE-MEDIATED SMAD7 GENE TRANSFER
ACKNOWLEDGMENTS
REFERENCES

Abstract

Renal inflammation and fibrosis are common features of acute and chronic kidney diseases. It is known that transforming growth factor-β (TGF-β) is upregulated in kidney disease and plays a diverse role in renal fibrosis by stimulating extracellular matrix (ECM) production, while inhibiting renal inflammation. Mice that overexpress active human TGF-β1 develop severe renal fibrosis, while TGF-β knockout mice exhibit widespread inflammation, including in the kidney. The accumulated knowledge gained through extensive studies in vitro and in vivo, including those of TGF-β knock-out and transgenic mice, supports the concept that TGF-β plays a distinct role in tissue scarring and inflammation.

Recent studies have identified TGF-β and its signaling molecules, Smads, as one of the major mechanisms responsible for renal fibrosis. It is now well accepted that TGF-β regulates fibrosis positively by receptor-associated Smads including Smad2 and Smad3, but negatively by an inhibitory Smad7. While it is clear that Smad7 exerts antifibrotic activity by blocking Smad2/3 activation, most recent findings from our laboratory also demonstrate that TGF-β induces Smad7 to exert its anti-inflammatory function. The major task of this chapter is to focus on the current understanding of the molecular basis and the role of TGF-β/Smad signaling and its crosstalk pathways in the pathophysiology of kidney diseases, particularly on the negative role and mechanisms of Smad7 in renal inflammation and fibrosis. A new gene therapy targeting the TGF-β/Smad signaling pathway using ultrasound-microbubble-mediated inducible Smad7 gene transfer in renal inflammation and fibrosis will be emphasized. It is our belief that the information provided in this chapter may be also applicable to a wide array of other disease conditions.

Key Words: Fibrosis; inflammation; ultrasound; Smad; kidney.

From: *Cancer Drug Discovery and Development: Transforming Growth Factor-β in Cancer Therapy, Vol I: Basic and Clinical Biology*
Edited by: Sonia B. Jakowlew © Humana Press Inc., Totowa, NJ

1. TGF-β AND ITS SIGNALING PATHWAY IN RENAL FIBROSIS

1.1. Role of TGF-β in Renal Fibrosis

TGF-β, a multifunctional cytokine with fibrogenic properties has long been considered as a key mediator of renal fibrosis in both experimental and human kidney diseases, including diabetic and hypertensive nephropathy *(1–3)*. TGF-β stimulates ECM deposition by increasing the synthesis of ECM proteins on the one hand, while acting to inhibit their degradation on the other *(3)*. In addition, TGF-β mediates renal fibrosis by inducing the transformation of tubular epithelial cells to ECM-producing myofibroblasts *(4)*. The important role of TGF-β in renal fibrosis is clearly demonstrated by the finding that renal fibrosis can be induced by the deliberate overexpression of TGF-β1 within the normal rat kidney *(3)*. Furthermore, renal fibrosis can be prevented or ameliorated by blockade of TGF-β with a neutralizing TGF-β Ab, decorin, and antisense strategies in a number of animal models *(3)*.

1.2. Activation of TGF-β Signaling in Renal Fibrosis

It is now clear that TGF-β signals through its signaling pathway, Smads, to mediate tissue fibrosis *(5,6)*. Many fibrogenic genes including COL1A1, COL1A2, COL3A1, COL5A1, COL6A1, COL6A3, COL7A1, and tissue inhibitor of matrix metalloproteinases-1 (MMP-1) genes have been shown to be Smad3-, but not Smad2-, dependent *(6,7)*. This is further supported by the recent finding that mice null for Smad3 are protected against renal fibrosis *(8)*, radiation-induced skin fibrosis *(9)*, and bleomycine-induced pulmonary fibrosis *(10)*. In the context of renal fibrosis, both Smad2 and Smad3 are strongly activated in experimental and human diabetic kidney *(11)*, in the rat with obstructive kidney disease *(12,13)*, and 5/6 nephrectomy *(14)*. Activation of Smad2 and Smad3 within the kidney contributes to the development of glomerulosclerosis, tubulointerstitial fibrosis, and vascular sclerosis, indicating that TGF-β signaling may be a critical pathway leading to tissue fibrosis. This is further documented in recent studies in vitro. Collagen matrix production by various kidney cells in response to TGF-β and other profibrogenic factors including advanced glycation end products (AGEs) and high glucose is mediated by activation of Smad2 and Smad3 *(1,2,11,15–17)*.

1.3. Smad Signaling Crosstalk Pathways in Renal Fibrosis

A novel and exciting finding from our recent studies is that Smad2 and Smad3 can also be activated by a non-TGF-β-dependent pathway. In the diabetic conditions, we find that AGEs, a key mediator in diabetic complications, are able to activate Smads directly and independently of TGF-β *(11)*. This is evident by the findings that AGEs can induce a rapid activation of Smad2 and Smad3 by tubular epithelial cells, mesangial cells, and vascular smooth muscle cells at 5 min with the peak at 30 min. AGE-induced a rapid activation of Smads occurs in the absence of TGF-β and its receptors and is mediated via the receptor for AGEs through the ERK/p38 MAP kinase-dependent crosstalk pathway *(11)*. A similar observation is also found in hypertension. We and other investigators have demonstrated that angiotensin II can also activate the Smad signaling pathway via TGF-β-dependent and MAP kinase-dependent mechanisms *(18,19)*. Thus, AGEs and angiotensin II mediate diabetic and hypertensive complications directly via the MAPK-Smad signaling pathway and indirectly through the classic TGF-β-Smad signaling pathway, which is outlined in Figure 1. Interestingly, substantial inhibition of AGE-induced Smad activation and collagen production by ERK/p38 MAP kinase inhibitors and, to a lesser extent, by an anti-TGF-β antibody, indicates that the MAP kinase-Smad signaling crosstalk pathway is a key mechanism in the pathogenesis of AGE-mediated diabetic scarring *(11)*. These new findings indicate that Smads act as signal integrators, forming the crosstalk pathways among the fibrogenic molecules to

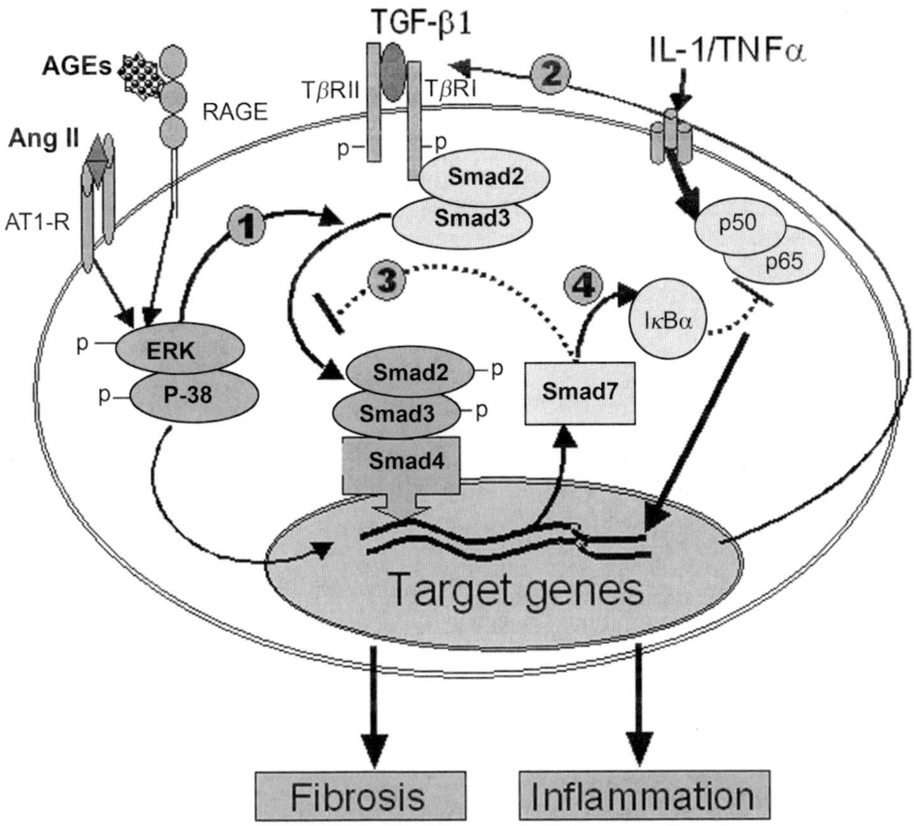

1. **The ERK/p38 MARK-Smad signaling crosstalk pathway**
2. **The TGF-β-dependent Smad signaling pathway**
3. **The Smad7-Smad2/3 negative feedback-loop**
4. **The Smad7-NF.κB negative feedback-loop**

Fig. 1. Smad signaling as a common pathway of fibrosis and inhibitory pathways of Smad7 in fibrosis and inflammation. While TGF-β signals through its receptors to activate Smad2/3 to mediate fibrosis, both AGEs and angiotensin II (Ang II) signal through their individual receptors to activate Smads via the ERK1/2 and p38 MAP kinases (MAPK)-Smad signaling crosstalk pathway (1), in addition to the classic TGF-β-dependent Smad signaling pathway (2). TGF-β induces an inhibitory Smad7 that negatively regulates Smad2/3 activation and fibrosis via the negative feedback-loop (3), and prevents NFκB activation by induction of IκBα, thereby inhibiting inflammatory responses.

mediate tissue scarring. These new data also imply that targeting Smad signaling, instead of anti-TGF-β therapy, may represent a better therapeutic strategy for diabetic and hypertensive complications.

2. TGF-β AND ITS SIGNALING PATHWAY IN RENAL INFLAMMATION

2.1. Role of Latent TGF-β1 in Immune and Inflammatory Diseases

Although the role of TGF-β and its signaling mechanisms in fibrosis has been firmly established, little attention has been paid to the role of TGF-β in inflammatory disease conditions. Furthermore, the underlying signaling mechanisms by which TGF-β exerts its anti-inflammatory function remain largely unclear. TGF-β1 is produced and secreted in vivo as a latent complex, consisting of mature dimeric TGF-β1, a latency-associated peptide (LAP), and a latent TGF-β binding protein (LTBP). LAP binds to the N-terminal of TGF-β, rendering TGF-β latent, thereby preventing TGF-β from binding to its receptors, while LTBP-1 binds the LAP/TGF-β complex and prevents TGF-β from interacting with local matrix proteins. TGF-β must be liberated from LAP to become activated. The factors involved in the liberation of TGF-β from LAP include plasmin, thrombospondin-1, reactive oxygen species and acid (pH). TGF-β functions in both autocrine and paracrine manners to regulate cell proliferation, apoptosis, differentiation, chemotaxis, ECM production, cell migration, and inflammatory/immune responses *(20)*. It is known that TGF-β exhibits its antiinflammation properties by inhibiting macrophage activation, cytokine (IL-1 and TNF-α) production, adhesion molecules (ICAM-1, ELAM-1) and chemokines (MCP-1, IL-8) expression *(21)*. TGF-β also exhibits chemotatic effect on neutrophils and macrophages *(21)*. Mice deficient in TGF-β1 develop lethal multiorgan inflammation at 3 wk of age *(22)*. Administration of TGF-β attenuates autoimmune diseases, including collagen-induced arthritis *(23)*, allergic encephalomyelitis *(24)*, and experimental colitis *(25)*. Moreover, a number of studies have convincingly shown the importance of LAP-TGF-β as therapeutic targets in inflammation and immune diseases. For example, T-cells engineered to produce latent TGF-β1 reverse allergen-induced airway hyperactivity and inflammation *(26)* and down regulate Th1-mediated autoimmune and Th2-mediated allergic inflammatory processes *(27)*. CD4+CD25- T-cells that express LAP on the surface suppress CD4+CD45RBhigh-induced colitis by a TGF-β-dependent mechanism *(28)*. All these studies suggest that over-expression of latent TGF-β1 on immune effector cells produces protective effects on inflammation and immune-mediated diseases.

2.2. Role of Active vs Latent TGF-β1 in Immune and Non-Immune-Mediated Kidney Diseases

Although the role of TGF-β in immune and inflammatory responses has been noted, its role in renal inflammation remains largely unknown. Based on the findings that immune cells engineered to produce latent TGF-β1 suppress immune and inflammatory response, while mice overexpressing the bioactive form of hepatic TGF-β1 develop progressive renal injury associated with highly increased levels of circulating active TGF-β1 *(29)*, We recently sought to test the hypothesis that increased circulating latent form of TGF-β1 may have a protective role in renal inflammation. This hypothesis was tested in a nonimmune-mediated progressive model of obstructive kidney disease. In contrast to a previous report that transgenic mice overexpressing an active form of TGF-β1 in the liver develop progressive renal injury *(29)*, mice that overexpress latent TGF-β1 in the skin show normal renal histology and function, and, importantly, are protected against progressive renal inflammation and fibrosis in obstructive kidney disease *(30)*. This is associated with a 10-fold increase in the circulating levels of latent form of TGF-β1. This novel observation suggests that, unlike active TGF-β1, latent TGF-β1 may have a unique role in antifibrosis and antiinflammation.

We also tested whether mice overexpressing latent TGF-β1 in the skin are protected against renal injury in an immunologically-induced rapid progressive crescentic glomerulonephritis. In wild-type mice, a severe crescentic glomerulonephritis is developed at 7 d after iv injection of a sheep antibody against mouse glomerular basement membrane (GBM). In contrast, mice overexpressing latent TGF-β1 in the skin did not develop crescentic glomerulonephritis. This was associated with inhibition of T-cell and macrophage-mediated renal injury *(31)*. All these findings further demonstrate that latent TGF-β1-derived from nonimmune cells also plays a protective role in inflammatory response in both immune and nonimmune-mediated kidney diseases.

2.3. Signaling Mechanisms of TGF-β in Anti-Renal Inflammation

While it is clear that TGF-β signals through Smad2/3 to mediate fibrosis, the signaling mechanisms by which TGF-β exerts its anti-inflammatory activity remain largely unclear. The most significant finding in our current studies is that inhibition of renal fibrosis and inflammation in both obstructive and anti-GBM glomerulonephritis is associated with a marked increase in renal Smad7 *(30,31)*. Because Smad7 plays a negative role in counter-regulation of TGF-β-mediated fibrosis by blocking Smad2/3 activation, we hypothesize that upregulation of renal Smad7 may be a central mechanism of TGF-β in antiinflammation under disease conditions. This is supported by the finding that overexpression of Smad7 on T-cells prevents the development of experimental glomerulonephritis *(32)*.

It is well accepted that NFκB is a key transcriptional factor that regulates inflammatory responses including those involved in the kidney diseases *(33)*. NFκB is composed of p50 and p65 subunits, of which p65 is a potent transcriptional activator. We found that a marked activation of p65 is closely correlated with the severity of renal inflammation in both obstructive and anti-GBM glomerulonephritis *(30,34)*. It is noted that the mouse Smad7 promoter contains a putative NFκB regulatory site, implying a functional link between the NFκB and Smad7 *(35)*. This was tested in a doxycycline-regulated Smad7-expressing tubular epithelial cell line. We were able to show that doxycycline-induced overexpression of Smad7 substantially suppresses NFκB activation as demonstrated by inhibition of NFκB/p65 nuclear translocation, NFκB transcriptional activity (reporter assay), NFκB DNA binding activity, and NFκB-dependent inflammatory responses induced by IL-1β and TNF-α *(30)*. These findings are consistent with the previous reports that induction of Smad7 inhibits transcriptional activity of NFκB in MDCK cells and in conditionally immortalized mouse podocytes *(36)*. Thus, we establish that TGF-β signals to induce Smad7 to inhibit renal inflammation by blocking NFκB activation.

NFκB activation is regulated by its inhibitor, IκBα. Normally, IκBα binds to NFκB p50/p65 subunits to prevent their activation from nuclear translocation. Once IκBα is phosphorylated or degraded, p50/p65 subunits become activated and translocate into the nucleus to activate the target genes. TGF-β is capable of inducing IκBα *(37)*. We found that Smad7 is able to directly induce expression of IκBα. This suggests that TGF-β may act by stimulating Smad7 to induce IκBα expression, thereby preventing NFκB activation. This in vitro finding is further supported by the in vivo finding that upregulation of renal Smad7 in TGF-β1 transgenic mice is associated with an increase in renal IκBα expression and prevention of κBα phosphorylation/degradation, resulting in suppression of NFκB activation and NFκB-driven inflammatory response in the diseased kidney *(30)*. Thus, inhibition of NFκB activation and NFκB-driven inflammatory response by induction of IκBα expression while preventing its phosphorylation/degradation by Smad7 may be a key signaling pathway whereby TGF-β1 exerts its anti-inflammatory properties.

3. GENE THERAPY FOR RENAL FIBROSIS AND INFLAMMATION USING ULTRASOUND-MICROBUBBLE-MEDIATED SMAD7 GENE TRANSFER

3.1. Smad2/3 as a Final and Common Pathway of Fibrosis and is a Novel Therapeutic Target for Tissue Scarring by Smad7

As described above, the TGF-β/Smad signaling pathway can be activated by TGF-β and many other profibrogenic factors including AGEs and angiotensin II via TGF-β-dependent and independent mechanisms. As outlined in Figure 1, Smad signaling may act as a central pathway leading to fibrosis regardless of the initial pathogenic causes in disease conditions such as hypertension and diabetes. Because Smad7 can act as a negative regulator of Smad signaling, overexpression of Smad7 may be able to inhibit Smad2 and Smad3 activation, thereby terminating Smad signaling and blocking Smad-mediated collagen matrix production. In vitro, we and other investigators have demonstrated that overexpression of Smad7 blocks fibrogenic effects of TGF-β, AGEs, angiotensin II, and high glucose on renal tubular epithelial cells, mesangial cells, and smooth muscle cells *(11,15–17)*. The ability of Smad7 to inhibit Smad2/3-mediated fibrosis in response to multiple fibrogenic factors provides strong evidence that Smad signaling may be a final and common pathway of fibrosis. This implies that specific targeting this pathway by overexpression of Smad7 may represent a novel therapeutic strategy for end stage disease. This hypothesis has been further tested in vivo. We and other investigators have shown that gene transfer of Smad7 using adenovirus or an ultrasound-mediated technique is able to inhibit Smad2 and Smad3 activation and fibrosis in multiple organs in different disease conditions including lung fibrosis included by bleomycin *(38)*, in rat models of obstructive nephropathy *(12,13)*, liver fibrosis *(39)*, and in a hypertensive-associated rat remnant kidney disease *(14)*. Results from these studies support the hypothesis of the central mechanism of Smad signaling in fibrosis and demonstrate that specific blockade of Smad signaling may represent a great therapeutic potential for chronic diseases associated with fibrosis.

3.2. Smad7 as a Novel Therapeutic Strategy for Renal Inflammation and Has a Unique Role in Anti-Fibrosis and Anti-Inflammation

Because targeted deletion of TGF-β and its signaling pathways result in massive inflammation *(20–22)*, it raises the fundamental question of whether blockade of TGF-β/Smad signaling with Smad7 enhances inflammation. An unexpected finding from our studies is that, unlike targeting TGF-β, inhibition of TGF-β signaling by overexpressing Smad7 is able to inhibit renal inflammation in both obstructive and remnant kidney diseases *(40)*. Indeed, Smad7 gene therapy results in substantial inhibition of renal inflammation including suppression of inflammatory cytokines (IL-1,TNF-α), adhesion molecules (ICAM-1and VCAM-1), chemotactic molecule (osteopontin), macrophage and T-cell accumulation, and cell proliferation *(40)*. As outlined in Figure 1, the inhibitory effect of Smad7 on renal inflammation is associated with blockade of NFκB activation *(40)*. Induction of IκBα expression and inhibition of its degradation may be a key mechanism by which Smad7 gene therapy inactivates the survival factor NFκB, resulting in cell apoptosis and suppression of NFκB-driven inflammation *(30)*. These new data demonstrate that Smad7 may have a unique role in antifibrosis and anti-inflammatory and immunosuppressive effects. Inhibition of Smad2 and Smad3 by Smad7 may terminate fibrosis, while inhibition of NFκB activation may be a key mechanism in the resolution of renal inflammation.

3.3. Ultrasound-Microbubble-Mediated Inducible Smad7 Gene Transfer is a Safe, Effective, and Controllable Gene Therapy for Kidney Diseases

It is belief that target gene therapy will ultimately have a major impact on clinical treatment of diseases in the future. Although biologic gene delivery using viral vectors or nonbiologic approaches, including gene gun, electroporation, liposome, and naked DNA injection, have been developed, the clinical use of gene therapy today remains problematic owing to the disadvantages in the aspects of sufficiency and safety. In addition, controlling the transgene expression at a therapeutic level without causing side effects is another important challenge.

It should be noted that extensive and uncontrollable overexpression of Smad7 also causes massive cell death through apoptosis *(12)*, which is consistent with previous reports that Smad7 is an inducer of cell apoptosis *(36)*. This may be associated with the inhibition of a survival factor NFκB and the activation of the JNK pathway *(41)*. Thus, it is critical to control the degree of Smad7 transgene expression within the diseased organs/tissues to maintain a physiological balance within the TGF-β, NF-κB, and JNK signaling crosstalk pathways when attempting to target Smad signaling with overexpression of Smad7. This implies that balancing the signaling pathways, rather than terminating the signals, is important when designing the therapeutic strategies targeting the signaling mechanisms.

To overcome these disadvantages and problems described above, we have recently developed a novel, safe, effective, and controllable gene therapy to specifically inhibit TGF-β/Smad-mediated renal fibrosis and NFκB-dependent renal inflammation using an ultrasound-microbubble-mediated inducible Smad7. This technique is safe because ultrasound itself is harmless to the body and is widely used clinically for many purposes such as physical therapy, diagnosis, guidance for deep organ biopsy, and local drug and genetic material delivery *(42)*. Optison is a contrast agent and is also safe and widely used clinically *(42–45)*. Optison is liquid at room temperature, but it becomes gas-filled microbubbles with an average of 3 μm in diameter at body temperature. Microbubbles can aid drug delivery in themselves and act as agents to carry drugs or genetic materials for site-specific treatment and gene therapy *(42–45)*. It is also effective because gene-bearing microbubbles can be injected intravenously or locally and ultrasound energy applied to the target region. As the microbubbles enter the region of insonation, they cavitate, locally releasing DNA. Cavitation also likely causes a local shockwave that increases cell permeability and thus improves cellular uptake of DNA *(42–45)*. The use of this technique substantially increases the gene transfection rate and transgene expression by up to a 1000-fold compared to the naked DNA strategy and results in Smad7 transgene expression in more than 90% of kidney cells *(12)*. Finally, controlling the transgene expression at the desired therapeutic levels, while minimizing the side effect is also fundamental when designing the gene therapy. To avoid the undesirable side effects caused by overexpression of Smad7, including severe inflammation resulting from oversuppression of TGF-β signaling and massive apoptosis owing to overinhibition of the NFκB signaling pathway, we transfected a doxycycline-regulated Smad7 gene into the kidney and induced Smad7 transgene expression at a therapeutic level by controlling the doses of doxycycline in the drinking water (200 μg/ml in the drinking water) *(12,14,40)*. By using this technique, we demonstrated that Dox-induced overexpression of Smad7 expression at the therapeutic level results in inhibition of renal fibrosis by controlling Smad2/3 activation at the normal level *(12,14)* and suppression of renal inflammation and cell proliferation by balancing NFκB signaling in obstructive and remnant kidney diseases *(30,40)*. As shown in Figures 2 and 3, severe renal histological damage is prevented and cell proliferation, as identified by positive for nucleated proliferative cell nuclear antigen (PCNA), is abolished in both obstructive and remnant kidney diseases. These changes are associated with the prevention of progressive renal injury *(12,14,40)*. Thus, ultrasound-microbubble-mediated inducible Smad7 gene therapy is a safe, effective, and controllable gene transfer method and may provide a novel therapeutic strategy for both inflammatory and fibrotic diseases.

In summary, the discovery of the TGF-β signaling pathway via Smads has markedly improved our understanding of the molecular mechanisms of tissue scarring and inflammation. Studies in individual Smads may allow better understanding of the specific role for each Smad in the pathophysiology of fibrosis, inflammation, and cancer. Identification of the Smad signaling crosstalk pathways will enable us to recognize in-depth the molecular mechanisms and the complicated processes in the pathogenesis of diseases. As outlined in Figure 1, in the context of fibrosis, Smad signaling may represent a common pathway of tissue scarring and overexpression of Smad7 by inducible gene therapy can balance this pathway by inactivating Smad2/3 activation, thereby inhibiting fibrosis in the disease conditions. TGF-β can also induce Smad7 to exert its anti-inflammatory activities by blocking

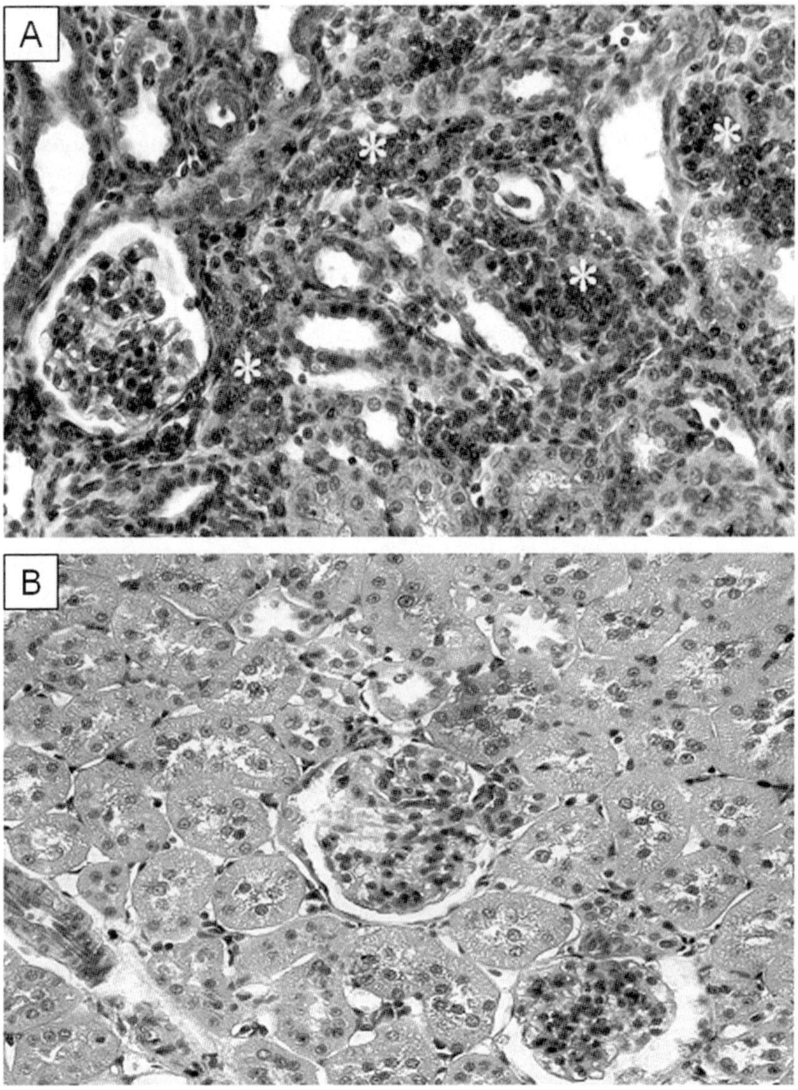

Fig. 2. Gene therapy using ultrasound-microbubble-mediated inducible Smad7 prevents severe kidney damage in remnant kidney disease in rats. (**A**) A rat remnant kidney treated with control empty vectors shows severe histology damage including numerous mononuclear cell infiltration (*) and fibrosis. (**B**) A rat remnant kidney treated with Dox-regulated Smad7 plasmid shows normal kidney histology. PAS-stained paraffin sections. ×250.

NFκB activation. The ability of Smad7 to block both pathways leading to fibrosis and inflammation demonstrates the unique role of Smad7 in anti-inflammation and antifibrosis and may have therapeutic potential for many diseases clinically.

ACKNOWLEDGMENTS

This work was supported by grants from Research Grants Council of Hong Kong (RGC, HKU7952/06M), Research Grants of HKU (CRCG/2006), and NIH (DK064233, DK062828, and HL076661).

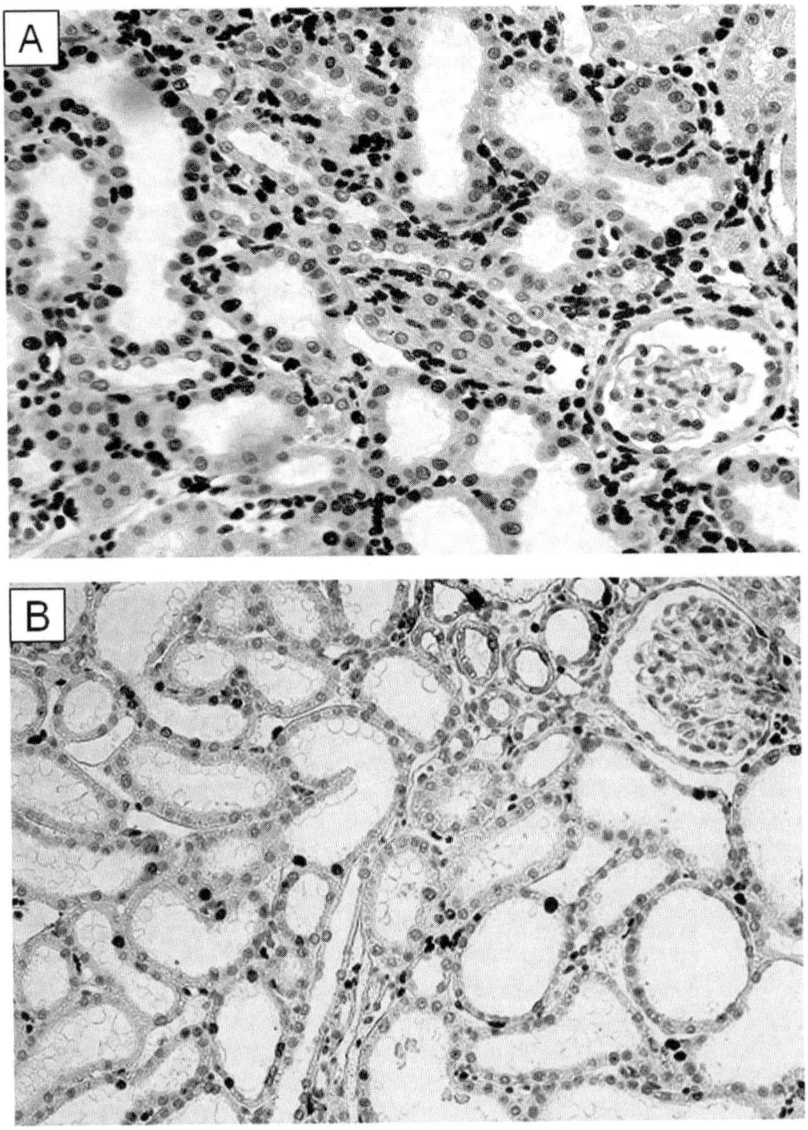

Fig. 3. Immunohistochemistry shows that ultrasound-microbubble-mediated inducible Smad7 gene therapy prevents cell proliferation in obstructive kidney disease in rats. (**A**) A rat obstructive kidney treated with control empty vectors shows a severe proliferative response as identified by numerous PCNA positive cells (dark-black nuclei) within the diseased kidney. (**B**) A rat obstructive kidney treated with Dox-regulated Smad7 plasmid shows a few proliferating cells labeled with PCNA. ×250.

REFERENCES

1. Wang W, Koka V, Lan HY. Transforming growth factor-beta and Smad signaling in kidney diseases. Nephrology (Carlton) 2005;10(1):48–56.
2. Bottinger EP, Bitzer M. TGF-beta signaling in renal disease. J Am Soc Nephrol 2002;13:2600–2610.
3. Border WA, Noble NA. Evidence that TGF-beta should be a therapeutic target in diabetic nephropathy. Kidney Int 1998;54:1390–1391.
4. Lan HY. Tubular epithelial-myofibroblast transdifferentiation mechanisms in proximal tubule cells. Curr Opin Nephrol Hypertens 2003;12:25–29.

5. Massagué J, Chen YG. Controlling TGF-β signaling. Genes Dev 2000;14:627–644.
6. Roberts AB, Piek E, Bottinger EP, Ashcroft G, Mitchell JB, Flanders KC. Is Smad3 a major player in signal transduction pathways leading to fibrogenesis? Chest 2001;120:S43–S47.
7. Verrecchia F, Chu ML, Mauviel A. Identification of novel TGF-beta /Smad gene targets in dermal fibroblasts using a combined cDNA microarray/promoter transactivation approach. J Biol Chem 2001;276:17,058–17,062.
8. Sato M, Muragaki Y, Saika S, Roberts AB, Ooshima A. Targeted disruption of TGF-beta1/Smad3 signaling protects against renal tubulointerstitial fibrosis induced by unilateral ureteral obstruction. J Clin Invest 2003;112(10):1486–1494.
9. Flanders KC, Sullivan CD, Fujii M, et al. Mice lacking Smad3 are protected against cutaneous injury induced by ionizing radiation. Am J Pathol 2002;160:1057–1068.
10. Bonniaud P, Kolb M, Galt T, et al. Smad3 null mice develop airspace enlargement and are resistant to TGF-beta-mediated pulmonary fibrosis. J Immunol 2004;173(3):2099–2108.
11. Li J, Hung XR, Zhu HJ, et al. Advanced glycation end products activate Smad signaling via TGF-β dependent and independent mechanisms: Implications for diabetic renal and vascular disease. FASEB J 2004;8(1):176–178.
12. Lan HY, Mu W, Tomita N, et al. Inhibition of renal fibrosis by gene transfer of inducible Smad7 using ultrasound-microbubble system in rat UUO model. J Am Soc Nephrol 2003;14(6):1535–1548.
13. Terada Y, Hanada S, Nakao A, Kuwahara M, Sasaki S, Marumo F. Gene transfer of Smad7 using electroporation of adenovirus prevents renal fibrosis in post-obstructed kidney. Kidney Int 2002;61: S94–S98.
14. Hou CC, Wang W, Huang XR, et al. Ultrasound-microbubble-mediated gene transfer of inducible Smad7 blocks transforming growth factor-beta signaling and fibrosis in rat remnant kidney. Am J Pathol 2005;166(3):761–771.
15. Li JH, Zhu HJ, Huang XR, Lai KN, Johnson RJ, Lan HY. Smad7 inhibits fibrotic effect of TGF-β on renal tubular epithelial cells by blocking Smad2 activation. J Am Soc Nephrol 2002;13:1464–1472.
16. Li JH, Huang XR, Zhu HJ, Johnson RJ, Lan HY. Role of TGF-β signaling in extracellular matrix production under high glucose conditions. Kidney Int 2003;63:2010–2019.
17. Schnaper HW, Hayashida T, Poncelet AC. It's a Smad world: regulation of TGF-beta signaling in the kidney. J Am Soc Nephrol 2002;13:1126–1128
18. Wang W, Huang XR, Canlas E, et al. Essential role of Smad3 in angiotensin II-induced vascular fibrosis. Circ Res 2006;98:1032–1039.
19. Rodriguez-Vita J, Sanchez-Lopez E, Esteban V, Ruperez M, Egido J, Ruiz-Ortega M. Angiotensin II activates the Smad pathway in vascular smooth muscle cells by a transforming growth factor-beta-independent mechanism. Circulation 2005;111(19):2509–2517.
20. Roberts A. Molecular and cell biology of TGF-beta. Miner Electrolyte Metab 1998;24:111–119.
21. Letterio JJ, Roberts AB. Regulation of immune responses by TGF-beta. Annu Rev Immunol 1998;16:137–161.
22. Yaswen L, Kulkarni AB, Fredrickson T, et al. Autoimmune manifestations in the transforming growth factor-beta 1 knockout mouse. Blood 1996;87:1439–1445.
23. Chernajovsky Y, Adams G, Triantaphyllopoulos K, Ledda MF, Podhajcer OL. Pathogenic lymphoid cells engineered to express TGF beta 1 ameliorate disease in a collagen-induced arthritis model. Gene Ther 1997;4:553–559.
24. Cautain B, Damoiseaux J, Bernard I, et al. Essential role of TGF-beta in the natural resistance to experimental allergic encephalomyelitis in rats. Eur J Immunol 2001;31:1132–1140.
25. Fuss IJ, Boirivant M, Lacy B, Strober W. The interrelated roles of TGF-beta and IL-10 in the regulation of experimental colitis. J Immunol 2002;168:900–908.
26. Hansen G, McIntire JJ, Yeung VP, et al. CD4(+) T helper cells engineered to produce latent TGF-beta1 reverse allergen-induced airway hyperreactivity and inflammation. J Clin Invest 2000;105:61–70.
27. Thorbecke GJ, Umetsu DT, deKruyff RH, Hansen G, Chen LZ, Hochwald GM. When engineered to produce latent TGF-beta1, antigen specific T cells down regulate Th1 cell-mediated autoimmune and Th2 cell-mediated allergic inflammatory processes. Cytokine Growth Factor Rev 2000;11:89–96.
28. Oida T, Zhang X, Goto M, et al. CD4+CD25- T cells that express latency-associated peptide on the surface suppress CD4+CD45RBhigh-induced colitis by a TGF-beta-dependent mechanism. J Immunol 2003;170:2516–2522.
29. Sanderson N, Factor V, Nagy P, et al. Hepatic expression of mature transforming growth factor beta 1 in transgenic mice results in multiple tissue lesions. Proc Natl Acad Sci USA 1995;92:2572–2576.

30. Wang W, Huang XR, Li AG, et al. Signaling mechanism of TGF-β1 in prevention of renal inflammation: role of Smad7. J Am Soc Nephrol 2005;16:1371–1383.
31. Huang XR, Li AG, Wang XJ, Lan HY. Mice overexpressing latent TGF-β1 are protected against crescentic glomerulonephritis. Inflamm Res (Sup2), 2005;S109.
32. Kanamaru Y, Nakao A, Mamura M, et al. Blockade of TGF-beta signaling in T cells prevents the development of experimental glomerulonephritis. J Immunol 2001;166:2818–2823.
33. Guijarro C, Egido J. Transcription factor-kappa B (NF-kappa B) and renal disease. Kidney Int 2001; 59:415–424.
34. Tomita N, Morishita R, Lan HY, et al. Molecular therapy with the nuclear transcription factor-kappaB decoy suppresses experimental crescentic glomerulonephritis. J Am Sci Nephrol 2000;11(7):1244–1252.
35. Nagarajan RP, Chen F, Li W, et al. Repression of transforming-growth-factor-beta-mediated transcription by nuclear factor kappaB. Biochem J 2000;348 (Pt 3):591–596.
36. Schiffer M, Bitzer M, Roberts IS, et al. Apoptosis in podocytes induced by TGF-beta and Smad7. J Clin Invest 2001;108:807–816.
37. Azuma M, Motegi K, Aota K, Yamashita T, Yoshida H, Sato M. TGF-beta1 inhibits NF-kappaB activity through induction of IkappaB-alpha expression in human salivary gland cells: a possible mechanism of growth suppression by TGF-beta1. Exp Cell Res 1999;250:213–222.
38. Nakao A, Fujii M, Matsumura R, et al. Transient gene transfer and expression of Smad7 prevents bleomycin-induced lung fibrosis in mice. J Clin Invest 1999;104:5–11.
39. Dooley S, Hamzavi J, Breitkopf K, et al. Smad7 prevents activation of hepatic stellate cells and liver fibrosis in rats. Gastroenterology 2003;125:178–191.
40. Ng YY, Hou CC, Wang W, Huang XR, Lan HY. Blockade of NFkappaB activation and renal inflammation by ultrasound-mediated gene transfer of Smad7 in rat remnant kidney. Kidney Int (Suppl) 2005;(94):S83–S91.
41. Mazars A, Lallemand F, Prunier C, et al. Evidence for a role of the JNK cascade in Smad7-mediated apoptosis. J Biol Chem 2002;276:36,797–36,803.
42. Newman CM, Lawrie A, Brisken AF, Cumberland DC. Ultrasound gene therapy: On the road from concept to reality. Echocardiography 2001;18:339–347.
43. Blomley MJK, Cooke JC, Unger EC, Monaghan MJ, Cosgrove DO. Science, medicine, and the future: Microbubble contrast agents: a new era in ultrasound. BMJ 2001;322:1222–1225.
44. Harvey CJ, Blomley MJ, Eckersley RJ, Cosgrove DO. Developments in ultrasound contrast media. Eur Radiol 2001;11:675–689.
45. Unger EC, Hersh E, Vannan M, McCreery T. Gene delivery using ultrasound contrast agents. Echocardiography 2001;18:355–361.

39 Transforming Growth Factor-β Peptide Signaling in Pulmonary Development, Bronchopulmonary Dysplasia, Fibrosis, and Emphysema

David Warburton, Wei Shi, Martin Kolb, and Jack Gauldie

CONTENTS

INTRODUCTION
THE TGF-β SUPERFAMILY
SPECIFIC FUNCTIONS OF TGF-β1, 2, 3
TGF-β MISEXPRESSION AND DYSREGULATION PHENOTYPES
DEVELOPMENTAL SPECIFICITY OF THE TGF-β1 OVEREXPRESSION PHENOTYPE
TGF-β FAMILY PEPTIDE SIGNALING IS TIGHTLY REGULATED IN MULTIPLE LAYERS
LUNG FIBROSIS IS INFLAMMATION INDEPENDENT AND MEDIATED BY EXCESSIVE TGF-β AND SMAD3 SIGNALING
TGF-β1 SIGNALING AND PROTEASE/ANTIPROTEASE BALANCE IN LUNG MORPHOGENESIS AND FIBROSIS
TOWARD TGF-β MODULATING THERAPIES FOR LUNG DISEASE
REFERENCES

Abstract

Transforming growth factor-β (TGF-β)1, 2 and 3 peptide superfamily signaling is not only essential for both prenatal and postnatal lung morphogenesis, but also plays a key role in the pathobiology of bronchopulmonary dysplasia, pulmonary fibrosis and emphysema. The respective null mutations of TGF-β1 reveals its function to protect against lung inflammation, of TGF-β2 in cardiopulmonary morphogenesis and of TGF-β3 in lung and palatal fusion. TGF-β signal transduction is tightly regulated at all levels from ligand bioavailability in the extracellular space to the nucleus. Protease-antiprotease balance, correct final assembly of lung matrix and hence completion of alveolarization are all important normal functions of the TGF-β signaling pathway. The consequences of excess TGF-β signaling depend on the developmental stage of the lung: alveolar hypoplasia and fibrosis in the growing lung, fibrosis in the adult lung. While inflammation can induce excessive TGF-β signaling, lung fibrosis per se is inflammation independent and mediated by excessive TGF-β and Smad3

From: *Cancer Drug Discovery and Development: Transforming Growth Factor-β in Cancer Therapy, Vol I: Basic and Clinical Biology*
Edited by: Sonia B. Jakowlew

signaling. Therapeutic manipulation of TGF-β-Smad3 function is therefore a rational target. However, its application to pulmonary medicine will not be easy because of the narrow therapeutic range and many-layered physiological regulation of this pathway.

Key Words: Pulmonary development; bronchopulmonary dysplasia; fibrosis; emphysema.

1. INTRODUCTION

The critical role of signaling initiated by the Transforming growth factor-β (TGF-β) peptide superfamily in pulmonary development and disease has become increasingly apparent over recent years. Misnamed originally for its ability to induce tumor-like growth of fibroblasts in soft agar, TGF-β signaling, is now recognized to play an essential physiological role in regulating extracellular matrix production and degradation as well as other critical cell behaviors. Moreover, the tight regulation of the bioactivity of this potent biomolecule family is critical to maintaining tissue homeostasis. This is illustrated by the fact that lack of TGF-β1 activity results in lethal defects of pulmonary cellular immunity, whereas excess TGF-β1 bioactivity and signaling underlies pulmonary fibrosis and bronchopulmonary dysplasia. In this brief review, we discuss the current biology of TGF-β superfamily signaling, together with the implications for devising possible rational therapies directed to the optimization of lung repair, healing and regeneration. Correctly adjusting the gain setting of TGF-β signaling in the lung will be critical to obtaining a satisfactory therapeutic outcome. Like the porridge in Goldilocks and the Three Bears, it will have to be "just right".

2. THE TGF-β SUPERFAMILY

The TGF-β superfamily can be divided into three subfamilies: activin, TGF-β, and BMP. There are three TGF-β peptide isoforms in mammals: TGF-β1, 2, 3. All of them have been detected in murine embryonic lungs *(1–4)*. In early mouse embryonic lungs (E11.5), TGF-β1 is expressed in the mesenchyme, particularly in the mesenchyme underlying distal epithelial branching points, while TGF-β2 is localized in distal epithelium, and TGF-β3 is expressed in proximal mesenchyme and mesothelium *(5)*. Finely regulated and correct physiologic concentrations and temporo-spatial distribution of TGF-β1, 2, 3 are essential for normal lung morphogenesis and defense against lung inflammation.

3. SPECIFIC FUNCTIONS OF TGF-β1, 2, 3

While the developing lung responds differently to the three TGF-β isoforms there appears to be significant overlap in their biological functions in the adult lung. Mice lacking TGF-β1 develop normally but die within the first month or two of life of aggressive pulmonary inflammation. When raised under pulmonary pathogen-free conditions these mice live somewhat longer but die of other forms of inflammation *(4)*. Thus, physiological concentrations of TGF-β1 appear to suppress the pulmonary inflammation that occurs in response to exogenous factors such as infection end endotoxin. TGF-β2 null mutants die in utero of severe cardiac malformations, while TGF-β3 mutants die neonatally of lung dysplasia and cleft palate *(5,6)*. Embryonic lung organ and cell cultures reveal that TGF-β2 plays a key role in branching morphogenesis, while TGF-β3 plays a key role in regulating alveolar epithelial cell proliferation during the injury repair response *(7)*. In the adult lung, TGF-β1 is one of the key mediators in wound healing and tissue repair processes, causing differentiation and accumulation of myofibroblasts and extracellular matrix. TGF-β2 seems to act in concert with TGF-β1, whereas recent experimental evidence suggests that TGF-β3 might be a counterbalance, opposing excessive fibrosis stimulated by TGF-β1 (Ask et al. manscript submitted).

4. TGF-β MISEXPRESSION AND DYSREGULATION PHENOTYPES

Overexpression of TGF-β1, driven by the *SP-C* promoter, in lung epithelium of transgenic mice causes hypoplastic phenotypes *(8)*. Similarly, addition of exogenous TGF-β to early embryonic mouse lungs in culture resulted in inhibition of lung branching morphogenesis, although each TGF-β isoform has a different IC_{50} (TGF-β2>1>3) *(9–11)*. In contrast, abrogation of TGF-β type II receptor stimulated embryonic lung branching through releasing cell cycle G1 arrest *(12)*. Moreover, overexpression of constitutively active TGF-β1, but not latent TGF-β1, in airway epithelium, is sufficient to have significant inhibitory effects on lung branching morphogenesis *(9)*. However, no inhibitory effect on lung branching was observed when TGF-β1 was over expressed in the pleura and subjacent mesenchymal cells. Furthermore, adenoviral overexpression of a TGF-β inhibitor, *Decorin*, in airway epithelium, completely abrogated exogenous TGF-β1-induced inhibition of embryonic lung growth in culture *(13)*. On the other hand, reduction of decorin expression by DNA antisense oligonucleotides was able to restore TGF-β1-mediated lung growth inhibition *(13)*. Therefore, TGF-β signaling in distal airway epithelium seems to be sufficient for its inhibitory function for embryonic lung growth. TGF-β specific signaling elements, such as Smad2/3/7, are expressed in distal airway epithelium *(13–16)*. Attenuation of Smad2/3 expression by a specific antisense oligonucleotide approach blocked the exogenous TGF-β1-induced inhibitory effects on lung growth. Moreover, expression of Smad7 in airway epithelium, which was induced by TGF-β, had negative regulatory functions for the TGF-β-Smad pathway in cultured cells, specifically blocking exogenous TGF-β-induced inhibitory effects on lung branching morphogenesis as well as on Smad2 phosphorylation in cultured lung explants. Because, blockade of TGF-β signaling not only stimulates lung morphogenesis in culture per se, but also potentiates the stimulatory effects of EGF and PDGF-A, it follows that TGF-β signaling functions downstream of or can over-ride tyrosine kinase receptor signaling.

5. DEVELOPMENTAL SPECIFICITY OF THE TGF-β1 OVEREXPRESSION PHENOTYPE

During embryonic and fetal life, epithelial misexpression of TGF-β1 results in hypoplastic branching and decreased epithelial cell proliferation. In contrast, neonatal misexpression of TGF-β1 using an adenoviral vector approach phenocopies bronchopulmonary dysplasia (BPD) with alveolar hypoplasia, some interstitial fibrosis and emphysema *(17,18)*. Adult misexpression of active TGF-β1, on the other hand, results in a chronic, progressive interstitial pulmonary fibrosis, resulting mainly from increased proliferation and matrix secretion by the mesenchyme; a process that depends on transduction through Smad3 *(19–22)*. Thus, the phenotype caused by excessive TGF-β1 production and signaling is always adverse, but the precise effect depends on the developmental stage of the lung: hypoplasia in embryonic, fetal and premature neonatal lung; fibrosis in premature and adult lung.

6. TGF-β FAMILY PEPTIDE SIGNALING IS TIGHTLY REGULATED IN MULTIPLE LAYERS

The TGF-β signaling system (Fig. 1) has recently been reviewed *(23–25)*. TGF-β signaling is finely regulated in multiple layers from latent peptide activation, through receptor mediated signaling to the nucleus and gene expression. Latent TGF-β ligands are bound to extracellular matrix elements such as Fibrillin *(26–27)* and require proteolytic activation prior to signal transduction by proteases such as plasmin and metalloproteinases. β6 integrin and thrombospondin also play key roles in TGF-β ligand activation *(28)*. Bioavailability of activated TGF-β ligand is further regulated by soluble binding proteins such as Decorin and Endoglin.

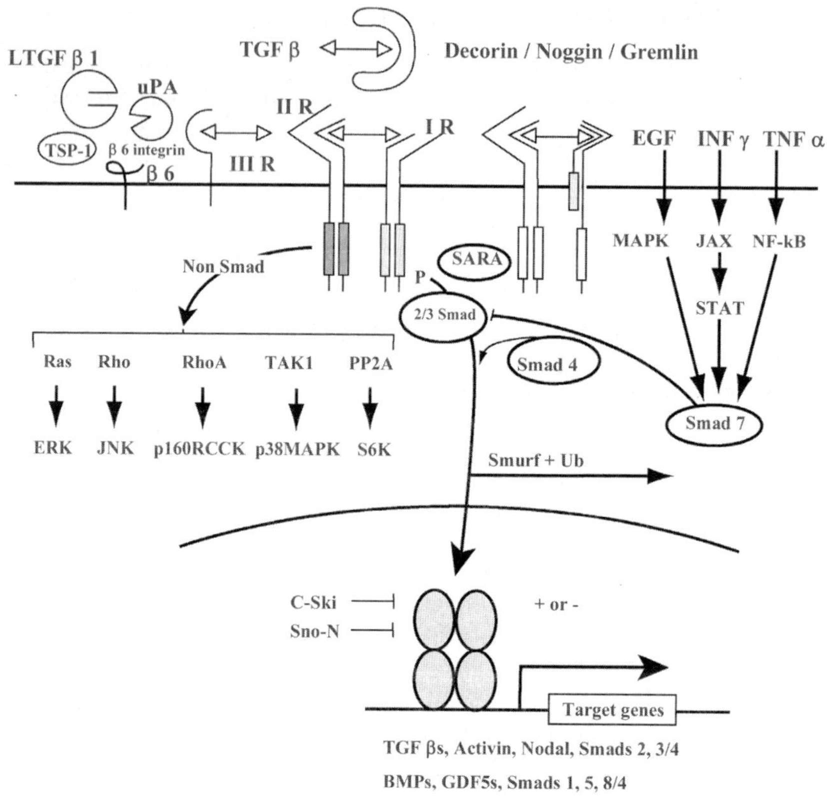

Fig. 1. The TGF-beta signaling pathway is complex and tightly regulated at many levels from outside the cell to the nucleus. This figure illustrates the elegant complexity of TGF-β signaling. Latent TGF-β1 (LTGF-β1) exists within the extracellular matrix, bound to such molecules as Fibrillin. Protease activity such as plasmin (uPA) and binding proteins such as β6 integrin are required to cleave the active peptide and/or to expose the active peptide sequence. Thrombospondin (TSP-1) also plays a role in the activation process. Dimeric TGF-β ligand then can be negatively regulated by binding proteins such as Decorin, Noggin or Gremlin. Free ligand, particularly TGF-β2, is then bound by TGF-β receptor III (IIIR), also called betaglycan), which increases the binding affinity and presents the ligand to a multimeric TGF-β receptor complex comprising pairs of TGF-β type I and II receptors (IR, IIR), assembled as tetramers. The TGF-β IIR has an intracellular serine/threonine kinase domain, which is constitutively active. It transphosphorylates and activates the TGF-βIR serine/threonine kinase. Subsequent signal transduction usually occurs through the canonical Smad pathway, but may also occur through non-Smad pathways linking to other tyrosine kinase receptors and signaling intermediates. The canonical Smad pathway involves phosphorylation and activation of Smads2 and 3 by the receptor kinase. These Smads in turn bind with the common Smad4. The Smad2 or 3 and 4 complex then enters the nucleus and up or down regulates transcription of a variety of genes important in cell cycle regulation and matrix production and turnover. Negative regulatory mechanisms active inside the cell include the inducible negative Smad7, Smurf, a ubiquitin ligase which degrades Smad complexes, and the intranuclear inhibitors Sno and Ski. Smad7 can also be phosphorylated downstream of several tyrosine kinase receptors.

Cognate receptor affinity for ligand binding may also be modulated by such factors as betaglycan, Endoglin or Decorin. Endoglin mutations provide the genetic basis for a subset of hereditary hemorrhagic pulmonary telangiectasia (Osler-Weber-Rendu disease) *(29)*. In the case of TGF-β2 ligand, betaglycan (TGF-β type III receptor) presents activated ligand to the signaling receptor complex and markedly increases ligand-receptor affinity. TGF-β receptors function predominantly as tetrameric transmembrane complexes, comprising pairs of TGF-β type I and II serine threonine kinase receptors. Following dimeric TGF-β ligand binding, the type I receptor kinase is phosphorylated and activated by the constitutively active TGF-β type II receptor kinase. The activated type I receptor serine/threonine kinase phosphorylates the receptor activated R-Smads2 and/or 3. However, this signal transduction step can be negatively modulated by inhibitory Smad7, which interferes with R-Smad activation by receptor. Phosphorylated R-Smads in turn form a complex with the common effector Smad4. This activated complex then becomes rapidly translocated to the nucleus and activates or represses transcription by binding to specific transcriptional complexes on certain gene promoters such as plasminogen activator inhibitor-1 (PAI-1) and cyclin A respectively. Smad complex stability is negatively regulated by Smurf 1, an ubiquitin ligase. Once in the nucleus, Smad complex mediated gene regulation is antagonized by the transcriptional regulators Sno and Ski.

7. LUNG FIBROSIS IS INFLAMMATION INDEPENDENT AND MEDIATED BY EXCESSIVE TGF-β AND SMAD3 SIGNALING

Adenoviral mediated overexpression of active TGF-β1 in the lung results in progressive fibrosis *(20)* and this process along with bleomycin induced fibrosis is totally blocked by administration of a TGF-βRII kinase inhibitor demonstrating the fibrogenic pathway proceeds through TGF-β mediated signaling *(30)*. The fibrosis induced by transient overexpression of IL-1β in rat lung is accompanied by persistent upregulation of TGF-β, suggesting a key role for TGF-β in this model *(31)*. Moreover, Smad3 null mutation substantially blocks pulmonary fibrosis induced by bleomycin and overexpression of TGF-β and IL-1β *(21,32, 33)*. Interestingly, the initial phase of lung inflammation induced by either bleomycin or overexpression of IL-1β were not blocked, neither was the induction of TGF-β1 expression. Rather, the key factor in blockade of experimental lung fibrosis was lack of Smad3 signaling, effectively linking the process of inflammation to fibrogenesis through the TGF-β/Smad3 pathway. Confirmation of this signal event is shown by total blockade of fibrogenesis induced by direct overexpression of active TGF-β1 in Smad3 null mice, thus separating the regulation of fibrogenesis through TGF-β1 from modulation of inflammation. Thus, Smad3 could act as a final common downstream target in the TGF-β-mediated pathobiologic sequence in the lung. Putative Ser/Thr phosphorylation in the link region of Smad2/3 by other non-Smad signaling pathways provides potential sites for crosstalk with other signaling pathways *(31)*. However, the TGF-β/Smad3 interaction is not the only pathway to be considered as therapeutic target. It has recently been demonstrated that TGF-β can signal through Abelson nonreceptor kinase (c-*Abl*), independently of Smad2/3 phosphorylation. Imatinib mesylate (gleevec) is an inhibitor of Abl family kinases, and has been shown to be an efficient antifibrotic agent in models of bleomycin induced lung fibrosis and renal fibrosis *(34,35)*. Thus there appear to be several necessary but not sufficient pathways activated by TGF-β leading to matrix activation and expression.

8. TGF-β1 SIGNALING AND PROTEASE/ANTIPROTEASE BALANCE IN LUNG MORPHOGENESIS AND FIBROSIS

Because, the accumulation of scar or matrix is the hallmark of fibrosis, this can develop through enhanced synthesis (fibrogenesis) or decreased metabolism (fibrolysis) or a combination thereof. Regulation of fibrolysis results from interplay between the matrix metabolizing enzymes, the matrix metalloproteinases (MMP) and their inhibitors, such as the tissue inhibitors of metalloproteinases (TIMP) and PAI-1. Blockade of fibrosis through the use of the TGF-βRII kinase inhibitor, or the use of Smad3 null mutation, is accompanied by a lack

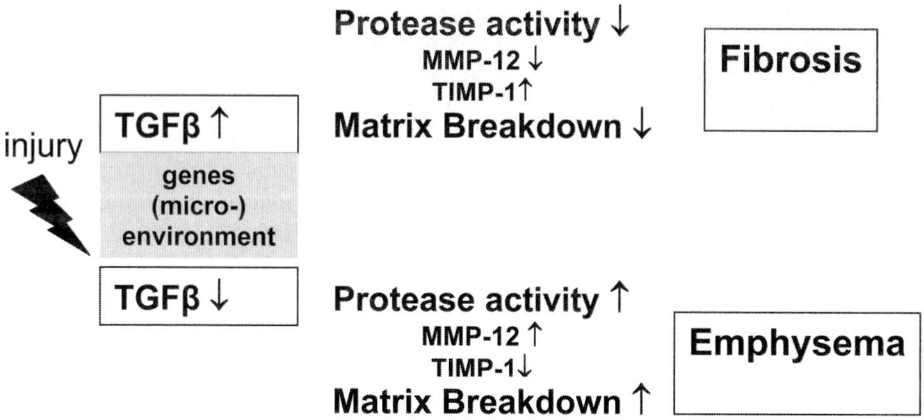

Fig. 2. TGFβ plays a critical role in defining the accumulation or metabolism of matrix in the tissue microenvironment.

of upregulation of TIMP and PAI-1, representing a ratio between MMP and their inhibitors, which creates a tissue microenvironment favoring matrix metabolism rather than accumulation (Fig. 2) *(22,36–38)*. In a similar manner, lung morphogenesis requires matrix metabolism and tissue regeneration and is also dependent on the microenvironmental balance between MMPs and TIMPs. Smad3 null mutant mice are not only resistant to lung fibrosis, caused by bleomycin or by adenoviral vector mediated TGF-β or IL1β overexpression, but also Smad3 null mutation leads to age dependent airspace enlargement associated with upregulation of MMP-12 and MMP-9. Similar findings of spontaneous airspace enlargement are seen in the αVβ6 null mouse, which has a defect in endogenous TGF-β activation *(39)*. Subsequently more detailed analysis shows that air space enlargement in Smad3 null mutant mice is preceded by a failure to correctly septate the peripheral alveolar spaces, with concomitant subtle defects in the spatiotemporal distribution of elastin and other matrix proteins *(40)*. Thus, we speculate that genes related to disruptions in the active TGF-β/Smad pathway could act as susceptibility factors for emphysema, while their overactivity may mediate the development of impaired tissue repair and lung fibrosis. One prominent gene in this cascade is Smad3. Smad3 mutation does protect against lung fibrosis, but not against inflammation, nor induction of TGF-β1 expression. In contrast, null mutation of TGF-β1 predisposes to lung inflammation. Therefore, exact regulation of TGF-β1 bioavailability and signaling actually is required to protect against lung damage.

9. TOWARD TGF-β MODULATING THERAPIES FOR LUNG DISEASE

The mechanistic studies briefly reviewed above strongly indicate the TGF-β pathway as a target for rational therapeutic manipulation to prevent or treat lung diseases ranging from bronchopulmonary dysplasia through tissue remodeling in asthma to pulmonary fibrosis and emphysema. The difficulty lies not only in the essential nature of the correct gain setting of TGF-β signaling to protect the lung against self-destruction owing to excessive inflammation, but on the other hand of the adverse effects of excessive TGF-β signaling. It is likely that the correct gain setting of the TGF-β signaling pathway plays a critical role in determining normal lung host defense and repair, vs adverse outcomes such as lung fibrosis and alveolar hypoplasia. Pretherapeutic approaches that have been tried include overexpression of ligand binding proteins such as betaglycan and Decorin, antiligand and/or anticognate receptor antibodies, overexpression of inhibitory Smad7 and blockade of TGF-βR kinase activity. All of these have shown some preliminary promise in inhibiting the bleomycin-induced model of fibrosis in mice or rats (reviewed in Giri, *[41]*). We have also tried to make interfering peptides that can interdict or modulate ligand-receptor interaction, with varying success in modulating TGF-β signaling in cell lines. The biggest problem confronting TGF-β therapeutics for lung disease, aside from selection of approach, formulation and delivery is pharmacodynamics and tissue localization. Correction of TGF-β ligand bioavailability and signaling to within a small window of therapeutic efficacy will not be trivial, but may provide an important therapeutic key to preserving lung function in lung injury recovery as well as in lung regeneration.

REFERENCES

1. Pelton RW, Johnson MD, Perkett EA, Gold LI, Moses HL. Expression of transforming growth factor-beta 1, -beta 2 and -beta 3 mRNA and protein in the murine lung. Am J Resp Cell Mol Biol 1991; 5:522–530.
2. Millan FA, Denhez F, Kondaiah P, Akhurst RJ. Embryonic gene expression patterns of TGF beta 1, beta 2 and beta 3 suggest different developmental functions in vivo. Development 1991;111:131–143.
3. Schmid P, Cox D, Bilbe G, Maier R, McMaster GK. Differential expression of TGF beta 1, beta 2 and beta 3 genes during mouse embryogenesis. Development 1991;111:117–130.
4. McLennan IS, Poussart Y, Koishi K. Development of skeletal muscles in transforming growth factor-beta 1 (TGF beta 1) null-mutant mice. Dev Dyn 2000;217:250–256.
5. Bartram U, Molin DG, Wisse LJ, et al. Double-outlet right ventricle and overriding tricuspid valve reflect disturbance of looping, myocardialization, endocardial cushion differentiation, and apoptosis in TGF-beta(2)-knockout mice. Circulation 2001;103:2745–2752.
6. Kaartinen V, Voncken JW, Shuler C, et al. Abnormal lung development and cleft palate in mice lacking TGF-beta 3 indicate defects of mesenchymal-epithelial interaction. Nat Genet 1995;11:415–421.
7. Buckley S, Bui KC, Hussain M, Warburton D. Dynamics of TGF-beta 3 peptide activity during rat alveolar epithelial cell proliferative recovery from acute hyperoxia. Am J Physiol 1996;271:L54–60.
8. Zhou L, Dey CR, Wert SE, Whitstett JA. Arrested lung morphogenesis in transgenic mice bearing an SP-C-TGF-Beta1 chimeric gene. Dev Biol 1996;175:227–228.
9. Zhao J, Sime PJ, Bringas P, Jr., et al. Spatial-specific TGF-beta 1 adenoviral expression determines morphogenetic phenotypes in embryonic lung. Eur J Cell Biol 1999;78:715–725.
10. Bragg AD, Moses HL. Serra R. Signaling to the epithelium is not sufficient to mediate all of the effects of transforming growth factor beat and bone morphogenetic protein 4 in murine embryonic lung development. Mech Dev 2001;109:13–26.
11. Serra R, Pelton RW, Moses HL. TGF beta 1 inhibits branching morphogenesis and N-myc expression in lung organ cultures. Development 1994;120:2153–2161.
12. ZhaoJ Crowe DL, Castillo C, Wuenschell C, Chai Y, Warburton D. Epithelium-specific adenoviral transfer of a dominant-negative mutant TGF-beta type II receptor stimulates embryonic lung branching morphogenesis in culture and potentiates EGF and PDGF-AA. Mech Dev 2000;93:71–81.
13. Zhao J, Sime PJ, Bringas P, Jr., Gauldie J, Warburton D. Adenovirus-mediated decorin gene transfer prevents TGF-beta-induced inhibition of lung morphogenesis. Am J Physiol 1999;277:L412–L422.
14. Zhao J, Lee M, Smith S, Warburton D. Abrogation of Smad3 and Smad2 or Smad4 gene expression positively regulates murine embryonic lung branching morphogenesis in culture. Dev Biol 1998;194: 182–195.
15. Zhao J, Crowe DL, Castillo C, Wuenschell C, Chai Y, Warburton D. Smad7 is a TGF-beta-inducible attenuator of Smad2/3-mediated inhibition of embryonic lung morphogenesis. Mech Dev 2000;93:71–81.
16. Zhao J, Shi W, Chen H, Warburton D. Smad7 and Smad6 differentially modulate transforming growth factor-beta-induced inhibition of embryonic lung morphogenesis. J Biol Chem 2000;275: 23,992–23,997.
17. Gauldie J, Galt T, Bonniaud P, Robbins C, Kelly M, Warburton D. Transfer of the active form of transforming growth factor-beat 1 gene to newborn rat lung induces changes consistent with bronchopulmonary dysplasia. Am J Pathol 2003;163:2575–2584.
18. Vicencio AG, Lee CG, Cho SJ, et al. Conditional overexpression of bioactive TGF-β1 in neonatal mouse lung: A new model for bronchopulmonary dysplasia? Am J Respir Cell Mol Biol 2004;31:650–656.
19. Lee CG, Cho SJ, Kang MJ, et al. Early growth response gene 1-mediated apoptosis is essential for transforming growth factor beta1-induced pulmonary fibrosis. J Exp Med 2004;200:377–389.

20. Sime PJ, Xing Z, Graham FL, Csaky KG, Gauldie J. Adenovector-mediated gene transfer of active transforming growth factor-beta induces prolonged severe fibrosis in rat lung. J Clin Invest 1997;100: 768–776.
21. Zhao J, Shi W, Wang YL, et al. Warburton. Smad3 deficiency attenuates bleomycin-induced fibrosis in mice. Am J Physiol Lung Cell Mol Physiol 2002;282:L585–L593.
22. Bonniaud P, Margetts PJ, Kolb M, et al. Adenoviral gene transfer of connective tissue growth factor in the lung induces transient fibrosis. Am J Resp Crit Care Med 2003;168:770–778.
23. Massagué J, Chen YG. Controlling TGF-beta signaling. Genes Dev 2000;14:627–644.
24. Derynck R, Akhurst RJ, Balmain A. TGF-beta signaling in tumor suppression and cancer progression. Nat Genet 2001;2:117–129.
25. Roberts AB, Derynck R. Meeting report: signaling schemes for TGF-beta. Sci STKE 2001;113:PE43.
26. Kaartinen V, Warburton D. Fibrillin controls TGF-beta activation. Nat Genet 2003;33:331–332.
27. Neptune ER, Frischmeyer PA, Arking DE, et al. Dysregulation of TGF-beta activation contributes to pathogenesis in Marfan syndrome. Nat Genet 2003;33:407–411.
28. Morris DG, Huang X, Kaminski N, et al. Loss of integrin alpha(v)beta6-mediated TGF-beta activation causes Mmp12-dependent emphysema. Nature 2003;422:169–173.
29. Shovlin CL, Hughes JM, Scott J, Seidman CE, Seidman JG. Characterization of endoglin and identification of novel mutations in hereditary hemorrhagic telangiectasia. Am J Hum Genet 1997;61:68–79.
30. Bonniaud P, Margetts PJ, Kolb M, et al. Progressive TGFβ induced lung fibrosis is blocked by an orally active ALK5 kinase inhibitor. Am J Respir Crit Care Med 2005;171:889–898.
31. Kolb M, Margetts PJ, Anthony DC, Pitossi F, Gauldie J. Transient expression of IL-1beta induces acute lung injury and chronic repair leading to pulmonary fibrosis. J Clin Invest 2001;107:1529–1536.
32. Bonniaud P, Kolb M, Galt T, et al. Smad3 null mutant mice develop airspace enlargement and are resistant to TGF-beta mediated pulmonary fibrosis. J Immunol 2004;173:2099–2108.
33. Bonniaud P, Margetts PJ, Ask K, Flanders K, Gauldie J, Kolb M. TGFbeta and Smad3 signaling link inflammation to chronic fibrogenesis. J Immunol 2005;175:5390–5395.
34. Daniels CE, Wilkes MC, Edens M, et al. Imatinib mesylate inhibits the profibrogenic activity of TGF-beta and prevents bleomycin-mediated lung fibrosis. J Clin Invest 2004;114(9):1308–1316.
35. Wang S, Wilkes MC, Leof EB, Hirschberg R. Imatinib mesylate blocks a non-Smad TGF-beta pathway and reduces renal fibrogenesis in vivo. FASEB J 2005;19(1):1–11.
36. Massagué J. Integration of Smad and MAPK pathways: a link and a linker revisited. Genes Dev 2003;17:2993–2997.
37. Kolb M, Bonniaud P, Galt T, et al. Differences in the fibrogenic response after transfer of active transforming growth factor-beta1 gene to lungs of fibrosis-prone and fibrosis-resistant mouse strains. Am J Resp Cell Mol Biol 2003;27:141–150.
38. Bonniaud P, Martin G, Margetts PJ, et al. Connective tissue growth factor is crucial to inducing a profibrotic environment in "fibrosis-resistant" BALB/c mouse lungs. Am J Resp Cell Mol Biol 2004; 31:510–516.
39. Munger JS, Huang X, Kawakatsu H, et al. The integrin alpha v beta 6 binds and activates latent TGF beta 1: a mechanism for regulating pulmonary inflammation and fibrosis. Cell 1999;96:319–328.
40. Chen H, Sun J, Buckley S, et al. Abnormal mouse lung alveolarization caused by Smad3 deficiency is a developmental antecedent of centrilobular emphysema. Am J Physiol 2004;288:L683–L691.
41. Giri SN. Novel pharmacological approaches to manage interstitial lung fibrosis in the twenty-first century. Ann Rev Pharmacol Tox 2003;43:73–95.

40 The Role of TGF-β in Radiation and Chemotherapy Induced Pulmonary Fibrosis: Inhibition of TGF-β as a Novel Therapeutic Strategy

Patricia J. Sime, R. Matthew Kottmann, Heather F. Lakatos, and Thomas H. Thatcher

CONTENTS

Introduction
Pulmonary Fibrosis in Cancer Therapy
The Role of TGF-β in Pulmonary Fibrosis
Anti-TGF-β Strategies as Therapy for Pulmonary Fibrosis
Conclusion
Acknowledgments
References

Abstract

Pulmonary fibrosis is a life-threatening side effect of cancer therapy, affecting 1–10% of patients receiving chemotherapy or thoracic radiation. Chemo- and radiation therapy may induce either an early phase acute pneumonitis or a late phase fibrotic process, or both. The acute phase can sometimes be treated effectively with corticosteroids, however, steroids are rarely effective in treating fibrosing disease and alternative therapies have proven elusive. Recent animal and human clinical data implicate the cytokine TGF-β as a key regulator of fibrosis. TGF-β promotes differentiation of fibroblasts to myofibroblasts, as well as fibroblast and myofibroblast proliferation and production of collagen, resulting in accumulation of fibrotic scar tissue and loss of lung function. It is proposed that interfering with TGF-β expression and/or signaling may be clinically useful strategies to block the progression of lung fibrosis. Several strategies in various stages of investigation, from the laboratory to phase III clinical trials, are discussed. These strategies fall into two general classes; direct inhibition of TGF-β itself (via soluble TGF-β decoy receptors, TGF-β antibodies, decorin, RNA interference, and the latency associated peptide) and interference with the downstream signaling cascades initiated by TGF-β (interferon-γ, pirfenidone, imatinib mesylate, and peroxisome proliferator activated receptor-γ [PPAR-γ] agonists). Concerns in implementing anti-TGF-β therapies in pulmonary fibrosis patients include possible systemic effects of inhibiting TGF-β, as well as possible effects on the underlying tumor of local or systemic inhibition of TGF-β.

Key Words: Pulmonary fibrosis; TGF-β; fibroblast; myofibroblast; bleomycin; radiation pneumonitis; decorin; pirfenidone; imatinib mesylate; peroxisome proliferator activated receptor-γ (PPAR-γ); interferon-γ (IFN-γ).

From: *Cancer Drug Discovery and Development: Transforming Growth Factor-β in Cancer Therapy, Vol I: Basic and Clinical Biology*
Edited by: Sonia B. Jakowlew © Humana Press Inc., Totowa, NJ

1. INTRODUCTION

In addition to the roles TGF-β plays in promoting or arresting malignant transformation and tumor growth, TGF is also implicated in a common and life-threatening side effect of cancer treatment, pulmonary fibrosis. Acute lung injury is a common side effect of some chemotherapeutic agents as well as thoracic irradiation. Acute pneumonitis is often followed by a late-phase, progressive fibrosing disease in which the lung accumulates fibrotic scar tissue, leading to impaired function and sometimes death. The acute phase can often be treated effectively with corticosteroids, however, steroids are rarely effective in treating fibrosing disease and alternative therapies have proven elusive. Recent animal and human clinical data implicate TGF-β as a key regulator of fibrosis. This chapter will reflect on specific cancer treatments known to cause fibrosis, and the mechanisms by which TGF-β directs fibrotic processes. A number of novel strategies aimed at interfering with the profibrotic effects of TGF-β, currently in preclinical animal studies and early clinical trials, will be discussed.

2. PULMONARY FIBROSIS IN CANCER THERAPY

Pulmonary fibrosis is a potentially fatal disorder characterized by proliferation of fibroblasts in the lung interstitium with increased synthesis and deposition of extracellular matrix proteins including collagen and fibronectin, which results in scarring of the lung with loss of alveolar function and respiratory distress *(1–3)*. This condition can arise from a number of insults to the lung including inhalation of fumes, dusts, and a variety of nuisance particles such as asbestos and silica. While malignancy itself generally does not cause fibrosis, pulmonary fibrosis can be caused by radiation therapy as well as several chemotherapeutic agents. Thus, pulmonary fibrosis is a significant and potentially life-threatening complication of many routine anticancer therapies, while concerns about fibrosis impose limits on the radiation dose that may be delivered to some patients.

2.1. Thoracic Irradiation

Thoracic irradiation is a well recognized cause of pulmonary inflammation and fibrosis *(4–8)*. The incidence of moderate to severe radiation pneumonitis ranges from 8–20% with the risk escalating with increased total radiation dose and increased volume of lung irradiated *(9,10)*. Pneumonitis is uncommon with total doses <25–30 Gray (Gy), and if the volume of lung irradiated is <25%, however it is almost universal with doses >40 Gy, and if >50% of the lung is irradiated *(8,9,11–14)*. Prior radiation therapy, age >65, and concomitant chemotherapy also increase the risk of developing radiation pneumonitis and subsequent fibrosis *(6,7,10,15–18)*.

Radiation pneumonitis follows a three phase time course. The acute or early phase occurs within the first month, the intermediate phase lasts from one to several months, and the late fibrotic phase typically develops over 6–24 mo, although patients may present with fibrosis without a history of acute radiation pneumonitis. There is some evidence that the early and intermediate phases are a distinct clinical entity that has been termed sporadic radiation pneumonitis, characterized by bilateral alveolitis sharing clinical features of hypersensitivity pneumonitis, that does not progress to fibrosis and is responsive to corticosteroid treatment *(5,19–22)*. Late phase radiation fibrosis is distinct in that it is confined to the direct field of irradiation, does not respond to corticosteroid therapy, and is associated with the release of cytokines and chemotactic factors involved in tissue injury and repair, including TGF-β *(5,6,8,19,21)*. Clinical studies have demonstrated elevated levels of TGF-β1 in the plasma and in irradiated lung tissue following thoracic irradiation for non-small cell lung carcinoma, and elevated plasma TGF-β1 is associated with the development of pulmonary toxicity and radiation pneumonitis *(23–28)*. Pathologic findings in the late phase response include

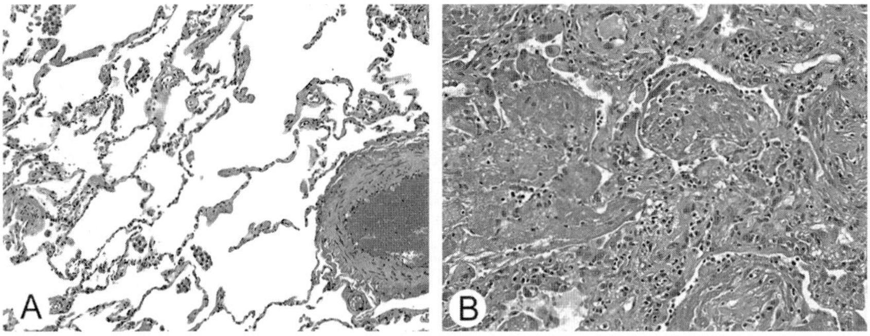

Fig. 1. Histologic appearance of radiation fibrosis in human lung. **(A)** Normal human lung. **(B)** Lung tissue from a 51 yr old male with prior diagnosis of adenocarcinoma, who received radiation prior to lobectomy. (**A**, author. **B**, courtesy of Dr. Haodong Xu, University of Rochester, and Dr. Konstantin Shiloh, Armed Forces Institute of Pathology.)

accumulation of fibroblasts, increased collagen deposition and hyaline membranes, thickening of alveolar septae and a loss of alveolar spaces and capillaries consistent with fibrosis *(4,5)* (Fig. 1).

2.2. Chemotherapy

Systemic administration of chemotherapeutic agents is also known to cause lung injury and fibrosis. Chemotherapeutic agents known to cause fibrosis include alkylating agents (busulfan, cyclophosphamide, chlorambucil, and melphalan), antimetabolics (methotrexate), nitrosourea (carmustine [BCNU], lomustine [CCNU]), vinca alkaloids (vinblastine), and antitumor antibiotics (bleomycin, mitomycin) *(29)*. Many of these agents are described in case reports that lack well-documented incidence rates, although BCNU and mitomycin-C have been documented to cause pulmonary fibrosis with incidence rates of 25%, and 3–12%, respectively *(29)*.

Bleomycin is the most well studied of the fibrosis-promoting antitumor agents. The reported incidence rates of pulmonary toxicity range from 2–40%, with the majority of reports around 10% *(29–31)*. Severe pulmonary toxicity can occur at any dose; however, the risk appears to be greatest when the cumulative dose is greater than 400–450 units *(31)*. The incidence of pulmonary toxicity is increased in patients over the age of 70 *(31)*, in patients receiving concomitant chest radiation or oxygen therapy *(32–39)*, and in patients with renal insufficiency *(40–42)*. There is no correlation between incidence and severity of toxicity and gender, route of administration, or tumor type *(31)*. The pulmonary effects of bleomycin are more pronounced at lower doses when used in combination with cisplatin *(41)*, and when combined with methotrexate, adriamycin, cyclophosphamide, vincristine, and dexamethasone in the M-BACOD regimen *(43)*, and in the BACOP program *(44)*.

Chemotherapy can cause acute lung injury, generally presenting immediately after the onset of chemotherapy or during the treatment regimen, characterized by inflammatory interstitial pneumonitis, edema, bronchospasm, and pleurisy. Pulmonary fibrosis is a late-onset disease, typically presenting more than 2 mo after the completion of chemotherapy *(6,7,30,43,45,46)*. Histopathologically, bleomycin injury is notable for diffuse alveolar wall damage, particularly in the periphery of the lung and at the bases, with hyperplasia and necrosis of pneumocytes, and septal widening with interstitial and alveolar fibrinous edema *(31,47–49)*. In advanced stages, the fibrotic pathology is indistinguishable from usual interstitial pneumonitis, with increased collagen deposition, thickened alveolar septae, and honeycombing in the periphery and bases of the lungs *(50)*. While there have been no reported studies associating chemotherapy with elevated TGF-β levels in fibrosis patients, there is clear evidence that bleomycin induces TGF-β in animal models of bleomycin toxicity (*see* Section 3.2.).

2.3. Current Therapies and Patient Outcomes

The incidence of pulmonary toxicity with bleomycin varies depending on risk factors as described above, with an overall mortality rate of 1–3% of all patients, increasing to 10% in patients receiving cumulative doses greater than 550 units *(30,31,46,51)*. Alkylating agents such as cyclophosphamide and busulfan have lower incidences of fibrosis (<1–4%) but mortality from fibrosis approaching 50% *(29)*. The majority of patients who develop symptomatic radiation pneumonitis will develop radiographic evidence of fibrosis *(4)*. Patients diagnosed with *severe* pneumonitis have significant mortality, and surviving patients are generally left with long-term impairment of lung function *(52,53)*. Corticosteroids are somewhat effective in treating acute lung injury but have no effect on the course of fibrosis *(5)*. At present, the only effective treatment for radiation and chemotherapy-associated fibrosis is prevention, i.e., limiting the dose and field of radiation. Effective antifibrosis treatments could reduce the morbidity and mortality associated with lung injury secondary to chemo- and radiation therapy while allowing pursuit of more aggressive treatment regimes.

3. THE ROLE OF TGF-β IN PULMONARY FIBROSIS

Like most forms of pulmonary fibrosis, the pathogenesis of radiation and chemotherapy-induced pulmonary fibrosis remains incompletely understood, but substantial data implicates the cytokine TGF-β1 (hereinafter referred to as TGF-β) as a critical effector.

3.1. TGF-β and the Myofibroblast

Fibroblasts and myofibroblasts are recognized as key effector cells in the development of fibrosis. Fibroblasts can differentiate to myofibroblasts following exposure to a variety of stimuli, particularly TGF-β *(54–56)*. Myofibroblasts are fibroblast-like cells with some characteristics of smooth muscle cells, including contractility and expression of α-smooth muscle actin (α-SMA) and calponin *(54,57)*. They are a major source of extracellular matrix proteins including collagen, as well as fibrogenic cytokines and chemokines. The differentiation of fibroblasts to myofibroblasts, along with proliferation of fibroblasts and myofibroblasts and increased deposition of extracellular matrix proteins, leads to the appearance of patchy fibroblastic foci in the lung parenchyma. The fibroblasts may be aligned in bundles similar to the contractile phase of normal wound healing. As fibrosis progresses these fibroblastic foci grow, merge and contract, leading to progressive destruction of normal lung architecture, and gas exchange abnormalities and loss of function *(3,56,58–61)*.

Extensive in vitro studies have demonstrated that TGF-β is a key stimulus of fibroblast to myofibroblast differentiation *(62,63)*, and that TGF-β also inhibits myofibroblast apoptosis *(55,64)*. TGF-β promotes expression of the matrix proteins collagen and fibronectin, and the tissue inhibitors of matrix metalloproteinases, which are also upregulated in human fibrosing diseases *(54,55,65–67)*. In vitro studies of cultured lung fibroblasts have demonstrated that bleomycin upregulates the profibrotic cytokines TGF-β, platelet derived growth factor (PDGF), and connective tissue growth factor, as well as the matrix proteins collagen and fibronectin, further implicating TGF-β as a key mediator of bleomycin induced fibrosis *(68,69)*.

3.2. TGF-β in Animal and Human Studies of Lung Fibrosis

Most of the data linking TGF-β to radiation and bleomycin fibrosis comes from animal models. Both bleomycin and thoracic irradiation lead to progressive fibrosis with upregulation

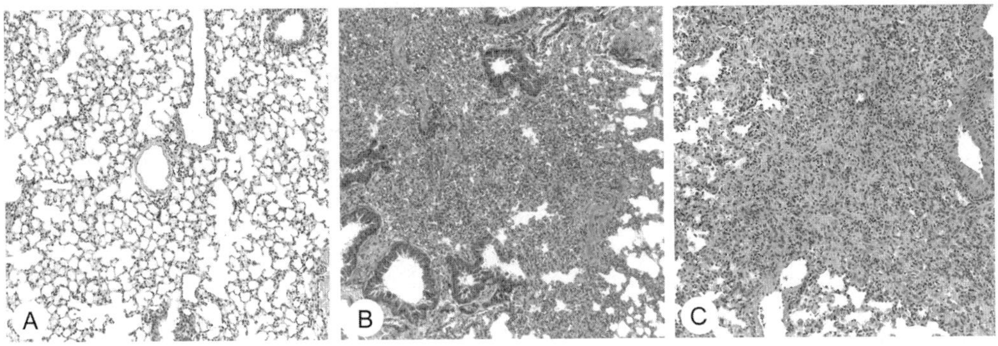

Fig. 2. Histologic appearance of fibrosis in the mouse model. **(A)** Normal (untreated) lung. **(B)** Mouse lung 12 wk after a single thoracic irradiation exposure of 15 Gy. **(C)** Mouse lung 28 d after a single intratracheal instillation of 3.75 units/kg of bleomycin. (**A** and **C**, authors. **B**, courtesy of Dr. Richard Phipps, University of Rochester.)

of TGF-β in mice *(65,70–73)*, rats *(74,75)*, and hamsters *(76)* (Fig. 2). Thoracic irradiation and bleomycin also promote the release of active TGF-β from its latency associated peptide (LAP) *(77–80)* and upregulate the TGF-β receptor *(81)*. TGF-β expression is localized to fibroblasts and myofibroblasts at sites of active fibrosis *(82,83)*. Support for the role of TGF-β in human fibrosis comes from observations of increased TGF-β expression and increased numbers of interstitial myofibroblasts expressing α-SMA and collagen in several human fibrosing diseases *(3,84–88)*. As discussed earlier, clinical studies indicate that plasma TGF-β levels increase following thoracic radiation, and elevated TGF-β levels are associated with radiation pneumonitis and fibrosis *(23–28)*.

However, these demonstrations of increased TGF-β expression in fibrosis do not directly prove that TGF-β is a key effector in fibrogenesis. To determine if overexpression of TGF-β could, in and of itself, lead directly to fibrogenesis, we utilized a replication deficient adenovirus vector to overexpress active TGF-β in rodent lung tissue. For the first time, we demonstrated that transient expression of a spontaneously active form of TGF-β in either rat or mouse lung induces rapid and severe fibrogenesis in the pulmonary interstitium with prominent induction of myofibroblasts that is histologically reminiscent of the pathological processes seen in radiation and bleomycin-associated fibrosis *(89–92)*. Strategies targeting TGF-β in lung fibrosis should therefore lead to decreased myofibroblast differentiation and proliferation, decreased production of extracellular matrix, contractility and lung remodeling, and increased apoptosis of lung fibroblasts, potentially arresting disease progression and even leading to recovery of lung function.

4. ANTI-TGF-β STRATEGIES AS THERAPY FOR PULMONARY FIBROSIS

As the key role of TGF-β in pulmonary fibrosis has emerged over the last few years, efforts have turned to investigating strategies to interfere with TGF-β as potential therapies for fibrosis *(3,84,93,94)*. These strategies fall into two general classes; direct inhibition of TGF-β itself (via soluble TGF-β decoy receptors, TGF-β antibodies, decorin, RNA interference, and LAP) and interference with the downstream signaling cascades initiated by TGF-β (interferon-γ, pirfenidone, imatinib mesylate, and peroxisome proliferator activated receptor-γ [PPAR-γ] agonists) (Table 1). The potential of these anti-TGF-β treatment strategies to directly impact tumor progression is discussed in Volume 2. The remainder of this chapter will review recent preclinical animal model and human trial data focusing on the specific application of these strategies to pulmonary fibrosis.

Table 1
Summary of Anti-TGF-β Strategies Under Investigation for Treatment of Pulmonary Fibrosis

Treatment strategy	*Mechanism of action*	*Stage of development*
TGF-β antibodies	Neutralizes TGF-β, prevents binding to receptor	Phase I and II
Soluble TGF-β receptors	Acts as a decoy receptor to reduce TGF-β signaling to cell-bound receptors	Laboratory
Decorin	Binds and neutralizes TGF-β	Laboratory
Respirable antisense oligonucleotides	Inhibits translation of TGF-β mRNA	Laboratory
Latency associated peptide (LAP)	Binds TGF-β	Laboratory
Interferon-γ	Possibly regulates TGF-β levels and TGF-β intracellular signaling	Phase III
Pirfenidone	Inhibits transcription of TGF-β and extracellular matrix proteins	Phase II/III
Imatinib Mesylate	Inhibits downstream signaling of TGF-β through c-Abl kinase	Phase III
PPAR-γ agonists	Unknown	Laboratory

4.1. Soluble TGF-β Receptors

One strategy for reducing the effects of TGF-β is to use the extracellular domain of the TGF-β receptor II (TβRII) as a decoy receptor. Soluble TβRII, usually expressed as a chimeric protein with the Fc region of IgG, competes with cell-bound TβRII for active TGF-β, resulting in reduced TGF-β signaling. Soluble TβRII inhibits TGF-β activity in vitro *(95–97)*, and inhibits liver fibrosis in a mouse model of chronic liver injury *(98)*. Soluble TGF-β receptors (STR) instilled into the lungs of hamsters via intratracheal instillation twice weekly reduced the development of lung fibrosis following a single dose of bleomycin *(99)*. STR also inhibited bleomycin fibrosis when given after the second dose of bleomycin in a three-dose model, as demonstrated by reductions in hydroxyproline content, histopathology score, and the activity of the TGF-β regulated enzyme prolyl hydroxylase *(100)*. STR treatment did not reduce the acute inflammation associated with early phase bleomycin pneumonitis, consistent with observations that while TGF-β drives fibrotic processes, it also has antiinflammatory effects *(101)*.

The use of gene therapy approaches to deliver STRs to target organs has also been investigated. Ueno and coworkers constructed a replication-deficient adenovirus vector expressing the extracellular domain of TβRII fused to the Fc portion of human IgG1 under the control of a strong constitutive promoter. This construct attenuated liver fibrosis when injected into the portal vein of rats treated with dimethylnitrosamine *(102)*. Direct organ targeting seems not to be required, as both liver fibrosis and corneal scarring were reduced when the adenovirus vector was injected into the femoral muscle prior to, or concomitant with, a fibrogenic insult *(103,104)*. However, given that TGF-β knockout mice develop spontaneous chronic inflammation and immune dysregulation *(105)*, systemic administration of STRs may have significant side effects. Because the lung is amenable to direct targeting via inhalation, development of inhaled forms of STRs or STR-expressing gene therapy vectors may prove successful.

4.2. TGF-β Antibodies

Neutralizing TGF-β antibodies represent another strategy to directly inhibit the profibrotic effects of TGF-β. TGF-β antibodies have been shown to inhibit experimental glomerulonephritis and diabetes-associated nephritic fibrosis in rats *(106–108)*. TGF-β antibodies also inhibited both dermal and pulmonary accumulation of extracellular matrix components in a mouse model of scleroderma *(109)*. Although, TGF-β antibodies have not been studied in models of chemotherapy or radiation induced lung fibrosis, systemic (intraperitoneal) injections of a TGF-β antibody were effective in attenuating lung fibrosis in mice repeatedly exposed to inhaled bacillus Calmette-Guerin *(110)*. While concerns have been raised about the possible side effects of systemic inhibition of TGF-β, mice receiving daily administration of therapeutic doses of TGF-β antibodies exhibited no adverse effects *(111)*. Phase I trials of TGF-β antibodies (GC1008 and CAT-192, Genzyme) for treatment of scleroderma and idiopathic pulmonary fibrosis (IPF) are underway (www.clinicaltrials.gov, Last accessed January 24, 2007). If trials of anti-TGF-β therapy demonstrate safety and efficacy for IPF, it will likely prove an effective therapy for other fibrosing diseases including chemotherapy and radiation-induced lung fibrosis.

4.3. Decorin

Decorin is a small, leucine-rich, chondroitin-dermatan sulfate proteoglycan that is a widely distributed component of extracellular matrix that may be involved in collagen stabilization *(112)*. Decorin is also a natural inhibitor of TGF-β, and may play a role in connective tissue homeostasis by countering the fibroblast chemotactic properties of collagen and fibronectin *(3,106,113)*. Decorin inhibits binding of TGF-β to mink lung cells and murine lung fibroblasts in vitro *(112,114)*, and also inhibits TGF-β-stimulated differentiation of human lung fibroblasts to myofibroblasts (Burgess and Sime, unpublished data). This molecule has previously been shown to significantly reduce experimental liver and kidney fibrosis *(85,115)*. In the lung, intratracheal administration of purified decorin attenuated bleomycin-induced lung fibrosis in hamsters *(116)*. However, the dose used in the hamster model (3.3 mg/kg, twice weekly) would be difficult to administer into the lungs of human patients by inhalation, and repeated intratracheal deposition (probably via bronchoscopy) would be unpleasant, impractical and costly. Gene therapy has the potential to confer long-lasting protein expression with a relatively small innoculum, and may therefore have greater clinical potential *(117,118)*.

To test the potential of gene therapy-based decorin treatment in lung fibrosis, we cloned the decorin core protein into a replication-deficient adenovirus vector. The expressed protein was an efficient inhibitor of TGF-β in vitro *(112)*. Intratracheal administration of the decorin expression vector significantly attenuated lung fibrosis in mice treated with a TGF-β overexpressing virus *(112)*, bleomycin *(119)*, or crystalline silica *(120)*. Interestingly, intravenous administration of a decorin-expressing viral vector resulted in decorin expression in the liver, but not the lung, and no reduction in lung fibrosis *(121)*. This is in contrast to the results seen with intramuscular injection of STR-expressing vectors and suggests that decorin must be present in the extracellular matrix at the site of fibrosis to be effective.

4.4. Latency Associated Peptide

TGF-β is secreted in an inactive, or latent form, bound to a LAP derived from the N-terminal end of the full-length TGF-β mRNA (reviewed in Chapter 3). Dissociation of LAP from TGF-β is required for biological activity. In vitro data indicates that both radiation and bleomycin stimulate the activation of latent TGF-β *(77,122)*. TGF-β is a high-affinity ligand for the integrins (including αvβ3, αvβ5, αvβ6, and αvβ8), and integrin-mediated activation of latent TGF-β has been demonstrated in fibroblast and epithelial cell cultures and in the mouse model of bleomycin fibrosis *(78,123–127)*. While antibodies to β6 and β8 inhibit TGF-β activation in vitro, their use in vivo has not been reported. However, LAP itself binds to TGF-β and can inhibit TGF-β function. For example, LAP reversed the differentiation of hepatic stellate cells to myofibroblasts in vitro *(128)*, and inhibited the production of collagen by pancreatic stellate cells *(129)*. Intravenous injection of LAP blocked skin thickening and collagen accumulation in a mouse model of scleroderma *(130)*. The best option for treating pulmonary disease with LAP might be to use an inhaled gene therapy vector to produce LAP at the site of TGF-β activation and fibrogenesis.

4.5. Respirable Antisense Oligonucleotides

Antisense oligonucleotides (AS-ONs) are short (15–20 bases) sequences complementary to an mRNA of interest that inhibit translation of the message by forming structures that either block translation by ribosomes or recruit RNAses to degrade the mRNA. AS-ONs have been a powerful tool for modulating gene expression in vitro for many years, but their use in vivo has been limited by difficulties with absorption, tissue distribution and excretion. To date, the potential of AS-ONs targeting TGF-β to modulate fibrotic disease has been investigated in the rat kidney. Injection of AS-ONs through the urethra into the kidney attenuated the development of renal fibrosis associated with unilateral ureteral obstruction, with reductions in TGF-β and collagen mRNA *(131)*. Recent technical advances have allowed the development of respirable antisense oligonucleotides (RASONs) that are delivered by inhalation and efficiently taken up by lung cells *(132,133)*. RASONs are being aggressively pursued as potential asthma therapies, targeting genes including the adenosine A_1 receptor, cytokines IL-4, IL-5, IL-13 and their receptors, TGF-β, and others *(133–138)*. Because one mRNA molecule can potentially be translated into thousands of protein molecules, RASONs targeting TGF-β mRNA may be much more effective in reducing interstitial TGF-β levels, and hence lung fibrosis, than STRs, decorin, LAP or TGF-β antibodies, which, as competitive inhibitors, must be present in relative excess to achieve therapeutic results.

4.6. Interferon-Gamma

Interferon-γ (IFN-γ) is a pleiotropic cytokine with potent immunomodulatory effects that is produced by T-cells, natural killer cells, macrophages and other immune cells. Bleomycin lung injury is attenuated in IFN-γ-knockout mice, suggesting that IFN-γ plays an important role in the initial acute inflammatory reaction *(139)*. However, there is considerable evidence that IFN-γ antagonizes the profibrotic properties of TGF-β *(3)*. In vitro, IFN-γ inhibits fibroblast proliferation and downregulates production of collagen as well as TGF-β itself *(140–144)*. IFN-γ has also been shown to reduce lung fibrosis in rats exposed to crystalline silica and in mice treated with bleomycin *(145,146)*.

IFN-γ has received FDA approval for treatment of chronic granulomatous disease and malignant osteoporosis (Actimmune®, InterMune). A small open label trial indicated that IFN-γ therapy restored lung function and vital capacity in patients with IPF, while a second small trial demonstrated no benefit of IFN-γ treatment *(147,148)*. A larger, placebo-controlled double-blind study found that IFN-γ therapy did not increase lung function or arrest progression of IPF, but did result in decreased mortality, particularly in subgroups with less severe disease at enrollment *(149–151)*. A phase III study (The INSPIRE Trial) is currently underway, which will assess survival time and other secondary outcomes over a minimum of 2 yr in patients with mild to moderate fibrosis. It would be interesting to determine whether patients with radiation and chemotherapy-associated fibrosis would be more amenable to IFN-γ treatment, owing either to the nature of the fibrogenic insult, or to the fact that fibrosis in cancer patients is probably identified at an earlier stage than in IPF.

4.7. Pirfenidone

Pirfenidone (5-methyl-1-phenyl-2-(1H)-pyridone, Deskar®, Marnac) is a small molecule orally active drug with antiinflammatory and antifibrotic effects in vivo and in vitro.

Pirfenidone downregulates the production of profibrotic cytokines including tumor necrosis factor-α (TNF-α), PDGF, and TGF-β *(152–154)*. Pirfenidone also inhibits synthesis of collagen and other extracellular matrix components *(155–157)*. Treatment with pirfenidone has shown promising results in animal models of kidney and liver fibrosis *(156,158–160)*, as well as lung fibrosis initiated by either bleomycin or cyclophosphamide *(161–163)*. Two small open-label trials of pirfenidone in IPF demonstrated some stabilization or improvement in lung function *(164,165)*. A double-blind, randomized, placebo-controlled phase II trial demonstrated improvements in lung function and a reduction in the number of acute exacerbations in IPF patients *(166)*. Additional clinical trials of pirfenidone in IPF and radiation fibrosis have been completed but not yet reported (www.clinicaltrials.gov, Last accessed January 24, 2007).

4.8. Imatinib Mesylate

Imatinib mesylate (Gleevec®, Novartis) was originally developed as a platelet derived growth factor receptor (PDGFR) inhibitor, and is a potent inhibitor of the Ableson family of protein tyrosine kinases including the PDGFR tyrosine kinase, c-*Abl* and c-*Kit (167)*. Imatinib also targets Bcr-Abl, the constitutively active kinase caused by the Philadelphia chromosome mutation in chronic myeloid leukemia (CML), and is FDA approved for the treatment of CML. PDGF is a potent profibrotic cytokine that is induced by TGF-β and is hypothesized to recruit fibroblasts to the site of injury while TGF-β promotes differentiation of fibroblasts to myofibroblasts *(168–172)*. In addition to upregulating PGDF expression, TGF-β directly activates the c-*Abl* kinase. Inhibition of c-*Abl* by imatinib blocked TGF-β-stimulated morphologic changes, cell proliferation, and production of extracellular matrix in lung fibroblasts in vitro *(173)*. Thus, imatinib has the potential to simultaneously block the effects of two strong profibrotic cytokines, TGF-β and PDGF, by inhibiting c-*Abl* and the PDGFR tyrosine kinase. Mice treated with imatinib and bleomycin developed less lung fibrosis, with reductions in histopathology score and collagen deposition, and fewer proliferating cells in the lung interstitium, compared to mice treated with bleomycin alone *(173,174)*. A phase II/III clinical trial of imatinib in IPF is ongoing. If successful in treating IPF, imatinib may also prove useful as a therapy for radiation and chemotherapy-induced pulmonary fibrosis.

4.9. PPAR-γ Agonists

PPAR-γ is a member of the ligand-activated nuclear receptor superfamily that includes the retinoic acid, steroid and thyroid hormone receptors *(175–177)*. PPAR-γ exists in the cytoplasm as a complex with the retinoid X receptor (RXR) and one or more corepressors *(178)*. PPAR-γ ligand binding (along with the binding of RXR to its ligand, 9-*cis*-retinoic acid) triggers shedding of the corepressors and nuclear translocation of the PPAR-γ/RXR heterodimer, leading to transcription of target genes carrying peroxisome proliferator receptor elements in their promoters. PPAR-γ is a key regulator of lipid metabolism and adipogenesis, the process by which fibroblast-like precursors differentiate into mature, fat-storing adipocytes *(178–180)*. Natural ligands for PPAR-γ include fatty acids (linoleic and arachidonic acid), phospholipids (lysophosphatidic acid), and eicosanoids (including 9/13-hydroxyoctadecadienoic acid and 15-hydroxyeicosatetraenoic acid), raising the possibility that PPAR-γ may act as a lipid sensor *(181)*. The thiazolidinedione (TZD) compounds (including ciglitazone, rosiglitazone, troglitazone, and pioglitazone), which were originally developed as insulin sensitizers for treatment of type 2 diabetes, are strong and selective PPAR-γ agonists and appear to mediate their insulin-sensitizing properties via PPAR-γ *(182–185)*. Pioglitazone (Actos®, Takeda) and rosiglitazone (Avandia®, GlaxoSmithKline) are currently approved for treatment of insulin resistant type 2 diabetes.

A growing body of evidence is emerging that PPAR-γ is involved in the regulation of inflammation and that PPAR-γ agonists may have potent antiinflammatory properties *(reviewed in [186])*. A metabolite of prostaglandin D_2, 15-deoxy-$\Delta^{12,14}$-prostaglandin J_2 (15d-PGJ_2), was identified as a potent natural ligand of PPAR-γ with antiinflammatory properties, and treatment with either 15d-PGJ_2 or selected TZDs suppresses inflammation in several animal models *(187–192)*. Furthermore, treatment with 15d-PGJ_2 differentiates fibroblasts into adipocyte-like cells *(193)*, suggesting that PPAR-γ agonists would antagonize the ability of TGF-β to differentiate fibroblasts to myofibroblasts *(143)*. This has led us and others to hypothesize that PPAR-γ agonists would also have antifibrotic properties.

The PPAR-γ agonists pioglitazone and troglitazone inhibit expression of α-SMA and collagen in hepatic and pancreatic stellate cells in vitro, and attenuate liver and pancreatic fibrosis in mice in vivo *(188, 194–197)*. To investigate the potential antifibrotic effects of the PPAR-γ agonists on lung fibrosis, we treated normal primary human lung fibroblasts with TGF-β and either 15d-PGJ_2 or rosiglitazone. Both 15d-PGJ_2 and rosiglitazone inhibited TGF-β-driven differentiation of fibroblasts to myofibroblasts, with reductions in the expression of α-SMA and collagen *(198)*. 15d-PGJ_2 and ciglitazone were effective in reducing lung fibrosis resulting from inhalation of crystalline silica *(199)*. Other laboratories have reported that 15d-PGJ_2, its precursor PGD_2, and rosiglitazone suppress bleomycin-mediated lung fibrosis in mice *(189,200)*.

The mechanism by which PPAR-γ agonists inhibit fibrogenesis is complex and not well understood. Several investigators have reported that some of the biological effects of 15d-PGJ_2 are mediated by a mechanism independent of PPAR-γ-dependent transcriptional activation *(201–203)*. We used a combination of genetic and pharmaceutical approaches to investigate the mechanism of PPAR-γ inhibition of fibrogenesis. The ability of 15d-PGJ_2 to inhibit TGF-β-stimulated myofibroblast differentiation was not affected by GW9662, an irreversible PPAR-γ antagonist, suggesting that the effects of 15d-PGJ_2 are PPAR-γ independent. On the other hand, overexpression of a dominant-negative PPAR-γ protein partially restored myofibroblast differentiation in fibroblasts treated with rosiglitazone and TGF-β, indicating that the effects of rosiglitazone may be partially PPAR-γ dependent *(198)*. Further research is necessary to understand the mechanisms by which PPAR-γ ligands inhibit fibrogenesis. However, it is clear that PPAR-γ is a promising target for therapy in fibrosing disease including bleomycin lung fibrosis. The availability in the clinic of the PPAR-γ agonists pioglitazone (Actos®, Takeda) and rosiglitazone (Avandia®, GlaxoSmithKline) should lead to rapid exploration in clinical trials for lung and other fibrosing diseases.

5. CONCLUSION

Pulmonary fibrosis has been a recognized side effect of radiation and chemotherapy for over six decades, and the disease remains almost completely refractory to traditional antiinflammatory therapies. The emerging understanding of fibrosis as a cytokine-mediated disease process has allowed for the development of several potentially effective antifibrotic therapies targeting a key profibrotic cytokine, TGF-β. Several of the treatments are currently licensed for clinical use, which should enable their usefulness in treating fibrosis to be rapidly evaluated. Although many of these clinical trials focus on IPF, the fact that TGF-β is a key regulator of fibrogenesis in both IPF and chemo/radiation therapy-induced fibrosis should allow the findings of these studies to be applied to cancer patients. However, several special circumstances have to be taken into account when projecting the use of anti-TGF-β therapies in cancer patients. As discussed above, chemo- and radiation therapy may elicit an acute lung inflammatory reaction that later develops into progressive fibrosis. As TGF-β appears to have important antiinflammatory properties, initiating anti-TGF-β therapy during this phase could result in exacerbation of pulmonary symptoms. It may be necessary to delay the start of therapy until a diagnosis of fibrosis (rather than inflammation) can be confirmed, or to first try a course of corticosteroid treatment, to differentiate between steroid-responsive inflammation and steroid-unresponsive fibrosis. On the other hand, prompt anti-TGF-β therapy might prevent the initiation of fibrosis, making it advantageous even if it caused a temporary increase in inflammation.

Additionally, the role of TGF-β in the underlying cancer will have to be considered. As discussed in Volume 2, many tumors secrete TGF-β which causes local immunosuppression, allowing the tumor to escape immune surveillance. Treatment with anti-TGF-β therapies following radiation or chemotherapy may have the dual effect of inhibiting fibrosis and restoring immune surveillance, thus inhibiting tumor regrowth. In these cases, a systemic therapy such as IFN-γ, imatinib, pirfenidone, or a TZD would be preferable over a lung-targeted therapy (except in the case of lung cancer, of course). However, there may also be tumors whose growth is inhibited by TGF-β, so that systemic inhibition of TGF-β would not be desirable. Such cases may benefit from targeting the anti-TGF-β therapy to the lung via inhalation of RASONs or gene therapy vectors expressing STRs, decorin, or LAP. The availability of multiple treatment strategies will enable optimal individualized treatment strategies taking into account the type of cancer and its response to TGF-β as well as the clinical presentation and history of radiation and chemotherapy-associated pulmonary disease.

ACKNOWLEDGMENTS

This work was supported in part by K08HL-04492, R01HL-754321, the James P. Wilmot Foundation, American Lung Association Grant DA004-N, The National Institute of Environmental Health Sciences Center Grant P30E80111247, and The Connor Fund. T. H. Thatcher was supported in part by a postdoctoral fellowship from Philip Morris, USA.

REFERENCES

1. Coultas DB, Zumwalt RE, Black WC, Sobonya RE. The epidemiology of interstitial lung diseases. Am J Respir Crit Care Med 1994;150:967–972.
2. Strieter RM. Mechanisms of pulmonary fibrosis: conference summary. Chest 2001;120:77S–85S.
3. Sime PJ, O'Reilly KM. Fibrosis of the lung and other tissues: new concepts in pathogenesis and treatment. Clin Immunol 2001;99:308–319.
4. Gross NJ. Pulmonary effects of radiation therapy. Ann Intern Med 1977;86:81–92.
5. Abratt RP, Morgan GW, Silvestri G, Willcox P. Pulmonary complications of radiation therapy. Clin Chest Med 2004;25:167–177.
6. Abid SH, Malhotra V, Perry MC. Radiation-induced and chemotherapy-induced pulmonary injury. Curr Opin Oncol 2001;13:242–248.
7. Ohe Y, Yamamoto S, Suzuki K, et al. Risk factors of treatment-related death in chemotherapy and thoracic radiotherapy for lung cancer. Eur J Cancer 2001;37:54–63.
8. Movsas B, Raffin TA, Epstein AH, Link CJ, Jr. Pulmonary radiation injury. Chest 1997;111:1061–1076.
9. Roach M 3rd, Gandara DR, Yuo HS, et al. Radiation pneumonitis following combined modality therapy for lung cancer: analysis of prognostic factors. J Clin Oncol 1995;13:2606–2612.
10. Mehta V. Radiation pneumonitis and pulmonary fibrosis in non-small-cell lung cancer: pulmonary function, prediction, and prevention. Int J Radiat Oncol Biol Phys 2005;63:5–24.
11. Byhardt RW, Martin L, Pajak TF, Shin KH, Emami B, Cox JD. The influence of field size and other treatment factors on pulmonary toxicity following hyperfractionated irradiation for inoperable non-small cell lung cancer (NSCLC)—analysis of a Radiation Therapy Oncology Group (RTOG) protocol. Int J Radiat Oncol Biol Phys 1993;27:537–544.
12. Libshitz HI, Southard ME. Complications of radiation therapy: the thorax. Semin Roentgenol 1974;9:41–49.
13. Keane TJ, Van Dyk J, Rider WD. Idiopathic interstitial pneumonia following bone marrow transplantation: the relationship with total body irradiation. Int J Radiat Oncol Biol Phys 1981;7:1365–1370.
14. Mah K, Van Dyk J, Keane T, Poon PY. Acute radiation-induced pulmonary damage: a clinical study on the response to fractionated radiation therapy. Int J Radiat Oncol Biol Phys 1987;13:179–188.
15. Mah K, Keane TJ, Van Dyk J, Braban LE, Poon PY, Hao Y. Quantitative effect of combined chemotherapy and fractionated radiotherapy on the incidence of radiation-induced lung damage: a prospective clinical study. Int J Radiat Oncol Biol Phys 1994;28:563–574.

16. Koc M, Polat P, Suma, S. Effects of tamoxifen on pulmonary fibrosis after cobalt-60 radiotherapy in breast cancer patients. Radiother Oncol 2002;64:171–175.
17. Bese NS, Umay C, Yildirim S, et al. The effects of tamoxifen on radiation-induced pulmonary fibrosis in Wistar albino rats: results of an experimental study. Breast 2006;15:456–460.
18. Freid JR, Goldberg H. Post-irradiation changes in the lung and thorax. Am J Roentgenol 1940;43: 877–895.
19. Morgan GW, Breit SN. Radiation and the lung: a reevaluation of the mechanisms mediating pulmonary injury. Int J Radiat Oncol Biol Phys 1995;31:361–369.
20. Roberts CM, Foulcher E, Zaunders JJ, et al. Radiation pneumonitis: a possible lymphocyte-mediated hypersensitivity reaction. Ann Intern Med 1993;118:696–700.
21. Abratt, RP, Morgan GW. Lung toxicity following chest irradiation in patients with lung cancer. Lung Cancer 2002;35:103–109.
22. Gibson PG, Bryant DH, Morgan GW, et al. Radiation-induced lung injury: a hypersensitivity pneumonitis? Ann Intern Med 1988;109:288–291.
23. De Jaeger K, Seppenwoolde Y, Kampinga HH, Boersma LJ, Belderbos JS, Lebesque JV. Significance of plasma transforming growth factor-beta levels in radiotherapy for non-small-cell lung cancer. Int J Radiat Oncol Biol Phys 2004;58:1378–1387.
24. Anscher MS, Marks LB, Shafman TD, et al. Risk of long-term complications after TFG-beta1-guided very-high-dose thoracic radiotherapy. Int J Radiat Oncol Biol Phys 2003;56:988–995.
25. Barthelemy-Brichant N, Bosquee L, Cataldo D, et al. Increased IL-6 and TGF-beta1 concentrations in bronchoalveolar lavage fluid associated with thoracic radiotherapy. Int J Radiat Oncol Biol Phys 2004;58:758–767.
26. Novakova-Jiresova A, Van Gameren MM, Coppes RP, Kampinga HH, Groen HJ. Transforming growth factor-beta plasma dynamics and post-irradiation lung injury in lung cancer patients. Radiother Oncol 2004;71:183–189.
27. Vujaskovic Z, Groen HJ. TGF-beta, radiation-induced pulmonary injury and lung cancer. Int J Radiat Biol 2000;76:511–516.
28. Anscher MS, Kong FM, Jirtle RL. The relevance of transforming growth factor beta 1 in pulmonary injury after radiation therapy. Lung Cancer 1998;19:109–120.
29. Twohig KJ, Matthay RA. Pulmonary effects of cytotoxic agents other than bleomycin. Clin Chest Med 1990;11:31–54.
30. Jules-Elysee K, White DA. Bleomycin-induced pulmonary toxicity. Clin Chest Med 1990;11:1–20.
31. Blum RH, Carter SK, Agre K. A clinical review of bleomycin—a new antineoplastic agent. Cancer 1973;31:903–914.
32. Douglas MJ, Coppin CM. Bleomycin and subsequent anaesthesia: a retrospective study at Vancouver General Hospital. Can Anaesth Soc J 1980;27:449–452.
33. Gilson AJ, Sahn SA. Reactivation of bleomycin lung toxicity following oxygen administration. A second response to corticosteroids. Chest 1985;88:304–306.
34. Goldiner PL, Carlon GC, Cvitkovic E, Schweizer O, Howland WS. Factors influencing postoperative morbidity and mortality in patients treated with bleomycin. Br Med J 1978;1:1664–1667.
35. Goldiner PL, Schweizer O. The hazards of anesthesia and surgery in bleomycin-treated patients. Semin Oncol 1979;6:121–124.
36. Samuels ML, Johnson DE, Holoye PY, Lanzotti VJ. Large-dose bleomycin therapy and pulmonary toxicity. A possible role of prior radiotherapy. JAMA 1976;235:1117–1120.
37. Einhorn L, Krause M, Hornback N, Furnas B. Enhanced pulmonary toxicity with bleomycin and radiotherapy in oat cell lung cancer. Cancer 1976;37:2414–2416.
38. Nygaard K, Smith-Erichsen N, Hatlevoll R, Refsum SB. Pulmonary complications after bleomycin, irradiation and surgery for esophageal cancer. Cancer 1978;41:17–22.
39. Hakkinen PJ, Whiteley JW, Witschi HR. Hyperoxia, but not thoracic X-irradiation, potentiates bleomycin- and cyclophosphamide-induced lung damage in mice. Am Rev Respir Dis 1982;126: 281–285.
40. Alliot C, Tabuteau S, Desablens B, Aubry P, Andrejak M. Fatal pulmonary fibrosis after a low cumulated dose of bleomycin: role of alpha1-antitrypsin deficiency? Am J Hematol 1999;62:198–199.
41. Rabinowits M, Souhami L, Gil RA, Andrade CA, Paiva HC. Increased pulmonary toxicity with bleomycin and cisplatin chemotherapy combinations. Am J Clin Oncol 1990;13:132–138.
42. McLeod BF, Lawrence HJ, Smith DW, Vogt PJ, Gandara DR. Fatal bleomycin toxicity from a low cumulative dose in a patient with renal insufficiency. Cancer 1987;60:2617–2620.

43. Bauer KA, Skarin AT, Balikian JP, Garnick MB, Rosenthal DS, Canellos GP. Pulmonary complications associated with combination chemotherapy programs containing bleomycin. Am J Med 1983;74: 557–563.
44. Skarin AT, Rosenthal DS, Moloney WC, Frei E 3rd. Combination chemotherapy of advanced non-Hodgkin lymphoma with bleomycin, adriamycin, cyclophosphamide, vincristine, and prednisone (BACOP). Blood 1977;49:759–770.
45. Van Barneveld PW, Sleijfer DT, van der Mark TW, et al. Natural course of bleomycin-induced pneumonitis. A follow-up study. Am Rev Respir Dis 1987;135:48–51.
46. White DA, Stover DE. Severe bleomycin-induced pneumonitis. Clinical features and response to corticosteroids. Chest 1984;86:723–728.
47. Brown WG, Hasan FM, Barbee RA. Reversibility of severe bleomycin-induced pneumonitis. JAMA 1978;239:2012–2014.
48. Bedrossian CW, Luna MA, Mackay B, Lichtiger B. Ultrastructure of pulmonary bleomycin toxicity. Cancer 1973;32:44–51.
49. Jones AW. Bleomycin lung damage: the pathology and nature of the lesion. Br J Dis Chest 1978;72: 321–326.
50. Luna MA, Bedrossian CW, Lichtiger B, Salem PA. Interstitial pneumonitis associated with bleomycin therapy. Am J Clin Pathol 1972;58:501–510.
51. Yagoda A, Mukherji B, Young C, et al. Bleomycin, an antitumor antibiotic. Clinical experience in 274 patients. Ann Intern Med 1972;77:861–870.
52. Inoue A, Kunitoh H, Sekine I, Sumi M, Tokuuye K, Saijo N. Radiation pneumonitis in lung cancer patients: a retrospective study of risk factors and the long-term prognosis. Int J Radiat Oncol Biol Phys 2001;49:649–655.
53. Postoperative radiotherapy in non-small-cell lung cancer: systematic review and meta-analysis of individual patient data from nine randomised controlled trials. PORT Meta-analysis Trialists Group. Lancet 1998;352:257–263.
54. Vaughan MB, Howard EW, Tomasek JJ. Transforming growth factor-beta1 promotes the morphological and functional differentiation of the myofibroblast. Exp Cell Res 2000;257:180–189.
55. Phan SH. The myofibroblast in pulmonary fibrosis. Chest 2002;122:286S–289S.
56. White ES, Lazar MH, Thannickal VJ. Pathogenetic mechanisms in usual interstitial pneumonia/ idiopathic pulmonary fibrosis. J Pathol 2003;201:343–354.
57. Chambers RC, Leoni P, Kaminski N, Laurent GJ, Heller RA. Global expression profiling of fibroblast responses to transforming growth factor-beta1 reveals the induction of inhibitor of differentiation-1 and provides evidence of smooth muscle cell phenotypic switching. Am J Pathol 2003;162:533–546.
58. Pardo A, Selman M. Molecular mechanisms of pulmonary fibrosis. Front Biosci 2002;7:d1743–d1761.
59. Selman M, King TE, Pardo A. Idiopathic pulmonary fibrosis: prevailing and evolving hypotheses about its pathogenesis and implications for therapy. Ann Intern Med 2001;134:136–151.
60. Rodemann HP, Bamberg M. Cellular basis of radiation-induced fibrosis. Radiother Oncol 1995;35: 83–90.
61. Kuhn C, McDonald JA. The roles of the myofibroblast in idiopathic pulmonary fibrosis. Ultrastructural and immunohistochemical features of sites of active extracellular matrix synthesis. Am J Pathol 1991;138:1257–1265.
62. Hu B, Wu Z, Phan SH. Smad3 mediates transforming growth factor-beta-induced alpha-smooth muscle actin expression. Am J Respir Cell Mol Biol 2003;29:397–404.
63. Desmouliere A, Geinoz A, Gabbiani F, Gabbiani G. Transforming growth factor-beta 1 induces alpha-smooth muscle actin expression in granulation tissue myofibroblasts and in quiescent and growing cultured fibroblasts. J Cell Biol 1993;122:103–111.
64. Zhang HY, Phan SH. Inhibition of myofibroblast apoptosis by transforming growth factor beta(1). Am J Respir Cell Mol Biol 1999;21:658–665.
65. Coker RK, Laurent GJ, Shahzeidi S, et al. Transforming growth factors-beta 1, -beta 2, and -beta 3 stimulate fibroblast procollagen production in vitro but are differentially expressed during bleomycin-induced lung fibrosis. Am J Pathol 1997;150:981–991.
66. Selman M, Ruiz V, Cabrera S, et al. TIMP-1, -2, -3, and -4 in idiopathic pulmonary fibrosis. A prevailing nondegradative lung microenvironment? Am J Physiol Lung Cell Mol Physiol 2000;279:L562–L574.
67. Ramos C, Montano M, Garcia-Alvarez J, et al. Fibroblasts from idiopathic pulmonary fibrosis and normal lungs differ in growth rate, apoptosis, and tissue inhibitor of metalloproteinases expression. Am J Respir Cell Mol Biol 2001;24:591–598.

68. Yamamoto T, Eckes B, Krieg T. Bleomycin increases steady-state levels of type I collagen, fibronectin and decorin mRNAs in human skin fibroblasts. Arch Dermatol Res 2000;292:556–561.
69. Yamamoto T, Katayama I, Nishioka K. Fibroblast proliferation by bleomycin stimulated peripheral blood mononuclear cell factors. J Rheumatol 1999;26:609–615.
70. Rube CE, Uthe D, Schmid KW, et al. Dose-dependent induction of transforming growth factor beta (TGF-beta) in the lung tissue of fibrosis-prone mice after thoracic irradiation. Int J Radiat Oncol Biol Phys 2000;47:1033–1042.
71. Phan SH, Kunkel SL. Lung cytokine production in bleomycin-induced pulmonary fibrosis. Exp Lung Res 1992;18:29–43.
72. Johnston CJ, Wright, TW, Rubin P, Finkelstein JN. Alterations in the expression of chemokine mRNA levels in fibrosis-resistant and -sensitive mice after thoracic irradiation. Exp Lung Res 1998;24: 321–337.
73. Finkelstein JN, Johnston CJ, Baggs R, Rubin P. Early alterations in extracellular matrix and transforming growth factor beta gene expression in mouse lung indicative of late radiation fibrosis. Int J Radiat Oncol Biol Phys 1994;28:621–631.
74. Breen E, Shull S, Burne S, et al. Bleomycin regulation of transforming growth factor-beta mRNA in rat lung fibroblasts. Am J Respir Cell Mol Biol 1992;6:146–152.
75. Williams AO, Flanders KC, Saffiotti U. Immunohistochemical localization of transforming growth factor-beta 1 in rats with experimental silicosis, alveolar type II hyperplasia, and lung cancer. Am J Pathol 1993;142:1831–1840.
76. Raghow B, Irish P, Kang AH. Coordinate regulation of transforming growth factor beta gene expression and cell proliferation in hamster lungs undergoing bleomycin-induced pulmonary fibrosis. J Clin Invest 1989;84:1836–1842.
77. Ehrhart EJ, Segarini P, Tsang ML, Carroll AG, Barcellos-Hoff MH. Latent transforming growth factor beta1 activation in situ: quantitative and functional evidence after low-dose gamma-irradiation. FASEB J 1997;11:991–1002.
78. Munger JS, Huang X, Kawakatsu H, et al. The integrin alpha v beta 6 binds and activates latent TGF beta 1: a mechanism for regulating pulmonary inflammation and fibrosis. Cell 1999;96:319–328.
79. Pittet JF, Griffiths MJ, Geiser T, et al. TGF-beta is a critical mediator of acute lung injury. J Clin Invest 2001;107:1537–1544.
80. Dhainaut JF, Charpentier J, Chiche JD. Transforming growth factor-beta: a mediator of cell regulation in acute respiratory distress syndrome. Crit Care Med 2003;31:S258–S264.
81. Zhao Y, Shah DU. Expression of transforming growth factor-beta type I and type II receptors is altered in rat lungs undergoing bleomycin-induced pulmonary fibrosis. Exp Mol Pathol 2000;69:67–78.
82. Franko AJ, Sharplin J, Ghahary A, Barcellos-Hoff MH. Immunohistochemical localization of transforming growth factor beta and tumor necrosis factor alpha in the lungs of fibrosis-prone and "non-fibrosing" mice during the latent period and early phase after irradiation. Radiat Res 1997;147:245–256.
83. Zhang K, Flanders KC, Phan SH. Cellular localization of transforming growth factor-beta expression in bleomycin-induced pulmonary fibrosis. Am J Pathol 1995;147:352–361.
84. Bartram U, Speer CP. The role of transforming growth factor beta in lung development and disease. Chest 2004;125:754–765.
85. Breitkopf K, Haas S, Wiercinska E, Singer MV, Dooley S. Anti-TGF-beta strategies for the treatment of chronic liver disease. Alcohol Clin Exp Res 2005;29:121S–131S.
86. Goumenos DS, Tsamandas AC, Oldroyd S, et al. Transforming growth factor-beta(1) and myofibroblasts: a potential pathway towards renal scarring in human glomerular disease. Nephron 2001;87:240–248.
87. Fireman E, Shahar I, Shoval S, Messer G, Dvash S, Grief J. Morphological and biochemical properties of alveolar fibroblasts in interstitial lung diseases. Lung 2001;179:105–117.
88. Pache JC, Christakos PG, Gannon DE, Mitchell JJ, Low RB, Leslie KO. Myofibroblasts in diffuse alveolar damage of the lung. Mod Pathol 1998;11:1064–1070.
89. Warshamana GS, Pociask DA, Fisher KJ, Liu JY, Sime PJ, Brody AR. Titration of non-replicating adenovirus as a vector for transducing active TGF-beta1 gene expression causing inflammation and fibrogenesis in the lungs of C57BL/6 mice. Int J Exp Pathol 2002;83:183–201.
90. Warshamana GS, Pociask DA, Sime P, Schwartz DA, Brody AR. Susceptibility to asbestos-induced and transforming growth factor-beta1-induced fibroproliferative lung disease in two strains of mice. Am J Respir Cell Mol Biol 2002;27:705–713.

91. Sime PJ, Xing Z, Graham FL, Csaky KG, Gauldie J. Adenovector-mediated gene transfer of active transforming growth factor-beta1 induces prolonged severe fibrosis in rat lung. J Clin Invest 1997; 100:768–776.
92. Kolb M, Bonniaud P, Galt T, et al. Differences in the fibrogenic response after transfer of active transforming growth factor-beta1 gene to lungs of "fibrosis-prone" and "fibrosis-resistant" mouse strains. Am J Respir Cell Mol Biol 2002;27:141–150.
93. Giri SN. Novel pharmacological approaches to manage interstitial lung fibrosis in the twenty-first century. Annu Rev Pharmacol Toxicol 2003;43:73–95.
94. Kelly M, Kolb M, Bonniaud P, Gauldie J. Re-evaluation of fibrogenic cytokines in lung fibrosis. Curr Pharm Des 2003;9:39–49.
95. Cui X, Shimizu I, Lu G, et al. Inhibitory effect of a soluble transforming growth factor beta type II receptor on the activation of rat hepatic stellate cells in primary culture. J Hepatol 2003;39:731–737.
96. Yamamoto H, Ueno H, Ooshima A, Takeshita A. Adenovirus-mediated transfer of a truncated transforming growth factor-beta (TGF-beta) type II receptor completely and specifically abolishes diverse signaling by TGF-beta in vascular wall cells in primary culture. J Biol Chem 1996;271: 16,253–16,259.
97. Rowland-Goldsmith MA, Maruyama H, Kusama T, Ralli S, Korc M. Soluble type II transforming growth factor-beta (TGF-beta) receptor inhibits TGF-beta signaling in COLO-357 pancreatic cancer cells in vitro and attenuates tumor formation. Clin Cancer Res 2001;7:2931–2940.
98. Yata Y, Gotwals P, Koteliansky V, Rockey DC. Dose-dependent inhibition of hepatic fibrosis in mice by a TGF-beta soluble receptor: implications for antifibrotic therapy. Hepatology 2002;35:1022–1030.
99. Wang Q, Wang Y, Hyde DM, et al. Reduction of bleomycin induced lung fibrosis by transforming growth factor beta soluble receptor in hamsters. Thorax 1999;54:805–812.
100. Wang Q, Hyde DM, Gotwals PJ, Giri SN. Effects of delayed treatment with transforming growth factor-beta soluble receptor in a three-dose bleomycin model of lung fibrosis in hamsters. Exp Lung Res 2002;28:405–417.
101. Schmidt-Weber CB, Blaser K. Regulation and role of transforming growth factor-beta in immune tolerance induction and inflammation. Curr Opin Immunol 2004;16:709–716.
102. Qi Z, Atsuchi N, Ooshima A, Takeshita A, Ueno H. Blockade of type beta transforming growth factor signaling prevents liver fibrosis and dysfunction in the rat. Proc Natl Acad Sci USA 1999;96: 2345–2349.
103. Sakamoto T, Ueno H, Sonoda K, et al. Blockade of TGF-beta by in vivo gene transfer of a soluble TGF-beta type II receptor in the muscle inhibits corneal opacification, edema and angiogenesis. Gene Ther 2000;7:1915–1924.
104. Ueno H, Sakamoto T, Nakamura T, et al. A soluble transforming growth factor beta receptor expressed in muscle prevents liver fibrogenesis and dysfunction in rats. Hum Gene Ther 2000;11:33–42.
105. Christ M, McCartney-Francis NL, Kulkarni AB, et al. Immune dysregulation in TGF-beta 1-deficient mice. J Immunol 1994;153:1936–1946.
106. Border WA, Noble NA, Yamamoto T, et al. Natural inhibitor of transforming growth factor-beta protects against scarring in experimental kidney disease. Nature 1992;360:361–364.
107. Border WA, Okuda S, Languino LR, Sporn MB, Ruoslahti E. Suppression of experimental glomerulonephritis by antiserum against transforming growth factor beta 1. Nature 1990;346:371–374.
108. Hill C, Flyvbjerg A, Rasch R, Bak M, Logan A. Transforming growth factor-beta2 antibody attenuates fibrosis in the experimental diabetic rat kidney. J Endocrinol 2001;170:647–651.
109. McCormick LL, Zhang Y, Tootell E, Gilliam AC. Anti-TGF-beta treatment prevents skin and lung fibrosis in murine sclerodermatous graft-versus-host disease: a model for human scleroderma. J Immunol 1999;163:5693–5699.
110. Denis M. Neutralization of transforming growth factor-beta 1 in a mouse model of immune-induced lung fibrosis. Immunology 1994;82:584–590.
111. Ruzek MC, Hawes M, Pratt B, et al. Minimal effects on immune parameters following chronic anti-TGF-beta monoclonal antibody administration to normal mice. Immunopharmacol Immunotoxicol 2003;25:235–257.
112. Kolb M, Margetts PJ, Sime PJ, Gauldie J. Proteoglycans decorin and biglycan differentially modulate TGF-beta-mediated fibrotic responses in the lung. Am J Physiol Lung Cell Mol Physiol 2001; 280:L1327–L1334.
113. Yamaguchi Y, Mann DM, Ruoslahti E. Negative regulation of transforming growth factor-beta by the proteoglycan decorin. Nature 1990;346:281–284.

114. Hildebrand A, Romaris M, Rasmussen LM, et al. Interaction of the small interstitial proteoglycans biglycan, decorin and fibromodulin with transforming growth factor beta. Biochem J 1994;302(Pt 2): 527–534.
115. Huijun W, Long C, Zhigang Z, Feng J, Muyi G. Ex vivo transfer of the decorin gene into rat glomerulus via a mesangial cell vector suppressed extracellular matrix accumulation in experimental glomerulonephritis. Exp Mol Pathol 2005;78:17–24.
116. Giri SN, Hyde DM, Braun RK, Gaarde W, Harper JR, Pierschbacher MD. Antifibrotic effect of decorin in a bleomycin hamster model of lung fibrosis. Biochem Pharmacol 1997;54:1205–1216.
117. Sime PJ, Xing Z, Foley R, Graham FL, Gauldie, J. Transient gene transfer and expression in the lung. Chest 1997;111:89S–94S.
118. Gauldie J, Graham F, Xing Z, Braciak T, Foley R, Sime PJ. Adenovirus-vector-mediated cytokine gene transfer to lung tissue. Ann N Y Acad Sci 1996;796:235–244.
119. Kolb M, Margetts PJ, Galt T, et al. Transient transgene expression of decorin in the lung reduces the fibrotic response to bleomycin. Am J Respir Crit Care Med 2001;163:770–777.
120. Thatcher TH, Burgess HA, Van Kirk JR, Redonnet M, Sime PJ, Gauldie J. Decorin gene therapy inhibits silica-induced pulmonary fibrosis in a pre-clinical animal model [Abstract]. Am J Respir Crit Care Med 2003;167:A583.
121. Shimizukawa M, Ebina M, Narumi K, Kikuchi T, Munakata H, Nukiwa T. Intratracheal gene transfer of decorin reduces subpleural fibroproliferation induced by bleomycin. Am J Physiol Lung Cell Mol Physiol 2003;284:L526–L532.
122. Barcellos-Hoff MH, Dix TA. Redox-mediated activation of latent transforming growth factor-beta 1. Mol Endocrinol 1996;10:1077–1083.
123. Asano Y, Ihn H, Yamane K, Jinnin M, Mimura Y, Tamaki K. Increased expression of integrin alpha(v)beta3 contributes to the establishment of autocrine TGF-beta signaling in scleroderma fibroblasts. J Immunol 2005;175:7708–7718.
124. Asano Y, Ihn H, Yamane K, Jinnin M, Mimura Y, Tamaki K. Involvement of alphavbeta5 integrin-mediated activation of latent transforming growth factor beta1 in autocrine transforming growth factor beta signaling in systemic sclerosis fibroblasts. Arthritis Rheum 2005;52:2897–2905.
125. Fjellbirkeland L, Cambier S, Broaddus VC, et al. Integrin alphavbeta8-mediated activation of transforming growth factor-beta inhibits human airway epithelial proliferation in intact bronchial tissue. Am J Pathol 2003;163:533–542.
126. Sheppard D. Integrin-mediated activation of transforming growth factor-beta(1) in pulmonary fibrosis. Chest 2001;120:49S–53S.
127. Khalil N, Corne S, Whitman C, Yacyshyn H. Plasmin regulates the activation of cell-associated latent TGF-beta 1 secreted by rat alveolar macrophages after in vivo bleomycin injury. Am J Respir Cell Mol Biol 1996;15:252–259.
128. Isono M, Soda M, Inoue A, Akiyoshi H, Sato K. Reverse transformation of hepatic myofibroblast-like cells by TGFbeta1/LAP. Biochem Biophys Res Commun 2003;311:959–965.
129. Luttenberger T, Schmid-Kotsas A, Menke A, et al. Platelet-derived growth factors stimulate proliferation and extracellular matrix synthesis of pancreatic stellate cells: implications in pathogenesis of pancreas fibrosis. Lab Invest 2000;80:47–55.
130. Zhang Y, McCormick LL, Gilliam AC. Latency-associated peptide prevents skin fibrosis in murine sclerodermatous graft-versus-host disease, a model for human scleroderma. J Invest Dermatol 2003;121:713–719.
131. Isaka Y, Tsujie M, Ando Y, et al. Transforming growth factor-beta 1 antisense oligodeoxynucleotides block interstitial fibrosis in unilateral ureteral obstruction. Kidney Int 2000;58:1885–1892.
132. Tanaka M, Nyce JW. Respirable antisense oligonucleotides: a new drug class for respiratory disease. Respir Res 2001;2:5–9.
133. Popescu FD. Antisense- and RNA interference-based therapeutic strategies in allergy. J Cell Mol Med 2005;9:840–853.
134. Nyce J. Respirable antisense oligonucleotides: a new, third drug class targeting respiratory disease. Curr Opin Allergy Clin Immunol 2002;2:533–536.
135. Ball HA, Sandrasagra A, Tang L, Van Scott M, Wild J, Nyce JW. Clinical potential of respirable antisense oligonucleotides (RASONs) in asthma. Am J Pharmacogenomics 2003;3:97–106.
136. Ali S, Leonard SA, Kukoly CA, et al. Absorption, distribution, metabolism, and excretion of a respirable antisense oligonucleotide for asthma. Am J Respir Crit Care Med 2001;163:989–993.
137. Dreyfus DH, Matczuk A, Fuleihan R. An RNA external guide sequence ribozyme targeting human interleukin-4 receptor alpha mRNA. Int Immunopharmacol 2004;4:1015–1027.

138. Duan W, Chan JH, McKay K, et al. Inhaled p38alpha mitogen-activated protein kinase antisense oligonucleotide attenuates asthma in mice. Am J Respir Crit Care Med 2005;171:571–578.
139. Chen ES, Greenlee BM, Wills-Karp M, Moller DR. Attenuation of lung inflammation and fibrosis in interferon-gamma-deficient mice after intratracheal bleomycin. Am J Respir Cell Mol Biol 2001;24: 545–555.
140. Sempowski GD, Derdak S, Phipps RP. Interleukin-4 and interferon-gamma discordantly regulate collagen biosynthesis by functionally distinct lung fibroblast subsets. J Cell Physio 1996;167:290–296.
141. Kaufman J, Sime PJ, Phipps RP. Expression of CD154 (CD40 ligand) by human lung fibroblasts: differential regulation by IFN-gamma and IL-13, and implications for fibrosis. J Immunol 2004; 172:1862–1871.
142. Tredget EE, Wang R, Shen Q, Scott PG, Ghahary A. Transforming growth factor-beta mRNA and protein in hypertrophic scar tissues and fibroblasts: antagonism by IFN-alpha and IFN-gamma in vitro and in vivo. J Interferon Cytokine Res 2000;20:143–151.
143. Ghosh AK, Yuan W, Mori Y, Chen S, Varga J. Antagonistic regulation of type I collagen gene expression by interferon-gamma and transforming growth factor-beta. Integration at the level of p300/CBP transcriptional coactivators. J Biol Chem 2001;276:11,041–11,048.
144. Higashi K, Inagaki Y, Fujimori K, Nakao A, Kaneko H, Nakatsuka I. Interferon-gamma interferes with transforming growth factor-beta signaling through direct interaction of YB-1 with Smad3. J Biol Chem 2003;278:43,470–43,479.
145. Chen Y, Chen J, Dong J, Liu W. Antifibrotic effect of interferon gamma in silicosis model of rat. Toxicol Lett 2005;155:353–360.
146. Gurujeyalakshmi G, Giri SN. Molecular mechanisms of antifibrotic effect of interferon gamma in bleomycin-mouse model of lung fibrosis: downregulation of TGF-beta and procollagen I and III gene expression. Exp Lung Res 1995;21:791–808.
147. Kalra S, Utz JP, Ryu JH. Interferon gamma-1b therapy for advanced idiopathic pulmonary fibrosis. Mayo Clin Proc 2003;78:1082–1087.
148. Ziesche R, Hofbauer E, Wittmann K, Petkov V, Block LH. A preliminary study of long-term treatment with interferon gamma-1b and low-dose prednisolone in patients with idiopathic pulmonary fibrosis. N Engl J Med 1999;341:1264–1269.
149. Raghu G, Brown KK, Bradford WZ, et al. A placebo-controlled trial of interferon gamma-1b in patients with idiopathic pulmonary fibrosis. N Engl J Med 2004;350:125–133.
150. King TE, Jr, Safrin S, Starko KM, et al. Analyses of efficacy end points in a controlled trial of interferon-gamma1b for idiopathic pulmonary fibrosis. Chest 2005;127:171–177.
151. Bajwa EK, Ayas NT, Schulzer M, Mak E, Ryu JH, Malhotra A. Interferon-gamma1b therapy in idiopathic pulmonary fibrosis: a metaanalysis. Chest 2005;128:203–206.
152. Iyer SN, Gurujeyalakshmi G, Giri SN. Effects of pirfenidone on transforming growth factor-beta gene expression at the transcriptional level in bleomycin hamster model of lung fibrosis. J Pharmacol Exp Ther 1999;291:367–373.
153. Cain WC, Stuart RW, Lefkowitz DL, Starnes JD, Margolin S, Lefkowitz SS. Inhibition of tumor necrosis factor and subsequent endotoxin shock by pirfenidone. Int J Immunopharmacol 1998;20: 685–695.
154. Gurujeyalakshmi G, Hollinger MA, Giri SN. Pirfenidone inhibits PDGF isoforms in bleomycin hamster model of lung fibrosis at the translational level. Am J Physiol 1999;276:L311–L318.
155. Shihab FS, Bennett WM, Yi H, Andoh TF. Pirfenidone treatment decreases transforming growth factor-beta1 and matrix proteins and ameliorates fibrosis in chronic cyclosporine nephrotoxicity. Am J Transplant 2002;2:111–119.
156. Di Sario A, Bendia E, Macarri G, et al. The anti-fibrotic effect of pirfenidone in rat liver fibrosis is mediated by downregulation of procollagen alpha1(I), TIMP-1 and MMP-2. Dig Liver Dis 2004; 36:744–751.
157. Iyer SN, Gurujeyalakshmi G, Giri SN. Effects of pirfenidone on procollagen gene expression at the transcriptional level in bleomycin hamster model of lung fibrosis. J Pharmacol Exp Ther 1999; 289:211–218.
158. Brook NR, Waller JR, Bicknell GR, Nicholson ML. The experimental agent pirfenidone reduces profibrotic gene expression in a model of tacrolimus-induced nephrotoxicity. J Surg Res 2005;125:137–143.
159. Garcia L, Hernandez I, Sandoval A, et al. Pirfenidone effectively reverses experimental liver fibrosis. J Hepatol 2002;37:797–805.
160. Al-Bayati MA, Xie Y, Mohr FC, Margolin SB, Giri SN. Effect of pirfenidone against vanadate-induced kidney fibrosis in rats. Biochem Pharmacol 2002;64:517–525.

161. Iyer SN, Wild JS, Schiedt MJ, et al. Dietary intake of pirfenidone ameliorates bleomycin-induced lung fibrosis in hamsters. J Lab Clin Med 1995;125:779–785.
162. Kehrer JP, Margolin SB. Pirfenidone diminishes cyclophosphamide-induced lung fibrosis in mice. Toxicol Lett 1997;90:125–132.
163. Iyer SN, Margolin SB, Hyde DM, Giri SN. Lung fibrosis is ameliorated by pirfenidone fed in diet after the second dose in a three-dose bleomycin-hamster model. Exp Lung Res 1998;24:119–132.
164. Raghu G, Johnson WC, Lockhart D, Mageto Y. Treatment of idiopathic pulmonary fibrosis with a new antifibrotic agent, pirfenidone: results of a prospective, open-label Phase II study. Am J Respir Crit Care Med 1999;159:1061–1069.
165. Nagai S, Hamada K, Shigematsu M, Taniyama M, Yamauchi S, Izumi T. Open-label compassionate use one year-treatment with pirfenidone to patients with chronic pulmonary fibrosis. Intern Med 2002;41:1118–1123.
166. Azuma A, Nukiwa T, Tsuboi E, et al. Double-blind, placebo-controlled trial of pirfenidone in patients with idiopathic pulmonary fibrosis. Am J Respir Crit Care Med 2005;171:1040–1047.
167. Savage DG, Antman KH. Imatinib mesylate—a new oral targeted therapy. N Engl J Med 2002;346: 683–693.
168. Heldin C-H, Westermark B. Mechanism of action and in vivo role of platelet-derived growth factor. Physiol Rev 1999;79:1283–1316.
169. Maeda A, Hiyama K, Yamakido H, Ishioka S, Yamakido M. Increased expression of platelet-derived growth factor A and insulin-like growth factor-I in BAL cells during the development of bleomycin-induced pulmonary fibrosis in mice. Chest 1996;109:780–786.
170. Yoshida M, Sakuma J, Hayashi S, et al. A histologically distinctive interstitial pneumonia induced by overexpression of the interleukin 6, transforming growth factor beta 1, or platelet-derived growth factor B gene. Proc Natl Acad Sci USA 1995;92:9570–9574.
171. Homma S, Nagaoka I, Abe H, et al. Localization of platelet-derived growth factor and insulin-like growth factor I in the fibrotic lung. Am J Respir Crit Care Med 1995;152:2084–2089.
172. Rice AB, Moomaw CR, Morgan DL, Bonner JC. Specific inhibitors of platelet-derived growth factor or epidermal growth factor receptor tyrosine kinase reduce pulmonary fibrosis in rats. Am J Pathol 1999;155:213–221.
173. Daniels CE, Wilkes MC, Edens M, et al. Imatinib mesylate inhibits the profibrogenic activity of TGF-beta and prevents bleomycin-mediated lung fibrosis. J Clin Invest 2004;114:1308–1316.
174. Aono Y, Nishioka Y, Inayama M, et al. Imatinib as a novel antifibrotic agent in bleomycin-induced pulmonary fibrosis in mice. Am J Respir Crit Care Med 2005;171:1279–1285.
175. Blumberg B, Evans RM. Orphan nuclear receptors—new ligands and new possibilities. Genes Dev 1998;12:3149–3155.
176. Willson TM, Brown PJ, Sternbach DD, Henke BR. The PPARs: from orphan receptors to drug discovery. J Med Chem 2000;43:527–550.
177. Kota BP, Huang TH, Roufogalis BD. An overview on biological mechanisms of PPARs. Pharmacol Res 2005;51:85–94.
178. Tontonoz P, Graves RA, Budavari AI, et al. Adipocyte-specific transcription factor ARF6 is a heterodimeric complex of two nuclear hormone receptors, PPAR gamma and RXR alpha. Nucleic Acids Res 1994;22:5628–5634.
179. Rosen ED, Sarraf P, Troy AE, et al. PPAR gamma is required for the differentiation of adipose tissue in vivo and in vitro. Mol Cell 1999;4:611–617.
180. Barak Y, Nelson MC, Ong ES, et al. PPAR gamma is required for placental, cardiac, and adipose tissue development. Mol Cell 1999;4:585–595.
181. Lehrke M, Lazar MA. The many faces of PPARgamma. Cell 2005;123:993–999.
182. Lehmann JM, Moore LB, Smith-Oliver TA, Wilkison WO, Willson TM, Kliewer SA. An antidiabetic thiazolidinedione is a high affinity ligand for peroxisome proliferator-activated receptor gamma (PPAR gamma). J Biol Chem 1995;270:12,953–12,956.
183. He W, Barak Y, Hevener A, et al. Adipose-specific peroxisome proliferator-activated receptor gamma knockout causes insulin resistance in fat and liver but not in muscle. Proc Natl Acad Sci USA 2003;100:15,712–15,717.
184. Nolan JJ, Ludvik B, Beerdsen P, Joyce M, Olefsky J. Improvement in glucose tolerance and insulin resistance in obese subjects treated with troglitazone. N Engl J Med 1994;331:1188–1193.
185. Rangwala SM, Lazar MA. Peroxisome proliferator-activated receptor gamma in diabetes and metabolism. Trends Pharmacol Sci 2004;25:331–336.

186. Standiford TJ, Keshamouni VG, Reddy RC. Peroxisome proliferator-activated receptor-γ as a regulator of lung inflammation and repair. Proc Am Thorac Soc 2005;2:226–231.
187. Alleva DG, Johnson EB, Lio FM, Boehme SA, Conlon PJ, Crowe PD. Regulation of murine macrophage proinflammatory and anti-inflammatory cytokines by ligands for peroxisome proliferator-activated receptor-gamma: counter-regulatory activity by IFN-gamma. J Leukoc Biol 2002; 71:677–685.
188. Kon K, Ikejima K, Hirose M, et al. Pioglitazone prevents early-phase hepatic fibrogenesis caused by carbon tetrachloride. Biochem Biophys Res Commun 2002;291:55–61.
189. Ando M, Murakami Y, Kojima F, et al. Retrovirally introduced prostaglandin D2 synthase suppresses lung injury induced by bleomycin. Am J Respir Cell Mol Biol 2003;28:582–591.
190. Cuzzocrea S, Wayman NS, Mazzon E, et al. The cyclopentenone prostaglandin 15-deoxy-Delta(12,14)-prostaglandin J(2) attenuates the development of acute and chronic inflammation. Mol Pharmacol 2002;61:997–1007.
191. Cuzzocrea S, Pisano B, Dugo L, et al. Rosiglitazone, a ligand of the peroxisome proliferator-activated receptor-gamma, reduces acute inflammation. Eur J Pharmacol 2004;483:79–93.
192. Cuzzocrea, S, Mazzon, E, Dugo, L, et al. Reduction in the evolution of murine type II collagen-induced arthritis by treatment with rosiglitazone, a ligand of the peroxisome proliferator-activated receptor gamma. Arthritis Rheum 2003;48:3544–3556.
193. Kliewer SA, Lenhard JM, Willson TM, Patel I, Morris DC, Lehmann JM. A prostaglandin J2 metabolite binds peroxisome proliferator-activated receptor gamma and promotes adipocyte differentiation. Cell 1995;83:813–819.
194. Shimizu K, Shiratori K, Kobayashi M, Kawamata H. Troglitazone inhibits the progression of chronic pancreatitis and the profibrogenic activity of pancreatic stellate cells via a PPARgamma-independent mechanism. Pancreas 2004;29:67–74.
195. Yuan GJ, Zhang ML, Gong ZJ. Effects of PPARg agonist pioglitazone on rat hepatic fibrosis. World J Gastroenterol 2004;10:1047–1051.
196. Jaster R, Lichte P, Fitzner B, et al. Peroxisome proliferator-activated receptor gamma overexpression inhibits pro-fibrogenic activities of immortalised rat pancreatic stellate cells. J Cell Mol Med 2005;9:670–682.
197. van Westerloo DJ, Florquin S, de Boer AM, et al. Therapeutic effects of troglitazone in experimental chronic pancreatitis in mice. Am J Pathol 2005;166:721–728.
198. Burgess HA, Daugherty LE, Thatcher TH, et al. PPARgamma agonists inhibit TGF-beta induced pulmonary myofibroblast differentiation and collagen production: implications for therapy of lung fibrosis. Am J Physiol Lung Cell Mol Physiol 2005;288:L1146–L1153.
199. Thatcher TH, Burgess HA, Daugherty LE, Redonnet M, Phipps RP, Sime PJ. PPAR-gamma agonists inhibit TGF-beta-induced myofibroblast differentiation in vitro and silica-induced pulmonary fibrosis in vivo. [Abstract]. Am J Respir Crit Care Med 2004;169:A301.
200. Genovese T, Cuzzocrea S, Di Paola R, et al. Effect of rosiglitazone and 15-deoxy-Delta12,14-prostaglandin J2 on bleomycin-induced lung injury. Eur Respir J 2005;25:225–234.
201. Harris SG, Phipps RP. Prostaglandin D(2), its metabolite 15-d-PGJ(2), and peroxisome proliferator activated receptor-gamma agonists induce apoptosis in transformed, but not normal, human T lineage cells. Immunology 2002;105:23–34.
202. Chawla A, Barak Y, Nagy L, Liao D, Tontonoz P, Evans RM. PPAR-gamma dependent and independent effects on macrophage-gene expression in lipid metabolism and inflammation. Nat Med 2001;7: 48–52.
203. Li L, Tao J, Davaille J, et al. 15-deoxy-Delta 12,14-prostaglandin J2 induces apoptosis of human hepatic myofibroblasts. A pathway involving oxidative stress independently of peroxisome-proliferator-activated receptors. J Biol Chem 2001;276:38,152–38,158.

41 Negative Regulation of the TGF-β Family Signal Pathway by Inhibitory Smads and Their Involvement in Cancer and Fibrosis

Susumu Itoh, Hiroyuki Suzuki, Fumiko Itoh, Kohei Miyazono, and Mitsuyasu Kato

CONTENTS

STRUCTURE OF INHIBITORY SMADS
REGULATION OF EXPRESSION AND SUBCELLULAR LOCALIZATION OF I-SMADS
REGULATION OF TGF-β SIGNALING BY I-SMADS
OTHER BIOLOGICAL FUNCTIONS OF I-SMADS
ROLE OF I-SMADS IN CANCER
I-SMADS IN OTHER DISEASES
GAIN OR LOSS OF FUNCTION OF I-SMAD IN MICE
PERSPECTIVES
REFERENCES

Abstract

Transforming growth factor-β (TGF-β) is a pivotal cytokine that regulates cell growth and differentiation of many different cell types *(1)*. TGF-β family signaling initiated by a specific receptor-ligand complex is mainly transmitted to the nucleus via intracellular signal transducing molecules, termed Smads. TGF-β family members, which include TGF-βs, activins, and bone morphogenetic proteins (BMPs), play roles in embryogenesis and maintenance of tissue homeostasis during adult life. In addition, TGF-β family signaling regulates physiological and pathological processes such as wound healing, immunomodulation, apoptosis, and fibrosis. Therefore, the deregulation of TGF-β family signaling is implicated in oncogenesis and heritable disorders *(2–4)*, and the Smad-mediated signaling by the TGF-β family must be tightly regulated. Inhibitory Smads (I-Smads) negatively act for regulation of duration and intensity of TGF-β family signaling. Whereas Smad7 acts as a general inhibitor for the TGF-β family signal pathway, Smad6 preferentially blocks BMP signaling. It has been reported that there are various mechanisms by which I-Smads inhibit the TGF-β family signal pathway (i.e., interaction with type I receptors, R-Smads or transcriptional factors). It has also been reported that several tumors express high levels of I-Smads, leading to blockage of TGF-β family signaling. However, the relevance between tumor development and elevated expression of I-Smads is not clear. In contrast, the expression (or function) of Smad7 is suppressed and augmented in scleroderma and gut inflammation, respectively. On the other hand,

From: *Cancer Drug Discovery and Development: Transforming Growth Factor-β in Cancer Therapy, Vol I: Basic and Clinical Biology*
Edited by: Sonia B. Jakowlew © Humana Press Inc., Totowa, NJ

there are some reports that I-Smads cooperate with the TGF-β family signal pathway. In this chapter, we focus on recent new insights into the functions of I-Smads (Smad6 and Smad7) and review how I-Smads regulate the TGF-β family signal pathway and are implicated in diseases.

Key Words: BMP; E3 ubiquitin ligase; I-Smad; scleroderma; Smad6; Smad7; TGF-β.

1. STRUCTURE OF INHIBITORY SMADS

Receptor-regulated Smads (R-Smads) and common Smads (Co-Smads) have highly conserved N- and C-terminal regions, termed Mad homology 1 and 2, (MH1) and MH2 domains, respectively. Whereas Inhibitory Smads (I-Smads) possess a conserved MH2 domain in their C- terminal region, their N-terminal regions show only weak similarity to the MH1 domain of R- and Co-Smads. I-Smads act in an opposing manner to R-Smads and antagonize the TGF-β family signal pathway. Smad7 inhibits all TGF-β family signaling, while Smad6 preferentially inhibits the BMP signal pathway. As a general concept, I-Smads compete with R-Smads for binding to activated type I receptors and inhibit the phosphorylation of R-Smads by activated type I receptor kinase. I-Smads bind to activated type I receptors through their MH2 domain and stably associate with these receptors without direct phosphorylation by type I receptor kinases *(5–9)*. Four basic amino acid residues (Lys-312, Lys-316, Lys-401, and Arg-409) in the MH2 domain of Smad7 play important roles in interaction with type I receptors. Among them, two amino acid residues within the basic surface groove (Lys-312 and Lys-316) of Smad7 are required for the specific inhibition of the TGF-β signal pathway, while two amino acids within the L3 loop (Lys-401 and Arg-409) are essential for the inhibition of both TGF-β and BMP signaling *(10)*. In swapping experiments using I-Smads, the N-terminal domain of Smad6 has less inhibitory activity than that of Smad7 *(11)*. There is a report that the MH2 domain of Smad6 forms a complex with phosphorylated R-Smads to compete with Co-Smad for their complex formation *(12)* (Fig. 1).

I-Smads are known to interact with HECT type E3-ubiquitin ligases, Smad ubiquitination regulatory factors (Smurfs) *(13–16)*, neural precursor cell expressed, developmentally down-regulated 4-2 (Nedd 4-2) *(17)*, WW domain-containing protein 1 (WWP1) *(18)* and atrophin 1-interacting protein 4 (AIP4) *(19)*. The WW motif in HECT type E3-ubiquitin ligases recognizes PY motifs in the linker domain of I-Smads. The interaction of HECT type E3-ubiquitin ligases with I-Smads potentiates nuclear export of I-Smads, type I receptor degradation and/or I-Smad degradation (*see* Section 2.2.). I-Smads also interact with a RING type E3-ubiquitin ligase, Arkadia, through the MH2 domain. The PLDLS motif, which was originally found as a binding motif of E1A to CtBP, binds to CtBP in the linker region of Smad6 to repress BMP-induced transcription *(20)* (Fig. 2).

2. REGULATION OF EXPRESSION AND SUBCELLULAR LOCALIZATION OF I-SMADS

2.1. Transcriptional Regulation of I-Smads

I-Smads are potently induced by the TGF-β family members, TGF-β, activin, BMP, Müllerian-inhibiting substance (MIS), and myostatin, in various cells and organs *(5,21–26)*. The induction of I-Smads is implicated in the negative feedback loop of the TGF-β family signal pathway. The promoter of the Smad7 gene contains Smad binding element (SBE), which is important for the binding of the complex between phosphorylated R-Smads and Smad4 and for the mediation of transcriptional activation of the Smad7 gene together with other transcriptional factors *(27–32)*. On the other hand, the GC-rich sequence in the promoter of the Smad6 gene was identified as a direct DNA binding element of BMP-responsive Smads *(33)*. Besides the GC-rich sequence in the Smad6 promoter, the CREB binding site

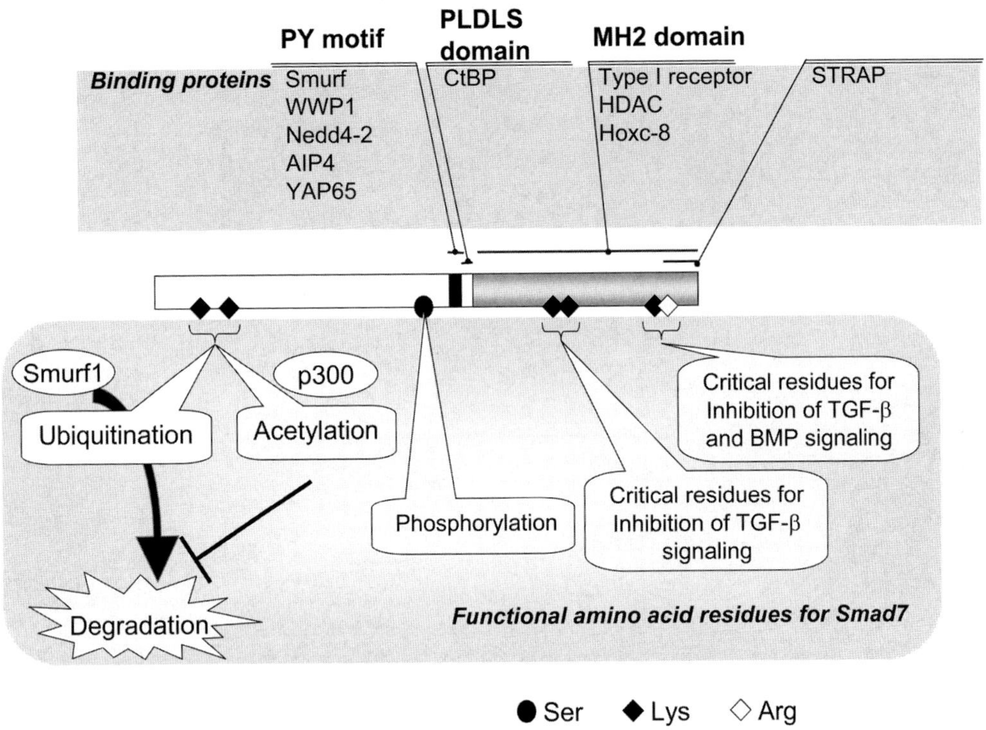

Fig. 1. Structure and functional domains of I-Smads. Among the binding partners to I-Smads reported, the proteins whose binding domains in I-Smads have already been determined are shown. CtBP and Hoxc-8 specifically interact with Smad6, whereas YAP65 and STRAP associate only with Smad7. The domains of I-Smad responsible for binding to WWP1, Nedd4-2 and AIP4 were analyzed using Smad7, while that of HDAC was determined using Smad6. Both I-Smads were used to determine their binding sites for Smurf and type I receptors.

close to the GC-rich sequence confers the transcriptional activity of the Smad6 gene *(34)*. It has been reported that the activation of the Smad7 gene by a low dose of BMP requires the enhancer in its first intron where GATA-5 or GATA-6 can make a complex with both Smad1 and Smad4 although a high dose of BMP is required for direct binding of the Smad1 and the Smad4 complex to the typical GC-rich sequence in the promoter of the Smad7 gene *(35)*. In contrast, Ski, a transcriptional corepressor, inhibits basal transcriptional activity of the Smad7 gene together with Smad4 *(36)*.

The induction of I-Smads is also seen via the JAK1/Stat1 pathway activated by interferon-γ *(37)*, NF-κB activation by LPS, TNF-α, IL-1β or norepinephrine *(30,38,39)*, and AP-1/JNK activity induced by stress *(40)*. Moreover, the activation of CD40 or ultraviolet irradiation can rapidly activate transcription of the Smad7 gene to block TGF-β/Smad signaling *(41–43)*. An oncoprotein, HER2/Neu, collaborates with ETS transcriptional factor ER81 to activate the Smad7 gene via the ERK pathway together with the TAK1 pathway *(44)*. Interestingly, treatment of IMR-90 cells, the human lung fibroblast cell line, with TPA revealed activation of the Smad7 gene via the PKCμ pathway bypassing MAPK/NF-κB activation and inhibition of the Smad6 gene through the PKCμ/ERK pathway *(45)*.

2.2. Posttranslational Regulation of I-Smads

Ubiquitination and proteasome-mediated degradation play key roles in the control of signal transduction. Ring finger domain-containing E3 ubiquitin ligases Arkadia and Jab1,

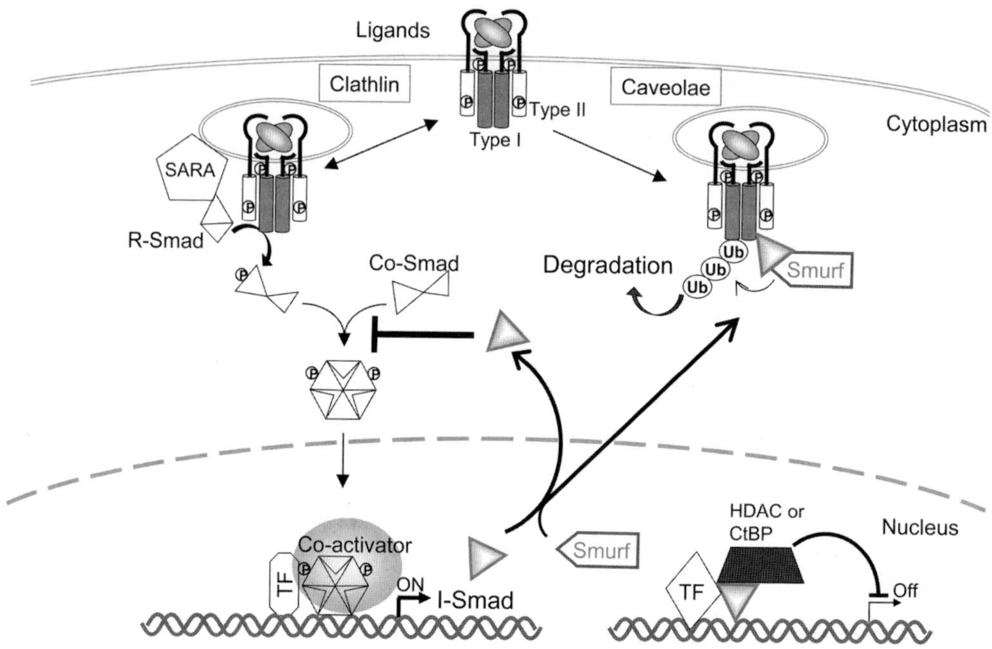

Fig. 2. Model of inhibitory mechanisms of I-Smads. Clathrin-dependent internalization into early endosomes promotes TGF-β family signaling. Subsequently, the Smad complex activates transcription of I-Smad genes. After I-Smads interacting with Smurfs are exported from the nucleus, Smurfs bind to the type I receptor indirectly through I-Smads. Then ubiquitination of the type I receptor is enhanced for its degradation. In addition, Smad6 competes with Smad4 for binding to phosphorylated R-Smads to block BMP signaling when BMP-Smads (Smad1, Smad5, and Smad8) are activated. As an inhibitory mechanism in the nucleus, Smad6 acts as a transcriptional corepressor. Smad6 binds to the promoter directly or indirectly via Hoxc-8 or GR together with CtBP or HDAC. Then, the transcription of genes is attenuated.

which is a component of the COP9 signalosome, promote ubiquitination of Smad7 to prolong TGF-β signaling, whereas AIP4 inhibits the TGF-β signal pathway despite its promotion of Smad7 ubiquitination *(19,46,47)*. Smurf1 and Smurf2 also catalyze the ubiquitination of Smad7 *(15)*. A lysine residue is known to be a target for E3 ubiquitin ligase, while protein acetylation is also catalyzed at the lysine residue. p300-mediated acetylation of Smad7 demonstrated the additional control of Smad7/Smurf1-dependent regulation of TGF-β signaling. Two lysine residues at 64 and 70 in Smad7 can be acetylated and ubiquitinated by p300 and Smurf1, respectively. p300-mediated acetylation of Smad7 serves to protect Smad7 from Smurf1-mediated ubiquitination *(48)*. In addition, histone deacetylase 1 (HDAC1)-mediated deacetylation of Smad7 decreases the stability of Smad7 by enhancing its ubiquitination *(49)*. Thus, the balance among acetylation, deacetylation and ubiquitination of two critical lysine residues in Smad7 determines its stability.

I-Smads are known to be phosphorylated proteins *(5,6,50)*. Serine 249 has been identified as a major phosphorylated amino acid residue in Smad7. Unlike wild-type Smad7 fused to Gal4 DNA binding domain (GBD), phosphorylation-defective Smad7 mutant fused to the GBD cannot inhibit Gal4-induced transcription. However, the context of phosphorylation in Smad7 for regulation of the TGF-β family signal pathway is not completely clear *(50)*.

We have recently detected the modification of I-Smads by protein arginine N-methyltransferase-1 (PRMT-1). PRMT-1-mediated arginine methylation is specific for I-Smads, but not for R- and Co-Smads (unpublished result). The functional significance of the methylation in I-Smads is under investigation.

2.3. Subcellular Localization of I-Smads

I-Smads are principally localized in the nucleus and rapidly exported to cytosols upon TGF-β family stimulation *(8,51–53)*. However, there are a few exceptions in respect to subcellular localization of I-Smads under different assay conditions *(11,54)*. Export of I-Smads upon TGF-β stimulation requires the MH2 domain of I-Smads *(8)*. In addition, overexpression of Smurf1 or Smurf2 in cells promotes nuclear export of Smad7 *(52,53)*. The nuclear export of Smad7 by Smurf1 is blocked by the inactivation of chromosomal region maintenance 1 *(53)*. However, the mechanism by which Smad7 is exported from the nucleus upon TGF-β stimulation remains to be elucidated.

The ubiquitin-proteasome pathway also controls TGF-β receptor trafficking machinery. When Smad7 interacts with the TGF-β type I receptor and promotes Smurf2-mediated ubiquitination of the TGF-β type I receptor for accelerated receptor turnover (*see* Section 3.2.), Smad7 associated with both the type I receptor and Smurf2 travel to caveolin-positive vesicles via lipid raft. On the other hand, the internalization of type I receptor through clathrin leads the type I receptor to early endosomal antigen 1 (EEA1)-positive early endosomes where TGF-β signals might be initiated together with Smad for anchor receptor activation (SARA) *(55–58)* (Fig. 2).

3. REGULATION OF TGF-β SIGNALING BY I-SMADS

The magnitude and duration of TGF-β signaling need to be tightly regulated under physiological conditions. I-Smads, early responsive genes induced by TGF-β family members, have the ability to inhibit TGF-β signaling by different mechanisms. Smad7 is a general inhibitor for the TGF-β family signal pathway, whereas Smad6 preferentially blocks the BMP signal pathway.

3.1. Inhibition of TGF-β Signaling Through the Binding of I-Smads to Type I Receptor

I-Smads are rapidly exported from the nucleus to cytoplasm by TGF-β stimulation, followed by formation of a stable complex with TGF-β type I receptors. The type I receptors that stably associate with I-Smads cannot phosphorylate R-Smads owing to competitive binding between I-Smads and R-Smads *(5–8)*. The phosphorylation of the type I receptor by the type II receptor kinase is required for I-Smads to bind to the type I receptor *(5)*. The MH2 domain of I-Smads is essential for I-Smads to block TGF-β signaling via binding to the type I receptors *(8,59)*. Four basic amino acids in the MH2 domain of Smad7 are important for interaction with type I receptors (*see* Section 1). STRAP recruits Smad7 to the activated type I receptor to make a stable complex between Smad7 and the activated type I receptor, resulting in efficient blockage of TGF-β signaling *(60,61)*. YAP65 promotes Smad7 to bind to the activated TGF-β type I receptor in order to inhibit TGF-β signaling *(62)*. On the other hand, Smad6 stably associates with BMP type I receptors and inhibits BMP response *(6,14)*. Tob protein enhances the interaction between Smad6 and BMP type I receptors, resulting in potent inhibition of BMP signaling compared with Smad6 alone *(63)*. Furthermore, I-Smads recruit various E3 ubiquitin ligases to the activated type I receptors and promote their degradation (*see* Section 3.2.) (Fig. 1).

3.2 Inhibition of TGF-β Signaling via Ubiquitination of Type I Receptors by E3 Ubiquitin Ligases

The degradation of the activated type I receptor is another inhibitory mechanism of the TGF-β signal pathway by I-Smads *(64)*. Smurf1 and Smurf2 were originally found to be recruited by I-Smads and to promote activated type I receptor degradation via the proteasomal pathway. Both Smurfs have WW domains which directly bind to the PY motif in I-Smads.

After the interaction of Smurfs with I-Smads, Smurfs indirectly associate with the activated type I receptors and ubiquitinate them, leading to efficient downregulation of TGF-β family signaling *(14–16)*. The N-terminal part of Smad7 presents a conjugating E2 enzyme, UbcH7, to the HECT domain of Smurf2, followed by enzymatic activation of Smurf2 *(65)*. Recently, it was found that other E3 ubiquitin ligases containing the WW domain, Nedd4-2 and WWP1, targeted the PY motif in Smad7 for its association with Smad7 and that they promoted the degradation of the TGF-β type I receptor via proteasome *(17,18)*. In contrast, a deubiquitinating enzyme, UCH37, interacts with Smad7 in a region distinct from its PY motif and promotes deubiquitination and stabilization of the TGF-β type I receptor *(66)* (Fig. 2).

3.3. Inhibition of BMP Signaling by Association Between Phosphorylated BMP-Smads and Smad6

Besides the inhibition of signaling via the association of type I receptors with I-Smads, Smad6 interacts with phosphorylated BMP-Smads. Thus, Smad4 can no longer make a complex with phosphorylated BMP-Smads because of competition with Smad6. Consequently, BMP signaling is not efficiently transduced to the nucleus *(12,51)* (Fig. 2).

3.4. Transcriptional Regulation by I-Smads

There are a few reports that Smad6 acts as a transcriptional corepressor. Hoxc-8 binds to the osteopontin promoter constitutively and inhibits the transcription of the osteopontin gene. Upon BMP stimulation, interaction between Smad1 and Hoxc-8 promotes dislodgement of Hoxc-8 from the osteopontin promoter, resulting in the activation of its promoter *(67,68)*. On the other hand, the expression of Smad6 inhibits BMP-induced osteopontin promoter activity because complex formation between Smad6 and Hoxc-8 prevents Smad1 from accessing Hoxc-8 *(69)*. Smad6 also makes a complex with HDAC1 in the nucleus together with Hoxc-8, resulting in the formation of the transcriptional repressor complex on the osteopontin promoter *(70)*. In addition to HDAC1, it was reported that HDAC3 attracted to the glucocorticoid receptor (GR) via Smad6 suppresses GR-induced transactivation *(71)*. As another function of nuclear Smad6, Smad6 is capable of inhibiting BMP-induced Id1 promoter activity through recruitment of CtBP, a transcriptional corepressor, in the Id1 promoter *(20)* (Fig. 2).

3.5. Cooperation of Other Interacting Partners with I-Smads

Smad7 is able to recruit GADD34, a regulatory subunit of the protein phosphatase 1 (PP1), to the TGF-β type I receptor upon TGF-β stimulation. Subsequently, GADD34 can interact with a catalytic subunit of PP1, followed by dephosphorylation of activated TGF-β type I receptor. Thus, the dephosphorylation of the TGF-β type I receptor via Smad7 effectively controls negative feedback in TGF-β signaling *(72)*. There are two reports that Smad6 can inhibit BMP signaling through the TAK1 pathway. BMP-induced apoptosis through the TAK1-p38 pathway in mouse hybridoma, MH60 cells, is blocked by direct binding of Smad6 to activated TAK1 *(73)*, whereas BMP-induced neurite outgrowth in PC12 cells is attenuated through the interaction of Smad6 with TAK1 or TAB1, or the interaction between Smad7 and TAB1 *(74)*. In contrast with negative regulation by cooperation between I-Smads and their interacting partners, AMSH prolongs BMP signaling because the interaction of AMSH with Smad6 prevents Smad6 from association with the BMP type I receptor and/or BMP-Smads *(51)*.

4. OTHER BIOLOGICAL FUNCTIONS OF I-SMADS

Although the main function of I-Smads is known to be the inhibition of the TGF-β family signal pathway, it has been demonstrated that in some cases I-Smads cooperate with the TGF-β family signal pathway. TGF-β-induced apoptosis is enhanced by overexpression of Smad7 via activation of p38, NF-κB or β-catenin *(75–78)*. On the other hand, Schiffer et al. demonstrated that cooperation of TGF-β with Smad7 in podocytes induces apoptosis owing to both TGF-β-induced p38 activation and Smad7-mediated NF-κB inactivation *(79)*. The activation of cdc42 by TGF-β is enhanced in Smad7-overexpressed prostate cancer PC-3U cells. Consequently, membrane ruffling is developed via the activation of p38 by Smad7 *(80,81)*. Adipocyte differentiation in 3T3-F442A cells is blocked by TGF-β. Interestingly, the expression of either Smad6 or Smad7 further promotes TGF-β-induced inhibition of adipocyte differentiation *(82)*.

A variant of Smad6, Smad6s, potentiates the promoter activity of the TGF-β1 gene in endothelial SVHA-1 cells, followed by enhancement of TGF-β-induced PAI-1 promoter activity *(83)*.

5. ROLE OF I-SMADS IN CANCER

TGF-β is well-known to inhibit tumorigenesis *(2,84,85)*. Because I-Smads block TGF-β family signaling, I-Smads are implicated in tumor development. The Smad7 gene is mapped in 18q21 *(86)* and located between the Smad2 and Smad4 genes within a 4-Mb gene cluster *(87)*. In colorectal cancer, either Smad2 or Smad4 gene is often inactivated by genetic alterations, the Smad7 gene is frequently amplified *(86)*. In pancreatic cancer, the expression of Smad6 and Smad7 is enhanced compared with that in normal tissue *(88,89)*. Endometrial cancer also expresses Smad7 at high levels, and this is associated with early recurrence *(90)*. In contrast, mutation analysis using samples from ovarian carcinomas, hepatocellular carcinomas and colorectal cancer indicated that there were no critical mutations in the Smad6 and Smad7 genes *(91–93)*. In addition, no mutations of the Smad7 gene could be found in juvenile polyposis in which Smad4 mutations have been reported *(94)*.

The high expression of Smad7 was seen in papilloma and squamous cell carcinoma in mouse skin treated with DMBA and TPA, whereas the expression of Smad1, Smad2, Smad3, Smad4, and Smad5 was lost during development of skin carcinoma *(95)*. The v-*ras*-induced squamous cell carcinoma model in mouse skin indicated that coexpression of Smad7 with oncogenic *ras* enhanced malignancy in skin carcinoma, but that of Smad6 did not *(96)*. Halder et al. demonstrated that injection of cells stably expressing Smad7 into nude mice resulted in tumor formation *(97)*. In contrast, inoculation of melanoma cells overexpressing Smad7 into nude mice resulted in delayed tumor formation compared with the control *(98)*. Recently, blockade of TGF-β superfamily signaling by systemic administration of adenovirus containing Smad7 cDNA has been shown to inhibit metastasis of mouse breast carcinoma to liver and lung in nude mice, possibly through regulating cell–cell adhesion *(99)*.

In spite of reports that the overexpression of I-Smads can be observed in some tumors, the relationship between tumor development and I-Smads is still inconclusive. It is possible that elevated expression of Smad7 in the early stages of cancer may result in acceleration of carcinogenesis owing to loss of inhibition of tumor growth, but that Smad7 may prevent progression of advanced cancer through inhibition of EMT, immunosupression, and angiogenesis induced by TGF-β.

6. I-SMADS IN OTHER DISEASES

TGF-β promotes fibroblast proliferation and synthesis of extracellular matrix (ECM) proteins. The enhanced TGF-β signaling in fibroblasts is known to promote fibrosis in various tissues.

6.1. Scleroderma

Scleroderma is characterized by vascular dysfunction and excessive accumulation of ECM proteins, resulting in progressive and widespread tissue fibrosis. The fibroblasts in the lesions of scleroderma may contribute to the fibrotic process *(100)*. It is known that TGF-β from activated fibroblasts is highly expressed in the lesions of scleroderma. Thus, TGF-β signaling is remarkably enhanced by activated fibroblasts in the lesions of scleroderma because of the autocrine loop of TGF-β signaling *(101)*, suggesting that TGF-β signaling might be involved in the pathogenesis of this disease *(102)*. Dong et al. found that expression of Smad7 decreases in skin biopsy samples from patients with scleroderma and that their fibroblasts possess a low response to induction of Smad7 mRNA by TGF-β, whereas Smad3 expression is increased in scleroderma skin. Thus, the process of scleroderma involves not only excessive TGF-β production but also marked Smad7 deficiency *(103)*. On the other hand, Asano et al. reported that fibroblasts from scleroderma are more potent than normal fibroblasts in the induction of Smad7 expression although Smad7 is scarcely able to inhibit TGF-β signaling in these tissues. They concluded that the impaired Smad7-Smurf-mediated inhibitory effect on TGF-β signaling might be involved in maintenance of the autocrine TGF-β loop in scleroderma fibroblasts *(104)*.

6.2. Other Fibrotic Diseases

In unilateral ureteral obstruction in mice, a model of progressive tubulointerstitial fibrosis, the expression of Smad7 decreased. Consistently, renal fibrosis was increased. It was demonstrated that the degradation of Smad7 via the ubiquitin-dependent pathway plays a critical role in the progression of tubulointerstitial fibrosis *(105)*. This evidence was further supported by reports that introduction of Smad7 to lesions of renal fibrosis improved this disease *(106,107)*. In lung and liver fibrosis, the introduction of Smad7 prevented progression of fibrosis owing to inhibition of TGF-β signaling *(108,109)*.

6.3. Gut Inflammation

TGF-β has been shown to effectively suppress inflammation in certain tissues, including those in inflammatory bowel diseases (IBD) *(110–112)*. IBD mucosa and purified mucosal T cells express Smad7 at high levels to inhibit TGF-β signaling. Reduction of Smad7 in cells isolated from patients with Crohn's disease restores TGF-β responsiveness and enables TGF-β to block proinflammatory cytokine production. Thus, blockade of TGF-β signaling by Smad7 promotes chronic production of inflammatory cytokines in gut mucosa, resulting in the inflammatory process in IBD *(113)*.

7. GAIN OR LOSS OF FUNCTION OF I-SMAD IN MICE

Targeted disruption and ectopic expression of I-Smads in mice have been reported. By targeted insertion of a *LacZ* reporter in the Smad6 locus, Smad6 expression is found to be largely restricted to the heart and blood vessels. Smad6$^{-/-}$ mice have multiple cardiovascular abnormalities including hyperplasia of the cardiac valves and outflow tract septation defects. In the adult cardiovascular system, the development of aortic ossification and elevated blood pressure are observed in viable Smad6$^{-/-}$ mice *(114)*.

Smad7 transgenic mice using the keratin 5 promoter exhibit severe defects in multiple epithelial tissues including delayed and aberrant hair follicle formation, a marked hyperplasia in the epidermis, aberrant development of the eyelid and cornea, and severe thymic atrophy. Transgenic skin exhibits a reduction in receptors of the TGF-β family and decreased phosphorylation of Smad1 and Smad2 *(115)*.

Transgenic mice expressing Smad7 selectively in mature T cells using the lck promoter exhibit enhancement of antigen-induced airway inflammation and airway reactivity associated with high production of both T helper (Th) cell type 1 (Th1) and 2 (Th2) cytokines. Peripheral T cells in Smad7 transgenic mice show resistance to TGF-β-induced growth inhibition *(116)*. These results suggest that TGF-β signaling negatively regulates the T-cell-mediated inflammatory response.

Transgenic mice expressing Smad6 in chondrocytes using the α2(XI) collagen chain gene promoter show postnatal dwarfism with osteopenia and inhibition of phosphorylation of BMP-Smads in chondrocytes. Endochondral ossification during development in these mice was associated with delayed chondrocyte hypertrophy and thin trabecular bone. Furthermore, double-transgenic mice expressing Smad6 and Smurf1 in chondrocytes exhibit more delayed endochondral ossification than do Smad6 transgenic mice *(117)*.

8. PERSPECTIVES

Genetic and biochemical studies have demonstrated the fine-tuning of the TGF-β family signal. In particular, I-Smads potently achieve negative regulation of the TGF-β family signal pathway. I-Smads stably interact with type I receptors to block TGF-β signaling. However, it remains unclear which region of the type I receptor I-Smads bind to and how I-Smads compete with R-Smads for association with type I receptors. Because I-Smads bind to a TGF-β type I receptor L45 mutant which does not phosphorylate Smad2 *(118)*, the binding site of the TGF-β type I receptor for I-Smads may not be the L45 loop where R-Smads can be recruited. Stoichiometric analysis of type I receptors and I-Smads is needed.

Recent studies demonstrated that Smad6 acts as a transcriptional corepressor in the nucleus without a direct binding to DNA. On the other hand, Bai and Cao showed that the N-terminal part of Smad6 binds to a Hoxc-8 binding site very weakly *(70)*. Interesting questions remain concerning whether Smad6 has a direct DNA binding ability and which DNA sequences it can bind to. Moreover, the function of Smad7 in the nucleus needs to be analyzed.

Although the loss of function for I-Smads has not demonstrated any enhancement of the TGF-β family signal, a gain of function showed the inhibitory effect of TGF-β signaling in in vitro experiments. It is highly important to know whether null mutations in the Smad6 or Smad7 genes lead mice to increased responsiveness to the TGF-β family in vivo. In addition, the overexpression of I-Smads has been reported in some cancers; whereas there is no evidence that transgenic mice carrying I-Smads develop tumors in vivo. Thus, whether the inhibition of TGF-β family signaling by I-Smads is required for development of certain cancers remains to be elucidated.

REFERENCES

1. Roberts AB, Sporn MB. The transforming growth factor-βs. In: Sporn MB, Roberts AB, editors. Peptide Growth Factors and Their Receptors, Part I. Berlin: Springer-Verlag 1990;95: pp. 419–472.
2. Massagué J, Blain SW, Lo RS. TGFβ signaling in growth control, cancer, and heritable disorders. Cell 2000;51:295–309.
3. ten Dijke P, Miyazono K, Heldin C-H. Signaling inputs converge on nuclear effectors in TGF-β signaling. Trends Biochem Sci 2000;25:64–70.
4. Derynck R, Zhang YE. Smad-dependent and Smad-independent pathways in TGF-β family signalling. Nature 2003;425:577–584.
5. Nakao A, Afrakhte M, Morén A, et al. Identification of Smad7, a TGF-β-inducible antagonist of TGF-β signalling. Nature 1997;389:631–635.
6. Imamura T, Takase M, Nishihara A, et al. Smad6 inhibits signalling by the TGF-β superfamily. Nature 1997;389:622–626.
7. Hayashi H, Abdollah S, Qiu Y, et al. The MAD-related protein Smad7 associates with the TGF-β receptor and functions as an antagonist of TGF-β signaling. Cell 1997;89:1165–1173.
8. Itoh S, Landström M, Hermansson A, et al. Transforming growth factor β1 induces nuclear export of inhibitory Smad7. J Biol Chem 1998;273:29,195–29,201.
9. Ishisaki A, Yamato K, Hashimoto S, et al. Differential inhibition of Smad6 and Smad7 on bone morphogenetic protein- and activin-mediated growth arrest and apoptosis in B cells. J Biol Chem 1999;274: 13,637–13,642.

10. Mochizuki T, Miyazaki H, Hara T, et al. Roles for the MH2 domain of Smad7 in the specific inhibition of transforming growth factor-β superfamily signaling. J Biol Chem 2004;279:31,568–31,574.
11. Hanyu A, Ishidou Y, Ebisawa T, Shimanuki T, Imamura T, Miyazono K. The N domain of Smad7 is essential for specific inhibition of transforming growth factor-β signaling. J Cell Biol 2001;155: 1017–1027.
12. Hata A, Lagna G, Massagué J, Hemmati-Brivanlou A. Smad6 inhibits BMP/Smad1 signaling by specifically competing with the Smad4 tumor suppressor. Genes Dev 1998;12:186–197.
13. Zhu H, Kavsak P, Abdollah S, Wrana JL, Thomsen GH. A SMAD ubiquitin ligase targets the BMP pathway and affects embryonic pattern formation. Nature 1999;400:687–693.
14. Murakami G, Watabe T, Takaoka K, Miyazono K, Imamura T. Cooperative inhibition of bone morphogenetic protein signaling by Smurf1 and inhibitory Smads. Mol Biol Cell 2003;14:2809–2817.
15. Ebisawa T, Fukuchi M, Murakami G, et al. Smurf1 interacts with transforming growth factor-β type I receptor through Smad7 and induces receptor degradation. J Biol Chem 2001;276:12,477–12,480.
16. Kavsak P, Rasmussen RK, Causing CG, et al. Smad7 binds to Smurf2 to form an E3 ubiquitin ligase that targets the TGF β receptor for degradation. Mol Cell 2000;6:1365–1375.
17. Kuratomi G, Komuro A, Goto K, et al. NEDD4-2 (neural precursor cell expressed, developmentally down-regulated 4-2) negatively regulates TGF-β (transforming growth factor-β) signalling by inducing ubiquitin-mediated degradation of Smad2 and TGF-β type I receptor. Biochem J 2005;386:461–470.
18. Komuro A, Imamura T, Saitoh M, et al. Negative regulation of transforming growth factor-β (TGF-β) signaling by WW domain-containing protein 1 (WWP1). Oncogene 2004;23:6914–6923.
19. Lallemand F, Seo SR, Ferrand N, et al. AIP4 restricts transforming growth factor-β signaling through a ubiquitination-independent mechanism. J Biol Chem 2005;280:27,645–27,653.
20. Lin X, Liang YY, Sun B, et al. Smad6 recruits transcription corepressor CtBP to repress bone morphogenetic protein-induced transcription. Mol Cell Biol 2003;23:9081–9093.
21. Takase M, Imamura T, Sampath TK, et al. Induction of Smad6 mRNA by bone morphogenetic proteins. Biochem Biophys Res Commun 1998;244:26–29.
22. Afrakhte M, Morén A, Jossan S, et al. Induction of inhibitory Smad6 and Smad7 mRNA by TGF-β family members. Biochem Biophys Res Commun 1998;249:505–511.
23. Bilezikjian LM, Corrigan AZ, Blount AL, Chen Y, Vale WW. Regulation and actions of Smad7 in the modulation of activin, inhibin, and transforming growth factor-β signaling in anterior pituitary cells. Endocrinology 2001;142:1065–1072.
24. Clarke TR, Hoshiya Y, Yi SE, Liu X, Lyons KM, Donahoe PK. Müllerian inhibiting substance signaling uses a bone morphogenetic protein (BMP)-like pathway mediated by ALK2 and induces SMAD6 expression. Mol Endocrinol 2001;15:946–959.
25. Li X, Ionescu AM, Schwarz EM, Zhang X, et al. Smad6 is induced by BMP-2 and modulates chondrocyte differentiation. J Orthop Res 2003;21:908–913.
26. Forbes D, Jackman M, Bishop A, Thomas M, Kambadur R, Sharma M. Myostatin auto-regulates its expression by feedback loop through Smad7 dependent mechanism. J Cell Physiol 2006;206:264–272.
27. von Gersdorff G, Susztak K, Rezvani F, Bitzer M, Liang D, Böttinger EP. Smad3 and Smad4 mediate transcriptional activation of the human Smad7 promoter by transforming growth factor β. J Biol Chem 2000;275:11,320–11,326.
28. Stopa M, Anhuf D, Terstegen L, Gatsios P, Gressner AM, Dooley S. Participation of Smad2, Smad3, and Smad4 in transforming growth factor β (TGF-β)-induced activation of Smad7. The TGF-β response element of the promoter requires functional Smad binding element and E-box sequences for transcriptional regulation. J Biol Chem 2000;275:29,308–29,317.
29. Brodin G, Åhgren A, ten Dijke P, Heldin C-H, Heuchel R. Efficient TGF-β induction of the Smad7 gene requires cooperation between AP-1, Sp1, and Smad proteins on the mouse Smad7 promoter. J Biol Chem 2000;275:29,023–29,030.
30. Nagarajan RP, Chen F, Li W, et al. Repression of transforming-growth-factor-β-mediated transcription by nuclear factor κB. Biochem J 2000;348:591–596.
31. Denissova NG, Pouponnot C, Long J, He D, Liu F. Transforming growth factor β-inducible independent binding of SMAD to the Smad7 promoter. Proc Natl Acad Sci USA 2000;97:6397–6402.
32. Nagarajan RP, Zhang J, Li W, Chen Y. Regulation of Smad7 promoter by direct association with Smad3 and Smad4. J Biol Chem 1999;274:33,412–33,418.
33. Ishida W, Hamamoto T, Kusanagi K, et al. Smad6 is a Smad1/5-induced Smad inhibitor. Characterization of bone morphogenetic protein-responsive element in the mouse Smad6 promoter. J Biol Chem 2000; 275:6075–6079.

34. Ionescu AM, Drissi H, Schwarz EM, et al. CREB Cooperates with BMP-stimulated Smad signaling to enhance transcription of the Smad6 promoter. J Cell Physiol 2004;198:428–440.
35. Benchabane H, Wrana JL. GATA- and Smad1-dependent enhancers in the Smad7 gene differentially interpret bone morphogenetic protein concentrations. Mol Cell Biol 2003;23:6646–6661.
36. Denissova NG, Liu F. Repression of endogenous Smad7 by Ski. J Biol Chem 2004;279:28,143–28,148.
37. Ulloa L, Doody J, Massagué J. Inhibition of transforming growth factor-β/SMAD signalling by the interferon-γ/STAT pathway. Nature 1999;397:710–713.
38. Kanamaru C, Yasuda H, Takeda M, et al. Smad7 is induced by norepinephrine and protects rat hepatocytes from activin A-induced growth inhibition. J Biol Chem 2001;276:45,636–45,641.
39. Bitzer M, von Gersdorff G, Liang D, et al. A mechanism of suppression of TGF-β/SMAD signaling by NF-κB/RelA. Genes Dev 2000;14:187–197.
40. Uchida K, Suzuki H, Ohashi T, Nitta K, Yumura W, Nihei H. Involvement of MAP kinase cascades in Smad7 transcriptional regulation. Biochem Biophys Res Commun 2001;289:376–381.
41. Werner F, Jain MK, Feinberg MW, et al. Transforming growth factor-β1 inhibition of macrophage activation is mediated via Smad3. J Biol Chem 2000;275:36,653–36,658.
42. Quan T, He T, Voorhees JJ, Fisher GJ. Ultraviolet irradiation blocks cellular responses to transforming growth factor-β by down-regulating its type-II receptor and inducing Smad7. J Biol Chem 2001;276:26,349–26,356.
43. Quan T, He T, Voorhees JJ, Fisher GJ. Ultraviolet irradiation induces Smad7 via induction of transcription factor AP-1 in human skin fibroblasts. J Biol Chem 2005;280:8079–8085.
44. Dowdy SC, Mariani A, Janknecht R. HER2/Neu- and TAK1-mediated up-regulation of the transforming growth factor β inhibitor Smad7 via the ETS protein ER81. J Biol Chem 2003;278:44,377–44,384.
45. Tsunobuchi H, Ishisaki A, Imamura T. Expressions of inhibitory Smads, Smad6 and Smad7, are differentially regulated by TPA in human lung fibroblast cells. Biochem Biophys Res Commun 2004;316:712–719.
46. Kim B-C, Lee H-J, Park SH, et al. Jab1/CSN5, a component of the COP9 signalosome, regulates transforming growth factor β signaling by binding to Smad7 and promoting its degradation. Mol Cell Biol 2004;24:2251–2262.
47. Koinuma D, Shinozaki M, Komuro A, et al. Arkadia amplifies TGF-β superfamily signalling through degradation of Smad7. EMBO J 2003;22:6458–6470.
48. Gronroos E, Hellman U, Heldin C-H, Ericsson J. Control of Smad7 stability by competition between acetylation and ubiquitination. Mol Cell 2002;10:483–493.
49. Simonsson M, Heldin C-H, Ericsson J, Gronroos E. The balance between acetylation and deacetylation controls Smad7 stability. J Biol Chem 2005;280:21,797–21,803.
50. Pulaski L, Landström M, Heldin C-H, Souchelnytskyi S. Phosphorylation of Smad7 at Ser-249 does not interfere with its inhibitory role in transforming growth factor-β-dependent signaling but affects Smad7-dependent transcriptional activation. J Biol Chem 2001;276:14,344–14,349.
51. Itoh F, Asao H, Sugamura K, Heldin C-H, ten Dijke P, Itoh S. Promoting bone morphogenetic protein signaling through negative regulation of inhibitory Smads. EMBO J 2001;20:4132–4142.
52. Suzuki C, Murakami G, Fukuchi M, et al. Smurf1 regulates the inhibitory activity of Smad7 by targeting Smad7 to the plasma membrane. J Biol Chem 2002;277:39,919–39,925.
53. Tajima Y, Goto K, Yoshida M, et al. Chromosomal region maintenance 1 (CRM1)-dependent nuclear export of Smad ubiquitin regulatory factor 1 (Smurf1) is essential for negative regulation of transforming growth factor-β signaling by Smad7. J Biol Chem 2003;278:10,716–10,721.
54. Zhu HJ, Iaria J, Sizeland AM. Smad7 differentially regulates transforming growth factor β-mediated signaling pathways. J Biol Chem 1999;274:32,258–32,264.
55. Di Guglielmo GM, Le Roy C, Goodfellow AF, Wrana JL. Distinct endocytic pathways regulate TGF-β receptor signalling and turnover. Nat Cell Biol 2003;5:410–421.
56. Itoh F, Divecha N, Brocks L, et al. The FYVE domain in Smad anchor for receptor activation (SARA) is sufficient for localization of SARA in early endosomes and regulates TGF-β/Smad signalling. Genes Cells 2002;7:321–331.
57. Penheiter SG, Mitchell H, Garamszegi N, Edens M, Doré Jr JJE, Leof EB. Internalization-dependent and -independent requirements for transforming growth factor β receptor signaling via the Smad pathway. Mol Cell Biol 2002;22:4750–4759.
58. Mitchell H, Choudhury A, Pagano RE, Leof EB. Ligand-dependent and -independent transforming growth factor-β receptor recycling regulated by clathrin-mediated endocytosis and Rab11. Mol Biol Cell 2004;15:4166–4178.

59. Souchelnytskyi S, Nakayama T, Nakao A, et al. Physical and functional interaction of murine and Xenopus Smad7 with bone morphogenetic protein receptors and transforming growth factor-β receptors. J Biol Chem 1998;273:25,364–25,370.
60. Datta PK, Chytil A, Gorska AE, Moses HL. Identification of STRAP, a novel WD domain protein in transforming growth factor-β signaling. J Biol Chem 1998;273:34,671–34,674.
61. Datta PK, Moses HL. STRAP and Smad7 synergize in the inhibition of transforming growth factor β signaling. Mol Cell Biol 2000;20:3157–3167.
62. Ferrigno O, Lallemand F, Verrecchia F, et al. Yes-associated protein (YAP65) interacts with Smad7 and potentiates its inhibitory activity against TGF-β/Smad signaling. Oncogene 2002;21: 4879–4884.
63. Yoshida Y, von Bubnoff A, Ikematsu N, et al. Tob proteins enhance inhibitory Smad-receptor interactions to repress BMP signaling. Mech Dev 2003;120:629–637.
64. Izzi L, Attisano L. Regulation of the TGFβ signalling pathway by ubiquitin-mediated degradation. Oncogene 2004;23:2071–2078.
65. Ogunjimi AA, Briant DJ, Pece-Barbara N, et al. Regulation of Smurf2 ubiquitin ligase activity by anchoring the E2 to the HECT domain. Mol Cell 2005;19:297–308.
66. Wicks SJ, Haros K, Maillard M, et al. The deubiquitinating enzyme UCH37 interacts with Smads and regulates TGF-β signalling. Oncogene 2005;24:8080–8084.
67. Shi X, Yang X, Chen D, Chang Z, Cao X. Smad1 interacts with homeobox DNA-binding proteins in bone morphogenetic protein signaling. J Biol Chem 1999;274:13,711–13,717.
68. Yang X, Ji X, Shi X, Cao X. Smad1 domains interacting with Hoxc-8 induce osteoblast differentiation. J Biol Chem 2000;275:1065–1072.
69. Bai S, Shi X, Yang X, Cao X. Smad6 as a transcriptional corepressor. J Biol Chem 2000;275: 8267–8270.
70. Bai S, Cao X. A nuclear antagonistic mechanism of inhibitory Smads in transforming growth factor-β signaling. J Biol Chem 2002;277:4176–4182.
71. Ichijo T, Voutetakis A, Cotrim AP, et al. The SMAD6/histone deacetylase 3 complex silences the transcriptional activity of the glucocorticoid receptor: potential clinical implications. J Biol Chem 2005;280:42,067–42,077.
72. Shi W, Sun C, He B, et al. GADD34-PP1c recruited by Smad7 dephosphorylates TGFβ type I receptor. J Cell Biol 2004;164:291–300.
73. Kimura N, Matsuo R, Shibuya H, Nakashima K, Taga T. BMP2-induced apoptosis is mediated by activation of the TAK1-p38 kinase pathway that is negatively regulated by Smad6. J Biol Chem 2000; 275:17,647–17,652.
74. Yanagisawa M, Nakashima K, Takeda K, et al. Inhibition of BMP2-induced, TAK1 kinase-mediated neurite outgrowth by Smad6 and Smad7. Genes Cells 2001;6:1091–1099.
75. Landström M, Heldin N-E, Bu S, et al. Smad7 mediates apoptosis induced by transforming growth factor β in prostatic carcinoma cells. Curr Biol 2000;10:535–538.
76. Edlund S, Bu S, Schuster N, Aspenström P, et al. Transforming growth factor-β1 (TGF-β)-induced apoptosis of prostate cancer cells involves Smad7-dependent activation of p38 by TGF-β-activated kinase 1 and mitogen-activated protein kinase kinase 3. Mol Biol Cell 2003;14:529–544.
77. Edlund S, Lee SY, Grimsby S, et al. Interaction between Smad7 and β-catenin: importance for transforming growth factor β-induced apoptosis. Mol Cell Biol 2005;25:1475–1488.
78. Lallemand F, Mazars A, Prunier C, et al. Smad7 inhibits the survival nuclear factor κB and potentiates apoptosis in epithelial cells. Oncogene. 2001;20:879–884.
79. Schiffer M, Bitzer M, Roberts IS, et al. Apoptosis in podocytes induced by TGF-β and Smad7. J Clin Invest 2001;108:807–816.
80. Edlund S, Landström M, Heldin C-H, Aspenström P. Transforming growth factor-β-induced mobilization of actin cytoskeleton requires signaling by small GTPases Cdc42 and RhoA. Mol Biol Cell 2002;13:902–914.
81. Edlund S, Landström M, Heldin C-H, Aspenström P. Smad7 is required for TGF-β-induced activation of the small GTPase Cdc42. J Cell Sci 2004;117:1835–1847.
82. Choy L, Skillington J, Derynck R. Roles of autocrine TGF-β receptor and Smad signaling in adipocyte differentiation. J Cell Biol 2000;149:667–682.
83. Berg DT, Myers LJ, Richardson MA, Sandusky G, Grinnell BW. Smad6s regulates plasminogen activator inhibitor-1 through a protein kinase C-beta-dependent up-regulation of transforming growth factor-β. J Biol Chem 2005;280:14,943–14,947.

84. Siegel PM, Massagué J. Cytostatic and apoptotic actions of TGF-β in homeostasis and cancer. Nat Rev Cancer 2003;3:807–820.
85. Derynck R, Akhurst RJ, Balmain A. TGF-β signaling in tumor suppression and cancer progression. Nat Genet 2001;29:117–129.
86. Boulay JL, Mild G, Reuter J, et al. Combined copy status of 18q21 genes in colorectal cancer shows frequent retention of SMAD7. Genes Chromosomes Cancer 2001;31:240–247.
87. Venter C, Adams MD, Myers EW, et al. The Sequence of the Human Genome. Science 2001;291: 1304–1345.
88. Kleeff J, Ishiwata T, Maruyama H, et al. The TGF-β signaling inhibitor Smad7 enhances tumorigenicity in pancreatic cancer. Oncogene 1999;18:5363–5372.
89. Kleeff J, Maruyama H, Friess H, Büchler MW, Falb D, Korc M. Smad6 suppresses TGF-β-induced growth inhibition in COLO-357 pancreatic cancer cells and is overexpressed in pancreatic cancer. Biochem Biophys Res Commun 1999;255:268–273.
90. Dowdy SC, Mariani A, Reinholz MM, et al. Overexpression of the TGF-β antagonist Smad7 in endometrial cancer. Gynecol Oncol 2005;96:368–373.
91. Wang D, Kanuma T, Mizumuma H, Ibuki Y, Takenoshita S. Mutation analysis of the Smad6 and Smad7 gene in human ovarian cancers. Int J Oncol 2000;17:1087–1091.
92. Kawate S, Ohwada S, Hamada K, et al. Mutational analysis of the Smad6 and Smad7 genes in hepatocellular carcinoma. Int J Mol Med 2001;8:49–52.
93. Fukushima T, Mashiko M, Takita K, et al. Mutational analysis of TGF-β type II receptor, Smad2, Smad3, Smad4, Smad6 and Smad7 genes in colorectal cancer. J Exp Clin Cancer Res 2003;22: 315–320.
94. Roth S, Sistonen P, Salovaara R, et al. SMAD genes in juvenile polyposis. Genes Chromosomes Cancer 1999;26:54–61.
95. He W, Cao T, Smith DA, Myers TE, Wang XJ. Smads mediate signaling of the TGFβ superfamily in normal keratinocytes but are lost during skin chemical carcinogenesis. Oncogene 2001;20:471–483.
96. Liu X, Lee J, Cooley M, Bhogte E, Hartley S, Glick A. Smad7 but not Smad6 cooperates with oncogenic ras to cause malignant conversion in a mouse model for squamous cell carcinoma. Cancer Res 2003; 63:7760–7768.
97. Halder SK, Beauchamp RD, Datta PK. Smad7 induces tumorigenicity by blocking TGF-β-induced growth inhibition and apoptosis. Exp Cell Res 2005;307:231–246.
98. Javelaud D, Delmas V, Möller M, et al. Stable overexpression of Smad7 in human melanoma cells inhibits their tumorigenicity in vitro and in vivo. Oncogene 2005;24:7624–7629.
99. Azuma H, Ehata S, Miyazaki H, et al. Effect of Smad7 expression on metastasis of mouse mammary carcinoma JygMC(A) cells. J Natl Cancer Inst 2005;97:1734–1746.
100. LeRoy EC. A brief overview of the pathogenesis of scleroderma (systemic sclerosis). Ann Rheum Dis 1992;51:286–288.
101. Ihn H, Yamane K, Kubo M, Tamaki K. Blockade of endogenous transforming growth factor β signaling prevents up-regulated collagen synthesis in scleroderma fibroblasts: association with increased expression of transforming growth factor β receptors. Arthritis Rheum 2001;44:474–480.
102. Rudnicka L, Varga J, Christiano AM, Iozzo RV, Jimenez SA, Uitto J. Elevated expression of type VII collagen in the skin of patients with systemic sclerosis. Regulation by transforming growth factor-β. J Clin Invest 1994;93:1709–1715.
103. Dong C, Zhu S, Wang T, et al. Deficient Smad7 expression: a putative molecular defect in scleroderma. Proc Natl Acad Sci USA 2002;99:3908–3913.
104. Asano Y, Ihn H, Yamane K, Kubo M, Tamaki K. Impaired Smad7-Smurf-mediated negative regulation of TGF-beta signaling in scleroderma fibroblasts. J Clin Invest 2004;113:253–264.
105. Fukasawa H, Yamamoto T, Togawa A, et al. Down-regulation of Smad7 expression by ubiquitin-dependent degradation contributes to renal fibrosis in obstructive nephropathy in mice. Proc Natl Acad Sci USA 2004;101:8687–8692.
106. Lan HY, Mu W, Tomita N, et al. Inhibition of renal fibrosis by gene transfer of inducible Smad7 using ultrasound-microbubble system in rat UUO model. J Am Soc Nephrol 2003;14:1535–1548.
107. Terada Y, Hanada S, Nakao A, Kuwahara M, Sasaki S, Marumo F. Gene transfer of Smad7 using electroporation of adenovirus prevents renal fibrosis in post-obstructed kidney. Kidney Int 2002;61: 94–98.
108. Nakao A, Fujii M, Matsumura R, et al. Transient gene transfer and expression of Smad7 prevents bleomycin-induced lung fibrosis in mice. J Clin Invest 1999;104:5–11.

109. Dooley S, Hamzavi J, Breitkopf K, et al. Smad7 prevents activation of hepatic stellate cells and liver fibrosis in rats. Gastroenterology 2003;125:178–191.
110. Wahl SM. Transforming growth factor β: the good, the bad, and the ugly. J Exp Med 1994;180: 1587–1590.
111. Kitani A, Fuss IJ, Nakamura K, Schwartz OM, Usui T, Strober W. Treatment of experimental (trinitrobenzene sulfonic acid) Ccolitis by intranasal administration of transforming growth factor (TGF)-β1 plasmid: TGF-β1–mediated suppression of T helper cell type 1 response occurs by interleukin (IL)-10 induction and IL-12 receptor β2 chain downregulation. J Exp Med 2000;192:41–52.
112. Hahm KB, Im YH, Parks TW, et al. Loss of transforming growth factor β signalling in the intestine contributes to tissue injury in inflammatory bowel disease. Gut 2001;49:190–198.
113. Monteleone G, Kumberova A, Croft NM, McKenzie C, Steer HW, MacDonald TT. Blocking Smad7 restores TGF-β1 signaling in chronic inflammatory bowel disease. J Clin Invest 200;108:601–609.
114. Galvin KM, Donovan MJ, Lynch CA, et al. A role for smad6 in development and homeostasis of the cardiovascular system. Nat Genet 2000;24:171–174.
115. He W, Li AG, Wang D, et al. Overexpression of Smad7 results in severe pathological alterations in multiple epithelial tissues. EMBO J 2002;21:2580–2590.
116. Nakao A, Miike S, Hatano M, et al. Blockade of transforming growth factor β/Smad signaling in T cells by overexpression of Smad7 enhances antigen-induced airway inflammation and airway reactivity. J Exp Med 2000;192:151–158.
117. Horiki M, Imamura T, Okamoto M, et al. Smad6/Smurf1 overexpression in cartilage delays chondrocyte hypertrophy and causes dwarfism with osteopenia. J Cell Biol 2004;165:433–445.
118. Itoh S, Thorikay M, Kowanetz M, et al. Elucidation of Smad requirement in transforming growth factor-β type I receptor-induced responses. J Biol Chem 2003;278:3751–3761.

42 TGF-β-Related Antifibrotic Strategies in the Eye

M. Francesca Cordeiro, Annelie Maass, Clive Migdal and Li Guo

CONTENTS

BACKGROUND
TGF-β AND FIBROSIS
TGF-β IN OCULAR FIBROSIS
TGF-β IN FIBROSIS ASSOCIATED WITH OCULAR SURGERY
TGF-β ANTIFIBROTIC STRATEGIES
CONCLUSIONS
REFERENCES

Abstract

The wound-healing response is involved in many diseases throughout the body, including the eye. It is a key factor in influencing results of surgery, particularly that occurring in the treatment of the blinding disease glaucoma, where the postoperative fibrotic response is the major determinant of outcome. Transforming growth factor-β (TGF-β) has been widely established as a target for post-operative antifibrotic therapy in glaucoma. Strategies against TGF-β have included antibodies, antisense phosphorothiate oligonucleotides, and naturally occurring antagonists. These have either reached preclinical or clinical trials in the eye, but offer potential widespread applications anywhere in the body where the wound-healing response requires modulation.

Key Words: TGF-β; ocular scarring; glaucoma; antifibrotic; antisense oligonucleotides; neutralizing antibodies; naturally occurring antagonists; siRNA.

1. BACKGROUND

Transforming growth factor-beta (TGF-β) is a multifunctional growth factor. It belongs to a large superfamily of polypeptide molecules that are highly conserved across and between species *(1)*.

There are three different mammalian isoforms of TGF-β—TGF-β1, TGF-β2, and TGF-β3, with different sites of gene expression. TGF-β1 mRNA is predominantly expressed in endothelial, hematopoietic, and connective tissue cells; TGF-β2 in epithelial and CNS cells; and TGF-β3 primarily in mesenchymal cells.

Most cell types simultaneously express the three main TGF-β receptors, Types I, II, and III (TBR-I, TBR-II, and TBR-III, respectively), although each is functionally very different

From: *Cancer Drug Discovery and Development: Transforming Growth Factor-β in Cancer Therapy, Vol I: Basic and Clinical Biology*
Edited by: Sonia B. Jakowlew © Humana Press Inc., Totowa, NJ

(2). TBR-I and TBR-II have been shown to be very important in signal transduction, unlike the Type III receptor betaglycan, which has an important function: it binds strongly to the ligand to form a stable ternary complex with TGF-β and TBR-II *(3)*. The biological significance of the Type IV receptor is still to be determined. Endoglin, another TGF-β receptor that is abundant in endothelial cells, is mutated in patients with hereditary hemorrhagic telangiectasia *(4)*.

Signaling pathways involve the Type I and II TGF-β receptors and Smads (derived from the Sma and MAD gene homologues in *Caenorhabditis elegans* and *Drosophila melanogaster*) with intracellular signaling specifically initiated by phosphorylation of Smad2 or Smad3, which then bind to Smad4. The resulting Smad complex then moves into the nucleus, where it interacts in a cell-specific manner with various transcription factors to regulate the transcription of many genes. Smad6 and Smad7 block the phosphorylation of Smad2 or Smad3, thus inhibiting TGF-β signaling *(5–7)*.

2. TGF-β AND FIBROSIS

TGF-β is intrinsically involved in wound healing, especially in the formation of scar tissue *(8–10)*. It has been demonstrated as affecting the scarring response not only in skin and skeletal processes but also throughout the body *(11)*.

TGF-β1 and -β2 are potent stimulators of the dermal scarring response *(8,9,12–15)*, and their activity has been shown to be blocked by the application of antibodies to adult cutaneous wounds *(16)*. The role of TGF-β3 in dermal wound healing, however, remains unclear *(9,17)*. Analysis of healing-impaired models including radiation-exposed, antiproliferative-agent, and glucocorticoid-treated animals have confirmed the stimulatory role of TGF-β *(18–21)*, with a suggestion that estrogen causes accelerated wound healing in females via its induction of dermal fibroblast production of TGF-β1 *(22)*. Abnormal fibrosis, as seen with keloids and hypertrophic scars characterized by excess accumulation of collagen within the wound, is also believed to be related to TGF-β *(23)*.

TGF-β is implicated in other pathological scarring conditions such as systemic sclerosis *(15,24)*, pulmonary *(19,25)*, renal *(26)*, and hepatic *(27)* scarring diseases. There is also increasing evidence of its association with vascular disease, hypertension, and atherosclerosis *(28–30)*.

3. TGF-β IN OCULAR FIBROSIS

Of the three isoforms, TGF-β2 is believed to be the predominant ocular isoform, as measured in vitreous and aqueous humor *(31,32)* and tears *(33)*. It is implicated in pathological and physiological conditions involving different ocular anatomical structures such as the cornea, lens, retina, and optic nerve head (ONH).

All three TGF-β isoforms *(34)* and TBR-I and -II receptors have been identified in the cornea *(35–37)*. TGF-β1 is implicated in the wound healing occurring after excimer laser photorefractive keratectomy *(38,39)*, and topical administration of an anti-TGF-β1/-β2/-β3 antibody *(40,41)* has been suggested as a strategy to reduce the induced haze *(42)*.

TGF-β has also been shown to be important in posterior capsular opacification (PCO), or secondary cataract *(43–46)*, and its modulation suggested as a treatment for inhibiting PCO formation *(45,47,48)*. TGF-β2 is also implicated as the most important isoform in cataract or lens pathology *(44,49–52)*. High levels of TGF-β2 in vitreous aspirates has been found to correlate with the degree of intraocular fibrosis and proliferative vitreoretinopathy (PVR) *(32,53)*, and anti-TGF-β antibodies have been found to inhibit in vitro models of PVR *(54)*. Conversely, exogenous TGF-β2 application has been advocated as a potential treatment strategy for the repair of macular holes *(55,56)* and retinal tears *(57)*. There is an association of increased levels of TGF-β2 with retinal laser photocoagulation

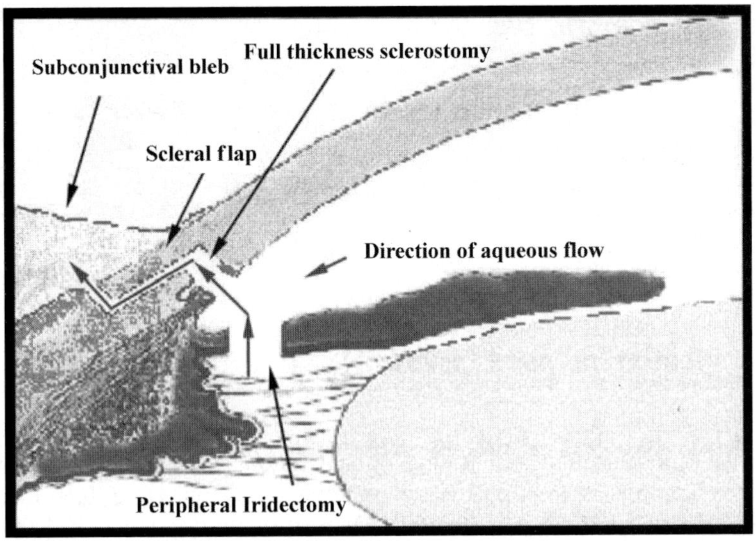

Fig. 1. The major determinant of glaucoma surgical success is the subconjunctival fibrosis response and the degree of scarring. The surgical procedure relies on the creation of a fistula allowing aqueous humor to flow from the anterior chamber of the eye, through a sclerotomy, and into a subconjunctival bleb (red-arrowed pathway), where it may be absorbed back into the circulatory system via aqueous veins, lymphatic channels, transconjunctivally, or via the suprachoroidal space.

(58–61), with TGF-β having been shown to be stimulated in diabetic retinopathy. Recently, using mice with a targeted deletion of Smad3, Roberts et al. *(62)* have suggested that Smad3 inhibition may have clinical applications in the prevention of TGF-β-associated retinal fibrosis.

TGF-β2 has been found to be elevated in the aqueous of patients suffering from the blinding disease glaucoma *(31,63,64)*. This has been suggested to be linked to the pathogenesis of the disease, which in part is attributed to the obstruction of aqueous humor outflow through the trabecular meshwork *(65)*. This is supported by recent findings in human trabecular meshwork cells *(66,67)*. Another mechanism in glaucoma in which TGF-β is implicated is in the process of extracellular matrix remodeling at the ONH *(68)*. TGF-β2 expression appears increased at the ONH in glaucomatous human and animal model eyes *(67,69–75)*, a finding which we have recently confirmed *(75)*.

4. TGF-β IN FIBROSIS ASSOCIATED WITH OCULAR SURGERY

TGF-β has also been shown to be key in the events associated with the wound-healing response following ocular surgery. This is particularly the case following glaucoma surgery *(68,76–80)*. Glaucoma is a disease that is the major cause of irreversible blindness in both the developed and the developing world, accounting for 15% of all blindness and over 500,000 new cases each year *(81–83)*. Compared to medical and laser treatments, glaucoma filtration surgery has been shown to be the most effective *(79,80)*. However, conjunctival fibrosis and the degree of scarring are major determinants of glaucoma surgical success (Fig. 1) *(68,78)*.

All three TGF-β isoforms have been shown to be expressed in the conjunctiva, although TGF-β2 is predominant *(84)*. We have previously demonstrated a spatial and temporal relationship between all three TGF-β isoforms in a model of conjunctival scarring, with an early peak in TGF-β2 mRNA expression followed by the protein *(85)*.

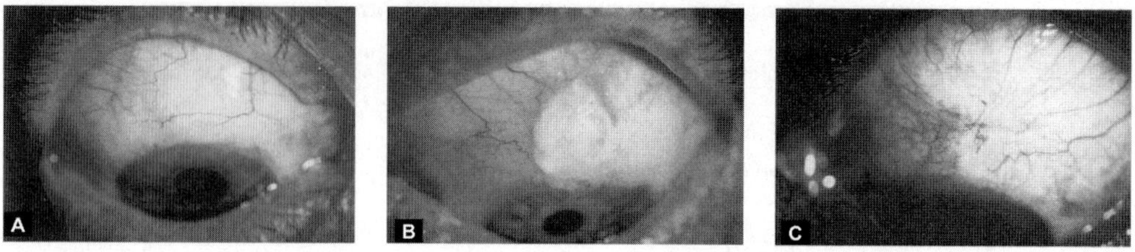

Fig. 2. A successful glaucoma surgical filtration procedure produces a visible, elevated conjunctival bleb (**A**). However, this may not always be possible even after the application of antimetabolites such as mitomycin **C**, which unfortunately can be associated with thin and cystic blebs at the risk of leakage and infection (**B**). The problem for glaucoma specialists therefore is how best to avoid conjunctival fibrosis associated with flat, scarred blebs and surgery failure (**C**).

Conjunctival fibrosis is mediated through human Tenon's fibroblast (HTF) cells. We have shown all three isoforms of TGF-β to stimulate in vitro HTF activity *(10,86)*, in a biphasic, concentration-dependent manner, with different peak activities associated with different fibroblast functions *(10)*. We have confirmed this finding in vivo, showing that all three isoforms, when applied exogenously and at the same concentration, produce a similar conjunctival scarring response *(87)*.

5. TGF-β ANTIFIBROTIC STRATEGIES

As TGF-β2 appears to be such an important component of conjunctival scarring, it is not surprising that its activity has made it an increasingly utilized target to improve glaucoma filtration surgery results (Fig. 2) *(76,88,89)*. Currently, mitomycin-C and 5-fluorouracil are antimetabolites used to reduce conjunctival scarring following glaucoma surgery, with application within the first week after surgery having been shown to improve outcome months to years later *(90–93)*. However, both treatments result in the production of thin avascular blebs, which can leak *(94,95)* and lead to the development of blinding infections *(96)* and hypotony *(97,98)*. The complications associated with their use can be partly attributed to the non-selective manner in which these agents work, causing widespread cellular destruction *(99)*.

A few years ago we began investigating more "physiological" ways of modulating the conjunctival scarring response with TGF-β as the target. We have looked at TGF-β2 antibodies and antisense oligonucleotides (OGN) *(71,76,89,100)*, but more recently other authors have assessed TGF-β-related strategies such as RNA interference molecules *(101)* and naturally occurring antagonists (e.g., tranilast and decorin) (Fig. 3) *(77,102)*. These antifibrotic strategies in glaucoma filtration surgery are described in more detail below.

5.1. TGF-β2-Neutralizing Antibodies

We previously investigated a humanized, monoclonal antibody to TGF-β2. In an in vitro HTF model, this antibody significantly inhibited TGF-β2-induced fibroblast proliferation, migration, and contraction *(71)*. We next investigated this molecule in an animal model of aggressive conjunctival scarring, where we showed it to significantly improve glaucoma filtration surgery outcome *(71)*. These successful preclinical studies were the basis of clinical trials. However, although a Phase I clinical trial of this antibody in glaucoma filtration surgery was very encouraging *(100)*; the European Phase II/III and international Phase III clinical trial results were not good, with no significant benefit demonstrated compared to control, and further development of this antibody as a glaucoma surgery modulator was curtailed in March 2005.

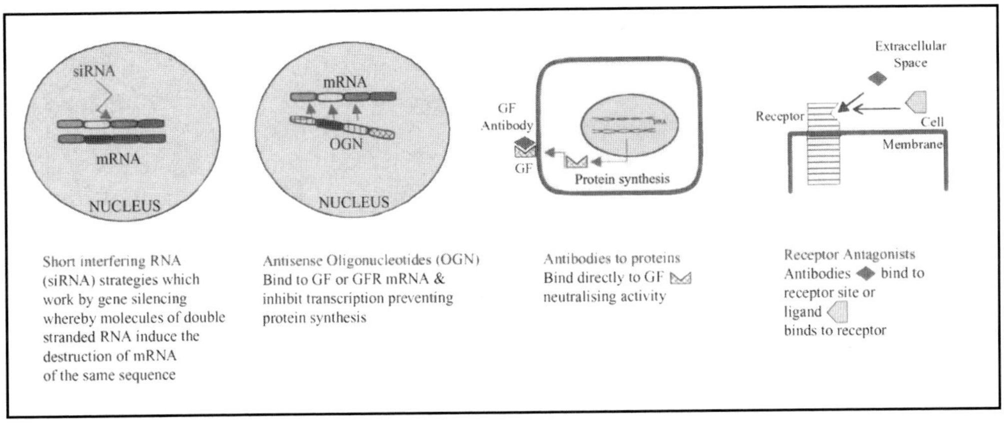

Fig. 3. Antifibrotic strategies involving TGF-β as the target have been investigated experimentally (and in some cases clinically) for the modulation of the conjunctival scarring response. This has been done at various levels, including at the mRNA level with siRNA and antisense oligonucleotides, at the protein level with neutralizing antibodies and natural antagonists.

5.2. TGF-β2 Antisense Oligonucleotides

We have also investigated the use of antisense OGN to inhibit gene expression of TGF-β, using an antisense OGN to TGF-β1 and -β2 (ISIS Pharmaceuticals, Carlsbad, CA) *(103)*. This was the first reported study of novel second-generation antisense phosphorothiate OGN against TGF-β in vivo.

Antisense OGN are synthetic molecules that bind to specific intracellular messenger RNA strands (mRNA), normally consisting of short DNA/RNA sequences (around 20 mer) *(104–106)*. They have been recognized as a promising therapeutic strategy in oncology, inflammatory and viral infective diseases *(107–110)*. Several possible mechanisms by which they produce decreased protein expression include the modulation of protein translation by disrupting ribosome assembly, RNase H mediated cleavage of targeted mRNA, and pretranslational modification of splicing *(107,111)*.

Although antisense molecules were described as long ago as 1978 *(112)*, there have been problems associated with their delivery into target cells or the cell nucleus, despite advances in viral and nonviral gene delivery systems.

Adjustment of their structures has led to improvements in pharmacokinetics and pharmacodynamics, with much better results obtained with the second-generation phosphorothioate OGN with nucleotide modification. The ISIS TGF-β OGNs we investigated contained both a phosphorothioate backbone and a 2′-methoxyethyl sugar modification to increase nuclease stability and antisense potency *(113,114)*.

In comparison to a TGF-β Type II receptor and TGF-β1, we demonstrated the TGF-β2 antisense OGN was more potent in inhibiting the conjunctival wound-healing response. We showed the TGF-β2 OGN to significantly reduce postoperative scarring ($p < 0.05$) and improve surgical outcome, in two different animal models of glaucoma filtration surgery. One of the most striking findings from the study was the long-lasting effects of the TGF-β OGN after only a single administration at the time of surgery. This was to be compared to our previous study using the TGF-β2-neutralizing antibody in the same model of aggressive conjunctival scarring, where the regimen consisted of repeated subconjunctival injections (5 applications in 7 days from the time of surgery) *(71)*.

5.3. TGF-β RNA Interference (RNAi) Strategies

A recent development in gene modulation is that of gene silencing whereby molecules of double-stranded RNA induce the destruction of mRNAs of the same sequence. Like antisense OGN technology, a major problem in achieving in vivo gene silencing by RNAi technologies has been delivery. However, chemical modifications of short interfering RNAs (siRNAs) have been shown recently to be successful, providing great potential in the treatment of diseases *(115,116)*.

A recent study by Nakamura et al. *(101)* used an siRNA to the TGF-β Type II receptor in models of ocular scarring both in vitro and in vivo. They demonstrated that the TGF-β siRNAs abrogated the Type II receptor transcript and protein expression in cultured corneal fibroblasts, with effects on fibronectin assembly and cell migration. They also showed that in an in vivo mouse model of conjunctival scarring, the siRNA significantly reduced the degree of fibrosis.

5.4. Naturally Occurring TGF-β Antagonists

Decorin is a naturally occurring proteoglycan *(117)*, which has been shown to bind to TGF-β and inhibit its activity. Using a model of glaucoma filtration surgery in rabbits, Grisanti et al. *(77)* have recently shown that perioperative subconjunctival decorin applications significantly reduced conjunctival scarring and improved surgical outcome.

In this group we include tranilast (*N*–(3,4-dimethoxycinnamoyl) anthranilic acid) because it is an antiallergic molecule that is known to produce an antiscarring effect by suppressing TGF-β activity. It has been shown in vitro to inhibit HTF proliferation and collagen synthesis, and a recent prospective randomized controlled trial of postoperative tranilast in trabeculectomy patients showed some benefit in its use, with efficacy still apparent at 2-yr follow-up *(118)*.

6. CONCLUSIONS

The identification of TGF-β as an important component of conjunctival scarring has made it a viable target for modulating the scarring response following glaucoma filtration surgery. Advances in molecular and cell therapy continue to offer the potential of more specific, safer, focal, and titratable treatment, with far-reaching clinical applications not only in the eye, but throughout the body.

REFERENCES

1. Roberts A, Sporn M. eds. The transforming growth factor-betas. In: Handbook of Experimental Pharmacology Number 1. Peptide Growth Factors and Their Receptors. Berlin: Springer-Verlag 1990; pp. 419–472.
2. Le Magueresse-Battistoni B, Morera A, Goddard I, Benahmed M. Expression of mRNA's for transforming growth factor-beta receptors in the rat testis. Endocrinology 1995;136:2788–2791.
3. Attisano L, Wrana J, Lopez-Casillas F, Massagué J. TGF-beta receptors and actions. Biochimica Biophysica Acta 1994;1222:71–80.
4. Blobe G, Schiemann W, Lodish H. Role of transforming growth factor β in human disease. New England Journal of Medicine 2000;342:1350–1358.
5. Nakao A, Imamura T, Souchelnytskyi S, et al. TGF-β receptor-mediated signalling through Smad2, Smad3 and Smad4. EMBO J 1997;16:5353–5362.
6. Piek E, Heldin C-H, ten Dijke P. Specificity, diversity, and regulation in TGF-β superfamily signaling. FASEB J 1999;13:2105–2124.
7. Massagué J, Chen Y. Controlling TGF-beta signaling. Genes Dev 2000;14:627–644.
8. Roberts AB, Sporn MB, Assoian RK, Smith JM, Roche NS. Transforming growth factor beta: Rapid induction of fibrosis and angiogenesis in vivo and stimulation of collagen formation in vitro. Proc Natl Acad Sci USA 1986;83:4167–4171.

9. Shah M, Foreman DM, Ferguson MW. Neutralisation of TGF-beta 1 and TGF-beta 2 or exogenous addition of TGF-beta 3 to cutaneous rat wounds reduces scarring. J Cell Sci 1995;108:985–1002.
10. Cordeiro MF, Bhattacharya SS, Schultz GS, Khaw PT. TGF-β1, -β2 & -β3 in vitro: Biphasic effects on Tenon's fibroblast contraction, proliferation & migration. Invest Ophthalmol Vis Sci 2000;41:756–763.
11. Critchlow MA, Bland YS, Ashhurst DE. The effect of exogenous transforming growth factor-beta 2 on healing fractures in the rabbit. Bone 1995;16:521–527.
12. Merwin JR, Roberts AB, Kondaiah P, Tucker A, Madri J. Vascular cell responses to TGF-β3 mimic those of TGF-β1 in vitro. Growth Factors 1991;5:149–158.
13. Levine JH, Moses HL, Gold LI, Nanney LB. Spatial and temporal patterns of immunoreactive transforming growth factor-beta-1, -beta-2 and -beta-3 during excisional wound repair. Am J Pathol 1993; 143:368–380.
14. Longaker MT, Bouhana KS, Harrison MR, Danielpour D, Roberts AB, Banda MJ. Wound healing in the fetus: Possible role for inflammatory macrophages and transforming growth factor-beta isoforms. Wound Rep Reg 1994;2:104–112.
15. Gruschwitz M, Muller PU, Sepp N, Hofer E, Fontana A, Wick G. Transcription and expression of transforming growth factor type beta in the skin of progressive systemic sclerosis: A mediator of fibrosis? J Invest Dermatol 1990;94:197–203.
16. Shah M, Foreman DM, Ferguson MW. Neutralising antibody to TGF-beta 1,2 reduces cutaneous scarring in adult rodents. J Cell Sci 1994;107:1137–1157.
17. Cox DA. Transforming growth factor-beta 3. Cell Biol Int 1995;19:357–371.
18. Wicke C, Halliday B, Allen D, et al. Effects of steroids and retinoids on wound healing. Arch Surg 2000;135:1265–1270.
19. Shukla A, Meisler N, Cutroneo K. Perspective article: transforming growth factor-beta: crossroad of glucocorticoid and bleomycin regulation of collagen synthesis in lung fibroblasts. Wound Repair Regen 1999;7:133–140.
20. Salomon G, Kasid A, Bernstein E, Buresh C, Director E, Norton J. Gene expression in normal and doxorubin impaired wounds: Importance of transforming growth factor-beta. Surgery 1990;108: 318–323.
21. Beck L, DeGuzman L, Lee W, Xu Y, Siegel M, Amento E. One systemic administration of transforming growth factor-beta 1 reverses age- or glucocorticoid-impaired wound healing. J Clin Invest 1993;92:2841–2849.
22. Ashcroft GS, Dodsworth J, van-Boxtel E, et al. Estrogen accelerates cutaneous wound healing associated with an increase in TGF-beta1 levels. Nat Med 1997;3:1209–1215.
23. Lee T, Chin G, Kim W, Chau D, Gittes G, Longaker M. Expression of transforming growth factor beta 1, 2, and 3 proteins in keloids. Ann Plast Surg 1999;43:179–184.
24. Cotton S, Herrick A, Jaysonm M, Freemont A. TGF beta – a role in systemic sclerosis? J Pathol 1998;184:4–6.
25. Sime P, O'Reilly K. Fibrosis of the lung and other tissues: new concepts in pathogenesis and treatment. Clin Immunol 2001;99:308–319.
26. Basile D. The transforming growth factor beta system in kidney disease and repair: recent progress and future directions. Curr Opin Nephrol Hypertens 1999;8:21–30.
27. Sanderson N, Factor V, Nagy P, et al. Hepatic expression of mature transforming growth factor beta1 in transgenic mice results in multiple tissue lesions. Proc Natl Acad Sci USA 1995;92: 2572–2576.
28. McCaffrey T. TGF-betas and TGF-beta receptors in atherosclerosis. Cytokine Growth Factor Rev 2000;11:103–114.
29. August P, Leventhal B, Suthanthiran M. Hypertension-induced organ damage in African Americans: transforming growth factor-beta(1) excess as a mechanism for increased prevalence. Curr Hypertens Rep 2000;2:184–191.
30. Suthanthiran M, Li B, Song J, et al. Transforming growth factor-beta 1 hyperexpression in African-American hypertensives: A novel mediator of hypertension and/or target organ damage. Proc Natl Acad Sci USA 2000;97:3479–3484.
31. Jampel HD, Roche N, Stark WJ, Roberts AB. Transforming growth factor-beta in human aqueous humor. Curr Eye Res 1990;9:963–969.
32. Connor TB, Roberts AB, Sporn MB, et al. Correlation of fibrosis and transforming growth factor-beta type 2 levels in the eye. J Clin Invest 1989;83:1661–1666.

33. Kokawa N, Sotozono C, Nishida K, Kinoshita S. High total TGF-β2 levels in normal human tears. Curr Eye Res 1996;15:341–342.
34. Imanishi J, Kamiyama K, Iguchi I, Kita M, Sotozono C, Kinoshita S. Growth factors: importance in wound healing and maintenance of transparency of the cornea. Prog Retin Eye Res 2000;19:113–129.
35. Nishida K, Sotozono C, Adachi W, Yamamoto S, Yokoi N, Kinoshita S. Transforming growth factor-beta 1, -beta 2 and -beta 3 mRNA expression in human cornea. Curr Eye Res 1995;14:235–241.
36. Zieske J, Hutcheon A, Guo X, Chung E, Joyce N. TGF-beta receptor types I and II are differentially expressed during corneal epithelial wound repair. Invest Ophthalmol Vis Sci 2001;7:1465–1471.
37. Hayashi K, Frangieh G, Wolf G, Kenyon K. Expression of transforming growth factor-beta in wound healing of vitamin A-deficient rat corneas. Invest Ophthalmol Vis Sci 1989;30:239–247.
38. Chen C, Michelini-Norris B, Stevens S, et al. Measurement of mRNAs for TGFss and extracellular matrix proteins in corneas of rats after PRK. Invest Ophthalmol Vis Sci 2000;41:4108–4116.
39. Blalock TD, Duncan MR, Varela JC, et al. Connective tissue growth factor expression and action in human corneal fibroblast cultures and rat corneas after photorefractive keratectomy. Invest Ophthalmol Vis Sci 2003;44:1879–1887.
40. Thom SB, Myers JS, Rapuano CJ, Eagle RC, Siepser SB, Gomes JA. Effect of topical anti-transforming growth factor-beta on corneal stromal haze after photorefractive keratectomy in rabbits. J Cataract Refract Surg 1997;23:1324–1330.
41. Myers JS, Gomes JA, Siepser SB, Rapuano CJ, Eagle RC, Thom SB. Effect of transforming growth factor beta 1 on stromal haze following excimer laser photorefractive keratectomy in rabbits. J Refract Surg 1997;13:356–361.
42. Jester JV, Barry Lane PA, Petroll WM, Olsen DR, Cavanagh HD. Inhibition of corneal fibrosis by topical application of blocking antibodies to TGF beta in the rabbit. Cornea 1997;16:177–187.
43. Saika S, Miyamoto T, Kawashima Y, et al. Immunolocalization of TGF-beta1, -beta2, and -beta3, and TGF-beta receptors in human lens capsules with lens implants. Graefes Arch Clin Exp Ophthalmol 2000;238:283–293.
44. Nishi O, Nishi K, Wada K, Ohmoto Y. Expression of transforming growth factor (TGF)-alpha, TGF-beta(2) and interleukin 8 messenger RNA in postsurgical and cultured lens epithelial cells obtained from patients with senile cataracts. Graefes Arch Clin Exp Ophthalmol 1999;237:806–811.
45. Lois N, Taylor J, McKinnon AD, Smith GC, van't Hof R, Forrester JV. Effect of TGF-beta2 and anti-TGF-beta2 antibody in a new in vivo rodent model of posterior capsule opacification. Invest Ophthalmol Vis Sci 2005;46:4260–4266.
46. Symonds JG, Lovicu FJ, Chamberlain CG. Posterior capsule opacification-like changes in rat lens explants cultured with TGFbeta and FGF: Effects of cell coverage and regional differences. Exp Eye Res 2005;13:13.
47. Meacock W, Spalton D, Stanford M. Role of cytokines in the pathogenesis of posterior capsule opacification. Br J Ophthalmol 2000;84:332–336.
48. Tobari I, Iwaki Y, Miyake K. Effect of tranilast eyedrops in preventing posterior capsule opacification: preliminary report. J Cataract Refract Surg 1999;25:1394–1399.
49. Lee E, Seomun Y, Hwang K, et al. Overexpression of the transforming growth factor-beta-inducible gene betaig-h3 in anterior polar cataracts. Invest Ophthalmol Vis Sci 2000;41:1840–1845.
50. Srinivasan Y, Lovicu F, Overbeek P. Lens-specific expression of transforming growth factor beta1 in transgenic mice causes anterior subcapsular cataracts. J Clin Invest 1998;101:625–634.
51. Ishida I, Saika S, Okada Y, Ohnishi Y. Growth factor deposition in anterior subcapsular cataract. J Cataract Refract Surg 2005;31:1219–1225.
52. Wormstone IM, Tamiya S, Eldred JA, et al. Characterisation of TGF-beta2 signalling and function in a human lens cell line. Exp Eye Res 2004;78:705–714.
53. Kon CH, Occleston NL, Aylward GW, Khaw PT. Expression of vitreous cytokines in proliferative vitreoretinopathy: a prospective study. Invest Ophthalmol Vis Sci 1999;40:705–712.
54. Carrington L, McLeod D, Boulton M. IL-10 and antibodies to TGF-beta2 and PDGF inhibit RPE-mediated retinal contraction. Invest Ophthalmol Vis Sci 2000;41:1210–1216.
55. Glaser BM, Michels RG, Kuppermann BD, Sjaarda RN, Pena RA. Transforming growth factor-β2 for the treatment of full-thickness macular holes. Ophthalmology 1992;99:1162–1173.
56. Kozy D, Maberley A. Closure of persistent macular holes with human recombinant transforming growth factor-beta 2. Can J Ophthalmol 1996;31:179–182.
57. Smiddy WE, Glaser BM, Green R, et al. Transforming growth factor beta. A biological chorioretinal glue. Arch Ophthalmol 1989;107:577–580.

58. Matsumoto M, Yoshimura N, Honda Y. Increased production of transforming growth factor-beta 2 from cultured human retinal pigment epithelial cells by photocoagulation. Invest Ophthalmol Vis Sci 1994;35:4245–4252.
59. Ie D, Gordon LW, Glaser BM, Pena RA. Transforming growth factor-beta 2 levels increase following retinal laser photocoagulation. Curr Eye Res 1994;13:743–746.
60. Ishida K, Yoshimura N, Yoshida M, Honda Y. Upregulation of transforming growth factor-beta after panretinal photocoagulation. Invest Ophthalmol Vis Sci 1998;39:801–807.
61. Yamamoto C, Ogata N, Yi X, et al. Immunolocalization of transforming growth factor beta during wound repair in rat retina after laser photocoagulation. Graefes Arch Clin Exp Ophthalmol 1998;236:41–46.
62. Roberts AB, Tian F, Byfield SD, et al. Smad3 is key to TGF-beta-mediated epithelial-to-mesenchymal transition, fibrosis, tumor suppression and metastasis. Cytokine Growth Factor Rev 2005;10:10.
63. Tripathi RC, Li J, Chan WF, Tripathi BJ. Aqueous humor in glaucomatous eyes contains an increased level of TGF-beta 2. Exp Eye Res 1994;59:723–727.
64. Inatani M, Tanihara H, Katsuta H, Honjo M, Kido N, Honda Y. Transforming growth factor-beta 2 levels in aqueous humor of glaucomatous eyes. Graefes Arch Clin Exp Ophthalmol 2001;239: 109–113.
65. Tripathi RC, Chan WF, Li J, Tripathi BJ. Trabecular cells express the TGF-beta 2 gene and secrete the cytokine. Exp Eye Res 1994;58:523–528.
66. Zhao X, Ramsey KE, Stephan DA, Russell P. Gene and protein expression changes in human trabecular meshwork cells treated with transforming growth factor-beta. Invest Ophthalmol Vis Sci 2004;45:4023–4034.
67. Lutjen-Drecoll E. Morphological changes in glaucomatous eyes and the role of TGFbeta2 for the pathogenesis of the disease. Exp Eye Res 2005;81:1–4.
68. Addicks EM, Quigley HA, Green WR, Robin AL. Histologic characteristics of filtering blebs in glaucomatous eyes. Arch Ophthalmol 1983;101:795–798.
69. Pena J, Taylor A, Ricard C, Vidal I, Hernandez M. Transforming growth factor isoforms in human optic nerve heads. Br J Ophthalmol 1999;83:209–218.
70. Fukuchi T, Ueda J, Hanyu T, Abe H, Sawaguchi S. Changes in transforming growth factor-beta and platelet-derived growth factor in the optic nerve head in monkey experimental glaucoma. Nippon Ganka Gakkai Zasshi 1999;103:193–200.
71. Cordeiro MF, Gay JA, Khaw PT. Human Anti-TGF-β2 monoclonal antibody: a new anti-scarring agent for glaucoma filtration surgery. Invest Ophthalmol Vis Sci 1999;40:2225–2234.
72. Ridet J, Malhotra S, Privat A, Gage F. Reactive astrocytes: cellular and molecular cues to biological function. Trends Neurosci 1997;20:570–577.
73. Cordeiro MF, Halfyard AS, Khaw PT, Fitzke FW, Keegan DJ. Objective assessment of changes at the optic nerve head in a rat model of glaucoma. Invest Ophthalmol Vis Sci 2001;42:S2750.
74. Yuan L, Neufeld AH. Activated microglia in the human glaucomatous optic nerve head. J Neurosci Res 2001;64:523–532.
75. Guo L, Moss SE, Alexander RA, Ali RR, Fitzke FW, Cordeiro MF. Retinal ganglion cell apoptosis in glaucoma is related to intraocular pressure (IOP) and IOP-induced effects on extracellular matrix. Invest Ophthalmol Vis Sci 2005;46:175–182.
76. Cordeiro MF. Beyond Mitomycin: TGF-beta and wound healing. Prog Retin Eye Res 2002;21:75–89.
77. Grisanti S, Szurman P, Warga M, et al. Decorin modulates wound healing in experimental glaucoma filtration surgery: a pilot study. Invest Ophthalmol Vis Sci 2005;46:191–196.
78. Hitchings RA, Grierson I. Clinico pathological correlation in eyes with failed fistulizing surgery. Trans Ophthalmol Soc UK 1983;103:84–88.
79. Jay JL. Rational choice of therapy in primary open angle glaucoma. Eye 1992;6:243–247.
80. Migdal C, Gregory W, Hitchings RA. Long-term functional outcome after early surgery compared with laser and medicine in open-angle glaucoma. Ophthalmology 1994;101:1651–1656.
81. Thylefors B, Negrel A. The global impact of glaucoma. Bulletin World Health Org 1994;72:323–326.
82. Foster A, Johnson GJ. Magnitude and causes of blindness in the developing world. Int Ophthalmol 1990;14:135–140.
83. Quigley HA. Number of people with glaucoma worldwide. Br J Ophthalmol 1996;80:389–393.
84. Cordeiro MF. The role of transforming growth factor-β in the conjunctival scarring process following glaucoma filtration surgery. In: Ophthalmology, University of London 1998.
85. Reichel ML, Cordeiro MF, Alexander RA, Bhattacharya SS, Schultz GS, Khaw PT. A new model of conjunctival scarring in the mouse eye. Br J Ophthalmol 1998;82:1072–1077.

86. Kay EP, Lee HK, Park KS, Lee SC. Indirect mitogenic effect of transforming growth factor-beta on cell proliferation of subconjunctival fibroblasts. Invest Ophthalmol Vis Sci 1998;39:481–486.
87. Cordeiro MF, Reichel MB, Gay JA, D'Esposita F, Alexander RA, Khaw PT. TGF-β1, -β2 & -β3 in vivo: Effects on normal and Mitomycin-C modulated conjunctival scarring. Invest Ophthalmol Vis Sci 1999;40:1975–1982.
88. Cordeiro MF, Siriwardena D, Chang L, Khaw PT. Wound healing modulation after glaucoma surgery. Curr Opin Ophthalmol 2000;11:121–126.
89. Cordeiro MF, Mead A, Ali RR, et al. Novel antisense oligonucleotides targeting TGF-beta inhibit in vivo scarring and improve surgical outcome. Gene Ther 2003;10:59–71.
90. The Fluorouracil Filtering Surgery Study Group. Fluorouracil filtering surgery study one-year follow-up. Am J Ophthalmol 1989;108:625–635.
91. Kitazawa Y, Kawase K, Matsushita H, Minobe M. Trabeculectomy with mitomycin: a comparative study with fluorouracil. Arch Ophthalmol 1991;109:1693–1698.
92. Katz GJ, Higginbotham EJ, Lichter PR, et al. Mitomycin C versus 5-fluorouracil in high-risk glaucoma filtering surgery. Extended follow-up. Ophthalmology 1995;102:1263–1269.
93. Lamping KA, Belkin JK. 5-Fluorouracil and mitomycin C in pseudophakic patients. Ophthalmology 1995;102:70–75.
94. Belyea DA, Dan JA, Stamper RL, Lieberman MF, Spencer WH. Late onset of sequential multifocal bleb leaks after glaucoma filtration surgery with 5-fluorauracil and mitomycin-C. Am J Ophthalmol 1997;124:40–45.
95. Greenfield DS, Liebmann JM, Jee J, Ritch R. Late-onset bleb leaks after glaucoma filtering surgery. Arch Ophthalmol 1998;116:443–447.
96. Parrish R, Minckler DS. "Late endophthalmitis" - filtering surgery time bomb? Ophthalmology 1996;103:1167–1168.
97. Stamper R, McMenemy M, Lieberman M. Hypotonous maculopathy after trabeculectomy with subconjunctival 5-fluorouracil. Am J Ophthalm 1992;114:544–553.
98. Kupin TH, Juzych MS, Shin DH, Khatana AK, Olivier MM. Adjunctive mitomycin C in primary trabeculectomy in phakic eyes. Am J Ophthalmol 1995;119:30–39.
99. Crowston JG, Akbar AN, Constable PH, Occleston NL, Daniels JT, Khaw PT. Antimetabolites-induced apoptosis in Tenon's capsule fibroblasts. Invest Ophthalmol Vis Sci 1998;39:449–454.
100. Siriwardena D, Khaw PT, King AJ, et al. Human antitransforming growth factor beta(2) monoclonal antibody—a new modulator of wound healing in trabeculectomy: a randomized placebo controlled clinical study. Ophthalmology 2002;109:427–431.
101. Nakamura H, Siddiqui SS, Shen X, et al. RNA interference targeting transforming growth factor-beta type II receptor suppresses ocular inflammation and fibrosis. Mol Vis 2004;10:703–711.
102. Chihara E, Dong J, Ochiai H. Effect of TGF-β1 suppressor Tranilast on filtering bleb and IOP after glaucoma surgery. Invest Ophthalmol Vis Sci 1999;40 (suppl):Abstract no. 5104.
103. Cordeiro MF, Balasubramanian L, Ali RR, et al. Effect and localization of a TGF-β1 antisense oligonucleotide in conjunctival scarring - A potential new anti-scarring strategy in glaucoma surgery. Invest Ophthalmol Vis Sci 2000;41:S3957.
104. Wu-Pong S, Weiss T, Hunt C. Antisense c-myc oligonucleotide cellular uptake and activity. Antisense Res Dev 1994;4:155–163.
105. Beardsley T. Making antisense. Drugs that turn off genes are entering human tests. Sci Am 1992;266: 107–108.
106. Fakler B, Herlitze S, Amthor B, Zenner H, Ruppersberg J. Short antisense oligonucleotide-mediated inhibition is strongly dependent on oligo length and concentration but almost independent of location of the target sequence. J Biol Chem 1994;269:16,187–16,194.
107. Tamm I, Dorken B, Hartmann G. Antisense therapy in oncology: new hope for an old idea? Lancet 2001;358:489–497.
108. Orr R. Technology evaluation: fomivirsen, Isis Pharmaceuticals Inc/CIBA vision. Curr Opin Mol Ther 2001;3:288–294.
109. Lang K, Peppercorn M. Promising new agents for the treatment of inflammatory bowel disorders. Drugs R D 1999;1:237–244.
110. Yacyshyn B, Bowen-Yacyshyn M, Jewell L, et al. A placebo-controlled trial of ICAM-1 antisense oligonucleotide in the treatment of Crohn's disease. Gastroenterology 1998;114:1133–1142.
111. Crooke S. Basic principles of antisense technology. In: Crooke ST, ed. Antisense Drug Technology: Principles, Strategies and Applications. New York: Marcel Dekker 2001; pp. 1–28.

112. Stephenson M, Zamecnik P. Inhibition of Rous sarcoma viral RNA translation by a specific oligodeoxyribonucleotide. Proc Natl Acad Sci USA 1978;75:285–288.
113. Dean N, Butler M, Monia B, Manoharan M. Pharmacology of 2′-O-(2-methoxy) ethyl - modified antisense oligonucleotides. In: Crooke ST, ed. Antisense Drug Technology: Principles, Strategies and Applications. New York: Marcel Dekker, 2001; pp. 319–338.
114. Dean NM, Griffey RH. Identification and charecterization of second-generation antisense oligonucleotides. Antisense Nucleic Acid Drug Develop 1997;7:229–233.
115. Soutschek J, Akinc A, Bramlage B, et al. Therapeutic silencing of an endogenous gene by systemic administration of modified siRNAs. Nature 2004;432:173–178.
116. Lingor P, Koeberle P, Kugler S, Bahr M. Down-regulation of apoptosis mediators by RNAi inhibits axotomy-induced retinal ganglion cell death in vivo. Brain 2005;128:550–558.
117. Tanihara H, Inatani M, Koga T, Yano T, Kimura A. Proteoglycans in the eye. Cornea 2002;21: S62–S69.
118. Chihara E, Dong J, Ochiai H, Hamada S. Effects of tranilast on filtering blebs: a pilot study. J Glaucoma 2002;11:127–133.

43 Transforming Growth Factor-β in Cutaneous Tissue Repair

Jisun Cha and Vincent Falanga

CONTENTS

INTRODUCTION
MATRIX DEPOSITION AND DEGRADATION
PROFIBROTIC ACTIVITY
GRANULATION TISSUE FORMATION
RE-EPITHELIALIZATION
ANIMAL STUDIES
HUMAN STUDIES AND EXPERIMENTS AIMED AT SCARRING
INTERACTIONS WITH OTHER CYTOKINES
ACKNOWLEDGMENTS
REFERENCES

Abstract

Transforming growth factor-βs and their family are important components of the injury response. By far, the best studied of these multifunctional peptides is TGF-β1. Evidence accumulated over the last two decades shows that TGF-βs have a profound stimulatory effect on overall extracellular matrix production, on the development of tensile strength after wounding, and on the formation of a stable scar. Some of the overall effects of the TGF-βs occur in concert with other cell types and often obscure the well-described inhibition these peptides have on endothelial cells and keratinocytes. In addition to their important effect on wound healing, TGF-βs have been shown to increase, modify, or even block the fibrotic response.

Key Words: Transforming growth factor-β; wound repair; tissue repair; matrix; fibroblast; keratinocyte.

1. INTRODUCTION

The body's ability to replace tissues after injury and inflammation is critical to survival. Upon injury, the host responds by setting in motion a series of highly interdependent events that serve to contain damage and prepare the wound for the repair process. The recognized events of wound healing specifically related to cutaneous injury include clotting of blood components, inflammation, granulation tissue formation, re-epithelialization, angiogenesis, fibroplasias, and connective tissue retraction *(1)*. These fundamental biological and molecular events, based on knowledge mainly derived from experimental wounds in animals, cannot be separated and categorized in a clear-cut way. However, it has been common to

From: *Cancer Drug Discovery and Development: Transforming Growth Factor-β in Cancer Therapy, Vol I: Basic and Clinical Biology*
Edited by: Sonia B. Jakowlew © Humana Press Inc., Totowa, NJ

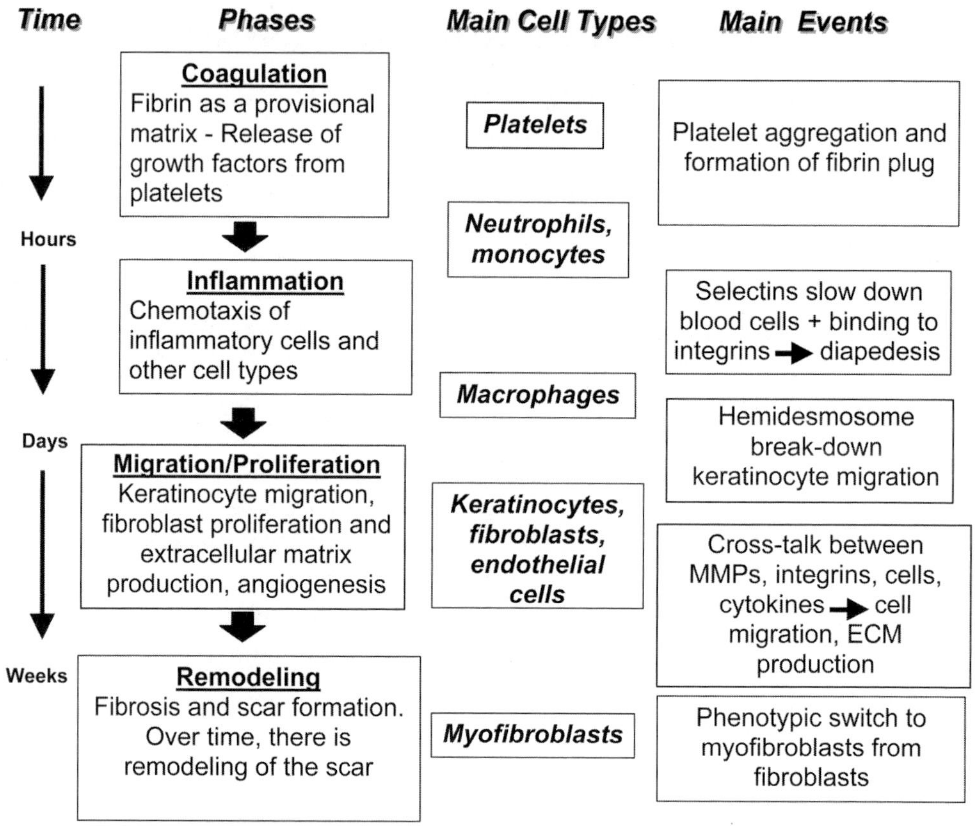

Fig. 1. A diagrammatic representation of the different phases of wound healing, the time line, the cells involved, and the main events (copyrighted V. Falanga, 2006).

divide the repair process into four overlapping phases of coagulation, inflammation, migration-proliferation (including matrix deposition), and remodeling (Fig. 1).

Multiple cytokines, growth factors, and other mediators are involved in these intricate processes, and TGF-β plays a fundamental role. TGF-β has multiple and often opposing effects on the response to cutaneous tissue injury. This cytokine belongs to a family of homologous polypeptides, which, at least in mammalian cells, includes three isoforms (TGF-β1 TGF-β2 and TGF-β3). The TGF-β1 family contains mediators with wide-ranging functions, such as bone morphogenic proteins, activins, inhibins, and mullerian-inhibiting substance *(2)*. The TGF-β1 isoform has the most widespread distribution in mammals and has been studied the most. Much of what will be discussed here has to do with TGF-β1.

TGF-β1 is a homodimeric protein produced by a variety of different cell types, including megakaryocytes (platelets), endothelial cells, lymphocytes, and macrophages. Native TGF-βs are synthesized as precursor proteins, which are secreted and then proteolytically cleaved to yield the biologically active growth factor and a second latent component. Active TGF-β binds to two cell surface receptors (Type I and Type II) with serine/threonine kinase activity, which trigger the phosphorylation of cytoplasmic transcription factors called Smads *(3)*. TGF-β first binds to the Type II receptor. After phosphorylation of the Type I receptor, a cascade of events occurs. There is phosphorylation of Smad2 and 3, which form heterodimers with Smad4, enter the nucleus, and associate with other DNA-binding proteins to activate or inhibit gene transcription (Fig. 2).

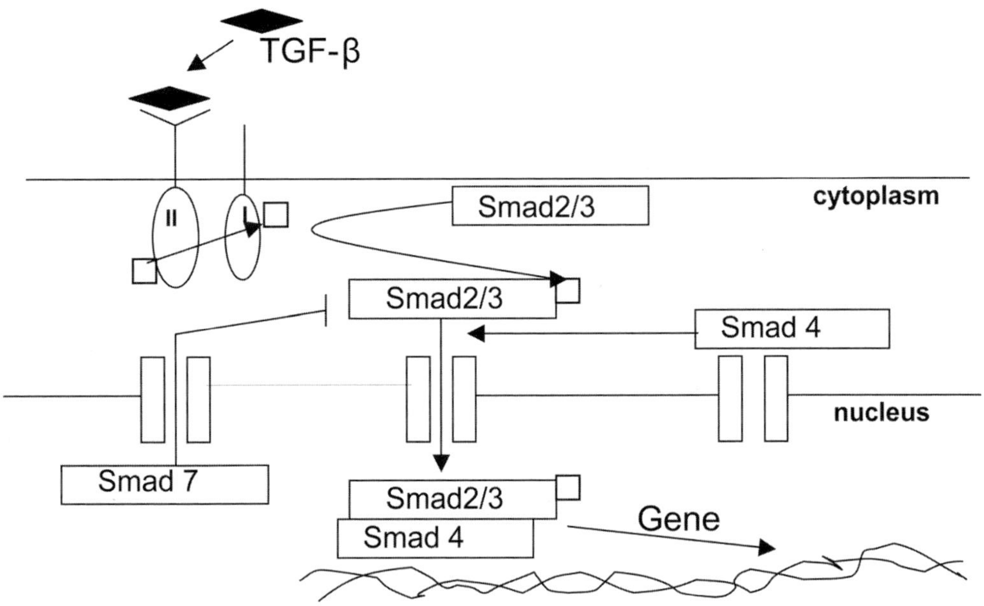

Fig. 2. A simplified representation of the TGF-β signaling pathway (copyrighted J. Cha, 2006).

Because of the large diversity of TGF-β effects, it has been said that TGF-β is extremely pleiotropic. As we will discuss, TGF-β is a growth inhibitor for endothelial cells, most epithelial cell types, and leukocytes. TGF-β blocks cell cycle by increasing the expression of cell-cycle inhibitors. For example, the protein p15 is an important mediator of the antiproliferative effects of TGF-β *(4,5)*. The increased expression of protein p15 in response to TGF-β is sufficient to cause cell-cycle arrest in certain epithelial cells *(6)*. The effects of TGF-β on mesenchymal cells depend on its concentration and culture conditions. Generally, TGF-β stimulates the proliferation of fibroblasts and smooth muscle cells and is a potent fibrogenic agent that stimulates fibroblast chemotaxis and enhances the production of collagen, fibronectin, and proteoglycans. These effects will be discussed in more detail later.

In the skin, TGF-β1 is expressed in the normal dermis and epidermis, and its expression is increased during wound healing *(7)*. Upon cutaneous injury, TGF-β is rapidly released and induced *(8)*. The initial release assists in the attraction of neutrophils, macrophages, and fibroblasts, which in turn release more TGF-β. Many cell types in the skin, including macrophages, fibroblasts, and keratinocytes, produce TGF-β during wound healing *(8)*; accordingly, TGF-β is consistently present in wound fluid throughout the repair process. Expression of TGF-β and TGF-β receptors are elevated in fibroblasts of human postburn hypertropic scars, in keloids, and in keloid-derived cultured fibroblasts *(9)*. The TGF-β activators found thus far are proteins intimately associated with the wound-healing response. For example, the plasmin proteases, matrix metalloproteinases (MMPs)-2 and -9, which promote matrix degradation *(10)*, activate TGF-β *(11)*. Another activator of TGF-β, thrombospondin-1(TSP-1), modulates cell adhesion, promotion of angiogenesis, and reconstruction of the matrix *(12)*. Yet another critical activator of TGF-β is the integrin αvβ6, which is normally expressed at low levels only in epithelia *(13)* but is induced during wounding or inflammation *(13)*. Studies using knockout mice have supported the notion that TSP-1 and αvβ6 integrin are involved in TGF-β activation. TSP-1 null mice demonstrate a partial phenotypic overlap with TGF-β null animals *(14)*.

TGF-β inhibits collagen degradation by decreasing matrix proteases and by increasing protease inhibitor activities. TGF-β is also involved in the development of fibrosis in a variety of conditions. At the same time, TGF-β has strong anti-inflammatory effects. Knockout mice lacking the TGF-β gene die from widespread inflammation and abundant lymphocyte proliferation, presumably because of unregulated T-cell proliferation and macrophage activation.

2. MATRIX DEPOSITION AND DEGRADATION

The balance between production and degradation of collagens plays critical roles in the development and maintenance of organ and tissue integrity. Type I collagen, the major component of extracellular matrix (ECM) in various organs, is produced predominantly in mesenchymal cells such as fibroblasts, osteoblasts, and myofibroblasts. Its chains are coordinately expressed but encoded by the distinct genes, *COL1A1* and *COL1A2*, respectively. TGF-β plays important roles in stimulating type I collagen gene expression mainly at the levels of transcription *(15)*. The *COL1A2* upstream sequence spanning from –313 to –183 is essential for basal transcription of the gene in the two cell types of mesenchymal origin and mediates the stimulatory effect of TGF-β on *COL1A2* transcription *(16–19)*. TGF-β upregulates *COL1A2* transcription by modifying the interaction between Box 3A-bound Sp1 and unknown nuclear factors bound to the neighboring Box B sequence. It has been show that Smad3, an intracellular mediator of TGF-β signal transduction, binds to the CAGACA sequence present in Box B and stimulates *COL1A2* transcription *(20,21)*. Smad3 and Smad4 have been shown to bind the CAGACA sequence present in the promoters of several TGF-β-inducible genes including plasminogen activator inhibitor (PAI)-1 *(22)*, *jun* B *(23)*, $p21^{WAF1/Cip1}$ *(24)*, Smad7 *(25)*. Box B of the *COL1A2* promoter also contains a CAGACA sequence (–265 to –260), and it has been shown that Smad3 binds to this sequence in vitro and stimulates *COL1A2* transcription *(20,21)*. Type VII collagen belongs to an extensive family of closely related proteins involved in cell anchoring to ECM and cartilage formation. Although some collagens have a widespread distribution, type VII collagen is found exclusively in the basement membrane of stratified squamous epithelia *(26,27)*. TGF-β in particular is a potent inducer of *COL7A1* expression in fibroblast and keratinocytes *(28–30)*. Smad proteins recognize the sequence CAGAC, commonly referred to as the Smad-binding element (SBE). To regulate specific target genes, however, activated Smad complexes must additionally interact or functionally cooperate with other transcription factors. In the case of *COL7A1* activation by TGF-β, previous studies have shown the involvement of Smad and AP1 transcription factors. The TGF-β-responsive region in the *COL7A1* contains a canonical SBE. Mutation of this element inhibits the activation of the *COL7A1* promoter by TGF-β in fibroblasts.

TGF-β's action on matrix is exerted through two complementary pathways, one that reduces matrix degradation and the other that stimulates matrix accumulation. TGF-β inhibits the synthesis of extracellular proteinases while upregulating the production of their inhibitors and that of structural ECM components. The combined action of TGF-β on the genes implicated in the formation and degradation of the ECM is mostly exerted at the transcriptional level through well-defined intracellular pathways *(31)*. Using a combined cDNA microarray/promoter transactivation approach, Verrecchia et al. have identified several new Smad gene targets, among which are *COL1A1, COL3A1, COL5A2, COL6A1, COL6A3*, and tissue inhibitor of metalloproteinases-1 (TIMP-1). Most notably, these data indicate that the Smad signaling pathways are crucial for simultaneous activation of several fibrillar collagen genes by TGF-β. About 60 other ECM-related genes were also identified as immediate-early gene targets downstream of TGF-β *(29)*. Ultimately, wound healing requires remodeling of the ECM. MMPs are essential to the ECM remodeling process as well as the removal of devitalized tissue and the re-epithelialization of cutaneous wounds *(32–34)*. There are three major classes of MMPs: collagenases, gelatinases, and stromelysins. The predominant collagenase that degrades the dermal collagen is MMP-1. MMP-2 (gelatinase) degrades damaged type I collagen, in addition to the basement membrane proteins, which are also degraded by the stromelysins *(35,36)*. TGF-β stimulates collagen and MMP-2 expression and inhibits MMP-1 expression in dermal fibroblasts. TGF-β is located either within the ECM or on the surface of cells in its latent form and is activated by proteases such as MMPs and plasmin. TGF-β-dependent synthesis of collagens 1, 3, 6, and 7 and TIMP-1 are Smad3 dependent *(29)*, as are the more complex processes of TGF-β-dependent chemotaxis and inhibition of epithelial migration *(28,37)*. Other important signaling pathways, including phosphoinositol-3 kinase and the mitogen-activated protein (MAP) kinases, also mediate effects of TGF-β and activin on cells *(38)*. Activation by TGF-β of other MAP kinases such as p38 MAP kinase *(39,40)* and JNK *(41,42)* has also been reported. The interaction with those MAP kinases could regulate Smad transcriptional activity in a positive *(43–45)* or a negative manner *(29,46,47)*. Some of the events initiated by TGF-β could be Smad independent, such as fibronectin induction through c-jun *(48)*.

3. PROFIBROTIC ACTIVITY

Formation and contraction of granulation tissues are essential in order to maintain tissue continuity, reduce the size of the wound, and produce a permanent scar. After incisional wounding, animals lacking Smad3 show accelerated wound healing, reduced granulation tissue formation, increased epithelialization, reduced inflammation possibly because of an impaired chemotactic response *(37)*, and are resistant to cutaneous fibrosis caused by radiation injury *(49)*. Giving further support of the role of Smad3 in the fibrotic response are observations that elevated levels of activated nuclear Smad3 are found in several situations and models, including bleomycin-induced fibrosis, hepatic stellate cell activation, and the leading edge of scleroderma lesions *(50,51)*. However, models for examining stages of established fibrosis do not show significant Smad3 activation, which suggests that over-expression of profibrotic markers is Smad independent *(52,53)*. Thus, although Smad3 plays a key role in the initiation of the fibrotic response, Smad3 *per se* might not be required for the maintenance of the fibrotic phenotype. Accordingly, TGF-β expression at fibrotic sites is limited to the early inflammatory stage in various fibrotic disorders in the skin, including systemic sclerosis and localized scleroderma *(54,55)*. However, the injection of CTGF after the initiation of fibrosis with TGF-β injection was found to be necessary for developing persistent fibrosis *(56)*. Persistent fibrosis by these cytokines in vivo is related to the sustained promoter activation of the gene for type I collagen, which is the most abundant component of the ECM in the skin.

4. GRANULATION TISSUE FORMATION

Wound fibroblasts are specialized, highly contractile collagen-producing cells. They are responsible for the production of many ECM proteins, including fibronectin, and proteoglycans. Gradually, however, wound fibroblasts assume a myofibroblast phenotype as the wound matures. Persistence of the myofibroblast in the wound is a feature of fibrosis. This cell type is responsible for the excessive contraction of matrix characteristic of scar tissue. During granulation tissue formation, fibroblasts evolve into the proto-myofibroblast, characterized by the formation of stress fibers expressing only cytoplasmic β- and γ-actins. Under the stimulation of growth factors and newly synthesized ECM components, the proto-myofibroblast differentiates into a myofibroblast, which is characterized by the expression of α-SM actin. TGF-β1 induces the epithelial mesenchymal transition (EMT). It is defined as the formation of mesenchymal cells from epithelia in different embryonic territories, although the delamination process is more precisely described in vitro in epithelial tissue explants *(57)*. The EMT results in the disruption of the polarized morphology of epithelial cells, formation of actin stress fibers, and enhancement of cell migration *(58,59)*. The transition between proto-myofibroblast and differentiated myofibroblast is essentially because of the combined action of the splice variant ED-A of cellular fibronectin and TGF-β *(60,61)*, particularly the TGF-β1 isoform. TGF-β1 induces fibroblasts synthesis of α-SM actin and stimulates the production of collagen type I. Accordingly, Gabbiani showed that TGF-β1 induced α-SM actin expression in growing and quiescent cultured fibroblasts. Neutralizing antibodies to TGF-β1 partly inhibit the expression of α-SM actin in fibroblasts cultured in the presence of whole blood serum but not of plasma-derived serum. The mechanical tension exerted by the myofibroblast is mediated through integrin-dependent signaling events including focal adhesion kinase, tyrosine kinase, and ERK activation *(62)*.

TGF-β induces α-smooth muscle actin (α-SMA), modulates the expression of adhesive receptors, and enhances the synthesis of ECM molecules including ED-A fibronectin (ED-A FN), an isoform of fibronectin arising from alternative splicing of the fibronectin mRNA *(48,63)*. ED-A FN is expressed during embryogenesis and is expressed *de novo* during wound healing and fibrosis.

ED-A FN deposition precedes α-SMA expression by fibroblasts during granulation tissue evolution in vivo and after TGF-β1 stimulation in vitro. Hence, TGF-β1 controls the two most important activities of granulation tissue, i.e., the capacity to modify tissue shape and the formation of ECM. In this respect, TGF-β1 is the key cytokine in the differentiation of fibroblasts to myofibroblasts.

Downstream of TGF-β, several proteins are required for the enhancement of a cellular response to TGF-β, and hence for prolongation of the wound healing and fibrotic response. These proteins either (1) enhance the contractile phenotype of the fibroblast or (2) prolong the induction of ECM by fibroblasts. Such peptides include connective tissue growth factor (CTGF) and the ED-A form of fibronectin *(64)*. In adult skin, CTGF is not normally expressed unless induced, for example, during the normal wound repair process *(65)*. TGF-β induces CTGF expression in dermal fibroblasts, but not in keratinocytes, via consensus Smad and transcription enhancer factor elements in the CTGF promoter *(52)*. In contrast to the situation in normal fibroblasts, CTGF is constitutively overexpressed in dermal fibrotic lesions such as in scleroderma *(65,66)*. These observations suggest that CTGF may act as a downstream effector of at least some of the profibrotic effects of TGF-β. CTGF promotes fibroblasts proliferation, matrix production, and granulation tissue formation *(64,66)* as well as cell adhesion and migration in a wide variety of cell types *(66)*. An expression vector encoding CTGF transfected into fibroblasts can activate a cotransfected reporter construct driven by the type I collagen promoter, suggesting that a CTGF response element exists in the type I collagen promoter. Experiments using recombinant CTGF and neutralizing antibodies targeting CTGF have suggested that CTGF mediates at least some of the effects of TGF-β on fibroblast proliferation, adhesion, and ECM production, including collagen and fibronectin *(67–70)*.

5. RE-EPITHELIALIZATION

The basal keratinocytes at the margins of the wound and within remaining skin appendages enter a migratory mode and migrate laterally across the provisional wound bed *(71,72)*. Sarret et al. *(73)* showed that TGF-β1 markedly inhibits the proliferation of human keratinocytes but not cell motility. Human keratinocyte motility is important in the re-epithelialization of human skin wounds. TGF-β1 plays a crucial role in the early phases of tissue repair, including regulation of the signal that drives keratinocyte migration during re-epithelialization *(8,74)*. Efficient migration is dependent on newly synthesized and deposited laminin 5 *(75–79)*, whose synthesis has been shown to be upregulated by TGF-β1 *(80,81)*.

TGF-β1-induced normal human keratinocyte migration is accompanied by a downregulation of interaction of both α3β1 and α6β4 integrins with laminin 5 *(82)* and increasing affinity of α3β1 integrin and increased βig-h3 *(83)*. In keratinocytes, Smad3 was identified as an essential receptor-activated Smad protein associated with TGF-β-induced biological responses. Knocking out Smad3 completely diminished TGF-β-induced chemotaxis and, partially, cell growth arrest of keratinocytes *(37,38)*.

6. ANIMAL STUDIES

A large number of animal studies have addressed the role of TGF-β in wound healing and fibrosis. Additional evidence supporting the profibrotic role of TGF-β in vivo are data which show that TGF-β1-deficient mice severely impaired late-stage wound repair, including decreased re-epithelialization and collagen deposition, compared with control mice. However, mice lacking TGF-β1 also show a severe wasting syndrome accompanied by a pronounced, generalized inflammatory response and tissue necrosis resulting in organ failure and death *(84,85)*. Gabbiani et al. reported that the local administration of TGF-β1 in vivo results in the formation of granulation tissue in which α-SM actin expressing myofibroblasts are particularly abundant. Amento et al. showed that a single intravenous administration of recombinant human TGF-β1 (rhTGF-β1) increased wound breaking strength in old or young rats with glucocorticoid-induced impaired healing, and to levels similar to normal control animals *(86)*. Intradermal injection of TGF-β1 gene accelerates wound closure *(87)* as well as wound contraction rate, ECM production, and tensile strength in vivo *(88,89)*. As discussed, TGF-β1 is a potent inhibitor of keratinocyte proliferation in vitro *(90,91)*. However, in a steroid-impaired rabbit wound model, application of TGF-β1 enhanced re-epithelialization *(86)*. Another study further supported that TGF-β1 promotes epidermal cell growth in a TGF-β1 transgenic model *(92)*. In an animal model of fascial wound healing, TGF-β3 increased fascia breaking strength, collagen deposition, and cellular proliferation *(93)*. Smad3-deficient mice showed accelerated wound healing associated with reduction of the influx of inflammatory cells and increased re-epithelialization *(37,47)*. However, exogenously expressed Smad3 by means of subcutaneous injection of adenovirus-containing Smad3 complementary DNA *(94)* showed some advantages over that of TGF-β peptide for treatment of cutaneous wounds. It has been difficult to reconcile these findings, and more work is needed in that area.

Recent evidence suggests that chronic wounds become unresponsive to growth factors, including PDGF and TGF-β1 *(95,96)* possibly resulting in delayed or incomplete wound healing. For instance, it was reported that fibroblasts from chronic wounds display decreased TGF-β Type II receptor expression, together with their lack of response to TGF-β1. These receptor abnormalities resulted in decreased phosphorylation of MAP kinases ERK2/1, as well as Smad2 and Smad3 *(97)*. The MAP kinase cascades appear to be profibrotic and interact with Smad signaling pathways in the control of matrix expression. It is of great interest to find out whether preferential switching to the MAPK pathway, away from the Smad signaling pathway, might stimulate wound healing.

Again, studies in animals have confirmed the effectiveness of TGF-β1 in wound healing. A single dose of TGF-β in a collagen vehicle restored the diabetes-related decrease in tensile strength of collagen *(98)*. Lee and Huang showed that electrophoretic delivery of TGF-β1 gene works synergistically with electric therapy to enhance diabetic wound healing in genetically diabetic mice (db/db mice). Simple local injection of TGF-β1 plasmid could induce re-epithelialization, wound closure, collagen deposition, and angiogenesis *(87)*. The effect of TGF-β on the ECM manifested as fibrotic tissue or scar because of increased collagen deposition in adult and neonatal wounds and was associated with overexpression of TGF-β, interleukin-6, and interleukin-8 *(99–102)*. Fetal wounds heal with minimal or no scarring, and this has been largely attributed to lower TGF-β activity in fetal wounds relative to neonatal or adult wounds *(100,101,103)*.

7. HUMAN STUDIES AND EXPERIMENTS AIMED AT SCARRING

Robson et al. evaluated TGF-β2 in acute incisional healing using a variety of vehicles. This isomer showed noticeably enhanced gaining of incisional strength in carboxymethylcellulose gel by day 3. Human therapeutic applications of exogenous TGF-β2 showed that it may normalize the distribution of TGF-β in venous ulcer *(104)* and diabetic neuropathic foot ulcer *(105)*.

TGF-β1 and TGF-β2 predominate in adult wounds, whereas TGF-β3 is the main isoform in fetal tissue, possibly promoting wound healing by regeneration with less scarring *(101)*. Ferguson used combined neutralization of TGF-β1 and -β2 or addition of TGF-β3 and showed reducing scar formation *(106)*. TGF-β1 is the principal isoform implicated in fibrotic disorders in most organ systems. In the eye, however, it is TGF-β2 rather than TGF-β1 that is believed to be the predominant isoform mediating fibrosis *(107)*. TGF-β2 is known to upregulate the expression and secretion of TGF-β1 by monocytes, macrophages, fibroblasts, and epithelial cells *(108)*. The role of TGF-β3 in tissue repair is less well documented, with a number of investigators demonstrating either a pro- or antifibrotic effect, depending on the body site investigated. Exogenous addition of TGF-β3 to rat incisional wounds led to an overall reduction in scarring, comparable to the effects observed after the combined neutralization of TGF-β1 and TGF-β2 *(106)*. TGF-β3 is markedly motogenic *(109)*, stimulating filopodia formation and rapid cell migration *(109)*. However, in animal models of wound healing, TGF-β3 increased fascia breaking strength, collagen deposition, and cellular proliferation *(93)*. Whether TGF-β3 is an antifibrotic isoform remains controversial.

8. INTERACTIONS WITH OTHER CYTOKINES

The proinflammatory cytokine TNF-α expressed by macrophages during the wound-healing response *(66)* has long been known to possess antifibrotic ability. It suppresses the expression of ECM genes *(51)* and the TGF-β-dependent induction of collagen and CTGF *(66,110)*. After wounding, mice lacking the TNF-α receptor p55 showed increased angiogenesis, collagen content, and re-epithelialization *(111)*. TNF-α has been proposed to suppress TGF-β signaling via the NF-κB induction of Smad7 *(112)*, but this appears to be a cell-type-specific effect and is not operative in dermal fibroblasts, because TNF-α does not induce Smad7 in this cell type *(113)*. Others have shown that TNF-α suppresses Smad3-dependent signaling in this cell type by elevating c-jun expression and activating JNK *(110)*. These effects result in suppression of the ability of Smad3 to activate gene expression because of the interaction of c-jun with Smad3, which leads to the off-DNA sequestering of Smad3 *(29,114)*. TNF-α suppresses the TGF-β induction of CTGF. However, overexpression of CTGF in scleroderma lesions, which occurs by a Smad-independent and Sp1-dependent mechanism *(52,114,115)* is not suppressed by TNF-α *(66)*. These latter findings suggest that the chronic fibrosis observed in scleroderma may be a consequence of the ability of lesional scleroderma fibroblasts to escape from the negative, antifibrotic effects of TNF-α.

Both TNF-α and TGF-β induce prostaglandin production in fibroblasts *(116,117)*. Prostaglandins, which are metabolites of arachidonic acid produced by the action of cyclooxygenase

1 and 2 (COX-1 and -2), are lipids used as signaling mediators by several pathways *(118)*. COX-1 expression is induced by TGF-β, whereas COX-2 is induced by TNF-α *(117)*. Prostaglandin (PGE2) and prostacyclin (PGI2) elevate cellular cAMP levels and activate protein kinase A (PKA) *(119)*. Some of the earliest known agents that suppressed collagen synthesis in cell culture and in vivo were those that activated PKA, including prostacyclin (PGI2), prostaglandin (PGE2), and stable cAMP analogs *(120–122)*. Recently 8-Br-cAMP was shown to block the TGF-β induction of granulation tissue in wound chambers, such as metal cylinders placed in the back of rats. The stable prostacyclin analog iloprost was shown to decrease the fibrotic response, including CTGF and collagen levels in scleroderma patients, and to reduce TGF-β induction of these proteins in vitro and in vivo *(52,123)*. The effect of iloprost was dependent on an elevation of cellular cAMP levels and resulted in antagonism of the ras/MEK/ERK signaling cascade necessary for induction of the profibrotic protein CTGF *(114,123)*. The potential utility of synthetic prostacyclins as antifibrotic therapies is heightened by observations that these compounds are well tolerated, already being used clinically to treat pulmonary hypertension *(124)*.

Interferon-γ (IFN-γ), a pleiotropic cytokine produced by T-cells and NK cells, plays fundamental roles in innate and acquired immune responses *(125)*. IFN-γ can negatively regulate the transcription of matrix *(126)*, but no common IFN-γ-specific inhibitory elements have yet been identified. Recently, it was shown that IFN-γ induces the Smad-presumed inhibitor Smad 7 in the fibrosarcoma-derived U4A cell line *(127)*; however, this mechanism does not seem to be generally applicable to normal fibroblasts *(128)*. Rather, in this cell type, IFN-γ seems to suppress ECM synthesis by the ability of nuclear Stat1α to compete with the transcriptional cofactor p300 *(128)*. More recently, an alternative IFN-γ signal transduction was shown to lead to transcriptional repression of *COL1A2*, namely, the Jak/CK2/YB-1 pathway *(129)*.

Experiments using antisense RNAs indicate that both YB-1 and STAT1 are necessary for transcriptional repression of *COL1A2* by IFN-γ through competition with the common Smad3 for the cellular p300 *(129)*. Thus, TGF-β and IFN-γ exert opposite effects on collagen synthesis. Because these two cytokines are secreted by inflammatory cells at sites of tissue injury, their antagonistic interactions regulating collagen synthesis are likely to be of great importance in the maintenance of connective tissue homeostasis.

Although TGF-β promotes angiogenesis in vivo *(130,131)*, it also inhibits the growth and proliferation of endothelial cell monolayers in vitro *(132–134)*. This apparent discrepancy between in vivo and in vitro activities may be attributable, in part, to the capacity of TGF-β in vivo to recruit and stimulate macrophages, which then produce other active angiogenesis factors *(135)*. An alternative or additional explanation is that TGF-β is a growth inhibitor for cultured endothelial cell monolayers, but a mitogen for cultured endothelial cells that have formed capillary-like tubes *(136)*. In fact, the types of TGF-β receptors on endothelial cells are altered when cultured endothelial cells form tubes *(137)*. Likewise, cultured monolayer endothelial cells produce PDGF-BB but have no receptor (PDGFR-β) for this ligand. In contrast, once the cultured cells form tubes, they express PDGFR-β and respond to the ligand that they no longer produce *(138)*. VEGF, a member of the PDGF family of growth factors, has potent angiogenetic effects as well as vasopermeable activity, which lead to its initial designation as vasopermeability factor *(139)*. This factor is produced in large quantities by the epidermis during wound healing *(140)*. Low oxygen tension, a common feature of tissue repair, is a major inducer of this growth factor *(141,142)* and its receptor *(143)*. Thus, cell disruption and hypoxia appear to be strong initial inducers of potent angiogenesis factors at the wound site. Recent data suggest that bFGF may set the stage for angiogenesis during the first 3 d of wound repair, whereas VEGF may be critical for angiogenesis during granulation tissue formation from days 4 through 7 *(144)*. Several additional members of the VEGF family have been found recently (VEGF-B, VEGF-C, and VEGF-D) *(145)*. Although their general role in angiogenesis is quickly being elucidated, their specific function in wound angiogenesis is not yet clear.

The angiopoietins have recently joined the members of the VEGF family as the only known growth factors largely specific for vascular endothelium. The angiopoietins include a naturally occurring agonist, angiopoietin-1, as well as a naturally occurring antagonist, angiopoietin-2, both of which act by means of the Tie2 receptor. Two new angiopoietins, angiopoietin-3 in mouse and angiopoietin-4 in human, have recently been identified but their function in angiogenesis is unknown *(146)*. Neither binds to the Tie2 receptor. The newly forming blood vessels first deposit a provisional matrix containing fibronectin and proteoglycans but, ultimately, form mature vascular basement membrane. TGF-β may induce endothelial cells to produce the fibronectin and proteoglycan provisional matrix as well as assume the correct phenotype for capillary tube formation. FGF, and other mitogens such as VEGF, stimulate endothelial cell proliferation, resulting in a continual supply of endothelial cells for capillary extension.

FGF-2 is a member of a large family of growth factors implicated in numerous biological processes, including cell proliferation, differentiation, migration, angiogenesis, and wound healing *(78,147)*. FGF-2 is an 18 kDa protein, and three high molecular mass forms have been described (22, 22.5, and 24 kDa) *(148–150)*. FGF-2 belongs to a family containing more than 20 heparin-binding proteins and its activity is mediated by binding to heparan sulfate proteoglycans and to high-affinity cell surface receptor tyrosine kinases *(78)*. The transient localization of FGF-2 in vivo 6 wk after injury and the coincident expression of heparan sulfate suggest that changes in growth factors in an avascular tissue may be used to mediate the crucial regulation of proteoglycans *(151,152)*.

Previously, it has been shown that TGF-β modulates the interaction of stromal cells with their ECM by inducing the synthesis of specific proteoglycan core proteins and their respective glycosaminoglycans *(153)*. More recently, Song et al. *(154)* demonstrated that injury rapidly increases TGF-β1 mRNA expression along the wound margin. Exogenous TGF-β1 also rapidly increases the level of its own mRNA. TGF-β1 also played a role in the increase in expression of FGF-2 following wound closure. In contrast, exogenous FGF-2 did not alter the expression of itself and inhibited that of TGF-β1. These results suggest a process whereby injury activates TGF-β and then primes cells to respond to other growth factors *(155)*.

ACKNOWLEDGMENTS

Work was supported by NIH grants P20RR018757, AR46557, and DK067836.

REFERENCES

1. Gailit J, Clark RA. Wound repair in the context of extracellular matrix. Curr Opin Cell Biol 1994; 6(5):717–725.
2. Massagué J. How cells read TGF-beta signals. Nat Rev Mol Cell Biol 2000;1(3):169–178.
3. Attisano L, Warana JL. Signal transduction by the TGF-beta superfamily. Science 2002;296:1646.
4. Hannon GJ, Beach D. p15INK4B is a potential effector of TGF-beta-induced cell cycle arrest. Nature 1994;371(6494):257–261.
5. Reynisdottir I, Polyak K, Iavarone A, Massagué J. Kip/Cip and Ink4 Cdk inhibitors cooperate to induce cell cycle arrest in response to TGF-beta. Genes Dev 1995;9(15):1831–1845.
6. Reynisdottir I, Massagué J. The subcellular locations of p15(Ink4b) and p27(Kip1) coordinate their inhibitory interactions with cdk4 and cdk2. Genes Dev 1997;11(4):492–503.
7. Levine JH, Moses HL, Gold LI, Nanney LB. Spatial and temporal patterns of immunoreactive transforming growth factor beta 1, beta 2, and beta 3 during excisional wound repair. Am J Pathol 1993; 143(2):368–380.
8. Kane CJ, Hebda PA, Mansbridge JN, Hanawalt PC. Direct evidence for spatial and temporal regulation of transforming growth factor beta 1 expression during cutaneous wound healing. J Cell Physiol 1991;148(1):157–173.

9. Coerper S, Sigloch E, Cox D, Starlinger M, Koveker G, Becker HD. Recombinant human transforming growth factor beta 3 accelerates gastric ulcer healing in rats. Scand J Gastroenterol 1997; 32(10):985–990.
10. Sato Y, Rifkin DB. Inhibition of endothelial cell movement by pericytes and smooth muscle cells: activation of a latent transforming growth factor-beta 1-like molecule by plasmin during co-culture. J Cell Biol 1989;109(1):309–315.
11. Schultz-Cherry S, Murphy-Ullrich JE. Thrombospondin causes activation of latent transforming growth factor-beta secreted by endothelial cells by a novel mechanism. J Cell Biol 1993;122(4):923–932.
12. Breuss JM, Gallo J, DeLisser HM, et al. Expression of the beta 6 integrin subunit in development, neoplasia and tissue repair suggests a role in epithelial remodeling. J Cell Sci 1995;108(Pt 6):2241–2251.
13. Miller LA, Barnett NL, Sheppard D, Hyde DM. Expression of the beta6 integrin subunit is associated with sites of neutrophil influx in lung epithelium. J Histochem Cytochem 2001;49(1):41–48.
14. Crawford SE, Stellmach V, Murphy-Ullrich JE, et al. Thrombospondin-1 is a major activator of TGF-beta1 in vivo. Cell 1998;93(7):1159–1170.
15. Ramirez F, Di Liberto M. Complex and diversified regulatory programs control the expression of vertebrate collagen genes. FASEB J 1990;4(6):1616–1623.
16. Boast S, Su MW, Ramirez F, Sanchez M, Avvedimento EV. Functional analysis of cis-acting DNA sequences controlling transcription of the human type I collagen genes. J Biol Chem 1990;265(22): 13,351–13,356.
17. Inagaki Y, Truter S, Ramirez F. Transforming growth factor-beta stimulates alpha 2(I) collagen gene expression through a cis-acting element that contains an Sp1-binding site. J Biol Chem 1994;269(20): 14,828–14,834.
18. Inagaki Y, Truter S, Greenwel P, et al. Regulation of the alpha 2(I) collagen gene transcription in fat-storing cells derived from a cirrhotic liver. Hepatology 1995;22(2):573–579.
19. Greenwel P, Inagaki Y, Hu W, Walsh M, Ramirez F. Sp1 is required for the early response of alpha2(I) collagen to transforming growth factor-beta1. J Biol Chem 1997;272(32):19,738–19,745.
20. Chen SJ, Yuan W, Lo S, Trojanowska M, Varga J. Interaction of smad3 with a proximal smad-binding element of the human alpha2(I) procollagen gene promoter required for transcriptional activation by TGF-beta. J Cell Physiol 2000;183(3):381–392.
21. Chen SJ, Yuan W, Mori Y, Levenson A, Trojanowska M, Varga J. Stimulation of type I collagen transcription in human skin fibroblasts by TGF-beta: involvement of Smad 3. J Invest Dermatol 1999;112(1):49–57.
22. Dennler S, Itoh S, Vivien D, ten Dijke P, Huet S, Gauthier JM. Direct binding of Smad3 and Smad4 to critical TGF beta-inducible elements in the promoter of human plasminogen activator inhibitor-type 1 gene. EMBO J 1998;17(11):3091–3100.
23. Jonk LJ, Itoh S, Heldin CH, ten Dijke P, Kruijer W. Identification and functional characterization of a Smad binding element (SBE) in the JunB promoter that acts as a transforming growth factor-beta, activin, and bone morphogenetic protein-inducible enhancer. J Biol Chem 1998;273(33):21,145–21,152.
24. Moustakas A, Kardassis D. Regulation of the human p21/WAF1/Cip1 promoter in hepatic cells by functional interactions between Sp1 and Smad family members. Proc Natl Acad Sci USA 1998; 95(12):6733–6738.
25. Nagarajan RP, Zhang J, Li W, Chen Y. Regulation of Smad7 promoter by direct association with Smad3 and Smad4. J Biol Chem 1999;274(47):33,412–33,418.
26. Sakai LY, Keene DR, Morris NP, Burgeson RE. Type VII collagen is a major structural component of anchoring fibrils. J Cell Biol 1986;103(4):1577–1586.
27. Uitto J, Chung-Honet LC, Christiano AM. Molecular biology and pathology of type VII collagen. Exp Dermatol 1992;1(1):2–11.
28. Vindevoghel L, Lechleider RJ, Kon A, et al. SMAD3/4-dependent transcriptional activation of the human type VII collagen gene (COL7A1) promoter by transforming growth factor beta. Proc Natl Acad Sci USA 1998;95(25):14,769–14,774.
29. Verrecchia F, Vindevoghel L, Lechleider RJ, Uitto J, Roberts AB, Mauviel A. Smad3/AP-1 interactions control transcriptional responses to TGF-beta in a promoter-specific manner. Oncogene 2001;20(26): 3332–3340.
30. Naso M, Uitto J, Klement JF. Transcriptional control of the mouse Col7a1 gene in keratinocytes: basal and transforming growth factor-beta regulated expression. J Invest Dermatol 2003;121(6):1469–1478.
31. Verrecchia F, Mauviel A. TGF-beta and TNF-alpha: antagonistic cytokines controlling type I collagen gene expression. Cell Signal 2004;16(8):873–880.
32. Agren MS. Matrix metalloproteinases (MMPs) are required for re-epithelialization of cutaneous wounds. Arch Dermatol Res 1999;291(11):583–590.

33. Makela M, Larjava H, Pirila E, et al. Matrix metalloproteinase 2 (gelatinase A) is related to migration of keratinocytes. Exp Cell Res 1999;251(1):67–78.
34. Pattison PS. The importance of reassessment to chronic wound care. Ostomy Wound Manage 2001; 47(8):24–25.
35. Westermarck J, Kahari VM. Regulation of matrix metalloproteinase expression in tumor invasion. FASEB J 1999;13(8):781–792.
36. Aimes RT, Quigley JP. Matrix metalloproteinase-2 is an interstitial collagenase. Inhibitor-free enzyme catalyzes the cleavage of collagen fibrils and soluble native type I collagen generating the specific 3/4- and 1/4-length fragments. J Biol Chem 1995;270(11):5872–5876.
37. Ashcroft GS, Yang X, Glick AB, et al. Mice lacking Smad3 show accelerated wound healing and an impaired local inflammatory response. Nat Cell Biol 1999;1(5):260–266.
38. Roberts AB, Piek E, Bottinger EP, Ashcroft G, Mitchell JB, Flanders KC. Is Smad3 a major player in signal transduction pathways leading to fibrogenesis? Chest 2001;120(1 Suppl):43S–47S.
39. Sano Y, Harada J, Tashiro S, Gotoh-Mandeville R, Maekawa T, Ishii S. ATF-2 is a common nuclear target of Smad and TAK1 pathways in transforming growth factor-beta signaling. J Biol Chem 1999; 274(13):8949–8957.
40. Bhowmick NA, Zent R, Ghiassi M, McDonnell M, Moses HL. Integrin beta 1 signaling is necessary for transforming growth factor-beta activation of p38MAPK and epithelial plasticity. J Biol Chem 2001;276(50):46,707–46,713.
41. Yamaguchi K, Shirakabe K, Shibuya H, et al. Identification of a member of the MAPKKK family as a potential mediator of TGF-beta signal transduction. Science 1995;270(5244):2008–2011.
42. Wang W, Zhou G, Hu MC, Yao Z, Tan TH. Activation of the hematopoietic progenitor kinase-1 (HPK1)-dependent, stress-activated c-Jun N-terminal kinase (JNK) pathway by transforming growth factor beta (TGF-beta)-activated kinase (TAK1), a kinase mediator of TGF beta signal transduction. J Biol Chem 1997;272(36):22,771–22,775.
43. Hanafusa H, Ninomiya-Tsuji J, Masuyama N, et al. Involvement of the p38 mitogen-activated protein kinase pathway in transforming growth factor-beta-induced gene expression. J Biol Chem 1999; 274(38):27,161–27,167.
44. Ravanti L, Toriseva M, Penttinen R, et al. Expression of human collagenase-3 (MMP-13) by fetal skin fibroblasts is induced by transforming growth factor beta via p38 mitogen-activated protein kinase. FASEB J 2001;15(6):1098–1100.
45. Zhang Y, Feng XH, Derynck R. Smad3 and Smad4 cooperate with c-Jun/c-Fos to mediate TGF-beta-induced transcription. Nature 1998;394(6696):909–913.
46. Dennler S, Prunier C, Ferrand N, Gauthier JM, Atfi A. c-Jun inhibits transforming growth factor beta-mediated transcription by repressing Smad3 transcriptional activity. J Biol Chem 2000;275(37): 28,858–28,865.
47. Falanga V, Schrayer D, Cha J, et al. Full-thickness wounding of the mouse tail as a model for delayed wound healing: accelerated wound closure in Smad3 knock-out mice. Wound Repair Regen 2004; 12(3):320–326.
48. Hocevar BA, Brown TL, Howe PH. TGF-beta induces fibronectin synthesis through a c-Jun N-terminal kinase-dependent, Smad4-independent pathway. EMBO J 1999;18(5):1345–1356.
49. Flanders KC, Major CD, Arabshahi A, et al. Interference with transforming growth factor-beta/Smad3 signaling results in accelerated healing of wounds in previously irradiated skin. Am J Pathol 2003;163(6):2247–2257.
50. Liu C, Gaca MD, Swenson ES, Vellucci VF, Reiss M, Wells RG. Smads 2 and 3 are differentially activated by transforming growth factor-beta (TGF-beta) in quiescent and activated hepatic stellate cells. Constitutive nuclear localization of Smads in activated cells is TGF-beta-independent. J Biol Chem 2003;278(13):11,721–11,728.
51. Mori Y, Chen SJ, Varga J. Expression and regulation of intracellular SMAD signaling in scleroderma skin fibroblasts. Arthritis Rheum 2003;48(7):1964–1978.
52. Holmes A, Abraham DJ, Sa S, Shiwen X, Black CM, Leask A. CTGF and SMADs, maintenance of scleroderma phenotype is independent of SMAD signaling. J Biol Chem 2001;276(14):10,594–10,601.
53. Dooley S, Delvoux B, Streckert M, et al. Transforming growth factor beta signal transduction in hepatic stellate cells via Smad2/3 phosphorylation, a pathway that is abrogated during in vitro progression to myofibroblasts. TGFbeta signal transduction during transdifferentiation of hepatic stellate cells. FEBS Lett 2001;502(1–2):4–10.
54. Falanga V, Gerhardt CO, Dasch JR, Takehara K, Ksander GA. Skin distribution and differential expression of transforming growth factor beta 1 and beta 2. J Dermatol Sci 1992;3(3):131–136.

55. Higley H, Persichitte K, Chu S, Waegell W, Vancheeswaran R, Black C. Immunocytochemical localization and serologic detection of transforming growth factor beta 1. Association with type I procollagen and inflammatory cell markers in diffuse and limited systemic sclerosis, morphea, and Raynaud's phenomenon. Arthritis Rheum 1994;37(2):278–288.
56. Mori T, Kawara S, Shinozaki M, et al. Role and interaction of connective tissue growth factor with transforming growth factor-beta in persistent fibrosis: a mouse fibrosis model. J Cell Physiol 1999; 181(1):153–159.
57. Thiery JP. Epithelial–mesenchymal transitions in development and pathologies. Curr Opin Cell Biol 2003;15(6):740–746.
58. Oft M, Heider KH, Beug H. TGFbeta signaling is necessary for carcinoma cell invasiveness and metastasis. Curr Biol 1998;8(23):1243–1252.
59. Oft M, Peli J, Rudaz C, Schwarz H, Beug H, Reichmann E. TGF-beta1 and Ha-Ras collaborate in modulating the phenotypic plasticity and invasiveness of epithelial tumor cells. Genes Dev 1996; 10(19):2462–2477.
60. Desmouliere A, Geinoz A, Gabbiani F, Gabbiani G. Transforming growth factor-beta 1 induces alpha-smooth muscle actin expression in granulation tissue myofibroblasts and in quiescent and growing cultured fibroblasts. J Cell Biol 1993;122(1):103–111.
61. Ronnov-Jessen L, Petersen OW. Induction of alpha-smooth muscle actin by transforming growth factor-beta 1 in quiescent human breast gland fibroblasts. Implications for myofibroblast generation in breast neoplasia. Lab Invest 1993;68(6):696–707.
62. Crowley E, Horwitz AF. Tyrosine phosphorylation and cytoskeletal tension regulate the release of fibroblast adhesions. J Cell Biol 1995;131(2):525–537.
63. Serini G, Bochaton-Piallat ML, Ropraz P, et al. The fibronectin domain ED-A is crucial for myofibroblastic phenotype induction by transforming growth factor-beta1. J Cell Biol 1998;142(3): 873–881.
64. Leask A, Denton CP, Abraham DJ. Insights into the molecular mechanism of chronic fibrosis: the role of connective tissue growth factor in scleroderma. J Invest Dermatol 2004;122(1):1–6.
65. Igarashi A, Nashiro K, Kikuchi K, et al. Connective tissue growth factor gene expression in tissue sections from localized scleroderma, keloid, and other fibrotic skin disorders. J Invest Dermatol 1996;106(4):729–733.
66. Abraham DJ, Shiwen X, Black CM, Sa S, Xu Y, Leask A. Tumor necrosis factor alpha suppresses the induction of connective tissue growth factor by transforming growth factor-beta in normal and scleroderma fibroblasts. J Biol Chem 2000;275(20):15,220–15,225.
67. Crean JK, Finlay D, Murphy M, et al. The role of p42/44 MAPK and protein kinase B in connective tissue growth factor induced extracellular matrix protein production, cell migration, and actin cytoskeletal rearrangement in human mesangial cells. J Biol Chem 2002;277(46):44,187–44,194.
68. Ito Y, Goldschmeding R, Bende R, et al. Kinetics of connective tissue growth factor expression during experimental proliferative glomerulonephritis. J Am Soc Nephrol 2001;12(3):472–484.
69. Wang S, Denichilo M, Brubaker C, Hirschberg R. Connective tissue growth factor in tubulointerstitial injury of diabetic nephropathy. Kidney Int 2001;60(1):96–105.
70. Grotendorst GR. Connective tissue growth factor: a mediator of TGF-beta action on fibroblasts. Cytokine Growth Factor Rev 1997;8(3):171–179.
71. Woodley DT. Re-epithelialization. In: Clark RAF, ed. The molecular and cellular biology of wound repair. 2nd ed. New York: Plenum Press, 1996;pp. 339–354.
72. Martin P. Wound healing—aiming for perfect skin regeneration. Science 1997;276(5309):75–81.
73. Sarret Y, Woodley DT, Grigsby K, Wynn K, O'Keefe EJ. Human keratinocyte locomotion: the effect of selected cytokines. J Invest Dermatol 1992;98(1):12–16.
74. Gailit J, Welch MP, Clark RA. TGF-beta 1 stimulates expression of keratinocyte integrins during re-epithelialization of cutaneous wounds. J Invest Dermatol 1994;103(2):221–227.
75. Zhang K, Kramer RH. Laminin 5 deposition promotes keratinocyte motility. Exp Cell Res 1996; 227(2):309–322.
76. Qin P, Kurpakus MA. The role of laminin-5 in TGF alpha/EGF-mediated corneal epithelial cell motility. Exp Eye Res 1998;66(5):569–579.
77. Goldfinger LE, Stack MS, Jones JC. Processing of laminin-5 and its functional consequences: role of plasmin and tissue-type plasminogen activator. J Cell Biol 1998;141(1):255–265.
78. Nugent MA, Iozzo RV. Fibroblast growth factor-2. Int J Biochem Cell Biol 2000;32(2):115–120.
79. Decline F, Rousselle P. Keratinocyte migration requires alpha2beta1 integrin-mediated interaction with the laminin 5 gamma2 chain. J Cell Sci 2001;114(Pt 4):811–823.

80. Korang K, Christiano AM, Uitto J, Mauviel A. Differential cytokine modulation of the genes LAMA3, LAMB3, and LAMC2, encoding the constitutive polypeptides, alpha 3, beta 3, and gamma 2, of human laminin 5 in epidermal keratinocytes. FEBS Lett 1995;368(3):556–558.
81. Kainulainen T, Hakkinen L, Hamidi S, et al. Laminin-5 expression is independent of the injury and the microenvironment during reepithelialization of wounds. J Histochem Cytochem 1998;46(3): 353–360.
82. Decline F, Okamoto O, Mallein-Gerin F, et al. Keratinocyte motility induced by TGF-beta1 is accompanied by dramatic changes in cellular interactions with laminin 5. Cell Motil Cytoskeleton 2003; 54(1):64–80.
83. Jeong HW, Kim IS. TGF-beta1 enhances betaig-h3-mediated keratinocyte cell migration through the alpha3beta1 integrin and PI3K. J Cell Biochem 2004;92(4):770–780.
84. Kulkarni AB, Huh CG, Becker D, et al. Transforming growth factor beta 1 null mutation in mice causes excessive inflammatory response and early death. Proc Natl Acad Sci USA 1993;90(2):770–774.
85. Bottinger EP, Letterio JJ, Roberts AB. Biology of TGF-beta in knockout and transgenic mouse models. Kidney Int 1997;51(5):1355–1360.
86. Beck LS, Deguzman L, Lee WP, Xu Y, McFatridge LA, Amento EP. TGF-beta 1 accelerates wound healing: reversal of steroid-impaired healing in rats and rabbits. Growth Factors 1991;5(4):295–304.
87. Chesnoy S, Lee PY, Huang L. Intradermal injection of transforming growth factor-beta1 gene enhances wound healing in genetically diabetic mice. Pharm Res 2003;20(3):345–350.
88. Pierce GF, Mustoe TA, Senior RM, et al. In vivo incisional wound healing augmented by platelet-derived growth factor and recombinant c-sis gene homodimeric proteins. J Exp Med 1988;167(3): 974–987.
89. Lanning DA, Diegelmann RF, Yager DR, Wallace ML, Bagwell CE, Haynes JH. Myofibroblast induction with transforming growth factor-beta1 and -beta3 in cutaneous fetal excisional wounds. J Pediatr Surg 2000;35(2):183–187; discussion 7–8.
90. Garlick JA, Taichman LB. Effect of TGF-beta 1 on re-epithelialization of human keratinocytes in vitro: an organotypic model. J Invest Dermatol 1994;103(4):554–559.
91. Garlick JA, Parks WC, Welgus HG, Taichman LB. Re-epithelialization of human oral keratinocytes in vitro. J Dent Res 1996;75(3):912–918.
92. Fowlis DJ, Cui W, Johnson SA, Balmain A, Akhurst RJ. Altered epidermal cell growth control in vivo by inducible expression of transforming growth factor beta 1 in the skin of transgenic mice. Cell Growth Differ 1996;7(5):679–687.
93. Tyrone JW, Marcus JR, Bonomo SR, Mogford JE, Xia Y, Mustoe TA. Transforming growth factor beta3 promotes fascial wound healing in a new animal model. Arch Surg 2000;135(10):1154–1159.
94. Sumiyoshi K, Nakao A, Setoguchi Y, Okumura K, Ogawa H. Exogenous Smad3 accelerates wound healing in a rabbit dermal ulcer model. J Invest Dermatol 2004;123(1):229–236.
95. Hasan A, Murata H, Falabella A, et al. Dermal fibroblasts from venous ulcers are unresponsive to the action of transforming growth factor-beta 1. J Dermatol Sci 1997;16(1):59–66.
96. Agren MS, Steenfos HH, Dabelsteen S, Hansen JB, Dabelsteen E. Proliferation and mitogenic response to PDGF-BB of fibroblasts isolated from chronic venous leg ulcers is ulcer-age dependent. J Invest Dermatol 1999;112(4):463–469.
97. Kim BC, Kim HT, Park SH, et al. Fibroblasts from chronic wounds show altered TGF-beta-signaling and decreased TGF-beta Type II receptor expression. J Cell Physiol 2003;195(3):331–336.
98. Bitar MS, Labbad ZN. Transforming growth factor-beta and insulin-like growth factor-I in relation to diabetes-induced impairment of wound healing. J Surg Res 1996;61(1):113–119.
99. Wang JF, Olson ME, Reno CR, Kulyk W, Wright JB, Hart DA. Molecular and cell biology of skin wound healing in a pig model. Connect Tissue Res 2000;41(3):195–211.
100. Gosiewska A, Yi CF, Brown LJ, Cullen B, Silcock D, Geesin JC. Differential expression and regulation of extracellular matrix-associated genes in fetal and neonatal fibroblasts. Wound Repair Regen 2001; 9(3):213–222.
101. Ellis IR, Schor SL. Differential effects of TGF-beta1 on hyaluronan synthesis by fetal and adult skin fibroblasts: implications for cell migration and wound healing. Exp Cell Res 1996;228(2):326–333.
102. Liechty KW, Nesbit M, Herlyn M, Radu A, Adzick NS, Crombleholme TM. Adenoviral-mediated overexpression of platelet-derived growth factor-B corrects ischemic impaired wound healing. J Invest Dermatol 1999;113(3):375–383.
103. Chegini N. The role of growth factors in peritoneal healing: transforming growth factor beta (TGF-beta). Eur J Surg Suppl 1997;163(577):17–23.

104. Wang X, Smith P, Pu LL, Kim YJ, Ko F, Robson MC. Exogenous transforming growth factor beta(2) modulates collagen I and collagen III synthesis in proliferative scar xenografts in nude rats. J Surg Res 1999;87(2):194–200.
105. Robson MC, Steed DL, McPherson JM, Pratt BM. Use of transforming growth factor beta-2(TGF-beta2) in the treatment of chronic foot ulcers in diabetic patients. Wound Repair Regen 1999;7(4): A266.
106. Gorvy DA, Herrick SE, Shah M, Ferguson MW. Experimental manipulation of transforming growth factor-beta isoforms significantly affects adhesion formation in a murine surgical model. Am J Pathol 2005;167(4):1005–1019.
107. Cordeiro MF. Beyond Mitomycin: TGF-beta and wound healing. Prog Retin Eye Res 2002;21(1):75–89.
108. O'Reilly MA, Danielpour D, Roberts AB, Sporn MB. Regulation of expression of transforming growth factor-beta 2 by transforming growth factor-beta isoforms is dependent upon cell type. Growth Factors 1992;6(3):193–201.
109. Taya Y, O'Kane S, Ferguson MW. Pathogenesis of cleft palate in TGF-beta3 knockout mice. Development 1999;126(17):3869–3879.
110. Verrecchia F, Tacheau C, Wagner EF, Mauviel A. A central role for the JNK pathway in mediating the antagonistic activity of pro-inflammatory cytokines against transforming growth factor-beta-driven SMAD3/4-specific gene expression. J Biol Chem 2003;278(3):1585–1593.
111. Mori R, Kondo T, Ohshima T, Ishida Y, Mukaida N. Accelerated wound healing in tumor necrosis factor receptor p55-deficient mice with reduced leukocyte infiltration. FASEB J 2002;16(9):963–974.
112. Bitzer M, von Gersdorff G, Liang D, et al. A mechanism of suppression of TGF-beta/SMAD signaling by NF-kappa B/RelA. Genes Dev 2000;14(2):187–197.
113. Verrecchia F, Pessah M, Atfi A, Mauviel A. Tumor necrosis factor-alpha inhibits transforming growth factor-beta /Smad signaling in human dermal fibroblasts via AP-1 activation. J Biol Chem 2000; 275(39):30,226–30,231.
114. Holmes A, Abraham DJ, Chen Y, et al. Constitutive connective tissue growth factor expression in scleroderma fibroblasts is dependent on Sp1. J Biol Chem 2003;278(43):41,728–41,733.
115. Leask A, Abraham DJ. The role of connective tissue growth factor, a multifunctional matricellular protein, in fibroblast biology. Biochem Cell Biol 2003;81(6):355–363.
116. Candela M, Barker SC, Ballou LR. Sphingosine synergistically stimulates tumor necrosis factor alpha-induced prostaglandin E2 production in human fibroblasts. J Exp Med 1991;174(6):1363–1369.
117. Diaz A, Chepenik KP, Korn JH, Reginato AM, Jimenez SA. Differential regulation of cyclooxygenases 1 and 2 by interleukin-1 beta, tumor necrosis factor-alpha, and transforming growth factor-beta 1 in human lung fibroblasts. Exp Cell Res 1998;241(1):222–229.
118. Lee JL, Mukhtar H, Bickers DR, Kopelovich L, Athar M. Cyclooxygenases in the skin: pharmacological and toxicological implications. Toxicol Appl Pharmacol 2003;192(3):294–306.
119. Balint GA. Experimental and clinical observations on the role of endogenous and on the effect of exogenous prostacyclin in gastric mucosa. A personal view. Exp Toxicol Pathol 1994;46(1):37–40.
120. Saltzman LE, Moss J, Berg RA, Hom B, Crystal RG. Modulation of collagen production by fibroblasts. Effects of chronic exposure to agonists that increase intracellular cyclic AMP. Biochem J 1982; 204(1):25–30.
121. Guller S, Wozniak R, Leibman MI, Lockwood CJ. Negative regulation of placental fibronectin expression by glucocorticoids and cyclic adenosine 3′,5′-monophosphate. Ann N Y Acad Sci 1994; 734:132–142.
122. Riquet FB, Lai WF, Birkhead JR, Suen LF, Karsenty G, Goldring MB. Suppression of type I collagen gene expression by prostaglandins in fibroblasts is mediated at the transcriptional level. Mol Med 2000;6(8):705–719.
123. Stratton R, Rajkumar V, Ponticos M, et al. Prostacyclin derivatives prevent the fibrotic response to TGF-beta by inhibiting the Ras/MEK/ERK pathway. FASEB J 2002;16(14):1949–1951.
124. Scorza R, Caronni M, Mascagni B, et al. Effects of long-term cyclic iloprost therapy in systemic sclerosis with Raynaud's phenomenon. A randomized, controlled study. Clin Exp Rheumatol 2001; 19(5):503–508.
125. Boehm U, Klamp T, Groot M, Howard JC. Cellular responses to interferon-gamma. Annu Rev Immunol 1997;15:749–795.
126. Gillery P, Serpier H, Polette M, et al. Gamma-interferon inhibits extracellular matrix synthesis and remodeling in collagen lattice cultures of normal and scleroderma skin fibroblasts. Eur J Cell Biol 1992;57(2):244–253.

127. Ulloa L, Doody J, Massagué J. Inhibition of transforming growth factor-beta/SMAD signalling by the interferon-gamma/STAT pathway. Nature 1999;397(6721):710–713.
128. Ghosh AK, Yuan W, Mori Y, Chen S, Varga J. Antagonistic regulation of type I collagen gene expression by interferon-gamma and transforming growth factor-beta. Integration at the level of p300/CBP transcriptional coactivators. J Biol Chem 2001;276(14):11,041–11,048.
129. Higashi K, Inagaki Y, Fujimori K, Nakao A, Kaneko H, Nakatsuka I. Interferon-gamma interferes with transforming growth factor-beta signaling through direct interaction of YB-1 with Smad3. J Biol Chem 2003;278(44):43,470–43,479.
130. Roberts AB, Sporn MB, Assoian RK, et al. Transforming growth factor type beta: rapid induction of fibrosis and angiogenesis in vivo and stimulation of collagen formation in vitro. Proc Natl Acad Sci USA 1986;83(12):4167–4171.
131. Yang EY, Moses HL. Transforming growth factor beta 1-induced changes in cell migration, proliferation, and angiogenesis in the chicken chorioallantoic membrane. J Cell Biol 1990;111(2):731–741.
132. Baird A, Durkin T. Inhibition of endothelial cell proliferation by type beta-transforming growth factor: interactions with acidic and basic fibroblast growth factors. Biochem Biophys Res Commun 1986; 138(1):476–482.
133. Frater-Schroder M, Muller G, Birchmeier W, Bohlen P. Transforming growth factor-beta inhibits endothelial cell proliferation. Biochem Biophys Res Commun 1986;137(1):295–302.
134. Heimark RL, Twardzik DR, Schwartz SM. Inhibition of endothelial regeneration by type-beta transforming growth factor from platelets. Science 1986;233(4768):1078–1080.
135. Weisman DM, Polverini PJ, Kamp DW, Leibovich SJ. Transforming growth factor-beta (TGF-beta) is chemotactic for human monocytes and induces their expression of angiogenic activity. Biochem Biophys Res Commun 1988;157:793–800.
136. Iruela-Arispe ML, Sage EH. Endothelial cells exhibiting angiogenesis in vitro proliferate in response to TGF-beta 1. J Cell Biochem 1993;52(4):414–430.
137. Sankar S, Mahooti-Brooks N, Bensen L, McCarthy TL, Centrella M, Madri JA. Modulation of transforming growth factor beta receptor levels on microvascular endothelial cells during in vitro angiogenesis. J Clin Invest 1996;97(6):1436–1446.
138. Battegay EJ, Rupp J, Iruela-Arispe L, Sage EH, Pech M. PDGF-BB modulates endothelial proliferation and angiogenesis in vitro via PDGF beta-receptors. J Cell Biol 1994;125(4):917–928.
139. Dvorak HF, Brown LF, Detmar M, Dvorak AM. Vascular permeability factor/vascular endothelial growth factor, microvascular hyperpermeability, and angiogenesis. Am J Pathol 1995;146(5):1029–1039.
140. Brown LF, Yeo KT, Berse B, et al. Expression of vascular permeability factor (vascular endothelial growth factor) by epidermal keratinocytes during wound healing. J Exp Med 1992;176(5):1375–1379.
141. Shweiki D, Itin A, Soffer D, Keshet E. Vascular endothelial growth factor induced by hypoxia may mediate hypoxia-initiated angiogenesis. Nature 1992;359(6398):843–845.
142. Detmar M, Brown LF, Berse B, et al. Hypoxia regulates the expression of vascular permeability factor/vascular endothelial growth factor (VPF/VEGF) and its receptors in human skin. J Invest Dermatol 1997;108(3):263–268.
143. Brogi E, Schatteman G, Wu T, et al. Hypoxia-induced paracrine regulation of vascular endothelial growth factor receptor expression. J Clin Invest 1996;97(2):469–476.
144. Nissen NN, Polverini PJ, Koch AE, Volin MV, Gamelli RL, DiPietro LA. Vascular endothelial growth factor mediates angiogenic activity during the proliferative phase of wound healing. Am J Pathol 1998;152(6):1445–1452.
145. Veikkola T, Alitalo K. VEGFs, receptors and angiogenesis. Semin Cancer Biol 1999;9(3):211–220.
146. Valenzuela DM, Griffiths JA, Rojas J, et al. Angiopoietins 3 and 4: diverging gene counterparts in mice and humans. Proc Natl Acad Sci USA 1999;96(5):1904–1909.
147. Richardson TP, Trinkaus-Randall V, Nugent MA. Regulation of basic fibroblast growth factor binding and activity by cell density and heparan sulfate. J Biol Chem 1999;274(19):13,534–13,540.
148. Arese M, Chen Y, Florkiewicz RZ, Gualandris A, Shen B, Rifkin DB. Nuclear activities of basic fibroblast growth factor: potentiation of low-serum growth mediated by natural or chimeric nuclear localization signals. Mol Biol Cell 1999;10(5):1429–1444.
149. Patry V, Bugler B, Maret A, Potier M, Prats H. Endogenous basic fibroblast growth factor isoforms involved in different intracellular protein complexes. Biochem J 1997;326(Pt 1):259–264.
150. Powell PP, Klagsbrun M. Three forms of rat basic fibroblast growth factor are made from a single mRNA and localize to the nucleus. J Cell Physiol 1991;148(2):202–210.
151. Brown CT, Vural M, Johnson M, Trinkaus-Randall V. Age-related changes of scleral hydration and sulfated glycosaminoglycans. Mech Ageing Dev 1994;77(2):97–107.

152. Trinkaus-Randall V, Nugent MA. Biological response to a synthetic cornea. J Control Release 1998;53(1–3):205–214.
153. Brown CT, Nugent MA, Lau FW, Trinkaus-Randall V. Characterization of proteoglycans synthesized by cultured corneal fibroblasts in response to transforming growth factor beta and fetal calf serum. J Biol Chem 1999;274(11):7111–7119.
154. Song QH, Singh RP, Richardson TP, Nugent MA, Trinkaus-Randall V. Transforming growth factor-beta1 expression in cultured corneal fibroblasts in response to injury. J Cell Biochem 2000; 77(2):186–199.
155. Song QH, Klepeis VE, Nugent MA, Trinkaus-Randall V. TGF-beta1 regulates TGF-beta1 and FGF-2 mRNA expression during fibroblast wound healing. Mol Pathol 2002;55(3):164–176.

44 Topical Application of TGF-β1 Peptide Inhibitors for the Therapy of Skin Fibrosis

Javier Dotor and José L. Pablos

CONTENTS

INTRODUCTION
TGF-β AND SCLERODERMA
TGF-β TARGETING STRATEGIES FOR THE THERAPY OF FIBROSIS
DEVELOPMENT OF SMALL TGF-β1 ANTAGONISTIC PEPTIDES
PRECLINICAL STUDIES OF P144
REFERENCES

Abstract

Transforming growth factor beta (TGF-β) plays a crucial role in the pathogenesis of skin fibrotic diseases by inducing extracellular matrix gene expression and sustaining fibroblast growth and differentiation. Systemic inhibition of TGF-β by different agents has been shown to effectively inhibit fibrosis in different animal models. However, systemic inhibition of TGF-β raises important safety issues because of the pleiotropic physiological effects of this factor.

Targeting of downstream factors specifically involved in TGF-β profibrotic signaling or local targeting of TGF-β represents potential alternatives to systemic inhibitors. Topical application of a short peptide derived from TGF-β1 type III receptor is effective in preventing or ameliorating established fibrosis in a model of bleomycin-induced scleroderma, suggesting that topical application of small anti-TGF-β peptides is a feasible strategy to treat pathological skin scarring and skin fibrotic diseases.

Key Words: Transforming growth factor-β; antagonists; fibrosis; skin; topical administration; scleroderma; therapy.

1. INTRODUCTION

Transforming growth factor beta (TGF-β) is a soluble glycoprotein corresponding to a large superfamily of structurally related regulatory proteins that includes TGF-β1, 2, and 3 isoforms, activins, and bone morphogenetic proteins. TGF-β structurally consists of two equal subunits linked by an interchain disulfide bond forming a homodimer of 25 kDa. Factors of this superfamily share structural and functional features, and all participate to some extent in widespread growth, development, differentiation, and reparative processes through their specific cellular membrane receptors *(1,2)*.

From: *Cancer Drug Discovery and Development: Transforming Growth Factor-β in Cancer Therapy, Vol I: Basic and Clinical Biology*
Edited by: Sonia B. Jakowlew © Humana Press Inc., Totowa, NJ

TGF-β role in tissue repair was first described by Roberts et al. in 1986 *(3)*. In this seminal work, subcutaneous injection of TGF-β was observed to induce the formation of a granulation tissue similar to what develops during wound healing and, in vitro, TGF-β was found to induce collagen synthesis in fibroblasts. Further studies have extensively confirmed the key role of TGF-β in normal and pathological tissue repair, through a variety of cellular effects such as increased synthesis and decreased degradation of multiple extracellular matrix (ECM) components and variable effects on mesenchymal cells recruitment, growth, and differentiation *(4)*. TGF-β binding to its cellular receptors induces the activation of several kinase signaling cascades leading to the phosphorylation of SMAD proteins as well as the activation of SMAD-independent kinases, which collectively activate ECM proteins expression and fibroblast growth and differentiation to an active reparative or profibrotic phenotype *(5,6)*. These effects are further enhanced by TGF-β inhibitory effects on ECM degradation, through both decreased synthesis of ECM proteases and increased synthesis of protease inhibitors *(7)*.

Excessive ECM accumulation in the connective tissue is the pathological hallmark of a large group of diseases, termed scarring or fibrotic diseases, which collectively represent a vast and heterogeneous chapter of human pathology *(8)*. Fibrosis is usually initiated by tissue injury of variable etiology, almost invariable accompanied by an inflammatory component, and followed by an abnormally strong and prolonged accumulation of a disorganized ECM that can lead to functional compromise and organ failure. Some examples are kidney, lung, or liver fibrosis, which represent the endpoint of a variety of toxic, infectious viral or autoimmune syndromes *(9,10)*. Whereas the role played by TGF-β in these conditions is clear *(11)*, the genetic and physiopathological bases explaining why this pathway becomes overfunctional leading to human fibrosis remain obscure.

2. TGF-β AND SCLERODERMA

A relevant group of scarring diseases is represented by abnormal skin wound healing conditions such as hypertrophic scars and keloids and two idiopathic skin fibrotic diseases: localized and systemic sclerosis (SSc) (or scleroderma). The first group represents an abnormal process of wound healing, with a strong individual and probably genetic component. SSc is paradigmatic of this group of diseases. It is classified as a connective tissue inflammatory and autoimmune disease, but its main feature is a progressive fibrosis of the skin and internal organs leading to severe disability and often to death *(12)*.

In both localized scleroderma (also called morphea) and SSc, pathological and in vitro cellular studies have clearly shown the imprint of TGF-β in their pathogenesis. Both diseases are characterized by variable inflammatory cell infiltration of the dermis with an important T-cell and monocyte–macrophage component during the early phases, accompanied by other pathological changes such as mast cell degranulation and, particularly in SSc, severe vascular damage *(13)*. Although the specific mechanisms that link these relatively nonspecific changes to the development of progressive fibrosis have not been fully elucidated, the local production of profibrotic mediators such as cytokines, chemokines, and growth factors, notably TGF-β, underlie the profound phenotypic changes observed in SSc fibroblasts that persist under tissue culture, a phenomenon that has greatly facilitated in vitro studies *(13,14)*. SSc lesional fibroblasts display both in vivo and in vitro enhanced ECM synthesis and a myofibroblast phenotype similar to that found in wounds and are abnormally resistant to apoptotic death, which is the physiological mechanism of myofibroblast elimination at the end of the wound repair process *(14–17)*. Therefore, abnormal growth and differentiation, together with increased ECM synthesis, are TGF-β-inducible features that characterize SSc fibroblasts.

Table 1
Evidences for the Role of TGF-β in the Pathogenesis of Scleroderma

Lesional TGF-β expression
Increased activation of latent TGF-β
Abnormal TGF-β type I and type II receptors expression and function
Constitutively increased SMAD2/3 activation
Deficitary inhibitory SMAD7 expression
Increased CTGF expression
Increased myofibroblasts development and survival

In animal models, different gain or loss of function approaches have shown that manipulation of TGF-β pathway at the ligand, receptor, or intracellular SMAD signaling levels are sufficient to induce or reduce skin fibrotic responses, respectively, underscoring the concept of TGF-β as the most relevant factor involved in this process *(18–22)*. Abundant evidences for the role of TGF-β in human scleroderma have come from skin molecular pathology studies or from in vitro studies focused on the phenotype and behavior of cultured scleroderma fibroblasts (Table 1) and can be summarized as follows.

In situ, increased expression of TGF-β and TGF-β type I and type II receptors has been demonstrated in scleroderma skin lesions compared with healthy controls using in situ hybridization or immunohistochemical analyses *(23,24)*. Furthermore, evidence for enhanced SMAD2/3 phosphorylation and decreased inhibitory SMAD7 expression or activity has been demonstrated in SSc skin, suggesting that increased TGF-β signaling occurs in vivo and is not properly counterbalanced by inhibitory SMAD signaling *(25,26)*.

In vitro, stimulation of healthy skin fibroblast by TGF-β1 reproduces many of the characteristics of cultured fibroblasts from scleroderma lesions. Spontaneously increased ECM synthesis characterizes the scleroderma fibroblast phenotype in vitro, and this process is paralleled by increased expression of TGF-β type I and type II receptors as well as increased TGF-β binding capacity *(27,28)*. In addition, constitutive activation of the intracellular TGF-β SMAD2/3 axis has also been demonstrated in vitro in cultured scleroderma fibroblasts *(25,26,29)*. Abnormalities in basal and TGF-β-induced expression of SMAD7 (a factor induced by TGF-β1 that downregulates TGF-β1 activity as part of a negative feedback loop) have been suggested to underlie excessive SMAD2/3 signaling in SSc, although this has not been uniformly confirmed in different studies *(25,26,29)*. Another factor potentially involved in the excessive autocrine TGF-β signaling of SSc fibroblasts is a higher activation rate of latent TGF-β that can be because of higher expression of $\alpha_v\beta3$ and $\alpha_v\beta5$ integrins and thrombospondin-1 *(30–32)*. Finally, increased and abnormally regulated expression of the specific TGF-β-inducible connective tissue growth factor (CTGF) has been found in SSc fibroblasts *(33)*. CTGF displays profibrotic effects by its own that synergize with TGF-β in the development of chronic skin fibrosis *(34)*.

Therefore, targeting TGF-β, its signaling cascade, or its downstream profibrotic effects in SSc fibroblasts appears as attractive therapeutic goal for the therapy of this still untreatable disease. In animal models, downregulation of fibroblasts profibrotic phenotype and improvement of skin fibrosis have been achieved by using small molecules, neutralizing peptides, or genetic approaches targeting the TGF-β signaling pathway at different levels. However, translating these approaches to the therapy of the human disease is still a challenge.

3. TGF-β TARGETING STRATEGIES FOR THE THERAPY OF FIBROSIS

Currently available options for the therapy of human fibrotic diseases are mostly symptomatic or oriented to downmodulate the inflammatory or autoimmune component frequently

Table 2
TGF-β-Based Antifibrotic Therapeutic Approaches

Systemic TGF-β inhibitors (antibodies, soluble receptors, LAP, short peptides)
TGF-β receptor associated kinases (s.m.i.*)
SMAD3 (s.m.i.)
Non-SMAD signaling kinases: p38, Abl, FAK, and so on (s.m.i.)
Distal TGF-β effectors inhibitors: CTGF inhibitors (antibodies)
TGF-β signaling antagonistic mediators: interferons, PPAR-γ agonists, PGI_2

(*) s.m.i.: small molecule inhibitors.

associated to these conditions. Traditional immunosuppressive drugs have shown a poor risk/benefit profile, particularly for SSc, challenging the physiopathological concept of fibrosis as the consequence of immune-mediated inflammation and suggesting that once the fibrotic process is established, it proceeds independently of the autoimmune component. Alternatively, the mechanisms of autoimmunity in SSc might not be efficiently or specifically targeted by current immunosuppressive drugs. In this regard, the efficacy and safety of more intensive regimes such as high-dose immunoablative therapy followed by hematopoietic support is currently being tested *(35)*. However, and in spite of the broad spectrum of human diseases where the fibrotic component is the main cause of organ failure, antifibrotic therapies directly targeting ECM accumulation are not yet available.

The lack of pathogenesis-based antifibrotic therapies has promoted the development of a variety of TGF-β-inhibitory strategies during the last decade (Table 2). As previously mentioned, satisfactory preclinical data have been obtained in several animal models, but so far, the very limited clinical data available do not permit to raise conclusions on their long-term efficacy or safety for human use in this group of chronic conditions.

In animal models of fibrosis, such as bleomycin injection or graft vs host disease, systemic therapy with neutralizing anti-TGF-β1 antibodies or soluble TGF-β latency-associated peptide (LAP) has shown to ameliorate skin fibrosis *(36,37)*. Preliminary data from a multicenter randomized trial of systemic TGF-β1-neutralizing mAb CAT-191 in SSc patients suggest that it can be safely administered in the short term, although biological or clinical evidences of efficacy were not detected *(38)*. However, this approach raises serious concerns in light of TGF-β's numerous homeostatic functions, including the potential development of tumors or autoimmune disease *(39,40)*. Furthermore, because intracellular TGF-β signaling is autonomously activated in scleroderma fibroblasts, it may be partially unresponsive to extracellular blockade of TGF-β *(41)*.

Alternatives to this approach are those specifically targeting intracellular pathways specifically involved in the fibrotic phenotype such as TGF-β type I receptor serine/threonine kinase, SMAD2/3 factors, or alternative non-SMAD TGF-β-activated kinases, such as MAPK (i.e., p38), PI3K, or c-Abl kinases. A vast array of small molecules is under development for this purpose *(17,42–46)*, but information regarding antifibrotic and safety profiles of these compounds is still very limited. A remarkable antifibrotic effect of the specific c-Abl kinase inhibitor imatinib has been observed in bleomycin-induced lung fibrosis, underscoring the potential role of non-SMAD-dependent TGF-β actions in fibrosis *(43)*.

Finally, targeting more distal factors downstream of TGF-β such as CTGF has been proposed as a potentially safer alternative because CTGF seems to partially mediate TGF-β profibrotic effects whereas sparing the beneficial aspects of TGF-β signaling that operate in normal physiology. However, CTGF-deficient mice have also revealed an important role for this factor in connective tissue and vascular development, again raising potential safety concerns *(47)*. Limited information from a phase I clinical trial with neutralizing

CTGF antibodies in lung fibrosis shows the safety of single dosing *(48)*, but efficacy or long-term safety data are not yet available.

4. DEVELOPMENT OF SMALL TGF-β1 ANTAGONISTIC PEPTIDES

Among the wide variety of approaches to develop TGF-β1 biological activity inhibitors, we have explored the feasibility of developing small TGF-β1 antagonistic peptides. As previously described by Ezquerro et al. *(49)*, potential peptide inhibitors of TGF-β1 can be successfully selected from the protein sequences of TGF-β1, its membrane receptors, and other TGF-β1 binding proteins using the two following strategies. The first was based on the synthesis of overlapping peptides encompassing the whole sequence of TGF-β1 or partial sequences of TGF-β1 binding proteins, including the extracellular region of TGF-β receptors. The second strategy was based on the generation of peptides from the complete sequence of TGF-β1 binding proteins (receptors, carrier proteins, and so on), with predicted TGF-β1 binding potential. Predictions were made using a specific program based on the assignation of potential interactions between two peptides, one from TGF-β1 and the other from a TGF-β1 interacting protein. The potential propensity interaction score takes into account the hydrophilicity/hydrophobicity and the net charge of the amino acid side chains from both peptides. Basically, the program calculates a putative score of interaction between sequential blocks of different length amino acid sequences from both proteins. To calculate the score of interaction between two blocks, the hydrophilicity of amino acid at each position is multiplied by the hydrophilicity of amino acid at the corresponding position second block, and so on until the amino acid at the last position. The sum of these products constitutes the score of interaction between the two partial protein sequences. These calculations were carried out using the amino acid hydrophilicity scale of Hopp and Woods *(50)*. Because in the scale of Hopp and Woods, the value of the hydrophilicity of hydrophilic hydrophobic amino acids is negative, higher scores are obtained when two amino acids of the same type are compared. However, because charged amino acid side chains may attract or repel depending on their charge, the products of the hydrophilicities between amino acid side chains having the same sign of charge were arbitrarily made negative to penalize the interaction between these repelling side chains. All other products were left unchanged.

Following this strategy, hundreds of peptides were selected, synthesized, and in vitro tested searching for peptides with TGF-β1 biological activity inhibitory effects. With this approach, only about 10 peptides gave significant in vitro TGF-β1 inhibitory activity (Table 3). Among them P144, encompassing amino acids 730–743 (accession number Q03167, SwissProt) from the sequence of human TGF-β1 type III receptor (betaglycan), showed a remarkable TGF-β1 inhibitory activity. P144 was able to prevent TGF-β1-dependent inhibition of Mv-1-Lu cell proliferation and, in cultured fibroblasts, it induced a dose-dependent decrease on TGF-β1-dependent stimulation of a reporter gene under the control of human α2(I) collagen promoter *(49)*.

5. PRECLINICAL STUDIES OF P144

The first studies demonstrating in vivo antifibrotic pharmacological activity of P144 were performed by intraperitoneal administration of P144 in CCl_4-induced liver fibrosis in rats *(49)*. Rats receiving CCl_4 inhalations for 11 wk were treated with 70 µg of P144 ip on alternate days during 6 wk starting at week 6. After this period, P144-treated rats displayed significantly less histological and biochemical collagen deposition. A significant decrease in the number of α-smooth muscle actin (α-SMA) positive activated HSCs, representing the most active mesenchymal cell involved in ECM deposition in fibrotic liver, was also observed in P144-treated rats.

Table 3
In Vitro Inhibition of Human TGF-β1 Activity by Synthetic Peptides Derived from TGF-β1 and TGF-β1 Type III Receptors

Peptide	*Sequence*	*Region of parent protein*	*Mv-1-Lu cell assay*
P11	HANFCLGPCPYIWSL**A**	(318–332) Hum TGF-β1	40%
P12	FCLGPCPYIWSLDT**A**	(321–334) Hum TGF-β1	96%
P106	SNPYSAFQVDIIVDI**A**	(245–259) Rat TGF-β1 RIII	40%
P54	TSLDATMIWTMM**A**	(730–742) Rat TGF-β1 RIII	97%
P144	TSLDASIIWAMMQN**A**	(730–743) Hum TGF-β1 RIII	80%

Numbers in brackets correspond to the amino acid region of the corresponding protein. Accession numbers through PubMed databases for Hum TGF-β1, Rat TGF-β1 type III receptor, and human type III receptor are: XP_008912, NP_058952, and Q03167, respectively. Inhibition of TGF-β1 by peptides was quantitated by comparing the growth of Mv-1-Lu cells cultured in the presence of 200 pg/mL of human TGF-β1 with or without 200 μg/mL of peptide. Last alanine (in bold) was added at the C-terminal end of all peptides for synthesis convenience. Only those peptides with an activity >40% are shown.

Local rather than systemic TGF-β1 inhibition represents a potentially safer strategy for the development of antifibrotic therapies. Local inhibition of TGF-β has previously been attempted by the direct application of neutralizing antibodies on skin or corneal open wounds, but the application of antibodies or large peptides through the epidermal barrier seems an unpractical approach *(51,52)*. Taking advantage of the small size and highly lipophilic character of P144, we explored the feasibility of its topical application for skin fibrosis therapy. This peptide was easily deliverable to the shaved mouse skin by its topical application at a concentration of 100 μg/mL on a lipogel vehicle. Among the described animal models of skin fibrosis, perhaps the best characterized is the model of repeated local injection of bleomycin. This model reproduces most of the features of human scleroderma such as skin inflammatory cell infiltration, vascular damage, mast cells activation, and prolonged skin fibrosis *(53)*. Furthermore, previous studies have demonstrated that either the administration of anti-TGF-β antibodies or genetic SMAD3 deficiency ameliorates fibrosis development, confirming the key role played by TGF-β through the SMAD pathway *(22,37)*. In this model, topical application of P144 along fibrosis development prevented dermal and hypodermal collagen accumulation and decreased skin-soluble collagen content compared to vehicle-treated mice (Fig. 1) *(54)*. Furthermore, and more importantly regarding human skin fibrotic diseases, established fibrosis was also significantly reduced following topical P144 application for 2 wk. Improvement of established skin fibrosis in this model by postonset therapy has been previously demonstrated with systemic interferon-γ *(55)* or superoxide dismutase therapy *(56)*, but not with systemic TGF-β inhibitors *(37)*.

Interestingly, changes in the density of inflammatory cell infiltration, mast cell infiltration, or morphological changes of the epidermis were not observed in the P144 or vehicle-treated mice compared to those receiving only bleomycin injections, pointing to a direct antifibrotic effect of this therapy *(54)*. Consistently, several surrogate markers for TGF-β effects on fibroblast were found downregulated by this therapeutic approach. First, α-SMA-positive myofibroblasts development was efficiently decreased by P144 therapy. Second, a decrease in the number of dermal fibroblasts displaying phosphorylated SMAD2/3, a direct consequence of TGF-β signaling in fibroblasts, was observed in P144-treated mice. Finally, CTGF, which was strongly expressed in fibroblasts and epidermal cells of bleomycin-treated mice, was also consistently reduced in P144-treated mice *(54)*.

Although systemic toxicity was not detected in short-term P144-treated mice, further studies are required to confirm its safety in the long term. Regardless of side effects, several

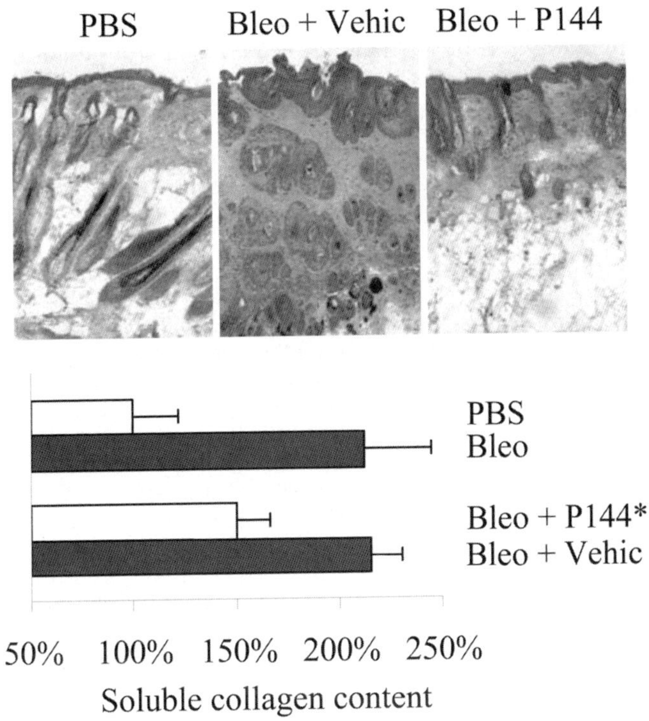

Fig. 1. Antifibrotic effect of topical P144 in bleomycin-induced skin fibrosis. Histopathological evaluation of bleomycin-induced skin sclerosis in P144-treated mice. C3H mice received subcutaneous injections of bleomycin (Bleo) and were simultaneously treated topically with TGF-β1 peptide inhibitor P144 on lipogel emulsion or vehicle (Vehic) for 4 wk. Control mice were daily injected with PBS. Skin sections show the dermal thickness. Percentual variation of soluble collagen content of the skin in P144-treated bleomycin-injected and control groups (representative of 10 mice per group). (*) $p < 0.05$.

issues regarding local vs systemic anti-TGF-β therapy emerge from these studies. First, a higher efficacy is suggested by the fact that, in contrast to systemic anti-TGF-β neutralizing antibodies *(37)*, topical P144 was able not only to prevent, but also to reduce established fibrosis in this model. Furthermore, the antifibrotic effect of topical therapy seems independent of inflammatory cell or mast cell infiltration, in contrast to systemic anti-TGF-β intervention *(37)*. An area of uncertainty is the cellular target of P144 delivered through the epidermal barrier. Preliminary data suggest limited penetration across in vitro cultured skin explants, but specific data on epidermal or dermal penetration are lacking. The lipophilic character of the peptide predicts a high rate of epidermal incorporation, whereas dermal absorption is uncertain. Interestingly, in P144-treated mice, we observed a higher reduction of CTGF expression in keratinocytes than in dermal fibroblasts, raising questions on the significance of dermal–epidermal crosstalk during fibrosis. Previous observations suggest that induced or enforced overexpression of different mediators, including TGF-β1, in the epidermal layer can mediate skin fibrosis *(20,21)*. Furthermore, profibrotic factors or markers for TGF-β signaling have also been detected in SSc epidermal keratinocytes *(25,57)*. Although the role of epidermal CTGF induction has not been established in fibrosis, previous studies demonstrate that it is expressed by normal keratinocytes in vivo and is involved in the regulation of procollagen synthesis in response to UV radiation *(58)*.

The demonstration of the effectiveness of topical application of a peptide inhibitor of TGF-β1 opens a potentially fruitful strategy for the therapy of pathological scarring and skin fibrotic diseases. Experiments are being carried out to determine to what extent P144 might be systemically absorbed through the skin. These experiments, together with a study of the potential toxicity of P144, will determine whether this peptide is suitable for human therapy.

REFERENCES

1. Massagué J. The transforming growth factor-beta family. Annu Rev Cell Biol 1990;6:597–641.
2. Roberts AB, Flanders KC, Heine UI, et al. Transforming growth factor-beta: multifunctional regulator of differentiation and development. Philos Trans R Soc Lond B Biol Sci 1990;327:145–154.
3. Roberts AB, Sporn MB, Assoian RK, et al. Transforming growth factor type beta: rapid induction of fibrosis and angiogenesis in vivo and stimulation of collagen formation in vitro. Proc Natl Acad Sci USA 1986;83:4167–4171.
4. Roberts AB, Heine UI, Flanders KC, Sporn MB. Transforming growth factor-beta. Major role in regulation of extracellular matrix. Ann N Y Acad Sci 1990;580:225–232.
5. Engel ME, McDonnell MA, Law BK, Moses HL. Interdependent SMAD and JNK signaling in transforming growth factor-beta-mediated transcription. J Biol Chem 1999;274:37,413–37,420.
6. Leask A, Abraham DJ. TGF-beta signaling and the fibrotic response. FASEB J 2004;18:816–827.
7. Verrecchia F, Mauviel A. Transforming growth factor-beta signaling through the Smad pathway: role in extracellular matrix gene expression and regulation. J Invest Dermatol 2002;118:211–215.
8. Blobe GC, Schiemann WP, Lodish HF. Role of transforming growth factor beta in human disease. N Engl J Med 2000;342:1350–1358.
9. Bitzer M, Sterzel RB, Bottinger EP. Transforming growth factor-beta in renal disease. Kidney Blood Press Res 1998;21:1–12.
10. Bedossa P, Paradis V. Transforming growth factor-beta (TGF-beta): a key-role in liver fibrogenesis. J Hepatol 1995;22(Suppl):37–42.
11. Border WA, Noble NA. Transforming growth factor beta in tissue fibrosis. N Engl J Med 1994;331: 1286–1292.
12. Silver RM. Clinical aspects of localized and systemic scleroderma. Curr Opin Rheumatol 1991;3: 973–978.
13. Abraham DJ, Varga J. Scleroderma: from cell and molecular mechanisms to disease models. Trends Immunol 2005;26:587–595.
14. Leroy EC. Connective tissue synthesis by scleroderma skin fibroblasts in cell culture. J Exp Med 1972; 135:1351–1362.
15. Desmouliere A, Redard M, Darby I, Gabbiani G. Apoptosis mediates the decrease in cellularity during the transition between granulation tissue and scar. Am J Pathol 1995;146:56–66.
16. Santiago B, Galindo M, Rivero M, Pablos JL. Decreased susceptibility to Fas-induced apoptosis of systemic sclerosis dermal fibroblasts. Arthritis Rheum 2001;44:1667–1676.
17. Mimura Y, Ihn H, Jinnin M, Asano Y, Yamane K, Tamaki K. Constitutive phosphorylation of focal adhesion kinase is involved in the myofibroblast differentiation of scleroderma fibroblasts. J Invest Dermatol 2005;124:886–892.
18. Clouthier DE, Comerford SA, Hammer RE. Hepatic fibrosis, glomerulosclerosis, and a lipodystrophy-like syndrome in PEPCK-TGF-beta1 transgenic mice. J Clin Invest 1997;100:2697–2713.
19. Munger JS, Huang X, Kawakatsu H, et al. The integrin alpha v beta 6 binds and activates latent TGF beta 1: a mechanism for regulating pulmonary inflammation and fibrosis. Cell 1999;96:319–328.
20. Ito Y, Sarkar P, Mi Q, et al. Overexpression of Smad2 reveals its concerted action with Smad4 in regulating TGF-beta-mediated epidermal homeostasis. Dev Biol 2001;236:181–194.
21. Chan T, Ghahary A, Demare J, et al. Development, characterization, and wound healing of the keratin 14 promoted transforming growth factor-beta1 transgenic mouse. Wound Repair Regen 2002;10: 177–187.
22. Lakos G, Takagawa S, Chen SJ, et al. Targeted disruption of TGF-beta/Smad3 signaling modulates skin fibrosis in a mouse model of scleroderma. Am J Pathol 2004;165:203–217.
23. Gruschwitz M, Muller PU, Sepp N, Hofer E, Fontana A, Wick G. Transcription and expression of transforming growth factor type beta in the skin of progressive systemic sclerosis: a mediator of fibrosis? J Invest Dermatol 1990;94:197–203.

24. Kubo M, Ihn H, Yamane K, Tamaki K. Upregulated expression of transforming growth factor-beta receptors in dermal fibroblasts of skin sections from patients with systemic sclerosis. J Rheumatol 2002;29:2558–2564.
25. Dong C, Zhu S, Wang T, et al. Deficient Smad7 expression: a putative molecular defect in scleroderma. Proc Natl Acad Sci USA 2002;99:3908–3913.
26. Asano Y, Ihn H, Yamane K, Kubo M, Tamaki K. Impaired Smad7-Smurf-mediated negative regulation of TGF-beta signaling in scleroderma fibroblasts. J Clin Invest 2004;113:253–264.
27. Kawakami T, Ihn H, Xu W, Smith E, LeRoy C, Trojanowska M. Increased expression of TGF-beta receptors by scleroderma fibroblasts: evidence for contribution of autocrine TGF-beta signaling to scleroderma phenotype. J Invest Dermatol 1998;110:47–51.
28. Pannu J, Gore-Hyer E, Yamanaka M, et al. An increased transforming growth factor beta receptor type I:type II ratio contributes to elevated collagen protein synthesis that is resistant to inhibition via a kinase-deficient transforming growth factor beta receptor type II in scleroderma. Arthritis Rheum 2004;50:1566–1577.
29. Mori Y, Chen SJ, Varga J. Expression and regulation of intracellular SMAD signaling in scleroderma skin fibroblasts. Arthritis Rheum 2003;48:1964–1978.
30. Asano Y, Ihn H, Yamane K, Kubo M, Tamaki K. Increased expression levels of integrin alphavbeta5 on scleroderma fibroblasts. Am J Pathol 2004;164:1275–1292.
31. Asano Y, Ihn H, Yamane K, Jinnin M, Mimura Y, Tamaki K. Increased expression of integrin alpha-v beta3 contributes to the establishment of autocrine TGF-beta signaling in scleroderma fibroblasts. J Immunol 2005;175:7708–7718.
32. Mimura Y, Ihn H, Jinnin M, Asano Y, Yamane K, Tamaki K. Constitutive thrombospondin-1 over-expression contributes to autocrine transforming growth factor-beta signaling in cultured scleroderma fibroblasts. Am J Pathol 2005;166:1451–1463.
33. Igarashi A, Nashiro K, Kikuchi K, et al. Significant correlation between connective tissue growth factor gene expression and skin sclerosis in tissue sections from patients with systemic sclerosis. J Invest Dermatol 1995;105:280–284.
34. Denton CP, Abraham DJ. Transforming growth factor-beta and connective tissue growth factor: key cytokines in scleroderma pathogenesis. Curr Opin Rheumatol 2001;13:505–511.
35. Farge D, Passweg J, van Laar JM, et al. Autologous stem cell transplantation in the treatment of systemic sclerosis: report from the EBMT/EULAR Registry. Ann Rheum Dis 2004;63:974–981.
36. Zhang Y, McCormick LL, Gilliam AC. Latency-associated peptide prevents skin fibrosis in murine sclerodermatous graft-versus-host disease, a model for human scleroderma. J Invest Dermatol 2003; 121:713–719.
37. Yamamoto T, Takagawa S, Katayama I, Nishioka K. Anti-sclerotic effect of transforming growth factor-beta antibody in a mouse model of bleomycin-induced scleroderma. Clin Immunol 1999;92:6–13.
38. Denton CP, Merkel PA, Furst DE, et al. Recombinant human anti-transforming growth factor-1 antibody therapy in systemic sclerosis: a multicenter, randomized placebo-controlled phase I/II trial of CAT-192. Arthritis Rheum 2004;56:323–333.
39. Bachman KE, Park BH. Duel nature of TGF-beta signaling: tumor suppressor vs. tumor promoter. Curr Opin Oncol 2005;17:49–54.
40. Li MO, Wan YY, Sanjabi S, Robertson AK, Flavell RA. Transforming growth factor-beta regulation of immune responses. Annu Rev Immunol 2005;8:8.
41. Varga J. Antifibrotic therapy in scleroderma: extracellular or intracellular targeting of activated fibroblasts? Curr Rheumatol Rep 2004;6:164–170.
42. Daniels CE, Wilkes MC, Edens M, et al. Imatinib mesylate inhibits the profibrogenic activity of TGF-beta and prevents bleomycin-mediated lung fibrosis. J Clin Invest 2004;114:1308–1316.
43. Mori Y, Ishida W, Bhattacharyya S, Li Y, Platanias LC, Varga J. Selective inhibition of activin receptor-like kinase 5 signaling blocks profibrotic transforming growth factor beta responses in skin fibroblasts. Arthritis Rheum 2004;50:4008–4021.
44. Ihn H, Yamane K, Tamaki K. Increased phosphorylation and activation of mitogen-activated protein kinase p38 in scleroderma fibroblasts. J Invest Dermatol 2005;125:247–255.
45. McGaha TL, Phelps RG, Spiera H, Bona C. Halofuginone, an inhibitor of type-I collagen synthesis and skin sclerosis, blocks transforming-growth-factor-beta-mediated Smad3 activation in fibroblasts. J Invest Dermatol 2002;118:461–470.
46. Asano Y, Ihn H, Yamane K, Jinnin M, Mimura Y, Tamaki K. Phosphatidylinositol 3-kinase is involved in alpha2(I) collagen gene expression in normal and scleroderma fibroblasts. J Immunol. 2004;172: 7123–7135.

47. Ivkovic S, Yoon BS, Popoff SN, et al. Connective tissue growth factor coordinates chondrogenesis and angiogenesis during skeletal development. Development 2003;130:2779–2791.
48. Mageto Y, FK, Brown K, Fong A, Raghu G. Safety and tolerability of human monoclonal antibody FG-3019, anti-connective tissue growth factor, in patients with idiopathic pulmonary fibrosis. Chest Meeting Abstracts 2004;126:S773.
49. Ezquerro IJ, Lasarte JJ, Dotor J, et al. A synthetic peptide from transforming growth factor beta type III receptor inhibits liver fibrogenesis in rats with carbon tetrachloride liver injury. Cytokine 2003;22:12–20.
50. Hopp TP, Woods KR. Prediction of protein antigenic determinants from amino acid sequences. Proc Natl Acad Sci USA 1981;78:3824–3828.
51. Jester JV, Barry-Lane PA, Petroll WM, Olsen DR, Cavanagh HD. Inhibition of corneal fibrosis by topical application of blocking antibodies to TGF beta in the rabbit. Cornea 1997;16:177–187.
52. Brahmatewari J, Serafini A, Serralta V, Mertz PM, Eaglstein WH. The effects of topical transforming growth factor-beta2 and anti-transforming growth factor-beta2,3 on scarring in pigs. J Cutan Med Surg 2000;4:126–131.
53. Yamamoto T, Takagawa S, Katayama I, et al. Animal model of sclerotic skin. I: Local injections of bleomycin induce sclerotic skin mimicking scleroderma. J Invest Dermatol 1999;112:456–462.
54. Santiago B, Gutierrez-Canas I, Dotor J, et al. Topical application of a peptide inhibitor of transforming growth factor-beta1 ameliorates bleomycin-induced skin fibrosis. J Invest Dermatol 2005;125:450–455.
55. Yamamoto T, Takagawa S, Katayama I, Mizushima Y, Nishioka K. Effect of superoxide dismutase on bleomycin-induced dermal sclerosis: implications for the treatment of systemic sclerosis. J Invest Dermatol 1999;113:843–847.
56. Yamamoto T, Takagawa S, Kuroda M, Nishioka K. Effect of interferon-gamma on experimental scleroderma induced by bleomycin. Arch Dermatol Res 2000;292:362–365.
57. Galindo M, Santiago B, Rivero M, Rullas J, Alcami J, Pablos JL. Chemokine expression by systemic sclerosis fibroblasts: abnormal regulation of monocyte chemoattractant protein 1 expression. Arthritis Rheum 2001;44:1382–1386.
58. Quan T, He T, Kang S, Voorhees JJ, Fisher GJ. Connective tissue growth factor: expression in human skin in vivo and inhibition by ultraviolet irradiation. J Invest Dermatol 2002;118:402–408.

45 Overexpressed Truncated TGF-β Type II Receptor Inhibits Fibrotic Behavior of Keloid Fibroblasts In Vitro and Experimental Scar Formation In Vivo

Wei Liu, Chekhau Chua, Zhen Gao,
Xiaoli Wu, and Yilin Cao

CONTENTS

INTRODUCTION
TGF-β AND FETAL SCARLESS WOUND HEALING
TGF-β AND THE FORMATION OF HYPERTROPHIC SCAR AND KELOID
CELLULAR AND MOLECULAR MECHANISM OF TGF-β PATHOLOGICAL ROLES IN HYPERTROPHIC SCAR AND KELOID FORMATION
OVEREXPRESSING TRUNCATED TGF-β RECEPTOR II FOR SCAR REDUCTION
REFERENCES

Abstract

Hypertrophic scar and keloid are common and difficult to treat diseases in plastic surgery, which cause patients' suffering both physiologically and psychologically. Results of wound-healing research over the past decades have demonstrated that transforming growth factor-β (TGF-β) plays an essential role in cutaneous scar formation. Recent studies also showed that TGF-β overfunction and its abnormal signaling are closely associated with wound scar formation. In addition, the dominant negative effect of overexpressed truncated TGF-β receptor II has been proven to be an effective anti-TGF-β strategy. This chapter intends to review the roles of TGF-β in scar formation as well as the molecular mechanism and to introduce our recent finding of inhibiting fibrotic behavior of keloid fibroblasts in vitro and experimental scar formation in vivo by the dominant negative effect strategy.

Key Words: TGF-β signaling; truncated TGF-β receptor II; wound scarring; keloid; scar gene therapy.

1. INTRODUCTION

Cancer and wound scarring are different diseases. Nevertheless, both cancer and pathological wound scarring (especially keloid) behave similarly in the features of unlimited growth and invasion to normal tissue. From this point, cancer may be considered as a wound that never heals.

From: *Cancer Drug Discovery and Development: Transforming Growth Factor-β in Cancer Therapy, Vol I: Basic and Clinical Biology*
Edited by: Sonia B. Jakowlew © Humana Press Inc., Totowa, NJ

It is well known that transforming growth factor-β (TGF-β) plays a dual role in cell growth and matrix production. It inhibits the growth of epithelial type of cells, but promotes growth and matrix production of mesenchymal type of cells. On the one hand, loss of TGF-β inhibitory effect leads to cancer formation of epithelial cell type *(1)*; on the other hand, TGF-β overfunction may lead to tumor formation of mesenchymal cell type *(2)*.

Pathological wound scarring is an example of uncontrolled growth of mesenchymal tissue, such as hypertrophic scar and keloids. In particular, keloid represents an aberrant scarring in human being, which usually extends beyond the boundary of an original wound and is difficult to treat clinically *(3)*. Thus, keloid is also considered as a benign tumor of dermis, in which TGF-β plays critical roles in its formation and growth. Because of the close relationship between sustained and high-level expression of TGF-β and the formation of keloid and hypertrophic scar *(4)*, this chapter will review the roles of TGF-β in scar formation, the molecular mechanism of its pathological effects, and also introduce the experimental finding that the dominant effect mediated by overexpressed truncated TGF-β receptor II can inhibit the fibrotic behavior of keloid fibroblasts in vitro and inhibit experimental scar formation in vivo.

2. TGF-β AND FETAL SCARLESS WOUND HEALING

It was conventionally believed that scarring in an adult wound may be inevitable; the discovery of scarless healing pattern in fetal wounds, however, indicates that inhibition of wound scarring might be possible if a healing pattern similar to that of a fetal wound can be mimicked in an adult wound. The mechanism of fetal scarless wound healing has been found quite complicated, which may involve the intrinsic function of fetal dermal fibroblasts *(5,6)*, the extracellular matrix (ECM) components *(7,8)*, the inflammatory response to tissue injury *(9,10)*, and the specific cytokine profiles *(11–13)*, and so on.

Despite the complexity of its mechanism, a lower level of TGF-β expression in fetal wounds clearly plays an essential role in scarless wound healing. Whitby and Ferguson *(7)* demonstrated in their study that TGF-β1 and TGF-β2 could be detected with immunolocalization in neonatal and adult wounds, but not in fetal wounds. Marin et al. *(14)* also found that both gene expression and protein production of TGF-β1 were rapid and transient in fetal wounds, whereas its expression and production appeared to be slower but sustained longer in adult wounds. Later, Krummel et al. *(15)* and Houghton et al. *(16)* found that fetal scarless wound healing actually could be converted into a scar-forming healing process if exogenous TGF-β was added into the wounds, indicating that the absence of TGF-β in fetal wounds is important for scarless wound healing.

Interestingly, no obvious platelets were found in the wounds of early stage fetuses, which might result in the lack of an initial dose of TGF-β that is usually released in adult wounds by degranulating platelets *(17)*. Additionally, a fetal wound is also characterized by its relatively weak inflammatory response to dermal wounding, a process that recruits wound infiltration of many TGF-β-producing cells, such as monocytes and macrophages *(9)*. In addition to low expression level, low bioactivity of TGF-β in a fetal wound may also contribute to scarless healing. It has been demonstrated that the expression level of fibromodulin, the TGF-β modulator, was much higher in fetal wounds than in adult wounds, which leads to a stronger inhibition of TGF-β activation in fetal than in adult wounds *(18)*. In addition, the gene expression level of TGF-β receptor II in rat fetuses was found to be much higher in the dermal fibroblast of later gestational stage (scar forming) than that of early stage (scar free) *(11)*, providing another way of modulating TGF-β function in fetal wounds. All these phenomena indicate that the diminished TGF-β effect plays an important role in fetal scarless wound healing. Therefore, reducing TGF-β production or attenuating its effect might be helpful in inhibiting scarring of an adult wound.

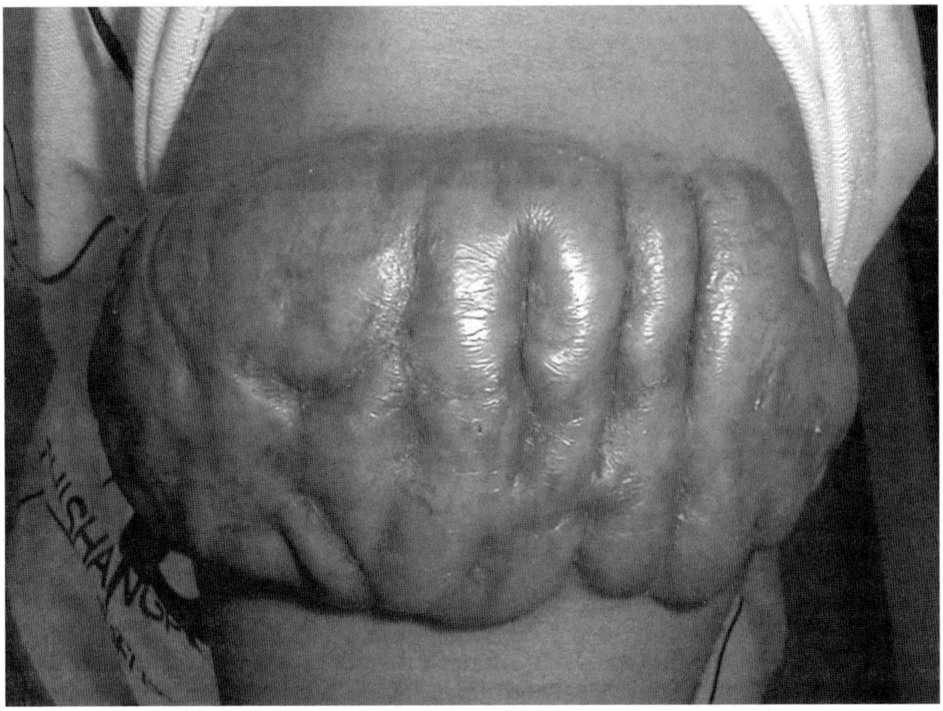

Fig. 1. A rapidly and unlimitedly growing keloid in a patient's arm.

3. TGF-β AND THE FORMATION OF HYPERTROPHIC SCAR AND KELOID

Accumulated evidence in numerous reports during the past two decades points to the fact that TGF-β overfunction is highly associated with pathological scar formation. For example, enhanced expression of TGF-β ligands and their receptors has been observed in hypertrophic scar and keloids, a striking contrast to fetal wound. Wang et al. reported that the hypertrophic scar tissue expressed five times as much TGF-β1 mRNA as normal skin tissue. Additionally, the in vitro cultured fibroblasts derived from the hypertrophic scar also expressed TGF-β1 mRNA in a level significantly higher than that of the normal fibroblasts *(19)*. In another report, Schmid et al. found that more TGF-β receptors I and II were expressed in the granulation tissues than in the normal skin at both mRNA and protein levels. However, the expression levels decreased gradually in the normal healing excisional wounds during the granulation tissue remodeling. By contrast, the receptor expression in the hypertrophic scar maintained at high levels up to 20 mo postinjury, suggesting that a persistent autocrine loop of TGF-β may exist and contribute to hypertrophic scar formation *(20)*.

Keloid is an extreme example of pathological scarring (Fig. 1) and is characterized by its uncontrolled growth and expansion beyond the original boundary through invading its surrounding normal tissues, a feature similar to that of cancer. Like other types of scarring, overexpression of TGF-β ligands and receptors has been observed in keloids. Lee et al. *(21)* reported that keloid fibroblasts produced more TGF-β1 and TGF-β2 protein, but not TGF-β3, than normal dermal fibroblasts. Chin et al. *(22)* later demonstrated an increased protein expression of TGF-β receptors I and II in the cultured keloid fibroblasts as compared with the normal dermal fibroblasts. We recently found that TGF-β receptor III is also overexpressed in keloid fibroblasts when compared with normal dermal fibroblasts (Fig. 2, preliminary data).

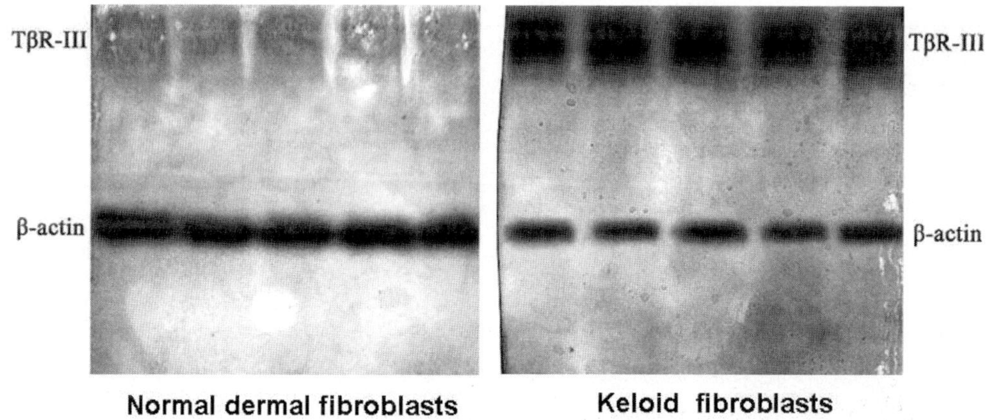

Fig. 2. Western blot analysis shows overexpression of TGF-β receptor III (betaglycan) in keloid fibroblasts with a level much higher than that of normal dermal fibroblasts (preliminary data).

All these results suggest that overexpressed TGF-β ligands and their receptors may contribute to the scar and keloid formation.

4. CELLULAR AND MOLECULAR MECHANISM OF TGF-β PATHOLOGICAL ROLES IN HYPERTROPHIC SCAR AND KELOID FORMATION

Although the mechanism of scar formation can be more complicated than the aberrant profiles of a few growth factors, a number of studies have shown that TGF-β might be the most important single factor that promotes scar formation. Additionally, TGF-β also involves multiple aspects of the pathological process that promotes scar formation, such as inducing inflammation, enhancing matrix production, reducing tissue remodeling, and abnormal regulation of cell proliferation and apoptosis *(23)*. Understanding the mechanism of these pathological roles played by TGF-β might be helpful in designing an antiscarring therapeutic strategy. During wounding, the initial dose of TGF-β appears to be an important factor to trigger the later cascade of TGF-β overproduction and scar formation. When a cutaneous wound is created, TGF-β released from the platelets can recruit monocytes and other types of cells into the wound and these cells further produce more TGF-β in the wound. Additionally, the initial dose of TGF-β in the early wound induces dermal fibroblasts, the key player of scar formation, to produce their own TGF-β in the later stage wound via autocrine regulation, leading to the overproduction of wound TGF-β and eventually to scar formation *(23,24)*.

The most distinctive pathological features of abnormal scarring include rapid fibroblast proliferation and collagen overproduction and there is sufficient evidence to indicate that TGF-β involves both. In vitro studies have shown that TGF-β1 promoted cell proliferation and collagen production of both keloid and normal dermal fibroblasts *(25,26)*, and the collagen production stimulated by TGF-β was mediated by Smad3 and Smad4 *(27)*. In vivo, overexpression of TGF-β1 gene has been observed at the peripheral area of keloid tissues, where highly proliferated cells reside. In addition, TGF-β1 gene expression is also colocalized to the area where collagen gene is actively expressed *(28)*, suggesting that TGF-β1 produced by keloid fibroblasts also promotes their own proliferation and collagen production in the local environment.

During hypertrophic scar or keloid formation, deposition of excessive collagen is also caused by inhibited proteolytic degradation of ECM in wounds. Plasminogen activator (PA) and matrix metalloproteinases (MMP) are the two major ECM-degrading enzymes, which are negatively regulated by plasminogen activator inhibitor 1 (PAI-1) and tissue inhibitor of metalloproteinases-1 (TIMP-1), respectively. Increased expression of TGF-β in wounds not only enhances collagen production, but may also inhibit collagen degradation by upregulating PAI-1 *(29)* and TIMP-1 *(30)* expression and by downregulating MMP expression *(31)*. In an in vitro fibroplasia assay, Tuan et al. found that PAI-1 expression was strikingly higher in keloid fibroblasts than in normal dermal fibroblasts. In contrast, the urokinase plasminogen activator (uPA) level was much lower in keloid fibroblasts, which led to the inhibited degradation of a fibrin gel when compared with the fibrin gel that contained normal fibroblasts *(32)*. A similar result has also been observed by Higgins et al. *(33)*, suggesting that TGF-β is involved in the altered ECM degradation process in keloids.

Wound tension is also known to promote scar formation *(34)*. In an in vitro assay, Peled et al. *(35)* have shown that mechanical strain could upregulate the gene expression of TGF-β1 and its receptors as well as type I procollagen in normal dermal fibroblasts, indicating that enhanced expression of TGF-β1 and its receptors by cellular strain may contribute to the tension-induced scarring.

Cell apoptosis is an important part of normal wound remodeling processes, and abnormal regulation of cell apoptosis may play an important role in the establishment and development of pathological scarring. Several studies have revealed the reduced apoptosis in keloid fibroblasts *(36–38)*. Mutation or downregulated expression of apoptosis-related genes has also been observed *(39,40)*. Interestingly, TGF-β seems to contribute to reduced apoptosis observed in keloid fibroblasts because a high apoptosis rate of normal fibroblasts can be reduced by adding TGF-β to the culture media *(37,38)*. Likewise, apoptosis rate of keloid fibroblasts can be increased by adding TGF-β-neutralizing antibody *(38)*.

The mechanism by which keloid fibroblasts regulate their TGF-β expression remains incompletely understood. The higher expression levels of TGF-β1, TGF-β2 *(21)*, and TGF-β receptors I and II *(22)* in keloid fibroblasts than in normal dermal fibroblasts suggest that a TGF-β autocrine loop may play a role. Recently, we have demonstrated that the TGF-β1 autocrine regulation existed in keloid fibroblasts (Fig. 3) (41) and TGF-β receptor III was overexpressed in keloid fibroblasts when compared to normal dermal fibroblasts (Fig. 2, preliminary data), which further support the autocrine role in keloid pathogenesis. In normal wound-healing process, TGF-β autocrine regulation is usually self-limited. This self-restriction is possibly mediated by the regulated expression of Smads. The study by Mori et al. *(42)* has shown that in normal dermal fibroblasts, Smad3 gene expression was downregulated by exogenous TGF-β1, whereas Smad7, an antagonistic signaling molecule, was upregulated by TGF-β1. Thus, an elevated expression level of TGF-β1 in normal cells may elicit a negative feedback, which limits the autoinduction process. Although a comparative study of TGF-β1 autocrine regulation between keloid and normal fibroblasts has not yet been performed, the sustained high expression levels of TGF-β ligands and their receptors in keloid cells imply that this autoinduction process may not be self-restricted, which might promote the unlimited growth of keloids clinically. In addition, the abnormal TGF-β signaling in keloid fibroblasts is likely to contribute to the abnormal TGF-β autocrine loop. This assumption is supported by the fact that Smad3 is overphosphorylated in keloid fibroblasts as compared with normal fibroblasts *(22)*. This abnormal signaling pathway may also explain why keloid fibroblasts exhibit an altered response to TGF-β treatment, such as increased cell proliferation and collagen production *(25,26)*. Babu et al. *(43)* also demonstrated that in the absence of serum, the TGF-β1-induced increase in fibronectin biosynthesis occurred more rapidly in keloid fibroblasts than in normal fibroblasts, indicating

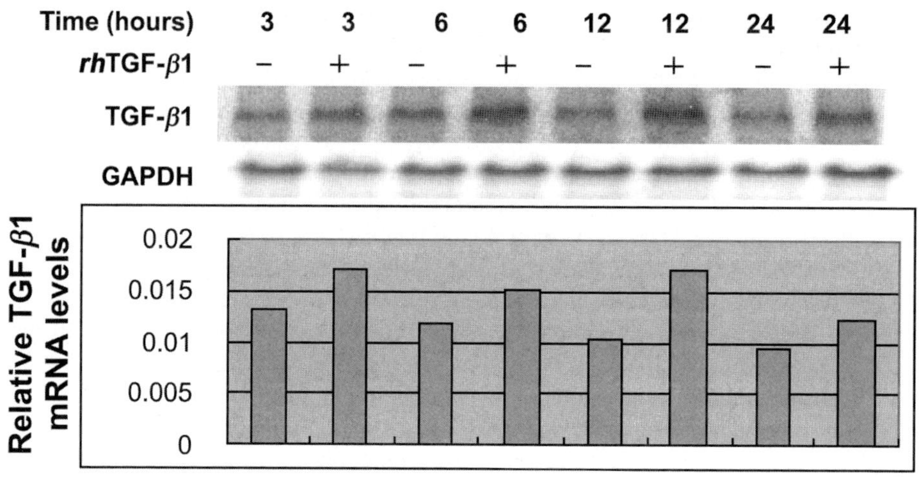

Fig. 3. rhTGF-β1 upregulates gene expression of TGF-β1 in keloid fibroblasts. Keloid cells were either not treated or treated with recombinant TGF-β1 (rhTGF-β1, 5 ng/mL) for 3, 6, 12, and 24 h, respectively. Nontreated and treated cells were harvested at various time points to isolate total RNA. A Northern blot assay was performed to analyze the gene expression of TGF-β1. The relative mRNA expression level was determined by dividing the expression level of TGF-β1 with GAPDH expression level *(41)*.

that keloid's overproduction of ECM components might be because of an inherent modification of the TGF-β regulatory program.

5. OVEREXPRESSING TRUNCATED TGF-β RECEPTOR II FOR SCAR REDUCTION

5.1. Dominant Negative Effect of Truncated TGF-β Receptor II

As described in this book, there are many different ways to disrupt TGF-β functions including using soluble TGF-β receptors, TGF-β receptor kinase inhibitor, and antisense compound. Use of dominant negative effect of truncated TGF-β receptor II to disrupt TGF-β function is also a widely used anti-TGF-β strategy, which has been proven to be efficient and effective by a number of reports *(44–46)*.

TGF-β binds all three types of receptors. However, it signals through TGF-β receptors I and II, the serine/threonine kinase receptors, and their downstream signaling molecules Smad *(47)*. The type III receptor, a membrane-anchored proteoglycan (betaglycan), contributes to the signaling by increasing the binding affinity between the ligands and their signaling receptors. The functional TGF-β receptor complex is presented at the cell surface in a tetramer form, which consists of two type II and two type I transmembrane serine/threonine kinase receptors *(48–51)*. Generally, type II and type I receptors exist as homodimers at the cell surface in the absence of ligand, and the binding between the ligand and the receptor II leads to the formation of a tetramer of receptor I/II complex. For TGF-β1 and TGF-β3, they can directly bind type II receptor without needing a type I receptor. However, TGF-β2 interacts only with type II–type I receptor combinations, suggesting that the heteromeric receptor complex form also exists even in the absence of ligand. Therefore, the low-affinity heteromeric receptor complex may provide a surface for ligand binding that conformationally stabilizes the complex *(48–51)*.

To activate TGF-β signaling, binding of the ligand to receptor II can first induce autophosphoration of receptor II at its intracellular part. The phosphorated cytoplasmic domain then transphosphorates the GS domain and activates the receptor kinase of receptor I, which further activates Smad molecules to transduce the signaling into the nucleus. Therefore, a truncated TGF-β receptor II can be formed by deleting its cytoplasmic domain. Importantly, the truncated receptor II has been shown to be able to form a molecular interaction with wild-type receptors I and II and thus disrupt the function of the TGF-β receptor complex and partially inhibit both the antiproliferative effect of TGF-β and the transcriptional activation by TGF-β *(44)*. However, another study showed that a similar truncated type II receptor abolished only the TGF-β-induced antiproliferative effect, without affecting the TGF-β-stimulated production of either PAI-I or fibronectin *(52)*.

To further demonstrate the dominant negative effect of mutated TGF-β receptor II, Yamamoto et al. constructed an adenoviral vector containing a truncated TGF-β receptor II gene, in which most of the cytoplasmic kinase coding region was deleted (except for seven amino acids in the intracellular region). Their study clearly demonstrated that the dominant negative effect, when the truncated receptor II was overexpressed to a sufficient amount in vascular wall cells, could specifically abolish diverse signaling mediated by TGF-β. These include complete inhibition of the TGF-β-induced antiproliferative effect on mink lung cells and endothelial cells, suppression of TGF-β-induced collagen production, and the inhibition of TGF-β-induced transcriptional activation of PAI-1 and fibronectin *(53)*. We used the same recombinant adenoviral vector provided by Yamamoto's group to examine the dominant negative effect on inhibiting fibrotic behavior of keloid fibroblasts in vitro and inhibiting experimental scar formation in vivo *(54)*.

5.2. Overexpression of Truncated TGF-β Receptor II Inhibits Fibrotic Behavior of Keloid Fibroblasts In Vitro

Clinically, keloid is characterized by its uncontrolled growth and invasion in normal skin tissue. Pathologically, it exhibits the features of rapid cell growth and overproduction and deposition of ECMs, which have been shown to be closely associated with TGF-β overfunction in many studies *(4)*. Overexpression of TGF-β ligands *(21)* and receptors *(22)* (Fig. 2) and overphosphoration of smad3 in keloid fibroblasts *(22)* suggest that abnormal TGF-β signaling may play important roles in keloid formation. Thus, we are interested in observing how disrupting TGF-β signaling affects keloid cell behavior by the dominant negative effect of truncated TGF-β receptor II.

First, we investigated the dominant negative effect on regulating TGF-β1 and its receptor expression. As previously described, overexpression of truncated TGF-β receptor II mediated by the recombinant adenovirus could significantly inhibit endogenous TGF-β1 mRNA expression (Fig. 4) *(41)*. Recently, we also demonstrated that TGF-β1 protein expression was significantly inhibited in keloid fibroblasts (Fig. 5) by the dominant negative effect (preliminary data). Interestingly, blocking TGF-β signaling did not downregulate the mRNA expression level of TGF-β receptors I and II, similar to what was observed in Yamamoto's study *(53)*. However, our recent finding demonstrated that such a dominant negative effect could significantly suppress TGF-β receptor III expression at both gene and protein levels (Fig. 6, preliminary data).

As reported, keloid fibroblasts proliferated much faster than normal dermal fibroblasts when treated with TGF-β *(25)* and were more resistant to Fas-mediated apoptosis, which could be abrogated with TGF-β-neutralizing antibody *(38)*. We thus investigate the relationship between disrupting TGF-β signaling and keloid fibroblast growth and cell-cycle change. Similar to previous reports *(4,41)*, keloid fibroblasts were seeded on six-well culture plates with a density of 5×10^3/well. After cell attachment and culture for certain time, cells were infected with the truncated receptor II adenovirus (experimental cells) or β-gal adenovirus (control cells). Both experimental and control cells from triplicate wells were harvested

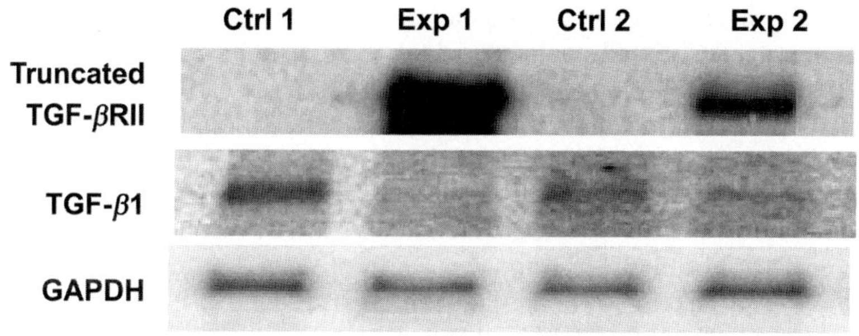

Fig. 4. Overexpressed truncated TGF-β receptor II downregulates TGF-β1 gene expression in keloid fibroblasts. Keloid cells were treated either with β-galactosidase adenovirus (Ctrl) or truncated TGF-βRII adenovirus (Exp). Cells were harvested at day 5 postviral infection for RNA extraction. A Northern blot assay was performed to analyze the gene expression of TGF-β1*(41)*.

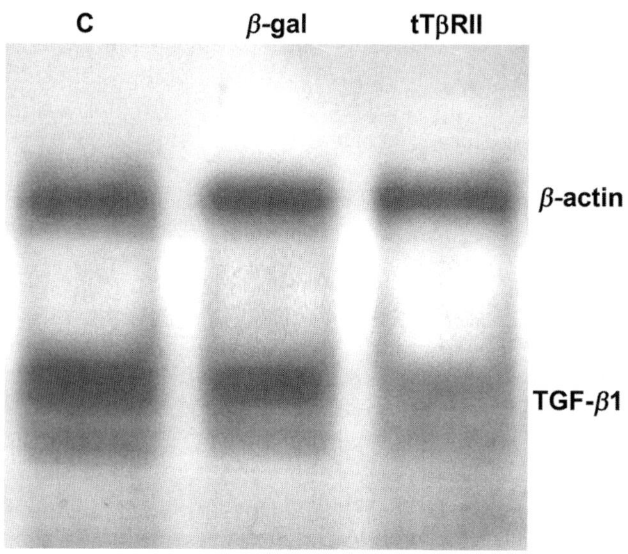

Fig. 5. Overexpressed truncated TGF-β receptor II inhibits TGF-β1 protein production in keloid fibroblasts. Keloid cells were treated either with truncated TGF-βRII adenovirus (tTβRII) or with β-galactosidase adenovirus (β-gal) or without viral infection as a blank control (C). Cells were harvested at day 5 postviral infection for Western blot analysis (preliminary data).

at the time points of days 1, 3, 5, and 7 postinfection and counted. In addition, cells without viral infection served as a blank control. As shown in Fig. 7, cell proliferation in the experimental group was significantly inhibited when compared to control cells ($p < 0.05$); however, no significant difference was found between vector control and blank control groups ($p > 0.05$).

To further explore the mechanism of cell proliferation inhibition by disrupting TGF-β signaling, we also examined cell-cycle distribution change between experimental and control keloid fibroblasts using a flow cytometry analysis. As shown in Fig. 8 and Table 1, blocking TGF-β signaling significantly reduced cell population in S phase and G2/M phase in the experimental group when compared to vector control group and blank control group. This result further confirms that the abnormal TGF-β signaling might be responsible for aberrant growth of keloid fibroblasts with the possible change of cell cycles, although the detailed mechanism needs to be investigated.

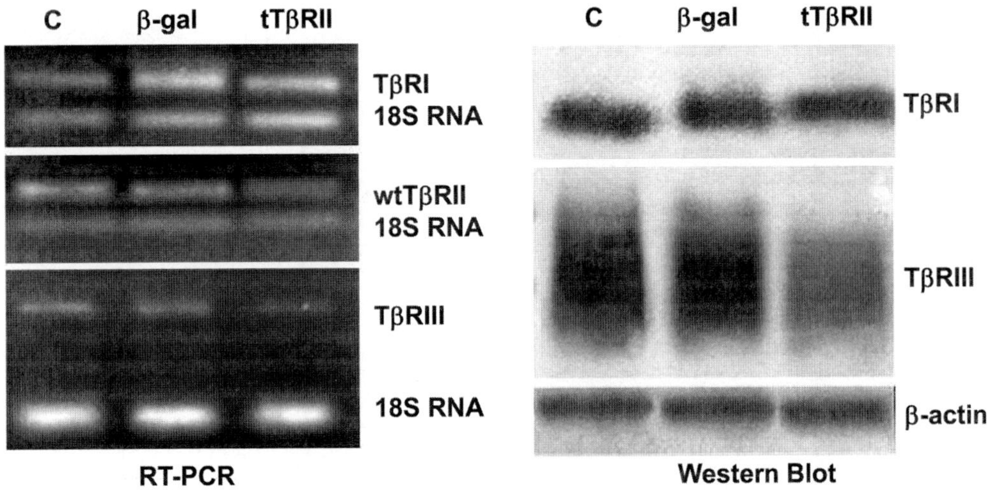

Fig. 6. Overexpressed truncated TGF-β receptor II downregulates TGF-β receptor III expression in keloid fibroblasts. Keloid cells were treated either with truncated TGF-βRII adenovirus (tTβRII) or with β-galactosidase adenovirus (β-gal) or without viral infection as a blank control (C). Cells were harvested at day 5 postviral infection for RT-PCR and Western blot analyses. Left: RT-PCR analysis shows that blocking the signaling can downregulate mRNA expression of TGF-β receptor III, but not type I and type II receptor mRNA expression; Right: Western blot shows significant downregulated protein expression of type III receptor by overexpressed truncated receptor II. It, however, does not affect type I receptor expression (preliminary data).

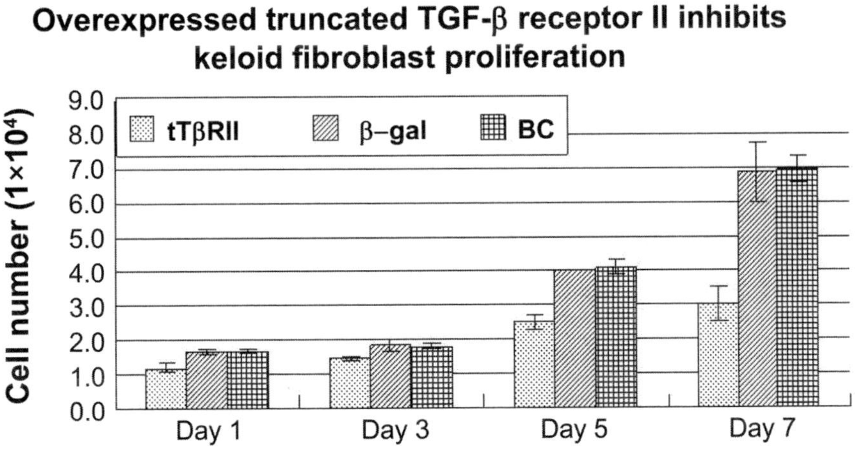

Fig. 7. Overexpressed truncated TGF-β receptor II significantly inhibits keloid fibroblast proliferation. Keloid cells were seeded on 6-well culture plates with a density of 5×10^3/well. After attachment and culture for certain time, cells were infected either with truncated TGF-βRII adenovirus (tTβRII) or with β-galactosidase adenovirus (β-gal) or without viral infection as a blank control (BC). Cells from triplicate wells were harvested at the time points of days 1, 3, 5, 7 postinfection for cell counting. The experiment was performed twice and statistical analysis shows a significant difference of cell numbers among different groups ($p < 0.05$), but not between vector control (β-gal) and blank control (C) groups (preliminary data).

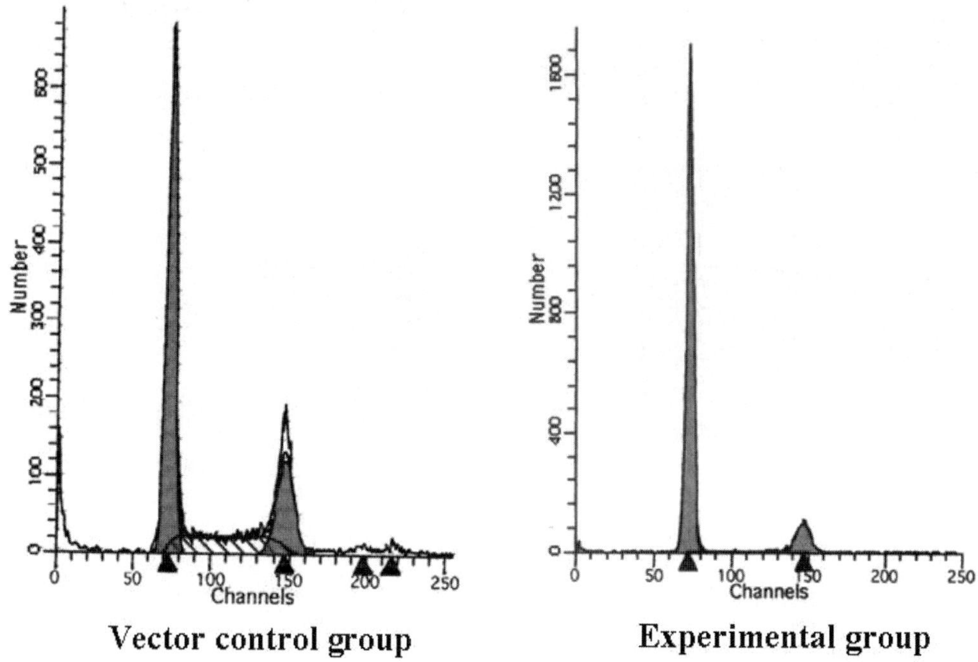

Fig. 8. Overexpressed truncated TGF-β receptor II decreases S phase and G2/M phase cell population in keloid fibroblasts. Keloid fibroblasts were infected either with truncated TGF-β receptor II adenovirus (experimental group) or with β-gal adenovirus (vector control group) and subjected to a flow cytometry analysis at day 5 postinfection (preliminary data).

Table 1
Blocking TGF-β Signaling Changes Keloid Fibroblast Cell-Cycle Distribution

	G0/G1 phase	*S phase*	*G2/M phase*
Experimental	92.62 ±0.78%	2.50 ±4.54%	4.88 ±0.31%
Vector control	67.78 ±3.45%	14.38 ±2.54%	17.84 ±1.10%
Blank control	66.01 ±3.26%	16.43 ±3.83%	17.56 ±1.66%
p value	<0.01	<0.01	<0.01

Because overproduction of ECM is another feature of keloid pathology, we also investigated how disrupting TGF-β signaling affects keloid cell matrix production. As mentioned above, experimental keloid fibroblasts and control keloid cells were infected with truncated receptor II virus and β-gal virus, respectively, and cells were harvested at day 5 postinfection for RT-PCR and Western blot analyses, respectively, to examine matrix production at both gene and protein levels; cells without viral infection served as a blank control. RT-PCR results showed that mRNA expression levels of procollagen I, procollagen III, and fibronectin were significantly downregulated in the experimental group when compared to vector control cells and blank control cells (Fig. 9, upper part), and the difference in gene expression levels between the experimental group and the control group was statistically significant ($p < 0.05$) when analyzed semiquantitatively. Furthermore, inhibition of matrix production in keloid fibroblasts by blocking TGF-β signaling was also confirmed at protein

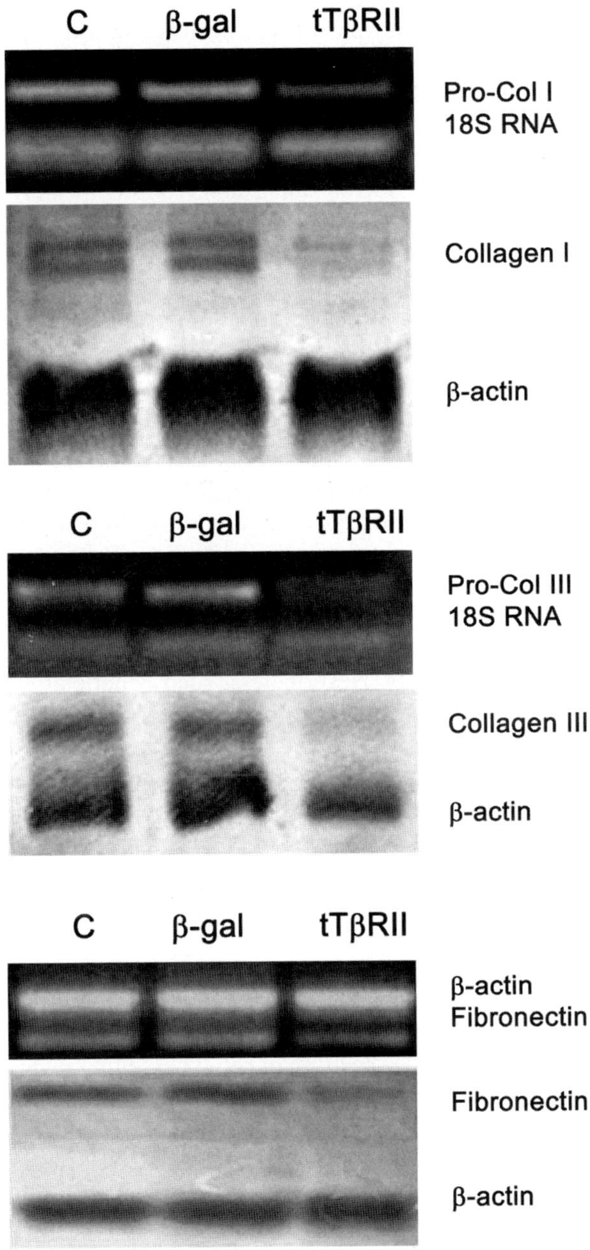

Fig. 9. Overexpressed truncated TGF-β receptor II significantly downregulates extracellular matrix expression at both gene and protein levels. Keloid cells were treated either with truncated TGF-βRII adenovirus (tTβRII) or with β-galactosidase adenovirus (β-gal) or without viral infection as a blank control (C) and were harvested at day 5 postviral infection for RT-PCR and Western blot analyses. Top panel: type I collagen; middle panel: type III collagen; bottom panel: fibronectin. In each panel, upper part is for RT-PCR and lower part for Western blot (preliminary data).

expression level. As demonstrated by Western blot analysis, protein expression levels of collagens type I and type III and fibronectin were all significantly lower in the experimental cells than in the control cells (Fig. 9, lower part) and semiquantification analysis also showed a statistically significant difference ($p < 0.05$) (preliminary data).

Although the association between keloid pathogenesis and TGF-β overproduction as well its aberrant TGF-β signaling has been suggested in reported studies *(21,22)*, this study clearly demonstrates that disrupting TGF-β signaling alone can significantly inhibit the fibrotic behaviors of keloid fibroblasts by suppressing their proliferation potential and matrix overproduction. This phenomenon further supports the hypothesis that TGF-β overfunction is a key player in keloid establishment and development. More importantly, the in vitro results of this study also provide insight into the development of an antikeloid therapeutic strategy.

5.3. Overexpression of Truncated TGF-β Receptor II Inhibits Wound Scar Formation In Vivo

Cutaneous scar formation is an integration of dynamic interactive processes, which involve soluble mediators, formed blood elements, ECM, inflammation, tissue formation, and tissue remodeling *(23,24)*. Because scar formation only happens in vivo, therefore in vitro results may not necessarily reflect the in vivo situation. Previous study by Ashcroft demonstrated that knock out of Smad3, a TGF-β signaling molecule, could gain certain beneficial effects on the healing of a linear cutaneous wound in Smad3-deficient mice, such as enhanced epithelialization, reduced inflammation and reduced wound width, which favor scar reduction in a healed wound *(55)*. Nevertheless, such beneficial effects will only be practically useful if TGF-β signaling is blocked in a local wound environment rather than disrupted systemically.

We thus performed an in vivo study to determine if the beneficial effects similar to those observed in Smad3-deficient mice *(55)* could be achieved by the dominant negative effect in a wound where truncated TGF-β receptor II will be overexpressed locally. In a previous study, we have shown that adenovirus-mediated gene transfer and incisional wound in newborn rat could serve as a proper model for manipulating cytokine profile in a wound *(56)*. Thus, the same model was applied in the in vivo study of scar manipulation.

Similar to the in vitro study, sufficient expression of truncated TGF-β receptor II is also required to achieve its dominant negative effect in vivo. Adenovirus is one of the most frequently used viral vectors in wound healing research because of its high gene transfer efficiency and its ability to target a wide spectrum of cell types, either dividing or nondividing cells *(4)*. To examine the transgene expression level mediated by the adenovirus, truncated receptor II adenoviruses were injected intradermally into the dorsal skin of 10-d-old newborn Sprague–Dawley rats and the injected skins were harvested at days 1, 2, 3, 5, 7 postinjection for immunohistochemical staining of TGF-β receptor II. As shown in Fig. 10, adenovirus-mediated expression of truncated TGF-β receptor II was observed as early as day 1 postinjection in the virus-injected dermis and the expression reached a very high level and maintained such a level at day 2. A high expression level remained observable at days 3 and 5 and the level declined at day 7. Immunohistochemical staining also demonstrated that the intradermally injected adenovirus mediated the gene expression mainly in dermal fibroblasts and partially in hair follicle (Fig. 10). Importantly, transgene expression was limited only to the injected area *(54)*.

For in vivo gene therapy study *(54)*, 1×10^9 plaque-forming units of truncated TGF-β receptor II adenovirus in phosphate-buffered saline was intradermally injected at the right-side dorsal skin of the newborn Sprague–Dawley rats as the experimental group ($n = 15$). At the left side of the same rat, either 1×10^9 plaque-forming units of β-gal adenovirus ($n = 11$) or an equal volume of phosphate-buffered saline ($n = 4$) was injected as the control group.

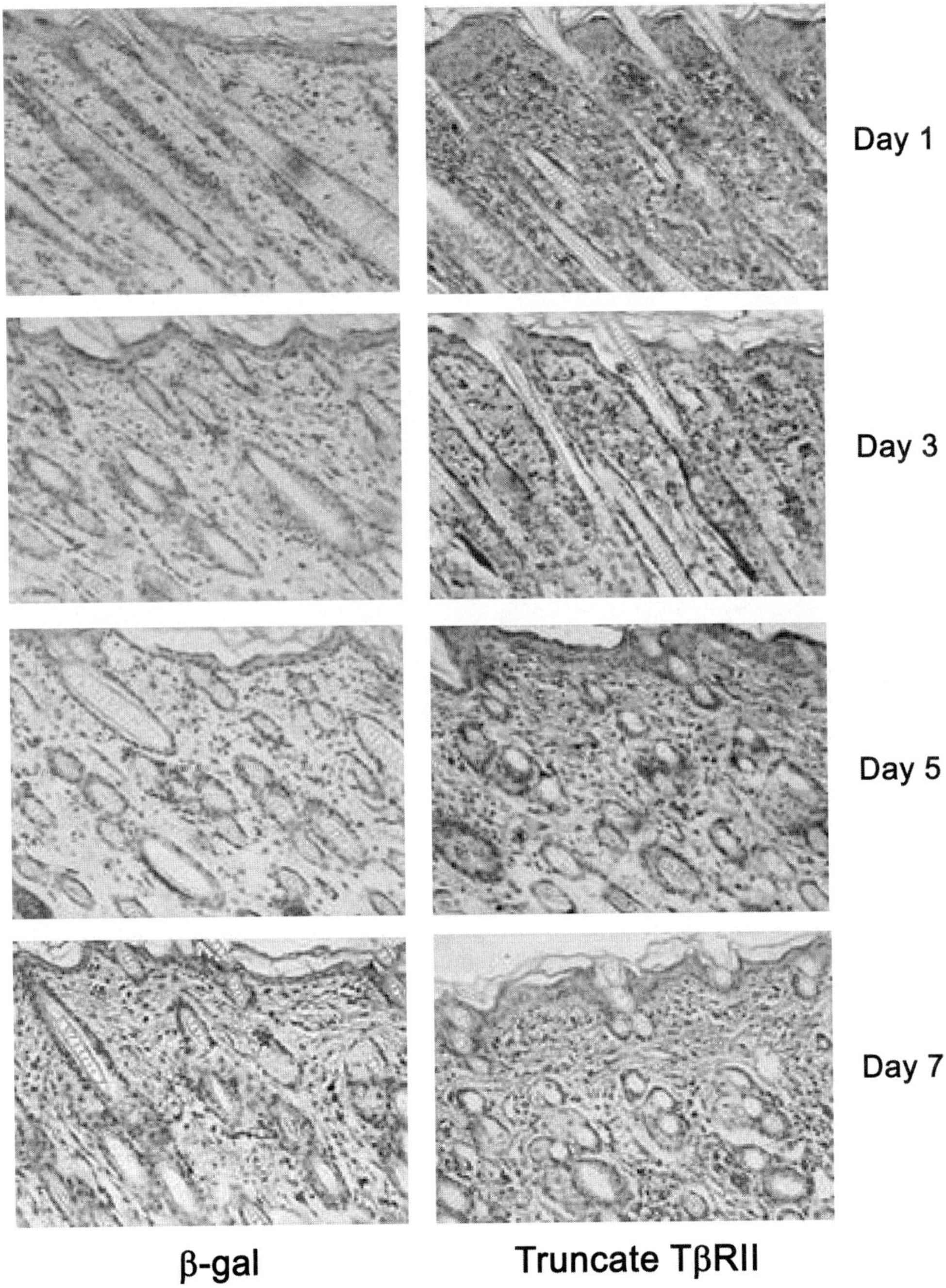

Fig. 10. Adenovirus-mediated overexpression of truncated TGF-β receptor II in newborn rat skin at days 1, 3, 5, 7 postinjection. Left: β-gal adenovirus injected skin shows a low-level expression of wild-type TGF-β receptor II; Right: tTGF-βRII virus injected skin demonstrates a high-level expression of truncated TGF-β receptor II from days 1 to 5 and the expression declines at day 7 postinjection. (Original magnification ×40) *(56)*.

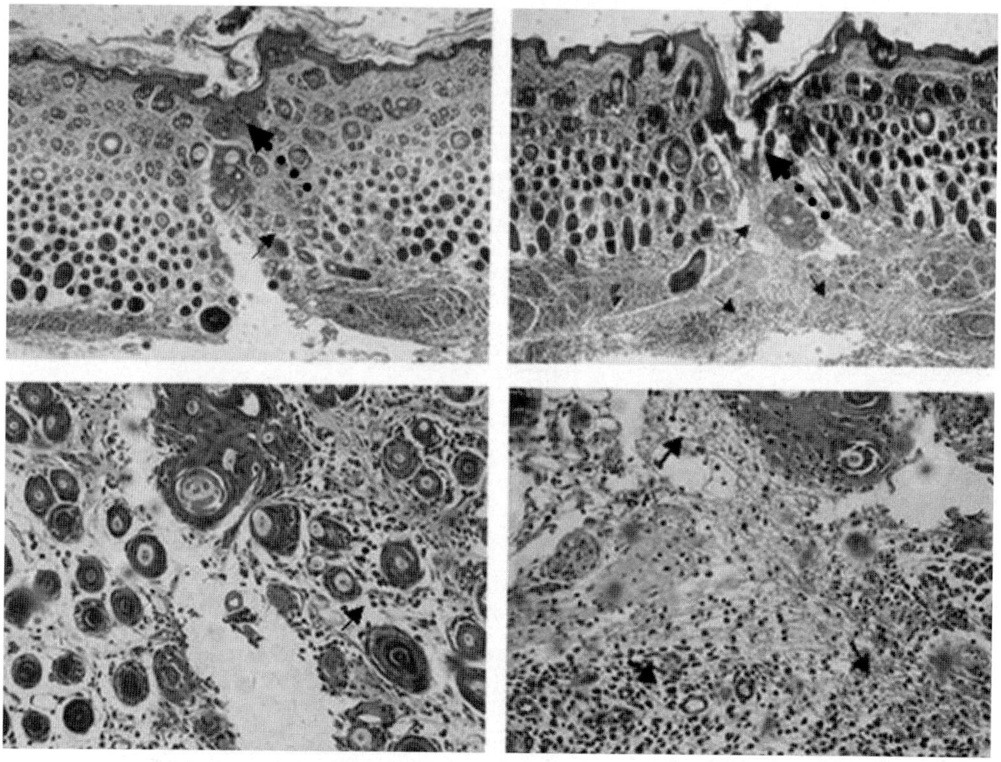

Fig. 11. Overexpressed truncated TGF-β receptor II enhances wound epithelialization and reduces wound inflammation at day 3 postwounding. Left: experimental wound; Right: β-gal control wound. Big dotted arrows indicate the wound epithelialization process, small solid arrows indicate the infiltrated inflammatory cells. (Original magnification: above, ×40; below, ×100) *(56)*.

At day 2 postinjection, a 5-mm-long full thickness incisional wound deep to and including panniculus muscle was made on the injected skin and the wound was left unrepaired as previously described *(56)*. To observe the wound-healing process in the early stage and the scar formation in the late stage, animals were sacrificed at days 3, 7, and 14 postwounding for histological examination.

In vivo results of this study demonstrated that blocking TGF-β signaling could promote wound healing by enhancing wound epithelialization. As shown in Fig. 11, at day 3 postwounding, epithelialization was completed in the experimental wound, but not in the control wound. In addition, less infiltrated inflammatory cells were observed in the experimental wound than in the control wound (Fig. 11) and quantitative analysis of infiltrated inflammatory cells showed a statistically significant difference ($p < 0.05$) between the experimental and the control wounds. At day 7, a wider wound width was observed in the control wound, which also contained abundant matrices. In contrast, the experimental wound was much narrower with less matrix deposition (Fig. 12). These results are similar to the reported beneficial effects observed in the early stage wound of Smad3-deficient mice *(55)*. Nevertheless, the TGF-β signaling was disrupted only in a local wound environment in this study.

Furthermore, at day 14 postwounding, the experimental wound healed with much less scarring than its control wound in each of 11 rats. Histology demonstrated that the experimental wound exhibited a much narrower fibrotic area when compared to that of the control wound (Fig. 13). To quantitatively analyze the difference of scar area between

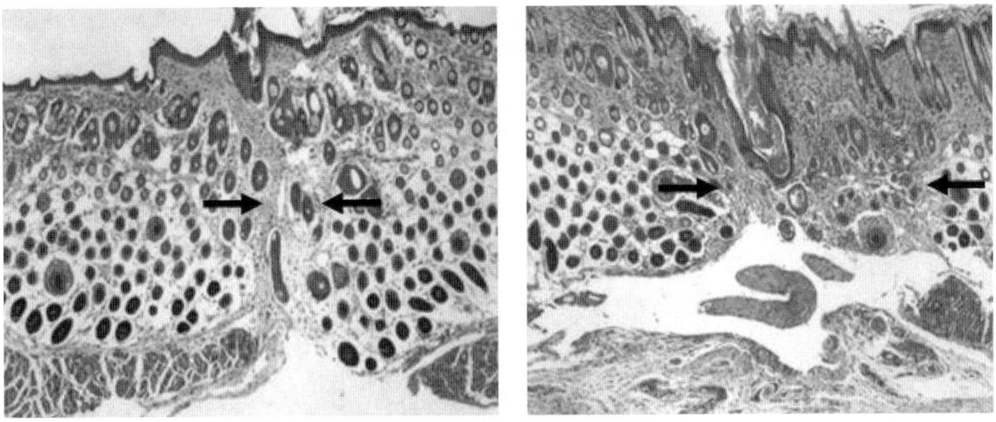

Fig. 12. Overexpressed truncated TGF-β receptor II reduces wound matrix deposition in newborn rats at day 7 postwounding. Left: less matrix production and a narrow wound width (between arrows) are observed in the experimental wound; Right: abundant matrix deposition with a wide wound width (between arrows) is shown in the control wound.

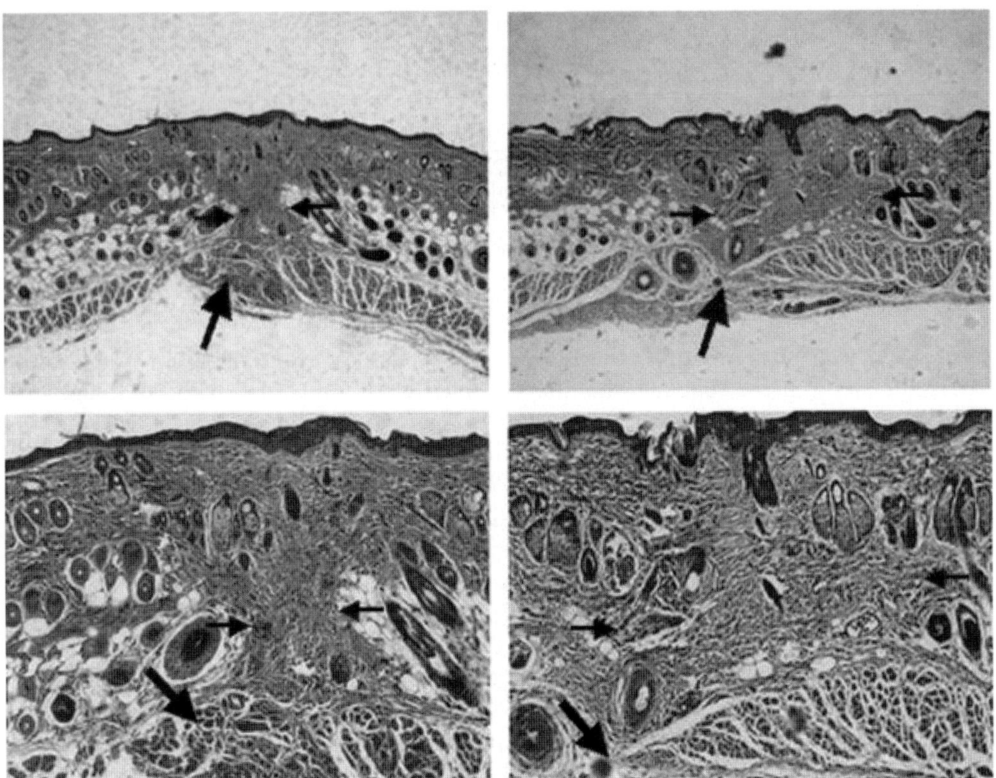

Fig. 13. Overexpressed truncated TGF-β receptor II inhibits wound scarring in newborn rats at day 14 postwounding. Left: reduced scar area (small arrows) and partially repaired panniculus muscle (large arrow) are observed in the experimental wound; Right: a wider scar area (small arrows) and a disrupted muscle layer (large arrow) are observed in the control wound. (Original magnification: above, ×40; below, ×100) *(56)*.

Table 2
Overexpressed Truncated TGF-β Receptor II Reduces Wound Scar Area

Group	*n*	*Relative area*	*SD*	*p-value*
Experimental	11	129,286.8	±36,234.6	<0.001
Control	11	252,101.1	±61,742.7	

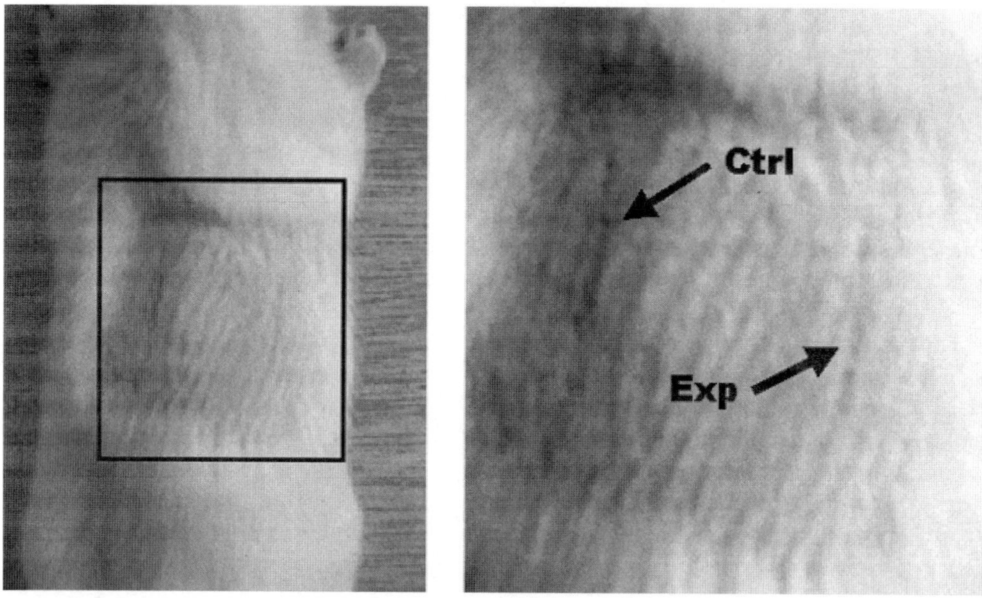

Fig. 14. Gross view shows that the treated wound (Exp) forms less visible scarring than the control wound (Ctrl) at day 14 after wounding *(56)*.

the experimental and the control wounds, H&E-stained tissue sections of day 14 wounds were observed under a microscope and the images were digitally recorded with the Image Pro-Plus System (Media Cybernetics, Silver Spring, MD). The relative area of wound scarring below the epidermis and above the panniculus muscle was measured and automatically recorded. Quantitative analysis of wound scarring using the Image Pro-Plus software demonstrated that an average of 49% reduction of wound scar area was achieved in the experimental wounds as compared with the control wounds, which was a statistically significant difference ($p < 0.001$) (Table 2). Interestingly, blocking TGF-β signaling in a wound also improved the tissue quality of repaired wounds. Among 11 experimental wounds, panniculus muscle was partially repaired in eight wounds and completely repaired in one wound. In contrast, only 2 out of 11 control wounds achieved partial repair of panniculus muscle. More importantly, by the gene therapy of an incisional wound with a strategy of the dominant negative effect mediated by overexpressed truncated TGF-β receptor II, the scar appearance of the wound has been apparently improved (Fig. 14). Thus, this approach may become an important application for future clinical treatment of wound scarring and patients can be benefited with such a strategy.

Based on the in vitro and in vivo data shown in this chapter, we would like to hypothesize the mechanism of this antiscarring gene therapy approach as the following. When a wound is created, the initial dose of TGF-β released by the platelets immediately binds the overexpressed truncated type II receptors and becomes unavailable to other cells.

Thus, its chemotactic effect is blocked, which leads to reduced infiltration of TGF-β-producing cells, such as monocytes and other cell types, and to significant reduction of wound TGF-β. The bound TGF-β also loses its ability to induce proliferation, collagen production, and TGF-β autoinduction of transduced wound cells because of the blocked signaling. All these effects eventually lead to the reduced scarring.

Nevertheless, further elucidation of the detailed mechanism and investigation of other issues, such as the strength of manipulated wound and long-term results, would be important for its translation to practical application. We hope that the results of this in vivo study can prove such a concept that wound scarring can be inhibited by disrupting wound TGF-β signaling using a gene therapy approach, which might provide insight into the designing of an antiscarring strategy for future clinical application.

REFERENCES

1. Paterson IC, Davies M, Stone A, et al. TGF-beta1 acts as a tumor suppressor of human malignant keratinocytes independently of Smad 4 expression and ligand-induced G(1) arrest. Oncogene 2002; 21:1616–1624.
2. Yamamoto T, Akisue T, Marui T, et al. Expression of transforming growth factor beta isoforms and their receptors in malignant fibrous histiocytoma of soft tissues. Clin Cancer Res 2004;10:5804–5807.
3. Tuan TL, Nichter LS. The molecular basis of keloid and hypertrophic scar formation. Mol Med Today 1998;4:19–24.
4. Liu W, Wang DR, Cao YL. TGF-beta: A fibrotic factor in wound scarring and a potential target for anti-scarring gene therapy. Curr Gene Ther 2004;4:123–136.
5. Chen WY, Grant ME, Schor AM, Schor SL. Differences between adult and foetal fibroblasts in the regulation of hyaluronate synthesis: correlation with migratory activity. J Cell Sci 1989;94:577–584.
6. Lorenz HP, Lin RY, Longaker MT, Whitby DJ, Adzick NS. The fetal fibroblast: the effector cell of scarless fetal skin repair. Plast Reconstr Surg 1995;96:1251–1259.
7. Whitby DJ, Ferguson MW. The extracellular matrix of lip wounds in fetal, neonatal and adult mice. Development 1991;112:651–668.
8. Longaker MT, Whitby DJ, Ferguson MW, et al. Studies in fetal wound healing: III. Early deposition of fibronectin distinguishes fetal from adult wound healing. J Pediatr Surg 1989;24:799–805.
9. Cowin AJ, Brosnan MP, Holmes TM, Ferguson MW. Endogenous inflammatory response to dermal wound healing in the fetal and adult mouse. Dev Dyn 1998;212:385–393.
10. Robinson BW, Goss AN. Intra-uterine healing of fetal rat cheek. Cleft Palate J 1981;18:251–255.
11. Hsu M, Peled ZM, Chin GS, Liu W, Longaker MT. Ontogeny of expression of transforming growth factor-beta 1 (TGF-beta 1), TGF-beta 3, and TGF-beta receptors I and II in fetal rat fibroblasts and skin. Plast Reconstr Surg 2001;107:1787–1794.
12. Liechty KW, Crombleholme TM, Cass DL, Martin B, Adzick NS. Diminished interleukin-8 (IL-8) production in the fetal wound healing response. J Surg Res 1998;77:80–84.
13. Haynes JH, Johnson DE, Mast BA, et al. Platelet-derived growth factor induces fetal wound fibrosis. J Pediatr Surg 1994;29:1405–1408.
14. Marin P, Dickson MC, Millan FA, Akhurst R. Rapid induction and clearance of TGF beta 1 is an early response to wounding in the mouse embryo. Dev Genet 1993;14:225–238.
15. Krummel TM, Michna BA, Thomas BL, et al. Transforming growth factor beta (TGF-beta) induces fibrosis in a fetal wound model. J Pediatr Surg 1988;23:647–652.
16. Houghton PE, Keefer KA, Krummel TM. The role of transforming growth factor beta (TGF-beta) in the conversion from scarless healing to healing with scar formation. Wound Repair Regen 1995;3:229–236.
17. Shaw AM. Recent advances in embryonic wound healing. In: Scarless Wound Healing, First Edition, Garg HG, Longaker MT, eds. New York: Marcel Dekker, 2000; pp. 227–237.
18. Soo CF, Hu Y, Zhang X, et al. Differential expression of fibromodulin, a transforming growth factor-beta modulator, in fetal skin development and scarless repair. Am J Pathol 2000;157:423–433.
19. Wang R, Ghahary A, Shen Q, Scott PG, Roy K, Tredget EE. Hypertrophic scar tissues and fibroblasts produce more transforming growth factor-beta1 mRNA and protein than normal skin and cells. Wound Repair Regen 2000;8:128–137.
20. Schmid P, Itin P, Cherry G, Bi C, Cox DA. Enhanced expression of transforming growth factor-beta type I and type II receptors in wound granulation tissue and hypertrophic scar. Am J Pathol 1998;152:485–493.

21. Lee TY, Chin GS, Kim WJ, Chau D, Gittes GK, Longaker MT. Expression of transforming growth factor beta 1, 2, and 3 proteins in keloids. Ann Plast Surg 1999;43:179–184.
22. Chin GS, Liu W, Peled Z, et al. Differential expression of transforming growth factor-beta receptors I and II and activation of Smad 3 in keloid fibroblasts. Plast Reconstr Surg 2001;108:423–429.
23. Roberts AB, Sporn MB. Transforming growth factor-beta. In: The Molecular and Cellular Biology of Wound Repair, Second Edition. Clark RAF, ed. New York: Plenum Press, 1996; pp. 275–308.
24. Singer AJ, Clark RA. Cutaneous wound healing. N Eng J Med 1999;341:738–746.
25. Bettinger DA, Yager DR, Diegelmann RF, Cohen IK. The effect of TGF-beta on keloid fibroblast proliferation and collagen synthesis. Plast Reconstr Surg 1996;98:827–833.
26. Younai S, Nichter LS, Wellisz T, Reinisch J, Nimni ME, Tuan TL. Modulation of collagen synthesis by transforming growth factor-beta in keloid and hypertrophic scar fibroblasts. Ann Plast Surg 1994; 33:148–151.
27. Chen SJ, Yuan W, Lo S, Trojanowska M, Varga J. Interaction of smad3 with a proximal smad-binding element of the human alpha2(I) procollagen gene promoter required for transcriptional activation by TGF-beta. J Cell Physiol 2000;183:381–392.
28. Peltonen J, Hsiao LL, Jaakkola S, et al. Activation of collagen gene expression in keloids: co-localization of type I and VI collagen and transforming growth factor-beta 1 mRNA. J Invest Dermatol 1991;97: 240–248.
29. Cao HJ, Hogg MG, Martino LJ, Smith TJ. Transforming growth factor-beta induces plasminogen activator inhibitor type-1 in cultured human orbital fibroblasts. Invest Ophthalmol Vis Sci 1995;36: 1411–1419.
30. Yang YY, Tsai HF, Lu SC, Huang YF, Chang YC. Regulation of tissue inhibitors of metalloproteinase-1 gene expression by cytokines in human gingival fibroblasts. J Endod 2002;28:803–805.
31. Eickelberg O, Kohler E, Reichenberger F, et al. Extracellular matrix deposition by primary human lung fibroblasts in response to TGF-beta1 and TGF-beta3. Am J Physiol 1999;276:L814–L824.
32. Tuan TL, Zhu JY, Sun B, Nichter LS, Nimni ME, Laug WE. Elevated levels of plasminogen activator inhibitor-1 may account for the altered fibrinolysis by keloid fibroblasts. J Invest Dermatol 1996;106: 1007–1011.
33. Higgins PJ, Slack JK, Diegelmann RF, Staiano-Coico L. Differential regulation of PAI-1 gene expression in human fibroblasts predisposed to a fibrotic phenotype. Exp Cell Res 1999;248:634–642.
34. McCarthy JG. Introduction to plastic surgery. In: Plastic Surgery, First Edition, McCarthy JG, May JW, Littler JW, eds. Philadelphia: W.B. Saunders, 1990; pp. 1–68.
35. Peled ZM, Liu W, Levinson H, et al. Cellular strain upregulates profibrotic growth factor and collagen gene expression. Surg Forum 2000;51:591–593.
36. Luo S, Benathan M, Raffoul W, Panizzon RG, Egloff DV. Abnormal balance between proliferation and apoptotic cell death in fibroblasts derived from keloid lesions. Plast Reconstr Surg 2001;107:87–96.
37. Chipev CC, Simman R, Hatch G, Katz AE, Siegel DM, Simon M. Myofibroblast phenotype and apoptosis in keloid and palmar fibroblasts in vitro. Cell Death Differ 2000;7:166–176.
38. Chodon T, Sugihara T, Igawa HH, Funayama E, Furukawa H. Keloid-derived fibroblasts are refractory to Fas-mediated apoptosis and neutralization of autocrine transforming growth factor-beta1 can abrogate this resistance. Am J Pathol 2000;157:1661–1669.
39. Sayah DN, Soo C, Shaw WW, et al. Downregulation of apoptosis-related genes in keloid tissues. J Surg Res 1999;87:209–216.
40. Saed GM, Ladin D, Olson J, Han X, Hou Z, Fivenson D. Analysis of p53 gene mutations in keloids using polymerase chain reaction-based single-strand conformational polymorphism and DNA sequencing. Arch Dermatol 1998;134:963–967.
41. Liu W, Cai Z, Wang D, et al. Blocking transforming growth factor-beta receptor signaling down-regulates transforming growth factor-beta1 autoproduction in keloid fibroblasts. Chin J Traumatol 2002;5:77–81.
42. Mori Y, Chen SJ, Varga J. Modulation of endogenous Smad expression in normal skin fibroblast by transforming growth factor-beta. Exp Cell Res 2000;258:374–383.
43. Babu M, Diegelmann R, Oliver N. Keloid fibroblasts exhibit an altered response to TGF-beta. J Invest Dermatol 1992;99:650–655.
44. Wieser R, Attisano L, Wrana JL, Massagué J. Signaling activity of transforming growth factor beta type II receptors lacking specific domains in the cytoplasmic region. Mol Cell Biol 1993; 13:7239–7247.
45. Choi ME, Ballermann BJ. Inhibition of capillary morphogenesis and associated apoptosis by dominant negative mutant transforming growth factor-beta receptors. J Biol Chem 1995;270:21,144–21,150.

46. Filvaroff EH, Ebner R, Derynck R. Inhibition of myogenic differentiation in myoblasts expressing a truncated type II TGF-beta receptor. Development 1994;120:1085–1095.
47. Massagué J, Watton D. Transcriptional control by the TGF-beta/Smad signaling. EMBO J 2000; 19:1745–1754.
48. Massagué J. How cells read TGF-β signals. Nat Rev Mol Cell Biol 2000;1:169–178.
49. Itoh S, Itoh F, Goumans MJ, ten Dijke P. Signaling of transforming growth factor-b family members through Smad proteins. Eur J Biochem 2000;267:6954–6967.
50. Moustakas A, Souchelnytskyi S, Heldin C-H. Smad regulation in TGF-b signal transduction. J Cell Sci 2001;114:4359–4369.
51. Derynck R, Feng XH. TGF-β receptor signaling. Biochim Biophys Acta 1997;1333:F105–F150.
52. Chen RH, Ebner R, Derynck R. Inactivation of the type II receptor reveals two receptor pathways for the diverse TGF-beta activities. Science 1993;260:1335–1338.
53. Yamamoto H, Ueno H, Ooshima A, Takeshita A. Adenovirus-mediated transfer of a truncated transforming growth factor-beta (TGF-beta) type II receptor completely and specifically abolishes diverse signaling by TGF-beta in vascular wall cells in primary culture. J Biol Chem 1996;271:16,253–16,259.
54. Liu W, Chua C, Wu X, et al. Inhibiting scar formation in rat wounds by adenovirus-mediated overexpression of truncated TGF-beta receptor II. Plast Reconstr Surg 2005;115:860–870.
55. Ashcroft GS, Yang X, Glick AB, et al. Mice lacking Smad3 show accelerated wound healing and an impaired local inflammatory response. Nat Cell Biol 1999;1:260–266.
56. Liu W, Mehrara BJ, Chin GS, Hsu M, Peled Z, Longaker MT. The use of newborn rats and an adenoviral gene delivery vector as a model system for wound-healing research. Ann Plast Surg 2000; 44:543–551.

Index

A

Activins
- angiogenesis effects, 416
- antagonists, *see* Follistatin; Inhibins
- dimerization, 491
- functional overview, 412
- gene structure, 484, 485
- glycosylation, 486, 487
- hepatocellular carcinoma expression, 498, 499
- immune modulation, 416, 417
- liver function, 493–497
- metastasis effects, 416
- processing, 489, 490
- prodomains, 485, 486
- prostate expression and function
  - cancer, 358–360
  - normal prostate, 357, 358
- receptors, 488, 489
- secretion, 490
- signaling, 412, 413
- structure, 412, 484, 485, 487, 488
- therapeutic targeting, 499, 500
- tumor progression activity, 415

Akt
- activation, 147
- isoforms, 147
- Smad repression, 147, 148
- transforming growth factor-β crosstalk, 146

Androgen receptor (AR), transforming growth factor-β/Smad system interactions, 529

Angiogenesis
- activin effects, 416
- bone morphogenetic protein-2 promotion, 376, 378, 379
- endoglin role, 399
- follistatin effects, 416
- transforming growth factor-β promotion, 31, 327

Anti-Mullerian hormone, *see* Mullerian inhibiting substance

Antisense oligonucleotides, transforming growth factor-β targeting
- ocular fibrosis, 666, 667
- pulmonary fibrosis, 636

Apoptosis
- Betaig-H3 induction, 476
- inhibitory Smad modulation, 654, 655
- transforming growth factor-β induction in cancer
  - endothelial cells, 204, 205
  - gastric cancer, 203, 204
  - hepatoma, 200–202
  - nervous system, 205, 206
  - overview, 26, 199, 200
  - prostate cancer, 202, 203
- transforming growth factor-β inhibition in lung injury model, 573

AR, *see* Androgen receptor

B

BAD, transforming growth factor-β induction of apoptosis in cancer, 201

BAMBI, *see* Bone morphogenetic protein and activin membrane-bound inhibitor

Benign prostate hyperplasia (BPH), features, 356

Betaglycan
- conservation between species, 453
- coreceptor activity, 452, 453
- ectodomain shedding as therapeutic target, 457
- function, 7
- gene, 453
- glycosaminoglycan chains, 455
- knockout mouse, 455, 456
- ligand-binding domains, 453, 454
- protein–protein interactions, 453
- structure, 453
- transforming growth factor-β-neutralizing activity of soluble protein, 456, 457
- transmembrane and cytoplasmic regions, 455
- type II receptor interactions, 454, 455

Betaig-H3
- apoptosis induction, 476
- cancer loss of expression, 477
- cell adhesion role, 476
- functional overview, 476
- promoter hypermethylation and silencing in cancer, 479, 480
- structure, 476
- transfection studies of tumor suppression, 478, 479

Bleomycin
- clinical utility, 581, 588–590
- DNA damage, 582, 583

pulmonary fibrosis induction
animal model
histopathology, 584–587
transforming growth factor-β inhibition studies, 588
prevention, 590
RNA damage, 583
structure, 582
BMP, *see* Bone morphogenetic protein
Bone morphogenetic protein (BMP)
angiogenesis promotion, 376
bone morphogenetic protein induction, 378, 379
endothelial cell activation, 376–378
functional overview, 373, 374
kidney
BMP7 and kidney disease, 464
regulation of local activities, 464
uterine sensitization-associated gene-1 antagonism, *see* Uterine sensitization-associated gene-1
ovary
cancer expression
granulosa cell tumor, 441, 442
ovarian epithelial tumor, 442
expression patterns, 427–430
functions
cumulus cell differentiation, 434
granulosa cell proliferation and differentiation, 431, 434
in vivo studies, 430–432
granulosa cell signaling pathways
ALK6, 440
alternative signaling pathways, 441
BMP15, 439, 440
miscellaneous BMPs, 440
prostate expression and function
normal prostate, 360–362
prostate cancer
BMP2, 362
BMP6, 362–365
BMP7, 365
GDF15, 365, 366
receptors, 366
tumor expression and promotion, 375, 376
vascular endothelial growth factor
vasculogenesis regulation by BMP2/4, 374, 375
Bone morphogenetic protein and activin membrane-bound inhibitor (BAMBI)
function, 413
Smad interactions, 12
Bone morphogenetic protein receptors
prostate cancer expression, 366
protein–protein interactions, 323, 324
signaling, 374, 384
types, 374
BPH, *see* Benign prostate hyperplasia
Breast cancer, Mullerian inhibiting substance therapy, 346, 348

C

E-Cadherin
embryonic liver fodrin interactions, 98, 99
epithelial-to-mesenchymal transition and transcriptomic reprogramming, 267–269
Calcium/calmodulin protein kinase II (CaMKII), Smad phosphorylation, 116
CaMKII, *see* Calcium/calmodulin protein kinase II
Camurati-Engelmann disease (CED), latency-associated peptide mutations, 41
β-Catenin, *see* Wnt signaling
Caveolin, transforming growth factor-β receptor interactions, 79
CBP, *see* CREB-binding protein
CD105, *see* Endoglin
CDK, *see* Cyclin-dependent kinase
C/EBP, Smad interactions, 12
CED, *see* Camurati-Engelmann disease
Cell cycle
endometrial carcinoma studies of transforming growth factor-β signaling effects on $p27^{Kip1}$
degradation prevention, 225–228
growth regulation in endometrium, 228, 229
transforming growth factor-β
dysregulation in cancer
aberrant signaling, 215, 216
cyclin-dependent kinase inhibitors, 218, 219
cyclin-dependent kinases, 217, 218
cyclins, 216, 217
c-Myc, 219, 220
$p27^{Kip1}$, 214, 215, 219–223
inhibition, 28, 214, 215
tumor evasion of antiproliferative effects, 28–31
Cervical cancer, Mullerian inhibiting substance therapy, 344–346
Chondrocyte
Smurf1 function, 163
transforming growth factor-β and differentiation, 325
Chromatin modification, *see* Histone methyl-transferase

Clathrin, transforming growth factor-β receptor interactions, 79
Collagen
Smad activation, 584
transforming growth factor-β induction
lung injury model, 573
wound healing, 678
Colon cancer
km23 and transforming growth factor-β production role cells, 176, 177
latent transforming growth factor-β-binding protein-4 deficiency, 64
Smad4 loss, 94
Common mediator Smads, *see* Smads
Connective-tissue growth factor (CTGF), transforming growth factor-β induction, 31
CREB-binding protein (CBP), Smad coactivation, 83, 84
Cripto
activin regulation, 492, 493
therapeutic targeting, 500
CTGF, *see* Connective-tissue growth factor
Cumulus cell, *see* Folliculogenesis
CUTL1, epithelial-to-mesenchymal transition mediation, 269
Cyclin, transforming growth factor-β regulation and cancer defects, 216, 217, 325
Cyclin-dependent kinase (CDK)
Smad phosphorylation, 114–116
transforming growth factor-β regulation and cancer defects, 217, 218
Cyclin-dependent kinase inhibitors, transforming growth factor-β regulation and cancer defects, 218, 219
Cysteine knot, transforming growth factor-β family proteins, 487, 488

D

Dab2, *see* Disabled-2
Dachshund, Ski interactions, 144
Daxx, transforming growth factor-β receptor interactions, 170
Decorin
ocular fibrosis and transforming growth factor-β targeting, 668
therapeutic potential in pulmonary fibrosis, 574, 635
transforming growth factor-β interactions, 282, 283
Deltanoids, transforming growth factor-β/Smad system regulation, 530–534
Disabled-2 (Dab2)
epithelial-to-mesenchymal transition mediation, 269
transforming growth factor-β receptor interactions, 170
DNA repair, transforming growth factor-β effects, 326
Dynein
function, 171
light chains
km23, *see* km23
overview, 171, 172
structure, 171, 172

E

ECM, *see* Extracellular matrix
ELF, *see* Embryonic liver fodrin
Embryonic liver fodrin (ELF)
function in cancer, 98, 99
isoforms, 95
knockout mice, 95–98
structure, 94
EMT, *see* Epithelial-to-mesenchymal transition
Endoglin (CD105)
angiogenesis role, 399
coreceptor activity, 452, 453
expression regulation, 398, 400
function, 7
prognostic value in cancer, 401, 402
structure, 396, 397
therapeutic targeting
ex vivo studies, 403
in vitro studies, 402, 403
in vivo studies, 403, 404
rationale, 401
vaccines, 404
tissue distribution, 398, 399
transforming growth factor-β interactions, 397, 398
tumor progression role, 399, 400
vasculogenesis role, 400, 401
Endometrial carcinoma
epidemiology, 223
management, 223
transforming growth factor-β effects in cell lines, 223, 224
transforming growth factor-β signaling effects on $p27^{Kip1}$
degradation prevention, 225–228
growth regulation in endometrium, 228, 229
Mullerian inhibiting substance therapy, 346
Endothelial cell
bone morphogenetic protein activation, 376–378
transforming growth factor-β induction of apoptosis, 204, 205
Epithelial-to-mesenchymal transition (EMT)

embryonic development, 186, 260
fibrosis, 261, 262
models, 262
Ras-transforming growth factor-β signaling crosstalk, 310, 311
Smurf1 role, 164
transforming growth factor-β induction
cancer, 31, 38, 260, 261
CUTL1 mediation, 269
Dab2 mediation, 269
lung injury model, 573
NET1 mediation, 269
Notch mediation, 269
Smad mediation, 263, 264
Smad-independent signaling, 264
transcriptomic reprogramming
E-cadherin, 267–269
Id proteins, 266, 267
overview, 265, 266
ER, *see* Estrogen receptor
Estrogen receptor (ER), transforming growth factor-β/Smad system interactions, 529, 536, 537
Evi1, Smad corepression, 86
Extracellular matrix (ECM), *see* also Matrix metalloproteinases
latent transforming growth factor-β-binding protein interactions, 44, 65–67
transforming growth factor-β modulation
matrix component expression, 584
wound healing, 678, 679
Eye, *see* Ocular fibrosis

F

FADD, transforming growth factor-β induction of apoptosis in cancer, 203
FGF, *see* Fibroblast growth factor
Fibrillin
knockout mouse, 68
latent transforming growth factor-β-binding protein interactions, 62
Fibroblast
matrix metalloproteinase expression in stromal cells, 285, 286
transforming growth factor-β sensitivity and cancer regulation in stroma
initiation, 189–192
metastasis, 193
progression, 192
Fibroblast growth factor (FGF), FGF-2 and transforming growth factor-β interactions in wound healing, 684
Fibrosis
epithelial-to-mesenchymal transition, 261, 262
eye, *see* Ocular fibrosis
fetal scarless wound healing, 704
keloid, *see* Keloid
kidney, *see* Renal fibrosis
latent transforming growth factor-β-binding protein expression, 67, 68
skin fibrosis, *see* also Scleroderma
therapeutic targeting of transforming growth factor-β
overview of strategies, 695–697
P144 antagonistic peptide development and preclinical studies, 697–700
transforming growth factor-β role, 694, 695
Smad3 studies using knockout mouse
eye, 602
kidney, 601
liver, 601
lung, 600, 601
skin, 600
therapeutic targeting, 602–604
thrombospondin-1 role
diabetes, 555
hypertension, 555
liver fibrosis, 556, 558
lung fibrosis, 555, 556
myocardial remodeling, 557
renal fibrosis, 554, 555
scleroderma, 557
truncated transforming growth factor-β receptor II overexpression and scar reduction
adenoviral vector, 709
dominant negative effect, 708, 709
fibroblast fibrotic behavior inhibition in vitro, 709, 710, 712, 714
scar formation inhibition in vivo, 714, 716, 718, 719
Fodrin, *see* Embryonic liver fodrin
Folliculogenesis
bone morphogenetic protein/growth differentiation factor
expression patterns, 427–430
functions
cumulus cell differentiation, 434
granulosa cell proliferation and differentiation, 431, 434
in vivo studies, 430–432
granulosa cell signaling pathways
ALK6, 440
alternative signaling pathways, 441
BMP15, 439, 440
GDF9, 438, 439

miscellaneous BMPs, 440
hormonal control, 426
Follistatin
activin binding, 489, 492
activin regulation, 357
angiogenesis effects, 416
isoforms, 492
prostate cancer expression and function, 358, 361
structure, 492
therapeutic targeting, 500

G

Gastric cancer, transforming growth factor-β induction of apoptosis, 203, 204
GDF, *see* Growth differentiation factor
Gene therapy, Smad7 ultrasound-microbubble-mediated gene transfer for renal fibrosis and inflammation
rationale, 613, 614
safety and efficacy, 614–616
Glucocorticoid receptor (GR), transforming growth factor-β/Smad system interactions, 529, 537, 538
GR, *see* Glucocorticoid receptor
Granulosa cell, *see* Folliculogenesis; Ovarian cancer
Growth differentiation factor (GDF)
GDF15 expression in prostate cancer, 365, 366
ovary
cancer expression
granulosa cell tumor, 441, 442
ovarian epithelial tumor, 442
expression patterns, 427–430
functions
cumulus cell differentiation, 434
granulosa cell proliferation and differentiation, 431, 434
in vivo studies, 430–432
granulosa cell signaling pathways
alternative signaling pathways, 441
GDF9, 438, 439

H

HCC, *see* Hepatocellular carcinoma
Hepatic fibrosis, Smad3 studies using knockout mouse, 601
Hepatocellular carcinoma (HCC), activin expression, 498, 499
Hepatocyte growth factor (HGF), transforming growth factor-β suppression, 574, 576
Hepatoma, transforming growth factor-β induction of apoptosis, 200–202
Heterochromatin protein 1 (HP1), histone methyltransferase recruitment, 384, 386–388
Histone methyltransferase
bone morphogenetic protein-induced transcriptional repression
carcinogenesis role, 391, 392
global changes, 390
specific genes, 388–390
heterochromatin protein 1 recruitment, 384, 386–388
Smad interactions in bone morphogenetic protein signaling, 384–388
trimethylation, 384
HP1, *see* Heterochromatin protein 1

I

IBD, *see* Inflammatory bowel disease
Id proteins
epithelial-to-mesenchymal transition and transcriptomic reprogramming, 266, 267
Id1 and bone morphogenetic protein-induced angiogenesis, 379
transforming growth factor-β effects, 28
IFN-γ, *see* Interferon-γ
IL-7, *see* Intereukin-7
Imatinib mesylate, pulmonary fibrosis management, 637
Inflammation
carcinogenesis correlates, 527, 528, 604, 605
Smad7 ultrasound-microbubble-mediated gene transfer for renal fibrosis and inflammation
rationale, 613, 614
safety and efficacy, 614–616
transforming growth factor-β/Smad signaling
overview, 527, 548
renal inflammation, 611–613
Inflammatory bowel disease (IBD), inhibitory Smad role, 656
Inhibins
discovery, 483, 484
functional overview, 412
glycosylation, 486, 487
processing, 489, 490
prodomains, 485, 486
prostate expression and function
normal prostate, 357, 358
prostate cancer, 360
secretion, 490
structure, 412, 485, 487, 488
α-subunit function, 417, 418
tumor suppressor activity, 413–415
Inhibitory Smads, *see* Smads

Integrins, transforming growth factor-β activation, 48–50, 61, 62
Intereukin-7 (IL-7), transforming growth factor-β suppression, 574
Interferon-γ (IFN-γ)
pulmonary fibrosis management, 636
transforming growth factor-β interactions in wound healing, 683

J

JPS, *see* Juvenile polyposis syndrome
Juvenile polyposis syndrome (JPS), Smad4 loss, 94

K

Keloid
features, 705
transforming growth factor-β system
expression of components, 705, 706
pathophysiological mechanisms, 706–708
truncated transforming growth factor-β receptor II overexpression and scar reduction
adenoviral vector, 709
dominant negative effect, 708, 709
fibroblast fibrotic behavior inhibition in vitro, 709, 710, 712, 714
scar formation inhibition in vivo, 714, 716, 718, 719
Keratinocyte, transforming growth factor-β response in wound healing, 680, 681
Kidney
bone morphogenetic proteins
BMP7 and kidney disease, 464
regulation of local activities, 464
fibrosis, *see* Renal fibrosis
uterine sensitization-associated gene-1
abundance, 469
expression, 468
knockout mouse resistance to renal tubular injury, 468, 469
km23
km23-1
ovarian cancer mutations, 174, 176
Smad2-dependent transforming growth factor-β signaling role, 174
structure, 173
transforming growth factor-β production role in colon carcinoma cells, 176, 177
transforming growth factor-β receptor interactions, 173, 174
tumor suppressor activity, 176
km23-2
Smad3-dependent transforming growth factor-β signaling role, 178
structure, 177
transforming growth factor-β signaling evidence, 177, 178
recruitment of transforming growth factor-β signaling endosomes for intracellular transport along microtubules, 178, 180
superfamily and homologies, 172, 173
therapeutic targeting in cancer, 180, 181

L

LAP, *see* Latency-associated peptide
Large latent complex, *see* Latent transforming growth factor-β-binding protein
Latency-associated peptide (LAP)
activation, *see* Transforming growth factor-β activation
glycosylation, 40
mutation in Camurati-Engelmann disease, 41
renal inflammation role, 612
small transforming growth factor-β latent complex, 5, 24, 38, 279, 550
therapeutic potential in pulmonary fibrosis, 635, 636
transforming growth factor-β interactions, 40, 41
Latent transforming growth factor-β-binding protein (LTBP)
activation, *see* Transforming growth factor-β activation
binding protein, 62
disease associations, 67, 68
extracellular matrix interactions, 44, 65–67
function studies
genetic approaches, 45, 46
in vitro approaches, 44, 45
large transforming growth factor-β latent complex, 5, 24, 38, 279, 550
structure, 41, 63
synthesis, 43
targeting in vivo, 67
tissue distribution, 43, 44, 62–64
transforming growth factor-β isoform specificity, 62, 64
tumor expression, 68, 69
types and functions
LTBP-1, 62, 63
LTBP-2, 64
LTBP-3, 64
LTBP-4, 64

overview, 5, 41, 65
Liver, *see* also Hepatocellular carcinoma; Hepatoma
activin function, 493–497
fibrosis and thrombospondin-1 role, 556, 558
transforming growth factor-β function, 493, 494
LMO4, Smad8 interactions, 323
LTBP, *see* Latent transforming growth factor-β-binding protein
Lung
fibrosis, *see* Pulmonary fibrosis
transforming growth factor-β
isoform functions, 622
overexpression phenotypes, 623
signaling layers, 623–625

M

α2-Macroglobulin, transforming growth factor-β interactions, 282, 283
MAD, Smad phosphatase, 111, 112
MAPK, *see* Mitogen-activated protein kinase
Marfan syndrome, latent transforming growth factor-β-binding protein role, 68
Matrix metalloproteinases (MMPs)
activation mechanisms, 277
betaglycan ectodomain shedding and therapeutic targeting, 457
cancer progression role, 276, 279
classification, 276
growth factor availability regulation, 279
pulmonary fibrosis protease/antiprotease balance, 625, 626
structure, 276, 278
transforming growth factor-β
interactions in tumors, 285, 86
regulation by matrix metalloproteinases
activation, 282
binding protein degradation, 283
regulation of matrix metalloproteinases
expression, 283–285
regulating protein expression regulation, 285
similarities, 281
Mesenchymal stem cell (MSC)
overview, 294
transforming growth factor-β signaling, 294–296
Wnt signaling
β-catenin nuclear accumulation
modulation of transforming growth factor-β effects, 299
response to transforming growth factor-β, 297, 298
Smad3 requirement, 298, 299
crosstalk with transforming growth factor-β signaling, 296, 297
prospects for study, 300, 301
regulation, 294–296
target gene cooperative regulation with β-catenin and Smad3, 299
Metabolism
myostatin regulation, 516, 517
transforming growth factor-β effects, 326, 327
Metastasis
activin effects, 416
fibroblast transforming growth factor-β sensitivity and cancer regulation in stroma, 193
transforming growth factor-β promotion, 31
MIS, *see* Mullerian inhibiting substance
Mitogen-activated protein kinase (MAPK)
epithelial-to-mesenchymal transition signaling, 264, 265
myostatin signaling, 513, 514
Ras-transforming growth factor-β signaling crosstalk, 310
Smad phosphorylation, 112–114
Smurf regulation of signaling, 160, 161
transforming growth factor-β
induction of apoptosis in cancer, 203
receptor signaling, 170, 437
MMPs, *see* Matrix metalloproteinases
MSC, *see* Mesenchymal stem cell
Mullerian inhibiting substance (MIS)
cancer therapy
breast cancer, 346, 348
cervical cancer, 344–346
endometrial cancer, 346
ovarian cancer, 340–343
prospects, 348
prostate cancer, 348
clinical utility, 339, 340
expression regulation, 335
function, 334, 335
history of study, 334–336
purification, 337
receptors, 338, 339
structure, 336, 337
Muscle, *see* Myostatin; Rhabdomyosarcoma
Myc
transforming growth factor-β effects, 28
transforming growth factor-β regulation of c-Myc and cancer defects, 219, 220
Myocardial remodeling, thrombospondin-1 role, 557
Myostatin
functional overview, 509, 51

metabolism regulation, 516, 517
muscle proliferation and differentiation inhibition, 515, 516
proteolytic processing, 510, 511
rhabdomyosarcoma expression, 517
signaling
mitogen-activated protein kinase, 513, 514
receptors, 511, 512, 517
Ski, 514, 515
Smads
Smad2–4, 512
Smad7 negative feedback, 512, 513
SnoN, 514, 515

N

NET1, epithelial-to-mesenchymal transition mediation, 269
NF-κB, *see* Nuclear factor-κB
Nkx2.1, Smad interactions, 12, 13
Notch, epithelial-to-mesenchymal transition mediation, 269
Nuclear factor-κB (NF-κB)
inhibitor expression, 613
renal inflammation role, 613

O

Ocular fibrosis
Smad3 knockout mouse studies, 602
transforming growth factor-β
expression, 664, 665
ocular surgery findings, 665
therapeutic targeting
antibodies, 666
antisense oligonucleotides, 666, 667
decorin, 668
RNA interference, 668
tranilast, 668
Osteoblast, Smurf1 function, 161–163
Ovarian cancer
bone morphogenetic protein/growth differentiation factor expression
granulosa cell tumor, 441, 442
ovarian epithelial tumor, 442
km23 mutations, 174, 176
Mullerian inhibiting substance therapy, 340–343
Ovarian folliculogenesis, *see* Folliculogenesis

P

p15Ink4b, transforming growth factor-β effects, 28, 218, 219
p21Cip1, transforming growth factor-β effects, 28, 218, 219
p27Kip1
cell cycle effects, 20
endometrial carcinoma studies of transforming growth factor-β signaling effects
degradation prevention, 225–228
growth regulation in endometrium, 228, 229
posttranslational regulation of levels and subcellular distribution, 220–223
reduced expression in cancer, 224
regulation in cancer, 214, 215, 218, 219
tumor suppression and progression functions, 229, 230
P144
development, 697
preclinical studies, 697–700
transforming growth factor-β antagonism rationale in fibrosis, 697
p300, Smad coactivation, 83, 84
Pancreatic cancer, Smad4 loss, 94
PDP, *see* Pyruvate dehydrogenase phosphatase
Peroxisome proliferator-activated receptor-γ (PPAR-γ), transforming growth factor-β/Smad system interactions, 529, 534–536, 538
Phosphatidylinositol 3-kinase (PI3K), epithelial-to-mesenchymal transition signaling, 264, 265
PI3K, *see* Phosphatidylinositol 3-kinase
Pioglitazone, pulmonary fibrosis management, 637, 638
Pirfenidone, pulmonary fibrosis management, 636, 637
PKC, *see* Protein kinase C
PPAR-γ, *see* Peroxisome proliferator-activated receptor-γ
PPM1A, Smad phosphatase, 110, 111
Prostate
development, 355, 356
hypertrophy, *see* Benign prostate hypertrophy
Prostate cancer
activin expression and function
normal prostate, 357, 358
prostate cancer, 358–360
bone morphogenetic protein expression and function
normal prostate, 360–362
prostate cancer
BMP2, 362
BMP6, 362–365
BMP7, 365
GDF-15, 365, 366
receptors, 366
inhibin expression and function
normal prostate, 357, 358

prostate cancer, 360
management, 356
metastasis, 356
Mullerian inhibiting substance therapy, 348
transforming growth factor-β induction of apoptosis, 202, 203
Proteasome system
regulatory Smads as adaptor proteins for E3 ligases, 249, 250
Smad ubiquitin regulating factors, *see* Smurfs
transforming growth factor-β signaling targets
cancer dysregulation, 252, 253
developmental regulation, 252
inhibitory Smad promotion of type I receptor destruction, 653, 654
overview, 248
regulatory Smads, 246, 247
Ski, 251
Smad4, 247–249
Smad7 as adaptor for destruction of transforming growth factor-β receptors and regulatory Smads, 250, 251
SnoN, 251
transforming growth factor-β, 246
ubiquitinating system, 156, 244–256
Protein kinase B, *see* Akt
Protein kinase C (PKC), Smad phosphorylation, 116
Proteomics
functional proteomics of transforming growth factor-β family signaling, 325–327
posttranslational modification of transforming growth factor-β receptors and Smads, 324, 325
prospects for transforming growth factor-β studies, 328
protein–protein interactions in transforming growth factor-β signaling, 321–324
systemic analysis of transforming growth factor-β family signaling, 327, 328
Pulmonary fibrosis
management, 573, 574, 576, 577
pathogenesis, 569, 570
protease/antiprotease balance, 625, 626
Smad3 studies using knockout mouse, 600, 601
therapy-induced pulmonary fibrosis
animal models, 632, 633
bleomycin-induced pulmonary fibrosis, *see* also Bleomycin
histopathology, 584–587
prevention, 590
transforming growth factor-β inhibition studies, 588
chemotherapy agents, 631, 632
management
imatinib mesylate, 637
interferon-γ, 636
peroxisome proliferator-activated receptor-γ antagonists, 637, 638
pirfenidone, 636, 637
prevention, 632
radiation therapy, 630, 631
thrombospondin-1 role, 555, 556
transforming growth factor-β
expression in idiopathic pulmonary fibrosis, 570, 571
lung injury animal model studies
apoptosis inhibition, 573
collagen synthesis induction, 573
cytokine response, 571, 572
epithelial-to-mesenchymal cell transition, 573
myofibroblast response, 632
signaling and inflammation mediation, 625
therapeutic targeting
antibodies, 634, 635
antisense oligonucleotides, 636
decorin, 574, 635
hepatocyte growth factor, 574, 576
interleukin-7, 574
latency-associated peptide, 635, 636
prospects, 626, 627
rationale, 573, 574
Smad7, 574
soluble receptors, 634
Pyruvate dehydrogenase phosphatase (PDP), Smad phosphatase, 111

R

RAR, *see* Retinoic acid receptor
Ras
mutation in cancer, 306
skin cancer mouse model studies
overview, 306, 307
transforming growth factor-β
expression and response changes, 307, 308
malignant phenotype enhancement, 309, 310
Ras carcinogenesis suppression mechanisms, 309
transgenic mice manipulation of signaling, 308, 309

transforming growth factor-β receptor
signaling, 170
signaling crosstalk, 310, 311
Regulatory Smads, *see* Smads
Renal fibrosis
Smad3 studies using knockout mouse
Smad7 ultrasound-microbubble-mediated gene transfer
rationale, 613, 614
safety and efficacy, 614–616
thrombospondin-1 role, 554, 555
transforming growth factor-β
inflammation signaling, 611–613
role, 610
signaling activation, 610
Smad signaling crosstalk, 610, 611
Retinoic acid receptor (RAR), transforming growth factor-β/Smad system interactions, 529, 537
Rhabdomyosarcoma, myostatin expression, 517
RhoA, Smurf regulation of signaling, 161
RNA interference
ocular fibrosis and transforming growth factor-β targeting, 668
Smad3 inhibition in fibrosis, 603
Rosiglitazone, pulmonary fibrosis management, 637
RUNX, transforming growth factor-β induction of apoptosis in cancer, 204, 205

S

SB431542
kinase specificity, 311
Smad3 inhibition in fibrosis, 603, 604
SB525344
kinase specificity, 311
Smad3 inhibition in fibrosis, 603
Scarring, *see* Fibrosis; Keloid
Scleroderma
inhibitory Smad role, 656
thrombospondin-1 role, 557
transforming growth factor-β role, 694, 695
Sclerostin
bone morphogenetic protein antagonism, 470
mutation and disease, 466
SD-208, preclinical studies, 311
SIP1, Smad corepression, 86
Ski
Dachshund interactions, 144
myostatin signaling, 514, 515
oncoprotein activity, 140–142
proteasomal degradation, 251
regulation
expression, 144
intracellular localization, 145
posttranslational modification, 145
Smad corepression, 85, 142–14
structure, 140–142
tumorigenesis role, 145, 146
Skin cancer, Ras signaling in mouse models
overview, 306, 307
transforming growth factor-β
expression and response changes, 307, 308
malignant phenotype enhancement, 309, 310
Ras carcinogenesis suppression mechanisms, 309
transgenic mice manipulation of signaling, 308, 309
Skin fibrosis, *see* Fibrosis; Scleroderma
Smads
coactivators, 83, 84
common mediator Smads, 78, 93, 437, 596
corepressors, 84–86
gene loci, 6
inhibitory Smads
apoptosis modulation, 654, 655
inflammatory bowel disease role, 656
overview, 11, 12, 78, 93, 437, 596
prospects for study, 657
regulation
posttranslational regulation, 651, 652
transcriptional regulation, 650, 651
scleroderma role, 656
structure, 650
subcellular localization, 653
transforming growth factor-β family signaling regulation mechanisms
bone morphogenetic protein signaling inhibition, 654
protein recruitment, 654
transcriptional regulation, 654
type I receptor binding, 653
type I receptor ubiquitination promotion, 653, 654
transgenic mouse studies, 656, 657
tumor expression, 655
modulators, 12, 13
myostatin signaling, 512, 513
nuclear translocation
dysregulation
cancer-causing mutations, 132, 133
signaling crosstalk, 133
export mechanisms, 128, 129
import mechanisms, 128
machinery, 127
nucleocytoplasmic shuffling versus retention, 129

overview, 25, 80, 81
regulatory Smad nuclear targeting, 131
Smad4
export, 130
import, 129, 130
nuclear targeting of activated complexes, 130, 131
transforming growth factor-β
regulation, 126
signal strength and duration of accumulation in nucleus, 133, 134
phosphorylative regulation
kinases
calcium/calmodulin protein kinase II, 116
cyclin-dependent kinase, 114–116
mitogen-activated protein kinase, 112–114
protein kinase C, 116
receptor-mediated phosphorylation, 106, 107, 109, 110
overview, 25, 78–80, 321
phosphatases, 110–112
prospects for study, 116, 117
sites in Smad2/3, 108
regulatory Smads, 9, 10, 25, 78, 93, 106, 437, 596
SARA modulation, 78, 79, 92, 109, 321
Smad2 knockout mouse, 597, 598
Smad3
β-catenin nuclear accumulation role, 298, 299
target gene cooperative regulation with β-catenin and Smad3, 299
Smad3
fibrosis studies using knockout mouse
eye, 602
kidney, 601
liver, 601
lung, 600, 601
skin, 600
therapeutic targeting, 602–604
knockout mouse, 597, 598, 600
wound healing role
excisional wounds, 598–600
incisional wounds, 598
Smad4
embryonic liver fodrin and tumor suppression, 95–98
loss in tumors, 94
tumor suppression, 93
Smad7 ultrasound-microbubble-mediated gene transfer for renal fibrosis and inflammation
rationale, 613, 614
safety and efficacy, 614–616
structure, 25, 78, 93, 106, 320
sumoylation, 249
transcriptional activation by Smads, 81–83
transforming growth factor-β receptor signaling mechanism, 8–10
trimer formation, 321
types, 24
ubiquitination, 26
ubiquitin regulation, *see* Proteasome system; Smurfs
Small latent complex, *see* Latency-associated peptide
Smurfs
cell polarity and transformed phenotype regulation, 163, 164
inhibitory Smads as adaptors, 159, 160
regulatory Smads
adaptors, 159, 160
regulation of protein levels, 158
Smad-independent signaling regulation
mitogen-activated protein kinase, 160, 161
RhoA, 161
Smad substrates, 11
Smurf1 in skeletal and bone homeostasis, 161–163
structure, 156–158
tumorigenesis and Smurf2, 164
ubiquitinating system, 156
SNIP1, Smad corepression, 86
SnoN
myostatin signaling, 514, 515
oncoprotein activity, 140–142
proteasomal degradation, 251
regulation
expression, 144
intracellular localization, 145
posttranslational modification, 145
Smad corepression, 85, 86, 142–144
structure, 140–142
tumorigenesis role, 145, 146
SREBP-2, Smad3 interactions, 323
Sumoylation, Smads, 249
Suv39h
bone morphogenetic protein-induced transcriptional repression
carcinogenesis role, 391, 392
global changes, 390
specific genes, 388–390
heterochromatin protein 1 recruitment, 384, 386–388
Smad interactions in bone morphogenetic protein signaling, 384–388
trimethylation, 384

T

TGF-β, *see* Transforming growth factor-β
TGIF, Smad corepression, 84, 85
Thrombospondin-1 (TSP-1)
developmental function, 554
fibrosis role
diabetes, 555
hypertension, 555
liver fibrosis, 556, 558
lung fibrosis, 555, 556
myocardial remodeling, 557
renal fibrosis, 554, 555
scleroderma, 557
immune function and infection response, 559, 560
prospects for study, 560, 561
transforming growth factor-β
activation, 47, 48, 61, 551–554
binding sequences, 552
tumorigenesis role, 558, 559
wound healing role, 557, 558
TNF-α, *see* Tumor necrosis factor-α
Tooth, uterine sensitization-associated gene-1 role in morphogenesis, 470
TRAIL, *see* Tumor necrosis factor-related apoptosis-inducing ligand
Tranilast, ocular fibrosis and transforming growth factor-β targeting, 668
Transforming growth factor-β (TGF-β)
discovery, 38
expression regulation, 38
gene loci, 6
isoforms, 4
latent complexes, *see* Latency-associated peptide; Latent transforming growth factor-β-binding protein
processing, 5, 38
sequence conservation between species, 39, 40
structure, 5
superfamily, 4, 426, 427
synthesis and secretion, 39
tumor progression activity, 26
tumor suppression in transgenic mice, 186
Transforming growth factor-β activation
integrins, 48–50, 61, 62
in vitro activation, 50, 58
matrix metalloproteinases, 282
overview, 46, 47, 279, 280, 550, 551
proteases, 47, 58, 59, 61
thrombospondin-1, 47, 48, 61, 551–664
Transforming growth factor-β receptors
accessory proteins, 7
coreceptors, *see* Betaglycan; Endoglin
gene loci, 6
heterotetramer formation, 320
ligand specificity, 6, 7
mutation in cancer, 29
phosphorylation, 320
protein--protein interaction, 170, 171
signaling, *see* km23; Mitogen-activated protein kinase; Smads
Smad-independent signaling, 26
Smad phosphorylation, 106, 107, 109, 110
stromal signaling
fibroblast sensitivity and cancer regulation
initiation, 189–192
metastasis, 193
progression, 192
overview, 187, 189
truncated transforming growth factor-β receptor II overexpression and scar reduction
adenoviral vector, 709
dominant negative effect, 708, 709
fibroblast fibrotic behavior inhibition in vitro, 709, 710, 712, 714
scar formation inhibition in vivo, 714, 716, 718, 719
types, 5, 6, 24, 434–436, 596
Triterpenoids, transforming growth factor-β/Smad system regulation, 534–536
Troglitazone, pulmonary fibrosis management, 638
TSP-1, *see* Thrombospondin-1
Tumor necrosis factor-α (TNF-α), transforming growth factor-β interactions in wound healing, 682, 683
Tumor necrosis factor-related apoptosis-inducing ligand (TRAIL), transforming growth factor-β induction of apoptosis in cancer, 202

U

Ubiquitinylation, *see* Proteasome system
USAG-1, *see* Uterine sensitization-associated gene-1
Uterine sensitization-associated gene-1 (USAG-1)
bone morphogenetic protein antagonism, 470
discovery, 464, 465
kidney
abundance, 469
expression, 468
knockout mouse resistance to renal tubular injury, 468, 469
protein–protein interactions, 467, 468
structure, 465–467
tooth morphogenesis role, 470
Wnt signaling modulation, 470, 471

V

Vascular endothelial growth factor (VEGF)
bone morphogenetic protein induction, 378, 379
transforming growth factor-β induction, 31
wound healing, 683, 684
Vasculogenesis
bone morphogenetic protein-2/4 regulation, 374, 375
endoglin role, 400, 401
VDR, *see* Vitamin D receptor
VEGF, *see* Vascular endothelial growth factor
Vitamin D receptor (VDR)
therapeutic targeting in cancer, 539
transforming growth factor-β/Smad system interactions, 528–534

W

Wnt signaling
mesenchymal stem cell
β-catenin nuclear accumulation
modulation of transforming growth factor-β effects, 299
response to transforming growth factor-β, 297, 298
Smad3 requirement, 298, 299
crosstalk with transforming growth factor-β signaling, 296, 297
prospects for study, 300, 301
regulation, 294–296
target gene cooperative regulation with β-catenin and Smad3, 299
overview, 294
uterine sensitization-associated gene-1 modulation, 470, 471
Wound healing
fetal scarless wound healing, 704
fibrosis, *see* Fibrosis
phases, 676
Smad3 role
excisional wounds, 598–600
incisional wounds, 598
thrombospondin-1 role, 557, 558
transforming growth factor-β
animal studies, 681, 682
cytokine interactions, 682–684
expression, 677, 694
extracellular matrix modulation, 678, 679
fibrosis induction, 679
granulation tissue formation, 679, 680
human studies, 682
re-epithelialization role, 680, 681